NUCLEAR
AND
PARTICLE PHYSICS

NUCLEAR AND PARTICLE PHYSICS

S. L. Kakani

M.Sc., Ph.D.

Shubhra Kakani

M. Sc.

Viva Books

New Delhi | Mumbai | Chennai | Kolkata | Bengaluru | Hyderabad | Kochi | Guwahati

First Published 2008
Reprinted 2011

VIVA BOOKS PRIVATE LIMITED

- 4737/23 Ansari Road, Daryaganj, New Delhi 110 002
  E-mail: vivadelhi@vivagroupindia.net, Tel. 42242200

- 76, Service Industries, Shirvane, Sector 1, Nerul, Navi Mumbai 400 706
  E-mail: vivamumbai@vivagroupindia.net, Tel. 27721273, 27721274

- Jamals Fazal Chambers, 26 Greams Road, Chennai 600 006
  E-mail: vivachennai@vivagroupindia.net, Tel. 28290304, 28294241

- B-103, Jindal Towers, 21/1A/3 Darga Road, Kolkata 700 017
  E-mail: vivakolkata@vivagroupindia.net, Tel. 22836381, 22816713

- 7, Sovereign Park Apartments, 56-58, K. R. Road, Basavanagudi, Bengaluru 560 004
  E-mail: vivabangalore@vivagroupindia.net, Tel. 26607409, 26607410

- 101-102 Mughal Marc Apartments, 3-4-637 to 641, Narayanguda, Hyderabad 500 029
  E-mail: vivahyderabad@vivagroupindia.net, Tel. 27564481, 27564482

- Beevi Towers, First Floor, SRM Road, Kaloor, Kochi 682 018, Kerala
  E-mail: vivakochi@vivagroupindia.net, Tel: 0484-2403055, 2403056

- 232, GNB Road, Beside UCO Bank, Silpukhuri, Guwahati 781 003
  E-mail: vivaguwahati@vivagroupindia.net, Tel: 0361-2666386

www.vivagroupindia.com

ISBN : 978-81-309-0040-7

Published by Vinod Vasishtha for Viva Books Private Limited, 4737/23, Ansari Road, Daryaganj, New Delhi 110 002. Printed and bound at Sanat Printers, Kundli - 131 028.

CONTENTS

PREFACE

This book has been designed to serve as a text book in *Nuclear and Particle Physics*, in accordance with the latest UGC syllabus for B.Sc. (Hons., pass), M.Sc. (Physics), M. Phil (Physics) students of all Indian and other universities. The aim of this text is to give students a thorough understanding of the principal features of atomic nuclei, nuclear decays, nuclear reactions, nuclear models, nuclear energy, experimental techniques, cosmic rays and elementary particle physics.

The subject matter has been selected and developed in a manner to bridge the gap between introductory and advanced level courses in Nuclear Physics. The salient features of the book are:

- Basic Principles and fundamental concepts are explained in a simple and lucid style.

- The presentation is quantitative and derivations of essential relations are given in full.

- A large number of typical worked out problems of different types have been given in each chapter. These worked out problems further supplement the text.

- A large number of self-explanatory accurate diagrams and tables have been provided to make the subject more interesting and explicit.

- At the end of each chapter good number of review questions and selected problems with answers are given.

To make book useful for UGC-CSIR; NET/SLET, GATE and other entrance examinations *Objective* and *Short Question Answers* type questions are also given at the end of each chapter.

We hope that with all its unique features, the book will definitely cater to the needs of both students and teachers as well as research students.

Honestly speaking, a text book of this kind cannot be a original work. We take this opportunity to place on record our indebtedness a large number of relevant books and journals that we have freely consulted. We heartily thank the Publishers, Viva Books Private Limited for all their efforts in publishing this book.

In spite of all precautions and care taken to avoid errors and misprints there might have crept some due to oversight. We will feel highly obliged to those fellow teachers and students who will bring them to their notice. Any suggestions for improvement will be thankfully acknowledged.

S. L. Kakani

Shubhra Kakani

Bhilwara

1

THE NUCLEUS
(Nuclear Model of the Atom and General Properties of the Nucleus)

1.1 INTRODUCTION

The Concept of an atom as of a minute indivisible particle of matter arose in the ancient times as an alternative to the concept of the continuous structure of matter. In modern times not only has the interest in the atoms of the ancient been retained but atomism has even acquired the feature of a scientific hypothesis. The discovery of real atoms is, however, associated with the name of Dalton, the British Chemist, who in 1803 was the first to substantiate a method for determining the relative masses of the atoms of simple substances. It took nearly a hundred years before the internal structure of the atom could be understood. The discovery of the electron by J.J. Thomson and the realisation that electrons were a common constituent of all matter provided the first significant insight into the structure of the atom. Electrons carry a negative charge ($-e$ = electronic charge). Atoms themselves are electrically neutral. Obviously, every atom must carry a positive charge to balance the negative charge on electrons. Further the electrons are very light compared to the atoms. This led J.J. Thomson to suggest that an atom is a uniform sphere of positively charged matter in which the electrons are embedded even as raisins are embedded in a plum pudding. Thomson's model of atom was put an experimental test in 1911 by Geiger and marsden to probe into the atoms and to find out what was inside them. From a study of the results of these experiments Rutherford concluded that Thomson's model of the atom had to be discarded. Rutherford developed a quantitative theory to account for the scattering of the alpha particles by the nucles of the thin foil of aluminium. The experimental results of Geiger and Marsden completely verified the correctness of Rutherford's conclusion.

1.2 RUTHERFORD'S THEORY OF ALPHA PARTICLE SCATTERING

Alpha particles (α) are highly energetic charged particles, emitted during the radioactive disintegration of certain heavy elements e.g. uranium, radium. The amount of positive charge on α particle is equal to two units of electronic charge and their mass is equal to that of a helium atom. When a beam of α-particles is passed through thin foils of metallic elements of high atomic mass, it is observed that, while the majority of α-particles are scattered through small angles, a significant number is deviated through large angles (greater than 90°) (Figure 1.1).

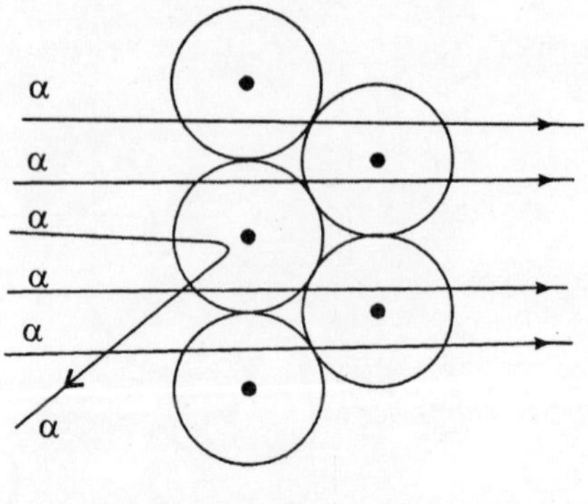

Geiger and Marsden observed that when a foil of gold of thickness 0.4×10^{-4} cm is used one α-particle in 20000 is deviated through an angle of $\dfrac{\pi}{2}$ or greater than

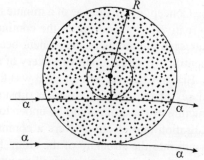

Fig. 1.1 Large and Small angle scattering of α-particle from an array of atoms

$\dfrac{\pi}{2}$ and few are turned back into the direction from which they come. Such large angle scattering cannot be explained on the basis of Thomson's model of the atom. Thomson model consists of a uniform distribution of positive and negative charge, sometimes called the plum-pudding or chemical model.

In traversing the homogeneous atoms of metal according to the Thomson model (Figure 2.2), the α-particles (helium ions) might be expected to be slightly deviated, according to the quantity of the positive charge traversed, the number of electrons encountered, and so forth. These effects would be small, and the alpha particles would be expected to suffer only minor deviations.

The experimental results were astonishingly different. In the first place, most of the α-particles were not deviated at all (Figure 1.1). Those that were deviated were turned through very large angles-some being even deflected back towards the source as shown in Figure 1.1.

To explain the observed large angle scattering of the α-particles, Rutherford in 1911 proposed a new model of the atom. He suggested that in a given atom the positive charge and the bulk of the mass is tightly concentrated in a "nucleus" of extremely small dimensions. The negatively charge, consisting of atom's electrons is distributed in the rest of the atom's volume which is thus largely empty

Fig. 1.2 Scattering of α-particle based on Thomson's model of the atom

space. Rutherford had discovered the nucleus which is a landmark in the history of the development of modern physics. Rutherford showed that the effect of the electrons in the atom outside the nucleus is negligible for the deflection of α-particles through angles greater than 1°. When an α-particle passes by very close to the nucleus, it experiences a strong electrostatic repulsive force due to the entire positive charge of the atom and hence scattered at a relatively large angle. At the suggestion of Rutherford experiments were conducted in 1911 by Geiger and Marsden to probe into the atoms and to find out what was inside the atom.

A schematic diagram of Rutherford and his colleagues experimental arrangement is shown in Figure 1.3. The apparatus obviously has to be enclosed in vacuum.

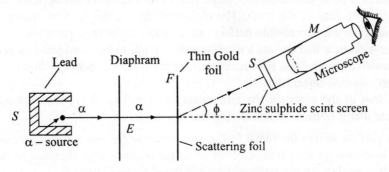

Fig. 1.3(a) A schematic diagram of the Rutherford scattering experiment. The system up to the scintillating screen, must be enclosed in a vaccuum.

During the experiment a narrow collimated beam of mono-energetic α-particles is allowed to fall on a very thin foil of gold ($Z = 79$) or silver ($Z = 47$) which can be rolled or beaten to thickness of the order of 10^{-7} m or less. The α-particles do not lose any appreciable energy in passing through such thin metal foils. The scattered α-particles were detected and counted by a fluorescent screen-microscope arrangement. Rutherford and his colleagues observed that many α-particles went straight through the foil or were deflected by a very small angle. However, a few α-particles were deflected through very large angles; some of them even got back scattered. This was contrary of the Thomson's atom model, i.e. the positive charge is uniformly smeared over the entire volume of the atom.

To explain the above observations of the experiment, Rutherford assumed that the positive massive part of the atom was concentrated in a very small volume at the centre of the atom. This core, now termed the *nucleus*, is surrounded by a cloud of electrons, which makes the entire atom electrically neutral. This nuclear model of the atom accounts for the large angle scattering of the α-particles in the following way. An α-particle approaching the centre of the atom experiences an increasingly large Coulomb repulsion. Since the atom is mostly empty space, most of the α-particles do not go sufficiently close to the nucleus to get sufficiently deflected and hence they pass the thin metal foil with practically no deviation. However, an α-particle that passes close to the nucleus is subjected to a very large Coulomb force exerted by massive positive core nucleus of charge $+Ze$ and is deflected at a large angle in a single encounter. The assumptions made by Rutherford during the explanation of the large angle α scattering are as follows:

(i) The positive charge of the atom is concentrated in an extremely small volume, known as 'nucleus' in the atom. The magnitude of this charge is an integer Z times the charge of an electron, where Z is an integer representing the atomic number of the element (Z is exactly equal to the number of orbital electrons in an atom). The negative charges (electrons) revolve around the nucleus in planetary orbits.

(ii) The whole mass of the atom (excluding the mass of the orbital electrons) is concentrated in the nucleus. Obviously, more than 99.99 % of the mass is concentrated within the small volume of the nucleus. Since the mass of the electron is negligible in comparison to that of the nucleon it follows that nuclei are very much denser. Obviously, the density of the nuclear matter is very high about 10^{15} times that of the density of the ordinary matter.

(iii) The scattering of α-particles takes place due to an encounter between the α-particle and the massive nucleus of the atom. The electrons outside the nucleus, owing to their very small mass, cannot appreciably deflect the far more massive α-particles.

(iv) Both the α-particle and the nucleus are small enough to be considered as point mass and point (positive) charge and the electrostatic repulsive force between them is governed by Coulomb's inverse square law.

(v) The nucleus of the atom (of the thin metal foil used) is so massive compared with the α-particle that it remains at rest during the encounter.

when an α-particle strikes the metal foil, it can penetrate the outer electron-cloud of an atom and approach closely the nucleus. It then moves under the action of a Coulombian force of repulsion and its path is hyperbola with the nucleus as the external focus (Figure 1.1).

1.2(a) Rutherford's Theory

Based on the above assumptions one can show with the help of classical mechanics that the path of the α-particle under the electrostatic repulsive force of the nucleus is a hyperbola with the scattering nucleus at the external focus S. In Figure 1.3 the path of the α-particle far off from the target nucleus is along AO. At α-particle comes nearer S, the repulsive force on it varies as $\dfrac{1}{r^2}$ where r is the instantaneous separation between the α-particle and the nucleus. The α-particle gets deflected and finally moves along OB, tracing the hyperbolic curve ADB. Let the angle of deflection (or scattering angle) of the α-particle be ϕ which is also the angle between the asymptotic direction of approach of the α-particle and the asymptotic direction along which α-particle recedes. If there were no force acting on the α-particle, it would pass by the nucleus at the minimum distance equal to p which is the length of the perpendicular from S on the asymptote AO. This perpendicular distance is called *impact parameter*.

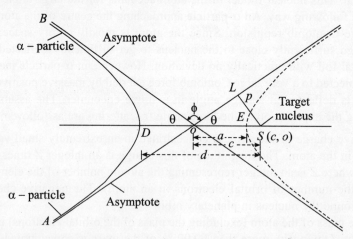

Fig. 1.3(b) Path of the α-particle during the scattering by a atomic nuclear.

Let us take the centre o of the hyperbola as the origin. Obviously, the vertex D of the hyperbola is at $x = -a$. Similarly the vertex E of the second branch of the hyperbola is at $x = a$. Let (c, o) be the coordinates of the target nucleus and d the distance of vertex D from the nucleus which is evidently the distance of the nearest approach of the α-particle to the nucleus. Further, let θ be the angle which the asymptotic direction AO of the α-particle makes with the x-axis. From Figure 1.3, the distance of nearest approach

$$d = c + a = c\left(1 + \frac{a}{c}\right)$$

But

$$\frac{a}{c} = \cos \theta^*$$

$\therefore$

$$d = c + c \cos \theta = c(1 + \cos \theta)$$

Also

$$\frac{p}{c} = \sin \theta \text{ or } c = \frac{p}{\sin \theta}$$

$\therefore$

$$d = \frac{p(1 + \cos \theta)}{\sin \theta} = \frac{p \, 2 \cos^2 \dfrac{\theta}{2}}{2 \sin \dfrac{\theta}{2} \cos \dfrac{\theta}{2}} \qquad \text{... (1.1)}$$

or

$$d = p \cot \frac{\theta}{2} \qquad \text{... (1.2)}$$

Let us now derive the relation between the impact parameter p and the angle of deflection ϕ. During the interaction between the α particle and the nucleus, the laws of conservation of kinetic energy and the angular momentum must hold good. When the α particle is at a large distance from the nucleus, its energy is only kinetic because the potential energy, being inversely proportional to

---

*The equation of hyperbola $\dfrac{x^2}{a^2} - \dfrac{y^2}{c^2 - a^2} = 1$, where c is the x-coordinate of the focus S and is equal to $a\,e$ (e being

the eccentricity of the hyperbola). The equation of the asymptote AO of the hyperbola is $y = \sqrt{c^2 - a^2} \, \dfrac{x}{a}$ or

$\sqrt{(c^2 - a^2)} \, x - ay = 0$. The general formula for the length of the perpendicular from a point (x_1, y_1) on the line $ax +$

$by + c = 0$ is $\dfrac{ax_1 + by_1 + c}{\sqrt{a^2 + b^2}}$. Obviously, the length of the perpendicular from the focus (c, o) to the asymptote AO

is

$$p = \frac{\left(\sqrt{c^2 - a^2}\right)c - a(0)}{\sqrt{(c^2 - a^2) + a^2}} = \sqrt{c^2 - a^2}$$

$$OL = \sqrt{c^2 - p^2} = \sqrt{c^2 - (c^2 - a^2)} = a$$

$\therefore$

$$\frac{OL}{OS} = \frac{a}{c} = \cos \theta$$

distance, can be taken as zero. If v is the velocity of the α-particle at this distance, the total energy is $\frac{1}{2} mv^2$. Let the velocity of the α-particle when it reaches the vertex D is v_0. Its kinetic energy at D is $\frac{1}{2} m v_0^2$ and potential energy is $\dfrac{(Ze)\,2e}{4\pi\,\varepsilon_0\,d} = \dfrac{Ze^2}{2\pi\,\varepsilon_0\,d}$, where Ze is the charge on the nucleus ($Z \to$ atomic number and $2e \to$ charge on the α-particle). According to law of conservation of energy

$$\frac{1}{2} mv^2 = \frac{1}{2} mv_0^2 + \frac{Ze^2}{2\pi\,\varepsilon_0\,d}$$

Dividing throughout by $\frac{1}{2} mv^2$, one obtains

$$1 = \frac{v_0^2}{v^2} + \frac{Ze^2}{\pi\,\varepsilon_0\,mv^2 d}$$

$$= \frac{v_0^2}{v^2} + \frac{b}{d} \qquad \left(b = \frac{Ze^2}{\pi\,\varepsilon_0\,mv^2} \right)$$

or

$$\frac{v_0^2}{v^2} = 1 - \frac{b}{d}$$

Substituting the value of d from Equation (1.1), one obtains

$$\frac{v_0^2}{v^2} = 1 - \frac{b \sin \theta}{b(1 + \cos \theta)} \qquad \qquad \dots (1.3)$$

From the law of conservation of angular momentum

$$mvp = mv_0 d$$

or

$$\frac{v_0}{v} = \frac{p}{d} = \frac{\sin \theta}{1 + \cos \theta} \qquad \text{(using Equation. (1.1))}$$

Squaring, one obtains

$$\frac{v_0^2}{v^2} = \frac{\sin^2 \theta}{(1 + \cos \theta)^2} = \frac{1 - \cos^2 \theta}{(1 + \cos \theta)^2} = \frac{1 - \cos \theta}{1 + \cos \theta} \qquad \qquad \dots (1.4)$$

Comparing (1.3) and (1.4), one obtains

$$\frac{1 - \cos \theta}{1 + \cos \theta} = 1 - \frac{b \sin \theta}{p(1 + \cos \theta)}$$

or

$$\frac{b \sin \theta}{p(1 + \cos \theta)} = 1 - \frac{1 - \cos \theta}{1 + \cos \theta} = \frac{2 \cos \theta}{1 + \cos \theta}$$

$\therefore$

$$\frac{b}{p} = 2 \cot \theta$$

But
$$\phi = \pi - 2\theta \quad \text{or} \quad \theta = \frac{\pi}{2} - \frac{\phi}{2}$$

∴
$$\frac{b}{p} = 2 \cot\left(\frac{\pi}{2} - \frac{\phi}{2}\right) = 2 \tan\frac{\phi}{2}$$

or
$$p = \frac{b}{2} \cot\frac{\phi}{2}$$

$$= \frac{Ze^2}{2\pi\,\varepsilon_0\,mv^2} \cot\frac{\phi}{2} \qquad \ldots (1.5)$$

$$= \frac{(Ze)(2e)}{4\pi\,\varepsilon_0\,mv^2} \cot\frac{\phi}{2} \qquad \ldots (1.6)$$

Equation (1.6) gives the relation between the impact parameter and the angle of scattering. When the impact parameter p becomes smaller, the angle of scattering ϕ increases (Figure 1.4). It further follows that an α-particle approaching a target nucleus with an impact parameter ranging from o to p will be scattered through an angle of ϕ or more. Obviously, an α-particle which is directed any where within an area πp^2 around the nucleus will be scattered through ϕ or more. The area πp^2 is called the *cross-section of interaction* and denoted by $\sigma\ (= \pi p^2)$.

Let us now consider beam of α-particles incident upon a thin foil of thickness t, containing n atoms per unit volume. Evidently, the number of target nuclei per unit area is nt so that an α-particle beam incident upon an area A faces ntA target nuclei. The total cross-section of interaction of the nuclei for a scattering angle of ϕ or more is equal to the number of target nuclei $nt\,A$ multiplied by the cross-section σ per nucleus for such scattering or equal to $n\,t\,A\,\sigma = nt\,A\,\pi p^2$. The fraction q of the α-particles scattered through an angle of ϕ or more is the ratio of the total cross-section of interaction and total target area A. Obviously,

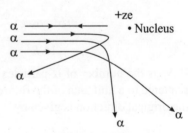

Fig. 1.4 α-particle scattering at different impact parameters

$$q = \frac{\text{Number of } \alpha\text{-particles scattered through } \phi \text{ or more}}{\text{Number of incident } \alpha\text{-particles}}$$

$$= \frac{\text{Total cross section of interaction}}{\text{Total target area}}$$

$$= \frac{nt\,A\,\pi p^2}{A} = \pi\,nt\,\pi^2$$

Substituting p from (1.5), one obtains

$$q = \pi\,nt\,\frac{b^2}{4} \cot^2\frac{\phi}{2} \qquad \ldots (1.6a)$$

The fraction of the incident α-particles which gets scattered through an angle ϕ and $(\phi + d\phi)$ can be obtained by differentiating (1.6) $w \cdot r \cdot t \cdot q$. Hence

$$dq = -\pi \, nt \, \frac{b^2}{4} \cot \frac{\phi}{2} \, \mathrm{cosec}^2 \frac{\phi}{2} \, d\phi \qquad \ldots (1.7)$$

The negative sign indicates that dq decreases as ϕ increases.

α-particles are detected by the scintillation which they produce on a fluorescent screen. Let a screen be placed at a distance R from a foil and suitably adjusted so as to receive the particles scattered through the angle between ϕ and $\phi + d\phi$ inside an annular ring of radius $R \sin \phi$ and width $R d\phi$ (Figure 1.5). The area of this ring is $2\pi R \sin \phi$

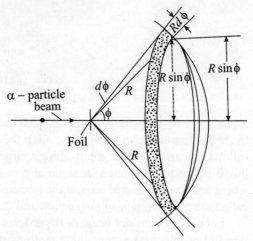

$$R d\phi = 2\pi^2 \sin \phi \, d\phi = 4\pi \, R^2 \sin \frac{\phi}{2} \, \cos \frac{\phi}{2} \, d\phi.$$

Thus the fraction of α-particles scattered between ϕ and $(\phi + d\phi)$ striking a unit area of the screen at a distance R from the scattering foil is given by

Fig. 1.5

$$\frac{dq}{dA} = \frac{\pi \, nt \, b^2 \cot \dfrac{\phi}{2} \, \mathrm{cosec}^2 \dfrac{\phi}{2}}{4 \times 4\pi \, R^2 \sin \dfrac{\phi}{2} \cos \dfrac{\phi}{2}} \frac{d\phi}{d\phi}$$

$$= \frac{nt \, b^2 \, \mathrm{cosec}^4 \dfrac{\phi}{2}}{16 \, R^2}$$

If N_0 is the number of α-particles actually incident normally on the foil per second, the number scattered to a unit area of the fluorescent screen at a distance R from the foil and at an angle ϕ with the original direction is given by

$$N = \frac{N_0 \, nt \, b^2 \, \mathrm{cosec}^4 \dfrac{\phi}{2}}{16 \, R^2}$$

Substituting for b, we have

$$N = N_0 \frac{nt(Ze)^2 \, e^2 \, \mathrm{cosec}^4 \dfrac{\phi}{2}}{16 \, \pi^2 \, \varepsilon_0^2 \, m^2 \, v^4 \, R^2} \qquad \ldots (1.8)$$

This is *Rutherford's scattering formula.*

1.2(b) Distance of Closest Approach

The distance d of the closest approach of the α-particle can also be calculated in terms of scattering and other known parameters. From Equation (1.1), we have

$$d = \frac{p(1 + \cos \theta)}{\sin \theta}$$

But $\qquad\qquad\qquad\qquad \phi = \pi - 2\theta$

$\therefore \qquad\qquad d = \dfrac{p\left[1 + \cos\left(\dfrac{\pi}{2} - \dfrac{\phi}{2}\right)\right]}{\sin\left(\dfrac{\pi}{2} - \dfrac{\phi}{2}\right)} = \dfrac{p\left(1 + \sin\dfrac{\phi}{2}\right)}{\cos\dfrac{\phi}{2}}$

$$= p\left[\dfrac{1}{\cos\dfrac{\phi}{2}} + \tan\dfrac{\phi}{2}\right]$$

Substituting p from (1.5), we have

$$d = \dfrac{b}{2}\cot\dfrac{\phi}{2}\left[\dfrac{1}{\cos\dfrac{\phi}{2}} + \tan\dfrac{\phi}{2}\right]$$

$$= \dfrac{b}{2}\left[\csc\dfrac{\phi}{2} + 1\right]$$

$$= \dfrac{Ze^2}{2\pi\,\varepsilon_0\,mv^2}\left(1 + \csc\dfrac{\phi}{2}\right) \qquad\qquad \dots (1.9)$$

1.2(c) Experimental Verification of Rutherford's Scattering Formula

From Equation (1.8) we see that the number of particles falling on a unit area of zinc sulphide screen at a given distance R from the screening foil should be proportional to

  (i) $\mathrm{Cosec}^4\,\dfrac{\phi}{2}$, where ϕ is the angle of scattering.

 (ii) t, thickness of scattering foil

(iii) $\dfrac{1}{E^2}$ or $\dfrac{1}{v^4}$, i.e. reciprocal of the fourth power of velocity of α-particles ($E = \dfrac{1}{2}\,mv^2$ is the energy of α-particles), and

(iv) $(Ze)^2$, i.e. the square of the nuclear charge of the scattering material.

The above conclusions of Rutherford's α-particle scattering theory were tested by a series of experiments perpormed by N.Geiger and E.Marsdon in 1913, two of Rutherford's close associates. The experimental arrangement is shown in Figure 1.6.

The thin foil F and source of α-particles 5 (small amount of radioactive radon gas 226 Rn emitting mono energetic α-particles were mounted on the pipe to the vacuum pump. During the experiments, this assembly was fixed in position. Foil and source were inside an evacuated cylindrical metal pot. Into the wall of this pot was inserted the microscope with a fluorescent screen of zinc sulphide S' mounted in the focal plane of objective O. The pot was supported on a platform P and could be rotated about the vertical axis through the plane of the scattering foil. Rotation without the necessity of breaking the vacuum was possible because of the greased cone joint in which the platform support was mounted. The platform P was provided with a graduated circle arranged so that the angle between the line from centre of F to the centre of S' and the line from the diaphragm D restricted the α-particles from the source S to a narrow beam, directed always normally on to F.

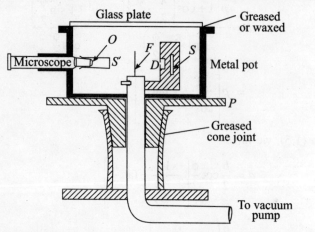

Fig. 1.6 Investigation of the scattering of α-particles by a thin metal foil

Investigation of the number of α-particles scattered to unit area of S' over the range of angles from $5°$ to $150°$ for a given number of incident α-particles provided a complete verification in the cases of silver, gold and platinum, of the $\operatorname{cosec}^4 \frac{1}{2} \phi$ law, where $\operatorname{cosec} \phi$ varied from 1 to 250,000.

Geiger and Marsden found that when α-particles of a definite velocity v are scattered from a foil of given thickness t, at different angles ϕ, the product $(N \times \sin^4 \frac{\phi}{2})$ remained constant within the limits of experimental errors. Table 1.1 supports the above conclusion.

Using the apparatus shown in Figure 1.7 Geiger and Marsden then studied the variation of scattering with (i) thickness of the foil, (ii) nature of the metallic element used for the foil, and (iii) the velocity of the incident α-particles.

The apparatus consisted of a cylindrical brass chamber of which each side was closed by a glass plate. The chamber was evacuated metallic foils F of various thickness were mounted on a disc G which could be rotated from outside the vacuum by turning the male part of greased cone joint J. Obviously, in this way, one foil after another could be presented to the normally incident beam of α-particles provided by a homogeneous source of $Ra\ B$ and $Ra\ C$ at S. S was just outside the chamber wall; the α-particles from it entered the chamber via the thin mica window and were collimated by the diaphragm D. Viewing of the scattered α-particles was at a fluorescent screen placed so that the angle of scattering was fixed at $25°$:

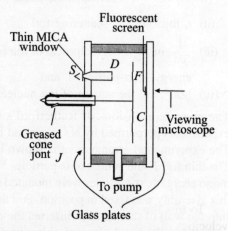

Fig. 1.7 Study of the effect of thickness and nature of metal foil, also velocity of α-particles on scattering.

Table 1.1

Scattering angle (φ)	Sin⁴ $\frac{\phi}{2}$	No. of α-particles falling on a unit area of the screen (N)	$N \times Sin^4 \frac{\phi}{2}$
15°	2.903×10^{-4}	132,000	38.4
30°	4.484×10^{-3}	7,800	35.0
45°	2.146×10^{-2}	1,435	30.8
60°	0.0625	477	29.8
75°	0.1379	211	29.1
105°	0.3952	69.5	27.5
120°	0.5586	51.9	29.0
135°	0.7245	43.0	31.2
150°	0.8695	33.1	28.8

Geiger and Marsden also measured the number of α-particles scattered through a given angle (φ = constant) using foils of different thickness. Figure 1.8 depicts the variation of N with t. Here the thickness of the foil is expressed in terms of the air-equivalent which is the thickness of air that would produce the same loss of energy as is actually produced by the given thickness of the foil. Geiger and Marsden used foils of different thicknesses and found that for a given material straight line graph confirmed N α t, there by confirming the conclusion (ii).

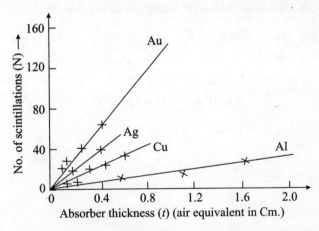

Fig. 1.8 Variation of number of scattered α-particles (N) with thickness of foil (t).

To test the conclusion (iii) Geiger and Marsden also performed experiment with different velocities of α-particle. When a collimated beam of mono-energetic α-particles of a given initial energy is allowed to pass through a thin mica foil, than the α-particles emerging from the foil lose a definite amount of kinetic energy, depending on the thickness of the foil. Obviously, the emergent α-particles form an approximately mono-energetic beam of reduced energy. When these α-particles of reduced energy are allowed to be scattered from a scattering foil then one can easily study the scattering of the α-particles at a reduced velocity. Following this Geiger and Marsden used mica foils of different thickness to reduce the energy of α-particles to different values and studies their scattering at a particular angle (keeping ϕ constant) by a scattering foil of given constant thickness. One can determine the velocity υ of α-particles by measuring the range R. They reported that the product $N\upsilon^4$ was a constant (Table 1.2)

Table 1.2 (From Geiger and marsden)

Range (R) of the incident α-particles (cm)	υ^{-4} (relative values)	No. of α-particles falling on a unit area of the screen (N)	Product $N\upsilon^4$
5.5	1.0	24.7	25
4.76	1.21	29.0	24
4.05	1.50	33.4	22
3.32	1.91	44	23
2.51	2.84	81.0	28
1.84	4.32	101	23
1.04	9.22	255.0	28

This verifies the law of velocities conclusion (iii) of Rutherford's theory. This is worthwhile to note that due to large errors in measurement Geiger and Marsden failed to verify the last conclusion of Rutherford's theory. However, this was done later by Chadwick.

Example 1.1 A 5 meV α-particle approaches a gold ($Z = 79$) nucleus with an impact parameter of 2.6×10^{-13} m. Through what angle will it be scattered? ($e = 1.6 \times 10^{-19}$ C, 1 eV $= 1.6 \times 10^{-19}$ J).

Sol. We have $\cot \dfrac{\phi}{2} = \dfrac{2p}{b} = \dfrac{(4\pi\varepsilon_0)\, p\, E_k}{Ze^2}$

$\therefore \qquad \cot \dfrac{\phi}{2} = \dfrac{2.6\times10^{-13}\times8\times10^{-13}}{9\times10^9\times79\times(1.6\times10^{-19})^2}$

$\qquad\qquad = 11.42$

$\therefore \qquad\qquad \dfrac{\phi}{2} = \cot^{-1}(11.42) = 5°$

or $\qquad\qquad \phi = 10°$

$E_K = 5$ meV
$\qquad = 8 \times 10^{-13}$ J
$p = 2.6 \times 10^{-13}$ J
$Z = 79$
$e = 1.6 \times 10^{-19}$ C
$\dfrac{1}{4\pi\varepsilon_o} = 9 \times 10^9$ N · m²/C²

Example 1.2 In Rutherford scattering experiment the number of particles observed at an angle of 10° is one million per minute. How many particles per minute will be observed at 90° and 180°?

Sol. $N(\phi) \propto \dfrac{1}{\sin^4\left(\dfrac{\phi}{2}\right)} = \dfrac{K}{\left(\sin^4\dfrac{\phi}{2}\right)}$

Here K is constant. Now $N(10°) = 10^6$/min.

$$\therefore \qquad 10^6 = K\,\dfrac{1}{\sin^4\left(\dfrac{10}{2}\right)^{\!\circ}} = \dfrac{K}{(0.0872)^4}$$

$$\therefore \qquad K = (0.872)^4 \times 10^6$$

Now
$$N(90^\circ) = K\,\dfrac{1}{\sin^4\left(\dfrac{90^\circ}{2}\right)} = (0.0872)^4 \times 10^6 \times \dfrac{1}{\left(\dfrac{1}{\sqrt{2}}\right)^4}$$

$$= 232$$

Similarly, one obtains $N(180^\circ) = 58$

Example 1.3 A beam of α-particles of energy 5.5 MeV is incident normally on a gold ($Z = 79$) foil at a rate of 10^4 particles per sec. An α-particles counter with an aperture 4 sq. cm is placed at a distance 0.25 m from the foil. Calculate the counts at an angle of 60°, if the thickness of the foil is 10^{-5} cm. The atomic weight of gold is 197 and its density is 19.3 gm/cm³. (Agvogadro number is 6.023×10^{23}).

Sol. we have

$$N(\phi) = \dfrac{N_0\, nt\, Z^2\, e^4}{4\,r^2\,(4\pi\varepsilon_0)^2\, E_k^2}\;\dfrac{1}{\sin^4\left(\dfrac{\phi}{2}\right)}$$

$$\therefore\; N(\phi) = \dfrac{10^4 \times 5.9 \times 10^{28} \times 10^{-7} \times (79)^2 \times (1.6 \times 10^{-19})^4}{4 \times (0.25)^2 \times \dfrac{1}{(9 \times 10^9)^2} \times (5.5 \times 1.6 \times 10^{-13})^2} \times \dfrac{1}{\left(\dfrac{1}{2}\right)^4}$$

$$= 1.6$$

The area of the aperature of the counter is 4 sq. cm
$$= 4 \times 10^{-4}\ \text{m}^2$$
$\therefore$ The number of α-particles entering the counter
$$= 1.6 \times 4 \times 10^{-4} = 6.4 \times 10^{-4}/\text{sec}$$

No = 10^4/sec.

n(number of atoms in the foil per unit volume)

$$= \dfrac{\text{Avogadro number}}{\text{Atomic weight}} \times \text{density}$$

$$= \dfrac{6.023 \times 10^{23}}{197} \times 19.3$$

$$= 5.9 \times 10^{22}\ \text{per cm}^3$$

$$= 5.9 \times 10^{28}/\text{m}^3$$

$t = 10^{-5}$ cm $= 10^{-7}$ m

r(distance of the counter from the foil) = 0.25 cm

$Z = 79$

$E_k = 5.5 \times 1.6 \times 10^{-13}$ J

$\phi = 60^\circ \quad \therefore\ \sin\dfrac{60^\circ}{2} = \dfrac{1}{2}$

Example 1.4 A thin (1 mg/cm²) target of ^{48}Ca is bombarded with a 10 nA beam of α-particles. A detector subtending a solid angle of 2×10^{-3} steradians, records 15 protons per second. If the angular distribution is measured to be isotropic, determine the total cross-section (in mb) for the ^{48}Ca (α, p) reaction. Take the atomic mass of ^{48}Ca = 48 u.

Sol. Since $\sigma(\theta)$ is isotropic, the total number of protons emitted into the full 4π solid angle is

$$\dfrac{15 \times 4\pi}{(2 \times 10^{-3})} = I. \text{ Using}$$

$$I = \dfrac{I_0\, \sigma\, m\, N_A}{A}$$

We have
$$\sigma = \frac{I_A}{(I_0 \, m \, N_A)}$$

where m is the mass of the target in units of mass per unit area. Substituting values, one obtains $\sigma = 241$ mb.

1.2(d) Rutherford's Model of the atom

The experimental verification of Rutherford's scattering of α-particles by Geiger and Marsden provide valuable support for Rutherford's model of the nuclear atom. According to Rutherford model of the nuclear atom, the entire positive charge and almost the entire mass of the atom are concentrated within a sphere of radius $R(\approx 10^{-10}$ m).

Rutherford's theory gives us some idea of nuclear dimensions. We have assumed Coulomb's inverse square law in deriving Rutherford's scattering formula and the formula has been confirmed experimentally for values of θ between 5° and 150°. It appears that the square law is valid for distance very much less than the diameter of an atom, which is of the order of 10^{-10} m. This means that the atomic nuclei are point particles, i.e. their dimensions are insignificant compared with the minimum or closest approach will be made by an α-particle approaching head-on to the nucleus. Let us estimate the distance of closest approach d when its impact parameter b is zero. At the instant of *closest approach* the initial kinetic energy (T) of the particle is entirely converted to electrostatic potential energy and so at that instant

$$K_\alpha = \frac{1}{4\pi\varepsilon_0} \frac{(Ze)(2e)}{d_m}$$

or
$$d_m = \frac{1}{4\pi\varepsilon_0} \frac{2Ze^2}{K_\alpha} \qquad \qquad \dots (1.10)$$

For α-particles from ^{222}Rn isotopes having the energy $K_\alpha = 5.486$ MeV, one obtains $d_m = 2.47 \times 10^{-14}$ m for silver ($Z = 47$) and $d_m = 4.15 \times 10^{-11}$ m for gold ($Z = 79$). Obviously, the radius of the atomic nucleus must be smaller than the above estimates, i.e. *nuclear radii* are smaller by factors of the order of 10^4 to 10^5 compared to the atomic radii. The above discussions clearly reveals that Coulomb's law of force between the scattering nucleus and the α-particle holds at distances of the order of 10^{-14} m. Obviously, the *nuclear radius* for the metals studied is of the order 10^{-14} m. It is appropriate at this point to ask what we really mean by "the radius" of a nucleus? The nuclear radius or *"nuclear force radius" represents the distance from the center of the nucleus at which an external uncharged nucleon first feels its nucleon. We must however, use the concept of nuclear radius with some care.*

In more recent years, with more accurate techniques it is found that the radius of an atomic nuclei lies within the range 10^{-14} to 10^{-15} m. The radius of lightest nucleus (hydrogen) is of the order of 1×10^{-15} m and that of the heaviest nucleus (U) is 7.7×10^{-15} m.

A nucleus has no clearly defined boundary. Hence nuclear radius has only an arbitrary boundary.

The experimental result is that the nuclear radius* is proportional to $A^{\frac{1}{3}}$, where A is the mass number of the nucleus, i.e.,

$$R = r_0 \, A^{\frac{1}{3}} \qquad \text{... (1.11)}$$

where r_0 is a constant, the same for all nuclei. Its accepted value is about 1.07×10^{-15} m.

One can easily see that the gravitational force is completely negligible inside the nucleus. Since a nucleus consists of protons and neutrons and since protons are positively charged, the question then arises as to how all these protons stay together in a nuclear size given by Equation (1.11)? Clearly inside the nucleus, over distances of about 10^{-14} m, there is specifically a *nuclear interaction* operating between protons and neutrons (nucleons) which is strong enough to overcome the Coulomb repulsion between the closely-packed protons. We shall study the nuclear forces later on.

Nuclear Density The volume of the nucleus is proportional to the number of nucleons it consists of. The density of matter is constant for all nuclei. Since the volume of a sphere is $\frac{4}{3}\pi \, R^3$, we conclude from (1.11) that the nuclear volume is

$$V = \frac{4}{3}\pi \, r_0^3 A = 1.12 \times 10^{-45} \, A/m^3 \qquad \text{... (1.12)}$$

Obviously, the volume of the nucleus is proportional to the number of nucleons A inside the nucleus. This suggests that the nucleons are maintained at fixed average distances, independent of the number of particles so that the *volume per nucleon is a constant quantity, the same for all nuclei.*

The mass of a nucleus of mass number A is approximately $M = 1.66 \times 10^{-27} A$ kg. Therefore the average density of nuclear matter is

$$\rho = \frac{M}{V} = \frac{1.66 \times 10^{-27} \, A \text{ kg}}{1.12 \times 10^{-45} \, A - m^3} = 1.49 \times 10^{18} \text{ kg}/m^3$$

Which is independent of A.* Obviously, nuclear density is about 10^{15} times greater than the density of matter in bulk, gives us an idea of the degree of compactness of the nucleons in a nucleus. It also shows that matter in bulk is essentially empty, since most of the mass is concentrated in the nuclei. However, we must note that the supposition, that the nucleus is spherical is not always true.

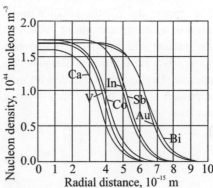

Fig. 1.8(a) Nuclear density as function of radial density.

*The density of nuclear matter varies with the distance from the centre of the nucleus as indicated in Figure 1.8(a). The density is approximately constant for a considerable distance from the center and then gradually decreases down to zero near the surface of the nucleus. The nuclear radius R may then be defined as the distance from the center of the nucleus at which the nuclear density has been reduced to one-half.

Example 1.5 Calculate the distance of closest approach of α-particles to the copper nucleus ($Z = 29$), when α particles of 5 MeV energy are scattered back by a thin sheet of copper. What would be the closest approach of a proton of the same energy?

Sol.
$$d = \frac{1}{4\pi\,\varepsilon_0}\,\frac{2Ze^2}{E_K}$$

or
$$d = \frac{2\times29\times(1.6\times10^{-19})^2}{8\times10^{-13}}$$

$$= 1.67\times10^{-14}\text{ m}$$

$$\frac{1}{4\pi\,\varepsilon_0} = 9\times10^9 \text{ N} - \text{m}^2/\text{C}^2$$

$$e = 1.6\times10^{-19}\text{ C}$$

$$E_k = 5\text{ MeV} = 8\times10^{-13}\text{ J}$$

$$Z = 29$$

(ii) The positive charge on the proton is e, i.e. half of that on α-particle. Therefore the electrostatic potential energy at the instant of closest energy is $\dfrac{1}{4\pi\,\varepsilon_0}\,\dfrac{(Ze)e}{d}$

$$\therefore \quad d = \frac{1}{4\pi\,\varepsilon_0}\,\frac{Ze^2}{E_K} = \frac{1.67\times10^{-14}}{2} = 0.835\times10^{-14}\text{ m}$$

Example 1.6 If the positive charge of the gold atom ($Z = 79$) is supposed to be spreaded uniformly over a spherical surface of diameter 1 A°, show that α-particles of energy greater than a certain value E will not be reflected back. Calculate E.

Sol. The distance of closest approach for an α-particle of initial kinetic energy E_K is

$$d = \frac{1}{4\pi\,\varepsilon_0}\,\frac{2Ze^2}{E_K}$$

The greater the E_K, the smaller the distance of closest approach d. If, however, the energy E_K exceeds a certain maximum value E, the α-particle approaches the nucleus so much closely that the nuclear size becomes comparable with the distance of the particle from it and the inverse-square law breaks down. Besides this, some short-range attractive nuclear forces also begin to operate. Hence under this condition the α-particle is not scattered back. The distance of closest approach of a particle of maximum kinetic energy E gives the upper limit of the radius of the nucleus.

$$\therefore \quad r_0 = \frac{1}{4\pi\,\varepsilon_0}\,\frac{2Ze^2}{E}$$

or
$$E = \frac{1}{4\pi\,\varepsilon_0}\,\frac{2Ze^2}{r_0} = \frac{9\times10^9\times2\times79\times(1.6\times10^{-19})}{0.5\times10^{-10}}$$

$$= 7.3\times10^{-16}\text{ J}$$

Example 1.7 A beam of protons and α-particles each of energy 15 meV is incident on a thin foil of tin ($Z = 50$). It is observed that the α-particles are scattered backward but not the protons.

Explain this observation and find the maximum possible value of the nuclear radius. Explain why electrons are not effective in scattering α-particles and protons ($e = 1.6 \times 10^{-19}$ C).

Sol.
$$d = \frac{1}{4\pi\,\varepsilon_0^2}\,\frac{2\,Ze^2}{E_K}$$

The particle on reaching at the distance of closest approach (d) from the nucleus is reflected back (scattered by 180°).

We have
$$d = 9 \times 10^9 \times \frac{2\times 50 \times (1.6\times 10^{-19})^2}{15\times 10^6 \times 1.6\times 10^{-19}}$$

$$= 9.6 \times 10^{-15}\ \text{m}$$

Obviously, 9.6×10^{-15} m is the maximum possible value of the radius of the tin nucleus. The charge on the proton is $+e$, i.e. half that of the charge on α-particle. Obviously, a proton of 15 meV energy will have $d = 4.8 \times 10^{-15}$ m. This means that the proton of energy 15 meV will penetrate the nucleus and will not be scattered back.

The electrons, because of their little mass are unable to deviate the α-particle and protons appreciably.

Example 1.8 Calculate the density of $^{12}_{6}$C nucleus. Given mass of $^{12}_{6}$C $= 12\ u$, $R = 2.7 \times 10^{-15}$ m.

Sol.
$$\rho = \frac{m}{\frac{4}{3}\pi\,R^3} = \frac{12\,u \times 1.66\times 10^{-27}\ \text{kg/u}}{\frac{4}{3}\pi\times (2.7\times 10^{-5})^3}$$

$$= 2.4 \times 10^{17}\ \text{kg/m}^3$$

The above figure is just equivalent to 4 billion tons per cubic inch and is essentially the same for all nuclei. The density of neutron stars is comparable with this figure.

1.3 CHADWICK'S METHOD OF MEASURING THE CHARGE ON THE NUCLEUS

Chadwick in 1920 adapted the experimental methods of Geiger and Marsden and used the Rutherford's theory of α-particle scattering to measure the charge Ze on the nucleus for the elements copper, silver and platinum. These experiments not only provided a further verification of the fact that the scattering was proportional to square of the atomic number Z, they also provided the only direct method of determining the nuclear charge. In the experiments, it was essential to measure the ratio of the number of α-particles scattered through a known angle ϕ, to the total number of incident particles at the scattering foil. To increase the number of α-particles deviated throught a certain angle, and so increase the accuracy of the determination, he used a metallic foil in the form of a ring, with the source of α-particles and the scintillation screen placed symmetrically on either side of this foil (Figure 1.9)

The annular scattering foil *FF* was mounted on a metal support. α-particles from the source *S* incident upon *FF* were confined to the region between cones of apex semi-angles $\frac{1}{2}\phi_1$ and $\frac{1}{2}\phi_2$ by the aperatures in the metallic diaphragm *D*. The fluorescent screen *S'* was placed beyond the foil F so that its perpendicular distance from the centre of the foil support was the same as the distance S from this centre. A lead screen L prevented the α-particles from direct falling on the detector from the source. The S' was just outside the vacuum chamber and α-particles reached it via a thin mica window in the chamber wall immediately opposite S'.

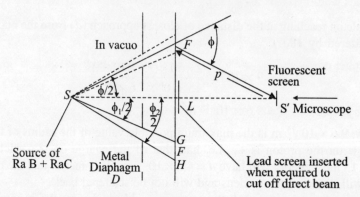

Fig. 1.9 Chadwick's experiment for the direct determination of Nuclear charg

Suppose N_0 is the number of α-particles emitted per second by the source S. Out of this, a fraction $\dfrac{\Omega}{4\pi}$ will reach the ring shaped foil F, where Ω is the solid angle subtended by a ring-shaped element at S by the ring.

The solid angle $d\Omega$ subtended by a ring shaped element at F is readily seen to be given by

$$d\Omega = 2\pi \sin \frac{\phi}{2} d\left(\frac{\phi}{2}\right)$$

where the angle $F\,S\,S'$ is $\dfrac{\phi}{2}$.

The number of α-particles incident per second on this elemental ring is therefore

$$N_0 \frac{d\Omega}{4\pi} = N_0 \frac{2\pi \sin \dfrac{\phi}{2}\, d\dfrac{\phi}{2}}{4\pi} = N_0 \frac{\sin \dfrac{\phi}{2}\, d\dfrac{\phi}{2}}{2} \qquad \ldots (1.12)$$

Using (1.8) the number of these particles scattering through the angle ϕ to reach S' can be calculated. In doing so, one must remember that Equation (1.8) was derived for normally incident particles. However, in the present case, the α-particles are incident at an angle $\dfrac{\phi}{2}$ to the normal. One can take this into account by replacing t in (1.8) by $t \sec \dfrac{\phi}{2}$, because the effective thickness of the foil is increased by the factor $\sec \dfrac{\phi}{2}$ on account of the off-normal incidence. Furthermore in deriving Equation (1.8) the fluorescent screen was considered to be perpendicular to the arriving α-particles.

In this experiment, the screen S' is at an angle $\frac{\phi}{2}$ to the particles arriving along r. The number of scintillations per unit area of the screen is observed. Unit area of a screen presented normally to incident α-particles would present an area of $\frac{\phi}{2}$ to particles incident at an angle of $\frac{\phi}{2}$. Obviously, the correction required is to multiply by $\cos\frac{\phi}{2}$.

Thus the number of α-particles dN scattered to unit area of the screen at S' in this experiment can be obtained by replacing N_0 in Equation (1.8) by $N_0\dfrac{\sin\frac{\phi}{2}\,d\left(\frac{\phi}{2}\right)}{2}$. Replacing t by $t\sec\frac{\phi}{2}$ and then multiplying by $\cos\left(\frac{\phi}{2}\right)$, one obtains

$$dN = \frac{\frac{1}{2}N_0 \sin\frac{\phi}{2}\,d\frac{\phi}{2}\cdot nt\sec\frac{\phi}{2}\cdot b^2\,\mathrm{cosec}^4\frac{\phi}{2}\cdot\cos\frac{\phi}{2}}{16\,R^2}$$

$$= \frac{N_0\,nt\,b^2\,\mathrm{cosec}^3\frac{\phi}{2}\,d\left(\frac{\phi}{2}\right)}{32\,R^2}$$

For the entire scattering foil F, the angular limits GSS' and HSS' equal to $\frac{1}{2}\phi_1$ and $\frac{1}{2}\phi_2$ respectively. Obviously, the total number of α-particles scattered to S' will be N, i.e.

$$N = \frac{N_0\,nt\,b^2}{32\,\bar{R}^2}\int_{\phi_1/2}^{\phi_2/2}\mathrm{cosec}^3\frac{\phi}{2}\,d\frac{\phi}{2}$$

Where $\bar{R}$ is the mean value of SG and SH. Evaluating the integral, one obtains

$$\frac{N}{N_0} = \frac{nt\,b^2}{64\,\bar{R}^2}\left(\log\tan\frac{\phi_2}{4} - \log\tan\frac{\phi_1}{2} + \cot\frac{\phi_1}{2}\,\mathrm{cosec}\frac{\phi_1}{2} - \cot\frac{\phi_2}{2}\,\mathrm{cosec}\frac{\phi_1}{2}\right) \quad \dots (1.13)$$

The number of α-particles which would fall directly on unit area of the fluorescent screen S' (i.e. lead screen L removed) would be $\dfrac{N_0}{4\pi l^2}$, where $l = SS'$. Hence both N and N_0 can be found, ϕ_1, ϕ_2 and $\bar{R}$ are known, $\dfrac{e}{m}$ for the α-particle is known, and v, the velocity of the α-particles from a RaB plus RaC source, can be found. Thus the charge on the nucleus (Ze) of the foil material concerned, can be evaluated.

Chadwick determined Z for a number of metallic elements. His results for copper, silver and platinum are given below and are compared with their atomic numbers.

Table 1.3 Results of Chadwick's Experiments

Element	Atomic number Z	Charge Z (determined experimentally by Chadwick)
Cu	29	29.3 ± 0.5
Ag	47	46.3 ± 0.7
Pt	78	77.4 ± 1

Chadwick's results are not precise enough to determine unique, integral values of Z, but they agree well with the values for three elements, obtained by an entirely independent method. Thus all four tests of Rutherford α-scattering theory were met successfully and constitute the earliest experimental evidence for the nuclear model of the atom.

1.4 GENERAL CHARACTERSTICS OF ATOMIC NUCLEI IN THEIR GROUND STATES

The atomic nucleus, like the atom itself, is a bound quantum system and hence can exist in different quantum states characterized by their energies, angular momenta etc. The lowest energy state is called as the ground state and normally the nuclei exist in this state. The properties of the nuclei corresponding to their *ground state* are usually termed as their *static* properties in contrast to the dynamic characterstics of the nuclei which are exhibited in the processes of nuclear decay, nuclear reactions, fission, fusion and nuclear excitation. The important static properties of the nuclei include their mass, electric charge, binding energy, size, shape, angular momentum, magnetic dipole moment, electric quadrupole moment, statistics, parity and ison-spin. We shall discuss these static characterstics of nuclei in this section.

1.4.1 Nuclear Mass and Binding Energy

The atomic nuclei are made of two types of elementary particles, *protons* and *neutrons*. Protons and neutrons are designated by the common name of *nucleons*. The hydrogen nucleus is called the *proton*. Protons (that is, ionized hydrogen) were observed as such in discharge tubes by Wilhelm Wien in 1898. In 1919, Ernest Rutherford observed protons coming out of the nitrogen nucleus. Proton carries one electronic unit of positive charge ($+e$) and has a mass about 1836 times the electronic mass (m_e). The neutron, another nuclear constituent is electrically neutral and slightly heavier than the proton (Table 1.4). The protons and the neutrons are held together inside the nucleus by very strong short range attractive force as *nuclear interaction*. The nuclear force is short range force and it is different from the more commonly forces, e.g. gravitational, electrical forces.

Table 1.4 Rest Masses of Particles

Particles	Rest mass (kg)	Rest mass (u)	Rest energy (eV)
Electron (m_e)	9.10953×10^{-31}	5.48580×10^{-4}	5.11003×10^{5}
Proton (M_p)	1.67265×10^{-27}	1.0072765	9.38280×10^{8}
Neutron (M_n)	1.67495×10^{-27}	1.0086650	9.39573×10^{8}

We identify a nucleus by the number of protons it has, or its *atomic number Z*, and by the total number of particles or *nucleons* it has, called its *mass number A*. Thus the number of neutrons is $N = A - Z$. The term *nuclide* has been introduced to designate in a general way, all nuclei having the same Z and N, and hence also the same A. In other words, in the same way that all atoms with the same Z belong to the same element, all nuclei of the same composition (the same Z and N) belong to the same nuclide. A nuclide is designated by the symbol of the chemical element to which it belongs, according to the value of Z, with a subscript to the left indicating the value of the mass number, such as ^{12}C, ^{23}Na, ^{107}Ag, ^{238}U. Some times it is convenient to write the atomic number Z explicityly, in which case a subscript to the left is added $^{12}_{6}C$, $^{23}_{11}Na$, $^{107}_{47}Ag$, $^{238}_{92}U$. In gereral a nucleus of an atom X of atomic number Z and mass number A is symbolically expressed as $^{A}_{Z}X$.

There are 92 natural chemical elements and 12 more have been produced artificially. There are about 1440 different known nuclides, some 340 existing in nature and about 1100 produced in the laboratory. Some of the nuclides (about 280 of them) are stable, but a large number are unstable or radioactive. Figure 1.10 indicates most of the known nuclides. The stable nuclides are indicated by the black squares and the unstable or radioactive nuclei by the other symbols.

One can observe from Figure 1.10 that, in light nuclei, neutrons and protons tend to be in equal numbers ($N \approx Z$), which indicates the independence of the nuclear interaction from the electric charge. However, in heavier nuclei the number of neutrons exceeds that of protons ($N > Z$). There must be an excess of neutrons to produce a stabilizing effect (through nuclear interactions); these extra neutrons balance the disrupting effect of the Coulomb repulsion between protons.

Because of great variety of nuclides, one can classify them in three categories: isotopes, isotones and isobars.

Isotopes   These are nuclides having the same number of protons but different number of neutrons. Obviously, they have the same atomic number Z but different neutron number N and different mass number A. Since a chemical element is identified by its atomic number Z, all isotopes corresponding to a given value of Z belong to the same element. For e.g. $^{1}_{1}H$ and $^{2}_{1}H$ are isotopes of hydrogen. Isotopes fall along vertical lines in Figure 1.10.

Isotones   These are nuclides having the same number of neutrons N but a different atomic number Z (or number of protons), and therefore also a different mass number A. Isotones having a given value of N obviously do not all correspond to the same chemical element. For examples, $^{13}_{6}C$ and $^{14}_{7}N$ are isotones. Both nuclei have $N = 7$. Isotones fall along horizontal lines in Figure 1.10.

Isobars   These are nuclides which have the same total number of nucleons (or same mass number A), but which differ in atomic number Z and also in neutron number N. $^{14}_{6}C$ and $^{14}_{7}N$ are examples of isobars. Isobars fall along the 45° lines shown in Figure 1.10.

Some Chemical elements occur in nature with one variety of nucleus or isotope. $^{19}_{9}F$ is an example. Others have several natural isotopes, such as tin which has ten, and carbon which has three: $^{12}_{6}C$, $^{13}_{6}C$ and $^{14}_{6}C$. In addition, four more isotopes of carbon have been produced artificially.

The analysis of the properties of isotopes, isotones, and isobars is improtant in that it discloses several features of nuclear structure. Such studies are helpful in predicting that what will happen to the stability of a nucleus when an extra neutron or proton is added to it, or to decide which configuration of neutrons and protons will be the most stable etc.

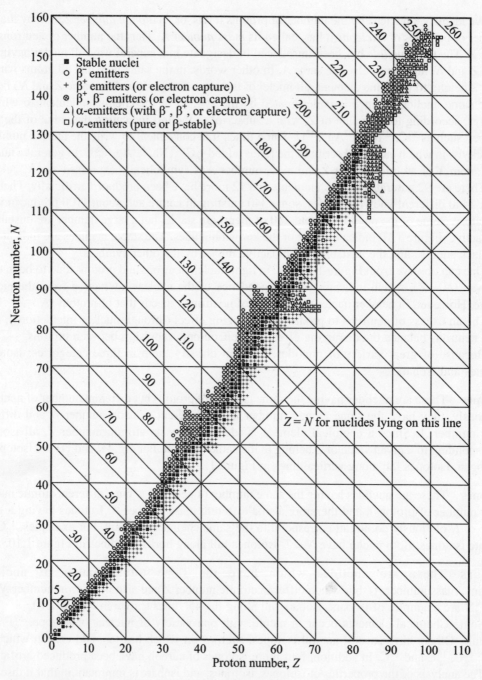

Fig. 1.10 Stable and radioactive nuclides. The 45° lines are lines of equal mass number A.

Atomic mass Unit (amu)

The atomic masses are measured in relative units or atomic mass units. The unit of atomic mass is defined to be $\frac{1}{12}$ of the mass of ^{12}C atom taken to be exactly 12 units and is designated by the symbol 'u' (unified atomic mass unit). This unit of atomic mass has been in use since 1961 by scientists (physicists and chemists) by international agreement. Prior to 1961, the atomic mass unit used by scientists were 1/16th of the mass of ^{16}O (taken to be exactly 16 units) and was called the atomic mass unit (amu). The conversion factor from unified atomic mass scale to the atomic mass unit scale is

$$1\,u : 1\,\text{amu} = 1.003172 : 1$$

The atomic mass unit used by chemists prior to 1961, was 1/16th of the average atomic weight of the natural oxygen consisting of three isotopes ^{16}O, ^{17}O and ^{18}O having the relative abundances 99.76%, 0.04% and 0.20% respectively.

One can easily obtain the absolute value of the unit of atomic mass in ^{12}C scale. We know that 1 mole of ^{12}C has the mass of 12 g or 12×10^{-3} kg. Now, 1 mole contains N atoms, where $N = 6.02205 \times 10^{23}$ is the Avogadro number. Obviously, the mass of each ^{12}C atom is

$$12 \times \frac{10^{-3}}{N} \text{ or } 12 \times 1.660566 \times 10^{-27} \text{ kg}$$

$\therefore$ The unit of atomic mass in ^{12}C scale is $1\,u = \frac{1}{12} \times \frac{12 \times 10^{-3}}{N} = 1.660566 \times 10^{-27}$ kg

By using the relation $E = mc^2$, one can express the energy equivalent of $1\,u$ in terms of Joules or electron volts. We have

$$\begin{aligned}1\,u &= 1.660566 \times 10^{-27}\,c^2 \\ &= 1.660566 \times 10^{-27} \times 3.98755 \times 10^{16} \\ &= 14.924427 \times 10^{-11} \text{ J} \\ &= \frac{14.924427 \times 10^{-11}}{1.60219 \times 10^{-13}} = 931.502 \text{ MeV}\end{aligned}$$

or about one billion eV (GeV).

The adoption of the ^{12}C scale for atomic masses put an end to the existence of the chemical and physical atomic mass units.

We must note that accurate determination of atomic masses is essential for the determination of nuclear binding energies and in the calculation of nuclear disintegration energies. Modern mass spectroscopies helps accuracies better than one part in a million. Let us consider an emample of α-disintegration of a heavy element like ^{226}Ra ($Z = 88$). These nuclei spontaneously disintegrate by the emission of α-particles of a few MeV energy (kinetic). This energy of α-particles is derived from the conversion of a part of the mass of the disintegrating parent nucleus into energy according to $E = mc^2$ relation. The α-disintegration of ^{226}Ra nucleus is expressed as

$$^{226}Ra \rightarrow ^{222}Rn + ^{4}He$$

where the masses of the different atoms taking part in the above process are

$$M(^{226}\text{Ra}) = 226.025436 \ u,$$

$$M(^{222}\text{Rn}) = 222.017608 \ u, \text{ and}$$

$$M(^{4}\text{He}) = 4.0026034 \ u$$

One can see that the α-disintegration energy in the above process is

$$Q_\alpha = [M(^{226}\text{Ra}) - M(^{222}\text{Rn}) - M(^{4}\text{He})]\,c^2$$

$$= [226.025436 - 222.017608 - 4.002603] \times 931.502$$

$$= 4.87 \ \text{MeV}$$

We can easily see that the disintegration energy in the above process is less than one part in 40,000 of the mass of the disintegrating nucleus. This clearly reveals that unless the masses of the atoms are determined with accuracies much better than the above, one can not correlate the measured disintegration energy with the change in mass due to the disintegration.

We must remember that the study of nuclear disintegration energies provides direct experimental evidence in support of $E = mc^2$ relation.

Mass Excess, Packing Fraction and Binding Energy

We have seen that accurate determination of atomic masses expressed on ^{12}C scale shows that these are very close to whole numbers, i.e. mass numbers of the atoms. This is also true when the atomic masses are expressed in 16O scale.

The masses of few atoms on ^{12}C scale are listed in Table 1.5. We must remember that the mass of ^{12}C on this scale is 12. The masses of all other atoms, though close to the corresponding mass numbers (integral), differ slightly from the latter.

From Table 1.5 it is clear that very light atoms with $A < 20$ and for very heavy atoms with $A > 180$, the masses are slightly greater than the corresponding mass numbers. In between the above values of A, we note that the atomic masses are slightly less than the corresponding mass numbers.

Table 1.5

Particle or Atom	Atomic mass (u)	Mass defect (u)	Packing Fraction (u)
Neutron (^{1}n)	1.008665	+0.008665	—
Proton (^{1}H)	1.007825	+0.007825	—
^{2}H	2.014102	+0.014102	+0.007051
^{4}He	4.002603	+0.002603	+0.00006507
^{12}C	12	0	0
^{16}O	15.994915	−0.005085	−0.0003178
^{31}P	30.973764	−0.026236	−0.008463
^{79}As	74.921597	−0.078403	−0.0010454
^{197}Au	196.96654	−0.03346	−0.0001698
^{238}U	238.05082	+0.05082	+0.0002135

The difference between the measured atomic mass $M(A, Z)$ from the mass number A is quite significant and is known as the *mass defect* or *mass excess* Δ:

$$\Delta = M(A, Z) - A \qquad \qquad \text{... (1.1)}$$

For example, let us consider the $_2^4$He. Its atomic mass 4.002603 u is slightly greater than the mass number 4. Obviously, its mass defect is +0.002603 u. On the other hand ^{127}I has the atomic mass 126.90497 u, which is slightly less than the mass number 127. Its mass defect is −0.09553 u. This clearly reveals that mass defect can be both positive and negative. From Table 1.5 it is evident that mass defect is positive for very light and very heavy atoms whereas it is negative in the intermediate region.

Aston in 1927, expressed the deviations of the masses from mass number in the form of a packing fraction for each nuclide, given by

$$\text{Packing fraction} \qquad f = \frac{M(A, Z) - A}{A}$$

$$= \frac{M(A, Z)}{A} - 1 \qquad \qquad \text{... (1.2)}$$

Also $\qquad M(A, Z) = A(1 + f) \qquad \qquad \text{... (1.3)}$

The packing fraction is defined as the mass defect per nucleon in the nucleus f is very small quantity. Aston multiplied the values of f by 10^4 to obtain figures which are more convenient to record. Thus usually the quoted values of packing fractions are expressed as

$$f = \frac{\Delta}{A} \times 10^4 \qquad \qquad \text{... (1.4)}$$

It is observed that the packing fraction f varies in a systematic manner with the mass number A. The variation of packing fraction with mass number for a number of nuclides is shown in Figure 1.11.

The packing fraction, with the exception of those for ^{4}He, ^{12}C and ^{16}O fall on or near the smooth curve. The values are high for elements of low mass numbers. From the graph it is evident that the values of packing fractions at first fall rapidly with increasing mass number pass through a relatively flat minimum and then rise, slowly but steadily, for the heavy elements beyond $A = 180$. The knowledge of packing fraction was very useful in the study of isotopic masses and also support the view that the atomic nucleus is built up from an integeral number of fundamental particles. However, packing fractions do not have a precise theoretical significance and precise physical meaning.

The systematic variation of f with A (Figure 1.11) can be understood from nuclear binding energy considerations. One can also get an idea of the strength of the nuclear interaction by

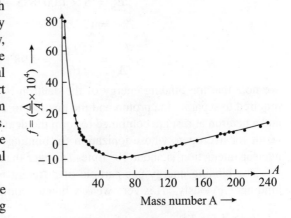

Fig. 1.11 Variation of packing fraction as a mass number.

studying the binding energy of nuclei. *Binding energy of a nucleus is the energy required to separate the nucleons composing a nucleus.* The mass of a nucleus is appreciably less than the sum of the masses of its component nucleons.

For a nucleus of mass M composed of A nucleons, of which Z are protons and $A - Z$ are neutrons, the binding energy (i.e., the energy required to separate all nucleons) is

$$E_b = [Z\,M_p + (A - Z)M_n - M(A, Z)]c^2 \qquad \ldots (1.5)$$
$$= 931.502[Z\,M_p + (A - Z)M_n - M(A, Z)] \text{ MeV}$$

A good indication of the stability of a nucleus is the average binding energy per nucleon, $\dfrac{E_b}{A}$. Let us estimate the values of $\dfrac{E_b}{A}$ for a few typical cases:

(a) *B.E. of Deutron* $(Z = 1, N = 1)$

$$E_b(^2\text{H}) = M_H + M_n - M(A, Z)$$
$$= (1.007825 + 1.008665 - 2.014102) \times 931.5$$
$$= 2.224 \text{ MeV}$$

$\therefore \qquad \dfrac{E_b}{A}(^2\text{H}) = 1.112$ MeV per nucleon.

(b) *B.E. of α-perticle* $(Z = 2, N = 2)$

$$E_b(^4\text{He}) = (2 \times 1.007825 + 2 \times 1.008665 - 4.002603) \times 931.5$$
$$= 28.3 \text{ MeV per nucleon}$$

$\therefore \qquad \dfrac{E_b}{A}(^4\text{He}) = \dfrac{28.3}{4} = 7.075$ MeV per nucleon.

(c) *B.E. of ^{16}O* $(N = 8, Z = 8)$

$$E_b = (8 \times 1.007825 + 8 \times 1.008665 - 15.994915)931.5$$
$$= 127.62 \text{ MeV}$$

$\therefore \qquad \dfrac{E_b}{A} = \dfrac{127.62}{16} = 7.98$ MeV per nucleon

We note that the binding energy of the deutron is $E_b = 2.224$ MeV. This is therefore the energy required to separate the proton and the neutron in a deutron, or the energy released when a proton and a neutron at rest are combined to form the deutron. Comparing this energy with the equivalent result for hydrogen, whose ionization (or binding) energy is only 13.6 eV, we conclude that the nuclear interaction is about 10^6 times stronger than electromagnetic interactions.

The binding energy per particle of different nuclei represent the relative strengths of their binding. Obviously, ^{2}H is very weakly bound, compared to ^{4}He or ^{16}O. The nature of variation of $\dfrac{E_b}{A}$ with A for different nuclei is shown graphically in Figure 1.12.

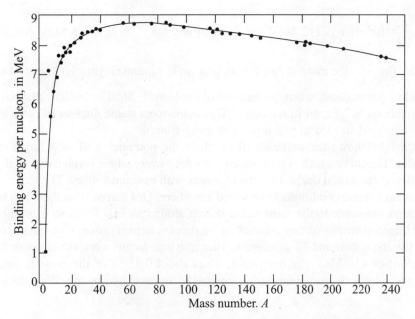

Fig. 1.12 Binding energy per nucleon as a function of mass number.

A careful analysis of graph in Figure 1.12 reveals the following facts:

(i) The binding energies of some of very light nuclides, such as ^{2}H are very small.

(ii) For nuclides having $A < 28$, there exists cylic recurrence of peaks corresponding to those nuclides whose mass numbers are multiple of four and contain equal number of protons and neutrons, i.e. the binding energies of ^{4}He, ^{8}Be, ^{12}C, ^{16}O etc. are considerably greater than those of their neighbours (Figure 1.13). This indicates that these nuclei are much more stable than their immediate neigbours. Similar, but less prominent peaks are observed at the values of Z or N equal to 20, 28, 50, 82 and 126. These are known as *magic numbers*.

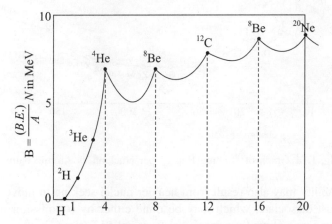

Fig. 1.13 Average binding energy for light elements in the periodic table

(iii) The $\dfrac{E_b}{A}$ for ^{2}H is 1.112 MeV per nucleon and for ^{209}Bi it is 7.8 MeV per nucleon. The

greater the $\dfrac{E_b}{A}$, the more stable the nucleus is. The graph (Figure 1.12) has its maximum at
8.8 MeV per nucleon, when the number of nucleons is 56. The nucleus that has 56 protons
and neutrons is $^{56}_{26}$Fe, an iron isotope. This is the most stable nucleus of them all, since the
most energy is needed to pull a nucleon away from it.

(iv) Figure 1.14 shows the variation of depth in the potential well as a function of atomic
number. The curve which is millions of volts deep every where is relatively shallow for light
and heavy nuclei and deepest for the elements with medium masses. The latter are therefore
in the most stable conditions to be found anywhere. One can see that there is a tendency for
nuclei to transmute to the most stable region around nickel. To do so from the side of the
light elements means further associating – a process termed *fusion*. One can see for example,
that for 26 protons and 30 neutrons to fuse into iron means a release of about 8.7 MeV per
nucleon, or 486 MeV per iron atom. Thus about 0.93 % of the mass is converted into
energy. Current nuclear technology involves the fusion of hydrogen or deuterium into tritium
or helium.

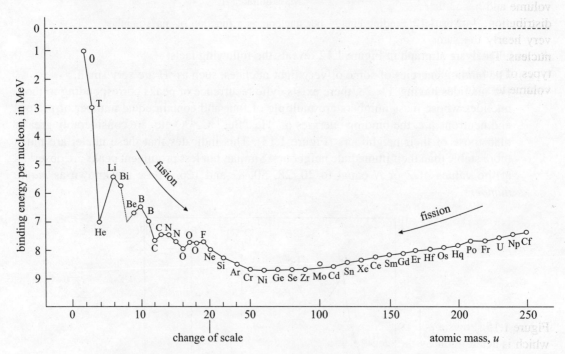

Fig. 1.14 Graph of Binding Energy per Nucleon vs Atomic Number.

Greater nuclear stability may also result from heavier nuclei seeking to move towards the center.
To do so, they must dissociate, which may be done either by α-emission or by breaking into
fragments including a pair of medium mass a process called *fission*. One can see that in a fission

reaction taking say uranium to the vicinity of medium masses (fission fragments usually are centered in two regions: about mass 100 and about mass 136), the binding energy per nucleon changes from about 7.9 MeV to 8.7 MeV. The energy release is about 0.8 MeV per nucleon. Thus about 0.09 percent of the mass is converted to energy. Although not efficient as fusion, the fission process is a remarkable source of energy. Recent theory indicates that just before fission the nucleus oscillates like a liquid drop, splitting into two droplets.

The binding energy curve has a good claim to being the most significant. The fact that binding energy exists at all means that nuclei more complex than the single proton of hydrogen can be stable. Such stability in turn accounts for the existence of the various elements and so far the existence of the many and diverse form of matter we see around us.

1.4.2 Nuclear size

In section 1.2(d) we have seen that when the shape of the nucleus is assumed to be spherical, its radius is given by Equation (1.11). We must note that the radius given by Equation (1.11) is the radius of nuclear mass distribution. One can also talk about the radius of *nuclear charge distribution*. It is found that the nuclear charge parameter (i.e., the atomic number) Z is almost linearly proportional to the mass number A and nuclear charge density ρ_c is approximately the same throughout the volume and hence the distribution of nuclear charge $+Ze$ also follow the pattern of nuclear mass distribution. *Obviously, the *charge radius* and *mass radius* of the nucleus may be expected to be very nearly the same. This type of behaviour is due to the strong attractive forces within the nucleus. There are strong evidences which clearly show that this is very nearly the same for both types of particles, viz the protons and neutrons. This makes their distributions within the nuclear volume to follow the same pattern.

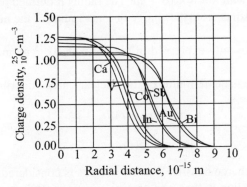

Fig. 1.14(a) Charge density as functions of radial distance

Figure 1.15 shows a potential energy diagram for a charged particle like a proton or an α-particle; which is acted upon by the electrostatic repulsive force of nuclear charge $+Ze$ when it is outside the nucleus ($r > R$), while inside the nucleus ($r < R$) a negative potential due to short range of the nuclear force acts upon it. Here O is the nuclear centre and r is the distance from it. One can

---

*Experiments with very fast electrons and muons suggest a charge distribution as indicated in Figure 1.14(a).

assume that nuclear force becomes zero at the nuclear sorce $r = R$ and electrostatic force is not effective inside the nucleus.

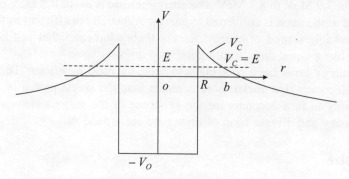

Fig. 1.15 Potential energy diagram for a nucleus.

It is evident from Figure 1.15 that the nucleus is surrounded by a Coulomb potential barrier $V_c = \dfrac{Z\,Z'\,e^2}{4\pi\,\varepsilon_0\,r}$ for an incident particle of charge $Z'e$ for $r > R$. One can see that the barrier height at the nuclear surface is

$$V_R = \frac{Z\,Z'\,e^2}{4\pi\,\varepsilon_0\,R} \qquad\qquad \text{... (1.6)}$$

If we take $r_o = 1.3 \times 10^{-15}$ m, then one obtains $V_R = 33.1$ MeV for an α-particle and 16.5 MeV for proton for the ^{92}U nucleus with $R = 8 \times 10^{-15}$ m. The radius R defined here is usually termed as the *potential radius*. This is distinct from the charge or mass radius of the nucleus and is slightly larger than the latter.

On can define the mean square radius of the nuclear charge distribution as

$$<r^2> = \frac{\displaystyle\int_0^\infty r^2\,4\pi\,r^2\,\rho(r)\,dr}{\displaystyle\int_0^\infty 4\pi\,r^2\,\rho(r)\,dr} \qquad\qquad \text{... (1.7)}$$

Where $\rho(r)$ is the nuclear charge density. When a nucleus is considered to be a uniformly charged sphere ($\rho = $ constant) of radius R, then one obtains (since $\rho = 0$ for $r > R$)

$$<r^2> = \frac{\displaystyle\int_0^R r^4\,dr}{\displaystyle\int_0^R r^2\,dr} = \frac{3}{5}\,R^2$$

Obviously,
$$R^2 = \frac{5}{3}<r^2> \qquad\qquad \text{... (1.8)}$$

1.4.3 Measurment of Nuclear Radius

The methods of measuring nuclear radius can be divided into two groups: (i) *Nuclear Methods*: Based on the study of the range of nuclear forces, where the nucleus is probed by a nucleon or light nucleus and (ii) *Electromagnetic methods*:

(i) Nuclear Methods:

    (a) Neutron scattering

    (b) α-decay

    (c) α-particle scattering

    (d) Isotope shift in line spectra

(ii) Electrical Methods:

    (a) Mesonic X-rays

    (b) Electron scattering

    (c) Coulomb energies of mirror nuclei

We shall restrict to the study of few methods only.

(a) Measurment of charge Radius

(i) **Electron scattering method**   Elastic scattering of high energy electrons by nuclei constitutes the most direct method of measuring the charge radius of the nucleus. This also helps to study the nature of the nuclear charge distribution. This is possible because there is no specifically nuclear force acting on the electrons. We know that only the Coulomb attractive force due to the nuclear charge acts on them. If the de-Broglie wave length of the electrons is reduced compared to the nuclear radius, then one can study many details form the electron scattering experiment.

The de Broglie wave length of a relativistic electron of rest mass m_0, having the total energy $E > m_0 c^2$ is given by

$$\lambda = \frac{ch}{e\left[V\left(V + \dfrac{2m_0 c^2}{e}\right)\right]^{\frac{1}{2}}}$$

Where e being the charge on the electron, $eV = E_K$ is the kinetic energy of the electron. Using the values of c, h, e and m_0, one obtains

$$\lambda = \frac{12.4 \times 10^3}{\left[V\left(V + 1.02 \times 10^6\right)\right]^{\frac{1}{2}}} \ A$$

where V is in Volts. For $E_K = 200$ MeV, $V = 200 \times 10^6$ Volts, one obtains

$$\lambda = 6.19 \times 10^{-5} \ A$$

and

$$\lambdabar = \frac{\lambda}{2\pi} = 10^{-15} \text{ m} = 1 \text{ fm}$$

Obviously, $\lambdabar$ is much smaller than the nuclear radius and hence such particle which interact strongly with electric charges are very useful for probing the nucleus.

R. Hofstadter and coworkers performed the pioneering experiments on the elastic scattering of electrons by nuclei, using the linear accelerator (SLAC), providing electron beam with energy upto 550 MeV. The experimental arrangement is shown in Figure 1.16.

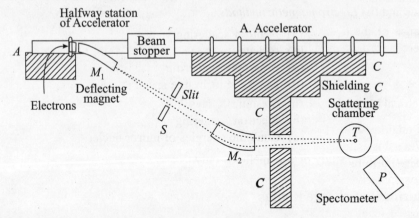

Fig. 1.16 High energy electron scattering experiment used by Hofstadter and coworkers.

The high energy electron beam from the linear accelerator A is deflected by means of deflecting magnet M_1 and collimated by the slit system S. The second deflecting magnet M_2 directs the beam on to the target inside the scattering chamber T. The large magnetic spectrometer P analyses the elastically scattered beam of electrons. The spectrometer can be rotated about an axis through the target so that the intensity of scattered electrons can be observed as a function of θ.

The quantum mechanical expression for the differential scattering cross section of a relativistic electron from a spin-less target at the centre of mass angle θ is expressed as

$$\sigma(\theta) = \sigma_M(\theta) \, [F(q)]^2 \qquad \qquad \text{... (1.1)}$$

Here $\sigma_M(\theta)$ is the mott cross section of elastic scattering from a point charge $+Ze$ and $\sigma(\theta)$ is the scattering cross section. Now,

$$\sigma_M(\theta) = \left(\frac{Ze^2}{8\pi \varepsilon_0 E} \right)^2 \frac{\cos^2 \dfrac{\phi}{2}}{\sin^4 \dfrac{\phi}{2}} \qquad \qquad \text{... (1.2)}$$

Where E is the energy of electrons in the C.M. system. We must note that Equation (1.2) is valid for low Z elements only. Using the Born approximation method, one can show that the *form factor*, $F(q)$ is given by

$$F(q) = \frac{1}{Ze} \int \rho(r) \exp(i\,\mathbf{q} \cdot \mathbf{r})\, dV \qquad \qquad \text{... (1.3)}$$

$$= \frac{4\pi}{Zeq} \int \rho(r) \, (\sin qr) \, r\, d\,r$$

Where
$$q = K - K' = \frac{1}{\hbar}(P - P') \qquad \qquad \dots (1.4)$$

The form factor gives the ratio by which the scattering cross-section is reduced when the charge $+Ze$ is spread out over finite volume. We must note that the form factor $F(q) < 1$, because of the destructive interference between the electron waves scattered from the different parts of the target nucleus.

The dependence of q on the angle of scattering is expressed by

$$|q| = \frac{2p}{\hbar} \sin \frac{\phi}{2} \qquad \qquad \dots (1.5)$$

The angular distribution of elastically scattered 200 MeV electrons is shown in Figure 1.17(a). It is assumed that nuclei have spherical uniform charge distribution (having uniform density up to radius r, shown in Figure 1.17(b)). The actual nuclear charge distribution is not exactly given by *step function*, but is shown by the dotted line.

Figure 1.17(a) clearly reveals that experiments do show diffraction maxima and minima as expected from a uniform charge distribution. Hofstadter and coworkers assumed that the density of charge in the nucleus for theoretical calculations is given by

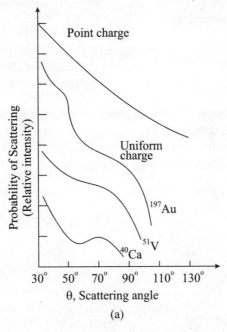

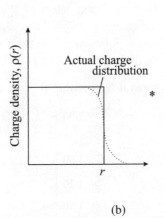

Fig. 1.17(a) Angular distribution (experimental) of scattered electrons from three difference nuclei. A curve for point charge assumption is also shown for the purpose of comparison.

Fig. 1.17(b) Step function exhibiting uniform charge distribution.

$$\rho(r) = \cfrac{\rho_0}{1 + \exp\left[\left(\cfrac{r - \cfrac{R_1}{2}}{a}\right)\right]} \qquad \dots (1.6)$$

This is known as *Fermi-distribution*. The parameters $\dfrac{R_1}{2}$ and a are adjusted to obtain the best fit with the experimental data. The density distribution given by (1.6) has the form shown in Figure 1.18. From Figure 1.18, we have

for $r = \dfrac{R_1}{2}$, $\rho = \dfrac{\rho_0}{2}$, where ρ_0 is the charge

density at the centre ($r = 0$). Obviously $\dfrac{R_1}{2}$ is the half-value radius. The parameter a determines the skin-thickness of the nucleus. Skin-thickness is the thickness in which $\rho(r)$ falls from 0.9 ρ_0 to 0.1 ρ_0 at the nuclear surface. One finds $t = 4.4a$. The charge distribution shown in Figure 1.17 was found to conveniently fit the experimental data involving many nuclei if one takes

$\rho_0 \sim 1.65 \times 10^{44}$ nucleons/m³, $a \sim 0.55$ fermi

and $R \sim 1.07\ A^{\frac{1}{3}}$ fermi.

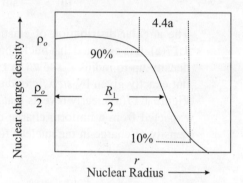

Fig. 1.18 Fermi distribution form for the nuclear charge density represented by Equation (1.6)

By varying the values of R and a, charge distribution could be obtained which would give the charge distribution as shown in Figure 1.17(a).

Figure 1.19 shows that the charge density is essentially same for almost all nuclei except for the lightest nuclei like H and He and decreases slowly for increasing A. It is found that the general relationship $R = r_0\ A^{\frac{1}{3}}$ holds good.

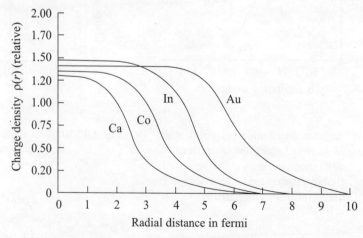

Fig. 1.19 Nuclear charge distribution obtained using high energy electrons as probes.

There are other alternative methods of determing the mean squared redius of the nuclear charge radius. One such method is *mesonic X-rays method*.

(ii) Muonic X-ray method   We will see in subsequent chapters that there is a large number of unstable fundamental particles (both charged and neutral) observed in nature (usually in the cosmic rays) or can be produced in the laboratory in high energy nuclear reactions. One of these particle is muon (previously called the μ-meson). When nuclei are bombarded by high energy particles or photons, then this may evoke the emission of π^+ or π^- mesons from the target nuclei. These π^+ mesons or pions are repalled by nuclei and decay into μ^+-mesons (nuons) as

$$\pi^+ \rightarrow \mu^+ + \nu \ (\tau = 0.02 \ \mu \ sec)$$

μ^+-mesons also decay into positrons as

$$\mu^+ \rightarrow e^+ + \bar{\nu} + \upsilon \ (t = 2.15 \ \mu \ sec)$$

The negative μ^--mesons are also produced in a similar manner. The muons carry one unit of electronic charge and both μ^+ and μ^- are known μ-meson has rest mass about 207 m_e where m_e is the electronic mass. These particles have the same spin and charge as an electron and subjected to the same type of interaction with the nuclei as the electrons. Obviously, only the electrostatic Coulomb force due to the nuclear charge acts on these muons. We must remember that strong nuclear force does not act on these muons.

When a beam of μ^--meson is allowed to pass through matter, some of them are readily captured in electron like Bohr type orbits round the nuclei of the capturing atoms forming a muonic atom (not for μ^+ atoms). The radii of the muonic orbits are much smaller than the electronic orbits, i.e. $\dfrac{m_e}{m_\mu} \sim \dfrac{1}{207}$.

We have the radius of nth electronic orbit as (Bohr's theory)

$$r_e = \frac{4\pi \varepsilon_0 \ n^2 \ \hbar^2}{Ze^2 \ m_e}$$

Where Ze is the charge on the nucleus and m_e is the mass of the electron. Similarly, one can write for the muonic orbit as

$$r_\mu = \frac{4\pi \varepsilon_0 \ n^2 \ \hbar^2}{Ze^2 \ m_\mu} \qquad \qquad ... (1.1)$$

Here, we have assumed that the nuclear charge Ze is concentrated at the centre. Let us calculate the radius of muonic K – orbit ($n = 1$),

$$r_\mu = \frac{m_e}{m_\mu \ Z} \times 0.529 \ A = 3.23 \times 10^{-5} \ A = 3.23 \times 10^{-15} \ m$$

Where the radius of the gold nucleus

$$R(Au) = r_0 \ A^{\frac{1}{3}} = 1.2 \times 10^{-15} \times (197)^{\frac{1}{3}} = 7 \times 10^{-15} \ m$$

Obviously, the muonic K – orbit is much smaller than the radius of gold nucleus. This means the muonic K – orbit may be expected to lie wholly inside the nucleus in the case of heavy atoms.

When the muon is captured by an atom, the μ-meson will fall to states of lower energy, first by means of non-radiatvie transitions (as all the mesonic atomic states are unoccupied) by collisions and by Auger electrons and then by radiative transitions. During the process, in the last few orbit transitions X-rays are radiated. The time for a μ-muon to slow down from about 2 keV and be captured in 1 s atomic orbit has been estimated to be about 10^{-13} sec in graphite. This is much smaller than the life time ($\tau = 2 \times 10^{-6}$ s) of the muon. Obviously, so many mesons may be capable of forming *mesonic-atoms*. Thus the measurement of energies of X-rays obtained from μ⁻-mesons while jumping from one orbit to the other, one can estimate the binding energies of the muon in different orbits. However the binding energy in a particular orbit will be greatly reduced if the nuclear charge is spread over a finite region so that a part of the captured μ⁻ meson wave function lies within the nucleus. Such type of behaviour is expected for the heavier nuclei. Rainwater and Fitch in 1953 allowed to pass a μ⁻ meson beam of well defined momentum from the cyclotron through a piece of 3″ thick Cu. Scintillation counter was used for the detection of X-rays produced in the detector. They used absorbers of different atomic numbers ranging from $Z = 13$ to $Z = 83$. The most probable transition could be a $2p \rightarrow 1$ s transition which give rise to a characterstic line. Figure 1.20 shows a typical X-rays spectrum from μ⁻ masonic atom of titanium. One can see that the characterstic line in the spectrum is at 0.955 MeV.

For Pb atom, the transition $2p_{\frac{3}{2}} \rightarrow 1s_{\frac{1}{2}}$ results in the emission of e.m-radiation of 6.02 MeV. While calculations based on point nucleus hypothesis yields 16.4 MeV. This difference is due to finite size of the nucleus. Muon spends about 50 % of its time inside the nuclear volume and therefore Coulomb field acting on the muon is reduced. This behaviour is similar to the gravitational intensity inside the earth's surface. The reduction of Coulomb forces reduces the binding of the muonic states and hence all the energy levels are raised.

One finds that this effect is most pronounced for 1 s orbit.

It is reported that the shift in the energy of the K-shell is small for light elements and approximately proportional to $R^2 Z^4$, where $R = r_0 A^{\frac{1}{3}}$. The shift is reported to be greatest for 1 s level, much less for the $2p_{\frac{1}{2}}$ level and still much smaller for $2p_{\frac{3}{2}}$ level (Table 1.6).

The nuclear radius parameter, r_0, estimated from muonic X-ray measurments obtained as $r_0 = (1.15 \pm 0.03) \times 10^{-15}$ m = 1.15 ± 0.03 fm are in reasonable agreement with the electron scattering experiments.

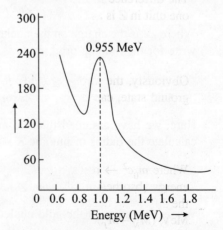

Fig. 1.20 X-ray spectrum from μ⁻-mesonic atom of titanium.

Table 1.6 Mesonic X-ray Transitions $\left(2p_{\frac{3}{2}} \rightarrow 1s_{\frac{1}{2}}\right)$

Element	Z	Energy calculation based on point nucleus (MeV)	Observed energy of X-ray (MeV)	Value of r_o based on "best fit" (fermi)
Al	13	0.363	0.35	–
Si	14	0.421	0.41	–
Ti	22	1.045	0.955	1.17
Cu	29	1.826	1.55	1.21
Zn	30	1.954	1.6	–
Hg	80	15.51	5.8	–
Pb	82	16.41	6.02	1.17

(iii) Mirror Nuclei Method   There are many light unstable nuclei in which Z differs from the neutron number N by 1. Among these nuclei there are pairs which consist of the same odd number of nucleons, say $(2Z + 1) = A$, where one of them has Z protons and $(Z + 1)$ protons and the other has $(Z + 1)$ protons and Z neutrons. Such nuclei are known as mirror nuclei. Pairs of isobaric nuclei, e.g. $^{11}_{6}C$ and $^{11}_{5}B$, $^{13}_{7}N$ and $^{13}_{6}C$ etc. are few known examples of mirror nuclei. The first member of the pair of mirror nuclei is usually β^+ active and undergoes β^+ transformation into the second, e.g.

$$^{13}_{7}N \rightarrow \,^{13}_{6}C + e^+ + v$$

Assuming that the charge is uniformly distributed in the nuclear volume of radius R, one obtains the total electrostatic energy,

$$E_c = \frac{3}{5}\left(\frac{Z^2 e^2}{4\pi\varepsilon_0 R}\right)$$

The difference in the electrostatic energies between the pair of mirror nuclei differing by one unit in Z is as follows:

$$E_c - E_c' = \Delta E_c = \frac{3}{5}\frac{e^2}{4\pi\varepsilon_0 R}[Z^2 - (Z\pm 1)^2] = \frac{3}{5}\frac{(2Z\pm 1)e^2}{4\pi\varepsilon_0 R} \qquad \dots (1.1)$$

Obviously, the maximum position energy, when the daughter nucleus is formed in the ground state, can be expressed as

$$E_{\beta^+} = \frac{3}{5}\left[\frac{(2Z-1)e^2}{4\pi\varepsilon_0 R}\right] - m_0 c^2 - (M_N - M_P)c^2 - \Delta B$$

Where $m_0 c^2 \rightarrow$ rest mass energy of the positron absorbed in creating it, $(M_N - M_P)c^2 \rightarrow$ energy absorbed in transforming a proton of mass M_P into a neutron of mass M_N and $\Delta B \rightarrow$ the difference between nuclear binding energies in the pair of mirror nuclei. For $m_0 c^2 = 0.51$ MeV, $(M_N - M_P)c^2 = 1.29$ MeV and $A = (2Z - 1)$, one obtains

$$E_{\beta^+} = \frac{3}{5}\frac{Ae^2}{4\pi\varepsilon_0 r_0 A^{\frac{1}{3}}} - 0.51 \text{ MeV} - 1.29 \text{ MeV} - \Delta B$$

$$= \frac{3}{5} \frac{e^2 A^{\frac{2}{3}}}{4\pi\,\varepsilon_0\,r_0} - 1.80 \text{ MeV} - \Delta B \qquad\qquad ... \ (1.2)$$

When the positron energy of the unstable members of the mirror nuclides is plotted against $A^{\frac{2}{3}}$, one obtains a straight line. The slope of this line $\dfrac{\frac{3}{5}e^2}{4\pi\,\varepsilon_0\,r_0}$ is a measure of the nuclear

unit radius r_0. The value obtained from it is about 1.45 fermi and it is about 20 % higher than the value obtained from electron scattering method. This difference reduced considerably when Pauli's principle is applied. Pauli's principle reduces effectively the probability of finding two protons close together and hence reduces Coulomb energy. Thus the estimated value of $r_0 = (1.28 \pm 0.05) \times 10^{-15}$ m = 1.28 $\pm$ 0.05 fm agree fairly well with those estimated by other method. The values of r_0 for several nuclei estimated in this are as below:

Nucleus	^{11}B	^{12}C	^{13}N	^{17}O	^{19}F	^{21}Ne	^{22}Na	^{25}Mg	^{27}Al
r_0(fm)	1.28	1.34	1.31	1.26	1.26	1.25	1.22	1.23	1.20

(iv) Isotope Shift Method   The potential $U(r)$ for a point nucleus is simply a Coulomb potential and is equal to $\dfrac{Ze}{4\pi\,\varepsilon_0\,r}$. The above relation is valid for $r > R$ for a finite nucleus, where R is the nuclear radius. When the nuclear charge is distributed uniformly throughout the nuclear volume, then the potential can be expressed as

$$U(r) = \frac{Ze}{4\pi\,\varepsilon_o\,R} \left[\frac{3}{2} - \frac{1}{2}\left(\frac{r}{R}\right)^2 \right] \quad r \le R \qquad\qquad ... \ (1.1)$$

If $\psi^2(r)$ represents the probability density of the electron at a distance r from the centre of the nucleus, them the potential energy of this electron in the central field $U(r)$ can be expressed as

$$e \int_0^\infty \psi^2(r)\, U(r)\, 4\pi\, r^2 dr$$

Obviously, the decrease of the atomic binding energy of an s-election is equal to

$$\Delta W = e \int_0^R \psi^2(r)\, [U_0(r) - U(r)]\, 4\pi\, r^2 dr \qquad\qquad ... \ (1.2)$$

It is reported that certain lines in the emission spectra of such elements having two or more stable isotopes consist of number of closely spaced components. These lines do not split further because the nuclear moments are zero valued. The isotope shift measurment provides the difference $\Delta W_1 - \Delta W_2$ and the finite value of this difference reveals that the protons in the nuclei of both isotopes move in the regions of different size.

Brix and Kopfermann compared the observed and theoretical values of the isotope shifts by taking $r_0 = 1.5$ fm and 1.1 fm and found that the data corresponding to the latter value are in agreement.

(b) Measurment of Potential Radius

Nuclear force is strong short range force and the potential from which this force is derived is oboviously a short range and has a steep slope at the edge of the nucleus. The potential owes its origin to the strong short raznge internucleon interaction. There are ample evidences which indicate that this is independent of the nature of the nucleons, i.e., charge state. This means that $p - p$ and $n - n$ forces are equal (charge symmetry). Moreover, the $p - n$ force is also the same in the same quantum state. This shows that for a complex nucleus, the specifically nuclear interaction will extend upto a distance of the same order of magnitude as the range of the inter nucleon interaction beyond the nuclear radius R of nuclear charge distribution (Figure 1.15). This radius is known as *potential radius* and it is slightly larger than r_0. The following methods can be used to estimate the potential radius:

(i) **Life time of α-emitters** The earliest method of estimating the potential radius of the nuclei was based on the study of α-disintegration of heavy nuclei, e.g. ^{238}W, ^{226}Ra etc. α-disintegration of heavy nuclei takes place due to the penetration of the Coulomb potential barrier surrounding the nucleus. According to α-disintegration theory, the barrier penetration probability or transmission coefficient, T is given by

$$T = \exp(-H) \qquad \qquad ...(1.1)$$

where
$$H = \frac{2}{\hbar}\left(\frac{M\,Ze^2\,b}{\pi\,\varepsilon_0}\right)^{\frac{1}{2}}\left[\cos^{-1}\sqrt{\frac{R}{b}} - \sqrt{\frac{R}{b} - \frac{R^2}{b^2}}\right] \qquad ...(1.2)$$

Where R is the potential radius of the nucleus and b is the distance from the centre to the point where the energy E of the α-particle is equal to the Coulomb potential energy, i.e. V_c

$= \dfrac{2Ze^2}{4\pi\,\varepsilon_0\,r}$. Here Z denotes the atomic number of the residual nucleus and M and $2e$ are the

mass and charge of the α-particle respectively. Here r is measured from the centre of the nucleus.

Let the frequency of collision of the α-particle against the nuclear wall inside the nucleus be n, then the probability of penetration through the potential barrier per second will be

$$P = nT$$

Obviously, the mean life of α-decay, which is the reciprocal of P is

$$\tau_m = \frac{1}{p} = \frac{1}{nT} \qquad \qquad ...(1.3)$$

Thus one can estimate the nuclear potential radius R by measuring the mean life. We have R

$= r_0\,A^{\frac{1}{3}}$. The potential radius parameter reported as $r_0 = 1.48 \times 10^{-15}$ m.

We must note that the above formula does not reproduce the α-decay life times accurately and may deviate by several orders of magnitude from the experimental result. However, it gives a much more pracise estimate of the nuclear radius R, even from a rough knowledge of the mean life τ_m. One obtains some what higher value of r_0 estimated by this method in comparison to charge or mass radius parameter. A correction due to the finite radius of the α-particle ($R_\alpha \sim 1.2 \times 10^{-15}$ m) yields the radius of the residual nucleus R_A.

$$R = R_A + R_\alpha \text{ and } R_A = r_{0A} A^{\frac{1}{3}} \qquad \text{... (1.4)}$$

Where $\qquad r_{0A} = 1.4 \times 10^{-15}$ m, is a new parameter.

(ii) **Neutron Scattering Experiments**   Neutrons possess no charge and hence the target cross section can be expected to be the geometrical cross section πR^2. Using neutron size as λ, we can estimate the total cross section σ_T as

$$\sigma_T = 2\pi(R + \lambda)^2 \approx 2\pi R^2 \qquad \text{... (1.1)}$$

One can consider neutron beam as a plane beam and hence at high energies the de Broglie wavelength $\lambda << R$, where $\lambda = \dfrac{\lambda}{2\pi}$. At such high energies the neutron scattering cross section is same as the absorption cross section, i.e.

$$\sigma_{SC} = \sigma_a \approx \pi(R + \lambda)^2 \approx \pi R^2 \qquad \text{... (1.2)}$$

$$\therefore \qquad \sigma_T \approx 2\pi(R + \lambda)^2 \approx 2\pi R^2$$

Obviously, total cross section is about twice the geometrical area of the nucleus. Here it is assumed that a perfectly black nucleus absorbs all the neutrons falling on it.

In these experiments monoenergetic beams of fast neutrons are allowed to be scattered by the target nuclei. Since neutrons interact mainly by the specifically strong nuclear interaction with the nucleus and hence this method actually detects the edge of the nuclear potential well.

The measurment of cross sections field a radius parameter

$$r_0 = 1.25 \times 10^{-15} \text{ m}$$

(iii) **Nuclear Reactions by charged Particles**   Neutron measurments are usually difficult and therefore using Bohr's compound nucleus model of nuclear reactions, one can estimate nuclear radius studying nuclear reactions by charged particles. When a target nucleus X is bombarded by the charged particle x(say), a compound nucleus ($X + x$) is formed. Usually α-particles or protons upto a few hundred MeV, which interact strongly with the nuclei at close range have been used. In the α-particle experiments, the critical angle (θ_c) of scattering at which deviations are observed from the Rugherford scattering is measured. One can correlate θ_c with the critical distance at which the effect of the specifically nuclear force begins to be felt.

In proton elastic scattering experiments range 5 to 300 MeV, diffraction patterns are observed due to the extension of the potential beyond the nuclear edge. Using Optical model, the following form of potentiall due to woods and saxon is usually employed to analyse the experimental data:

$$V(r) = \frac{V_0}{1 + \exp\left[\dfrac{\left(r - \dfrac{R_1}{2}\right)}{a}\right]} \qquad \text{... (1.1)}$$

Here $\dfrac{R_1}{2}$ and a have the usual meanings specified earlier. $V(r)$ has a radial dependence similar to Fermi charge distribution. A value $r_0 = 1.33 \times 10^{-15}$ m is obtained from this study. It is worth while to mention that the potential radius is about 0.7 fm greater than the charge radius which may be taken to be the measure of the range of the nuclear force.

The results of the different types of measurments are summarized in Table 1.7.

Table 1.7 Nuclear Radius Parameter (r_0)

Measurment	symbol	Value (m)
Mass distribution	r_{om}	1.1×10^{-15} m
Charge distribution (equivalent square well)	r_{oc}	$(1.2 \text{ to } 1.3) \times 10^{-15}$
Optical potential	r_{ov}	1.25×10^{-15}

1.4.4 Nuclear Spin, Parity, Moments and Statistics

So far we have seen that electrons cannot be within a nucleus and constitutents of nuclei are protons and neutrons. In this section, we will study some important quantam mechanical properties of nuclei.

(a) Angular Momentum Elementary mechanics tells us that the angular momentum of an isolated system is conserved. Obviously, the angular momentum of a nucleus is a constant quantity because nucleus being an isolated system, i.e. the net external torque equal to zero. This situation is analogous to an atom. Following this analogy, let I represents the total angular momentum quantum number, then the eigen values of the square of the total angular momentum operator are $I(I + 1)$ in the units of h. One can write the eigen value equation as

$$\hat{I}^2 \psi = I(I + 1)h^2 \psi \qquad \ldots (1.1)$$

Where $\hat{I}^2$ is the operator belonging to the square of the total angular momentum I and ψ is the common eigen function of $\hat{I}^2$ and I_Z. Here I_Z is the Z-compoment off total angular momentum I. We have for operator $\hat{I}_Z$,

$$\hat{I}_Z \psi = m \hbar \psi \qquad \ldots (1.2)$$

It is worth while to note that (1.1) and (1.2) are satisfied by a set of common eigen functions because the $\hat{I}^2$ and $\hat{I}_Z$ commute, i.e.

$$[\hat{I}^2, I_Z] = O \qquad \ldots (1.3)$$

Obviously, (1.1) reveals that the magnitude of the total angular momentum I is quantized and is restricted to values.

Total angular momentum $(I) = \sqrt{I(I+1)}\hbar \qquad \ldots (1.4)$

One can see from (1.2) that the Z-component of the angular momentum can take only values $M\hbar$. We know that the quantum numbers I and M are integers and for a given value of I, M can take $2I + 1$ values from $-I$ to $+I$. One can express the space quantization in vector diagram as shown in Figure 1.21.

We must note that in the above example, allowed values of I are not only integers, but also half integers, i.e.

$$I = 0, \frac{1}{2}, 1, \frac{3}{2}, 2, \frac{5}{2}, ... \qquad ... (1.5)$$

Equation (1.5) clearly reveals that half integer values of I are a result of Pauli's proposal of *'intrinsic spin'*. Obviously, the total angular momentum is the sum of orbital angular momentum (L) and spin angular momentum, i.e.

$$\mathbf{I} = \mathbf{L} + \mathbf{S} \qquad ... (1.6)$$

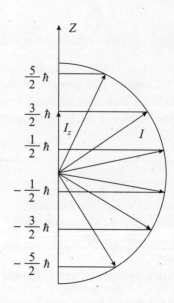

Fig. 1.21 Vector diagram exhibiting possible orientation of **I** for $I = \frac{5}{2}$. M can take values $\pm\frac{5}{2}\hbar$, $\pm\frac{3}{2}\hbar$ and $\pm\frac{1}{2}\hbar$.

Experimentally the values of integral spins are integers (if mass numbers A is even) or half integers (if A is odd) ranging from zero, as in ^{4}He and ^{12}C, upto 7, as in ^{176}Lu. Table 1.8 presents some nuclear spins. It has been noted that practically all even-even nuclei (i.e., nuclei that have an even number of neutrons and protons i.e. both N and Z even) have zero spin. *Spin has no analog in classical mechanics.*

Quantum mechanically, one can express the total orbital and spin angular momenta of the nucleus as follows

$$p_I^2 = I(I + 1)\hbar^2$$

$$p_L^2 = L(L + 1)\hbar^2$$

$$p_S^2 = S(S + 1)\hbar^2$$

The magnitudes in the three cases stated above are I, L and S respectively in the units of $\hbar$. The resultant spin angular momentum of a nucleus is obtained by the vector addition of the spins of individual nucleons, i.e. $S = \Sigma s_i$. Similarly, the resultant orbital angular momentum is given by $L = \Sigma l_i$. Now, $s_i = \frac{1}{2}$ and hence S can be either integral or half-integral, depending upon whether the number of nucleons (A) in the nucleus is even or odd. Since l_i is integer (0, 1, 2, 3, ...) and hence L is integer or zero. Obviously the total angular momentum I (also usually referred as the nuclear spin) can be either integral (for A even) or half odd integral (for A even) or half odd integral (for A odd). It has been noted that practically all even-even nuclei (i.e., nuclei that have an even number of neutrons and protons have $I = O$, which

Table 1.8 Properties of Some Nuclides

Nuclide	Atomic mass, amu	Spin	Magnetic dipole moment, nm	Electric quadrupole moment, 10^{-28} m^2
$^1_0$n	1.008665	$\frac{1}{2}$	−1.9131	0
$^1_1$H	1.007825	$\frac{1}{2}$	2.7927	0
$^2_1$H	2.014102	1	0.8574	0.00282
$^3_2$He	3.016989	$\frac{1}{2}$	−2.1275	0
$^3_1$H	3.016050	$\frac{1}{2}$	2.9789	0
$^4_2$He	4.002603	0	0	0
$^7_3$Li	7.016004	$\frac{3}{2}$	3.2563	−0.045
$^{12}_6$C	12.00000	0	0	0
$^{13}_6$C	13.00335	$\frac{1}{2}$	0.7024	0
$^{14}_7$N	14.00307	1	0.4036	0.007
$^{15}_7$N	15.00011	$\frac{1}{2}$	−0.2831	
$^{16}_8$O	15.99492	0	0	0
$^{17}_8$O	16.99913	$\frac{5}{2}$	−1.8937	−0.026
$^{21}_{10}$Na	20.99395	$\frac{3}{2}$	−0.6618	0.093
$^{27}_{13}$Al	26.98153	$\frac{5}{2}$	3.6414	0.15
$^{35}_{17}$Cl	34.96885	$\frac{3}{2}$	0.8218	−0.080
$^{40}_{20}$Ca	39.96260	0	0	0
$^{43}_{20}$Ca	42.95878	$\frac{7}{2}$	−1.3172	
$^{56}_{26}$Fe	55.9349	0	1.06	
$^{57}_{26}$Fe	56.9354	$\frac{1}{2}$	0.0905	
$^{60}_{27}$Co	59.9338	5	3.8100	
$^{63}_{29}$Cu	62.9296	$\frac{3}{2}$	2.2260	−0.180
$^{79}_{35}$Br	78.9183	$\frac{3}{2}$	2.1060	0.310
$^{88}_{38}$Sr	87.0564	0	0	
$^{93}_{41}$Nb	92.9064	$\frac{9}{2}$	6.1670	−0.220
$^{103}_{45}$Rh	102.9048	$\frac{1}{2}$	0	0
$^{114}_{48}$Cd	113.9034	0	0	0
$^{127}_{53}$I	126.9045	$\frac{5}{2}$	2.8080	−0.790
$^{155}_{64}$Gd	154.9277	$\frac{3}{2}$	−0.2700	1.300
$^{175}_{71}$Lu	174.9409	$\frac{7}{2}$	2.2300	5.600
$^{176}_{71}$Lu	175.9427	7	3.1800	8.000
$^{177}_{71}$Lu	176.9439	$\frac{7}{2}$	2.2400	5.400
$^{180}_{72}$Hf	179.9468	0	0	
$^{185}_{75}$Re	184.9501	$\frac{5}{2}$	3.1716	2.6
$^{208}_{82}$Pb	205.9892	0	0	0
$^{209}_{93}$Bi	208.9804	$\frac{9}{2}$	4.0802	−0.340
$^{227}_{83}$Ac	227.0278	$\frac{3}{2}$	1.1	1.7
$^{233}_{92}$U	233.0395	$\frac{5}{2}$	0.54	3.5
$^{235}_{92}$U	235.0439	$\frac{7}{2}$	0.35	4.1
$^{238}_{92}$U	238.0508	0	0	0
$^{241}_{94}$Pu	240.4236	$\frac{5}{2}$	−0.730	5.600

indicates that identical nucleons tend to pair their angular momenta in opposite directions. This is called the pairing effect. Even-odd nuclei (i.e., nuclei that have an odd number of either protons or neutrons) all have half integral angular momenta, and it is reasonable to assume that the nuclear spin coincides with the angular momentum of the last or unpaired nucleon, a result which seems to hold in many cases. Odd-odd nuclei have two unpaired nucleons (one neutron and one proton) and the experimental results are a little more difficult to predict, but their angular momenta are integers, as one would expect, since there is an even total number of particles. Any theory of nuclear forces, to be satisfactory, must account for the experimental values of *I*.

We must note that the measured values of ground state spins of the nuclei are small integers or half odd integers. The highest measured value is $\dfrac{9}{2}$, which is small compared to the sum of the absolute values of l_i and s_i of all the individual nucleons within the nucleus.

This is a further confirmation of earlier statement of pair formation within the nuclei. It seems that majority of nucleons of either type within the nucleus form a core in which even numbers of protons and neutrons are grouped in pairs of zero spin and orbital angular momenta so that the core itself has zero angular momentum. Obviously, the few remaining nucleons outside the core within the nucleus determine the nuclear spin which is a small number, integral or half odd integral.

We shall discuss the methods of measurment of the ground state spins of nuclei later on. One can deduce the spins of excited states of nuclei from nuclear disintegration and nuclear reaction data.

(b) Parity   This refers to the behaviour of the wave function when evaluated at two symmetric positions. One may say that for problems having a centre of symmetry, the stationary states are described by wave functions having a well defined parity, even or odd. To make it more clear, let us change the coordinate system at the origin from *x*, *y*, *z* to −*x*, −*y*, −*z*, then the wave function of the physical system changes from ψ(*x*, *y*, *z*) to ψ(−*x*, −*y*, −*z*). Now, if the Hamiltonian of the physical system remains invariant under space inversion, then the changed wave function can be related to the originas wave function in following two different ways:

$$\psi(-x, -y, -z) = +\psi(x, y, z) \qquad\qquad \text{... (1.1)}$$

or
$$\psi(-x, -y, -z) = -\psi(x, y, z) \qquad\qquad \text{... (1.2)}$$

In the first case, we say that the wave function has *even parity* while in the second case, the wave function has *odd parity*.

The concept of parity, as defined above, refers to the operation of space reflection, either in a plane or through a point, such as the origin of coordinates. In the case of a particle acted upon by a central force, the parity is determined by the azimuthal quantum number *l*, being even for *l* = 0 or even and odd for *l* = odd.

Apart from the orbital property, elementary particles may possess intrinsic parity, which is related to the inversion of some internal axis of the particle. Truly speaking, it is defined in

---

*Total parity mean the product of the orbital and intrinsic parities.

a relative manner. By convention, one takes the parity of nucleons even. It is then fixed for the other particles in such a way that in an interaction between the particles involving strong nuclear or electromagnetic force, total parity of the system be conserved.* The law of conservation of parity states that *parity is conserved in a process if the mirror image of the process is also a process which can occur in nature.*
We shall discuss it in detail later on.

(c) **Statistical Properties of Nuclei** The concept of statistics is related to the behaviour of an assemblage of a large number of particles, e.g. the distribution of energies of velocities among a large number of molecules of a gas can be described by the classical Maxwell Boltzmann statistics. However, when we are dealing with subatomic particles, then quantum statistics comes into picture. The wave function of a quantum mechanical system is either symmetric or anti-symmetric in the interchange of the coordinates of the two particles. When by such an interchange the sign of the wave function does not change, one obtains a *symmetric* wave function and the resulting quantum statistics is known as *Bose-Einstein Statistics*. In the reverse case, when the sign of the wave function changes as a results of the interchange, one obtains an *antisymmetric wave function* and the resulting quantum statistics is called *Fermi-Dirac statistics*. These can be summarized as follows:

$$\psi(x_1, x_2) = \pm \psi(x_1, x_2) \qquad \qquad ...(1.1)$$

The sign we choose depends on the type of subatomic particle we are considering. One must choose the negative sign, when electron, proton etc. are considered. The eigen function for the two electrons is antisymmetric for an exchange of co-ordinates. Such subatomic particles are said to satisfy Fermi-Dirac (FD) statistics and are called *fermions*. These particles have half-integral spins $\left(\dfrac{1}{2}, \dfrac{3}{2}, ...\right)$. On the other hand for the class of particles having spin O or integral obey *Bose-Einstein (BE)* statistics and are known as *Bosons*. Photon is an example of boson. For bosons, one must choose the positive sign so that the eigen function of two bosons is symmetric for an exchange of co-ordinates.

For two electrons, one can write the antisymmetric state function as

$$\psi(x_1, x_2) = \sqrt{\frac{1}{2}} \left[\psi_\alpha(x_1)\,\psi_\beta(x_2) - \psi_\beta(x_1)\,\psi_\alpha(x_2)\right] \qquad \qquad ...(1.2)$$

If the two electrons are in the same quantum state the R.H.S. of (1.2) vanishes and the state function becomes equal to zero. Thus $\psi_\alpha \neq \psi_\beta$ (where α and β denotes quantum states) and no two electrons can be in the same quantum state. Obviously, the particles obey Pauli's exclusion principal. For bosons, the particles do not obey the exclusion principle and number of bosons can occupy the same quantum state.
We must note that statistics is conserved in nuclear reactions.
Pauli was the first scientist to realize the connection between spin and symmetry of the wave function on exchange of coordinates. Table 1.9 summarizes the spin-symmetry relationship.

Table 1.9 Spin-Symmetry Relationship for subatomic particles

Spin (I)	Particles	Wave function on exchange of coordinates
Zero or Integer	Bosons	Symmetric
Half-Integer	Fermions	Antisymmetric

(d) Magnetic dipole moment   Figure 1.22 shows that the magnetic dipole moment μ associated with a current loop of area A and carrying a current i. we have

$$\mu = i\,A \qquad \ldots (1.1)$$

Let us consider an electron moving in a circular orbit of radius r with a period T. We have

$$i = \frac{e}{T} = \frac{e}{2\pi\dfrac{r}{\upsilon}} \text{ and } A = \pi\,r^2$$

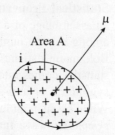

Fig. 1.21 Magnetic dipole moment (μ) given rise to by a circular current loop of area A and carrying a current i

$$\therefore \qquad \mu = \frac{e\upsilon}{2\pi r} \times \pi r^2 = \frac{e}{2}\upsilon r$$

or

$$\mu = \frac{e}{2m} m\upsilon r = \frac{e}{2m}L \qquad \ldots (1.2)$$

Where L is the angular momentum. Obviously a moving charge possesses an orbital magnetic dipole moment μ_L which is proportional to its orbital angular momentum. Equation (1.2) tells us that the angular momentum L gives rise to a magnetic dipole moment μ for a charged particle. For the case of proton, charge = e and hence

$$\mu_L = \left(\frac{e}{2M_P}\right)L \qquad \ldots (1.3)$$

Where M_P is the mass of the proton. Equation (1.2) can be generalized as

$$\mu = \frac{e}{2m}Lg \qquad \ldots (1.4)$$

Where g is called the gyromagnetic ratio or *g-factor*. For a nucleus, one can put (1.4) as

$$\mu = \frac{e}{2m}Ig \qquad \ldots (1.5)$$

Usually e is taken positive and the sign of the magnetic dipole moment μ is governed by the sign of the g-factor. I, the total angular momentum of the nucleus has the units of $\hbar$ and hence $\dfrac{1}{\hbar}$ is dimensionless. One can write (1.5) as

$$\mu = g\,\mu_o\,\frac{I}{\hbar} \qquad \ldots (1.6)$$

where
$$\mu_0 = \frac{e\hbar}{2m} \qquad \qquad \qquad \qquad \qquad ...\ (1.7)$$

The constant m_0 is called a *magneton*. μ_0 is taken as the unit to measure magnetic moments. Taking $m = m_e$, the electron mass, one finds

$$\mu_B = \frac{e\hbar}{2m_e} = 0.927 \times 10^{-23}\ \text{J/Weber/m}^2 \qquad \qquad ...\ (1.8)$$

Where μ_B is called Bohr magneton.

For nuclei, magnetic moments are expressed in terms of nuclear magneton (μ_N), defined as

$$\mu_N = \frac{e\hbar}{2m_p} = \frac{\mu_B}{1836} = 5.05 \times 10^{-27}\ \text{J/Weber/m}^2 \qquad ...\ (1.9)$$

Obviously, μ_N is much smaller than μ_B, being only $\dfrac{1}{1836}$ part of the latter. The measured values of the magnetic moments of the proton and the neutron are

$$\mu_p = +2.7927\ \mu_N$$

and
$$\mu_n = -1.9131\ \mu_N \qquad \qquad \qquad \qquad ...\ (1.10)$$

The above values reveals that the proton and neutron magnetic moments are of the order of 10^{-3} times the electronic magnetic moment, which is equal to the Bohr magneton $(\mu_e = \mu_B)$. The actual measured values of nuclear magnetic moments are broadly between $-3\ \mu_N$ and $+10\ \mu_N$. The positive sign indicates that μ is parallel to the spin where as the negative sign gives their directions antiparallel. However the values of magnetic moment of proton and neutron given by (1.10) are unexpected one and quite different from the expected values, i.e. one nuclear magneton for proton and zero for neutron (no charge). This indicates that the proton and neutron both have a non uniform charge distribution and in case of neutron the direcrtion of the magnetic moment is opposite to that of its intrinsic spin angular momentum vector. These anamalous values of the magnetic moments of both proton and neutron can be understood, qualitatively, on the basis of *meson theory*. π-mesons may have either sign of electric charge and are constantly exchanged between nuclear particles. This helps the neutron to contribute to a non-zero orbital magnetic moment.

The magnetic moments of the proton and the neutron are intimately related to their intrinsic spin angular momenta as

$$p_p = s_p \hbar\ \text{and}\ p_n = s_n \hbar \qquad \qquad \qquad ...\ (1.11)$$

Where $s_p = s_n = \dfrac{1}{2}$. One can easily show that the ratio of the magnetic moment μ_e to the spin angular momentum p_e of the electron is given by

$$\frac{\mu_e}{p_e} = g_e\ \frac{e}{2m_e} \qquad \qquad \qquad \qquad ...\ (1.12)$$

Where $p_e = s_e \hbar = \dfrac{\hbar}{2}$ and g_e is Lange factor. g_e has the value -2. The r.h.s. of Equation (1.12) represent the gyromagnetic ratio for the spin motion of the electron. Goudsmit and Uhlenback first introduced the factor $g = -2$ on ad hoc basis but later found justification from the Dirac electron theory.

In analogy with (1.12), one can write form proton and neutron as

$$\frac{\mu_p}{p_p} = g_p \frac{e}{2M_p} \qquad \qquad \text{... (1.13)}$$

and

$$\frac{\mu_n}{p_n} = g_n \frac{e}{2M_p} \qquad \qquad \text{... (1.14)}$$

Using (1.11) , Equations (1.13) and (1.14) becomes

$$\mu_p = g_p \frac{e}{2M_p} s_p \hbar = \frac{g_p}{2} \mu_N \qquad \qquad \text{... (1.15)}$$

$$\mu_n = g_n \frac{e}{2M_n} s_n \hbar = \frac{g_n}{2} \mu_N \qquad \qquad \text{... (1.16)}$$

Comparing with Equation (1.9), one obtains

$$g_p = +2 \times 2.7927 = +5.5855 \text{ and } g_n = -2 \times 1.9131 = -3.8263 \qquad \text{... (1.17)}$$

These results suggest that the proton and the neutron have a complex structure which we have not yet been able to determine exactly.

Equations (1.15) and (1.16) can also be expressed in vector forms as

$$\mu_p = \frac{1}{2} g_p \sigma_p \qquad \qquad \text{... (1.18)}$$

and

$$\mu_n = \frac{1}{2} g_n \sigma_n \qquad \qquad \text{... (1.19)}$$

Where σ_p and σ_n are the Pauli spin matrials.

For a complex nucleus one has to add vectorially intrinsic magnetic moments of all the protons to obtain the resultant $\sum \mu_{pi}$. Similarly, one has to add vectorially the intrinsic magnetic moments of all the neutrons to obtain the resultant $\sum \mu_{ni}$. Here must not that the orbital rotations of the protons will also contribute to the net magnetic moment of the nucleus equal to $\sum (\mu_{ip})_i$. One can define mlp in the same way as in the case of the orbital motion of the electron. If the resultant orbital angular momentum due to orbital motion of the protons be pL then one can write

$$\frac{\mu_L}{p_L} = g_L \frac{e}{2M_P}$$

Now, $p_L = L\hbar$; therefore

$$\mu_L = g_L \frac{e}{2M_p} \cdot L\hbar = g_L L \mu_N \qquad \qquad ... (1.20)$$

Where L is the orbital angular momentum quantum number. $g_L = I_L$ for the orbital motion of the proton and hence

$$\mu_L = L \mu_N \qquad \qquad ... (1.21)$$

L can be zero or can have only integral values. It is to be noted that no contribution to the magnetic moment of the nucleus comes from the orbital motion of the neutrons, i.e. $g_L = 0$ for neutrons.

Obviously, one can obtain the resultant magnetic moment of the nucleus by the vector addition of $\Sigma \mu_{pi}$, $\Sigma \mu_{ni}$ and $\Sigma \mu_{ip}$.

We have seen that protons and neutrons tend to form pairs with oppositely aligned spins, giving a resultant spin O. This means such pairs of protons and neutrons will also have zero magnetic moments. Thus the net magnetic moment of the nucleus will be determined by the nucleons outside the even Z-even N core for which the net magnetic moment of the nucleus will be determined by the nucleons outside the even Z-even N core for which the net magnetic moment is zero. In the case of nuclear spin, this makes the value of the magnetic moment of the nucleus comparable to the magnetic moments or proton of neutron.

Table 1.8 gives some values of magnetic dipole moment. A satisfactory theory of nuclear interactions must account for the observed values of magnetic dipole moment of nuclei.

We shall discuss the experimental methods for the determination of the dipole moments in subsequent chapters.

(e) Electric moments of nuclei We have seen that the atomic nucleus is a positively charged body of finite dimensions. We all know that any distribution of electric charge produces an electric potential $\phi(r, \theta)$ at a distance r in the Z direction. This potential $\phi(r, \theta)$ due to an azimuthally symmetric distribution of electric charges can be expanded in ascending powers of $\frac{1}{r}$, where r is the distance of the point wher the potential is measured from the origin of the coordinate system.

$$\phi(r, \theta) = \frac{1}{r} \sum_{n=0}^{\infty} \frac{a_n}{r^n} P_n(\cos \theta) \qquad \qquad ... (1.1)$$

Where P_n's are the Legendre Polynomials of different orders. The different terms in the expansion of (1.1) correspond to potentials due to electric multipoles of different orders located at the origin. The first term in the expansion corresponds to the potential due to *electric multipoles* of different orders located at the origin. The first term in the expansion corresponds to the potential due to an *electric* momopole which is a point charge $+Ze$, Z being the atomic number. The second term in the above expansion corresponds to the potential due to an electric dipole.

It turns out that a nucleus must have a zero electric dipole moment. A nucleus is made up of positively charged protons (each of charge $+e$) and neutrons which are electrically neutral.

A displacement of centres of mass of two types of particles will cause an electric dipole moment to appear within the nucleus, given by $p = Zed$, where Z is the atomic number of the nucleus. Since protons are distributed throughout the nuclear volume, one can easily show that $p = 0$, i.e. *electric dipole moment of a nucleus in its ground state vanishes*. This is also true for all non-degenerate excited states of the nucleus. This result also holds good for static electric moments of all odd orders (e.g., octupole moment) which are obviously all zero for the nucleus. The third term in the expansion is called the *quadrupole moment*.

(f) Electric quadrupole moment   The third term in (1.1) corresponds to the potential due to electric quadrupole moment Q of the given cylindrically symmetric charge distribution located at the origin of the coordinate system. The four charge (quadrupole) shown in Figure 1.23 has net charge and electric dipole moment zero and hence the entire electric field is produced by the electric quadrupole moment. Atomic nuclei, the spin of which is ≥ 1, have quadrupole moments other than zero. It shows that their form is not strictly spherical. The quadrupole moment has a plus sign if

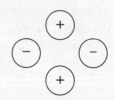

Fig. 1.23 Charge distribution having an electric quadrupole moment

the nucleus is extended along the spin axis (a spindle-shaped body) and minus sign if the nucleus is extended in a plane perpendicular to the spin axis (a lenticular body). Nuclei with positive and negative quadrupole moments are known (Table 1.8). A nucleus possessing a quadrupole moment produces a nonspherically symmetrical field. This results in the formation of additional energy levels of atomic electrons and gives rise to lines of superfine structure in the atomic spectra, the distances between which depend on the value of the quadrupole moment.

One can generate a simplest system of charges which has electric quadrupole moment by the displacement of an electric dipole w.r.t. itself with the sign of the dipole moment p reversed (Figure 1.24). The quadrupole moment of the system is given by $Q = 2pd'$, where d' is the displacement obviously, $Q = 2\theta\, dd'$. The units of both d and d' are of length and hence dd' has the dimensions of area (m^2) which is also taken to be the unit of Q.

Fig. 1.24 Two electric dipoles oppositely aligned.

Let Q_0 be the intrinsic quadrupole moment and Q be the observed quadrupole moment of a nucleus. Then

$$Q_0 = \frac{1}{e}\int (3Z'^2 - r'^2)\,\rho(r')\,d\tau' \qquad \qquad \text{... (1.1)}$$

Where the integeration is carried out over the entire volume of the nucleus. $r'(x', y', z')$ is the distance measured from the centre of mass of the nucleus. It is assumed that the nucleus have a symmetry axis along z' and e is the charge on the proton. Normally the expression for nuclear quadrupole Q_0 is divided by e so that its unit is (length)^2. In MKSC system, Q_0 is expressed in m^2. Sometimes a unit called the *barn* is used; one barn is equal to 10^{-28} m^2.

We have for a spherically symmetric charge distribution

$$\int \rho(r')\, x'^2\, d\tau' = \int \rho(r')\, y'^2\, d\tau'$$

$$= \int \rho(r')z'^2\, d\tau' = \frac{1}{3} \int \rho(r')\, r'^2\, d\tau' \qquad \dots (1.2)$$

For a spherical nucleus for which $I = 0$, one finds $Q_0 = 0$ from Equation 1.1 (Figure 1.25(a)). This menas non-zero quadrupole moment of a nucleus gives a meausre of the departure of the shape of the nucleus from sphericity.

One can always distinguish between intrinsic quadrupole moment Q_0 and observed quadrupole moment Q for a non-spherical nuclei. A reference frame fixed to the nucleus is considered and Q_0 is measured in it. Q is measured in a reference frame fixed in the laboratory. Q and Q_0 are related by the following relation.

$$Q = \frac{I(2I - 1)}{(I + 1)(2I + 3)}\, Q_0 \qquad \dots (1.3)$$

Equations 1.1 an 1.2 reveals that for a nucleus in the shape of a prelate spheroid, elongated along the Z'-axis (cigar-shaped), $Q_0 > 0$ (Figure 1.25(b)) as

$$\int \rho(r')z'^2\, d\tau' > \frac{1}{3} \int \rho(r')\, r'^2\, d\tau' \qquad \dots (1.4)$$

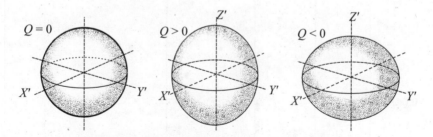

Fig. 1.25 Charge distribution, symmetric w.r.t. Z' axis (a) sphere (b) prolate ellipsoid and (c) oblate ellipsoid.

When the shape of the nucleus is an oblate spheroid (Pancake-shaped) as shown in Figure 1.25(c), we have

$$\int \rho(r')z'^2\, d\tau' < \frac{1}{3} \int \rho(r')\, r'^2\, d\tau' \qquad \dots (1.5)$$

Obviously $Q_0 < 0$ for such nuclei.

Figure 1.26 shows the observed electric quadrupole moments of a few nuclei. We note that some nuclei have abnormally large electric quadrupole moments, indicating very large deformation. A satisfactory theory of nuclear interactions must account for the observed values of Q.

In an external inhomogeneous electric field, a nucleus with quadrupole moment acquires an energy that depends on the orientation of the nucleus with respect to the field gradient. The interaction of the nuclear electric quadrupole moment with the external homogeneous electric field due to the atomic electron distribution produces additional hyperfine splitting. This departs from the interval rule followed by the normal hyperfine splitting due to the interaction between the nuclear magnetic moment and the atomic magnetic moment. This interaction enables one to determine Q. As an example, the deutron quadrupole moment measured from observations on the hyperfine structure of the atomic spectral lines is,

$$Q_{\text{deutron}} = 0.282 \text{ fm}^2$$

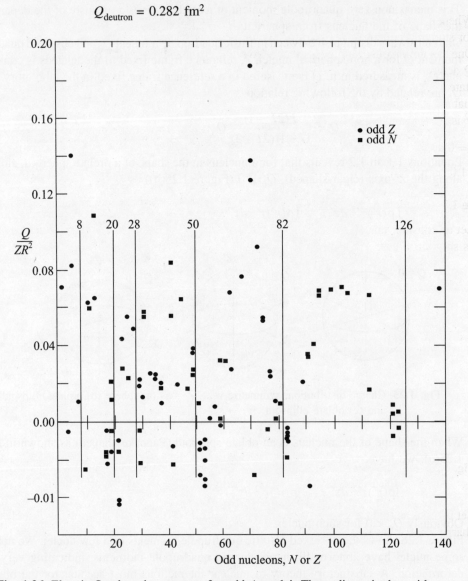

Fig. 1.26 Electric Quadrupole moments for odd-A nuclei. The ordinate is the unitless quantity $\dfrac{Q}{ZR^2}$, which is a better indicator of the deviation from a spherical shape.

This result has important implications as regards to our understanding of nuclear interactions. There is also another aspect of quadrupole moment measurment. It makes possible to determine *nuclear deformation*. The relation between the quadrupole moment and the nuclear deformation parameter *d* is given by

$$Q = \frac{4}{5} Z r_0^2 \delta \left(1 + \frac{1}{2}\delta + ...\right) \qquad ... (1.6)$$

Where r_0 is the mean nuclear radius and Z is the atomic number. Knowledge of δ is important for studying collective behaviour of subatomic particles.

One can define quadrupole moment Q quantum mechanically as the expectation value Q_I of Q determined by expectation value of $(3Z'^2 - r'^2)$ for a given charge distribution for the state $M_I = I$, where M_I is the projection of I along a given direction Z in space. One can see that $M_I = I$ is the maximum projection of I along the given direction. Since the magnitude of I^2 is $I(I + 1)$ and hence vector I can never align along Z. The above value of Q_I vanishes for $I = 0$ and $I = \frac{1}{2}$. It is reported that measured values of Q_I vary from -8 barns for $^{123}_{71}$Lu to -1.0 barn for $^{123}_{51}$Sb.

Example 1.8 Calculate quadrupole moment for an ellipsoidal charge distribution.

Sol. Let us consider that Z' axis be the axis of symmetry of the ellipsoid with semi axes a and b as show in Figure 1.27(a).

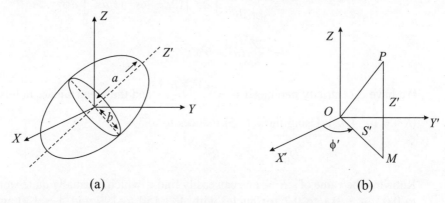

(a)   (b)

Fig. 1.27 **(a)** The charge distribution with a body fixed symmetry axis (Z') for an ellipsoid.   **(b)** Transformation to cylindrical polar coordinates

Let us assume that the charge distribution is uniform with density ρ. Now for a nucleus of charge Ze, we have

$$\rho = \frac{Ze}{\frac{4}{3}\pi ab^2} \qquad ... (1.1)$$

We have the equation of ellipsoid as

$$\frac{x'^2}{b^2} + \frac{y'^2}{b^2} + \frac{z'^2}{a^2} = 1 \qquad \qquad \dots (1.2)$$

With the help of Figure 1.27(b), Equation (1.2) reduces to in cylindrical polar coordinates (s', ϕ', z') as

$$\frac{s'^2}{b^2} + \frac{z'^2}{a^2} = 1 \qquad \qquad \dots (1.3)$$

where

$$s'^2 = b^2 \left(1 - \frac{z'^2}{a^2}\right) \qquad \qquad \dots (1.4)$$

The quadrupole moment of the nucleus in the body fixed coordinate system is

$$\begin{aligned}
Q_0 = Q_{z'z'} &= \frac{1}{e} \int_{\tau'} \rho(3z'^2 - r'^2) d\tau' \\[2mm]
&= \frac{3Z}{4\pi ab^2} \int_{\tau'} (3z'^2 - x'^2 - y'^2 - z'^2) d\tau' \\[2mm]
&= \frac{3Z}{4\pi ab^2} \int_{\tau'} (2z'^2 - s'^2) dz' \, d\phi' \, s' \, ds' \\[2mm]
&= \frac{3Z}{4\pi ab^2} \int_{-a}^{a} dz' \int_{0}^{b} (2z'^2 - s'^2) s' \, ds' \int_{0}^{2\pi} d\phi' \\[2mm]
&= \frac{2Z}{5} (a^2 - b^2) \qquad \qquad \dots (1.5)
\end{aligned}$$

We have eccentricity parameter $\in = \dfrac{(a^2 - b^2)}{(a^2 + b^2)}$ and the mean square radius of the nucleus $\langle R^2 \rangle = (a^2 + b^2)$. Using these (1.5) reduces to

$$Q_0 = \frac{4}{5} Z \in \langle R^2 \rangle \qquad \qquad \dots (1.6)$$

Knowing the value of $\langle R^2 \rangle$, one can easily find $\in$ which is usually quite small, being -0.01 to 0.02. $\in = 0.1$ to 0.2 for nuclei with $A = 150$ to 190 and $A > 120$ and considerable deformation is observed for these nuclei.

Example 1.9 Quantum mechanical expression for the observed quadrupole moment Q corresponding to $M_I = I$ is

$$Q = \frac{3K^2 - I(I+1)}{(I+1)(2I+3)} Q_0$$

Where the symbols have usual meanings. Discuss various situations.

Sol. The given expression reveals that

(i) Q is always less than Q_0 (ii) For I large, $I(I + 1) > 3K^2$, i.e. the sign of Q is always opposite to that of Q_0. (iii) Nuclear rotational angular momentum $R = 0$ for the nuclear ground state. This means $I = K$. Using this one obtains from the given expression

$$Q = \frac{I(2I-1)}{(I+1)(2I+3)} Q_0$$

Obviously, for $I = 0$ (even-even nuclei) or for $I = \frac{1}{2}$, $Q = 0$ even if $Q_0 \neq 0$.

Also. $I > 1$, Q is non zero, if $Q \neq Q_0$.

When $I = 1$, $\frac{Q}{Q_0} = 0.1$. For $I = \frac{3}{2}$, $\frac{Q}{Q_0} = 0.2$ Concluding, one can say that for $I \gg 1$, $\frac{Q}{Q_0}$ approaches 1.

We must note that the intrinsic quadrupole moment Q_0 can be determined from the cross-section of Coulomb excitation of the rotational levels of deformed nuclei and the probability of γ-transitions between such levels.

(g) **Isotopic Spin** The charge symmetry and charge independence of nuclear forces and the almost equality of the masses of the neutron and the proton are strongly suggestive that the neutron and proton are the same particles in two different charge states. Obviously, one can treat them mathematically on an equal footing. One can do it conveniently by introducing a mathematical three dimensional *"charge space"*, which is called the *isotopic spin* or *isospin* space (or *isospace*). The neutron and proton are described by an isotopic spinor (isospinor) field in the isospace, the two isospiner components corresponding to the two charge states of the particle. Particles with very close mass values and, correspondingly, similar properties are united into families of one, two or three particles, characterized by the same *isotopic spin* value equal to 0, $\frac{1}{2}$ or 1, respectively. Analogous to an ordinary spin, the number of isotopic spin "projections", gives the number of particles in a family, from which originates the name 'isotopic spin'.

The ordinary spin has two states. Following the same the former development of the isospiner algebra is exactly the same as for the spin $s = \left(\frac{\hbar}{2}\right)\sigma$ whose Z components are $s_Z = \pm\frac{\hbar}{2}$. In the absence of any other interaction, these two spin states having the spin components $s_Z = +\frac{\hbar}{2}$ and $-\frac{\hbar}{2}$ must have the same energy, i.e., the Hamiltonian of the system is invariant under the rotation of the spin vector **s**. However, this symmetry is broken by the presence of the magnetic field. Magnetic field splits up the two states which have different energies in the magnetic field.

Neutron and proton together is given the common name nucleon. A nucleon is a particle with isospin $t = \frac{1}{2}\tau$, whose Z-components in the isospin space

$$t_3 = \frac{1}{2}\, \tau_3 = +\frac{1}{2} \text{ and } -\frac{1}{2} \qquad\qquad \dots (1.1)$$

Correspond respectively to the proton $\left(t_3 = +\dfrac{1}{2}\right)$ and neutron $\left(t_3 = -\dfrac{1}{2}\right)$. We must remember that sometimes opposite convension is also used. One can write the three components of the matrix τ as

$$\tau_1 = \begin{pmatrix} 0 & 1 \\ 1 & 0 \end{pmatrix}, \ \tau_2 = \begin{pmatrix} 0 & -i \\ i & 0 \end{pmatrix} \text{ and } \tau_3 = \begin{pmatrix} 1 & 0 \\ 0 & -1 \end{pmatrix} \qquad\qquad \dots (1.2)$$

One can write the wave functions for the proton and neutron as

$$\eta(p) = \eta\left(t_3 = +\frac{1}{2}\right) = \begin{pmatrix} 1 \\ 0 \end{pmatrix} \qquad\qquad \dots (1.3)$$

and

$$\eta(n) = \eta\left(t_3 = -\frac{1}{2}\right) = \begin{pmatrix} 0 \\ 1 \end{pmatrix} \qquad\qquad \dots (1.4)$$

Obviously $t_3\, \eta(p) = \dfrac{1}{2}\, \eta(p)$ and $t_3\, \eta(n) = \dfrac{1}{2}\, \eta(n)$ $\qquad\qquad \dots (1.5)$

Relations (1.5) clearly reveals that the eigenvalues of the operator t_3 and $+\dfrac{1}{2}$ and $-\dfrac{1}{2}$ for proton and neutron respectively.

One can write the charge operator Q whose eigen values are +1 for the proton state and O for the neutron state as

$$Q = \frac{1}{2}(1 + t_3) = \begin{pmatrix} 1 & 0 \\ 0 & 0 \end{pmatrix} \qquad\qquad \dots (1.6a)$$

with $\qquad\qquad Q\, \eta(p) = \eta(p) \text{ and } Q\, \eta(n) = 0 \qquad\qquad \dots (1.6b)$

We must note that the concept of isospin is not merely a formal analogy but has deep physical significance based on solid experimental observations.

The operators τ_+ and τ_- which, respectively, increase and decrease charge of the nucleon by one unit can be expressed as

$$\tau_+ = \frac{1}{2}(\tau_1 + i\, \tau_2), \ \tau_- = \frac{1}{2}(\tau_1 - i\, \tau_2) \qquad\qquad \dots (1.7)$$

Obviously $\qquad \tau_+\, \eta(p) = 0, \qquad\qquad \tau_-\, \eta(p) = \eta(n) \qquad\qquad \dots (1.8a)$

$$\tau_+\, \eta(n) = \eta(p), \qquad\qquad \tau_-\, \eta(n) = 0 \qquad\qquad \dots (1.8b)$$

One can generalize the concept of isospin to the case of a complex nucleus with Z protons and N neutrons $(A = Z + N)$. To obtain the resultant third component T_3 of the nucleus one will have to add algebraically the third component of the isospin analogous to the Z-component of ordinary spin for the individual nucleons, i.e.

$$T_3 = \sum_i t_{i3} = \sum_i (t_{p3} + t_{n3}) = \frac{Z}{2} - \frac{N}{2}$$

$$= \frac{1}{2}(N - Z) = -\frac{1}{2}(A - 2Z) \qquad \text{... (1.9)}$$

Since the component cannot exceed the magnitude of the vector and hence one must have for the magnitude of the vector T as

$$T \ge \left| \frac{N - Z}{2} \right| \qquad \text{... (1.10)}$$

There may be different combinations of N and Z values for a given A (given isobaric multiplets. This means T_3 will be different for the different nuclei for a given value of A (= $N + Z$). Examples are the Isospin multiplets $^{14}_6C_8$, $^{14}_7N_7$ and $^{14}_8O_6$ at $A = 14$. One can have other possible combinations N and Z for this particular A (= 14) but these are the only three nuclei known to exist in nature or are artificially produced. Obviously, the values of T_3 in this case are -1, 0 and $+1$ respectively for ^{14}C, ^{14}N and ^{14}O. Thus are can write $T_3 = 1$. The Isospin multiplicity = $2T + 1$. Obviously, in the present case isospin multiplicity is 3 and so we have the above triplet.

Example 1.10 What do you understand by isotopic invariance? Explain by considering a system of two nucleons.

Sol. We know that a nucleon consists of one proton and one neutron. Obviously, for a two nucleon system, there are three possibilities: (p, p) (n, n) and (p, n). The possible values of T_3 are $+1$, -1 and O. Thus one can write $T = 1$. If t_1 and t_2 are two isospin vectors for the two nucleons, then the vector addition of the two can yield a resultant $\mathbf{T} = \mathbf{t}_1 + \mathbf{t}_2$ such that T can have two values, i.e. $T = 1$ for parallel alignment of t_1 and t_2 and $T = 0$ for antiparallel alignment of t_1 and t_2. In the first case T_3 can have values 1, 0, -1 corresponding to the cases given above. In the second case we have $T = 0$ and therefore $T_3 = 0$, i.e. it is a singlet. This corresponds to a different isospin state for the (p, n) system. One can see that the two groups of states, the isospin triplet and the isospin singlet, correspond to the ordinary spin singlet (^1S) and triplet (^3S) respectively. Thus the isospin triplet ($T = 1$) for the two nucleon system can be expressed as

$$^1S : (p, p) \text{ with } T_3 = +1, (p, n) \text{ with } T_3 = 0 \text{ and}$$

$$(n, n) \text{ with } T_3 = -1.$$

Pauli exclusion pronciple allows only anti-parallel spin alignment with $S = 0$ for the ground state ($L = 0$) for two identical nucleons (p, p or n, n). However, in the case of the (p, n) system, both anti-parallel and parallel spin alignments are possible with resultant spins $S = 0$ and $S = 1$ respectively. The first of these is analogous to the (p, p) and (n, n) systems where as second one with $S = 1$ and $L = 0$ is the ground state of the deutron, which is a bound state with the binding energy 2.226 MeV. We must note that the other state of the (p, n) system is unbound, just as the (p, p) and (n, n) systems are

Concluding we can say that the nature of nuclear interaction (excluding e.m. interaction) is independent of the type of the nucleon, i.e. of the projection T_3. Obviously, T_3 characterizes the difference in e.m. properties and the nuclear interaction is determined by the value of vector T. Obviously, the nuclear interaction is invariant under rotation in *isospin space*. This is *isotopic invariance*. This implies that the isospin must be conserved in nuclear interaction.

Example 1.11 Calculate the repulsive potential energy due to Coulomb interaction among the protons in a nucleus. Using the result, explain the stability of the nucleus.

Sol. The electric energy E of a homogeneously charged sphere of total charge Q and radius R, is

$$E = \frac{3}{5} \frac{Q^2}{4\pi \varepsilon_o R}$$

Assuming that the Z protons in a nucleus are distributed uniformly, we may take $Q = Ze$ and $R = r_0 A^{\frac{1}{3}}$, then

$$E = \frac{3}{5} \left(\frac{e^2}{4\pi \varepsilon_0 r_0} \right) \frac{Z^2}{A^{\frac{1}{3}}}$$

Substituting numerical values, one obtains

$$E = 9.87 \times 10^{-14} Z^2 A^{-\frac{1}{3}} \text{ J} = 0.617 Z^2 A^{-\frac{1}{3}} \text{ MeV}$$

We must note that electric charge is not spread uniformly over the nuclear volume but is concentrated at the protons and hence one should take $Z(Z-1)$ instead of Z^2 because each of Z-protons interacts only with the remaining $(Z-1)$ protons. This correction is especially important for nuclei with small Z, such as He. For example, for $^4_2$He, with $Z = 2$ and $A = 4$, we have $E_p = 0.776$ MeV (using the corrected formula) and for $^{238}_{92}$U, with $Z = 92$ and $A = 238$, we have $E_p = 879$ MeV. This reveals that because of the Z^2 term, the Coulomb repulsion energy increases very rapidly with the number of protons. This means that unless the nuclear interaction if strong enough and attractive, the nucleus can not be stable.

Example 1.12 Show that the quadrupole moment of a system in which a proton is circling a spherical nucleus having equal number of protons (Z) and neutrons (N) is negative.

Sol. For spheroid, quadrupole moment is zero and $<x^2> = <y^2> = <z^2> = \frac{1}{3} R^2$. For the given system, proton is circling in $x - y$ plane, hence change will be also in $x - y$ plane. Obviously, the distribution of charge is of oblate shape and hence the quadrupole moment of the given system is negative.

Example 1.13 Show that the maximum energy shift that can be observed for a body whose quadrupole moment is Q, is given by

$$\frac{1}{8}\ e\ Q\left(\frac{\partial^2\phi}{\partial z^2}\right)_o \frac{2I-1}{I+1}$$

Where the symbols have usual meanings.

Sol. When the nucleus is under the influence of the potential ϕ, then one can write the interaction energy as

$$U = \int_\tau \phi(r)\ \rho(r)\ d\tau \qquad \dots (1.1)$$

Here ρ is the proton density in the nucleus. Expanding the electric potential ϕ, we have the quadrupole interaction energy

$$\Delta U_2 = \frac{1}{4}\left(\frac{\partial^2\phi}{\partial z^2}\right)_o \int_\tau (3z^2 - r^2)\rho\ d\tau = \frac{1}{4}eQ\left(\frac{\partial^2\phi}{\partial z^2}\right)_o$$

When the nuclear symmetry axis is the I axis, the energy shift is maximum. Obviously, Q to the precession axis is

$$Q_{pre} = \frac{1}{2}\left(\frac{2I-1}{I+1}\right)Q_{intrinsic}$$

Therefore the maximum energy shift for a system of $Q_{intrinsic}$ is

$$(\Delta U_2)_{max} = \frac{1}{8}\ e\ Q\left(\frac{\partial^2\phi}{\partial z^2}\right)_o\left(\frac{2I-1}{I+1}\right)$$

Example 1.14 Calculate the atomic number of the most stable nucleus for a given mass number A.

Sol. The most stable nucleus with a given mass number A is that which has the maximum value of the binding energy. Obviously, one have to compute $\frac{\partial E_b}{\partial Z}$ with A constant, and equate it to zero. The expression for binding energy is

$$E_b = a_1 A - a_2 A^{\frac{2}{3}} - a_3 z^2 A^{-\frac{1}{3}}$$

Where a_1, a_2 and a_3 are constants. Using $N - Z = A - 2Z$, one finds

$$\frac{\partial E_b}{2Z} = -2a_3 Z\ A^{-\frac{1}{3}} + 4a_4(A - 2Z)A^{-1} = 0$$

Substituting the numerical values of a_3 and a_4, one obtains

$$Z = \frac{A}{Z + 0.0157\ A^{\frac{2}{3}}}$$

For light nuclei, i.e. for small A, one can neglect the second term in the denominator, and obtains approximately $Z \approx \frac{1}{2} A$. This result is confirmed experimentally.

SUGGESTED READINGS

1. R.D. Evans: The Atomic Nucleus, McGraw Hill (1955)
2. Y.M. Shirokov and N.P. Yudin, Vol. I, Mir Publishers, Moscow, (1982).
3. W.E. Burcham, Elements of Nuclear Physics, (ELBS) Longman (1979).
4. K.N. Mukhin, Experimental Nuclear Physics, Mir Publishers, Moscow (1987).
5. J.M. Blatt and W.F. Weisskoff, John Wiley and Sons, Newyork (1952).
6. H.A. Bethe and P. Morrison, Elementary Nuclear Theory, John Wiley and Sons (1956)
7. M.A. Preston, Physics of the Nucleus, Addison-Wesley Pub. Co. (1962).
8. L.R.B. Eliton, Introductory Nuclear Theory, Pitman and Sons (1959).
9. I. Kaplan, Nuclear Physics, Addison-Wesley Pub. Co. (1962).
10. R.R. Roy and B.P. Nigam, Nuclear Physics, John Wiley and Sons., New York (1997).
11. David Halliday, Introductry Nuclear Physics, (2nd ed.), John Wiley and Sons. (1955).
12. J. Lilley, Nuclear Physics, John Wiley (2001).
13. K. Heyde, Basics Ideas and concepts in Nuclear Physics, I P O, Bristol (1999).

REVIEW QUESTIONS

1. State the assumptions employed in the Rutherford's α-ray scattering experiment. Obtain an expression for the number of α-particles scattered from a metallic foil of a given thickness.
2. Two individual α-particles with the same energy are observed to be scattered through different angles. Considering only scattering by gold nucleus, which particle does come closer the nucleus the one scattered through a small angle or the one scattered through a large angle.
3. Calculate the distance of closest approach of α-particle of a given initial kinetic energy to an atomic nucleus. Explain why there is an upper limit for the energy of the incident particle beyond which the particle cannot be reflected back (i.e., 180° scattering) by a nucleus. What is the significance of this observation?
4. If we assume that the charge $+Ze$ of a nucleus is spread over a sphere of radius R, show that the fastest α-particles (charge $2e$, mass $4m_p$) which can suffer 180° scattering will have the speed given by

$$v = \sqrt{\frac{Ze^2}{(4\pi\,\varepsilon_0)\,M_p\,R}}$$

5. Describe how nuclear size is estimated from Rutherford scattering.
6. What do you know about isotopes, isobars, and isotones and how they differ from each other?
7. What is mass defect and how is it related to nuclear binding energy? Explain with example.
8. Explain the meaning of angular momentum, magnetic dipole moment and parity of a nucleus. What is meant by the electric quadrupole moment of a nucleus? Obtain an expression for the same.
9. Define mass defect and binding energy. Draw a curve showing variation of binding energy per nucleon with mass number A of the elements and explain its features.
10. How will you calculate the size of the nucleus by α-scattering? Give in short about nuclear density.
11. Define the following (i) Packing fraction (ii) Mass defect. How stability of nucleus is related with binding energy of nucleus?

12. What is meant by electric quadrupole moment of a nucleus? Prove that

$$Q = \frac{e}{Z}(3Z^2 - r^2)$$

If two protons are $x = \pm R$ on the equator, what will be the electric quadrupole moment?
13. Write short notes on
    (i) Parity
    (ii) Isospin
    (iii) Electric quadrupole moment and its importance

SHORT QUESTIONS

1. State the assumptions made by Rutherford in his theory of the scattering of α-particles.
2. What is impact parameter?
3. Write the relation between the impact parameter and the angle of scattering?
4. What are constitutents of an atomic nucleus?
5. Differentiate between isotopes and isobars?
6. Explain the significance of binding energy
7. Show that 1 amu $= 931$ MeV
8. What are magic numbers?
9. Draw the curve showing the variation of binding energy per nucleon as a function of the mass number of stable nuclei. Discuss the interesting conclusions from this curve.
10. Explains the terms binding energy and packing fraction.
11. Explain how nuclear stability depends on binding energy.
12. Enumerate the basic properties of atomic nuclei.
13. Is the use of the classical expression for kinetic energy justified in the derivation of Rutherford's scattering formula? If so why?
14. What do you understand by the magnetic moment of a neutron.
15. What holds the nucleons in the nucleus together.
16. Discuss the important properties of nucleus
17. Explain how some of the important properties of the nucleus could be established experimentally?
18. Write the form of charge density assumed by Hofstader and his colleagues.
19. Draw a vector diagram showing possible orientations of the total angular momentum vector **I** for $I = \frac{5}{2}$.
20. How the quadrupole moment Q is related to the nuclear deformation parameter δ?
21. Explain spin-symmetry relationship for a nucleus.

PROBLEMS

1. Calculate the distance of closest approach of a 5 MeV alpha paritcles to a gold nucleus of atomic number $Z = 79$. (**Ans.** 6.8×10^{-12} m)

2. What is the mass number A of nucleus whose radius $A = 2.71$ fm? Given $R = 1.50 \times 10^{-15} \, A^{\frac{1}{3}}$ m. [**Ans.** 9]

3. Calculate the fraction of α-particles of energy of 8 MeV which is scattered by less than 1° when incident upon a copper foil of thickness 0.25 μm   (**Ans.** 0.9987)

4. In an experiment on α-scattering from gold foil, the observer counted 400 counts per minute at a scattering angle of 30°. How many counts per minute would be observed if the gold foil is replaced by an aluminium foil of same thickness.   (**Ans.** 75 Counts/minute)

5. A beam of α-particle energy 5.5 MeV is incident normally on a gold foil at a rate of 10^4 particles/sec. An α-particle counter with an aperature 4 sq. cm. is placed at a distance 0.25 m from the foil. Calculate the counts at an angle 60°, if the thickness of the foil is 10^{-5} m.   [**Ans.** 7 Counts/hour]

6. Calculate the nuclear radius of ^{16}O, ^{120}Sn and ^{208}Pb.

   [**Ans.** 3.62×10^{-15} m; 6.90×10^{-15} m and 8.28×10^{-15} m]

7. What nuclei have a radius equal to one-half the radius of ^{236}U?   [**Ans.** ^{28}Si, ^{29}Si, ^{30}Si]

8. Estimate the Coulomb repulsion energy of the two protons in ^{3}He (assuming that they are 1.7×10^{-15} m apart). Compare this energy with the difference in the binding energies of ^{3}H and ^{3}He. Is the result compatible with the assumption that nuclear forces are charge independent?

9. The binding energy of $^{35}_{17}$Cl is 289 MeV. Calculate its mass in amu   [**Ans.** 34.98 amu]

10. Mass of α-particle = 4.002603 amu, mass of proton = 1.007825 amu, mass of neutron = 1.008664 amu. Taking 1 amu = 982 MeV, calculate the binding energy of an α-particle.   [**Ans.** 28.28 MeV]

11. Assuming a uniform charge distribution within a nucleus of radius $R = r_0 \, A^{\frac{1}{3}}$. Show that for high energy electron scattering the form factor $F(q)$ is given by

$$F(q) = \frac{3}{(q^2 \, r_0^3 A)} \left[\frac{\sin q \, R}{q} - R \cos q \, R \right]$$

    where $q = \left(\frac{2p}{\hbar} \right) \sin \frac{\phi}{2}$.

12. An aluminium foil scatters 103 α-particles per second in a given direction. If the aluminium foil is replaced by a gold foil of identical thickness, how many α-particles will be scattered per second in the same direction?   (**Ans.** 3.62×10^4 particles/s).

13. Taking nuclear radius $R = 4 \times 10^{-15}$ m, find the numerical value of classical electric quadrupole moment of a nucleus having a finite angular momentum and containing one proton at the nuclear equator in addition to a spherically symmetric distribution of charge. Also find the quantum mechanical value for the same quadrupole moment.

    [**Ans.** -0.08×10^{-28} m²/electron; -0.16×10^{-28} m²/electron]

OBJECTIVE QUESTIONS

1. Isobars are nuclides which have
   (a) same A but different Z
   (b) same A and Z
   (c) same Z but different A
   (d) same Z and same N

2. $^{19}_9$F is an example of chemical element occuring in nature
   (a) with no natural isotope
   (b) with three natural isotopes
   (c) with seven natural isotopes
   (d) with several natural isotopes

3. Nuclear radius is proportional to

    (a) $A^{\frac{2}{3}}$ (b) $A^{\frac{1}{3}}$

    (c) A (d) $\dfrac{3}{4}$

4. Nuclear volume is proportional to

    (a) $A^{\frac{1}{3}}$ (b) A^2

    (c) A (d) $A^{\frac{1}{2}}$

5. The resultant angular momentum of a nucleus is called

    (a) nuclear radius   (b) nuclear mass

    (c) nuclear spin   (d) nuclear qudrupole moment

6. Practically all even-even nuclei have resultant angular momentum, i.e. nuclear spin $I = 0$, which indicates that identical nucleons tend to pair their angular momenta in opposite directions. This effect is called

    (a) isotopic invariance   (b) Parity

    (c) Pairing effect   (d) Gamow effect

7. Nuclear magneton is smaller than the Bohr magneton by a factro about

    (a) 20   (b) 200

    (c) 2000   (d) 20000

8. A nucleus must have

    (a) Zero dipole moment

    (b) non zero but finite and positive dipole moment

    (c) infinite dipole moment

    (d) negative dipole moment

9. The quadrupole moment is related to nuclear deformation by a relation

    (a) $Q = \dfrac{4}{5} Z R_0{}^3 \delta$ (b) $Q = \dfrac{4}{5} Z R_0{}^3 \delta \left(1 + \dfrac{\delta}{2} + ... \right)$

    (c) $Q = \dfrac{4}{5} Z R_0{}^3 \delta^{\frac{2}{3}}$ (d) $Q = \dfrac{4}{5} Z R_0{}^3 \delta^2$

10. For a spherical symmetry, the quadrupole moment

    (a) −negative   (b) +positive

    (c) 0   (d) none of the above

11. Quadrupole moment is exhibited by

    (a) spherical nuclei   (b) ellipsoidal nuclei

    (c) neither (a) nor (b)   (d) both (a) and (b)

12. In an atom there are p-protons, n-neutrons and e electrons. For atoms of small mass the relation between p, n and e is

    (a) $p = n \approx ne$ (b) $p \approx n \approx e$

    (c) $p = e \approx n$ (d) $n = p - 1 = e$

13. Assuming that a nucleus is a sphere of nuclear matter of radius $1.2 \times A^{\frac{1}{3}}$ fm, the average density in SI units is

(a) $2.29 \times 10^{17}\,\text{kg/m}^3$ (b) $2.29 \times 10^{13}\,\text{kg/m}^3$

(c) $2.29 \times 10^{11}\,\text{kg/m}^3$ (d) $2.29 \times 10^{9}\,\text{kg/m}^3$

ANSWERS

1. (a)	**2.** (a)	**3.** (c)	**4.** (c)	**5.** (c)
6. (c)	**7.** (c)	**8.** (a)	**9.** (b)	**10.** (c)
11. (b)	**12.** (c)	**13.** (a)		

$$\boxed{2}$$

RADIOACTIVITY

2.1 INTRODUCTION

Henry Becquerel in 1986 during his experiments on the fluorescence of uranium salts noticed that a photographic plate wrapped in black paper was affected by the salts kept outside the paper. He showed that this is due to the emission of some new type of radiation by the uranium salts independent of any external influence. The radiation from the uranium salts is found to capable of producing ionization in gases and thereby rendering gases conducting. The phenomenon discovered by becquerel is called *radioactivity*. Becquerel's discovery was followed by intensive research by Curies who found that thorium salts too are radioactive. Pierre and Madan Curie discovered two new radioactive elements which they named polonium and radium. The radiations from radioactive substances were classified as α-and β-rays by Rutherford from his experiments on the penetrating powers of these radiations. To these two types of radiations was added a third-type, named γ-rays, by villard. Radioactivity is exhibited by heaviest elements in periodic table. About 40 elements are known to be naturally radioactive. No single phenomenon has played so significant a role in the development of nuclear physics as radioactivity. Following three aspects of radioactivity are extraordinary from the perpective of classial physics:

(i) When a nucleus undergoes α or β decays its atomic number Z changes and it becomes the nucleus of a differnt atom. Obviously, the elements are not immutable.

(ii) The energy liberated during radioactive decay comes from within individual nuclei without external excitation, unlike the case of atomic nuclei.

(iii) Radioactive decay is a statistical process.

We must note that radioactivity is observed not only amongst a few naturally occuring elements. There are large number of radioactive isotopes of elements which are not normally radioactive, have been produced in the laboratory. These are termed as artifical radioactive substances.

2.2 PROPETIES OF RADIOACTIVE RAYS

(i) Radioactive rays ionize the surrounding air and affect photographic plates
(ii) Curie found that one gm of radium liberates 140 calories in one hour. Although this is small but this energy is released continuously over a very long period of time.
(iii) Fluorescence is produced in substances like *ZnS*, barium platinocynide, etc. When minute quantities of radium is added to, say *ZnS*, one finds that the compound obtained is continuously luminuous in the dark.
(iv) These rays act differently on different cells and tissues. These rays destroy most readily those cells which multiply rapidly. This made radium an invaluable aid to physicians to fight tumours, particularly cancerous growths.
(v) Radioactivity is a nuclear phenomenon. The nuclei of atoms spontaneously disintegrate to produce other nuclei by the emission of corpuscular radiations. Rutherford found that a beam of radioactive rays from a radium sample into three components (α, β or γ rays) in strong magnetic or electric field (Figure 2.1).

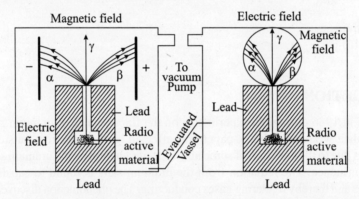

Fig. 2.1 Splitting or radiations emitted by radioactive substance when subjected to electric or magnetic field.

(a) The α-rays (particles) Rutherford and Royds in 1909 proved that the α-rays were nothing but the nuclei of the helium atoms $^4_2$He having the mass number $A = 4$ and the atomic number $Z = 2$. Obviously, the emission of an α-particle from a nucleus $^A_Z X$ (parent nucleus) reduces the mass number by 4 units and the atomic number by 2 units, i.e. a new nucleus $^{A-4}_{Z-2} Y$ (daughter nucleus) is formed, as a result of α-disintegration. For example of radon, one can write the radioactive transformation equation as

$$^{222}_{86}\text{Rn} \rightarrow {}^4_2\text{He} + {}^{218}_{34}\text{Po} \qquad \qquad \text{... (2.1)}$$

We must note that during a radioactive transformation, the mass number and total charge is conserved. Few more examples of α-decay are

$$^{238}_{88}\text{Ra} \rightarrow {}^4_2\text{He} + {}^{234}_{90}\text{Th} \qquad \qquad \text{... (2.2)}$$

$$^{238}_{92}\text{U} \rightarrow {}^4_2\text{He} + {}^{234}_{90}\text{Th} \qquad \qquad \text{... (2.3)}$$

$$^{218}_{84}\text{PO} \rightarrow {}^4_2\text{He} + {}^{214}_{82}\text{Pb} \qquad \qquad \text{... (2.4)}$$

α-rays cause intense ionization in air and can be stopped by a thin sheet of paper. Usually the velocities of emitted α particles are found between $\sim 1.5 \times 10^7$ m/s and $\sim 2.2 \times 10^7$ m/s. Any group of α-particles emitted from the same type of nuclei are found to possess a definite velocity and hence a definite energy.

α-particles are found to cover a definite distance in a given material, practically without any loss of intensity and then suddenly in a small distance are absorbed completely. The definte distance travelled within a given material is called the *range* of α-particles in that material. For example, α-particles from $^{226}_{88}$Ra have a range in air equal to 3.4 cm at O^oC and 76 cm of pressure.

(b) The β-rays (particles) Various experiments have established that the β-rays are identical with electrons. β-particles are very high energy electrons and therefore a β-particle has a mass $\left(\dfrac{1}{1836}\right)$ of mass of a proton. Since β-particle carry one unit of nagative charge the positive charge of the nucleus is increased by one unit as a result of β-particle emission. Obviously, the atomic number Z of the parent nucleus is increased $Z + 1$ in the daughter nucleus in this case. Mass number A remains unchanged. In general β-particle unchanged. In general β-particle disintegration can be represented as,

$$^{A}_{Z}X \underline{\;\beta\;} ^{A}_{+1}Y \qquad \qquad \qquad ... (2.5)$$

A few examples of β-decay are

$$^{14}_{6}C \rightarrow ^{14}_{7}N + ^{0}_{-1}e \qquad \qquad \qquad ... (2.6)$$

$$^{210}_{83}Bi \rightarrow ^{210}_{84}Po + ^{0}_{-1}e \qquad \qquad \qquad ... (2.7)$$

$$^{234}_{90}Th \rightarrow ^{234}_{91}Pa + ^{0}_{-1}e \qquad \qquad \qquad ... (2.8)$$

β-rays cause much less ionization in air. β-rays are about 100 times more penetrating than α-rays. It is found that β-rays can penetrate a sheet of aluminium a few mm thick.

The velocities of β-particles upto 0.99 c. A particular β-active nucleus emits β-particles with energy varying between zero and certain maximum. This maximum energy is termed as *end-point energy*.

(c) γ-rays These are very high electromagnetic radiations with very short wavelenghts. In many cases wavelenghts of γ-rays are even shorter than those of X-rays. Thus, usually γ-rays photons are more energetic than the X-ray photons and are even more penetrating than X-rays. α or β disintegration of nucleus is usually followed by the emission of one or more γ-ray photons. γ-rays are about 100 times more penetrating than β-rays.

The emission of γ-rays does not produce any fundemental change in the nature of the nucleus. γ-rays are also emitted during various types of nuclear transmutations induced by artificial means. During the γ-ray emission a nucleus makes a transition from one energy state to another.

In general γ-decay can be expressed as

$$^{A}_{Z}X^* \rightarrow ^{A}_{Z}X + \gamma$$

An example of γ decay is

$$^{87}_{38}Sr^* \rightarrow ^{38}_{38}Sr + \gamma$$

Where * denotes an excited nuclear state

γ-rays have much greater penetrating power and hence their interaction with matter is much different than that of α-and β-rays. The absorption of γ-rays by matter is accompanied by mainly three processes: (i) *photo-electric absorption* (ii) *Compton scattering and* (iii) *pair production*. We will study α, β and γ decays in detail later on.

2.3 RADIOACTIVE DECAY

Rutherford and Soddy, as a result of their extensive studies on radioactivity, particularly of uranium and actinium, put forward the following conclusions:

(i) The radioactive emission is characterstic of the isotope, it varies from one isotope to another of the same element.

(ii) On emission of α-or β-rays which are usually, but not always, accompanied by the emission of γ-rays, the emitting parent nuclides transformed into those of new daughters elements; the daughter nuclides themselves are radioactive, so that the process of radioactive disintegration continue, until the original active parent nuclide gets transformed into a stable one, usually lead, through a series of successive disintegrations.

(iii) The emission occurs spontaneously and cannot be speeded up or slowed down by physical means such as change of pressure or temperature, magnetic or electric fields, or non relativistic speeds. A seeming exception to this statement is the slight effect of chemical form on a nuclear process known as *K*-capture.

(iv) The disintegration occurs at random and which will disintegrate first is simply a matter of chance, i.e. the nature of radioactive decay is *statistical*.

(v) The rate of disintegration of a particular substance (i.e., number of atoms disintegrating per second) at any instant is proportional to the number of atoms persent at that time.

Let us consider a radioactive sample containing. N_0 nuclei at time $t = 0$, i.e., at the begining. Our aim is to calculate the number N of undecayed nuclei left after time t.

The number of nuclei of a given radioactive sample disintegrating per second is called the activity of a radioactive sample. The rate of change $\dfrac{dN}{dt}$ of N with time, which is a measure of the activity of the sample, is proportional to N, i.e.

$$\frac{dN}{dt} = \text{Rate if decrease of nuclei with time or activity at time } t$$

or
$$-\frac{dN}{dt} = \propto N \qquad\qquad \text{... (2.1)}$$

or
$$-\frac{dN}{dt} = \lambda N \qquad\qquad \text{... (2.2)}$$

where $\lambda > 0$, is the proportionality constant and known as *disintegration constant* or *decay constant* and it depends in no way on the amount of substance presents. The negative sign in the L.H.S. of (2.2) is due to diminution in the number of atoms N with time. For unit time (2.2) can be expressed as

$$\lambda = \left(-\frac{dN}{N} \right) \qquad \qquad \text{... (2.3)}$$

Obviously, λ is fractional change in N per sec and one can see that λ is not a mere proportionality constant, but it gives the probability of decay per unit interval of time. Therefore, λ is also called *probability constant.*

Integrating (2.2), one obtains

$$\int_{N_0}^{N} \frac{dN}{N} = \int_{0}^{t} \lambda \, dt$$

or $\qquad \qquad \log_e N = -\lambda t + k \qquad \qquad \text{... (2.4)}$

Where k is the integration constant. We have at $t = o$, $N = N_0$. Thus

$$k = \log_e N_0 \qquad \qquad \text{... (2.5)}$$

using (2.5), Equation (2.4) becomes

$$\log_e N = -\lambda t + \log_e N_0$$

or $\qquad \qquad \log_e \frac{N}{N_0} = -\lambda t$

or $\qquad \qquad N = N_0 \, e^{-\lambda t} \qquad \qquad \text{... (2.6)}$

Equation (2.6) show that the number of active nuclides decreases exponentially with time (i.e., more rapidly at first and slowly afterwords). Equation (2.6) is represented in Figure 2.2.

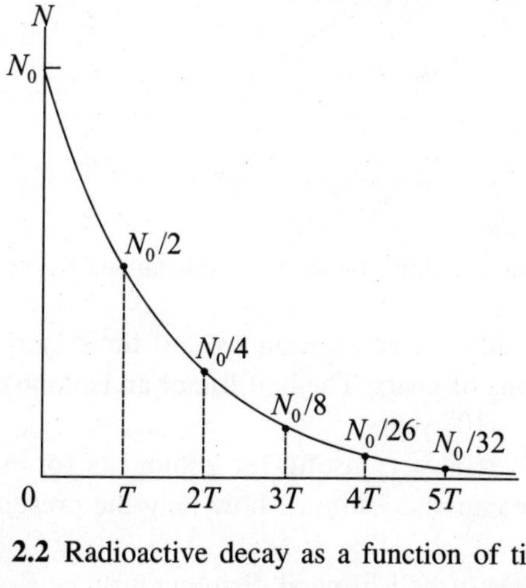

Fig. 2.2 Radioactive decay as a function of time.

Half Life One can define a time interval (T) during which half of a given sample of radioactive substance decays. This interval of time T which is fixed for a given element is called the *half life* or *half value period* of the radioactive substance. The half life varies widely from one substance to another. For example for radium its value is 1620 years while for radon it is only 3.80 days. Obviously, if one starts with one milligram of radon gas then at the end of 3.82 days only 0.5

milligram will remain and at the end of next 3.82 days 0.25 mg will remain and at the end of the next 3.82 days only $\frac{1}{8}$ mg will remain.

From Equation (2.6), at $t = T$ (half life period),

$$N = \frac{1}{2} N_0. \text{ Therefore}$$

$$\frac{N_0}{2} = N_0 \, e^{-\lambda T}$$

or $$e^{-\lambda T} = \frac{1}{2}$$

or $$\lambda T = \log_e 2$$

or $$T = \frac{\log_e 2}{\lambda} = \frac{2.303 \times \log_{10} 2}{\lambda} = \frac{2.303 \times 0.3010}{\lambda}$$

or $$T = \frac{0.693}{\lambda} \qquad\qquad ... (2.7)$$

As λ incrases T decreases and vice versa. The half life (T) of any radioactive element is absolutely invariant and changes of pressure, temperature, electric and magnetic fields have no effect on it. Every radioactive element has its own characterstic value of λ and T (Figure 2.3)

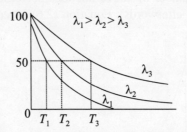

Fig. 2.3 Decay curves for three radioactive substances having different half lives

The half life of natural radioactive substances and their isotopes vary from a fraction of a second to hundreds of millions of years. The half life of an isotope of polonium ($^{214}_{84}$Po) is 10^{-5} s, where half life of $^{238}_{92}$U is 4.5×10^9 years.

The knowledge of half lives is very useful for geologists for estimating the ages of mineral deposits, rocks and earth. One can also estimate how long the present stock of natural radioactive substances will last.

Table 2.1 depicts the values of half life and disintegration constant for the uranium series.

Mean Life or Average Life: The atoms of a radioactive substance are continuously disintegrating. Which atom will disintegrate first is a matter of chance. Obviously, atoms which disintegrate in the begining have a very short life and those which disintegrate at the end have the longest life. Thus individual radioactive atoms may have life spans between zero and infinity. Therefore, it is meaningful

Table 2.1 The Uranium Series ($4n + 2$)

Radioactive species	Nuclide	Type of disinte-gration	Half-life (T)	Disintegration constant (λ) s^{-1}	Particle energy MeV
Uranium I (UI)	$^{238}_{92}U$	α	4.51×10^9 y	4.88×10^{-18}	4.19
Uranium X_1 (UX$_1$)	$^{234}_{90}Th$	β	24.1 d	3.33×10^{-7}	0.19
Uranium X_2 (UX$_2$)	$^{234}_{91}Pa$	β	1.18 m	9.77×10^{-3}	2.31
Uranium Z (UZ)	$^{234}_{91}Pa$	β	6.66 h	2.88×10^{-5}	0.5
Uranium II (UII)	$^{234}_{92}U$	α	2.48×10^5 y	8.80×10^{-14}	4.768
Thorium (Th)	$^{230}_{90}Th$	α	8.0×10^4 y	2.75×10^{-13}	4.68
Radium (Ra)	$^{226}_{88}Ra$	α	1620 y	1.36×10^{-11}	4.777
Radon (Rn)	$^{222}_{86}Rn$	α	3.82 d	2.10×10^{-6}	5.486
Radium A (RaA)	$^{218}_{84}Po$	α, β	3.05 m	3.78×10^{-3}	α:5.998
Radium B (RaB)	$^{214}_{82}Pb$	β	26.8 m	4.31×10^{-4}	0.7
Astatine-218 (^{218}At)	$^{218}_{85}At$	α	1.3 s	0.4	6.70
Radium C (RaC)	$^{214}_{83}Bi$	α, β	19.7 m	5.86×10^{-4}	α: 5.51
					β: 3.17
Radium C′ (RaC′)	$^{214}_{84}Po$	α	1.64×10^{-4} s	4.23×10^3	7.683
Radium C″ (RaC″)	$^{210}_{81}Tl$	β	1.32 m	8.75×10^{-4}	1.96
Radium D (RaD)	$^{210}_{82}Pb$	β	21 y	1.13×10^{-9}	0.0185
Radium E (RaE)	$^{210}_{83}Bi$	β	5.0 d	1.60×10^{-6}	1.155
Radium F (RaF)	$^{210}_{84}Po$	α	138.4 d	5.80×10^{-8}	5.300
Thallium-206 (^{206}Tl)	$^{206}_{81}Tl$	β	4.2 m	2.75×10^{-3}	1.51
Radium G (RaG)	$^{206}_{82}Pb$	Stable			

to talk about the mean or average life (τ). From the curve of Figure 2.4, we note that each of the dN number of radioactive nuclei has lived a life of t seconds, i.e. the total life span of dN nuclei is ($dN \cdot t$) seconds. Thus one obains,

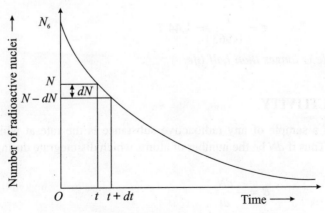

Fig. 2.4 Decay of dN number of nuclei in time dt.

$$\tau = \frac{\text{Total life time of all nuclei in a given sample}}{\text{Total number of nuclei in that sample}}$$

or

$$\tau = \frac{\int_0^\infty t \, dN}{\int_{N_0}^0 dN} = \frac{-N_0 \lambda \int_0^\infty t \, e^{-\lambda t} \, dt}{-N_0}$$

$$(\because \ dN = -\lambda N \, dt = -\lambda \, N_0 \, e^{-\lambda t} \, dt)$$

$$= \lambda \int_0^\infty t \, e^{-\lambda t} \, dt$$

Integrating by parts, one obtains

$$\tau = \lambda \left[\frac{t \, e^{-\lambda t}}{-\lambda} - \int \frac{e^{-\lambda t}}{-\lambda} \, dt \right]_0^\infty$$

The first term of R.H.S. on substituting the limits become zero. Therefore

$$\tau = \int_0^\infty e^{-\lambda t} \, dt = \left[\frac{e^{-\lambda t}}{-\lambda} \right]_0^\infty$$

$$= \left[o - \frac{1}{-\lambda} \right] = \frac{1}{\lambda} \qquad \qquad \ldots (2.8)$$

Obviously, the mean life is the reciprocal of the transformation or decay constant.

Relation between Mean life (τ) *and Half life* (*T*)

We have from (2.8)

$$\tau = \frac{1}{\lambda}$$

and from (2.7), we have

$$T = \frac{0.693}{\lambda}$$

$$\therefore \qquad \qquad \tau = \frac{T}{0.963} = 1.44 \, T \qquad \qquad \ldots (2.9)$$

Obviously, *mean life is longer than half life.*

2.4 UNIT OF ACITIVITY

The activity (R) of a sample of any radioactive substance is the rate at which the constitutent atoms disintegrate. Thus if dN be the number of atoms which disintegrate during a time interval dt, we have

$$R = -\frac{dN}{dt} \qquad \qquad \ldots (2.1)$$

The negative sign in (2.1) indicates that the number of atoms is decreasing with time.

If at any time the number of atoms in the sample is N, then it is found experimentally that the rate of disintegration of atoms is proportional to N, i.e.

$$-\frac{dN}{dt} \propto N = \lambda N \qquad \text{... (2.2)}$$

Where λ is the decay constant for the particular atom (rather the particular isotope). From Equations (2.1) and (2.2)

$$R = \lambda N \qquad \text{... (2.3)}$$

Obviously, the activity of a sample depends upon the number of atoms in the sample, i.e. upon the mass of the sample and upon the type of atom. The traditional unit of activity is *Curie* (Ci). The Curie was originally defined as the activity of 1 gm of radium. With the improvement in measurment techniques, the value of Curie varied. Now it is arbitrarily defined as

1 Curie (Ci) = 3.7×10^{10} disintegrations/sec. The activity of 1 gm of radium is very nearly equal to 1 Curie. The smaller units are 'millicurie (mCi) and 'microcurie' (μCi)

$$1 \text{ m Ci} = 10^{-3} \text{ Ci} = 3.7 \times 10^{7} \text{ disintegrations/s}$$

$$1 \text{ } \mu \text{ Ci} = 10^{-6} \text{ C}; = 3.7 \times 10^{4} \text{ disintegrations/s}$$

Sometimes one uses another unit of activity, called *rutherford* (*rd*)

$$1 \text{ rutherford } (rd) = 10^{6} \frac{\text{disintegrations}}{\text{sec}}$$

$$1 \text{ m } rd = 10^{-3} \text{ } rd$$

$$1 \text{ } \mu \text{ } rd = 10^{-6} \text{ } rd$$

The *SI* unit of activity is the '*becquerel*' (*Bq*) named after the discoverer of radioactivity:

$$1 \text{ Bq} = 1 \text{ disintegration/s}$$

$$\therefore \qquad 1 \text{ Ci} = 3.7 \times 10^{10} \text{ Bq} = 37 \text{ G Bq}.$$

The larger units such as M Bq (= 10^{6} Bq) and G Bq (= 10^{9} Bq) are often used because the activities that we come across in practice are very much larger.

2.5 SUCCESSIVE DISINTEGRATIONS (RADIOACTIVE GROWTH AND DECAY)

It is observed that a radioactive element undergoes a long series of transformations till the final stable element is obtained. Obviously, it is useful to know the number of atoms of various generations at any specified time, in a mixture.

Let us suppose that a substance A decay into a substance B which in turn decays into C. For the sake of simplicity let us suppose that initally (t = o) the number of A atoms is N_o and that of B is zero whereas after time t, the number of A atoms reduces to N, and that of B becomes N_2. We must note that every disintegration of A atom increases the number of B atoms and every disintegration of B atom reduces the number of B atoms. Let the decay constants for these two transformations are λ_1 and λ_2, then

$$A \xrightarrow{\lambda_1} B \xrightarrow{\lambda_2} C$$

The rate of decay of the element $A = \dfrac{dN_1}{dt} = \lambda_1 N_1$ and the rate of decay of element $B = \lambda_2 N_2$

$$\qquad \text{... (2.1)}$$

Rate of formation of $B = (\lambda_1 N_1 - \lambda_2 N_2)$ $\qquad\qquad$ ... (2.2)

But $\qquad\qquad\qquad N_1 = N_0\, e^{-\lambda_1 t}$ $\qquad\qquad\qquad\qquad$ ... (2.3)

Using (2.3), one obtains from (2.2)

$$\frac{dN_2}{dt} + \lambda_2 N_2 = \lambda_1 N_0\, e^{-\lambda_1 t}$$

Multiplying throughout by the factor $e^{\lambda_2 t}\, dt$, one obtains

$$\left(\frac{dN_2}{dt} + \lambda_2 N_2\right) e^{\lambda_2 t}\, dt = \lambda_1 N_0\, e^{-\lambda_1 t}\, e^{\lambda_2 t}\, dt$$

or $\qquad\qquad\qquad \dfrac{d}{dt}(N_2\, e^{\lambda_2 t}) = \lambda_1 N_0\, e^{(\lambda_2 - \lambda_1)}\, dt$

Integrating, one obtains

$$N_2\, e^{\lambda_2 t} = \frac{\lambda_1}{\lambda_2 - \lambda_1}\, N_0\, e^{(\lambda_2 - \lambda_1)t} + K \qquad\qquad\qquad \text{... (2.4)}$$

Where K is a constant of integration. But we have supposed that $\dfrac{\lambda_1}{\lambda_2 - \lambda_1}\, N_2 = 0$ at $t = 0$.

$$\therefore \qquad\qquad\qquad o = \frac{\lambda_1 N_0}{(\lambda_2 - \lambda_1)} + K \text{ or } K = -\frac{\lambda_1 N_0}{\lambda_2 - \lambda_1} \qquad\qquad \text{... (2.5)}$$

Substituting in (2.4), one obtains

$$N_2\, e^{\lambda_2 t} = \frac{\lambda_1}{\lambda_2 - \lambda_1}\, N_0(e^{(\lambda_2 - \lambda_1)t} - 1)$$

or $\qquad\qquad\qquad N_2 = \dfrac{\lambda_1}{(\lambda_2 - \lambda_1)}\, N_0(e^{-\lambda_1 t} - e^{-\lambda_2 t})$ $\qquad\qquad$ ... (2.6)

Figure 2.5 shows the decay of parent and growth of the daughters atoms with times. Only the parent $A(T \approx 1 \text{ hr})$ is initially present. The daughter B, has a half life of 5 hr and the third member, C, is stable.

Equation (2.6) gives the variation of the number of B atoms N_2 with time. One can see from (2.6) that $N_2 = 0$ at $t = 0$. It increases with increasing t and attains a maximum at some time t_m (say). One can determine t_m by differentiating N_2 w.r.t. time:

$$\left(\frac{d N_2}{dt}\right)_{t_m} = \frac{\lambda_1 N_0}{\lambda_1 - \lambda_2}\, \{-\lambda_2 \exp(-\lambda_2 t) + \lambda_1 \exp(-\lambda_1 t)\} t_m = 0$$

or $\qquad\qquad\qquad \lambda_2 \exp(-\lambda_2 t_m) = \lambda_1 \exp(-\lambda_1 t_m)$

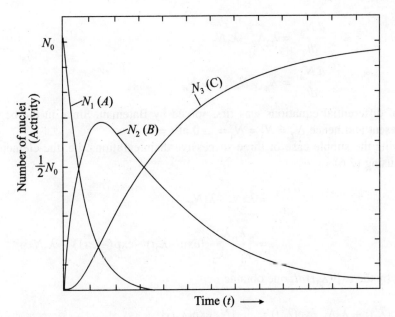

Fig. 2.5 Radioactive decay and growth with three members A, B and C

or
$$\exp\{(\lambda_1 - \lambda_2)t_m\} = \frac{\lambda_1}{\lambda_2}$$

$\therefore$
$$t_m = \frac{1}{\lambda_1 - \lambda_2} \, ln\left(\frac{\lambda_1}{\lambda_2}\right) \qquad \qquad \dots (2.7)$$

One can easily check that N_2 becomes maximum at $t = t_m$ by a second differentiation of N_2. One finds that $\dfrac{d^2N}{dt^2}$ is negative at $t = t_m$.

Sometimes one finds that the daughter product C in the radiodisintegration of B is also radioactive. Again the daughter product D formed from the disintegration of C may be radioactive. One can write such a series of successive disintegrations as

$$A \xrightarrow{\lambda_1} B \xrightarrow{\lambda_2} C \xrightarrow{\lambda_3} D \xrightarrow{\lambda_4} \dots\dots$$

We must note that the last arrow after D indicates that D is also radioactive, i.e. radioactive chain continues after D. Obviously the chain terminates after a few steps till a stable end product is formed.

Let λ_1, λ_2, λ_3, λ_4, ... etc. represent the decay constants and N_1, N_2, N_3, N_4, ... etc., are the number of atoms at the instant t of the elements A, B, C, D, etc. then one finds.

$$\frac{dN_1}{dt} = -\lambda_1 N_1$$

$$\frac{dN_2}{dt} = \lambda_1 N_1 - \lambda_2 N_2$$

$$\frac{dN_3}{dt} = \lambda_2 N_2 - \lambda_3 N_3$$

$$\frac{dN_n}{dt} = \lambda_{n-1} N_{n-1} - \lambda_n N_n$$

This series of differential equations was first solved by Bateman. Since in the begining only A atoms are present and hence $N_2 = N_3 = N_4 = ...\, 0$ at $t = 0$.

Considering the simple case of three successive disintegrations, i.e. the element D is stable, one obtains, using (2.6)

$$\frac{dN_3}{dt} = \lambda_2 N_2 - \lambda_3 N_3$$

$$= \frac{\lambda_1 \lambda_2}{\lambda_1 - \lambda_2} \{\exp(-\lambda_2 t) - \exp(-\lambda_1 t)\} - \lambda_3 N_3$$

Multiplaying by the $\exp(\lambda_3 t)\, dt$, one obtains

$$\frac{dN_3}{dt} \exp(\lambda_3 t) + \lambda_3 N_3 \exp(\lambda_3 t) = \frac{d}{dt} \{N_3 \exp(\lambda_3 t)\}$$

$$= \frac{\lambda_1 \lambda_2}{\lambda_1 - \lambda_2} \{\exp(-\lambda_2 t) - \exp(-\lambda_1 t)\} \exp(\lambda_3 t)$$

Integrating one obtains

$$N_3 \exp(\lambda_3 t) = \frac{\lambda_1 \lambda_2}{\lambda_1 - \lambda_2} N_0 \left\{ \frac{\exp(\lambda_3 - \lambda_2)t}{\lambda_3 - \lambda_2} - \frac{\exp(\lambda_3 - \lambda_1)t}{\lambda_3 - \lambda_1} \right\} + K$$

Where K is constant of integration. knowing $N_3 = 0$ at $t = 0$, one obtains

$$\frac{\lambda_1 \lambda_2 N_0}{\lambda_1 - \lambda_2} \frac{\lambda_2 - \lambda_1}{(\lambda_3 - \lambda_2)(\lambda_3 - \lambda_1)} + K = 0$$

or

$$K = \frac{\lambda_1 \lambda_2 N_0}{(\lambda_3 - \lambda_1)(\lambda_3 - \lambda_2)}$$

Thus one obtains

$$N_3 = \lambda_1 \lambda_2 N_0 \{K_1 \exp(-\lambda_1 t) + K_2 \exp(-\lambda_2 t) + K_3 \exp(-\lambda_3 t) \quad ... \;(2.8)$$

where

$$K_1 = \frac{1}{(\lambda_2 - \lambda_1)(\lambda_3 - \lambda_1)}$$

$$K_2 = \frac{1}{(\lambda_1 - \lambda_2)(\lambda_3 - \lambda_1)}$$

$$K_3 = \frac{1}{(\lambda_1 - \lambda_3)(\lambda_2 - \lambda_3)}$$

Generalising the above for n successive disintegrations, one finds the number of atoms of the element in the nth step as

$$N_n = N_0\{K_1 \exp(-\lambda_1 t) + K_2 \exp(-\lambda_2 t) + K_3 \exp(-\lambda_3 t) + \ldots + K_n \exp(-\lambda_n t)\} \ldots (2.9)$$

where

$$K_1 = \frac{\lambda_1 \lambda_2 \lambda_3 \ldots \lambda_{n-1}}{(\lambda_2 - \lambda_1)(\lambda_3 - \lambda_1) \ldots (\lambda_n - \lambda_1)}$$

$$K_2 = \frac{\lambda_1 \lambda_2 \lambda_3 \ldots \lambda_{n-1}}{(\lambda_1 - \lambda_2)(\lambda_3 - \lambda_2) \ldots (\lambda_n - \lambda_2)}$$

$$K_n = \frac{\lambda_1 \lambda_2 \lambda_3 \ldots \lambda_{n-1}}{(\lambda_1 - \lambda_n)(\lambda_2 - \lambda_n) \ldots (\lambda_{n-1} - \lambda_n)}$$

and so on

2.6 RADIOACTIVE EQUILIBRIUM

We have already mentioned that each parent nuclei in a radioactive series decays into its daughter which decays in turn till a stable end product is formed. one can derive an expression for the number of atoms of any member in this radioactive chain as a function of time as derived in the previous section. The number of atoms of any member changes with time as a result of its birth by the previous member and its decay into the next. However, under certain specific conditions, the number of atoms of any member in the series may become constant, i.e. the rate of its formation may become, equal to the rate of its decay. Obviously, *when the number of atoms of any member in a radioactive chain becomes constant, one can say that it is in radioactive equiliblium with its parent.*

During successive integration, it is observed that the number of any particular member does not change, its derivative w.r.t. time is zero. Let us consider a radioactive series which contain n members and let N_1, N_2, N_3, ... N_n be the number of the atoms of the members at any time and λ_1, λ_2, λ_3, ... λ_n be their respective decay constants. let us suppose that the members to be in a state of radioactive equilibrium, the number of atoms in any member becomes constant obviously,

$$\frac{dN_1}{dt} = -\lambda_1 N_1 = 0 \qquad \ldots (2.1)$$

$$\frac{dN_2}{dt} = (\lambda_1 N_1 - \lambda_2 N_2) = 0 \text{ or } \lambda_1 N_1 = \lambda_2 N_2$$

Similarly, one obtains

$$\lambda_2 N_2 = \lambda_3 N_3$$

$$- \quad - \quad -$$

$$\lambda_{n-1} N_{n-1} = \lambda_n N_n$$

One can easily group all the equations in the above series except the first, into a single relation,

$$\lambda_1 N_1 = \lambda_2 N_2 = \lambda_3 N_3 = \ldots \lambda_n N_n \qquad \ldots (2.2)$$

(2.1) and (2.2) gives the conditions for radioactive equilibrium. Equations (2.1) implies that $\lambda_1 = 0$, i.e., parent radioactive sunstance should not decay at all. However, this condition is contradictory and one cannot evidently achieve a radioactive series without decay of parent atom. One can approach very close to this condition for equilibrium if the rate of decay of the parent substance is very small as compared to the other members in the decay chain, i.e. the half life of parent atom is very long as compared to any member.Some of the naturally occuring radioactive chains found to satisfy this condition. For e.g., the half life of uranium I(UI) is 4.5×10^9 years. Obviously, the decay constant (λ) for uranium is, therefore, much smaller than any other member of the series. The fraction of uranium-I atoms which decays in a given time is, therefore, negligible and hence the number of atoms N, can be considered to be constant. This means that the condition given by (2.1) is nearly satisfied. The condition given by (2.2) is strictly valid and an equilibrium satisfying this condition is known as *secular equilibrium*. In terms of ther mean lives, one can express (2.2) as

$$\frac{N_1}{T_1} = \frac{N_2}{T_2} = \frac{N_3}{T_3} = \dots \frac{N_n}{T_n} \qquad \dots (2.3)$$

Obviously, *when a state of secular equilibrium is reached, each member element is present in an amount proportional to its averge life.*

A long lived parent and a short lived daughter ultimately attain a state of secular equilibrium. This is maintained as long as the parent element A and the different products remain together. If anyone of the daughter products is separated from the parent, then the equilibrium is disturbed. The number of the product atoms then beings to grow again and attains a saturation value. Obviously, one can separate a product element of relatively short half life from a very long-lived parent element repeatedly.

Transient Equilibrium: Now, we consider a case where parent is longer-lived than the daughter but the half life of the parent is not very long. Obviously, $\lambda_2 \gg \lambda_1$, the term $e^{-\lambda_2 t}$ approaches zero faster. The number of daughter atoms shall eventually be given by (Equation (2.6) of section 5)

$$N_2 = N_1 \frac{\lambda_1}{\lambda_2 - \lambda_1} \exp(-\lambda_1 t) \qquad \dots (2.4)$$

Thus the daughter eventually starts decaying with the half life of the parent.
Now, $N_0 \exp(-\lambda_1 t) = N_1$ and hence one obtains

$$N_2 = N_1 \frac{\lambda_1}{\lambda_2 - \lambda_1}$$

or $$\frac{N_2}{N_1} = \frac{\lambda_1}{\lambda_2 - \lambda_1} \qquad \dots (2.5)$$

Obviously, the ratio $\frac{N_2}{N_1}$, becomes constant. When this state has been reached the sample is said to be in *transient equilibrium.*

When $\frac{\lambda_1}{\lambda_2}$, i.e. mean life of the parent is shorter than that of the daughter, Equation (2.6) of section 5 reduces to

$$N_2 = - N_0 \; \frac{\lambda_1}{\lambda_2 - \lambda_1} \; \exp(-\lambda_2 t)$$

Thus the daughter ultimately decays with its own rate of decay. The parent in this case practically disappears after a certain time and only daughter decaying at its own rate is left and hence no radioactive equilibrium is possible.

2.7 RADIOACTIVE SERIES

Almost all natural radioactive elements live in the range of atomic numbers from $Z = 83$ to $Z = 92$. The nuclei of these elements are unstable and disintegrate by ejecting either an α-particle or *a* β particle. The process of disintegration continues through a series of elements ending up finally with an atom which is stable and non-radioactive. The disintegration amongst almost all naturally occuring heavier elements of periodic table end up with stable lead atoms, $Z = 82$. Thus a series of new radioactive elements is produced by successive disintegrations until a stable element is obtained.

There are four main series of radioactive elements: the *uranium, thorium, actinium* and *neptunium* series (Figure 2.6)

(i) **Uranium Series**   In this series the parent element is $^{238}_{92}U$. This is known as uranium 1 or U 1, because other isotopes of uranium exist. U1 emits α-particles; in doing so it gives the immediate by product U X_1 with $Z = 90$ and $A = 234$. U X_1 then, itself emits a β-particle; the atomic number $Z = 90$ therefore increases to $Z + 1 = 91$, whilst mass number A remains unchanged at 234. The product is $^{234}_{91}U X_2$, an *isobar* of U X_1. U X_2 then emits a β-particle, resulting in an increase of atomic number to $Z + 2$, i.e. 92. The mass number A remains unchanged at 234. The product is $^{234}_{92}U$; it is an isotope of U1 because it has the same atomic number of 92, though a different mass number. $^{234}_{92}U$ is known as Uranium 11, or U11.

One can represent this series of transformation as

$$U1 \xrightarrow{\alpha} U X_1 \xrightarrow{\beta} U X_2 \xrightarrow{\beta} U11$$

i.e. $^{238}_{92}U \xrightarrow{\alpha} {}^{234}_{90}U X_1 \xrightarrow{\beta} {}^{234}_{91}U X_2 \xrightarrow{\beta} {}^{234}_{92}U11$

These are only the first four stages in the transformation series. The U11 is radioactive and emits α-particles to give ionium, $^{234}_{90}Io$. Ionium then emits an α-particle to produce radium, $^{226}_{88}Ra$. The subsequent decay scheme of radium is

$$^{226}_{88}Ra \xrightarrow{\alpha} {}^{222}_{86}Rn \xrightarrow{\alpha} {}^{218}_{84}Ra A \xrightarrow{\alpha} {}^{214}_{82}Ra B \xrightarrow{\beta} {}^{214}_{83}Ra C$$

The series is still not compelte. Ra C is radioactive and transforms on the emission of *either* an α-or β-particle. In fact, there are still six further radioactive by-products terminating with lead, which is stable (see U-238 series in Figure 2.6)

In Table 2.1 the details of the whole of the uranium-238 series are given, including the values of half life periods.

In Table 2.1, we have

*.   U X_1 exhibits a *branching effect*: either it emits β-particles to form U X_2 or it emits β-particles to form a further radioactive element, uranium Z. The transformations are as follows:

99.65 % of atoms of U X_1 disintegrate as shown in Table 2.1, i.e.,

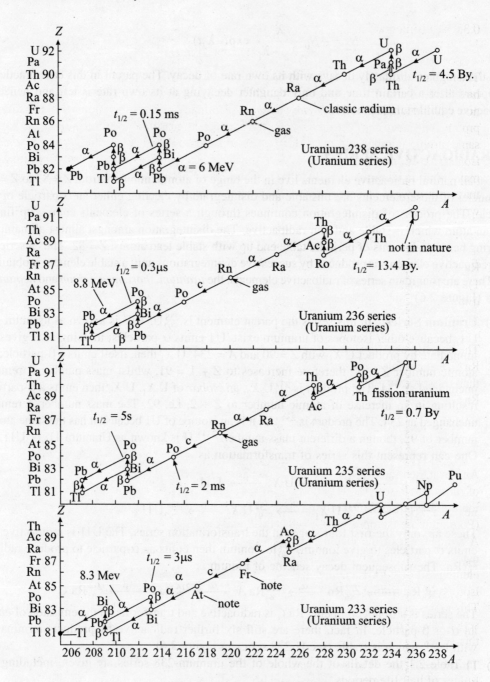

Fig. 2.6 Radioactive Families

$$U X_1 \ (99.65\ \%) \xrightarrow{\ \beta\ } U X_2 \xrightarrow{\ \beta\ } U11 \qquad \qquad \dots (a)$$

0.33 %, disintegrate as

$$\text{U } X_1 \; (0.35 \text{ %}) \xrightarrow{\;\beta\;} \text{UZ} \xrightarrow{\;\beta\;} \text{U11} \qquad\qquad \text{... (b)}$$

The half life of U X_2 is 1.17 min, whereas the half life of UZ is 6.7 hr. These two elements have same Z and A, i.e. they are represented by $^{234}_{91}\text{U } X_2$ and $^{234}_{91}\text{UZ}$ respectively. We must note that though the nuclei of these atoms are identical as regards their number of constitutent protons and neutrons, yet their nuclei have different energy levels. Such elements having same atomic number and mass number but different nuclear levels are known as *nuclear isomers*. U X_2 nucleus has 0.394 MeV greater energy than the UZ nucleus. Indeed, U X_2 may either disintegrate directly to U11 on the emission of a β-particle as showm in (a), or it may emit a γ-ray of energy 0.394 MeV first; it then becomes UZ. Obviously, (b) can be written as

$$\text{U } X_1 \xrightarrow{\;\beta\;} \text{U } X_2 \xrightarrow[\text{0.394 MeV}]{\;\beta\;} \text{UZ} \xrightarrow{\;\beta\;} \text{U11}$$

†. Ra A exhibits *branching*. 99.96 % of Ra A atoms emit α-particles to become Ra B (Table 1). The remaining 0.04 %, of Ra A atoms emit β-particles to give the radioactive element *astatine* (At), which then emits α-particles in disintegrating to Ra C, as

$$^{218}_{84}\text{Ra } A \xrightarrow{\;\beta\;} ^{218}_{85}\text{At} \xrightarrow{\;\alpha\;} ^{214}_{83}\text{Ra } C$$

The half life of At is 2 sec.

‡. Ra C also exhibit *branching*. Either the atom emits first an α-particle to produce Ra C' and then this product emits β-particle to give RaD, or the RaC emits a β-particle first to give Ra C' and then Ra C' emits an α-particle to produce RaD. We have

$$\text{Ra } C \, (0.04 \text{ %}) \xrightarrow{\;\alpha\;} \text{Ra } C'' \xrightarrow{\;\beta\;} \text{Ra } D$$

$$\text{Ra } C \, (99.96 \text{ %}) \xrightarrow{\;\beta\;} \text{Ra } C' \xrightarrow{\;\alpha\;} \text{Ra } D$$

Amongst the radioactive elements of the U-series (Table 2.1) there occurs serveral examples of *radioactive isotopes*. U1 and U11 are isotopic forms of uranium having Z = 92. U X_1 and ionium are isotopes of thorium (Z = 90), itself a radioactive element. Ra A, Ra C' and Ra F (Polonium are isotopes with Z = 84. Ra B and Ra D are isotopes of lead.

A comparison of the elements of Table 2.1 with a Table of the 92 elements occuring in nature exhibit that some natural radioactive elements of this series are isotopes of stable elements. For e.g., Ra C and Ra E are isotopes of bismuth (Z = 83) whilst Ra C'' is an isotope of tellurium (Z = 81).

This series is known as (4n + 2) series, because the atomic weight of the element of this series can be represented by (4n + 2), where n is some interger. A = 238 = 4 × 59 + 2 for this series

(ii) **Thorium Series** This series starts with $^{232}_{90}\text{Th}$. it undergoes α-disintegration having a half life of 1.39×10^{10} years. After a number of successive α and β-disintegrations ends with stable isotope of lead $^{208}_{82}\text{Pb}$ ($^{208}_{82}\text{Th } D$). This is known as 4n series (Figure 2.6). The properties of the various memebers of this series are listed in Table 2.2. For this series A = 232 = 4 × 58.

Table 2.2 Thorium series $(4n)$

Radioactive Element	Atoms number Z	Mass number A	Symbol	Radiation emitted	Half life (T)
Thorium	90	232	^{232}Th	α, γ	1.39×10^{10} y
Meso-thorium I	88	228	^{228}Ra	β⁻	5.7 y
Meso-thorium II	89	228	^{228}Ac	β⁻, γ	6.13 h
Radio-thorium	90	228	^{228}Th	α, γ	1.90 d
Thorium-X	88	224	^{224}Ra	α, γ	3.64 d
Thorium Emanation	86	220	^{220}Rn	α	55.5 s
**Thorium-A	84	216	^{216}Po	α(> 99 %) β⁻(0.014 %)	0.16 s
Thorium-B	82	212	^{212}Pb	β⁻, γ	10.6 h
**Thorium-C	83	212	^{212}Bi	β⁻(66.3 %) α(33.7 %)	60.5 min
Thorium-C′	84	212	^{212}Po	α	3×10^{-7} s
Thorium-C″	81	208	^{208}Tl	β⁻, γ	3.1 min
Thorium-D	82	208	^{208}Pb	—	Stable

[** Th A and Th C exhibit branching effect. The scheme of branching in Th C is as below.

$$\text{Th } C \ (33.7 \ \%) \xrightarrow{\alpha} \text{Th } C'' \xrightarrow{\beta} \text{Th } D$$

$$\text{Th } C \ (66.3 \ \%) \xrightarrow{\beta} \text{Th } C' \xrightarrow{\alpha} \text{Th } D.]$$

(iii) Actinium Series This series starts from the rarer isotope $^{255}_{92}$U which undergoes α-disintrgration with a half life 7.1×10^8 y. The relative abundance in natural uranium is only 0.7 %. It is usually called actino-uranium. After a succession of α and β disintegrations the series terminates at the stable isotope $^{207}_{82}$Pb ($^{207}_{82}$Ac D). This is known as $(4n + 3)$ series. (Figure 2.6). The properties of various members of this series are listed in Table 2.3. For this serves $A = 235 = 4 \times 58 + 3$.

Table 2.3 Actinium series $(4n + 3)$

Radioactive Element	Atomic Number	Mass number	Symbol	Radiation emitted	Half life (T)
Actino-uranium	92	235	^{235}U	α, γ	7.1×10^8 y
Uranium-Y	90	231	^{231}Th	β⁻, γ	25.6 h
Proto-actinium	91	231	^{231}Pa	α, γ	3.23×10^4 y
**Actinium	89	227	^{227}Ac	β⁻(98.8 %) α(1.2 %), γ	22.0 y
Radio-actinium	90	227	^{227}Th	α, γ	18.7 d
Actinium-K	87	223	^{223}Fr	β⁻, γ	21 min
Actinium-X	88	223	^{223}Ra	α, γ	11.2 d
Actinium-Emanation	86	219	^{219}Ra	α, γ	3.92 s

(Contd.)

Radioactive Element	Atomic Number Z	Mass number A	Symbol	Radiation emitted	Half life (T)
**Actinium-A	84	215	^{215}Po	$\alpha(>99\%)$ $\beta^-(5 \times 10^{-4}\%)$	1.8×10^{-3} s
Actinium-B	82	211	^{211}Pb	β^-, γ	36.1 m
Astatin-215	85	215	^{215}At	α	10^{-4} s
**Actinium-C	83	211	^{211}Bi	$\beta^-(0.32\%)$ $\alpha(99.68\%)$, γ	2.16 min
Actinium-C'	84	211	^{211}Po	α, γ	0.52 s
Actinium-C''	81	207	^{207}Tl	β^-, γ	4.79 min
Actinium-D	82	207	^{207}Pb	—	Stable

(iv) Neptunium Series   There is no naturally radioactive series corresponding to $A = 4n + 1$. However, this series has been identified with the transuranic element $^{241}_{94}$Pu with half life of 15 y. The isotope of the transuranic element neptunium, $^{237}_{93}$Np is the longest lived member of the series. The series ends with the stable isotope $^{209}_{83}$Bi. This series is known as $(4n + 1)$ series. None of the member nuclides of this series has a half life comparable to the life of the universe and hence this series does not occur in nature. One can produce all its members by artificial means in the laboratory. The series is shown in Figure 2.6 and the properties of the various members of this series are listed in Table 2.4.

Table 2.4 Neptunium series $(4n + 1)$

Radioactive Element	Atomic number Z	Mass number A	Symbol	Radiation emitted	Half life (T)
**Plutonium-241	94	241	^{241}Pu	α, β^-, γ	14.7 y
Uranium-237	92	237	^{237}U	β^-, γ	6.75 d
Americium-241	95	241	^{241}Am	α, γ	432 y
Neptunium-237	93	237	^{237}Np	α, γ	2.14×10^6 y
Proto-actinium-233	91	233	^{233}U	β^-, γ	27.0 d
Uranium-233	92	233	^{233}U	α, γ	1.6×10^5 y
Thorium-229	90	229	^{229}Th	α, γ	7.34×10^3 y
Radium-225	88	225	^{255}Ra	β^-, γ	14.8 d
Actinium-225	89	225	^{225}Ac	α, γ	10.0 d
Francium-221	87	221	^{221}Fr	α, γ	4.8 min
Astatine-217	85	217	^{217}At	α, γ	0.032 s
**Bismuth-213	83	213	^{213}Bi	α, β^-, γ	45.6 min
Thallium-209	81	209	^{209}Tl	β^-, γ	2.2 min
Polonium-213	84	213	^{213}Po	α	4.2×10^{-6} s
Lead-209	82	209	^{209}Pb	β^-	3.3 h
Bismuth-209	83	209	^{209}Bi	—	Stable

2.8 RADIOACTIVE BRANCHING

We have seen that there are few cases where the radioactive substances decay by two or more different mechanisms. This phenomenon is called as *'branching'*. For e.g. 99.96 % of the isotope of $^{214}_{83}$Bi undergo β-decay to form the $^{214}_{84}$Po while 0.04 % of the nuclei undergo α-decay to form $^{210}_{81}$Tl during the same time interval.

Let us consider the α and β branching and let λ_α and λ_β are partial decay constants. Obviously, the probability of α-emission by one atom in a short time dt will be $\lambda_\alpha\,dt$ and probabilityof β-emission by the same atom in the same time will be $\lambda_\beta\,dt$. Thus the total probability of the decay of the atom in time dt is $(\lambda_\alpha + \lambda_\beta)dt$

$$\therefore \qquad \frac{dN}{dt} = -(\lambda_\alpha + \lambda_\beta)N = -\lambda N$$

Where $\lambda = \lambda_\alpha + \lambda_\beta$. Therefore the mean life T of the substance is given by $T = \dfrac{1}{\lambda}$. Expressing in terms of partial mean life times T_α and T_β, one obtains

$$\frac{1}{T} = \frac{1}{T_\alpha} + \frac{1}{T_\beta}$$

The ratio of the number of α to β particles emitted during a certain time is equal to $\dfrac{\lambda_\alpha}{\lambda_\beta}$ and this is called the *branching ratio R*. The ratios $\dfrac{\lambda_\alpha}{\lambda}$ and $\dfrac{\lambda_\beta}{\lambda}$ give the branching ratios for the two brances.

2.9 RADON GAS

In each naturally occuring radioactive series a radioactive isotope of the element $Z = 86$ is produced which is gaseous at ordinary room temperature and heaviest known inert gas. In the begining of 19th century, when it was discovered, different names had been proposed for it. In the case of U-Ra series it was called radium-emanation (^{222}Em). In the cases of thorium and actinium sereis, this was called as thorium-emanation (^{220}Em) and actinium-emanation (^{219}Em) respectively. Later the common name *radon* (Rn) was given to it. The element radon is similar to the inert gases neon, organ, krypton etc. and its outermost 6 p shell is completely filled with 6 electrons. It falls in the last column of the periodic table of elements. Rutherford and Soddy were able to liquify radon gas by colling it to $-150°$ C which established its true nature. One can separate the radon gas from its parent element (and other products) by means of an air current sent through a glass tube.

2.10 RADIOACTIVE ISOTOPES OF LIGHTER ELEMENTS

It is found that the lowest atomic number among the naturally occurring radioactive series is 81. However, radioactive isotopes of lower atomic numbers have also been reported in nature. the half lives of these isotopes is very long and have very low abundances. Obviously, the activity of samples is extremely small. It is found that continuous and extensive testing of nuclear weapons by various countries has added a new radioactive isotopes to atmosphere and polluting it. ^{90}Sr is an example of this pollution. The half life of 27.7 years and hence it is a serious hazard to atmosphere. Table 2.5 gives some of the prominent radio isotopes of lighter elements.

Table 2.5 Prominent Radioa Isotopes of Lighter Elements

Atomic Number Z	Element	Mass Number A	Half Life (T) yrs.	Type of radioactivity
1	H	3	12.26	β
6	C	14	5730	β
19	K	40	1.3×10^9	β
38	Sr	90	27.7	β
57	La	138	1.1×10^{11}	β
60	Nd	144	2.4×10^{15}	α
62	Sm	147	1.06×10^{11}	α
72	Hf	174	2×10^{15}	α
78	Pt	190	6×10^{11}	α
82	Pb	204	1.4×10^{17}	α

We must note that the nuclei of ^{14}C are being constantly produced by cosmic ray neutron bombardment of ^{14}N. Obviously, its activity is continuously replenished.

2.11 STATISTICAL NATURE OF RADIOACTIVITY

While obtaining the law of disintegration, $N = N_0 \exp(-\lambda t)$, we have treated N as a continuous variable. One is justified in doing this as long as N is very large compared to dN, the number of nuclei decaying between times t and $t + dt$. This is usually the situation. One can easily see that even in a minute quantity of radioactive material there are very large number of nuclei, e.g. one microgram of radium contains about 2.7×10^{15} radium nuclei.

Truly speaking N varies discontinuously and the smallest value of dN is one, corresponding to decay of a single nucleus. Obviously, one can say that the law of radioactive decay is valid only for a sample of radioactive material which is sufficiently large to enable one to treat dN as a differential. if a sample contains only a few nuclei then the concept of half life and mean life becomes meaningless. Suppose you are given 200 active nuclei on a sample holder and the half life of the material is 1 hour, then it does not mean that one will be left with 100 nuclei after 1 hour. As the number of nuclei in the sample is appreciably increased – through one cannot say which nuclei will decay at a given instant – but one cay say that about half of the nuclei will decay after one hour. Obviously, the phenomenon of radioactivity is statistical in nature. One cannot say which radioactive nucleus in the sample will decay at a given instant of time, though one can predict what portion of the given sample will decay in an interval of time, provided the sample contains sufficiently large number of nuclei. One can say that this is the characterstic of all quantum mechanical phenomena e.g. emission of light when atoms *de* excite.

One can obtain the law of radioactive decay purely on the basis of statistical basis.

Let us assume that the disintegration of nucleus depends only on the law of chance and the probability P for a nucleus to decay is the same for all nuclei. Let P be independent of age or past history of the nucleus.

Obviously, P depands only on time interval and for short intervals, P is proportional to the interval, i.e.

$$P = \lambda \, \Delta t \qquad \qquad \text{... (2.1)}$$

The probability Q_1, that a nucleus will not decay during the time interval Δt is given by

$$Q_1 = 1 - P = 1 - \lambda \, \Delta t \qquad \qquad \text{... (2.2)}$$

Similarly, the probability Q_2, that a nucleus will not decay during time $2 \, \Delta t$ is

$$Q_2 = (1 - P)(1 - P) = (1 - \lambda \, \Delta t)^2 \qquad \qquad \text{... (2.3)}$$

Generalizing, the probability Q_n, that a nucleus survives n such intervals, is

$$Q_n = (1 - \lambda \, \Delta t)^n \qquad \qquad \text{... (2.4)}$$

Let us consider that a finite interval t, made up of n intervals of Δt, i.e.

$$\Delta t = \frac{t}{n}$$

$$\therefore \qquad Q_n = \left(1 - \frac{\lambda t}{n}\right)^n$$

We know that the limit of this quantity Q_n, as $n \to \infty$ is $\dfrac{N}{N_0}$, the surviving fraction after time t

$$\therefore \qquad Q_n = 1 - n \frac{\lambda t}{n} + \frac{n(n-1)}{\underline{|2}} \frac{\lambda^2 \, t^2}{n^2} - \frac{n(n-1)(n-2)}{\underline{|3}} \frac{\lambda^3 \, t^3}{n^3} + \cdots$$

$$= 1 - \lambda t + \left(1 - \frac{1}{n}\right) \frac{\lambda^2 \, t^2}{\underline{|2}} - \left(1 - \frac{1}{n}\right)\left(1 - \frac{2}{n}\right) \frac{\lambda^3 \, t^3}{\underline{|3}} + \cdots$$

$$\text{or} \qquad \lim_{n \to \infty} Q_n = \frac{N}{N_0} = 1 - \lambda t + \frac{\lambda^2 \, t^2}{\underline{|2}} - \frac{\lambda^3 \, t^3}{\underline{|3}} + \cdots$$

$$= e^{-\lambda t}$$

Thus one obtains the law of radioactive decay as

$$N = N_0 \, e^{-\lambda t} \qquad \qquad \text{... (2.5)}$$

2.12 STATISTICAL ERRORS OF NUCLEAR PHYSICS

One can always count the radioactive particles with the help of particle detectors, e.g. Geiger Counter. The time t necessary for the observation of any finite number of counts N is subject to statistical fluctuations. This causes an error in the observed counting rate, $n = \dfrac{N}{t}$.

We must note that this is not a property of the instrument and this is fundamental to the phenomenon of radioactivity. Now, we can see that the number of counts observed in a given time obeys a *Poisson Distribution*.

Let us consider that the probability that N particles are observed in time t be P_N. Let us suppose that the interval t is divided into n equal intervals so small that the probability of emission of two particles within an interval is negligible. Obviously, the probability of the emission of one particle in a given interval is, $\dfrac{N}{n}$, where $N \rightarrow$ average number.

Now, we consider the fluctuations or differences between the actual number of decaying nuclei N and $<N>$, where $<N>$ denotes average number.

Now, one can write the probability of emission of N particles in the first N intervals and none in the remaining $(n - N)$ intervals is:

$$\left(\frac{\bar{N}}{n} \right)^N \left(1 - \frac{\bar{N}}{n} \right)^{n - N} \qquad \qquad \ldots (2.6)$$

This is only one possible way of obtaining N particles in the total time t. The first particle could have been in any one of the n intervals, the second one in any one of the remaining $(n - 1)$ and so on. Thus the Nth in any of the remaining $(n + 1 - N)$. The Nth particle can choose anyone of the intervals between the one it occupies and all the $(n - N)$ unoccupied intervals. Thus one finds the number of ways of distributing the N particles in the n intervals as

$$n(n - 1)(n - 2) \ldots (n - N + 1)$$

We must note that all these ways are not independent, since the particles may be interchanged without influencing the result. Obviously, the number of essentially different ways

$$= \frac{n(n - 1)(n - 2) \ldots (n - N + 1)}{\text{Number of ways of interchanging particles}}$$

One can choose any one of the N particles as the first, any one of the remaining $(N - 1)$ as the second and so on.

$\therefore$ Number of interchanging particles $= \lfloor N$

Therefore, one obtains tha probability of N counts as

$$P_N = \frac{n(n - 1)(n - 2) \ldots (n - N + 1)}{\lfloor N} \left(\frac{\bar{N}}{n} \right)^N \left(1 - \frac{\bar{N}}{n} \right)^{n - N} \qquad \ldots (2.7)$$

This is known as *binomial distribution law*. When $n \rightarrow \infty$, (2.7) becomes

$$P_N = \frac{n^N}{\lfloor N} \left(\frac{\bar{N}}{n} \right)^N e^{-\bar{N}}$$

$$= \frac{\bar{N}^N \, e^{-\bar{N}}}{\lfloor N} \qquad \qquad \ldots (2.8)$$

(2.8) is known as *Poisson distribution.* Figure 2.7 shows the Posson's series.

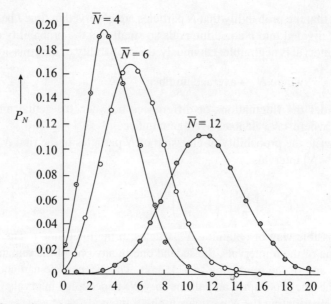

Fig. 2.7 Illustration of Poisson's distribution, i.e. Poisson's series. As N becomes larger the curve becomes more symmetrical about the maximum.

We must note that (2.8) is valid for only integral values of N. One finds that a continuous curve of P_N against N plotted through the points given by (2.8) yields a curve with an asymmetrical maximum for small values of N. For N becomes large the curve becomes symmetrical about the maximum and one can show analytically that the Poisson distribution approaches a Gauss error curve. Then

$$P_N = \frac{1}{\sqrt{2\pi\,\bar{N}}} \exp\left[-\left(\frac{(N-\bar{N})^2}{2N}\right)\right] \qquad \text{... (2.9)}$$

This is Gauss's law.

Suppose that on an average N counts are observed in a given time t, the standard in the number of counts

$$S = \lfloor N \qquad\qquad\qquad \text{... (2.10)}$$

($\because$ the precision index $h = \dfrac{1}{\sqrt{2}\,\bar{N}}$ and $S = \dfrac{\sqrt{2}}{h}$)

Usually a single reading is taken and it is assumed to be close enough to N.

Thus
$$S = \sqrt{N}$$

2.13 MEASUREMENT OF DECAY CONSTANT

Various methods have been developed for the measurement of λ of radioactive substance. We shall now consider some of the more important techniques.

(i) For half lives from a few. Seconds to about 10^{-15} years, the direct measurements are practicable. One can measure the decay constant by following the decrease of the intensity of radiation emitted by a radioactive sunstance, i.e. activity with time. One obtains a straight line with a slope (λ) from the plot of the natural lograrithm of the activity with time. When the activity of the sample is very feeble, one finds that the accuracy of the method is not very good. Geiger-Muller counter or scintillation counter is usually used for the measurment of intensity of the radiation.

(ii) **Weighing Method** When T (half life) is in the range 10 to 10^{18} y, we have from the relation $A = \dfrac{dN}{dt} = -\lambda N$, λ can be measured if the number of N (can be calculated by weighing and using the atomic weight and Avogadro's number) in the source and $\dfrac{dN}{dt}$ is measured. The knowledge of the fraction of emitted particles entering the detector and the detector efficiency is essential for this method.

(iii) One can use the *condition of secular equilibrium* between two genetically related substances for the determination of the ratio of their decay constants. Using this method the ratio for radium and ^{238}U has been estimated.

(iv) In the case of α-emitting substances one can determine α by measuring the ranges (R) of emitted α-particles Using the empirical Geiger-Nuttall law $\lambda = a + b \log R$. This method provides an estimation of λ and the accuracy of the method is very low.

(v) In the case of certain very long lived α-emitters, decay constant (λ) has been determined with the help of *nuclear emulsion photographic* plates. The plate is stored for several months and after developing it the number of α-tracks of the correct range are counted carefully. Using the relation $\Delta N = -\lambda N \, \Delta t$, one can determine λ.

(vi) **Calorimetry** One can determine the half lifes of the order of thousand years with the help of calorimetry. It is reported that the rate of emission of α-radiation by a known weight of ^{239}Pu has been determined by measuring the rate of production of heat due to self absorption of α-particles. Using the relation $\Delta N = -\lambda N \, \Delta t$, one can determine λ.

(vii) For very short half lives of the order of 1 second, one can used the mechanical methods. if the substance is in gaseous from, then for the interval of this duration, flow methods are effective. The gas is allowed to flow through a tube with a definite velocity v. With the help of suitable radiation detectors placed by the side of the tube, separted by known distance (d), one finds the time taken by the gas to flow from one detector to the next $= \dfrac{d}{v}$. The count rate recorded by the stream detector is $\exp\left(-\lambda \dfrac{d}{v}\right)$ times that of up stream detector. With the help of this method the half life of ^{12}B has been reported to be 0.204 s.

(viii) **The Electronic Method** To measure the life times of excited nuclear states ($\tau \approx 10^{-9}$ s), electronic methods have been used. The methods requires that the decay to be studied be preceded by another event, which is actually an electronic gate and which stays open for a time t. Obviously

$$K \int_0^t \lambda \, \exp(-\lambda x) \, dx = K(1 - e^{-\lambda t})$$

The apparatus is arranged in such a manner so that one can count the number of gates during which a particle is detected. One can obtain λ by measuring the ratio between the number of times a particle is detected while the gate is open and the total number of gates formed as a function of t.

Making use of *delayed coincidence measurement* procedure, keeping t constant and less than τ, the time T is varied after the first triggering pulse at which the gate opens. We have the number of decay coincidences/disintegration proportional to $e^{-\lambda t} (1 - e^{-\lambda t})$ or to $\lambda t\, e^{-\lambda t}$ for $t << \lambda$. Using this method a value of $T = 8.7 \times 10^{-8}$ s for the 0.197 MeV ^{19}Fe state have been reported.

(ix) Rotating Disc Method In few cases, it is observed that the recoiling short half lived decay product is deposited at one point on a rotating disc, along the circumference of which are placed a succession of suitable detectors (Figure 2.8). When the disc is rotated at a uniform speed, the deposited radioctive material passes by the window of the successive detectors at regular intervals of time.

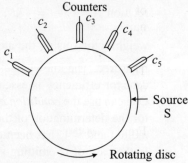

One finds that the intensity of the radiation emitted by the radioactive material decreases as the deposited source passes from one detector to the next. The half of ^{216}Po (Th A) obtained from method is found to 0.158 s.

Fig. 2.8 Rotating disc method

(x) Recoil Distance Method A recoil of daughter atom is observed when the parent atom emits some radiation, e.g. an α or a β-particle. If the daughter nuclei is radiactive, then the recoil atoms disintegrate in the course of travel from the source situated at one end within a tube (evacuated) to the other end. One can measure the diminution in the intensity of radiation emitted by the recoil atoms with the help of two successive detectors placed along of the tube (Figure 2.9), provided the half life is very short. The half life of the recoiling atom ^{214}Po (Ra C') produced in the β-decay of ^{214}Bi (Ra C) found with this method is 1.64×10^{-4} s.

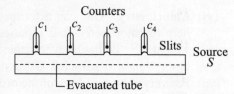

Fig. 2.9 Recoil distance method

2.14 RADIOACTIVE DATING: THE AGE OF MINERALS AND ROCKS

The decay of radioactive elements and its complete independance from physical and chemical conditions provides a method for estimating the ages of old minerals rocks and the earth. There are strong evidences, geophysical and astronomical, that the elements present in the universe were created about 10^9 years ago and the solidfication of the earth s crust took place about 3.5 billions years ago. A comparison of the half lives of uranium and thorium is of the same order of magnitude as the age of the universe ($\sim 10^9$ yrs.).

One can determine the geological ages of rocks, minerals containing U and Th from the study of their radioactivity. Some of these methods are as follows.

(a) Lead Method We have seen that the different natural radioactive series start from long lived radioisotopes ^{238}U, ^{235}U or ^{232}Th and after a number of successive α- and β-disintegrations attain the stable end products, i.e. isotopes of lead

$$^{238}U \rightarrow {}^{206}Pb + 8\alpha + 6\beta + 1 \qquad \text{... (a)}$$

$$^{235}U \rightarrow {}^{207}Pb + 7\alpha + 4\beta + 1 \qquad \text{... (b)}$$

Obviously, the successive disintegration in the case of the U-Ra series result in the formation of stable isotope of ^{206}Pb. Let the time which has elapsed since the solidification of the minerals bearing U in the earth's crest uptil the present date is t, and if N_{oU} was the number of atoms of ^{238}U which were present intially in the given sample of the mineral, then the number of ^{238}U atoms present in the sample at present time is

$$N_U = N_{oU} \exp(-\lambda_U t) \qquad \text{... (2.1)}$$

The number of ^{238}U atoms which have undergone radioactive transformation during number of ^{206}Pb atoms formed due to these radioactive transformation is

$$N_{Pb} = N_{oU} - N_U = N_U \left[\exp(\lambda_U t) - 1\right] \qquad \text{... (2.2)}$$

From (2.1) and (2.2), one obtains

$$t = \frac{1}{\lambda_U} \ln\left(\frac{N_{Pb} + N_U}{N_0}\right) \qquad \text{... (2.3)}$$

One can determine N_{Pb} and N_U by chemical and mass spectroscopic methods. Thus one can determine t, i.e. the geological age of the mineral. Following the same procedure, one can determine the age of the ^{232}Th bearing mineral.

The above method is based upon the following assumptions.

(a) The sample of the mineral under study contain only radiogenic lead isotope (^{206}Pb) originating due to the disintegration of ^{238}U and no other lead.

(b) The entire amount of lead (^{206}Pb) so produced in this process has been retained in this mineral.

The above assumptions seems to be reasonable. However, some loss of ^{206}Pb due to leaching cannot be ruled out.

(b) Helium Method The α-particles have quite small ranges in solid matter and it may be trapped inside the rock specimen provided the rock is quite fine grained. Obviously, the α-particles emitted during the successive disintagration starting from ^{238}U turn into neutral helium atoms by acquiring orbital electrons and there by accumulating as helium gas within the mineral containing Uranium. During the long period that has lapsed since the solidification of the mineral, sufficient quantity of helium gas is produced. By measuring the quantity of helium, one can estimate the age of the mineral.

During the transformation of ^{238}U to ^{206}Pb, one finds that 8 α-particles are emitted. Therefore the number of helium atoms formed is

$$N_{He} = 8(N_{oU} - N_U) = 8 N_U \left[\exp(\lambda_U t) - 1\right] \qquad \text{... (2.4)}$$

one obtains,
$$t = \frac{1}{\lambda_U} \ln\left(\frac{8 N_U + N_{He}}{8 N_U}\right) \qquad \text{... (2.5)}$$

Obviously, by measuring the amount of helium gas accumulated within a given sample containing the known amount of ^{238}U, one can find the lime time t of the mineral.

We must remember that α-particles i.e., helium gas is produced due to the successive α-disintegrations of both the isotopes of ^{238}U and ^{235}U. We can see that an atom of ^{235}U emits 7 α-particles before it is transformed into the stable ^{207}Pb isotope. Obviously 7 helium atoms are produced for each atom of ^{235}U transformed. Taking this into account, the number of atoms of helium comes out to be

$$N_{He} = 8\, N_U\, (\exp(\lambda_U t) - 1] + 7\, N'_U\, [\exp(\lambda'_U t) - 1] \qquad \dots (2.6)$$

where λ_U and λ'_U are the decay constants of ^{238}U and ^{235}U respectively and N_U and N'_U are the number of their atoms in the given amount of the material.

Following assumptions have been made

(i) All the helium, accumulated within the sample since the begining of its formation, is retained within the mineral.

(ii) No helium gas from any other source is present in the mineral

The aboe assumptions are not correct, because some helium gas is likely to diffuse in or out of porous structures of the rocks.

(c) Potassium Method This method utilizes stable isotope of ^{40}Ar $(Z = 18)$, one of the decay product of a long lived radioactive isotope ^{40}K of naturally occuring potassium $(Z = 19)$. There are two competing decay modes

$$^{40}_{19}\text{K} \Big\langle \begin{array}{l} ^{40}_{18}\text{Ar by orbital electron capture, } 11.7\ \%,\ T = 1.2 \times 10^{10} \text{ years} \\[6pt] ^{40}_{20}\text{Ca with } \beta^- - \text{emission, } 88.3\ \%,\ T = 1.48 \times 10^9 \text{ years} \end{array}$$

Nature Argon has three stable isotopes ^{36}Ar, ^{38}Ar and ^{40}Ar with the relative abundances 0.337 %, 0.063 % and 99.6 % respectively. The normal ratio of ^{40}Ar to ^{36}Ar is known.

If radiogenic ^{40}Ar is present, the mass spectrometer will show an abnormal $\dfrac{^{40}\text{Ar}}{^{36}\text{Ar}}$ ratio and one can attribute the excess to the ^{40}K decays. The ages of the samples of the order 2×10^6 y has been determined successfully.

(d) Isotopic abudance method One can estimate the age of the mineral by measuring the ratio of the relative abudances of the lead isotopes ^{206}Pb and ^{207}Pb of radiogenic origin by mass spectroscopic method. This is considered to be the best method for estimating the age of minerals. This method is less sensitive to the chemical and mechanical losses of lead from the mineral. The absence of any ^{204}Pb isotope ensures radiogenic origin of the two isotopes ^{206}Pb and ^{207}Pb of lead in the given mineral.

(e) Carbon dating The decay of ^{14}C is used to determine more recent ages. ^{14}C is unstable and decays as

$$^{14}_{6}\text{C} \rightarrow {}^{14}_{7}\text{N} + \beta^-;\ T = 5568\ y$$

There is a trace of ^{14}C in the atmosphere due to bombardment of cosmic neutron with $^{14}_{7}$N. We have

$$^{14}_{6}\text{C} + {}^{1}_{0}\text{n} \rightarrow {}^{14}_{6}\text{C} + {}^{1}_{1}\text{p}$$

$^{14}_{6}$C is a nuetron-rich (6 p and 8 n) isotope of carbon, which is β^- active.

$$^{14}_6\text{C} \rightarrow {}^{14}_6\text{N} + \beta^- + \text{antineutrino}$$

Half Life of $^{14}_6\text{C}$ is 5730 years.

Obviously $^{14}_6\text{C}$ would not have been present in the atmosphere, had it not been continuously replenished. $^{14}_6\text{C}$ gets combined with hydrogen and oxygen and thus find its way in all organic matter. Since $^{14}_6\text{C}$ has a half life of 5730 and hence it is an ideal radioactive isotope for studying the age of civilization and one uses it extensively in archaeology and anthropology. When an living thing, i.e. animal or plant dies, its intake of carbon stops and from that moment decay of ^{14}C is the only process that continues. Obviously, at death, equilibrium is ended and the exponential radioactive decay takes place. Coal and petroleum are organic in nature, but they are so old that there is no trace of ^{14}C in them. This has been verified experimentally. ^{14}C in a young tree or animal has the same activity as atmospheric carbon. Obviously, one can date an object containing organic carbon by measuring the specific activity of Carbon. Thus ^{14}C provides a radioactive clock for anothropologists. By carefully measuring the ^{14}C activity of a fossil or dead tree, one can estimate its age. The radioactivity property of ^{14}C has been used in the age determination of archaeological objects upto 25000 yrs.

(f) **Glaciology Studies**   The radioactive isotopes occuring in nature which have relatively much shorter half lifes, e.g. tritium ^3H $(T = 12 + y)$ is useful for glaciology studies

$$^3_1\text{H} \rightarrow {}^3_2\text{He}$$

(g) **Photometric examination** of pleochroic halos in maca due to α-radiation is also a very useful method for radioactive dating.

2.15 ARTIFICIAL RADIOACTIVITY

The radioactive elements, in which natural disintegration of nucleus occurs are the heavier elements. Rutherford was the first scientist to accomplish in 1919, the artificial transmutation of one element into another. He succeeded in disintegrating nitrogen nuclei by bombaring ordinary nitrogen gas with α-particles emitted from Ra C'.

The apparatus used by rutherford is schematically shown in Figure 2.10. The apparatus consisted of a long chamber with side opening covered by silver foil F. A zinc sulphide (Zn S) screen was placed just outside the opening and a microscope M was placed for observing any scintillations occuring on the screen. The source of α-particles, S (^{212}Po), was deposited on a metal plate placed inside the chamber. One could vary the distance of S from Zn S screen. The chamber could be filled with different gases through the side-tubes T_1 and T_2.

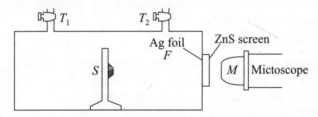

Fig. 2.10 Rutherford's apparatus to study the first successful artificial disintegrate of nuclei.

When the chamber was filled with oxygen or CO_2 gas, no scintillations were observed on the screen. But when the chamber was filled with nitrogen, scintillations were observed on the screen, even though, this screen waas shielded from the α-particle source S by the silver foil F which was thick enough to absorb all the α-particles. Rutherford concluded that the scintillation were produced by particles which were ejected from nitrogen nuclei as a result of α-bombardment. Since the scintillations were observable even when the distance of S from the screen was as long as 40 cm. Magnetic deflection method indicated that these particles had a positive charge equal to the charge of an electron and a mass equal to that of a hydrogen nucleus (1.672×10^{-27} kg). Later these particles were indentified as 'protons'.

The reaction can be written as

$$^{14}_{7}N + {}^{4}_{2}He \rightarrow {}^{17}_{8}O + {}^{1}_{1}H \qquad ...(2.1)$$

Obviously, Rutherford transformed ordinary nitrogen into a rare isotope of oxygen. This was the first artifical nuclear transformation. Nuclear disintegrations of this kind are now referred to as α-particle proton nuclear reactions.

(a) α-p reactions

Following Ruherford's successful experiment on artificial nuclear disinteration Rutherford and Chadwick disintegrated other light elements by bombarding them with α-particles. They found that protons could be ejected out of the nuclei of all the light elelments from boron to potassium (with the exception of carbon and oxygen). The reaction involved can be represented by the general equation

$$^{A}_{Z}X + {}^{4}_{2}He \rightarrow \left({}^{A+4}_{Z+2}C \right) \rightarrow {}^{A+3}_{Z+1}Y + {}^{1}_{1}H + Q \qquad ...(2.2)$$

Here ${}^{A+4}_{Z+2}C$ is a compound nucleus, postulated by Bohr, which disintegrates almost immediately to give the element ${}^{A+3}_{Z+1}Y$ and a proton. Q is the energy evolved or absorbed during the nuclear reaction. In some cases the energy of the ejected protons was even greater than that of the bombarding α-particles. This result further verified that the protons were emitted due to nuclear disintegration, the extra energy being acquired due to the nuclear disintegration.

(b) Discovery of Neutrons

In 1930, Bothe and Becker found that when beryllium (or boron) was bombarded with α-particles a highly penetrating but very poorly ionizing radiation was emitted. As this radiation was shown to be uncharged it was reasonable at that time to assume that it was heghly energetic γ-radiation, and produced in accordance with the nuclear reaction

$$^{9}_{4}Be + {}^{4}_{2}He \rightarrow ({}^{13}_{6}C) \rightarrow {}^{13}_{6}C + \gamma \qquad ...(2.3)$$

This supposition led, however, to difficulties. The measurement of absorption of the radiation in lead showed that if it was a γ-radiation then its energy should be about 7 MeV. This value was greater than the energy of any γ-radiation was known at that time.

In 1932, Curie and Joliot measured the intensity of this radiation, using ionisation chambers, and showed that the ionization increased markedly if the radiation were first passed through paraffin wax or other material containing hydrogen. The ionisation increase was shown to be due to protons ejected from the paraffin wax. Furthermore, these protons had a maximum range of 26 cm in air as measured by a cloud chamber, corresponding to an energy of 4.5 MeV. Calculations based on this basis revealed that each incident γ-ray photon must have had an energy of 55 MeV, a value about 8 times higher than that deduced from the absorption measurments. Obviously, there was a serious difference between the values of the energy of the supposed γ-radiation given the said two methods.

In 1932, chadwick studies this serious anomaly. He showed that the radiation from beryllium undergoing the impact of α-particles had to have if it were assumed to be γ-radiation, an energy of 55 MeV to accounts for the protons ejected from the paranffin wax, and an energy of 90 MeV to eject nitrogen ions of the range found when using targets containing nitrogen atoms. In these experiments, α-particles from polonium set up in a vacuum chamber impinged on beryllium. The radiation produced was then passed through a paraffin-wax block and the maximum range of the protons produced was measured by interposing various thickness of aluminium between the paraffin wax and the ionisation chamber used as a detector. They replaced then the paraffin was by a slab of nitrogenous material (Figure 2.11)

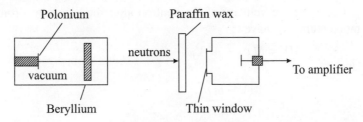

Fig. 2.11 Chadwick's apparatus used for the discovery of neutron

These results quite inconsistent with the laws of conservation of energy and momentum, were explained by Chadwick when he abandoned the suggestion that the radiation from the beryllium was in the form of γ-rays. Instead he asserted that a new type of particle was being produced which had zero charge, and was therefore undeflected in magnetic and electric fields and left no track in an ionization chamber. He called this particle as *neutron* and the correct nuclear reaction on bombarding baryllium with α-particles was not represented by (2.3) but by

$$_{4}^{9}\text{Be} + _{2}^{4}\text{He} \rightarrow (_{6}^{13}\text{C}) \rightarrow _{6}^{12}\text{C} + _{0}n^{1} \qquad \text{... (2.4)}$$

Most of the light elements upto fluorine can be made to yield neutrons on the impact of α-particles of suitable energies. The exceptions are the elements hydrogen, helium, carbon and oxygen. A mixture of beryllium powder with a polonium in a cavity in a block of paraffin wax is a useful simple source of neutrons.

Nuclear reactions in which the impact of α-particles produces neutrons are called α-*n* (*alpha neutron*) reactions. One can represent them by general reaction:

$$_{Z}^{A}X + _{2}^{4}\text{He} \rightarrow \left(_{Z+2}^{A+4}Y\right) \rightarrow _{Z+2}^{A+3}Y + _{0}n^{1} + Q \qquad \text{... (2.5)}$$

A few typical α-*n* reactions are

$$_{3}^{7}\text{Li} + _{2}^{4}\text{He} \rightarrow _{5}^{10}\text{B} + _{0}^{1}\text{n}$$

$$\,^{14}_{7}\text{N} + \,^{4}_{2}\text{He} \rightarrow \,^{17}_{9}\text{F} + \,^{1}_{0}\text{n}$$

$$\,^{27}_{13}\text{Al} + \,^{4}_{2}\text{He} \rightarrow \,^{20}_{15}\text{P} + \,^{1}_{0}\text{n}$$

Bombardment of these elements with α-particles can also yield protons. In fact, the probability of production of either protons or neutrons depends on the energy of α-particles.

Neutrons can also be produced by bombarding deuterium (heavy water) or beryllium with γ-rays obtained from artificial radioactive atoms like $\,^{24}_{11}\text{Na}$ or $\,^{124}_{51}\text{Sr}$. Such sources are named as '*photoneutron*' sources and give practically monoenergetic neutrons.

Nowadays, the nuclear reactor is the most powerful source of neutrons. During the fission of heavy nuclei in nuclear reactors, neutrons are produced.

(c) Determination of Mass of the Neutron

Neutron is not a charged particle, its mass cannot be determined directly by deflecting it in electric or magnetic field. Chadwick determined the mass of the neutron by measuring the maximum velocities of recoiling nuclei of hydrogen and nitrogen struck by neutrons. Suppose a nutron of mass M and maximum velocity V impinges on a nucleus of mass m and imparts to it a maximum recoil velocity v_R. Let V_1 be the velocity of the neutron after the collision. From the law of the conservation of momentum, we have

$$MV = M v_R + M V_1 \qquad \qquad \qquad \text{... (2.7)}$$

From (2.7) we have

$$V_1 = \frac{MV - m v_R}{M} \qquad \qquad \qquad \text{... (2.9)}$$

Using (2.9), Equation (2.8) becomes

$$M V^2 = M v_R^2 + \frac{MV - m v_R}{M}$$

$$\therefore \qquad M^2 V^2 = mM v_R^2 + M^2 V^2 - 2 MV m v_R + m^2 v_R^2$$

$$\therefore \qquad M v_R - 2 MV + m v_R = 0$$

or

$$v_R = \frac{2MV}{M + m} \qquad \qquad \qquad \text{... (2.10)}$$

If the neutrons impinge upon hydrogen atoms in paraffin wax, let the maximum recoil velocity v_R be v_H, the velocity of the protons each of mass m_H, Equation (2.10) then becomes

$$v_H = \frac{2 MV}{M + m_H} \qquad \qquad \qquad \text{... (2.11)}$$

If nitrogen atoms are concerned in a nitrogenous subject to the neutron impact, then

$$v_N = \frac{2 MV}{M + m_N} \qquad \qquad \qquad \text{... (2.12)}$$

Where v_N is the maximum recoil velocity of the nitrogen nuclei each of mass m_N. From (2.11) and (2.12), one obtains

$$\frac{v_H}{v_N} = \frac{M + m_N}{M + m_H} \qquad\qquad \text{... (2.13)}$$

m_N and m_H are known. If, therefore, v_H and v_N are measured in recoil experiments, M the mass of the neutron can be found.

Chadwick obtained v_H to be 3.3×10^7 m/s from the measurment of the range of the recoil protons in air. The mass of the neutron is found to be 1.00898 amu or 1.675×10^{-25} kg, which slightly greater than that of a proton.

Properties

(i) Neutrons is an uncharged particle. Therefore, it cannot be accelerated to high velocities by means of electric fields. For the same reasons they cannot be focussed by means of magnetic fields.

(ii) Neutron is a highly penetrating particle and can pass through thick sheets of lead.

(iii) Being chargeless particle, it produces practically no ionisation in a gas, and hence no track in a cloud chamber.

(iv) Being chargeless, it can easily penetrate the nucleus of an atom. Obviously, it can produce a nuclear excitation or nuclear disintegration far more readily than almost any other fundamental particle. We must note that charge particles have to overcome the strong electrostatic repulsion offered by the nucleus.

(v) When the probability of nuclear excitation or nuclear disintegration is small, the neutrons, on striking matter, are simply scattered by the nuclei of the atom. When neutrons collide with heavy nuclei, the neutrons are scattered with very little loss of energy. When neutrons Collide with light nuclei, they are slowed down in a few collisions. Heavy water (D_2O), light water (H_2O), carbon and paraffin wax are very effective in slowing down neutrons. These substances are called '*moderators*'. A slow neutron is found more efficient in producing nuclear disintegration because it spends more time near a nucleus than a fast neutron and stands a greater chance of being captured.

(vi) Neutrons possessing energies of 1 MeV or more are termed as *fast* neutrons. The neutrons possessing energy below 1 eV are described as *slow* neutrons. The neutrons in thermal equilibrium with a moderator at normal temperature and pressure are called thermal neutrons and these neutrons have energies of the order of ~ 0.03 eV.

(vii) The neutron, being a non-ionising particle, does not produce a track in a cloud chamber. Obviously, G.M. tube or a cloud chamber cannot be used to detect a neutron. Indirect methods are to be used for its detection.

Slow neutrons can be detected by means of a G.M tube or ionisation chamber filled with boron trifluoride gas containing $^{10}_{5}B$. The neutrons passing through the gas disintegrate boron nuclei which thus emit α-particles

$$^{10}_{5}B + ^{1}_{0}n = ^{7}_{3}Li + ^{4}_{2}He \qquad\qquad \text{... (2.14)}$$

The α-particles so produced cause ionisation of the gas and are thus detected.

A G.M. tube or ionisation chamber containing hydrogen can detect fast neutrons. When a fast neutron collides with a hydrogen nucleus which in turn produces ionisation in the surrounding gas and is detected.

(viii) A free neutron outside an atomic nucleus is *unstable* and decays into a proton, emitting an electron and an antineutrno.

$$\begin{smallmatrix}1\\0\end{smallmatrix}n \rightarrow \begin{smallmatrix}1\\1\end{smallmatrix}H + \begin{smallmatrix}0\\-1\end{smallmatrix}e + \bar{\upsilon} \qquad \text{... (2.15)}$$

The half life of the neutron is estimated to be 12.8 min.

(ix) Neutrons find wider applications. These are used in medicine, specially in the treatment of cancer. Slow neutrons are used in nuclear fission. Fast and alow neutrons are used for artificial disintegration of nuclei and producing radio-isotopes.

(x) The neutron flux in a beam of monoenergetic neutrons is the number n of neutron/cc multiplied by the velocity V of the neutrons in cm/sec. The product $n\upsilon$ then represents the number of neutrons per second which traverse an area of 1 sq. cm. drawn perpendicular to the beam. Obviously, $\sigma\, n\upsilon$ decides the cross-section σ of a nucleus for neutron interactions, σ is decided by two types of process: scattering and capture, i.e.

$$\sigma = \sigma_s + \sigma_c \qquad \text{... (2.16)}$$

Where σ_s is the scatter cross-section and σ_c is the capture cross-reaction. σ depends, the type of interaction concerned, and the velocity of the incident neutrons. In several cases, the phenomenon of *resonance capture* occurs where the collision cross-section of a nucleus for neutrons reaches peak values at certain discrete incident neutron energies.

(d) Artificial Transmulation of Elements

The experiments on the artificial disintegration of nuclei led to the discovery of induced or artificial radioactivy. In 1934, *M.* and *Mme* Curie-Joliot observed that when certain light elements like boron and aluminium, were bombarded by α-particles, the resulting products of disintegration emitted positrons, and the emission persisted even after bombardment by α-particles was stopped. The positron activity decayed exponentially with time just in the same way as natural radioactivity. This showed that as a result of the α-particle bombardment, the stable element were converted into unstable radioactive isotopes. *This phenomenon in which a stable element is converted into a radioactive isotope by an artificial disintegration is called artificial radioactivity.* These artificially produced radioactive isotopes have comparatively much shorter half lives than the natural radioactive elements. Few examples of artificial radioactivity are

$$\begin{smallmatrix}10\\5\end{smallmatrix}B + \begin{smallmatrix}4\\2\end{smallmatrix}He \rightarrow (\begin{smallmatrix}14\\7\end{smallmatrix}N) \rightarrow \begin{smallmatrix}13\\7\end{smallmatrix}N + \begin{smallmatrix}1\\0\end{smallmatrix}n$$

Compound
nucleus

proton rich residual nucleus decay by positron emission

$$\begin{smallmatrix}13\\7\end{smallmatrix}N \rightarrow \begin{smallmatrix}13\\6\end{smallmatrix}C + \begin{smallmatrix}0\\+1\end{smallmatrix}e + \text{neutrino} \ (T = 9.96 \text{ min}) \qquad \text{... (2.17)}$$

$$\begin{smallmatrix}27\\13\end{smallmatrix}Al + \begin{smallmatrix}4\\2\end{smallmatrix}He \rightarrow (\begin{smallmatrix}31\\15\end{smallmatrix}P) \rightarrow \begin{smallmatrix}30\\15\end{smallmatrix}P + \begin{smallmatrix}1\\0\end{smallmatrix}n$$

Compound
nucleus

followed by

$$\begin{smallmatrix}30\\15\end{smallmatrix}P \rightarrow \begin{smallmatrix}30\\14\end{smallmatrix}Si + \begin{smallmatrix}0\\+1\end{smallmatrix}e + \text{neutrino} \ (T = 2 \text{ smins}) \qquad \text{... (2.18)}$$

$$\begin{smallmatrix}11\\5\end{smallmatrix}B + \begin{smallmatrix}4\\2\end{smallmatrix}He \rightarrow (\begin{smallmatrix}15\\7\end{smallmatrix}N) \rightarrow \begin{smallmatrix}14\\6\end{smallmatrix}C + \begin{smallmatrix}1\\1\end{smallmatrix}H$$

followed by

$$^{14}_{6}\text{C} \rightarrow {}^{14}_{7}\text{N} + {}^{0}_{-1}\text{e} \ (T = 5730 \text{ yrs}) \qquad \ldots (2.19)$$

$$^{34}_{17}\text{Cl} \rightarrow {}^{34}_{16}\text{S} + {}^{0}_{+1}\text{e} + \text{neutrino} \ (T = 32 \text{ mins}) \qquad \ldots (2.20)$$

$$^{7}_{3}\text{Li} + {}^{4}_{2}\text{He} \rightarrow ({}^{11}_{5}\text{B}) \rightarrow {}^{11}_{5}\text{B} + \gamma \qquad \ldots (2.21)$$

$$^{2}_{1}\text{H} + {}^{2}_{1}\text{H} \rightarrow ({}^{4}_{2}\text{He}) \rightarrow {}^{3}_{1}\text{H} + {}^{1}_{1}\text{H}$$

followed by

$$^{3}_{1}\text{H} \rightarrow {}^{3}_{2}\text{He} + {}^{0}_{-1}\text{e} + \bar{v} \ (T = 12.26 \text{ hrs}) \qquad \ldots (2.22)$$

$$^{27}_{13}\text{Al} + {}^{1}_{0}\text{n} \rightarrow ({}^{28}_{13}\text{Al}) \rightarrow {}^{24}_{11}\text{Na} + {}^{4}_{2}\text{H}$$

followed by

$$^{24}_{11}\text{Na} \rightarrow {}^{24}_{12}\text{Mg} + {}^{0}_{-1}\text{e} + \bar{v} \ (T = 15 \text{ hrs})$$

$$^{14}_{7}\text{N} + {}^{1}_{0}\text{n} \rightarrow ({}^{15}_{7}\text{N}) \rightarrow {}^{14}_{6}\text{C} + {}^{1}_{1}\text{H}$$

followed by

$$^{14}_{6}\text{C} \rightarrow {}^{14}_{7}\text{N} + {}^{0}_{-1}\text{e} + v$$

In general, artificial radioactive isotopes are produced by the bombardment of stable elements with accelerated charged particles ($^{1}_{1}\text{p}$, $^{2}_{1}\text{D}$, α). Most of the artificially produced radioactive isotopes have small half lives. This can be a valuable aid in the identification of an element by chemical analysis. By far the most useful particles for producing artifical radioactivity are the neutrons which are available in large number in a nuclear reactor. Normally, now all the elements can be made artificially radioactive.

It is observed that some radioactive isotopes emit positrons, while some others emit electrons. In few cases γ-ray emission also takes place together with positrons or electrons. Artificially radioactive isotopes which emit α-particles are much less common and occur mostly in transuranic elements ($Z > 92$).

(e) Transuranic Elements

The naturally occuring heaviest element is uranium ($_{92}\text{U}$). Since 1940, elements having an atomic number (Z) greater than 92 have been *artificially* produced in the laboratory by the bombardment of certain heavy nuclei with appropriate particles. These *elements* with $Z > 92$ are called '*transuranic elements*'. Many of the transuranic elements have been produced in isotopic forms and these isotopic nuclides and their daughter products are generally radioactive. The transuranic elements ranging from $Z = 93$ to $Z = 106$ are as follows:

(i) **Neptunium** Np ($Z = 93$): This is produced by the resonance capture of slow neutrons of certain discrete energies by $^{238}_{92}\text{U}$:

$$^{238}_{92}\text{U} + {}^{1}_{0}\text{n} \longrightarrow ({}^{239}_{92}\text{U}) \rightarrow {}^{239}_{92}\text{U} + \gamma$$

followed by

$$^{238}_{92}\text{U} \xrightarrow{\ \ } {}^{239}_{93}\text{Np} + {}^{0}_{-1}\text{e} + \bar{v}_{e}$$
$$\text{23.5 minutes}$$

One can also produce it by the action of fast neutrons on $^{238}_{92}U$;

$$^{238}_{92}U + ^{1}_{0}n \longrightarrow ^{237}_{92}U + ^{1}_{0}n + ^{1}_{0}n$$

followed by

$$^{237}_{92}U \longrightarrow ^{237}_{93}Np + ^{0}_{-1}e + \bar{v}_e$$
$$\text{6.7 days}$$

and

$$^{237}_{93}Np \longrightarrow ^{233}_{91}Pa + ^{4}_{2}He$$

(ii) Plutonimu, Pu (Z = 94)   The $^{239}_{93}Np$ isotope is itself radioactive. It decays from the spontaneous β^- activity into $^{239}_{94}Pu$, an isotope of the second transuranic element plutonium.

$$^{239}_{93}Np, \longrightarrow ^{239}_{94}Pu + ^{0}_{-1}e + \bar{v}_e$$
$$\text{2.33 days}$$

followed by

$$^{239}_{94}Pu \longrightarrow ^{235}_{92}U + ^{4}_{2}He$$
$$\text{24400 years}$$

one can also produce by bombarding $^{238}_{92}U$ with deutrons;

$$^{238}_{92}U + ^{2}_{1}H \longrightarrow (^{240}_{93}Np) \longrightarrow ^{238}_{93}Np + ^{1}_{0}n + ^{1}_{0}n$$

followed by

$$^{238}_{93}Np \longrightarrow ^{238}_{94}Pu + ^{0}_{-1}e + \bar{v}_e$$
$$\text{2.1 days}$$

and

$$^{238}_{94}Pu \longrightarrow ^{234}_{92}U + ^{4}_{2}He$$
$$\text{90 yrs}$$

(iii) Americium, Am (Z = 95)   It is produced from α-particle bombardment of ^{238}U and from deuteron bombardment of ^{239}Pu;

$$^{238}_{92}U + ^{4}_{2}He \longrightarrow (^{242}_{92}Pu) \longrightarrow ^{241}_{94}Pu + _{0}n^1$$

followed by

$$^{241}_{94}Pu \longrightarrow ^{241}_{95}Am + ^{0}_{-1}e + \bar{v}_e$$
$$\text{14 yrs}$$

$$^{239}_{94}Pu + ^{2}_{1}H \longrightarrow (^{241}_{95}Am) \longrightarrow ^{240}_{95}Am + ^{1}_{0}n$$

(iv) Curium, Cm (Z = 96)   This is produced from α-particle bombardment of $^{238}_{94}Pu$

$$^{239}_{94}Pu + ^{4}_{2}He \longrightarrow (^{243}_{96}Cm) \longrightarrow ^{242}_{96}Cm + _{0}n^1$$

followed

$$^{242}_{96}Cm \longrightarrow ^{238}_{94}Pu + ^{4}_{2}He$$
$$\text{162 days}$$

(v) **Berkelium, Bk (Z = 97)** This is produced from α-particle bombardment of ^{241}Am;

$$^{241}_{95}\text{Am} + ^4_2\text{He} \longrightarrow ^{245}_{97}\text{Bk}$$

followed by $\quad ^{245}_{97}\text{Bk} \underset{\text{5 days}}{\longrightarrow} ^{241}_{95}\text{Am} + ^4_2\text{He}$

(vi) **Californium, Cf (Z = 98)** This is produced from α-particle bombardment of ^{242}Cm,

$$^{242}_{96}\text{Cm} + ^4_2\text{He} \longrightarrow ^{246}_{98}\text{Cf}$$

followed by $\quad ^{246}_{98}\text{Cf} \underset{\text{36 hrs}}{\longrightarrow} ^{242}_{98}\text{Cm} + ^4_2\text{He}$

(vii) **Einsteinium, Es (Z = 99)** This is produced by bombarding ^{238}U by synchrocyclotron-accelerated ^{14}N ion

$$^{238}_{92}\text{U} + ^{14}_7\text{N} \longrightarrow (^{252}_{99}\text{Es}) \longrightarrow ^{246}_{99}\text{Es} + n + n + n + n + n + n$$

(viii) **Fermium, Fm (Z = 100)** It is produced by bombarding ^{238}U with synchrocyclotron-accelerated ^{14}N ion

$$^{238}_{92}\text{U} + ^{16}_8\text{O} \longrightarrow (^{254}_{100}\text{Fm}) \longrightarrow ^{250}_{100}\text{Fm} + n + n + n + n$$

(ix) **Mendelevium, Mv (Z = 101)** This is produced from α-particle bombardment of ^{253}Es:

$$^{253}_{99}\text{Es} + ^4_2\text{He} \longrightarrow (^{257}_{101}\text{Mv}) \longrightarrow ^{256}_{101}\text{Mv} + _0n^1$$

(x) **Nobelium, No (Z = 102)** This is produced by bombarding ^{244}Cm with synchrocyclotron-accelerated ^{13}C ions

$$^{244}_{96}\text{Cm} + ^{13}_6\text{C} \longrightarrow (^{257}_{102}\text{No}) \longrightarrow ^{253}_{102}\text{No} + n + n + n + n$$

(xi) **Lawrencium, Lw (Z = 103)** This is produced by bombarding ^{246}Cf with high energy ^{11}B ions;

$$^{246}_{98}\text{Cf} + ^{11}_5\text{B} \longrightarrow ^{257}_{103}\text{Lw}$$

followed by $\quad ^{257}_{103}\text{Lw} \underset{\text{8.6 s}}{\longrightarrow} ^{253}_{101}\text{Mv} + ^4_2\text{He}$

(xii) **Kurchatovium, Z = 104** This is produced by bombarding ^{242}Pu with high energy ^{22}Ne ions.

(xiii) **Hahnium, Z = 105** This element is produced by bombarding ^{249}Cf with high energy ^{15}N ions.

(xiv) Z = 106 element is produced by bombarding ^{207}Pb or ^{208}Pb with high energy ^{54}Cr ions, or ^{249}Cf with high energy ^{18}O ions.

Transuranic elelemts form a series begining with actinium and this is referred as *actinide series*. This series is analogous to the *rare earth* series. Except neptunium and plutonium other transuranic elements are not found in nature. This is possibly due to their short half lives compared with the age of the earth.

2.16 USES OF RADIO-ISOTOPES

We have read that in addition to the naturally occuring radio-isotopes such as radium, hundreds of radio-isotopes have been made artificially. These radio-isotopes finds numerous applications in agriculture, medicine, industry and research. Many applications employ a special technique called as '*tracer technique*'

Tracer Techniques   To study the dynamics of a particular system, a radio-isotope is added and its path is traced by means of a G.M. Counter. For intance $^{15}_{8}O$ mixed with normal breathed by a patient, enables radio-detection of the path of inhaled oxygen in lungs. Phosphorus uptake by plants from the soil is studied using a phosphate fertilizer containing the radioisotope ^{32}P. The mechanism of plant photosynthesis can be understood by growing plants in an atmosphere of $^{14}CO_2$. Constrictions in blood vessels are detected using the soluble $^{24}NaCl$ as an indicator injected in the blood stream. I-131 is used for detection of brain and thyroid tumors.

α, β, γ radiations emanating from radio-isotopes are utilized in fundamental research, e.g. routine laboratory measurments and testing of materials. The material under test is interposed between the radio-isotope (source) and a radiation detector. The change in counting of the physical properties and structure of the material.

Radio-isotopes find wider applications in molecular biology for producing destructive effects. To kill bacteria and insects, large doses of radiations are needed. One can produce genetic changes in grain seeds and thus obtain mutations from the radiations emerging from radio-isotopes. New strains are developed which can resist bad weather, disease and pests, and provide higher yields and mature early.

Certain perishable creals exposed to radiations remain fresh beyond their normal life span. One can delay the ripening of fruits and vegetable and their deterioration can be checked by exposing them to radio-isotopes under controlled conditions. One can enhance the storage time of the food-item by prior-irradiation. Very small doses of radiation precent sprouting and spoilage of onions, potatoes and gram. Higher doses (25 to 50 kilorad) prolong the shell life of fruits like bananas and oranges.

Radio-isotopes are being used in many diverse fields in engineering works. In order to locate a blockage in an underground sewage pipe, a rubber ball having ^{24}Na is introduced into the pipe. A G.M. Counter above ground give us the position of the ball when it has come to rest. Radio-isotopes are also useful in transporting different oils through underground pipe to distant places. When the type of the oil flowing through the pipe is changed, a small quantity of radio-isotope is mixed exactly at the position where the change takes place. Near the other end of the pipe line, a G.M. Counter is placed which gives a signal when the radio-isotope passes. Radioactivated sand deposited in harbours helps study the movement of silt. Studies carried in the sea bed enable selection of possible dumping sites for dredged silt. To detect the flaws and cracks in castings and welded joints, Iridium-192 is used. Cobalt-60 emits gamma ray of energy 1.3 MeV which have a much higher penetration in metals than X-rays. Using gamma rays from this isotope sheet thicknesses are routinely ganged.

The tracer technique is also used in research to study the exchange of atoms between various molecules, and to investigate the solubility and vaporisation of materials.

The tracer technique is used for testing the uniformity of mixtures, e.g. for testing a chocolate mixture, a small quantity of short-lived radio-isotope such as ^{24}Na or ^{56}Mn is added to the primary ingradients. Several different samples of the final product are then tested for radioactivity by means of a G.M. Counter. If one obtains the same counting rate, then the mixing is said to be uniform. One can also use this method in mixing processes occuring in the manfacture of soap, cement, chocolate, fertilisers, paints cattle food and medical tablets.

Besides the uses employing the tracer technique there are several other uses of radio-isotopes in various fields. A very promising application of radio-isotopes is found in nuclear medicine.

Nuclear Medicine Uses   The radio-isotopes which are used for medical purposes are categorised as radiopharmaceuticals. They are classified under two categories; (i) where the primary element is radioactive, e.g. radioactive iodine, iron, chromium, etc. and (ii) Where a radioactive label is introduced into a complex chemical molecule like radioiodine labelled albumin, where iodine is not a natural constitutent of the albumin.

Scintigraphy   This is a technique, in which the patient is injected with a radiopharmaceutical which concentrates in a specific organ of the body. One finds that a focus of disease in that organ stands out as a void, a non-functioning and non-concentrating area. It is reported that there are some radiopharmacenticals which do not concentrate in healthy cells of an organ but would only concentrate in the disease part because of the pathological defect in that part, e.g. radiopharmaceuticals used for brain scanning do not concentrate in the normal brain because of the blood-brain barrier which allows only for few selected compounds to penetrate and concentrate in the normal brain tissue. In brain scanning radiopharmaceuticals primarily localise at the site of the disease like brain tumour.

One can get the *imaging* of an organ of the body with the help of two different kinds of instruments:

(i) **Rectilinear Scanner**   In this instrument, a scintillation detector systematically and automatically moves over an organ point by print detecting the presence and amount of radioactivity at each point during its tranverse and displaying it on paper of an X-ray film so that a composite map of all the points results in an image of distribution of radioactivity in the organ, and

(ii) **Scintillation Camera**   In this the deetector is so large that it views the entire organ at one and the same time and then electronically discerns the amount of radioactivity arising from each point. We must note that the camera is quicker than the scanner and permits one to visualise the dynamic function of an organ.

A commonly used tracer in nuclear medicine is radioactive iodine. Radioactive iodine traces iodine metabolism in the body and while doing so, provides useful information about the function of the throid gland. One can determine the percentage of the administered dose accumulating in the gland at various times. Increased trapping indicates hyperactivity, while decreased trapping indicates hyperactivity of the gland.

Some dynamic function studies cover the kidneys and the lever of the body. Hippuran (radioiodine) is rapidly excreted through the kidneys after an intravenous injection and one can monitor its passage through two externally placed detectors over the kidneys. This is the simplest known method of simutaneous assessment of the functions of both the kidneys. This study is helpful in excluding the rental origin of hypertension in the patients suffering from this disease due to kidney infections. One can visulize the passage of radioiodine labelled dye Rose Bengal by a series of detectors placed at suitable sites over the body. A dectector placed over the heart exhibit blood clearance, a detector over the liver region exhibit liver uptake and liver clearance, and a third detector over the intestinal region exhibit excretion of radioisotope from the gall bladder into the intestine in response to a fatty meal.

There are two traces namely radioiodine labelled albumin and transferrin bound indium. These tracers confines itseld to the blood and just remains in circulation. Obviously, it can provide information about the rate of blood flow through several organs. These flow studies include peripheral blood flow in a limb, blood flow in a brain, output of heart, etc.

For the kinetic studies, the movement to a tracer through various parts of the body can be traced to obtain useful physiological information, e.g. the rate of plasma disappearance of radioactive iron informs how avidly the iron stores in the blood and how bone marrow picks up the injected iron. In the case of anaemia due to iron deficiency this iron would disappear rapidly from the circulation to get into the bone marrow. One can do this sort of kinetic study with the help of many tracers, e.g. labelled albumin, calcium, vitamin B-12, etc.

In vitro assays provide another unique technique where no radioactivity is administered to the patient. One introduces the radioisotope only in the test-tube containing blood samples from the patient for estimation of the amount of the harmone present in his blood. One can made bind most of the harmones in the blood specific antibodies.When a known amount of radioactive harmone is added to the test-tube, then how much of the harmone now remains bound to the anti-body and how much is displaced depends on how much of the harmone was originally present in the patient's blood.

There are very few clinical situations wherre radio-isotopes can be used in large amount to destroy tissues where they would concentrate without affecting the surrounding healthy organs one exception is radioactive iodine. Which achieves selective localization in the thyroid and this property is utilised to destroy overactive thyroid or a thyroid ridden with cancer. Similarly, radioactive phosphorus is used to destroy excessive production of red blood cells in the bone marrow in polycycnthemia. Technetium 99 m (Tc) is another widely used radiopharmaceutical and is obtained from Mo-99 produced in nuclear reactors when stable Mo-98 is subjected (n, γ) reactions. Chromium-51 is taken up by the red blood cells and is used in location of hemorrhage. Similarly, Ba-137 $(T = 127$ s) is used in radiocardiography. Electronic counters are used to detect the radiations emitted by these tracers.

About 400 different radioisotopes are being manfactured at BARC (Bhaba Atomic Research Centre), Bombay. These are used indigenously and are also exported.

Example 2.1 An accident occurs in a laboratory in which a large amount of radioactive material with a known half life of 20 days gets embedded in floor and walls etc. Tests show that the level of radiation is 32 times the permissible level of normal occupancy of the room. Assuming that the last statement is correct, find after how many days can the laboratory be safely used?

Sol. We have the initial level of radiation as 32 times that of the permissible limit. Let the activity drops to $\dfrac{1}{32}$ of its initial value in time t, then

$$\frac{N}{N_0} = \frac{1}{32}$$

Half life T of the sample = 20 days

$$\therefore \quad \lambda = \frac{0.693}{20} \text{ (days)}^{-1}$$

Let the laboratory become safe for use after t days, then

$$\frac{N}{N_0} = e^{-\lambda t}$$

or

$$\frac{1}{32} = \exp\left(-\frac{0.693 \, t}{20}\right)$$

or

$$t = \frac{2.303 \times 1.5051 \times 20}{0.693} = 100 \text{ days.}$$

Example 2.2 The activity of $^{200}_{92}\text{Au}$ of mass 20×10^{-9} kg is 60 Curie. Determine the half life of $^{200}_{92}\text{Au}$.

Sol. Let the number of atoms in 20×10^{-9} kg of $^{200}_{79}\text{Au}$ is N. Then

$$N = (20 \times 10^{-9} \text{ kg}) \left(\frac{1 \text{ k mole}}{200 \text{ kg}}\right)\left(6.025 \times 10^{23} \frac{\text{atoms}}{\text{mole}}\right)$$

$$= 6.025 \times 10^{16} \text{ atoms}$$

The acitivity is given by (60 Curie) $\dfrac{3.7 \times 10^{10} \text{ disintegration/sec}}{1 \text{ Curie}}$

$$= 2.22 \times 10^{12} \text{ disintegrations/s}$$

$$\therefore \quad \lambda = \frac{\text{Activity}}{N} = \frac{2.22 \times 10^{12}}{6.025 \times 10^{16}} = 0.368 \times 10^{-4}\text{/s}$$

$$\therefore \quad \text{Half life } T = \frac{0.693}{\lambda} = \frac{0.693}{0.368 \times 10^{-4}} \text{ s} = 1.883 \times 10^{-4} \text{ s}$$

Example 2.3 Compute the mass of 1.00 Ci of ^{14}C. The half life of ^{14}C is 5570 years.

Sol. $T = 5570 \text{ y} = (5.570 \times 10^3 \text{ y}) \times (2.156 \times 10^7 \text{ s/y})$

$$= 1.758 \times 10^{11} \text{ s}$$

$$\lambda = \frac{0.693}{T} = 3.94 \times 10^{-12}\text{/s}$$

Now $\left|\dfrac{dN}{dt}\right| = 1 \text{ Ci} = 3.70 \times 10^7\text{/s}$

$$\therefore \quad N = \frac{1}{\lambda}\left|\frac{dN}{dt}\right| = 9.38 \times 10^{18} \text{ nuclei of } ^{14}\text{C,}$$

Which is also the number of carbon atoms present. The atomic mass of $^{14}C = 14.0077$ amu.

$\therefore$ Mass of the 9.38×10^{18} atoms of ^{14}C

$$= (14.0077 \times 1.6604 \times 10^{-27} \text{ kg/atom}) (9.38 \times 10^{18} \text{ atoms})$$

$$= 2.18 \times 10^{-7} \text{ kg}$$

Example 2.4 Using particle accelerators particular type of nucleus with decay constant λ is produced at a steady rate of P nuclei per second. Show that the number of nuclei $N(t)$ present t seconds after the production starts is given by $N(t) = \dfrac{P}{\lambda} (1 - e^{-\lambda t})$.

Sol.   Rate of formation of nuclei = P

Rate of declay of nuclei = $\lambda N(t)$ at time t

At time $t = 0$, $N(t) = 0$

and $\qquad\qquad \dfrac{dN(t)}{dt} = P - \lambda N$

Multiplying by $e^{\lambda t} dt$ on both sides, we have

$$e^{\lambda t} dN(t) = D\ e^{\lambda t}\ dt - \lambda\ N(t)\ e^{\lambda t}\ dt$$

Integerating $e^{\lambda t}\ dN(t)$ by parts and R.H.S in a straight forward manner, one obtains

$$N(t)\, e^{\lambda t} = \int N(t) \lambda\ e^{\lambda t}\ dt = \frac{P}{\lambda} e^{\lambda t} - \int N(t) \lambda\ e^{\lambda t}\ dt + C$$

$\therefore \qquad\qquad$ At $t = 0$, $N(t) = 0$ $\therefore C = -\dfrac{P}{\lambda}$

or $\qquad\qquad N(t)\, e^{\lambda t} = \dfrac{P}{\lambda} (e^{\lambda t} - 1)$

or $\qquad\qquad N(t) = \dfrac{P}{\lambda} (1 - e^{-\lambda t})$

Example 2.5 $^{212}_{83}Bi$ decays to $^{208}_{81}Tl$ by α-emission in 34 % of disintegrations and to $^{212}_{84}Po$ by β-emission in 66 % of disintegrations. If the total half-value period is 60.5 minutes, find the decay constants for α and β emissions.

Sol.   The probability of disintegration

$$\lambda = \lambda_\alpha + \lambda_\beta$$

$$T = \frac{0.693}{\lambda} = \frac{0.693}{\lambda_\alpha + \lambda_\beta} \qquad\qquad\qquad \ldots (2.1)$$

$\therefore \qquad\qquad \lambda_\alpha + \lambda_\beta = \dfrac{0.693}{T} = \dfrac{0.693}{60.5 \text{ min}} = 0.01145/\text{min} \qquad \ldots (2.2)$

Now $$\frac{dN}{dt} = -\lambda N$$

$\therefore$ $$0.34 = -\lambda_\alpha N \qquad \qquad \qquad \dots (2.3)$$

and $$0.66 = -\lambda_\beta N \qquad \qquad \qquad \dots (2.4)$$

(2.3) and (2.4) give

$$\frac{0.34}{0.66} = \frac{\lambda_\alpha}{0.01145 - \lambda_\alpha} \qquad (\because \lambda_\alpha + \lambda_\beta = 0.01145)$$

Solving one obtains $\lambda_\alpha = 0.003893/\text{min}$

and $\lambda_\beta = 0.007557/\text{min}.$

Example 2.6 The mean lives of a radioactive substance are 1620 and 405 years for α-emissions and β-emissions respectively. Find out the time during which $\frac{3}{4}$th of a sample will decay if it is decaying both by α-emission and β-emission simultaneously.

Sol. Total decay constant $\lambda = \dfrac{1}{1620} + \dfrac{1}{405} = \dfrac{1}{324} /\text{year}$

We have $$N = N_0 e^{-\lambda t}$$

$\therefore$ $$\frac{1}{4} = e^{-\lambda t} \text{ or } t = \frac{1}{4} \log e^4 = 449 \text{ years}$$

Example 2.7 The isotopes U^{238} and U^{235} occur in nature in the ratio 140:1. Assuming that at the time of earth's formation they were present in equal ratio, make an estimation of the age of the earth. The half-lives of U^{238} and U^{235} are 4.5×10^9 years and 7.13×10^8 years respectively. ($\text{Log}_{10}^{140} = 2.1461$, $\log_{10}^2 = 0.3010$)

Sol. Let N_1 and N_2 be the number of atoms of U^{238} and U^{235} and T_1 and T_2 be their half lives. Now

$$N = N_0 e^{-\lambda t}$$

$\therefore$ $$\frac{N_1}{N_2} = e(\lambda_2 - \lambda_1)t$$

where t is the elasped time. From this one ontains

$$t = \frac{\log_e \left(\dfrac{N_1}{N_2} \right)}{\lambda_2 - \lambda_2} = \frac{\log_e \left(\dfrac{140}{1} \right)}{\log_e 2 \left(\dfrac{1}{T_2} - \dfrac{1}{T_1} \right)} \qquad \left(\because \lambda = \log_e 2/T \right)$$

$$= \frac{\log_{10}^{140}}{\log_{10}^2} \left(\frac{T_2\, T_1}{T_2 - T_1} \right)$$

$$= \frac{2.1461}{0.3010} \times \frac{7.13 \times 10^8 \times 4.5 \times 10^9}{37.87 \times 10^8} = 6 \times 10^9 \text{ yrs.}$$

Example 2.8 10 mg of carbon from living matter produce 200 counts per minute due to a small proportion of the radioactive isotope ^{14}C. A piece of ancient wood of mass 10 mg is found to give 50 counts per minute. Determine the age of the wood assuming that ^{14}C content of the atmosphere has remained unchanged. The half life of ^{14}C is 5700 years.

Sol. The amount of ^{14}C in the atmosphere remains constant and a living matter represents the amount of ^{14}C present in the atmosphere. This means the ancient wood, which has now 50 counts per minute from 10 mg. In the begining, must have had the same count rate as from living matter. Obviously, the age of the ancient wood is the time in which ^{14}C count rate has decreased from the initial value of 200 counts per minute from 10 mg to the final value of 50 counts per minute from 10 mg. Now, the disintegration constant for ^{14}C

$$\lambda = \frac{0.693}{T} = \frac{0.693}{5700} \text{ (yr)}^{-1}$$

Let t represent the age of the wood, then

$$50 = 200 \exp\left(-\frac{0.693 t}{5700}\right)$$

or $\qquad t = \dfrac{\ln e^4 \times 5700}{0.693} \text{ years} = 1.41 \times 10^4 \text{ years}$

Example 2.9 The ratio of the mass of ^{208}Pb to the mass of ^{238}U in a certain rock specimen is found to be 0.5. Assuming that the rock originally contained no lead, show that its age is 2.63×10^9 years. Given half life $^{238}U = 4.5 \times 10^9$ yrs.

Sol. $\qquad \lambda = \dfrac{0.693}{T} = \dfrac{0.693}{4.5 \times 10^9} \text{ (yr)}^{-1}$

Age of rock $\qquad t = \dfrac{1}{\lambda} \times \log\left(\dfrac{N_{Pb} + N_U}{N_U}\right) = \dfrac{1}{\lambda} \log_e\left(\dfrac{N_{Pb}}{N_U} + 1\right)$

$$= \frac{4.5 \times 10^9}{0.693} \log_e (0.5 + 1) = 2.63 \times 10^9 \text{ yrs.}$$

Example 2.10 Show that, if N_1^0 be the number of parent atoms initially and λ_1, λ_2 the decay constants of the parent and daughter respectively, the time at which the number of daughter atoms is a maximum is given by

$$t_m = \frac{\ln\left(\dfrac{\lambda_2}{\lambda_1}\right)}{\lambda_2 - \lambda_1}$$

(Delhi Hons.)

Sol. We have from the relation

$$(N_2)_t = \frac{\lambda_1}{\lambda_2 - \lambda_1} N_{10}(e^{-\lambda_1 t} - e^{-\lambda_2 t})$$

where $(N_2)_t$ denotes the number of daughter stoms at any time t, one obtains,

$$\frac{d(N_2)_t}{dt} = \frac{\lambda_1}{\lambda_2 - \lambda_1} N_{10}(-\lambda_1 e^{-\lambda_1 t} + \lambda_2 e^{-\lambda_2 t})$$

For $(N_2)_t$ to become a maximum, $\dfrac{d(N_2)_t}{dt} = 0$, which implies

$$\therefore \quad -\lambda_1 e^{-\lambda_1 t} + \lambda_2 e^{-\lambda_2 t} = 0$$

or, $$\frac{\lambda_2}{\lambda_1} = \frac{e^{-\lambda_1 t}}{e^{-\lambda_2 t}} = e^{(\lambda_2 - \lambda_1)t}$$

Calling the time t as t_m, one obtains

$$\ln \frac{\lambda_2}{\lambda_1} = (\lambda_2 - \lambda_1)t_m; \quad \therefore t_m = \frac{\ln\left(\dfrac{\lambda_2}{\lambda_1}\right)}{\lambda_2 - \lambda_1}$$

We may note that $\dfrac{d^2(N_2)_t}{dt^2}$ is negative at t_m satisfying fully the condition of maximum.

Example 2.11 Calculate the age of the earth given that the isotopic abundances of U-238 and U-235 today are 99.28 % and 0.72 % respectively. Originally the two isotopes were present in equal abudance and there has been no isotopic separation physically or chemiclaly. Given $T_{\frac{1}{2}}$ of U-238 = 4.5×10^9 years and $T_{\frac{1}{2}}$ of U-235 = 7.1×10^8 years. [Punjab]

Sol. We have, $$N_{235} = (N_0)_{235}\, e^{-\lambda_{235} t}$$
$$N_{238} = (N_0)_{238}\, e^{-\lambda_{238} t}$$

Where t is the decay time. But, from the problem,

$$(N_0)_{238} = (N_0)_{235}$$

$$\therefore \quad \frac{e^{-\lambda_{238} t}}{e^{-\lambda_{235} t}} = \frac{N_{238}}{N_{235}} = \frac{99.28}{0.72} = 138$$

$$\therefore \quad e^{(\lambda_{235} - \lambda_{238})t} = 138$$

$$\therefore \quad (\lambda_{235} - \lambda_{238})t = 2.303 \log_{10} 138 = 4.91$$

But, $$\lambda_{238} = \frac{0.693}{4.5 \times 10^9} = 1.54 \times 10^{-10}/\text{yr}$$

$$\lambda_{235} = \frac{0.693}{7.1 \times 10^8} = 9.76 \times 10^{-10} \,/\text{yr}$$

$\therefore$ We have, $\qquad t = \dfrac{4.91}{\lambda_{235} - \lambda_{238}} = \dfrac{4.91}{(9.76 - 1.54) \times 10^{-10}} = 6 \times 10^9$ years.

Thus the age of the earth is 6 billion years. (This is not the exact age but this is the upper limit).

Example 2.12 The half lives of radium of radon are 1622 years and 3.825 days. What is the volume of radon gas at N. T. P. equivalent to one curie? [Raj, MDS]

Sol. Using the condition for equilibrium between radium and its emanation radon

$$\lambda_{Rn} N_{Rn} = \lambda_{Ra} N_{Ra} \qquad (N = \text{no. of atoms per g})$$

$$\therefore \qquad N_{Rn} = \frac{\lambda_{Ra}}{\lambda_{Rn}} N_{Ra} = \frac{\left(T_{\frac{1}{2}}\right)_{Rn}}{\left(T_{\frac{1}{2}}\right)_{Ra}} N_{Ra}$$

$$= \frac{3.825}{1622 \times 365} N_{Ra}$$

Now, $\qquad N_{Ra} = \dfrac{N_A}{226}$, where N_A = Avogadro number. So, from above

$$N_{Rn} = \frac{3.825 \, N_A}{1622 \times 365 \times 226}$$

Since N_A atoms of radon gas occuples 22.4 litres at N. T. P., the volume occupied by N_{Rn} atoms of radon at N.T.P. is

$$V = \frac{N_{Rn}}{N_A} \times 22.4 \text{ litres}$$

$$= \frac{3.825 \, N_A \times 22.4 \times 10^3}{N_A \times 1622 \times 365 \times 226} \text{ cc}$$

$$= 6.4 \times 10^{-4} \text{ cc}$$

$$= 0.64 \text{ cubic mm}$$

Example 2.13 The range in standard air of α-particles from Ra ($T_{\frac{1}{2}}$ = 1622 yrs) is 3.36 cm and that of α's from Po ($T_{\frac{1}{2}}$ = 138 days) is 3.85 cm. If the range of α-particles from Ra C' be 6.97 cm, find the half life of Ra C' using Geiger Nuttal law. [Vikram]

Sol. We have $\lambda_{Ra} = \dfrac{0.693}{1622 \times 365}$ day^{-1} and $R_{Ra} = 3.36$ cm

$$\therefore \quad \ln \dfrac{0.693}{1622 \times 365} = a + b \ln 3.36 \qquad \text{(i)}$$

Using the Geiger-Nuttal relation, one finds

$$\lambda_{Po} = \dfrac{0.693}{138} \text{ day}^{-1} \text{ and } R_{Po} = 3.85 \text{ cm}$$

$$\therefore \quad \ln \dfrac{0.693}{138} = a + b \text{ in } 3.85 \qquad \text{(ii)}$$

From (i) and (ii) $\quad b \ln \dfrac{3.82}{3.36} = \ln \dfrac{1622 \times 365}{138}$

$$\therefore \qquad b = 61.43$$

Substituting this value of b in (ii),

$$\ln \dfrac{0.693}{138} = a + 61.43 \ln 3.85$$

$$\therefore \qquad a = -38.3$$

$\therefore$ For the given series (radioactive), the Geiger-Nuttal relation is

$$\ln \lambda = -38.3 + 61.43 \ln R$$

$\therefore$ For Ra C', $\qquad \ln \lambda = -38.3 + 61.43$ In 6.97

$$\therefore \qquad \lambda = 2.51 \times 10^{13} \text{ days}^{-1}$$

$$\therefore \quad \text{Half life of Ra } C' = \dfrac{0.693}{2.51 \times 10^{13}} \text{ days} = \dfrac{0.693 \times 24 \times 3600}{2.51 \times 10^{13}}$$

$$= 2.4 \times 10^{-9} \text{ s}$$

Example 2.14 The half life of ^{131}Te is 30 hrs and that of ^{131}I is 193 hrs. They are the respective parent and daughter products for successive growth in a radioactive series. Find the time when the intermediate daughter product would be maximum in its decay rate. [Punjab]

Sol. If t_m be the time for maximum, then t_m is given by

$$t_m = \dfrac{\ln \left(\dfrac{\lambda_2}{\lambda_1} \right)}{\lambda_2 - \lambda_1}$$

$$= \dfrac{\ln \left(\dfrac{193}{30} \right)}{0.69 \left(\dfrac{1}{30} - \dfrac{1}{193} \right)}$$

$$= \frac{1.86}{0.69 \times 0.028} = 95.4 \text{ hr}$$

Example 2.15 The maximum permissible dose using γ-radiation is 6.25 millioentgen/hr. Find the safe working distance from a 20 Ci source having a dose rate of 25 roentgen per hour at a distance of 1 m. If the source is enclosed in a lead shield, the γ-radiation is reduced to 2 %; how much closer then can one go safely?   [Raj, Udaipur]

Sol. Let $I(r_0)$ be the intensity at a distnce r_0 and $I(r)$ that at r, i.e., the safe distance.

We have

$$\therefore \qquad \frac{I(r)}{I(r_0)} = \frac{r_0^2}{r^2} = \frac{6.25 \times 10^{-3}}{25} = \frac{(1.0)^2}{r^2}$$

$$[\therefore \quad I(r) = 6.25 \times 10^{-3} \text{ roentgen/hr and } r_0 = 1.0 \text{ m}]$$

$$\therefore \qquad \text{Safe distance, } r = 1.0 \times \sqrt{\frac{25}{(6.25 \times 10^{-3})}} = 63.25 \text{ m}$$

With Pb-shield the intensity at 1.0 m $= 25 \times \dfrac{2}{100} = 0.5$ roentgen/h

$$\therefore \qquad \text{Safe distance, } r = 1.0 \sqrt{\frac{0.5}{6.25 \times 10^{-3}}} = 8.95 \text{ m}$$

REFERENCES

1. Irving Kaplan, Nuclear Physics, Addition Wesley Publishing Co. Inc. (1963)
2. R.E.Lapp and H.L.Andrews, Nuclear physics (3rd ed.), Prentice Hall Inc. (1963)
3. W. E. Burcham and M. Jobes, Nuclear and Particle Physics, Addision Wesley Longman Ltd. England (1995)
4. Samuel S.M. Wong, Introductory Nuclear Physics, Prentice Hall International, Inc., New Jersy (USA) (1990).
5. J. S. Rasey and N. J. Nelson, 'Radiation Research' 85, 69 (1981)

REVIEW QUESTIONS

1. What is natural radioactivity? Explain the terms (a) decay constant (b) half life and (c) average life.
2. What do you understand by radioactive disintegration? State the laws of radioactive disintegration and derive Rutherford soddy law. Discuss the satistical nature of radioactive disintegration.
3. What is meant by the half life time of a radioactive substance? Establish a relation between the half life time and decay constnt.
4. Explain the term mean life time of a radioactive substance. Show that the mean life time of an atom of a radioactive sunstance is the reciprocal of its decay constant. Hence obtain the relation between mean life time and half life time of a radioactive substance.

5. Deduce expressions for the half life and mean life of a radio active substance in terms of a decay constant.
6. What is secular equilibrium? Obtain the condition for such an equilibrium
7. Give a short account of the various radioactive transformation series. Indicate the various stages by which ^{238}U is converted into an isotope of lead in the course of natural radioactivity.
8. Describe the general equations of radioactive series growth and decay. Explain secular and transistent radioactive equilibrium
9. A nucleus A decays into B with half life T_1 and B decays into C with half life T_2. Discuss the composite process of decay and distinguish between the situations
   (i) $T_1 > T_2$ (ii) $T_1 = T_2$ and   (iii) $T_1 < T_2$
10. Define the activity of a radioactive sample and state the units in which it is measured. Define Curie and becquerel and write relation between these.
11. Explain how will you measure the half life (or decay constant) of a radioactive substance.
12. Explain how the age of rocks or earth can be inferred from a measurment of their intensity of radioactivity
13. What is a radon gas? How it can be separated from its parent element?
14. Give a brief account of various methods for the measurment of decay constant
15. Give a brief account of the methods used for the determination of the geological ages of minerals and rocks.
16. Give an account of the uses of radioisotopes
17. What is artificial transmutation of elements?
    Explain with example.
18. What is artificial radioactivity? How can stable nuclei be made radioactive?
19. How the neutron was discovered? Discuss its nature, Characterstic properties, detection and uses. Is a free neutron stable particle?

PROBLEMS

1. The half life of ^{90}Sr is 28 years. Determine (a) the disintegration constant for ^{90}Sr (b) the activity of 1 mg of ^{90}Sr in Curie and as nuclei per second, (c) the time for the 1 mg to reduce to $^{250}\mu g$, (d) the acitivity in this letter time.
   [**Ans.** (a) $7.86 \times 10^{-8}\,s^{-1}$ (b) $5.27 \times 10^{11}\,s^{-1}$, 112 yr (d) $1.32 \times 10^4\,s^{-1}$]
2. A sample contains 10^{-2} kg each of two substances A and B with half lives 4 s and 8 s respectively. Their atomic weights are in the ratio 1 : 2. Find the amounts of A and B after an interval of 16 seconds.
   [**Ans.** $m_B = 2.5 \times 10^{-3}$ kg, $m_A = 7.5 \times 10^{-3}$ kg]
3. A radioactive element disintegrates for an interval of time equal to its means life.
   (i) What fraction of element remains?   (ii) What fraction has disintegrated?
   [**Ans.** (i) $\dfrac{1}{e}$ (ii) $\dfrac{(e-1)}{e}$]
4. The half life of ^{24}Na is 15 hr. How long does it take for 87.5 % of this isotope to decay?
   [**Ans.** 45 hours]
5. A radioactive substance at a given instant emits 4750 particles per minute. Find the decay constant and half life of the substance.
   [**Ans.** $\lambda = 0.113$/min (b) $T = 6.13$ min]

6. The atomic ratio between the uranium isotopes U^{238} and U^{234} in a mineral sample is found to be 1.8×10^4. The half life of U^{238} is 2.5×10^8 yrs. Find the half life of ^{238}U.

[**Ans.** 4.5×10^9 years]

7. Consider the decay scheme $A \to B \to C$, with $\lambda_A < \lambda_B$. after transient equilibrium is established between A and B, show that the interval of time Δt, such that the activity of A at $t - \Delta t$ is equal to the activity of B at t, is given by

$$\Delta t = T_A \log_e \left(\frac{\lambda_B}{\lambda_B - \lambda_A} \right)$$

and that this approaches to T_B as $\dfrac{T_B}{T_A}$ approaches zero.

8. The nucleus of $^{238}_{92}$U decays into a stable isotope of some element through successive emission of 8 α-particles and 6 β-particles. Identify the stable element, and give its Z and A values

[**Ans.** $Z = 82$, $A = 206$ Lead-82]

9. One milligram of thorium emits 22 α-particles per unit solid angle per minute. Calculate the half life of thorium. Given atomic mass of thorium = 232 and Avogadro's number = 6.02×10^{23} per mole.

[Hint $N = \dfrac{22 \times 4\pi}{60} = \dfrac{2\pi}{15}$,   **Ans.** $T = 1.238 \times 10^{10}$ s]

10. Radioactive isotopes ^{32}p and ^{33}p are mixed in the ratio 2 : 1 by the number of atoms. The activity is 2 micro Curies. Find the activity after 30 days. Given half life of ^{32}p = 14 days and that of ^{33}p = 25 days

[**Ans.** 0.598 Ci]

11. Two consecutive radiations from a long lived parent have deay constants λ_1, λ_2. Show that the apparent mean life of second radiation is $\left(\dfrac{1}{\lambda_1} + \dfrac{1}{\lambda_2} \right)$.

12. Assuming that the average disintegration rate of radioactive carbon $^{14}_{6}$C in a living organism is 1.53×10^4 disintegrations per minute for each kg of carbon, estimate the age of an archaelogical specimen of some dung for kg. The half life of radioactive carbon is 5600 years.

[**Ans.** 2680 years]

There are no physical cases known for which $T_A = T_B$, where A is present and B is daughter nuclei. Considering $T_A \simeq T_B$ show that

$$t_{\max} = \sqrt{\tau_A \, \tau_B}$$

[Hint: use $T_A = T_B (1 + \delta)$ and $\delta \ll 1$]

SHORT QUESTION ANSWERS

1. If a radioactive substance is placed in a vaccum, what will be the effect on its rate of disintegration?
Ans. No effect.

2. When a nucleus emits a γ-ray photon what happens to its atomic number and atomic mass?
Ans. Remains unchanged.

3. What happens when a radioactive nucleus disintegrates by emitting and α-particle?
Ans. It gets displaced by two places to the left in the periodic table of elements

4. what happens when a radioactive nucleus disintegrates emitting a β-particles?

Ans. It gets displaced by one place to the right in the periodic table of elements.

5. How is the rate of disintegration affected when a solution of radioactive material is heated?

Ans. The rate of disintegration is not affected by change in temperature.

6. What is the final product of all four natural occuring radioactive series?

Ans. Stable lead.

7. What are the conservation laws that are obeyed during radioactive decay?

Ans. (i) conservation of charge and   (ii) conservation of mass

8. Why in spontaneous radioactive disintegration the rest mass of the parent nucleus be equal or more than the combined rest mass of the daughter and the particles emitted?

Ans. If no energy is to be released the mass should be equal but as the energy is always released and hence the rest mass of the parent should be greater than the combined mass of the daughter and the particle emitted.

OBJECTIVE QUESTIONS

1. The radioactivity of a sample is x at a time t_1 and y at a time t_2. If the mean life of the specimen is T, the number of atoms that have disintegrated in the time interval $(t_2 - t_1)$ is

   (a) $x - y$

   (b) $\dfrac{(x - y)}{T}$

   (c) $(x - y)T$

   (d) $xt_1 - yt_2$

2. The half life of ^{218}Pa is 3 minutes. The fraction of 10 gm sample of ^{218}Pa left after 15 minutes is

   (a) $\dfrac{1}{32}$

   (b) $\dfrac{1}{64}$

   (c) $\dfrac{1}{25}$

   (d) $\dfrac{1}{5}$

3. The end A of a metallic wire is irradiated with α-rays and the end B is irradiated with β-rays, then

   (a) a current will flow from end B to end A

   (b) a current will flow from end A to end B

   (c) there will be no current in the wire

   (d) None of the above statement is correct

4. The radioactive constant for an element whose half life period is 20 years is about

   (a) 3.4×10^{-2} yrs

   (b) 6.93×10^{-2} yrs

   (c) 10 yrs

   (d) 40 yrs

5. The half life of a radioactive nucleus is 2.5 days. The percentage of the original substance which will disintegrate in 7.5 days is

   (a) 8.75 %

   (b) 87.5 %

   (c) 12.5 %

   (d) 100 %

6. The half life of radon 3.8 days. After 16.5 days $\dfrac{N}{N_o}$ will be

   (a) 1 : 10

   (b) 1 : 20

   (c) 1 : 30

   (d) 1 : 40

7. The half life of radium is 1600 yr. The time taken for $\dfrac{15}{16}$ of a sample of radium to decay is

   (a) 1600 yr

   (b) 3200 yr

   (c) 800 yr

   (d) 6400 yr

8. A nuclide with mass number m and atomic number n disintegrates emitting an α-particle and a negative β-particle. The resulting nuclide has mass number and atomic number respectively equal to
   (a) $m-2, n+1$ (b) $m-2, n$
   (c) $m-4, n-2$ (d) $m-4, n-1$

9. Which of the following radioactive decay emit α-particle
   (a) $^{234}_{90}\text{Th} \rightarrow ^{234}_{91}\text{Pa} + ...$ (b) $^{214}_{82}\text{Pb} \rightarrow ^{214}_{83}\text{Bi} + ...$
   (c) $^{238}_{92}\text{U} \rightarrow ^{234}_{90}\text{Th} + ...$ (d) $^{234}_{91}\text{Pa} \rightarrow ^{234}_{92}\text{U} + ...$

10. ^{210}Bi has half life of 5 days. The time taken for $\frac{7}{8}$th of a sample to decay is
    (a) 20 days   (b) 15 days
    (c) 10 days   (d) 5 days

11. A sample contains 16 g of a radioactive material. The half life is 2 days. After 32 days the amount of radioactive material left in the sample is
    (a) 0.5 g   (b) 1 g
    (c) 0.75 g   (d) less than 1 mg

12. If 10 % of radioactive material decays in 5 days, then the amount of original material left after 20 days is approximately.
    (a) 37 %   (b) 65 %
    (c) 78 %   (d) 83 %

13. As a result of decay of $^{238}_{92}\text{U}$, $^{234}_{91}\text{Pa}$ nucleus is obtained. The particles emitted during this decay are
    (a) two β and one n (b) two β and one p
    (c) one p and two n (d) one α and one β

ANSWERS

1. (a)	2. (a)	3. (b)	4. (a)	5. (c)
6. (b)	7. (d)	8. (d)	9. (c)	10. (b)
11. (d)	12. (b)	13. (d)		

3

NUCLEAR DECAY
[Decay of Unstable Nuclei: α-Decay, β-decay, γ-decay and Electromagnetic Transitions, Mossbauer Effect]

INTRODUCTION

3.1 NUCLEAR DECAY

This divides itself into three categories. One is α-decay – the spontaneous emission of an α-particle from a nucleus of large atomic number. This process and also the closely related process of spontaneous fission, is responsible for setting an upper limit on the atomic numbers of chemical elements occurring in nature. A second type of nuclear decay is β-decay – the spontaneous emission or absorption of an electron or positron by a nucleus. This is particularly interesting because it tell us much about the β-decay interaction, or forces of nature. A third type of nuclear decay is γ-decay – the spontaneous emission of high-energy photons when a nucleus makes transitions from an excited state to its ground state. γ-decay gives detailed information about the excited states of nuclei that can be used to improve the nuclear models. γ-decay is used in the *Mossbaner effect* to make extremely high-resolution energy measurments in many different fields of physics.

3.2 ALPHA(α)-DECAY

The main properties of spontaneous α-decay were established early in the history of nuclear physics by the work of Rutherford and his coworkers on the naturally occurring radioactive elements. It is now known, following the development of the semi-empirical mass formula and the actual measurment of atomic masses throughout the periodic system, that most nuclei with A > 150 are actually unstable with respect to the tightly bound $^{4}_{2}$He (α) particle causing a change of atomic number $\Delta Z = 2$. Usually the α-particle emitted by radioactive substances have kinetic energies range from 8.9 MeV for $^{212}_{84}$Po to 4.1 MeV for $^{232}_{90}$Th corresponding to velocities of the order of 10^7 m/s. The $\frac{q}{m}$ value for the α-particles as obtained by Lord Rutherford and coworkers is

$$\frac{q}{m} = 4.82 \times 10^7 \text{ C/kg}$$

Rutherford and Royds showed experimentally, that α-particles were doubly ionised helium atoms or helium nuclei. Rutherford and Geiger measured the charge (q) or α-particles and found it to be twice the charge of the electron. The value of $\frac{q}{m}$ calculated for doubly ionised helium, came out to be very close to the value obtained experimentally. The numerical values of charge (q) and mass (m) of the α-particle are as follows:

$$q = 3.19 \times 10^{-19} \text{ C}$$

$$m = 6.62 \times 10^{-27} \text{ kg}$$

The fact that this form of decay, and also the process known as spontaneous fission, are relatively rare phenomena compared with instability against α-particle emission or against fission is due to the fact that these processes are always retarded by the repulsive Coulomb barrier surrounding the nucleus. For the lighter unstable nuclei, with one or two exceptions (e.g. ^{147}Sm), the Q value for α-decay is so low that the process is effectively inhibited. For the heavier nuclei the possibility of quantum mechanical penetration explains the early difficulty that observed α-particle energies were often considerably less than the barrier energy inferred from Rutherford's elastic scattering experiments. In α-decay, an α-particle is contained by a barrier, called the Coulomb barrier, formed by the strong attractive potential and electrostatic repulsive $\frac{1}{r}$ potential between the α-particle and the rest of the nucleus. The strong dependence of the probability of tunnelling this barrier (Figure 3.1) on the α-particle's energy is responsible for the observed systematics of α-decay.

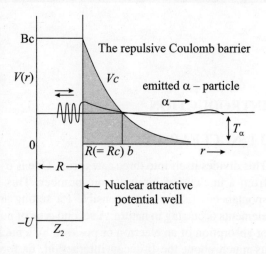

Fig. 3.1 Mechanism of α-decay. The wave representing the α-particle has large amplitude and short wave length within the nuclear well and is attenuated exponentially in the region of negative kinetic energy ($R < r < b$). The kinetic energy of emitted particle is T_α; this is determined by internal excitation in the parent nucleus, but is manifested as a result of repulsion of the particle in the Coulomb field of the final nucleus.

One of the most important properties of α-particles is to produce *ionisation* when they pass through matter. The extent of ionisation depends on the energy of α-particles emitted by a certain nuclei. This property of α-particle helped in their detection in most of the nuclear detectors, e.g. counters, ionisation chambers, cloud chambers, scintillation counters, etc.

3.3 RANGE OF α-PARTICLES

The α-particles exhibit a definite *range* when they traverse a material medium. When α-particles travel through matter, they lose energy through collision with the particles of matter (air or gas). Usually 34 eV of energy is required to produce one ion-pair in air. Due to large mass of α-particle, the loss of energy per collision is relatively small. Obviously, a large number of collisions, therefore, (~ 10^6) take place along the path till the energy of α-particle falls below the ionisation energy of the medium concerned. α-particles are absorbed after travelling a certain distance in medium. *This distance upto which α-particles travel in a medium prior to being absorbed in the medium is known as range of α-particles.* Figure 3.2 shows the range of α-particles emitted by ^{214}Po. We may note that towards the end of the range, ionization increases due to slow movement of α-particles.

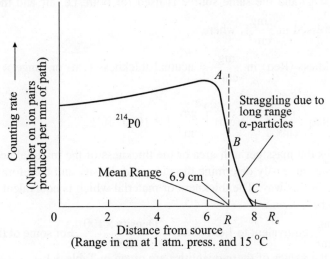

Fig. 3.2 Range of α-particles in air from ^{214}Po.

Due to this, α-particles need more time to interact with surrounding atoms and molecules, thereby increasing the number of ion pairs so produced. One normally measures the extra polated range R_e, which can be obtained by drawing a tangent to the curve at its point of inflection. When the linear portion *ABC* of the graph is extrapolated to zero intensity, one obtains the extrapolated range (R_e). On the other hand, the distance from the source at which the intensity, is half the initial intensity is known as the *mean range R* which is a few mm shorter than R_e. The mean range has a value such that 50 % of α-particles in the incident beam have ranges greater than R and 50 % have ranges less than R. The distance at which the counting rate drops to half its original value is called as range of α-particles. The range of α-particles in air is given at a temperature of 15°C and 760 mm of mercury. The *straggling* is due to the long-range α-particles. The mean range of α-particles in air from ^{214}Po is about 6.9 cm corresponds to an energy of 7.68 MeV. For ^{210}Po α-rays ($E_\alpha = 5.3007$ MeV) the mean range obtained is $R = 3.842$ cm. For this the ionisation extrapolated range obtained is, $R_i = 3.870$ cm and extrapolated range $R_e = 3.897$ cm.

3.3.1 Stopping Power (S(E))

The stopping power is defined as the energy lost by α-particle per unit distance as it travels through the medium, i.e.

$$S(E) = \frac{dE_\alpha}{dx} \qquad \qquad ... (3.1)$$

The use of relative stopping power (RSP) is more convenient for practical purposes. For any material, we have

$$RSP = \frac{\text{Range of α-particle in air}}{\text{Range of α-particle in material}}$$

We may note that in defining RSP, air is considered as a standard at 15°C and 1 atm pressure of mercury (76 cm of Hg) and the same source is used for both, i.e. air and the material. STP is sometimes also expressed in $\frac{mg}{cm^2}$, where

Equivalent thickness (Req) in $\frac{mg}{cm^2}$ = acutual thickness (cm) or range × density $\left(\frac{gm}{cm^3}\right)$ × 1000. Thus

or $$Req\left(\frac{mg}{cm^2}\right) = R(cm)\left(\frac{gm}{cm^2}\right) \times 1000 \qquad ... (3.3)$$

Obviously, Req gives the mass per unit area or the thickness of the material needed to absorb the α-particles. Since one can easily determine the mass and density and therefore it is advantageous to express the range in this way. The thickness of material which is equivalent in stopping power to 1 cm is found to be,

Thickness in $\frac{mg}{cm^2}$ equivalent to 1 cm of air = $\frac{\text{Density} \times 100}{RSP}$. For some of the most commonly used foil materials, the values of these quantities are given in Table 3.1.

Table 3.1

Substance	Extapolated range (cm)	RSP	Density (ρ) $\left(\frac{gm}{cm^3}\right)$	Equivalent thickness $\left(\frac{mg}{cm^2}\right)$	Thickness $\left(\frac{mg}{cm^2}\right)$ equivalent to 1 cm of air
Gold	0.0014	4950	19.33	27.1	3.89
Mica	0.0036	1930	2.8	10.1	1.45
Copper	0.00183	1800	8.93	16.3	2.35
Aluminium	0.00406	1700	2.7	11.0	1.57

The stopping power divided by density (r) of the absorber is called *mass stopping power*. Mass stopping power divided by the number of atoms per c.c. of the material is known as the *atomic stopping power*.

Bethe in 1930 deduced an approximate quantum mechanical relationship for stopping power. Bloch in 1933 derived a more exact expression based on quantum mechanics. The Bethe relation taking relativistic effects into account at high particle speeds is

$$-\frac{dE}{dx} = \frac{Z^2 e^2 Z_n}{4\pi \varepsilon_0^2 m_0 v^2}\left[\log_e \frac{2m_0 v^2}{hv} - \log_e\left(1 - \frac{v^2}{c^2}\right) - \frac{v^2}{c^2}\right] \qquad \ldots (3.3)$$

Where Z is the atomic number of the medium, Ze is the charge of α-particle moving with velocity v. For all except the highest speeds the last two terms within the square bracket in (3.3) almost cancel each other. Since the term $\log_e\left(\dfrac{2m_0 v^2}{hv}\right)$ is reasonably slowly varying with v and therefore the main variation of stopping power with v is with the term outside the square bracket. The observed rate of energy loss of α-particles (charge particles) is having good agreement with (3.3). However, perfect agreement is not expected for several important reasons, e.g. $v \sim c$, electric field of charged particle temporarily polarize, etc.

3.3.2 Measurment of Range of α-particles

Bragg and Kleeman were the first to determine the α-range in gases by measuring the ionisation produced by α-particles at different distances from the sources along their path within the medium. The experimental arrangement is shown in Figure 3.3.

The radioactive source (S) of α-particles was placed on a movable platform placed within a cylindrical tube (T). At the other end of T, an ionisation chamber, consisting of a plate (P) and a grid (G) was placed. The ionisation produced by α-particles at different distances from S was determined by the ionisation chamber. An electrometer was connected in series to read the ionisation current. For noting down the ionisation current a potential difference was maintained between the grid and the plate for this purpose. As the α-particles collide with the atoms or molecules of the gas, a very larg number ($\sim 10^6$) of collisions along its path take place till the energy of α-particles falls below the ionisation energy of the gas. At various distances from the source, specific ionisation, i.e. ion-pairs or ionisation per unit length is determined. Figure 3.4 shows the plot of specific ionisation against the distance from the source (S). The curve so obtained is called *Bragg's curve*.

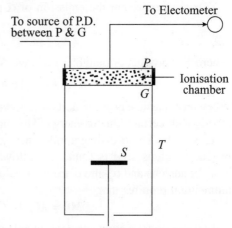

Fig. 3.3 Bragg and Kleeman apparatus for the determination of the range of α-particles.

From the curve it is clear that the specific ionisation remains almost contant upto a certain distance and then it attains first a maximum and afterwards gradually falls to zero. We may note that the range of α-particles is not precise, very close to the end of curve, i.e. the path. We note that the exhibits a 'tail' instead of vertically going to zero on the distance axis. As the α-particles lose their energy during the path due to collisions with the particles of the medium and therefore the velocity of α-particles decreases. This causes α-particles to able to stay for more time towards the end of their journey causing more ionisation. This is why the curve exhibit maximum prior to attaining zero. As stated above, we can explain the tail due to the phenomenon of *straggling*. It is also possible that near the end of their path, the α-particles may start neutralised, first some of α-particles may absorb one electron, then He⁺ may still continue to travel producing

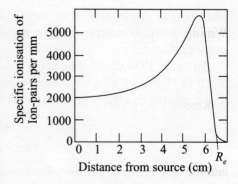

Fig. 3.4 Bragg's curve

ionisation and ultimately it may absorb one more electron to become a neutral He atom.

Another method of determining the range of α-particles is to take the photographs of their tracks in Wilson cloud chamber. Geiger and Nuttall, and Holloway and Livingston used different methods for the determination of the range of α-particles.

3.4 α-DISINTEGRATION ENERGY

When an α-particle is emitted, the product or residual nucleus recoils, carrying with it a certain amount of energy. The nucleus emitting an α-particle is normally at rest in *L*-system and having zero momentum. Therefore, we may regard that in the process of α-emission, both energy and momentum are conserved.

One can represent the emission of α-particle from a parent nucleus (say X) as

$$_Z^A X \rightarrow {}_{Z-2}^{A-4}Y + {}_2^4\text{He} \ (\alpha\text{-particle})$$

Where Y is product or daughter nucleus. A familiar α-emission reaction is

$$_{92}^{238}\text{U} \rightarrow {}_{90}^{234}\text{Th} + {}_2^4\text{He} + E_\alpha$$

When an α-particle is emitted, it carries certain kinetic energy and momentum and product nucleus recoils with certain kinetic energy. The momentum it carries is equal and opposite to that carried by α-particle. The α-disintegration energy is the sum of the kinetic energies of α-particle and the product nucleus, and is found as follows. If M_α and M are the masses of α-particle and the product nucleus and v_α and v are their velocities respectively, then the principle of conservation of momentum requires that

$$Mv = M_\alpha v_\alpha \qquad \qquad ... (3.1)$$

and energy conservation (neglecting relativistic effects) requires that the α-disintegration energy (E_α) is given by

$$E_\alpha = \frac{1}{2}M_\alpha v_\alpha^2 + \frac{1}{2}Mv^2 \qquad \qquad ... (3.2)$$

Using (3.1) and (3.2), one obtains

$$E_\alpha = \frac{1}{2}M_\alpha v_\alpha^2\left(1 + \frac{M_\alpha}{M}\right)$$

Writing $\frac{1}{2} M_\alpha v_\alpha^2 = K_\alpha$ (kinetic energy of α-particle, we have

$$E_\alpha = K_\alpha \left(1 + \frac{M_\alpha}{M} \right)$$

$\therefore$ Kinetic energy of α-particle is

$$K_\alpha = \left(\frac{M}{M + M_\alpha} \right) E_\alpha \qquad \ldots (3.4)$$

and the kinetic energy of the product nucleus (K) is

$$K = \left(\frac{M_\alpha}{M_\alpha + M} \right) E_\alpha \qquad \ldots (3.5)$$

(3.4) and (3.5) can also be expressed in terms of mass number. Mass number of parent nucleus is A (amu), that of product nucleus is $(A - 4)$ amu. Thus

$$K_\alpha = \left(\frac{A - 4}{A} \right) E_\alpha \qquad \ldots (3.6)$$

Since $M_\alpha + M \simeq A$ and $M \simeq A - M_\alpha = A - 4$

$\therefore$
$$E_\alpha = (M_i - M_f - M_\alpha) c^2 \qquad \ldots (3.7)$$

Where M_i is the initial mass and M_f is the final mass (expressed in amu)

$\therefore$
$$E_\alpha = 931 \, (M_i - M_f - M_\alpha) \text{ MeV} \qquad \ldots (3.8)$$

For the spontaneous emission of α-particles, it is necessary that E_α must be positive for $M_i >$ $(M_f + M_\alpha)$. In Equation (3.8), $M_i c^2$, $M_f c^2$ and $M_\alpha c^2$ are the rest mass energies of parent nucleus, daughter nucleus and the α-particle respecively. The experimental value of E_α for the α-particles from a number of nuclei show that the maximum value of the energy always agrees with Equation (3.8). In the decay of ^{222}Rn, the disintegration energy $E_\alpha = 5.587$ MeV, whereas the observed kinetic energy of α-particle = 5.486 MeV. We may note that for almost all the α-emitters in natural radioactive nuclides, the E_α value for the emission of any other nuclear particle like a neutron, a proton or a deutron comes out to negative. This means that α-emission is always preferred over emission of other nuclear particles, even though a nucleus contains only neutrons and protons, e.g. for the decay of $^{235}_{92}$U $\rightarrow$ $^{234}_{90}$Th $+$ $^{4}_{2}$He (α), one finds $E_\alpha = 5.40$ MeV from Equation (3.8), whereas for the emission of n or p, $E_\alpha = -7.16$ MeV or -6.05 MeV respectively. Obviously, emission of α-particle is easier due to its large binding energy. The quantity E_α is the total energy released in α-decay process and is called *disintegration energy*.

Most of the α-emitter nuclei have mass numbers around more than 200 and the disintegration energy mostly appears as the kinetic energy of α-particle, as the residual or daughter nucleus being heavy compared to α-particles, carries very small amount of kinetic energy.

3.5 RANGE-ENERGY RELATIONSHIP FOR α-PARTICLES

The measured values of the ranges and energies of α-particles shows that the empirical relationship between the two is

$$R = a \ E^{\frac{3}{2}} \qquad \qquad \ldots (3.1)$$

If E in MeV and R in meter, the constant $a = 0.315 \times 10^{-2}$
Equation (3.1) can also be expressed as

$$E = b \ R^{\frac{2}{3}} = 2.12 \times 10^2 \ R^{\frac{2}{3}} \qquad \qquad \ldots (3.2)$$

If v be the velocity of α-particle, then since $v \propto \sqrt{E}$, one can write

$$R = cv^3$$

or $\qquad \qquad v^3 = 1.03 \times 10^{27} \ R$. The constant C has same value in all cases.
In case of a solid the range R_s (in meters) is having following relationship with R in air:

$$R_s = \frac{0.312 \ R \ A^{\frac{1}{2}}}{\rho} \qquad \qquad \ldots (3.4)$$

Where ρ is the density of the solid of mass number A.

Knowing the range (R) of α-particle, one can calculate the energy (E) or velocity (v) of α-particle emitted by a source

We may note that the Geiger empirical law is found to be applicable only for α-particles having

range (R) between 3 to 7 cms in air. The range R is found to be approximately proportional to $v_0^{\frac{3}{2}}$

for lower ranges, whereas at higher energies it is proportional to v^4. The value $\dfrac{R}{E^{3/2}}$ for α-particles

emitted from different sources having varying kinetic energies and mean range (R) in air is almost constant. This establishes the validity of Geiger formula.

3.6 GEIGER-NUTTALL LAW

An important *quantitative (empirical) relation* between the range R of α-particles and the decay constant λ of the emitting nuclei was experimentally by Geiger and Nuttall in 1911. This law is called as Geiger-Nuttall law and expressed as

$$\log \lambda = A + B \log R \qquad \qquad \ldots (3.1)$$

Where A and B are constants having values different for different radioactive series. The constant A has almost the same value for all the three radioactive series and constant B has different value for each series. This law reveals that α-particles emitted by substances having larger λ (or shorter half lives) have longer ranges and vice versa.

The plot of $\log \lambda$ against $\log R$ would thus give a straight line, the slope of which gives the value B. For different radioactive series, one obtains different straight lines essentially parallel to each other (Figure 3.5).

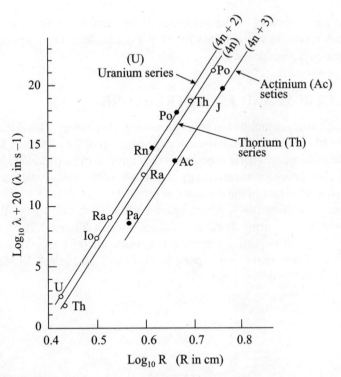

Fig. 3.5 The Geiger-Nuttall rule for α-emitters in three different radioactive series

The Geiger-Nuttall acquired prominence because this enabled one to estimates half lives of α-emitters which could not be determined easily by experimental measurments. We have half life $T_{\frac{1}{2}} = \dfrac{\log e^2}{\lambda}$. One can express the Geiger-Nuttall law by relating the variation of log $T_{\frac{1}{2}}$ with log R or log E. In this also one will obtain straight lines with negative slopes. We have the Geiger-Nuttall law as expressed between decay constant λ and energy E as

$$\log \lambda = C \log E + D \qquad \ldots (3.2)$$

$(\because R \propto v^3 \text{ or } R \propto E^{\frac{3}{2}})$.

Where C and D are constants. Since $\lambda = \dfrac{0.693}{T}$, the variation of log T with log E is shown in Figure 3.6.

We may note that the Geiger-Nuttall rule is not very exact. More accurate relations have been obtained subsequently, e.g., the log λ – values of different isotopes of a given element

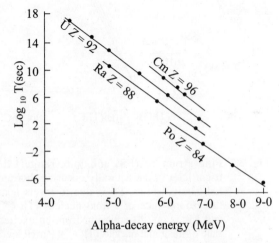

Fig. 3.6 Variation of log T with log E

(Z = constant) and the reciprocal of the velocities of the particles emitted from their nuclei are directly related. They represent straight lines for even-even nucleus. One can obtain these relations with the help of quantum mechanics.

3.7 THE α-SPECTRUM AND FINE STRUCTURE

So far we have said that every α-emitter has only one associated α-energy. This is true (experimentally) for many α-emitters where one finds that the velocity spectrum of the α-particle form these isotopes is always a *sharp-line* spectrum. This is expected as the emission of an α-particle is a result of energy transition between two definite nuclear energy states, where the *initial state* is that of the parent nucleus and the final state is that of the daughter or product nucleus.

However, there are some α-emitters, which exhibit a *fine structure* in the line spectrum, e.g. α-particles in the naturally occuring thorium radioactive series (Figure 3.7), $^{214}_{84}$Po decay to $^{210}_{82}$Pb by emitting four groups of α-particles having ranges in air 6.91 cm, 7.79 cm, 9.04 cm and 11.51 cm (Figure 3.8). These ranges correspond to energies 7.68 McV, 8.28 MeV, 9.07 MeV and 10.51 MeV respectively. The decay scheme is

$$^{238}_{90}\text{Th} \rightarrow {}^{224}_{88}\text{Ra} + \alpha$$

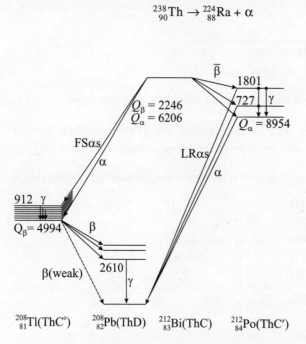

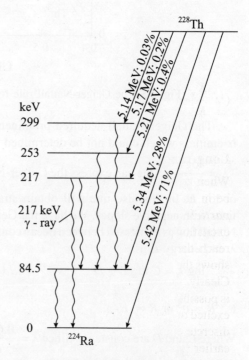

Fig. 3.7 Fine structure (FS) and long range(LR) α-particles in the naturally occuring thorium radioactive series. FS is due to levels near the ground state of a final nucleus (Th *C″*) and LR is due to emission from an excited state of the initial nucleus (Th *C″*). Partial level schemes only are shown and level energies when entered, and *Q* in keV.

Fig. 3.8 Five groups of α-particles shown by slanting lines emitted by ^{228}Th. Intensities of α-lines are shown as percentages.

We note that there are five groups of α-particles with different energies. Out of these five groups, of α-particles, four leave the daughter nucleus in the excited state and fifth group of 5.42 MeV α-rays takes one to the ground state of the daughter nuclei. The excited states of nuclei reach the ground state by emitting γ-rays. Obviously, the fine structure of the α-spectrum tells us about the energy levels in the daughter nucleus. One can easily say that the existence of these different α-energy groups and γ-rays proves the existence of *discrete nuclear energy states* within the nucleus. Obviously, the final structure in α-particle energy spectrum depends on the nuclear energy levels of the parent and daughter nuclei involved in the α-particle disintegration. There exist also a strong correlation between the differences in the energies of the groups of α-particles emitted by the parent nucleus (X) and the energies of the γ-photons emitted by the daughter [${}^A_Z X \rightarrow {}^{A-4}_{Z-2} Y + {}^4_2 \text{He}$].

Therefore the energy of the γ-ray,

$$E_\gamma = E_\alpha - (K_\alpha + K_y)$$

$$= E_\alpha - K_\alpha \left(1 + \frac{K_y}{K_\alpha} \right)$$

$$= E_\alpha - K_\alpha \left(1 + \frac{M_\alpha}{M_y} \right)$$

$$\left(\because \frac{K_y}{K_\alpha} = \frac{\frac{1}{2} M_y v_y^2}{\frac{1}{2} M_\alpha v_\alpha^2} = \frac{M_y M_\alpha^2}{M_\alpha M_y^2} = \frac{M_\alpha}{M_y} \text{ for } M_\alpha v_\alpha = M_y v_y, \text{ from the conservation of momentum} \right)$$

Long-range α-particles

When a parent α-emitter emits α-particles, it is said to be in an excited state, then one obtains *long-range α-particles*. This is due to the fact that the energy of excitation becomes available to the α-particles as they reach the ground state of daughter nuclei. Figure 3.9 shows the emission of long range α-particles from ${}^{212}_{84}\text{Po}$. Clearly, this type of emission of long-range α-particles is possible only if transitions from different energy states, excited or ground, of the parent-nuclide occur to different discrete energy states of the product nuclide. As stated earlier ${}^{212}_{84}\text{Po}$, α-emitter have a type of fine structure in which there are a few α-particles with much greater energies than those of the main group. The half-life ${}^{212}_{84}\text{Po}$ is very short (~ 10^{-6} s). The main group has an energy of about 9.95 MeV and in addition, there are two groups of energies 9.67 MeV and 10.75 MeV. We may note that the intensities of these groups are very small, i.e. about

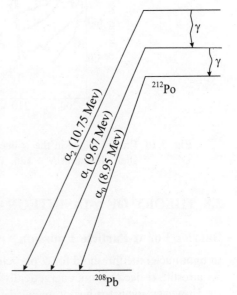

Fig. 3.9 Emission of long-range α-particles from ^{212}Po.

100 in 10^6 of the main group. The reason is not far to seek. Excited parent nuclides usually go down to their respective ground state by γ-emission of different energies, either directly or stepwise, and then under go α-disintegration. Thus the probability of γ-transitions are much higher than that of α-tansitions from the excited state directly to the ground state of the product nucleus. This is why the intensities of the main group particles are comparatively much higher.

This type of α-emission occurs only if the half-lives of decay are very short, e.g. Th C'. In the case of long-range particles the excited state will usually have been formed by a preceding β-decay, and the 'β-delayed' α-decay competes with γ-emission to the ground state (Figure 3.7).

In Figure 3.9, $α_0$ is the normal α-group which corresponds to transition between ground states of ^{212}Po and ^{208}Pb. The other two groups $α_1$ and $α_2$ originate from the excited states of the parent ^{212}Po and directly go the ground state of daughter ^{208}Pb. Obviously, nuclear spectroscopic studies of long range α-emitters provide information about nuclear energy levels of the parent nucleus.

α-decay energies are determined by the detailed structure of the mass surface and are sensitive to shell effects. Figure 3.10 shows the dramatic effect for $A \approx 210$ of the closure of the $N = 126$ neutron shell. Extra energy is available just beyond shell closure, since two loosely bound neutrons are then removed by the emission of a tightly bound α-particle. The same effect occurs for $N = 82$ and results in the α-activity among the rare earths.

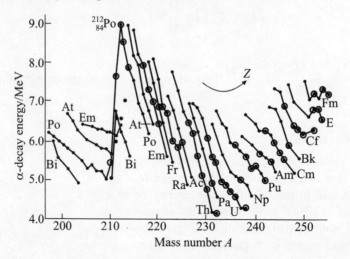

Fig. 3.10 Energy states in the α-decay of the heavy elements, showing the effect of neutron shell closure at $N = 126$. The nuclei ringed are stable against β-decay.

3.8 THEORY OF α-DISINTEGRATION

Barrier For α-Particle Emission

In light nuclei the threshold for α-particle emission is comparable with those for nuclear emission. As a result, α-decay is not energetically possible for the low-lying excited states until $A > 150$ or so. However, even for heavy nuclei, life-times are very long by strong interaction time scales. Further more, the kinetic energies of α-particles emitted are very similar, usually in the range of 4 to 9 MeV, but the half lives associated with these emissions differ by a wide range. For this reason, α-decay by natural occurring radioactive nuclei was a puzzle in early days of modern physics, one which eventually led to the discovery of *quantum mechanical tunnelling*.

It is perhaps easier to visualize the barrier an α-particle must tunnel through by considering first the inverse process of an α-particle, approaching a heavy nucleus from large distances. Outside the nuclear surface, the interaction is purely Coulomb and one can represent the repulsive potential in the form,

$$V_c(r) = \frac{1}{4\pi\varepsilon_0}\frac{2Ze^2}{r} = \frac{2Z\alpha\hbar c}{r} = \frac{2.88\,Z}{r}\ \text{MeV} \qquad \dots (3.1)$$

Where r is measured in *femo meters* in the last equality.

Once the α-particle is inside the nuclear surface, short range nuclear forces become effective. Some particles are bound to the nucleus at distances $r < R$ (Figure 3.1), the combined result of Coulomb and nuclear forces must produce a minimum in the potential at these short distances. For simplicity, we shall thake the shape of this part of the potential to be an attractive square well, as shown schematically in Figure 3.1. In this approximation, the height of the barrier that retains the α-particle inside the nucleus may be estimated from the amount of work required to overcome Coulomb repulsion in bringing an α-particle with two units of charge to the surface of a heavy nucleus such as $^{238}_{92}\text{U}$,

$$E_c = \frac{1}{4\pi\varepsilon_0}\frac{2Ze^2}{R} = \frac{2Z\alpha\hbar c}{r_0\,A^{\frac{1}{3}}} \approx 35\ \text{MeV} \qquad \dots (3.2)$$

Better estimates put the Coulomb energy of an α-particle in this region of mass number to be just below 30 MeV.

Classically, in order for an α-particle to be emitted from a nucleus, it must some how acquire enough energy to reach the top of the potential barrier. Once it is there, it can leave the nucleus carrying with it all the energy it acquired. Consequently, when the α-particle is far away from the nucleus, it should have a kinetic energy at least equal to the barrier height. The observed kinetic energies of α-particles emitted are, however much smaller than this value and, consequently a different mechanism must be operating here. The quantum mechanical explanation is that the α-particle does not have to go over the top of the potential barrier before being emitted: instead, it *tunnels* through it. The basic reason for such a possibility comes from the fact that the amplitude of the α-particle wave function does not vanish inside a barrier of finite height and, as a result, there is a finite probability of finding the particle outside the nucleus, as shown schematically in Figure 3.1.

The reason that α-particle emission is an important channel of decay for heavy nuclei comes from a combination of two reasons. The first is the saturation of nuclear force which turns α-clusters into tightly bound groups of four nucleons inside a nucleus. As we have seen earlier, the average binding energy per nucleus between α-clusters is much less than that between nucleons inside such a cluster. This reduces the energy required for an α-cluster to be separated from the rest of the nucleus. The second reason is the increase in Coulomb repulsion in heavy nuclei due to the large number of protons present. The combined effect of both reasons enables the Q-value for α-emission to become positive (negative in terms of the separation energy) for $A > 150$.

Decay Probability

Even before any theoretical models constructed to explain the decay probability, the Geiger-Nuttall empirical law,

$$\log_{10} W = C - \frac{D}{\sqrt{E_\alpha}} \qquad \dots (3.3)$$

Where W is the decay probability and parameters C and D are weakly dependent on Z but not on neutron number, revealed that the large range of α-decay half lives, from μs to 10^{17} years, may be related to the square root of the kinetic energy E_α of the α-particle as given by (3.3). A simple, one body theory of α-particles provides the foundation for this observed relation.

We do not expect α-particles to exist as entities that can be readily identified inside a heavy nucleus. On the other hand there is a finite probability of finding two protons and two neutrons to be correlated in such a way that they have an α-particle like structure. Let us call such an object an α-cluster. In this way, the probability W for α-particle emission from a heavy nucleus by tunnelling may be expressed as a product of three factors,

$$W = p_\alpha \nu T \qquad \dots (3.4)$$

Where p_α is the probability of finding an α-cluster inside a heavy nucleus, ν is the frequency of the α-cluster appearing at the inside edge of the potential barrier where it can leak out, and T is the transmission coefficient for the α-cluster to tunnel through the barrier.

Once an α-cluster is formed inside a nucleus, the frequency ν with which it appears at the edge of the potential well depends on the velocity v it travels and the size of the potential well. A reasonable way to estimate ν is to take the well size as twice the nuclear radius R. With this assumption, one obtains the result

$$\nu = \frac{v}{2R} = \frac{\sqrt{\dfrac{2K}{M_\alpha}}}{2R} \qquad \dots (3.5)$$

Where K is the kinetic energy of the α-cluster inside the well and M_α is the mass. The precise value of K depends on the depth of the potential well and is not well known. It is reasonable to expect that K is of the same order of magnitude as E_α, the kinetic energy of the α-particle emitted. We take $K = E_\alpha$. This leads to

$$\nu = \frac{\sqrt{\dfrac{2E_\alpha}{M_\alpha}}}{2R} = \frac{\sqrt{E_\alpha}}{A^{\frac{1}{3}}} \times 2.9 \times 10^{21} \ s^{-1} \qquad \dots (3.6)$$

Where E_α is measured in MeV and $R = r_0 A^{\frac{1}{3}}$, where $r_0 = 1.2$ fm. Using these, one obtains, $\nu = 10^{21} \ s^{-1}$ for ^{238}U with $E_\alpha = 5.6$ MeV. It is an order of magnitude larger than the best values deduced from measurments.

Part of the reason for poor agreement seems from the fact that heavy nuclei do not have the simple spherical shape assumed here. Moreover, the replacement of K by E_α may also have caused some loss of accuracy.

3.8.1 Quantum Mechanical Theory of α-Decay and Transmission Coefficient

George Gamow in 1928 and also independently, R. Gurney and E. Condon applied wave mechanics to the problem of α-decay and were able to successfully resolve the α-decay paradox.

Through the actual potential barrier in the case of α-disintegration has a complex shape (Figure 3.1), but one can replace it by a rectangular potential barrier (Figure 3.11). For a given energy E_α of the α-particle, one can divide whole of one dimensional space into following three region

Region I, $x < 0; V = 0$

Region II, $0 < x < a; V = V_0$

Region III, $x > a; V = 0$

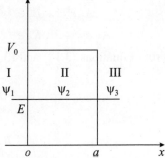

Fig. 3.11 A rectangular potential barrier

Assuming ψ_1, ψ_2 and ψ_3 be the wave functions in regions I, II and III respectively, one can write corresponding Schrodinger's equations for the said three regions as

$$\frac{d^2 \psi_1}{dx^2} + \frac{2M_\alpha}{\hbar^2} \psi_1 = 0 \qquad \dots (3.1)$$

$$\frac{d^2 \psi_2}{dx^2} - \frac{2M_\alpha}{\hbar^2}(V_0 - E)\psi_2 = 0 \qquad \dots (3.2)$$

$$\frac{d^2 \psi_3}{dx^2} + \frac{2M_\alpha}{\hbar^2} \psi_3 = 0 \qquad \dots (3.3)$$

Since the region I has both incident and reflectes α-waves, hence the solution of (3.1) in region I is

$$\psi_1 = A_1\, e^{ik_1 x} + B_1\, e^{-ik_1 x} \qquad \dots (3.4)$$

Where

$$k_1 = \sqrt{\frac{2M_\alpha E}{\hbar^2}} \qquad \dots (3.5)$$

Since the region II has both forward moving and reflected α-waves from the other side of the barrier, hence the solution of Eauation (3.2) in region II is

$$\psi_2 = A_2\, e^{k_2 r} + B_2\, e^{-k_2 r} \qquad \dots (3.6)$$

Where

$$k_2 = \frac{\sqrt{[2M_\alpha(V_0 - E)]}}{\hbar} \qquad \dots (3.7)$$

Since region III has only forward moving transmitted wave hence the solution of Equation (3.1) in this region is

$$\psi_3 = A_3\, e^{ik_1 r} \qquad \dots (3.8)$$

One can determine the constants A_1, A_2, A_3, B_1 and B_2 with the help of following boundary conditions

$$\psi_1 = \psi_2 \quad \text{and} \quad \frac{\partial \psi_1}{\partial x} = \frac{\partial \psi_2}{\partial x} \quad \text{at } x = 0$$

and
$$\psi_2 = \psi_3 \text{ and } \frac{\partial \psi_2}{\partial x} = \frac{\partial \psi_3}{\partial x} \text{ at } x = a$$

Now, substituting the values of ψ_1, ψ_2 and ψ_3 in the above relations, one obtains

$$A_1 + B_1 = A_2 + B_2 \qquad \qquad \text{... (3.10)}$$

$$ik_1 A_1 - ik_1 B_1 = k_2 A_2 - k_2 B_2 \qquad \qquad \text{... (3.11)}$$

$$A_2 e^{k_2 a} + B_2 e^{-k_2 a} = A_3 e^{ik_1 a} \qquad \qquad \text{... (3.12)}$$

$$A_2 e^{k_2 a} - B_2 K_2 e^{-k_2 a} = ik_1 A_3 e^{ik_1 a} \qquad \qquad \text{... (3.13)}$$

From Equations (3.12) and (3.13), one obtains

$$A_2 = \frac{1}{2} A_3 \left(1 + \frac{i k_1}{k_2} \right) e^{(i k_1 - k_2)a} \qquad \qquad \text{... (3.14)}$$

$$B_2 = \frac{1}{2} A_3 \left(1 - \frac{i k_1}{k_2} \right) e^{(i k_1 + k_2)a} \qquad \qquad \text{... (3.15)}$$

From Equations (3.10) and (3.11), one obtains

$$A_1 = \frac{1}{2} A_2 \left(1 + \frac{k_2}{i k_2} \right) + \frac{1}{2} B_2 \left(1 - \frac{k_2}{i k_1} \right) \qquad \qquad \text{... (3.16)}$$

Substituting the values of A_2 and B_2, one obtains

$$A_1 = \frac{1}{4} A_3 \left(1 + \frac{i k_1}{k_2} \right) \left(1 + \frac{k_2}{i k_1} \right) e^{(i k_1 - k_2)a} + \frac{1}{4} A_3 \left(1 - \frac{i k_1}{k_2} \right) \left(1 - \frac{k_2}{i k_1} \right) e^{(i k_1 + k_2)a} \qquad \text{... (3.17)}$$

The velocity of α-particle in region I is same as in region III, hence transmission probability of incident α-particle

$$T = \frac{\text{Transmitted flux}}{\text{Incident flux}} = \frac{|A_3|^2 \, \upsilon}{|A_1|^2 \, \upsilon} = \frac{|A_3|^2}{|A_1|^2} \qquad \qquad \text{... (3.18)}$$

Since in practice $k_2 a \gg 1$, hence first term of Equation (3.17) can be easily neglected in comparision to the second term. Thus, one finds

$$\frac{|A_1|^2}{|A_3|^2} = \left(\frac{A_1}{A_3} \right) \left(\frac{A_1}{A_3} \right)^* = \frac{1}{16} \left(1 - \frac{i k_1}{k_2} \right) \left(1 + \frac{i k_1}{k_2} \right) \left(1 - \frac{k_2}{i k_1} \right) \left(1 + \frac{k_2}{i k_1} \right) e^{2 k_2 a}$$

$$= \frac{(k_1^2 + k_2^2)^2}{16 \, k_1^2 \, k_2^2} \, e^{2 k_2 a}$$

$\therefore$ Transmissivity of the barrier

$$T = \frac{16 \, k_1^2 \, k_2^2}{(k_1^2 + k_2^2)^2} \, e^{-2 k_2 a}$$

$$= \frac{16 \, E (V_0 - E)}{V_0^2} \, \exp(-2 k_2 a) \qquad \qquad \text{... (3.19)}$$

We note that T is strongly influenced by the barrier thickness a and the energy E of the particle for a given barrier height V_0, as a and E both are occuring in the exponent of the exponential. When $2k_2a \gg 1$, the most important factor in (3.19) is the exponential, which then will be very small. The factor in front of the exponential part is usually of the order of magnitude of unity (maximum value is 4). For order of magnitude calculations, one can take

$$T = e^{-2k_2a}$$

Equation (3.20) represents the fraction of α-particles that will penetrate the barrier of width a and height V_0 (> E). If the potential is not constant in the region $o < x < a$, one can always approximate it with a series of small steps, each with a constant potential. As the total probability is the product of individual probabilities, and therefore one obtains a sum in exponent of the exponential. By making the intervals smaller and smaller, the sum goes over to into the integral. Thus, we have

$$T = e^{-2 \int k_2 \, dx} \qquad \qquad \dots (3.21)$$

The integral is taken through the whole region between R and b (Figure 3.1), where the Coulomb repulsion $V(r)$ is greater than the energy E of the α-particle. Now, substituting the value of k_2 in Equation (3.21), one finds

$$T = \exp\left[-\frac{2\sqrt{2M_\alpha}}{\hbar} \int_R^b [V(r) - E]^{\frac{1}{2}} \, dr \right] \qquad \qquad \dots (3.22)$$

We note that changes in either a or E or both changes the value of T by a very large factor which, shows qualitatively the reason for very large variation of the α-disintegration probability due to small variation of the α-particle energy.

Now, we take a numerical example.

We take $\qquad\qquad\qquad V_0 = 15$ MeV, $a = 2 \times 10^{-14}$ m and $E = 5$ MeV.

Then for $\qquad\qquad\qquad M_\alpha = 4 \times 1.66 \times 10^{-27}$ kg, we have

$$q^2 = \frac{2M}{\hbar^2} (V_0 - E)$$

$$= \frac{2 \times 4 \times 1.66 \times 10^{-27}}{(1.05 \times 10^{-34})^2} \times 10 \times 1.6 \times 10^{-13}$$

$$= 1.927 \times 10^{30}$$

$$\therefore \qquad\qquad\qquad q = 1.388 \times 10^{15}$$

and $\qquad\qquad\qquad 2qa = 55.53$

and $\qquad\qquad\qquad \exp(-2qa) = 7.649 \times 10^{-25}$

$$\therefore \text{Transmission coefficient } (T) = \frac{16 \times 5 \times 10 \times 7.649 \times 10^{-25}}{15 \times 15} = 2.72 \times 10^{-24}$$

This reveals that the particle has the probability of about 2.7×10^{24} collisions against the potential barrier wall to penetrate through the potential barrier.

Now, we apply the above to the α-disintegration. An α-particle of velocity v inside the nucleus of radius $R = 10^{-14}$ m takes a time $t = \dfrac{2R}{V}$ to cross the nucleus from one end to the other. We may note that the velocity of α-particle v inside the nucleus is not the same as its velocity at infinity though they are of the same order or magnitude. Now

$$n = \frac{V}{2R} = \frac{1.552 \times 10^7}{2 \times 10^{-14}} = 7.76 \times 10^{20}/s$$

Thus the probability of escape of the α-particle per second, which is equal to the decay constant λ is $P = \lambda = nT$. Thus

$$\lambda = 7.76 \times 10^{20} \times 2.72 \times 10^{-24} = 2.11 \times 10^{-3}$$

$\therefore$ Mean life time against α-decay is

$$T_m = \frac{1}{\lambda} = \frac{10^3}{2.11} = 474 \text{ s} = 7 \text{ min } 54 \text{ sec}$$

Now, if we increase the barrier width (a) by 20 %, i.e. $a = 2.4 \times 10^{-14}$ m, then $2\,qa = 66.63$.

This gives mean life as

$$T_m' = 3.13 \times 10^7 \text{ s}$$

T_m' is about 6.6×10^4 larger than T_m. Now, if the energy of the particle is 20 % less, then one obtains $T_m'' = 9.018 \times 10^3$ s $= 2\,h$, 30.3 m which is about 19 times as large.

These calculations reveals that slight change either in the width of the barrier or the energy of the particle, make large changes in the value of T_m and hence of λ (decay constant). It is further noted that in the case of more realistic barrier (Figure 3.1), change in the energy of the α-particle also affects the barrier thickness (right hand side of Figure). Concluding, we can say that the mean life of α-disintegration is enormously changed by slight change in the α-particle energy in agreement with Geiger-Nuttall law.

Now, we discuss the more realistic theory of α-disintegeration and calculate the transmission coefficient.

3.8.2 Realistic theory of α-disintegration

In the case of α-disintegration, the rectangular potential barrier is an idealized barrier, where as the actual barrier is complex. The appropriate shape is as shown in Figure 3.12. Prior to the consideration of the penetration of potential barrier, we first consider the penetration of a barrier of an arbitrary shape as shown in Figure 3.12.

One can treat this problem by the W.K.B. method (semi-classical approximation). This method is applicable to the problems in which the potential varies very slowly, e.g. the change in the de Broglie wavelength (λ) is small as compared to λ itself the

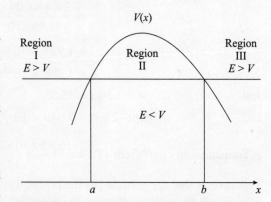

Fig. 3.12 Potential barrier for disintegration process having arbitrary shape

distances of the order of the de-Broglie wave length. One can easily show that the probability of penetration of a potential barrier, extending from $x = a$ to $x = b$ of arbitrary shape is given by

$$T = \exp(-D) \qquad \qquad \ldots (3.1)$$

Where

$$D = \left[-\frac{2}{\hbar} \int_a^b \sqrt{2 M_\alpha \{V(x) - E\}} \right] \qquad \qquad \ldots (3.2)$$

Now, replacing $x = r$ for the present case, one finds

$$V = V(r) = \frac{(Z-2)e^2}{2\pi \varepsilon_0 r}$$

Where Ze is the charge of the parent nucleus undergoing by the α-disintegration. Then, one finds

$$D = \frac{2}{\hbar} \left[\sqrt{2 M_\alpha} \left\{ \frac{(Z-2)e^2}{2\pi \varepsilon_0 r} - E \right\}^{\frac{1}{2}} dr \right]$$

$$= \frac{2}{\hbar} \left\{ \frac{M_\alpha (Z-2)e^2}{\pi \varepsilon_0} \int_R^b \left[\frac{1}{r} - \frac{2\pi \varepsilon_0 E}{(Z-2)e^2} \right] \right. \qquad \qquad \ldots (3.3)$$

The limits of integeration are for $r = R$ to $r = b$. At these limits, the Coulomb energy is equal to the α-energy. Thus, we have

$$\frac{2(Z-2)e^2}{4\pi \varepsilon_0 b} = E$$

or

$$\frac{1}{b} = \frac{2\pi \varepsilon_0 E}{(Z-2)e^2}$$

$\therefore$

$$D = \frac{2}{\hbar} \left\{ \frac{M_\alpha (Z-2)e^2}{\pi \varepsilon_0} \right\}^{\frac{1}{2}} \int_R^b \left(\frac{1}{r} - \frac{1}{b} \right) dr$$

$$= \frac{2}{\hbar} \left\{ \frac{M_\alpha (Z-2)e^2 b}{\pi \varepsilon_0} \left[\cos^{-1} \left(\frac{R}{b} \right)^{\frac{1}{2}} - \left(\frac{R}{b} - \frac{R^2}{b^2} \right)^{\frac{1}{2}} - \left(\frac{R}{b} - \frac{R^2}{b^2} \right)^{\frac{1}{2}} \right] \right. \qquad \ldots (3.4)$$

Usually $b \gg R$ and therefore $\dfrac{R}{b} \ll 1$. Thus, one can neglect the term $\left(\dfrac{R}{b} - \dfrac{R^2}{b^2} \right)^{\frac{1}{2}}$ on the r.h.s. of

(3.4). Further $\cos^{-1} \left(\dfrac{R}{b} \right)^{\frac{1}{2}} = \dfrac{\pi}{2}$, One obtains

$$D = \frac{2}{\hbar} \left\{ \frac{M_\alpha (Z-2)e^2 b}{\pi \varepsilon_0} \right\}^{\frac{1}{2}} \frac{\pi}{2} = \frac{\pi}{\hbar} \left\{ \frac{M_\alpha (Z-2)e^2 b}{\pi \varepsilon_0} \right\}^{\frac{1}{2}} \qquad \ldots (3.5)$$

If υ be the velocity of α-particle, we have

$$E = \frac{1}{2} M_\alpha \upsilon^2 = \frac{(Z-2)e^2}{2\pi\varepsilon_0 b}$$

or

$$b = \frac{(Z-2)e^2}{\pi\varepsilon_0 M_\alpha \upsilon^2}$$

and

$$D = \frac{\pi}{\hbar} \left\{ \frac{M_\alpha (Z-2)e^2}{\pi\varepsilon_0} \right\}^{\frac{1}{2}} \left\{ \frac{(Z-2)e^2}{\pi\varepsilon_0 M_\alpha \upsilon^2} \right\}$$

$$= \frac{\pi}{\hbar} \frac{(Z-2)e^2}{\pi\varepsilon_0 \upsilon} = \frac{(Z-2)e^2}{\varepsilon_0 \hbar \upsilon} \qquad \qquad \text{... (3.6)}$$

Now, the probability or barrier penetration becomes

$$T = \exp(-D) = \exp\left\{ \frac{(Z-2)e^2}{\varepsilon_0 \hbar \upsilon} \right\} \qquad \qquad \text{... (3.7)}$$

If $E = 5$ MeV, one finds

$$\upsilon = \sqrt{\frac{2E}{M_\alpha}} = 1.55 \times 10^7 \text{ m/s}$$

and $Z - 2 = 86$, one obtains

$$D = \frac{86 \times 1.6 \times 10^{-19} \times 36\pi}{10^{-9} \times 1.05 \times 10^{-34} \times 1.55 \times 10^7} = 153$$

Now, we take $E = 10$ MeV. For this, one obtains

$$\upsilon = \sqrt{2} \times 1.55 \times 10^7 = 2.19 \times 10^7 \text{ m/s}$$

and

$$D' = \frac{153}{\sqrt{2}} = 108$$

Thus the ratio of α-emission probabilities in the two cases is

$$\frac{T_{10}}{T_5} = \frac{\exp(-D')}{\exp(-D)} = \exp(D - D')$$

$$\exp(45) = 3.49 \times 10^9$$

These calculations clearly reveal that if we double the α-particle energy then the α-emission probability increases by a factor of about 3.5×10^{19}. We can see that this agrees qualitatively with Geiger-Nuttall law.

No doubt, the expression for barrier penetration explains qualitatively the obsevered dependence of the α-decay half life on the energy of the α-particles, but quantitative calculations in actual cases do not reproduce the observed half lives.

Let us again consider the case of the rectangular potential penetration by α-particle discussed earlier. One can write the α-disintegration constant as

$$\lambda = \frac{v_{\text{in}}}{2R} \exp(-D)$$

Using (3.6), we have

$$\therefore \quad \log \lambda = \log\left(\frac{v_{\text{in}}}{2R}\right) - D = \log \frac{v_{\text{in}}}{2R} - \frac{(Z-2)e^2}{\varepsilon_0 \hbar v} \quad \ldots (3.8)$$

For α-emitters, V_{in} and R do not vary much among the naturally radioactive series and therefore $\log\left(\dfrac{v_{\text{in}}}{2R}\right)$ may be taken to be almost a constant. Putting $A = \log\left(\dfrac{v_{\text{in}}}{2R}\right)$, one can write (3.8) as

$$\log \lambda = A - \frac{B(Z-2)}{v} \quad \ldots (3.9)$$

Where $\quad B = \dfrac{e^2}{\hbar}$ and ε_0 is constant.

We can regard (3.9) as a theoretical form of Geiger-Nuttall rule. Using (3.9), we can easily see that

$$\log \lambda = A' + \frac{B'}{\sqrt{E}} \quad \ldots (3.10)$$

Where A' and B' are constants and E is the α-disintegration energy (here we have assumed $v_{\text{in}} = v$).

It is very difficult to make a simple estimate for p_α, the probability of finding an α-cluster inside a nucleus. Since this factor depends on the wave function of a nuclear state, one can expect it to be somewhat different from nucleus to nucleus. However, the value of p_α must be essentially of the same order of magnitude for all heavy nuclei, as there are only small fractional differences in their masses. For an estimate, let us take $p_\alpha = 0.1$ for all the heavy, α-radioactive nuclei. In view of the uncertainties in calculating v and T, the assumption of a constant p_α for all heavy nuclei is justified.

Energy and Mass Dependence

We now have reasonable estimates for all three factors in $W = p_\alpha\, v\, T$. Since the Geiger-Nuttall law is a relation between the logarithm of the rate of α-particle emission as a function of the α-particle kinetic energy E_α, one need to convert the expression for W derived earlier to logarithm in the base 10, and one obtains

$$\log_{10} W = \log_{10} p_\alpha + \log_{10} v + \log_{10} T$$

$$= 20.46 + \log_{10} \frac{\sqrt{E_\alpha}}{A^{\frac{1}{3}}} + 1.42 \sqrt{Z\, A^{\frac{1}{3}}} - 1.72 \frac{Z}{\sqrt{E_\alpha}} \quad \ldots (3.11)$$

The dominant energy dependence comes from the last term, in agreement with the empirical result of Geiger-Nuttall law.

In addition to energy dependence, there is also a dependence of $\log_{10} W$ on Z, and to a lesser extent on A in Equation (3.11). To show the Z-dependence, one can plot $\log_{10} W$ as a function of $\sqrt{E_\alpha}$ separately for each element. The experimental values are now clustered much closer to straight lines. The curves belonging to different Z values run parallel to each other.

We have implicitly assumed in the above discussion that there is a single kinetic energy E_α for all the α-particles emerging from a nucleus. In fact, it is common to find several different groups of α-particles emitted by the same parent nucleus. ^{212}Bi has more than 14 different decays known and each one leaves the residual nucleus in a differnet state. In addition to naturally occurring α-radioactive nuclei, there are also man made α-activities to enrich the variety of samples that can be used in the study.

In the above discussion of α-disintegration. We have neglected some factors, which are significant.

(i) We have assumed the shape of the potential well as square, where as it has a diffused boundary. For heavy nuclei, deformations may set in, and boundary may not even be spherically symmetrical. However, this introduces an error of only a few percent.

(ii) **Effect of angular momentum change**   We have only considered $l = 0$, i.e. where there is no change in angular momentum takes place. This is particularly true in the case of even-even nuclei. However, there are great discrepancies for even-odd, odd-even and odd nuclei, e.g. in the case of an even-odd nucleus, the half life in many cases is 100 to 1000 times larger thatn that for an even-even nucleus of the same Z and energy. In no case is the half-life shorter than that for the even-even nuclei.

The ratio of the observed half life to the calculated half life of even-even nuclei is termed as hinderance factor:

$$\text{Hinderance factor (H.F.)} = \left(\frac{T_{1/2,\,0}}{T_{1/2,\,\text{Cal}}} \right) = \frac{\lambda_{\text{Cal}}}{\lambda_0}$$

The patterns for the odd-odd and odd-even nuclei are found similar. The average hinderance factor for the former is perhaps greater than for the latter. The most important reason for hindered α-disintegration is due to the non-spherical nature of the nuclei.

In the case of α-disintegration of an even-even nucleus in the ground state leading to the formation of the even-even daughter nucleus in the ground state, the emitted α-particle has $\lambda = 0$, i.e. it carry no angular momentum. Obviously, in this case both the parent and product nuclei have $I = 0$ and same parity (even). However, in the cases, where α-particles are emitted with non-zero l, they are subjected to the action of the centrifugal potential (V_l) in addition to Coulomb potential (V_c), where

$$V_l = \frac{\hbar^2}{2\,M_\alpha} \frac{l(l+1)}{r^2} \quad \text{and} \quad V_c = \frac{2(Z-2)e^2}{4\pi\varepsilon_0\,r}$$

The Coulomb potential which, then is effective outside the range of nuclear force is modified and expressed as

$$V_{\text{eff}} = \frac{\hbar^2}{2\,M_\alpha} \frac{l(l+1)}{r^2} + V_c = V_l + V_c \qquad \qquad \dots (3.12)$$

The centrifugal potential term reduces the value of transmission coefficient. However, for a typical heavy nucleus, the effect of centrifugal potential (V_l) on the transmission coefficient is small as compared to the effects of energy E or even the radius b. Usually, $V_l \ll V_c$ as can be checked by finding the ratio,

$$\sigma = \frac{V_l}{V_c} = \frac{\pi \hbar^2 \, \varepsilon_0 l (l+1)}{M_\alpha \, e^2 \, r (Z-2)} \qquad \dots (3.13)$$

Taking $Z = 88$ and $r \sim 10^{-14}$ m, one obtains

$$\sigma \simeq 0 : 0021 \; l(l+1)$$

The effect of centrifugal barrier in heavy nuclei ($Z = 80$ to 92) is rather small, the decay constant or the life time does not depend strongly on l-values. For $Z = 88$, $E_\alpha = 4.88$ MeV and $r \simeq 10^{-14}$ m, the numerical values for the ratio $\dfrac{\lambda_l}{\lambda_0}$ for certain values of l are as follows:

1	0	1	2	3	4	5	6
$\dfrac{\lambda_l}{\lambda_0} \left(\dfrac{T_{1/2,0}}{T_{1/2,1}} \right)$	1	0.699	0.37	0.137	0.037	0.0709	0.0011

The above values show that for l-small ($l = 1, 2, 3$) the decay constants, calculated and observed are equal to one another within the likely error of theoretical calculation.

In order to obtain the probability of α-disintegration, one will have to take the total potential $V = V_c + V_l$. We have for barrier penetration probability T:

$$T = \exp(-D) = \exp\left\{ -\frac{2}{\hbar} \int \left[2 M_\alpha (V_c + V_l - E) \right]^{\frac{1}{2}} dr \right\}$$

$$\therefore \qquad D = \frac{2}{\hbar} \sqrt{2 M_\alpha} \int \left[\frac{2(Z-2) e^2}{4 \pi \varepsilon_0} \frac{1}{r} + \frac{\hbar^2}{2M} \frac{l(l+1)}{r^2} - E \right]^{\frac{1}{2}} dr \qquad \dots (3.14)$$

(iii) Mechanism of α-particle formation inside the nucleus We have mentioned that it is certain that α-particles are not permanent constitutents of the nuclei. In any given state, we can say that the configuration is continuously changing. Tolhock and Brussard in 1955 considered the mechanism for α-particle formation inside the nucleus. They considered the α-particles as formed from nucleus in outer orbits, with the inner part acting only as the origin or potential-well, without exchanging energy with the α-particle. They also calculated the probability p_α, of 2 neutrons and 2 protons, combining together to form an α-particle by taking the component of the wave function, which represents the wave function of an α-particle with the same total energy as total K.E. of 4 nucleons. Obviously, $E = 2 E_p + 2 E_n + E_x$, where E_x is the binding energy of the α-paticle inside the nucleus. They estimated the probability (p_α) of α-particle by considering that α-particle is formed when four nucleons ($2p$ and $2n$) are within the α-particle radius r_α. Assuming that nucleon wave functions are constant over the nuclear volume (a sphere of radius R), they used the final value for the probability of α-formation to be $n_\alpha \, p_\alpha$, with

$$p_\alpha = 64 \left(\frac{r_\alpha}{R} \right)^9 \qquad \qquad \ldots (3.15)$$

Here n_α is the number of ways in which an α-particle can be formed from all the nucleons in outer orbits. They estimated the value of n_α to be 3 and $p_\alpha = 1.4 \times 10^{-4}$ from the decay data and electron scattering from Po^{214} with $r_\alpha = 1.6 \times 10^{-15}$ m.

3.9 BETA PARTICLES

The early experiments showed that in β^--decay a nucleus decays to another nucleus of the same mass number with the emission of an electron with a varying amount of energy. The simplest example of such a decay is β^--spectrum for ^{211}Bi shown in Figure 3.13.

For the radioactive series, β^--decay involves a nucleus Z, A emitting a negatively charged electron and being transformed into the nucleus $Z + 1$, A.

If just an electron were emitted in the β^--decay, then energy and momentum conservation would require all the electrons to have the same energy determined by E_0, the Q-value for the decay which is equal to the difference in rest mass energies of the initial and final state. The shape of the spectrum

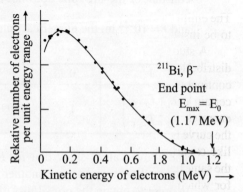

Fig. 3.13 Electron energy spectrum in the β^--decay of ^{211}Bi

therefore implies that another particle is also emitted which shares the available decay energy. By 1934 the production of positron emitters in (α, n) reactions had been demonstrated by Curie and Joliot and Fermi had formulated the theory of β-decay, using Pauli's concept of neutrino. The β^--decay of neutron therefore corresponds to:

$$n \rightarrow p + e^- + \bar{\nu}$$

where $\bar{\nu}$ represent antineutrino.

There are two closely related processes to β^--decay, which are β^+-decay where a proton bound in a nucleus decays to a bound neutron with the emission of a positron (e^+) and a neutrino (ν), and *electron capture* (EC) where a proton bound in a nucleus changes to a bound neutron through interacting with one of the atomic electrons and a neutrino is emitted. Examples of these processes are:

(i) ^{22}Na $\rightarrow$ ^{22}Na $+ e^+ + \nu$ (β^+-decay)

The energy spectrum of e^+ emitted in the β^+ decay of ^{13}N is shown in Figure 3.14

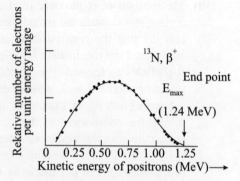

Fig. 3.14 Positron energy spectrum in the β^+ decay of ^{13}N

and (ii) $e^- + {}^7Be \rightarrow {}^7Li + \nu$ (Electron Capture)

and when erergetically possible β^+-decay and EC compete.

Fermi's theory, in which an electron and neutrino are created by an interaction which has a zero range, accounts for the observed electron energy spectra and spread of β-decay life times. The modern theory of weak interaction, as arising through the exchange of heavy intermediate vector bosons ($\sim 80\ GeV/C^2$), predicts a very short range for the weak interaction $\sim 10^{-3}\ F$. The observation of parity violation showed that the Fermi theory was incomplete and that the weak matrix element must contain a scalar and pseudoscalar terms. Now, we will discuss also these issues.

3.9.1 β-Decay

The emitted electron in the β-decay is not an orbital electron. Uncertainty principle forbids electrons to be inside the nucleus. The emitted electron is the one created within the nucleus itself.

A study of energy (momentum) of β-particles from a parent beta active source reveals that the distribution is quite unlike to that of α-particle energy distribution. β-particle energy distribution is continuous spectrum extending from a minimum, attaining maximum and then it falls to zero at a certain energy called as *end-point energy*, whereas α-particles show discrete group of different energies. The end point energy, E_{max}, is the characterstic of a beta emitter. Although the shape of the curve is similar, but the end-point energy of different β-emitters is differnet. However, β-particles, like α-particles, do have characterstic half-lives and all parts of continuous spectrum decay with the same half life. Figure 3.13 shows the energy spectrum of the β^--particles emitted from *Ra E* ($_{83}^{211}Bi$) for which $E_{max} = 1.17$ MeV. Figure 3.14 shows the energy spectrum of β^+ (positron) emitted in the β^+-decay of ^{13}N with $E_{max} = 1.24$ MeV. For β-emitters, the values of E_{max} very from 0.25 MeV to 2.15 MeV.

There are some β-emitters whose experimental momentum distribution shows sharp peaks superimposed over the continuous spectrum (Figure 3.15). The curve is due to the electrons of definite energies and the peaks in the curve are due to the internal conversion of γ-rays emitted by the daughter nucleus formed in β-decay process.

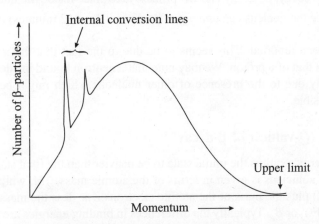

Fig. 3.15 Momentum distribution of β-particles from a radioactive source. The peaks in the curve correspond to internal conversion lines.

As stated earlier, there are three β-decay processes, all isobaric transition. After the decay, the mass number A of the parent nucleus remains the same, whereas the atomic number Z, i.e. charge, increases or decreases by one. We have $A = Z + N$, where N is neutron number. Obviously, whenever β⁻ (electron) emission takes place, Z-increases by one to $(Z + 1)$, i.e. decreasing the N by one, i.e. $(N - 1)$ and A remains unchanged. Likewise in β⁺ (positron) process Z decreases by one, i.e. $(N - 1)$. Again A remains unchanged. One can represent these decay processes as:

β⁻-decay $^A_Z Y \rightarrow \,_{Z+1}^{\quad A}Y + \beta^-$(electron)

β⁺-decay $^A_Z X \rightarrow \,_{Z-1}^{\quad A}Y + \beta^+$(positron)

Obviously, in β-decay processes, nucleon inside the nucleus gets transformed. In β⁻ (electron) emission a neutron transforms into a proton,

β⁻-decay, $n \rightarrow p + \beta^-$

thereby increasing the Z of daughter nucleus Y by one. On the other hand, in β⁺ (positron) emission, a proton transforms into a neutron,

β⁺-decay, $p \rightarrow n + \beta^+$

thereby decreasing the Z of daughter nucleus by one. Obviously in β⁻-decay, the proton number increases by one, whereas in β⁺-decay, the neutron number increases by one.

In electron capture (EC) process, mostly taking place in the K-shell along with positron emission,

EC process, $p + \beta^- \rightarrow n$

the parnet nucleus, instead of emitting a positron captures an electron from one of the innermost atomic orbits. We know that the K-electrons are nearest to the nucleus, therefore, the probability of their being captured by a proton is relatively large and hence the process is called K-electron capture. The vacancy, thus created is filled by the electrons from the higher energy states, e.g. L, M, N... etc., resulting into the X-ray emission. The wave length of such X-ray photons are found to be characterstics of daughter atom. One can represent K-capture process as

$$^A_Z X_N + \beta^- \rightarrow \,_{Z-1}^{\quad A}Y_{N+1}$$

We may note that β⁻ decay, β⁺ decay and *EC* process take place inside the nucleus. A free neutron, i.e. a neutron outside the nucleus can undergo a β⁻-decay $(T_{\frac{1}{2}} \simeq 12$ minutes) but so far no decay of a free proton has been reported. This seems to be due to the fact that the mass of the neutron is slightly greater than that of a proton. We may note that a neutron bound inside the nucleus does not decay spontaneously due to the presence of other nucleons which might be making the process energetically impossible.

3.9.2 Energetics (Q-values) of β-decay

Conservation of energy requires the initial state to be heavier than the final state before a decay can occur. The mass of a nucleus is given in terms of the atomic mass, $^A_Z M$, which equals the mass of the nucleus $M(A, Z)$ plus the mass of the atomic electrons $Z m_e$ less the mass equivalence of their binding energy (B.E.)$_e$ or B_e. Typically the differences in binding energies are small compared with the Q-value for a decay and can be neglected. Also the mass of neutrino m_ν (which equals that of its anti-particles $m_{\bar{\nu}}$) can be set equal to zero to a very good approximation as its value is known to be very small (if not zero) (≤ 30 eV/c^2). With these approximations the conditions for β^-, β^+-decay and *EC* to be energetically possible are:

$$_Z^A M > {}_{Z+1}^{A} M \qquad \beta^+\text{-decay} \qquad \text{... (3.1)}$$

$$_Z^A M > {}_{Z-1}^{A} M + 2\,m_e \qquad \beta^+\text{-decay} \qquad \text{... (3.2)}$$

$$_Z^A M > {}_{Z-1}^{A} M \qquad EC \qquad \text{... (3.3)}$$

The energy expressions in terms of atomic masses can be expressed as follows:

$$Q_{\beta^-} = [M(A, Z) - M(A, Z + 1)]c^2 \qquad \beta^-\text{-decay} \qquad \text{... (3.4)}$$

$$Q_{\beta^+} = [M(A, Z) - M(A, Z - 1) - 2\,m_e]c^2 \qquad \text{... (3.5)}$$

$$Q_{EC} = [M(A, Z) - M(A, Z - 1)]c^2 - B(K) \qquad \text{... (3.6)}$$

Where $B(K)$ is the binding energy of the K-electron in the parent atom (if electron from any other shell, i.e. L, M, N, ... is captured, then $B(K)$ will correspond to the B.E. of that electron).

From the above expressions for the Q-values of the three β-decay processes we may note:

(i) β^--emission will take place if the mass of the parent atom is greater than the mass of the daughter atom. Example is

$$^{14}_{6}\text{C} \rightarrow {}^{14}_{7}\text{N} + \beta^- + \bar{\nu} \text{ (antineutrino)}$$

$$M_{Z+1} \,(^{14}_{7}\text{N}) = 14.007515 \ u \text{ and } M_Z \,(^{14}_{6}\text{C}) = 14.007682 \ u$$

Clearly $M_{Z+1} > M_Z$ and hence β^--decay is possible.

(ii) For β^+-emission, the difference in the masses of parent and daughter atoms should be greater than $2\,m_e$, i.e. twice the mass of an electron. The rest mass of an electron is 0.511 MeV, and therefore the difference in the masses of parent and daughter atoms (energy units) should be greater than $2 \times 0.511 = 1.022$ MeV. In other words, the energy that is required to create two electrons (~ 1.022 MeV) have to be supplied by decay process, half of it for positron (β^+) emission and other half for the atomic electron ($_{-1}e^0$) that must be ejected in going from a neutral atom $^A_Z X$ to one with $_{Z-1}^{A} Y$ atom. The familiar example of β^+-decay is

$$^{11}_{6}\text{C} \rightarrow {}^{11}_{5}\text{B} + \beta^+ + \nu \text{ (neutrino)}$$

Here $\qquad [M(^{11}_{6}\text{C}) - M(^{11}_{5}\text{B})]c^2 \simeq 1.985$ MeV

Obviously, 1.985 MeV is greater than 1.022 MeV and hence positron emission can take place. Now,

$$Q_{\beta^+} = (1.985 - 1.022) \text{ MeV} = 0.963 \text{ MeV}$$

(iii) In *EC*, for K-electron capture, it is required that the difference in the atomic masses of parent and daughter atoms should be greater than the binding energy of K-electron in atom. We may note that in the lighter nuclides, small Z-value elements, the B.E. $B_e(K)$ is small and one can ignore it. Obviously, in this case the condition for electron capture just reduces to $M(A, Z)$ being greater than $M(A, Z - 1)$ and both electron capture and positron emission can happen in the same nucleus. On the other hand, in heavy nuclides, large Z-values elements, electron capture occurs more often than positron emission.

Decay of ^7_4Be into ^7_3Li by capturing an K-electron is an example of electron capture

$$^7_4\text{Be} + \beta^- \rightarrow ^7_3\text{Li} + \nu \text{ (neutrino)}$$

$$M(^7_4\text{Be}) - M(^7_3\text{Li}) = 0.866 \text{ MeV}$$

This mass difference is less than $2m_e c^2$ or 1.022 MeV. This means positron emission is not possible, i.e. decay occurs only through electron capturing.

We may note that electron capturing process is competitive with positron emission. There are nuclei which can undergo decay through both the processes, with different Q values, e.g. the decay of $^{80}_{35}\text{Br}$ to $^{80}_{34}\text{Se}$ as

$$^{80}_{35}\text{Br} \rightarrow ^{80}_{34}\text{Se} + \beta^+ + \nu$$

and

$$^{80}_{35}\text{Br} + \beta^- \rightarrow ^{80}_{34}\text{Se} + \nu$$

We may see that the atomic mass difference between the parent and daughter nuclei is around 2.66 MeV (> 1.022 MeV = $2m_e$).

In *EC* it is energetically possible for *K*-capture to be forbidden but *L*-capture to be allowed as the binding energy of a *K*-shell electron is greater than of an *L*-shell electron. Double β-decay is energetically possible for several nuclei but is very improbble, as it is a *second-order weak* process, with a half life of ~ 10^{22} yr.

3.9.3   Continuous β-ray spectrum and Pauli's Neutrino Hypothesis

The continuous nature of β-ray spectrum (Figures 3.13 and 3.14) is unexpected, as β-transition connects two states of definite energy (Figure 3.16). Most of the electrons, corresponding to peaks in the curve (Figure 3.13) are emitted with only about $\frac{1}{3}$ rd of the maximum energy, E_{max}.

What happens to about $\frac{2}{3}$ rd of E_{max} (missing) energy.

What happens to this missing energy? However, this missing must be acconted for to uphold the energy conservation principle. Several attempts were made to detect the missing energy, for instance by placing the β-decaying material inside a calorimeter with very thick lead walls, but they were fruitless. The situation was grave enough that some physicists were begining to seriously consider abandonig the law of conservation of relativistic energy, when Pauli proposed a less repugnant alternative.

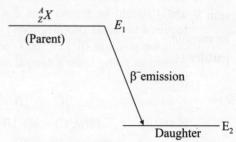

Fig. 3.16 β⁻-decay as a transition between two definite energy states E_1 and E_2

In 1931 Pauli postulated that a particle, now called the *antineutrino* $\bar{\nu}$, is also emitted in the electron emission process, but it is not normally detected because its *interaction with matter is extremely weak*. He also posulated that the antineutrino has (i) *zero charge*, (ii) *intrinsic spin* $s = \frac{1}{2}$, and (iii) *zero rest mass*. The first property permits charge conservation to be maintained in electron emission. The second property allows angular momentum to be conserved. Consider the nucleus Z, A emitting an electron to become the nucleus $Z + 1$, A and assume, for example that A is even. Then the nuclear spin i is an integer for both the initial and final nuclei. If only the electron with intrinsic spin $s = \dfrac{1}{2}$ were emitted, it would be impossible to conserve angular momentum, because the sum of a half-integral angular momentum (the electron) and an integral angular momentrum (the final nucleus) can only be half integral. If an anti-neutrino ($\bar{\nu}$) with $s = \dfrac{1}{2}$ is also emitted, the difficulty is removed. The third property was postulated to agree with measurments showing that the end point $E_e^{\max}$ (or $K_e^{\max}$) of the electron spectrum equals the decay energy, to the accuracy of the measurments. When an electron happens to be emitted at the end point, it carries away all the decay energy and none is left for rest mass energy of antineutrino. In positron (e^+) emission and electron capture (EC), the particle that is emitted, but very difficult to detect, is called the *neutrino*, ν. It has the same zero charge, spin $\dfrac{1}{2}$, and zero rest mass as the antineutrino $\bar{\nu}$. The spin of ($\beta + \nu$) system would be 1 or 0.

The relation between ν and $\bar{\nu}$ is explained by Dirac's relativistic quantum mechanics. This theory shows that every particle with intrinsic spin $s = \dfrac{1}{2}$ has its antiparticle, e.g. electron and its antiparticle positron, proton and antiproton, neutron and antineutron, etc. Electron and positron are produced in pairs. This is also found in the three β-decay processes. In electron emission a particle (electron) is produced with an antiparticle (anti neutrino), while in positron emission a particle (neutrino) is produced with an antipartilce (positron). Electron capture fits into this scheme since in the Dirac theory the destruction of an electron is identical to the creation of a positron.

It is possible to distinguish between ν and its antiparticle $\bar{\nu}$, although both of them have the same properties. Experiments show that ν exhibits a 'handedness' similar to a screw. A ν has its spin S_ν always *anti-parallel* to its momentum p_ν while spin $S_{\bar{\nu}}$ of an $\bar{\nu}$ is always *parallel* to its momentum $p_{\bar{\nu}}$. Obviously ν is *left-handed* particle, while its anti-particle $\bar{\nu}$ is a *right-handed* partilce (Figure 3.17). Neutrino is a *fermion*.

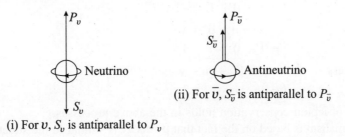

(i) For υ, S_υ is antiparallel to P_υ

(ii) For $\bar{\upsilon}$, $S_{\bar{\upsilon}}$ is antiparallel to $P_{\bar{\upsilon}}$

Fig. 3.17 Handedness in neutrino (ν) and its antiparticle $\bar{\nu}$ (antineutrino).

W. Pauli in 1931 proposed that in a beta decay process, if E_β is the energy of the beta particle and E_0 is the end-point energy, then the energy taken away by the neutrino is $E_n = E_0 - E_\beta$, i.e., in each β-decay, the reaction disintegration energy is shared in a continuous manner by ν, β-particle and the recoil nucleus.

Fermi postulated that the electron and neutrino are *created* at the time of β-particle emission. This situation is similar to the situation where a photon is created at the time of atomic or nuclear *de-excitation*. Obviously, β-decay is now a three-body process in the C.M. frame of reference. Clearly, the linear momenta of three particles, i.e. β, ν and recoil nucleus should add to zero (Figure 3.18).

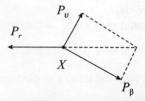

Fig. 3.18 Conservation of momentum in β-decay process. p_β, p_ν and p_r are the momentum of β-particle, neutrino (ν) and recoil nuclus respectively.

With the discovery of neutrino, one can express the β-decay process as

β⁻-decay $^A_Z X_N \rightarrow\ ^A_{Z+1} Y_{N-1} + \beta^- + \bar{\nu}$

β⁺-decay $^A_Z X_N \rightarrow\ ^A_{Z-1} Y_{N+1} + \beta^+ + \nu$

and Electron capture (*EC*)

$$^A_Z X_N + \beta^- \rightarrow\ ^A_{Z-1} Y_{N+1} + \nu$$

Now, the examples of β-decay reads as

$$^3_1 H_2 \rightarrow\ ^3_2 He_1 + \beta^- + \bar{\nu}$$
$$^{11}_6 C_5 \rightarrow\ ^{11}_5 B_6 + \beta^+ + \nu$$
$$^7_4 Be_3 + \beta^- \rightarrow\ ^7_3 Li_4 + \nu$$

3.9.4 Fermi's Theory of β-decay

Based on Pauli's neutrino hypotheses, E. Fermi in 1934 developed a theory for β-decay to obtain the continuous energy spectrum as well as the decay constant for β-emission. Fermi postulated that the electron and the neutrino are created at the time of β-particle emission. β⁻-particle and neutrino belong to *Leptons* (light particles) in the scheme of classification of elementary particles. If we denote the leptonic number for leptons by *l*, then for antilepton it will be – *l*. For nucleons *l* = 0, for electron and neutrino each, *l* = 1. Likewise for positron and anti-neutrino it is –1. According to law of conservation of leptons, one can see the β-decay processes as

β⁻-decay $n \rightarrow p + \beta^- + \bar{\nu}$
                  $l = 0 \quad\ 0 \quad\ 1 \quad -1$

β⁺-decay $p \rightarrow n + \beta^+ + \nu$
                  $l = 0 \quad\ 0 \quad -1 \quad\ 1$

EC process $p + \beta^- \rightarrow n + \nu$
                  $l = 0 \quad\ 1 \quad\ 0 \quad\ 1$

Obviously, law of lepton conservation holds in the above processes.

Fermi's formalism is based on the fact that β⁻-decay is similar to the situation where a photon is created at the time of atomic or nuclear de-excitation. Fermi assumed that the interaction responsible for β⁻-decay is *weak*, so that the perturbation theory of quantum mechanics can be applied. Fermi used the result of Dirac's time dependent perturbation theory, according to which, the transition probability per unit time for the transition is given by

$$W = \frac{2\pi}{\hbar}\ |H_\beta|^2_{fi}\ \rho_f(E)$$

$$= \frac{2\pi}{\hbar}\ |<f|H_\beta|\ i>|^2\ \rho_f(E) \qquad\qquad ...\ (3.1)$$

Where $|H_\beta|_{fi}$ is the *matrix element* of the perturbating interaction causing the transition and between the initial state i and final state f is given by

$$|H_\beta|_{fi} = \int \psi_f^* \, H_\beta \, \psi_i \, d\tau \qquad \qquad \dots (3.2)$$

$\rho_f(E)$ is the density of the final state, i.e. the number of momentum states in the final system per unit energy range and is equal to

$$\rho_f(E) = \frac{d\,n_f}{d\,E_0} \qquad \qquad \dots (3.3)$$

Where n_f is the number of states of particles involved in the decay that can be put in a given volume and E_0 is the total energy of transformation (same as E_{max}). Fermi called Equation (3.1) the *Golden rule.*

The nature and form of the interaction H_β being not known, Fermi assumed a very simple form for it, $H_\beta - gQ$, where g is Fermi coupling constant, which determines the strength of the interaction. This universal constant g has the value 0.9×10^{-4} MeV fm^3 and this is analogous to the electron charge e in the photon decay theory. Further, the emission of a neutrino and the absorption of an anti-neutrino of opposite momentum are equivalent and may replace $\psi_{\bar{\nu}}^*$ by ψ_ν to make the equations more symmetrial. Thus one can write the matrix element as

$$H_{if} = g \int \left[\psi_f^* \, \psi_e^* \, \psi_\nu \right] M \, \psi_i \, d\tau \qquad \qquad \dots (3.4)$$

Where M is a dimensionless matrix element and this is an operator.

The neutrino interact weakly with nucleons and therefore one can reasonably use a time independent wave function for neutrinos, which characterizes a free particle with the propagation constant $k = \dfrac{p_\nu}{\hbar}$, as

$$\psi_\nu = V^{-\frac{1}{2}} \exp\left[-\left(\frac{i}{\hbar} \right) \mathbf{p}_\nu \cdot \mathbf{r} \right] \qquad \qquad \dots (3.5)$$

One may ignore the electrostatic effect of the nucleus upon the ejected electron for high velocity electrons and use the wave function

$$\psi_e^* = V^{-\frac{1}{2}} \exp\left[-\left(\frac{i}{\hbar} \right) \mathbf{p}_e \cdot \mathbf{r} \right] \qquad \qquad \dots (3.6)$$

Here V is the entire volume in which one encloses the system for normalization purposes, $\mathbf{r}$ is the position coordinate and $\mathbf{p}_\nu$ and $\mathbf{p}_e$ are the momenta of the neutrino and electron respectively. As the magnitudes of wave functions ψ_ν and ψ_e at the position of the nucleus are certainly much larger for S-wave neutrinos and electrons than for these particles with larger angular momenta and hence one can neglect the spins. Assuming plane wave form for the wave functions of the electron and neutrino, one may neglect the possible interactions of these particles with the nucleus. Thus the matrix element takes the form,

$$H_{if} = g \int \psi_f^* \, \frac{1}{V} \left\{ \exp\left[-\frac{i}{\hbar} (\mathbf{p}_e + \mathbf{p}_\nu) \cdot \mathbf{r} \right] \right\} M \psi_i \, d\tau \qquad \qquad \dots (3.7)$$

One can write the exponential factor as

$$\exp\left[-\frac{i}{\hbar}(\mathbf{p}_e + \mathbf{p}_\nu) \cdot \mathbf{r}\right] = 1 - \frac{i}{\hbar}(\mathbf{p}_e + \mathbf{p}_\nu) \cdot \mathbf{r} - \frac{1}{2\hbar^2}[(\mathbf{p}_e + \mathbf{p}_\nu) \cdot \mathbf{r}]^2 + ... \qquad \text{... (3.8)}$$

The nuclear wave functions have appreciable values only in the regions of the order of nuclear dimensions, and therefore, the significant values of r should be no greater than the nuclear radius R. Now, if p_e and p_ν are both of the order of m_c, as is usually the case, one finds the exponent will

be of the order of $\dfrac{2mcR}{\hbar} \simeq \dfrac{1}{50}$. This smallvalue, clearly suggests that one need to retain only the

highest term, unity in the power series. When we insert this term in (3.7), the matrix element is nonvanishing and no longer it is dependent upon the energies or momenta of the light particles. This gives the allowed transitions. Matrix elements of forbidden transitions will be energy dependent as they contain terms of the form $(\mathbf{p}_e + \mathbf{p}_\nu) \cdot \mathbf{r}$. Thus for allowed transitions, one can write Equation (3.7) as

$$H_{if} = \frac{g}{V} \int \psi_f^* M \psi_i \, d\tau = \frac{g}{V} |M_{if}| \qquad \text{... (3.9)}$$

Where $|M_{if}|$ is the *overlap integral* or the nuclear matrix element of the final and initial wave functions of the nucleus, i.e. M_{if} is the nuclear part of the matrix element excluding the stength g and normalization volume V. One can compute this only in few cases where the structure of the nuclei is reasonably known.

3.9.4.1 Computation of Density of states or the Statistical Factor $\rho_f(E)$

One can represent the position and momentum of electron or neutrino by a point in *phase space, the space containing three spatial and three momentum dimensions.* The uncertainty principle prevents one from representing a moving particle by a single vector. This is due to the fact that such a representation would amount to specifying both the position and momentum exactly. This means, one will have to divide the phase space into cells of volume

$$\Delta x \, \Delta y \, \Delta z \, \Delta p_x \, \Delta p_y \, \Delta p_z \approx (2\pi\hbar)^3$$

One can easily show that the number of states of a particle restricted to a volume V in actual space and whose momentum lies between the limits p and $p + dp$ is given by

$$dN = \frac{V \, 4\pi \, p^2 \, dp}{(2\pi\hbar)^3} \qquad \text{... (3.10)}$$

The number of states of the neutrino corresponding to the appearance in volume V with the momentum in the range p_ν to $p_\nu + d\,p_\nu$ is

$$d N_\nu = \frac{4\pi V \, p_\nu^2 \, d p_\nu}{(2\pi\hbar)^3} \qquad \text{... (3.11)}$$

Similarly $\qquad\qquad d N_e = \dfrac{4\pi V \, p_e^2 \, d p_e}{(2\pi\hbar)^3} \qquad\qquad\qquad\qquad$ ... (3.12)

Thus the total number of states available for the β(electron) and ν in the above momentum ranges, when they are confined within the volume V is obtained as

$$dN = \left[\frac{(4\pi V)\, p_\nu^2\, dp_\nu}{(2\pi\hbar)^3} \right]\left[\frac{4\pi V\, p_e^2\, dp_e}{(2\pi\hbar)^3} \right]$$

Thus the number of states per unit energy of the electron is

$$\frac{dN}{dE_0} = \frac{16\,\pi^2 V^2}{(2\pi\hbar)^6}\, p_e^2\, p_\nu^2\, dp_e\, \frac{dp_\nu}{dE_0} \qquad \ldots (3.13)$$

But

$$E_0 = E_\nu + E_e \qquad \ldots (3.14)$$

Where E_0 is the total available energy. Now, for a fixed electron energy E_e, one obtains

$$dE_0 = dE_\nu \qquad \ldots (3.15)$$

The momenta p_e and p_ν are related to the energy of electron (E_0) and neutrino energy (E_ν) respectively by the following equations

$$E_0^2 = p_e^2\, c^2 + m^2\, c^4 \qquad \ldots (3.16a)$$

and

$$E_\nu = c\, p_\nu \qquad \ldots (3.16b)$$

Here, we have assumed zero rest mass.

Making use of Equations (3.14) to (3.16) in Equation (3.13), one obtains

$$\frac{dN}{dE_0} = \frac{16\pi^2\, V^2}{(2\pi\hbar)^6}\, p_e^2 \left(\frac{E_0 - E_e}{c} \right)^2 \frac{dp_e}{c} \qquad \ldots (3.17)$$

From Dirac's expression for transition probability per unit time of an atomic system to exit photon, using time dependent perturbation theory, the probability that an electron of momentum p_e and $p_e + dp_e$ is emitted per unit time may be expressed as

$$P(p_e)dp_e = \frac{2\pi}{\hbar}\, |H_{if}|\, \frac{dN}{dE_0} \qquad \ldots (3.18)$$

Inserting $\dfrac{dN}{dE_0}$ from Equation (3.17) and H_{if} from Equation (3.8) in (3.18), one obtains

$$P(p_e)\, dp_e = \frac{2\pi}{\hbar}\, \frac{g^2}{V^2}\, |M_{if}|^2\, \frac{16\pi^2\, V^2}{c^3\,(2\pi\hbar)^6}\, p_e^2(E_0 - E_e)^2\, dp_e$$

$$= \frac{g^2\, |M_{if}|^2}{2\pi^3\, c^3\, \hbar^7}\, (E_0 - E_e)^2\, p_e^2\, dp_e \qquad \ldots (3.19)$$

Expression (3.19) is independent of V as it should be. This expression gives the Fermi distribution for the emitted β-particles in allowed β-decay. This equation gives the momentum spectrum for electrons emitted in β-decay using the energy and momentum relationship for the electrons (neglecting the effect of Coulomb field of the nucleus) and has a shape very similar to that obtained experimentally. Using

$$E_0 = \sqrt{m_0^2 c^4 + c^2 p_{max}^2} \quad \text{to define } p_{max}, \text{ one obtains, } \lambda \text{ decay probability}$$
$$\text{per unit time} = P(p_e)d\, p_e$$

or

$$\lambda = \frac{g^2 |M_{if}|^2}{2\pi^3 \hbar^7 c^3} \left(\sqrt{m_0^2 c^4 + c^2 p_{max}^2} - \sqrt{m_0^2 c^4 + p^2 c^2} \right)^2 p^2 dp$$

The mean life $\tau \left(= \dfrac{1}{\lambda} \right)$ is found by integration over all possible values of p, i.e.

$$\lambda = \frac{1}{\tau} = \frac{g^2 |M_{if}|^2}{2\pi^3 \hbar^7 c^3} \int_0^{p_{max}} \left(\sqrt{m_0^2 c^4 - c^2 p_{max}^2} - \sqrt{m_0^2 c^4 + p^2 c^2} \right)^2 p^2 dp$$

$$= \frac{g^2 |M_{if}|^2}{2\pi^3 \hbar^7} m_0^5 c^4 F(\eta_0) \qquad\qquad [p_{max} \text{ is also } p_0]$$

Wher $F(\eta_0)$ is a *Fermi function*. The final expression for τ is

$$\lambda = \frac{0.693}{T} = \frac{1}{\tau} = \frac{g^2 m_0^5 c^4}{2\pi^3 \hbar^7} |M_{if}|^2 F(z_0, \eta_0)$$

If $|M_{if}|^2$ does not change, the quantity $FT = $ constant. The FT values are found to be a measure of the forbiddenness of the transition. From the observed rates for simple β-decays it is found that $\gamma \approx 1.4 \times 10^{-62}$ J m³. One can also express this quantity in dimensionless form by using $\hbar c$ and $\dfrac{\hbar}{m_p c}$, where m_p is the proton mass, as convenient units of energy × length and of length, one obtains

$$g \approx 1.4 \times 10^{-62} \text{ J} - \text{m}^3 = 10^{-5}\, \hbar c \left(\frac{\hbar}{m_p c} \right)^2$$

The quantity FT is sometimes called the *comparative lifetime*. It can be used to compare β decays of different decay energy, and rank them according to the lifetimes they would have if they all had the same decay energy. That it, multiplying T by F removes the energy dependence, and so produces a quantity whose value depends only on a collection of universal constants and on the value of nuclear matrix element $|M_{if}|$. Since the matrix element contains the eigen functions for the nuclear states involved in a β-decay, it is apparent that the FT values for the decay can provide information those nuclear states. One can obtain the energy spectrum for the emitted electrons using

$$d\, p_e = \frac{E_e\, d\, E_e}{c(E_e^2 - m_0^2 c^4)^{1/2}} \qquad\qquad \text{... (3.20)}$$

The probability for the emission of electrons with energy E_e and $E_e + d\, E_e$ is given by

$$P(E_e)d\, E_e = \frac{g^2 |M_{if}|^2}{2\pi^3 c^6 \hbar^7} E_e (E_0^2 - m_0^2 c^4)^{\frac{1}{2}} (E_0 - E_e)^2\, d\, E_e \qquad \text{... (3.21)}$$

Fermi assumed that the operator g is a scalar of order 1, the matrix element $|M_{if}|$ becomes of the order or unity as such $|M_{if}|^2 \approx 1$ in (3.19) and (3.21). Thus, under Fermi's approximation, the number of electrons emitted per second in the energy range E_e and $E_e + dE_e$ becomes

$$P(E_e)dE_e = C^2 E_e \ (E_0^2 - m_0^2 c^4)^{\frac{1}{2}} \ (E_0 - E_e)^2 \ dE_e \qquad \qquad \text{... (3.22)}$$

Where
$$C^2 = \frac{g^2}{2\pi^3 \, \hbar^7 \, c^6} \qquad \text{... (3.23)}$$

These transitions for the β-emission under Fermi approximations are called *allowed transitions*. A plot of Equation 3.22) gives the energy spectrum for allowed transitions. Figure 3.19 shows a plot of $\frac{P}{C^2}$ as a function of E_e.

From the graph it is evident that the distribution falls to zero for both high and low electron energies. This is found to be contrary to what is actually observed, where more low energy electrons are found. We will obtain a similar curve for positrons also.

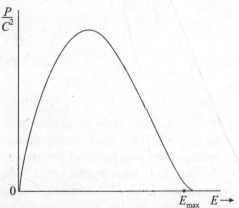

Fig. 3.19 Energy spectrum of β-particles for allowed transitions obtained by plotting $\frac{P}{C^2}$ as a function of E_e.

3.9.4.2 Coulomb Correction

The disagreement of Fermi distribution for β-spectrum as stated above was partly because in the derivation of the above relation no account has been taken of the Coulomb interaction which can be neglected only for the lighter nuclei $(Z < 10)$ and sufficiently high electron energies. If we take account fo the Coulomb field of the nucleus, then the electronic wave function is affected and it cannot be a plane wave as assumed above. In other words, the plane wave will have to be replaced by a distorted Coulomb wave function.

In order to take account of the Coulomb interaction, one will have to multiply $|\psi_e|^2$ with a factor $F(Z, E_e)$. This factor is sometimes called a *Coulomb factor* and also called as *Fermi Function*. Its sign changes, depending on whether it is applied for electrons or positrons. It is essentially a barrier penetration factor and in the non-relativistic limit, i.e. when the velocity (v) of β-particle is much less than the velocity of light (c), i.e. $v \ll c$. The Fermi function is the ratio of electron density at the daughter nucleus to the density at infinity, i.e.

$$F(Z, E_e) = \frac{|\psi_e(0)|^2_{\text{Coulomb}}}{|\psi_e(0)|_{\text{free}}} \qquad \qquad \text{... (3.24)}$$

In the non-relativistic approximation, Fermi function has the approximate form

$$F(Z, E_e) = \frac{2\pi\eta}{(1 - e^{-2\pi\eta})} \qquad \qquad \text{... (3.25)}$$

Where $\eta = \dfrac{Ze^2}{4\pi\varepsilon_0\hbar v}$ for positrons and $\eta = -\dfrac{Ze^2}{4\pi\varepsilon_0\hbar v}$ for positrons, Z is the atomic number of the product nucleus and v is the velocity of the electrons far away from the nucleus. Tables of the Fermi functions have been prepared. When consideration is given to this effect, Equation (3.19) takes the form

$$P(p_e)dp_e = \frac{g^2|M_{if}|^2}{2\pi^3 c^3 \hbar^7} F(Z, E_e)(E_0 - E_e)^2 p_e^2 \, dp_e$$

$$= C^2 F(Z, E_e)(E_0 - E_e)^2 p_e^2 \, dp_e$$

Where
$$C = g|M_{if}| \, (2\pi^3 c^3 \hbar^7)^{-\frac{1}{2}} \qquad \qquad \text{... (3.27)}$$

One can directly obtain the behaviour of β-spectrum at very low electron or positron energies with the help of (3.26). The Coulomb correction enhances the probability of the electron emission and decreases the probability of positron emission, especially at low energies. At high energies, the Coulomb force loses its effect on the shape of the spectrum and spectrum approaches that computed without the Coulomb correction. One can easily explain this as: the Coulomb field accelerates the positive electron and decelerates the negative electron. This means, the positron has fewer slow particles and the electron spectrum has more slow particles, than they would have in the absence of the Coulomb correction (Figure 3.20).

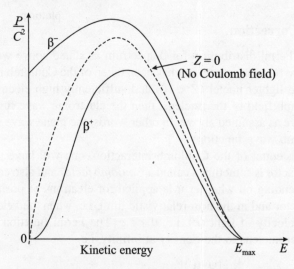

Fig. 3.20 Effect of Coulomb interaction on β-energy spectrum (Fermi theory). The dotted line curve corresponds to the spectrum without Coulomb field.

The decay of $^{64}_{29}\text{Cu}$ which emits both electrons and positrons of almost same and point energy demonstrates nicely the effect of Coulomb interactions. We have,

$$^{64}_{29}\text{Cu} \rightarrow {}^{64}_{30}\text{Zn} + e^- \ (\beta^-), \ E_0 \sim 0.57 \text{ MeV}$$

and
$$^{64}_{29}\text{Cu} \rightarrow {}^{64}_{28}\text{Ni} + e^+ \ (\beta^+), \ E_0 \sim 0.66 \text{ MeV}$$

Figure 3.21 shows the energy distribution of the electrons and positrons from $^{64}_{29}Cu$

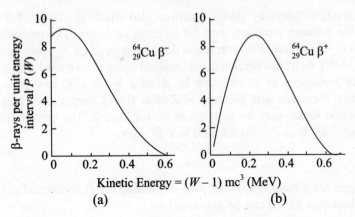

Fig. 3.21 Energy distribution of β^- (a) and β^+ (b) from the decay of $^{64}_{29}Cu$. In Fig. $\left(W = \dfrac{E_e}{m_0 c^2}\right)$

Figure 3.22 shows the corresponding momentum distributions for β^- and β^+ by the decay of $^{64}_{29}Cu$.

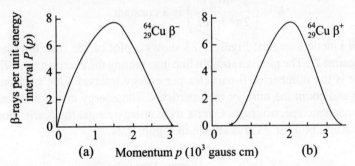

Fig. 3.22 Momentum distribution of β^- (a) and β^+ (b) particles from the decay of $^{64}_{29}Cu$.

Classically, no positron would be expected to emerge below the barrier energy $\dfrac{Ze^2}{4\pi\,\varepsilon_0\,R}$, but in actual spectrum some low energy positrons as a result of the tunnel effect appear. One finds the relation identical to that for α-decay, as for low energy β^+-emission.

$$F(Z, E_e) = -\frac{2\pi\,Ze^2}{4\pi\,\varepsilon_0\,\hbar\nu}\frac{1}{\left[1 - e^{\frac{2\pi\,Ze^2}{4\pi\,\varepsilon_0\,\hbar\nu}}\right]}$$

$$\approx \frac{Ze^2}{2\varepsilon_0\,\hbar\nu}\,e^{\frac{-Ze^2}{2\varepsilon_0\,\hbar\nu}} \qquad\qquad \dots (3.28)$$

3.9.4.3 Screening by Atomic Electrons

Screening of the nuclear charge by atomic electrons also affects the shape of β-spectra, which is more important for positron emission than for electron emission. The maximum correction for electron emission is ~ 5 %. Screening correction does not depend on E_β appreciably, except at very low energy (< 35 MeV). Reitz has tabulated and obtained corrections from 1.5 to 400 keV. Screening effect reduces the probability of β⁻-emission by about 2 % for a 50 keV β⁻-ray at Z ~ 50. The screening correction increases with increase of Z or as β⁻-ray energy decreases. For low energy β⁺ ray, the correction factor may be as large as 10 for high Z. The probability of β⁺ emission increases by about ~ 37 % at Z ~ 50 for a 50 keV β⁺-rays.

Kurie Plot

Kurie in 1936 suggested a more convenient way of plotting the experimental results and checking with the theory. Equation (3.26) can be expressed as

$$\left[\frac{P(p_e)}{p_e^2 \, F(Z, E_e)} \right]^{\frac{1}{2}} = K(E_0 - E_e) = K(T_0 - T_e) \qquad \ldots (3.29)$$

Where $$K = \left(\frac{g^2 |M_{if}|^2}{2\pi^3 \, \hbar^7 \, c^3} \right) \text{ is a constant} \qquad \ldots (3.30)$$

and T_0 and T_e are kinetic energies. Figure 3.23 shows a plot of the function on the LHS side of Equation (3.29) against T_e. The plot is a straight line intersecting the energy axis at T_0. Experimentally the transition rate is the number of β-particles per energy interval. One may employ the Geiger Counter to detect and count the number of β-particles. The energy or momentum interval may be determined by a magnetic spectrograph. Over a wide energy range, the Kurie plot is found to be a straight line. The Kurie plot for ⁶⁰Co is shown in Figure 3.24.

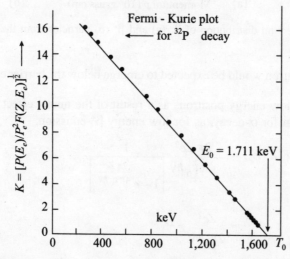

Fig. 3.23 Kuriet plot or Fermi-Kurie plot for ³²P (a single transition).

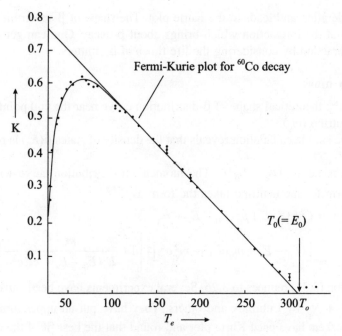

Fig. 3.24 Fermi-Kurie plot for ^{60}C-decay

Kurie-plot provides a good means of determining the maximum energy of the β-spectrum. For ^{64}Cu – β-decay, the maximum energies of electron and positron obtained by Kurie plots are 571 and 657 keV respectively.

Sometimes deviations are reported in straight line Kurie plots. Probably, this may be due to following reasons.

(i) A forbidden transition usually result in concave-up shape type graph. In this case, one has to modify the simple theory for allowed transition.

(ii) Complex β-specturm is another reason for the deviation. Due to this the transitions to two or more states of daughter nucleus takes place. This situation is quite analogous to emission of α-rays with a fine structure. The transitions to the upper states are usually followed by γ-rays emission or internal conversion electrons (Figure 3.25).

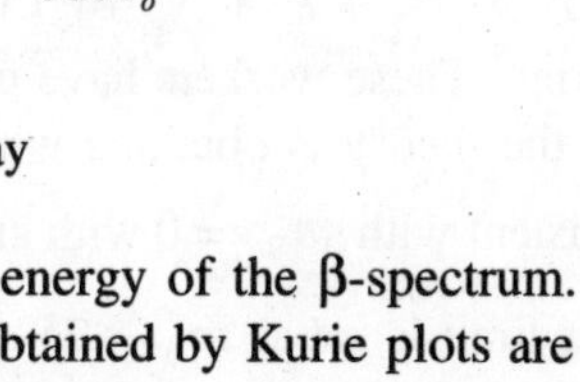

We may note that Fermi-Kurie plots show non-linearity below 200 keV energy due to scattering of electrons in the source. The best one can go, therefore is about 50 keV. If one takes into account a relativistic mass correction term for the neutrino, then one finds that Fermi plot near the end point energy is not a straight line but turns sharply to intersect the energy axis at a point smaller than the value for zero mass neutrino.

Fig. 3.25 Complex β-spectrum (transitions in ^{187}W → ^{187}Re).

We may also note tat the *Matrix element* $|M_{if}|$ in the Fermi theory being momentum dependent (for most β-decay) does not influence the shape of β-spectrum. This shape is mainly due to the phase space consideration and leads to the Kurie plot. The shape of β-spectrum thus can not give us information about the interaction which brings about β-decay. One can get an idea about the strength of the interaction by considering the life times of β-emitters.

3.9.4.4 Neutrino mass

We may note that the theoretical shape of β-distribution curve near the end point $E_\beta = E_0$ depends on the mass of neutrino (m_ν).

In this case, the simple calculation reveals that the density of states $\rho(E_0)$ is proportional to the neutrino momentum, i.e. to $(E_o - E_\beta)^{\frac{1}{2}}$. The momentum distribution curve with the relativistic mass correction term for the neutrino takes the form as

$$P(p_e)\,dp_e = C^2 F(Z, E_e)\, p_e^2 (E_0 - E_e + m_\nu c^2)$$

$$\left[(E_0 - E_e + m_\nu c^2)^2 - m_\nu^2 c^4\right]^{\frac{1}{2}} \left[1 + \frac{m_\nu c^2}{E_e(E_0 - E_e + m_\nu c^2)}\right] dp_e \quad ...\ (3.31)$$

For $m_\nu = 0$, Equation (3.31) reduces to 3.26. Several experiments have been carried on β-spectrum of ^{3}H $\rightarrow$ ^{3}He $+ e^- + \bar{\nu}_e$ by Lubimov and others. They have put an upper limit for the mass of neutrino. These workers have used Kurie plot and found that the best fit of the experimental data with the theory is obtained with $m_\nu = 7$ eV/c^2. It is almost zero, i.e. the current best results are consistent with $m_{\bar\nu_e} = 0$ with an upper limit of the order of 7 eV at the 90 % confidence level. The upper limit is set at $m_{\bar\nu_e} < 25$ eV. In a similar fassion the limits on muon (m_{ν_μ}) type and tau-type (m_{ν_τ}) have been obtained from a study of muon spectrum in the decay $\pi \rightarrow \mu + \nu_\mu$ and from the study of pion spectrum in the decay $\tau = 5\pi + \nu_\tau$ as $m_{\nu_\mu} < 270$ keV and $m_{\nu_\tau} < 35$ MeV respectively.

3.9.4.5 Allowed and Forbidden Transitions

We have read that β-decay is the result of interaction between β-active nucleus and the field of leptons and there exists a relation between decay constant λ of the nucleus and the maximum beta energy $(E_0 = E_{max})$. During the disintegration of a nuclei, the emission of the electron and neutrino are regarded as free particles. Neglecting Coulomb effects, the wave functions for electron (ψ_e) and neutrino (ψ_ν) are defined as plane waves, i.e.

$$\psi_e \sim \exp\left[\frac{i}{\hbar}(\mathbf{p}_e \cdot \mathbf{r})\right]$$

We have also read that exponentials in the plane waves can be expanded since for beta particles of energy around 1 MeV, $\dfrac{\mathbf{p} \cdot \mathbf{R}}{\hbar} = \dfrac{R}{\lambda}$, for medium size nucleus ~ 0.01, where R is nuclear radius and $\lambda = 2\pi\lambdabar$ is the wavelength of electron or neutrino. Now, p_r is a measure of angular momentum carried by leptons, whose permissible values are $\sqrt{l(l+1)}\hbar$ (according quantum mechanics), where l is the orbital angular momentum quantum number of emitted leptons. We have

$$\phi = \exp\left[\frac{i}{\hbar}(\mathbf{p}\cdot\mathbf{r})\right] = 1 + \frac{i}{\hbar}(\mathbf{p}\cdot\mathbf{r}) + \left(\frac{i}{\hbar}\right)^2\left(\frac{\mathbf{p}\cdot\mathbf{r}}{2}\right)^2$$

Retaining only the first term in the expansion, i.e. only considering $l = 0$ terms for leptons, as the electron and the neutrino have, on an average approximately equal and opposite momenta, the inclusion of higher order terms means, one is considering $l = 1$, $l = 2$, ... etc. terms. However, the rate of emission, is largest when the total angular momentum, L of electron-neutrino pair is zero. β-decay or β-transitions with $L = 0$ are called as *allowed transitions* and those with $L > 0$ are referred as *forbidden transitions*. The order of magnitude $\dfrac{pR}{\hbar} \approx 10^{-2}$ also support it. Since then $\dfrac{pR}{\hbar} \ll 1$ and hence $L = 0$ is the most probable β-decay. However, if parity changes ($\Delta\pi = $ yes) due to decay, then the first term in the expansion of $e^{ikr} = e^{\frac{ipr}{\hbar}}$ becomes zero and therefore higher order terms will have to be considered $l = 1$ refers to first forbidden transition and so on. Clearly, the allowed transitions are most intense and in forbidden transitions the intensity decreases by a factor $\left(\dfrac{R}{\lambda}\right)^2$ as compared to allowed transitions. We know that besides orbital angular momentum (L), electron and neutrino, both have spin angular momentum of $\dfrac{1}{2}$ each in unit of $\hbar$. Depending on their spins being anti-parallel ($\uparrow\downarrow$ singlet state) or parallel ($\uparrow\uparrow$ triplet state) their total spin S can be 0 or 1. The transitions are said to be *Fermi transitions* when spins are anti-parallel, i.e. $S = 0$, and *Gammow-Teller Transitions* when spins are parallel, i.e. $S = 1$. For both types of transitions, there are allowed and forbidden transitions.

In the expression for transition probability, the interaction matrix $|M_{if}|$ is referred as *Fermi interaction* $|M_F|$ where the interaction operator Q which connects the initial nuclear state (parent nucleus) and the final nuclear state (daughter nucleus) is a unit and at the same time it is a scalar. When $I_i = I_f$, $|M_F| \approx 1$. When the interaction operator Q is spin operator σ, the interaction is called *Gammow-Teller (GT) interaction* $|M_{GT}|$. One cannot evaluate the matrix elements for GT operator unless the wave functions for the parent nucleus (initial state u_i) and that of daughter nucleus (final state u_f) are known. Properties of operators itself helps one to find that the initial and final spins of the nucleus are coupled vectorially by a unit vector. However, σ being an axial vector, i.e. pseudo vector, cannot change the parity between initial and final states. The β-decay theory then predicts that the shapes of all allowed spectra should be same and the shapes of forbidden spectra should differ according to the order of forbiddeness and the type of assumed interaction.

3.9.4.6 Selection Rules

The β-decay transitions are governed by certain selection rules, depending on whether β-decay is allowed or forbidden. The rules connecting the characterstics of the transition in terms of change in angular momentum and parity, with the transition order, are termed as *selection rules*. The origin of selection rules is the incorporation of the laws of conservation of angular momentum and parity into the β-decay theory, and is governed by the nature of interaction. By comparing the experimentally determined order of β-transition with that of theoretically expected, one can know about the type of interaction responsible for β-decay.

There are two selection rules in practice. A nuclear state is characterized by the total angular momentum **I** and parity **π**. The total angular momentum or the "spin" of the nucleus is the vector sum of the orbital and spin angular momenta, i.e.

$$\mathbf{I} = \mathbf{L} + \mathbf{S}$$

When all the transitions take place from an initial state I_i to a final state I_f, one can write the change in I as

$$\Delta\mathbf{I} = \mathbf{I}_f - \mathbf{I}_i \qquad \qquad \text{... (3.1)}$$

such that

$$\Delta\mathbf{I} = \Delta\mathbf{L} + \Delta\mathbf{S} \qquad \qquad \text{... (3.2)}$$

In the case of allowed transitions, $L = 0$ as such ΔI must be decided by the spin angular momentum of electron-neutrino pair. For Fermi transitions ($\mathbf{S} = 0$) and hence one finds,

$$\Delta\mathbf{I} = \mathbf{I}_f - \mathbf{I}_i = 0 \qquad \qquad \text{... (3.3)}$$

Here $\mathbf{I}_i$ and $\mathbf{I}_f$ both can be zero, i.e. $\mathbf{I}_i = \mathbf{I}_f = 0$ is possible. Such transitions are known as Fermi transitions or $\Delta\mathbf{I} = 0$.

In the second care, the vector difference between the initial and final angular momentum of the nucleus must be 1, i.e.

$$\Delta I = \pm 1 \text{ or } 0, \text{ but no } 0 \to 0 \qquad \qquad \text{... (3.4)}$$

The selection rule is due to Gamow and Teller ($L = 0$) and in this case

$$\mathbf{I}_f = \mathbf{I}_i + 1$$
$$|\Delta I| = |\mathbf{I}_f - \mathbf{I}_i| = \pm 1, 0$$

i.e.,

$$I_f = I_i \pm 1, \text{ and } I_f = I_i$$

Obviously, in $G - T$ allowed transitions, there can not be a transition between $I_f = I_i = 0$, as these transitions can not satisfy the condition of $\mathbf{S} = 1$ for the total spin angular momentum. We may remember that $I_f = I_i = 0$ or $0^+ \to 0^+$ is a pure Fermi allowed transition. In accordance with the conservation of parity, the initial wave function (of parent nucleus) must be same as the product of the parities of the wave functions of the daughter nucleus, electron and neutrino, i.e.

$$\pi_i = \pi_f (-1)^L \qquad \qquad \text{... (3.5)}$$

Where π_i and π_f denotes the parities of the wave functions corresponding to parent and daughter nuclei respectively, and $(-1)^L$ is the parity of the electron-neutrino pair. We may note that for Fermi allowed transitions $L = 0$ and hence $\pi_i = \pi_f$ and thus there is no change in the parity. Also in case of GT transition there is no change in parity. Summarising, we have two selection rules for the allowed transitions

(i) $\Delta I = 0$, Parity 'No' Fermi selection rule

(ii) $\Delta I = 0$ or ± 1, Parity 'No' GT selection rule (except $0 \to 0$)

Example of a pure Fermi allowed transition is

$$\ce{^{14}_{8}O} \to \ce{^{14}_{7}N^*} + \beta^+ + \nu, \quad \genfrac{}{}{0pt}{}{(0^+ \to 0^+)}{\text{Parily \quad No}}$$

The ground state of daughter N^{14} – has a spin 1, but 2.31 MeV excited state $^{14}N^*$ has a spin O. Both have even parities and for ^{14}O, $I_i = O^+$ and for $^{14}N^*$, $I_f = O^+$. Thus $I_i - I_f = 0$ and there is no change in parity.

Experiments shows that the allowed transitions of the type $\Delta I = 1$, obeying $G - T$ selection rule, are forbidden by Fermi (F) selection rule, as in the decay

$$^6_2\text{He} \rightarrow \, ^6_3\text{Li} + e^- + \bar{\nu}, \qquad (0^+ \rightarrow 1^+, \text{No})$$

We can see that in the above, there is no change in parity, as both the nuclei have even parities. However, ^6Li has spin angular momentum 1, i.e. $I_f = 1^+$ and for ^6He, $I_i = 0$ and thus $\Delta I = I_f - I_i = 1$.

We may note that even parity denoted by (+) over the angular momentum I, (I^+) and for odd parity (–) is used (I^-).

There are also allowed transitions of the $0 \rightarrow 0$ type. These are allowed by F-selection rules but forbidden by $G - T$ selection rules, for e.g.

$$^{14}\text{O} \rightarrow \, ^{14}\text{N}^* + e^+ + \nu \qquad (0^+ \rightarrow 0^+, \text{No})$$

The ground state ^{14}N has a spin 1 but 2.31 MeV excited state $^{14}\text{N}^*$ has a spin O.

There are several transitions which are allowed by both the selection rules. This is always possible in the allowed decays in which $I_i = I_f \neq 0$ ($i \rightarrow$ initial and $f \rightarrow$ final states). Some examples of such decays are

$$^1_0 n \rightarrow \, ^1_1\text{H} + \beta^- + \bar{\nu}, \qquad \left(\frac{1^+}{2} \rightarrow \frac{1^+}{2} , \text{No} \right)$$

$$^3_1\text{H} \rightarrow \, ^3_2\text{He} + \beta^- + \bar{\nu}, \qquad \left(\frac{1^+}{2} \rightarrow \frac{1^+}{2} , \text{No} \right)$$

$$^{35}_{16}\text{S} \rightarrow \, ^{35}_{17}\text{Cl} + \beta^- + \bar{\nu}, \qquad \left(\frac{3^+}{2} \rightarrow \frac{3^+}{2} , \text{No} \right)$$

The appkicability of the selection rules is also confirmed by ft values. The allowed transitions are furhter classified as *favoured* (*super allowed*) and *unfavoured transitions*. An allowed transition if said to be favoured if the nucleon which changes its charge remains in the same level and it is unfavoured if the nucleon changes its level. We find that most of the allowed β-transitions are unfavoured. We may note that for the β^--decay, the change of a neutron (n) into a proton (p) without the change of level would increase the total energy of the nucleus and thus it would not lead to a spontaneous decay. For the β^+-decay, the surplus of neutrons makes the allowed transitions unfavoured. There are exceptions also, e.g.,

$$\text{H}^3 \rightarrow \text{He}^3, \text{He}^6 \rightarrow \text{Li}^6, \text{C}^{11} \rightarrow \text{B}^{11} \text{ and } \text{F}^{18} \rightarrow \text{O}^{18}$$

We may note that the conditions for super-allowed transitions are same as that for allowed transitions. The matrix element $|M_{if}|$ is also energy independent. The main difference between these two transitions is that the allowed transitions cannot be between mirror nuclei.

We now consider the case when the transition from the initial to final nucleus does not take place by the emission of S-wave electron and neutrino. The emission of electron and neutrino with orbital angular moment other than zero is also possible due to the finite size of the nucleus. With increasing orbital angular momentum, the magnitudes of the wave functions ψ_e and ψ_ν for p-wave, d-wave etc. over the nuclear volume decrease rapidly. β-transitions with angular momentum, carried off by two light particles together $l_\beta = 1, 2, 3, 4, \ldots$ etc. are termed as first, second, third, fourth etc. *forbidden transitions*.

Initial and final nucleus must have opposite particles if l_β is odd, i.e. parity changes in these transitions. However, for even l_β, the initial and final nucleus must have same parity, i.e. there is no change in parity. Moreover, the emision of leptons (electron and neutrino) in the singlet state (Fermi-selection rule) like allowed transitions demands $\Delta I \leq l_\beta$, whereas triplet state ($G-T$ selection rule) emission demands $\Delta I \leq l_\beta + 1$. One can write the selection rules for forbidden transitions as

First Forbidden Transition: $l_\beta = 1$ and parity changes

Fermi-selection rules: $\Delta I = \pm 1, 0$ (except $0 \to 0$)

G.T. Selection rules: $\Delta I = \pm 2, \pm 1, 0$ (except $0 \to 0, \dfrac{1}{2} \to \dfrac{1}{2}, 0 \leftrightarrow 1$)

Few examples are:

$$^{37}S \to {}^{37}Cl + \beta^- \qquad \left(\frac{7}{2} \to \frac{3}{2}\right)$$

$$^{85}Kr \to {}^{85}Rb + \beta^- \qquad \left(\frac{9}{2} \to \frac{5}{2}\right)$$

$$^{87}Kr \to {}^{87}Rb + \beta^- \qquad \left(\frac{5}{2} \to \frac{3}{2}\right)$$

$$^{111}Ag \to {}^{111}Cd + \beta^- \qquad \left(\frac{1}{2} \to \frac{1}{2}\right)$$

$$^{137}Cs \to {}^{137}Ba^* + \beta^- \qquad \left(\frac{7}{2} \to \frac{11}{2}\right)$$

$$^{141}Ce \to {}^{141}Pr + \beta^- \qquad \left(\frac{7}{2} \to \frac{5}{2}\right)$$

Second Forbidden Transitions: $l_\beta = 2$, Parity: no change.

Fermi selection rules: $\Delta I = \pm 2, \pm 1$ (except $0 \leftrightarrow 1$)

G.T. selection rules: $\Delta I = \pm 3, \pm 2, 0 \to 0$ (except $0 \leftrightarrow 2$)

Few examples are:

$$^{10}Be \to {}^{10}B + \beta^- \qquad (0 \to 3)$$
$$^{22}Na \to {}^{22}Ne + \beta^+ \qquad (3 \to 0)$$
$$^{135}Cs \to {}^{135}Ba + \beta^- \qquad \left(\frac{7}{2} \to \frac{3}{2}\right)$$

Third Forbidden Transitions:
Change in parity. An example is

$$^{87}Rb \to {}^{87}Sr + \beta^- \qquad \left(\frac{3}{2} \to \frac{9}{2}\right)$$

Fourth Forbidden Transitions:
No change in parity. An example is

$$^{115}\text{In} \rightarrow {}^{115}\text{Sn} + \beta^- \qquad \left(\frac{9}{2} \rightarrow \frac{1}{2}\right)$$

nth Forbidden Transitions

Fermi selection rules: $\Delta I = \pm n, \pm(n-1)$; change in parity for *n*-odd

GT selection rules: $\Delta I = \pm n, \pm(n+1)$; Parity remains unchanged for *n* even.

We see from these selection rules that a β^--decay between two $I = 0$ states of the same parity can occur only through a Fermi transition ($S = 0$). There are also many situations when decays go in lowest order, only through GT transitions, including all cases where $\Delta I = l + 1$. Such transitions are referred to as *unique*, the shape of the electron spectrum from them can be calculated accurately. In cases, where there is a mixture of Fermi and G. T. transitions, the shape of the electron spectrum in forbidden transitions depends on the details of this mixture and, hence, can not be calculated accurately.

3.9.4.7 Total Decay Rate for β-Disintegration

One can obtain the total transition probability or the decay rate λ for β^--decay, i.e. electron emission in a β-decay by integrating

$$P(p_e)dp_e = \frac{g^2|M_{if}|^2}{2\pi^3 \hbar^7 c^3} F(Z, p_e)(E_0 - E_e)^2 p_e^2 dp_e$$

Over the electron momentum distribution from zero to a maximum momentum p_0, i.e.

$$\lambda = \int_0^{p_0} P(p_e)dp_e$$

$$= \int_0^{p_0} \frac{g^2}{2\pi^3} \frac{|M_{if}|^2}{\hbar^7 c^3} F(Z, E_e)(E_0 - E_e)^2 p_e^2 dp_e \qquad \ldots (3.1)$$

Now, assuming that matrix element $|M_{if}|$ is independent of energy, one finds

$$\lambda = \frac{1}{\tau} \frac{g^2|M_{if}|^2}{2\pi^3 \hbar^3 c^3} \int_o^{p_o} F(Z, E_e)(E_0 - E_e)^2 p_e^2 dp_e \qquad \ldots (3.2)$$

Here τ is the mean life for β^--decay, i.e. electron emission. We have

$$\lambda = \frac{\log 2}{T_{\frac{1}{2}}} = \frac{1}{\tau}$$

Where $T_{\frac{1}{2}}$ is half life for β^--decay.

or
$$\lambda = \frac{g^2 |M_{if}|^2 m_e^5 c^4}{2\pi^3 \hbar^7} \int\limits_0^{p_0/m_e c} F(Z, E_e) \left\{\frac{E_0 - E_e}{m_e c^2}\right\}^2 \frac{p_e^2}{m_e c^2} \frac{d p_e}{m_e c}$$

$$= \frac{g^2 |M_{if}|^2 m_e^5 c^4}{2\pi^3 \hbar^7} f(Z, E_0) \qquad \qquad \dots (3.3)$$

Where
$$f(Z, E_0) = \frac{1}{me^4 c^7} \int\limits_0^{p_0/m_e c} F(Z, E_0)(E_0 - E_e)^2 p_e^2 d p_e \qquad \dots (3.4)$$

is the dimensionless function. The function $f(Z, E_0)$ is usually referred as *Fermi integral*. One has to evaluate the integral numerically except in the case $Z = 1$ for the daughter nucleus.

One can obtain β-decay half life from the transition probability as $T_{\frac{1}{2}} = \frac{\log 2}{\lambda}$. Making use of Equation (3.3), one obtains

$$T_{\frac{1}{2}} = \frac{\log 2}{\lambda} = \frac{2\pi^3 \hbar^7}{g^2 |M_{if}|^2 m_e^5 c^4} \frac{\log 2}{f(Z, E_0)} \qquad \dots (3.5)$$

The lifetime, as mean life τ or half life $T_{\frac{1}{2}}$, ranges from a fraction of a second to 10^6 a and is determined by conventional timing techniques, using pulsed accelerators to produce the actual nuclei in the case of very short-lived bodies. This extensive range of lifetimes may be roughly ordered by calculating a quantity known as the *comparative half life* $ft_{\frac{1}{2}}$ often called the *ft* value.

The numerical factor *f* is determined mainly by the energy release in the decay since this defines the accessible region of phase space for the emitted particle, but calculable Coulomb correction factors are essential. One ontains

$$ft = f(Z, E_0)\frac{T_{\frac{1}{2}}}{2} = \frac{2\pi^3 \hbar^7 \log 2}{g^2 |M_{if}|^2 m_e^5 c^4}$$

or
$$ft = \frac{\tau_0 \log 2}{|M_{if}|^2} \qquad \qquad \dots (3.6)$$

The measured *ft* values provide an inverse measure of the trasition matrix element. It is expressed in seconds and covers a wide range of values, viz 10^3 to 10^{23} sec, most of which cluster around $10^4 - 10^9$ sec. It is often more connenient to use $\log_{10}$ ft values. Since *ft* is inversely proportional to the square of $|M_{if}|^2$, as such for all allowed transitions $|M_{if}|$ being almost same and successively smaller for forbidden transitions of increasing order.

If one calculates the values of $\log_{10}$ ft for different β-emitters from their known values of E_0 (end point energy) and $T_{\frac{1}{2}}$ (half lives), the values so obtained are expected to fall into groups in accordance with allowed, first, second, etc. forbidden transitions Figure 3.26 shows such groups for odd mass nuclei. We can see that the different groups of values of $\log_{10}$ ft help one to categorise the β-emitters. As the ft value give its order of for biddenness and therefore it must also correspond to ΔI values (selection rules) which govern β-decay.

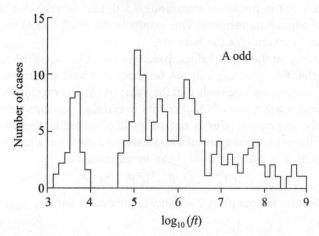

Fig. 3.26 Distribution of log ft values for β-transitions between ground states of odd mass (A-odd) nuclei

The smallest values of ft are found for a group of light nuclei particularty mirror nuclei for which $\log_{10}$ ft = 2.7 to 3.7. In these nuclei the ground state wave functions of the parent (initial) and daughter nuclei (final) are alike. There is a high probability for occurring these decays. They are called as *allowed* and *favoured* or *super allowed* transitions. Few examples of such transitions are as follows:

Decay	Half life $T_{\frac{1}{2}}$	log ft
$^{23}\text{Mg} \rightarrow {}^{23}\text{Na} + \beta^+ + \nu$	11.6 sec	3.5
$^1_0\text{n} \rightarrow {}^1_1\text{p} + \beta^- + \bar{\nu}$	750 sec	3
$^{11}\text{C} \rightarrow {}^{11}\text{B} + \beta^+ + \nu$	1230 sec	3.6
$^3_1\text{H} \rightarrow {}^3_2\text{He} + \beta^- + \bar{\nu}$	319×10^8 sec	3

The above examples reveal that though the half life of decays have a wide range but log ft values are almost constant.

One can group the next range of log ft value between 4 to 6. However, the wave functions ψ_i and ψ_f may not be strictly indentical, such that $|M_{if}| < 1$. These transitions are allowed but not favoured like super allowed transitions. Few examples of such types of transitions are:

Decay	Half life $\left(T_{\frac{1}{2}} \text{ sec}\right)$	log ft
$^{35}\text{S} \rightarrow {}^{35}\text{Cl} + \beta^- + \bar{\nu}$	7.5×10^6	5
$^{24}\text{Na} \rightarrow {}^{24}\text{Mg} + \beta^+ + \bar{\nu}$	5.4×10^4	6

Log ft values between 6 to 10 correspond to first forbidden transitions, whereas values between 10 to 14 correspond to the second forbidden transitions. Obviously, the higher order transitions will have higher values for log ft. In case of forbidden transitions, the large values of log ft values for β-decays is for the fact that in forbidden transitions, $l > 0$, i.e., the emission of β-particles have to face an barrier due to angular momentum. This results in the small size of nuclear matrix element $|M_{if}|$ and therefore an in increase in the ft-values.

Although, depending on the log ft values, β-decays could be classified in various groups, but Figure 3.26 reveals that this grouping can not be sharp and well separated. In the case of pure Fermi and pure G. T. transitions, one finds that the value of $|M_{if}|$ is maximum. However when one considers ft values such a behaviour of $|M_{if}|$ occurs in certain cases categorised as *super-allowed transitions*. Obviously, in gerneral, this is not so, as is observed by the spread of ft values for allowed transitions. However, when allowed transitions are mixed, i.e. where Fermi and $G - T$ both transitions are fovoured, it is found that $|M_{if}|$ can be expressed as

$$|M_{if}|^2 = |C|_F^2 + |M_{if}|_F^2 + |C|_{GT}^2 |M_{if}|_{GT}^2 \qquad \ldots (3.7)$$

Where C_F and C_{GT} denotes the coupling constants expressed in units of g, and bears a relation,

$$|C|_F^2 + |C|_{GT}^2 = 1 \qquad \ldots (3.8)$$

and $|M_{if}|_F$ and $|M_{if}|_{GT}$ are matrix elements for Fermi and G-T interactions respectively. Now, one can express (3.6) as

$$ft = \frac{K}{(1-x)|M|_F^2 + x|M|_{GT}^2} \qquad \ldots (3.9)$$

Where K is constant given by

$$K = \log 2 \cdot \tau_0 \ \frac{2\pi^3 \ \hbar^3 \ \log 2}{g^2 c^4 \ m_e^5} \qquad \ldots (3.10)$$

and $x |C|_{GT}^2$. Experimentally it is found that K is 2787 ± 79 sec and $x = 0.56$. Knowing K, one obtains the coupling constant for Fermi interaction, $g_F = 1.435 \times 10^{-62}$ J – m^3.

3.9.4.8 Energy and Half life Relationships: Sargent curves

B. W. Sargent in 1933 reported that when log λ is plotted against log E_{max} for the naturally occurring β-emitters, most of the points fell on or near two straight lines (Figure 3.27). These lines are called *Sargent curves or lines* and represent empirical rules, analogous to Geiger-Nuttall law in α-decay. These lines or curves can be represented by an expression of the form,

$$\lambda = K E_0^5 \qquad \ldots (3.1)$$

Where K is a constant and differs by a factor of 100 for the two Sargent curves for the same maximum energy E_0. Obviously, the decay constant or the half lives for the naturally occurring radioactive elements also differ by a factor of about 100. Gamow explained this difference in behaviour of decay constants in terms of transitions. Accordingly, the short lived or large l (upper curve) corresponds to the allowed transitions and the small l or longlived (lower curve) radioactive elements correspond to forbidden transitions. The transitions are now referred as first, second, third, ... or *n*-forbidden. However on the availability of large number of artificially produced β-emitter radioactive elements such a sharp distinction is not possible and the points scatter around such almost parallel lines.

One can establish Sargent's fifth power law using Fermi's theory of β-decay. The total transition probability or the decay constant for allowed transition is given by

$$\lambda = \text{cosntant} \times f(E_0) \qquad ... (3.2)$$

Where we have neglected Coulomb effects. The Fermi integral in (3.2) is

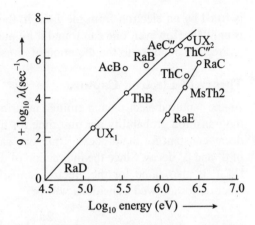

Fig. 3.27 Sargent curves for the naturally occurring β-emitters

$$f(E_0) = \frac{1}{m_e^5 c^7} \int_0^{p_0} (E_0 - E_e)^2 \, p_e^2 \, dp_e \qquad ... (3.3)$$

One can evaluate the Fermi integral as $E_0 >> m_0 c^2$, $p_0 >> m_e c$ for all radio-active nuclides. One then obtains (3.1). Obviously, Sargent's observations agree with theory particularly in the case of low Z-nuclei, i.e. light nuclei. However, for heavy nuclei, Coulomb effects causes certain distortions, as such the theory is more complicated.

3.9.4.9 Orbital Electron Capture Process

We have read that the electron capture type of β-decay takes place when the positron emission is not energetically possible, i.e. there is a competition between positron emission and capture of orbital electrons (*K* or *L* electron, etc.) by the nucleus. Moreover, during the transformation of a proton into a neutron as a result of weak interaction, inside a β-active nucleus (high-Z-values) if the difference between the atomic mass between the parent nucleus and daughter nucleus is less than $2\,m_e c^2$, capture of an orbital electron is fovoured. In this case the rest mass energy of the captured electron is then added to the released energy. The probability of *K*-capture depends on the nature of the electron-cloud around the nucleus. Since the energy of the electron (*K*- or *L*) is fixed, the released energy must be carried, by the neutrino, and thus all the neutrinos come out with the same energy. The capture of orbital electron by nucleus can be expressed as

$$p^+ + e^- \rightarrow {}_0 n^1 + \nu$$

An example of *K*-capture is ^{4}Be nucleus,

$$^4\text{Be} + e_K^- \rightarrow {}^7\text{Li} + \nu$$

K-Capture effect is expected to be appreciable for light elements.

Once an orbital electron is captured, a vacancy is caused in the corresponding orbit. This vacancy is filled by an electron from the next higher orbit and characterstic *X*-ray photon is emitted. The emission of the characterstic *X*-ray or of Auger electrons following *K*-capture merely indicates that such capture has actually taken place, e.g. if *K*-electron is captured and the vacancy is filled by an electron from the *L*-orbit, then characterstic $K - X$-rays from the daughter nucleus is emitted. However, one can found it by measuring the energy of the product nucleus recoiling in a direction opposite to the direction of emission of the neutrino.

Theory of Electron Capture

As the captured electron and emitted neutrino both have definite energies in β-decay and therefore the transition probability per unit time for the electron capture process would be just equal to the decay constant for the process. Thus, one can use the same expression as, used for the probability of β⁻ and β⁺ decay. Since the interaction of the field, i.e. weak interaction being same here and thus $|M_{if}|^2$ is also same for the electron emission. The decay constant for the *K*-electron capture electron for allowed transitions will be

$$\lambda = \frac{2\pi}{\hbar} \left| \int \psi_i^* H \psi_J \, d\tau \right|^2 \frac{dN}{dE_\nu} \qquad \text{... (3.1)}$$

Where $\dfrac{dN}{dE_\nu}$ refers to the density of neutrino states only. One can express the neutrino field as a plane wave.

$$\psi_\nu = \frac{e^{i\mathbf{K}\cdot\mathbf{r}}}{\sqrt{V}} = \frac{1}{\sqrt{V}} \, e^{\frac{1}{\hbar}(\mathbf{p}\cdot\mathbf{r})}$$

$$= \frac{1}{\sqrt{V}} \left\{ 1 + \frac{1}{\hbar} \mathbf{p}\cdot\mathbf{r} + ... \right\}$$

Thus, at $r = 0$, we have

$$|\psi_\nu(0)|^2 = \frac{1}{\sqrt{V}} \qquad \text{... (3.2)}$$

Now, the electron wave function should characterize the bound state of the initial system. One may approximately use for the *K*-electron the 15 wave function of a hydrogenic system, with the charge number of the parent nuclide. One finds,

$$\psi_{e,\,K} = \psi_s = \frac{1}{\sqrt{\pi}} \left(\frac{Z}{a_B} \right)^{\frac{3}{2}} \exp\left(-\frac{Zr}{a_B} \right) \qquad \text{... (3.3)}$$

Where $a_B \left(= \dfrac{4\pi \varepsilon_0 \hbar^2}{m_e c^2} \right)$ is the Bohr radius. We know that the nuclear wave functions are appreciable over only at a small distance, i.e. smaller than $\left(\dfrac{a_B}{Z} \right)$, hence one may expand the neutrino and electron wave functions about $r = 0$ and retain only the constant terms. Approximately, for $r = 0$

$$\psi_e(0) = \psi_K(0) = \frac{1}{\sqrt{\pi}} \left(\frac{Z}{a_B} \right)^{\frac{3}{2}}$$

or $\qquad \psi_e(0) = \frac{1}{\sqrt{\pi}} \left\{ \frac{Z\, m_e\, e^2}{4\pi\, \varepsilon_0\, \hbar^2} \right\}^{\frac{3}{2}} \qquad \qquad \ldots (3.4)$

The statistical factor or the density of states $\frac{dN}{dE_\nu} = \rho_\nu$ for the neutrino is to be given by the number of neutrino states in the momentum range p_ν and $p_\nu + dp_\nu$. Therefore, one finds

$$\rho_\nu = \frac{dn_\nu}{dE_\nu} = \frac{4\pi V}{(2\pi \hbar)^3} \frac{p_\nu^2\, dp_\nu}{dE_\nu} \qquad \qquad \ldots (3.5)$$

The energy E_ν of the neutrino and its momentum p_ν bears a relation $E_\nu = c p_\nu$. We have

$$dE_\nu = cd p_\nu \qquad \qquad \ldots (3.6)$$

Moreover, the neutrino energy in terms of binding energy B_K of the K-electron and the maximum available energy E_0 is given by, (where E_0 is the difference of nuclear masses of parent and daughter nucleus)

$$E_\nu = E_0 + m_e c^2 - B_K \qquad \qquad \ldots (3.7)$$

Using (3.6) in (3.5), one obtains

$$\rho_\nu(E) = \frac{4\pi V E_\nu^2}{(2\pi \hbar)^3 c^2} \frac{1}{c} \qquad \qquad \ldots (3.8)$$

Now, making use of Equations (3.2), (3.4) and (3.8) in (3.1), one obtains for decay constant λ_K for the K-electron capture as

$$\lambda_K = \frac{2\pi}{\hbar} g^2 |\psi_e(0)|^2 |\psi_\nu(0)|^2 |M_{if}| \frac{dN}{dE_\nu}$$

$$= \frac{2\pi}{\hbar} g^2 |M_{if}|^2 \frac{1}{V} \frac{1}{\pi} \left\{ \frac{Z\, m_e\, c^2}{4\pi\, \varepsilon_0\, \hbar^2} \right\}^3 \frac{4\pi V E_\nu^2}{2\pi \hbar^3\, c^3}$$

or $\qquad \lambda_K = \frac{2\pi}{\hbar} g^2 |M_{if}|^2 \frac{4 E_\nu^2}{(2\pi \hbar^3) c^3} \left\{ \frac{Z\, m_e\, e^2}{4\pi\, \varepsilon_0\, \hbar^2} \right\} \qquad \ldots (3.9)$

Using (3.7), one obtains

$$\lambda_K = \frac{g^2\, m_e^3\, e^6\, Z^3 |M_{if}|^2}{64\, \pi^2\, \varepsilon_0^3\, \hbar^{10}\, c^3} (E_0 - m_e c^2 - B_K)^2 \qquad \ldots (3.10)$$

We may note that to take into account for the presence of two electrons in K-shell, one may multiply λ_K in (3.10) by a factor 2.

Now, putting $W_0 = \dfrac{E_0}{m_e c^2}$ and $W_B = \dfrac{B_K}{m_e c^2}$, one obtains

$$\lambda_K = \frac{g^2 \, m_e^5 \, c^4}{\pi^2 \, \hbar^7} \left\{ \frac{Ze^2}{4\pi \, \varepsilon_0 \, \hbar c} \right\}^3 |M_{if}|^2 \, (W_0 + 1 - W_B) \qquad \ldots (3.10)$$

Now, putting

$$f_K = 2\pi (\alpha Z)^3 \, (W_0 + 1 - W_B)^2 \qquad \ldots (3.11)$$

Where $\alpha = \dfrac{e^2}{4\pi \, \varepsilon_0 \, \hbar c} = \dfrac{1}{137}$ is *Sommerfeld's fine structure constant*. One obtains,

$$\lambda_K = \frac{g^2 \, m_e^2 \, c^4}{2\pi^3 \, \hbar^7} \, |M_{if}|^2 \, f_K \qquad \ldots (3.12)$$

Writing $\tau_0 = \dfrac{2\pi^3 \, \hbar^3}{g^2 \, m_e^5 \, c^4}$ in (3.12), where τ_0 is a universal time constant, one obtains

$$\lambda_K = \frac{|M_{if}|^2 \, f_K}{\tau_0} \qquad \ldots (3.13)$$

If t_k be the half life for K-electron capture, then

$$\lambda_K = \frac{\log 2}{t_K} = \frac{|M_{if}|^2 \, f_K}{\tau_0}$$

or

$$f_K \, t_K = \frac{\tau_0 \log 2}{|M_{if}|^2} \qquad \ldots (3.14)$$

The expression (3.10) is corrected for the screening of the nuclear charge by the atomic nucleus. The effect of the screening is to reduce the value of Z. For K-electrons Z has to be replaced by $Z - 0.35$ and for L-electrons by $Z - 4.15$.

There is competition between β^+ emission and K-capture in certain β-active nuclei. Therefore, one can calculate the branching ratio $\dfrac{\lambda_K}{\lambda_{\beta^+}}$ independent of $|M_{if}|^2$, since it is same for the two competing processes. From Equation (3.10) and

$$\lambda = \frac{g^2 \, |M_{if}|^2 \, m_e^5 \, c4}{2\pi^3 \, \hbar^7} \, f(Z, E_0) \qquad \ldots (3.15)$$

One obtains

$$\frac{\lambda_K}{\lambda_{\beta^+}} = \frac{(2\pi \, \alpha \, Z)^3 \, (W_0 + 1 - W_\beta)^2}{f(Z, E_0)} \qquad \ldots (3.15a)$$

Equation (3.15) reveals that λ_K increases rapidly with increasing Z (as Z^3) and for high Z, the positron has to penetrate a Coulomb barrier and so decreases with increasing Z. The Coulomb effect is there in the Fermi integral $f(Z, E_0)$ in Equation (3.15). Obviously, electron capture becomes more probable in high Z-elements. For low Z-value, i.e. lighter elements, β^+-emission predominates. β^+-emission in light nuclei is favourable for high-end point energies. In a number of β-active nuclei, the experimental observation of branching ratio is in good agreement with the theoretical results.

We have the binding energy of K-electron,

$$B_K = \frac{1}{2}\frac{m_e\, e^4\, Z^2}{(4\pi\,\varepsilon_0\,\hbar)^2} \quad \text{and} \quad W_B = \frac{B_K}{m_e\, e^2} = \frac{1}{2}\frac{m_e\, e^4\, Z^2}{m_e\, e^2\,(4\pi\,\varepsilon_0\,\hbar)^2}$$

$$\therefore \qquad W_B = \frac{1}{2}\frac{e^4\, Z^2}{(4\pi\,\varepsilon_0\,\hbar c)^2} = \frac{1}{2}\alpha\, Z^2 \qquad \qquad \text{... (3.16)}$$

Where α is fine structure constant $\left(= \dfrac{e^2}{4\pi\,\varepsilon_0\,\hbar c} = \dfrac{1}{137}\right)$. Using Equation (3.16) in Equation (3.11), one obtains

$$f_K = 2\pi(\alpha Z)^3 \left[W_0 + 1 - \frac{1}{2}(\alpha Z)^3 \right] \qquad \qquad \text{... (3.17)}$$

The *ft* values agree fairly well with those of β^+-transitions, *ft* determines the forbiddenness of a transition. Obviously, one can easily see whether in a given K-capture process, whether K-capture transition is allowed trasition or a forbidden transition. Let us consider an example of decay of ^{7}Be by K-capture,

$$^7_4\text{Be} + e^-_K \rightarrow\ ^7_3\text{Li} + \nu$$

The maximum energy available for the transition $E_0 = [M(^7_4\text{Be}) - M(^7_3\text{Li})]c^2 = 0.35$ MeV. This gives

$$W_0 = \frac{0.35}{0.51} = 0.68, \ m_e c^2 = 0.51 \text{ MeV}$$

For the process half life $t_K = 53.4$ days. In Be $(Z = 4)$, the binding energy of K electron is about 212 eV. Obviously, this is negligible compared to E_0. Thus, one obtains

$$f_K = 4\pi\,(\alpha Z)^3\,(W_0 + 1) \qquad \qquad \text{... (3.18)}$$

$$= 4\pi\left(\frac{4}{137}\right)^3 (1.68)2 = 8.6 \times 10^{-4}$$

and $\qquad\qquad t_K = 53.4 \text{ days} = 4.6 \times 10^6 \text{ s}$

Thus $\qquad\qquad f_K\, t_K = 8.6 \times 10^{-4} \times 4.6 \times 10^6 = 3956$

and $\qquad\qquad \log f_K\, t_K = 3.596 \approx 3.6$

Clearly, the Be decay through K-electron capture is a *super allowed transition*. Since it is a case of $\left(\dfrac{3}{2}\right)^-$ to $\left(\dfrac{3}{2}\right)^-$ transition, it is a mixed Fermi and GT super allowed transition. Moreover, this shows that the strength of the interaction must be approximately the same in both the processes. The various degrees of forbiddness also apply to EC process.

Neutrino can not be detected. Thus, the measurment of daughter nucleus recoiling in a direction opposite to the direction of emitted neutrino, gives an idea of change of energy involved in the process. This is achieved by applying retarding potential to stop the recoiled nucleus.

One can apply the same theory to calculate, if instead of K-capture, an electron from L-orbit is captured by a radioactive, β-active nucleus. The probability of L-electron capture is smaller than for K-electron, as L-electron is farther from the nucleus as compared to L-electron. However, the probability for the electrons from the successive higher orbits will decrease.

3.9.4.10 The Beta-Decay Interaction ($V - A$ interaction)

The β-decay interaction is the least familiar of four interactions (nuclear, electromagnetic, β-decay, gravitational) that govern the operation of every thing in the universe. All the particles involved in β-decay are fermions $\left(\text{spin} \dfrac{1}{2} \hbar \right)$ and the interaction giving rise to β-decay takes place between a nucleon and combined field of electron-antineutrino or positron-neutrino. Fermi assumed that the interaction was like the electromagnetic interaction and was described by Lorent Z-invariant scalar matrix element formed by the dot product of two vectors, i.e. $\mathbf{V} \cdot \mathbf{V}$ interaction. This interaction is not complete, however, as it does not account for GT (Gamow-Teller) transitions and in fact there are five possible combinations that give a Lorent Z-invariant scalar matrix: $\mathbf{S} \cdot \mathbf{S}, \mathbf{P} \cdot \mathbf{P}, \mathbf{V} \cdot \mathbf{V}, \mathbf{A} \cdot \mathbf{A}$, and $\mathbf{T} \cdot \mathbf{T}$, where S, P, V, A and T stand for *scalar, pseudo scalar, vector axial vector* and *tensor operators*, respectively. These five covariants are:

			parity change
Scalar	S	$\Delta I = 0$	No
Vector (Polar)	V	$\Delta I = 0$	No
Tensor (anti-symmetri)	T	$\Delta I = 0, \pm 1$	No
Axial Vector	A	$\Delta I = 0, \pm 1$	No
Pseudo scalar	P	$\Delta I = 0$	Yes

The $\mathbf{A} \cdot \mathbf{A}$ and $\mathbf{T} \cdot \mathbf{T}$ combinations would give rise to GT transitions. Each of these interactions has a definite form of the perturbation operator H_i' which conserves parity and angular momentum in weak interaction. In general, one can write $H' = \sum C_i \, H_i'$, where i takes on the five values corresponding to the five possible forms of the interaction.

The discovery of parity violation showed that the matrix element $|M_{if}|$ must be a mixture of scalar and pseudo scalar terms so it would not be invariant under the parity operation. Feynmann and coworkers postulated that the interaction was an equal mixture of V and A operators. They considered the following wave equation for a spin $\dfrac{1}{2}$ massive particle.

$$\frac{1}{2} \left(\frac{i\hbar}{c} \frac{\partial}{\partial t} - \boldsymbol{\sigma} \cdot \mathbf{p} \right) \left(\frac{i\hbar}{c} \frac{\partial}{\partial t} + \boldsymbol{\sigma} \cdot \mathbf{p} \right) \phi = m^2 c^2 \phi \qquad \qquad ... (3.1)$$

Where ϕ is a two component wave function and $\boldsymbol{\sigma}$ is the Pauli spin operator. One can deduce it from the energy-momentum relation

$$\frac{E^2}{c^2} - p^2 = m^2 c^2 \qquad \dots (3.2)$$

substituting the operator quantity $(\boldsymbol{\sigma} \cdot \mathbf{p})^2 = \mathbf{p}^2$. For a massless particle, then either $\dfrac{E \phi_R}{c} = \boldsymbol{\sigma} \cdot \mathbf{p} \, \phi_R$

or $\dfrac{E \phi_L}{c} = \boldsymbol{\sigma} \cdot \mathbf{p} \, \phi_L$. These equations were first proposed by Weyl in 1929, i.e. there are just two solutions corresponding to left (L) and right (R) handed *helicity* and these describe a massless neutrino (ϕ_L) and anti-neutrino (ϕ_R). The wave functions ϕ_L and ϕ_R are not eigenstates of parity as under parity $\phi_L \to \phi_R$ and $\phi_R \to \phi_L$; however, they are eigenstates under the combined operation of charge conjugation and parity (CP).

If the mass is not zero, then Equation (3.1) is equivalent to following two equations:

$$\left.\begin{aligned}
\left(\frac{i\hbar}{c} \frac{\partial}{\partial t} + \boldsymbol{\sigma} \cdot \mathbf{p}\right)\phi_L &= m_c \phi_R \\[2mm]
\left(\frac{i\hbar}{c} \frac{\partial}{\partial t} - \boldsymbol{\sigma} \cdot \mathbf{p}\right)\phi_R &= m_c \phi_L
\end{aligned}\right\} \qquad \dots (3.3)$$

Now, in β-decay only left handed neutrinos are involved, i.e. ϕ_L. Feynmann and Gell-Mann assumed that the electron must also be described by only ϕ_L, with ϕ_L satisfying Equation (3.3). With this assumption the only combination of S, P, V, A and T operators which give a non-zero matrix element is $(\mathbf{V} - \mathbf{A}) \cdot (\mathbf{V} + \mathbf{A})$ (If the neutrino had right-handed helicity then it would be $(\mathbf{V} + \mathbf{A}) \cdot (\mathbf{V} + \mathbf{A})$. The angular distributions of the leptons emitted in a β-decay predicted by this interaction are shown in Figure 3.28. Also shown are those for a scalar $(\mathbf{S} \cdot \mathbf{S})$ and tensor $(\mathbf{T} \cdot \mathbf{T})$ interaction, which would also give rise to Fermi and GT transitions respectively.

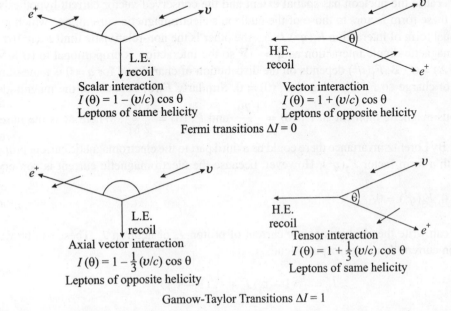

Fig. 3.28 Angular distribution, $I(\theta)$, in β-decay predicted by various types of interactions.

Experimentally the electron-neutrino correlation can be inferred from measuring both the electron momentum and the nuclear recoil e.g. in the decay ^{6}He $(O^+) \rightarrow {}^6$Li $(1^+) + e^- + \bar{\nu}$. The pseudo-scalar interaction P is insignificant in nuclear β-decay as the nucleon velocities are much less than c. The theory also predicts the helicities of the emitted leptons: $e^- = -\dfrac{v}{c}$, $e^+ = +\dfrac{v}{c}$, $\bar{\nu} = +1$ and $\nu = -1$.

The electron helicity can be measured by observing the circular polarisation of γ-rays produced when the electrons are stopped in an absorber. The sign of the helicity of the forward γ-rays is the same as that of the electrons and can be determined by measuring the γ-ray absorption in magnetised iron as in the experiment on helicity of the neutrino.

Incidentanlly, the observed asymmetry in the electron distribution from the decay of polarised ^{60}Co nuclei can be seen as a direct consequence of the negative helicity of the electrons and the leptons' spins coupling to spin 1 in a GT transition. The transition is from a $I = 5$ state in ^{60}Co to a $I = 4$ state in ^{60}Ni, so to conserve angular momentum they must tend to be emitted in the opposite direction to the initial orientation of the $J = 5$ ^{60}Co nuclei.

The conserved-vector current (CVC) Hypothesis

For a purely leptonic process, such as:

$$\mu^- \rightarrow e^- + \bar{\nu}_e + \nu_\mu$$

the weak interaction matrix element is $G_F (\mathbf{V} - \mathbf{A}) \cdot (\mathbf{V} - \mathbf{A})$, where G_F is called the Fermi coupling constant. However, nuclear β-decay is a semi leptonic weak interaction and the nucleon weak current is proportional to $(\mathbf{V} - \lambda\mathbf{A})$, where λ is the ratio of GT (axial) to Fermi (vector) coupling constants, $\lambda = 1.24$, and is the ratio of the axial to vector form factors for q^2 small. There are form factors because the nucleon has spatial extent and the conserved-vector current hypothesis relates some of these form factors to those of the nucleon's electromagnetic current: one which gives rise to the usual form of interaction $F_1(q^2) J_\mu A_\mu$, the other is the non-relativistic limit $F_2(q^2)(\boldsymbol{\sigma} \cdot \mathbf{q}) \cdot \mathbf{A}$. This is magnetic dipole interaction as $\mathbf{q} = -i\nabla$ so the interaction is proportional to $(\boldsymbol{\sigma} \times \nabla) \cdot \mathbf{A} = \boldsymbol{\sigma} \cdot (\nabla \times \mathbf{A}) = \boldsymbol{\sigma} \cdot \mathbf{B}$. $F_1(q^2)$ depends on the distribution of charge and for $q = 0$ is a measure of the amount of charge so $F_1^P(0) = 1$ and $F_1^n(0) = 0$. Similarly $F_2(0)$ is equal to the magnitude of the anamalous magnetic moment so $F_2^P(0) = \dfrac{1.79}{2\,M}$ and $F_2^n(0) = \dfrac{1.91}{2\,M}$, Where M is the mass of the nucleon. By Lorentz invariance there could be a third part to the electromagnetic current proportional to q_μ with a form factor $F_3(q^2)$. However, because the electromagnetic current is conserved, i.e.

$$\dfrac{\partial I_\mu}{\partial x_\mu} = 0, \; F_3(q^2) = 0.$$

one can write the electromagnetic current of proton J_p or neutron J_n. These can be written as a nucleon current J_N using isospin. Sicne:

$$J_N = \frac{1}{2}(1 - \tau_3)J_p + \frac{1}{2}(1 + \tau_3)J_n$$

$$= \frac{1}{2}(J_p + J_n) + \frac{1}{2}\tau_3(J_n - J_p)$$

$$= J_N^S + \tau_3 J_N^V \qquad .$$

Where $\dfrac{\tau_3}{2}$ is the nucleon's third component of isospin. There are therefore isoscalar J_N^S and isovector J_N^V contributions to the electromagnetic current. Likewise the weak nucleon current can be expressed as

$$J_W = \tau^{\pm}(V_N - \lambda A_N)$$

Where $\tau^{\pm}$ cause a change in τ_3 of $+1$ and -1 and so correspond to $p \rightarrow n$ or $n \rightarrow p$, respectively. This is an isovector current and the term giving rise to Fermi transitions is $\tau^{\pm}\,V_N$, which is isovector weak vector current. By Lorentz invariance there could be three parts to this current with associated form factors $f_1(q^2)$, $f_2(q^2)$ and $f_3(q^2)$, just like there could be to the electromagnetic vector current. In allowed Fermi transitions only the first part of this current contributes and the matrix element is multiplied by the form factor $f_1(q^2)$, i.e. the Fermi constant G_β includes $f_1(q^2)$ while the muon weak coupling constant G_F does not as the muon behaves as a point-like particle.

The constants G_β and G_F are very similar in magnitude and this similarity led Feynman and Gell-Mann to propose that the isovector weak vector current and the isovector electromagnetic vector current are members of an isotriplet of currents all of which are conserved. This idea is called the conserved-vector current (CVC) hypothesis and it means that

$$f_1(q^2) = f_{V_1}(q^2),\, f_2(q^2) = F_{V_2}(q^2) \text{ and } f_3(q^2) = 0,$$

where:

$$F_V(q^2) = \frac{[F^n(q^2) - F^p(q^2)]}{2} \text{ as } J_N^V = \frac{(J_n - J_p)}{2}$$

The prediction that $f_1(q^2) = F_{V_1}(q^2)$ would also predict that $G_\beta = G_F$ as $F_V(q^2)$ is essentially 1 at the q^2 found in nuclear β-decay. The lack of exact quality between G_β and G_F is not accounted for by radiative corrections and is now understood in terms of Cabbibo's hypothesis that the weak interaction is shared between $\Delta S = 0$ and $\Delta S = 1$ (S = strangeness) transitions such that the coupling strength in nuclear β-decay ($\Delta S = 0$) is $G_F \cos \theta_C$ and in $\Delta S = 1$ transitions is $G_F \sin \theta_C$. This means that $G_\beta = G_F \cos \theta_C$ and the experimental value for $\cos \theta_C = 0.975$.

A striking test of the CVC hypothesis is given in a comparison of the β-decays: $^{12}_{5}\text{B}(1^+)$ $\xrightarrow{\beta^-}$ $^{12}_{6}\text{C}(0^+)$ and $^{12}_{7}\text{N}(1^+)$ $\xrightarrow{\beta^-}$ $^{12}_{6}\text{C}(0^+)$, which are allowed G-T transitions. The dominant nuclear term is $\lambda(\sigma)$ but the term involving $f_2(q^2)$ in the vector current can also contribute and its effect is called *weak magnetism* as the analogous term in the electromagnetic current gives rise to a magnetic dipole interaction. In the non-relativestic limit the weak magnetic interaction is of the form $(\sigma \times \mathbf{q}) \cdot \mathbf{L}$, where $\mathbf{L}$ is the lepton current so its contribution to the nuclear part of the matrix element is $f_2(q^2) <\sigma> \cdot \mathbf{q}$. The CVC hypothesis is that $f_2(q^2) = F_{V2}(q^2)$, which for nuclear β-decay means $f_2(q^2) = F_{V2}(0) = \dfrac{[F_2^p(0) - F_2^n(0)]}{2} = \dfrac{3.70}{2\,M}$. Allowing for this term alters the shape of the spectrum of emitted positrons and electrons in different ways and the agreement with experiment is good.

3.9.5 Nuetrino(ν) and Anti neutrino ($\overline{\nu}$)

The neutrino has no electric charge. Its magnetic moment and rest mass probably equal zero. As it interacts very weakly with the substance this particle is extremely difficult to observe. Therefore, originally, the existence of the neutrino was postulated only on the basis of indirect data. The neutrino produces in a number of decay processes under the action of weak forces. Among such processes the β-decay of atomic nuclei is the best known. This decay is a spontaneous transformation of a neutron into a proton or of a proton into a neutron into a proton or of a proton into a neutron in the interior of the atomic nucleus:

$$n \to p + e^- + \overline{\nu} \qquad\qquad \text{... (3.1)}$$

$$p \to n + e^+ + \nu \qquad\qquad \text{... (3.2)}$$

Such transformations are possible in those nuclei the charge of the $p - n$ composition of which due to decays, (3.1) or (3.2), is profitable from the stand point of energy, i.e. results in the formation of nuclei of lower mass. As the sum of the masses of the proton and electron is somewhat lower than the mass of the neutron, process (3.1), as opposed to (3.2), proceeds simultaneously also in the case of a free neutron not bound in the nucleus.

It was found that there are two kinds of neutral particles, almost identical, associated with β-decay. One the neutrino (ν) in emitted in β⁺-decay and electron capture (inverse β-decay) $[A_Z^X + e^- \to {}_{Z-1}^{A}Y + \nu;\ p + e^- \to n + \nu]$, while the particle emitted in β⁻-decay (Equation 3.1) is called antineutrino ($\overline{\nu}$). The emission of $\overline{\nu}$ along with the emission of an electron is in accordance with the law of conservation of leptons. From the 'reactor associated' counting rate of 36 ± 4 events/hour the cross-section for the inverse β-decay was found to be $(11 \pm 2.6) \times 10^{-48}$ m².

Neutrinos were generated in copious quantities during the formation of the universe and are thought today to be, by far, the most abudant of all of the known particles of physics. The earth is totally transparent for them. $\overline{\nu}_e$ and ν_e are antiparticle and particle respectively in the same way that $(e^+\ e^-)$ form an antiparticle-particle pair.

The properties required of the neutrino to interpret the experimental evidence so far discussed are that it shall have

(i) Zero electric charge

(ii) Zero or nearly zero rest mass

(iii) Half integral spin angular momentum (being fermion)

(iv) Extremely small interaction with matter, because of the failure of experiments to show even the feeblest ionisation caused by neutrinos passing through absorbers or track chambers. This means it has zero magnetic moment.

(v) Stable, i.e. it has infinite mean life

(vi) Parity non-conserving behaviour in β-decay, a general property of the fundamental weak interaction demands that this particle must have a definite *helicity*. This is a correlation between the spin direction of a particle with its linear momentum which makes it move like a screw. This implies that the intrinsic spin angular momentum of a particle is parallel or anti parallel to its direction of motion.

Fifty years after its suggestion the neutrino has become a familiar particle, with its own reactor and accelerator technology for production and massive track chambers, counters or radio chamical assemblies for detection.

Stars are neutrino sources. The solar neutro flux to the earth is determined by the sun luminescence that has a value $10^{14} - 10^{15}$ $1/(m^2 s)$. Cosmic antineutrinos are of some other origin, and if antistars exist, antineutrinos are emitted by them in thermo nuclear synthesis like the neutrinos of ordinary stars.

Neutrinos produce not only in β-decay but in some other decay processes as well. Neutrinos emitted in the π-meson decay turned out to be not identical to β-decay neutrinos. Neutrinos forming μ-mesons in the interaction with nucleons were called μ-meson neutrinos to differentiate them from electron neutrinos forming in the β-decay. It follows from the experiment that electron and μ-meson neutrinos differ in the character of their interaction with other elementary particles, while the other properties of all the neutrinos are similar. After this discovery, however, the lepton charges of electrons and electron neutrinos should be considered differing from the lepton charges of μ-mesons and μ-meson neutrinos to the same extent as the lepton charges differ from the barion charges.

Majorana theory for charge less elementary particles favours ν and ν̄ to be same, but there are definite evidences that ν and ν̄ are two different particles. The phenomenon of $β^+$-disintegration itself, is a most convincing proof for the existence of ν and ν̄.

3.9.6 Double β-decay

Double β-decay is the process by which two electrons or two positrons are emitted,

$$A(Z, N) \rightarrow A(Z + 2, N - 2) + 2e^- + 2\bar{\nu}_e$$

$$A(Z, N) \rightarrow A(Z - 2, N + 2) + 2e^+ + 2\nu_e$$

Such processes are caused by second-order perturbations induced by weak interactions and are therefore much slower than β-decays in which only a single charged lepton is emitted. As a result, double β-decays are expected to be long-lived, with typical half-lives of the order of 10^{20} years. Processes with such long half-lives may be observed only in nuclei where ordinary β-decay and other faster reactions are forbidden by Q-value considerations. In spite of this limitation, a number of such cases are known as we can see by comparing the binding energies of neighbouring nuclei, e.g. $^{82}_{34}Se$ is stable against $β^-$-decay to $^{82}_{35}Br$ since the Q value is -0.90 MeV. However, it is unstable against double $β^-$-decay to $^{82}_{36}Kr$ with a Q value of $+3.00$ MeV.

It is not surprising that a large number of nuclei can, in principle, undergo double β-decay. In general, these are even-even nuclei near, but not at, the bottom of the valley of stability. Because of pairing energy, they are more tightly bound compared with neighbouring odd-odd nuclei. On the otherhand, a neighbouring even-even nucleus with two more protons and two less neutrons may be even more tightly bound because of a larger symmetry energy. This term is proportional to $(N - Z)^2$. Since most nuclei in the medium to heavy range have a neutron excess, an isobars with neutrons less can often be more tightly bound as a result. For this reason, more nuclei are known to be capable of double β^--decay than the number of nuclei which can decay by emitting two positrons. Double β^+-decay is possible, for e.g., in the decay of $^{106}_{48}$Cd to $^{106}_{46}$Pd: the Q-value of 0.7 MeV is however, smaller than the typical double β^--decay values of two to three MeV.

One of the primary interests in nuclear double β-decay is the question of whether the process can take place without emitting neutrinos. If neutrinos are Majorana fermions, with no distinction between particles and antiparticles, one can imagine that the neutrino from the first β-decay in a double β-decay proces is absorbed in the intermediate state and that this absorption induces the emission of the second charged lepton. In such cases, no neutrino will emerge from the decay. On the other hand such "neutrinoless" double β-decay processes are strictly forbidden if neutrinos are Dirac particles with particles distinct from their anti particles.

The fact that neutrinos and anti neutrinos are different particles was confirmed by an experiment due to Davis in 1955 — using the reaction,

$$\text{neutrino} + {}^{37}\text{Cl} \rightarrow e^- + {}^{37}\text{Ar}$$

The source of the neutrino for this classic experiment was a reactor which produces mainly $\bar{\nu}_e$. The observed cross-section for this reaction was much smaller than one expected for Majorama neutrinos. The observation forms a proof that neutrinos are Dirac particles.

However, in a neutrinoless double β-decay the neutrinos are virtual particles and may be different from real neutrinos observed in the said experiment. If virtual neutrinos are Majorana particles, then double β-decay can take place without emitting any physical neutrinos and can therefore proceed on a much faster scale, perhaps by as much as six orders of magnitude. An important fact in support of the faster rate, among others, is that the phase space available for the final states of a neutrinoless double β-decay is, much larger than the competing two-neutrino mode.

One way to distinguish between the two types of double β-decay is the spectrum of the electrons emitted: If two neutrinos are emitted, the sum of the energies of the two electrons is equal to the Q-value of the decay (again ignoring the small amount of energy associated with nuclear recoil). On the other hand, if two neutrinos are emitted, the sum of the energies of the two electrons will have a continuous distribution given by energy momentum conservation of the five-body final state. One of the recent measurements of the double β^--decay of ^{82}Se to ^{82}Kr by Elliot and Coworkers [1987] $^{82}_{34}$Se $\rightarrow$ $^{82}_{36}$Kr $+ 2\beta^- + 2\bar{\nu}$ gives a limit of the half life of the decay to be 4.4 $\times$ 10^{20} years and an energy spectrum of the two electrons emitted consistent with the two neutrino-mode.

A similar experiment on double β-decay was in the case of decay of ^{130}Te to ^{130}Xe:

$$^{130}_{52}\text{Te} \rightarrow {}^{130}_{54}\text{Xe} + 2\beta^- + 2\bar{\nu}$$

The calculated value for half life is around 10^{21} years. If $\nu = \bar{\nu}$, then the decay scheme would be

$$^{130}_{52}\text{Te} \rightarrow {}^{130}_{54}\text{Xe} + 2\beta^-$$

and the half life is expected to be around 10^{16} years while the experimental value is about 10^{21} years.

Measuring long liftimes alone do not necessarily rule out the possibility of neutrinoless double β-decay. We have seen in the case of single β-decays that there is a large spread in the log ft values, especially for the allowed decays. Such a divergence in the rate is due primarily to the differences in the value of nuclear transition matrix element involved. The same may also be true for double β-decays. If the nuclear matrix elements in double β-decays are much smaller than expected, the life times of 10^{20} years could even be an underestimate of the two neutrino mode. Consequently, long measured half-lives by themselves do not rule out the neutrinoless mode. In this sense, the energy spectrum of the experiment of Elliot and coworkers provide more conclusive evidence against the neutrinoless mode than life time measurments.

Since we are considering very slow processes, there are also other possibilities for double β-decay in addition to two neutrino and neutrinoless modes. One is the weak decay of a Δ-particle to a nucleon with the emission of two charged leptons. The normal decay mode of Δ is to a pion plus a nucleon via strong interaction, but a weak branch involving leptons can not be ruled out, especially when the Δ-particle is a part of a nucleus. The other possibility is that instead of two neutrinos, a boson, given the name '*Majaron*' may be emitted. The detection of any such events requires measurments involving half-lives of the order of 10^{20} year, and these are not easy experiments to carry out.

3.9.7 The Detection of Neutrino

The zero charge and almost zero rest mass make the neutrino the most difficult partilce to detect experimentally. Beta decay is a three-body problem and conservation of momentum demands that after emission of β-particle and neutrino, the daughter nucleus recoil in a direction not exactly opposite to the emitted β-particle (Figure 3.29). If one can detect this recoil of the daughter, then it would form an indirect evidence of the existence of neutrino.

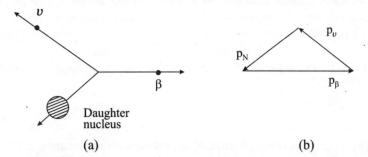

Fig. 3.29 (a) Recoil of a daughter nucleus after β-emission. (b) The momentum triangle: $p_\beta \rightarrow$ momentum of β-particle, $p_v \rightarrow$ momentum of neutrino and $p_N \rightarrow$ momentum of daughter nucleus

One can estimate the recoil energy for the daughter nucleus. If the neutrino carries all momentum is

$$(p_v)_{max} = \frac{(E_v)_{max}}{c} \qquad \qquad \dots (3.1)$$

Where $(E_v)_{max} = E_0$, the end point energy of the β-spectrum. The light nucleus ^{14}O decays the process

$$^{14}_{8}O \rightarrow ^{14}_{7}O + \beta^+ + v \qquad \qquad ...(3.2)$$

The end point energy $E_0 = (E_v)_{max} = 1.812$ MeV

$$\therefore \qquad \qquad (p_v)_{max} = \frac{1.812}{c} = p_N \text{ (Momentum Conservation principle)}$$

Where p_N is momentum of daughter nucleus. Obviously, this momentum corresponds to daughter

nucleus recoil energy, $E_N = \dfrac{p_N^2}{2M}$, where M is the mass of the daughter nucleus $^{14}_{7}$N.

$$\therefore \qquad \qquad E_N = \frac{(1.812 \text{ MeV})^2/c2}{(26087.73 \text{ MeV})^2} = 0.000125 \text{ MeV}$$

$$= 125 \text{ eV} \qquad \qquad ...(3.3)$$

We note that even for a very favourable case of a light nucleus the recoil energy is very small ≈ 125 eV. This low value of E_N makes it extremely difficult to make experimental measurments.

The recoil experiments have been performed with several β-emitters, e.g. n, ^{6}He, ^{7}Be, ^{19}Ne, ^{35}A, etc. The nuclear recoil in ^{6}He has been observed in low pressure cloud chamber. A cloud chamber photograph reveals that the daughter recoils in a direction not opposite to the path of emitted β-particle, which clearly demonstrates that β-decay is a three-body problem and provide indirect proof of the existence of the neutrino. However, these measurment demands difficult techniques involving the further acceleration of the recoil daughter nucleus in an external magnetic field.

Inspite of these difficulties Cowan and Reines in 1953 succeeded in observing the direct interaction of free neutrino. They repeated their experiment with better equipment in 1957. As stated earlier, Pauli postulated the neutrino to satisfy the various conservation laws. However, the verification of these conservation laws does not provide any independent proof the existence of neutrino. Cowan and Reines experiment provides us with direct, independent proof of the existence of the neutrino.

We have

$$^{1}_{0}n \rightarrow ^{1}_{1}p + ^{0}_{-1}e + \bar{v} \qquad \qquad ...(3.4)$$

One may invert the above reaction by bombarding protons with anti-neutrinos. The inverse reaction is

$$\bar{v} + ^{1}_{1}p \rightarrow ^{1}_{0}n + ^{0}_{+1}e \qquad \qquad ...(3.5)$$

Obviously, when protons are bombarded with antineutrinos, one expect that they emit positrons to become neutrons. To observe this reaction, Cowan and Reines utilized the large flux of anti neutrinos existing in the neighbourhood of a nuclear reactor. Although neutrinos interact extremely weakly with matter, because of the very large flux of anti-neutrinos near a nuclear reactor there is a possibility of observing the reaction (S). Cowan and Reines observed reaction (S) with the help of experimental arrangement shown in Figure 3.30.

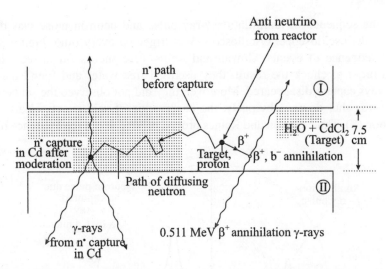

Fig. 3.30 Schematic diagram of Cowan and Reines neutrino detection experimental arrangement.

The protons for the reaction (*S*) are provided by water in a large plastic tank having about 200 litres of water. In the figure, I and II are liquid scintillation counters in the form of tanks. Each tank contains about 400 litres of liquid scintillation solution.

Cd has a large cross section (probability) of capturing thermal (very slow) neutrons having energy ~ eV. $CdCl_2$ was dissolved in water to provide Cd nuclei for neutron detection. On the two sides of the water tank two large tanks I and II (Figure 3.30) of liquid scintillator were placed. In order to restablish the reaction (3.5), the following sequence of events have to be identified:

(i) In the water tank, an anti-neutrino from the reactor interacts with a proton and produces a neutron and a positron according to reaction (3.5).

(ii) The positron comes to rest very quickly and it annihilates with an electron, resulting the emission of two gamma rays in opposite directions. Each of these γ-rays can have an energy equal to the electron rest mass (= 0.511 MeV). Liquid scintillators marked I and II can detect these γ-rays. No doubt, a large number of photo multiplier tubes are required to collect the light emitted by the scintillator. Cowan and Reines observed this radiation which follows very quickly (time interval ~ 10^{-9} s) the emission of positron.

(iii) The neutron diffused in the water containing dissolved $CdCl_2$ and after many collisions with protons slows down to thermal energies and eventually get captured (absorbed) by cadmium (Cd) nucleus. However, this capture of neutron by Cd nucleus takes place much later than the emission of anni-hilation γ-rays. The time gap is about 10^{-5} s. This capture results in the emission of γ-rays by the (*n*, γ) reaction (usually three γ-rays in upper detector I and one γ-ray in the lower detector II. On an average, a neutron requires about 6 micro seconds to be captured. Cowan and Reines succeeded in establishing the above cited three characterstic sequence of events. However they delayed the process of annihilation so that it could be displayed at the begining of the oscilloscope trace. For this purpose a delay of 30 μs was desired. An electronic circuit was *gated* for about 20 μs triggered by capture γ-rays, so that it was able to accept a *neutron pulse*.

Each time the sequence of annihilation γ-ray pulse and neutron pulse was displayed and photographed by the oscilloscope. Oscilloscope was triggered every time. Figure 3.31(a) shows this acceptable sequence of events. Cowan and Reins were successful in recording about 30 events/hour. In order to check the result, they used Cd free water and found that the neutron pluses due to γ-rays capture disappeared. Moreover, they did not observed the said events after the nuclear reactor was shut down. Obviously, the observation of said events clearly establish the nuclear interaction (*S*). Thus, Cowan and Reines proved beyond doubt the existence of the neutrino experimentally.

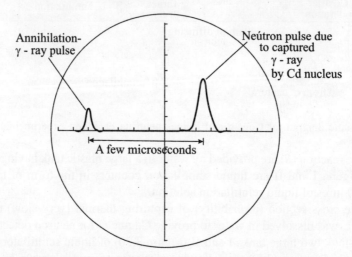

Fig. 3.31(a) Triggered pulses as a result of positron annihilation and neutron capture this sequence establishes direct neutrino interaction in accordacne with $\bar{v} + {}_1^1p \rightarrow {}_0^1n + {}_{+1}^0e$

Cowan and Reins experiment provides a clear proof in favour of the two particles, i.e. v and $\bar{v}$ being different.

The experimental demonstration of non-conservation of parity in β-decay assigned a new property called *Helicity* to the neutrinos. Helicity is a two valued quantity (±1), which implies that the intrinsic spin momentum of a particle is parallel or anti-parallel to the direction of motion. Neutrino has helicity –1 (or left handed), whereas anti-neutrino has +1 (or righ handed) as shown in Figure 3.31(b).

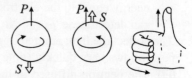

Neutrino, (v); Antineutrino, ($\bar{v}$)

Fig. 3.31(b) Helicities of neutrino (v) and antineutrino ($\bar{v}$)

Solar Neutrino

In the sun or a star, when a sufficient amount of ^{4}He is accumulated, the reaction

$$^3\text{He} + {}^4\text{He} \rightarrow {}^7\text{Be} + \gamma \qquad \qquad ... (3.6)$$

through which a ^{3}He is captured by a ^{4}He nucleus, becomes probable. ^{7}Be may capture a proton to form ^{8}B which, in turn, β⁺-decays to ^{8}Be. Since ^{8}Be is unstable toward decay into two ^{4}He. We have a conversion of four protons into a ^{4}He nucleus together with two positrons and two electron neutrinos

$$p + {}^{7}_{4}\text{Be} \rightarrow {}^{8}_{5}\text{B} + \gamma$$

$${}^{8}_{5}\text{B} \rightarrow {}^{8}_{4}\text{Be}^{*} + e^{+} + \nu_{e} + 14.02 \text{ MeV}$$

$${}^{8}_{4}\text{Be} \rightarrow {}^{4}_{2}\text{He} + {}^{4}_{2}\text{He}$$

The interest of this reaction is in the high energy (7 MeV on the average) neutrino released in the β^{+}-decay of B. The energy is sufficient to initiate the reaction.

$$\nu_{e} + {}^{37}\text{Cl} \rightarrow {}^{37}\text{Ar} + e^{-} \qquad \qquad ... (3.7)$$

This process is used in the study of solar neutrinos and solar energy production.

Davis et al. Obtained a surprising experimental result. They observed flux of solar neutrinos approximately a quarter of the calculated flux $(6 \times 10^{10} \text{ m}^{-2} \text{ s}^{-1})$. This calculated value depends on a model of the solar interior but this is believed to be understood sufficiently well that there is a very significant discrepancy, which is now known as *solar neutrino problem*.

There is currently speculation that the reduced flux of solar neutrinos might arise through the phenomenon of neutrino oscillations in which the flux of electron neutrinos is reduced to about one third of their initial value while the remaining flux becomes made up of muon and tan neutrinos. Several experiments have been planned to measure the solar neutrino flux.

3.9.8 Parity Violation (non-conservation) in β-decay

Prior to 1956, it has been assumed that parity was conserved in the weak interaction like in the electro-magnetic interaction.

One may define a parity operator P. For the odd parity wave function a change in sign on reflection reads,

$$\psi(x, y, z) = -\psi(-x, -y, -z) \qquad \qquad ... (3.1)$$

so that parity operator $\qquad P = -1$ $\qquad \qquad ... (3.2)$

For even parity wave function, on reflection one finds

$$\psi(x, y, z) = \psi(-x, -y, -z) \qquad \qquad ... (3.3)$$

so that parity operator $\qquad P = 1$ $\qquad \qquad ... (3.4)$

The parity of a system of particles is obtained by the product of the parties of the individual parities. Thus $P^2 \psi(\mathbf{r}) = \psi(\mathbf{r})$, since two parity Operations result in a return to the initial state. Therefore $P^2 = 1$ and the eiganvalues of $P = +1$ parity and -1 odd parity. The invariance of a Hamiltonian (H) under a parity operation is equivalent to the commutator of P and H being zero, i.e. $[P, H] = 0$.

In 1956, Lee and Yang pointed out that parity conservation in weak interaction had not been experimentally tested and shortly afterwards it was demonstrated that parity was not conserved in the weak interaction.

To develop the consequences, we proceed as

$$[P, H] = 0$$

This means $\dfrac{d<P>}{dt} = 0$, i.e. the rate of change of the expectation value of P with time is zero. If H is written as $H = T + V$, where T is the kinetic energy and V the potential energy operator, then:

$$[P, H] = 0 \Rightarrow [P, V] = 0 \qquad \qquad \text{... (3.5)}$$

This means that in a transition from an initial state ψ_i of parity p_i to final state ψ_f of parity p_f caused by an interaction described by the potential V, then $p_f = p_i$ as can be seen from the following. The transition matrix element M_{if} is given by the volume integral:

$$
\begin{aligned}
M_{if} &= \int \psi_f^* \, V \, \psi_i \, dt \\
&= \int \psi_f^* \, V P^2 \psi_i \, dt & (\because P^2 = 1) \\
&= \int \psi_f^* \, P V \psi_i \, dt & [P, V] = 0 \\
&= \int (P \, \psi_f)^* \, V P \, \psi_i \, dt & P \to \text{Hermitian} \\
&= p_f \, p_i \, M_{if} & \text{... (3.6)}
\end{aligned}
$$

So either $M_{if} = 0$ or $p_f = p_i$

Since the final state after a transition brough about by a parity conservating interaction has a definite parity ($p_f = p_i$) there is an important experimental consequence, which is that the expectation value in the final state of any pseudoscalar will be zero. A pseudoscalar quantity is one whose value changes sign under the parity operation. An example is $\mathbf{s} \cdot \mathbf{p}$, where $\mathbf{s}$ is the spin operator of the particle and $\mathbf{p}$ is the momentum of the particle under parity $\mathbf{p} \to \mathbf{p}$ while $\mathbf{s} \to \mathbf{s}$ so $\mathbf{s} \cdot \mathbf{p} = \mathbf{s} \cdot \mathbf{p}$. The quantity $\mathbf{s} \cdot \mathbf{p}$ divided by $|\mathbf{p}|\,|\mathbf{s}|$ is called the *helicity* or handedness of a particle.

$$H = \frac{\mathbf{s} \cdot \mathbf{p}}{(|\mathbf{s}|\,|\mathbf{p}|)} \qquad \qquad \text{... (3.7)}$$

For example, if an electron has its spin vector pointing along its direction of motion then H = +1 and the electron is called right-handed, and if H = −1 then it is called left handed.

To show why the expectation value E of a pseudoscalar operator O is zero in a state ψ_f whose parity eigenvalue is p_f consider:

$$
\begin{aligned}
E &= \int \psi_f^* \, O \, \psi_f \, dt = <f\,|O|\,f> \\
&= <f\,|OP^2|\,f> & (\because P^2 = 1) \\
&= -<f\,|POP|\,f> & (\because PO = OP) \\
&= -p_f^2 <f\,|O|\,f> & (\because P \text{ is Hermition}) \\
&= -E
\end{aligned}
$$

So E is zero. The expectation value would not necessarily be zero if the state ψ_f did not have good parity. Therefore *if in a transition form a state ψ_i of good parity p_i to a state ψ_f the expectation value of a pseudo scalar quantity in the final state is not zero then parity has been violated in the transition.*

Another important experimental consequence concerns the angular distribution of particles emitted in a transition. Consider an initial state with a definite parity and a parity-conserving interaction which brings about a transition to a final state containing two or more particles, e.g. $^{16}O^* \to {}^{12}C + \alpha$, $^{24}Mg^* \to {}^{24}Mg + \gamma$ or $^{16}O^* \to {}^{12}C + \alpha + \gamma$, then parity conservation means that the parity of the final state is the same as that of the initial state. The parity of the final state is the product of the intrinsic parities of the particles and the parities of wave functions describing the relative motions of the particles.

If the wave function describing one of the particles in the finall state is $\phi(\mathbf{r})$ with $P\phi(\mathbf{r}) = \phi(-\mathbf{r}) = p\,\phi(\mathbf{r})$, then:

$$|\phi(-\mathbf{r})|^2 = p^2|\phi(\mathbf{r})|^2 = |\phi(\mathbf{r})|^2$$

so the probability of the particle being at $\mathbf{r}$ is the same as its being $-\mathbf{r}$. This means that in a parity conserving decay an emitted particle is as likely to be travelling in the direction $\mathbf{r}$ as $-\mathbf{r}$. Figure 3.32 shows the γ-ray angular distribution produced by $1^- \to 0^+$ decay when the initial state is polarised along the Z-axis, i.e. only the $m = 1$ substate is populated. As can be seen any direction $\mathbf{r}$ is as equally likely as $-\mathbf{r}$, which is equivalent to saying the distribution is symmetric about any plane through the origin. Such an anisotropic distribution will not be seen if there is a random orientation of initial nuclear orientations as these will produce a random superposition of γ-ray intensity patterns which will give an isotropic intensity distribution.

Therefore if the distribution of an emitted particle from a polarised initial state of good parity is not symmetric about any plane through the origin of the particle then the transition has been brought about by an interaction which violates parity. It was such an observation that provided the first evidence that parity is violated in the weak interaction.

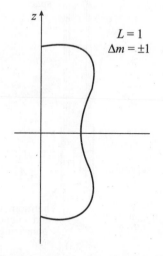

Fig. 3.32 Polar representation of the dipole angular distribution for a $1^- (m = 1) \to 0^+$–decay.

Experimental Evidence for Parity Violation

The first experiment (to test the theory of Lee and Yang, who proposed for the first time in 1956, and 1957; that in weak interactions in particle physics ($\theta \to 2\pi$, $\tau \to 3\pi$), (called $\theta - \tau$ *puzzle*) or in β-decay in nuclear physics, e.g. ($n \to p + e^- + \bar{\nu}$); parity is not conserved) that showed that parity was not conserved in the weak interaction was carried out by Wu et al. in 1957 at National Bureau of standards, who looked at the β-deacy of $^{60}\text{Co}(5^+) \to {}^{60}\text{Ni}(4^+)$, which is an allowed GT transition (Figure 3.33).

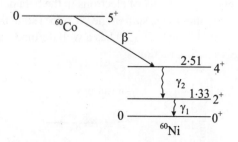

Fig. 3.33 The $^{60}\text{Co} \xrightarrow{\beta^-} {}^{60}\text{Ni}$ β-decay scheme.

$$^{60}_{27}\text{Co} \to {}^{60}_{28}\text{Ni} + \beta^- + \bar{\nu} \qquad \qquad \dots (3.7)$$

The experimental arrangement is shown in Figure 3.34.

They polarised the ^{60}Co nuclei by first cooling the ^{60}Co nuclei to 4.2 K using liquid ^4He and then to ~ 0.01 K by adiabatic demagnetisation. An external magnetic field was then applied which polarised the atomic magnetic moments. The resulting internal field polarised the nuclear magnetic moments and hence the spins of the ^{60}Co nuclei. The degree of polarisation was measured by looking at the anisotropy of the γ-rays from the decay of the 4th state in ^{60}Ni.

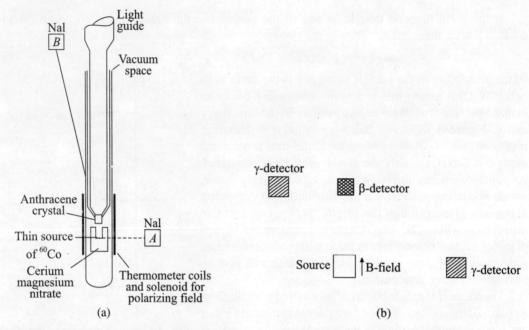

Fig. 3.34(a) The experimental arrangement used by Wu et al. for the verification of violation of parity in β-decay.

(b) Schematic diagram of the experimental arrangement to look for parity violation in ⁶⁰Co β-decay.

The ratio of counts in the two γ-detectors is a measure of the degree of polarisation (Figure 3.35a). Changing the direction of the magnetic field **B** and hence the direction of polarisation significantly altered the intensity of electrons in the beta detector (Figure 3.35b), with more emitted in a direction anti-parallel to the spin of the ⁶⁰Co (5⁺) state than parallel, i.e. more in a direction **r** than in −**r**. Since the initial state ⁶⁰Co (5⁺) has a well defined parity, this is a clear indication that parity is violated in the weak interaction.

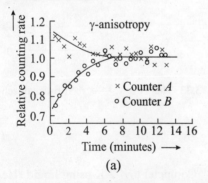

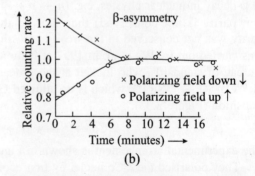

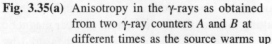

Fig. 3.35(a) Anisotropy in the γ-rays as obtained from two γ-ray counters A and B at different times as the source warms up

(b) Anisotropy in β-intensity for two opposite polarizing magnetic fields

The actual angular distribution of the electrons in the pure GT beta-decay of ^{60}Co is given by the relation

$$I(\theta) = 1 - \frac{1}{3}\frac{\upsilon}{c}\cos\theta \qquad\qquad ...\,(3.8)$$

Where υ is the velocity of the electron. For such a distribution $I(\theta) \neq I(180 - \theta)$ as the electrons are in a state of mixed polarity, the state being a mixture of $l = 0$ and $l = 1$ wave functions.

In β-decay there is in gereral a non-zero expectation value for the electron helicity and this is present whether the source is polarised or not. The helicity of electrons produced in β-decay has been measured by Wu et al. and found to be $-\dfrac{\upsilon}{c}$ in agreement with $V - A$ theory.

P and CP Invariance

One may also express the consequences of parity conservation by the statement:

The mirror image of a process is also an observable process if parity is conserved.

For ^{60}Co β-decay, the above statement is illustrated in Figure 3.36. The electrons have on average negative helicity. The mirror image shows that if parity were conserved then as many electrons should have been emitted in the direction of the spin of ^{60}Co as opposed to it. Also as many electrons should have positive as negative helicity. We may note that although in a γ-ray transition photons of definite helicity, i.e. circularly polarised, are emitted there are as many right-hand as left-hand circularly polarised photons emitted in a transition

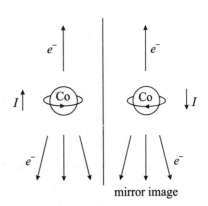

Fig. 3.36 The violation of mirror symmetry in ^{60}Co β-decay.

since the electromagnetic interaction conserves parity. This is an example of the statement that a zero expectation value of a pseudoscalar quantity does not mean that every measurment must be zero, only that the average must be zero.

Although the weak interaction is not invariant under parity, it is invariant (to a very good approximation) under the combined operation, called CP of charge conjugation and parity. Charge conjugation changes a particle into its anti-particle or vice versa so CP invariance is equivalent to saying that:

The charge conjugated mirror image of a process is also an observable process if CP is conserved.

Figure 3.37 illustrates the above statement for weak decay: $\mu^- \rightarrow e^- + \bar{\nu}_e + \nu_\mu$, which predicts that in μ^+ decay positive helicity positrons will be emitted and this is what is found. There is however, a very small violation of CP invariance observed in the weak decay of the strange particle K°, the magnitude of which is still unexplained.

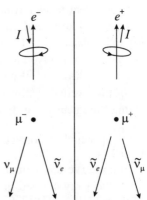

Fig. 3.37 The charge-conjugate mirror symmetry of $\mu^\pm$ β-decay.

3.9.9 Helicity of Neutrino

The discovery of parity violation in the weak interaction indicated definite helicities for anti neutrino and neutrino and in 1958 the helicity of the neutrino was measured in a most elegant experiment by Goldhaber et al.

The metable state of ^{152}Eu decays about 24 % of the time by K-electron capture to an excited state of ^{152}Sm which in trun decays to the ground state of samarium by emission of a 961 keV γ-ray. (The life time of the excited state is extremely short, $\approx 10^{-14}$ s). Furthermore, since K-capture is a two-body process, the emitted neutrino has a unique energy, $E_\nu \approx 900$ keV in this case, i.e. close to the γ-ray energy, an important consideration it this case. The relevant portion of the decay scheme is shown in Figure 3.38. We may note in particular the spin-parities of the levels involved.

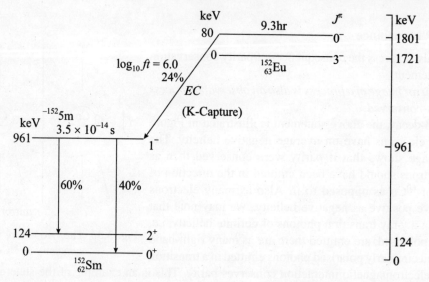

Fig. 3.38 A portion of β-decay scheme of ^{152}Eu.

The experiment takes advantage of the near equality of the neutrino energy in the electron-capture (E_ν = 840 keV) and γ-ray energy of $1^- \rightarrow 0^+$ decay (E_γ = 960 keV), to enable resonant scattering to be observed of $1^- \rightarrow 0^+$ γ-ray when its helicity is the same as that of the neutrino. The spins and momenta involved in the experiment are illustrated in Figure 3.39.

In Figure 3.39(a) the $^{152}_{63}$Eu captures an electron essentially with zero momentum. To conserve momentum the excited $^{152}_{62}$Sm nucleus with $J^\pi = 1^-$ recoils with a momentum $\mathbf{R}$ equal to the momentum of the neutrino $\mathbf{q}$. In this figure the helicity of the neutrino has been assumed negative $\left(H = \dfrac{(\mathbf{s} \cdot \mathbf{p})}{(s \cdot p)} \right)$, which is correct, so the direction of the spin of the 1^- state in $^{152}_{62}$Sm is as shown in order to conserve the angular momentum. In Figure 3.39(b) the γ-decay of 1^- $^{152}_{62}$Sm state is shown with the γ-ray emitted in the same direction as the $^{152}_{62}$Sm nucleus is recoiling. In the rest frame of the recoiling $^{152}_{62}$Sm the energy E_0 of the $1^- \rightarrow O^+$ ray is given by:

$$E_0 = E_x \left(1 - \frac{E_x}{2Mc^2} \right) \quad \text{conservation of momentum}$$

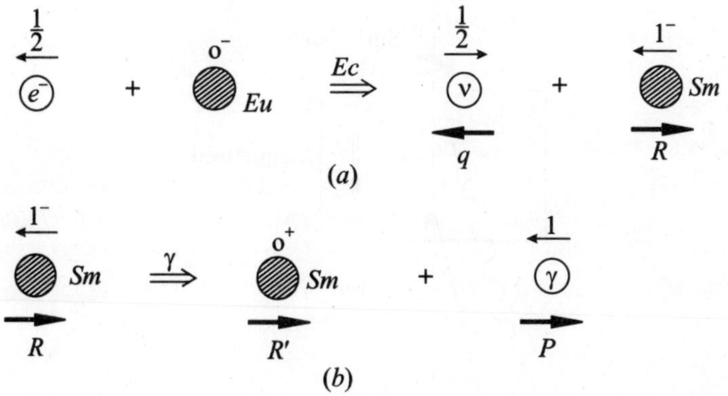

Fig. 3.39 Representation of (a) the β-decay, and (b) the γ-decays in the ^{152}Eu
neutrino helicity experiment

Where M is the mass of the $^{152}_{62}$Sm nucleus and E_x is the excitation energy of 1^- state. In the laboratory frame the energy of the γ-ray is Doppler shifted because the $^{152}_{62}$Sm nucleus is moving, and its energy E_γ is related to E_0 by:

$$E_\gamma = E_0\left(1+\frac{v}{c}\right) = E_0\left(1+\frac{E_v}{Mc^2}\right)$$

Since $\mathbf{R} = \mathbf{q}$ so $v = \dfrac{q}{M} = \dfrac{E_v}{Mc}$. Therefore

$$E_\gamma = \left(1-\frac{E_x}{2Mc^2}\right)\left(1+\frac{E_v}{Mc^2}\right) \approx E_x\left(1+\frac{E_x}{2Mc^2}\right) \quad (\because E_v \approx E_x) \qquad \text{... (3.1)}$$

Obviously, the γ-ray has the required energy for absorption by $^{152}_{62}$Sm from the 0^+ ground state to the 1^- excited state and this is the condition for resonant scattering of the γ-ray by a piece of $^{152}_{62}$Sm. The energy E_γ does not have to exactly satisfy Equation (3.1) because of the thermal motion of the $^{152}_{62}$Sm nuclei. Figure 3.40 shows the experimental arrangement. The scatterer was in the form of a ring of samarium oxide $(Sm_2 O_3)$ and the γ-rays were detected with an NaI(Tl) detector which was shielded from the direct radiation by 12 inch of lead. The helicity of the γ-rays was analysed by transmission through fully magnesized iron.

To measure the polarisation of the γ-rays an analysing magnet was used to magnetise an iron absorber as shown in figure 3.40, which changes the absorption coefficient of the γ-rays depending on their polarisation measuring the effect of reversing the magnetic field enabled the polarisation the γ-rays to be measured. The result was that neutrino has a negative helicity and was consistent with complete polarisation (i.e. $H_v = -1$).

The helicity of the antineutrinos has been inferred to be positive from the asymmetry measurments in the decay of polarized neutrons, a result which is supported by later measurments. The v and $\bar{v}$ are thus distinguished from each other by their helicity; the neutrino has helicity -1 while the antineutrino has helicity $+1$: the parity violation is said to be maximal.

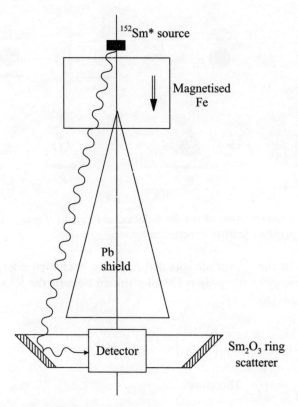

^{152}Sm* source

Magnetised Fe

Pb shield

Detector

Sm_2O_3 ring scatterer

Fig. 3.40 Experimental arrangement used by Goldhaber et al. to determine the helicity of the neutrino.

3.10 ELECTROMAGNETIC TRANSITIONS

In nuclear physics the absorption or emission of radiation normally connects two nuclear levels. Both of these may be unstable but often the lower state is the ground state of the nucleus and then the radiative processes can be discussed in terms of Figure 3.41 which indicates the relevant energies and quantum numbers.

The probability of absorption or of emission of a photon of correct energy E_γ depends on (a) the energy E_γ itself, (b) the number of units $\hbar$ of angular momentum transferred in the process, also known as the *multipolarity*, and (c) the matrix element which is constructed from the wave functions for the two levels using the appropriate electro-magnetic multiple operator. Experimentally, one can determine the multipolarity L by conversion or angular correlation phenomena and the transition polarity $T(L)$ may then be predicated from detailed nuclear models.

The multipolarity possible for a radiative transition is limited by selection *rules*, based on the conservation of angular momentum. These rules are:

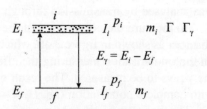

Fig. 3.41 Emission of a radiation from a level of width Γ and radiative width Γ_γ. The converse excitation between the levels is also shown in the Figure.

$$I_i + I_f \geq L \geq |I_i - I_f|$$
$$m_i = M + m_f$$

Where $M\hbar$ is the resolved part of the photon total angular momentum L. Parity is also conserved in the electro magnetic interaction to a high degree of accuracy, but a given multipole field may be of even or odd parity (Table 3.2)

Table 3.2 Classification of Radiative Transitions

	E1	E2 M1	E3 M2
Name	Electricdipole	Electric quadrupole Magnetic dipole	Electric octupole Magnetic quadrupole
Multipolarity	1	2, 1	3, 2
Parity change	yes	No	yes

The names electric and magnetic derive from the semi classical theory of radiation, in which a radiation field arises because of the time variation of a charge and current distribution. Monopole transitions, e.g. E_0, do not exist in the real radiation field; the E_0 *giant resonance* is excited or by 'virtual' photons with longitudional field components. If $I_i = I_f = 0$ radiation as such is strictly prohibited.

Electromagnetic transitions form the dominant mode of decay for low-lying excited states in nuclei, particularly for the light ones. The main reason is that nucleon emission, a much faster process than γ-decay is forbidden until the excitatin energy is above nucleon seperation energies. Neutron separation energies are of the order of 8 to 10 MeV and the corresponding values for protons are some what lower because of Coulomb repulsion. Other possible decay modes are β-decay, α-particle emission, and fission. Gererally speaking, these are slower processes than γ-decay and their Q-value considerations, i.e., the energy involved in the decay, are also different.

Besides γ-ray emission, nuclei can also de-excite through *electromagnetic interaction* by internal conversion whereby one of the atomic electrons is ejected. This is usually more important for heavy nuclei, where the nuclear electromagnetic fields are strong and the orbits of the inner shell electrons are close to the nucleus. Instead of a γ-ray, an *electron-positron pair* may also be created as a result of de excitation process. The probability for such internal pair creation is, in gereral, much smaller than γ-ray emission and becomes important where γ-ray emission is forbidden by angular momentum considerations. This happens in the case of transitions from an initial state with $I^\pi = 0^+$ to a final state that is also 0^+.

The first step in a discussion on electromagnetic transitions is to establish a connection between the transition probability W and the nuclear matrix element M_{if}. One can achieve this using first-order time-dependent perturbation theory. The gereral result, known as Fermi's golden

$$W = \frac{2\pi}{\hbar} \ |<\phi_k(\mathbf{r})|H'|\phi_0(\mathbf{r})<|^2 \ \rho(E_f) \qquad \dots (3.1)$$

Where H' represents the perturbation due to coupling between nuclear and electromegnetic fields. The density of final statese (E_f) here is a product of the number of nuclear and magnetic states per energy interval at E_f. Similarly, the initial and final wave functions, $\phi_0(\mathbf{r})$ and $\phi_k(\mathbf{r})$, respectively, are products of nuclear and electromagnetic parts.

The photon (quantum of electromagnetic radiation) has a spin angular momentum quantum $S = 1$. An electromagnetic field can have oly transverse components and hence along a fixed direction (say Z-direction) the component of spin angular momentum $S\hbar = \hbar$. Obviously, when a nucleus emits a photon, its total angular momentum along the Z-direction must change by at least one unit. This means that the emitted photon has to take away an angular momentum equal to the difference between the angular momenta of the nucleus in the two states between which the transition takes place. However, the angular momentum carried away by a photon defines the multipolarity of the transition. If the angular momentum (along the fixed direction, i.e. Z-direction) carried by the photon $L = 1$, then the transition is said to be *dipole transition* and the radiation is termed as *dipole radiation*. Similarly, for $L = 2$ the transition is said to be *quadrupole transition* and the radiation is termed as quadrupole radiation and for $L = 3$ it is *octopole*. The *multipolarity* of γ-transition for angular momentum L of radiation field is $2L$ and the transition may (or may not) cause a change in the *parity* (π) of the nuclear state. If π_i and π_f be the particles of initial and final states respectively in a γ-transition with $L = 1$, then if $\Delta\pi$ = yes or $\dfrac{\pi_i}{\pi_f} = (-1)^1$, then the emitted radiation is called an electric *dipole radiation*. However, if $\Delta\pi$ = No, i.e. $\dfrac{\pi_i}{\pi_f} = (-1)^{\circ}$, the nucleus emits a *magnetic dipole radiation*. One can define the electric or magnetic nature of a transition in which an angular momentum L is carried away be the radiation field as (Table 3.3)

Table 3.3

Transition	$L =$	1	2	3	4	$\dfrac{\pi_i}{\pi_f}$
Electric	$\Delta\pi =$	Yes	No	Yes	No	$(-1)^L$
Magnetic	$\Delta\pi =$	No	Yes	No	Yes	$(-1)^{L+1}$

If in the initial and final states, the total angular momentum quantum number or spin of a nucleus is I_i and I_f respectively, then the radiation field can carry away any angular momentum L between $|I_i + I_f|$ and $|I_i - I_f|$. Thus, one finds

$$|I_i - I_f| \le L \le |I_i + I_f|$$

For e.g., if $I_i = 1$ and $I_f = 2$, one finds the possible values of $L = 1, 2$ and 3. Since the change in parity is fixed (yes or no), the γ-transition is a mixture of alternate electric and magnetic multipoles. When $\Delta\pi$ = yes, i.e. there is change in parity, the γ-transition is a mixture of electric dipole ($E1$), magnetic quadrupole ($M2$) and electric octapole ($E3$) transitions, and if $\Delta\pi = 0$, i.e. there is no change in parity, the transition is a mixture of $M1$, $E2$ and $M3$ (Table 3.2).

The probability of γ-ray transition varies as $\left(\dfrac{R}{\hbar}\right)^{2L}$ where $\lambda = 2\pi\hbar$ is the wavelength of γ-radiations and R is the radius of the nucleus emitting γ-ray. On finds,

$$\frac{R}{\lambda} = \frac{R\hbar\omega}{\hbar c} = \frac{R E_\gamma}{\hbar c} \qquad \qquad \dots (3.2)$$

Where E_γ is the energy of the γ-ray photon. We may note that $E_\gamma << \dfrac{\hbar c}{R}$ $(= 10 \text{ MeV})$, $\dfrac{R}{\lambda} << 1$, one finds the transitions probability for higher values of L is very small. This is why that one observes only the lowest multipole transitions.

We may note that the electric multipole transitions are visualised due to vibrating charges, whereas the magnetic transitions are visualised as due to fluctuating currents inside the nucleus. Moreover, the probability of magnetic transition is less than that of an electric transition of the same multipolarity by a factor of about $\dfrac{v^2}{c^2}$ (≈ 0.05), where v is the velocity of the nucleons within the nucleus. Obviously, when the lowest multipole transition is electric, the higher multipole transitions magnetic or electric are seldom observed. When the lowest multipole transition is magnetic, the next higher electric multipole transition may sometimes be mixed with it. This means, one may observe a mixed M_1 and E_2 transition, but never a M_2 transition mixed with E_1 transition. As stated earlier monopole transitions E_0 do not exist in real field. This means that when initial and final spins of the excited nucleus are both zero $(I_i - I_f - 0)$, it is not possible to have a γ-transition. We may remember that the minimum angular momentum carried away by a photon field is $L = 1$.

3.10.1 Multipolarity of Radiation Field

In order to obtain an expression for the multipole expansion of a radiation field one usually employs quantum-electrodynamics. However, for a good gain insight, about the e.m. transitions one can also use a semi-classical theory.

We have Maxwell's equations in vacuum as

$$\nabla \times \mathbf{H} = \varepsilon_0 \frac{\partial \mathbf{E}}{\partial t} \left(\text{or } \nabla \times \mathbf{B} = \mu_0 \varepsilon_0 \frac{\partial \mathbf{E}}{\partial t} \right) \qquad \text{... (3.1)}$$

$$\nabla \times \mathbf{E} = -\mu_0 \frac{\partial \mathbf{H}}{\partial t} \qquad \text{... (3.2)}$$

$$\nabla \cdot \mathbf{H} = 0 \ (\text{or } \nabla \cdot \mathbf{B} = 0) \qquad \text{... (3.3)}$$

$$\nabla \cdot \mathbf{E} = 0 \qquad \text{... (3.4)}$$

The solutions of Equations (3.1) to (3.4) describes the electromagnetic waves with superposition of the incoming and outgoing waves of different frequencies. One can write the electric and magnetic fields satisfying Equations (3.1) to (3.4) as

$$\mathbf{E} = \int_0^\infty \left[\mathbf{E}_r (r, \omega) e^{i\omega t} + \mathbf{E}_r^* (r, \omega) e^{-i\omega t} \right] d\omega \qquad \text{... (3.5)}$$

and

$$\mathbf{H} = \int_0^\infty \left[\mathbf{H}_r (r, \omega) e^{i\omega t} + \mathbf{H}_r^* (r, \omega) e^{-i\omega t} \right] d\omega \qquad \text{... (3.6)}$$

Where $\mathbf{E}_r$ and $\mathbf{H}_r$ are respectively the electric and magnetic fields of the radiations of different frequencies. Now, substituting the value of $\mathbf{E}$ from (3.5) and $\mathbf{H}$ from (3.6) in (3.1) and (3.2) respectively, one obtains

$$\nabla \times \mathbf{H} = -i\omega\,\varepsilon_0\,\mathbf{E} \qquad \qquad \dots (3.7)$$

$$\nabla \times \mathbf{E} = i\omega\,m_0\,\mathbf{H} \qquad \qquad \dots (3.8)$$

Taking cure of (3.7) and (3.8), one obtains

$$\nabla \times \nabla \times \mathbf{H} = \omega^2\,e_0\mu_0\,\mathbf{H} \qquad \qquad \dots (3.9)$$

$$\nabla \times \nabla \times \mathbf{E} = \omega^2\,\varepsilon_0\,\mu_0\,\mathbf{E} \qquad \qquad \dots (3.10)$$

Let us put $\omega^2\,\varepsilon_0\,\mu_0 = K^2$, we obtains

$$[\nabla \times \nabla \times -k^2]\mathbf{H} = 0 \qquad \qquad \dots (3.11)$$

and $\qquad\qquad [\nabla \times \nabla \times -k^2]\mathbf{E} = 0 \qquad \qquad \dots (3.12)$

Since the photon carries away momentum, the electric and magnetic fields $\mathbf{E}$ and $\mathbf{H}$ should be angular momenturm eigen functions. Now, we have to find some angular momentum operator which commutes with the operator $\nabla \times \nabla$. We see that eigen function of the operator $\nabla \times \nabla$ can represent the magnetic and electric fields of equations (3.11) and (3.12).

　　Angular momentum operator $\mathbf{L}$ is defined as

$$\mathbf{L} = -i\hbar\,\mathbf{r} \times \nabla \qquad \qquad \dots (3.13)$$

We have the Z-component of $\mathbf{L}$ as

$$L_z = -i\hbar\left(x\frac{\partial}{\partial y} - y\frac{\partial}{\partial x} \right) \qquad \qquad \dots (3.14)$$

and $\qquad\qquad L^2 = -\hbar^2(\mathbf{r} \times \nabla)^2 \qquad \qquad \dots (3.12)$

We can easily see that the operator L^2 does not commute with $\nabla \times \nabla$. However, if we define an operator $S_z\,(= -i\hbar\,\mathbf{k}\,x)$ and add to L_z, then we find that operator $L_z + S_z$ commutes with $\nabla \times \nabla$. Now, putting

$$J_z = L_z + S_z \qquad \qquad \dots (3.13)$$

One finds that eigen function of the operator J_z should be able to represent the field $\mathbf{H}$ and $\mathbf{E}$ given by Equations (3.9) to (3.10). Now, We define the operator J^2 as

$$J^2 = J_x^2 + J_y^2 + J_z^2 \qquad \qquad \dots (3.14)$$

(3.14) is also found to commute with $\nabla \times \nabla$. Let $\mathbf{B}$ is an eigen function defined as

$$\mathbf{B} = L\,R_L(r)\,Y_{LM}(\theta,\,\phi) \qquad \qquad \dots (3.15)$$

We find $\qquad\qquad J_z\,\mathbf{B} = M\hbar\,\mathbf{B}$

where $\qquad\qquad L_z = M\hbar$ and $J^2\,\mathbf{B} = \hbar^2\,L(L+1)\mathbf{B} \qquad \qquad \dots (3.17)$

Obviously $\mathbf{B}$ is an eigen function of the operators J_z and J^2 and can represent the electric and magnetic fields $\mathbf{E}$ and $\mathbf{H}$ respectively.

　　Now, one can write Equations (3.9) to (3.10) as

$$(\nabla \times \nabla \times -k^2)\mathbf{B} = 0$$

or $\qquad\qquad \nabla(\nabla \cdot \mathbf{B}) - \nabla^2\,\mathbf{B} - k^2\,\mathbf{B} = 0 \qquad \qquad \dots (3.18)$

Since in free space $\nabla \cdot \mathbf{H} = 0$ and $\nabla \cdot \mathbf{E}$ and therefore $\mathbf{B}$ can represent either $\mathbf{H}$ or $\mathbf{E}$. One can write the above equation as

$$(\nabla^2 + k^2)\mathbf{B} = 0 \qquad \qquad \dots (3.19)$$

We note that (3.19) is just similar to Schrodinger equation. One can write the eigen function in polar coordinates as $R(r) \, y_{LM}(\theta, \phi)$ where the equation satisfied by the radial part is

$$\left(\frac{d^2}{dr^2} - \frac{L(L+1)}{r^2} + k^2 \right) R_L(r) = 0 \qquad \text{... (3.20)}$$

The solution of (3.20) are the spherical Bersel functions $J_L(kr)$. One can write eigen function satisfying the Equations (3.9) and (3.10) as

$$\mathbf{B} = C \, \mathbf{L} \, j_L(kr) \, Y_{LM}(\theta, \phi) \qquad \text{... (3.21)}$$

Where C is a constant.

Now, if $\mathbf{B}$ represents the magnetic field vector satisfying Equation (3.9), the corresponding electric field vector $\mathbf{E}$ can be obtained from Equation (3.7) and if $\mathbf{B}$ represents the electric field vector of Equation (3.10) then one can obtain the corresponding magnetic field vector using Equation (3.8). Let us chose that $\mathbf{B}$ represent the magnetic field vector $\mathbf{H}_{LM}^e$, then

$$\mathbf{H}_{LM}^e = C^e \, L_{JL}(kr) \, Y_{LM}(\theta, \phi) \qquad \text{... (3.22)}$$

One finds that the radial component of $\mathbf{H}_{LM}^e$ is zero as $L_r = 0$ ($\mathbf{L} = -i\hbar \, \mathbf{r} \times \mathbf{\nabla}$). The field in which the radial component of the magnetic field is zero are termed as *electric multipole fields* and are denoted by a superscript e. One obtains the corresponding electric field E_{LM}^e from Equation (3.7) as,

$$E_{LM}^e = \frac{1}{i\omega\varepsilon_0} \mathbf{\nabla} \times \mathbf{H}_{LM}^e = -\frac{C^e}{i\omega\varepsilon_0} \mathbf{\nabla} \times L \, j_L(kr) \, Y_{LM}(\theta, \phi) \qquad \text{... (3.23)}$$

Now, if $\mathbf{B}$ represents the electric field vector E_{LM}^m of Equation (3.10), then one obtains E_{LM}^m as,

$$E_{LM}^m = C^m \, L \, j_r(kr) \, Y_{LM}(\theta, \phi) \qquad \text{... (3.24)}$$

We see that the radial component of the electric field is zero. Such fields are termed as *magnetic multipole fields* and are denoted with the superscript m.

One obtains the corresponding magnetic field vector for a magnetic multipole field as,

$$H_{LM}^m = \frac{C^m}{i\omega\mu_0} \mathbf{\nabla} \times \mathbf{L} \, j_L(kr) \, Y_{LM}(\theta, \phi) \qquad \text{... (3.25)}$$

One can easily evaluate the constants C^e and C^m by normalising the radiation fields such that after the emission of a photon in which a radiation field of multipolarity L and frequency ω appears, the field energy is $\hbar\omega$. We may note that in an electromagnetic field half the average energy is in electric fields and half in magnetic field. Thus, one finds

$$\varepsilon_0 |E|^2 = \mu_0 |H|^2 = \frac{\hbar c k}{2} \qquad \text{... (3.26)}$$

If the wave is enclosed in a large sphere of radius R_0 ($R_0 \to \infty$) so that $k R_0 \gg 1$, one obtains the single frequency of the radiation. The radiation field is zero at $r = R_0$ and therefore,

$$j_L(k R_0) = 0 \qquad \text{... (3.27)}$$

Now, putting the value of $\mathbf{H}$ in Equation (3.26), one obtains

$$\frac{\hbar c k}{2\mu_0} = (C^e)2 \int\limits_0^{R_0} |j_L(kr)|^2 \ r^2 \ dr \int (\mathbf{L} \, Y_{LM})^* \ \mathbf{L} \, Y_{LM} \ d\Omega$$

L being Hermitian operator, one finds

$$\int (\mathbf{L} \, Y_{LM})^* \ \mathbf{L} \, Y_{LM} \ d\Omega = \int Y_{LM}^* \ \mathbf{L} \mathbf{L} \, Y_{LM} \ d\Omega 1$$

$$= \int Y_{LM}^* \ L^2 \, Y_{LM} \ d\Omega$$

$$= L(L+1)\hbar^2 \qquad \qquad \dots (3.28)$$

For large values of r, one finds

$$j_L(kr) = \frac{1}{r} \cos\left(kr - \frac{\pi}{2}(L+1) \right)$$

$$\therefore \quad \int\limits_0^{R_o} |j_L(kr)|^2 \ r^2 \ dr \approx \frac{1}{k^2} \int\limits_0^{R_o} \left| \cos\left(kr - \frac{\pi}{2}(L+1) \right) \right|^2 \ dr$$

$$\approx \frac{R_0}{2k^2} \qquad \qquad \dots (3.29)$$

Using (3.28) and (3.29), one finds

$$\frac{\hbar c k}{2\mu_0} = (C^e)^2 \, L(L+1)\hbar^2 \, \frac{R_0}{2k^2}$$

$$\therefore \quad (C^e)^2 = \frac{c k^3}{\mu_0 \, \hbar \, R_0 \, L(L+1)} \qquad \qquad \dots (3.30)$$

Similarly, one can obtain the value of $(C^m)^2$ by putting the value of $\mathbf{E}$ in Equation (3.26). Thus,

$$(C^m)^2 = \frac{c k^3}{\varepsilon_0 \, \hbar \, R_0 \, L(L+1)} \qquad \qquad \dots (3.31)$$

Now, by substitution of the values of C and C^m, one obtains the multipole expansion of the radiation field as,

$$E_{LM}^e \doteq i\left[\frac{ck}{\varepsilon_0 \, \hbar \, R_0 \, L(L+1)} \right]^2 \nabla \times \mathbf{L} \, j_L(kr) \, Y_{LM}(\theta, \phi) \qquad \dots (3.32)$$

$$H_{LM}^e = k\left[\frac{ck}{\mu_0 \, \hbar \, R_0 \, L(L+1)} \right]^2 \mathbf{L} \, j_L(kr) \, Y_{LM}(\theta, \phi) \qquad \dots (3.33)$$

$$E_{LM}^m = k\left[\frac{ck}{\varepsilon_0 \, \hbar \, R_0 \, L(L+1)} \right]^2 \mathbf{L} \, j_L(kr) \, Y_{LM}(\theta, \phi) \qquad \dots (3.34)$$

$$H_{LM}^m = -i \left[\frac{ck}{\mu_0 \, \hbar \, R_0 \, L(L+1)} \right]^2 \nabla \times \mathbf{L} \, j_L(kr) \, Y_{LM}(\theta, \phi) \qquad \text{... (3.35)}$$

One can show that the operator

$$\nabla \times \mathbf{L} = \frac{\hbar}{i} \left[\mathbf{r} \, \nabla^2 - \nabla \left(1 + r \frac{\partial}{\partial r} \right) \right] \qquad \text{... (3.36)}$$

One can evaluate the above expression with the use of the properties of Bessel's function. One finds,

$$\nabla^2 \, j_L(kr) \, Y_{LM}(\theta, \phi) = -k^2 \, j_L(kr) \, Y_{LM}(\theta, \phi) \qquad \text{... (3.37)}$$

and

$$r \frac{\partial}{\partial r} \, j_L(kr) = kr \, j_{L+1}(kr) - (L+1) \, j_L(kr)$$

$$= L \, j_L(kr) - kr \, J_{L+1}(kr) \qquad \text{... (3.38)}$$

Making use of (3.37) and (3.38), one obtains

$$\nabla \times \mathbf{L} = j_L(kr) \, Y_{LM}(\theta, \phi) = \frac{\hbar}{i} \left[-k^2 r - (L+1)\nabla \right] j_L(kr)$$

$$Y_{LM}(\theta, \phi) + \frac{\hbar}{i} \nabla(kr) \, j_{L+1} \, Y_{LM}(\theta, \phi) \qquad \text{... (3.39)}$$

Outside the nucleus, i.e. beyond nuclear radius, the wave function $R(r)$ is insignificant. Since r is very small and therefore $kr \to 0$. Now, the relations takes the form

$$\nabla \times \mathbf{L} = -\frac{\hbar}{i} (L+1) \nabla \, j_L(kr) \, Y_{LM}(\theta, \phi) \qquad \text{... (3.40)}$$

The parity of the function Y_{LM} is $(-1)^L$, whereas the parity of the operator L is $+1$ and that of the operator ∇x is -1. Obviously, the parity of the functions E_{LM}^e and E_{LM}^m is $(-1)^L$ and that of the functions H_{LM}^m and H_{LM}^e is $(-1)^{L+1}$. Moreover, as $Y_{0,0} = 0$ and $\nabla \, Y_{0,0} = 0$ and therefore there is no radiation field with $L = 0$. This result is in conformity with the transverse nature of the electromagnetic waves. With $L \geq 1$, the angular momentum carried away by the electromagnetic field is $\hbar \sqrt{l(L+1)}$. One can build up any electromagnetic field satisfying Maxwell's equations by the mixture of various multipole fields. In general, one finds,

$$\mathbf{E} = \sum_{M=-L}^{L} \sum_{L=1}^{\infty} (a_{LM}^e \, \mathbf{E}_{LM}^e + a_{LM}^m \, \mathbf{E}_{LM}^m) \qquad \text{... (3.41)}$$

and

$$\mathbf{H} = \sum_{M=-L}^{L} \sum_{L=1}^{\infty} (b_{LM}^e \, \mathbf{H}_{LM}^e + b_{LM}^m \, \mathbf{H}_{LM}^m) \qquad \text{... (3.42)}$$

Where a and b denotes the relative amplitudes of the corresponding fields. The parity of the system is conserved during the emission of electromagnetic radiations. However, the parity of the initial and the final state of the nucleus emitting radiation as well as the parity of the electromagnetic field is definite. Obviously, the parities of the radiation fields mixing together must, therefore be same.

This means that one may able to mix the fields $\mathbf{E}^e_{LM}$ and $\mathbf{E}^m_{L+1}$, M as they both have the same parity. Similarly, one may mix the field $\mathbf{H}^e_{LM}$ with $\mathbf{H}^m_{L+1}$, M.

We know that one can describe the electric and magnetic fields in terms or a vector potential ($\mathbf{A}$) and a scalar potential (ϕ). Thus,

$$\mathbf{H} = \nabla \times \mathbf{A} \qquad \ldots (3.43)$$

Making use of Equation (3.8), one finds

$$\mathbf{A} = \frac{i}{i\omega\mu_0} - \nabla\phi \qquad \ldots (3.44)$$

If we choose $\mathbf{A}$ such that $\nabla \cdot \mathbf{A} = 0$, the scalar potential ϕ may be taken as zero at $r \to \infty$. Moreover, $\nabla^2 \phi = 0$, i.e. the scalar potential vanishes everywhere and the vector potential is then given as,

$$\mathbf{A} = \frac{i}{i\omega\mu_0} \mathbf{E} \qquad \ldots (3.45)$$

3.11 γ-DECAY AND γ-RAY TRANSITION PROBABILITY

Excited states if bound to particle emission nearly always decay through the electromagnetic interaction usually with the emission of a γ-ray, though if the nucleus has a large Z and excitation is low then decay by internal conversion in which an electron is ejected is more likely. Very occasionally when the angular momentum change in a low-energy γ-transition is very large, β-decay competes with γ-decay. For unbound states well above threshold, particle decay predominates. Near threshold, charge particle decay is hindered by the Coulomb barrier and neutron decay by the transmission across the potential discontiunity at the nuclear surface, with the result that γ-decay can be comparable to particle decay for excited states near threshold.

3.11.1 γ-ray Transition Probability

We know that nucleus is a system composed of neutrons and protons and any excited state of nucleus is a state of entire nucleus, i.e. one can justify a nuclear excited state due to the excitation of a single nucleon. However, to simplify the semi-classical calculations of γ-ray transition probability, one may assume that the excited state of the nucleus is due to a proton. A photon is emitted, when the proton interacts with the electromagnetic field. The Hamiltonian of a proton in an electromagnetic field can be expressed as

$$\mathcal{H} = \frac{1}{2M_p} (\mathbf{p} - e\mu_0 \mathbf{A})^2 + u(r) + \frac{\mu_N}{\hbar} g_{sp}(\mathbf{S}_p \cdot \mathbf{H}) \qquad \ldots (3.1)$$

Where $u(r)$ is the spherically symmetric nuclear potential, μ_N nuclear magneton, S_p the spin of the proton and $\mathbf{H}$ the magnetic component of the electromagnetic field. The last term in (3.1) represents the energy of interaction of the intrinric magnetic moment of the proton with the magnetic field of the radiations. One can expand the first term in (3.1) as

$$\frac{1}{2M_p} (\mathbf{p} - e\mu_0 \mathbf{A})^2 = \frac{1}{2M_p} [p^2 + e^2\mu_0^2 A^2 - e\mu_0(\mathbf{p} \cdot \mathbf{A} + \mathbf{A} \cdot \mathbf{p})]$$

One can neglect the term A^2 as it is a second order term and represents emission of two photons. Now the Hamiltonian (3.1) reduces to

$$\mathcal{H} = \left[\frac{p^2}{2 M_p} + u(r) \right] + \left[\frac{e \mu_0}{2 M_p} (\mathbf{p} \cdot \mathbf{A} + \mathbf{A} \cdot \mathbf{p}) \right] + \frac{\mu_N}{\hbar} g_{sp} (\mathbf{S}_p \cdot \mathbf{H}) \quad \dots (3.2)$$

$$= H_0 + H_1 \qquad \dots (3.2a)$$

Where H_0 is the Hamiltonian of proton in the absence of the electromagnetic field and H_1 is the perturbation caused due to the presence of the radiation field. Now, the eigen value equation is

$$H_0 \psi = E_n \psi_n \qquad \dots (3.3)$$

Where E_n represent the energy levels of the proton and n is an integral number.

We may note that the transition from one energy state of the proton to another state is due to the perturbation H_1. One can write

$$H_1 = H_1 e^{iwt} + H_1^* e^{-i\omega t}$$

$$= \frac{e \mu_0}{2 M_p} (\mathbf{p} \cdot \mathbf{A} + \mathbf{A} \cdot \mathbf{p}) + \frac{\mu_N}{\hbar} g_{sp} (\mathbf{S}_p \cdot \mathbf{H}) \qquad \dots (3.4)$$

We have the transition probability from the state ψ_i to the state ψ_f as

$$P = \frac{2\pi}{\hbar} \left| \int \psi_f^* H_1 \psi_i \, dV \right|^2 \frac{dn}{dE} \qquad \dots (3.5)$$

Where $\dfrac{dn}{dE}$ is the density of final states.

The operator $H_1 e^{i\omega t}$ takes care of the transition from a higher energy state to a lower energy state with the emission of radiation, whereas the operator $H_1^* e^{-i\omega t}$ gives rise to an absorption of radiation from ψ_f to ψ_i with transition. In order to calculate the energy density of final states, we consider the sruface of a large sphere of radius R_0, such that $k R_0 \gg 1$. Now, the asymptotic form of the Bessel function is

$$j_L(kr) \approx \frac{1}{kr} \operatorname{Cos}\left(kr - \frac{k}{2}(L+1) \right)$$

$$= \frac{1}{kr} \operatorname{Sin}\left(kr - \frac{L\pi}{2} \right)$$

Since $r \rightarrow R_0$, $\operatorname{Sin}\left(k r_0 - \dfrac{L\pi}{2} \right) \rightarrow 0$

$$\therefore \qquad k R_0 - \frac{L\pi}{2} = n\pi \qquad \dots (3.6)$$

Where n is an integer. Since $k = \dfrac{E_\gamma}{\hbar c}$, we have

$$E_\gamma = \frac{\pi \hbar c}{R_0} \left(n + \frac{L}{2} \right)$$

on differentiation, this yields

$$\frac{dn}{dE} = \frac{R_0}{\pi \hbar c} \qquad \dots (3.7)$$

Outside the nuclear radius r_0, the proton wave functions ψ_i and ψ_f are zero, where $k\, r_0 \ll 1$. The interaction of proton with the radiation field is significant only for $r \ll r_0$. For $kr \ll 1$, we have

$$j_L(kr) = \frac{2^L\, L!}{(2L+1)!}\, (kr)^L = \frac{(kr)_L}{(2L+1)!!} \qquad \dots (3.8)$$

Where $\qquad (2L+1)!! = 1, 3, 5, \dots (2L+1)$

Now, substituting the value of E^e_{LM} (Equation 3.32, section 3.9) in Equation (3.45) (section 3.9), one obtains

$$\mathbf{A}^e_{LM} = \frac{1}{\mu_0\, ck} \left[\frac{c_k}{\varepsilon_0\, R_0 \hbar\, L(L+1)} \right]^{\frac{1}{2}} \nabla \times \mathbf{L}\, j_L(kr)\, Y_{LM}(\theta, \phi)$$

After putting the values of $\nabla \cdot \mathbf{L}$ (Equation 3.40, section 3.9) and $j_L(kr)$ (Equation 3.8), one obtains

$$A^e_{LM} = \frac{1}{\mu_0\, ck} \left[\frac{c_k}{\varepsilon_0\, R_0 \hbar\, L(L+1)} \right]^{\frac{1}{2}} \frac{1}{(2L+1)!!}\, \frac{\hbar}{i}\, \nabla F(kr, \theta, \phi) \quad \dots (3.9)$$

Here $\qquad F(kr, \theta, \phi) = (kr)^L\, Y_{LM}(\theta, \phi) \qquad \dots (3.10)$

Now, to obtain the matrix element of Equation (3.5), we have

$$\int \psi_f^* \,(\mathbf{p}\cdot\mathbf{A} + \mathbf{A}\cdot\mathbf{p})\, \psi_i\, dV = \int \psi_f^* \,(\mathbf{p}\cdot\nabla F^* + \nabla F^* \cdot \mathbf{p})\, \psi_i\, dV$$

$$= \frac{i}{\hbar} \int \psi^* \,(\mathbf{p}\cdot\mathbf{p}\, F^* + \mathbf{p}\, F^*\, \mathbf{p})\, dV$$

Since $\nabla = \dfrac{i}{\hbar}\,\mathbf{p}$, one obtains after absorbing $\dfrac{i}{\hbar}$ in the constant,

$$\int \psi^* \,(\mathbf{p}\cdot\mathbf{A} + \mathbf{A}\cdot\mathbf{p})\psi_i\, dV = \int \mathbf{p}\,\psi^*\, p\, F^*\, \psi_i\, dV + \int \psi_f^* \,(\mathbf{p}\, F^*)\, p\, \psi_i\, dV \qquad \dots (3.11)$$

We have the following relations which are generally valid

$$(\mathbf{p}\, F^*)\psi = \mathbf{p}(F^*\,\psi) - F^*\, \mathbf{p}\,\psi$$

and $\qquad (\mathbf{p}\, F^*)p\psi = \mathbf{p}(F^*\, p\psi) - F^*\, p^2\psi$

Using the above relations, r.h.s. of (3.11) gives

$$\text{R.H.S.} = \int \mathbf{p}\,\psi_f^*\, [\mathbf{p}\, F^*\, \psi_i - F^*\, p\, \psi_i]\, dV + \int \psi^*\, \mathbf{p}\, (F^*\, \mathbf{p}\, \psi)\, dV - \int \psi^*\, F^*\, p^2\, \psi_i\, dV$$

$$= \int p^2\, \psi_f^*\, F^*\, \psi_i\, dV - \int \mathbf{p}\,\psi_f^*\, F^*\, (\mathbf{p}\,\psi_i)\, dV + \int \mathbf{p}\,\psi_f^*\, F^*\, \psi_i\, dV - \int \psi_f^*\, p^2\, \psi_i\, F^*\, dV$$

$$= \int p^2\, \psi_f^*\, F^*\, dV - \int \psi_f^* \,(p^2\, \psi_i)\, F^*\, dV$$

Now, adding and subtracting the term

$$2 M_p \int \psi_f^* u_r F^* \psi_i \, dV, \text{ one obtains}$$

$$\text{R.H.S.} = \int 2 M_p \left(\frac{p^2}{2 M_p} + u(r) \right) F^* \psi_f^* \psi_i \, dV - \int \psi^* F^* 2 M_p \left(\frac{p^2}{2 M_p} + u(r) \right) \psi_i \, dV$$

$$= 2 M_p (E_i - E_f) \int \psi_f^* F^* \psi_i \, dV$$

$$= 2 M_p \hbar\omega \int \psi_f^* F^* \psi_i \, dV \qquad \qquad \dots (3.12)$$

Since $\mathbf{A}^* = -A$, $p^* = -p$, and $\mu_N = \dfrac{e\hbar}{2 M_p}$, one obtains

$$-\frac{e\mu_0}{2 M_p} \int \psi_f^* (\mathbf{p} \cdot \mathbf{A} + \mathbf{A} \cdot \mathbf{p}) \psi_i \, dV = e\hbar \left[\frac{c_k}{\varepsilon_0 \hbar R_0 L(L+1)} \right]^{\frac{1}{2}} \left(\frac{L+1}{(2L+1)!!} \right) \int \psi_f^* F^* \psi_i \, dV \dots (3.13)$$

and

$$\int \psi^* \frac{\mu_N g_{sp}}{\hbar} (\mathbf{S}_p \cdot \mathbf{H}) \, dV = \frac{e\mu_0}{2 M_p} g_{sp} k \left[\frac{c_k}{\mu_0 \hbar R_0 L(L+1)} \right]^{\frac{1}{2}}$$

$$\frac{1}{(2L+1)!!} \hbar \int \psi_f^* \left(\mathbf{S}_p \cdot \frac{\mathbf{L} F}{\hbar^2} \right)^* \psi_i \, dV \quad \dots (3.14)$$

Now, substituting in (3.5), one obtains the γ–ray transition probability as,

$$P = \frac{2\pi}{\hbar} \frac{R_0}{\pi \hbar c} \frac{e^2 \hbar^2 ck}{\hbar \varepsilon_0 R_0 L(L+1)} \frac{1}{[(2L+1)!!]^2}$$

$$\times (L+1) \int \psi_f^* F^* \psi_i \, dV + g_{sp} \frac{\hbar\omega}{2 M_p c^2} \int \psi_f^* \frac{\mathbf{S}_p \cdot \mathbf{L} F^*}{\hbar^2} \psi_i \, dV \quad \dots (3.15)$$

Neglecting the term $\dfrac{\hbar\omega}{2 M_p c^2}$ being small, one obtains

$$P_{LM}^l = \frac{2k e^2 \hbar^2}{\hbar^3 \varepsilon_0} \frac{1}{[(2L+1)!!]^2} \left(\frac{L+1}{L} \right) \left| \int \psi_f^* F^* \psi_i \, dV \right|^2$$

$$= \frac{2 e^2 \omega}{\hbar c \varepsilon_0} \left(\frac{L+1}{L} \right) \frac{1}{[(2L+1)!!]^2} \left| \int \psi_f^* F^* \psi_i \, dV \right|^2$$

$$= \left(\frac{L+1}{L} \right) \frac{8\pi \alpha \, d\omega}{[(2L+1)!!]^2} \left| \int \psi_f^* (kr)^L Y_{LM}^* (\theta, \phi) \psi_i \, dV \right|^2 \quad \dots (3.16)$$

Where

$$\alpha = \frac{e^2}{4\pi \varepsilon_0 \hbar c} \qquad \qquad \dots (3.17)$$

$r^L \dfrac{1}{LM} (\theta, \phi)$ term in (3.16) is the *multipole* operator. The term represents the dynamic multipole moment giving rise to the transition.

One can write the nuclear wave functions ψ_i and ψ_f for a spherical symmetric potential as product of radial wave function $\dfrac{u(r)}{r}$, spin wave function x_s and angular wave function. Thus, we have

$$\psi = \frac{u_l(r)}{r} \phi_{jl\,mj}$$

$$= \frac{u_l(r)}{r} \sum_{m_s = -\frac{1}{2}}^{m_s = \frac{1}{2}} C\left(\frac{1}{2}, l, j; m_s\, m_j\right) x_{\frac{1}{2} m_s} Y_{l,\, m_l} \qquad \dots (3.18)$$

We can specify the proton state in the nucleus by its total angular momentum j, orbital angular momentum l and spin angular momentum $S = \dfrac{1}{2}$, with the corresponding quantum numbers m_j, m_l and m_s. Obviously, we can specify the state as $\psi(j, l, m_j)$ and the initial and final states as $\psi_i = \psi(j', l', m'_j)$ and $\psi_f = \psi(j, l, m_j)$ respectively. Since, there is no preference of direction of emission of the photon, the initial and final states are $(2j' + 1)$ and $(2j + 1)$ fold degenerate respectively summing over all the states $(2j' + 1)$ and $(2j + 1)$, the transition probability is obtained as

$$P^e(j', l' \to j, l) = 8\pi\omega\alpha\left(\frac{L+1}{L}\right)\frac{1}{[(2L+1)!!]^2}\left|\int u_l\, u_{l'}\,(kr)^L\, dr\right|^2 S_{j'\,l'\,L\,jl} \qquad \dots (3.19)$$

Where $\qquad S_{j'\,l'\,L\,jl} = \dfrac{1}{2j+1}\left|\displaystyle\sum_{m_j\, m_{j'}} |\phi^*_{jl\,m_j}\, Y^*_{LM}\, \phi_{j'\,l'}\, m'_j\, d\Omega\, dS|^2\right. \qquad \dots (3.20)$

The integral in (3.20) can be written as

$$\int \phi^*_{jl\,mj}\, Y_{j'\,l'\,m'\,Lj}\, d\Omega\, dS = \sum_{m_s\, m_{s'}} C\left(\frac{1}{2}\,lj, m_s\, m_j\right) C(l'\, j'\, m_{s'}\, m_{j'})$$

$$\int \chi_{\frac{1}{2} m_s}\, \chi_{\frac{1}{2} m_{s'}}\, dS \int Y^*_{l\,m_l}\, Y^*_{LM}\, Y_{l'\,m'_l}\, d\Omega \qquad \dots (3.21)$$

If $\qquad m_s = m_{s'},\ m_{l'} - m_l = M$ and $(l - L) < l' < (l + L),\ L \neq 0,$ $\qquad \dots (3.21a)$

the above integral gives non-vanishing value.

Unless $l + L^* + l'$ is even, the integral over the spherical wave function in Equation (3.20) is zero. This means that for electric radiation with $L = 0$ (dipole) $l + l'$ must be odd. Since the parity of a state is given by $(-1)^l$, then if $(l + l')$ is odd, then the parities of the two states must be different. For electric quadrupole radiation, i.e. $L = 2$, $l + l'$ is even and hence both l and l' must either be odd or even. Clearly, the parity of both the initial and final states is similar, i.e. parity is either odd or even.

We may note that in the absence of the exact knowledge of nuclear potential, the wave function $u_l(r)$ depends upon the model employed. In a simplest model, one may assume that both initial and final states, the proton is uniformly distributed over the nuclear volume, say spherical with radius R. In this model, one finds that

$$u_l = u_{l'} = r \sqrt{4\pi \rho_p} \qquad (0 \le r \le R)$$

$$= 0 \qquad (r > R)$$

Where ρ_p is the proton density in the nucleus. For a single proton $\rho_p = \dfrac{3}{4\pi R^3}$. Now, the integral over radial wave function in Equation (3.19) gives

$$4\pi \int_0^R \rho \, (kr)^L \, r^2 \, dr = (kr)^L \, \frac{3}{L+3} \qquad \qquad \dots (3.22)$$

Now, if we put the integral $S_{j' l' L' jl}$ equal to unity, then the transition probability comes out to be as Weisskopf's estimates. If we put the value of $k = \dfrac{\omega}{c}$, R in units of Fermi, i.e. 10^{-15} m and γ-ray transition energy in MeV, then we obtain the transition probability for electric multipole transition as

$$P_L^e = \eta^2 \, \frac{4.4(L+1)}{L[(2L+1)!!]^2} \left(\frac{3}{L+3} \right)^2 \left[\frac{\hbar\omega(\text{MeV})}{197} \right]^{2L+1} R^{2L} \times 10^{21}/\text{sec} \quad \dots (3.23)$$

Similarly, one can obtain the transition probability for a magnetic multipole radiation as

$$P_L^m = \eta^2 \, \frac{1.9(L+1)}{L[(2L+1)!!]^2} \left(\frac{3}{L+3} \right)^2 \left[\frac{\hbar\omega(\text{MeV})}{197} \right]^{2L+1} R^{2L-2} \times 10^{21}/\text{sec} \ \dots (3.24)$$

η^2 appearing in (3.23) and (3.24) gives the effective charge of the proton in the nucleus and is given by

$$\eta^2 = \left[\varepsilon \left(\frac{A-1}{A} \right)^2 + (-1)^L \left(\frac{Z-\varepsilon}{A^2} \right) \right] \qquad \qquad \dots (3.25)$$

Where $\varepsilon = 1$ is for a proton and $\varepsilon = 0$ is for neutron. A is the mass number and Z the atomic number of the nucleus emitting γ-rays.

We have obtained the transition probability for a proton making a transition. However, a neutron can also lead to the transition except that its effective charge is given by (3.25) we may remember that the transition takes place between two nuclear states and not individual nucleon, i.e. proton or neutron states. This clearly indicates that one must employ a more realistic wave function $u(r)$. We may note that the Weisskopf's estimate of transition probability describes the γ-ray life times only within a factor of about 10 and realistic nuclear wave functions are also available only for very light nuclei.

We have seen that γ-ray transition probability varies as $\left(\dfrac{E_r}{197} \right)^{2L+1}$. Since $E_r < 1$ MeV, the transition probability strongly reduces with the increase of the multipolarity L and the transition energy decreases. This indicates clearly the mixing of higher multipoles. This effect of the effective charge correction indicates that transitions in heavy nuclei with $L > 1$ take place preferably due to protons. One can obtain the reduced width of the transition by dividing the transition probability by the r.h.s. of Equation (3.23). According to Equation (3.19), the reduced width provides the value of $S_{j' l' L jl}$ and correction to the radial matrix element due to proper wave functions $u(r)$. One can equate the reduced width to a matrix element $|m|^2$ which takes into account of both the above mentioned factors. Weisskopf has taken this matrix element to be unity in his estimates. One can obtain the value of matrix element $|m|^2$ by comparing the experimental results of the γ-ray life times with that of Weirskopf's estimates.

The γ-ray transition takes place only by the lowest allowed multipole due to the strong dependence on L. As stated earlier, the electric multipole transitions are depicted as E_1 (electric dipole), E_2 (electric quadrupole), E_3 (electric octopole) and so on. Similarly, magnetic transitions are depicted as M_1, M_2, M_3, M_4, etc. Parity considerations allows the mixing of the multipoles as $E1$, $M2$, $E3$ and $M4$ transitions or $M1$, $E2$, $M3$, $E4$ transitions. However, one never finds mixing of $E1$ with $M2$. The transition probability of $E2$ transition is normally $10^{-4} - 10^{-5}$ of that of $M1$ transition. There are number of cases reported where $E2$ is mixed with $E1$. As we know that $E2$ transition takes place due to quadrupole moment operator. In the case of a deformed nucleus the quadrupole moment is large which makes the matrix element $|m|^2$ very large. This permits a comparable transition probability for $M1$ and $E2$ transitions. Generally in a decay more than one type of transition is possible, e.g. for a $3^+ \rightarrow 2^+$ decay $M1$, $E2$, $M3$, $E4$ and $M5$ transitions are all possible. But because of the strong dependence on multipole order only the lowest or the two lowest multi polarities are usually important, i.e. for a $3^+ \rightarrow 2^+$ transition $M1$ and $E2$. The multipole mixing ratio 3.8, for example given by

$$\delta^2 = \frac{\Gamma_\gamma (E2)}{\Gamma_\gamma (M1)} \qquad \qquad \ldots (3.26)$$

Where Γ_γ is the radiative width, can be obtained from conversion coefficients or angular correlation experiments. We will study this in latter sections.

3.11.2 Selection Rules

The radiation patterns of oscillating dipoles, quadrupoles, octupoles, etc. are different. An important aspect of the classification of the types of γ-emission is that it gives information concerning the nuclear spin of the excited state. We have already studied that a distinction is made between electric and magnetic multipoles according to the parity associated with the electromagnetic radiation. This in turn is directly determined by whether or not there is a difference in parities of the nuclear states between which the transition occurs. We have seen that electric multipole radiation of order L has opposite parity to that of magnetic radiation of the same multipolarity L in accordance with the parity rule

$$P^E = (-1)^L \qquad \qquad \ldots (3.27)$$

Magnetic multipole radiation of order L has parity

$$P^M = -(-1)^L = (-1)^{L + 1} \qquad \qquad \ldots (3.28)$$

The main selection rules governing the γ-transitions are given in 3.21(a), are

$$m_S = m_{S'}$$

$$|m_{l'} - m_l| = M$$
$$|l' - l| \leq L \leq |l' + l|$$

Where M is the projection of L along the Z-direction.

In addition to the above mentioned selection rules, there are selection rules involving the isotopic spins of the states involved. According to these selection rules, One finds that γ-transition takes place if

$$\Delta T_3 = 0 \text{ and } \Delta T = 0 \pm 1$$

Where ΔT_3 denote the change in the charge of the nucleus, which obviously does not happen. The selection rules for the isotopic spin are applicable to the very light nuclei where states of different isotopic spins are observed. However, in light nuclei too, the excited states do not differ in isotopic spin T by more than 1. Obviously, the selection rule for isotopic spin is of little practical importance.

When the number of neutrons and protons in a nuclei are equal, the isotopic selection rule is $\Delta T = \pm 1$. However, the isotopic spin is not a good quantum number and one can have transition with $\Delta T = 0$ and the reduced with for such transition is quite less than for $\Delta T = \pm 1$ transitions.

We have enough data on the life time of γ-ray transitions of different multipoles. One can compare this with the Weisskopf's estimates to obtain the reduced width or the matrix element $|m|^2$ of the transition.

In the case of light nuclei the calculated value of the matrix element is ~ 0.05 whereas the measured value is 0.044. There is a good agreement between the two. The agreement for $M1$ transition is also fairly good with $|m|^2 \simeq 0.15$. The experimental value for $E2$ transition give $|m|^2 \simeq 5$. $|m|^2 \simeq 30$ for the only known $E3$ transition from the 6.14 MeV level of ^{16}O.

In heavy nuclei for $E1$ transitions, $|m|^2$; 10^{-5}, whereas for deformed nuclei $|m|^2 \simeq 10^{-2}$. In other nuclei for $E2$, $|m|^2$; 10^{-2}. $|m|^2$; 10^{-2} in heavy nuclei for $E3$, $E4$, $M1$ and $M2$ transitions. $|m|^2 \sim 1$ for $M3$ and $|m|^2 \sim 5$ for $M4$ transitions.

The measurment of lifetime of the nuclear states helps one to obtain information about the multipolarity of the transition. We may note that in deformed nuclei it gives approximately the quadrupole moment.

3.11.3 Isomerism

Figure 3.42 shows that electromagnetic transitions of high multipolarity and low energy are relatively slow processes. Excited levels of nuclei which can only decay by such transitions may therefore have a long life on a nuclear time

scale, say half life $\left(\dfrac{t_1}{2}\right) > 10^{-6}$ s and nuclei excited to these

levels are said to be *isomeric* with respect to their ground state. From the Weisskopf's estimates, one finds that most of the $E2$, $M3$, $E4$ and $M4$ transitions have life times greater than 10^{-6} s and therefore they form isomeric states.

Nuclear isomerism is found in groups of nuclei located just below the major shell closures at Z, $N = 50$, 82 and 126 as shown in Figure 3.43. This can be understood early on

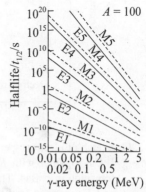

Fig. 3.42 Lifetime-energy relations for γ radiation according to the single-particle formula of Weisskopf, for a nucleus of mass number $A = 100$.

the single-particle shell model. Isomerism in $^{137}_{56}\mathrm{Ba}^m$ is shown in Figure 3.44. Isomeric decay usually shows strong internal conversion.

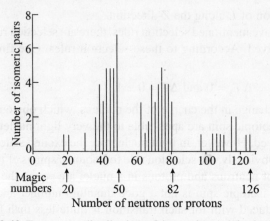

Fig. 3.43 Frequency distribution of isomeric nuclei of odd A.

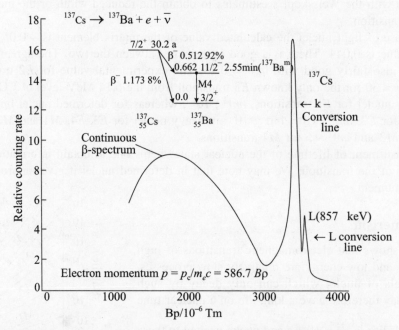

Fig. 3.44 Sketch of electron spectrum seen in a simple 180° spectrometer of the β-decay electrons from $^{137}\mathrm{Cs}$, together with K and L internal conversion electrons from the 0.662 MeV isomeric transition in $^{137}\mathrm{Ba}^m$. The decay scheme is shown, with energies in mega-electron volts. The K and L vacancies lead to X-ray and Auger electron emission from the Ba atom.

3.11.4 Internal Conversion of γ-rays

Besides γ-ray emission, electromagnetic decay of an excited nuclear state can also be accomplished through *internal conversion* and *internal pair production* processes. In *internal conversion*, an atomic electron is ejected in place of a γ-ray being emitted from the nucleus, Thus internal conversion is the electromagnetic process in which an excited nuclear state de-excites with the ejection of an atomic electron. The interaction is the electromagnetic interaction between the nucleus and the bound electron. The kinetic energy of the electron is equal to the de-excitation energy $E_i - E_f (= E_\gamma)$ minus the (atomic) binding energy of the electron. As a result electrons emitted from internal conversion processes have discrete energies and can therefore be distinguished form the continuous spectrum of electrons emitted in β⁻-decays. Since both types of decay can take palce from the same excited state in medium and heavy nuclei, the difference makes it possible to distinguish between them.

One can visualize the internal conversion in the following way. When a nucleus de-excites, say, either by a nucleon jumping from one single-particle orbit to another or by a change of the angular momentum of the nucleus as a whole, a sudden disturbance is sent to the surrounding electromagnetic field. Atomic electrons, especially those in the inner most orbits, such as the $K -$ and $L -$ orbits, spend a large fraction of time in the vicinity of the nucleus, the source of the electromagnetic field of interest here. It is therefore probable for the disturbance in the electromagnetic field to transfer all the excess energy in the nucleus to one of the electrons and eject it from the atomic orbit. This is similar to the automic *Auger effect*, where, instead of emitting a photon when an atomic electron de-excites from a higher to a lower energy orbit, one of the atomic electrons is ejected.

Internal conversion electrons appear in magnetic spectrum as homogeneous lines of kinetic energy

$$E_e = E_\gamma - B \qquad \qquad \dots (3.1)$$

Where B is the binding energy of the electron ejected, and are often superimposed upon the continuous distribution arising from an associated β-decay. Figure 3.44 shows two internal conversion lines from ^{137}Cs.

$L -$ Conversion line is actually due to L_I, L_{II}, L_{III} etc. subshells. Figure 3.45 shows the conversion spectrum of ^{165}Ho, 94.6 transition.

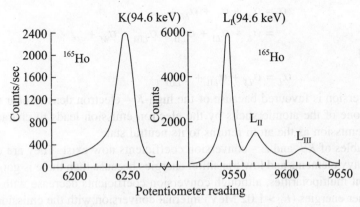

Fig. 3.45 $K -$ and $L -$ subshell conversion spectrum of 94.6 keV transition in ^{165}Ho.

The *internal conversion coefficient* (α) for a particular transition is defined by the ratio of the probability λ_e of emission of an electron to that of emission of a photon, i.e. λ_γ.

$$\alpha = \frac{\lambda_e}{\lambda_\gamma} \qquad \qquad \dots (3.2)$$

If the number of internal conversion electrons and γ-ray photons emitted per second from a radioactive source are N_e and N_γ respectively, then we have

$$\alpha = \frac{\lambda_e}{\lambda_\gamma} = \frac{N_e}{N_\gamma} \qquad \qquad \dots (3.3)$$

The total probability (λ) of the decay of the excited state is the sum of the probabilities for decay by emission of an electron (λ_e) and a photon (λ_γ) per second. Thus,

$$\lambda = \lambda_e + \lambda_\gamma = \lambda_\gamma \left(1 + \frac{\lambda_e}{\lambda_\gamma} \right) = \lambda_\gamma (1 + \alpha) \qquad \dots (3.4)$$

The value of α – is found to depend on transition energy, multipole character of transition and the atomic number Z. One can experimentally determine the value of a by measuring the intensities of γ-lines and corresponding conversion lines.

The life time T of an excited state and the partical life times for emission of a conversion electron (T_e) and that for emission of a photon (T_γ) are related as

$$\frac{1}{T} = \frac{1}{T_e} + \frac{1}{T_\gamma} \qquad \qquad \dots (3.5)$$

Further, the partial life time for time for γ-ray emission (T_γ) is related to life time T of the excited state as

$$T_\gamma = (1 + \alpha) T \qquad \qquad \dots (3.6)$$

Internal conversion is also possible (though less, as compared to the K-shell) for higher atomic shells, L, M, etc. Thus

$$\alpha = \alpha_K + \alpha_L + \alpha_M + \dots$$
$$= \alpha_K + \alpha_{LI} + \alpha_{LII} + \alpha_{LIII} + \alpha_M + \dots \qquad (3.7)$$

For the L – shell

$$\alpha_L = \alpha_{LI} + \alpha_{LII} + \alpha_{LIII} \qquad \qquad \dots (3.8)$$

Usually, K-conversion is favoured bacause of the high-K – electron density near the nucleus. The vacancy left in one of the atomic shells by the electron emission leads to characterstic X-ray or Auger electron emission as the atom returns to its neutral state.

Extensive tables of K – and L – conversion coefficients now exist. They are especially useful in the case of heavy elements and low transition energies because then there is good discrimination between different multipolarities, although conversion coefficients decrease with increasing γ-ray energy. For higher energies ($E_\gamma > 1.02$ MeV) internal conversion with the emission of a positron – electron pair is an additional mode of decay and the probability of this process *increases* with transition process.

Internal conversion is the main mode of decay for the *EO* monopole transition between states of zero spin, and in such cases (e.g. in *Ra C'*, 1.414 MeV) we speak of *total internal conversion*.

O → O Transitions   If the initial and final nuclear states each have zero, γ-emission is impossible as the γ-ray photon must carry at least one unit of angular momentum. However, the $O \rightarrow O$ transition is possible by internal conversion provided the electrons wave function does not vanish at the origin. This means that the electron has a finite chances of being within the nuclear volume. The $O \rightarrow O$ transitions are usually distinguished experimentally by the emission of conversion electrons and the complete absence of γ-ray emission. Obviously, the internal conversion coefficient (α) is infinite.

3.11.5 Probability of Internal conversion of γ-rays

It is found that internal conversion coefficient α depends on: (i) E_γ, the energy of γ-ray transition, (ii) Multipolarity L of the γ-ray transition, (iii) Atomic number Z of the atom, and (iv) The shell or subshell of the atom in which the internal conversion is taking place. If the γ-ray transition has the mixed polarity and α_L, the internal conversion coefficient in a particular shell for the transition multipolarity L, then α in the shell is expressed as

$$\alpha = \Sigma a_L \, \alpha_L \qquad \qquad ... (3.1)$$

Where a_L denotes the intensity of the γ-transition of multipolarity L. One can calculate α with the help of follwing assumptions:

(i) Internal conversion is taking place in the K-shell of the atom.
(ii) The wave function of the K-shell electron is hydrogen like atom, i.e. we are not considering the effects of the electrons in the outer shells.
(iii) The interaction between the excited nucleus and the electron in the K-shell is purely electrostatic.
(iv) After ejection, the electrons behaves as a plane wave and does not experience any electromagnetic interaction.
(v) The kinetic energy of the electron is much larger than its binding energy.

In the light of above assumptions, one can calculate the probability for the internal conversion. On dividing this probability by the probability for γ-ray emission, one obtains the internal conversion coefficient of the γ-ray.

Let ψ_i and ψ_f are the initial and final states wave functions of the nucleus respectively. These depend upon the nuclear coordinates. Let ϕ_i and ϕ_f are the wave functions of K-shell electron and free electron respectively. We can write the wave function for hydrogen like K-electron as

$$\phi_i = (\pi a^3)^{-\frac{1}{2}} e^{-\frac{R}{a}} \qquad \qquad ... (3.1)$$

Where a_0 is Bohr radius, and $a = \dfrac{a_0}{Z}$. Z is the atomic number of the atom. The wave function ϕ_f of plane wave emitted electron is

$$\phi_f = V^{-\frac{1}{2}} e^{i\,\mathbf{K}_e \cdot \mathbf{R}} \qquad \qquad ... (3.2)$$

Where $\mathbf{K}_e$ is the wave vector and $\mathbf{R}$ the positron vector of the electron enclosed in the volume V.

We have the transition probability for the nucleus going from the state ψ_i to ψ_f and one of the two K-shell electrons going from the state ϕ_i to ϕ_f as

$$P = 2 \cdot \frac{2\pi}{\hbar} \, |H_{if}|^2 \, \frac{dn}{dE} \, d\Omega \qquad \qquad ...(3.3)$$

Where $\dfrac{dn}{dE} \, d\Omega$ represents the number of states of the ejected electrons per unit energy range for the electron directions within the solid angle $d\Omega$. Now, the number of states in volume V and with momentum p and $p + dp$ moving in solid angle $d\Omega$ is given by

$$dn = \frac{p^2 \, dp}{(2\pi\hbar)^3} \, d\Omega$$

Since

$$E = \frac{p^2}{2m} \qquad \therefore \quad dE = \frac{p \, dp}{m}$$

$$\therefore \qquad \frac{dn}{dE} = \frac{V \, mp}{(2\pi\hbar)^3} = \frac{V \, m\hbar \, K_e}{(2\pi\hbar)^3} \qquad \qquad ...(3.4)$$

One can write the matrix element as

$$H_{if} = \sum_{i=1}^{Z} \int \phi_f^* \, \psi_f^* \, \frac{e^2}{4\pi\,\varepsilon_0 \, |\mathbf{R} - \mathbf{r}_i|} \, \phi_i \, \psi_i \, dV \qquad \qquad ...(3.5)$$

We may note that the electrostatic interaction is between a proton at position r_i and the electron at positron R and the summation is over all the protons within the nucleus. Making use of (3.1) and (3.2), (3.5) takes the form

$$H_{if} = \frac{(\pi\,a^3\,V)^{-\frac{1}{2}}}{4\pi\,\varepsilon_0} \sum_{i=1}^{Z} \int d^3R \int d^3R \, e^{-i\,\mathbf{K}_e \cdot \mathbf{R}} \, \psi_f^* \left| \frac{e^2}{|\mathbf{R} - \mathbf{r}_i|} \right| e^{-\frac{R}{a}} \, \psi_i \qquad ...(3.6)$$

The integration in (3.6) is over the volume of the nucleus (d^3r) and the volume of the electron (d^3R). Since the major contribution to the integral is when $\mathbf{R} > \mathbf{r}_i$, and hence

$$\frac{1}{|\mathbf{R} - \mathbf{r}_i|} = \sum_{L=0}^{\infty} \sum_{m=-L}^{L} \left(\frac{4\pi}{2L+1} \right) \left(\frac{r_i^L}{R^{L+1}} \right) Y_{LM}(\theta, \phi) \, Y_{LM}(\theta_i, \phi_i) \qquad ...(3.7)$$

Where θ, ϕ corresponds to polar angles of $\mathbf{R}$ and θ_i and ϕ_i to the polar angles of r_i. Now, making use of (3.7), (3.6) reduces to

$$H_{if} = \sum_{L=0}^{M} \sum_{m=-L}^{L} \frac{4\pi e}{(\pi\,a^3\,V)^{\frac{1}{2}}} \frac{1}{4\pi\,\varepsilon_0\,(2L+1)} \, H_{LM} \, J_{LM} \qquad ...(3.8)$$

Where H_{LM} is the electric multipole matrix element for the transition from the nuclear state ψ_i to ψ_f and J_{LM} is the matrix element for the change of the state of the electron. Thus

$$H_{LM} = \sum_{Z=1}^{Z} \sum_{L=1}^{\infty} \sum_{m=-L}^{L} e \int \psi_f^* \, r_i^L \, Y_{LM}(\theta_i, \phi_i) \psi_i \, d^3 r \qquad \text{... (3.9)}$$

and

$$J_{LM} = \sum_{L=1}^{\infty} \sum_{M=-L}^{L} e^{i\,\mathbf{K}_e \cdot \mathbf{R}} Y_{LM}(\theta, \phi) e^{-\frac{R}{a}} R^{-(L+1)} \, d^3 r \qquad \text{... (3.10)}$$

Since the energy imparted to the electron is quite large in comparison to its binding energy in the K-shell and therefore the wavelength $(k_e)^{-1}$ of ejected electron is quite smaller than the radius of K-shell. This means $k_a \gg 1$. In the approximation $e^{-\frac{R}{a}} \approx 1$. Now, one can evaluate the integral J_{LM} with the help of addition theorem of spherical harmonics. Thus,

$$J_{LM} = 4\pi (i)^{-L} \frac{(ke)^{L-2}}{(2L-1)!!} \, Y_{LM}(\theta, \phi) \qquad \text{... (3.11)}$$

Where θ and ϕ are the polar angles made by vector $\mathbf{k}_e$ w.r.t. an arbitrary Z-axis. Now, we have the matrix element for emission for an internally converted K-electron as

$$H_{if} = \frac{(4\pi)^2 \, e(i)^{-L} (k_e)^{L-2}}{4\pi \, \varepsilon_0 \, (\pi \, a^3 \, V)^{\frac{1}{2}} \, (2L+1)(2L-1)!!} \, Y_{LM}(\theta, \phi) \, H_{LM}$$

or

$$H_{if} = \frac{(4\pi)^2 \, e(i)^L}{4\pi \, \varepsilon_0 \, (\pi \, a^3 \, V)^{\frac{1}{2}} \, [(2L+1)!!]} \, Y_{LM}(\theta, \phi) \, H_{LM}(\theta_i \, \phi_i)$$

or

$$|H_{if}|^2 = \frac{256\pi^4 \, e^2}{(4\pi \, \varepsilon_0)^2 \, \pi \, a^3 \, V \, [(2L+1)!!]^2} \, |H_{LM}|^2 \int Y_{LM}^* \, Y_{LM} \, d\Omega \qquad \text{... (3.12)}$$

Thus we obtain the probability for internal conversion of γ-ray as

$$P_e = 2 \frac{2\pi}{2} \frac{dn}{dW} \int |H_{if}|^2 \, d\Omega$$

$$= \frac{4\pi}{\hbar} \left(\frac{m\hbar \, \text{keV}}{(2\pi\hbar)^3} \right) \left(\frac{256\pi^4 \, e^2}{\pi \, a^3 \, V} \right) \frac{k_c^{2L-4}}{[(2L+1)!!]^2} |H_{LM}|^2$$

$$= \frac{128\pi \, m^2 \, z^2 \, k_c^{2L-3}}{\hbar^3 \, a_0^3 \, [(2L+1)!!]^2} |H_{LM}|^2 \qquad \text{... (3.13)}$$

Where

$$a = \frac{a_0}{Z}.$$

In the previous section we have obtained the probability of emission of a γ-ray photon as

$$P_\gamma = 8\pi \, \alpha \, \omega \, \frac{L+1}{L} \, \frac{1}{[(2L+1)!!]^2} \left| \int \psi_f \, (k_r \, r)^L \, Y_{LM}(\theta, \phi) \psi_i \, d^3 r \right|^2 \qquad \text{... (3.14)}$$

Dividing (3.13) by (3.14) and putting $a_0 = \dfrac{4\pi\,\varepsilon\,\hbar^2}{me^2}$ and $\alpha = \dfrac{c^2}{4\pi\,\varepsilon\,\hbar^2}$, one obtains the internal conversion coefficient for multipolarity L as,

$$\alpha_K^L = \frac{P_L(k)}{P_L(h\nu_0)} = \frac{16\,Z^3}{a_0^{\,4}}\,\frac{L}{L+1}\,\frac{k_e^{\,2L-3}}{k^{2L+1}} \qquad \ldots (3.15)$$

Since the wave vector k for photon $= \dfrac{\omega}{c}$ and the wave vector k_e for electron is given by

$$\frac{\hbar^2\,k_e^{\,2}}{2m} = \hbar\omega - \text{B.E.} \approx \hbar\omega \qquad \ldots (3.16)$$

$$k_e = \sqrt{\frac{2\,\omega m}{\hbar}} \qquad \ldots (3.17)$$

Substituting the values of k and k_e in (3.15), one obtains

$$\alpha_K^L = \frac{16\,Z^3}{a_0^{\,4}}\left(\frac{L}{L+1}\right)\left(\frac{2m\omega}{\hbar}\right)^{\frac{2L-3}{2}}\left(\frac{c}{\omega}\right)^{2L+1}$$

$$= 16\,Z^3\left(\frac{L}{L+1}\right)\left(\frac{me^2}{4\hbar^2\,\pi\,\varepsilon_0}\right)^{4}\left(\frac{2mc^2}{\hbar\omega}\right)^{L+\frac{5}{2}}\frac{\hbar^4}{2^4\,m^4\,c^4}$$

$$= Z^3\left(\frac{e^2}{4\pi\,\varepsilon_0\,\hbar c}\right)^{4}\left(\frac{L}{L+1}\right)\left(\frac{e\,mc^2}{\hbar\omega}\right)^{L+\frac{5}{2}}$$

$$= Z^3\left(\frac{L}{L+1}\right)\alpha^4\left(\frac{2mc^2}{E_\gamma}\right)^{L+\frac{5}{2}} \qquad \ldots (3.18)$$

We may note that (3.18) is an approximate expression for internal conversion coefficient for electric multipole transitions in the K-shell of an atom as the electron wave function should be treated relativistically. We may also see that α_k^L increases strongly with the increase in the atomic number Z (since $\propto Z^3$) and multipolarity L of the γ-ray transition, but decreases with the increasing transition energy $\hbar\omega$. The *screening effect* of the outer shell electrons is quite significant for high atomic number (Z) atoms.

In order to obtain the internal conversion coefficient in the magnetic multipole transitions, we have to take the electromagnetic interaction between the nucleus and the atomic electrons.

The internal conversion coefficient for different energy γ-rays, in different shells of almost all the elements for different electric and magnetic multipole radiations have been calculated by Rose. These values have been tabulated. Rose used relativistic Dirac equations for the description of electron state and suitable theory to take account the finite size of the nucleus and screening effects.

The variation of *K*-shell internal conversion coefficients with transition energy and multipole order in atoms with $Z = 40$ is shown in Figure 3.46. We can see that the internal conversion coefficient for magnetic transition is about 10 times of the internal conversion coefficient of an electric transition of the same with polarity.

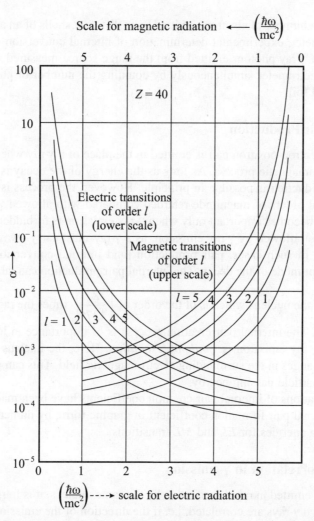

Scale for magnetic radiation $\longleftarrow \left(\dfrac{\hbar\omega}{mc^2}\right)$

$Z = 40$

Electric transitions of order l (lower scale)

Magnetic transitions of order l (upper scale)

$l = 5 \quad 4 \quad 3 \quad 2 \quad 1$

$l = 1 \quad 2 \quad 3 \quad 4 \quad 5$

$\alpha \longrightarrow$

$\left(\dfrac{\hbar\omega}{mc^2}\right) \dashrightarrow$ scale for electric radiation

Fig. 3.46 *K*-shell internal conversion coefficients for an element ($Z = 40$). Figure shows the lower scale for electric multipole radiations and upper scale for magnetic multipole radiation l indicates multipolarity 2^l.

We may also note that the internal converion coefficient is large for transitions for which γ-emission probability is low ($\hbar\omega$ small and L large).

Internal conversion is important in heavier nuclei for two reasons. In the first place, the radii of atomic electron orbits are smaller because of the stronger Coulomb fields provided by heavy nuclei. The probability of transition is increased as a result of the large overlap between the wave functions of the nucleus and inner shell atomic electrons. For this reason, the electrons ejected come mainly from the innermost shells. In the second place, the stronger Coulomb field in heavy nuclei exerts a larger influence on the surrounding. For these reasons, the importance of internal conversion increases roughly as Z^3 and becomes competitive with γ-ray emission for medium and heavy nuclei.

By measuring the number of electrons emitted from different shells of an atom with the help of a β-particle spectrometer, experimental determination of internal conversion coefficient has been made. The number of γ-ray photons emitted from the range source measured many times with the help of β-particle spectrometer simultaneously by counting the number of photoelectrons ejected from a very thin gold foil.

3.11.6 Internal Pair Production

In this process, an electron-positron pair is emitted in the place of a γ-ray when an excited nucleus decay through electromagnetic process. As long as the energy of the decay is greater than $2\,m_e c^2$ ≈ 1.02 MeV, pair production is possible in principle. However, the process is not an efficient one and is usually several orders of magnitude retarded compared with allowed γ-ray emissions. Pair production therefore becomes important only when γ-ray emission is forbidden. For example, a O^+ $\rightarrow O^+$ transition is not allowed by γ-ray emission, as a γ-ray must carry away at least one unit of angular momentum. In such cases, pair production (and internal conversion for heavy nuclei) becomes the dominant mode of the decay. The internal pair production can take place anywhere in

the Coulomb field of the nucleus which is of the order of $\left(\dfrac{Z}{137}\right)^2$ times the radius of the K-shell of

the atomic electrons. The internal pair creation process is of importance in low Z-elements.

The inverse of γ-ray emission is *Coulomb excitation*. Here, the nucleus is excited to higher state as a result of changes in the surrounding electromagnetic field. This can place, for e.g., as the result of a charged particle passing nearby.

Extensive calculations of internal pair creation coefficients have been made. Rose was able to present the total internal pair formation coefficient in graphic form, by numerical integration, as a function of transition energies for *EL* and *ML* transitions.

3.11.7 Angular Correlation in γ-emission

When two γ-rays are emitted in rapid succession by the same nucleus, it is found that the directions of emission of the two γ-rays are correlated, i.e. if the direction of the emission of the first gamma (γ) ray is taken as Z-axis, the probability of the second γ-ray falling into the element of solid angle $d\omega$ is not constant but depends mainly upon the angle θ between the directions. This type of correlation is known as angular correlation in γ-rays or $\gamma - \gamma$ correlation. There are also similar angular correlations for other pairs of successive radiations, e.g. $\alpha - \gamma$, $\beta - \gamma$, $\beta - e^{-1}$, $\gamma - e^{-1}$, ... etc. The measurment of angular distribution of γ-radiations or of the angular correlation between two radiations can yield valuable information on multipole mixing ratios or spins of nuclear levels. $\gamma - \gamma$ correlation is especially observed when the excited states (Figure 3.47) has a spin which differs by several units of $\hbar$ from that of the ground state and the level of intermediate spin lies between the edcited state and the ground state.

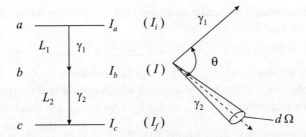

Fig. 3.47 Two successive γ-ray transitions decay scheme. L_1 and L_2 are multipolarities of γ_1 and γ_2 respectively.

We see that the nuclear excited state (spin I_c) de-excites by a cascade of γ-rays γ_1 and γ_2 through an intermediate state (spin I_b) to the final state (spin I_c). The multipolarity of γ-rays γ_1 and γ_2 are L_1 and L_2 respectively. A typical angular correlation experiment is shown in Figure 3.48(a) and correlation pattern is shown in Figure 3.48(b).

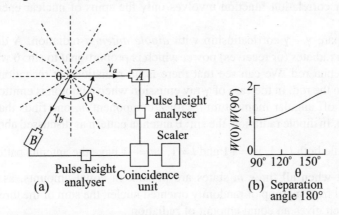

Fig. 3.48 Angular correlation of successive radiations from a radioactive nucleus at O. (a) Counters; (b) Correlation pattern

If the intermediate state (Figure 3.47) is sufficiently short-lived, coincidences between counts in the two counters (to accept signals γ_1 and γ_2) yield the angular correlation between the two γ-rays. The counting (coincidence) rate is measured (between γ_1 and γ_2) as a function of θ between the directions of emission of γ_1 and γ_2. One can define the *directional correlation*, $W(\theta)\,d\Omega$ as the relative probabiltiy that γ_2 ray is emitted into the solid angle $d\Omega$ at an angle θ w.r.t. γ_1 and expressed as

$$W(\theta) = 1 + A_2 P_2(\cos\theta) + A_4 P_4(\cos\theta) + \cdots = \sum_{i=0}^{L} A_{2i}\, P_{2i}(\cos\theta) \qquad \ldots (3.1)$$

Where A_{2i} are the coefficients which depends on the spins of levels a and b (Figure 3.47) and on L_1 and L_2, and $P_{2i}(\cos\theta)$ are the Legendre Polynomials having values:

$$P_2(\cos\theta) = \frac{1}{2}(3\cos^2\theta - 1),\ P_4(\cos\theta) = \frac{1}{8}(35\cos^4\theta - 30\cos^2\theta + 3),\ \ldots \qquad \ldots (3.2)$$

An equivalent and more common is usually employed for comparison with experiment is the power series in even powers of cos θ

$$W(\theta) = 1 + a_2 \cos^2\theta + a_4 \cos^4\theta + \dots a_{2^L} \cos^{2^L} \theta \qquad \dots (3.3)$$

Where a^{2^L} coefficients are the functions of the angular momenta I_a, I_b, I_c, L_1 and L_2 but not of relative parities of levels. The highest power of cos θ which appear in the expansion is governed by the smallest of the three quantities $2\,I_b$, $2\,L_1$ and $2\,L_2$. Obviously, for quadrupole radiation, $\cos^4\theta$ is the highest power encountered. This means that for $I_b = 0$ or $\dfrac{1}{2}$, $A_2 = 0$, $W(\theta) = 1$ and the angular correlation distribution will be *isotropic*. Clearly, if an angular correlation is to be observed, the intermediate state must, therefore, have a spin of at least 1.

The complexity of the angular distribution is expressed by the term *anisotropy*, defined as

$$\text{Anisotropy} = \frac{W(180°)}{W(90°) - 1} = a_2 + a_4 + \dots a_{2^L}$$

Obviously, angular correlation function involves only the spins of nuclear energy levels and not their parities.

One can compare γ – γ corrleatioship with *dipole antenna* radiation. A dipole antenna is a straight rod, which radiates (or receives) power which is proportional to $\sin^2\theta$ where θ is the angle relative to the antenna rod. We can see that there is no radiation (or reception) of power in the direction parallel to the rod. In the case of γ-ray emission when a photon is emitted during a nuclear transition carrying off angular momentum l with Z component m, one finds that (l, m) determine the antenna pattern. In dipole radiation, the $\sin^2\theta$ antenna pattern as discussed above is characterstic of $l = 1$, $m = 0$; while both $l = 1$, $m = +1$ and $l = 1$, $m = -1$ have the antenna pattern $\dfrac{1}{2}(1 + \cos^2\theta)$.

We may note that when all three m states are present in equal amounts, as is the case when radiation is emitted from a group of randomly oriented nuclei, the sum of the three antenna patterns is isotropic and each gives an equal amount of radiation.

Let us consider a γ-ray transition from an excited state of nucleus characterised by spin $\mathbf{I}_i$ to another state with spin $\mathbf{I}_f$ emitting a photon having angular momentum $\mathbf{L}$. We have

$$\mathbf{I}_i = \mathbf{I}_f = \mathbf{L} \qquad \dots (3.4)$$

or $\qquad |I_i - I_f| \le L \le |I_i + I_f|$

The component of $\mathbf{L}$ along some fixed direction Z is $L_Z = M\hbar$, where $L^2 = L(L + 1)\hbar^2$. We know that each nuclear state with spin I is composed of $(2I + 1)$ magnetic states which may or may not be degenerate depending upon the enviornment of the nucleus. In the initial and final states, the magnetic subleveles have magnetic quantum number m_i and m_f respectively. For γ-transition, the selection rules demand that $m_i - m_f = M$.

The photons emitted in a transition from m_i to m_f substates possesses a characterstic angular distribution $F_L^M(\theta)$ which is independent of spins I_i and I_f. Here θ is the angle between the directions of Z-axis and emitted γ-ray photon. For a dipole radiation with $L = 1$ and $M = 0$ or ± 1, we have the angular distribution functions as

$$F_1^0(\theta) = 3 \sin^2\theta = 2 - 2\,P_2(\cos\theta)$$

$$F_1^{\pm 1}(\theta) = \frac{3}{2}(1 + \cos^2 \theta) = 2 + 2P_2(\cos \theta)$$

Here P_2 are the second order Legendre polynomials.

For calculating the angular distribution of γ-rays emitted during the transition from state I_i to I_f, one will have to take into account the angular distribution for ech transition $m_i - m_f$ and its relative intensity as the energy separation between the various magnetic sublevels of a nuclear state is very small and one cannot observe it by any γ-ray spectrometer. Let the relative population of each of m_i states is $P(m_i)$ and the transition probability from m_i to m_f is $G(m_i\, m_f)$, then the angular distribution of the γ-rays for state I_i to state I_f is,

$$F_L(\theta) = \sum_{m_i, m_f} P(m_i)\, G(m_i\, m_f)\, F_L^M(\theta) \qquad \text{... (3.5)}$$

$m_i \rightarrow m_f$ transition probability is a product of a *nuclear factor* and *geometrical factor*, where the nuclear factor is independent of m_i and m_f and is a constant and geometrical factor gives the transition probability $G(m_i,\, m_f)$. Using Clebsch-Gordon coefficients, the state $|I_i,\, m_i>$ can be expended in terms of the state $|I_f,\, m_f> |L,\, M>$ as

$$|I_i,\, m_i> = \sum_{m_f,\, M} |I_f,\, m_f> |L,\, M> <I_f,\, m_f;\, L,\, M|\, I_i,\, m_i> \qquad \text{... (3.6)}$$

The last term in (3.6) is Clebsch-Gorden coefficient. One obtains the transition probability for a given values of I_i, I_f, m_i, m_f and L as

$$G(m_i,\, m_f) = <I_f,\, m_f;\, L,\, M|\, I_i,\, m_i> \qquad \text{... (3.7)}$$

We may note that the relative population of $P(m_i)$ of the substates depends upon the energies of the substates and on the manner in which the state I_i was created. The angular distribution $F_2(\theta)$ is isotropic if states mi are all equally populated. If a nucleus is cooled below 1 K in a magnetic field, then only one value of m_i is possible, and the distribution of the emitted γ-rays is anisotric in such a case. We have

$$F_L(\theta) = G(m_i,\, m_f)\, F_L^M(\theta) \qquad \text{... (3.8)}$$

3.11.7.1 Angular Correlation in a $\gamma - \gamma$ Cascade

Let us consider a nucleus in an excited state I_a (or I_i) making a transition to state I_b (or I) with the emission of γ_1 ray of multipolarity (L_2) as shown in Figure 3.47. Let the direction of emission of γ_1 be along Z-direction. The angular distribution $F_L(\theta)$ of the γ_2-ray w.r.t. Z-axis represents the angular correlation $\omega(\theta)$ between γ_1 and γ_2. Thus

$$\omega(\theta) = F_L(\theta) \qquad \text{... (3.1)}$$

Let us assume that when the excited nucleus in state I_a (or I_i) is formed, all the substates m_i are equally populated. The probability of forming a substate m of the state I after transition from state I_i is as

$$P(m) = \sum_{m_i} G(m_i,\, m)\, F_{L_i}^M(\theta = 0) \qquad \text{... (3.2)}$$

Since γ_1 is emitted in Z-direction, the effective components of L_1 is only $L_{1Z} = M_l = \pm 1$ and the angular correlation function $\omega(\theta)$ is given as

$$\omega(\theta) \propto \sum_{m_f\, mm_i} <\mathrm{Im}\, L_1 \pm 1>^2 F_{L1}^{\pm 1}(\theta) <I_f\, m_f\, L_Z\, m_Z | I_m>^2 F_{LZ}^M(\theta) \qquad \ldots (3.3)$$

$L_{1Z} = \pm$ represents the left and the right circularly polarisation of γ_1-ray. We can see that there is no interference term for the various transitions $m_i \rightarrow m \rightarrow m_f$. However, (3.3) gives $\omega(\theta)$ only for pure multipole transitions where L_1 and L_2 have only one value.

A rigorous derivation of the angular correlation of two cascade γ-rays is as

$$\omega(\theta) = \sum_K 1 + A_{KK}\, P_K(\cos \theta) \qquad \ldots (3.4)$$

Where K_{max} = minimum value of $2I$, $2L_1$ and $2L_2$.

One can write the coefficient A_{KK} as two factors ecah depending upon one γ-ray transition only. Thus

$$A_{KK} = F_K(L_1\, L\, I_i\, I)\, F_K(L_2\, L\, I_f\, I) \qquad \ldots (3.5)$$

Where

$$F_K(L_1\, L\, I_i\, I) = (-1)^{I_i + I - 1}\, (2L_1 + 1)(2I + 1) \begin{pmatrix} L_1 & L_1 & k \\ 1 & -1 & 0 \end{pmatrix} \begin{Bmatrix} L_1 & L_1 & k \\ I & I & I_i \end{Bmatrix} \qquad \ldots (3.6)$$

The last two factors in (3.6) represent the Clebsch-Gordan and Racah coefficients. For different values of $L\, I_i$ and I, the values of F_K are tabulated. One can obtain the angular correlation function $\omega(\theta)$ from these values.

In case if one or both the γ-ray transitions are a, mixed multipole of L_1 and L_1' and L_2 and L_2', then the angular correlation function takes the form as

$$\omega(\theta) = 1 + A_{kk}\, P_k(\cos \theta)$$

Where

$$A_{kk} = A_k(L_1\, L_1'\, I_i\, I)\, A_k(L_2\, L_2'\, I_f\, I)$$

and

$$A_k(L_1\, L_2'\, I_2\, I) = \frac{F_k(L_1\, L_1\, I_i\, I) + 2\delta_1\, F_k(L_1\, L_1',\, I_1\, I) + \delta_1^2\, F_k(L_1'\, L_1'\, I_1\, I)}{1 + \delta_1^2} \qquad \ldots (3.7)$$

Here δ_1 is the mixing ratio of L_1' and L_1 multipoles in γ_1. Similar expression holds good for the second factor.

Brady and Deutsch in 1948 made the first measurment of the angular correlation of the cascade γ-rays of ^{60}Ni observed in the decay of ^{60}Co (Figure 3.49). The radioisotope ^{60}Co decays to the 4 MeV excited state of ^{60}Ni (Figure 3.49). As shown in the figure, the state decays with the emission of two γ-rays in cascade to the ground state of ^{60}Ni. We note that the spins of the ground, first and second excited states are 0, 2 and 4 respectively and each of the γ-ray transition is pure multipole with $L = 2$. One expects the angular correlation function for spin sequence as

$$\omega(\theta) = 1 + \frac{1}{8}\, P_2(\cos \theta) + \frac{1}{24}\, P_4(\cos \theta) \qquad \ldots (3.8)$$

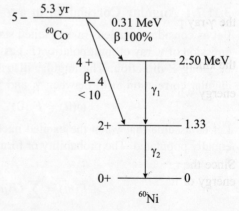

Fig. 3.49 Cascade γ-rays of ^{60}Ni observed in the decay of ^{60}Co.

Figure 3.48(b) shows the angular correlation function $W(\theta)$ for ^{60}Co γ-ray cascade, which is in conformity to the theory.

3.11.8 Resonance Scattering and Absorptions of γ-rays

We know that an oscillating electron emits electomagnetic radiations and the electron experiences a radiation resistant due to which electron oscillations are damped. The radiations have distribution for such an oscillator in frequency (ω). We have the intensity of radiations of frequency ω as

$$I(\omega) = \frac{A^2 \, \Gamma}{(\omega - \omega_0)^2 + \left(\dfrac{\Gamma}{2\hbar}\right)^2} \qquad \text{... (3.1)}$$

Where ω_0 is the frequency of the undamped oscillator, T the width in the frequency due to damping and A is a constant. One finds that

$$I(\omega) = \frac{1}{2I(\omega_0)} \text{ for } \omega = \omega_0 \pm \frac{\Gamma}{2}$$

Clearly, Γ is the frequency width of the radiations.

Rayleigh in 1980 predicted resonance absorption of atomic radiations based on the above concept. Wood in 1904 reported the resonance absorption of sodium light by sodium vapours. Weisskopf in 1930 explained the emission and resonance absorption of atomic radiations with the help of Dirac theory of electron.

Quantum electrodynamics also predicted similar results in a system where energy levels are quantised. Kuhn in 1929 made efforts to observe resonance absorption of nuclear γ-rays but could not succeed.

Let us consider a nucleus in a state a making transition to state b (Figure 3.50). Let

$$E_a - E_b = E_0$$

i.e. energy difference between levels a and b is E_0 and the energy of the γ-ray photon emitted is E_γ. The momentum of photon $= \dfrac{E_\gamma}{c}$ and

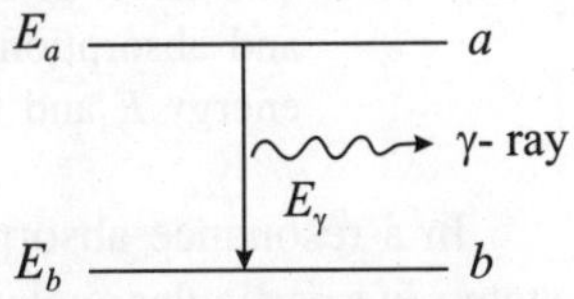

Fig. 3.50 γ-ray transition

the kinetic energy of the recoil atom $= \dfrac{p^2}{2M} = \dfrac{E_\gamma^2}{2Mc^2}$, where M is the mass of the atom. The recoil energy which is very small can be expressed as

$$R = \frac{E_\gamma}{2Mc^2} \approx \frac{E_0^2}{2Mc^2}$$

Since the recoil energy R has to come out of the excitation energy E_0 of the atom and hence the energy of the emitted photon is

$$E_\gamma = E_0 - R = E_0 - \frac{E_0^2}{2Mc^2} = E_0 - R \qquad \text{... (3.2)}$$

Obvisouly, when a photon is absorbed by a nucleus, it has to provide the recoil energy $=$ $\left(E_0 - \dfrac{E_0^{\,2}}{2Mc^2} \right)$ to the nucleus. Clearly, for resonance absorption of the photon, the energy of the incident photon should be $= E_0 + \dfrac{E_0^{\,2}}{2Mc^2}$. Thus the photon emitted from the nucleus falls short of the resonance energy by an amount $R = \dfrac{E_0^{\,2}}{2Mc^2}$ and this short fall of energy from resonacne increases as square of the transition energy. If the atom containing the nucleus is moving, the energy of the γ-ray will be altered by the Doppler effect and for thermal motion this gives rise to Doppler broadening of the γ-ray line shape as shown in Figure 3.51. Also shown is the resonance absorption spectrum corresponding to transitions from the ground state to the state x which is centred about $E_0 + R$.

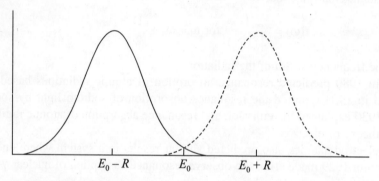

Fig. 3.51 The effect of recoil and Doppler broadening on the γ-ray emission spectrum (solid line) and absorption cross-section (dotted line) for transitions between a state at an excitation energy E and the ground state.

In a resonance absorption experiment the intensity of γ-rays from the γ-decay of an excited state x in a particular nucleus is measured after the γ-rays have passed through an obsorber containing the same nuclei in their ground state. The γ-rays are provided by a source which feeds the level x, e.g. a ^{57}Co source which β-decays to the ^{57}Fe (14.4 keV) level.

The transmission depends on the amount of overlap of the emission and absorption spectra and on the natural line width T_γ of the excited state. By varying the temperature of the source and absorber, and if necessary their relative velocity to increase the Doppler effect and enhance the overlap, T_γ can be measured.

Cross Sections

We know that scattering and absorption are compeimentary processes, and whenever there is absorption there is scattering as well. A photon of energy E and wave length $\lambda (= 2\pi\lambdabar)$ is absorbed and scattered by a absorber with cross-sections σ_a and σ_s respectively. σ_a and σ_s are given by

$$\sigma_a(E) = \sigma_0 \frac{\Gamma \, \Gamma_\gamma}{4(E - E_0)^2 + \Gamma^2} \qquad \qquad \text{... (3.3)}$$

and
$$\sigma_s(E) = \sigma_0 \frac{\Gamma_\gamma^2}{4(E - E_0)^2 + \Gamma^2} \qquad \qquad \text{... (3.4)}$$

Where Γ_γ is the partial γ-ray width of the level and Γ is the total width of the excited state. Now, if a is the internal conversion coefficient of γ-ray, then

$$\Gamma_\gamma = \frac{\Gamma}{1 + \alpha} \text{ and } \sigma_s = \frac{\sigma_a}{1 + \alpha}$$

and
$$\sigma_0 = \frac{2 I_a + 1}{2 I_b + 1} (2\pi \lambdabar^2) \qquad \qquad \text{... (3.5)}$$

Where I_a and I_b denotes the spins of the ground state and excited states respectively. The distribution of the energy of the γ-rays emitted from a nuclear source or scattered from an absorber has an *Lorenzian shape* represented by

$$I(E) = \frac{T}{2\pi} \frac{A^2}{\left[(E - E_0)^2 + \left(\frac{T}{2}\right)^2 \right]} \qquad \qquad \text{... (3.6)}$$

Where A is the normalisation factor. The intensity $I(E)$ is normalised, i.e.

$$\int_0^\infty I(E)\, dE = 1 \qquad \qquad \text{... (3.7)}$$

For absorption and scattering cross-sections, the expressions (3.3) and (3.4) holds good for a definite energy E of the photon. If the intensity distribution of the photon is governed by (3.1), the observed cross section for absorption and scattering will be as

$$\sigma_{Obs} = \frac{\int_0^\infty I(E)\sigma(E)\, dE}{\int_0^\infty I(E)\, dE} \qquad \qquad \text{... (3.8)}$$

We may note that in the case of Doppler broadening the photon energy distribution $I(E)$ is no longer Lorenzian but Maxwellian and one will have to calculate the observed cross-section accordingly. One can observe the maximum absorption coefficient when the photon energy is E_0. We have

$$\sigma_a = \frac{\sigma_0}{T} \Gamma_\gamma = \frac{\sigma_0}{1 + \alpha} \qquad \qquad \text{... (3.9)}$$

In case, when the incoming photons have a Doppler broadening having a width D and centred at the resonance energy E_0, the observed cross sections is as

$$\sigma_a = \sigma_0 \frac{\Gamma_\gamma}{\Gamma} \frac{\Gamma}{D} = \frac{\sigma_0 \, \Gamma_\gamma}{D} \qquad\qquad \dots (3.10)$$

We have in optical transitions at room temperature $\dfrac{\Gamma_\gamma}{D} \approx 10^{-1}$ and one can observe the resonance

absorption. However, in the case of γ-ray, we have $\dfrac{\Gamma_\gamma}{D} \approx 10^{-4} - 10^{-7}$. Obviously, even if the γ-ray

line is centred at E_0, even though the effective cross section is very low.

3.11.9 Coulomb Excitation

When we expose stable nuclei to a beam of charged particles, at energy insufficient for the penetration

of the electrostatic potential barrier $\left(E_{in} < \dfrac{Z \, ze^2}{4\pi \, \varepsilon_0 \, r} \right)$, γ-rays are observed. One can assume these

γ-rays as due to the emission from the excited nucleus. Obviously, Coulomb excitation is an important example of an inelastic scattering reaction caused by the Coulomb interaction. In this process the charged bombarding particle brings about an excitation of the target nucleus from the ground state to a higher state.

There are several reasons why Coulomb excitation is of interest in nuclear physics. In the first place, the reaction mechanism is well known and may be regarded essentially as the inverse of electromagnetic decay. Experimentally very strong Coulomb fields can be created by bombarding nuclei with a beam of heavy ions. When this advantage is coupled with the precision that can be achieved in charged particle experiments, we have a powerful tool for investigating some of the detailed properties of nuclei.

Coulomb excitation is in principle possible using electrons but in practice it is difficult to observe as an emission process because of the presence of *bremsstrahlung quanta*. The *inelastically scattered electrons*, however, may be observed with good discrimination background at large angles of scattering and peaks in their spectrum corresponding with the excitation of the of the lower excited levels of the target nuclei may be inentified. Inelastic electron scattering, in contrast with Coulomb excitation, is able to excite monopole transitions; it is also used in particle physics to excite nucleon resonances.

The theoretical treatment of Coulomb excitation is similar to that of internal conversion. In this process nucleus absorbs energy from the incident particle where as in internal conversion a charged particle (electron) absorbs the nuclear excitation energy. Corresponding to Rutherford scattering trajectories, the states of the incident charged particle in Coulomb excitation process are distorted plane waves.

One can calculate the cross-section for such a reaction approximately by assuming the incident particle is on a classical Rutherford trajectory and integrating the probability of a transition along the trajectory. The differential cross-section can be written as,

$$\frac{d\sigma}{d\Omega} = \frac{d\sigma}{d\Omega} \text{ (Rutherford) } P(\text{trajectory}) \qquad\qquad \dots (3.1)$$

Where P (trajectory) is the integrated transition probability. If the perturbing interaction between the incident charged particle q and the nucles is V then the probability amplitude for a transition from the ground state to an excited state f is given by

$$a_{if} = \frac{1}{i\hbar} \int V_{if} \exp(i\omega t)\, dt \qquad \qquad ...\ (3.2)$$

Where $\hbar\omega = E_f - E_i$, P(trajectory) $= |a_{if}|^2$ and

$$V_{if} = \int \psi_f^* \, V \, \psi_i \, d\tau \qquad \qquad ...\ (3.3)$$

Coulomb excitation has been observed for good number of nuclei from low to high Z. Bombarding particles include α-particles and also heavior ions, e.g. ^{14}N, ^{40}A, etc. Several Coulomb excitation processes involving $E2$ type transitions have been reported, which permit the study of the low lying rotational levels in deformed nuclei. Such nuclei have internal quadrupole moment Q_0 which can easily be measured with the help of Coulomb excitation cross-section, since $B(E2)$ in this case depends on $Q_0^2 \left[B(E) = \left(\dfrac{5}{16\pi} \right) Q_0^2 \right]$. Q_0 for ^{114}Cd had been determined in this way $= -0.15$ barn.

3.12 MOSSBAUER EFFECT

In 1958 Mossbauer made a discovery that allows the extremely small width to energy ratio of low-lying excited states to be used in many different applications as an energy spectrometer of extremely good resolution. The basic idea of the Mossbauer effect is illustrated in Figure 3.52. A source nucleus in an excited state makes a transition to its ground state, emitting a γ-ray. The γ-ray is subsequently caught by an unexcited absorber nucleus of the same species, which ends up in the same excited state. The potentialities as an energy spectrometer becomes clear when it is realized that changes in the source energy, the absorber energy, or the energy of the γ-ray in flight, will destroy the "resonant" absorption – even if the energy change is only a few parts in 10^{11}. For some years physicists had been attempting to utilize these potentialities, but with little success. The problem had to do with recoil of the nuclei upon emission and absorption of the γ-ray. Mossbauer effect has been observed in many nuclei, but the limitation to low transition enrgies is rather restrictive. In nuclear physics the high resolution available $\left(\dfrac{\Gamma_\gamma}{E_\gamma} \approx 10^{-13} \right)$ has been explored mainly in the measurment of nuclear moments and *isomer shifts* which arise because of the difference in mean square radius of nuclear ground and excited states. These shifts may be seen in resonant absorption experiment if the electron density at the mossbauer nuclei differs between source and absorber. In Mossbauer resonant scattering exact resonance can sometimes be destroyed by the Doppler shift due to motion of the source with velocities as low as a few millimeters per second.

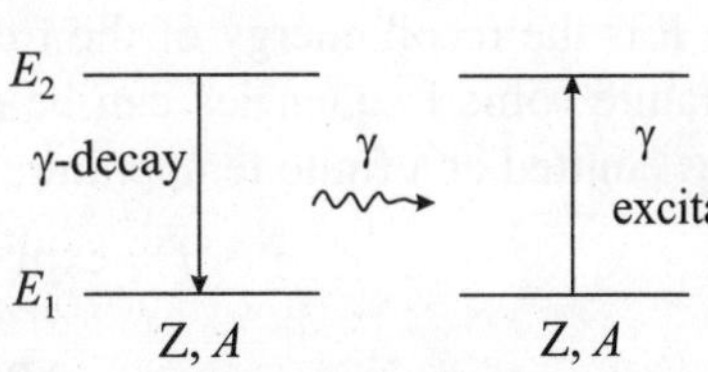

Fig. 3.52 Resonant absorption, the basis of the Mossbauer effect.

Recoilless emission and absorption can take place if the nucleus is bound in a crystal for then the possibility exists that the recoil can be taken up by the whole lattice, in which case effice M in Equation (3.2) (Section 10.8) is vastly increased with $E_\gamma = E_0$ to a very good approximation indeed for both emission and absorption what determines the probability of no recoil or recoil free is the strength of the binding of the nucleus and the temperature of the crystal. The number of atoms which do not show recoil, increases as the temperature of the source or the absorber is lowered; however, it is appreciable at liquid nitrogen temperature (77 K). At low temperature the whole crystal takes up the recoil momentum and there is no loss of energy of emitted photon. It is therefore only when both source and absorber are lowered in temperature that an increased absorption is expected and this is what Mossbauer discovered.

Similar phenomena were known in other branches of physics but the similarity was not appreciated. In X-ray diffraction there are scattered X-rays with the same wavelength as the incident X-rays but going in a different direction, there is therefore a momentum transfer with no loss of energy and same thing occurs in neutron scattering in crystals.

We know that the atoms in a crystal lattice can vibrate about their mean position with some frequency, say ω. Now, the frequency ν of the emitted γ-ray should be modulated in order to give the possible frequencies of the γ-ray as ν, $\nu \pm \omega$, $\nu \pm 2\omega$, $\nu \pm 3\omega$, ... etc. Einstein explained the specific heat of solids by assuming that in a crystal the atoms vibrate with only one frequency. Debye explained the T^3 term appearing in the specific heat relation by assuming that all frequencies upto a maximum frequency ω_D (Debye frequency) obeys the distribution law

$$c(\omega) = A\, \omega^2$$

Where A is the constant. The frequency distribution has a sharp cut off at frequency ω_D. We have

$$\hbar\, \omega_D = k\, \theta_D \qquad \qquad \text{... (3.1)}$$

Where θ_D is Debye temperature of the crystal and k is Boltzmann constant.

We have the wave length of the vibrations corresponding to the Debye frequency ω_D is $\lambda = 2d$, where d is the lattice spacing. We may note that λ is shortest wave length. There in the crystal lower ω is also present, i.e. there are larger wavelengths too. After emission of a photon, recoil of the radioactive atom can excite these vibrations. However, its probability is small as excitation of such vibrations involves moving of a large number of atoms in the lattice. Debye theory gives us the number of photons emitted without any energy loss by recoil. According to Debye theory, the recoil free fraction (f) is given by

$$f = \exp\left(\frac{-3R}{2k\,\theta_D}\right) \qquad \qquad \text{... (3.2)}$$

Where R is the recoil energy of the free atom. Equation (3.2) is true at 0 K. However, at a finite temperature some frequencies can be excited and the photon can loose energy. The fractions of photons emitted at a finite temperature without loss of energy is given by

$$f = \exp(-2W) \qquad \qquad \text{... (3.3)}$$

Where

$$W = \frac{3R}{k\,\Theta_D}\left(\frac{T}{\Theta_D} + \frac{1}{4}\right)^2 \int_0^{\Theta_D/T} \frac{x}{e^x - 1}\,dx \qquad \qquad \text{... (3.4)}$$

f is known as *Lamb-Mossbauer factor*. A large value of f would hence the recoil free emission and absorption of γ-rays. Obviously, one can realise this if R is small, Debye temperature Θ_D is large and the temperature of the source and the absorber is low.

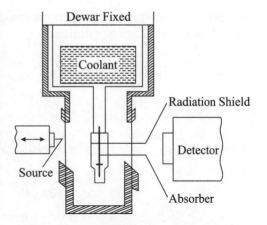

Mossbauer, in his original experiment measured the emission of 0.129 MeV gamma radiations from radioactive ^{191}Ir. They were allowed to pass through a metallic iridium absorber (39 % of ^{191}Ir) then on to a detector. ^{191}Ir (γ-source) was formed in an isomeric state in the β-decay of ^{191}Os. The natural width and life time of the 0.129 MeV level of ^{191}Ir are 5×10^{-6} eV and 1.3×10^{-10} s respectively. Figure 3.53 shows an experimental set up for the demonstration of Mossbauer effect. Many more substances which could be used have been discovered. The substance which has been used most extensively is the 14.4 keV excited state of ^{57}Fe formed in the decay of 270 day activity of ^{57}Co. Decay scheme of ^{57}Co is shown in Figure 3.54. The positive and the negative signs after the quantum numbers refer to the parity of the states. The mean life time of 14.4×10^{-7} s and the corresponding line width is 4.6×10^{-9} eV. This mean life time is long enough for the excited ^{57}Fe ions to occupy suitable sites in the iron crystal lattice prior to decay. The 14.4 keV γ-rays can be passed through an iron absorber which can be enriched in ^{57}Fe so as to enhance the probability of recoilless absorption. The energy of recoil of atom on emitting 14.4 keV photon is 0.002 eV.

Fig. 3.53 Set up for the demonstration of Mossbauer Effect. The source emits a monochromatic line which may be Doppler shifted by a mechanical motion.

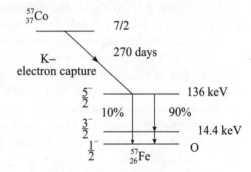

Fig. 3.54 Decay scheme of ^{57}Co to 14.4 keV ^{57}Fe.

Mossbauer soruce is usually prepared by incorporating radioactive ^{57}Co in the lattice of rhodium metal having f-factor = 0.7 at room temperature. The line width from the source is very close to the natural line width of the 14.4 keV γ-ray line and therefore, one can study the absorption of 14.4 keV γ-ray emitted from the Mossbauer source in any absorber which has iron as one of its constitutents. The γ-rays can be detected by a proportional counter or scintillation counter.

When one uses an iron foil as an absorber, one observes the Zeeman splitting of the ground state and the 14.4 keV excited state as there is an internal field (magnetic) of about 330 kilogauss in iron metal.

In the Mossbauer experiment, there was arrangement for rotating the crystal containing the source. When rotated the crystal in one direction, the source moves towards the absorber while for rotation in the opposite direction, the source moves away from the absorber. The frequency of the emitted radiation changes when the source moves due to Doppler effect, which destroys the condition of resonance because of the very high degree of mono-chromaticity of the radiation. The velocity (v) of the source at which the Doppler energy shift (ΔE) becomes equal to the natural width Γ of the level can be estimated

$$\Delta E = E_\gamma \frac{v}{c} \qquad \qquad ... (3.5)$$

or

$$\frac{v}{c} = \frac{T}{E_\gamma} \qquad \qquad ... (3.6)$$

In our case $\qquad \Gamma = 5 \times 10^{-6}$ eV. We obtain

$$\frac{v}{c} = \frac{5 \times 10^{-6}}{129 \times 10^3} = 4 \times 10^{-11}$$

or $\qquad v = 0.012$ m/s

The Mossbauer experiment is performed by noting down the count rate at the detector at different relative velocities of the sources. The count rate is plotted as a function of relative velocity. Figure 3.55(a) shows a typical Mossbauer spectrum. The remarkable sharpness of the spectrum is evident. Figure 3.55(b) shows the determination of the natural line width of the 129 keV excited state in ^{191}Ir in a resonance absorption experiment, under conditions of significant recoilless emission and absorption. It was observed that at velocities of a few cm/s, for which the Doppler shift is of the order of 10^{-5} eV or less, the resonance condition is disrupted. This confirms that a recoiless trasition giving rise to a γ–ray line of intrinsic width $\sim 5 \times 10^{-6}$ eV must have been produced in Mossbauer experiment.

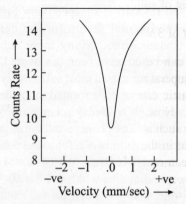

Fig. 3.55(a) A typical Mossbauer spectrum observed in transmission geometry. Sharpness of the resonance is remarkable.

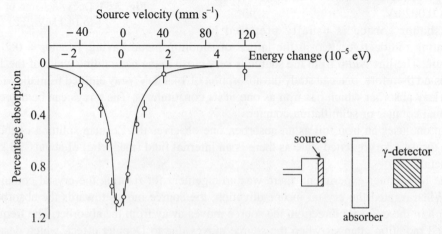

Fig. 3.55(b) Determination of the natural line width of the 129 keV excited state in ^{191}Ir in a resonance absorption experiment, under condition of significant recoilless emission and absorption.

Other than ^{57}Fe, the substance used most frequently is probably ^{117}Sn isotope of tin for which E_γ = 23.8 keV, τ = 2.6 × 10^{-8} s and Γ = 2.5 × 10^{-8} eV $\left(\dfrac{\Gamma}{E_\gamma} = 10^{-12} \right)$. In addition to ^{117}Sn, fair amount of work has also been done with some rare-earth elements.

The effect of recoilless emission is greater when, (i) the energy (E) γ-ray is small; (^{57}Fe with 14.4 keV transition is the best example), (ii) the temperature of the crystal source is lower, and (iii) high Debye temperature of the crystal lattice.

3.12.1 Theortical Considerations

Theoretical explanation of Mossbauer effect is based on Einstein's quantum theory of the specific heat of solids modified by Debye. According to Einstein's theory, a solid consists of N atoms is a set of $3N$ harmonic oscillators having same frequency, $\nu = \dfrac{\omega}{2\pi}$, where ω is the circular frequency ω can be determined from $\hbar\omega = k\,T_E$, where k is Boltzmann constant and T_E is the Einstein's temperature which is characterstic of a solid. T_E can be determined by its atomic mass and the elastic constants.

Now, if a solid consists of atoms with excited nucleus, then the emission of a γ-ray quantum by such a nucleus will lead to the recoil of the latter. In general, the recoil energy is different from that in the case of emission from a nucleus in a free atom. However, calculation reveals that the mean recoil energy even in this case is equal to the recoil energy of a free nucleus, i.e. $<E_r> = \dfrac{E_\gamma^2}{2Mc^2}$, where M is the mass of the nucleus. Now, if $<E_r> < \hbar\omega$, then some of the γ-ray transitions will be recoilless, i.e. there will be no excitation of the oscillator since the minimum energy transfer into the oscillators is $\hbar\omega$. Obviously, in some cases the emission of a photon will not lead to any energy transfer to the emitting nucleus and therefore fail to excite the oscillators. One can express the condition $<E_r> < \hbar\omega$ as

$$<E_r> = (1-f)\hbar\omega \qquad \ldots (3.7)$$

$$\therefore \qquad f = 1 - \frac{<E_r>}{\hbar\omega} \qquad \ldots (3.8)$$

Here f is called as Mossbauer coefficient and gives the fraction of recoiless transitions.

We know that the total energy of an oscillator is twice the mean kinetic energy E_K and also at 0 K the zero point energy, $E_{0sc} = \dfrac{\hbar\omega}{2}$. Thus, one finds

$$\hbar\omega = 2\,E_{0sc} = 2 \times 2 <E_K> = \frac{<p^2>}{2M} \qquad \ldots (3.9)$$

Where p is momentum. We have

$$E_r = \frac{p_r^{\,2}}{2M} = \frac{p_v^{\,2}}{2M} = \frac{E_v^{\,2}}{2Mc^2}$$

$$\therefore \qquad <E_r> = \frac{p_v^2}{2M} = \frac{\hbar^2}{2M\lambda^2} \qquad \qquad \text{... (3.10)}$$

Where $\lambda(= 2\pi\lambda)$ is the wave length of the radiation. From Equations (3.9) and (3.10), one obtains

$$\frac{<E_r>}{\hbar\omega} = \frac{\hbar^2}{<p^2>\lambda^2} = \frac{<x^2>}{\lambda^2} \qquad \qquad \text{... (3.11)}$$

Where $<x^2>$ respresents the mean square position of the oscillator. We have the uncertainty relation $\Delta x\,\Delta p \geq \dfrac{\hbar}{2}$. Thus for an oscillator

$$\Delta x^2 = <x^2> - <x>^2 = <x^2>$$

and
$$\Delta p^2 = <p^2> - <p>^2 = <p^2>$$

One finally obtains the Mossbauer coefficient (f) as

$$f = 1 - \frac{<E_r>}{\hbar\omega} = \frac{<x^2>}{\lambda^2} \qquad \qquad \text{... (3.12)}$$

For the general case $(T \neq 0)$, a more rigorous treatment gives

$$f = \exp\left[\frac{<x^2>}{\lambda^2}\right] \qquad \qquad \text{... (3.13)}$$

We can see that in the limit $T \to 0$, (3.13) reduces to the approximate expression (3.12). Equation (3.12) reveals that the recoilless transition become more possible, if

  (i) T is low (for which Δx^2 and hence $<x^2>$ are quite small)
 (ii) The emitter is a solid for which Δx^2 and hence $<x^2>$ are quite small.
(iii) Transition energy is low. This makes λ large.
 (iv) Elastic constants of the material are large. These makes the frequency ω large so that T_E is correspondingly large.

Debye proposed a more rigorous theory of specific heat. According to this theory the solid oscillate with a whole range of frequencies upto a maximum ω_m, where $\omega_m = k T_D$, T_D is the characterstic Debye temperature of the solid and can be determined by elastic constants of the crystal.

Theory has been modified by replacing T_D with T_E. For a mono atomic crystal, one obtains

$$f = \exp\left[\frac{-6<E_r>}{k T_D}\left\{\frac{1}{4} + \left(\frac{T}{T_D}\right)^2 \int_0^{T_D} \frac{y}{e^y - 1}\,dy\right\}\right] \qquad \qquad \text{... (3.14)}$$

(3.14) reduces in the limit $T \to 0$ as

$$f = \exp\left(-\frac{3}{2}\frac{<E_r>}{k T_D}\right) \to 1 - \frac{3}{2}\frac{<E_r>}{2\hbar\,\omega_m} \qquad \qquad \text{... (3.15)}$$

Usually, the following simplified expression is used for obtain f (Lamb-Mossbauer factor)

$$f \simeq \left[\frac{-3E_r}{2kT_D} \left\{ 1 + \frac{2}{3}\left(\frac{\pi T}{T_D} \right)^2 \right\} \right]$$ ... (3.16)

In order to use the Mossbauer effect, we must consider transitions in which:

(i) f is adequately large
(ii) for a good precision E_0 must be fairly large
(iii) there should be small livelihood of internal conversion.
(iv) as far as possible, fine structure splitting should not occur.

When these requirements are satisfied, one should observe following precautions:

(i) On emission and absorption of γ-rays, the changes in nuclear mass. This causes the γ-ray to change,
(ii) the differences in the average isotopic mass number,
(iii) between source and absorber, the differences in the Debye and absolute temperatures
(iv) the differences in chemical constitution of presence or lattice defects if any.
(v) relative velocity between emitter and absorber.

3.12.2 Importance of Mossbauer Effect

Moossbauer effect not only demonstrated for the first time the feasibility of observing resonance fluorescence of γ-rays but it also opened out possibility of studying a variety of hyperfine interactions between the nucleus (+ very charged) and surrounding electron cloud. When the emitting and absorbing atoms are embedded in a well-bound crystalline lattices, then there is a difinite probability for emission and absorption of γ-rays without changing the phonon occupation number of the lattice. Such processes are termed zero phonon processes. Under such circumstances, the emitted γ-ray carries with it the full transition energy. Much more important than recoilless emission of γ-ray is the fact that the γ-ray emitted under conditions appropriate to Mossbauer effect has a natural line width determined entirely by the life time of the nucleus in the excited state. Considering Mossbauer isotope ^{57}Fe m, 14.4 KeV γ-ray emitted as a result of transition from the $\frac{3}{2}$ state to $\frac{1}{2}$ state has a line width of 4.19×10^{-9} eV determined entirely by the life time of the nucleus ^{57}Fe m in the excited state ($T \cong 9.77 \times 10^{-8}$ sec). This is about six orders of magnitude sharper than the line width of the γ-ray emitted when the lattice is excited. In other words we can say that the fractional line whdth of 14.4 KeV γ-ray emitted by ^{57}Fe m nucleus $\left(\cong \frac{4.19 \times 10^{-9}}{14.4 \times 10^3} \right) \cong 10^{-13}$, i.e. we can measure the energy of the γ-ray and in turn the nuclear levels is an accuracy of one part in 10^{13}. Using isotopes such as ^{67}Zn, one can increase the accuracy to even one part in 10^{15}. This makes Mossbauer effect the most precisely determined electromagnetic radiation available for physical measurements. With such sharp lines it is possible to observe and study a variety of hyperfine interaction e.g. (a) electrostatic monopole interaction between the nucleus and the s electrons density at the nucleus, (b) electric quadrupole interaction between the quadrupole moment of the nucleus and the electric field gradient at the nucleus, and (c) magnetic dipole interaction between the magnetic dipole moment of the nucleus and the magnetic field at the nucleus. It is worthwhile to mention that the electromagnetic radiation with Mossbauer's comparable stability and line width has not yet been obtained by other means. Even the optical maser, which is the best source of narrow-line infrared and visible radiation has failed to attain the said limit. With the help of Mossbauer effect not only it has been possible to study hyperfine interactions but it has made it possible to observe gravitational red shift in a laboratory and test the equivalence principle.

3.12.3 Applications

Mossbauer effect have important novel applications in various fields. We will mention few of them over here:

(i) **Chemical or Isomer shift**   Most application of the Mossbauer effect deal with situation for which the emitter and absorber are in different enviornments, so that the emission and absorption peaks do not occur at precisely the same energy. The relative velocity required to obtain maximum absorption is measured and the results used to study the enviornment of the emitter or absorber.

We know that chemical states of an atom are associated with differences in the electron distribution around the nucleus and contribute to the energy of transition

$$E_\gamma = \Delta E_{nuc} + \Delta E_{ele} \qquad \qquad ... (3.14)$$

Here ΔE_{nuc} denotes the change in nuclear binding energy (B.E.) and ΔE_{ele} denotes the change in B.E. of the atomic electrons. Now, if the emitting nucleus (excited state) and the absorbing nucleus (ground state) are in different chemical compounds, then the distribution in atomic electrons will be different. This will make differences in ΔE_{elec} and thus in E_γ and obviously change in the position of the Mossbauer peak. This is known as *chemical or isomer shift*.

By placing the emitter in various solids and measuring the chemical shift for each situation, it is possible to obtain information about the charge state of an ion and about changes in the electron distribution brought about by changes in bonding. Even if it is chiefly the distribution of electrons in p and d subshells which change, as an covalent or partially covalent bonds, these influence the s-subshell electron distribution and the chemical shift.

(ii) **Hyperfine Splitting of Nuclear Energy Levels**   We know that hyperfine splitting of atomic energy levels arises due to magnetic interactions of the nuclear magnetic moment with the magnetic field due to the orbital electrons at the nucleus. Nuclear magnetic moments are of the order of nuclear magneton $(\mu_N) = 5.05 \times 10^{-27}$ J/T. Due to the electronic current in the atomic shell the average magnetic field at the nucleus is

$$B = \mu_0 \frac{q}{4\pi r^3} \mathbf{r} \times \mathbf{v} \qquad \qquad ... (3.15)$$

But $v = \dfrac{l\hbar}{mr}$, where λ is the *azimuthal quantum number*, one obtains

$$B = \frac{\mu_0 \, q \, \hbar l}{4\pi \, m \, r^3} \approx 2T \qquad \qquad ... (3.16)$$

Let us assume that q = electronic charge (1.6×10^{-19} C) and $l = 1$. Thus, one obtains the hyperfine splitting as

$$\Delta E \sim \mu_N B = 10^{-7} \text{ to } 10^{-8} \text{ eV}$$

Now, the atomic levels have energies ~ 1 eV and therefore the relative magnitude of hyperfine splitting of the atomic energy levels is

$$\frac{\Delta E}{E} \sim 10^{-7} \text{ to } 10^{-8}$$

Clearly, the hyperfine structures of the atomic spectral lines can be resolved well by optical spectroscopic methods.

However, the nuclear levels have energies of the order of 10^4 to 10^5 eV or even more. The value of $\frac{\Delta E}{E}$ in this case is

$$\frac{\Delta E}{E} \sim 10^{-11} \text{ to } 10^{-13}$$

Which is much smaller. Due to very small magnitude, nuclear splitting of levels can only be studied with the help of Mossbauer effect. For this purpose, one of the most widely studied nuclei is $^{57}_{26}$Fe. Hyperfine splitting of a nuclear level was first observed in the case 14.4 keV γ-ray line of Fe. Unstable $^{57}_{27}$Co nuclei, implanted in the sample, decay by means of electron capture to the first excited state of $^{57}_{26}$Fe and many of the iron nuclei decay to the ground state by γ-emission. The two $^{57}_{26}$Fe states of interest are separated in energy by 14.4 keV and the width of the excited state is on the order of 10^{-9} eV. The two levels $\frac{3^-}{2}$ and $\frac{1^-}{2}$ which are involved in this transition split up in magnetic field (Figure 3.56). This is also called *nuclear zeeman effect*. The ground state splits into 2 levels and the excited state splits into 4 levels. The splitting is proportional to $\mu \cdot \mathbf{B}$, where μ is the magnetic dipole moment of the nucleus. The transitions between there levels are governed by the selection rules $\Delta m = 0, \pm 1$.

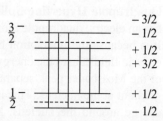

The levels in both the emitting and the absorbind nuclei undergo hyperfine splitting and therefore the observed Mossbauer pattern would be very complicated. In order to avoid this, the emitting nuclei are usually mounted on a stainless steel

Fig. 3.56 Hyperfine splitting of two nuclear levels $\frac{3^-}{2}$ and $\frac{1^-}{2}$ and the possible transitions between in $^{57}_{26}$Fe.

grating which is diamagnetic. Thus there is no hyperfine splitting in them and the observed pattern corresponds to the splitting of the absorber nuclei.

The exciting line from the emitter nuclei is obviously a single line, which can produce the six different absorption lines of slightly different frequencies through resonance fluorescence by moving the source with different velocities, which is of the order of 1 mm/s or even less (Figure 3.57).

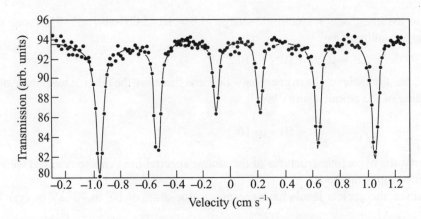

Fig. 3.57 The effect of the hyperfine interactions within an iron absorber on the transmission of 14.4 keV γ-rays from the decay of the $J = \frac{3}{2}$ 14.4 keV eccited to $J = \frac{1}{2}$ ground state of ^{57}Fe. (The difference in environment of emitting and absorbing nuclei accounts for the transmission pattern not being centred on zero speed).

Both the local magnetic field at the site of nucleus and the ratio of the magnetic dipole moments of the excited and ground states can be calculated from the positions of the Mossbauer peaks. The Mossbauer effect is particularly useful for the study of the magnetic field in ferromagnetic materials. For example, the transition to a paramagnetic state can be investigated. The effect is also used to study the enviornement of Fe atoms in *biological materials*.

Due to orbital electrons the magnetic field at the nucleus of $^{57}_{26}$Fe has been found to be $B = 33\ T$ (assuming $\mu = 0.09\ \mu_N$ in the ground state). From this experiment, the magnetic moment of the excited ^{57}Fe nucleus comes out to be, $\mu_{exc} = -0.153\ \mu_N$.

(iii) **Quadrupole Hyperfine Splitting** Splitting of nuclear levels also occurs if the nucleus has an electric quadrupole moment ($I \geq 1$) and is situated in a spatially varying electric field. It is because the interaction between nuclear quadrupole moment and the inhomogeneous electric field causes the energy of the nucleus to depend on its orientation. The measurments of the Mossbauer peak separation can be used to obtain information about the electric field gradient at the nucleus. This information, in turn, provides knowledge of the distribution of charge around the nucleus. Mossbever studies have been used to study (determine) the number of bonds formed by atoms in solids.

An example of quadrupole splitting is the Mossbaver effect in the salt sodium nitroprusside (Figure 3.58). We note that quadrupole splitting of the ^{57}Fe give rise to two line pattern. Moreover, the molecular fields in Sodium Nitroprusside (Na$_2$ Fe(CN)$_5$ No) 2H$_2$O give rise to strong electric quadrupole field.

Figure 3.59 shows how in the case of ^{57}Fe the $\dfrac{3}{2}$ level splits up into a quadrupole *doublet* due to the electric field gradient originating from the nuclear electric quadrupole moment. The energy difference between the two lines is expected to be

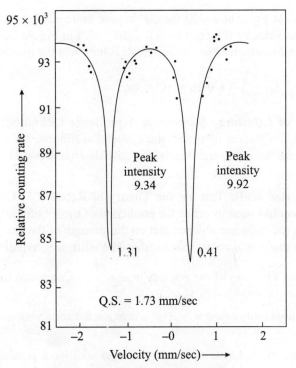

Absorber: Sodium Nitroprusside

Source: Co^{57}/Pd at room temperature

Fig. 3.58 Mossbauer absorption spectrum of sodium Nitroprusside exhibiting quadrupole splitting of the ^{57}Fe states giving rise to two line pattern.

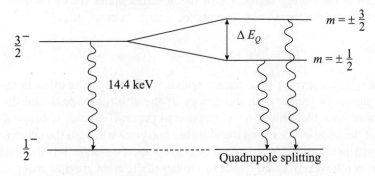

Fig. 3.59 Quadrupole doublet splitting of $I = \dfrac{3}{2}$ level in ^{57}Fe as a result of an electric field gradient.

$$\Delta E_Q = \frac{1}{2}\, e\, V_{ZZ}\, Q \qquad \qquad \ldots (3.17)$$

Where V_{ZZ} is field gradient and Q the quadrupole moment. For the 14.4 keV transition in ^{57}Fe, the source velocity 0.1 cm/s ($= 4.8 \times 10^{-8}$ eV) in Figure 3.59. This is the value of ΔE_Q, which the Mossbauer experiment yields. One finds the quadrupole moment as

$$Q\left(\frac{3}{2}\right) \sim + 0.28 \times 10^{-24} \text{ cm}^2$$

(iv) **Measurment of Lifetime**   Mossbaver experiment technique is quite good for the measurment of fairly short lifetimes, since when the lifetime exceeds about 10^{-10} s, extra nuclear fields can cause line broadening. For such determination the suitable range is 10^{-13} s $\leq t \leq 10^{-10}$ s.

(v) **Gravitational Red Shift: Test for the Theory of Relativity**   One important use of the Mossbaver effect has been to verify the prediction of theory of relativity that the frequency of electromagnetic radiation is dependent on the strength of the gravitational field. Suppose the emitter is a distance d above the absorber in a uniform gravitational field. When it is in the ground state, the mass of the nucleus is $m = \dfrac{E_0}{c^2}$. Compared to the absorber, it has an additional potential energy $mgd = \dfrac{E_0\, gd}{c^2}$, where g is the acceleration due to gravity. Similarly, when it is in the excited state the nucleus has an additional potential energy $\dfrac{E_1\, gd}{c^2}$. The energy difference of the emitter states is now

$$E_1\left(1+\frac{gd}{c^2}\right) - E_0\left(1-\frac{gd}{c^2}\right) = \Delta E\left(1+\frac{gd}{c^2}\right) \qquad \text{... (3.18)}$$

Where ΔE is the energy difference of the absorber states (or of the emitter states in the absence of a gravitational field. The photon energy is now

$$h\nu = h\nu_0\left(1+\frac{gd}{c^2}\right) \qquad \text{... (3.19)}$$

Where $h\nu_o$ is the energy of a photon which will cause a transition in the absorber. The photon energy is greater than the energy of the absorption peak and the absorber must move away from the emitter for absorption to occur. If emitter is below the absorber, the energy of the photon is less and the absorber must move toward the emitter. General theory of relativity predicted the phenomena of *gravitational red shift*. It is the change in energy of a photon as it travels from one place to another of different gravitational potential. Equation (3.19) gives this fractional energy change as $\dfrac{gd}{c^2}$. This is very minute. Thus the frequency of light will decrease by

$$\Delta\nu_{gr} = \frac{\Delta E_{gr}}{h} = \frac{E_\gamma\, gd}{c^2 h} \qquad \text{... (3.20)}$$

This decrease causes the light of shorter wavelength to shift towards the longer wave length, an effect known as the gravitational red shift. We may note that in the reverse case, if light moves downwards against the force of gravity, it will suffer a *blue shift*. The experiment-s were first performed by Pound and Rebka around 1960, and an excellent agreement with theory was obtained. In their experiment $d = 22.6$ m, which gives

$$\frac{\Delta E_{gr}}{E_\gamma} = \frac{gd}{c^2} = \frac{9.8 \times 22.6}{(3 \times 10^8)^2} = 2.5 \times 10^{-15}$$

In 1965 they performed the experiment with improved accuracy.

Hay et at. in 1960 provided experimental verification of *time dilatation* based on Mossbaver method.

Example 3.1 $^{212}_{83}$Bi decays with a half life of 60.5 min. by emitting 5 groups of α-particles having energies 6.08 MeV, 6.04 MeV, 5.76 MeV, 5.62 MeV and 5.60 MeV. Show that the α-disintegration energies are 6.20 MeV, 6.16 MeV, 5.87 MeV, 5.73 MeV and 5.71 MeV. Also show that the daughter nucleus is $^{208}_{81}$Tl. Sketch the energy level scheme for the daughter nucleus.

Sol. α-disintegration energy $(E_\alpha) = \left(\dfrac{A}{A - u}\right) K_\alpha$

$$\therefore \qquad E_{\alpha 1} = \left(\frac{212}{208}\right) \times 6.08 \text{ MeV} = 6.20 \text{ MeV}$$

$$E_{\alpha 2} = \left(\frac{212}{208}\right) \times 6.04 \text{ MeV} = 6.16 \text{ MeV}$$

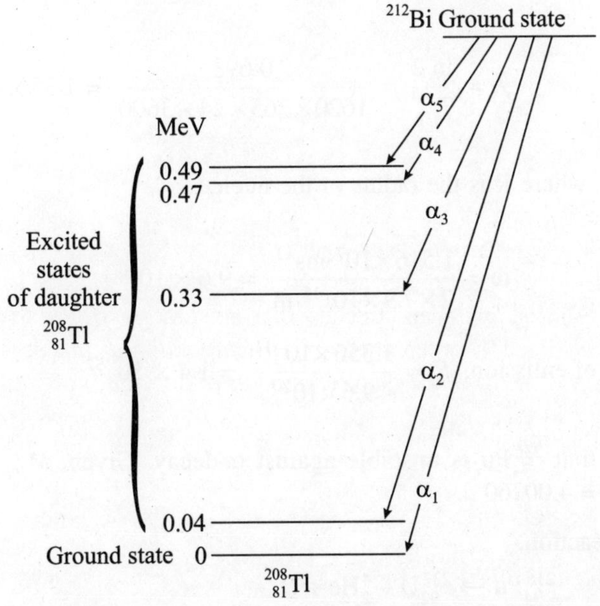

Fig. 3.60 Energy level (excited states of $^{208}_{81}$Tl fed by the α-ray gorups

$$E_{\alpha3} = \left(\frac{212}{208}\right) \times 5.76 \text{ MeV} = 5.87 \text{ MeV}$$

$$E_{\alpha4} = \left(\frac{212}{208}\right) \times 5.62 \text{ MeV} = 5.73 \text{ MeV}$$

$$E_{\alpha5} = \left(\frac{212}{208}\right) \times 5.60 \text{ MeV} = 5.71 \text{ MeV}$$

We can easily see that the daughter nucleus is $^{208}_{81}\text{Tl}$ ($^{212}_{83}\text{Bi} \rightarrow {}^{208}_{81}\text{Tl} + {}^{4}_{2}\text{He} + E_{\alpha}$). The energy levels.

Example 3.2 A radio-nuclide emits α-particles of energy 4.8 MeV and has a half-life 1620 years. Compute the velocity of α-particles and the probability of α-emission (Mass of α = 4.0026 u, radius of residual nucleus = 7.9×10^{-15} m)

Sol. The energy E_{α} of the α-particle (non-relativistically) is given by

$$E_{\alpha} = \frac{1}{2}mv^2 \Rightarrow v = \sqrt{\frac{2E_{\alpha}}{m}}$$

$\therefore$ velocity, $v = \left(\dfrac{2 \times 4.8 \times 1.6 \times 10^{-13}}{4.0026 \times 1.66 \times 10^{-27}}\right)^{\frac{1}{2}} = 1.516 \times 10^7 \text{ ms}^{-1}$

Probability of α-emission $P_{\alpha} = \dfrac{\lambda}{\omega}$, where λ is the decay constant and ω, the frequency of hitting the barrier of the nucleus.
Now,

$$\lambda = \frac{\ln 2}{T_{1/2}} = \frac{0.693}{1620 \times 365 \times 24 \times 3600} = 1.356 \times 10^{-11} \text{ s}^{-1}$$

and, $\omega = \dfrac{v}{2R}$, where R is the radius of the nucleus.

$$\therefore \qquad \omega = \frac{1.516 \times 10^7 \text{ ms}^{-1}}{2 \times 7.9 \times 10^{-15} \text{ m}} = 9.6 \times 10^{20} \text{ s}^{-1}$$

$\therefore$ Probability of emission, $P_{\alpha} = \dfrac{1.356 \times 10^{-11}}{9.6 \times 10^{20}} = 1.4 \times 10^{-32}$

Example 3.3 Show that $^{236}_{94}\text{Pu}$ is unstable against α-decay. Given, $M_{Pu} = 236.0460\ u$, $M_U = 232.03717\ U$ and $M_{\alpha} = 4.00260\ u$. [Punjab]

Sol. We have the reaction,

$$^{236}_{94}\text{Pu} \rightarrow {}^{232}_{92}\text{U} + {}^{4}_{2}\text{He} + Q$$

$$\therefore \qquad Q = (236.04607\ u - 232.03717\ u - 4.00260\ u)\,(931.5)$$
$$= 5.87\ \text{MeV}$$

Obviously, Q is positive and therefore ^{236}Pu will decay by α-emission.

Example 3.4 Find the maximum height of the potential barrier for α-penetration through ^{238}U nucleus. The radius of the residual nucleus is 9.3×10^{-13} Cm. [Bangalore]

Sol. The Coulomb repulsion energy, $U(r) = \dfrac{Zze^2}{4\pi\varepsilon_0\,r}$. The maximum value corresponds to a distance $r = R$, the nuclear radius.

$$\therefore \qquad \text{Barrier, } B = \frac{1}{4\pi\varepsilon_0}\frac{Zze^2}{R}$$

$$= \frac{92\times2\times(1.6\times10^{-19})^2}{4\times3.14\times8.85\times10^{-12}\times9.3\times10^{-15}}\ \text{J}$$

$$= \frac{92\times2\times2.56\times10^{-38}}{4\times3.14\times8.85\times10^{-12}\times9.3\times10^{-15}\times1.6\times10^{-13}}\ \text{MeV}$$

$$\simeq 28\ \text{MeV}$$

Example 3.5 Calculate, for an α-particle of energy 5 MeV, the order of magnitude of probability for leakage through a potential barrier of width 10^{-14} m and height 10 MeV [Kerala]

Sol. We have for a rectangular potential barrier,

$$T = \frac{16\,e^{-2k_2a}}{\left[1+\left(\dfrac{k_1}{k_2}\right)^2\right]\left[1+\left(\dfrac{k_2}{k_1}\right)^2\right]}$$

Here $\qquad k_2 = \sqrt{\left[\dfrac{2M_\alpha(V_0-E)}{\hbar^2}\right]}$ and $k_1 = \sqrt{\left[\dfrac{2M_\alpha E}{\hbar^2}\right]}$

$$\therefore \qquad \frac{k_1}{k_2} = \left[\frac{E}{(V_0-E)}\right]^{\frac12} = \left[\frac{5}{(10-5)}\right]^{\frac12} = 1$$

Hence $\qquad T = 4\,e^{-2k_2a}$

Example 3.6 Show that the degree of hindrance for the 4.2 MeV α-particles from $^{238}_{92}$U is about 1.6. Assume that the daughter nucleus acquires angular momentum l = 2. [Raj]

Sol. If we ignore the recoil effects, we have

$$V = \sqrt{\frac{2E}{M_\alpha}} = \sqrt{\frac{2\times4.2\times1.6\times10^{-13}}{4\times1.65\times10^{-27}}}$$

$$= 1.43 \times 10^7\ \text{m/s}$$

Now, the radius of the nucleus

$$R = r_0\, A^{\frac{1}{3}} = 1.4 \times 10^{-15} \times (234)^{\frac{1}{3}}$$
$$= 8.6 \times 10^{-15} \text{ m}$$

We have for heavy nuclei, an approximate analytical form due to Gamow is

$$T_l = \exp\left[\frac{Z_1\, e^2}{\hbar\, \varepsilon_0}\left(\frac{M_\alpha}{2}\right)^{\frac{1}{2}} E^{-\frac{1}{2}} - \frac{4e}{\hbar}\left(\frac{M_\alpha}{\pi\, \varepsilon_0}\right)^{\frac{1}{2}} Z_1^{\frac{1}{2}} R^{\frac{1}{2}}\left(1 - \frac{\sigma}{2}\right)\right]$$

Where Z_1 is mass number of daughter nucleus and σ is the centrifugal barrier to the Coulomb barrier,

$$\sigma = \frac{\dfrac{l(l+1)\hbar^2}{2\, m_\alpha\, R^2}}{\dfrac{2\, Z_1\, e^2}{4\pi\, \varepsilon_0\, R}}$$

$\therefore$ Hindrance factor $\dfrac{T_l}{T_o} = \left[\dfrac{2e}{\hbar}\left(\dfrac{M_\alpha}{\pi\, \varepsilon_0}\right)^{\frac{1}{2}} Z, R^{\frac{1}{2}} \times \dfrac{l(l+1)\hbar^2\,\pi\,\varepsilon_0}{Z_1\, e^2\, R\, M_\alpha}\right]$

Substituting the proper values and simplifying, one obtains

$$\frac{T_l}{T_0} \approx 1.6.$$

Example 3.7 An α–particle is emitted by the parent nucleus $^{212}_{84}$Po. Estimate the Coulomb potential if feels at the nuclear surface, and then make an approximate plot of the sum of the Coulomb and nuclear potentials acting on the Coulomb and nuclear potentials acting on the α-particles in various locations. [Punjab]

Sol. Let us assume that the daughter nucleus and the α-particle are uniformly charged spheres, then the Coulomb repulsion potential energy when they are just touching will be

$$V_0 = +\frac{2\, Z_1\, e^2}{4\pi\, \varepsilon_0\, r'}$$

Where $Z_1\, e$ is the charge on daughter nucleus and r' is the sum of the radii of the α-particle and daughter nucleus. One can estimate these radii by using the charge density half value radii a of the actual charge distributions found in the electron scattering measurments

$$a = 1.07\, A^{\frac{1}{3}} \text{ F}$$

One obtains the sum of radii

$$r' = \left[(4)^{\frac{1}{3}} + (208)^{\frac{1}{3}}\right] 1.07 \text{ F} = 8 \text{ F}$$

$$\therefore \quad V_0 = \frac{2 \times 82 \times (1.6 \times 10^{-19}\ C)^2}{1.1 \times 10^{-10} \times \dfrac{C^2}{N-m^2} \times 8 \times 10^{-15}\ m} = 4.8 \times 10^{-12}\ J$$

$$= 30\ \text{MeV}$$

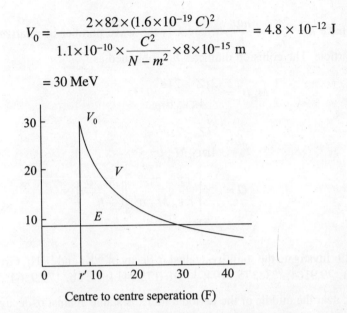

Fig. 3.61 Approximate represantation of the Coulomb + nuclear potential V acting on a α-particle from a $^{212}_{84}$Po nucleus, and the total energy E of the α-particle.

Figure indicates the total (Coulomb + nuclear) potential acting on the α-particle. As it approaches the nucleus, it feels the repulsive Coulomb potential increasing in inverse proportion to the distance between the centres of α-particle and nucleus, and reaching the value of V_0 when this distance equals r_1. Inside the surface it feels a rapid onset of the strong attractive nuclear potential, which soon dominates. (The onset is, of course, not quite as rapid as shown in Figure). Also indicated is the $^{212}_{84}$Po α-decay energy $E = 8.9$ MeV, which is the energy of the emitted α-particle. We may note that it is much less than V_0, the height of the Coulomb barrier.

Example 3.8 Show that the Gamow factor can be expressed as $G \simeq 3^{-\pi k b}$, where k is the wave number and b is the collision diameter. [Ujjain]

Sol. We have the probability of transmission of S-wave,

$$T = e^{-\gamma}$$

Where

$$\gamma = \frac{2\sqrt{2\,M_\alpha}}{\hbar} \int_R^b [V(r) - E]^{\frac{1}{2}}\, dr$$

$$= \frac{e^2\,(Z-2)}{\varepsilon_0\,\hbar} \left(\frac{M_\alpha}{2}\right)^{\frac{1}{2}} E^{-\frac{1}{2}} - \frac{4e}{\hbar}\left(\frac{M_\alpha}{\varepsilon_0\,\pi}\right)^{\frac{1}{2}} (Z-2)^{\frac{1}{2}}\,R^{\frac{1}{2}} \qquad \dots (3.1)$$

The first term in (3.1) is larger than the second. The first term is sometimes called the *Gamow exponent* and the corresponding approximate value of the barrier transparency is termed as Gamow factor.

We have $k = \dfrac{2\pi}{\lambda} = \dfrac{mv}{\hbar}$, where k is the wave number of relative motion for the emitted α-particle. The collision diameter b is obtained as

$$\frac{1}{2} M_\alpha v^2 = \frac{2(Z-2)e^2}{4\pi \varepsilon_0 b}$$

or

$$b = \frac{(Z-2)e^2}{4\pi \varepsilon_0 M_\alpha v^2} \qquad \qquad \dots (3.2)$$

$\therefore$

$$G = e^{-\pi \left[\frac{(Z-2)e^2}{\pi \varepsilon_0 M_\alpha v^2}\right]\left(\frac{M_\alpha v}{\hbar}\right)}$$

$$= e^{-\pi b k}$$

Example 3.9 Invstigate the stability against α-decay of ^{80}Kr and ^{176}Hf, Given the following atomic masses ^{80}Kr, 79.9164; ^{76}Se, 75.9194; ^{176}Hf, 175.9414; ^{172}Yb, 171.9364; ^{4}He, 4.0026.

Sol. ^{80}Kr, near the middle of the periodic table, is stable against α-decay but ^{176}Hf, with a much larger Coulomb energy, is unstable, though a decay is impeded by the barrier.

Example 3.10 The Q value for the α-decay of $Ra\,C$ (^{214}Po) is 7.83 MeV. Show that the energy of emitted α-particle is 7.68 MeV.

Sol. The Q-value relates to the centre of mass system. Allowing for the energy taken by the residual heavy nucleus the α-particle energy is

$$7.83 \times \frac{210}{214} = 7.68 \text{ MeV}$$

Example 3.11 If the range in standard air of the α-particles from sodium (half life, $T_{\frac{1}{2}}$ = 1622 years) is 3.36 cm, whereas from polonium ($T_{\frac{1}{2}}$ = 138 days) this range is 3.85 cm. Assuming the Geiger Nuttell law, is valid, show that the half life of $Ra\,C'$ for which α-particle range is 6.97 cm is 2.4×10^{-9} s. [Bhopal]

Sol. Geiger – Nuttell law is

$$\text{Log}_e \lambda = a + b \log_e R$$

For radium, we have $\lambda = \dfrac{0.693}{T_{\frac{1}{2}}} = \dfrac{0.693}{1622 \times 365}$

and

$$R = 3.36 \text{ cm}$$

$\therefore$ $\text{Log}_e\left(\dfrac{0.693}{1622 \times 365}\right) = a + b \log_e 3.36 \qquad \qquad \dots (3.1)$

For Polonium,

$$\lambda = \frac{0.693}{138}, R = 3.85 \text{ cm}$$

$$\lambda = \frac{0.693}{138} = a + b \log_e 3.85 \text{ cm} \qquad \ldots (3.2)$$

Subtracting (3.1) from (3.2), one obtains

$$b \log_e \frac{3.85}{3.36} = \log_e \frac{1622 \times 365}{138}$$

$$\therefore \qquad b = 61.43$$

Substituting this value of b in (3.2), one obtains

$$\log_e \frac{0.693}{138} = a + 61.43 \text{ loge } 3.85$$

$$\therefore \qquad a = -38.3$$

One can write the Geiger-Nuttell law for this radioactive series as

$$\log_e \lambda = -38.3 + 61.43 \log_e R$$

For *Ra C'*, $\qquad R = 6.97 \text{ Cm}$

$$\therefore \qquad \log_e \lambda = -38.3 + 61.43 \log_e 6.97 = 2.51 \times 10^{13}/\text{day}$$

$$\therefore \qquad T_{\frac{1}{2}} \text{ of } Ra C' \text{ is}$$

$$T_{\frac{1}{2}} = \frac{0.693 \times 24 \times 3600}{2.51 \times 10^{13}} \text{ sec} = 2.4 \times 10^{-9} \text{ s.}$$

Example 3.12 Tritium emits negative β-particles. Represent by an equation the decay process and calculate the end-point energy of the emitted particles. Given: $M(^3_1H)$, $M(^3_2He)$ and $M(^0_{-1}e)$ as 3.01695 amu, 3.01693 amu and 0.00055 amu respectively. [Delhi]

Sol. The decay process may be represented as

$$^3_1H \rightarrow ^3_2He + ^0_{-1}e + \nu$$

The emitted β-particle has its energy maximum when it is not shared at all by the accompanying neutrino. This is the end-point energy E, which is the total energy of disintegration.

$$\therefore \qquad E = (\text{mass of } ^3_1H - \text{mass of } ^3_2He) \text{ amu}$$

$$= (3.01695 - 3.01693) \text{ amu}$$

$$= 2 \times 10^{-5} \text{ amu}$$

$$= 2 \times 10^{-5} \times 931.2 \text{ MeV} = 0.01862 \text{ MeV}$$

Example 3.13 A sample of *RaE* contains 4 mg. If the half-life is 5 days and the average energy of β-particles emitted is 0.34 MeV, at what rate does the sample emit energy? At. wt. or *RaE* = 210.

[Raj]

Sol. Here the decay constant, $\lambda = \dfrac{0.693}{T_{\frac{1}{2}}} = \dfrac{0.693}{5 \times 24 \times 60 \times 60}$ s^{-1}.

Also, the number of atoms, $N = \dfrac{6.02 \times 10^{23}}{210} \times 4 \times 10^{-3}$.

From the relation $\dfrac{dN}{dt} = \lambda N$; one obtains

$$\frac{dN}{dt} = \frac{0.693}{5 \times 24 \times 60 \times 60} \times \frac{6.02 \times 10^{23}}{210} \times 4 \times 10^{-3} = 1.83 \times 10^{13}.$$

These are the number of particles emitted per second. If E_β be the average energy per particle, then the rate of emission of energy,

$$P = E_\beta \frac{dN}{dt} = 0.34 \times 1.83 \times 10^{13} \text{ Mev/s}$$

$$= 0.34 \times 1.83 \times 10^{13} \times 1.602 \times 10^{-13} \text{ J/s} = 1 \text{ Watt}.$$

Example 3.14 Tritium emits electrons and Mg-23 positrons. Represent the two decay processes by equations and calculate in each case the end-point energy of the particles emitted. Given the atomic masses: $M(^3\text{H}) = 3.01695$, $M(_{\pm1}^{0}e) = 0.00055$; $M(^3\text{He}) = 3.01693$, $M(^{23}\text{Na}) = 22.99618$ and $M(^{23}\text{Mg}) = 23.0002$, all in amu.

[Kerala]

Sol. The first decay process is: $_1^3\text{H} \rightarrow {_2^3}\text{He} + {_{-1}^{0}}e + \nu$. When the energy of the neutrino is zero, the emitted electron has the maximum energy—the total disintegration energy, or the end-point energy, E_d

$$E_d = M(_1^3\text{H}) - M(_2^3\text{He})$$

$$= 3.01695 - 3.01693 = 2 \times 10^{-5} \text{ amu}$$

$$= 2 \times 10^{-5} \times 931 \text{ MeV} = 0.0186 \text{ MeV}$$

The second decay process is: $_{12}^{22}\text{Mg} \rightarrow {_{11}^{23}}\text{Na} + {_{+1}^{0}}e + \nu$

∴ The end-point energy E_d of positron is,

$$E_d = M(_{12}^{23}\text{Mg}) - M(_{11}^{23}\text{Na}) - 2M(_{+1}^{0}e)$$

$$= 23.0002 - 22.99618 - 2 \times 0.00055$$

$$= (23.0002 - 22.99618 - 0.0011) \text{ amu}$$

$$= 0.00292 \times 931 \text{ MeV} = 2.72 \text{ MeV}$$

Example 3.15 Calculate the energy of γ-rays emitted in the β-decay of ^{28}Al. Given $E_{max} = 2.86$ MeV.

[MDS]

Sol.

$$^{28}\text{Al} \xrightarrow{\ _{-1}^{0}e\ } {}^{28}\text{Si} + Q$$
(daughter)

Since γ-rays are emitted during the above β decay, the β-transformation should be leading to an excited state of a nuclei, ^{28}Si (Figure).

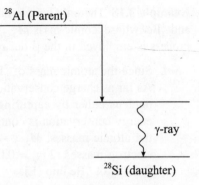

We have ^{28}Al = 27.981908 u,

^{28}Si = 27.976929 u

∴ $Q = (27.981908 - 27.976929)$ u

= $4.979 \times 10^{-3} \times 931.5$ MeV

= 4.638 MeV

Example 3.16 Show that the relativistic esepression for the ratio of nuclear recoil energy to maximum electron energy in β-decay is $\dfrac{T_R}{T_e} = \dfrac{(Q + 2m_e)}{(Q + 2m_R)}$, where Q is the energy available for decay products and m_R is the mass of the recoil nucleus.   [Mysore]

Sol. The electron energy is a maximum when the neutrino energy is zero and the energy and momentum equations are then

$$Q = T_e + T_R \qquad \qquad \text{... (3.1)}$$

$$P_e = P_R \qquad \qquad \text{... (3.2)}$$

The relativistic expression for P^2 is

$$E_e^2 - M_e^2 = E_R^2 - M_R^2 \qquad \qquad \text{... (3.3)}$$

But for each particle, $E = T + m$, therefore

$$T_e^2 - T_R^2 = 2 m_R T_R - m_e T_e$$

or $(T_e + T_R)(T_e - T_R) = 2 m_R T_R - 2 m_e T_e$

or $Q T_e + 2 m_e T_e = 2 m_R T_R + Q T_R$

or $\dfrac{T_R}{T_e} = \dfrac{(Q + 2m_e)}{(Q + 2m_R)}$

Example 3.17 In a hight energy β spectrum, with the maximum energy $E_0 \gg m_e c^2$, show that decay constant λ is approximately proportional to E_0^5.

Sol. At high energies, we assume $p_e \approx E_e$ (taking $c = 1$). We have the decay probability for hte emission of β particles with momentum between p_e and $p_e + d p_e$ is

$$d\,\lambda(P_e) = \frac{1}{2\pi^3} \frac{G^2 |M_{fi}|^2}{\hbar^2 c^4} \, p_e^2 \, (E_0 - E_e)^2 \, d p_e$$

Now, using $p_e \approx E_e$, we obtain

$$\lambda \propto E_0^5$$

Example 3.18 The only known nuclei with $A = 7$ are $_3^7\text{Li}$, whose atomic mass $M_{3,\,7} = 7.01600$ u, and $_4^7\text{Be}$, whose atomic mass $M_{4,\,7} = 7.01693$ u. Which of these nuclei is stable to β-decay? What process is employed in the β-decay of the unstable nucleus to the stable nucleus?   [Kerala]

Sol. Since the atomic mass of $_3^7\text{Li}$ is the lowest, it is the nucleus which is β-stable.

As far as charge conservation is concerned, the β-unstable $_4^7\text{Be}$ could decay into the stable nucleus either by capturing an atomic electron or by emitting a positron. But as far as energy conservation is concerned, only electron capture is possible since the difference in the atomic masses, $M_{4,\,7} - M_{3,\,7} = 7.016934 - 7.01600$ u $= 0.00093$ u, is less than two electron masses, $2\,m_e = 0.00110$ u. Thus electron capture is the process employed in the β decay of $_4^7\text{Be}$ into $_3^7\text{Li}$.

Example 3.19 One of the simplest β-decays is

$$_1^3\text{H} \rightarrow {}_2^3\text{He} + e + \bar{\nu}$$

The measured values of the decay energy and half life are $E = 0.0186$ MeV and $T_{\frac{1}{2}} = 12.3$ yr.

Calculate the value of FT.

Sol. Since Z is very small, one can evaluate F from a plot end point energy $K_e^{\max}$ (MeV) vs × log F. Using $K_e^{\max} = E = 0.0186$ MeV, one finds

$$\log F \simeq -5.7$$

or

$$F \simeq 2.1 \times 10^{-6}$$

Now life time $$T = \dfrac{T_{\frac{1}{2}}}{0.693} = \dfrac{12.3 \text{ yr} \times 365\dfrac{\text{days}}{\text{yr}} \times 24\dfrac{\text{hr}}{\text{day}} \times 60\dfrac{\text{min}}{\text{hr}} \times 60\dfrac{\text{sec}}{\text{min}}}{0.693}$$

$$\simeq 5.6 \times 10^8 \text{ s}$$

so $FT \simeq 2.1 \times 10^{-6} \times 5.6 \times 10^8 \text{ s} = 1.2 \times 10^3 \text{ s}$

Example 3.20 A free neutron decays into a proton, an electron and an antineutrino. If $M(n) = 1.00898$ u, $M(p) = 1.00759$ and $M(e) = 0.00055$ u, find the kinetic energy shared by the electron and the antineutrino.   [MDS]

Sol. The neutron decay may be repressented as

$$n \rightarrow p = e^- + \bar{\nu}$$

Since the total energy is to be conserved, we get

$$M(n)c^2 = M(p)c^2 + E_p + M(e)c^2 + E_e + M(\bar{\nu})c^2 + E_\nu$$

Where E's represent the kinetic energies of the respective particles.

Now, $M(\bar{\nu}) = 0$ and $E_p = 0$, the proton being at rest. We have,

$$E_e + E_\nu = M(n)c^2 - M(p)c^2 - M(e)c^2$$

$$= \{M(n) - M(p) - M(e)\}c^2$$

$$= (1.00898 - 1.00759 - 0.00055)c^2$$

$$= \frac{(0.00084 \times 1.00 \times 10^{-87} \times 9 \times 10^{16})}{1.6 \times 10^{-13}} \text{ MeV} = 0.78 \text{ MeV}$$

Example 3.21 ^{40}K decays to ^{40}Ar by electron capture. Assuming that the initial kinetic energy of the electron and the recoil energy of the nucleus are zero, show that the kinetic energy of the neutrino is 1.504 MeV. [Udaipur]

Given: Mass of ^{40}K = 39.96399 u, Mass of ^{40}Ar = 39.962384 u.

Sol. We have for electron capture

$$^{40}\text{K} + e^- \rightarrow {}^{40}\text{Ar} + \nu$$

$$\therefore \qquad Q = [M(^{40}\text{K}) - M(^{40}\text{Ar})]c^2$$

$$= (39.963999 \text{ u} - 39.9623844) \, 931.5 \, \frac{\text{MeV}}{\text{u}}$$

$$= 1.504 \text{ MeV}$$

∴ The kinetic energy of neutrino = 1.504 MeV

Example 3.22 $^{11}_{6}$C decays to $^{11}_{5}$B by positive β-emission. What are the maximum and minimum energies of neutrino? Given: mass of ^{11}C = 11.011433 u, mass of ^{11}B = 11.009305. [Bhopal]

Sol. We have

$$Q = [M(^AX) - M(^AX') - 2m_e]c^2$$

$$= [11.0114334 - 11.009305 \text{ u}] \, 931.5 \left(\frac{\text{MeV}}{\text{u}}\right) - 2 \times 0.511 \text{ MeV}$$

$$= 0.96 \text{ MeV}$$

This is the Q-value of decay. Now, the energy of neutrino is maximum, when the positron energy is minimum. Clearly, the maximum energy the neutrino can have is 0.96 MeV and the minimum energy is zero.

Example 3.23 A beam of mononergetic γ-rays is incident on an *Al*-sheet of thickness 10 cm. The sheet reduces the intensity of the beam to 21 % of the original. Calculate the linear and mass absorption coefficients, given density of *Al* = 2700 kg.m^{-3}.

Sol. Let I_0 be the original intensity and I the intensity on absorption by a thickness t.

$$\therefore \quad \frac{I}{I_0} = e^{-\mu t}. \text{ where } \mu = \text{linear absorption coefficient. Here } \frac{I}{I_0} = \frac{21}{100} \, ; t = 0.1 \text{ m}.$$

$$\therefore \qquad 0.21 = e^{-0.1 \, \mu}$$

$$\therefore \qquad -0.1 \, \mu = \ln(0.21) = -1.56$$

$$\therefore \qquad \mu = 15.6 \text{ m}^{-1}$$

$\therefore$ Mass absorption coefficient, $\mu_m = \dfrac{\mu}{\rho}$

$$\mu_m = \frac{15.6}{2700}$$

$$= 5.78 \times 10^{-3}\ m^2.kg^{-1}$$

Example 3.24 The isotope ^{12}N beca decays (positron) to an excited state of ^{12}C which decays to the ground state with the emission of a 4.43 MeV gamma rays. Given mass of $^{12}N = 12.018613$ u, show that the maximum energy of the β-particle (positron) is 16.32 MeV. [Bhopal, MDS]

Sol. We have the process

$$^{12}_{7}N \rightarrow {}^{12}_{6}C + {}_{+1}e^0 + \nu$$

The Q-value for the above process is

$$Q = [m(^{12}N) - m(^{12}C) - 2\,m_e]c^2$$

$$= (12.0186134 - 12.000\ u - 2 \times 0.0005494)\ 931.5\ \frac{MeV}{u}$$

$$= 16.32\ MeV$$

Example 3.25 The β-decay of ^{141}Ce leads to ^{141}Pr. The daughter nuclei ^{141}Pr are found to have been left in an excited state. They reach the ground state by emission of γ-rays or conversion electrons. If for the conversion of K-electrons, $B\rho = 1135$ gauss-cm and the binding energy of the K-electrons is equal to 42 keV, show that the excitation energy of a ^{141}Pr nucleus is ^{145}keV.

[Kerala, Jodhpur]

Sol. We have

$$K_e = \Delta E - \text{B.E.}$$

Here binding energy (B.E.) of K_e-electron = 42 MeV. Now, We want to find ΔE, the excitation energy of a ^{141}Pr nucleus. For this purpose, we must know the kinetic energy of the K-conversion electrons. Now, $B\rho$ value is a measure of electron Momentum. Electrons being always relativistic. We have the relation between the kinetic energy of electrons in MeV and $B\rho$ value in gauss-cm, i.e.

$$\left(\frac{K_e}{0.511} + 1 \right)^2 = \left(\frac{B_\rho}{1704} \right)^2 + 1 = \left(\frac{1135}{1704} \right)^2 - 1 = 1.444$$

$\therefore \qquad\qquad K_e = 0.103\ MeV = 103\ keV = \text{K.E. or the } K_e\text{-Conversion}$

$\therefore \qquad\qquad \Delta E = 103\ keV + 42\ keV = 145\ keV$

Obviously, 145 keV is the excitation energy of a ^{141}Pr nucleus.

Example 3.26 From Figure determine the types of radiation emitted by $^{38}_{18}A$ in the γ-decays between its three lowest energy states. [Mysore, Kerala]

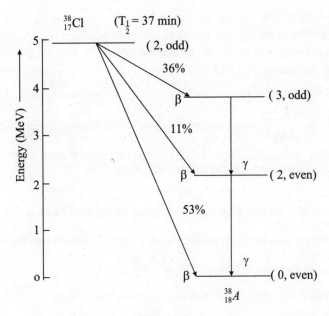

Fig. 3. 62

Sol. In the decay scheme shown between the states of $i = 2$, even parity, and $i = 0$, even parity, we have $|i_i - i_f| = 2 = L$. Since this value is even, and hence the nuclear parity does not change, the radiation is electric quadrupole.

Example 3.27 ^{203}Tl atoms resulting from β-decay of ^{203}Hg atoms emit 4 groups of conversion electrons with kinetic energies of 266.3, 264.2, 263.6 and 193.3 keV. To what shell of Tl atom, K, L_I, L_{II}, L_{III} does each group correspond? The electron binding energies (B.E.) in the shells are 87.7, 15.4, 14.8 and 12.7 keV respectively. Show that the energies of γ-quanta are concurrent with that decay.   [Goa, Mumbai]

Sol. We have

$$K_e = \Delta E - \text{B.E.}$$

Obviously, the maximum kinetic energy group corresponds to L_{III} conversion electrons at they are least tightly bound. We have

$$266.3 \text{ keV} = \Delta E - 12.7 \text{ keV}$$

$$\therefore \qquad \Delta E = 279 \text{ keV}$$

Similarly, one obtains

$$264.2 \text{ keV} = \Delta E - 14.8 \text{ keV}$$

$$263.6 \text{ keV} = \Delta E - 15.4 \text{ keV}$$

and $\qquad 193.3 \text{ keV} = \Delta E - 87.7 \text{ keV for } K\text{-shell}$

For all these, we obtain $\Delta E = 279$ keV.

This clearly reveals that the γ-energy concurrent with the β-decay is 279 keV.

Example 3.28 The 0.014 MeV gamma say that follows electron capture in ^{57}Co is used in a study of Mossbauer effect. Assuming that the recoil momentum is effectively distributed between 1000 neighbouring atoms in the crystal lattice, show that the relative source absorption velocitity required to obtain resonance absorption is 0.079 m/s. [MDS]

Sol. Let M is the mass of the nucleus and E is energy of emitted photon. The energy of recoil,

$$E_{re} = \frac{Ee^2}{2Mc^2} = \frac{E_0^2}{2Mc^2}.$$ This energy is distributed among N neighbouring atoms in the

crystal lattice, then energy imparted to each atom $E = \dfrac{E_{re}}{N} = \dfrac{E_0^2}{M\,Nc^2}$. Thus the energy

difference between centers of the emission and absorption lines is $\Delta E = \dfrac{E_0^2}{M\,Nc^2}$.

When a source of radiation is travelling with velocity v relative to the observer, the change

in energy due to Doppler effect is $\Delta E_D = \dfrac{E_0\, v}{C}$.

∴ For resonance absorption $\Delta E = \Delta E_D$

or
$$v = \frac{E_0}{M\,Nc} = \frac{0.014 \times 1.6 \times 10^{-13}}{57 \times 1.67 \times 10^{-27} \times 1000 \times 3 \times 10^8}$$

$$= 0.079 \text{ m/s}$$

Example 3.29 Mossbauer's original resonant absorption used γ-rays emitted in transitions from the 0.129 MeV first excited state to the ground state of $^{191}_{79}$Ir. (a) consider the recoil of the nucleus, assumed to be free, when it emits the γ-ray, and determine the down ward shift in the energy of the γ-ray that results from the energy taken by the nuclear recoil. (b) Then compare this energy shift to the width of the first excited state of $^{191}_{77}$Ir, which has a measured lifetime $T = 1.4 \times 10^{-10}$ s. [Indore]

Sol. Since the total linear momentum of the decaying nucleus is zero before emitting the γ-ray, the magnitude of the nuclear recoil momentum p_n after the emission must equal the magnitude of the momentum p_γ carried by the emitted γ-ray. As the nuclear mass M is quite high, its recoil velocity is low, so one may use the classical expression

$$p_n = \sqrt{2MK} \qquad \qquad \text{... (3.1)}$$

Now, the γ-ray momentum p_γ is related to its enegy E by the relativistic expression

$$p_\gamma = \frac{E}{c} \qquad \qquad \text{... (3.2)}$$

∴
$$p_\gamma = \frac{E}{c} = p_n = \sqrt{2MK}$$

or
$$\frac{E^2}{c^2} = 2\,MK$$

$$\therefore \qquad K = \frac{E^2}{2Mc^2} \qquad \qquad \dots (3.3)$$

Since the sum of the γ-ray energy E and the nuclear recoil energy K must equal the energy available in the γ-decay, i.e., the 0.129 MeV energy of the first excited state of the decaying nucleus. One finds that E is less than the energy of the first excited state by an amount K. This is the downward shift ΔE in the energy of the γ-ray due to nuclear recoil, i.e.

$$\Delta E = -K = -\frac{E^2}{2Mc^2} \qquad \qquad \dots (3.4)$$

Since M is quite large, ΔE is very small compared to E, and one may evaluate it approximately by setting $E = 0.129$ MeV.

$$\therefore \qquad \Delta E \simeq -\frac{(0.129)^2 \ (\text{MeV})^2}{2 \times 191 \times 931 \ \text{MeV}}$$

$$= -4.7 \times 10^{-8} \ \text{MeV} = -4.7 \times 10^{-2} \ \text{eV}.$$

One can also obtain the above result by considering the γ–ray to be emitted from a moving source, the recoiling nucleus, and using the longitudional Doppler shift formula to evaluate the downward shift in its frequency or energy.

(b) If the lifetime of the first excited state of $^{191}_{77}$Ir is $T = 1.4 \times 10^{-10}$ s, its width is

$$\Gamma = \frac{\hbar}{T} = \frac{6.6 \times 10^{-16} \ \text{eV-s}}{1.4 \times 10^{-10} \ \text{s}} = 4.7 \times 10^{-6} \ \text{eV}.$$

Clearly, the γ-ray emitted by the decay from the first excited state of the $^{191}_{77}$Ir source nucleus cannot excite a $^{191}_{77}$Ir absorber nucleus from its ground state to its first excited state. The nuclear recoil shift of the γ-ray is larger by a factor of 104 than the width of the state it is supposed to excite. So the γ-ray is thrown completely out of resonance, and the resonant absorption is destroyed. (If there actually were an absorption, there would be two sources to the total recoil shift, one due to recoil of the emitting nucleus and the other due to recoil of the absorbing nucleus. This is because to be absorbed by the free nucleus, the γ-ray must have an energy that is greater than the energy difference of the nuclear states by the amount $\Delta E = +K$. There would also be two sources of the total width of the resonance, one due to the width of the state emitting the γ-ray and the other due to the width of the state absorbing it).

SUGGESTED READINGS

1. K. Siegbah, 'Alpha, Beta and Gamma ray spectroscopy', North-Molland (reprinted 1979)
2. J.M. Pearson, 'Nuclear Physics: Energy and Matter', Adam Hilger, Ch. III (1986)
3. J.B. Blatt and V.F. Weisskof, 'Theoretical Nuclear Physics', Wiley (1952)
4. W.D. Hamilton (ed), 'The Electromagnetic Interaction in Nuclear Spectroscopy' North-Holland (1975).
5. M.Morita, 'Beta decay and Muon capture', Benjamin (1973)

6. T.E. Cranshaw, E.W. Dale, G.O. Longworth and C.E. Johnson, 'Mossbauer Spectroscopy and Its Applications' Cambridge University Press (1985).
7. S. L. Kakani, Solid State Physics (4th ed.) Sultan Chand New Delhi-2 (2007)
8. J. Lilley, Nuclear Physics, John Wiley (2001)

REWIEW QUESTIONS

1. Explain Gamow's theory of α-decay. How is Geiger-Nuttel law dervied from it.
2. What is meant by the range of α-particle? How it is determined experimentally? Explain the 'straggling' of the range of α-particles.
3. Explain with the help of Gamow's theory how α-particles with energies less than the height of the potential barrier are emitted from a radioactive nucleus. Write the special features of α-ray spectra.
4. How, one can determine accurately the range of an α-particle emitted by a natural radioactive element? Discuss the relationship between
   (i) range and energy of an α-particle and
   (ii) range of an α-particle and the half-life of the radioactiv element emitting it.
5. Why some nuclei emit α-particles of more than a single energy?
6. Describe the theory of α-decay. Explain with the help of energy level diagram the fine structure of α-ray spectrum.
7. Discuss the salient features of β-ray spectra and explain how Pauli's hypothesis of neutrino –β emission solved the anomalies in the β-ray spectra. Mention the properties of neutrino.
8. Give complete theory of β-decay due to Fermi. How far this theory has been verified experimentally?
9. List the conservation laws obeyed in β-decay. Which conservation law is violated? How has this been experimentally verified.
10. When electron is not an integral part of a nucleus, how do we justify the emission of electrons from nuclei in β-decay?
11. The continuous β-ray spectrum of radioactive substances presented a peculiar difficulty. How it was solved?
12. What led Pauli to put forward the neutrino hypothesis? Give the elementary theory of β-decay. Explain how the distortions in Kurie plots of the β-ray energy spectrum were accounted for.
13. Give Fermi's theory of β-decay. How far this theory has been verified experimentally?
14. Give a general account of the β-ray spectrum and indicate its importance in relation to the conservation of energy in nuclear processes.
15. A nucleus emits an α-particle followed by two β-particles. Show that the final nucleus is an isotope of the original one.
16. What is a neutrino? Explain qualitatively how the hypothesis of a neutrino solves the apparent breakdown of conservation of momentum and energy in β-decay.
17. Write the decay process for (i) electron emission (ii) positron emission, and (iii) electron capture.
18. The three β-decay processes involve the emission of neutrinos or antineutrinos. In which decay process do the neutrinos have a continuous energy spectrum? In which system is the neutrino monoenergetic?
19. Derive the momentum spectrum for electrons emitted in an allowed β-transition where a β-particle and neutrino are emitted with zero angular momentum with respect to the nucleus. Assume that

$$\lambda = \frac{2\lambda}{\hbar} |H \rightarrow f|^2 \frac{dn}{dE}$$

Where the symbols have their usual significance. Take the matrix element of the interaction as some constant.

20. The natural decay chain $^{238}_{92}U \to {}^{206}_{82}Pb$ consists of several α-decays and β-decays. (i) How many α-decays must occur in the chain?

    (ii) For that number of α-decays, how many β-decays must occur to make the end product?

21. Discuss briefly the different mechanisms through which γ-rays are absorbed in matter. Describe how γ-ray energy is measured in a scintillation counter.

22. Write Klein and Nishina for the probability of compton scattering process to take place when γ-rays pass through matter. Explain the significance of each term involved.

23. Examine critically the different physical processes resulting from the interaction of γ-rays with matter and the relative importance of three processes at different energies of radiation.

24. Describe briefly the methods for measuring the energies of γ-rays.

25. Write short notes on
    (i) Nuclear isomerism
    (ii) Pair production
    (iii) Internal conversion
    (iv) Mossbauer effect

26. In γ-decays from states of excitation energy around 1 MeV, or less, to ground states, electric dipole radiation is almost never observed? Use the shell model to explain this.

27. What are the differences between single particle states and many particle states? How are they related? What about γ-decaying states?

28. Why it is easier for an incident nucleon to enter a nucleus than it is for either of the nucleons, resulting from its first collision, to escape?

29. Electric dipole radiation is emitted with a characterstic spatial pattern. Does this suggest an experimental technique for determining the type of radiation emitted in a γ-decay? What would be the difficulty in using such a technique?

30. Predict, from the shell model, the regions of the periodic table in which the first excited states of nuclei have particularly long lifetimes for γ-decay.

PROBLEMS

1. Determine the velocity and the kinetic energy of the daughter nucleus formed in α-decay (of energy 4.19 MeV) of the parent nucleus U-238. **[Ans. 0.072 MeV, 2.44×105 m/s]**

2. The α-particles emitted when ^{238}U decays has a kinetic energy of 4.2 MeV. Calculate the disintegration energy (Q) and recoil kinetic energy of the daughter nucleus.
   [Ans. 4.27 MeV, 0.0718 MeV]

3. Show that in α-decay, the nucleus carries away a fraction $\dfrac{4}{(4 + A_D)}$ of the total energy available, where A_D is the mass number of daughter nucleus.

4. The α-particles from *Ra* have a range 3.36 cm in air whereas those from Po have 3.85 cm. Assuming Geiger-Nuttall law, find the half-life of *Ra C′* for which the range of α-particles is 6.97 cm. Given half-lives of *Ra* and *Po* as 1622 yrs and 132 days respectively. **[Ans. 2.4×10⁻⁹ s]**

5. At what distance from the centre of a ^{238}U nucleus is the α-particle of its radioactive decay released with zero kinetic energy, the disintegration energy is 4.27 MeV **[Ans. 6.1×10⁻¹⁴ m]**

6. Show that *s*-wave barrier transmission exponents can be expressed as the converging series

$$\gamma = \pi \frac{4Z}{137\beta}\left[1 - \frac{4}{\pi}\left(\frac{E}{B}\right)^{\frac{1}{2}} + \frac{2}{3\pi}\left(\frac{E}{B}\right)^{\frac{3}{2}} + \frac{1}{10\pi}\left(\frac{E}{B}\right)^{\frac{5}{2}} + \dots\right]$$

where $\dfrac{E}{B} = \dfrac{mv^2 \, \pi \varepsilon_0 \, R}{Z_D \, e^2}$ [Raj, MDS]

7. Protons of energy 1.2 MeV are scattered by $^{197}_{79}$Au nucleus. Assuming a Rutherford type of scattering, estimate the distance of closest approach. Also show that deutrons with the same kinetic energy as the above protons may come within the same distance of a gold nucleus.

[**Ans.** 9.47×10^{-14} m]

8. The α-particles form *Ra* have a range 3.36 cm in air whereas those from *Po* have 3.85 cm. Assuming Geiger-Nuttall law, estimate the half life of *Ra C'* for which the range of α-particles is 6.97 cm. Given half lives of *Ra* and *Po* are 1622 yrs and 132 days respectively. [**Ans.** 2.4×10^{-9} s]

9. *Ra E'* ($A = 120$) containing 4 mg emits β-particles and radiates energy at the rate of 1 Watt. If half life of *Ra E'* = 5 days, then show that the average energy of β-particles is 0.34 MeV.

10. The nucleus $^{21}_{11}$Na decays to $^{21}_{10}$Ne with the emission of a positron. The radius of the nuclei of mass number 21 is 3.6 F; estimate the maximum energy of the positron. (A spherical uniform distribution of charge Q has a Coulomb potential energy of $\dfrac{3}{5}\left(\dfrac{Q^2}{4\pi\varepsilon_0 R}\right)$. [**Ans.** 3.226 MeV]

11. The three leptons e, μ and τ may be assumed to have the same weak interaction coupling constant. The τ lepton (mass 1784 MeV/c^2) and the muon (mass 105 MeV/c^2) have decay modes and branching values as follows:

$$\mu = e + \nu + \overline{\nu} \qquad (100\%)$$

$$\tau = e + \nu + \overline{\nu} \qquad (17\%)$$

Given the mean lifetime of muon is 2.2×10^{-6} s make an estimate of the mean lifetime of the τ-lepton.

If lepton with momentum 5 GeV/c are produced in an $e^+ e^-$ collider, calculate the mean flight path before decay in the laboratory system. [**Ans.** 2.64×10^{-13} s; 222 μm]

12. Calculate the fraction of electrons emitted within 100 eV of the end point in tritium β-decay, assuming the mass of the neutrino is zero. ($^3_1$H $\rightarrow$ $^3_2$He $+ e^- + \overline{\nu}_e + 18.60$ keV) [**Ans.** 3.4×10^{-7}]

13. The positron decay of ^{8}B to ^{8}Be has an end-point energy of 14.09 MeV. What fraction of the neutrinos prodiced in the decay of ^{8}B have sufficient energy to induce the reaction $\nu + {}^{37}$Cl $\rightarrow$ ^{37}A $+ e^-$? ($Q = -0.81$ MeV). [**Ans.** 99.84 %]

14. Show that the energy released in electron capture by beryllium represented by

$$^{74}\text{Be} + e^- \rightarrow {}^7_3\text{Li} + \nu$$

is 0.861 MeV.

15. $^{23}_{10}$Ne decays to $^{23}_{11}$Na by negative β^- emission.
   (i) Find the maximum kinetic energy of the electron
   (ii) What will be its minimum energy?
   (iii) Also find the energy of the antineutrino in each case.

[**Ans.** (i) 4.4 MeV, (ii) zero, (iii) zero, $4-4$ MeV]

16. ^{14}C undergoes β-decay transition to form $^{14}_{7}$N. If the end point energy of the transition is 3.94 MeV and the mass of the initial atom is 14.007685 u, show that the mass of final atom is 14.007517 u.

17. Show that for $E_m << me\,c^2$, the mean kinetic energy of the β-particles is equal to $\dfrac{E_m}{3}$.

18. Show that in the β-transformation

$$^{A}_{Z}X \rightarrow \,^{A}_{Z+1}Y + \beta^{-} + \bar{\nu}$$

the kinetic energy of the recoil nucleus is given by

$$E_y = \frac{(Q + 2m_0 c^2)\,E_m}{2\,M_y\,c^2}$$

Where Q is the β-disintegration energy and E_m is the maximum kinetic energy of the β-particle. You may assume the motion of the recoil nucleus to be non-relativistic

19. Show that the recoil energy of the nucleus undergoing electron capture type of β-decay is given by

$$E_R = \frac{E^2}{2Mc^2}$$

Where E is the total released energy.

20. In an experiment on γ-rays from ^{161}Ba, the rays are allowed to fall on a lead radiator. Four groups of photoelectrons are observed corresponding to Br-values 1250, 1445, 2050 and 2520 gauss-cm respectively. The binding energy of the K-shell electrons in lead is 0.0891 MeV. Show that the energies of γ-rays are 0.212 MeV, 0.248 MeV, 0.377 MeV and 0.490 MeV.

21. A heterogeneous beam of γ-rays from a radioactive source is allowed to fall on an aluminium radiator and the electrons ejected are analyzed in a magnetic spectrograph. The electron fall into three groups with maxima at 11960, 5845 and 2752 gauss-cm respectively. Show that the energies of the three γ-rays are 3.35 MeV, 1.53 MeV, and 0.64 MeV.

22. Show that the electron-positron pair creation by a high energy photon cannot take place in vacuum.

23. Show that a free electron cannot absorb the entire energy of a photon incident on it.

24. If the maximum energy of the recoil electrons be 1 MeV in the compton the energy of the photon is 1.234 MeV.

25. In a Mossbauer experiment the source, which emit photons with frequency ν, is raised to a height l above the absorber and detector. By equating the photon's initial mass $\dfrac{h\nu}{c^2}$ with its gravitational mass, show that the frequency ν' of the photon at the absorber is given by $\nu' = \nu\left(1 + \dfrac{gl}{c^2}\right)$, where g is the acceleration due to gravity. For $l = 0.3$ m calculate the fractional change in frequency and the source velocity required to give resonance. **[Ans. 3.3×10^{-15}; 1.0 mm/s]**

SHORT QUESTION-ANSWERS

1. What do you understand by α, β and γ decays?

Ans. α-decay is the spontaneous emission of an α-particle from a nucleus of large atomic number. This and spontaneous fission is responsible for setting an upper limit on the atomic numbers of the chemical elements occuring in nature. β-decay is the spontaneous emission or absorption of an electron or positron by a nucleus. It tells us much about β-decay interaction, which is one of the fundamental interactions, or forces, of nature. γ-decay is the spontaneous emission of high energy photons when a nucleus makes transitios from an excited state to its ground state. This gives detailed information about the excited states of nuclei that can be used to improve the nuclear models. γ-decays is also used in the Mossbauer effect to make extremely high-resolution energy measurments in many different fields.

2. How α-particle decay illustrate the importance of the phenomenon of quantum mechanical tunnelling?

Ans. In α-decay, an α-particle is contained by a barrier, called the Coulomb barrier, formed by the strong attractive potential and the electrostatic repulsive $\dfrac{1}{r}$ potential between the α-particle and the rest of the nucleus. Since every decay energy is far less than the height of the Coulomb barriers, which is $\sim 30\,\text{MeV}$ for all α-decays the α-particle tends to be trapped by the barrier in every decay. It can escape only by the quantum mechanical process of *barrier penetration*.

3. How α-decaying nuclei with very short life-times can be found in nature?

Ans. The short lifetime α-emitters are in equilibrium in decay families with long life-time parents, called radioactive series.

4. Why nuclei of large Z spontaneously emit α-particles (^4_2He), but do not spontaneously emit any of the particles ^3_2He, ^2_1H, or ^1_1H, even though emitting any of these particles reduces the Coulomb energy of the nucleus?

Ans. For the particles other than ^4_2He the binding energy per nucleon $\dfrac{\Delta E}{A}$, is much smaller than it is for a typical nucleus. Thus their emission is not energetically favourable.

5. What are long-range α-particles?

Ans. α-particles may be emitted from an excited state of the initial nucleus as well as from its ground state and the particles of increased energy are known for historical reasons as long-range α-particles.

6. What do you understand by fine structure?

Ans. In some decays, several low-lying states of final nucleus are accessible with reasonable probability and the α-particle spectrum then shows a branching leading to fine structure.

7. What are two closely related processes to β⁻-decay?

Ans. The two closely related processes to β⁻-decay are β⁺-decay where a proton bound in a nucleus decays to a bound neutron with the emission of a positron and and a neutrino, ν, and electron capture (EC) where a proton bound in a nucleus changes to a bound neutron through interacting with one of the atomic electrons and a neutrino is emitted. Examples of these processes are:

$$^{22}\text{Na} \rightarrow {}^{22}\text{Ne} + e^+ + \nu \qquad \text{(β⁺-decay)}$$
$$e^- + {}^7\text{Be} \rightarrow {}^7\text{Li} + \nu \qquad \text{(Electron capture)}$$

8. Give the simplest example of β⁻-decay

Ans. β⁻-decay of a neutron to a proton

$$n \rightarrow p + e^- + \bar{\nu}$$

9. What is Fermi theory of β⁻-decay?

Ans. Fermi's theory, in which an electron and neutrino are created by an interaction which has a zero range, accounts for the observed electron energy spectra snd spread of β-decay lifetimes.

10. What is modern theory of Beta-decay?

Ans. Beta-decay is an example of a weak interaction and this interaction, like the strong, electromagnetic and gravitational interactions, can be understood as arising through the exchange of particles. In beta-decay the exchanged particles are charged intermediate vector bosons W^- and W^+, whose masses are $\sim 80 \text{ GeV/c}^2$ (mass of a proton $\sim 1 \text{ GeV/c}^2$)

11. What is the relation between the range of an interaction (R) and the mass (m) of the particle?

Ans. $R \approx \dfrac{\hbar}{mc}$

As an example the force between two neutrons arising from the exchange of a pion, which has a mass of $\sim 140 \text{ MeV/c}^2$, would be predicted to have a range $R \approx 1.4$ F, which is in good agreement with the observed range of the nuclear force between the nucleons.

12. What is parity violation?

Ans. It had been assumed prior to 1956 that parity was conserved in the weak interaction like in the electromagnetic interaction. Lee and Yang pointed out in 1956 that this statement has not been experimentally tested and shortly afterward it was demonstrated that parity was not conserved in the weak interaction. The observation of parity violation showed that the Fermi theory was incomplete and that the weak matrix element must contain a mixture of scalar and pseudoscalar terms. This is provided by the $V - A$ theory, which accounts for the observed helicities of the emitted neutrinos and electrons.

13. What are the expected types of gamma ray transitions between the following states of odd A nuclei; $g_{\frac{9}{2}} \to p_{\frac{1}{2}}$; $f_{\frac{5}{2}} \to n_{\frac{3}{2}}$; $h_{\frac{11}{2}} \to d_{\frac{5}{2}}$; $h_{\frac{11}{2}} \to d_{\frac{3}{2}}$?

Ans. (i) For transition $g_{\frac{9}{2}} \to p_{\frac{1}{2}}$, we have $J_a = \dfrac{9^+}{2}$ and $J_b = \dfrac{1^-}{2}$. This means L-values are 4 and 5. Since the parity changes, the possible transitions are $M4$ and $E5$.

(ii) $f_{\frac{5}{2}} \to p_{\frac{3}{2}}$ transition. We have $J_a = \dfrac{5^-}{2}$ and $J_b = \dfrac{3^-}{2}$. Thus, possible L-values are 1, 2, 3, 4. Now, the parity does not change and therefore the possible transition are $M1$, $E2$, $M3$ and $E4$.

(iii) $h_{\frac{11}{2}} \to d_{\frac{5}{2}}$ transition. We have $J_a = \dfrac{11^-}{2}$ and $J_b = \dfrac{5^+}{2}$. Clearly L-values are 3, 4, 5, 6, 7, 8. The parity changes and hence the possible transitions are $E3$, $M4$, $E5$, $M6$, $E7$ and $M8$.

(iv) $h_{\frac{11}{2}} \to d_{\frac{3}{2}}$ transition. Here $J_a = \dfrac{11^-}{2}$ and $J_b = \dfrac{3^+}{2}$. Possible L-values are 4, 5, 6, 7 Parity changes. The possible transitions are $M4$, $E5$, $M6$ and $E7$.

In the above four transitions, the transition probability decreases rapidly with the increase of multipole order L. Obviously, the prominently decay modes are $M4$, $M1$, $E3$ and $M4$ in the above transitions.

14. What are γ-rays?

Ans. γ-rays are emitted from many of the nuclei of the radioactive series. These are *photons* of electromagnetic radiation that carry away the excess energy when nuclei make γ-decay transitions from excited states to lower energy states.

15. What is the energy of γ-rays?

Ans. The energy difference in nuclear excited states range upwards from $\sim 10^{-3}$ MeV, as such γ-rays have energies greater than $\sim 10^{-3}$ MeV.

16. What is the most accurate technique for measuring the energy of γ-rays?

Ans. To study their diffraction from a crystal lattice of known lattice spacing.

17. What is the cause of nuclear decay through the emission of a γ-ray?

Ans. The cause of such transitions is the interaction between the nucleus and an external electromagnetic field.

18. Why electromagnetic transitions form the dominant mode of decay for low-lying excited states in nuclei, Particularly for lighter ones?

Ans. The main reason is that nuclear emission a much faster process than γ-decay, is forbidden until the excitation energy is above nucleon separation energies.

19. What do you understand by internal conversion?

Ans. Besides γ-ray emission, nuclei can also de-excite through electromagnetic interaction by *internal conversion* where by one of the atomic electrons is ejected. This is usually more important for heavy nuclei, where the nuclear electromagnetic fields are strong and the orbits of the inner shell electrons are closed to the nucleus. Instead of a – γ-ray, an electron-positron pair may also be created as a result of the de-excitation process. The probability for such internal pair creation is, in general, much smaller than γ-ray emission and becomes important where γ-ray emission is forbidden by angular momentum considerations.

20. Does internal conversion compete with γ-ray emission?

Ans. Internal conversion does not compete with γ-ray emission in the sense that one process inhibits the other. The processes are independent alternatives.

21. What will be the full recoil shift between emission and absorption lines for the 411 keV level in ^{198}Hg and the source velocity necessary to give complete overlap?

Ans. The full recoil shift is $\dfrac{2\,E_\gamma^2}{mc^2}$, where m is the absolute mass of the atom and the source velocity for the overlap is obtained Doppler shift of γ-ray energy for velocity V. Numerical evaluation gives recoilshift = 0.91 eV and the source velocity, $V = 667$ m/s.

22. The 14.4 keV γ-ray transition in ^{57}Fe has been extensively studied using the Mossbauer effect, and it is found that $J = \dfrac{1}{2}$ ground state is split with an energy difference corresponding with a source velocity of 3.96 mm/s. If the ground state magnetic moment is 0.0955 μ_N, what will be the internal magnetic field at ^{57}Fe nucleus?

Ans. The splitting is due to the up and down alignment of the nuclear spin and magnetic moment in the internal field B. The Doppler shift of the 14.4 keV radiation due to the given source velocity. On numerical evaluation, one obtains the value of the internal field = 31.5 T.

OBJECTIVE QUESTIONS

1. Geiger-Nuttall law expression is (A and B are constants)
   - (a) $\log \lambda = A \log R + B$
   - (b) $\log \lambda = AR + B$
   - (c) $\log \lambda = AR^2 + B$
   - (d) $\log \lambda = AR^3 + B$

2. The relation between range (R) and energy (E) of α-particle due to Geiger is
   (a) $R = KE^3$
   (b) $R = KE^{1.5}$
   (c) $R = KE$
   (d) $R = KE^5$
3. For thick rectangular barriers, the particle transmission coefficient of the barrier (T) due to Gamow is

   (a) $T \sim \dfrac{e^{-2bm V^{\frac{1}{2}}}}{\hbar}$
   (b) $T \sim \dfrac{e^{-2bm E^{\frac{1}{2}}}}{\hbar}$

   (c) $T \sim \dfrac{e^{-2b[2m(V-E)]^{\frac{1}{2}}}}{\hbar}$
   (d) $T \sim \dfrac{e^{-2bm E^{\frac{1}{2}} V^{\frac{1}{2}}}}{\hbar}$

4. In Contrast with β emission, α-decay is
   (a) One-body process
   (b) three-body process
   (c) two-body process
   (d) five body process
5. α-decay energies are sensitive to
   (a) shell effects
   (b) temperature
   (c) velocity of the nucleus
   (d) temperature as well as velocity
6. α-decay illustrate the importance of the phenomenon of
   (a) quantum-mechanical tunnelling
   (b) fine structure of α-particles
   (c) long-range α-particles
   (d) None of the above
7. The following is the example of: $n \rightarrow p + e^- + \bar{v}$
   (a) γ-decay
   (b) α-decay
   (c) β-decay
   (d) α, β and γ decays
8. The condition for β⁺-decay energetically possible is
   (a) $_Z^A M >\ _{Z+1}^A M$
   (b) $_Z^A M >\ _{Z-1}^A M + 2m_e$
   (c) $_Z^A M =\ _{Z-1}^A M$
   (d) $_Z^A M >\ _{Z-1}^A M + p$
9. Pauli postulated that in the electron emission process a particle is also emitted, which has
   (a) zero charge only
   (b) zero rest masss only
   (c) intrinsic spin $s = \dfrac{1}{2}$ only
   (d) zero charge, zero mass and intrinsic spin $s = \dfrac{1}{2}$
10. According to Pauli, the emitted particle in the electron emission process is
    (a) neutrino
    (b) anineutrino
    (c) proton
    (d) alpha particle
11. The β-decay matrix element (M) is

$$M = \beta \int \psi_f^* \psi_i \, d\tau = \beta M'$$

   The M' is so called
   (a) electric dipole moment matrix element
   (b) nuclear matrix element
   (c) density of states
   (d) none of the above

12. β-decay is a consequence of
    (a) nuclear force
    (c) gravitational force
    (b) electromagnetic force
    (d) weak interaction force

13. β-decay coupling constant has the value
    (a) 10^{-14} J - m^3
    (c) 10^{-36} J - m^3
    (b) 10^{-24} J - m^3
    (d) 10^{-62} J - m^3

14. β-decay interaction is weaker than the nuclear interaction by a factor of
    (a) 10^{-3}
    (c) 10^{-9}
    (b) 10^{-5}
    (d] 10^{-14}

15. Which one of the follwing is not conserved in β$^-$-decay
    (a) isospin
    (c) baryon number
    (b) parity
    (d) strangeness

16. $V - A$ theory accounts for the observed helicities in the β$^-$-decay process of the emitted
    (a) proton and electrons
    (c) electrons and neutrons
    (b) neutrinos and electrons
    (d) neutrons and protons

17. The multipolarity possible for a radiative transition is limited by *selection rules*, based on
    (a) spin
    (c) Pauli principle
    (b) angular momentum
    (d) parity

18. The relation between the mean life (τ_γ) of the level conce.ned for the emission of radiation and radiative width of upper level (T_γ) is
    (a) $\tau_\gamma \Gamma_\gamma = 1$
    (c) $\tau_\gamma \Gamma_\gamma^{-1} = 1$
    (b) $\tau_\gamma \Gamma_\gamma = h$
    (d) $\tau_\gamma^{-1} \Gamma_\gamma = h$

19. Nuclear isomerism if found in groups of nuclei located just below the major shell closures at
    (a) $Z, N = 50, 82$ and 126
    (c) $Z, N = 2, 8$ and 18
    (b) $Z, N = 6, 16$ and 32
    (d) None of the above

ANSWERS

1. (a)	2. (b)	3. (c)	4. (c)	5. (a)
6. (a)	7. (c)	8. (b)	9. (d)	10. (b)
11. (b)	12. (d)	13. (d)	14. (d)	15. (b)
16. (b)	17. (b)	18. (b)	19. (a)	

4

NUCLEAR FORCES AND TWO
BODY PROBLEM

4.1 INTRODUCTION

The nature of forces holding nucleons in nuclei has not yet fully understood. At the same time, much data have been obtained on the physical properties of nuclei, as well as on interaction of free nucleons in collisions over a very wide range of kinetic energies, from 10^{-4} to 10^{10} eV. The analysis of the observed phenomena allows the following conclusions to be drawn about the forces between nucleons. Nuclear forces are strong attractive forces between nucleons that binds them together. That nuclear force in turn has its origin in the quark-gluon constitution of the nucleon, i.e. this force or interaction between quarks is arising from the exchange of spin 1 particles called *gluons*.

Nucleons are made up of three quarks: a neutron of two down quarks each of charge $-\frac{1}{3}$ and an up quark of charge $+\frac{2}{3}$, a proton of one down and two up quarks. The gluon exchange force has the property of increasing with increasing quark separation with the result that quarks are confined. The nuclear force between nucleons arises from the exchange of mesons, which are made up of quark-antiquark pair. It therefore bears somewhat of the same relationship of the strong interaction as the Van der Waals force between molecules does to the electrostatic interaction. However, this does not mean that a good description is not possible of the nuclear force using a meson exchange picture but only its connection with the fundamental quark-gluon interaction is extremely complicated. The analysis of the observed phenomena allows one to draw following conclusions about the nuclear forces between nucleons. Nuclear forces are strong attractive forces, acting only at short distances. They posses the property of saturation, due to which nuclear forces are attributed exchange character (exchange forces), nuclear forces depend on spin, not on electric charge, and are not central forces. A comparison of nuclear interaction with other basic interactions is made in Table 4.1

Table 4.1 Four Basic Interactions

Type of Interaction	Relative Strengths	Range Particle	Mediating Time	Characterstic	Typical particles
Strong (nuclear) Interaction	1	10^{-14} m ~ 10^{-5} m	gluon	$< 10^{-22}$ s	π, n, p, K
Electromagenetic	~10^{-2}	Infinite	photon	$10^{-14} - 10^{-20}$ s	e, π, n, p, μ, K
Weak interaction (β-decay)	~10^{-13}	Almost zero	$W^{\pm}, Z^0$	$10^{-8} - 10^{-13}$ s	all
Gravitational interaction	~10^{-39}	Infinite (long range)	graviton	years	all

The simplest system to study the nuclear force, i.e, the internucleon force is that of two nucleons, i.e. the two body system. Some understanding of the nuclear force has materially aided the construction of nuclear models that attempt approximation to the complete nuclear Hamiltonian.

4.2 GROUND STATE OF DEUTRON ($^2_1$D)

The deuterium nucleus ($A = 2, Z = N = 1$) is a bound state of the neutron-proton system, into which it may be disintegrated by irradiation with γ-rays of energy above the binding energy, E_{Bd} of 2.2245 ± 0.0002 MeV. Deutron is the only two-nucleon bound system. The two other possible two-nucleon systems, namely the diproton (^{2}He) and the dineutron do not exit as bound systems.

The experimental determined results about the deutron nucleus are as follows:

(i) **Binding Energy (E_{Bd})** The binding energy of deutron is 2.2245 ± 0.0002 MeV. Thus the binding energy per nucleon in the deutron ($f_{Bd} = $ B. E./A) is 1.1122 MeV. Obviously, this is much smaller than the mean value of binding fraction ($f_B = B/A$) for the nuclei with $A = 4$ or more. Clearly, deutron is a rather weakly bound structure as compared to most of the other nuclei. One can contain the binding energy of the deutron using the radiactive capture of a neutron by hydrogen nucleus forming a deutron ($n + p \rightarrow d + \gamma$). Clearly, the energy or the emitted γ-ray in the said reaction gives E_{Bd}. One can also obtain E_{Bd} by measuring the γ-ray energy sufficient to break the between proton and neutron within the deutron ($d + \gamma \rightarrow n + p$), or by measuring the masses of neutron, proton and deutron.

(ii) *The spin I (total angular momentum) of the deutron in the ground state in the unit of $\hbar$ is I_d = 1. The parity of the deutron ground state is even (+) . One can understand it as, the wave function of the state is separted into a product of three wave functions: intrinsic wave function each of proton and neutron and orbital wave function for the relative motion between the two nucleons. As the neutron and proton are just two states of a nucleon and hence the intrinsic wave functions of both has same parity. As a consequence, the product of their intrinsic wave function has even (+) parity. Obviously, the parity of the deutron has to be determined by the relative motion between the neutron and the proton. If L is the* angular momentum wave function is given by $(-1)^L$. The value of the L must be even as the parity of the deutron is even.

We have $I = L + S$. The total angular momentum (I) of the ground state of deutron is 1. The possible value of S are 0 or 1 as $S = \left|\dfrac{1}{2} + \dfrac{1}{2}\right|$ 0 or 1. One cannot combine $S = 0$. with any of the allowed even values of L to form $I = 1$ state and hence one can eliminate $S = 0$. Likewise, even values of L greater are also not possible. Obviously, the only possible values of L and S are $L = 0$, $S = 1$ and $L = 2$, $S = 1$, part of the ground state wave function is $L = 0$, $S = 1$ component. Further, one finds that there is a small but significant admixture of $L = 2$ state also. Obviously, for the ground state of deutron $L = 0$, $I = L + S$ and $I = S = 1$.

(iii) The measured magnetic moment of the deutron (μ_d) in nuclear magneton (μ_N) unit as measured by magnetic resonance method is $\mu_d = (0.857414 \pm 0.000019)\mu_N$.

(iv) The deutron also possesses a small but finite electric quadrupole moment (Q_d). The measure value of $Q_d = 2.82 \times 10^{-31}$ m^2.

(v) High energy electron scattering yield the radius of the deuteron $(r_d) \approx 4.2$ fm, which corresponds to root mean square distance $\left(\sqrt{<r^2>}\right)$ between the neutron and proton. Thus

$$r_d = \left(\sqrt{<r^2>}\right) = 4.2 \text{ fm} \qquad\qquad (1 \text{ fm} = 10^{-15} \text{ m})$$

The quantum mechanical theory of a system acted upon by a central force clearly reveals that the orbital angular movement L is a constant of motion, i.e. each energy eigen-state of the system is characterized by a definite value of L. In particular, the ground state is an $L = 0$ state. Let us assume that this is true for the deutron. This means that the potential of the interaction between the neutron and the proton is a function of r, i.e. $V = V(r)$, where r is the scalar distance between the two nucleons. This is a very strong short-range interaction having the general appearence as shown in Figure 4.1.

The exact mathematical form of the interaction potential is not known. One can approximate such a strong short range, spherically symmetric potential by one of the following expressions.

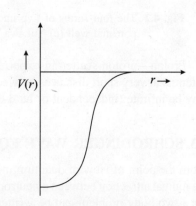

Fig. 4.1 General form of the internucleon rotential (central)

(i) **Rectangular-well potential** (Figure 4.2a)

$$V = -V_0 \text{ for } r \leq b$$
$$= 0 \text{ for } r > b \qquad\qquad\qquad ... (4.1)$$

Often used in calculation, where $V_0 > 0$ and b is the radius of the well.

(ii) **Expotential Well Potential** (Figure 2b)

$$V = -V_0 \exp\left(-\frac{r}{b}\right) \qquad \ldots (2.2)$$

(iii) **Yukuwa Potetial** (Figure 2.2 (c))

$$V = -\left\{\frac{V_0}{r/b}\right\} \exp\left(-\frac{r}{b}\right) \qquad \ldots (2.3)$$

(iv) **Wooden -Saxon Potential**

$$V = -\frac{V_0}{\left[1 + \exp\left\{-\frac{(r-b)}{c}\right\}\right]} \qquad \ldots (2.4)$$

Where c is constant and representing the skin depth.

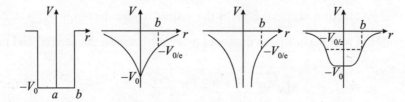

Fig. 4.2 The four forms of two nucleons potential: (a) Rectangular potential well (b) Exponential potential well (c) Yukawa potential (d) Wools-saxon potential well.

Proton—proton scattering experiments shows that there is a repulsive core of the two-body potential at very short distances between the two protons. However, such a repulsive core potential may be infinite (independent or hard core).

4.3 SCHRODINGER WAVE EQUATION FOR DEUTRON AND ITS SOLUTION

From the point of view of quantum mechanics deutron is regarded as a stationary state formed by the mutual attraction between a neutron and proton. In the C-system, the schrodinger wave equation for a two body problem can be written as

$$\nabla^2 \psi + \frac{2\mu}{\hbar^2} [E - V(r)] \psi = 0 \qquad \ldots (4.1)$$

Where $\psi = \psi(r, \theta, \phi)$. To obtain a wave function $\psi(r)$ for the deutron we use the spherically symmetrical square well potential. μ is the reduced mass of the neutron-proton system, given by

$$\mu = \frac{M_p M_n}{M_p + M_n} \qquad \ldots (4.2)$$

Since $M_p \approx M_n = M$ (say), so the $\qquad \ldots (4.2)$

$$\mu = \frac{M}{2} \qquad \qquad \qquad \text{... (4.3)}$$

Usually, M is taken as proton mass (M_p). Writing the Schrodinger equation (4.1) in spherical polar coordinates as

$$\left[\frac{1}{r^2} \frac{\partial}{\partial r} \left(r^2 \frac{\partial}{\partial r} \right) + \frac{1}{r^2 + \sin\theta} \frac{\partial}{\partial\theta} \left(\sin\theta \frac{\partial}{\partial\theta} \right) + \frac{1}{r^2 \sin^2\theta} \frac{\partial^2}{\partial\phi^2} \right] \psi + \frac{2\mu}{\hbar^2} [E - V(r)] \psi = 0 \qquad \text{... (4.4)}$$

If the lowest state is stable, i.e. the ground state of deutron is taken to be purely as ^3_1S state, the force is *central,* i.e. the force depends only upon r (the seperation between the nucleon spin orientation with respect to the line joining the nucleous, i.e. upon θ and ϕ. We know that the central force is conservative and hence we can express it as the gradient of a potential function $V(r)$, spherically symmetric. This means the wave function is also symmetrical in θ and ϕ. The potential $V(r)$ is spherically symmetric being independent of θ and ϕ. Equation (4.4) can be separted into angular and radial parts. We get the radial equation as

$$\frac{d^2\psi(r)}{dr^2} + \frac{2}{r} \frac{d\psi(r)}{dr} + \frac{2\mu}{\hbar^2} \left[E - V(r) - \frac{l(l+1)}{2\mu\, r^2} \right] \psi(r) = 0 \qquad \qquad \text{... (4.5)}$$

The last term, viz $\dfrac{l(l+1)\hbar^2}{2\mu\, r^2}$ is called the *centrifugal potential* and it has the effect of tearing the neutrin-proton apart. For binding the proton and neutron together as in the deutron, $V(r)$ must be attractive at least over a limited range r and should have a value at least to compensate for the repulsive centrifugal potential. One can easily achieve this in $l = 0$ state. Clearly, for a two body sytem, e.g. deutron, $l = 0$ or and S-state is always the lowest quantum mechanical state. For $l = 0$, i.e. for an S-state of deutron, Schrodinger equation (4.5) becomes

$$\frac{d^2\psi(r)}{dr^2} + \frac{2}{r} \frac{d\psi(r)}{dr} + \frac{2\mu}{\hbar^2} [E - V(r)] \psi(r) = 0 \qquad \qquad \text{... (4.6)}$$

One can simplify (4.6) by introducing a new wave function $u(r)$ related to $\psi(r)$ as

$$\psi(r) = \frac{u(r)}{r} \qquad \qquad \qquad \text{... (4.7)}$$

Using (4.7), Equation (4.6) takes the form

$$\frac{d^2u(r)}{dr^2} + \frac{M}{\hbar^2} [E - V(r)]\, u(r) = 0 \qquad \qquad \text{... (4.8)}$$

Assuming the rectangular well potential (or square-well potential) (Figure 4.2a), which is the simplest and can be handled mathematically, we obtain the solution of (4.8). Here, we may note that the solution of (4.8) for deutron have been obtained with different types of potentials and results are found not to depend much on the shape of the potential assumed. Writing Equation (4.8) for $l = 0$ (ground state), we have

For $r < b$
$$\frac{d^2u_{in}}{dr^2} + \frac{M}{\hbar^2} (E + V_0)\, U_{in} = 0 \qquad \qquad \text{... (4.9a)}$$

For $r > b$
$$\frac{d^2 u_{out}}{dr^2} + \frac{ME}{\hbar^2}\, u_{out} = 0 \qquad \qquad \text{... (4.9b)}$$

Here u_{in} and u_{out} denotes the radial functions inside ($r < b$) and outside ($r > b$) regions of the potential well respectively. Since the deutron ground state is a bound state. Its energy is negative, i.e., $E = -E_d = -2.226$ MeV. For $E_d > 0$, putting $E = -E_d$ in 4.9(a) and 4.9(b), one obtains

$$\frac{d^2 u_{out}}{dr^2} + \frac{M}{\hbar^2}\, (V_0 - E_d)\, u_{in} = 0 \qquad \qquad \text{... (4.10a)}$$

$$\frac{d^2 u_{out}}{dr^2} - \frac{M}{\hbar^2}\, E_d\, u_{out} = 0 \qquad \qquad \text{... (4.10b)}$$

Now, writing
$$k_1^2 = \frac{M}{\hbar^2}\, (V_0 - E_d) \qquad \qquad \text{... (4.11a)}$$

$$k_0^2 = \frac{M}{\hbar^2}\, V_0 \qquad \qquad \text{... (4.11b)}$$

and
$$\alpha^2 = \frac{M}{\hbar^2}\, E_d \qquad \qquad \text{... (4.11c)}$$

We have
$$k_1^2 = k_0^2 - \alpha^2 \qquad \qquad \text{... (4.12)}$$

Using (4.11) and (4.12), wave equations (4.10) becomes

$$u''_{in} + k_1^2\, u_{in} = 0 \qquad \qquad \text{... (4.13a)}$$

$$u''_{out} - \alpha^2\, u_{out} = 0 \qquad \qquad \text{... (4.13b)}$$

Where $u'' \left(= \dfrac{d^2 u}{dr^2} \right)$

The solutions of (4.13) are of the forms:

$$u_{in} = A \sin k_1 r + A' \cos k_1 r \qquad \qquad \text{... (4.14a)}$$

$$u_{out} = B \exp(-\alpha r) + B' \exp(\alpha r) \qquad \qquad \text{... (4.14b)}$$

We have the radial function $R(r) = \dfrac{u(r)}{r}$. One must put $A' = 0$ in Equation (4.14(a)) which would otherwise make $U_{in} \to \infty$ as $r \to 0$. Similarly, one must put $B' = 0$ in 4.14(b) which would otherwise make $u_{out} \to \infty$ as $r \to 0$. Thus, one obtains finally

$$u_{in} = A \sin k_1 r \qquad \qquad \text{... (4.15a)}$$

and
$$u_{out} = B \exp(-\alpha r) \qquad \qquad \text{... (4.15b)}$$

Now, the two solutions given above must be matched at the boundary, i.e. $r = b$. This requires

$$\left(\frac{u'_{in}}{u_{in}} \right)_b = \left(\frac{u'_{out}}{u_{out}} \right)_b \qquad \qquad \text{... (4.16)}$$

Where $u' = \dfrac{\partial u}{\partial r}$. The condition (4.16) requires that

$$k_1 \cot k_1 b = -\alpha \qquad \qquad ...(4.17)$$

Now, substituting the values of k_1 and α_1, one obtains that transcendental equation

$$\cot k_1 b = -\frac{\alpha}{k_1} = \sqrt{\frac{E_d}{V_0 - E_d}} \qquad \qquad ...(4.18)$$

One cannot solve Equation (4.18) analytically and therefore, one has to solve it graphically. If V_0 is in the range 30 – 60 MeV, the quantity represented in (4.18) is small, At this stage, one can obtain an idea about the minimum depth V_{om} of the potential. This will give just a bound $n - p$ system. For this, we can put $E_d = O$ (being small). One obtains, this (4.18) reduces to

$$\cos k_0 b = 0 \qquad \qquad ...(4.19)$$

This gives
$$k_0 b = \sqrt{\frac{M V_{om}}{\hbar^2}} \qquad b = \frac{\pi}{2} \qquad \qquad ...(4.20)$$

or
$$V_{0m} b^2 = \frac{\pi^2 \hbar^2}{4 M} = 10^{-28} \text{ MeV} - \text{m}^2 \qquad \qquad ...(4.21)$$

Which is a relation between the range of the *central force* and the depth of the triplet well. For a well depth of 30 MeV the well radius 1.83 fm = 1.83×10^{-15} m, which is the mean internucleon separation in the nuclei.

In the actual case, we have $E_d = 2.226$ MeV. The value of the minimum depth V_{om} given above $E_{Bd} \ll V_0$. Now, putting $E_d = E_{Bd}$ in Equation (4.11), one ontains

$$k_1 = \sqrt{\frac{M (V_0 - \varepsilon d)}{\hbar}} \quad \text{and} \quad \alpha = \sqrt{\frac{M \varepsilon d}{\hbar}} \qquad \qquad ...(4.22)$$

Now, one can write the transcendental Equation (4.18) as

$$\cot k_1 b = \frac{\alpha b}{k_1 b} \qquad \qquad ...(4.23)$$

Writing $x = k_1 b$, one then obtains the following two equations

$$y = \cot x \qquad \qquad ...(4.24a)$$

$$y = -\frac{\alpha b}{x} \qquad \qquad ...(4.24b)$$

Now, substituting the proper numerical values, one obtains

$$\alpha^2 = \frac{M \varepsilon_d}{\hbar^2} = \frac{1.67 \times 10^{-27} \times 2.226 \times 1.6 \times 10^{-13}}{(1.054 \times 10^{-34})^2}$$

or
$$\alpha = 2.314 \times 10^{14} = 0.2314 \text{ fm}^{-1}$$

Since $\alpha b = 0.463$, for $b = 2$ fm. Using this value Equation (4.24b) gives

$$y = -\frac{0.463}{x} \qquad \text{... (4.24c)}$$

One can solve Equations 4.24(a) to 4.24(c) graphically. Figure 4.3 shows the two graphs. The points of intersection of these graphs give the possible solutions. We note from the figure that the points of intersection of Equation 4.24(c) with the set of graphs $y = \cot x$ (Equation 4.24a) have small negative values. Further, the successive values of $k_1 b$ must then be slightly greater ($\geq$) than $\frac{\pi}{2}$, $\frac{3\pi}{2}$, $\frac{5\pi}{2}$ etc. or in general, we have

$$k_1 b \geq (2n+1)\frac{\pi}{2} \qquad \text{... (4.25)}$$

Where $n = 0, 1, 2, 3$, etc. $n = 0$, i.e. the smallest of these values corresponds to the ground state for which $k_1 b \geq \frac{\pi}{2}$. Now, writing

$$k_1 b = \frac{\pi}{2} + \varepsilon_0 \qquad \text{... (4.26)}$$

Where ε_0 is a small number. One obtains

$$\cot k_1 b = \cot\left(\frac{\pi}{2} + \varepsilon_0\right) = -\tan \varepsilon_0 \geq -\varepsilon_0 \qquad \text{... (4.27)}$$

Hence from Equation (4.23), one obtains

$$k_1 b \cot k_1 b = -\left(\frac{\pi}{2} + \varepsilon_0\right)\varepsilon_0 = -\alpha b = -0.0463 \qquad \text{... (4.28)}$$

Neglecting ε_0^2, one finds

$$\varepsilon_0 \approx \frac{2\pi b}{\pi} = 0.295 \qquad \text{... (4.29)}$$

$$\therefore \qquad k_1 b = \frac{\pi}{2} + \varepsilon_0 = 1.87 \qquad \text{... (4.30)}$$

Now, putting $b = 2$ fm $= 2 \times 10^{-15}$ m, one obtains using (4.22)

$$V_0 - E_d = 36 \text{ MeV}$$

or $\qquad V_0 = 38$ MeV $\qquad$ ... (4.31)

Figure 4.4 shows the relative magnitudes of the depth V_0 of the rectangular potential and the binding energy E_{Bd} ($= E_d$) of the deuteron for $b = 2$ fm $= 2 \times 10^{-15}$ m.

Fig. 4.3 Graphical solutions of transcendental equation: $\cot x = -\dfrac{\alpha b}{x}$, where $x = k_1 b$.

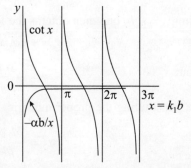

Fig. 4.4 For $b = 2 \times 10^{-15}$ m, the ground state of deuteron in the rectangular potential well of depth, $V_0 = 38$ MeV

4.3.1 Nature of the Deutron Wave functions

Figure 4.5(a) and 4.5(b) shows the nature of the wave functions of the deutron.

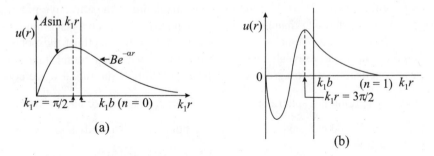

Fig. 4.5 For $r < b$ and $r > b$, the nature of deutron wave function: (a) for $n = 0$, and (b) for $n = 1$.

We have the internal function u_{in} is sinusoidal, whereas the external function u_{out} is a decreasing expotential. These two functions can join smoothly at the boundary $r = b$ only if u_{in} has negative slope at this point. This is only possibly if Equation (4.25) is satisfied for different values of n. In the ground state, i.e. for $n = 0$, this condition reduces to $k_1 b \geq \dfrac{\pi}{2}$. This means that the portion of $\sin k_1 r$ graph contained within the region $0 < r < b$ is slightly greater than half a loop of the sine curve (Figure 4.5a). We note that there is no node of the wave function. In the case of $n = 1$, the internal region $0 < r < b$ contains one full loop and slightly more than half a loop of the sine curve (Figure 4.5b). Obviously, the condition $k_1 b > \dfrac{3\pi}{2}$ is fulfilled. In this case, there is one node of the wave function.

We further note that the number of nodes of the wave function increases with increasing n, being equal to the value of n. One can interpret this as being equivalent to the radial quantum number.

4.3.2 Excited states of deutron

Let us investigate the possibility of excited bound state of deutron. Experimentally there is no evidence for the existence of any excited states. One can consider two possibilities: the existence of a state with a higher value of n and that with a higher value of n.

We first consider the case of the $n = 1$ state. For this, the condition (4.25) gives for a just bound state $(E_d = 0)$, $k_0 b = \dfrac{3\pi}{2}$ or $b = 2$ fm. Thus

$$V_0 = \frac{9\pi^2 \, \hbar^2}{4 M \, b^2} = 9 \, V_{om} = 225 \text{ MeV}$$

(Use of Equation (4.21) has been made)

However, this value of potentoal depth is quite different that obtained from the $n - p$ potential depth obtained from the deutron ground state energy. Moreover, for higher value of n, the disagreement would be still more. Clearly, no excited bound state of the deutron exists with $n > 0$ value.

Let us now consider the case for $l > 0$. The repulsive centrifugal potential in Equation (4.25) tends to diminish the strength of binding the deutron. Moreover, the effect increases with increasing value of l and is the least for $l = 1$. One finds the magnitude of the centrifugal potential at the boundary, $r = b$ for this case ~ 21 MeV, which reduces the depth of the attractive potential to only 17 MeV. This value is much lower than the value that required to produce a just bound $n - p$ system ($V_{om} \approx 25$ MeV). In the interior regions of the well, i.e. $r < b$, the depth of the potential well is reduced even more. This clearly reveals that there cannot be any excited bound state of the deutron for $l = 1$. These conclusions are in agreement with the solution of the radial wave equation with $l = 1$ which gives as solution the spherical Bessel function $J_{3/2} (k_0 r)$ of order $\frac{3}{2}$. We may note that matching of the internal and external solutions at the boundary also leads to a contradiction with the bound state, $l = 0$ theory.

4.3.3 Normalization of deutron wave function

Following the usual procedure, one can normalize the wave functions u_{in} and u_{out} as:

$$\int_0^\infty |u|^2 \, dr = 1 \qquad \qquad \dots (4.1)$$

or

$$\int_0^b |u_{in}|^2 \, dr + \int_b^\infty |u_{out}| \, dr = 1 \qquad \qquad \dots (4.2)$$

Prior to the evaluation of integrals appearing in (4.2), we have to find the relation between A and B appearing in 4.15(a) and 4.15(b) of section 3. Applying boundary conditions $(u_{in})_b = (u_{out})_b$ and $(u'_{in})_b = (u'_{out})_b$, one obtains

$$A \sin k_1 b = B \exp(-\alpha b)$$

$$k_1 A \cos k_1 b = -\alpha B \exp(-\alpha b)$$

or

$$A \cos k_1 b = -\left(\frac{\alpha}{k_1}\right) B \exp(-\alpha b)$$

Squaring and adding, one obtains

$$A^2 = B^2 \left(1 + \frac{\alpha^2}{k_1^2}\right) \exp(-2\alpha b)$$

$$= B^2 \left(\frac{k_0^2}{k_1^2}\right) \exp(-2\alpha b)$$

or

$$B^2 = \left(\frac{k_1^2}{k_0^2}\right) A^2 \exp(-2\alpha b)$$

and

$$B = \left(\frac{k_1}{k_0}\right) A \exp(\alpha b) \qquad \qquad \dots (4.3)$$

Now, from (4.2), we have

$$\int_0^b |u_{in}|^2\, dr + \int_b^\infty |u_{out}|^2\, dr = A^2 \int_0^b \sin^2 k_1 r\, dr + B^2 \int_b^\infty \exp(-2\alpha r)\, dr$$

$$\approx A^2 \left(\frac{b}{2} + \frac{k_1^2}{k_0^2} \frac{1}{2\alpha} \right)$$

$$\approx \frac{A^2}{2\alpha}\ (1 + \alpha b) \qquad\qquad ...\ (4.4)$$

In obtaining (4.4), we have assumed $k_1 b \approx \dfrac{\pi}{2}$ and $k_1^2 \approx k_0^2$. Now, one obtains

$$\left(\frac{A^2}{2\alpha} \right)(1 + \alpha B) = 1$$

or
$$A = \left(\frac{2\alpha}{1+\alpha b} \right)^{\frac{1}{2}} = \sqrt{2\alpha}\left(1 - \frac{\alpha b}{2} \right) \qquad\qquad ...\ (4.5a)$$

and
$$B = \left(\frac{2\alpha}{1+\alpha b} \right)^{\frac{1}{2}} \exp(\alpha b) = \sqrt{2\alpha}\left(1 + \frac{\alpha b}{2} \right) \qquad\qquad ...\ (4.5b)$$

again, we have assumed $k_1^2 \approx k_0^2$.

Hence the radial wave functions for the deuteron are

$$u_{in} = \left(\frac{2\alpha}{1+\alpha b} \right)^{\frac{1}{2}} \sin k_1 r \qquad\qquad ...\ (4.6a)$$

$$u_{out} = \left(\frac{2\alpha}{1+\alpha b} \right)^{\frac{1}{2}} \exp[\, +\alpha(b + r)] \qquad\qquad ...\ (4.6b)$$

The probability of finding the neutron and proton within the range of the nuclear potential is given by

$$P = \int_0^b |u_{in}|^2\, dr = A^2 \int_0^b \sin^2 kr\, dr \approx \frac{\alpha b}{1+\alpha b}$$

where
$$\alpha = \sqrt{\frac{M\,\varepsilon_d}{\hbar}} = 0.2314 \text{ fm}^{-1} \qquad\qquad ...\ (4.7)$$

Using the above value of α, one obtains

$$P = 0.315$$

or
$$P = 31.5 \ \% \qquad\qquad ...\ (4.8)$$

This clearly reveals that the probability of finding the neutron and proton within the range of nuclear force is only about 30 %, i.e. small. This means that the neutron and proton stay outside the range of the $n - p$ internucleon force nearly70 %. This indicates that the deutron is a loosely bound structure (low binding energy, 1.113 MeV per nucleon). From classical mechanics of point of view, However, the nucleous cannot stay outside the potential well. However, quantum mechanically, one finds that they do stay outside the well. This quantum mechanical phenomena is to be associated with tunnelling, i.e. wave penetration through the barrier.

4.3.3 Size or radius of the deutron

Usually, the distance R_d at twhich the exponential function $u_{out} = B \exp(-\alpha r)$ extended backward upto $r = 0$, then the distance at which u_{out} falls to $\dfrac{1}{e}$ of its value at $r = 0$ may be taken to be a measure of the deutron radius R_d. This gives

$$\alpha\, R_d = 1 \text{ or } R_d = \frac{1}{a}, \text{ but } \alpha = \sqrt{\frac{M}{\hbar^2}\varepsilon_B} = 0.237 \times 10^{15} \text{ m}^{-1}$$

$$\therefore \qquad\qquad R_d = 4.3 \times 10^{-15} \text{ m} = 4.3 \text{ fm}$$

Obviously, R_d is much larger than b.

4.3.4 Mixing of orbitals in deutron

The above discussion reveals that the observed size and binding energy of deutron are consistent with the quantum mechanical description of the nucleus, based on the strong attractive centre potential of specified depth and width. However, the assumption of central potential cannot explain the electric quadrupole moment and magnetic dipole moment of the deutron.

We have read that the ground state of the deutron is predominanly an s-state ($l = 0$). The total intrinsic spin of deutron is $S = 1$. This reveals that the spin vectors of neutron and proton are parallel ($\uparrow\uparrow$) and one can write the state of the deutron in spectroscopic notation as a triplet s-state ${}^3_1 S$. Now, if, the nuclear forces were independent of spin directions, one may expect a spin singlet state ${}^1_0 S$ also with the same energy. In this situation, the two spin vectors will be antiparallel ($\uparrow\downarrow$). However, no bound state with a neutron and a proton having $I = 0$ ($L = 0$, $S = 0$) has been observed. Obviously, this means that the mutual force between the neutron and a proton inside the nucleus is strong enough when their spin vectors are parallel compared to when spins are antiparallel. However, this does not mean that nuclear forces do not exist for antiparallel spin vectors.

Let us consider a simple picture of a pure ${}^3_1 S$ state. The charge distribution is spherically symmetric. This suggests that the electric quadrupole moment (Q), should be zero. Further, the magnetic dipole moment, μ_d should be the algebraic sum of the magnetic moments of neutron (μ_n) and proton (μ_p). In fact, the deutron is found to have a finite, but small electric quadrupole moment, $Q = 2.88 \times 10^{-31}$ m^2. Moreover, the observed magnetic moment μ_d is not equal to ($\mu_n + \mu_d$), i.e. the observed value of $\mu_d = 0.8573\ \mu_N$, whereas $\mu_n + \mu_p = -1.9132 + 2.7928 = 0.87\ \mu_N$. The small difference between the expected value and observed value of $\mu_d = 0.0223\ \mu_N$ and the non-zero quadrupole moment suggest that the deutron ground state is not a pure ${}^3_1 S$ state.

In order to explain the disagreement between the experimental results and corresponding predictions from the central potential, Rarita and schwinger in 1941 suggested that the nuclear force between a neutron and a proton is not purely central but one will have to add a term to the central potential which produces a non-central or a tensor force. Obviously, the nuclear force between nucleons not only depends on the radial distance between them (central force) but also depends on the relative orientation of their spin directions (non-central). It is found that the non-central or the tensor component depends on the angle which their spin axes make with the line joining the two nucleons. In vector notation, the tensor force is therfore a function of $(\mathbf{S} \cdot \mathbf{r})$. However, for the same separation distance the tensor force is different for different relative orientation of spins.

Under the effect of non-central force, the orbital motion of the nucleons is such that the charge distribution acquires a spheroidal shape (we may remember that for spherically symmetric charge distribution, the orbital angular momentum of their relative motion around the centre of mass is zero). When the probability density of particles is more along the spin axis (axis of symmetry) its shape is prolated or cigar type. The quadrupole moment (Q) is positive and mutual interaction between the neutron and proton is attractive. However, when the probability density of the charge distribution is more along the line perpendicular to the spin axis, the shape is oblate and the mutual interaction is repulsive and Q is negative (we may note that the attractive and repulsive nature of the mutual interaction between n and p in the said two situations of the spin orientation resembles similar behaviour of magnetic dipoles when the two lie along the line parallel to their magnetic axis, resulting into attraction. When the magnetic axis is perpendicular to the line joining them, two repel each other). The above two situations are depicted in Figure 4.6

We may note that the orbital motion contributes towards the magnetic moment which must be combined with the intrinsic magnetic moment of n and p. Obviously, this accounts for the difference between the observed μ_d and $(\mu_n + \mu_p)$

From the above discussion it is clear that besides a predominant s state (i.e. $l = 0$) there has to be another state $(l \neq 0)$ with $I = 1$ when the deuteron stays in the ground state. When the wave functions of such states are combined, one finds that the resultant wave function must have definite parity and $I = 1$. We have read that states with $(1, S)$ values $(0, 1)$ and $(2, 1)$

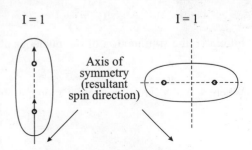

Fig. 4.6 (a) Q positive, prolate:
(b) Q negative, oblate

each have even parity, and on combination can give $I = 1$, While $l = 1$ and $S = 0$ and 1 can also be combined vectorially to give $I = 1$, but these P states (^1_1P and ^3_1P) have odd parity (Figure 4.7). Conservation of parity demands that only states of same parity can co-exist as such the wavefunction could be a linear combination of same parity wave functions. Clearly, the ground state of deutron due to tensor force is therefore not a pure ^3_1S state but a mixture of ^3_1S and ^3_1D states. This picture of non-central force giving rise to the mixture of states, i.e. ($^3_1S + ^3_1D$) states explain satisfactorily the magnetic moment and quadrupole moment of deutron.

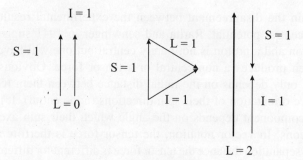

Fig. 4.7 In the ground state of deutron the effect of tensor force on charge distributions

4.3.5 Magnetic moment of deutron

The deutron magnetic moment is due to the orbital motion of the proton having positive electric charge and due to the intrinsic spin of each nucleon. However, neutron being electrically neutral, its orbital motion does not contribute towards the magnetic moment.

In the units of nuclear magneton (μ_N), the magnetic moment (μ_l) arising from orbital motion is

$$\mu_l = g_l \, l \qquad \dots (4.1)$$

Where g_l is gyromagnetic ratio-and l is orbital angular momentum quantum number of a nucleon,

$$g_l = 1 \text{ for proton}$$
$$= 0 \text{ for neutron} \qquad \dots (4.2)$$

$$\therefore \qquad \mu_l(p) = l_p \qquad \dots (4.3)$$

Similarly, the spin motion of the nucleons also contributes towards magnetic moment. We have

$$\mu_s = g_s \, \mathbf{s}_p + g_n \, \mathbf{s}_n \qquad \dots (4.4)$$

Where $\mathbf{s}_p$ and $\mathbf{s}_n$ are spin angular momenta of proton and neutron respectively in units of $\hbar$, i.e.

$$s_p = s_n = \frac{1}{2}$$

$$g_p = 2\,\mu_p = 2 \times 2.7928 = 5.5856 \,\mu_N$$

$$g_n = 2\,\mu_n = -2 \times 1.9132 = -3.8264 \,\mu_N$$

Therefore, the magnite moment of deutron is

$$\mu_d = g_p \, \mathbf{s}_p + g_n \, \mathbf{s}_n + l_p \qquad \dots (4.3)$$

Since the masses of the nucleons are nearly equal, one may assume that each one carries one half of the orbital angular momentum associated with the relative motion between them. Now, if L is the orbital angular momentum of deutron, then

$$\mu_d = g_p \, \mathbf{s}_p + g_n \, \mathbf{s}_n + \frac{1}{2}\,\mathbf{L}$$

$$= \frac{1}{2}\,(g_p + g_n)\,(\mathbf{s}_p + \mathbf{s}_n)(\mathbf{s}_p - \mathbf{s}_n) + \frac{1}{2}\,\mathbf{L} \quad (s_p = s_n = \frac{1}{2}) \qquad \dots (4.6)$$

For a triplet state, the eigen value of $(s_p - s_n)$ is zero. $s_p - s_n = \dfrac{1}{2}(\sigma_p - \sigma_n)$, where σ are Pauli matrices and $\mathbf{S} = s_p + s_n$, is the total angular momentum. We have

$$\mu_d = \frac{1}{2}(g_p + g_n)\,\mathbf{S} + \frac{1}{2}\mathbf{L} \qquad \qquad \text{... (4.7)}$$

The expectation value, $<\mu_d>$ depends on the vlaue of I_z, (given by the projection of spin I on Z-axis). However, by connection, we calculate the expectation value of the Z-component of the magnetic moment.

Now, taking the projection of μ_d along Z-axis, one finds the expectation value of magnetic moment in Dirac notation as

$$\mu_d <\mu_z> \ = \ <I, I_z = I \,|\mu_z|\, I, I_z = I>$$

Now, using the formalism of spherical tensor and rotation matrix for I, I_z and μ_z, along with C. G. Corfficients and Wigner-Eckart theorm, one can obtain the expectation value of μ_z as

$$<I, I_z \,|\mu_z|\, I, I_z> \ = \ \frac{I_z}{I(I+1)} <I, I_z|\mu_d \cdot \mathbf{I}|I, I_z> \qquad \qquad \text{... (4.9)}$$

$$\mu_d = <\mu_z> \ = \ \frac{I_z}{I(I+1)} <I, I_z|\frac{1}{2}(g_p + g_n)\mathbf{S}, \mathbf{I} + \frac{1}{2}\mathbf{L}, \mathbf{I}|I, I_z> \qquad \text{... (4.10)}$$

Now, $\qquad\qquad\qquad \mathbf{I} = \mathbf{L} + \mathbf{S} \text{ or } I^2 = L^2 + S^2 + 2\,\mathbf{L}\cdot\mathbf{S} \qquad \qquad \text{... (4.11)}$

Further, $\qquad\qquad\quad \mathbf{I} = \mathbf{L} + \mathbf{S}$

$$\therefore \qquad \mathbf{S}\cdot\mathbf{I} = \mathbf{S}\cdot\mathbf{L} + \mathbf{S}\cdot\mathbf{S} + S^2 + \frac{1}{2}(I^2 - L^2 - S^2)$$

$$= \frac{1}{2}(I^2 - L^2 + S^2) \qquad\qquad \text{... (4.12)}$$

and , similarly, one obtains

$$\mathbf{L}\cdot\mathbf{I} = L^2 + \mathbf{L}\cdot\mathbf{S} = L^2 + \frac{1}{2}(I^2 - L^2 - S^2)$$

$$= \frac{1}{2}(I^2 + L^2 - S^2)$$

Now, using Equations (4.12) and (4.13) with Equations (4.10), one obtains

$$\mu_d = <\mu_z> \ = \ \frac{I_z}{I(I+1)}\,[<I, I_z\,|\,\frac{1}{2}(g_p + g_n), \frac{1}{2}(I^2 - L^2 + S^2)|I, I_z>$$

$$+ <I, I_z|\frac{1}{2}\cdot\frac{1}{2}(I^2 + L^2 - S^2)|I, I_z> \qquad \text{... (4.14)}$$

and the value of the magnetic moment in a state I, L, S and $I = I_z$ reduces to

$$\mu_d = \frac{1}{4(I+1)}\,[(g_p + g_n)\,\{I(I+1) - 1(L+1) + S(S+1)\}$$

$$+ \{I(I+1) + L(L+1) - S(S+1)\}] \quad \text{... (4.15)}$$

Now, for a triplet *s*-sate, ^3_1S state, $L = 0$, $I = 1$ and $S = 1$. One finds

$$\mu_d\,(^3_1S) = \frac{1}{2}\,(g_p + g_n) = \mu_p + \mu_n = 0.8796\;\mu_N$$

For a ^3_1D-state, $L = 2$, $I = 1$, $S = 1$, the magnetic moment is obtained as

$$\mu_d\,(^3_1D) = \frac{1}{8}\,[(g_p + g_n)\,(-2) + 6]$$

$$= 0.3102\;\mu_N$$

Since $\mu_d\,(^3_1D)$ is smaller than $\mu_d\,(^3_1S)$ and therefore a suitable mixture of the two can be obtained. One can find the respective proportion of mixture from the experimental value of μ_d and claculated value of μ_d and claculated values of $\mu_d\,(^3_1S)$ and $\mu_d\,(^3_1D)$.

Let us take the ground state wave function of the deutron, as

$$\psi = a\,\psi_s + b\,\psi_D \qquad\qquad\text{... (4.16)}$$

Where ψ_s and ψ_D represents the wave functions for S and D states respectively and a and b are the corresponding wave amplitudes. The normalisation constant being

$$a^2 + b^2 = 1 \qquad\qquad\text{... (4.17)}$$

as there are no cross-terms for the magnetic moment between the states ^3_1S and ^3_1D. Therefore the deutron magnetic dipole moment is obtained as

$$\mu_d = a^2\,\mu_d(^3_1S) + b^2\,\mu_d\,(^3_1\mu) \qquad\qquad\text{... (4.18)}$$

$$0.8573 = a^2(0.8796) + b^2(0.3102) \qquad\qquad\text{... (4.19)}$$

Where, experimentally known value of μ_d is used in (4.18).

Now, one obtains from the two simultaneous equations (4.17) and (4.19): $b^2 = 0.04$ and $a^2 = 0.96$. Obviously, there is 4 % mixture of ^3_1D state in the ^3_1S ground state. Clearly, the deutron spends about 4 % of its time in ^3_1D state.

4.3.6 Quadrupole moment (*Q*) of deutron

We known that the quadrupole moment is the departure from the spherical charge distribution in a nucleus and expressed as

$$Q = (3Z^2 - r^2) = r^2(3\cos^2\theta - 1)$$

In the C. M. system, the distance of proton and neutron from CM is $\dfrac{r}{2}$ and only the proton contributes to the quadrupole moment, (neutron being chargeless), we have

$$Q_d = \frac{r^2}{4}\,(3\cos^2\theta - 1)$$

$$= \frac{r^2}{2}\,P_{2,\,0}\,(\cos\theta)$$

Where $P_{2,\,0}\,(\cos\theta)$ is legendre polynomial.

Now, the expectation value of Q is given by

$$<Q_d> = \int \psi_d^* \, Q_d \, \psi_d \, d^3r = \int \psi_d^* \, \frac{r^2}{2} \, P_{2,0} \, (\cos \theta) \, \psi_d \, d^3r$$

We have already shown in the previous section that the ground state wave function, ψ_D, of the deutron is an admixture of $_1^3S$ and $_1^3D$ states, i.e.

$$\psi = a \, \psi_s + b \, \psi_D$$

Where ψ_s and ψ_D are wave functions of $_1^3S$ and $_1^3D$ states respectively and a and b are normalisation constants.

$$\therefore \qquad Q_d = \int \frac{r^2}{2} \, P_{2,0} \, (\cos \theta) \, |\psi_D|^2 \, d^3r$$

$$= \int \frac{r^2}{2} \, P_{2,0} \, (\cos \theta) \, |a \, \psi_s + b \, \psi_D|^2 \, r \sin \theta \, d\theta \, d\phi \, dr \qquad \text{... (4.13)}$$

or

$$Q_d = \int \frac{r^2}{2} \, P_{2,0} \, (\cos \theta)\{|a \, \psi_s|^2 + |b \, \psi_D|^2 + 2ab \, \psi_s \, \psi_D\} d^3r \quad \text{... (4.14)}$$

Now, the pure $_1^3S$ state cannot contribute to the quadrupole moment and hence

$$\int \frac{r^2}{2} \, P_{2,0} \, (\cos \theta) a^2 \, |\psi_s|^2 \, d^3r = 0$$

The second term corresponds to the contribution to the quadrupole moment from $_1^3D$ state. Now, using

$$\psi_D = \frac{u_2 \, (r)}{r} \left[Y_{2,2} \, \chi_{1,1} - \sqrt{\frac{3}{10}} \, Y_2, \chi_{1,0} + \frac{1}{10} \, Y_{2,0} \, \chi_{1,1} \right]$$

Where $Y's$ are spherical harmonics and $\chi's$ are spin wave functions.

We may note that Q is a function of space part only, and does not depend on spin. Due to orthogonality of spin wave function, it drops out.

Now, expressing the spherical harmonics in terms of *Associated Legendre Polynomials* and by making use of the orthogonality of the polynomials the second term on integration yield

$$-\frac{b^2}{20} \int_0^\infty r^2 \, u_2^2 \, (r) \, dr \qquad \text{... (4.15)}$$

as the contribution from $_1^3D$ state. Similarly, one can write the third term as

$$2ab \int_0^r r_2^2 \, P_{2,0} \, (\cos \theta) \, \psi_s \, \psi_D \, d^3r = \frac{2ab}{4\pi} \int_0^\infty \frac{1}{2} r^2 \, u_0(r) \, u_2(r) dr$$

$$\int_0^\pi P_{2,0} \, (\cos \theta) \, \frac{1}{\sqrt{2}} \, P_{(2,0)} \, (\cos \theta) \sin \theta \, dQ \int_0^{2\pi} d\phi$$

$$= \frac{1}{\sqrt{50}} \int_0^\infty r^2 \, u_0(r) \, u_2(r) dr \qquad \text{... (4.16)}$$

In order to estimate the magnitude of Q_d, one will have to know the radial wave functions, which depends on the form and nature of nuclear potential.

We have seen that magnetic moment and positive quadrupole moment establishes the fact that the ground state of deutron ($l = 0$) is an admixture of predominantly ^3_1S state (about 96 %) and a small fraction of ^3_1D state (about 0.04 %). We have already mentioned that experimentally no excited state has been observed for the deutron so far. One can understand this as follows:

Let us consider the general solution of the Schrodinger wave equation. For any angular mommentum, the radial equation is

$$\frac{d^2 u_l}{dr^2} + \frac{M}{\hbar^2}(E - V)u_l - \frac{l(l+1)}{r^2}u_l = 0 \qquad \text{... (4.17)}$$

We can easily see that equation (4.17) is equivalent to an S-wave radial equation with the potential

$$V_{\text{eff}} = V(r) + \frac{\hbar^2 l(l+1)}{M r^2} \qquad \text{... (4.18)}$$

The second term on the r.h.s. of (4.18) is due to orbital motion, i.e., the motion in the θ-direction gives an effective force in the r-direction, namely the centrifugal force which gives this potential. It is positive, therefore, repulsive and as it increases with l, the binding energy or the total energy of the lowest state for a given l decreases with the increasing l. One uses this fact to show that no bound *P*-state ($l = 1$) exists for deutron.

One can write the schrodinger equation for $l = 1$ in the two regions by putting $l = 1$ in Equation (4.17)

$$\frac{d^2 u_{1,i}}{dr^2} + k^2 u_{1,i} - \frac{2}{r^2}u_{1,i} \qquad \text{for } r < b \qquad \text{... (4.9)}$$

$$\frac{d^2 u_{1,i}}{dr^2} + \gamma^2 u_{1,0} - \frac{2}{r^2}u_{1,0} \qquad \text{for } r > b \qquad \text{... (4.10)}$$

Where $k^2 = \frac{M}{\hbar^2}(V_0 - E_B)$ and $\gamma^2 = \frac{M}{\hbar^2}E_B \cdot u_{1,i}$ and $u_{2,0}$ are the wave functions inside and outside the potential well respectively for $l = 1$ state. In the limit $E_B \to 0$, the solutions of Equations (4.9) and (4.9) can be obtained with the help of Neumann and Bessel's equations. We may note that while solving these equations in terms of the Bessel and Neumann functions one has to match the wave functions and their derivatives at the boundary b of the potential well. One finds that this leads to the condition that kb should be equal to π. But

$$kb = \frac{M}{\hbar^2}(V_0 - E_B)^{\frac{1}{2}}b$$

In the limit $E_B \to 0$, the above reduces to

$$kb = \left(\frac{M}{\hbar^2}V_0\right)^{\frac{1}{2}}b$$

Now, in accordance with the condition $kb = \pi$, one obtains

$$\frac{M}{\hbar^2} V_0 b^2 = \pi^2$$

or
$$V_0 = \frac{\pi^2 \hbar^2}{M b^2} \qquad \qquad \text{... (4.11)}$$

For $b = 2 \times 10^{-15}$ m, one finds the depth of the potential ≈ 144 MeV, which is about the four times of value of the potential in the ground state. However, for large l values, the potential needed to produce a bound state, would be even deeper clearly, no bound state exists for $l > 0$.

We may note that even no excited s-state exists. If one assume that there is a bound first excited state, then the value of $kb = \dfrac{3\pi}{2}$, which would give the corresponding value of potential.

However, when $kb = \dfrac{3\pi}{2}$, one finds that a wave function produces a node inside the potential well at $kb - \pi$, but the wave function for the bound state should not produce any node. Obviously, for $l = 0$ also there can be no excited state.

Thus, in deutron no excited bound state exists and this is in agreement with experimental observations.

4.4 NUCLEON-NUCLEON SCATTERING

In principal there are four types of scattering measurments involving two nucleons that can be carried out: *Proton-proton* (pp) *scattering, np-scattering, pn scattering* and *nn scattering*. The scattering of neutrons off proton targets (*np* scattering) is an important source of information on two nucleon systems. A system of two protons, or two neutrons, can only be coupled together to isospin $T = 1$, whereas a neutron-proton system can be either $T = 0$ or 1. Hence *np*-scattering and the corresponding *pn*-scattering, contain information that cannot be obtained from *pp*-and *nn*-scattering.

The information obtained from *pn*-and *nn*-scatting may not be any different from that obtained from *np*-and *pp*-scattering, e.g. the only difference between *pn*-and *np*-scattering is whether the neutron or the proton is the target. Under time-reversal invariance, these two arrangements should give identical results.

Both *pp*-and *nn*-are isospin $(T) = 1$ systems. If nuclear force is charge independent, the results of *pp*-and *nn*-scattering can only be different by the contribution made by Coulomb interaction. Since the latter is well known, a comparison of *pp*-and *nn*-scattering results can, in principle, test the charge independence of nuclear interaction. However, the accuracy that can be achieved with nn-scattering is still inadequate for such a task. Now, we will study n-p scattering

4.4.1 Low energy neutron-proton (*np*) scattering

One can study the relative motion of two particles of masses M_1 and M_2 with the help of Schrodinger wave equation

$$-\frac{\hbar^2}{2\mu} \nabla^2 \psi + V(r) \psi = E \psi \qquad \qquad \text{... (4.1)}$$

wherre $\mu = \dfrac{M_1 M_2}{M_1 + M_2}$ is the reduced mass. $E = (E_L - E_C)$ is the internal energy of the system, E_L is the energy in the L-system and E_C is the kinetic energy of C_M given by

$$E_C = \frac{M_1}{M_1 + M_2} E_L \qquad \qquad \text{... (4.2)}$$

Now, for n-p scattering, $M_1 \approx M_2 = M$ (say)

so that
$$E_C = \frac{E_L}{2}$$

Obviously, only half the energy is available for scattering the C-system, i.e.

$$E = E_L - E_C = \frac{E_L}{2} \qquad \qquad \text{... (4.4)}$$

We have already discussed the kinematics of the collision in earlier chapter, where we have shown that the angle of scattering θ_L is related to that in the C-system, i.e. θ_C for n-p scattering as

$$\theta_C = 2\,\theta_L \qquad \qquad \text{... (4.5)}$$

Also the angle between the neutron and proton after scattering in the L-system is 90°.

Since, $\mu = \dfrac{M}{2}$, Equation (4.1) can be expressed as

$$\nabla^2 \psi + \frac{M}{\hbar^2}\,[E - V(r)]\,\psi = 0 \qquad \qquad \text{... (4.6)}$$

Here $\psi = \psi(r, \theta, \phi) \cdot \theta$ and ϕ are the angles in CM system and r is the distance between the neutron and the proton. For scattering $E > 0$.

Partial wave treatment of *n*-*p* scattering

We will discuss this in nuclear reaction theory.

We have read from the theory of scattering that an incident wave can be expanded in terms of a series of spherical outgoing and spherical incoming waves in the limit of larger ($kr \ll 1$), where r is range of interaction (~ 1.4 fm), we have

$$\psi_{inc} = \frac{1}{2ikr} \sum_{l=0}^{\infty} (2l+1)\, i^l \left[\exp\left\{ i\left(kr - \frac{l\pi}{2} \right) \right\} - \exp\left\{ -i\left(kr - \frac{l\pi}{2} \right) \right\} \right] P_l (\cos\theta) \qquad \text{... (4.7)}$$

Where P_l (cos θ) is the Legendre polynomial of the order l; $k^2 = \dfrac{2ME}{\hbar^2}$

Spherical outgoing waves in the above expression are affected either in phase or in amplitude or in both from the scatterer. When only elastic scattering is taking place, i.e. no reaction, then only phase is affected. Since the scattered wave is a spherical outgoing wave with amplitude dependent on the angle of scattering, one can write $\psi_{SC} = \left[\dfrac{f(\theta)}{r} \right] \exp(ikr)$. Thus, in the presence of scatter, the total wave function is

$$\psi(r) = \psi_{inc} + \psi_{sc} = \psi_{inc} + \frac{f(\theta)}{r} \exp(ikr) \qquad \dots (4.8)$$

In analogy with Equations (4.7), (4.8) takes the form

$$\psi(r) = \frac{1}{2\pi \, ikr} \sum_{l=0}^{\infty} (2l+1)i^l \left[\eta_l \exp\left\{ i\left(kr - \frac{l\pi}{2} \right) \right\} - \exp\left\{ -i\left(kr - \frac{l\pi}{2} \right) \right\} \right] P_l(\cos\theta) \qquad \dots (4.9)$$

where $\eta_l = \exp(2i\delta_l)$ with δ_l real as the phases and not the amplitudes of outgoing waves are affected. This means $|\eta_l|^2 = 1$.

In the actual scattering experiment, an incident beam of monoergetic particle is scattered from an infinitely heavy scattering centre (in the C-system) at an angle θ. Then the differential scattering cross-section is given by

$$\sigma(\theta) = |f(\theta)|^2 \qquad \dots (4.10)$$

Comparing Equations (4.7) and (4.9) and making use of Equation (4.8), one finds

$$f(\theta) = \frac{1}{2ik} \sum_{l=0}^{\infty} (2l+1)(\eta_l - 1) P_l(\cos\theta)$$

$$= \frac{1}{2ik} \sum_{l=0}^{\infty} (2l+1)[\exp(2i\delta_l) - 1] P_l(\cos\theta)$$

$$= \frac{1}{k} \sum_{l=0}^{\infty} (2l+1) \exp(i\delta_l) \sin\delta_l \, P_l(\cos\theta) \qquad \dots (4.11)$$

Using (4.10), one obtains

$$\sigma(\theta) = |f(\theta)|^2 = \frac{1}{k} \left| \sum_{l=0}^{\infty} (2l+1) \exp(i\delta_l) \sin\delta_l \, P_l(\cos\theta) \right|^2 \qquad \dots (4.12)$$

Now, the total scattering cross section is obtained by integrating (4.12) over the entire 4π solid angle. One obtains

$$\sigma_{tot} = \int \sigma(\theta) \, d\Omega = \int_0^\pi |f(\theta)|^2 \, 2\pi \sin\theta \, d\theta$$

$$= \frac{2\pi}{k^2} \int_0^\pi \left| \sum_{l=0}^{\infty} (2l+1) \exp(i\delta_l) \sin\delta_l \cdot P_l(\cos\theta) \right|^2 \sin\theta \, d\theta$$

Because of the orthogonality of the Legendre's polynomials, one finally obtains,

$$\sigma_{tot} = \frac{2\pi}{k^2} \sum_{l=0}^{\infty} (2l+1)^2 \sin^2\delta_l \left(\frac{2}{2l+1} \right)$$

$$= \frac{4\pi}{k^2} \sum_{l=0}^{\infty} (2l + 1) \sin^2 \delta_l \qquad \qquad ... (4.13)$$

Equation (4.11) reveals that one can express the scattering ampletude $f(\theta)$ as a sum over the amplitudes $f_l(\theta)$ of scattering of the different partial waves

$$f(\theta) = \sum_{l=0}^{\infty} f_l(\theta) \qquad \qquad ... (4.14)$$

where
$$f_l(\theta) = \frac{(2l+1)}{2ik} [\exp(2i\delta_l) - 1] P_l(\cos \theta)$$

$$= \frac{2l+1}{k} \exp(i\delta_l) \sin \delta_l \cdot P_l (\cos \theta) \qquad \qquad ... (4.15)$$

In order to calculate the cross sections, the phase shifts δ_l must be known for different sight, it appears that the calculation of the cross sections would be difficult due to involvement of the large number of partial waves of different l. However, in practice, only a few l values are involved in these calculations, depending on the energy.

Engery limits for the scattering of different partial waves

Let us consider a neutron in the C-system having linear momentment p and incident with an impact parameter q. Obviously, scattering will take place only if $q < b$. For the case $q > b$, clearly the incident neutrons will not feel the interaction potential, which has a range b. Obviously, the largest impact parameter q, for which scattering occurs is $q_m = b$. Now, for the neutrons having this impact parameter, the angular momentum will be $p q_m = pb$. According to quantum mechanics this can only be an integral multiple of $\hbar$.

Let us suppose that for a given momentum, pb has a value such that

$$l\hbar \le p_b \le (l + 1)\hbar \qquad \qquad ... (4.16)$$

Where l is an integer. Obviously, the angular momentum of these particles will be $l\hbar$. The maximum value p_{max} for which the said condition is satisfied is

$$p_{max} b = (l + 1)\hbar$$

This yields

$$p_{max} = \frac{(l+1)\hbar}{b}$$

and
$$E_{max} = \frac{p_{max}^2}{M} = \frac{(l+1)^2 \hbar^2}{M b^2} \qquad \qquad ... (4.17)$$

Thus for S-wave n-p scattering, i.e. $l = 0$, one obtains the maximum energy in the C-system as

$$E_{max} = \frac{\hbar^2}{M b^2} = 10 \text{ MeV} \qquad \qquad ... (4.17a)$$

From Equation (4.4) the maximum energy for S-wave n-p scattering in the L-system will be double that of in C-system. Thus in L-system, the maximum energy for S-wave n-p scattering is 20 MeV. The limiting values are usually taken to be half those mentioned above, i.e. $E_{max} = 5$ MeV in C-system and $(E_L)_{max} = 10$ MeV for L-system for S-wave n-p scattering.

As mentioned earlier, one can divide the whole space surrounding the scattering centre into coaxial cylindrical zones with the axis parallel to the direction of the incident beam having radii λ, 2λ, 3λ, etc.

Phase Shift

We note that the above analysis is qualitative in nature and one cannot expect to give a correct quantitative picture. Under suitable approximations from the solutions of the radial equations with and without the scatter, one can determine the nature of variation of the phase shift with energy.

We make a separation of variables in the wave Equation (4.6). In the region $r < b$, One can obtain the radial equation in the presence of the scattering potential $V(r)$ as

$$u_l'' + \left[k^2 - u(r) - \frac{l(l+1)}{r^2} \right] u_l = 0$$

Where $u_l(r)$ is the radial function for the lth partial wave, $k^2 = \dfrac{M}{\hbar^2} E$ and $U(r) = \dfrac{M}{\hbar^2} V(r)$.

In the absence of scatterer, one can write the radial wave equation as

$$v_l'' + \left[k^2 - \frac{l(l+1)}{r^2} \right] v_l = 0$$

The asymptotive solution for the above radial equations are of the form

$$u_l \sim \sin\left(kr - \frac{l\pi}{2} + \delta_l \right) \qquad \text{... (4.18)}$$

$$v_l \sim \sin\left(kr - \frac{l\pi}{2} \right) \qquad \text{... (4.19)}$$

The sign of δ_l in (4.18) shows whether the wave function $u_l(r)$ is ahead or lagging in phase with respect to the function $V_l(r)$.

For attractive (negative) potential, $U(r) < 0$ and therefore u_l''. This means that the wave function u_l (solid curve in Figure (4.8a)) has a greater curvature in the interior region ($r > b$) than v_l (dashed curve in Figure (4.8a)). Figure (4.8a) illustrates $\delta_l > 0$ for an attractive potential. For a repulsive potential, Figure (4.8b) illustrates $\delta_l < 0$.

For low energy scattering, quantitative estimate reveals that potential $V(r) \propto \dfrac{1}{r^n}$. One can write δ_l as $\delta_l \sim k^{2l+1}$ for $2l < (n-3)$.

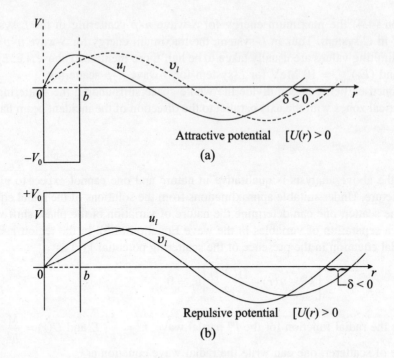

Fig. 4.8 Diagram illustrating the nature of the radial solution for a given *l*: (a) attractive potential, $U(r) < 0$, $(\delta_l > 0)$ and (b) repulsive potential, $U(r) > 0$, $(\delta_l < 0)$.

Interestengily, the measured phases δ_l completely determine the potential $V(r)$. However, the determination of phases from the cross sections is very complicated and the problem has not been solved in practice for any case. One can obtain fairly reliable information about the phases for those cases in which only a small number of partial waves (*l*) are involved.

4.5 SCATTERING CROSS SECTION

One of the useful information obtained in scattering experiments in the scattering cross-section. Let $\sigma_c(\theta_c)$ and $\sigma_L(\theta_L)$ denotes the differential scattering cross-sections in the *C* and *L* systems respectively. The number of particles scattered into the solid angle $d\Omega$ per unit incident flux by a target of unit surface density is given by $\sigma(\theta)\, d\Omega$.

One obtains the total scattering cross section by integrating differential scattering cross section over all directions

$$\sigma = \int \sigma(\theta)\, d\Omega = \int_0^{2\pi} d\phi \int_0^\pi \sigma(\theta)\, \sin\theta\, d\theta$$

Since $d\theta$ is independent of $d\phi$, we have

$$\sigma = 2\pi \int \sigma(\theta)\, \sin\theta\, \delta\theta$$

Since by definition of the scattering cross-section it is required that the number of particles scattered into an element of solid angle $d\Omega_L$ about θ_L in laboratory system (L) should be same as scattered into an element of solid angle must be independent of the coordinate chosen, i.e.

Fig. 4.9 Solid angles $d\Omega_C$ and $d\Omega_L$ in C and L systems respectively

$$Nn \; \sigma_L(\theta_L) \, d\Omega_L = Nn \; \sigma_C(\theta_C) \, d\Omega_C \qquad \text{... (4.1)}$$

Since scattering is independent of ϕ, $\phi_L = \phi_C$

$$d\phi_L = d\phi_C$$

$\therefore$
$$Nn \; \sigma_L(\theta_L) \, 2\pi \sin\theta_L \, d\theta_L = Nn \; \sigma_C(\theta_C) \, 2\pi \sin\theta_C \, d\theta_C$$

or
$$\sigma_L(\theta_L) \sin\theta_L \, d\theta_L = \sigma_C(\theta_C) \sin\theta_C \, d\theta_C \qquad \text{... (4.2)}$$

or
$$\sigma_L(\theta_L) = \frac{\sin\theta_C}{\sin\theta_L} \frac{d\theta_c}{d\theta_L} \; \sigma_C(\theta_C) \qquad \text{... (4.3)}$$

The velocity u_1 of particle m_1 after scattering in L-system is obtained by adding vectorially the velocity V_{CM} of CM in L system to V_1' of m_1 in CM system after scattering (Figure 4.10).
we have

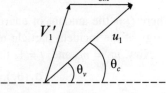

$$\mathbf{u}_1 = \mathbf{V}_1' + \mathbf{V}_{CM} \qquad \text{... (4.4)}$$

or
$$u_1 \cos\theta_L = V_1' \cos\theta_C + V_{CM} \qquad \text{... (4.5)}$$

and
$$u_1 \sin\theta_L = V_1' \sin\theta_C \qquad \text{... (4.6)}$$

Fig. 4.10 Relation between θ_L and θ_C

Since the scattering is in the same plane, the azimuthal angle in the two systems remain the same. Dividing (4.6) by (4.5), one obtains

$$\tan\theta_L = \frac{V_1^1 \sin\theta_C}{V_1^1 \cos\theta_C + V_{CM}} \qquad \text{... (4.7)}$$

$$= \frac{\sin\theta_C}{\cos\theta_C + \dfrac{V_{CM}}{V_1'}}$$

Now, defining
$$\gamma = \frac{V_{CM}}{V_1'} = \frac{\text{speed of CM in } L\text{-system}}{\text{speed of the scattered particle in CM-system}}$$

or
$$\tan\theta_L = \frac{\sin\theta_C}{\cos\theta_C + \gamma} \qquad \text{... (4.8)}$$

Obviously, (4.8) gives a relation between the angle θ_L at which m_1 is scattered in the L-system as a function of θ_C in the CM system. From Equation (4.8)

$$\frac{d\theta_C}{d\theta_L} = \frac{(\gamma + \cos\theta_C)^2}{(1 + \gamma\cos\theta_C)^2} \frac{1}{\cos^2\theta_L} \qquad \dots (4.9)$$

and

$$\sin\theta_L = \frac{\sin\theta_C}{(1 + \gamma^2 + 2\gamma\cos\theta_C)^{\frac{1}{2}}} \qquad \dots (4.10)$$

$$\cos^2\theta_L = \frac{(\gamma + \cos\theta_C)^2}{(1 + \gamma^2 + 2\gamma\cos\theta_C)} \qquad \dots (4.11)$$

Using Equations (4.9), (4.10) and (4.11) in Equation (4.8), one obtains

$$\sigma_L(\theta_L) = \frac{(1 + \gamma^2 + 2\gamma\cos\theta)^{\frac{3}{2}}}{1 + \gamma + \cos\theta_C} \sigma_c(\theta_c) \qquad \dots (4.12)$$

Where $\gamma = \dfrac{m_1}{m_2}$, for elastic scattering and for inelastic scattering

$$\gamma = \frac{m_1}{m_2}\left[1 - \frac{Q}{E}\left(1 + \frac{m_1}{m_2}\right)\right]^{-\frac{1}{2}} \qquad \dots (4.13)$$

Where Q is the amount of energy that is lost or gained as a result of impact and E is the kinetic energy of the incident particle measured in the L-system.

Now, $m_1 = m_2$ and $\gamma = 1$, then (4.12) reduces to

$$\sigma_L(\theta_L) = 4\cos\theta_L\,\sigma_C(\theta_C) \qquad (\because \theta_C = 2\theta_L) \qquad \dots (4.14)$$

We may note that the cross-section given by (4.14) is in the C-system. However, the measurment of the cross-section is made in the L-system. With the help of (4.14), the measured cross-section $\sigma_L(\theta_L)$ must be transformed into $\sigma_C(\theta_C)$ for comparison with theoretical results.

4.6 SCATTERING LENGTH AND EFFECTIVE RANGE

If we make a partial wave expansion of the scattering wave function, and substitute the results into the Schrodinger equation, we obtain an equation for the radial wave function of the lth partial wave

$$-\frac{\hbar^2}{2\mu}\left\{\frac{1}{r^2}\frac{d}{dr}r^2\frac{d}{dr} - \frac{l(l+1)}{r^2}\right\}R_e(k,r) + V(r)R_l(k,r) = E R_l(k,r) \qquad \dots (4.1)$$

The l-dependent term, $\dfrac{l(l+1)}{r^2}$, comes from the angular part of the kinetic energy and is sometimes referred to as a centrifugal barrier since it is repulsive to an incoming particle. The effective potential experienced by the scattering particle is then

$$V'(r) = V(r) + \frac{\hbar^2}{2\mu} \frac{l(l+1)}{r^2} \qquad \qquad \text{... (4.2)}$$

Because of the barrier, scattering at low energies (particularly for $E < 10$ MeV) is dominated by low order partial waves. For $E < 10$ MeV, nucleon-nucleon scattering is given by s-wave ($l = 0$) alone, as can be seen from the fact that, among the observed values, only $l = 0$ phase shifts δ_0 are significantly different from zero.

When kinetic energy $E \to 0$, the total cross section

$$\sigma_{\text{tot}} = \int \sigma(\theta) \, d\Omega = \frac{4\pi}{k^2 \, \text{cosec}^2 \, \delta_0} = \frac{4\pi}{k^2 + k^2 \cot^2 \delta_0} \qquad \text{... (4.3)}$$

where $$\sigma(\theta) = |f(\theta)|^2 = \frac{\text{Sin}^2 \, \delta_0}{k^2} \qquad \qquad \text{... (4.3a)}$$

and $$\sigma_{\text{tot}} = \int \sigma(\theta) \, d\Omega = \frac{4\pi \, \text{Sin}^2 \, \delta_0}{k^2}, \text{ which is transformed as (4.3)} \quad \text{... (4.3b)}$$

remains finite for nucleon-nucleon scattering. As the energy (E) decreases, k also decreases, which means that both $\sigma(\theta)$ and σ_{tot} increases. Equations (4.3a) and (4.3b) would make $\sigma(\theta)$ and stot go to infinity in the limit $E \to 0$, i.e. $K \to 0$. However, that would make the number of scattered particles infinitaly large, which is quite impossible. Therefore, we stipulate that the limiting value of $k \cot \delta_0$ in the denominator of (4.3) should remain finite as $k \to 0$. Let us write

$$\frac{1}{a_k} = -k \cot \delta_0 \qquad \qquad ..(4.4)$$

then one obtains from (4.3)

$$\sigma_{\text{tot}} = \frac{4\pi}{k^2 + \dfrac{1}{ak^2}} \qquad \qquad \text{... (4.5)}$$

The quantity a_k is known as the *scattering length* for the energy E. In order that σ_{tot} may remain finite as $k \to 0$, one stipulate that

$$\lim_{k \to 0} k \cot \delta_0 = \lim_{k \to 0} \left(-\frac{1}{a_k} \right) = -\frac{1}{a} \qquad \qquad \text{... (4.6)}$$

where a is known as *Fermi scattering length*. One then obtains for zero energy neutrons

$$\lim_{k \to 0} \sigma_{\text{tot}} = 4\pi a^2 = \sigma_0 \text{ (say)} \qquad \qquad \text{... (4.7)}$$

The s-wave scattered amplitude at low energies ($E_L < 10$ MeV),

$$f(\theta) = \frac{\exp(i\delta_0)}{k} \, \text{Sin} \, \delta_0 \qquad \qquad \text{... (4.8)}$$

shows that $f(\theta)$ remains finite as $k \to 0$ as δ_0 also goes to zero in this limit, i.e. $\lim_{k \to 0} \delta_0 = 0$. Since $\sin \delta_0 = \delta_0$ as $\delta_0 \to 0$, one obtains in this case

$$\lim_{k \to 0} f(\theta) = \lim_{k \to 0} \frac{\delta_0}{k} = -a \text{ (say)} \qquad \ldots (4.9)$$

We note that the amplitude of the scattered wave in the limit of zero energy neutrons is thus equal to the negative of the Fermi scattering length.

We may also note that it is often convenient to discuss extermely low energy scattering in terms of scattering length instead of *s*-wave phase shift. In general a and δ_0 bears the following relationship

$$a = \lim_{k \to 0} R_e \left\{ -\frac{1}{k} e^{ik\delta_0} \sin \delta_0 \right\} \qquad \ldots (4.10)$$

where $k^2 = \dfrac{2\mu_e}{\hbar^2}$. The energy dependence of δ_0 at low energies is given by the *effective range parameter*, r_e, defined by the relation

$$k \cot \delta_0 = -\frac{1}{a} + \frac{1}{2} r_e k^2 \qquad \ldots (4.11)$$

or

$$\cot \delta_0 = -\frac{1}{ka} + \frac{1}{2} kr_e \qquad \ldots (4.11a)$$

Scattering length and effective range provide a useful way to parametrize information on low energy nucleon-nucleon scattering furthermore, these parameters may be related to observations other than *NN* scattering, such as deutron binding energy. Substituting equation (4.11a) in

$$\sigma_0 = \frac{4\pi}{k^2} \sin^2 \delta_0 \qquad \ldots (4.12)$$

one obtains,

$$\sigma_0 = 4\pi a^2 \left[\left(1 - \frac{1}{2} a r_e k^2 \right)^2 + a^2 k^2 \right]^{-1} \qquad \ldots (4.13)$$

For a bound state (simple square well) of energy $-E_B$, one can easily show that

$$\frac{1}{a} = \beta - \frac{1}{2} \beta^2 r_e \qquad \ldots (4.14)$$

where

$$\beta^2 = \frac{2m E_B}{\hbar^2}$$

Obviously, (4.14) relates the binding energy E_B to the scattering length (a) and effetive range (r_e). For the deutron, in which the neutron and proton are in a triplet state, the value of $\beta = 0.2316$ fm^{-1}, and $a_t = 5.425$ fm, so $r_t = 1.76$ fm.

In addition to the above, very accurate results can be obtained for the *np* system by scattering slow neutrons off protons in hydrogen atoms bound in H_2 molecules. This is why, a great deal of attention is devoted to the measurement and understanding of these parameters.

4.7 RADIAL EQUATION AT LOW ENERGY

The radial equation for $l = 0$ is

$$\frac{d^2u}{dr^2} + \frac{M}{\hbar^2}[E - V(r)]u = 0 \qquad \text{... (4.1)}$$

Here $E > 0$. Now, assuming a rectangular potential well, Equation (4.1) reduces to

$$u''_{in} + k_2^2\, u_{in} = 0 \qquad \text{... (4.2a)}$$
$$u''_{out} + k^2\, u_{out} = 0 \qquad \text{... (4.2b)}$$

Where $u_{in}(r)$ and $u_{out}(r)$ represents the radial functions inside and outside the well, and

$$k_2^2 = \frac{M}{\hbar^2}(E + V_0), \qquad k^2 = \frac{M}{\hbar^2}E \qquad \text{... (4.3)}$$

One obtains the solutions as

$$u_{in} = A \sin k^2 r \qquad \text{... (4.4a)}$$
$$u_{out} = B \sin (kr + \eta_0) \qquad \text{... (4.4b)}$$

One can also find the nature of the radial function $u(r)$ at large r (asymptotic form) both with and without the scatterer by expansion of Equations (4.18) and (4.19) of sec 4.1 respectively for $l = 0$ partial waves. These are:

(i) without scatterer: $V_0(r) \sim \dfrac{\text{Sin } kr}{r}$ (4.5a)

(ii) with scatterer: $V_0(r) \sim \dfrac{\text{Sin } (kr + \delta_0)}{k}$ (4.5b)

Comparing the Equations (4.4b) and (4.5b), one finds the phase of the external wave to be $\eta_0 = \delta_0$.

One can use Equation (4.4) to calculate δ_0, applying the boundary condition at $r = b$, one obtains

$$k_2 \cot k_2 b = k \cot(kb + \delta_0) \qquad \text{... (4.6)}$$

After simplification, one obtains

$$\sin \delta_0 = \frac{\text{Sin } kb\,(k \text{ Cot } kb - k_2 \text{ Cot } k_2\, b)}{(k^2 + k_2^2 \text{ Cot}^2\, k_2\, b)} \qquad \text{... (4.7)}$$

Obviously Equation (4.7) gives δ_0 as a function of E, b and V_0.

Now, we find *solutions for neutrons of zero energy*. Obviously, in this case $E = 0$ and hence $k = 0$ and

$$k_2^2 = k_2^0 = \frac{M\, V_0}{\hbar^2}$$

The radial equations are

$$u''_{in} + k_0^2\, u_{in} = 0 \qquad \text{for } r < b \qquad \text{... (4.8a)}$$
$$u''_{out} = 0 \qquad \text{for } r > b \qquad \text{... (4.8b)}$$

The solutions of the above equations are

$$u_{\text{in}} = A \sin k_0 r \qquad \qquad \text{... (4.9a)}$$

$$u_{\text{out}} = B(r - a') \qquad \qquad \text{... (4.9b)}$$

But from Equation (4.4b), one obtains in the zero energy limit (for which k and δ_0 go to zero).

$$\lim_{k \to 0} u_{\text{out}} = \lim_{k \to 0} B \sin (kr + \delta_0)$$

$$= \lim_{k \to 0} B (kr + \delta_0)$$

$$= \lim_{k \to 0} Bk \left(r + \frac{\delta_0}{k} \right) = B'(r - a)$$

Where $-a$ is substituted for $\lim_{k \to 0} \left(\dfrac{\delta_0}{k} \right)$ from Equation (4.9) of sec. 6. Now, comparing Equations (4.9b) and (4.9c), one finds $a' = a = $ Fermi scattering length. Clearly, the external solution for zero energy neutrons reduces to

$$u_{\text{out}} = B(r - a) \qquad \qquad \text{... (4.9d)}$$

Where we have put $B = B'$.

Let us study the *nature of wave functions* $k^2 \ll k_0^2$, which means that k_2^2 in Equations (4.2) is only slightly greater than k_0^2, i.e. $k_2^2 \geq k_0^2$. Further, we have $k_0^2 \geq k_1^2$, where $k_1 = \dfrac{\sqrt{M (V_0 - \varepsilon_d)}}{\hbar}$. This follows from the fact that the deuteron binding energy $E_{Bd} \ll V_0$. Now, we have from the nature of the solution in the deuteron ground state $k_0 b \geq \dfrac{\pi}{2}$. This reveals that as in the case of the deuteron, the portion of the $\sin k_2 r$ graph for the internal solution of Equations (4.4a) contained in the range $0 < r < b$ is slightly greater than half a loop of the sine curve (Figure 4.11). Obviously, the graph has a negative slope at $r = b$ so that the external solution (Equation 4.4b) must also have a negative slope at this point. Further, since $k \ll k_2$, the wavelength of the external solution $\lambda \gg \lambda_2$ the wavelength of the internal solution. Figure 4.11 clearly illustrates these features of the solutions for low energy $n - p$ scattering in relation to the sqaure well potential.

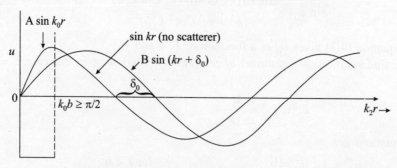

Fig. 4.11 Wave functional of the scattered wave for $0 < r < b$ and for $r > b$ in relation to square wave potential

We now consider the zero energy of neutrons ($k = 0$) for which $k_2^2 = k_0^2$. From Figure 4.12, we note that the portion of the graph of the internal solution $\sin k_0 r$ between $0 < r < b$ is again slightly larger than half a loop of the sine curve. At $r = b$, this also has a negative slope and therefore the external solution which is the straight line given by Equation (4.9d) has a negative slope while intersects r-axis at $r = a > 0$. Obviously, the Fermi scattering length a is positive

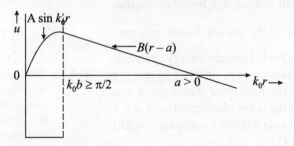

Fig. 4.12 Diagram illustrating the nature of wave functions for zero energy neutrons ($k = 0$). The function outside the potential well is a straight line having negative slope.

One can calculate Fermi scattering length a from the boundary conditions applied to the solutions (4.9a) and (4.9b) at $r = b$:

$$\left(\frac{u'_{in}}{u_{in}}\right)_b = \left(\frac{u'_{out}}{u_{out}}\right)_b$$

This yields

$$k_0 \cot k_0 b = \frac{1}{b - a} \qquad \text{... (4.10)}$$

or

$$a = b - \frac{\tan k_0 b}{k_0} \qquad \text{... (4.11)}$$

For $V_0 = 38$ MeV and $b = 2$ fm, one finds $k_0 = 0.963$ fm^{-1}. This gives $k_0 b = 1.93$ radians and $\tan k_0 b = -2.69$. One obtains Fermi scattering length $a = 4.8$ fm. This gives $\sigma_0 = 4\pi a^2 = 2.88$ barns.

Sign of the Fermi scattering length (a) and its importance

We have seen that an attractive $n - p$ potential gives rise to bound state of the deutron ($E = E_{Bd} < 0$) and the Fermi scattering length (a) for this must be postive this is due to the fact that the internal wave functions (u_{in}) both for deutron and for low energy ($E < V_0$) scattering including $E = 0$ have negative slopes at $r = b$ (boundary). This means both $k_1 b$ and $k_0 b$ are slightly greater than $\frac{\pi}{2}$.

Clearly, for zero energy $n - p$ scattering, the external linear function must intersect the r - axis at a point $r = a > 0$.

Now, suppose that V_0 is gradually reduced so that k_0 is also reduced till $k_0 b$ becomes just equal to $\frac{\pi}{2}$. Then the slope of the internal function is zero at $r = b$, i.e. $\cot k_0 b = \cot \frac{\pi}{2} = 0$. For zero

energy neutrons, the outside function is a straight line
with zero slope, i.e. in this case it is parallel to the r -
axis extending to infinity. Obviously this corresponds
to the just the bound case for which $a = \infty$ as illustrated
in Figure 4.13.

Now, reduce V_0 still further, $k_0 b$ becomes smaller

than $\dfrac{\pi}{2}$. Thus the slope of the internal function is now

postive ($\cot k_0 b > 0$) at $r = b$. For zero energy neutrons,
the outside linear function has a positive slope at $r = b$
and it is produced in the backward direction, it can
intersect the r-axis. At the point of intersection $r = a <$
0, i.e. for this unbound case a (Fermi scattering length)
is negative (Figure 4.14).

For a negative (attractive) n-p potential, one can
conclude as:

(i) Fermi scattering length, $a > 0$ for scattering of
     zero energy neutrons form a potential giving rise
     to a bound state of $n - p$ system.
(ii) Fermi scattering length, $a = \infty$ for scattering of
     zero energy neutrons from a potential giving rise
     to a just bound n-p system.
(iii) Fermi scattering length, $a < 0$ for scattering of
     zero energy neutrons from a weak attractive
     potential for which no bound state of the n-p
     system is possible.

Concluding, one can say that the knowledge of
the sign of the Fermi scattering length (a) helps
to get an idea about the nature of a particular
state of the n-p system is bound or unbound.

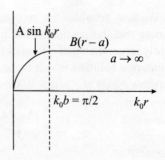

Fig. 4.13 Illustration of the nature of the
wave functions for the just bound
case. Clearly, the external function
is a straight line with zero slope
and corresponds to just bound
case for which $a = \infty$.

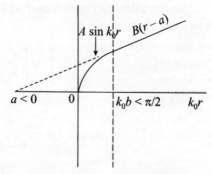

Fig. 4.14 Illustration of the nature of wave
functions for the unbound case.

4.8 SPIN DEPENDENCE OF n-p INTERACTION

In section 7, assuming a rectangular potential well, we calculated low, i.e. zero energy n-p cross-
section, σ_0. The value of σ_0 obtained to be about 3 barns. However, this value is quite different
from the experimental value of zero energy incoherent scatting cross section $\sigma_0 = 20.38$ barns.
What is the reason behind such a large discrepancy? E. P. Wigner was first to clear up this
discrepancy. He assumed that the n-p interaction potential is dependent on the relative spin

orientations of the neutron and proton. Since, neutron and proton are $\dfrac{1}{2}$ spin particles, therefore in

n-p scattering the neutron and proton spins way either be parallel or anti-parallel. In deutron-the
bound state of the n-p system the neutron and proton spins arre parallel. The resultant spin angular
momentum is $S = 1$, which has a statistical weigh $2S + 1 = 3$. For antiparallel spin orientations, the

resultant spin-angular momentum of the n-p interaction system is $S = 0$ which has a statistical weigth $2S + 1 = 1$.

Let us consider that an unpolarized beam of neutrons is incident on a target containing protons with random orientiations of spins. Some of the neutrons will be scattered by protons with parallel spins orientations (triplet scattering) and the others will be scattered by protons with anti parallel spin orientations (singlet scattering). Wigner stated that the potentials of interaction in the above two cases are different. Let them be V_t and V_s. Both these potentials are strong, short-range and attractive and may be assumed to be rectangular with depths V_{ot} and V_{os} having range b_t and b_s respectively.

In n-p scattering experiment, the singlet and triplet state will occur in proportion to the statistical weight factors for these states which are $\dfrac{1}{4}$ and $\dfrac{3}{4}$ respectively. The total scattering cross-section therefore shall be made up of two parts, σ_t (the cross section for scattering in the triplet state) and σ_s (the scattering crossection in singlet spin) as follows:

$$\sigma = \frac{3}{4}\,\sigma_t + \frac{1}{4}\,\sigma_s \qquad\qquad \text{... (4.1)}$$

Obviously, the probability of forming a singlet state is only one in four. We know that under zero energy neutrons, the triplet scattering cross section can be written as $\sigma_t = 4\pi a_t^2$, where $a_t \approx +4.8$ fm is the triplet scattering length. Here the + sign with a_t indicates that the n-p interaction potential in the triplet case can produce a bound state of n-p system, i.e. the deuteron ground state. We can use calculated value of $\sigma_t \approx 2.33$ barns for zero energy and experimental information on σ_0 (zero energy cross-section) ≈ 20.36 barns to calculate σ_s. One obtains from (4.1)

$$20.36 = \frac{3}{4}\,(2.33) + \frac{1}{4}\,\sigma_{s,0}$$

or
$$\sigma_{s,0} = 74.45 \text{ barns} \qquad\qquad \text{... (4.1)}$$

which shows that $\sigma_{s,0}$ is very large as compared to $\sigma_{t,0}$ for slow neutrons, and although the singlet occurs once in four, it still contributes for most of the observed scattering cross section.

In analogy with the triplet case, one can always write

$$\sigma_s = 4\pi a_s^2 \qquad\qquad \text{... (4.2)}$$

where a_s is the singlet Fermi scattering length. One can easily estimate a_s, using the relation

$$\frac{a_s^2}{a_t^2} = \frac{\sigma_s}{\sigma_t} \approx \frac{74.45}{2.33}$$

$$\therefore \qquad \frac{a_s}{a_t} = \sqrt{\frac{74.45}{2.33}}$$

or
$$a_s = \sqrt{\frac{74.45}{2.33}} \times 4.8 \approx 24.3 \text{ fm} \qquad\qquad \text{... (4.3)}$$

One can not determine the sign of the singlet scattering length from the above considerations. If $a_s > 0$, then one may expect that a singlet bound state of n-p system (i.e., of deuteron) exist.

However, as is evident, its energy will be quite small for the relatively large value of a_s. We may note that for $a_s < 0$, the singlet state will not be bound state.

Singlet state energy

Let us obtian the energy expression for the low energy n-p scattering cross-section. This can be used for calculation of σ_t and σ_s.

In the previous section, we have obtained the conditions of matching the solutions of the radial wave equations for the deutron ground state as also for low energy n-p scattering as:

For the deutron: $k_1 \cot k_1 b = -\alpha t$

and for the scattering: $k_2 \cot k_2 b = k \cot (k_b + \delta_0)$

$$\alpha_t^2 = \frac{M}{\hbar^2} \, \varepsilon_t, \; k^2 = \frac{M}{\hbar^2} \, E$$

$$k_1 = \frac{M}{\hbar^2} \, (V_0 - \varepsilon_t) \text{ and } k_2^2 = \frac{M}{\hbar^2} \, (V_0 + E)$$

Here, we have $E_{Bd} = \varepsilon_t = \varepsilon_d$

Now, both ε_t and $E \ll V_0$, one can write

$$k_1^2 \approx k_2^2 \approx k_0^2 = \frac{M}{\hbar^2} \, V_0$$

From the above equations, one obtains

$$-\alpha_t = k \cot (k_b + \delta_0) \qquad \qquad \text{... (4.4)}$$

or $$\cot (k_b + \delta_0) = -\frac{\alpha_t}{k} \qquad \qquad \text{... (4.5)}$$

Now, considering the scattering of neutrons of energy $(E) = 1$ eV, one obtains

$$\frac{\alpha_t^2}{k^2} = \frac{E_{Bd}}{E} = 2.245 \times 10^6$$

or $$\frac{\alpha_t}{k} \approx 1.5 \times 10^3 \approx 1 \qquad \qquad \text{..(4.6)}$$

Obviously, $\cot(k_b + \delta_0)$ in (4.5) has a large negative value. This means

$$k_b + \delta_0 \approx \pi \qquad \qquad \text{... (4.7)}$$

For $b \approx 2$ fm, $\alpha_t b = 0.464$, one finds

$$k_b = \frac{\alpha_t b}{1.5 \times 10^3} = \frac{0.464}{1.5 \times 10^3} \qquad \qquad \text{... (4.8)}$$

Obviously $k_b \ll 1$. Comparing with (4.7), one can conclude that $kb \ll \delta_0$. This yields

$$\cot (k_b + \delta_0) = \cot \delta_0 = -\frac{\alpha_t}{k} \qquad \qquad \text{... (4.9)}$$

One then obtain the low energy triplet scattering cross section as

$$\sigma_t = \frac{4\pi}{k^2 + k^2 \cot^2 \delta_0} = \frac{4\pi}{k^2 + \alpha_t^2} = \frac{4\pi \hbar^2}{M} \left(\frac{1}{E + \varepsilon_d} \right) \qquad \dots (4.10)$$

Similar expression can also be wrtitten for singlet state cross-section σ_s for low energy neutrons:

$$\sigma_s = \frac{4\pi}{k^2 + \alpha_s^2} = \frac{4\pi \hbar^2}{M} \left(\frac{1}{E + \varepsilon_s} \right) \qquad \dots (4.11)$$

Here $\alpha_s^2 = \left(\dfrac{M}{\hbar^2} \right) \varepsilon_s$, ε_s being the energy of the singlet state of the n-p system. This is analogous to the triplet state energy $\varepsilon_d = E_{Bd}$.

Now, one can write the low energy n-p scattering cross-section as

$$\sigma = \frac{3}{4} \sigma_t + \frac{1}{4} \sigma_s = \frac{\pi \hbar^2}{M} \left(\frac{3}{E + \varepsilon_d} + \frac{1}{E + \varepsilon_s} \right) \qquad \dots (4.12)$$

From (4.12), one finds zero nergy scattering cross-section as

$$\sigma_0 = \frac{\pi \hbar^2}{M} \left(\frac{3}{\varepsilon_d} + \frac{1}{\varepsilon_s} \right) \qquad \dots (4.13)$$

Now, using the measured value of zero energy n-p scattering cross section, one can estimate the energy of the singlet state, ε_s. Taking $\varepsilon_d = 2.226$ MeV, and $\sigma_0 = 20$b in (4.13), one obtains

$$\varepsilon_s = 90 \text{ keV} \qquad \dots (7.14)$$

One can also obtain an estimate of the depth V_{os} of the single state potential. Obviously, V_{os} must be shallower than the triplet potential depth V_{ot}. The estimated values of the parameter for the rectangular triplet an singlet potentials are summarized in Table 4.1.

Table 4.1 Estimated values of the parameters for the rectangular triplet and singlet potentials

State	Potential depth (Rectangular)	Range of the potential
Triple ^3_1S	$V_{ot} \approx 38.5$ MeV	$b_t \approx 1.93$ fm
Single ^1_0S	$V_{os} \approx 14.3$ MeV	$b_s \approx 2.50$ fm

4.9 EFFECTIVE RANGE THEORY

Let us formulate the theory of scattering in a manner, that the range over which the potential is especially effective, comes explicit in the expression of the scattering cross-section along with the phase shift and scattering length. We may note that the theory is applicable only for low energies.

The low energy n-p scattering cross-section can be expressed as

$$\sigma = \frac{4\pi}{k^2 + \dfrac{1}{a_k^2}}$$

Where a_k is the scattering length for the energy E and $k^2 = \left(\dfrac{ME}{\hbar^2}\right)$. If one can express a_k as a function of k, then the energy dependence of the low energy scattering cross-section can be obtined.

Let us consider the radial equations for low energy n-p scattering at kinetic energies E_1 and E_2. Then the Schrodinger equation at these two energies will be

$$u_1'' + [k_1^2 - U(r)]\, u_1(r) = 0 \qquad\qquad \text{... (4.1a)}$$

$$u_2'' + [k_2^2 - U(r)]\, u_2(r) = 0 \qquad\qquad \text{... (4.1b)}$$

where $$k_1^2 = \frac{M\,E_1}{\hbar^2},\ k_2^2 = \frac{M\,E_2}{\hbar^2}\ \text{and}\ U(r) = \frac{MY(r)}{\hbar^2} \qquad\qquad \text{... (4.2)}$$

The subscripts 1 and 2 corresponding to E_1 and E_2 respectively. Multiplying (4.1a) by u_2'' and (4.1b) by u_1', one obtians

$$u_2\, u_1'' + k_1^2\, u_1 u_2 = U(r)\, u_1 u_2 \qquad\qquad \text{... (4.3a)}$$

$$u_1\, u_2'' + k_2^2\, u_1 u_2 = U(r)\, u_1 u_2 \qquad\qquad \text{... (4.3b)}$$

Subtracting (4.3b) from (4.3a), one obtains

$$u_2\, u_1'' - u_1\, u_2'' - \frac{d}{dr}(u_2\, u_1' - u_1\, u_2') = (k_2^2 - k_1^2) u_1 u_2 \qquad\qquad \text{... (4.4)}$$

where u_1' and u_2' denote the first order differentiation w.r.t. time and u_1'' and u_2'' denote the second order differentiation. Integrating the two sides, one obtains

$$\int_0^r \frac{d}{dr}(u_2\, u_1' - u_1\, u_2')\,dr = (k_2^2 - k_1^2)\int_0^r u_1 u_2\, dr \qquad\qquad \text{... (4.5)}$$

Since $$u_1(0) = u_2(0) = 0,\ \text{one obtains}$$

$$(u_2\, u_1' - u_1\, u_2')_r = (k_2^2 - k_1^2)\int_0^r u_1 u_2\, dr \qquad\qquad \text{... (4.6)}$$

Let us introduce an-auxillary function $V_1(r)$ which represents the asymptotic behaviour of u_1 at large distance, given by

$$V_1 = \frac{\operatorname{Sin}(k_1 r + \delta_1)}{\operatorname{Sin}\delta_1},\ V_2(r) = \frac{\operatorname{Sin}(k_2 r + \delta_2)}{\operatorname{Sin}\delta_2} \qquad\qquad \text{... (4.7)}$$

We must remember that actually the internal wave function u_1, will never be 1 at $r = 0$. We have only fixed the amplitude of external wave-function in such a manner that if extended back, its value is 1 at $r = 0$. The valur of $u_1(r)$ will be on the other hand, be zero at $r = 0$. V_1 and V_2 are normalized to unity at $r = 0$ so that $V_1(0) = 1$ and $V_2(0) = 1$, i.e., V_1 being only special case of u_1 at $r \to \infty$ where the potential $U(r) = o$. Equation (4.6) may be applied to V's also so that one can also write

$$V_2\, V_1'' - V_1 V_2'' = \frac{d}{dr}(V_2\, V_1' - V_1\, V_2') = (k_2^2 - k_1^2)\, V_1 V_2 \qquad\qquad \text{... (4.8)}$$

Integration yield

$$(V_2 - V_1' - V_1 V_2')_r - (V_2 V_1' - V_1 V_2')_0 = (k_2^2 - k_1^2) \int_0^r V_1 V_2 \, dr \qquad \ldots (4.9)$$

From (4.7), one obtains

$$V_1' = \frac{k_1 \cos(k_1 r + \delta_1)}{\sin \delta_1}, \quad V_2' = \frac{k_2 \cos(k_2 r + \delta_2)}{\sin \delta_2}$$

Then
$$V_1'(0) = k_1 \cos \delta_1 \text{ and } V_2'(0) = k_2 \cot \delta_2 \qquad \ldots (4.10a)$$

Now, one obtains

$$(V_2 V_1' - V_1 V_2')_r - (k_1 \cot \delta_1 - k_2 \cot \delta_2) = (k_2^2 - k_1^2) \int_0^r V_1 V_2 \, dr \qquad \ldots (4.11)$$

Pushing up the upper limits of intergration in (4.6) and (4.11) to $r = \infty$, one obtains

$$(u_2 u_1' - u_1 u_2')_\infty = (k_2^2 - k_1^2) \int_0^\infty u_1 u_2 \, dr \qquad \ldots (4.12a)$$

$$(V_2 V_1' - V_1 V_2')_\infty - (k_1 \cot \delta_1 - k_2 \cot \delta_2) = (k_2^2 - k_1^2) \int_0^\infty V_1 V_2 \, dr \qquad \ldots (4.12b)$$

At large r, $u_1(r)$ and $u_2(r)$ reduce to the asymptotic forms $V_1(r)$ and $V_2(r)$ respectively. Now, one obtains on substraction of (4.12a) from (4.12b).

$$(-k_1 \cot \delta_1 - k_2 \cot \delta_2) = (k_2^2 - k_1^2) \int_0^\infty (V_1 V_2 - u_1 u_2) dr \qquad \ldots (4.13)$$

Here $\rho(E_1, E_2) = \int_0^\infty (V_1 V_2 - u_1 u_2) dr$ is called the '*effective range*' for energies E_1 and E_2. The quantity $\rho(E_1, E_2)$ has the dimension of length, and is of the order of the effective range of nuclear potential, because outside the nuclear range $V_1, V_2 = u_1, u_2$ and the integral in (4.13) vanishes so that the integral is only effective inside the nuclear range. However, the integral is a difference quantity and quite sensitive to the properties of nuclear potential. We must remember that $v's$ also depend on the potential. Now, choosing the energy $E_1 = 0$ so that $k_1 = 0$. Also let $E_2 = E$ and $k_2 = k$. Then,

$$\lim_{k_1 \to 0} (k_1 \cot \delta_1) = -\frac{1}{a} \qquad \ldots (4.14)$$

where a is the Fermi scattering length. Moreover

$$k_2 \cot \delta_2 = k \cot \delta = -\frac{1}{a_k}$$

One then obtains

$$\frac{1}{a} - \frac{1}{a_k} = k^2 \int_0^\infty (V_0 V - u_0 u)\, dr \qquad \ldots (4.15)$$

or

$$\frac{1}{a_k} = \frac{1}{a} - k^2 \int_0^\infty (V_0 V - u_0 u)\, dr \qquad \ldots (4.16)$$

Both $u(r)$ and $V(r)$ depend on E and hence on k. Clearly, r.h.s. of (4.16) is in general energy dependent. Now, writing

$$\int_0^\infty (V_0 V - u_0 u)\, dr = \frac{1}{2} \rho(0, E) \qquad \ldots (4.17)$$

Where $\rho(0, E)$ depends on the energies $E_1 (= 0)$ and $E_2 (= E)$. Thus, one obtains

$$\frac{1}{a_k} = \frac{1}{a} - \frac{1}{2} \rho(0, E) k^2 \qquad \ldots (4.18)$$

Since for small r, $V(r) > E$, the wave functions u_0 and V_0 for zero energy will differ very little from u and V respectively for the energy E. Thus for small r, one can write $u(r) \approx u_0(r)$ and $V(r) \approx V_0(r)$. On the other hand for large r, i.e. beyond the range of n-p interation potential, one finds $u(r) \to V(r)$ and $u_0(r) \to V_0(r)$. Thus for some arbitrarily chosen range a of the potential, one finds

$$\int_0^a (V_0 V - u_0 u)\, dr = \int_0^a (V_0^2 - u_0^2)\, dr \qquad \ldots (4.19a)$$

and

$$\int_0^\infty (V_0 V - u_0 u)\, dr = 0 \qquad ..(4.19b)$$

Thus, one finally obtains

$$\int_0^\infty (V_0 V - u_0 u)\, dr = \int_0^\infty (V_0^2 - u_0^2)\, dr = \frac{1}{2} \rho(0, 0) = \frac{1}{2} r_0 \qquad \ldots (4.20)$$

Where r_0 is a constant and known as the *effective range of interaction* between n and p.

Thus from (4.18), one obtains

$$\frac{1}{a_k} = \frac{1}{a} - \frac{1}{2} r_0 k^2 \qquad \ldots (4.21)$$

Relation (4.21) is an important relation between the phase shift, scattering length and the effective range. We may note that effective range contains the properties of the potential $U(r)$; so also δ is related to $U(r)$. For higher approximations, one expands u and V around $k^2 = 0$ so that

$$u(r) = u_0(r) + k^2 \left[\frac{du(r)}{dk}\right] + \ldots$$

$$V(r) = V_0(r) + k^2 \left[\frac{dV(r)}{dk} \right] + \ldots \qquad \ldots (4.22)$$

Remebering that at $r = 0$

$$u_0(0) = 0, \quad \frac{du}{dk} = 0; \quad V(0) = 1 \text{ and } \frac{dV}{dk} = 0 \qquad \ldots (4.23)$$

One can write

$$V(r) \underset{k \to 0}{=} \frac{\text{Sin}(kr + \delta)}{\text{Sin}\,\delta} = 1 + kr \cot \delta$$

$$= 1 + \left[-\frac{1}{a} + \frac{1}{2} k^2 \, \rho(0, E) \right] \qquad \ldots (4.24)$$

where
$$\rho(0, E) = 2 \int_0^\infty [V_0(r)\,V(r) - u_0(r)\,u(r)]dr$$

$$= \int_0^\infty [V_0^2(r) - u_0^2(r)]dr + 2k \int_0^\infty \left[V_0(r) \left(\frac{dV}{dk} \right)_{k=0} - u_0 \left(\frac{du}{dk} \right)_{k=0} \right] dr \ \ldots (4.25)$$

Now, defining P as:

$$P = -\frac{1}{r_0^3} \int_0^\infty [V_0(r)V_0'(r) - u_0(r)\,u_0'(r)]dr \qquad \ldots (4.26)$$

$$\left[\text{Where } V_0'(r) = \left(\frac{dV}{dk} \right)_{k=0} \text{ and } u_0'(r) = \left(\frac{du}{dk} \right)_{k=0} \right]$$

one obtains

$$\rho(0, E) = r_0 - 2P\,k^2\,r_0^3 \qquad \ldots (4.27a)$$

and
$$k \cot \delta = \frac{1}{\alpha} + \frac{1}{2}\,k^2\,r_0 - P\,k^4\,r_0^3 \qquad \ldots (4.27b)$$

We may note that Equation (4.21), which does not involve P is less sensitive to the exact shape of the potential chosen, than Equation (4.27b). This is why equation (4.21) is called *shape independent equation.*

Using (4.21), one can write (4.6) for low energy n-p scattering cross-section for triplet and singlet scattering as

$$\sigma_t = \frac{4\pi}{k^2 + \left(\dfrac{1}{a_t} - \dfrac{1}{2} r_{ot}\, k^2 \right)^2} \qquad \ldots (4.28a)$$

$$\sigma_s = \frac{4\pi}{k^2 + \left(\dfrac{1}{a_s} - \dfrac{1}{2}r_{os}k^2\right)^2} \qquad \text{... (4.28b)}$$

Where r_{ot} and r_{os} are the effective ranges for the triplet and singlet interactions respectively. One obtians the total cross-section for low energy n-p scattering from Equation (4.1) of sec. 8 as:

$$\sigma = \pi \left[\frac{3}{k^2 + \left(\dfrac{1}{a_t} - \dfrac{1}{2}r_{ot}k^2\right)^2} + \frac{1}{k^2 + \left(\dfrac{1}{a_s} - \dfrac{1}{2}r_{os}k^2\right)^2} \right] \qquad \text{... (4.29)}$$

Equation (4.29) has been used to analyse the scattering data upto neutron energy of 10 MeV. The parameters a_t, a_s, r_{ot} and r_{os} needed for the analysis have been determined accurately in various experiments.

Effective Range Theory and Deutron Problem

One can relate the above equations with the parameters of deutron problem. The radial equation of the ground state of the deutron is

$$U_0'' - \left[\frac{ME}{\hbar^2} + \frac{M}{\hbar^2} U(r) \right] u_0(r) = 0 \qquad \text{... (4.30)}$$

writing

$$k^2 = \frac{M}{\hbar^2} E = \frac{M}{\hbar^2}(-\varepsilon_d) = -\alpha^2 \qquad \text{... (4.31)}$$

$$\therefore \qquad k = i\alpha = i\sqrt{\frac{M\,\varepsilon_d}{\hbar^2}} \qquad \text{... (4.32)}$$

The wave equation for the outside region then becomes

$$u_0'' + k^2 u_0 = u_0'' - a^2 u_0 = 0 \qquad \text{... (4.33)}$$

The solution of (4.33) is

$$u_0 = B \left\{ \frac{\exp[i(kr + \delta)] - \exp[-i(kr + \delta)]}{2i} \right\}$$

$$= \frac{B}{2i} \{ \exp(i\delta)\exp(-\alpha r) - \exp(-i\delta)\exp(\alpha r) \} \qquad \text{... (4.34)}$$

For the bound state the coefficient of $\exp(\alpha r)$ must be zero. Thus $\exp(-i\delta) = 0$. This means that $(i\delta)$ must be large real positive number. Clearly, this would make $B\exp(i\delta)$ to be large, unless B is made small. Clearly $B\exp(i\delta)$ remains finite. One obtains in this case $\cos\delta = i$ and

$$k\cot\delta = ik = -\alpha \qquad \text{... (4.35)}$$

Equation (4.13) with $k_1 = 0$ and $k_2 = k = i\alpha$ then gives

$$\frac{1}{a} + k\cot\delta = \frac{1}{a} - \alpha = -\alpha^2 \int\limits_0^\infty (V_0 V - u_0 u)\,dr \qquad \text{... (4.36)}$$

In the approximation of shape independence, the above reduces to

$$\alpha = \frac{1}{a} + \frac{1}{2} \, r_0 \, \alpha^2 \qquad\qquad \dots (4.37)$$

To a first approximation, if one neglects $\frac{1}{2} \, r_0 \, \alpha^2$ in (4.37) and $\frac{1}{2} \, r_0 \, k^2$ in (4.21), one finds

$$\alpha \approx \frac{1}{a_k} \approx \frac{1}{a}$$

This gives the total (triplet) scattering cross-section as

$$\sigma_t = \frac{4\pi}{k^2 + \alpha^2}$$

Which is the same as obtained in the previous section.

4.10 EFFECT OF CHEMICAL BINDING ON n-p SCATTERING

So far we have discussed n-p scattering by free protons. However, protons are not found free in nature and in all experiments on n-p scattering, the target protons are always bound in molecules, either in H_2 or in paraffin. In such cases, a number of different kinds of effects arise. We will discuss them:

(i) **Reduced Mass effect** When neutrons of sufficiently high energy are incident on a hydrogenous substance usually the chemical bond in the molecule is broken and the scattering is essentially from free protons. However, when the neutron energy is very low, such bond rupture may not be possible and scattering takes place from the entire molecule. Thus, one has to take into account the reduced mass of the neutron-molecule system for calculating scattering cross-section.

In order to calculate scattering cross-section, Fermi applied the quantum mechanical Born approximation. Usually the Born approximation is applicable when the interaction potential is small compared to the energy. For low n-p scattering ($E < 10$ MeV), this condition is obviously not satisfied as the interaction potential is of the order of 38 MeV for triplet and about 14 MeV for singlet scattering (when the potential is rectangular well). However, Bethe has shown that one can apply Born approximation in this case also since the actual potential may be replaced by another much shallower potential well with a relatively longer range. This gives the same result as the former.

Using Born approximation, the differential scattering cross-section obtained is

$$\sigma(\theta) = \text{constant} \times \mu^2 \left| \int \psi_f^* \, V \, \psi_i \, d\tau \right|^2 \qquad\qquad \dots (4.1)$$

Here μ is reduced mass, V is interaction potential, ψ_i and ψ_f are the wave functions for the initial and final states respectively. The quantity within the sign of modulus represent the matrix element of the transition between the initial (i) and the final (f) states. However, this is independent of the mass of the scattering molecule. We may note that the reduced mass μ is different for different molecules. In case of scattering from free protons

$$\mu_f = \frac{M}{2} \qquad \qquad \text{... (4.2)}$$

and scattering from a molecule of mass M'

$$\mu_m = \frac{MM'}{M + M'} \qquad \qquad \text{... (4.3)}$$

In case of scattering by H_2 molecules, $M' = 2M$, $\therefore \mu_m(H_2) = \frac{2M}{3}$. In case of scattering by a heavy paraffin molecule $M' \gg M$. Thus $\mu_m = M$. Obviously, the ratio of the cross-section for scattering from a heavy paraffin molecule to that by a free proton is

$$\frac{\sigma_m}{\sigma_f} = \left(\frac{\mu_m}{\mu_f} \right)^2 = 4 \qquad \qquad \text{... (4.4)}$$

The above effect is observed is scattering experiments. The extrapolated value of the zero energy n-p scattering cross-section is reported to be about 80 barns when the target protons are embedded in heavier paraffin incoherent scattering cross-section by free protons, $\sigma_f \approx 20$ barns.

The criterion for deciding when a proton may be regarded as free and when bound, is that for $E < h\nu$, the proton may be considered as bound. Here ν is the vibrational frequency of the proton in the subgroup of the molecule. For the case of the C H bond in paraffin $h\nu \approx 0.4$ eV.

The variation of the ratio $\dfrac{\sigma_m}{\sigma_f}$ with neutron energy is shown in Figure 4.15. The neutron will not be able to excite the vibrational levels of the molecule for $E < h\nu$. Thus, there will be no energy loss in the collision between the neutron and the molecule

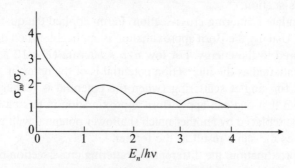

Fig. 4.15 Variation of the ratio of the scattering cross section of neutrons from protons bound in a molecule (σ_m) to that from free protons (σ_f), i.e. $\dfrac{\sigma_m}{\sigma_f}$ with neutron energy.

We note that the ratio $\dfrac{\sigma_m}{\sigma_f}$ goes down as E increases above zero upto $E = h\nu$, when an abrupt rise in the ratio is observed. In this case, the incident neutron loses one quantum of energy to excite the first vibrational state within the molecule. Further, similar breaks occur at energies $2h\nu$, $3h\nu$,... etc. In the case for E much greater than the highest vibrational energy of the proton in the molecule, i.e. above the dissociation energy the proton may be considered as free, so that $\dfrac{\sigma_m}{\sigma_f} \to 1$.

Neutrons with $E < h\nu$ cannot be easily slowed down by collisions with protons within molecules, as neutrons cannot transfer any energy to the latter by inelastic collisions to excite the first vibrational level. However, they can lose energy to vibrations in the whole CH_2 groups which have smaller quantum energies. Concluding, one may say that the neutrons are easily slowed down, i.e. cooled to room temperature $\left(\sim \dfrac{1}{40} \text{ eV} \right)$ but cannot be easily cooled to 20 K or lower temperature.

(ii) **Scattering of Neutrons by ortho and para hydrogen** A hydrogen molecule can exist in two possible spin states, one in which the spins of two protons are parallel ($S = 1$) is known as *ortho hydrogen*, which have the statistical weight $2S + 1 = 3$. The other is known as para hydrogen, where the spins of the protons are antiparallel ($S = 0$). The statistical weight of the *para hydrogen* molecule is $2S + 1 = 1$. At normal temperatures, say around room temperatures, the ordinary hydrogen gas consists of an equilibrium mixture of ortho and para hydrogen in the ratio of 3 : 1. Though the para-hydrogen molecules are in the lower energy state than that for the ortho hydrogen the transition from ortho to para state is very ·slow. However, at very low temperatures ($T < 20$ K), most of the ortho hydrogen molecules go over to para state in the presence of suitable catalyst and one finds that all ortho-molecules are converted into para-hydrogen molecules and almost pure para-hydrogen is obtained.

In a hydrogen molecule, the scattering of low energy neutrons is coherent and one could observe the interference effects in the scattered beam because of large wavelength of incident neutrons compared to the distance between the two protons. However, the process of neutron scattering is analogous to the diffraction of light from a double slit where the width is much smaller than the wavelength of light.

The total wave function of a hydrogen molecule is written as

$$\psi = \psi_{el} \times \psi_{\upsilon} \times \psi_{rot} \times \psi_{s} \qquad \qquad \text{... (4.1)}$$

wavefunction ψ must be antisymmetric in the interchange of two protons as these obey F. D. statistics. For hydrogen molecule, the electronic wave function is symmetric. Moreover, vibrational wave function ψ_{υ} is always symmetric. Clearly, the symmetry of ψ is determined by the symmetry properties of the nuclear spin function, ψ_{s} and the rotational function ψ_{rot}. One can easily prove that ψ_{s} is symmetric for $S = 1$ (ortho-molecules) and anti-symmetric for $S = 0$ (para-molecules). Obviously, ψ_{s} for the ortho-molecules must be combined with antisymmetric rotational functions for which the rotational quantum number j is odd, i.e. $j = 1, 3, 5, ...$ etc. Conversely, one must combine ψ_{s} for the para molecules with symmetric rotational functions for which j is even, i.e. $j = 0, 2, 4, ...$ etc. Clearly, the lowest energy level thus belongs to para-hydrogen, the rotational energy being proportional to $\dfrac{j(j+1)}{2I}$, where I denotes the moment of inertia.

At normal room temperature, the ordinary hydrogen is an equilibrium mixture of ortho and para molecules in the ratio 3 : 1. Although the para-molecules are in the lower energy state, there is normally no transition from ortho to para state. In the presence of a suitable catalyst at very low temperatures ($T < 20$ K) such transitions may occur and in this case all ortho-molecules are converted into para-hydrogen.

The experiments on scattering of slow neutrons by pure para hydrogen and also by an equilibrium mixture of ortho and para hydrogen (in the ratio of 3 : 1) would give scattering cross sections for ortho and para hydrogen, from which one can calculate the corresponding scattering lengths.

In a hydrogen molecule, the scattering of low energy neutrons is coherent and one could observe the interference effects in the scattered bem because of the large wavelength of incident neutrons compared to the distance between the two protons. We may note that the process of neutron scattering is analogous to the diffraction of light from a double slit where the width is much smaller than the wave length of light. If the neutron kinetic energy is E, then the de Broglie wavelength is

$$\lambda = \frac{h}{p} = \frac{h}{\sqrt{2ME}} = \frac{0.288}{\sqrt{E}}$$

where λ is in $A^\bullet$ and E is in eV. Now, if the neutrons are in thermal equilibrium with the scatterer at the emperature T, their mean kinetic energy in eV is

$$E = kT = 8.625 \times 10^{-5} \, T$$

Thus, one finds

$$\lambda = \frac{30.93}{\sqrt{T}}$$

The values of neutron wavelengths for different values of E and T are given in Table 4.2

Table 4.2 Values of neutron wave lengths for different E and T values

$T(k)$	$E(eV)$	$\lambda(A)$
–	100	0.0288
–	1	0.288
300	0.026	1.8
20	17.25×10^{-4}	–7

In the hydrogen molecule, the separation between the two protons is 0.74 A. Obviously, if the scattering experiment is performed by cooling the hydrogen to 20 K or less, the de Broglie wave length of the incident neutron (7A) is about 10 times larger than the distance between the two protons in the molecule. Clearly, the situation is analogous to the diffraction of light from a double slit in which the slit separation is small compared to the wavelength. Thus the resultant scattered wave in this case can be obtained by simply adding up the amplitudes of the waves scattered from the two protons without consideration of the phases. In the limit of zero energy the scattered amplitude is $\lim_{k \to 0} f(\theta) = -a$, where a is the scattered of a low energy neutrons from the two protons in hydrogen molecule can be expressed as

$$A = a_1 + a_2 \qquad\qquad\qquad ...\,(4.2)$$

Where a_1 and a_2 are scattering lengths for scattering from protons designated as 1 and 2 respectively. a_1 and a_2 can be either a_t or a_s depending on whether the neutron spin s_n is parallel or antiparallel to the proton spin s_p. Thus in the case of scattering from ortho and para H_2 molecules, one finds that the following situations may arise (Figure 4.16). From Figure (4.16) it is clear that in the case of ortho hydrogen scattering the incident neutron spin s_n is either parallel to both proton spins (s_{p_1} and s_{p_2}) as in (a) or antiparallel to both, as in (b). In the first case, one finds that there is triplet scattering from both, with the scattering lengths at a_{t_1} and a_{t_2} repectively. In the second case, one finds that there is singlet scattering from both the protons with the scattering lengths a_{s_1} and a_{s_2} respectively.

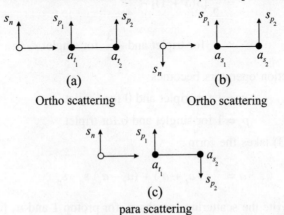

(a) Ortho scattering
(b) Ortho scattering
(c) para scattering

Figure 4.16 Scattering of neutrons from ortho and para hydrogen.

In the case of para scattering as shown in Figure (c) there is triplet scattering from one proton and singlet scattering from the other, with the scattering lengths a_{s_1} and a_{s_2} respectively. One can combine the triplet and singlet scattering lengths in a single expression in terms of projection operators as

$$a = a_t \, p_t + a_s \, p_s \qquad \qquad \ldots (4.3)$$

Where

$$p_t = \frac{3}{4} + s_n \cdot s_p \qquad \qquad \ldots (4.4)$$

and

$$p_s = \frac{1}{4} - s_n \cdot s_p \qquad \qquad \ldots (4.5)$$

For a n-p scattering, the total spin is algebraic sum, i.e. $s = s_n + s_p$. Thus

$$S^2 = s_n^2 + s_p^2 + 2 s_n \cdot s_p \qquad \qquad \ldots (4.6)$$

Since $S = 1$ for triplet and $S = 0$ for singlet state. We have the eigen (expectation) values of S^2 are

$$<S^2> = S(S + 1) = 2 \text{ for triplet}$$

$$= 0 \text{ for singlet}$$

Similary, one obtains

$$<s_n^2> = <s_p^2> = \frac{1}{2}\left(\frac{1}{2}+1\right) = \frac{3}{4} \qquad ... (4.7)$$

From Equation (4.6), one obtains

$$<s_n \cdot s_p> = \frac{1}{2}[<S^2> - <s_n^2> - <s_p^2>]$$

$$= \frac{1}{2}[S(S+1)] - \frac{3}{4}$$

$$= \frac{1}{4} \text{ for triplet and } -\frac{3}{4} \text{ for singlet}$$

Now, the projection operators becomes $\qquad ... (8.8)$

$$p_t = 1 \text{ for triplet and } 0 \text{ for singlet}$$

and $\qquad p_s = 1$ for singlet and 0 for triplet

Using (4.9), (4.3) takes the form $\qquad ..(4.9)$

$$a = \frac{1}{4}(3a_t + a_s) + (a_t - a_s)\, s_n \cdot s_p \qquad ..(4.10)$$

Now, one can write the scattering lengths a_1 for proton 1 and a_2 for proton 2 as

$$a_1 = \frac{1}{4}(3a_t + a_s) + (a_t - a_s)\, s_n \cdot s_{p_1}$$

$$a_2 = \frac{1}{4}(3a_t + a_s) + (a_t - a_s)\, s_n \cdot s_{p_2}$$

and $\qquad A = a_1 + a_2 = \frac{1}{2}(3a_t + a_s) + (a_t - a_s)\, s_n \cdot (s_{p_1} + s_{p_2})$

$$= \frac{3a_t + a_s}{2} + (a_t - a_s)\, s_n \cdot s_H \qquad ... (4.11)$$

where $\qquad s_H = s_{p_1} + s_{p_2} \qquad ... (4.12)$

In case of para hydrogen, the spins of the two protons are antiparallel. Thus

$$s_H = s_{p_1} + s_{p_2} = 0 \qquad ..(4.13)$$

$$\therefore \qquad A_{\text{para}} = \frac{3a_t + a_s}{2} \qquad ..(4.14)$$

The scattering cross-section for para hydrogen molecule is obtained as

$$\sigma_{\text{para}} = 4\pi A^2 = \frac{4\pi(3a_t + a_s)^2}{4} = \pi(3a_t + a_s) \qquad ... (4.15)$$

The scattering cross section for ortho hydrogen molecule is obtained as

$$\sigma_{ortho} = 4\pi A^2 = 4\pi \left[\frac{(3a_t + a_s)^2}{2} + (a_t - a_s)(\mathbf{s}_n \cdot \mathbf{s}_H) \right]^2$$

$$= 4\pi \left[\left(\frac{3a_t + a_s}{2} \right)^2 + (a_t - a_s)^2 (\mathbf{s}_n \cdot \mathbf{s}_H)^2 + (3a_t + a_s)(a_t - a_s)(\mathbf{s}_n \cdot \mathbf{s}_H) \right] \quad \dots (4.16)$$

Since the incident neutron beam being unpolarised, the neutron spin must be averaged over all possible polarisation of the incident beam. Making use of the Pauli matrices, averaging over all possible orientations of the intrinsic spin of the neutron yield

$$(\mathbf{s}_n \cdot \mathbf{s}_m)_{av} = 0$$

Now, $$(\mathbf{s}_n \cdot \mathbf{s}_m)^2_{av} = [(s_{nx} + s_{ny} + s_{nz})(s_{Hx} + s_{Hy} + s_{Hz})]^2_{av}$$

$$= (s^2_{nx} s^2_{Hx} + s^2_{ny} s^2_{Hy} + s^2_{nx} s^2_{Hz}) + \text{cross terms}$$

According to the properties of Pauli matrices each of the cross terms will average to zero, e.g. terms of the type:

$$(s_{nx} s_{ny} s_{Hx} s_{Hy})_{av} = 0$$

and $$(s^2_{nx})_{av} = (s^2_{ny})_{av} = (s^2_{nz}) = \frac{1}{4}$$

$$(\mathbf{s}_n \cdot \mathbf{s}_H)^2_{av} = \frac{1}{4}(s^2_{Hx} + s^2_{Hy} + s^2_{Hz})$$

$$= \frac{1}{4}(s^2_H)_{av} = \frac{1}{4} s_H(s_H + 1)$$

$$= \frac{1}{4} \times 2 = \frac{1}{2} \quad \dots (4.17)$$

Similarly, for ortho-hydrogen,

$$s_H = s_{p_1} + s_{p_2} = 1 \quad \dots (4.18)$$

Using Equation (4.17) with the result $(\mathbf{s}_n \cdot \mathbf{s}_H)_{av} = 0$, the equation (4.16) becomes

$$\sigma_{ortho} = 4\pi \left[\frac{1}{4}(3a_t + a_s)^2 + \frac{1}{2}(a_t - a_s)^2 \right] \quad \dots (4.19)$$

where $$A^2_{ortho} = \frac{1}{4}(3a_t + a_s)^2 + \frac{1}{2}(a_t - a_s)^2 \quad \dots (4.20)$$

Using Equation (4.15) in Equation (4.19), one obtains

$$\sigma_{ortho} = \sigma_{para} + 2\pi(a_t - a_s)^2 \quad \dots (4.21)$$

The second term in σ_{ortho} Equation (4.21) arises due to interference. Taking the ratio of σ_{ortho} to σ_{para}, one obtains

$$\frac{\sigma_{ortho}}{\sigma_{para}} = 1 + \frac{2(a_t - a_s)^2}{(3a_t + a_s)^2} \quad \dots (2.22)$$

The value of this ratio depends on the sign of a_s, the sign of a_t being positive. The experimental values of the scattering cross sections are, $\sigma_{ortho} = 125$ barns and $\sigma_{para} = 4$ barns. If $a_s > 0$, the second term on the r.h.s. of Equation (4.22) will be very small, then $\sigma_{ortho} \approx \sigma_{para}$. Even if $a_s = a_t$, $\sigma_{ortho} = \sigma_{para}$. In either situations, one finds that the results obtained are contradictory to experimental values. Now, if $a_s < 0$, i.e. if singlet scattering length is negative, $|3a_t + a_s|$ will be much smaller than $|a_t - a_s|$, since a_t is positive obviously, σ_{ortho} will then be much greater than σ_{para}, as required by the experimental values as well. This means that singlet scattering length a_s should be negative. The small experimental value of σ_{para} also reveals that a_t and a_s are of opposite signs. Clearly, the negative scattering length a_s confirms the Wigner's hypothesis that n-p interactions are spin dependent. The singlet state of deutron is a virtual state, i.e. unbound state. The values of scattering lengths obtianed experimentally are:

$$a_t = 5.37 \text{ F} \approx 5 \text{ F}$$

and

$$a_s = -23.73 \text{ F} \approx -24 \text{ F}$$

Using these values of a_t and a_s, one obtains the scattering lengths for ortho and para hydrogen molecules as

$$A_{ortho} = 20.92 \text{ F and } A_{para} = -3.81 \text{ F}$$

We may note that the said calculations of the scattering lengths of para and ortho hydrogen are based on the assumption that the protons are free. However, in the actual molecule, the protons are bound as such for close comparison with the experimental results. This means that the calculations have to be modified, because of the changed reduced mass. For the n-p system, the reduced mass is twice when the protons are bound than when the protons are free. The thermal motion of the molecules and the presence of certain amount of the ortho component also introduces inaccuracy in the experimental results. We may note that the more accurate experiments for the scattering lengths for singlet are: (i) the coherent scattering of slow neutrons by hydrogen atoms in crystals and (ii) the reflection of slow neutrons by liquid hydrocarbon mirrors. The value of A_{para} obtained from the reflection method is

$$A_{para} = -3.78 \text{ F}$$

We may note that the comparison of ortho and para-hydrogen cross sections for neutron also confirm that the neutron spin $s_n = \dfrac{1}{2}$.

Reflection of neutrons from a liquid mirror

One can determine the coherent scattering length a_{para} given by Equation (4.14) quite accurately by experiments on the total reflection of a neutron beam from a liquid mirror. The idea has been dervied from refraction of x-rays. The refractive index n in accordance with Sellmeier's equation is given by

$$n^2 = 1 - \frac{Ne^2}{4\pi^2 \, \varepsilon_0 \, m} \frac{1}{v^2 - v_0^2} \qquad \qquad \text{... (4.23)}$$

Where ν is the frequency of the radiation and ν_0 is the frequency of the classical oscillators in the medium, which are set into vibration by the electric vector of the incident e.m. wave, and ε_0 is the permittivity of empty space. e and m are charge and mass of the electron respectively. We have $\lambda = \dfrac{c}{\nu}$, we find, since $\nu >> \nu_0$ for x-rays

$$n^2 = 1 - \frac{Ne^2}{4\pi^2 \, \varepsilon_0 \, m\nu^2} = 1 - \frac{Ne^2 \, \lambda^2}{4\pi^2 \, \varepsilon_0 \, mc^2}$$

$$= 1 - \frac{N \, r_0 \, \lambda^2}{\pi}$$

where $r_0 = \dfrac{e^2}{4\pi \, \varepsilon_0 \, mc^2}$ is called the classical radius of the electron. On substituting the numerical values, the second term of the above expression gives value $\sim 10^{-6}$ Thus, one finds

$$n \approx 1 - \frac{N \, r_0 \, \lambda^2}{2\pi} \qquad \qquad ... (4.24)$$

In analogy with the above, one can write the expression for the refractive index of the neutron waves of de Broglie wavelength λ incident on a medium for which the scattering length is a as

$$n = 1 - \frac{N \, a \, \lambda^2}{2\pi} \qquad \qquad ... (4.25)$$

Since $a \sim 10^{-14}$ m, $\lambda \sim 10^{-10}$ m and $N \sim \dfrac{10^{28}}{m^3}$, and therefore

$$\frac{Na \, \lambda^2}{2\pi} \sim \frac{10^{-6}}{2\pi} << 1$$

So $n^2 \sim 1$. Since n is slightly less than 1 and thus the above gives rise to the possibility of *total external reflection*, as in the case of X-rays. Now, if θ_c be the critical angle of reflection, then we have

$$n = \sin \theta_c = \cos \varepsilon_c$$

where $\varepsilon_c = \dfrac{\pi}{2} - \theta_c$ is the grazing angle of the incidence for total reflection. Now, writing

$$n = 1 - \frac{\varepsilon_c^2}{2} \qquad \qquad ... (4.26)$$

one obtains

$$\varepsilon_c^2 = \frac{Na \, \lambda^2}{\pi} = \frac{N \left(\dfrac{\sigma}{4\pi} \right)^{\frac{1}{2}} \lambda^2}{\pi} \qquad \qquad ... (4.27)$$

Here σ is the coherent scattering cross-section for a liquid hydrogen mirror containing pure para-hydrogen ($T = 20.4$ K).

One can write for the product Na,

$$\sum_i N_i \, a_i = N_s \, a_s + N_t \, a_t \qquad \qquad \dots (4.28)$$

where $a_s \approx -24$ fm, $a_t \approx 5$ fm and $\dfrac{N_s}{N_t} \approx \dfrac{1}{3}$. Hence $\sum_i N_i \, a_i < 0$ which makes $n > 0$ so that no total reflection takes place from the liquid hydrogen mirror. There are good number of heavier nuclei for which $a > 0$. Clearly, by suitable choice of an organic liquid containing large percentage of such nuclei mixed with hydrogen, one can make $\sum_i N_i \, a_i > 0$ and $n < 1$. This means, for such a liquid total reflection may take place. In actual experiments, liquid hydrocarbons like cyclohexane ($C_6 H_{10}$), triethyl benzene ($C_{12} H_8$) and a mixture of two with a variable ratio was used. It is found that the critical grazing angle ε_c varies with this ratio. By measuring the ratio for which ε_c becomes zero it is possible for one to determine the coherent scattering length which come out to be $a_{coh} = a_{para} = -3.78$ F.

Scattering parameters from low energy data

We have seen that there are four parameters a_t, a_s, r_{ot} and r_{os} sufficient to analyze the low energy data-one can determine these parameters by a number of measurments involving the coherent scattering length (a_{para}), zero energy incoherent scattering cross section (σ_0), deutron binding energy (E_{Bd}) and the scattering cross section at a known low energy. Then using appropriate relations derived earlier, these four parameters can be determined accurately. Table 4.3 summarizes the latest values in 10^{-15} m

Table 4.3

Two nucleon system	a_t	a_s	r_{ot}	r_{os}
np	5.245 ± 0.0014	-23.714 ± 0.013	1.749 ± 0.008	2.73 ± 0.03
pp	—	$-7.82 \pm 0.04*$	—	2.830 ± 0.017
nn	—	-17.4 ± 1.8	—	2.4 ± 1.5

* When Coulomb correction is taken into account, it becomes about -17 F

4.11 NON-CENTRAL FORCE (TENSOR FORCE)

In the earlier sections we have pointed out that the theoretical description of the deutron based on central force characterstic of the two body nuclear interaction failed to produce its experimentally observed magnetic moment and a quadrupole moment. One can remove this discrepancy by adding a non-central (tensor) component to the central force. We have used the similar consideration to explain the difference in scattering of a neutron by para-hydrogen (anti-parallel spin) and ortho-hydrogen (parrallel spins). Clearly, the nuclear force should not only depend on the relative position vector **r** but also on the relative orientation of spins, i.e. on the angle between the spin directions and the line joining the two particles. If we have to construct a non-central potential, then it must satisfy the following conditions:

(i) Under rotation of coordinates system used to describe the particles, it should be invariant.

(ii) Under reflection of coordinates system used to describe the particles (parity transformation), it should be invariant. Obviously, the potential must be scalar. It has, however to retain the basic requirement that its derivative should yield nuclear force and further the force should be independent of velocity.

Let us construct a tensor operator satisfying the said restrictions. let the relative position vector be **r** and the spin vectors of the two particles are σ_1 and σ_2 which satisfy Pauli matrix algebra. Moreover, under reflection σ_1 and σ_2 are invariant (space inversion but not invariant under rotation). $\sigma_1 \cdot \sigma_2$, however is invariant under rotation. Under reflection, the position vector **r** is not invariant. Thus in the tensor potential r should occur an even number of times. Clearly, σ_1 and σ_2 should appear in a bilinear combination. By using spin identities one can use any polynomial in σ to an expression linear in σ. Thus the only linearly independent quantities which one can construct from r, σ_1 and σ_2 are $(\sigma_1 \cdot \sigma_2)$ and $(\sigma_1 \cdot \mathbf{r})(\sigma_2 \cdot r)$. It is found that the tensor component of the potential, then have the form

$$V_T = V(r)\, S_{12}(r) \qquad \qquad \dots (4.1)$$

Where S_{12} denotes the dependence on the angle between the vector **r** and the spin directions of the nucleons. The tensor operator S_{12} should be such that its average over all directions vanishes. S_{12} is defined as

$$S_{12} = \frac{3}{r^2}(\sigma_1 \cdot \mathbf{r})(\sigma_2 \cdot \mathbf{r}) - \sigma_1 \cdot \sigma_2 \qquad \qquad \dots (4.2)$$

$$= 3(\mathbf{s}_1 \times \hat{r})(\sigma_2 \cdot \hat{r}) - (\sigma_1 \cdot \sigma_2) \qquad \qquad \dots (4.3)$$

Where $\hat{r}\left(= \dfrac{\mathbf{r}}{|\mathbf{r}|}\right)$ is the unit vector.

In Equations (4.2) and (4.3), the second term is used so that the average of S_{12} over all directions of r becomes zero.

We have $\qquad \dfrac{1}{4\pi}\displaystyle\int (\sigma_1 \cdot \mathbf{r})(\sigma_2 \cdot \mathbf{r})\, d\Omega = \dfrac{1}{3}(\sigma_1 \cdot \sigma_2) \qquad \qquad \dots (4.4)$

Hence the average of S_{12} over all directions of r reduces to zero, i.e.

$$<S_{12}> = \frac{1}{3}(\sigma_1 \cdot \sigma_2) \cdot 3 - (\sigma_1 \cdot \sigma_2) = 0$$

Wigner in 1941 showed that if the n-p interaction is invariant under displacement, rotation and inversion of the observer's coordinate system as well as independent of the velocities of the particles, the most general form of two body nuclear potential satisfying the above requirements can be expressed as

$$V(r) = V_R(r) + V_\sigma(r)\, \sigma_1 \cdot \sigma_2 + V_T(r)\, S_{12} \qquad \qquad \dots (4.5)$$

The first two terms correspond to central potential while the last term the non-central potential. We can see that the second term depends only on the directions of spins of the two particles $\sigma_1 \cdot \sigma_2 = +1$ for triplet state (spins antiparallel), assuming rectangular well. Accordingly, one can express the central potential for the two states as

For triplet state $V_T(r) = V_R(r) + V_\sigma(r)$

and for singlet state $V_S(r) = V_R(r) - 3V_\sigma(r)$

one can estimate $V_R(r)$ and $V_\sigma(r)$ for rectangular potential well (n-p) scattering) from the theoretically obtained values of $V(r)$ as

$$V_T(r) = V_R(r) + V_\sigma(r) = 38 \text{ MeV}$$

$$V_\sigma(r) = V_R(r) - 3\,V_\sigma(r) = 14 \text{ MeV}$$

This gives $V_R(r) = 32$ MeV and $V_\sigma(r) = 6$ MeV. Since different shapes of potentials fit the experimentally known parameters, e.g., binding energy, magnetic moment and quadrupole moment of deutron and also low energy n-p and p-p scattering. However, one cannot made the uniquely the estimation of the coefficients $V_R(r)$ and $V_\sigma(r)$.

One may easily see qualitatively by noting that in the single spin state, there is no preferred direction, hence we expect S_{12} in the singlet state to be zero. Obviously, in the singlet state, the nuclear forces are only central and tensor force is zero.

For triplet state, i.e. for $S = 1$, there are three states possible, i.e. $3S_1$, $3P_1$ and $3D_1$.

We know that the orbital and spin angular momenta L and S correspond to infinitesimal rotation operators for space and spin coordinates. These operators commute with the Hamiltonian H containing the central potential. Since L_z and S_z commute with H, both m_L and m_S are good quantum numbers. Moreover, L^2 commutes with both H and L_z which reveals that each energy eigenstate has definite values of the orbital angular momentum $\hbar\sqrt{L(L+1)}$ and its Z-component $m_{L\hbar}$. Similarly, one finds that the spin angular momentum and its Z-component have definite values $\hbar\sqrt{S(S+1)}$ and $m_s\hbar$ for each eigenstate. Clearly, the quantum numbers specifying the different energy eigenstates are L, S, m_L and m_S.

When the non-central interaction is present along with the central, the Hamilton H is invariant only under combined rotation of space and spin Co-ordinates. This means, in general, $L^2 + S^2$ may not commute with H but J^2 will commute with H where $\mathbf{J} = \mathbf{L} + \mathbf{S}$ is the total angular momentum. This means J and m_J are good quantum numbers in this case.

In general, when non-central interaction is present S is not a good quantum number. However, S can be shown to be a good quantum number in the case of a system of two spin $\frac{1}{2}$ particles like the neutron and proton.

4.12 EXCHANGE INTERACTION AND SATURATION OF NUCLEAR FORCES

While studying the nucleus, we have noted that each nucleon inside the nucleus interacts only with a limited number of nucleons, i.e. the nuclear force is satured. This is supported by two important facts: (i) the density of nucleons is roughly equal for all nuclei (saturation of binding energy), i.e. binding energy $(E)\,\alpha\,A$. The saturation of the nuclear force also make the nuclear volume linearly proportional to A.W. Heisenberg in 1932 suggested that the saturation of the nuclear force can be explained by assuming the existence of *exhange interaction* between nucleon pairs. Exchange forces are well known in chemistry, e.g. sharing of an electron between two protons in a hydrogen molecule. Such exchange force gives attraction in some states, but repulsion in other states, which results in overall reduction in the binding energy and density of packing, compared to what would be observed with the so called ordinary force. We may note that the latter gives attraction in all states.

Exchange Forces

Let us first investigate the properties of different possible types of exchange forces between two nucleons.

One can write the wave equation in the C-system as

$$\frac{\hbar^2}{2\mu} \nabla^2 \psi + E\psi = V\psi$$

Where μ is reduced mass, $\psi = \psi(\mathbf{r}_1, \mathbf{r}_2) \chi (\sigma_1, \sigma_2) = \psi_{12} \chi_{12}$ is a function of the position and spin coordinates of the two particles. We have seen that space and spin coordinates of nucleons play an important role in characterising the nuclear force. Consequently, different types of exchange forces have been proposed, e.g. Wigner force, Heisenberg charge exchange force, Majorana space exchange force, Bartlett interaction or spin exchange force. As stated earlier, the nuclear force is charge independent and charge symmetric, the exchange force acts between pairs of identical nucleons. Now, we will discuss the various proposed exchange forces:

(i) **Ordinary or Winger Force**

$$V = V(r)P_w$$

Where $P_w = 1$ for all states and it is called as Wigner operator. This would make wave equation

$$\left(\frac{\hbar^2}{2\mu} \nabla^2 + E\right) \psi_{12} \chi_{12} = V(r) P_w \psi_{12} \chi_{12} \qquad \text{... (4.1)}$$

$$= V(r) \psi_{12} \chi_{12} \qquad \text{... (4.1a)}$$

Obviously, the interaction potential has the same sign in all states. Moreover, it is attractive in the ^3S –state ($L = 0$) and hence $V(r)$ is always attractive. Clearly, in this type of force no exchange is involved and such a force does not produce saturation

(ii) **Majorana Interaction or space exchange force** This is a space exchange force where exchange of space coordinates only takes place. The exchange potential is defined by

$$V = V(r) P_M$$

Where P_M is the Majorana or space exchange operator.

$$P_M \psi_{12} = P_M \psi(r_1 \sigma_1, r_2 \sigma_2) = P_M \psi(r_1, r_2) \chi (\sigma_1, \sigma_2)$$

$$= \psi_{21} \chi_{12} \qquad \text{... (4.2)}$$

We note that the Majoana operator is same as the parity operator. Obviously, one can express majorana operator as

$$P_M \psi_{12} \chi_{12} = (-1)^L \psi_{12} \chi_{12}$$

and $\qquad V(r) = + V_M(r)$ for $L = 0$, even

$$= -V_M(r) \text{ for } L = \text{Odd} \qquad \text{... (4.3)}$$

Wave equation becomes

$$\left(\frac{\hbar^2}{2\mu} \nabla^2 + E\right) \psi_{12} \chi_{12} = V(r) P_M \psi_{12} \chi_{12} = V(r) \psi_{21} \chi_{12}$$

$$= (-1)^L V(r) \psi_{12} \chi_{12}$$

$$= + V(r) \, \psi_{12} \, \chi_{12} \text{ for } L = 0, \text{ even}$$

$$= - V(r) \, \psi_{12} \, \chi_{12} \text{ for } L \text{ odd} \qquad \text{... (4.3a)}$$

The n-p interaction potential is attractive in the ^3S-state ($L = 0$) and therefore it must be attractive in all L-even states (D, G, etc. with $L = 2$, 4, etc.) and repulsive in all L-odd states (P, F, etc. with $L = 1$, 3, etc.). We may note that such a force could provide saturation.

(iii) **Bartlett Interaction or spin exchange force**

we write $\qquad\qquad V = V(r) \, P_B$

Where P_B is Bartlett or spin exchange operater having the property

$$P_B \, \chi_{12} = P_B \, \chi(\sigma_1, \sigma_2) = \chi(\sigma_2, \sigma_1) = \chi_{21}$$

Obviously, P_B interchanges the spins of two nucleons. For triplet states ($S = 1$) χ_{12} is symmetric, i.e. $\chi_{12} = + \chi_{21}$ and for singlet state ($S = 0$), χ_{12} is antisymmetric, i.e. $\chi_{12} = - \chi_{21}$. In general, we can write

$$\chi_{21} = \chi(\sigma_2, \sigma_1) = (-1)^{S+1} \, \chi(\sigma_1, \sigma_2)$$

$$= (-1)^{S+1} \, \chi^{12} \qquad \text{... (4.3)}$$

The wave equation becomes

$$\left(\frac{\hbar^2}{2\mu} \nabla^2 + E \right) \psi_{12} \, \chi_{12} = V(r) \, P_B \, \psi_{12} \, \chi_{12}$$

$$= V(r) \, \psi_{12} \, \chi_{21}$$

$$= (-1)^{S+1} \, V(r) \, \psi_{12} \, \chi_{12}$$

$$= V(r) \, \psi_{12} \, \chi_{12} \text{ for } S = 1 \text{ (triplet)}$$

$$= - V(r) \, \psi_{12} \, \chi_{12} \text{ for } S = 0 \text{ (singlet)} \qquad \text{... (4.3)}$$

Obviously, all triplet state forces (^3P, ^3D, ... etc.) are attractive as ^3S force is known to be attractive, where as all singlet state forces are repulsive (^1S, ^1P, ^1D, etc.). Experiments at low energies n-p scattering show that ^1S force is attractive, which clearly contradicts the above conclusion. Clearly, nuclear force cannot be totally of Bartlett type, since we have seen that it is attractive for both singlet and triplet states.

(iv) **Heisenberg Interaction: Space plus spin or charge Exchange Force**   In this type of interaction, one finds that the exchange of charge between the nucleons is the same as the exchange of both the space and spin coordinates. We write.

$$V = V(r) \, P_H \qquad \text{... (4.1)}$$

Where the Heisenberg interaction operator P_H has the property of interchanging both space and spin coordinates:

$$P_H \, \psi_{12} \, \chi_{12} = P_H \, (r_1, r_2) \, \chi(\sigma_1, \sigma_2)$$

$$= \psi(r_2, r_1) \, \chi(\sigma_2, \sigma_1)$$

$$= \psi_{21} \, \chi_{21} \qquad \text{... (4.2)}$$

The space exchange or the parity is governed by the orbital angular momentum L and spin exchange by the spin angular momentum quantum number S. As stated earlier, space exchange is symmetric for even parity states (*L*-even) and antisymmetric for odd-parity states (*L*-odd) and spin triplet states, i.e. $S = 0$ are antisymmetric. When one combine these characterstics to determine the Heisenberg exchange force,

$$P_H \psi_{12} \chi_{12} = (-1)^L \psi_{12} (-1)^{S+1} \chi_{12}$$
$$= (-1)^{L+S+1} \psi_{12} \chi_{12} \qquad \ldots (4.3)$$

The wave equation now takes the form

$$\left(\frac{\hbar^2}{2\mu} \nabla^2 + E \right) \psi_{12} \chi_{12} = V(r) P_H \psi_{12} \chi_{12}$$
$$= (-1)^{L+S+1} V(r) \psi_{12} \chi_{12}$$
$$= +V(r) \psi_{12} \chi_{12} \text{ for } L = 0, \text{ even}; S = 1; \text{ or } L = \text{ odd}; S = 0;$$
$$= -V(r) \psi_{12} \chi_{12} \text{ for } L = 0, \text{ even}; S = 0; \text{ or } L = \text{ odd}; S = 1, \ldots (4.4)$$

Clearly, the force is attractive in ^3S, ^3D, etc. states as also in ^1P, ^1F, etc. states. The force is repulsive in ^1S, ^1D, etc. states and also in ^3P, ^3F, etc. states. We see that this again contradicts the experimental observation, since ^3S and ^1S interactions should have the same sign. Obviously, Heisenberg charge exchange interaction cannot also be the only type of exchange force in the n-p system.

It is found that the estimated strengths of the ^3S and ^1S forces for the n-p scattering can be explained by postulating a mixture of about 25 % Heisenberg or Barlett interaction with about 75 % Wigner or Majorana interaction for a range ~ 2.8 F.

The effect of the Wigner (W), Bartlett (B), Heisenberg (H) and majorana (M) operators on two nucleon state is shown in Figure 4.17.

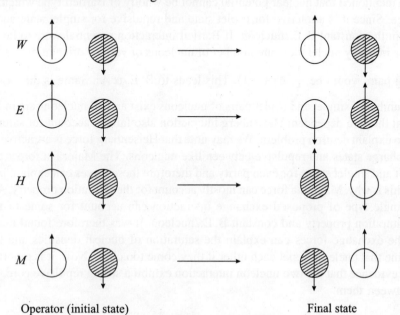

Operator (initial state)   Final state

Fig. 4.17 Effect of Wigner (W), Bratlet (B), Heidenberg (H), and Majorana (M) Operators on two nuclear states.

The characteristics of Wigner, Bartlett, Heisenberg and Majorana operators are summarized in Table 4.4.

Table 4.4 Characteristics of nuclear force exchange operators

	Even parity states		odd parity states	
	Triplet ($S=1$)	*Singlet* ($S=0$)	*Triplet* ($S=1$)	*Singlet* ($S=0$)
Exchange operator				
Wigner (P_W)	+1	+1	+1	+1
Heisenberg (P_H)	+1	−1	−1	+1
Majorana (P_M)	+1	+1	−1	−1
Bartlett (P_B)	+1	−1	+1	−1

4.13 EXCHANGE AND SATURATION OF NUCLEAR FORCES

We have already studied the two important properties of nuclei namely, $R \propto A^{\frac{1}{3}}$, i.e. a relation between nuclear radius (R) and mass number (A) and secondly, the binding energy of nuclei being proportional to the total number of nucleons inside a nuclei indicating that B. E./ nucleon is almost constant for most of the nuclei. These two important properties of nuclei indicate the saturation properties of the nuclear force. We now examine, how the above mentioned four exchange forces are able to explain the saturation properties.

We have mentioned that nuclear potential cannot be wholly of Bartlett type which accounts for spin exchange. Since it is attractive for triplet state and repulsive for singlet state and therefore it cannot explain the constant B.E./nucleon. If Bartlett interaction potential were to hold, then B.E. $\propto A^2$, because in heavy nuclei there are number of nucleons of various orientations and the number of interacting pairs would be $\frac{1}{2} A(A-1)$. This leads to B. E. $\propto A^2$. Same is the case for $^4_2$He. In $^4_2$He, equal number of singlet and triplet pairs of nucleons exist and all nucleons are in S-states. One also finds that the spin dependent Heisenberg interaction also fails to account for saturation and is, yet adquate to explain deutron problem. We may note that Heisenberg force is attractive for nucleons of different charge states and repulsive between like nucleons. The Majorana forces are attractive in both triplet and singlet states for even parity and therefore these forces can explain large B.E. for α-particle. This is why, Majorana force can mostly account for the saturation property. As mentioned earlier, no single type of proposed exchange interaction can account for some of the observed facts like. Saturation property and constant B. E./nucleon. It was therefore found that a judicious mixture of the exchange forces can explain the saturation of nucleat densities and there was no need to assume that nucleons repel each other if they come too close. Now, it is reported that there is very good evidence that the two nucleon interaction exhibit a strong repulsive core, at very close distances between them.

4.14 ISOSPIN FORMATION

One can express the exchange operators discussed in section 4.12 in terms of Pauli spin operators σ_1, σ_2 and isotopic spin operators $\tau^{(1)}$ and $\tau^{(2)}$ for the two particles.

We have seen that for two particles of spin $\dfrac{1}{2}$ each.

$$\sigma_1 \cdot \sigma_2 = +1 \text{ for triplet } (S = 1) \qquad \qquad \text{... (4.1)}$$
$$= -3 \text{ for singlet } (S = 0) \qquad \qquad \text{... (4.2)}$$

In analogy of the above, for the two isospin operators $\tau^{(1)}$ and $\tau^{(2)}$ for the two nucleon system, we have

$$\tau^{(1)} \cdot \tau^{(2)} = +1 \text{ for isospin triplet } (T = 1) \qquad \text{... (4.2a)}$$
$$= -3 \text{ for singlet } (S = 0) \qquad \qquad \text{... (4.1b)}$$

In analogy of the above, for the two isospin operators $\tau^{(1)}$ and $\tau^{(2)}$ for the two nucleon system, we have

$$\tau^{(1)} \cdot \tau^{(2)} = +1 \text{ for isospin triplet } (T = 1) \qquad \text{... (4.2a)}$$
$$= -3 \text{ for isospin singlet } (T = 0) \qquad \text{... (4.2b)}$$

Obviously, the total wave function ψ of a two nucleon system depends on the space coordinates r_1, r_2; spin coordinates σ_1, σ_2 and the charge coordinates $\tau^{(1)}$, $\tau^{(2)}$ of the two nucleons. We may note that the last pair of coordinates determine the charge state (whether proton or neutron) of the nucleon. It is the component τ (analogous to Z-component of σ) which gives the charge state of the nucleon. We can write for two protons $\tau_3^{(1)} = +1$ and $\tau_3^{(2)} = -1$ ($T_3 = t_{13} + t_{23} = +1$); for two neutrons $\tau_3^{(1)} = -1$; $\tau_3^{(2)} = -1$ ($T_3 = -1$); for a neutron and a proton $\tau_3^{(1)} = -1$, $\tau_3^{(2)} = +1$ ($T = 0$). Clearly, the isospin triplet p-p ($T_3 = +1$), n-n ($T_3 = -1$) and p-n ($T_3 = 0$) for which $T = 0$. Clearly, the isospin multiplicity is $(2T + 1)$.

We now introduce iso-spin-space. In analogy with the spin function $\chi(\sigma_1, \sigma_2)$, the isospin function for the two nucleon system can be expressed as $\xi(\tau^{(1)}, \tau^{(2)})$, where ξ is the charge state of the system. The isospin 'up' represents a proton state and written as γ and isospin 'down' represents the neutron state and written as δ. In Pauli spin formalism, one can write γ and δ as

$$\gamma = \begin{pmatrix} 1 \\ 0 \end{pmatrix} \qquad \delta = \begin{pmatrix} 0 \\ 1 \end{pmatrix} \qquad \text{... (4.3)}$$

In the case of two nucleon system, we have $T = 1$ (*symmetric triplet*):

$$\xi(\tau^{(1)}, \tau^{(2)}) = \xi_{12} = \gamma_1 \gamma_2 \qquad (p\text{-}p \text{ or } {}^2He) \qquad T_3 = +1$$
$$= \frac{\gamma_1 \delta_2 + \delta_1 \gamma_2}{\sqrt{2}} \qquad (p\text{-}n \text{ system}) \qquad T_3 = 0$$
$$= \delta_1 \delta_2 \qquad (n\text{-}n \text{ or deuteron}) \qquad T_3 = -1 \qquad \text{... (4.4)}$$

$T = 0$ (*Antisymmetric singlet*): $T = 0$

$$\xi_{12} = \frac{\gamma_1 \delta_1 - \delta_1 \gamma_2}{\sqrt{2}} \qquad (p\text{-}n \text{ system})$$

One can write the total wave function as

$$\psi = \psi(\mathbf{r}_1, \mathbf{r}_2) \, \chi(\sigma_1, \sigma_2) \, \xi(\tau^{(1)}, \tau^{(2)}) \qquad \ldots (4.5)$$

The two nucleons obey Fermi-Dirac statistics, the total wave function must be antisymmetric in the interchange of all the coordinate pairs r_1, r_2; σ_1, σ_2; and $\tau^{(1)}$, $\tau^{(2)}$.

As discussed above, the interchange of space coordinates multiples $\psi(\mathbf{r}_1, \mathbf{r}_2)$ by $(-1)^L$.

One then finds $\psi(\mathbf{r}_2, \mathbf{r}_1) = +\psi(\mathbf{r}_1, \mathbf{r}_2)$, i.e. $\psi_{21} = \psi_{12}$. Clearly in this case the symmetry of the total wave function is determined by the symmetry properties of $\chi(\sigma_1, \sigma_2)$ and $\xi(\tau^{(1)}, \tau^{(2)})$. If $\chi(\sigma_1, \sigma_2)$ is symmetric (spin triplet with $S = 1$), then $\xi(\tau^{(1)}, \tau^{(2)})$ should be antisymmetric (isospin singlet with $T = 0$). However, if $\chi(\sigma_1, \sigma_2)$ is antisymmetric (spin singlet with $S = 0$), then $\xi(\tau^{(1)}, \tau^{(2)})$ must be symmetric (isospin triplet with $T = 1$).

The isospin function for the two nucleon system (in analogy with the case of the spin) on interchange of $\tau^{(1)}$ and $\tau^{(2)}$ becomes

$$\xi_{21} = \xi(\tau^{(2)}, \tau^{(1)}) = (-1)^{T+1} \, \xi(\tau^{(1)}, \tau^{(2)})$$
$$= (-1)^{T+1} \, \xi_{12} \qquad \ldots (4.6)$$

Obviously, the above bears out the symmetry properties of $\xi(\tau^{(1)}, \tau^{(2)})$.

Thus the total wave function, as determined by the interchange of the pairs of coordinates of the two particles for $L = 0$ gives

$$\psi_{21} = \psi_{21} \, \chi_{21} \, \xi_{21}$$
$$= (-1)^L \, \psi_{12} (-1)^{S+1} \, \chi_{12} (-1)^{T+1} \, \xi_{12}$$
$$= (-1)^{L+S+T+2} \, \psi_{12} \, \chi_{12} \, \xi_{12}$$
$$= (-1)^{L+S+T+2} \, \psi_{12} \qquad \ldots (4.7)$$

Obviously, the nature of the wave function is decided by the L, S and T values of the state. This represents the *generalized Pauli exclusion principle* for the two particle system. Since $\psi_{21} = -\psi_{12}$, for $S = 1$ (spin triplet), one finds $T = 0$ (isospin singlet; p-n system). On the otherhand for $S = 0$ (spin singlet) one finds $T = 1$ (isospin triplet; p-p, p-n or n-n systems).

As an example, we consider deutron (n-p system) in the ground state, a mixture of a S-state ($L = 0$) and a D-state ($L = 2$) on the inter-change of space coordinates, the space part of the wave function is symmetric. The spin wave function is symmetric, the two spins are parallel and the total spin is $1(S = 1)$. $\chi_{21} = + \chi_{12}$ and to have ψ_{total} to be antisymmetric, the wave function of iso-spin should be anti-symmetric, i.e. $\xi_{21} = (-1) \, \xi_{12}$, i.e. it should be an iso-spin singlet ($T = 0$) state. Clearly, the ground state of the deutron (mixture of $L = 0$ and $L = 2$) is an ordinary triplet ($S = 1$) and iso-spin singlet ($T = 0$) state. Now, if the spins of neutron and proton in the deutron are anti-parallel, then it is an ordinary spin singlet state, i.e. $S = 0$, the spin wave function χ being antisymmetric, the iso-spin wave function ξ must be summetric ($T = 1$). Clearly, the singlet state of the deutron is an isospin triplet.

4.15 EXCHANGE OPERATORS

Let us consider the operator

$$P_\sigma = \frac{1}{2} (1 + \sigma_1 \cdot \sigma_2) \qquad \ldots (4.1)$$

But
$$\sigma_1 \cdot \sigma_2 = +1 \text{ for triplet } (S = 1)$$
$$= -3 \text{ for singlet } (S = 0) \qquad \ldots (4.2)$$

Using (2), one obtains $P_\sigma = +1$ for triplet $(S = 1)$ spin state and $P_\sigma = -1$ for singlet $(S = 0)$ spin state. In general, one can write

$$P_\sigma = (-1)^{S+1}$$

One then obtains

$$P_\sigma \psi_{12} = P_\sigma \psi_{12} \chi_{12} \xi_{12} = \psi_{12} (P_\sigma \chi_{12}) \xi_{12}$$
$$= (-1)^{S+1} \psi_{12} \chi_{12} \xi_{12} = (-1)^{S+1} \psi_{12} \qquad \ldots (4.3)$$

We note that this property of P_σ is identical with that of Bartlett spin exchange operator P_B. Let us identify P_B with P_σ.

$$P_B = P_\sigma = \frac{1}{2} (1 + \sigma_1 \cdot \sigma_2) \qquad \ldots (4.4)$$

Now, we consider the operator

$$P_\tau = \frac{1}{2} (1 + \tau^{(1)} \cdot \tau^{(2)}) \qquad \ldots (4.5)$$

This operator operates on the "charge function" $\xi(\tau^{(1)} \cdot \tau^{(2)})$. We have

$$\tau^{(1)} \cdot \tau^{(2)} = +1 \text{ for isospin triplet } (T = 1)$$
$$= -3 \text{ for isospin singlet } (T = 0) \qquad \ldots (4.6)$$

Using (4.6), we obtain

$$P_\tau = +1 \text{ for isospin triple } (T = 1)$$
$$P_\tau = -1 \text{ for isospin singlet } (T = 0)$$

Thus, in general, one can write

$$P_\tau = (-1)^{T+1}$$

$$\therefore \qquad P_\tau \xi(\tau^{(1)}, \tau^{(2)}) = (-1)^{T+1} \xi(\tau^{(1)}, \tau^{(2)}) \qquad \ldots (4.7)$$

However, this is exactly the effect of charge exchange operation

$$P_\tau \xi_{12} = \xi_{21} = (-1)^{T+1} \xi_{12} \qquad \ldots (4.8)$$

Let us now consider the Heisenberg interaction operator P_H. Obviously, $P_H P_\tau$ operation interchanges all the three pairs of coordinates $(\mathbf{r}_1, \mathbf{r}_2)$, (σ_1, σ_2) and $(\tau^{(1)}, \tau^{(2)})$.

$$P_H P_\tau \psi_{12} = P_H P_\tau \psi_{12} \chi_{12} \xi_{12}$$
$$= P_H \psi_{12} \chi_{12} P_\tau \xi_{12}$$
$$= \psi_{12} \chi_{21} \xi_{21} = \psi_{21} = -\psi_{12} \qquad \ldots (4.9)$$

This follows from the generalized Pauli's exclusion principle. Accordingly, the interchange of all the coordinates must produce the original wave function with the sign changed. This gives

$$P_H P_\tau = -1 \qquad \ldots (4.10)$$

One obtains from the above equation

$$P_H \, P_\tau \, P_\tau = -P_\tau$$

But, we have

$$P_\tau^2 = P_\tau \, P_\tau = +1 \qquad \qquad \text{... (4.11)}$$

$$\therefore \qquad P_H = -P_\tau = -\frac{1}{2}(1 + \tau^{(1)} \, \tau^{(2)}) \qquad \qquad \text{... (4.12)}$$

We have seen that Majorana operator P_M interchanges the space coordinates and it operates on $\psi_{12} = \psi(\mathbf{r}_1, \mathbf{r}_2)$ only. The Bartlett operator P_B operates on the spin part $\chi_{12} = \chi(\boldsymbol{\sigma}_1, \boldsymbol{\sigma}_2)$ of the wave function only and interchanges σ_1 and σ_2. This means that the operator $P_M \, P_B$ interchanges both the space and spin coordinates and this is the property of the Heisenberg operator. Obviously,

$$P_H = P_M \, P_B$$

or
$$P_H \, P_B = P_M \, P_B \, P_B = P_M \, P_B^2 = P_M \qquad (\therefore \, P_B^2 = +1) \qquad \text{... (4.13)}$$

This shows that P_H and P_B can be interchanged on the left side, since they operate on the charge part (ξ_{12}) and spin part (χ_{12}) of the wave function ψ_{12} respectively. Using (4.4) and (4.12), one obtains

$$P_M = P_B \, P_H = -\frac{1}{4}(1 + \boldsymbol{\sigma}_1 \cdot \boldsymbol{\sigma}_2)(1 + \tau^{(1)} \cdot \tau^{(2)}) \qquad \text{... (4.14)}$$

4.16 ISOBARIC ANALOGUE STATES

The neutron and the proton are regarded as two isospin states with different values of τ_3 of the nucleon. Here τ_3 is the 2×2 Pauli matrix, identical with σ_z. The isospin vector t has the third component $t_3 \left(= \dfrac{\tau_3}{2} \right)$. For a complex nucleus having z-protons and n-neutrons, one can obtain the isospin vector $\mathbf{T}$ by the vector addition of the isospin vectors of all the nucleons, i.e. $\mathbf{T} = \sum\limits_i \mathbf{t}_i$ obviously, the third component $T_3 = -\dfrac{1}{2}(N - Z)$.

One can express the charge independence of the nuclear forces in terms of the isospin formalism by requiring the isospin T to be conserved in strong interaction; which is responsible for the binding force. We may note that the conservation of T implies that the specifically nuclear force is independent of the charge state of the nucleons, i.e. independent of the value of the third component T_3 of $\mathbf{T}$. In case of two nucleon system, $T = 1$ in the spin singlet $(S = 0)$ state, the force is the same for the n-n $(T_3 = -1)$, p-p $(T_3 = +1)$ and n-p $(T_3 = 0)$. Obviously, if we do not take into account the Coulomb force, the energies (masses) of the said three two body systems should be equal. However, the presence of the charge on the proton produces the difference in the energies (masses) of the three states. One can express this by saying that the symmetry between the three states required by isospin conservation is broken by the electomagnetic interaction. This means that the isospin is not conserved in electromagnetic interaction.

In the above formalism, the n-p system can also have $T = 0$, $T_3 = 0$ state which is nothing but the spin triplet state with $S = 1$. Obviously, this is deutron ground state.

Isospin conservation is expressed quantum mechanically as

$$[H, T] = 0$$

Where H is Hamiltonian operator.

Let us operate the isospin function by the operators T_2 and T_3. We have

$$T^2 \xi = T(T + 1)\xi$$

$$T^3 \xi = T_3 \xi$$

For the isospin *singlet* state, we have $T = 0$, $T_3 = 0$ and for the isospin *triplet*, $T = 1$, $T_3 = 0, \pm 1$.

One can extend the isospin formalism to the case of the excited states of the complex nuclei. Let us consider isobaric nuclei with $A = N + Z$. The different components in it differ in T_3 such that $T > |T_3|$. For a particular case, the values of T to be chosen depends on the number of nuclei (stable and unstable) actually observed, e.g. in the case of $^{13}_6C$ and $^{13}_7N$ (mirror nuclei) at $A = 13$, we take $T = \dfrac{1}{2}$ since the multiplicity $M = 2T + 1 = 2$. The two components of the isospin multiplet have $T_3 = \pm \dfrac{1}{2}$. However, if one takes into account the more nuclei actually existing in nature, e.g.

$^{13}_5B$ and $^{13}_8O$ at the same $A(=13)$, the multiplicity becomes 4. In this case, we have to take $T = \dfrac{3}{2}$ with T_3 ranging from $-\dfrac{3}{2}$ to $+\dfrac{3}{2}$. We may note that the concept of isospin is quite useful in classifying the levels of the system of nuclei for a given A.

We have already mentioned that in the isospin formalism it is assumed that the only difference between the protons and neutrons arises due to the electric charge carried by the protons, which makes the energies (masses) of the different given A different from each other. However, due to the electrostatic repulsive energy of the protons, the nuclei with greater Z have usually higher energy. As an example, we condider the isobaric doublet constituted by the mirror nuclei $^{13}_6C$ and $^{13}_7N$. Clearly, they are of the type $^A_Z X$ and $^A_{Z+1} Y$ such that $A = 2Z + 1$. We note that N and Z are interchanged in the two nuclei and they differ by unity in both. Obviously, the two nuclei differ in having a neutron replaced by a proton in the second member w.r.t the first. Figure 4.18 shows the arrangement of neutrons and protons in the said two nuclei. Pauli exclusion principle permits only four nucleons ($2p$ and $2n$) to be accomodated in each level. This is why that the three lowermost levels in both mirror nuclei are fully occupied by a pair of protons (with opposite spins) and a pair of neutrons (with opposite spins) each. Obviously, the 13th nucleon which is a neutron in the case of ^{13}C and a proton in the case of ^{13}N goes to the fourth, i.e. topmost level. This is the only difference in these two mirror nuclei, which are otherwise identical.

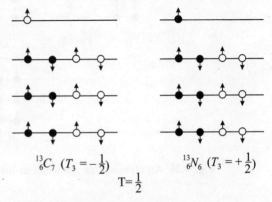

$$^{13}_6C_7 \ (T_3 = -\tfrac{1}{2}) \qquad \qquad ^{13}_7N_6 \ (T_3 = +\tfrac{1}{2})$$
$$T = \tfrac{1}{2}$$

Fig. 4.18 Nucleon arrangement in the mirror nuclei $^{13}_6C$ and $^{13}_7N$.

Figure 4.19 shows the energy level diagram of the said two mirror nuclei. Figure shows the remarkable similarity between the two except for the Coulomb energy difference of about 3 MeV stated above. These are known as *isobaric analogue states* (IAS). The other examples of isobaric analogue states are ^{28}Mg and ^{28}Si. These analogue states have also been studied as resonanaces in reactions such as ^{89}Y(p, n) ^{89}Zr, ^{51}V(p, n) ^{51}Cr. We may note that for heavier isobaric multiplets the analogue states can be at an excitation greater than the binding energy. They may then be seen, for instance in proton reactions as *formation* (or *isobaric analogue*) *resonances* rather than as final states. In this case they are analogues not to states of the target nucleus but to those of the nucleus formed by adding one neutron to it, since they exist in a compound nucleus of mass number one greater than that of the target.

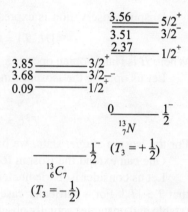

Fig. 4.19 Isobaric analogue states in two mirror nuclei ^{13}N and ^{13}C

Experimentally for a large number of nuclei it is found that the calculated excitation energy E_c of the analogue state, which equals the Coulomb energy difference less the neutron H atom mass difference, is equal to the measured eacitation energy (E_x) plus the ground state mass difference (E_Δ), showing that analogue states do exist in heavy nuclei. this is mainly for heavy nuclei because the degree to which the protons contribute to isospin mixing is reduced by a factor $\dfrac{1}{(N - Z - 2)}$ compared to simple estimates.

Now, we consider the case of the *isotriplet* $^{12}_{5}$B ($T_3 = 1$), $^{12}_{6}$C ($T_3 = 0$) and $^{12}_{7}$N ($T_3 = + 1$) for which $T = 1$. Figure 4.20 shows the arrangement of nucleons in different levels of these three nuclei and Figure 4.21 shows the energy levels of these nuclei.

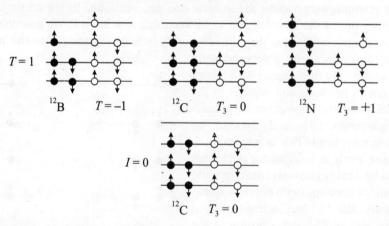

Fig. 4.20 Arrangement of nucleons in the different levels of isobars ^{12}B, ^{12}C and ^{12}N.

2.62 1^-	17.23 1^- $T=1$	1.65 1^-
1.674 2^-	16.58 2^- $T=1$	1.20 2^-
0.953 2^+	16.11 2^+ $T=1$	0.969 2^+
0 $\quad 1^+$	15.11 1^+ $T=1$	0 $\quad 1^+$
$^{12}_{5}$B	12.71 1^+ $T=0$	$^{12}_{7}$N
$(T_3=-1)$	9.638 3^- $T=0$	$(T_3=+1)$
	7.653 0^+ $T=0$	
	4.439 2^+ $T=0$	
	0 $\quad 0^+$ $T=0$	
	$^{12}_{6}$C	
	$(T_3=0)$	

Fig. 4.21 Energy levels in ^{12}B, ^{12}C and ^{12}N exhibiting analogue states.

Amongst the three nuclei, ^{12}C has the lowest mass energy in the ground state (≈ 12.6 MeV), which is lower than that of ^{12}B. Figure 4.20 shows in the lower part the sequence of level filling of ^{12}C ground state pairs of protons and neutrons according to Pauli principle. We can see that in the three lowermost levels, the three pairs of protons and the three pairs of neutrons are accomodated. On the otherhand in $^{12}_{5}$B and $^{12}_{7}$N the two lowermost levels are filled with a pair of protons and a pair of neutrons in each of the four remaining nucleons three are identical in both so that only two of them can go into the third level from the bottom. Obviously, in ^{12}B there are two neutrons and a proton in this level while the last, i.e. 7th neutron goes into the 4th level from the bottom. In ^{12}N, there are $2p$ and a neutron in the 3rd level with the 7th proton (p) going into the 4th level from the bottom. However, for the difference between the proton and the neutron, the two would have the same energy (mass) in their ground states. One can also have a similar type of level-filling in ^{12}C (upper central part of Figure 4.20). In ^{12}B and ^{12}N, the two lowermost levels are filled with a pair of protons and a pair of neutrons in each of the remaining 4 nucleons two of the same type (p or n) along with one of the different type (n or p) go into the 3rd level. The remaining last nucleon (p or n) is now placed in the 4th level. Obviously, the configuration is analogous to the other two so that the energy of ^{12}C in this excited state would be the same as those of the other two except for the Coulomb energy difference. This means that the three nuclei in this configuration have the same value of the isospin, $T=1$ with $T_3=-1$, 0 and $+1$ for them. We may note that the ground state configuration of ^{12}C is however different from $T=0$ and $T_3=0$.

Figure 4.21 shows the energy levels of these three nuclei. From figure, we can see that for ^{12}C the levels below about 15 MeV excitation all belong to the $T=0$ configuration while those below 15 MeV belong to the $T=1$ configuration. Figure also shows the corresponding levels with their energies of ^{12}B and ^{12}N ($T=1$). We note that all those states are analogous to the $T=1$ levels of ^{12}C and constitute isobaric states.

In the case of isobaric doublet ^{13}C – ^{13}N, all the levels upto about 15 MeV have $T=\dfrac{1}{2}$, $T_3=\pm\dfrac{1}{2}$. An analogue state is observed in ^{13}N at about 15.07 MeV which corresponds to the ground state of $^{13}_5\text{B}$ and $^{13}_8\text{O}$ and has $T = \dfrac{3}{2}$ and $T_3 = \pm \dfrac{1}{2}$. The ground state of latter two nuclei

having $T = \dfrac{3}{2}$ and $T_3 = -\dfrac{3}{2}$ and $+\dfrac{3}{2}$ respectively. The states of both ^{13}C and ^{13}N above 15.07 MeV

are seen of which some have $T = \dfrac{3}{2}$ while others have $T = \dfrac{1}{2}$.

We have already mentioned that isospin is conserved in strong interaction. The experimental results on the production of 1.74 MeV, $T = 1$ excited state of the ^{10}B nucleus provide the validity of said assertion. It is reported that though this state $^{10}\text{B}^*$ is produced by inelastic scattering of protons of ^{10}B, it cannot be produced by inelastic duetron sacttering. The reactions involving strong interaction must conserve T. One can represent the two reactions as

$$^{10}\text{B} \; + \; p \; \rightarrow \; ^{10}\text{B}^* \; + \; p'$$
$$(T = 0 \qquad \dfrac{1}{2} \qquad 1 \qquad \dfrac{1}{2})$$

$$^{10}\text{B} \; + \; d \; \rightarrow \; ^{10}\text{B}^* \; + \; d'$$
$$(T = 0 \qquad 0 \qquad 1 \qquad 0)$$

One can easily see from the values of T for the different particles that conservation of T is possible in pp' scattering (first case) whereas in the second case (dd' scattering) it is not possible.

4.17 HIGH ENERGY n-p AND p-p SCATTERING

One can use the high energy scattering experiments to investigate the details of the potential shape and to learn about the two nucleon forces in states other than S-state. With the development of high energy accelerators, it is now possible to extend the energy range of n-p and p-p scattering to energies at which the de-Broglie wavelength of the relative motion is considerably smaller than the range of nuclear forces. At high energy, the presence of higher angular momentum states results in a large number of parameters, making it difficult to interpret the data. The available data are strange and unexpected. The analysis for neutrons is still more complicated as a mono-energetic neutron beam is not available. The experimentally observed cross-section is an average over a rather large energy region and may consequently fail to show rapid energy dependence. In comparision to high energy n-p scattering data, it is easy to interpret high energy p-p sacttering data because the Pauli exclusion principle cuts the number of phase shifts in half and one can obtain the experimantal cross-sections with mono-energetic protons.

(i) **High energy n-p scattering**   We have mentioned earlier that high energy n-p scattering involve partial waves of higher values of l in the expression for the scattered amplitude $f(\theta)$. Numerical values of E_{max} in C-system for the first few partial waves are as follows

L	0	1	2	3
E_{max} (MeV)	10	40	90	160

We may note that the corresponding values of the energy in the L-system are twice the values cited above.

We now consider scattering involving S, P and D waves. One can write

$$f(\theta) = \frac{1}{2ik} [\{\exp(2i\delta_0) - 1\} + 3\{\exp(2i\delta_0) - 1\}\cos\theta + \frac{5}{2}\{\exp(2i\delta_2) - 1\}(3\cos^2\theta - 1)]$$

$$= \frac{1}{k} [\exp(i\delta_0)\sin\delta_0 + 3\exp(i\delta_1)\sin\delta_1\cos\theta + \frac{5}{2}\exp(i\delta_2)\sin\delta_2(3\cos^2\theta - 1)]$$

Assuming δ_1 and δ_2 to be small as compared to δ_0, one can neglect small quantities of second order, i.e. δ_1^2, δ_2^2 and $\delta_1\delta_2$.

$$\therefore \quad \sigma(\theta) = |f(\theta)|^2$$

$$= \frac{1}{k^2} [\sin^2\delta_0 + 6\sin\delta_0\sin\delta_1\cos(\delta_0 - \delta_1)\cos\theta$$

$$+ 5\sin\delta_0\sin\delta_2\cos(\delta_0 - \delta_2)(3\cos^2\theta - 1)]$$

$$= \frac{1}{k^2} [\sin^2\delta_0 + 3\sin 2\delta_0\sin\delta_1\cos\theta + \frac{5}{2}\sin 2\delta_0\sin\delta_2(3\cos^2\theta - 1)] \quad ... (4.1)$$

In obtaining (1), we have neglected δ_1 and δ_2 in the terms containing $\cos(\delta_0 - \delta_1)$ and $\cos(\delta_0 - \delta_2)$. Now, writing $\sin\delta_1 \approx \delta_1$ and $\sin\delta_2 \approx \delta_2$, one obtains

$$\sigma(\theta) = \frac{1}{k^2} [\sin^2\delta_0 + 3\delta_1\sin 2\delta_0\cos\theta + \frac{5}{2}\delta_2\sin 2\delta_0(3\cos^2\theta - 1)] \quad ... (4.2)$$

One can write the above expression as

$$\sigma(\theta) = A + B\cos\theta + C\cos^2\theta \quad\quad\quad ... (4.3a)$$

We may note that the coefficient C is small. For an ordinary or Wingner type n-p scattering δ_0, δ_1 and δ_2 all have the same sign (positive for an attractive potential) and A, B and C are of the same sign. Thus $\sigma(\theta)$ is peaked forward ($\theta = 0$) [Figure 4.22a]. Figure 4.22a shows the contribution of the various terms in equations (4.2). If however, Majorana type space exchange force acts between n-p, i.e. there is repulsion in the p-state ($l = 1$) then the sign of δ_1 would be opposite to that of δ_0 and δ_2. Due to this the coefficient B will be negative. Then $\sigma(\theta)$ given by equation (4.2) would show a backward peak (at $\theta = \pi$ in C-system). This is shown in Figure 4.22b).

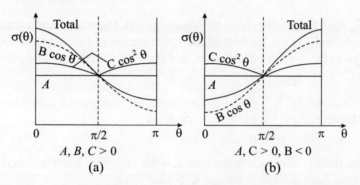

Fig. 4.22 For high energy n-p scattering the variation of $\sigma(\theta)$ with θ: (a) by an ordinary potential and (b) by space exchange potential, i.e. Majorana type.

Figure 4.23 shows the variation of $\sigma(\theta)$ with θ obtained experimentally. Figure shows large n-p scattering cross-sections in both forward and backward directions with a dip around $\theta = 90°$. This reveals that n-p scattering is neither wholly ordinary and nor wholly of Majorana type.

One can explain qualitatively the experimental results, if we assume that the n-p interaction is a mixture of ordinary and space-exchange type in approximately equal proportions. Serber proposed an interaction potential given by

$$V = [(1 - f)\, P_W + f\, P_W]\, V(r) \qquad \qquad ... (4.3)$$

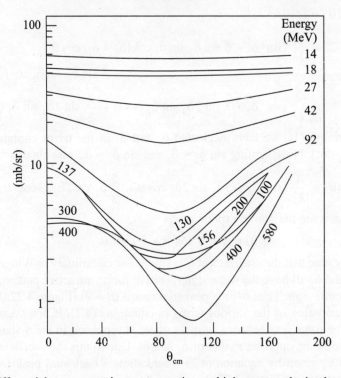

Fig. 4.23 Differential n-p scattering cross-section at high energy obtained experimentally.

Where P_W and P_M are the Wigner (ordinary) and Majorana (space-exchange) operators respectively and f is the fraction of the space-exchange force. Serber took $f = 0.5$.

Since $P_M = (-1)^L = +1$ for $L = 0$, even, and $P_W = +1$ for all states, one finds for the even L states

$$V_e = V(r) \qquad \qquad ... (4.4)$$

For odd L states, $P_M = -1$ and hence one finds

$$V_0 = 0 \qquad \qquad ... (4.5)$$

Obviously, there is attraction in the even L states including $L = 0$ and there is no force in the odd L states. Such a force is termed as *Serber force*.

Figure 4.23 shows slight anisotropy in the angular distribution about $\theta = \dfrac{\pi}{2}$ and thus showing higher values of $\sigma(\theta)$ for $\theta > \dfrac{\pi}{2}$ over these for $\theta < \dfrac{\pi}{2}$. This clearly reveals that the space exchange force is stronger than ordinary force. Obviously, this shows the deficiency of the Serber force for $E > 100$ MeV. We may note that Serber force cannot also explain the saturation of the nuclear force which requires attraction in some states and repulsion in other states. It is reported that saturation in complex nuclei would require $|V_M| > 4\,|V_W|$, where V_M and V_W are the Majorana and wigner forces reapectively.

At still higher energies, one will have to use Born approximation (see Quantum Mechanics by Kakani, Sultan Chand, 4th ed.)

Born Approximation

This method is applicable to the case of perturbation of a system with a set of continuous eigenfunctions.

One can write the wave function as

$$\psi(\mathbf{r}) = \exp(ikz) + V(\mathbf{r}) \qquad \qquad \text{... (4.1)}$$

The first term represents the incident wave along Z. One can write the wave equations as

$$\nabla^2 \psi + k^2 \psi = U(r)\psi \qquad \qquad \text{... (4.2)}$$

Where $k^2 = \dfrac{2\mu E}{\hbar^2}$, $U(r) = \dfrac{2\mu V(r)}{\hbar^2}$ and μ is reduced mass. Assuming the scattering wave function $|V(r)| \ll 1$, Equation (4.2) takes the form

$$(\nabla^2 + k^2)\, V(\mathbf{r}) = U(r) \exp(ikz) \qquad \qquad \text{... (4.3)}$$

Assuming that the energy to be high, i.e $E \gg V(r)$ so that $k^2 \gg U(r)$. Let $u_k(r)$ be a complete set or orthonomal functions with real eigen values k^2, one finds

$$\nabla^2 u_k(\mathbf{r}) = k^2\, u_k(\mathbf{r}) \qquad \qquad \text{... (4.4)}$$

Expanding the function $V(\mathbf{r})$ in terms of $u_k(\mathbf{r})$, one obtains

$$V(\mathbf{r}) = \int A_k\, u_k(\mathbf{r})dk \qquad \qquad \text{... (4.5)}$$

We have

$$\int u_{k'}^{*}(\mathbf{r})\, u_k(\mathbf{r})\, d\mathbf{r} = \delta(k - k') \qquad \qquad \text{...: (4.6a)}$$

and

$$\int u_k^{*}(\mathbf{r})\, u_k(\mathbf{r}')dk = \delta(\mathbf{r} - \mathbf{r}') \qquad \qquad \text{... (4.6b)}$$

Where $\delta(k, k')$ and $\delta(\mathbf{r}, \mathbf{r}')$ are delta functions.

Then

$$\nabla^2 V(\mathbf{r}) = \int A_k k^2\, u_k(\mathbf{r})dk \qquad \qquad \text{... (4.7)}$$

Writing the inhomogeneous equations (4.3) as

$$(\nabla^2 - k_0^2)\, V(\mathbf{r}) = \int A_k (k^2 - k_0^2)\, u_k(\mathbf{r})dk = F(\mathbf{r})$$

Where
$$F(\mathbf{r}) = -U(\mathbf{r}) \exp(ik_0 z) \qquad \ldots (4.7a)$$

One can slove (4.7a) by introducing the Green's finction for ∇^2 and k_0^2.

$$G_{k_0}(\mathbf{r}, \mathbf{r}') = \frac{u_k^*(\mathbf{r}') u_k(\mathbf{r})}{k^2 - k_0^2} d\mathbf{k} \qquad \ldots (4.8)$$

Then one obtains
$$V(\mathbf{r}) = \int G_{k_0}(\mathbf{r}, -\mathbf{r}')\, F(\mathbf{r}')\, d\mathbf{r}' \qquad \ldots (4.9)$$

We have for a free particle, a suitably normalized eigenfunction of $-\nabla^2$ with the eigen value $-k'^2$ as

$$u_{k'}(\mathbf{r}) = \frac{1}{(2\pi)^{\frac{3}{2}}} \exp(i\mathbf{k}' \cdot \mathbf{r}) \qquad \ldots (4.10)$$

Now, the Green function takes the form

$$G_k(\mathbf{r}, \mathbf{r}') = \frac{1}{4\pi^2 \rho} \int_0^\infty \frac{K \sin K}{K^2 - \sigma^2} \cdot dK \qquad \ldots (4.11)$$

Here
$$\rho = r - r',\ K = k'\rho \text{ and } \sigma = k_0\rho$$

on perfoming the intergral in (4.11) by the method of residue, one obtains

$$G_k(\mathbf{r}, \mathbf{r}') = \frac{\exp(ik\rho)}{4\pi\rho} \qquad \ldots (4.12)$$

Now, we apply the above result for calculating the *differential cross-section for scattering*.

Using (4.9) and (4.12), we obtain

$$V(\mathbf{r}) = \int G_k(\mathbf{r}, \mathbf{r}')\, \{-U(\mathbf{r}') \exp(ikz')\} d\mathbf{r}' \qquad \ldots (4.13)$$

Now, we can write the total wave function as

$$\psi(\mathbf{r}) = \exp(ikz) - \frac{1}{4\pi} \int \frac{U(\mathbf{r}') \exp(ikz') \exp(ik\rho)}{\rho} d\mathbf{r}'$$

$$= \exp(ikz) - \frac{1}{4\pi} \int U(\mathbf{r}') \exp(ikz') \exp(ik|\mathbf{r} - \mathbf{r}'|) d\mathbf{r}' \qquad \ldots (4.14)$$

One can interpret (4.14) with the help of Figure 4.24. A spherical wave starting from $P(\mathbf{r}')$ and arriving at $S(\mathbf{r})$ will have the form

$$\frac{\exp(ik|\mathbf{r} - \mathbf{r}'|)}{|\mathbf{r} - \mathbf{r}'|} = \frac{\exp(ik\rho)}{\rho}$$

The amplitude of the above spherical wave is proportional to $U(\mathbf{r}') \exp(ikz')$ or to $V(\mathbf{r}') \exp(ikz')$. Integrating over the entire scatterer and if the scatterer is small, then $r' \ll r$, we have

$$\rho = |\mathbf{r} - \mathbf{r}'| \approx r - r' \cos \beta$$
$$= r - wr'$$

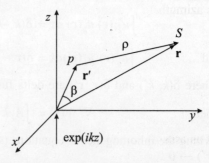

Fig. 4.24 Illustration of scattering from different points of the scatter.

$$\therefore \qquad \frac{1}{\rho} = \frac{1}{r}\left(1 + \frac{wr'}{r}\right) \approx \frac{1}{r}$$

Thus the asymptotic form of ψ is

$$\psi(\mathbf{r}) = \exp(ikz) - \frac{\exp(ikz)}{4\pi r} \int U(\mathbf{r}') \exp\{ik(z' - wr')\}d\mathbf{r}'$$

Now, writing

$$\psi(\mathbf{r}) = \exp(ikz) + f(\theta, \phi)\frac{\exp(ikz)}{r}$$

One obtains the scattering amplitude as

$$f(\theta, \phi) = \frac{1}{4\pi} \int U(\mathbf{r}') \exp\{ik(z' - wr')\}dr' \qquad \dots (4.15)$$

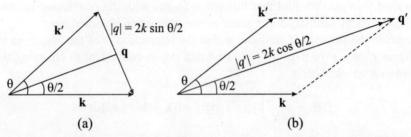

(a)   (b)

Fig. 4.25 Illustration of change of wave vector in elastic scattering: (a) under the influence of an ordinary force, and (b) under the influence of Majorana space exchange force.

Change of the wave vector on scattering at the angle θ (Figure 4.25a) is $q = 2k \sin\dfrac{\theta}{2}$, then

$$\mathbf{q} \cdot \mathbf{r}' = (\mathbf{k} - \mathbf{k}') \cdot \mathbf{r}' = k(z' - r'w)$$

$$f(\theta, \phi) = -\frac{1}{4\pi} \int U(\mathbf{r}') \exp(i\mathbf{q} \cdot r')dr' \qquad \dots (4.16)$$

Thus $\qquad\qquad \sigma(\theta, \phi) = |f(\theta, \phi)|^2 \qquad\qquad\qquad\qquad \dots (4.17)$

Now, if the scatteing potential is spherically symmetric, one have $U(\mathbf{r}') = U(r')$. Since the scattering has azimuthal symmetry, one obtains the scattering amplitude as

$$f(\theta) = -\frac{2\pi}{4\pi} \int_0^\infty r'^2 \, U(r')dr' \int_{-1}^1 \exp(iqr' \cos\alpha)\, d(\cos\alpha)$$

$$= -\frac{1}{q} \int_0^\infty U(r')r' \sin(qr')dr' \qquad \dots (4.18)$$

Let us assume that the potential giving rise to the scattering has a *short range R*. This means that $V(r') \to 0$ for $r' > R$. We have $qR \ll 1$, for low, energy. In this case, we can write $\sin qr' \approx qr'$, as the integrand in (4.5) has the non-vanishing value for $r' < R$, then

$$f(\theta) = -U(r')\, r'^2\, dr' \qquad\qquad \text{... (4.19)}$$

Obviously, the scattering is spherical symmetric, i.e. there is no θ-dependence.

At higher energies $qr \gg 1$ except at very small angles. One finds in this case,

$$f(\theta) = -\int_0^{\infty} U(r')\, \frac{\mathrm{Sin}\, qr'}{qr'}\, r'^2\, dr' \qquad\qquad \text{... (4.20)}$$

We may note that the integrand has significant values only within the range of interaction, i.e. $r' < R'$. Within this range qr' is usually large. Thus $\dfrac{\mathrm{Sin}\,(qr')}{qr'}$ has a small value and therefore it does not contribute much to the integral except for very small angles $\theta \approx 0$ of scattering for which $\theta = 2k\sin\dfrac{\theta}{2}$. Thus the scattering amplitude $f(\theta)$ has large value at small angles of scattering only, i.e. $\sigma(\theta)$ is peaked forward. We find that this result agrees with the conclusion drawn earlier for high energy scattering by ordinary force same as the interaction assumed here.

Let us examine the situation if we assume that the interaction is of the Majorana type. In this case it transforms $\mathbf{r}'$ to $-\mathbf{r}'$ or z' to $-z'$. We find that this is equivalent to changing the scattered wave vector from $\mathbf{k}$ to $-\mathbf{k}$. Using

$$f(\theta) = \frac{1}{4\pi} \int U(r')\, \exp\{\, -i(\mathbf{k} + \mathbf{k}') \cdot \mathbf{r}'\}dr' \qquad\qquad \text{... (4.21)}$$

From Figure (4.25b), we have

$$q' = |\mathbf{k} + \mathbf{k}'| = 2k\cos\frac{\theta}{2} \qquad\qquad \text{... (4.22)}$$

Obviously, the scattered amplitude has significant values only for small values of q' which occurs for $\theta \to p$, i.e. the scattering cross section is peaked backwards.

The expected variation of $\sigma(\theta)$ with θ for two cases discussed above is illustrated in Figure 4.26.

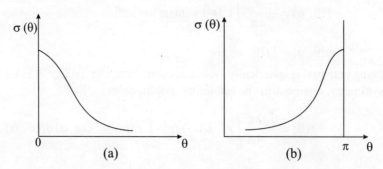

Fig. 4.26 Variation of scattering cross section $\sigma(\theta)$ with θ according to Born approximation (a) ordinary force (b) Majorana force.

(ii) **High Energy p-p scattering**   Just as in the case of n-p scattering we can expect to obtain more information concerning the shape and exchange character of the potential from high energy p-p scattering Figure 4.27 shows the experimental differential cross section for high energy (>10 MeV) p-p scattering. The experimental observations are quite puzzling. For energies about 500 MeV, the angular distribution is flat except at small angles, revealing that the dependence of the differential cross section on θ is almost very small. Such spherically symmeric angular distribution seems to be only possible for S-wave, i.e $l = 0$ scattering. However, this is impossible at these high energies. Further, we note that p-p total cross section is almost constant (~ 23 millibarns) between 150 MeV to 600 MeV whereas n-p total cross section shows $\dfrac{1}{E}$ variation.

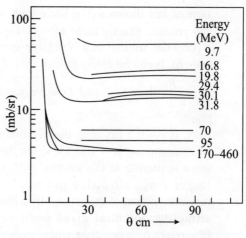

Fig. 4.27 Experimental curves for the variation of $\sigma_{pp}(\theta)$ with θ at high energies.

Clearly, the hypothesis of no p-p force for vlaues of $l > 0$ is not tenable as the calculated value of total S-wave cross section comes out to be 11 millibarns agains the experimental value of 23 millibarns. Obviously, the balance must be due to higher l-values.

If one assumes that only S, P and D waves are present then there should be following terms in the expression for $\sigma(\theta)$:

(i) a term proportional to $\sin^2 \delta$ (due to ^1S scattering)
(ii) a term in $\sin^2 \delta_2 [P_2(\cos \theta)]^2$ (due to ^1D scattering)
(iii) a term in $\sin \delta_0 \sin \delta_2 [P_2(\cos \theta)]$ (due to interference between S and D scattering)
(iv) terms due to ^3P wave scattering with different phase shifts δ_{10}, δ_{11} and δ_{12} for ^3_0P, ^3_1P substates respectively due to *tensor force*.
(v) Other interference terms which are propotional harmonics.

One can then explain the nearly isotropic angular distribution of p-p scattering using the following two hypotheses: (i) there exists a force which makes the coefficients of S and D interference terms negative, and (ii) there exists a tensor force in the triplet state which builds up the angular distribution in the middle region. There is a possibility that such a combination of forces may produce a flat distribution at one particulr energy. However, such a mixture of forces seems not able to produce flat distribution at all energies.

For this to happen, as the energy rises the D-scattering should become more prominent and to compensate this the S-D interference term must change by the right amount. Actually at high energy the coefficient of this interference term involving $\sin \delta_0 \sin \delta_1$ should be negative and should decrease with rise in energy as shown in Figure 4.28.

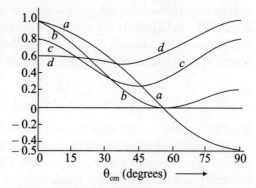

Fig. 4.28 Contributions from S and D waves to p-p scattering.

Jastrow has shown a p-p force with a repulsive core can produce exactly these effects. Figure 4.29 shows a potential similar to a repulsive core having range $c(<<b)$. It can be easily shown that if E_m is such that

$\lambda_m = \dfrac{R}{3}$, then S, P and D wave scattering will take place.

Now, if $c \sim 0.5$ fm, then the maximum energy for S-wave scattering is ~ 100 MeV. On the hand P and D wave scattering at this energy will occur due to the longer range attractive potential. We know that repulsive potential gives negative phase shifts while attractive potential gives positive phase shifts. Obviously, δ_0 is negative, where as δ_2 is positive. Thus $\sin \delta_0 \sin \delta_2$ will be negative as required to make the angular distribution flat even at high energies.

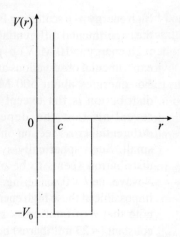

Fig. 4.29 Repulsive core potential with the range c (<< b).

The energies at which the phase shift for a given l changes sign depends entirely on the details of potential shape. In agreement with the experiment, the potential shape reveals that the ^1S phase shift (δ_0) become maximum between 10 to 20 MeV and negative at around 150 MeV. Obviously, the observed flat angular distribution between 150 to 500 MeV can be explained in terms of a uniformly changing phase shift in a appropriately chosen potential with a repulsive core.

We may note that the repulsive core is postulated for p-p system only which is a singlet state.

We may note that at higher energies in n-p and p-p interactions, tensor forces are important. very recently, a promising technique involving the scattering of polarized protons has been developed to obtain more detailed informations about the nucleon interaction.

4.18 POLARIZATION: HIGH ENERGY NUCLEON SCATTERING

The spin orientations of nucleons in a beam from an accelator are randomly distributed, i.e. unpolarized. However, if the spins somehow orient themself in some preferred direction, the beam is said to be polarized. If the number of nucleons in the beam with the spin component parallel to this preferred direction is N_+ and the number with anti parallel spin component is N_-, then the degree of polarization or *polarizing power*, is defined as

$$P = \frac{N^+ - N^-}{N^+ + N^-} \qquad \qquad \dots (4.1)$$

If either N_+ or N_- is zero then the beam is said to be completely polarized.

Nucleons may be polarized by elastic scattering on a spinless centre provided that the interaction producing the scattering is spin dependent. Figure 4.30 shows the effect of spin-orbit force in the scattering of a nucleon from a spinless target T. It is assumed that the force is attractive when **L** and **S** are parallel and repulsive for **L** and **S** antiparallel. In Figure (4.30a) the incident nucleons have spin 'in', i.e. the **S** goes into the paper. Those nucleons going to the left of T have **L** and **S** parallel and hence nucleons are scattered to the right due to attraction. On the otherhand, nucleons going to the right of T suffer repulsion as **L** and **S** are antiparallel. Obviously, in both the cases the beam is deflected to the right. In Figure (4.30b), the incident nucleons have spin 'out', i.e. out of the paper. In this case nucleons are deflected to the left in both the cases, i.e. L and S parallel and antiparallel.

One may scatter a beam of protons on a target and select the beam scattered under angle θ_1. By scattering this beam second time on a target under angle θ_2, one finds that the scattered intensity depends not only on θ_2, but also on angle ϕ, where ϕ is the angle between the two planes of scattering we have.

$$I = A(\theta_2) + B(\theta_2) \cos \phi \qquad \dots (4.2)$$

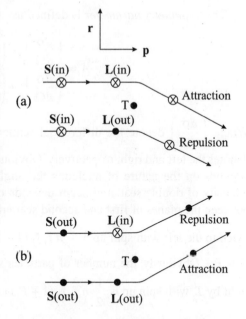

Fig. 4.30 Effect of spin-orbit interaction $(\mathbf{L} \cdot \mathbf{S})$ in the scattering of a nucleon from a spinless target T.

Double scattering

Let n be the number of nucleons in the initial unpolarised beam. One can assume that in this beam there are $\frac{1}{2} n$ nucleons with spins of each sign (+ or –).

After scattering by target T_1 to the left, the number of nucleons reaching the target T_2 with polarization P_1 and spin up (+) is $\frac{1}{2} n f_1 (1 + P_1)$ and with spin down (–) is $\frac{1}{2} n (1 - P_1)$, where f_1 is the fraction of nucleons reaching T_2. Similarly, those nucleons scattered to the right by T_1, $\frac{1}{2} n f_1 (1 + P_1)$ will reach T_2 with spin up and $\frac{1}{2} n f_1 (1 + P_1)$ will reach T_2 with spin down. Here P_2 and f_2 are the polarisation with spin up and fraction of all nucleons scattered by the target T_2 respectively. Figure (4.31) depicts the double scattering.

In an actual experiment, an asymmetry is observed in the intensities of the beam scattered to the left and right after the second screening. The asymmetry is a simple function of the polarization at the two scatterings.

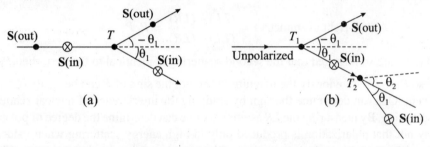

Fig. 4.31 (a) Scattering of an unpolarised beam from a target T and (b) Double scattering from target T_1 and T_2

The *asymmetry parameter* is defined as

$$A = \frac{\left(\dfrac{d\sigma}{d\Omega}\right)_L - \left(\dfrac{d\sigma}{d\Omega}\right)_R}{\left(\dfrac{d\sigma}{d\Omega}\right)_L + \left(\dfrac{d\sigma}{d\Omega}\right)_R} \qquad \text{... (4.3)}$$

Where $\left(\dfrac{d\sigma}{d\Omega}\right)$ denote the differential scattering cross section after second scattering. L and R denotes the left and right respectively. Obviously, the value of A denotes on P_1 and P_2 which in turn depends on the nature of nucleons the angle of scattering (θ_1 and θ_2) and the energy E. The intensity of double scattered beam depends on the angle θ_2 as well as on the azimuthal angle ϕ between the planes of first and second scattering as shown in Figure (4.32). The number scattered twice to the left with spin up = $\dfrac{1}{2} n f_1 f_2 (1 + P_1)(1 + P_2)$ and with spin down = $\dfrac{1}{2} n f_1 f_2 (1 - P_1)$ $(1 - P_2)$. Obviously, the number of particles scattered twice, first to the left by T_1 and then to the right by T_2 with spin up is $\dfrac{1}{2} n f_1 f_2 (1 + P_1)(1 - P_2)$. Thus the number of particles scattered twice, first to the left and then to the right (LR) = $\dfrac{1}{2} n f_1 f_2 (2 - 2 P_1 P_2)$. So the left-right asymmetry (A) becomes

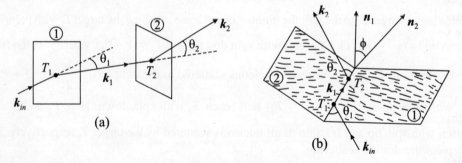

Fig. 4.32 Double scattering experiment. The angle between the normals $\mathbf{n}_1$ and $\mathbf{n}_2$ of the two scattering planes shown by (4.1) and (4.2) is ϕ.

$$A \text{ (left-right asymmetry)} = \frac{(LL) - (LR)}{(LL) + (LR)} = P_1 P_2 \qquad \text{... (4.4)}$$

If the conditions are so arranged that the second scattering is identical to the first, then $P_1 = P_2 = P$ and $A = P^2$ so $P = \sqrt{A}$. Obviously the magnitude and not the sign of P can be easily obtained from the experiment. One can determine the sign by studying the interference of nuclear scattering with Coulomb scattering. By measuring the LR asymmetry one can determine the degree of polarization P.

We may not that polarization is produced only in high energy scattering when values of $l > 0$ are present in the scattering. At low energies S-wave scattering with $l = 0$ occurs. Clearly, the polarization takes place due to **L · S** (spin-orbit interaction). Polarization in scattering becomes significant only at high energies ~ 100 MeV.

This asymmetry can be explained if the interaction has a tensor character with the spin-orbit coupling. As stated above, the asymmetry is large if l is large.

Triple scattering experiments have also been performed which determines completely the scattering matrix.

4.19 MESON THEORY OF NUCLEAR FORCE

We have described the nuclear force between two nucleons is due to a potential which is mainly a function of distance between the nucleons. The force is propagated instantaneously and also the description is non-relativistic. However, this description of nuclear force was found inadequate to explain many of the experimental results, e.g. magnetic moment and quadrupole moment of deutron. In order to overcome these difficulties, Yukawa in 1935 proposed that nucleon is a source of a force field in which the other nucleon finds itself. Yukawa further argued that the Coulomb force between electric charges is of infinite range and electromagnetic interaction between charges acts through the exchange of photons. In the similar way, one can describe the nuclear interaction between nucleons in terms of force field and force may be conveyed through the exchange of certain nuclear particle between the nucleons. We may note that there is a marked difference between the nuclear interaction and electromagnetic interaction, which lies in their range. As stated earlier, nuclear force is a short (finite) range force, whereas electromagnetic force is a long range force. Obviously, when developing on the lines of electromagnetic interaction, one must include a parameter which may take care for the finite range.

The wave equation derived from Maxwell's field equations is

$$\nabla^2 V - \frac{1}{c^2} \frac{\partial^2 V}{\partial t^2} = 0 \qquad \qquad ...\ (4.1)$$

Where V is a scalar potential. Equations (4.1) reduces to the Laplace's equations in free space for static charges.

$$\nabla^2 V = 0 \qquad \qquad ...\ (4.2)$$

If a point source with charge q is located at the origin, the equation takes the form of Poisson's equation.

$$\nabla^2 V = - \frac{1}{4\pi \varepsilon_0} 4\pi q\, \delta(r) \qquad \qquad ...\ (4.3)$$

Here $\delta(r)$ is familiar Dirac delta function. The solution of (4.3) is the Coulomb potential

$$V = - \frac{q}{4\pi \varepsilon_0 r} \qquad \qquad ...\ (4.4)$$

Now, if there is a charge q at a point distance r from the origin, its potential energy $U = qV = \frac{q^2}{4\pi \varepsilon_0 r}$

In order to account for short range nuclear interaction, Yukawa modified the wave equation (4.1) as

$$\nabla^2 \phi - \frac{1}{c^2} \frac{\partial^2 \phi}{\partial t^2} - \mu^2 \phi = 0 \qquad \text{... (4.5)}$$

where ϕ is nuclear potential function, μ is a constant parameter having the dimension of recipocal length. We will see that μ is related to the range (b) of the internucleon. Putting $M = 273 \, m_e$, one

obtains $b = \dfrac{1}{\mu} = 1.41$ F. Now, using a wave like solution for (5), one obtains

$$-k^2 + \frac{\omega^2}{c^2} - \mu^2 = 0 \qquad \text{... (4.6)}$$

de Broglie relations for energy (E) and momentum (p) are,

$$E = \hbar \omega \text{ and } p = \hbar k \qquad \text{... (4.7)}$$

Where k is wave vector and ω is angular frquency. Using (4.7), Equation (4.6) reduces to

$$E^2 = c^2 p^2 + M^2 c^4 \qquad \text{... (4.8)}$$

Where

$$M = \frac{\hbar \mu}{c} \qquad \text{... (4.9)}$$

Equation (4.8) is a familiar relativistic equation showing relationship between energy and momentum of a particle whose rest mass is M, given by equation (4.9). Obviously, the constant parameter μ defines the rest mass of the particle. The time independent equation as obtained from equation (4.5) is

$$\nabla^2 \phi - \mu^2 \phi = 0 \qquad \text{... (4.10)}$$

If the mass of the particle is zero, i.e. $\mu = 0$, then equation (4.10) reduces to equation similar to the Laplace's equation, i.e. $\nabla^2 \phi = 0$.

One can obtain the equation corresponding to Possion's equation by introducing a point source strength g located at the origin. One can write such equation as

$$\nabla^2 \phi - \mu^2 \phi = g \, \delta(r) \qquad \text{... (4.11)}$$

The solution of equation (4.11) is

$$\phi(r) = -\frac{g \, e^{-\mu r}}{4 \pi r} \qquad \text{... (4.12)}$$

Here g is an arbitrary constant. One finds that with proper choice of μ, i.e. mass μ of the particle, one can make ϕ to describe the short range of nuclear interaction. The constant g behaves as a strength of a point source situated at the origin. If another source g is situated at r, then the interaction energy, i.e potential energy between two such sources is given by

$$U = g \, \phi(r) = -\frac{g^2 (r)}{4 \pi} \frac{e^{-\mu r}}{r} \qquad \text{... (4.13)}$$

Equation (4.13) clearly shows that unlike Coulomb interaction, the nuclear interaction decreases more rapidly as one moves away from the source due to the exponential factor. This takes care of the short range of nuclear interaction. This is well known Yukawa potential. If we put $\mu = 0$ (or $M = 0$) and $g = \dfrac{g}{\varepsilon_0}$, equation (4.13) reduces to (4.4). In terms of range, we have $g^2 = V_0 b$. For the singlet p-p potential, $V_0 = 43$ MeV, $b = 1.18$ F. One obtains $g^2 \sim 8 \times 10^{-27}$ and $\dfrac{g^2}{\hbar c} = 0.3$. One must compare it with the value of the fine structure constant $\alpha = \dfrac{1}{137}$, which is the coupling constant for the electromagnetic interaction. The strength of the meonic charge is thus much larger than that of the electric charge. One of the consequences of this large coupling constants that the exchange of two or more mesons may have appreciable probability.

Particles of mass around $280\ m_e$ were observed in cosmic ray experiments in 1947. These particles were named as π-mesons or simply *pions*. There are three types of pions: electrically charged pions ($\pi^\pm$) and electrically netural pion (π^0). These pions are "carriers" of nuclear force field. Exchange of pions to provide interaction between two nucleons (say 1 and 2) can be expressed as

$$p \xleftrightarrow{\ \pi^+\ } n + \pi^+$$

$$n \xleftrightarrow{\ \pi^-\ } p + \pi^-$$

$$p \xleftrightarrow{\ \pi^0\ } p + \pi^0$$

$$n \xleftrightarrow{\ \pi^0\ } n + \pi^0$$

We may note that such exchange processes are not observable for free nucleons. A free neutron, for e.g. decays with a half life of about 12 minutes

$$n \to p + e^- + \bar{\nu}$$

A nucleon within the nucleus cannot provide the essential energy for the production of pions and hence the processes are forbidden classically. Obviously, the law of conservation of energy is violated. Quantum mechnically, one can made these processes possible provided that the energy $\Delta E = M C^2$, required for the production of pions may be obtained from the nucleon and return the same within a time Ñt given by the uncertainty relation

$$\Delta E\ \Delta t \approx \hbar$$

or
$$\Delta t \approx \frac{\hbar}{\Delta E} = \frac{\hbar}{M C^2}$$

If we assume that the pion travels with a velocity of light (c) within the nucleus, the maximum distance it can travel in time $\Delta t = c\Delta t$. Let this distance is the range of nuclear interaction, then

$$b = c\Delta t = \frac{c\,\hbar}{Mc^2} = \frac{\hbar}{Mc} = \frac{1}{\mu}$$

Obviously, this is same as indicated by (4.12). We may note that the pions exist for very short time 10^{-23} s, and therefore it cannot be observed. This is why such a pion is termed as virtual pion and the excahnge process is also termed as virtual process. In a virtual process, for e.g. a proton emits a π^+ gets transformed into a neutron and this pion (π^+) is absorbed by a neutron and transformed into a proton. This characterises strong interaction between a proton (p) and a neutron (n). Similary a π^- is emitted by *a n*, transferred into *a p* and this π^- is absored by *a p* and transformed into *a n*. Similarly, the neutral pion (π^0) is exchanged between like nucleons, i.e. $n \leftrightarrow n$ or $p \leftrightarrow p$.

In order to take into account the excahnge interaction, one will have to introduce excahnge operators which can be constructed using the isospin operators τ_1, τ_2, τ_3 identical with the Pauli spin operators σ_x, σ_y, σ_z. If $\gamma = \binom{1}{0}$ and $\delta = \binom{0}{1}$ represent the two possible isospin states (the p and the n respectively), one finds

$$\tau_1 \gamma = \delta, \qquad \tau_1 \delta = \gamma$$
$$\tau_2 \gamma = i\delta, \qquad \tau_2 \delta = -i\gamma$$
$$\tau_3 \gamma = \gamma, \qquad \tau_3 \delta = -\delta \qquad \text{... (4.14)}$$

One obtains the exchange operators τ_+ and τ_- as

$$\tau_+ \gamma = \frac{1}{\sqrt{2}}(\tau_1 + i\tau_2)\gamma = 0, \; \tau_+ \delta = \sqrt{2}\,\gamma \qquad \text{... (4.15a)}$$

$$\tau_- \gamma = \frac{1}{\sqrt{2}}(\tau_1 - i\tau_2)\gamma = \sqrt{2}\delta, \; \tau_- \delta = 0 \qquad \text{... (4.15b)}$$

Thus τ_+ transforms a neutron (δ) into a proton(γ) and τ_- transforms a proton (γ) into a neutron (δ). However, τ_3 operating on the proton (neutron) wave function gives the same wave function, though with the changed sign in the case of the neutron (δ). Obviously, it is different from the "neutral field" of Yukawa's meson theory.

Now, the source term in the wave function (4.5) [written as $\nabla^2 \phi - \frac{1}{c^2}\frac{\partial^2 \phi}{\partial t^2}\,\mu^2 \phi = 4\pi\,\eta(\mathbf{r})$, where $\eta(\mathbf{r})$ is the source strength], analogous to the charge density for the em field has to expressed as $\eta(\mathbf{r})\,\tau_n$, where $\tau_n = \tau_\pm$ in the 'charged theory' $\tau_n = \tau_\pm$, τ_3 in symmetry theory' and $\tau_n = 1$ in the neutral theory due to Yukawa. Both $\pi^\pm$ and π^0 are exchanged in this symmetric theory. Then the static wave equation can be expressed as

$$\nabla^2 \phi - \beta^2 \phi = 4\pi\,\tau_n\,\eta(\mathbf{r}) \qquad \text{... (4.16)}$$

For a print nucleon at $\mathbf{r}_1$, the solution of (4.16) is

$$\phi(\mathbf{r}) = -g_n \tau_n \frac{\exp[-\beta|\mathbf{r} - \mathbf{r}_1|]}{|\mathbf{r} - \mathbf{r}_1|} \qquad \text{... (4.17)}$$

The potential energy of a second nucleon having strength $g_n'\,\tau_n'$ is thus obtained as

$$U = -g_n\,g_n'\,\tau_n\,\tau_n'\,\frac{\exp[-\beta|\mathbf{r}_2 - \mathbf{r}_1|]}{|\mathbf{r}_2 - \mathbf{r}_1|} \qquad \text{... (4.18)}$$

Here τ_n and τ_n' act on the wave functions of emitting and absorbing nucleons respectively.

We may note that the meson field can either be scalar or pseudoscalar. Further, a scalar field does not change sign on inversion while a pseudoscalar changes sign on inversion.

If $\tau^{(1)}$ and $\tau^{(2)}$ are the isospin operators for the two nucleons, then one can see that

$$\tau^{(1)} \cdot \tau^{(2)} = \tau^{(1)} \tau^{(2)} + \tau_2^{(1)} \tau_2^{(2)} + \tau_3^{(1)} \tau_3^{(2)}$$

$$= \tau_+^{(1)} \tau_-^{(2)} + \tau_-^{(1)} \tau_+^{(2)} + \tau_3^{(1)} \tau_3^{(2)} \qquad \dots (4.19)$$

For different *cases*, one obtains different results.

The Yukawa's meson field theory has been a useful step towards the understanding of strong nuclear forces. If a π-meson is actually exchanged between interacting nucleons, most of the properties of nuclear forces follow directly. The repulsive core, however, requires the exchange of heavier mesons and spin-orbit force vector meson with spin-parity I^-. This theory quantitatively account for the 1.4 Fm range part of the nuclear force.

SUGGESTED READINGS

1. J. M. Blatt and V. F. Weisskoff, Theoretical Nuclear Physics, John Wiley (1952).
2. D. M. Brink, Nuclear Forces, Pergamon (1965).
3. J. M. Pearson, Nuclear Physics; Energy and Matter, Adam Hilger, ch-II (1986).
4. A. de Shalit and H. Feshback, Theoretical Nuclear Physics, Wiley (1974).
5. K. Wildermuth and Y.C. Tang, A Unified Theory of Nucleus, Vieweg, Braunschweig, Germany (1977).
6. R. G. Sachs, Nuclear Theory, Addition Wesley (1955).
7. W. Magie, A Source Book in Physics, Cambridge (1963).
8. R. R. Roy and B. P. Nigam, Nuclear Physics, John Wiley (1967).
9. G. E. Brown, Unified Theory of Nuclear Models and Nuclear Forces, North Holland (1968).
10. S. L. Kakani and H. M. Chandalia, Quantum Mechanics, (4th ed. 2005), Sultan Chand, New Delhi-2
11. L. I. Schiff, Quantum Mechanics, 3rd ed. McGraw Hill, (1968).

Example 4.1 The experimentally measured mass of π-meson is 140 MeV/c². Show that the range of the nuclear force is ~ 1.4 fm.

Sol. The range of nuclear force is

$$r = \frac{\hbar}{m_\pi c} = \frac{\hbar c}{m_\pi c^2}$$

$$= \frac{1.05 \times 3 \times 10^{-26} \text{ J} - \text{m}}{140 \times 1.6 \times 10^{-13} \text{ J}}$$

$$= 1.4 \times 10^{-15} \text{ m} = 1.4 \text{ fm}$$

$c = 3 \times 10^8$ m s⁻¹

$\hbar = 1.05 \times 10^{-34}$ J-s

$m_\pi = 140$ MeV/c²

$\quad = 140$ MeV/c² $\times 1.6 \times 10^{-13}$ J/MeV/c²

$\quad = 140 \times 1.6 \times 10^{-13}$ J

Example 4.2 Assuming that the zero range wave function $u(r) = r\psi(r) = C \exp(-\alpha r)$ is valid for the deutron from $r = 0$ to $r = \infty$, obtain an expression for the normalization constant C. If $\alpha = 0.232$ fm⁻¹ find the probability that the separation of the two particles in the deutron exceeds a value of 2 fm. Find also the average distance of interaction for this wave function. [Bangalore]

Sol. The normalization constant is given by

$$4\pi \int_0^\infty |\psi|^2 r^2 \, dr = 1 \qquad \dots (4.1)$$

We have
$$r\,\psi(r) = C\,\exp(-\alpha r) \qquad\qquad \ldots (4.2)$$

$$\therefore \qquad 4\pi \int_0^\infty C^2 \exp(-2\alpha r) = 1 \text{ or } C = \sqrt{\frac{\alpha}{2\pi}}$$

Using (4.2), (4.1) yield $C = \sqrt{\dfrac{\alpha}{2\pi}}$

The probability of a seperation greater than R is

$$4\pi\, C^2 \int_R^\infty \exp(-2\alpha r)dr = \exp(-2\alpha r)$$

$$= 0.395 \ (\because \alpha = 0.232 \text{ fm}^{-1})$$

The average distance of interaction

$$<r> = \frac{\int r\,|\psi|^2\, r^2\, dr}{\int |\psi|^2\, r^2\, dr} = \frac{1}{2\alpha} = 2.2 \text{ fm.}$$

Example 4.3 Given the magnetic moment of proton, $\mu_p = 1.41061 \times 10^{-26}$ J T^{-1} and that of neutron, $\mu_n = -0.96624 \times 10^{-26}$ J T^{-1}, calcualate the force between these two particles in a triplet state at a separation of 3×10^{-15} m and the work required, on account of this force, to bring the neutron from infinity to this distance from the proton. Assume that the spins always point along the line joining the particles. [Calcutta, Allahabad]

Sol. The force between the two magnetic dipoles in the end on position is $\left(\dfrac{\mu_0}{4\pi}\right)\left(\dfrac{6\mu_1\mu_2}{r^4}\right)N$

and if this a triplet state it is repulsive because of the opposite sign of the two moments. The work is

$$\therefore \qquad W = -\int_\infty^r \frac{\mu_0}{4\pi}\frac{6\mu_1\mu_2}{r^4}\,dr = \frac{\mu_0}{4\pi}\frac{2\mu_1\mu_2}{r^3}\,\text{J}$$

Substituting the values and simplifying, one obtains

$$F = 1 \text{ N}$$

and

$$W = 6267 \text{ eV.}$$

$\mu_0 = 4\pi \times 10^{-7}$ NA^{-2}
$\mu_1 = \mu_p = 1.41 \times 10^{-26}$ JT^{-1}
$\mu_2 = \mu_n = -0.966 \times 10^{-26}$ JT^{-1}
$r = 3 \times 10^{-15}$ m

Example 4.4 Assuming that the deutron-internucleon potential is of rectangular well type with depth V_0 and range r_0, show that the radius of deutron is given by $r_d = \dfrac{2\,r_0\,V_0^{\frac{1}{2}}}{\left(\pi\,E_B^{\frac{1}{2}}\right)}$, where E_B is the binding energy of the deutron. [Raj., Jodhpur]

Sol. When we assume deutron-internucleon potential is of rectangular type, the relation that holds good in its ground is

$kr_0 = \dfrac{\pi}{2}$. The deutron radius is given by

$$r_d = \frac{1}{r}\sqrt{\frac{\hbar^2}{M\,E_B}} \qquad \qquad \text{... (4.1)}$$

Now

$$k = \sqrt{\frac{M}{\hbar^2}(V_0 - E_B)} \qquad \qquad \text{... (4.2)}$$

We have

$$kr_0 = \frac{\pi}{2} \quad \therefore\ k = \frac{\pi}{2r_0} \qquad \qquad \text{... (4.3)}$$

From (4.2) and (4.3), we have

$$\frac{\pi}{2r_0} = \sqrt{\left[\frac{M}{\hbar}(V_0 - E_B)\right]} \qquad \qquad \text{... (4.4)}$$

(4.1) and (4.4) gives

$$r_d = \frac{2r_0}{\pi}\sqrt{\frac{V_0 - E_B}{E_B}} \qquad \qquad \text{... (4.5)}$$

Since the binding energy of deutron, E_B is very small, i.e. $E_B \ll V_0$ and hence

$$V_0 - E_B \approx V_0$$

$$\therefore \qquad \qquad r_d = \frac{2r_0}{\pi}\sqrt{\frac{V_0}{E_B}} \qquad \qquad \text{... (4.6)}$$

Example 4.5 The meson theory of nuclear force assumes the virtual exchange of pions. If a nucleon emits a virtual pion of rest mass $270\,m_e$, show that the range of nuclear force is 1.43 fm.

[Kerala, Punjab]

Sol. We have

$$\Delta E = \Delta m\,c^2 \qquad \qquad \text{... (4.1)}$$

From uncertainty relation

$$\Delta E\,\Delta t \approx \hbar$$

$$\therefore \qquad \qquad \Delta t = \frac{\hbar}{\Delta E} = \frac{\hbar}{\Delta m c^2}$$

Assuming tha the emitted pion travels at the speed of light, distance travelled by it during this time is given by

$$r_0 = c\,\Delta t = \frac{c\hbar}{\Delta m c^2} = \frac{\hbar}{\Delta m c}$$

$$= \frac{1.05 \times 10^{-34} \text{ J-s}}{270 \times (9.1 \times 10^{-31} \text{ kg}) \times (3 \times 10^8 \text{ m/s})}$$

$$= 1.43 \times 10^{-15} \text{ m} = 1.43 \text{ fm}$$

Example 4.6 Given, $\sigma_{para} = 4.19$ and $\delta_{ortho} = 128$ barns, calculate the scattering lenghts a_t and a_s.

[MS, Raj.]

Sol. We have the theoretical expressions for the total cross sections are

$$\sigma_{para} = 6.69 \, (3a_t + a_s)^2$$

$$\sigma_{ortho} = 6.69 \, [(3a_t + a_s)^2 + 2(a_t - a_s)^2] + 1.74 \, (a_t - a_s) \qquad \text{... (4.2)}$$

Now, substituting the values of σ_{para} and σ_{ortho} in (4.1) and (4.2) and solving these equations, one obtains

$$
\begin{array}{c|c|c|c|c}
a_t = & 0.91 & 0.52 & -0.52 & -0.91 \\
a_s = & -1.95 & -2.34 & -3.38 & -3.77
\end{array} \times 10^{-14} \text{ m}
$$

One can immediately discard the last two pairs as they have negative a_t values. One can also discard the first pair because it is not consistent with the n-p scattering data. The remaining values are:

$$a_t = 0.52 \times 10^{-14} \text{ m and } a_s = -2.34 \times 10^{-14} \text{ m}$$

Example 4.7 Given the density of Cadium (Cd) as 8.67×10^3 kg/m^3 and its scattering cross-section for neutrons involved to be 20800 barns, show that the attenuation coefficient and thickness of Cd^{113} required to reduce the intensity of neutron beam by 1 % of its value are 962×10^2 m^{-1} and 0.48×10^{-4} m respectively

[Punjab]

Sol. (i)

$$\mu = \frac{\sigma \, N_A \, \sigma}{A}$$

$$= \frac{8.67}{113} \times 6.022 \times 10^{23} \times 20800 \times 10^{-22} \text{ m}^{-1}$$

$$= 962 \times 10^2 \text{ m}^{-1}$$

(ii)

$$N\sigma t = \mu t = \text{Log}_e \frac{N_0}{N_t}$$

$$= \text{Log}_e \, 100 = 2.303 \times \text{Log}_{10}(2 \times 2.303)$$

$$= 4.606$$

$$\therefore \qquad t = \frac{4.66}{\mu} = \frac{4.606}{962 \times 10^2 \text{ m}^{-1}} = 0.48 \times 10^{-4} \text{ m}^{-1}$$

Example 4.8 Given $a_t = 5.38$ fm, $a_s = -23.7$ fm, $r_{ot} = 1.70$ fm and $r_{os} = 2.40$ fm, show that total cross section for n-p scattering at neutron energy 2 MeV (Lab.frame) is 2.904 barns. [Jodhpur]

Sol. Total cross section

$$\sigma = \frac{3}{4}\,\sigma_t + \frac{1}{4}\,\sigma_s$$

or

$$\sigma = \frac{3\pi\,a_t^2}{\left[1 - \frac{1}{2}\,a_t\,r_{ot}\,k^2\right]^2 + a_t^2\,k^2} + \frac{\pi\,a_s^2}{\left[1 - \frac{1}{2}\,a_s\,r_{os}\,k^2\right]^2 + a_s^2\,k^2}$$

In CM frame system, $E_{CM} = \dfrac{1}{2}\,E_{Lab} = 1$ MeV

$$\therefore \qquad k^2 = \frac{ME}{\hbar^2} = \frac{1.6748 \times 10^{-27} \times 1.6 \times 10^{-13}}{(1.0549 \times 10^{-34})^2}$$

$$= 2.413 \times 10^{28}$$

$$\therefore \qquad \sigma = 1.83 \times 10^{-28} + 1.074 \times 10^{-28} = 2.904 \text{ barns}$$

REVIEW QUESTIONS

1. Define scattering cross section and scattering length. Give an account of effective range theory of n-p scattering at low energies [Kerala, Raj.]

2. Show that for square well type of inter-nucleon potential of depth $V_0 \ll E_B$ and range r_0, the scattering length a for a spinless neutron can be express as: $a = b - \left(\dfrac{1}{\cot k\,r_0}\right)$.

   [Mysore, Jodhpur]

3. Discuss the neutron-proton scattering at low energy. Show how the assumption of spin-dependence of nuclear force can explain the experimental results. [MDS, Punjab]

4. Give a theory of the deutron. Obtain and plot the wave function for the deutron ground state taken as an S-state [Delhi]

5. Give an account of meson theory of nuclear forces. [MDS]

6. Give an account of n-p scattering at energies below 0 MeV. Using experimental datas, can you determine the shape of n-p potential? [Mysore]

7. Deutron forms a bound state of a neutron and proton whereas a system consisting of two neutrons does not form a bound state. What informations about the nature of a nuclear force can be derived from it? Explain.

8. What are the general characterstics of a nuclear force? Make a comparison of four basic interactions? [MDS]

9. Assuming a square well type of nuclear potential, give the theory of the deutron problem and obtain a relation between the depth and width of the well and deutron binding energy.

   [Lucknow, Meerut]

10. Set the Schrodinger equation for the deutron for $l = 0$ $(s-)$ state-assuming a square-well potential and show that no excited state is possible for the deutron. [Allahabad, Meerut]

11. What are exchange forces? How will you interpret high energy n-p scattering in terms of these forces?

12. Assuming rectangular well type of nuclear potential give an account of the phase shift analysis of n-p scattering at low energies. Also that the descrepancy between theoretical and experimental results can be resolved if we assume that nuclear force is spin-dependent.   [MDS, Mysore]

13. Give an account of the effective range theory of n-p scattering at low energies and show that the scattering cross section besides being a function of energy, depends upon the scattering length and the effective range.

14. Give an account of tensor forces. How the study of deutron problem indicates the spin-dependent tensor character of nuclear forces? Explain.   [Punjab, Vikram]

15. Explain differential and total scattering cross sections. For n-p scattering at low energies, obtain an expression for the total cross section.

16. Write short notes
    (a) General characterstics of nuclear force
    (b) Meson theory of nuclear forces
    (c) Exchange forces
    (d) Tensor forces
    (e) Deutron problem

17. The nucleon-nucleon force is usually described as (a) strong, (b) short range, (c) charge symmetric (d) charge independent (e) containing both ordinary and exchange terms, (f) spin-independent, (g) non-central, (h) saturating in nuclei. Explain the meaning of these and discuss briefly the experimental evidence for them.

18. What is meant by the statement that strong interactions conserve isospin? Why does isospin conservation imply the charge independence of nucleon-nucleon force, while pion-nucleon interactions are not generally charge independent? The reaction $d + d \rightarrow {}^4\text{He} + \pi^0$ has never been observed? Why is this? Can the reaction occur at all?   [Bangalore]

19. Discuss why ${}^8\text{He}$ is stable against nucleon emission but ${}^5\text{He}$ is not.

20. The life times of the charged and neutral pions are respectively 2.6×10^{-8} s and 8×10^{-17} s. Why, in spite of the difference between these two numbers, it is useful to regard the pions as members of an isospin triplet? Are the pion lifetimes are relevant to nuclear forces?

21. Why is it that the proton field cannot contain a π^--meson and the neutron field cannot contain a π^+-meson?

22. In (a) neutron-proton scattering and (b) in proton-proton scattering, which species of π-mesons are excahnged?

23. Show that meson-theory, leads not only to the attractive radial depending part, but also to a repulsive core.   [Allahabad]

PROBLEMS

1. Assuming a simple square well potential of depth V_0 and radius R for the potential between the neutron and proton in a deutron, show that the matching condition $\tan KR = -\dfrac{K}{\beta}$ gives the condition $V_0 R^2$ approximately constant and that the value of the constant is ~ 1 MeV barn.

2. The potential energy operator V acting on the wavefunction ψ is $V(\mathbf{r})\,\psi(\mathbf{r})$ for a local potential and $\int V(\mathbf{r}, \mathbf{r}')\,\psi(r')\,d^3 r'$ for a non-local potential. Show that such a non-local potential is equivalent to momentum dependent potential by showing that

$$\int V(\mathbf{r}, \mathbf{r}') \, \psi(\mathbf{r}') \, d^3 r' = \int U(\mathbf{r}, \mathbf{p}) \phi(\mathbf{p}) \, \exp\left(\frac{i\mathbf{p} \cdot \mathbf{r}}{\hbar}\right) d^3 p$$

Where $\qquad \psi(\mathbf{r}) = (2\pi)^{-\frac{3}{2}} \int \exp\left(\frac{i\mathbf{p} \cdot \mathbf{r}}{\hbar}\right) \phi(\mathbf{p}) \, d^3 p$

and the momentum dependent potential

$$U(\mathbf{r}, \mathbf{p}) = (2\pi)^{-\frac{3}{2}} \int \exp\left(\frac{i\mathbf{p} \cdot \mathbf{x}}{\hbar}\right) V(\mathbf{r}, \mathbf{r} + \mathbf{x}) \, d^3 x.$$

3. Use the parameters $a_t = 5.38$ F, $r_{ot} = 1.70$ fm, and $a_s = -23.7$ fm and determine r_{os} from the experimental data: $\sigma = 1.69$ barns at energy 4.75 MeV (*L*- system). [2.4 fm]

4. Calculate the total cross section for neutrons of energy 4 MeV in the CM system for *n-p* scattering given: $a_t = 5$ fm, $a_s = 24$ fm, $r_{ot} = 1.6$ fm and $r_{os} = 2.4$ fm. [0.8134 barns]

5. Show that Heisenberg exchange potential is equivalent to an ordinary potential, which changes sign according to whether **L** + **S** is even or odd, where **L** is the total two nucleon orbital angular momentum and **S** is the total spin.

6. A 6 MeV γ-ray is absorbed and dissociates a deutron into a proton and a neutron. If it makes an angle of 20° with the direction of the γ-ray determine the kinetic energies of the proton and the neutron. [Raj., MDS] [$K_p = 1.91$ MeV, $K_n = 1.86$ MeV]

7. Show that *p-p* scattering can take place only in ^1S, ^3P, ^1D, ... states. What about the total isospin of the *n-p* system in these states? [Punjab]

8. Show that ^3S, ^1P, ^3P and ^3D states of a neutron-proton system are the only one compatible with a nuclear spin equal to one. [Vikram]

9. Show that the low energy *n-p* scattering parameters have the following values $a_t = 5.38$ fm, $a_s = -23.7$ fm, $r_{ot} = 1.718$ fm and $r_{os} = 2.41$ fm by using the following data:
   Zero energy incoherent *n-p* scattering cross section $\sigma_0 = 20.38$ b, para-hydrogen scattering length $a_{para} = -3.78$ fm, deutron binding energy $\varepsilon_d = 2.226$ MeV which gives $\alpha_t = 0.232$ fm^{-1}, cross section for scattering of 2.375 MeV (C-system) neutrons is $\sigma = 1.69$ b. [Goa]

SHORT QUESTION ANSWERS

1. Is nuclear force between nucleons is to a good approximation both charge symmetric and charge independence?

Ans. Yes

2. What conclusion you can draw about nuclear froce from the analysis of both α-particle scattering and α-particle emission?

Ans. These studies showed that there is a nuclear force, which is attractive, acting between the particle and the nucleus, in addition to the repulsive Coulomb force acting between the two. They indicated that the nuclear force is of very *short-range*, i.e. that it extends only for a distance apprecialy less than 10 fm. The analysis also indicated that the nuclear force is *strong*, compared to the Coulomb force, since it dominates the latter, which is repulsive, to produce an overall attraction on the α-paricle when it is very close to the nucleus.

3. What conclusions you can draw about nuclear forces from modern experiments involving the scattering of protons from protons and protons from neutrons?

Ans. Scattering of protons from protons show that the range of nuclear force is $\simeq 2$ fm, and that the magnitude of the negative energy associated with the attractive force is larger than their Coulomb energy, when the two protons are separated by that distance, by roughly a factor of 10^2. Scattering of protons from neutrons indicate that the nuclear force is *charge independent,* i.e. the nuclear force between p-p, p-n and n-n is same (except for exclusion principle effects that apply in first and last cases only).

4. Can meson-exchange theory of nuclear forces help to understand qualitatively the spin, tensor, spin-orbit and exchange components of nuclear forces?

Ans. Yes

5. What inference you can draw about nuclear forces within nuclei from Pauli exclusion principle together with the tensor and exchange forces?

Ans. This can account for the *saturation* of the nuclear force. Moreover, this also implies that the effective one-body potential is momentum dependent.

6. What the saturation property of the nuclear force indicate?

Ans. This shows that a nucleon within a nuclei only interacts with a limited number, but not all, of the other nucleons within the nucleus.

7. What is a tensor force? How it differs from spin-orbit force?

Ans. This is velocity independent and cannot account for the polarization effects, which depend on velocity. They can be understood if there is a coupling between the spin and orbital motion of a nucleon in a potential field. This *spin-orbit force* in a complex nucleus may be derived from a two body spin-orbit force between a pair of nucleons and is velocity dependent, in contrast with the tensor force.

8. What do you understand by spin-dependency of nuclear force?

Ans. The nuclear force keeping two nucleon system bound, depends on whether the spins of two nucleons are parallel or antiparallel, i.e. where the total spin $S = 1$ or 0. This feature of nuclear force is quite different than, for e.g., the Coulomb force which goes as $\dfrac{1}{r^2}$ and so is just a function of r.

9. Is tensor force a non-central force?

Ans. Yes, the value of the tensor force depends on the angle θ which is measured from the direction of the spin vector **S**. Obviously, a tensor force is a function of **S**, **r**. This force is just similar to the force law for two magners, which depends on the way they are oriented.

10. Can a tensor force explain the deutron quadrupole moment?

Ans. Yes.

11. Write the three types of exchange forces

Ans. (i) The space exchange or Majorana force
 (ii) The spin exchange or Bartlett force
 (iii) The space-spin exchange or Heisenberg force.

OBJECTIVE QUESTIONS

1. Yukawa potential is

 (a) $V = -V_0 \exp\left(-\dfrac{r}{R}\right)$
 (b) $V = -V_0 \dfrac{\exp\left(-\dfrac{r}{R}\right)}{\dfrac{r}{R}}$

(c) $V = -V_0 \dfrac{R}{r}$

(d) $V = -V_0 R \exp\left(+\dfrac{r}{R}\right)$

2. Which one of the following represents exponential well potential:

(a) $V = -V_0 \dfrac{\exp\left(-\dfrac{r}{R}\right)}{\dfrac{r}{R}}$

(b) $V = -V_0$ for $r \le b = 0$ for $r > b$

(c) $V = -V_0 \exp\left(-\dfrac{r}{b}\right)$

(d) $V = -\dfrac{V_0}{\left[1 + \exp\left\{\dfrac{-(r-b)}{c}\right\}\right]}$

3. Schrodinger wave equation for the deutron of reduced mass μ in the C-system is

(a) $\nabla^2 \psi + \dfrac{2\mu}{\hbar^2} V(r) \psi = 0$

(b) $\nabla^2 \psi + \dfrac{2\mu}{\hbar^2} E\psi = 0$

(c) $\nabla^2 \psi + \dfrac{2\mu}{\hbar^2} [E - V(r)] \psi = 0$

(d) $\nabla^2 \psi + \dfrac{2\mu}{\hbar^2} [E - V(r)]\psi = 0$

4. The resultant cross section of scattering of an unpolarized beam of neutrons from protons can be written as

(a) $\sigma = \dfrac{3}{4} \sigma_t (\sigma_t = 4\pi a_t^2) + \dfrac{1}{4} \sigma_s (\sigma_s = 4\pi a_s^2)$

(b) $\sigma = \dfrac{1}{4} \sigma_t + \dfrac{3}{4} \sigma_s$

(c) $\sigma = \dfrac{\sigma_t}{2} + \dfrac{\sigma_s}{2}$

(d) $\sigma = \dfrac{\sigma_t}{3} + \dfrac{2}{3} \sigma_s$

($t \rightarrow$ triplet scattering, $s \rightarrow$ singlet scattering)

5. The wave function for a scalar neutral meson in the presence of a source of strength g_s satisfies:
(a) $\nabla^2 \phi - \mu^2 \phi = 4\pi g_s \, \delta(x)$
(b) $\nabla^2 \phi - \mu^2 \phi = 4\pi g_s \, \delta(x) \, \sigma \cdot \nabla$
(c) $\nabla^2 \phi - \mu^2 \phi = 0$
(d) $\nabla^2 \phi = 4\pi g_s$

6. The wave function for a pseudoscalar neutral meson in the presence of a source of strength satisfies:
(a) $\nabla^2 \phi - \mu^2 \phi = 4\pi g_s \, \delta(x)$
(b) $\nabla^2 \phi = 4\pi g_s \, \delta(x)$
(c) $\nabla^2 \phi - \mu^2 \phi = 4\pi g_s \, \delta(x) \, \sigma \cdot \nabla$
(d) $\nabla^2 \phi - \mu^2 \phi = 0$

7. For a saturing nuclear force:
(a) the separation energy equals the binding energy per nucleon
(b) the separation energy is one half of the binding energy per nucleon.
(c) the separation energy is one fourth of the binding energy per nucleon.
(d) the separation energy is one third of the binding energy per nucleon

8. If a_k is the scattering length for the energy E and $k^2 = \dfrac{ME}{\hbar^2}$, the low energy scattering cross section is

(a) $\sigma = \dfrac{4\pi}{k^2 + a_2^{-2}}$

(b) $\sigma = \dfrac{4\pi}{k^2}$

(c) $\sigma = \dfrac{4\pi}{a_k^2}$

(d) $\sigma = \dfrac{4\pi}{k^2 + a_k^2}$

9. Neutron-proton capture is inverse process of
   - (a) pair production
   - (b) photo electric effect
   - (c) photo-disintegration
   - (d) none of the above

10. If μ_p and μ_n are magnetic moments of proton and neutron respectively, the condition for which there would be no reaction is
    - (a) $\mu_p > \mu_n$
    - (b) $\mu_p < \mu_n$
    - (c) $\mu_p = \mu_n$
    - (d) none of the above

11. The mass of virtual meson is about
    - (a) $\sim 50\, m_e$
    - (b) $\sim 100\, m_e$
    - (c) $\sim 200\, m_e$
    - (d) $\sim 500\, m_e$
    $(m_e \rightarrow \text{electron mass})$

12. The Yale potential has the form
    - (a) $V = V_{OPEP} + V_C + V_T\, S_{12} + V_{LS}\, \mathbf{L} \cdot \mathbf{S} + V_Q \{Q_{12} - (\mathbf{L} \cdot \mathbf{S})^2\}$
    - (b) $V = -\dfrac{V_0}{\left(\dfrac{r}{b}\right)} \exp\left(-\dfrac{r}{b}\right)$
    - (c) $V = -\dfrac{V_0}{\left[1 + \exp\left\{-\dfrac{(r-b)}{c}\right\}\right]}$
    - (d) $V = -\dfrac{g^2 \exp(-\beta |\mathbf{r}_2 - \mathbf{r}_1|)}{|\mathbf{r}_2 - \mathbf{r}_1|}$

ANSWERS

1. (b)	**2.** (c)	**3.** (d)	**4.** (a)	**5.** (a)
6. (c)	**7.** (a)	**8.** (a)	**9.** (c)	**10.** (c)
11. (c)	**12.** (a)			

5

NUCLEAR STRUCTURE AND MODELS

5.1 INTRODUCTION

We know much about the arrangement of the orbital electrons in an atom. This is largely because the force between the electron and the nucleus as well as between the electrons themselves is electrical in nature and mathematical treatment of such forces is well established. Unfortunately, there is no satisfactory explanation for treating the strong nuclear interaction based on the concept of pion transfer. Consequently there is no simple theory of the detailed structure of the nucleus.

An atomic nucleus exhibit a number of properties, e.g. (i) a radius proportional to $A^{\frac{1}{3}}$ ($A \rightarrow$ mass number), (ii) a well-defined and uniform density, (iii) a binding energy per nucleon which is almost constant (except for very small A) and (iv) a neutron to proton ratio which is close to unity for small A but increases progressively with increasing A. It also exhibit many types of nuclear reactions, radioactive decay, nuclear fission and nuclear fusion.

What physicists have done so far is to propose "models" which can be used to interpret certain aspects of the behaviour of nuclei. There are several isolated facts which require explanation when one adopts any nuclear model. Few of them are (a) why do nuclei emit α and β particles when these are known to contain only protons and neutrons (b) why is binding energy per nucleon almost constant? (c) why are the 4n nuclei particularly stable? (d) How one can explain the excited states of nuclei (e) How one can explain Geiger-Nuttall law? (f) How one can interpret the special properties of nuclei, e.g. spin, stability, magnetic moment, etc.

Various models which have been proposed for the nucleus are the different collective models (of which the liquid drop model is one), Fermi gas model and the shell models with different types of coupling. The characteristics of nucleon-nucleon forces in conjuction with the Pauli exclusion principle cause nucleus of various elements to exhibit apparently contradictory behaviour. One finds that the macroscopic properties, e.g. constant density and constant binding energy per nucleon, resemble those of a drop of liquid. On the otherhand, the microscopic properties, e.g., nuclear wave functions and particle motions resemble those of a weakly interacting gas. One can see that the resemblance to a drop of liquid serves as basis for the *liquid drop model* and *collective model* and the resemblance to a weakly interacting gas serves as the basis for the *Fermi gas* model and the *shell model*. A neutron-proton assembly is now the basis of all nuclear models which attempt to interpret overall nuclear properties in terms of individual particle states. These states are essentially those occupied by fermions in a spherical nucleus represented by a potential well of depth ~ 50 MeV [idealized form in Figure 5.1(a) for a nucleus of $A \approx 100$, $R_C \approx 5.2$ fm, (which is sufficiently large to avoid dominance of surface effects)].

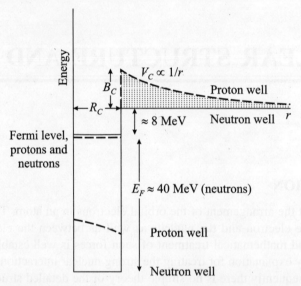

Fig. 5.1(a) The nuclear potential well drawn to show separate the wells for protons and neutrons. The Fermi levels for the two-component fermi systems are at about the same energy. The radius of the wells of R_C and the Coulomb barrier height (idealized) is B_C.

The search for a single nuclear model that accounts for all the properties of a nucleus is there. The more or less extensively studied nuclear models are briefly reviewed in this chapter.

5.2 FERMI GAS MODEL

It is a statistical model of the nucleus. This model pictures the nucleus as a degenerate gas of protons and neutrons much like the free electron gas in metals. The gas is considered degenerate because all the particles are crowed into the lowest possible states in a manner consistent with the requirement of Pauli exclusion principle. In this case the nature of the microscopic particles is fully reflected in its effect on the ensemble as a whole.

Nucleons are fermions having spin $\frac{1}{2}$. Thus the behaviour of the neutron or the proton gas will be determined by Fermi-Dirac statistics. In such a gas at 0 K, all the energy levels upto a maximum, known as Fermi energy E_F, are occupied by the particles, each level being occupied by two particles with opposite spins.

Neglecting for the moment, the electrostatic charge of the protons and supposing that the nucleus has $N = Z = \dfrac{A}{2}$. The nucleons move freely within a spherical potential well of the proper diameter with depth adjusted so that the Fermi energy raises the highest lying nucleons upto the observed binding energies. The potential well is filled separately with nucleons of each type, allowing just two particles of a given type with opposite spin to each cell in phase space of volume h^3. According to Fermi-Dirac statistics, the number of neutron states per unit momentum interval is

$$\frac{dN}{dp} = \frac{2 \times 4\pi \, p^2 V}{(2\pi\hbar)^2} = \frac{V p^2}{\pi^2 \, \hbar^3} \qquad \ldots (5.1)$$

Where V is the volume of the nucleus. If $p_f \left[= (2M \, E_F)^{\frac{1}{2}} \right]$ is the limiting momentum below which all the states all filled. Obviously, the number of neutrons occupying momentum states upto this maximum momentum is obtained as

$$N = \int_0^{p_f} \frac{dN}{dp} \, dp = \frac{V}{3\pi^2 \, \hbar^2} \, p_f^3 \qquad \ldots (5.2)$$

We have

$$p_f = (3\pi^2)^{\frac{1}{3}} \, \hbar \left(\frac{N}{V} \right)^{\frac{1}{3}}$$

$$\therefore \qquad E_f = \frac{p_f^2}{2M} = (3\pi)^{\frac{2}{3}} \frac{\hbar^2}{2M} \left(\frac{N}{V} \right)^{\frac{2}{3}} \qquad \ldots (5.3)$$

$$= \frac{\hbar^2}{2M} \left(\frac{3\pi^2 N}{V} \right)^{\frac{2}{3}} \qquad \ldots (5.3a)$$

Where $V = \dfrac{4}{3} \pi \, r_0^3 A$ is the nuclear volume which contain N particles (fermions) and M is the nucleonic mass.

We have two different types of Fermi gas in the nucleus (i) the proton gas and (ii) the neutron gas. The respective numbers of protons and neutrons are Z and $A - Z$. Now, assuming that the number of nucleonic states to be equal to the nucleon number in each case, one obtains the density of states for the two gases as

$$n_p = \frac{Z}{V} = \frac{Z}{\frac{4}{3} \pi \, r_0^3 A} = \frac{3Z}{4\pi \, r_0 \, A} \qquad \ldots (5.4)$$

$$n_n = \frac{A - Z}{V} = \frac{3(A - Z)}{4\pi \, r_0 \, A} \qquad \ldots (5.5)$$

We have $r_0 \approx 1.2$ fm. One obtains after assuming $N = A - Z = \dfrac{A}{2}$

$$n_p = n_n = \frac{\dfrac{3}{2}}{4\pi(1.2)^3} = 0.069 \ \frac{\text{nucleons}}{\text{m}^3}$$

$\therefore$ Nucleon density $\qquad N = n_p + n_n = 0.138 \ \dfrac{\text{nucleons}}{\text{m}^3}$

Remembering that each state can be occupied by nucleons of opposite spins and substituting the above in (5.3), one obtains

$$E_f = \frac{\hbar^2}{2M}\left(\frac{3}{2}\pi^2 n_p\right)^{\frac{2}{3}} = 21 \ \text{MeV}$$

However, the number of protons (Z) and neutrons $(A - Z)$ in an actual nucleus are not equal and hence N being somewhat greater than Z. Obviously, the Fermi energies of the two types of nucleons are different. Now $N > Z$, $(E_f)_n > (E_f)_p$ and hence the potential wells for the protons and neutrons have different depths, i.e. the former being less deep than the latter. The depth of the potential well is obtained as

$$V_0 = E_f + \frac{E_B}{A} = E_f + f_B \qquad\qquad \ldots (3.6)$$

Here $f_B = \dfrac{E_B}{A}$ is called the mean binding energy per nucleon (binding fraction) and is of the order of 8 MeV/nucleon for both protons and neutrons. Figure 5.1(b) exhibit Fermi gas model of nucleonic potential wells. Figure exhibit the difference in the depths of wells for neutron and proton.

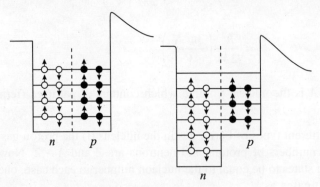

Fig. 5.1(b)

From Figure 5.1(b), we note that the Fermi energies for both protons and neutrons are represented by the same horizontal line, corresponding to about 8 MeV below the top of the potential well (Coulomb effect is neglected). One can visualize this that if these are at different depths below the top of the well, then the nucleons of one type from the higher Fermi level (say, neutrons) would make spontaneous transitions to the lower Fermi level for the other type (protons) by β-transformations. Obviously, the levels would ultimately equalize.

Thus one finds the depth of the potential well approximately,

$$V_0 \approx 21 + 8 = 29 \text{ MeV}$$

We must note that the neutron depth is slightly greater than the proton depth.

Let us now consider a hypothetical infinite medium of nuclear matter of uniform density in which the numbers of neutrons and protons are equal, i.e. $N = Z$ and the Coulomb interaction of the proton is considered negligible. In this situation, one obtains from the semi-empirical binding energy Bethe-Weizsackar relation

$$\left(\frac{E_B}{A} \right) = \left(\frac{E_v}{A} \right) = 15.9 \text{ MeV/nucleon}$$

Where E_B is binding energy. Adding this to the depth of the potential well below the Fermi level, one obtains the depth of the potential well as

$$V_0 = 21 + 15.9 = 36.9 \text{ MeV/nucleon}$$

One expect that any successful theory of nuclear matter should be able to correlate the above value of V_0 to the nature of internucleon nuclear force.

We have so far assumed nuclear temperature to be $T = 0$ K corresponding to the ground state. When some excitation energy is supplied to the nucleus, then the thermal energy of the nucleus corresponds to $T > 0$ K. In this case, one can easily show that the total excitation energy is

$$E_t = E_p + E_n = 11(kT)^2 \text{ MeV}$$

Since $kT \approx 1$ and hence $E_t \approx 11$ MeV.

One can also obtain the energy density of the nuclear levels for a given excitation energy using the entropy relation: $S = k \ln W$ and thermodynamic expression for entropy.

One cannot predict detailed properties of low lying states of nuclei observed in the radioactive decay processes from this model. This model is particularly useful in describing phenomena which are sensitive to the high momentum part of the nucleon spectrum. The model suggests that nucleon collisions often do not transfer small amount of momentum to the nucleus, because the nucleon momentum states near the origin are filled. However, this limitation does not affect collisions in which large momentum transfer takes place. Obviously, this statistical model helps to explain the properties of the nucleus in excited states. One can also treat the unbound states of heavy and medium nuclei with the help of this model.

5.3 LIQUID DROP MODEL

The binding energies and volumes of the nuclei are proportional to the number of nucleons present in nuclei. The proportionality, however, indicates the short range and saturation characters of the nuclear forces. This follows that there is strong interaction amongst all the neighbouring inside the nucleus. These properties of a nucleus are analogous to the properties of the forces which hold a liquid drop together. Hence, a nucleus may be considered to be analogous to a drop of incompressible fluid of very high density ($\sim 10^{17} \text{ kg/m}^3$). This analogy led Niels Bohr and Wheeler and independently Frenkel to propose the liquid drop drop model of the nucleus. This model is not considered about the individual characterstics of the nucleons and hence this is statistical model. According to this model, the nucleus is regarded as an incompressible and uniformly charged liquid drop.

(a) Similarities Between a Nucleus and Liquid Drop

   (i) There are a large number of particles in a nucleus as in a drop of liquid, protons and neutrons (except $^1_1$H which contains one proton only) in the former and molecules or atoms in the latter). For example, considering the volume of a drop of water to be 0.05×10^{-6} m^3, it would contain $\approx 10^{21}$ molecules, Whereas the same volume of nuclear substance would contain $\approx 10^{37}$ nucleons, the density being 0.165 nucleon per fm^3.

   (ii) Both the liquid drop and nucleus exhibit homogeneity and incompressibility. This implies that the charge density ($\approx 10^{17}$ m^{-3} for the nucleus) and all other properties are uniform throughout the drop and the nucleus, except at the surface boundary. In both the cases, the density is independent of dimensions.

   (iii) Each nucleus in a nucleus interacts strongly with a small number of adjacent nucleons just as do the molecules in a liquid drop. This follows that nucleon forces are of short-range and of saturation character.

   (iv) Both the liquid drop and nucleus show surface tension effect, i.e. the surface energy of the nucleus is analogous to the surface tension of liquids.

   (v) Apart from the Coulomb repulsion, the forces between the n-n, n-p, and p-p are the same in the nucleus, just as the intermolecular forces (between the solute-solute, solute-solvent and solvent-solvent molecules) in an ideal solution.

   (vi) Because of internucleon forces in a nucleus or the intermolecular forces in a liquid drop, any excess energy associated with an individual particle is rapidly shared by all the nucleons in the nucleus or the particles in the drop.

   (vii) The energy levels of the nucleus are regarded was the quantum states of the nucleus as a whole rather than of a single nucleion like various molecules as in a liquid drop.

   (viii) Evaporation from a liquid drop is analogous to the loss of nucleons from a nucleus in the nuclear reaction.

   (ix) The fusion of small drops into a bigger one, and the breaking up of large drop into small droplets are quite analogous to the fusion of light nuclei into a heavy (or compound) nucleus and the fission of a heavy nucleus into light nuclei, respectively. The fusion and fission processes in both the cases are exoergic.

   (x) The thermal agitation of the molecules in a drop is quit analogous to the kinetic energy (movement of the nucleons) of the nucleons in the nucleus.

   Note: The nuclear reactions, accompanied by the emission of nucleons from the nucleus, can be interpreted on the basis of liquid drop nuclear model. If the thermal agitation in the drop of a liquid is increased by heating, evaporation of molecules takes place. In a similar way, when an energetic particle is captured by a nucleus, a compound nucleus is formed which emits almost immediately. The emission of nucleons from a compound nucleus is thus analogous to the evaporation of molecules from a hot liquid-drop.

(b) Merits of the Liquid Drop Model   The liquid drop model has been employed with great success in the interpretation of intra-nuclear forces and of nuclear transformations. The merits of this model in the satisfactory explanation. The merits of this model in the satisfactory explanation of various nuclear phenomena are given below.

(i) Unlike the shell model, the liquid drop model can be applied to the nuclei in excited states. Thus it provides a mechanism for low energy nuclear reactions, for which Niels Bohr used this model as the basis of the compound nucleus theory.

(ii) The phenomenon of nuclear fission has been explained satisfactorily with the help of this model.

(iii) The calculation of the nuclear binding energies and the study of the properties of isobars have been made successfully on the basis of the liquid drop model of the nucleus.

(c) Bethe-Weizsacker Semi-Empirical Mass Formula On the basis of the liquid drop model, Weiszacker, and several others, have attempted to express the masses of nuclei in terms of nuclear characterstics in connection with their binding energy and stability. This formula is known as semi-empirical mass formula.

Let $M(A, Z)$ be the atomic mass of the isotope of an element X of atomic number Z and mass number A, then

$$M(A, Z) = ZM_H + NM_n - E_B \qquad \qquad ... (5.1)$$

Where M_H and M_n are the masses of the hydrogen atom and netron respectively and E_B is the nuclear binding energy. $N = A - Z$ is the number of neutrons in the nucleus. One can express the binding energy E_B as the sum of a number of terms as below:

(i) **Volume energy** Nucleons are bound by short-range force, i.e. each nucleon is held strongly by other nucleons in its immediate vicinity. These short range forces are responsible for the binding energy of the nucleus. The binding energy is proportional to the number of nucleons, (i.e. mass number A). Since the nuclear volume is proportional to A, as a first approximation, there is an attraction or volume energy, proportional to A. This volume energy (E_v) is positive and is given by

$$E_v = a_1 A \qquad \qquad ... (5.2)$$

Where a_1 is constant.

(ii) **Surface Energy** The nucleus is made up of protons and neutrons and the nuclear radius is proportional to the cube root of the nuclear mass number, i.e. $R = r_0 A^{\frac{1}{3}}$. Obviously, the nucleus is assumed to resemble a spherical liquid drop of radius R. As in case of a liquid drop, the nucleons on the surface of the nucleus are not surrounded by as many neighbours as those in the interior. In other words, the nucleons are less strongly bound on the surface, and consequently, the nucleon forces are unsaturated. Since the binding energy is proportional to the number of nucleons (i.e., A), and the number of nucleons is reduced on the surface, the binding energy is, therefore lowered by an amount which varies as the surface area of the nucleus. As the nuclear radius is proportional to $A^{\frac{1}{3}}$, the area of the nuclear surface is related to $A^{\frac{2}{3}}$. The energy term corresponding to this surface tension effect is called the surface energy, and is represented by

$$E_S = -a_2 A^{\frac{2}{3}} \qquad \qquad ... (5.3)$$

Where a_2 is the proportionality constant.

(iii) **Coulomb energy**   The protons present in the nuclear volume experience a Coulomb repulsion, (i.e. long range electrostatic force of repulsion) which tends to lower the binding energy. As each proton is repelled by $(Z-1)$ protons, the total Coulomb energy is proportional to $\dfrac{Z(Z-1)e^2}{R}$. The effect of this Coulomb repulsion on the binding energy is termed as the Coulomb energy and can be calculated as follows:

Let us assume that the total nuclear charge $Q = +Ze$ be distributed uniformly, then the charge density is obtained as

$$\rho_c = \frac{3Ze}{4\pi R^3} = \frac{3Ze}{4\pi r_0^3 A} \qquad \qquad \dots (5.4)$$

One can calculate the potential energy of the charged sphere of radius R by considering that the sphere is built up layer by layer and making use of the fact that the field outside a spherical distribution of charge is the same as if all the charge is concentrated at the centre. Assuming that the sphere has been built up to a radius r, one finds the total charge in it as

$$q = \frac{4}{3}\pi r^3 \rho_c$$

Now, we add an infinitesimal shell of charge of thickness dr to this. The work done in bringing the charge $dq = 4\pi r^2 \rho_c\, dr$ in the shell from infinity to the radius r against the field of charged sphere of radius r, as above is

$$dW = \frac{q\,dq}{4\pi \varepsilon_0 r} = \frac{4}{3}\pi r^3 \rho_c \times \frac{4\pi r^2 \rho_c\, dr}{4\pi \varepsilon_0 r}$$

$$= \frac{1}{4\pi \varepsilon_0 r} \times \frac{16\pi^2}{3}\rho_c^2\, r^4\, dr$$

$\left(\varepsilon_0 = \dfrac{10^{-9}}{36\pi}\dfrac{F}{m}\right.$ is the permitivity of the free space$)$

∴ Total work done in forming the charged sphere of radius R is obtained as

$$W = \frac{1}{4\pi \varepsilon_0}\frac{16\pi^2}{3}\rho_c^2 \int_0^R r^4\, dr = \frac{1}{4\pi \varepsilon_0}\frac{16\pi^2}{3}\rho_c^2 \frac{R^5}{5} = \frac{1}{4\pi \varepsilon_0}\frac{3}{5}\frac{Q^2}{R}$$

We know that the potential energy is the negative of work done and hence the Coulomb energy of the charged sphere is

$$E_c = -\frac{3}{5}\times\frac{Q^2}{4\pi \varepsilon_0 R} = -\frac{3}{5}\times\frac{(Ze)^2}{4\pi \varepsilon_0 R}$$

$$= -\frac{3}{5}\left(\frac{(Ze)^2}{4\pi \varepsilon_0\, r_0\, A^{\frac{1}{3}}}\right) = -a_3\frac{Z^2}{A^{\frac{1}{3}}} \qquad \qquad \dots (5.5)$$

Where $a_3 = \dfrac{3}{5}\left(\dfrac{e^2}{4\pi\varepsilon_0\, r_0}\right)$ is a constant $\qquad$ ... (5.6)

We must note Equation (5.5) for the Coulomb energy is not exact. One has apply corrections due to (i) non uniformity of the nuclear charge distribution; (ii) effect of the uncertainty on the localization of the protons; (iii) the requirement of the discrete arrangement of the charges on the protons; (iv) corrections of the positions of the protons and (v) non-sphericity of the nucleus.

(iv) **Asymmetry energy** We have seen that the maximum stability occurs when $Z = \dfrac{A}{2}$, i.e. the numbers of protons and neutrons are equal in the most stable (light) nuclides. But when the number of neutrons exceeds, i.e. $(A - Z) > Z$, instability of the nuclide appears. This is know as the *asymmetry* or *composition effect*. Since the excess neutrons occupy the higher quantum states than the other nucleons, they contribute a smaller amount (per neutron) to the total binding energy. Obviously, with the introduction of asymmetry, the binding energy decreases. The lowering of binding energy, called the *asymmetry* or *neutron* excess energy is proportional to the square of the neutron excess, $= (A - 2Z)^2$ and inversely to A, i.e.

$$E_a = -a_4\,\frac{(A - 2Z)^2}{A} \qquad\qquad ... (5.7)$$

Where a_4 is the proportionality constant.

(v) **Pairing energy** It has been observed that even-Z, even-N nuclides are the most abundant amongst the stable nuclides. Nuclides with odd-Z, odd-N are least the least stable, while nuclides with even-Z, odd-N or vice versa, are of intermediate stability. This may be attributed to the pairing of nucleons of the same type or pairing of the nucleon spins. In case of even-even nuclei, all the spins are paired, and so there is a positive contribution to the binding energy for nuclei with odd-Z, odd-N because of the presence of unpaired proton and unpaired neutron spins. There is no contribution for nuclei with even-Z, odd-N or the reverse. This contribution to the binding energy, arising from the spin or odd-even effect is known as the pairing energy. The pairing energy term $E_\delta(A, Z)$ depends only on A and is taken to be zero for odd A nuclei, positive for even-even nuclei and negative for odd-odd nuclei and given by

$$E_\delta = \pm a_5\, A^{-\frac{3}{4}} \qquad\qquad ... (5.8)$$

Where a_5 is the proportionality constant, the positive sign implies an increase in the binding energy for the even-even nuclei, and the negative sign for the odd-odd nuclei, which results in a decrease in the binding energy. $E_\delta = 0$ for odd a nuclei.

This pairing interaction arises from the short-range and attractive nature of the nuclear force and leads to greater binding between like nucleons if their angular momenta are coupled to zero spin. When coupled like this the two nucleons are on average closer than when coupled to non-zero spin and hence are more bound. This is illustrated in figure 5.1(a) and (b). Two like nucleons cannot both be in the same magnetic substate because of the Pauli exclusion principle but if they are in opposite substates (i.e. m_j and $-m_j$) then their combined spin is zero and the spatial overlap of their wave functions is high (see figure 5.1a); however, if they are not coupled to zero spin then the overlap is less and the nucleons are less bound (see figure 5.1b). The overlap (figure 5.1a) increases with increasing orbital angular momentum l, as the orbitals become more localised, with the result that the pairing interaction increases.

This pairing interaction between like nucleons is responsible for the last term, the pairing term δ, in the semi-empirical mass formula. The function δ has been parametrised by (as stated above)

$$+a_p A^{-\frac{3}{4}} \quad \text{for odd-odd nuclei}$$
$$\delta = 0 \quad \text{for even-odd nuclei}$$
$$-a_p A^{-\frac{3}{4}} \quad \text{for even-even nuclei}$$

and the term accounts for the increased stability observed for nuclei which have an even number of like nucleons. The reason the pairing interaction is not so effective between neutrons and protons in stable nuclei with $A > 40$ is because such nuclei have $N > Z$ because of the Coulomb repulsion of the protons which displaces the neutron and proton potential wells as shown in figure 5.1(c)

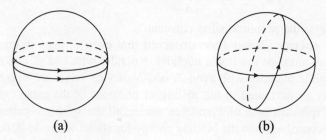

(a)   (b)

Fig. 5.1(c) Illustration of the grater of overlap when two nucleons (a) are coupled up to spin zero, than (b) when they are not.

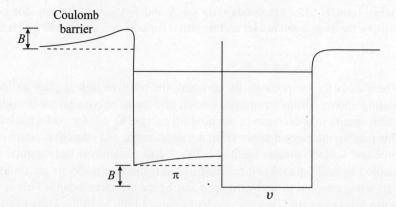

Fig. 5.1(d) The effect of the Coulomb energy of the protons on the relative depth and shape of the proton and neutron potential wells.

The total binding energy of a nuclide of mass number A proton number Z is obtained by combining all the energy terms, i.e.

$$E_b(A, Z) = E_v - E_s - E_c - E_a \pm E_\delta$$

$$= a_1 A - a_2 A^{\frac{2}{3}} - a_3 z^2 A^{-\frac{1}{3}} - a_4 (A - 2Z)^2 A^{-1} \pm E_\delta \qquad \dots (5.9)$$

In Figure 5.2(a) some terms on the right is plotted against A, together with their sum, $\dfrac{\text{B.E.}}{A}$. The curve representing the total binding energy per nucleon is fairly close to the experimental curve.

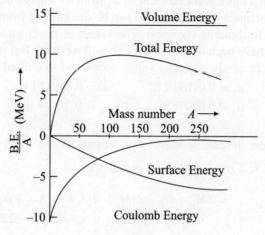

Fig. 5.2(a)

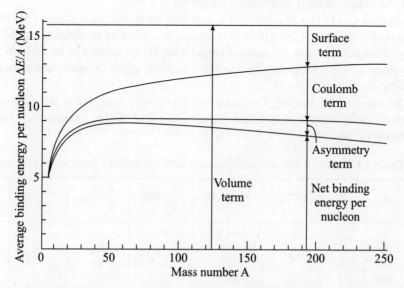

Fig. 5.2(b) Illustrating how the volume, surface, Coulomb, and asymmetry terms of semiempirical mass formula combine to yield the average binding energy per nucleon.

Obviously, the analogy of a nucleus with a liquid drop has some validity at least. Equation (5.9) is known as semi-empirical binding energy formula. For the Calculation of the binding energy and the atomic mass, the values of constants apprearing in (5.9) must be known. The constant a_3 determing the Coulomb energy is calculated as (by taking $r_0 = 1.22$ fm)

$$a_3 = \frac{3}{5} \frac{e^2}{4\pi \varepsilon_0 \, r_0} = \frac{3}{5} \frac{(1.602 \times 10^{-19})^2}{4\pi \times \left(\dfrac{10^{-9}}{36\pi}\right) \times 1.22 \times 10^{-15}}$$

$$= 0.000761 \text{ u}$$

One can obtain a more accurate value of a_3 from the energy difference of mirror nuclei. The other constants are empirical and can be determined from the experimental values of the masses (or binding energies). The values of the constants in the semi-empirical mass formula have been determined by several workers. But their values are somewhat different. The currently accepted values expressed in unified atomic mass scale are

$$a_1 = 0.016919 \text{ u}; \qquad a_2 = 0.019114 \text{ u};$$
$$a_3 = 0.0007626 \text{ u}; \qquad a_4 = 0.02544 \text{ u};$$
$$a_5 = 0.036 \text{ u}$$

One can express the atomic mass of an isotope as

$$M(A, Z) = ZM_H + NM_n - E_B(A, Z)$$

$$= ZM_H + (A - Z)M_n - a_1 A + a_2 A^{\frac{2}{3}} + a_3 Z^2 \, A^{-\frac{1}{3}}$$
$$+ a_4(A - 2Z)^2 \, A^{-1} \pm E_\delta \qquad \dots (5.11)$$

This is the semi empirical mass formula which gives binding energies within about 1 % of the experimentally determined values for $A > 40$.

Equation (5.11) is a quadratic in Z and obviously, for each value of A, there is some particular value of Z for which M is minimum. A family of different nuclides for which the mass number A is the same is called a family of *isobars* or an *isobaric family*. The Z value which corresponds to minimum mass M gives the most stable member of this particular isobaric family.

The ratios of the surface, Coulomb and asymmetry energies to the volume energy for some typical nuclei in different mass regions, i.e. light, intermediate and heavy are shown in Table 5.1.

Table 5.1 Ratios of the surface, Coulomb and Asymmetry Energies to the Volume energy

A	Z	$A - 2Z$	E_v (MeV)	$\dfrac{E_s}{E_v}$ (%)	$\dfrac{E_c}{E_v}$ (%)	$\dfrac{E_a}{E_v}$ (%)	$\dfrac{E_B}{A}$ (MeV)
6	3	0	94.8	62	3.7	0	5.4
20	10	0	316	41	8.2	0	7.95
50	24	2	790	30.5	14	0.24	8.71
160	64	32	2528	20.8	21.2	6	8.22
200	80	40	3160	19.3	24.6	6	7.92
250	100	50	3950	18	28.5	6	7.52

One finds from Table 5.1 that for very light nuclei, the main contributory factor reducing the volume energy is the surface energy term. This is because of the fact that the surface area to volume ratios for these are relatively much larger than for the heavy nuclei. The effect of the surface energy term diminishes rapidly as A increases in the case of very light nuclei which is responsible for the rapid increase in the value of $\dfrac{E_v}{A}$ with the increase of A.

One also finds from table 5.1 that the Coulomb energy effect $\left(\dfrac{E_c}{E_v}\right)$, which is small for lighter nuclei, becomes quite dominant for the heavier nuclei containing a large number of protons.

The asymmetry effect is important for heavier nuclei and insignificant at low A values for which N and Z are almost equal.

(d) Applications of the Semi-empirical binding energy relation

(i) α-decat of $^{A}_{Z}X$ nucleus into the nucleus $^{A-4}_{Z-2}Y$ can be written as

$$^{A}_{Z}X \rightarrow {}^{A-4}_{Z-2}Y + {}^{4}_{2}He \qquad \text{... (5.1)}$$

The α-disintegration energy is

$$Q_\alpha = M(A, Z) - M(A-4, Z-2) - M(^{4}_{2}He) \qquad \text{... (5.2)}$$

Expressing in terms of binding energies (E_B) of the nuclei involved, Equation (5.2) becomes for A and Z large

$$Q(\alpha, Z) = E_B(A-4, Z-2) + E_B(^{4}_{2}He) - E_B(A, Z)$$

$$= a_1(A-4) - a_2(a-4)^{\frac{2}{3}} - a_3 \frac{(Z-2)^2}{(A-4)^{\frac{1}{3}}} - a_4 \frac{(A-2Z)^2}{A-4}$$

$$- a_1 A + a_2 A^{\frac{2}{3}} + a_3 Z^2 A^{-\frac{1}{3}} + a_4(A-2Z)^2 A^{-1} + E_b(^{4}_{2}He)$$

$$= E_B(^{4}_{2}He) - 4\,a_1 + a_2\left[\left\{A^{\frac{2}{3}} - (A-4)^{\frac{2}{3}}\right\}\right] + a_3\left\{\frac{Z^2}{A^{\frac{1}{3}}} - \frac{(Z-2)^2}{(A-4)^{\frac{1}{3}}}\right\}$$

$$+ a_4(A-2Z)^2\left\{\frac{1}{A} - \frac{1}{A-4}\right\}$$

$$= 28.3 - 4a_1 + \frac{8}{3}a_2 A^{-\frac{1}{3}} + 4a_3 Z\left(1 - \frac{Z}{3A}\right)A^{-\frac{1}{3}} - \frac{4a_4(A-2Z)^2}{A(A-4)} \qquad \text{... (5.3)}$$

We have neglected the pairing energy term. The binding energy of the α-particle has been taken to be 28.3 MeV. Taking the values of constants a_1, a_2, a_3 and a_4 obtained earlier, one finds $Q_\alpha > 0$ for $A > 160$. Obviously, nuclei with $A > 160$ should be α-disintegrating. α-disintegration is actually observed primarily in the region $A > 200$. For $A < 200$, i.e. lighter nuclei the energy release is so small that barrier penetration probability becomes very small.

(ii) **Stability Properties of β-decaying nuclei: Mass Parabola**   The semi-empirical mass formula accounts for the stability properties of nuclei, especially, the β-activity and stability properties of isobars.

The nuclei undergoing β-decay give rise to the product (daughter) nuclear which are isobaric with the initial (parent) nuclei. Thus isobars although have the same mass number, they contain different numbers of protons and neutrons, for which their binding energies will be different. Applying the semi-empirical mass formula to nuclides of constant A, one can rewrite it without the δ term as

$$M(A, Z) = f_A + pz + qZ^2 \qquad \qquad \text{... (5.1)}$$

Where
$$f_A = A(M_n - a_1 + a_4) + a_2\, A^{\frac{2}{3}}$$
$$p = -4a_4 - (M_n - M_H)$$

and
$$q = \frac{1}{A}\left(a_3\, A^{\frac{2}{3}} + 4a_4\right) \qquad \qquad \text{... (5.2)}$$

Equation (5.1) is the equation of a *parabola*. Differentiating $M(A, Z)$ with respect to Z for a given A(i.e. a given isobaric line), one obtains

$$\left(\frac{\partial M}{\partial Z}\right)_A = p + 2qZ \qquad \qquad \text{... (5.3)}$$

Putting $\left(\dfrac{\partial M}{\partial Z}\right)_A = 0$, one can obtain the lowest point of the parobola $Z = Z_A$. This gives

$$p + 2qZ_A = 0$$

or
$$Z_A = -\frac{p}{2q} = \frac{(M_n + M_H + 4a_4)\,A}{2\left(a_3\, A^{\frac{2}{3}} + 4a_4\right)} \qquad \qquad \text{... (5.4)}$$

Putting $Z = Z_A$ in (5.1), one obtains

$$M(A, Z_A) = f_A + pZ_A + qZ_A^2$$

$$= f_A + p\left(-\frac{p}{2q}\right) + q\left(\frac{p^2}{4q^2}\right)$$

$$= f_A - \frac{p^2}{4q} \qquad \qquad \text{... (5.5)}$$

$$\therefore \qquad M(A, Z) - M(A, Z_A) = \frac{p^2}{4q} + pZ + qZ^2$$

$$= q(Z - Z_A)^2 \qquad \qquad \text{... (5.6)}$$

Obviously, $Z = Z_A$ is the lowest point of the mass parabola for a given isobar (A = Constant) as the r.h.s. of (5.6) is positive. We note that $M(A, Z)$ has the smallest value for a given A when $Z = Z_A$, this means that this nucleus would have the largest binding energy amongst the isobars for the given A, i.e. Z_A would give the value of Z for the most stable isobar. Making use of the numerical values of M_n, M_H, a_3 and a_4 in Equation (5.4), one obtains

$$Z_A = \frac{A}{1.98 + 0.015\, A^{\frac{2}{3}}} \qquad \text{... (5.7)}$$

One finds that using Equation (5.7) calculations for Z_A do not usually yield an integral value, i.e. one obtains the fractional value. It is found that in most cases, the value of Z is nearest to Z_A corresponds to actual stable nucleus for a given A. Let us make it clear by taking an example for $A = 63$. Equation (5.7) gives $Z_A = 28.4$. Actually at $Z = 29$, the isotope ^{63}Cu is found to be stable. Similarly, for $A = 109$, one obtains from (5.7), $Z_A = 46.94$. The isotope ^{109}Ag with $Z = 47$ is found stable.

When pairing energy term is taken into account, the mass parabolas for the different isobars defined by (5.1) fall into two different categories according to whether A is odd or even. It is obvious that the pairing energy term keeps the mass parabola for odd-mass-nuclei unaltered ($\delta = 0$) where as it produces two parabolas for even A nuclei, one for even $Z(e-e)$ and other for odd $Z(0-0)$. We have seen that the pairing energy term is subtracted, i.e. negative for $e-e$ nuclei and is added for $0-0$ nuclei and therefore the parabola for the $0-0$ nuclei lies above that for the $e-e$ nuclei for the same A by an amount 2δ. In Figure 5.3, mass parabolas are drawn for $A = 135$, $A = 136$ and $A = 200$.

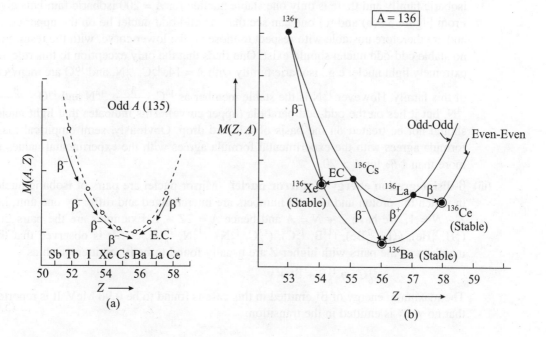

(a)

(b)

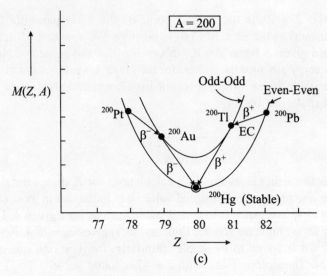

Fig. 5.3 Mass parabolas for $A = 135$, $A = 136$ and $A = 200$

The lowest point for odd A isobars with $A = 135$ (dashed line) is at $Z_A = 56.85$. The stable isobar that is actually observed at this mass number is ^{135}Ba. for which $Z = 56$. We note that nuclides falling on either side of the stable isobar are all unstable. We further note that nuclides on lower Z side ($Z < 56$) are β^- active while those on the higher Z side ($Z > 56$) are β^+ active or electron capturing. Each of these nuclei undergoes β^- transformation till the stable end product is reached.

We note from Figure 5.3(b) and (c) that there are three stable nuclides for $A = 136$ isobaric family and there is only one stable nuclide for $A = 200$ isobaric family.

From Figure 5.3(b) and (c) one can see that the odd-odd nuclei lie on the upper curve and are therefore unstable with respect to those on the lower curve, with the result that no stable odd-odd nuclei should exist. One finds that the only exception to this rule are extremely light nuclei e.g., isobaric family with $A = 14$, $^{14}_{6}$C, $^{14}_{7}$N, and $^{16}_{8}$O are members

of this family. However $^{14}_{7}$N is the stable member as ^{14}C $\xrightarrow{\beta^-}$ ^{14}N and ^{14}O $\xrightarrow{\beta^-}$ ^{14}N, but it lies on the odd-odd parabola (upper curve). This indicates that light nuclei should not be treated on the basis of a liqued drop. Obviously, semi empirical mass formula agrees with the experimental formula agrees with the experimental values to more than 1 % for $A > 20$.

(iii) **β-disintegration energy of mirror nuclei**   Mirror nuclei are pairs of isobaric nuclei in which the proton and neutron numbers are interchanged and differ by one unit, i.e. $Z - N = 1$. We have $Z + N = A$ and hence $A = 2Z - 1$. Examples are the pairs like $(^{3}_{1}$H, $^{3}_{2}$He$)$, $(^{7}_{3}$Li, $^{7}_{4}$Be$)$, $(^{11}_{5}$B, $^{11}_{6}$C$)$, $(^{13}_{6}$C, $^{13}_{7}$N$)$, $(^{15}_{7}$N, $^{16}_{8}$O$)$, etc. It is observed that the members of the pairs with higher Z are usually found to be β^+ emitters such as

$$^{11}_{6}\text{C} \rightarrow {}^{11}_{5}\text{B} + \beta^+ + \nu$$

The maximum energy of β^+ emitted in this case is found to be 0.96 MeV. It is reported that no γ-ray is emitted in the transition.

Writing semi-empirical mass formula for odd A nuclei ($\delta = 0$)

$$M(A, Z) = Z M_H + N H_n - a_1 A + a_2 A^{\frac{2}{3}} + a_3 Z^2 A^{-\frac{1}{3}} + a_4 (A - 2Z)^2 A^{-1}$$

$$= Z M_H + (Z - 1) M_n - a_1 A + a_2 A^{\frac{2}{3}} + a_3 Z^2 A^{-\frac{1}{3}} + a_4 A^{-1} \quad \dots (5.1)$$

For the daughter atom, we have

$$M(A, Z - 1) = (Z - 1) M_H + Z M_n - a_1 A + a_2 A^{\frac{2}{3}} + a_2 A^{\frac{2}{3}}$$
$$+ a_3 (Z - 1)^2 A^{-1} + a_4 A^{-1} \quad \dots (5.2)$$

One obtains

$$M(A, Z) - M(A, Z - 1) = M_H - M_n + a_3 (2Z - 1) A^{-\frac{1}{3}}$$

$$\therefore \qquad Q_{\beta^+} - M(A, Z) - M(A, Z \quad 1) \quad 2 m_e$$

$$= a_3 A^{\frac{2}{3}} - (M_n - M_H + 2 m_e)$$

$$= a_3 A^{\frac{2}{3}} + 1.804 \text{ MeV} \qquad \dots (5.3)$$

Where $a_3 = \dfrac{3e^2}{20 \pi \varepsilon_0 r_0}$ and expressed in MeV. The plot of disintegration energy Q_{β^+}

against $A^{\frac{2}{3}}$ is a straight line with a slope a_3. The estimation of nuclear radius parameter r_0 from this plot comes out to be about 1.44×10^{-15} m. This is a higher value because of limitations. Using more correct formula, one obtains $r_0 = 1.27 \times 10^{-15}$ m.

We shall take the discussion of the application of the semi-empirical formula to explain the phenomenon of fission later.

Limitations of Liquid Drop Model   In preceding sections we came across with different important applications of the liquid drop model of the nucleus in connection with the nuclear properties. The liquid drop model finds its importance. Nevertheless, this model is associated with the following limitations:

(i) The model does not consider the independent behaviour of the nucleons in the nucleus with respect to their motion, spin, parity and magnetic moment effects. In other words, the nucleon forces are charge and spin independent.

(ii) The binding energy or mass formula ignores the closed shell effects as revealed by the discontinuities of the atomic masses at the neutron or proton magic numbers (it may be noted that the mass of any isobar at the nucleon magic numbers obtained form β-decay energetics lies about 1 or 2 MeV below the value predicted by the smoothly varying mass formula).

(iii) The enhanced binding energy of even Z, and N nuclei due to the positive contribution of the odd-even effect to the total binding energy is consistent with the p-p and n-n pairing effects.

(iv) This model applies well to the medium mass nuclei, (i.e., masses in the region of 100), as for light nuclei surface effect predominates over the Coulomb effect and, for heavy nuclei, the reverse is the case.

The liquid drop model gives a good account of the average behaviour of nuclei in regard to mass, or binding energy, i.e. their stability. Figure 5.4 shows a plot of the difference between the value of E_n (the minimum energy required to separate a neutron) measured for a number of nuclei, and the value predicted by the semi empirical formula. Except for the effect of the pairing term, the predicted value is a smooth function that decreases slowly from around 8 MeV for intermediate value of N to around 6 MeV for large values of N. There is unusual stability of nuclei with $N = 28, 50, 82, 126$ indicated by the exceptionally large energy requirement to remove their last neutron.

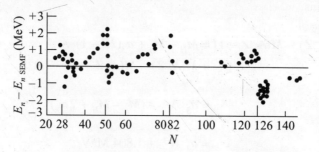

Fig. 5.4 The difference between the binding energy of the last neutron and the prediction of the semiempirical mass formula, as a function of the number of neutrons in the nucleus. These data provide clear evidence for the magic numbers 28, 50, 82, and 126, for neutrons. Similar evidence shows that 20, 28, 50 and 82 are also magic numbers for protons. But there is no concrete evidence, pro or con, concerning 126 for protons since nuclei with such large Z values have not yet been detected.

When the nuclear magic numbers were first being discussed seriously, around 1948, it seemed very difficult to understand how nucleons could move independently in a nucleus. The reason was that the liquid drop model had been dominant for a number of years, and it seemed basic to this model that a nucleon in a nucleus (of density ($\sim 10^{18}$ kg/m^3) would constantly interact with its neighbours through the strong nuclear force. If so, the nucleon would be repeatedly scattered in travelling through the nucleus, and it would follow an erratic path, resembling Brownian motion much more than the motion of an electron moving independently through its orbit in an atom.

The application of the semi-empirical mass formular to explain the phenomenon of nuclear fission will be discussed in chapter on nuclear energy.

The limits of stability predicted by the semi-empirical mass formula in which the ratio $\dfrac{N}{Z}$ is plotted against A is shown in Figure 5.5. The line of maximum stability predicted by Equation (5.3) is surrounded by a region of stable isobars.

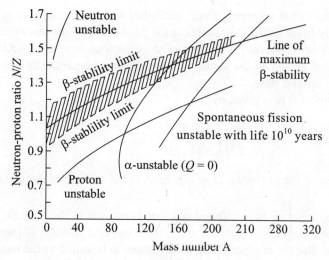

Fig. 5.5 The limits of stability predicted by the semi-empirical mass formula.

5.4 THE FERMI GAS MODEL

Weisskopf first pointed out that there is a similar explanation of how nucleons can move independently through a nucleus in its ground state. The explanation is based on the *Fermi gas model of the nucleus*, which is a statistical model and pictures the nucleus as a degenerate gas of protons and neutrons much like the free electron gas in metals. This model assumes that each nucleon of the nucleus moves in an attractive net potential, that represents with other nucleons in the nucleus. The net potential has a constant depth inside the nucleus since the distribution of nucleons is constant in this region; outside the nucleus it goes to zero within a distance equal to the range of nuclear forces. Thus the net potential is approximately like a three-dimensional finite square well of a radius a little larger than the nuclear radius and depth ~ 50 MeV.

Since the nucleons in a nucleus are all spin $\frac{1}{2}$ particles and they are fermions. Obviously, the behaviour of the proton or the neutron gas will be determined by Fermi-Dirac statistics as in the case of electron gas in the metal. In such a gas at 0 K, all the energy levels upto a maximum, known as the Fermi energy, E_f are occupied by the nucleons, each level being occupied by two nucleons with opposite spins.

Figure 5.6 indicates the quantum states filled by neutrons in the ground state of a nucleus. Since protons are distinguishable from neutrons, the exclusion principle operates independently on the two types of nucleons, and we must imagine a separate and independent diagram representing the quantum states filled by the protons. This reveals that why the Pauli *exclusion principle* prevents

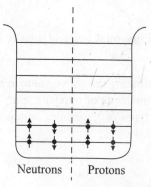

Fig. 5.6 A schematic representation of independent Fermi gases of neutrons and protons in the minimum energy state of a nucleus of very small Z, which is indicated by a square well with rounded edges.

almost all the nucleons from scattering from each other when the nucleus is in its ground state. If there is a set of partly filled degenerate startes at the Fermi energy, the few nucleons in these states can collide with each other, but only a small fraction of the total number of nucleons can be in such states. This means that almost all the nucleons that compose a nucleus can move freely within the nucleus if it is in its ground state.

The Fermi energy at 0 K is given by

$$E_f = \frac{\hbar^2}{2M} \left(\frac{3\pi^2 N}{\Omega} \right)^{\frac{2}{3}} \qquad \dots (5.1)$$

Where M is the mass of the nucleons, $\Omega = \frac{4}{3}\pi r_0^3 A$ is the volume of the nucleus which contains N particles.

As states above, there are two different types of Fermi gas in the nucleus, the proton gas and the neutron gas. Their respective numbers are Z (protons) and $A - Z = N$ (neutrons).

We may assume that the number of nucleonic states to be equal to the number of nucleons in each case, one obtains the density of states for the two types of gases as

$$n_p = \frac{Z}{\Omega} = \frac{Z}{\frac{4}{3}\pi r_0^3 A} = \frac{3Z}{4\pi r_0^3 A} \qquad \dots (5.2)$$

and

$$n_n = \frac{A-Z}{\Omega} = \frac{3(A-Z)}{4\pi r_0^3 A} = \frac{3N}{4\pi r_0^3 A} \qquad \dots (5.3)$$

Now, taking $r_0 = 1.2$ fm and assuming $N = A - Z = \frac{A}{2}$, one obtains

$$n_p = n_n = \frac{\frac{3}{2}}{4\pi (1.2)^3} = 0.069 \text{ nucleons/m}^3$$

Thus, the total nucleon density is

$$n_t = n_p + n_n = 0.139 \text{ nucleons/m}^3$$

Substituting the above value in Equation (5.1) and remembering that each state can be occupied by two nucleons of opposite spins (Figure 5.6), one obtains the Fermi energy for each of the above two types of gases, i.e. proton and neutron as

$$E_F = \frac{\hbar^2}{2M} \left(\frac{3}{2}\pi^2 n_p \right)^{\frac{2}{3}} = 43 \text{ MeV} \qquad \dots (5.4)$$

We may note that in an actual nucleus, Z and $N(= A - Z)$ are not equal, N being somewhat greater than Z. Obviously, the Fermi energies of two types of nucleons are different.

The density of nuclear states increases rapidly with excitation energy E. By treating nucleons in a nucleus as non-interacting fermions, Bethe in 1937 derived the relation,

$$\rho_A(E) = \frac{1}{12a^{\frac{1}{4}} E^{\frac{5}{4}}} e^{2\sqrt{aE}} \qquad \dots (5.5)$$

known as the Fermi-gas model formula. Where a is the level density parameter and often obtained empirically by fitting (5.5) to the measured state densities. The relation (5.5) gives an acceptable estimates of the state density at various energies and is used widely in many nuclear reaction calculations.

A better form of the nuclear state density is given by the back-shifted Fermi gas model formula,

$$\rho_A(E) = \frac{1}{12\, a^{\frac{1}{4}}\, (E - \Delta)^{\frac{5}{4}}}\, e^{2\sqrt{a(E-\Delta)}} \qquad \text{... (5.6)}$$

Where a and Δ are considered as adjustable parameters to be determined by fitting to known data.

Since in a nucleus $N > Z$ and hence $(E_f)_n > (E_f)_p$. This clearly reveals that the potential wells for the protons and neutrons have different depths, i.e. the former being less deep than the latter. Actually, the depth of the potential well is given by

$$V_0 = E_f + \frac{E_B}{A} = E_f + f_B \qquad \text{... (5.7)}$$

Where $f_B = \dfrac{E_B}{A}$ is the mean binding energy per nucleon (usually called as binding fraction) which is ~ 8 MeV per nucleon for both protons and neutrons. Figure 5.7 illustrates this.

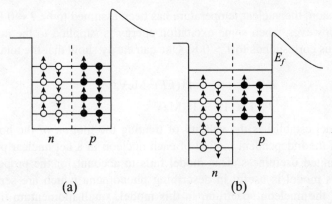

Fig. 5.7 Fermi gas model picture of nucleonic potential wells. The difference in the depth of the neutron and potential wells is clearly visible.

Fermi energies for both protons and neutrons are represented by the same horizontal line in Figure 5.7, which is about 8 MeV below the top of the potential well (Coulomb effect is neglected). This can be understood from the fact that if these are at different depths below the top of the well, then the nucleons of type from higher Fermi level (say, the neutrons) would make spontaneous transitions to the lower Fermi level for the other type the lower Fermi level for the other type (for protons) through β-transformations.

This means that the levels will ultimately equalize.

The depth of the potential well is approximately

$$V_0 \approx 21 + 8 = 29 \text{ MeV}$$

We may note that the neutron depth being slightly greater than the proton depth.

Let us now consider a hypothetical infinite medium of nuclear matter of uniform density in which the numbers of neutrons and protons are equal, i.e. $N = Z$. Neglecting the Coulomb interaction of the protons, the semi-empirical binding energy formula yield

$$\frac{E_B}{A} = \frac{E_v}{A} \simeq 15.9 \text{ MeV/nucleon} \qquad \text{... (5.8)}$$

Adding this to the depth of the potential well below the Fermi level, one obtains the depth of the potential well as

$$V_0 = 21 + 15.9 = 36.9 \text{ MeV} \qquad \text{... (5.9)}$$

Any successful theory of nuclear matter should be able to correlate the value of V_0 given by (5.9) with the nature of inter nucleon force.

The level density at the Fermi level $\rho = \left(\dfrac{dn}{dE}\right)_{E_f}$ can be estimated from this model. Noting that

$\dfrac{dp}{p} = \dfrac{dE}{E_f}$ gives:

$$\rho = \frac{3A}{8E_f} = 0.011 \text{ A(MeV)}^{-1}$$

In the above discussion, the nuclear temperature has been assumed to be $T = 0$ K corresponding to the ground state. However, when some excitation energy is supplied to the nucleus, the thermal energy of the nucleus corresponds to $T > 0$ K. One can easily show that the total excitation energy comes out to be

$$E_t = E_p + E_n = 11(kT)^2 \text{ MeV}$$

$$11 \times (1)^2 = 11 \text{ MeV} \qquad\qquad (\because kT \simeq 1 \text{ MeV})$$

The Fermi gas model establishes the validity of treating the motion of the bound nucleons in a nucleus in terms of the independent motion of each nucleon in a net nuclear potential. However, this model is of limited usefulness. This model fails to account for the properties of low lying nuclear states. This model is useful in describing phenomena which are sensitive to the high momentum part of the nucleon spectrum. In this model, small momentum transfers in nuclear collisions are not permitted as all the low lying states are filled up. However, this limitation of the model does not affect collisions in which large momentum transfer takes place. Obviously, one can describe the properties of the nucleus in excited states with the help of this model. One can also treat the unbounded states of the heavy nuclei with the help of this model.

5.5 THE SHELL MODEL

The liquid drop model can explain only some limited features of the nucleus, e.g. the observed variation of the nuclear binding energy with the mass number and fission of heavy nuclei. However, the model predicts very closely spaced energy levels in nuclei which is contrary to the observation at low energies. The low lying excited states in nuclei are quite widely spaced and liquid drop model fails to explain this and certain other properties of the nucleus. This requires to solve the Schrodinger equation for that potential, and to obtain a detailed description of the behaviour of the nucleons. This procedure is adopted in the shell model of the nucleus. The shell model plays a role in nuclear physics comparable to that played by the Hartree theory in atomic physics. But the shell model is cruder since the exact form of the net atomic potential is internally determined by the self-consistent atomic theory, while the exact form of the net nuclear potential must be inserted into the nuclear model. We will have to consider the motion of the individual nucleons in a potential well which would give rise to the existence of a nuclear shell structure, similar to the electronic shells in the atoms.

The existence of shell structure in atoms arises from the atomic electrons moving in a good approximation independently in a central potential. Taking the clue from the chemical stability of closed electron sub-shells and shells in the atoms, the physicists enquired if nucleons too form similar closed sub-shell and shells in nuclei, i.e. if protons and neutrons in a nucleus are arranged in some type of shell structure.

The properties of electron shells are well known. If the number of electrons in the atom equals 2, 10, 18, 54 or 86, then the outer shell is closed and is of such a stability that the electron exchange with other atoms is never energetically advantageous. Therefore, neither chemical reactions nor the formation of homopolar molecules are feasble. The above numbers are the atomic numbers of inert gases: $He(Z = 2)$, $Ne(Z = 10)$, $Ar(Z = 18)$, $Kr(Z = 36)$, $Xe(Z = 54)$ and $Rn(Z = 86)$. Analogous numbers of protons and neutrons, to which particular properties also correspond, were found in atomic nuclei as well. But these numbers are different:

$$Z = 2, 8, 20, 28, 50, 82 \text{ and } 126$$

They have been called *magic numbers*. There are also characterstic jumps in the single particle separation energies S_p and S_n at the magic numbers as shown for S_n in Figure 5.8. The separation energies S_p and S_n are given by

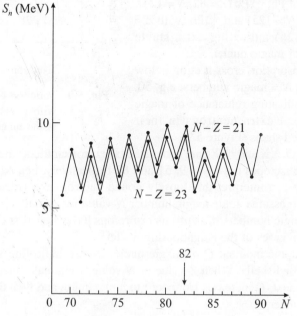

Fig. 5.8 A plot of neutron separation energies, S_n exhibiting the discontinuity at $N = 82$.

$$S_p = B(Z, N) - B(Z - 1, N)$$

and

$$S_n = B(Z, N) - B(Z, N - 1)$$

These changes are similar to those found to occur in atoms of intert gases at closed electronic shells and support strongly the shell structure of nuclei. The following are the main evidences to show the existence of shell structure within the nuclei:

(i) Just as intert gases, with $Z = 2, 10, 18, 36, 54, ...$ electrons having closed shells show high chemical stability, nuclei with $Z = 2, 8, 20, 28, 50, 82$, and 126 nucleons – the so-called *magic numbers* of the same kind (either proton or neutron) are particularly stable. (Nuclei with $Z = N$, a magic number are said to be doubly magic and show exceptionally high stability).

(ii) The number of stable *isotopes* (Z = const.) and isotones (N = const.) is larger with respective number of protons and neutrons and equal to either of magic numbers, e.g. $S_n(Z = 50)$ has 10 stable isotopes, $Ca(Z = 20)$ has 6; the biggest group of isotone is at $N = 82$, then at $N = 50$ and $N = 20$. The relative abundances of naturally occurring isotopes whose nuclei contain magic numbers of protons or neutrons, have greater relative abundances, e.g. the isotopes ^{88}Sr ($N = 50$), ^{138}Ba ($N = 82$) and ^{140}Ce ($N = 82$) have relative abundances of 82.56 %, 71.66 % and 88.48 % respectively.

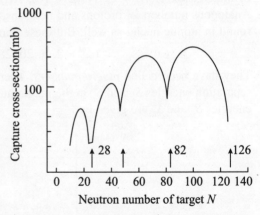

Fig. 5.9 Variation of neutron crose-section with N exhibiting discontinuities at the magic numbers.

(iii) The three naturally occurring radioactive series decay to the stable end product the three isotopes of Pb; $^{206}_{82}Pb$ ($Z = 82, N = 124$), $^{207}_{82}Pb$ ($Z = 82, N = 125$) and $^{208}_{82}Pb$ (with $Z = 82$ and $N = 126$) indicating extra stable configuration of magic nuclei.

(iv) The neutron absorption cross-section is low for nuclei with N = magic numbers, e.g. 50, 82 and 126, indicating reluctance of magic nuclei to accept extra neutrons in their completely filled shells (Figure 5.9).

(v) Isotopes like $^{17}_8O$, $^{87}_{36}K$ and $^{137}_{54}Xe$ are spontaneous neutron emitters when excited by preceding β-decay. These isotopes have $N = 9, 51$ and 83 respectively, i.e., $N(8 + 1), N(50 + 1)$ and $N(82 + 1)$. We can interpret this loosely bound neutron as a *valence neutron* which the isotopes emit to assume some magic number N-value for stability.

(vi) Nuclei with magic numbers of neutrons or protons have their first excited states at higher energies than in cases of the neighbouring nuclei.

(vii) Electric quadrupole moment Q of magic nuclei is zero indicating spherical symmetry of nucleus for closed shells. When Z-value or N-value is gradually increases from one magic number to the next, Q increases from zero to a maximum and then decreases to zero at the next magic number.

(viii)   The energy of α– or β– particles emitted by magic radioactive nuclei is larger. Figure 3.10 shows the plot of the α-disintegration energies of the heavy nuclei as functions of the mass number A for a given Z. The curve shows a regular variation till the magic neutron number N = 126 is reached when there is sudden discontinuity. Obviously, this confirms the magic character of the neutron number 126. Similar discontinuities are reported amongst the β-emitters as the magic neutron or proton numbers.

(ix)   The asymmetry of the fission of the uranium nucleus could involve the sub-structure of nuclei, which is expressed in the existence of the magic numbers.

All these experimental facts lend strong support to the shell structure of nuclei.

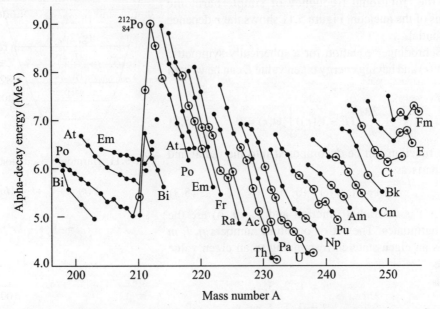

Fig. 5.10 Discontinuities in the α-disintegration energies at N = 126 for heavy nuclides. This confirms the magic character of the neutron number 126.

The procedure of the shell model involves first finding the neutron and proton energy levels for an assumed form of the net potential of a particular nucleus. That is, if each nucleon is treated as moving independently in a net nuclear potential V(r), which are found by solving the Schrodinger equation for that potential. The only forms for the net potential considered are spherically symmetrical functions, V(r), where r is the distance from a nucleon to the centre of the nucleus; other forms would greatly increase the difficulty of solving the Schrodinger equation. Just as in the Hartree theory of atoms, it is found that the energy of a nucleon energy level of net nuclear potential V(r) depends on quantum numbers n and l, which specify the radial and angular behaviour of a nucleon in the level. The quantum number l is just the same as the one we encounter throughout atomic physics when dealing with any spherically potential like V(r). The quantum number n used in nuclear physics is related to, but not the same as, the quantum number of atomic physics that is symbolized by the same letter. Because of the approximate square well form of the net potential V(r) which arises in nuclear physics, it is more convenient in that field to use what is called the radial node quantum number n.

Singe Particle States in nucleus

We assume an infinite three dimensional harmonic oscillator potential of the form (Figure 5.11).

$$V(r) = -V_0 + \frac{1}{2} M\omega^2 r^2 \qquad ... (5.1)$$

Where M is the nucleon mass, V_0 is the potential well depth and ω is the rotational frequency of simple harmonic oscillations of the nucleon. Figure 5.11 shows the r-depence of the potential.

The Schrodinger equation for a spherically symmetric potential $V(r)$ and having energy eigen value E can be written as

$$\left[\nabla^2 + \frac{2M}{\hbar^2}(E - V(r)) \right] \psi(r) = 0 \qquad ... (5.2)$$

The wave function can be decomposed into the radial and spherical parts as

$$\psi(\mathbf{r}) = R_{nl}(r)\, Y_{lm}(\theta, \phi) \qquad ... (5.3)$$

Where $R_{nl}(r)$ is the radial function and $Y_{lm}(\theta, \phi)$ are the spherical harmonics. The set of quantum numbers n, l, m determines an eigen-state corresponding to an eigen value E_{nl}, where

$$n = 1, 2, 3, ...$$

and

$$l = 0, 1, 2, ..., n$$

The radial wave function R_{nl} is a solution of the equation

$$\frac{1}{r^2}\frac{d}{dr}\left(r^2 \frac{dR_{nl}}{dr} \right) + \frac{2M}{\hbar^2}\left[E_{nl} - V(r) - \frac{l(l+1)\hbar^2}{2Mr^2} \right] R_{nl} = 0 \qquad ... (5.4)$$

The term in $l(l + 1)$ is the centrifugal potential.

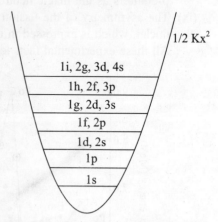

$1/2\, Kx^2$

| 1i, 2g, 3d, 4s |
| 1h, 2f, 3p |
| 1g, 2d, 3s |
| 1f, 2p |
| 1d, 2s |
| 1p |
| 1s |

Fig. 5.11 Harmonic oscillator potential

$$V(r) = -V_0 + \frac{1}{2} M\omega^2 r^2$$

$$= -V_0 + \frac{1}{2} kr^2 \ (K = M\omega^2)$$

(i) **Square-well of Infinite Depth** Let us try to calculate the position of energy levels for the simplest case of an infinitely deep square well of radius r. Let us assume that the potential is zero inside the well and infinite outside. The radial wave function $R_{nl}(r)$ vanishes at the boundary of the well. The radial solutions are regular at the origin and inside the potential well are the well known *spherical Bessel functions* $J_l(k_{nl}r)$. Thus

$$R_{nl}(r) = J_l(k_{nl}r) = \frac{A}{\sqrt{k_{nl}\, r}}\, J_{l+\frac{1}{2}}(k_{nl}r) \qquad ... (5.5)$$

Where A is constant, $J_{l+\frac{1}{2}}(k_{nl}r)$ is a Bessel function and k_{nl} is the wave number can be easily defined by the following equation

$$k_{nl}^2 = \frac{2M}{\hbar^2}[E_{nl} - V(r)] \qquad \qquad \text{... (5.6)}$$

Where E'_{nl} is the total negative energy and $V(= -U)$ is the potential well depth. Measuring the energies from the bottom of the well is always convenient. We have

$$K_{nl}^2 = \left(\frac{2M}{\hbar^2}\right)E'_{nl} \qquad \qquad \text{... (5.7)}$$

Where E'_{nl} is positive and measured from the bottom of the potential well.

The boundary condition restrict the permitted values of k_{nl}. For an well of infinite depth, the wave function has to vanish at the nuclear boundary. Thus at $r = R$, we have

$$U_{nl}(R) = J_l(k_{nl}R) = 0$$

We find that each l-value has a set of zeros and each of them corresponds to a value of k_{nl} given by $k_{nl}R = x$. Obviously, the eigen-value $k_{nl}R$ is the nth zero of the lth *spherical Bessel function*. The number $n(= 1, 2, 3, ...)$, giving the number of zeros of the radial part of the wave function (not counting the origin), is called as the *radial quantum number*. We may note that this differs from the principal quantum number of atomic spectroscopy, since the latter counts all the total wave function (angular as well as radial) and is of major importance for specifying the energy of the corresponding state.

Spherical bessel function for $l = 0$, 1, and 2 is shown in Figure 5.12. One finds that there is succession of zeros at $k_{nl}R = x$ numbered serially $n = 1, 2, ...$ and moreover these values differ for different l. The order of energy levels in a spherical potential well of infinite depth is determined by the order of the zeros in the Bessel functions one can indicate such levels in order of increasing energy. The following relation represent the level of energies

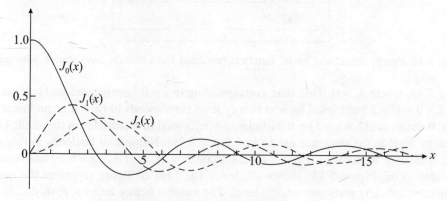

Fig. 5.12 Spherical Bessel functions for $l = 0$, 1, 2

$$E'_{nl} = k_{nl}^2 \frac{\hbar^2}{2M} = \left(\frac{\hbar^2}{2MR^2}\right)x^2$$

For $R = 8 \times 10^{-15}$ m, the quantity $\dfrac{\hbar^2}{2MR^2} = 0.34$ MeV. The symbols (spectroscopic) represent the states $l = 0, 1, 2, \ldots$ with a different angular wave function $Y_{l,\,m}(\theta, \phi)$. Due to two different orientations of the nucleon spin, a level can have $2(2l + 1)$ protons. As neutron and proton are two different particles, hence this level can also accomodate $2(2l + 1)$ neutrons in addition to $2(2l + 1)$ protons.

Figure 5.13 shows the energy levels and magic numbers predicted from this model, i.e. infinite deep square well. The first energy level corresponds to

l^2	Level	No of Particles	Total No of Particles
118.90	3p	6	132
110.52	1i	26	132
100.54	3p	14	106
88.83	3s	2	82
87.53	1d	22	90
12.72	2d	10	68
66.99	15	18	58
59.68	1p	6	40
40.84	1f	14	34
32.51	2s	2	20
33.2	1d	10	18
42.18	4p	6	8
1.87	1f	2	2

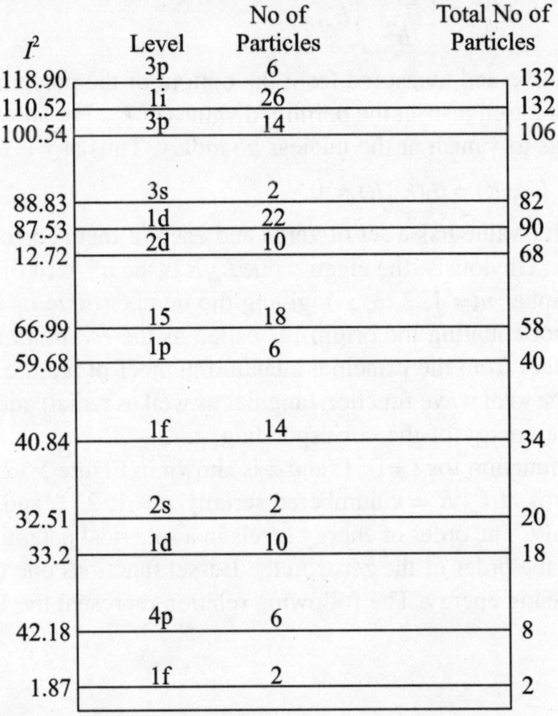

Fig. 5.13 Energy levels and magic numbers predicted from infinite deep square potential weel.

$x = 3.14$, where $J_0 = 0$. This state corresponding to $l = 0$ is expressed as $1s$. It consists of $2(2 \times 0 + 1) = 2$ particles. The next energy level corresponds to $l = 1$ state and expressed as $1p$. It contains $2(2 \times 1 \pm 1) = 6$ particles, giving a total of 8 particles in the shell. The other energy level in the sequence are $1d$ [$l = 2$, $x = 5.76$]. Number of nucleons in this state are 10. Similarly, we have $2s$ ($l = 0$, $x = 6.28$, number $= 2$), $1f$ [$l = 3$, $x = 6.99$ and number 14], ... and so on. Figure 5.13 shows all these. The LHS of figure represent the number of particles upto any particular energy level. The number before terms $s, p, d, f, \ldots$ represents the order of zero in the Bessel function, i.e. $1s$ for first zero in the Bessel function $J_0(x)$, $2s$ for second zero in this function. We note that shell closes at total particle numbers 2, 8, 18, 20, 34, 40, 58, ... We may see that these are not the nuclear magic numbers.

If we now consider the level for finite rather than infinite well. We note that the level order is not essentially different although the excitation alters. This explains the pronounced stability of $^4_2$He, $^{16}_8$O, $^{40}_{20}$Ca. However, there is no indication of gaps occuring at the magic numbers 28, 50, 82, 126.

(ii) **Harmonic Oscillator Potential**   Now, We assume that the particle is moving with simple harmonic motion isotropically bound in three dimensions. A particle of mass M, held to a fixed point by a restoring force kr, where k is a force constant, will vibrate about this fixed point with frequency $v = \dfrac{1}{\sqrt{2\pi}}\sqrt{\dfrac{k}{M}}$ or angular velocity $\omega = \sqrt{\dfrac{k}{M}}$. Obviously, the potential function of the oscillating particle is

$$V(r) = \int_0^r kr\, dr = \frac{1}{2}kr^2 = \frac{1}{2}M\omega^2 r^2 \qquad \text{... (5.9)}$$

The energy eigen value corresponding to eigen function

$$\psi_{n,\,l,\,m}(r,\,\theta,\,\phi) = u_{nl}(r)\,\psi_{l,\,m}(\theta,\,\phi) \qquad \text{... (5.10)}$$

is given by

$$E_{n,\,l} = \hbar\omega\left(2n + l - \frac{1}{2}\right); \quad n = 1, 2, 3, \dots \quad l = 0, 1, 2, \dots$$

$$= \hbar\omega\left(\eta + \frac{3}{2}\right) = E_\eta \qquad \text{... (5.11)}$$

Where $\qquad \eta = 2n + 1 - 2.$

When we examine the angular dependence of the wave function, we find that for each η value, there is a degenerate group of levels with different l values such that $l \le \eta$ and even (odd) corresponds to even (odd) values of η. Clearly, the sequence of single particle levels for the harmonic oscillator well consists of bands of degenerate levels, each band separated by energy $\hbar\omega$ from the next. Thus the degeneracy for each l is $2(2l + 1)$ as before. However, the eigenstates corresponding to the same value of $2n + 1$, (i.e. the same η) are also degenerate. So for different values of η, the eigen values of energies are given by

$$E_{nl} = \hbar\omega\left(\eta + \frac{3}{2}\right) = E_\eta$$

But for each η one can have different combinations of n and l. Since $2n = \eta - l + 2$ is even for η even or odd, the degenerate eigenstates are

$$(\eta,\, l) = \left[\frac{1}{2}(\eta + 2), 0\right], \left[\frac{\eta}{2}, 2\right], \dots [2, \eta - 2], [1, \eta] \dots \text{ for } \eta \text{ even}$$

$$= \left[\frac{1}{2}(\eta + 1), 1\right], \left[\frac{1}{2}(\eta - 1), 3\right], \dots [2, \eta - 2], [1 - \eta] \dots \text{ for } \eta \text{ odd}$$

$\therefore$ Number of particles, i.e. number of neutrons or protons with the eigen value E_η

$$N_\eta = \sum_{k=0}^{\eta/2} 2[(2k)+1] \qquad\qquad \text{for even } \eta$$

$$= \sum_{k=0}^{(\eta-1)/2} 2[2(2k+1)+1] \qquad\qquad \text{for odd } \eta$$

$$= (\eta+1)(\eta+2) \qquad \text{in either case} \qquad\qquad ... (5.12)$$

and the total number of particles for all levels upto h is

$$\sum_\eta N_\eta = \frac{1}{3}(\eta+1)(\eta+2)(\eta+3) \qquad\qquad ... (5.13)$$

The *single particle level scheme* and the spectroscopic notation of the states are given in Table 5.2.

Table 5.2 Order of levels in an oscillator potential and number of particles of each kind (protons or neutrons) in a shell.

η	E_η	n	l	nl for each η	l_j	Number of particles in a shell $(\eta+1)(\eta+2)$	Number of particles upto and including the shell $\frac{1}{3}(\eta+1)(\eta+2)(\eta+3)$
0	$\frac{3}{2}\hbar\omega$	1	0	$1s$	$s_{\frac{1}{2}}$	2	2
1	$\frac{5}{2}\hbar\omega$	1	1	$1p$	$p_{\frac{3}{2}}, p_{\frac{1}{2}}$	6	8
2	$\frac{7}{2}\hbar\omega$	1,2	2,0	$1d, 2s$	$d_{\frac{5}{2}}, d_{\frac{3}{2}}, s_{\frac{1}{2}}$	12	20
3	$\frac{9}{2}\hbar\omega$	1,2	3,1	$1f, 2p$	$f_{\frac{7}{2}}, f_{\frac{5}{2}}, p_{\frac{3}{2}}$	20	40
4	$\frac{11}{2}\hbar\omega$	1,2,3	4,2,0	$1g, 2d, 3s$	$g_{\frac{9}{2}}, g_{\frac{7}{2}}, d_{\frac{6}{2}}$ $d_{\frac{3}{2}}, s_{\frac{1}{2}}$	30	70
5	$\frac{13}{2}\hbar\omega$	1,2,3	5,3,1	$1h, 2f, 3p$	$h_{\frac{11}{2}}, h_{\frac{9}{2}}, f_{\frac{7}{2}}$ $f_{\frac{5}{2}}, p_{\frac{3}{2}}, p_{\frac{1}{2}}$	42	112
6	$\frac{15}{2}\hbar\omega$	1,2,3,4	6,4,2,0	$1i, 2g, 3d, 4s$	$i_{\frac{13}{2}}, i_{\frac{11}{2}}, g_{\frac{9}{2}}$ $g_{\frac{7}{2}}, d_{\frac{5}{2}}, s_{\frac{1}{2}}$	56	168

We can see that in the harmonic oscillator potential, the levels appear in group such as 1s; 1p; 1d, 2s; 1f, 2p; etc. We may note that these grouped levels are degenerate, occupying the same energy state, whereas in the infinite square well potential, the degeneracy is split. The energy levels are of even parity when the oscillator number η is even and of odd parity when η is odd. One can conclude from the above results that the closure of a shell for harmonic oscillator potential occurs corresponding to neutron or proton numbers 2, 8, 20, 40, 112 and 168, whereas the magic numbers suggested by the square well potential are 2, 8, 18, 20, 34, 40, 58, 68, 70, 92, 106 112, 138 and 156. Experimentally observed values of magic numbers are 2, 8, 20, 50, 82 and 126. Obviously, the truth may lie in between these two potentials.

Nuclear Spin-orbit interaction

The mystery of magic numbers was solved in 1949 by Meyer, and independently by Jenson, who introduced the idea of a nuclear spin-orbit interaction. They proposed that each nucleon in a nuclear potential, a *strong inverted spin-orbit interaction* proportional to $\mathbf{s} \cdot \mathbf{l}$, the dot product of its spin and orbital angular momentum vectors. Strong means that the interaction energy is much (about 20 times) larger than would be predicted by using the atomic spin-orbit relation,

$$\mathbf{s} \cdot \mathbf{l} = \frac{\hbar^2}{2} [j(j+1) - l(l+1) - s(s+1)]$$

equating $V(r)$ to the nuclear potential and m to the nucleon mass. Inverted means that the energy of the nucleon is decreased when $\mathbf{s} \cdot \mathbf{l}$ is positive, and increased when it is negative. Thus the sign of the interaction is opposite to the sign of the magnetic spin-orbit interaction experienced by an electron in an atom, i.e. the interaction energy is *negative* when the total angular momentum of the nucleion $\mathbf{j} = \mathbf{s} + \mathbf{l}$ has its *maximum* possible magnitude (i.e. when s and l are as parallel as possible, and $\mathbf{s} \cdot \mathbf{l}$ is positive). However, as the magnitude of spin-orbit interaction is proportional to $\mathbf{s} \cdot \mathbf{l}$ just as it is for an atomic electron, the magnitude of spin-orbit splitting of the nucleon energy levels will be approximately proportional to the value of the quantum number l, just as it is for the electron energy levels. Although, there are similarities between the atomic and nuclear spin-orbit interactions, their differences make it clear that the latter is not in magnetic origin. Instead, it is an attribute of the nuclear force.

The spin-orbit potential V_{ls}, which in non-central, can be written as

$$V_{ls} = -\phi(r) \, \mathbf{l} \cdot \mathbf{s} \qquad \qquad ... (5.14)$$

Where
$$\phi(r) = b \frac{1}{r}\left(\frac{\partial f}{\partial r}\right) \qquad \qquad ... (5.15)$$

Here $l\hbar$ and $s\hbar$ are the azimuthal and spin angular momenta of the nucleon under consideration. $f(r)$ is a spherically symmetric potential function. Thus the potential which determines the single particle wave function will be $\{V(r) - f(r) \, \mathbf{s} \cdot \mathbf{l}\}$, where $V(r)$ and $f(r)$ depend only on the radial distance and the size of the nucleus. Assuming strong coupling between the spin and orbital angular momenta of each individual nucleon giving rise to a total angular momentum j for each so that one can write

$$j = l + s \qquad \qquad ... (5.16)$$

Since $s = \dfrac{1}{2}$ for each nucleon, the two possible values of j are $j = l + \dfrac{1}{2}$ and $j = l - \dfrac{1}{2}$. Due to strong spin-orbit coupling, these two levels now have different energies. One can calculate the splitting of the two levels by computing the expectation values of spin-orbit potential (Equation 5.14) in the two states of j. One can easily, see that

$$2\mathbf{l} \cdot \mathbf{s} = \hbar^2 \{ j(j + 1) - l(l + 1) - s(s + 1) \} \qquad \text{... (5.17)}$$

(5.17) gives the following two values of $\mathbf{l} \cdot \mathbf{s}$:

$$j = l + \frac{1}{2} \qquad : \qquad \mathbf{l} \cdot \mathbf{s} = \frac{l}{2} \qquad \text{... (5.18)}$$

$$j = l - \frac{1}{2} \qquad : \qquad \mathbf{l} \cdot \mathbf{s} = \frac{-(l+1)}{2} \qquad \text{... (5.19)}$$

Now, the expectation value of the spin-orbit interaction potential is

$$<V_{ls}> = E_{ls} = -(\mathbf{l} \cdot \mathbf{s}) \int_0^\infty \psi_{nl}^*(r) \, \phi(r) \, \psi_{nl}(r) dr$$

$$= -(\mathbf{l} \cdot \mathbf{s}) <\phi(r)> \qquad \text{... (5.20)}$$

Where $<\phi(r)>$ is the expection value of $\phi(r)$ appearing in Equation (5.14). Now, for the two states, we have

$$j = l + \frac{1}{2} \qquad : \qquad \varepsilon_{ls} = -\frac{l}{2} <\phi(r)> \qquad \text{... (5.21)}$$

$$j = l - \frac{1}{2} \qquad : \qquad \varepsilon_{ls} = \frac{l+1}{2} <\phi(r)> \qquad \text{... (5.22)}$$

Thus the spin-orbit splitting of the two levels is then

$$\Delta\varepsilon_{ls} = \varepsilon_{ls}\left(l - \frac{1}{2}\right) - \varepsilon_{ls}\left(l + \frac{1}{2}\right)$$

$$= \left(l + \frac{1}{2}\right) <\phi(r)> \qquad \text{... (5.23)}$$

The observed level-spacing due to spin-orbit interaction is given by the following empirical relation:

$$\Delta\varepsilon_{ls} = 10(2l + 1) \ A^{-\frac{2}{3}} \ \text{MeV} \qquad \text{... (5.24)}$$

R.H.S. of Equation (5.23) is positive and therefore the state with $j = l + \dfrac{1}{2}$ lies below the state $j = 1 - \dfrac{1}{2}$ and is thus more tightly bound than the term with $j = 1 - \dfrac{1}{2}$. The splitting which is of the order of a few MeV increases with l and can become so large that for a given n the term with the largest l value and $j = l + \dfrac{1}{2}$ can slide down to energies as low as those of the multiplet with a quantum number $(n-1)$.

The assumed spin-orbit interaction potential resembles that which would arise due to a simple magnetic effect that causes the fine-structure splitting in atomic physics, but such effects are much too weak to give necessary splitting. There is intimate relation between central potential, $V(r)$ and $\phi(r)$ as

$$\phi(r) = \frac{\beta}{r}\left(\frac{\partial V}{\partial r}\right) \qquad \qquad \dots (5.25)$$

Where β is spin-orbit constant.

There is an evidence for the existence of a strong spin-orbit force between nucleons from high energy polarisation experiments.

Figure 5.14 shows the sequence of energy levels, taking spin-orbit interaction into consideration. The LHS part of Figure 5.14 shows the ordering and approximate spacing of the energy levels which nucleons are filling in the nuclei with potentials $V(r)$ in the form of square wells with sounded edges. The right-hand part of Figure 5.14 shows the nucleon energy levels are split by the nuclear spin-orbit interaction. In the presence of spin-orbit interaction, m_l and m_s are no longer useful quantum numbers because the Z-components of the orbital and intrinsic spin angular momenta of a nucleon are no longer constants when these angular momenta are coupled by the interaction. Thus n, l, j, m_j must be used to label the split energy levels. The quantum number j specifies the magnitude of the total angular momentum, J, of a nucleon, which is the sum of its spin and orbital angular momenta; and m_j is the quantum number specifying the Z-component of its total angular momentum J_z. As a result of the spin-orbit interaction, the energies of the levels depend on j as well as on n and l, with the larger j (corresponding to the larger value of J, or $\mathbf{s} \cdot \mathbf{l}$) yielding the smaller energy since the sign of the nuclear spin-orbit interaction is inverted. According to the exclusion principle, each of these levels has a capacity of $(2j+1)$, which is equal to the number of possible values of m_j. In accordance with the three quantum numbers n, l, and j, the levels are designated as follows: $1s_{\frac{1}{2}}$; $1p_{\frac{3}{2}}$, $1p_{\frac{1}{2}}$, $1d_{\frac{3}{2}}$; ... $2s_{\frac{1}{2}}$; $2p_{\frac{3}{2}}$, $2p_{\frac{1}{2}}$; ... $2f_{\frac{7}{2}}$, $2f_{\frac{5}{2}}$; etc.

The lowest state $2s_{\frac{1}{2}}$ holds two nucleons and thus there is a closed shell at 2. In the next state the interaction is yet not strong enough to separate the $1p_{\frac{3}{2}}$ and $1p_{\frac{1}{2}}$ levels by an amount comparable with well spacing so that the next closed shell occurs, when all the $6p$ states are occupied, at $2 + 6 = 8$. Now, the state $1d_{\frac{5}{2}}$ is definitely lower than $1d_{\frac{3}{2}}$, but it still forms part of the same shell.

The states $1d_{\frac{5}{2}}$, $1d_{\frac{3}{2}}$ and $2s_{\frac{1}{2}}$ are sufficiently close to constitute a single shell. Obviously, the next shell closes at $8 + 12 = 20$.

For the state $1f_{\frac{7}{2}}$, we have the l value, $l = 3$, is quite high enough to make it to join the group from $N = 2$. Obviously, this shell closes at $20 + 8 = 28$. The next shell comparises $1f_{\frac{5}{2}}$, $2p_{\frac{3}{2}}$, $2p_{\frac{1}{2}}$

and $1g_{\frac{9}{2}}$. The last coming down due to spin-orbit coupling from the next higher group of levels.

This contains 22 sub-levels. This shell closes at $28 + 22 = 50$.

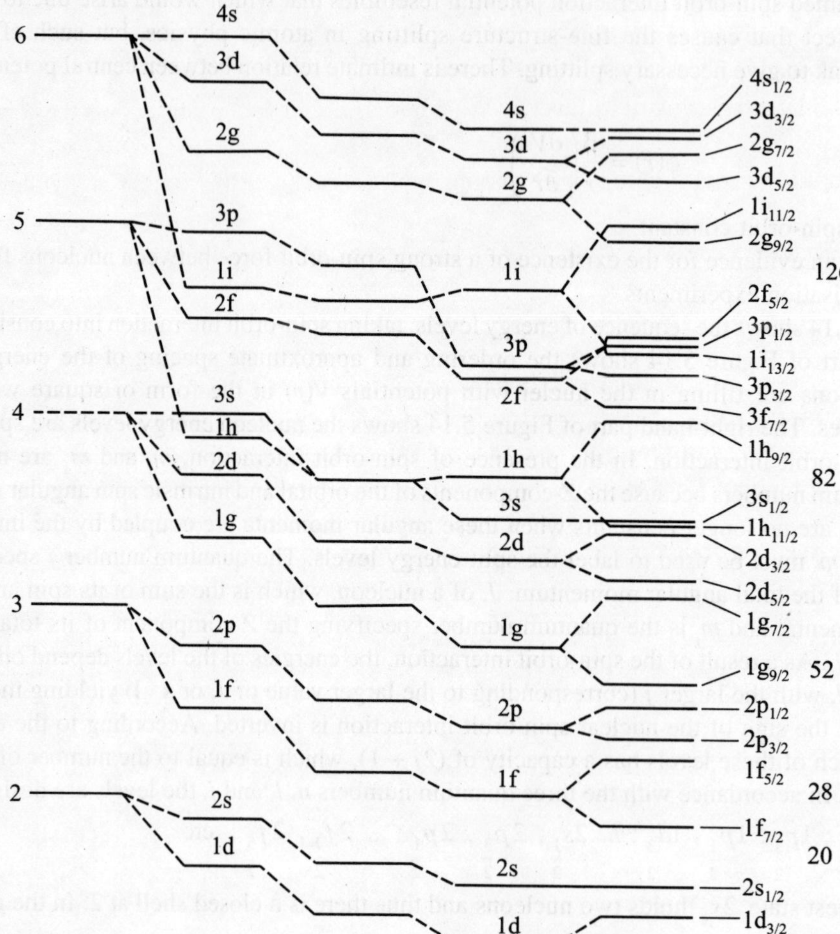

Fig. 5.14 Ordering of energy states according to shell model, using various potentials.

The next shell is made up of 32 sub-levels of $1g_{\frac{7}{2}}$, $2d_{\frac{5}{2}}$, $2d_{\frac{3}{2}}$, $3s_{\frac{1}{2}}$ and $1h_{\frac{11}{2}}$ and closes at 50 + 32 = 82 and one after that will 44 sub-shells of $1h_{\frac{9}{2}}$, $2f_{\frac{7}{2}}$, $2f_{\frac{5}{2}}$, $3p_{\frac{3}{2}}$, $3p_{\frac{1}{2}}$ and $1i_{\frac{13}{2}}$ at 82 + 44 = 126.

Finally the large splitting of $1i$ level at $N = 6$ into $1i_{\frac{13}{2}}$ and $1i_{\frac{11}{2}}$ pushes down the former level to the vicinity of the $N = 5$ group of remaining sublevels. The new group of six sublevels $1h_{\frac{9}{2}}$, $2f_{\frac{7}{2}}$, $2f_{\frac{5}{2}}$, $3p_{\frac{3}{2}}$, $3p_{\frac{1}{2}}$ and $1i_{\frac{13}{2}}$ can accomodate a maximum of 10 + 8 + 6 + 4 + 2 + 14 = 44 nucleons, which when added to the number 82 accounts for occurrence of the particle numbers 126.

In this way, one can see that the shell closures occur at particle numbers 2, 8, 20, 28, 50, 82, 126, 184 exactly as required by experiments.

Table 5.3 shows the orbitals, listed in order of increasing energy and each subshell can accomodate upto $(2j + 1)$ protons or neutrons. We see that *extreme single particle shell model with strong inverted spin-orbit interaction* predicts precisely the magic numbers.

If all inter-nucleon coupling are ignored, the model is then called *single particle shell model*. If however, couplings are considered it is known as *independent particle shell model*.

Table 5.3 Level scheme after inclusion of spin-orbit interaction in the oscillator potential

Shell number	State in a shell	Number of particles in a shell	Total particles upto the shell closure
1	$1s_{\frac{1}{2}}$	2	2
2	$1p_{\frac{3}{2}}$, $1p_{\frac{1}{2}}$	6	8
3	$1d_{\frac{5}{2}}$, $2s_{\frac{1}{2}}$, $1d_{\frac{3}{2}}$	12	20
3a	$1f_{\frac{7}{2}}$	8	28
4	$2p_{\frac{3}{2}}$, $1f_{\frac{5}{2}}$, $2p_{\frac{1}{2}}$, $1g_{\frac{9}{2}}$	22	50
5	$1g_{\frac{7}{2}}$, $2d_{\frac{5}{2}}$, $2d_{\frac{3}{2}}$, $3s_{\frac{1}{2}}$, $1h_{\frac{11}{2}}$	32	82
6	$1h_{\frac{9}{2}}$, $2f_{\frac{7}{2}}$, $2f_{\frac{5}{2}}$, $3p_{\frac{3}{2}}$, $3p_{\frac{1}{2}}$, $1i_{\frac{13}{2}}$	44	126
7	$2g_{\frac{9}{2}}$, $3d_{\frac{5}{2}}$, $1i_{\frac{11}{2}}$, $2g_{\frac{7}{2}}$, $4s_{\frac{1}{2}}$, $3d_{\frac{3}{2}}$, $1j_{\frac{15}{2}}$	58	184

Figure 5.14 is so frequently used by the nuclear scientists that many of them have it memorized. An easier way is to construct it by using the acrostic.

spuds if pug dish of pig

Which means: (eat) potatoes if the pork is bad. Deletion of all vowels, except the last, yields

s p d s f p g d s h f p i g

This is the ordering of *l* for all the unsplit levels, through those leading to the magic number 126. The values of *n* are assigned easily since the first *s* level is 1*s*, the second is 2*s*, etc. The remainder of the figure is constructed by applying an inverted spin-orbit splitting, proportional to *l*.

We may note that Figure 5.14 is not an energy level diagram for any particular nucleus: instead it gives the order in which the nuclear levels appear below the Fermi energy as the radius of the nuclear potential increases in proportion to $A^{\frac{1}{3}}$, i.e. it gives the order in which the highest energy levels of various nuclei fill. It also gives an indication of the relative magnitudes of the separation between adjacent levels as they are filling. Obviously, it is analogous to the diagram that could be constructed for atoms by using only the left hand side of Figure. 5.14.

There are some recent experimental and theoretical evidence showing that there may be small but important changes from Figure 5.14 in the filling order or the highest levels in the case of protons.

Predictions of the Shell Model

The shell model with its spin-orbit coupling acconts for the magic nucleon number. Mayer and Jensen were awarded *Nobel prize in* 1963 for their contributions to the development of nuclear shell structure. However, shell model can do much more than predict the magic numbers. Now, we shell examine these predictions:

(i) **Ground state Spin and Parity**   In the single-particle shell model the remaining residual interaction is assumed to be just the pairing interaction. With this assumption, one can understood the spins and parities (J^π) of many ground states especially spins of odd *A*-nuclei. Let us use the term nuclear 'spin' to denote the angular momentum *I* of the nucleus which is the vector sum $\mathbf{I} = \mathbf{L} + \mathbf{S}$.

We know that the spin s of each nucleon couples with its orbital angular momentum *l* to form the total angular momentum $j = l + s$. One can then obtain the total angular momentum *I* of the system of nucleons by the coupling of the *j* vectors of the individual nucleons within the nucleus, i.e. $\mathbf{I} = \sum \mathbf{j}_i$. This coupling is called as $j - j$ coupling. We may note that in nuclear physics it is customary to denote the total angular momentum by the symbol *I* in place of *J*. There are following rules for the angular momenta and parities of nuclear ground states (we may remember that according to shell model neutron and proton levels fill independently):

(i) Even-even nuclei of any one kind have total ground state angular momentum $I = O^+$ and parity even. Thus, the total angular momentum of any shell with an even number of nucleons is zero, while the angular momentum with an odd number of nucleons is equal to the angular momentum of the last odd unpaired nucleon.

The level configurations for the simple odd proton nuclei $^7_3\text{Li}_4$ and $^{15}_7\text{N}_8$ in their ground states is as follows:

$$_3^7\text{Li}_4 : \left(1s_{\frac{1}{2}}\right)^2, \left(1p_{\frac{3}{2}}\right)^1$$

$$_7^{15}\text{N}_8 : \left(1s_{\frac{1}{2}}\right)^2, \left(1p_{\frac{3}{2}}\right)^4, \left(1p_{\frac{1}{2}}\right)^1$$

After putting pairs of particles in the earlier shells, one finds that they give rise to spins of lass odd protons as $\frac{3}{2}$ and $\frac{1}{2}$ respectively. One obtains the parities of both states as $(-1)^l$ and is odd (–) since $l = 1$ for the last odd protons in both the examples. Thus the states can be specified as I^π equals to $\frac{3^-}{2}$ and $\frac{1^-}{2}$ respectively.

We now consider another example of odd neutron nuclei, $^{33}\text{S}_{17}$ and $^{29}\text{Si}_{15}$ having following neutron level configurations:

$$_{16}^{33}\text{S}_{17} : \left(1s_{\frac{1}{2}}\right)^2 \left|\left(1p_{\frac{3}{2}}\right)^4 \left(1p_{\frac{1}{2}}\right)^2\right| \left(1d_{\frac{5}{2}}\right)^6 \left(2s_{\frac{1}{2}}\right)^2 \left(id_{\frac{3}{2}}\right)^1$$

$$_{14}^{29}\text{Si}_{15} : \left(1s_{\frac{1}{2}}\right)^2 \left|\left(1p_{\frac{3}{2}}\right)^4 \left(1p_{\frac{1}{2}}\right)^2\right| \left(1d_{\frac{5}{2}}\right)^6 \left(2s_{\frac{1}{2}}\right)^1$$

and have spins $\frac{3}{2}$ and $\frac{1}{2}$ respectively.

Let us now take the example of mirror nuclei, $^{13}\text{C}_7$ and $^{13}\text{N}_6$. The level configuration for 7 odd protons in $^{13}\text{N}_6$ or 7 odd neutrons in $^{13}\text{C}_7$ is the same and is expressed as

$$\left(1s_{\frac{1}{2}}\right)^2, \left(1p_{\frac{3}{2}}\right)^4, \left(1p_{\frac{1}{2}}\right)^1$$

Both have same spin $\frac{1^-}{2}$ as predicted by the shell model and confirmed experimentally.

There is another example of mirror nuclei $^{17}_8\text{O}_9$ and $^{17}_9\text{F}_8$. The configuration for both these for the odd neutron or odd proton as

$$\left(1s_{\frac{1}{2}}\right)^2 \left|\left(1p_{\frac{3}{2}}\right)^4 \left(1p_{\frac{1}{2}}\right)^2\right| \left(1d_{\frac{5}{2}}\right)^1$$

Shell model predicts spin $\frac{5^+}{2}$ for both these mirror nuclei, which is in accordance with the experiment.

There is no known exception to this rule. We can summarize it as

- For N or Z even: $I = 0$
- For N even, Z odd or for N odd, Z even: I depends on the number of p of odd nucleons. $p = Z$ in the first case and $p = N$ in the second case.

Nordheim's Rule for odd Z-odd N nuclei

For finding the spins of odd Z-odd N nuclei, Nordheim proposed following two empirical coupling rules. They predict the ground state spins, J, of odd Z-odd N nuclei. They are:

(i) $I = |J_1 - J_2|$ for $J_1 = l_1 \pm \dfrac{1}{2}$ and $J_2 = l_2 \mp \dfrac{1}{2}$

(ii) $|J_1 - J_2| \le I \le (J_1 + J_2)$ for $J_1 = l_1 \pm \dfrac{1}{2}$ and $J_2 = l_2 \pm \dfrac{1}{2}$

Where l_1, l_2 and J_1, J_2 are orbital and total angular momenta respectively of odd proton and odd neutron. I is the total angular momentum of the odd Z-odd N nucleus

Modified Nordheim's Rules

Brennen and Bernstein after comparing with experimental results, modified Nordtheim's rules as follows:

(i) $I = |J_1 - J_2|$ for $J_1 = l_1 \pm \dfrac{1}{2}$ and $J_2 = l_2 \mp \dfrac{1}{2}$

(ii) $I = |J_1 \pm |$ for $J_1 = l_1 \pm \dfrac{1}{2}$ and $J_2 = l_2 \pm \dfrac{1}{2}$

(iii) $I = J_1 + J_2 - 1$

With the help of above modified rules the angular momentum of the deutron which contains one protons and one neutron can be calculated to be either $1\hbar$ or 0. Experimentally it was found to be $1\hbar$.

However, there are few exceptions to this simple model. The level configuration of nuclei $^{47}_{22}\text{Ti}_{25}$ and $^{55}_{25}\text{Mn}_{30}$ is

$$\left(1S_{\frac{1}{2}}\right)^2 \left|\left(1p_{\frac{3}{2}}\right)^4 \left(1p_{\frac{1}{2}}\right)^2\right| \left(1d_{\frac{5}{2}}\right)^6 \left(2s_{\frac{1}{2}}\right)^2 \left(1d_{\frac{3}{2}}\right)^4 \left(1f_{\frac{7}{2}}\right)^5$$

The 25th proton or neutron in these nuclei predict $\dfrac{7}{2}$ value instead of the experimentally observed $\dfrac{5}{2}$. Now, we consider nuclei $^{75}_{33}\text{As}_{42}$ and $^{61}_{28}\text{Ni}_{33}$. The odd particle configuration of these nuclei are

$$\left(1s_{\frac{1}{2}}\right)^2 \left(1p_{\frac{3}{2}}\right)^4 \left(1p_{\frac{1}{2}}\right)^2 \left(1d_{\frac{5}{2}}\right)^6 \left(2s_{\frac{1}{2}}\right)^2 \left(1d_{\frac{3}{2}}\right)^4 \left(1f_{\frac{7}{2}}\right)^8 \left(2p_{\frac{3}{2}}\right)^4 \left(1f_{\frac{5}{2}}\right)^1$$

These two nuclei with 33 protons and 33 neutrons respectively are expected to have spin $\dfrac{5}{2}^-$, whereas experimentally reported value is $\dfrac{3}{2}^-$ spin. Moreover, exceptions have also been reported for neutron numbers 57, 59, and 61. These exceptions were explained by modifying the rules and stating that if the *high spin shell comes after low spin shell the high spin shell fills faster, pairing its particles before the low spin shell can be filled completely.* According to this rule the '*pairing energy*' (Pairing interaction is a residual nuclear interaction, i.e., a part of the total nuclear interaction experienced by the nucleons that is not described by the spherically symmetrical net potential $V(r)$ of the shell model, or by the spin-orbit interaction. Although not described by these attributes of the shell model, pairing interaction can be predicted by them. The net potential $V(r)$ represents the interactions experienced by a nucleon on the *average*. The pairing interaction represents a *departure from the average interaction* described by $V(r)$, that arises when the nucleon is particularly close to another nucleon with which it can have an individual interaction. It involves the collision of nucleons in degenerate states of a partly filled subshell. A pair of nucleons having the same value of j but the opposite values of m_j (e.g. $j = \frac{5}{2}$, $m_j = \frac{5}{2}$, $j = \frac{5}{2}$, $m_j = -\frac{5}{2}$) collide with each other in such an interaction, and after the collision enter previously empty states that have different but still opposite values of m_j (e.g. $j - \frac{5}{2}$, $m_j - \frac{3}{2}$; $j = \frac{5}{2}$, $m_j = -\frac{3}{2}$) It is clear that angular momentum is conserved in such collisions, and the collisions are not inhibited by the exclusion principle. The energy of the system is reduced because, when colliding, the nucleons are particularly close together, and the exclusion principle does not prevent them from exerting on each other the strongly attractive short range nuclear force) which is a negative potential energy connected with double occupancy of a level, and it is larger, the higher the l-value of the level. This means in $^{75}_{33}$As and $^{61}_{33}$Ni nuclei, the pairing energy favours the occupancy of the $f_{\frac{5}{2}}$ level ($l = 3$) by the proton or neutron pairs leaving $p_{\frac{3}{2}}$ level for the unpaired proton or neutron respectively. Thus the spin for these two nuclei in their ground states is $\frac{3}{2}^-$.

The pairing energy also explains why high spins such as $\frac{11}{2}$ and $\frac{13}{2}$ are not found in the ground states. It is because the high spin levels are filled prior to the low spin levels start to fill as a result of pairing energy.

The assumption that like nucleon outside the closed shells pair off so as to cancel their angular momenta is really not at all convincing and self evident. It is reported that the above assumption is correct except when the $1d_{\frac{5}{2}}$, $1f_{\frac{7}{2}}$ and $1g_{\frac{9}{2}}$ levels are filled to the extent of three nucleons/holes. In these three cases the angular momenta would be $\frac{3}{2}$, $\frac{5}{2}$, $\frac{7}{2}$ instead of the reported (predicted) values $\frac{5}{2}$, $\frac{7}{2}$, $\frac{9}{2}$ respectively, e.g. $^{23}_{11}$Na has angular momentum $\frac{3}{2}$ instead of $\frac{5}{2}$, $^{55}_{25}$Mn has $\frac{5}{2}$ instead of $\frac{7}{2}$ and $^{79}_{34}$Se has $\frac{7}{2}$ instead of $\frac{9}{2}$. Clearly, in these three nuclides the total angular momentum is due to the three nucleons outside the closed shell and one can explain it on the basis of single particle model.

We may note that the source of pairing energy between two neutrons (or protons) in the same shell over and above the individual neutron binding energies is not accounted by the extreme single particle model. As stated above, it arises due to the residual internucleon interaction and increases with increasing j.

By taking account of residual internucleon interaction in a shell in the single particle model, one can explain the observed I values of most of the nuclei. Apart from two or three isolated cases, e.g. $^{79}_{34}$Se and $^{201}_{80}$Hg, the exceptional I values all occur for $155 \leq A \leq 180$ and $N > 140$. In addition to these, there are certain other regions, where the nuclei are considerably deformed from sphericity and due to this the sequence of levels of a spherical potential cannot be expected to occur in their cases.

Because the nuclear force exerted between two nucleons is strong and short range, the departures from the average described by the pairing interaction are pronounced. Thus pairing interaction is fairly strong, although it is less strong than the spin-orbit interaction. It is short range, just like the nuclear force leading to the fluctuation it represents. It is *attractive* because that force is attractive. A similar interaction resulting from a departure from the average, called the *residual Coulomb interaction*, arises in the treatment of atoms.

(ii) **Magnetic Moments of Nuclei**   The shell model is not so successful in predicting the dipole moments of nuclei.

The magnetic moment of a nucleus is the vector sum of the spin magnetic moment, μ_s and orbital magnetic moment, μ_L. Thus

$$\mu = \mu_s + \mu_L \qquad \qquad ... (5.1)$$

Where μ_s is the vector sum of the intrinsic magnetic moments of the individual nucleons in the nucleus. We have intrinsic moments for protons and neutrons are

$$\mu_p = g_p \, \mu_{\frac{N}{2}} \text{ and } \mu_n = g_n \, \mu_{\frac{N}{2}} \qquad \qquad ... (5.2)$$

Where $\mu_N = \dfrac{e\hbar}{2M_p}$ is the *nuclear magneton* and M_p being the mass of proton. g_n and g_p are the gyromagnetic ratios for the neutron and proton respectively and their numerical values are

$$g_n = -2 \times 1.9131 \text{ and } g_p = 2 \times 2.7927 \qquad \qquad ... (5.3)$$

One can calculate the magnetic moment of an odd nucleus with the help of *extreme single particle shell model*. We can write the magnetic moment of a nucleus of spin I, i.e. total angular momentum as

$$\mu_I = g_I \sqrt{I(I+1)} \, \mu_N \qquad \qquad ... (5.4)$$

A magnetic field has to be applied while measuring the magnetic moments and it is the component of μ_I in the field direction Z which is determined. Applying space quantization rule, (5.4) takes the form

$$\mu_z = \mu_I \cos(I \cdot B) = \frac{\mu_I \, m_I}{\sqrt{I(I+1)}} \qquad \qquad ... (5.5)$$

Where μ_I is the magnetic quantum number and can take up the values $\mu_I = I, I-1, ..., -I$, and B is the magnetic induction field. The component corresponding to $\mu_I = I$ is the largest and gives the measurmed magnetic moment,

$$\mu_z = \frac{\mu_I \, I}{\sqrt{I(I+1)}} = g_I \, I \, \mu_N \qquad ... (5.6)$$

From the above, it is clear that in the extreme single particle model, any one kind always gives the resultant spin ($I = 0$). Obviously, the magnetic moment of an even-even nucleus will be zero, i.e.

$$(\mu_I)_{ee} = 0$$

This means that in an odd A nucleus, it is the last odd nucleon (proton or neutron) determines the magnetic moment. Obviously, for such a nucleus, $I = J$, where J is the total angular momentum of last unpaired nucleon. In order to get the total magnetic moment, we have to add both the intrinsic magnetic moment (μ_s) and the magnetic moment due to its orbital motion (μ_l), i.e.

$$\mu_J = \mu_l + \mu_s$$

Where $\mu_s = \mu_p$ for the proton and $\mu_s = \mu_n$ for the neutron.
The orbital motion of a nucleon having azimuthal angular momentum produces a magnetic moment,

$$\mu_l = g_e \, \mu_N \, \sqrt{l(l+1)} \qquad ... (5.7)$$

Now, since the neutron is an uncharged particle, its orbital motion does not produce any magnetic moment, i.e. $g_l = 0$, Thus

$$(\mu_l)_n = 0$$

One can write $g_l = 1$ in the case of proton. Thus the orbital contribution is

$$(\mu_{lp}) = \mu_N \, \sqrt{l(l+1)} \qquad ... (5.8)$$

Nucleons are spin $\dfrac{1}{2}$ particles and hence one can write the quantum mechanical values of intrinsic magnetic moment as

$$\mu_s = g_s \, \mu_N \, \sqrt{s(s+1)} \qquad ... (5.9)$$

Where $s = \dfrac{1}{2}$ is the spin quantum number, $g_s = g_p$ for the proton and $g_s = g_n$ for the neutron.
In the direction of J, the total magnetic moment is

$$m_I = m_J = m_l \cos(l, J) + \mu_s \cos(s, j)$$

$$= \mu_N \left[g_l \, \sqrt{l(l+1)} \cos(l, J) + g_s \, \sqrt{s(s+1)} \cos(s, j) \right] \qquad ... (5.10)$$

Using (5.4), one can write

$$\mu_j = g_j \, \sqrt{j(j+1)} \, \mu_N \qquad ... (5.11)$$

Form the cosine law, we have

$$\cos(l, j) = \frac{j(j+1) + l(l+1) - s(s+1)}{2\sqrt{j(j+1)}\ \sqrt{l(l+1)}}$$

and

$$\cos(s, j) = \frac{j(j+1) + s(s+1) - l(l+1)}{2\sqrt{j(j+1)}\ \sqrt{s(s+1)}}$$

Thus, we have

$$\mu_j = g_l\,\mu_N \frac{j(j+1) + l(l+1) - s(s+1)}{2\sqrt{j(j+1)}} + g_s\,\mu_N \frac{j(j+1) + s(s+1) - l(l+1)}{2\sqrt{j(j+1)}} \quad \dots (5.12)$$

For a spin $\frac{1}{2}$ particle, j can have two values, $j = l \pm \frac{1}{2}$. Thus for a given j, l, one can have the following two values:

For $\quad j = l + \frac{1}{2}, -l = j - \frac{1}{2}$; for $j = l - \frac{1}{2}, l = j + \frac{1}{2}$

For the above two cases, one obtains two different values of μ_j from (5.12). Using (5.11), one obtains

$$g_j = g_l \frac{j(j+1) + l(l+1) - s(s+1)}{2j(j+1)} + g_s \frac{j(j+1) + s(s+1) - l(l+1)}{2j(j+1)} \quad \dots (5.13)$$

As stated earlier, the measured magnetic moment μ_z is the largest possible component of μ_j in the magnetic field direction and its given by (5.5). Now, replacing I by j, one obtains

$$\mu_z = g_j\, j\, \mu_N$$

$$= \left\{ g_l \frac{[j(j+1) + l(l+1) - s(s+1)]}{2[j(j+1)]} + g_s \frac{[j(j+1) + s(s+1) - l(l+1)]}{2j(j+1)} \right\} \mu_N \quad \dots (5.14)$$

For $l = j - \frac{1}{2}$, we have

$$\mu_z = \left[g_l \frac{j - \frac{1}{2}}{j} + \frac{g_s}{2j} \right] j\mu_N \quad \dots (5.15)$$

and $\quad l = j + \frac{1}{2}$, we have

$$\mu_z = \left[g_l \frac{j + \frac{3}{2}}{j+1} - \frac{g_s}{2(j+1)} \right] j\mu_N \quad \dots (5.16)$$

For odd A, either the proton number is odd (in the $o-e$ nucleus) or the neutron number is odd (in the $e-o$ nucleus). Thus, we have the following possibilities:

Odd proton ($g_l = 1$, $g_s = g_p$):

$$l = j - \frac{1}{2} \quad : \quad \mu_z = \left(j - \frac{1}{2} + \frac{g_p}{2} \right)\mu_N \qquad \ldots (5.17a)$$

$$l = j + \frac{1}{2} \quad : \quad \mu_z = \frac{j}{j+1}\left(j + \frac{3}{2} - \frac{g_p}{2} \right)\mu_N \qquad \ldots (5.17b)$$

Odd neutron ($g_l = 0$, $g_l = g_n$):

$$l = j - \frac{1}{2} \quad : \quad \mu_z = \frac{g_n\,\mu_N}{2} \qquad \ldots (5.18a)$$

$$l = j + \frac{1}{2} \quad : \quad \mu_z = -\frac{j}{j+1}\frac{g_n\,\mu_N}{2} \qquad \ldots (5.18b)$$

The numerical values of g_p and g_n are given by Equation (5.3). Magnetic moments of odd A nuclei as functions of nuclear spin I which is taken to be equal to the j value of the last odd nucleon are given by Equation (5.17) and (5.18). We may note that the above values of the nuclear magnetic moments are known as *Schmidt limits*. These limits are shown joined by lines in Figure 5.15 together with the measured magnetic moments of odd-A nuclei. In this figure Schmidt values are plotted as functions of $I = j$ for the four cases discussed above. these plots are known as *Schmidt diagrams*.

From Figure 5.15 it is evident that there are a few nuclei in good agreement, e.g. $^{17}_{8}$O and $^{39}_{19}$K, most nuclei fall between these limits. Obviously, the shell model is not so successful in predicting the magnetic dipole moments of nuclei. The data show only a barely recognizable tendency to follow qualitatively the predictions of the shell model.

The failure of the shell model is due to its assumption that the nuclear magnetic dipole moment is due entirely to the single odd nucleon. It is not true that all the other nucleons are *always* paired off with total angular momenta and magnetic dipole moments that strictly cancel. The assumption is good enough to lead to the prediction of correct magitude for the total angular momentum, since this quantity is quantized. If occasionally the pairs have a non zero total angular momentum, then at that time the odd nucleon *must* have exactly the right total angular momentum to compensate and keep the magnitude of the total angular momentum constant. This kind of compensation cannot also take place for the magnetic dipole moments since the g factors, which relate the magnitudes of the magnetic dipole moments to the magnitude of the angular momenta, change as the angular momentum couplings change. And since the nuclear magnetic dipole moment does not have not have a *quantized magnitude*, there is nothing to enforce such a compensation.

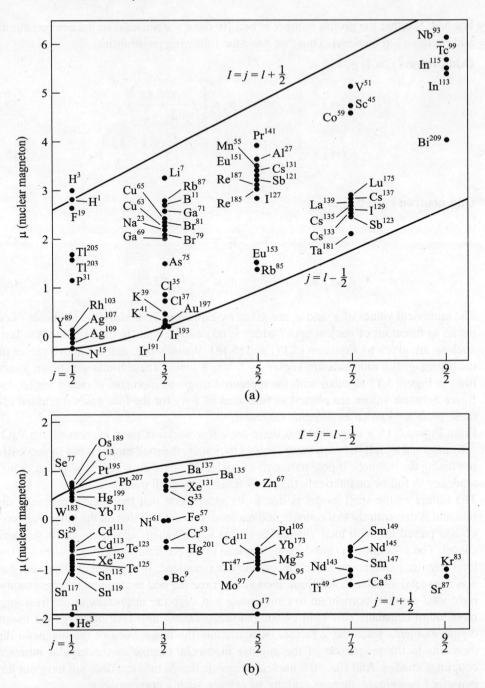

Fig. 5.15 (a) Measured magnetic dipole moments of even N, odd Z nuclei and the shell model predictions. The upper line is the prediction if the spin and orbital angular momenta of the odd proton are assumed to be essentially parallel, and the lower line is the prediction if they are assumed to be essentially anti parallel, **(b)** The same for odd N, even Z. Here the lower line is for the "parallel" assumption and the upper line is for the "anti parallel" assumption.

(iii) **Islands of Isomerism** We have mentioned about the long lived nuclear excited states, known as isomeric states in some nuclei. These states occur when there is a large difference in angular momenta between the excited and the ground states, specially when the energy difference between the two states is relatively small.

Usually, isomeric states are found amongst nuclei for which either N or Z is near to the end of the shell. These regions are called as *islands of isomerism* (Figure 5.16). The grouping of the nuclei with isomeric states is found most prominent just below the major closed shells at Z or N equal to 50 and 82. However, the grouping is less marked near Z or $N = 20$ and near $N = 126$.

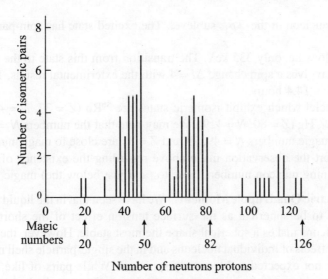

Fig. 5.16 Islands of isomerism

One can explain the existence of isomerism in terms of single particle shell model. The various sublevels in one shell within the nucleus do not differ much in energy. When a nucleus is excited from the ground state, it usually involves the transition of an odd nucleon in an unfilled sublevel to an upper sublevel. There will be a large change in the nuclear spin I when the nucleus goes down to the lower state by the emission of electromagnetic radiation, provided there is a large difference in the j values between these two states of the nucleus. This is the basic condition for isomerism. From Figure 5.14 it is evident that such large changes in the j-values occur near the magnetic numbers 50 and 82. In the first case for magic number 50 the last sublevel is $1g_{\frac{9}{2}}$ ($l = 4$, $j = \frac{9}{2}$),

while those just below it are $2p_{\frac{1}{2}}$, $1f_{\frac{7}{2}}$ and $2p_{\frac{3}{2}}$. In case of magic number 82, the last three sublevels are $1h_{\frac{11}{2}}$, $2d_{\frac{3}{2}}$ and $3s_{\frac{1}{2}}$.

We now consider nucleus ^{115}In ($Z = 49$, $N = 66$). Its ground state spin and parity are $\frac{9^+}{2}$ exhibiting that the last odd upaired proton occupies the $1g_{\frac{9}{2}}$ sublevel while the rest of the protons fill up all the lower sublevels upto $2p_{\frac{1}{2}}$ in pairs. When excited, one proton from $2p_{\frac{1}{2}}$ is raised to the $1g_{\frac{9}{2}}$ sublevel and form a pair with the odd proton in the latter leaving

an unpaired nucleon in the $2p_{\frac{1}{2}}$ sublevel. The excited state has spin-parity $\frac{1^-}{2}$. Its energy is relatively low, i.e. only 335 keV. The transition from this state to the ground state is the M4 since it involves a spin change $\Delta I = 4$ with the exterimental results. The life time of the excited state is 14.4 hours.

The other nuclei which exhibit isomeric states are ^{86}Rb ($Z = 37$, $N = 49$), ^{131}Te ($Z = 52$, $N = 79$) and ^{199}Hg ($Z = 80$, $N = 119$). We may note that the numbers $N = 49$, 79 and $Z = 80$ are close to magic numbers $N = 49$, 79 and $Z = 80$ are close to magic numbers 50, 82. This further support the observation made above regarding the existence of isomeric states in nuclei containing nucleon numbers close to (and just below the) magic numbers.

(iv) **Nuclear Electric Quadrupole Moment**   We have seen that in the liquid drop model, nuclei are assumed to be spherical as the surface tension effect of the short-range interaction between nucleons makes a spherical shape the most stable. However, the liquid drop model neglects the effects of individual nucleons and in the single-particle shell model odd-A nuclei are generally not expected to be exactly spherical. While pairs of like nucleons coupling upto $J = 0$ give rise to a spherical mass distribution and hence spherical potential for the odd nucleon to move in, the distribution of the odd nucleon is not spherically symmetric unless its spin is a half. The resultant deviation from a sphere, though, is expected to be small as only one nucleon is involved.

A measure of the deviation of a charge distribution from a spherical shape is the electric quadrupole moment (Q) of the distribution. The Q of a nucleus is the average of quantity $(3Z^2 - r^2)$ for the charge distribution in the nucleus. For a spherically symmetric charge distribution this average is zero and obviously $Q = 0$ for even-even nuclei which have ground state spin $I = 0$. In the case of a single odd proton nucleus in the state j this averaging gives the quadrupole moment as

$$Q_{sp} = -<r^2> \frac{(2j-1)}{(2j+1)} \qquad \ldots (5.19)$$

Where $<r^2>$ is the average square radius of the charge distribution. In the present case this is equal to the mean square distance of the proton from the nuclear centre. Figure 5.17 shows the measured quadrupole moments Q for odd proton nuclei versus the proton number Z and for odd neutron nuclei *vs.* the neutron number N. The values are generally much larger than the values as predicted by the shell model.

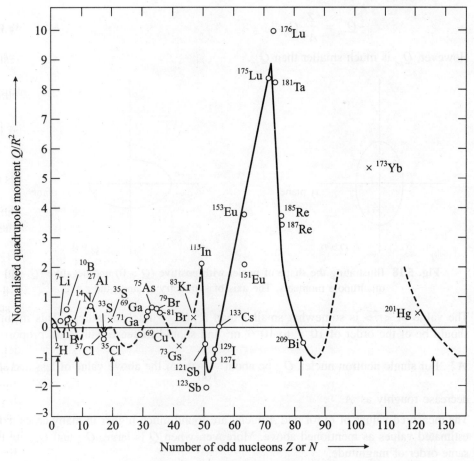

Fig. 5.17 Measured quadrupole moments Q (expressed in units of the square of the nuclear radius, $R^2 \equiv (1.5 \times 10^{-13} \, A^{\frac{1}{3}})^2$ cm^2) of odd-proton nuclei (excepting ^{6}Li and ^{36}Cl) plotted as circles against the number of protons Z, and odd-neutron nuclei, plotted as crosses against the number of neutrons N. Arrows indicate the closure of major shells. The solid curve represents the regions where the quadrupole data appear to be well established, the dashed part indicates more doubtful regions.

The negative sign on the r.h.s. of Equation (5.19) shows that orbital motion of the proton in the equitorial plane makes the charge distribution an oblate spheroid, whereas an odd hole in the state j would make the charge distribution a prolate spheroid for which $Q > 0$ (Figure 5.18).

We may normally not expect any quadrupole moment in the case of a single odd neutron nucleus. However, the orbital motion of the neutron gives rise to the recoil motion of the rest of the nucleus. One may take this to be a charge Z at a distance $\dfrac{r_n}{A}$, (where r_n is the radius of the odd neutron orbit) from the centre of the mass. This means, a small quadrupole moment Q_{sn} may be expected as

$$Q_{sn} = \frac{Z}{A^2} Q_{sp}$$
... (5.20)

However, Q_{sn} is much smaller than Q_{sp}.

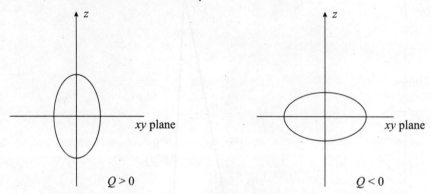

Fig. 5.18 Illustrating the shape of nuclei with positive ($Q > 0$) and negative ($Q < 0$) quadrupole moments. The axis of symmetry is along the z axis.

The value of $<r^2>$ is somewhat smaller than R^2 ($R \rightarrow$ nuclear radius). This means Q_{sp} should be of the order of 10^{-28} to 10^{-29} m^2 and should increase with A, in proportion to $A^{\frac{2}{3}}$. For single neutron nuclei, Q_{sn} be about $\frac{1}{100}$ th of the above value of less and should decrease roughly as $A^{-\frac{1}{3}}$.

The measured values of Q for odd A nuclei are found much higher in many cases than the estimated values as mentioned above. Moreover, when Q is large, Q_{sp} and Q_{sn} are of the same order of magnitude.

Obviously, single particle shell model cannot explain the very large values of Q in many nuclei. It seems that these nuclei acquire permanent deformation.

(v) **Configuration Mixing**   It is assumed that the single particle potentials are not distributed by the interaction between the nucleons. However, there is a possibility that a strong inter particle interaction can modify the shell model wave function and the modified wave functions can be produced by the superposition of two or more configurations. This is termed as the configuration mixing, e.g. the ground state of $^{18}_{7}$F is found to show an admixture of $1d_{\frac{5}{2}}$, $2s_{\frac{1}{2}}$ and $1d_{\frac{3}{2}}$ states.

5.6 SINGLE PARTICLE SHELL MODEL

In the previous section, we have already mentioned about this. Unlike in the extreme single particle model, one has to consider the couping of the angular momenta of all the odd nucleons in the last occupied sublevel outside the core of closed shells within the nucleus. One can assume that the residual interaction between these loose nucleons in the outermost sublevel does not cause any perturbation of the single particle levels determined by (n, l, j).

We know that nucleons are fermions. The total wave function of a pair of nucleons must be antisymmetric in the exchange of all the coordinates of a pair of nucleons. However, this condition restricts the states which can be formed for a number of nucleons in a state of given j, e.g. for two nucleons of the same time in a given (n, l, j) state (equivalent nucleons) the interchange of the particles multiplies the wave function by $(-1)^{J+1}$ and thus only even values of $I = J$ (where $J = J_1 + J_2$ for the two particles with $J_1 = J_2$) are possible. In the case of two non-equivalent nucleons for which one or more of the quantum numbers (n, l, j) are different, the wave function is multiplied by $(-1)^{j_1 + j_2 - J}$ when the two particles are interchanged. Obviously, the process of anti-symmetrization is more involved in this case. We may note that all possible values of J lying between $|j_1 - j_2|$ and $|j_1 + j_2|$ are allowed. There is no need of any distinction between equivalent and non-equivalent cases, i.e. for two *non-indentical* nucleons (i.e., a proton and a neutron), all values of J between $|j_1 - j_2|$ and $|j_1 + j_2|$ are permissible.

In the case of more than two nucleons ($k > 2$), the procedure for antisymmetrization involves the choice of suitable antisymmerizing operator which is allowed to operate on the product of an antisymmetric wave function of the last neutron. Following this procedure, one can show that for the three nucleons there can not be any state with $I = \dfrac{1}{2}$.

For k-identical nucleons in a state j (configuration j_k), the possible states are listed in Table 5.4

Table 5.4 Possible antisymmetric state for k identical nucleons in a state j (configuration j_k).

j	k	l
$\frac{1}{2}$	1	$\frac{1}{2}$
	2	0
$\frac{3}{2}$	1	$\frac{3}{2}$
	2	$0, 2$
$\frac{5}{2}$	1	$\frac{5}{2}$
	2	$0, 2, 4$
	3	$\frac{3}{2}, \frac{5}{2}, \frac{9}{2}$
$\frac{7}{2}$	1	$\frac{7}{2}$
	2	$0, 2, 4, 6$
	3	$\frac{3}{2}, \frac{5}{2}, \frac{7}{2}, \frac{9}{2}, \frac{11}{2}, \frac{15}{2}$
	4	$0, 22, 42, 5, 6, 8$
$\frac{9}{2}$	1	$\frac{9}{2}$
	2	$0, 2, 4, 6, 8$
	3	$\frac{3}{2}, \frac{5}{2}, \frac{7}{2}, \left(\frac{9}{2}\right)^2, \frac{11}{2}, \frac{13}{2}, \frac{15}{2}, \frac{17}{2}, \frac{21}{2}$
	4	$0^2, 2^2, 3, 4^3, 5, 6^3, 7, 8^2, 9, 10, 12$
	5	$\frac{1}{2}, \frac{3}{2}, \left(\frac{5}{2}\right)^2, \left(\frac{7}{2}\right)^2, \left(\frac{9}{2}\right)^3, \left(\frac{11}{2}\right)^2, \left(\frac{13}{2}\right)^2, \left(\frac{15}{2}\right)^2, \left(\frac{17}{2}\right)^2, \frac{19}{2}, \frac{21}{2}, \frac{25}{2}$

The results of the single particle model are as follows:

(i) For even Z and even N nuclei, the spin $I = 0$;

(ii) For odd A nuclei, one can determine I by the j value of the last odd nucleon

(iii) For odd Z, and odd N nuclei, the value of I can be determined by the Nordheim rules.

5.7 INDIVIDUAL (INDEPENDENT) PARTICLE MODEL

In the previous two sections we have discussed the various aspects of the shell model. We have seen that a common feature of all of them was the existence of an inert core of even-even nucleons. Moreover, in all these cases, the inert core belong to magic number particles and it was assumed that the core is not excited. Now, the question arises, what happens, if there is no inert core and particles are excited? This is the situation, where the individual particle model is invoked. In this model all the A nucleons are assumed to move independently of one another in a common potential field. One can use this model to obtain the wave function of the system which can be quite accurate in principle, though in practice it depends on the approximate validity of the shell model to some extent.

The wave function $\psi^v(r)$, for configuration is a Slator determinant of the single particle wave function ψ^{v_i} for all A particles, where v_i specifies the quantum state.

$$\psi_{IM}^n (r) = \left(\underline{n}\right)^{-\frac{1}{2}} \begin{vmatrix} \phi_{J_1 m_1}(1) & \phi_{J_2 m_2}(1) & \cdots & \phi_{J_n m_n}(1) \\ \phi_{J_1 m_2}(2) & \phi_{J_2 m_2}(2) & \cdots & \phi_{J_n m_n}(2) \\ \vdots & & & \\ \phi_{J_1 m_1}(n) & \phi_{J_2 m_2}(n) & \cdots & \phi_{J_n m_n}(n) \end{vmatrix}$$

$$\equiv |\phi_{J_1 m_1} \quad \phi_{J_2 m_2} \quad \cdots \quad \phi_{J_n m_n}| \qquad \qquad \ldots (5.1)$$

Which is antisymmetric (from the properties of the determinant), to the interchange of any two particles and vanishes when two particles occupy the same quantum state. Due to this condition, the only possible M state, for the configuration $I = (j)^{2I+1}$ (when particles completely fill the orbital), is

$$M = \sum_i m_i = 0 \qquad \qquad \ldots (5.12)$$

where $$m_i = j_{m_i}.$$

However, now there is no central common potential and ψ^v are the solutions of Schrodinger equation:

$$H_1 \psi^v = E_1 \psi^v \qquad \qquad \ldots (5.3)$$

where $$H_1 = \sum_{i=1}^{A} \left[-\frac{\hbar^2}{2M} \nabla_i^2 + V_1(r_i) \right] \qquad \qquad \ldots (5.4)$$

Where $V_1(r_i)$ is suitably chosen single particle potential in which each nucleon moves and one can obtain it by certain self-consistent procedures, e.g. Hartree Fock method or the wave function is chosen by an insight into the physical situation. By this process, one obtains a set of wave functions

ψ^{v_i}, which is complete and orthogonal. After this, one will have to set up energy matrix, i.e. the matrix formed by evaluating $(\psi^v|H|\psi^v)$, where

$$H = -\frac{\hbar^2}{2M}\sum_{i=1}^{A}\nabla_i^2 + \sum_{i<j=2}V_{ij}(r_{ij}) \qquad \ldots (5.5)$$

Here $V_{ij}(r_{ij})$ is the two-nucleon interaction between nucleons i and j. This matrix is then diagonalised. The diagonal elements are the eigen values E_k of the energy of the actual system. In this way, one obtains the eigen functions and eigen energies of the Schrodinger equation

$$H\psi = E\psi \qquad \ldots (5.6)$$

$$\psi = \sum_v a_v \psi^v \qquad \ldots (5.7)$$

We may note that for a realistic case, the number of elements ψ^v are very large or infinite, though in practice a few values of v are sufficient to determine ψ.

One chooses either $L - S$ coupling or so-called Russel Saunders coupling while writing the wave functions (generally for light nuclei), or $j - j$ coupling for intermediate nuclei.

The energy level configuration and binding energies predicted by this model are in fair agreement with the observed experimental values. Figure 5.19 shows the energy levels of $_3\text{Li}^7$ and $_4^7\text{Be}$ predicted by this model and actually observed.

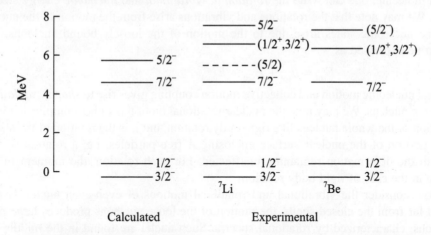

Fig. 5.19 Energy levels of nuclides with $A = 7$. (experimental and calculated).

5.8 COLLECTIVE MODEL

The shell model is based upon the idea that the constitutents parts of a nucleus move independently. The liquid drop model implies just the opposite, since in a drop of incompressible liquid the motion of any constitutent part is correlated with the motion of all the neighbouring parts. The conflict between these ideas emphasizes that a model provides a description of only a *limited set* of phenomena, without regard to the existence of contrary models used for the description of very *large set* of phenomena. At the border lines between its own set of phenomena and other set of phenomena, a theory fuses without conflict into the theories used for the description of other sets.

As nuclear physics evolves, attempts are made to remove conflicts between various models and unify them into more comprehensive models. The most successful and most important example is the *collective model of the nucleus*, which combines certain features of the shell and liquid drop models. It is partly the work of Aage Bohr (son of Niels Bohr) and Mottelson. The collective model assumes that the nucleons in the unfilled subshells of a nucleus move independently in a net nuclear potential produced by a *core* of filled subshells, as in the shell model. However, the net potential due to the core is not static spherically symmetrical potential $V(r)$ used in the shell model; instead it is a potential capable of undergoing deformations in shape. These deformations represent the correlated, or collective, motion of the nucleons in the core of the nucleus that are associated with the liquid drop model.

In nearly spherical nuclei, the coupling between the collective motion of the nucleons in the core and the motion of the loose nucleons outside the core is weak. On the otherhand, for strong coupling the surface is distored and the potential experienced by the loose particles is not spherically symmetric. These particles, moving in a non-spherically symmetric shell model potential, maintains the deformed nuclear shape. Large quadrupole moments are found in some cases, mainly among the rare earth elements (Figure 5.20) with major axis 30 % larger than the minor. The situation resemble with the linear molecules corresponding to the *rotation*, *vibration* and *electronic* energy states of a molecule, one can write the *rotational*, *vibrational* and *nucleonic* energy states within the nuclei. We may note that the rotations and vibrations arise from the motion of the nuclear core, whereas the nucleonic states arise due to the motion of the loosely bound nucleons. The total energy can be expressed as

$$E_{tot} = E_{rot} + E_{vib} + E_{nucl}. \qquad \qquad ... (5.1)$$

The external nucleonic motion and collective motion coupling gives rise to *shape-oscillations* at the surface of the nucleus. We may note the nuclear rotational motion is rather complicated in that it is not a rotation of the whole nucleus like rigid body rotation, but it is the rotation of the shape of the deformed portion of the nuclear surface enclosing A free particles, i.e. a rotation of the shape occurs with the deformation remaining unaffected. For such rotation, the moment of inertia is lower than in the case of rigid body rotation.

We now consider the vibrational and rotational motions of even-even nuclei. Experiments reveal that far from the closed shells, the motion of the loose nucleons produces large permanent deformations, characterized by rotational spectra. Such nuclei are found in the middle of $1d$, $2s$ shells in the range of $145 < A < 185$ and for $A > 226$. The reported value of energy difference between the O^+ ground state and 2^+ first excited state is of the order of 100 keV between them. Far from the deformed regions and nearer the closed shells, the equilibrium shape is spherical. Characterstic vibrational spectra are produced by low energy excitations. At the closed shells, excited states can be produced by the break up of the core. This gives rise to new particle states.

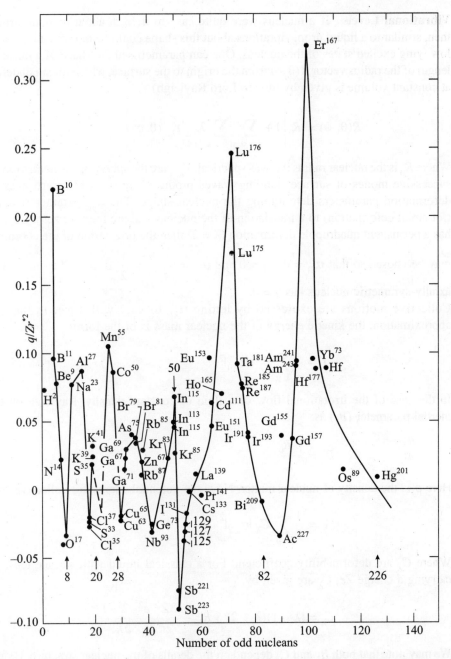

Fig. 5.20 The nuclear electric quadrupole moment, q, about the symmetry axis, divided by Zr'^2, for odd-A nuclei. The distance r' is the average from the center of the ellipsoidal distribution, of charge $+Ze$, to the surface. The quantity $1 + \dfrac{q}{Zr'^2}$ is approximately equal to the ratio of the distances from the center to the surface measured parallel to, and perpendicular to, the symmetry axis.

(i) **Vibrational Levels**   If a nucleus were quite easy to deform about its equilibrium shape then, similar to a liquid drop, vibrations about this shape could be important in describing in low lying excited states of the nucleus. One can parametrised the shape of a nucleus by the length of the radius vector $R(\theta, \phi)$ from the origin to the surface, which for small deformation at constant volume is given by (due to Lord Rayleigh)

$$R(\theta, \phi) = R_0 \left[1 + \sum_{\lambda=0}^{\infty} \sum_{\mu=-\lambda}^{\lambda} \lambda_{\lambda\mu} Y_{\lambda\mu} (\theta, \phi) \right] \qquad \text{... (5.2)}$$

Where R_0 is the nuclear radius if it was spherical. $Y_{\lambda, \mu}$ are the spherical harmonics representing successive modes of surface standing waves produced by surface disturbances, $\alpha_{\lambda\mu}$ are deformation parameters determining the nuclear shape. The $\alpha_{1\mu}$ parameters correspond (for small deformation) to a translation of the nucleus and are therfore zero. If the nucleus has a permanent quadrupole deformation ($\lambda = 2$) then the orientation of the coordinate axes may be chosen so that $\alpha_{21} = \alpha_{2-1} = 0$ and $\alpha_{22} = \alpha_{2-2} = \dfrac{1}{2} \beta \sin \gamma$. We may note that an axially-symmetric nucleus has $\gamma = 0$.

Collective motions are expressed by letting $\alpha_{\lambda\mu}$ to vary with time. In the quadratic approximation, the kinetic energy of the nuclear mass is of the form

$$T = \frac{1}{2} \sum_{\lambda, \mu} B_\lambda \, |\alpha_{\lambda\mu}|^2 \qquad \text{... (5.3)}$$

In the case of the irrotational flow of a fluid of constant density, Rayleigh obtained for inertial parameter (B_λ) as,

$$B_\lambda = \frac{\rho \, R_0^5}{\lambda} \qquad \text{... (5.4)}$$

Here ρ is the density of nuclear matter. Now, the potential energy for collective motion is

$$V = \frac{1}{2} \sum_{\lambda, \mu} C_\lambda \, |\alpha_{\lambda\mu}|^2 \qquad \text{... (5.5)}$$

Where C_λ are deformability coefficient. For a classical liquid with surface tension S and carrying a charge Ze, C_λ are given by,

$$C_\lambda = (\lambda - 1)(\lambda + 2) \, S \, R_0^2 - \frac{3Z^2 \, e^2}{2\pi \, R_0} \frac{\lambda - 1}{2\lambda + 1} \qquad \text{... (5.6)}$$

We may note that both B_λ and C_λ depend on the details of the nuclear structure. Hamiltonian H for the system is

$$H = T + V = E_0 + \sum_{\lambda, \mu} \left[B_\lambda |\alpha_{\lambda\mu}|^2 + \frac{1}{2} \sum_{\lambda, \mu} C_\lambda |\alpha_{\lambda\mu}|^2 \right] \qquad \text{... (5.7)}$$

Hence the classical circular frequency of oscillation associated with the variable $\alpha_{\lambda\mu}$ is given by

$$\omega_\lambda = \left(\frac{C_\lambda}{B_\lambda}\right)^{\frac{1}{2}} \qquad \ldots (5.8)$$

Equation (5.4) gives $B_\lambda \to \infty$ for $\lambda = 0$ and hence $\omega_0 = 0$. Further, Equations (5.5) and (5.6) gives $C_\lambda = 0$ for $\lambda = 1$ and therefore $\omega_1 = 0$. This means $\omega_\lambda = 0$ for both $\lambda = 0$ and 1 which represent respectively the oscillations of the charge density and translational vibrational of the entire drop. We may note that $\lambda = 0$ motion implies change of density and not an oscillator. These considerations are classical.

For the harmonic oscillator, the quantized result for the energy eigenvalues corresponding to Hamiltonian H (Equation 5.8) are given by

$$E = E_0 + \sum_{\lambda,\,\mu}\left(n\,\lambda_\mu + \frac{1}{2}\right)\hbar\,\omega_\lambda \qquad \ldots (5.9)$$

or

$$E_\lambda = E_0 + \sum_\mu\left(n_\mu + \frac{1}{2}\right)\hbar\,\omega_\lambda = E_0 + \left(n_\lambda + \frac{2\lambda+1}{2}\right)\hbar\,\omega_\lambda \qquad \ldots 5.9a)$$

Where the integer $n_\lambda = \sum_\mu n_\mu = 0, 1, 2, 3, \ldots$ gives the number of phonons of order λ in the excited state. Moreover, the same value of n_λ is associated with each of the coordinates $\alpha_{\lambda\mu}$ for the same λ but different μ. This reveals that the states are degenerate. This reveals that the states are degenerate. The with n_λ has a degeneracy of $(2\lambda + 1)$ and angular momentum $\lambda\hbar$. For irrotational motion, the phonon of the type λ_μ carries an angular momentum quantum number λ, with Z-Component m and parity $(-1)^\lambda$. We know that phonons like photons obey BE statistics.

The energy $\hbar\omega_\lambda$ increases rapidly with λ. Now, neglecting the charge effect and making use of Equation (5.4) and (5.6), one obtains

$$\frac{\omega_3}{\omega_2} = \sqrt{\frac{30}{8}} \approx 2 \text{ and } \frac{\omega_4}{\omega_2} = \sqrt{\frac{72}{8}} = 3$$

Here ω_2, ω_3, ω_4 ... etc. correspond to the frequencies of quadrupole, octupole, 2^4 pole phonons respectively. Thus $\omega_3 = 2\omega_2$ and $\omega_4 = 3\omega_2$. Clearly, for examining the low lying nuclear states, we have to consider only concerned low values of λ.

Let us consider that there are collective vibrations about a spherical shape ($\lambda = 0$), the first excited state will be a one quadrupole phonon state with $\lambda = 2$ and thus it would be a 2^+ state. Again since a single octupole phonon with $\lambda = 3$ has twice the energy of the quadrupole phonon with $\lambda = 2$. The $\lambda = 3$ single phonon state (3^- state) has about the same energy as two $\lambda = 2$ phonons. Obviously, this give rise to the three states 0^+, 2^+, and 4^+ by the coupling of the two angular momenta of 2 units. The degeneracies of these states are removed by perturbations involving an harmonic terms. We have not considered the perturbations.

A good example of a nucleus with an excited state spectrum looking like a quadrupole vibrational one is $^{106}_{46}$Pd even-even nucleus, whose spectrum is shown in Figure 5.21. Figure shows the 2$^+$ single photon vibrational states at an energy about 0.5 MeV and the triplet of levels 0$^+$, 2$^+$, and 4$^+$, almost degenerate and approximately at twice the excitation of the first excited state which is a 2$^+$. Side by side, the expected vibrational levels are also shown. Such a sequence of low lying levels is observed amongst many even-even nuclei. This demonstrates the validity of assumption concerning collective vibrational states of even-even nuclei. We may note that the photon energies of 0.5 MeV are much lower than the very much larger energies ($\sim$ 10 MeV) associated with the single-particle oscillator level spacing of the shell model. However, the excitation of the surface waves as discussed above do not involve any change of the single particle oscillator number.

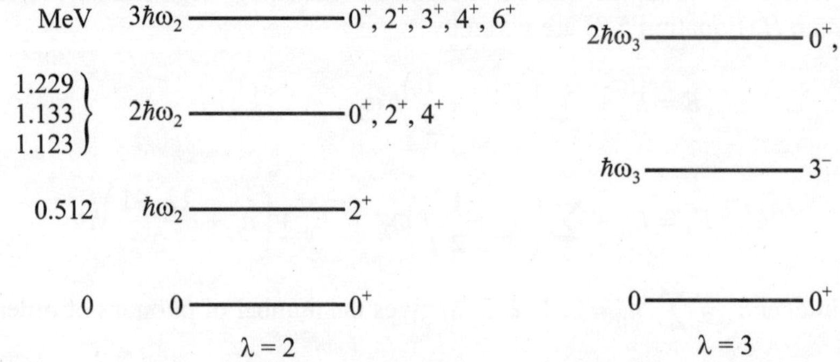

Fig. 5.21 Harmonic energy spectra for the quadrupole ($\lambda = 2$) and octupole ($\lambda = 3$) surface oscillations of $^{106}_{46}$Pd. Figure also shows the energies (in MeV) and the spin-parity values of the levels. The degeneracies of the two phonon triplet states (0$^+$, 2$^+$, 4$^+$) are removed by perturbation.

It is reported that there are also quadrupole excitations of much higher energy, corresponding to particle excitations across two major shells. They give rise to *quadrupole resonance* in the region of unbound states near giant dipole resonaces.

(ii) **Vibrations in a Permanently Deformed Nucleus** When a nucleus has a spherical equilibrium shape, for convenience one uses a different set of coordinates. In this case $\alpha_{\lambda\mu}$ describe changes in the shape w.r.t. a set of coordinates (x, y, z) fixed in space. If (x', y', z') represent the coordinates fixed in the nucleus, then one can carried out the transformation between (x, y, z) and (x', y', z') with the help of *Eulerian angles* $(\theta_1, \theta_2, \theta_3)$ of the principal axes of the nucleus shown in Figure 5.22(a).

Let us confine ourself to quadrupole equilibrium shape with $\lambda = 2$. In this case m can take the value $\mu = 2, 1, 0, -1, -2$. The equation of nuclear surface is

$$R = R_0\left[1 + \sum_{\mu=-2}^{2} a_{2\mu} Y_2^{\mu}(\theta', \phi')\right] \qquad \ldots (5.10)$$

and

$$\sum_{\mu} a_{2\mu} Y_2^{\mu}(\theta, \phi) = \sum_{\nu} a_{2\nu} \qquad \ldots (5.11)$$

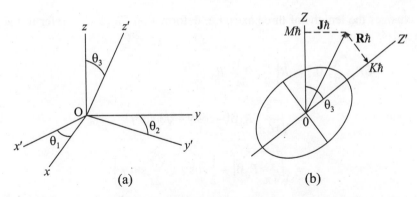

Fig. 5.22 **(a)** Transformation between the *body fixed* and space fixed axes of a deformed
nucleus with the help of Eulerian angles $(\theta_1, \theta_2, \theta_3)$ of the principal axes,
(b) The angular momentum coupling for a deformed nucleus.

One can carryout the transformation from the body fixed axes to space fixed axes using the
rotational matrices $D_{\mu\nu}$, which are unitary matrices and used in the study of angular
momentum[1].

Since the body axes are principal axes, the products of inertia are zero. This means that

$$a_{20} = \beta \cos \gamma; \qquad a_{22} = \left(\frac{\beta}{\sqrt{2}}\right) \sin \gamma \qquad \qquad \text{... (5.12)}$$

so that $$\beta^2 = |a_{20}|^2 + 2|a_{22}|^2 \text{ and } \tan \gamma = \sqrt{2}\,\frac{a_{22}}{a_{20}}$$

Where β and γ are new parameters. These parameters along with the Eulerian angles
$(\theta_1, \theta_2, \theta_3)$ which replace the five variables $\alpha_{2\mu}$ discribe the system completely.

Now, we obtain the deformations δR along the principal axes $j = 1, 2, 3$. Substituting the

values of spherical harmonics $y_2^0(\theta', \phi') = \sqrt{\dfrac{5}{16\pi}}\,(3 \cos^2 \theta' - 1)$ and $y_2^{\pm 2}(\theta', \phi') = \sqrt{\dfrac{15}{32\pi}}$

$\sin^2 \theta' \times \exp(\pm 2i\phi')$, one obtains from Equation (5.10)

$$R - R_0 = \sqrt{\frac{5}{16\pi}}\,R_0\beta\,[\cos \gamma(3 \cos^2 \theta' - 1) + \sqrt{3}\,\sin \gamma \sin^2 \theta' \cos 2\phi'] \text{ ... (5.13)}$$

Where β gives the deformation. We have

$$\beta^2 = \sum_\mu |\alpha_{2\mu}|^2 = \sum_\mu |\alpha_{2\mu}|^2 \qquad \qquad \text{... (5.14)}$$

Now, the potential energy is given by

$$V = \frac{1}{2}\,C|\alpha_{2\mu}|^2 = \frac{1}{2}\,C\beta^2 \qquad \qquad \text{... (5.15)}$$

[1]M. A. Perston, Physics of the nucleus

The changes the lengths of three axes, i.e. deformation δR (1, 2, 3 refer to the body-fixed axes x', y', z') are

$$\delta R_1\left(\frac{\pi}{2}, 0\right) = R\left(\frac{\pi}{2}, 0\right) - R_0$$

$$= \sqrt{\frac{5}{16\pi}} \, R_0 \beta \left(-\cos\gamma + \sqrt{3}\sin\gamma\right)$$

$$= \sqrt{\frac{5}{4\pi}} \, R_0 \beta \left(-\frac{1}{2}\cos\gamma + \frac{\sqrt{3}}{2}\sin\gamma\right)$$

$$\therefore \qquad \delta R_1 = \left(\frac{5}{4\pi}\right)^{\frac{1}{2}} \beta R_0 \cos\left(\gamma - \frac{2\pi}{3}\right) \qquad \qquad \dots (5.16)$$

Similarly, one obtains

$$\delta R_2\left(\frac{\pi}{2}, \frac{\pi}{2}\right) = R\left(\frac{\pi}{2}, \frac{\pi}{2}\right) - R_0$$

$$= \left(\frac{5}{4\pi}\right)^{\frac{1}{2}} \beta R_0 \cos\left(\gamma - \frac{4\pi}{3}\right) \qquad \qquad \dots (5.17)$$

and $\qquad \delta R_3(0, \phi) = R(0, \phi') - R_0 = \left(\frac{5}{4\pi}\right)^{\frac{1}{2}} R_0 \beta \cos\gamma \qquad \dots (5.18)$

In general, $\qquad \delta R_j = \left(\frac{5}{4\pi}\right)^{\frac{1}{2}} \beta R_0 \cos\left(\gamma - \frac{2\pi}{3}\right) \qquad \qquad \dots (5.19)$

For $\gamma = 0$, one obtains,

$$\delta R_3 = \left(\frac{5}{4\pi}\right)^{\frac{1}{2}} \beta R_0 > 0$$

and $\qquad \delta R_1 = \delta R_2 = -\left(\frac{5}{4\pi}\right)^{\frac{1}{2}} < 0$

This reveals that the ellipsoid is elongated along Z' axis. So the nucleus would be *prolate* (cigar shaped) spheroid with the axis-3 as its axis of symmetry. For $\gamma = \frac{2\pi}{3}$ and $\frac{4\pi}{3}$, the nuclei would be prolate spheroids with axis-1 and axis-2 as its symmetry axis. However, for $\gamma = \pi$, $\frac{\pi}{3}$, $\frac{5\pi}{3}$, the nuclei are oblate spheroids. If $\gamma \neq \frac{n\pi}{3}$, the shape of nucleus is that of an ellipsoid whose three axes are unequal.

If β changes with time, we have β-vibrations. If γ remains constant in this case, the nucleus preserves its symmetry. However, the eccentricity of the ellipse changes. The nucleus loses its axial symmetry for γ-changing with time, i.e. γ-vibrations.

(iii) **Nuclear Rotation and Rotational States**   One can obtain the transformed expression for the kinetic energy to the new coordinates with the help of the unitary matrices $D_{\mu\nu}$ as

$$T = \frac{1}{2}\, B(\dot{\beta}^2 + \beta^2\, \dot{\gamma}^2) + \frac{1}{2}\sum_{K=1}^{3} I_k\, \omega_k^2 \qquad\qquad ... \ (5.1)$$

Here $\omega_k's$ are the angular velocities of the principal axes w.r.t. the space fixed axes. The last term in Equation (5.1) is of the form of rotational kinetic energy with the moment of inertia,

$$I_k = 4B\, \beta^2\, \sin^2\!\left(\gamma - \frac{2\pi k}{3}\right) \qquad\qquad ... \ (5.2)$$

Now, if there is an axis of symmetry of the nucleus which corresponds to $\gamma - 0$ or π, we have $I_3 = 0$. On the other hand $I_1 = I_2 = 3B\, \beta^2$.

Since β is small, very little nuclear matter is actually taking part in the effective rotation.

Rotational States

One finds that the observable rotational motion is possible if the nucleus is considered to be a *fluid drop* or to have any form with a definite surface. This rotational effect can be either *rigid* in which case particles actually move in circles around the axis of rotation, or *wave like* in which case particles execute oscillatory motions and consequently the geometrical shape of the drop changes only. If we consider the shell model as a reasonable good picture of independent motions of individual nucleons within a nucleus, then for us it is difficult to picture the nucleus as a rigid body rotating around an axis. These oscillations actually give rise to the rotation of the shape of the nucleus, but not of the nucleons, i.e. particles. These wave like oscillations around the nuclear surface leads to an irrotational flow of the loose nucleons (Figure 5.23).

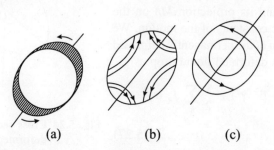

$$\text{(a)} \qquad\qquad \text{(b)} \qquad\qquad \text{(c)}$$

Fig. 5.23 Irrotational flow of the loose nucleons as a result of wave travelling around the nuclear surface: (a) rigid (b) wave like rotational motion of a deformed nucleus, i.e. irrotational flow (c) rigid rotations.

In this case, the moment of inertia is

$$I = \frac{3}{5}\, MA\, (\delta R)^2 \qquad\qquad ... \ (5.23)$$

Where
$$\delta R^2 = \delta R_1^2 + \delta R_2^2 + \delta R_2^3 = \frac{15}{8\pi} R_0^2 \beta^2 \qquad \text{...(5.23a)}$$

and M is the nucleon– mass. Thus, we have

$$I = \frac{9}{8\pi} \, MA \, R_0^2 \, \beta^2 \qquad \text{...(5.24)}$$

For the rigid sphere of radius R_0, the moment of inertia is

$$I_{\text{rig.sph.}} = \frac{2}{5} \, MA \, R_0^2 \qquad \text{...(5.25)}$$

From (5.24) and (5.25), we have

$$I = \left(\frac{45}{16\pi} \beta^2 \right) I_{\text{rig.sph.}} \qquad \text{...(5.26)}$$

The experimental observed values of I lie between I calculated above and $I_{\text{rig.}}$ being about 30 – 50 % of the latter. Obviously, this reveals that there is only partial breakdown of the shell model. At low excitation, the absence of any rotational states in spherical nuclei, clearly shows that there is no quasi-rigid rotation about the axis of symmetry. This means, I_3, if not zero, must be small and rotation about the symmetry axis, if any, can be said to be fairly at high excitation only.

We mentioned above that the actual moment of inertia (I) of the nucleus lies between I and $I_{\text{rig.sph.}}$ reveals that the nuclear matter is a mixture of a superfluid and viscous fluid, i.e. nuclear matter exhibits partial *superfluidity*.

The Rotational spectrum

Let us consider an ellipsoidally deformed nucleus (Figure 5.24) in which the body-fixed symmetry axis is shown by Oz'. If the Hamiltonian of the whole system is independent of spatial orientation, then the total angular momentum $\mathbf{J}$ and its projection $M\hbar$ on the spaced fixed axis Oz are constants of the motion. We can write the Hamiltonian of the permanently deformed nucleus, considered as a deformed liquid drop as

$$H = \frac{1}{2} B(\beta^2 + \beta^2 \, \dot{\gamma}^2) + \frac{1}{2} \sum_K I_K \, \omega_K^2$$

$$+ \frac{1}{2} C\beta^2 \qquad \text{...(5.27)}$$

$$= H_{\text{vib}} + H_{\text{rot}} \qquad \text{...(5.27a)}$$

Where $H_{\text{rot}} = \dfrac{1}{2} \sum_K I_K \, \omega_K^2 = \dfrac{1}{2} \sum_K \dfrac{L_K^2}{2 I_K} \qquad \text{...(5.28)}$

is the Hamiltonian for rotational motion. H_{vib} is the Hamiltonian for the vibrational motion considered

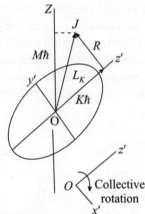

Fig. 5.24 Cross section of axially symmetric deformed nucleus showing body-fixed axis Oz' and space-fixed axis Oz. The total angular momentum vector $\mathbf{J}$ is indicated; $M\hbar$ is the resolved part of $\mathbf{J}$ along Oz and $K\hbar$ is the resolved part along Oz'. $\mathbf{R}$ represents collective rotation which adds vectorially to the intrinsic angular momentum $K\hbar$ to form $\mathbf{J}$.

earlier. L_K ($= K\hbar$) is the component of the angular momentum along the z'-axis (fixed to the body of the nucleus). If there is an axis of symmetry, the angular momentum about that axis is zero. We have read that rotations involves only the nucleons in the surface region of the nucleus and is intermediate between rigid rotation and wave like surface motion.

If only intrinsic and rotational motions are taking place, the total angular momentum **J** is a vector sum of the corresponding momenta and this addition is constrained in two ways by the circumstance of axial symmetry. First, there is no collective rotation about the symmetry axis so that (classically) the rotation **R** is perpendicular to this axis. Secondly, because the Hamiltonian is invariant to the orientation of the nucleus about the symmetry axis, the component of **J** along the symmetry axis is the third constant of motion. This component must be provided by the intrinsic motion of the valence nucleons about the symmetry axis and we call its resultant value $K\hbar$; it is the sum of the corresponding components: $K_i\hbar$ for all the individual valence nucleons. We may note that the individual particle angular momenta are *not* constants of motion in a non-spherical field. Concluding there are three constants of motion:

$$(Jh)^2 \; (= J(J + 1)h)^2, \; K\hbar \text{ and } M\hbar.$$

If each individual nucleon has angular momentum j with a projection Ω_j along the axis of symmetry, then, $\mathbf{J} = \sum j_i$ and $K = \sum \Omega_{jj}$. R is the angular momentum of collective rotation perpendicular to the symmetry axis.

There is symmetry about the 3-axis and hence $I_1 = I_2 = I$, where I_1 and I_2 are moments of inertia about 1– and 2– axes. Thus the Hamiltonian for the rotational motion becomes:

$$H_{\text{rot}} = \sum_K \frac{\hbar^2}{2I_K} J_k^2 \; = \; \frac{\hbar^2}{2I}(J_1^2 + J_2^2) + \frac{\hbar^2}{2I_3} J_3^2 \qquad \dots (5.29)$$

$$= \frac{\hbar^2}{2I}(J^2 + J_3^2) + \frac{\hbar^2}{2I_3} J_3^2 \qquad \dots (5.30)$$

As stated earlier, J^2 and J_3 are both consants of motion and their eigen values are as

$$J^2 = J(J + 1) \text{ and } J_3 = K \qquad \dots (5.31)$$

Thus, one obtains the energy of rotational motion as

$$E_r = \frac{\hbar^2}{2I}\{J (J + 1) - K^2\} + \frac{\hbar^2}{2I_3} K^2 \qquad \dots (5.32)$$

$$= \frac{\hbar^2}{2I} J (J + 1) + \hbar^2 \left(\frac{1}{2I_3} - \frac{1}{2I} \right) K^2 \qquad \dots (5.33)$$

We have read that for a nucleus with axis of symmetry, the angular momentum about the axis of symmetry is small, if not zero. In other words we can say that if there is no angular momentum due to the intrinsic motion of the valence nucleons ($K = 0$), the spin J is due solely to the collective rotation. Thus neglecting the terms containing K^2 in (5.33), we have

$$E_r = \frac{\hbar^2}{2I} J (J + 1) \qquad \dots (5.34)$$

If ω is angular velocity of rotation, then

$$I\omega = |\mathbf{J}| = \sqrt{J(J+1)}\ \hbar \qquad \ldots (5.35)$$

We may note that the rotational eigen functions are the functions $D_{\lambda\mu}$ used in the transformation of the spherical harmonics.

For *even-even deformed nuclei*, we have axis of symmetry and Equation (5.34) is applicable. For such types of nuclei, we have pairs of identical particles with $+\Omega_j$ and $-\Omega_j$ added for each pair which make $K = 0$ with even parity. Because of the symmetry about a plane perpendicular to Oz', only even-J values occur and the spon-parity values for the rotational bare

$$(K^P = 0^+) \qquad I^P = 0^+,\ 2^+,\ 4^+,\ 6^+,\ \ldots$$

We may note that the wave function must be invariant under reflection through $180°$. Such level sequence is actually observed for many even-even nuclei in the strongly deformed regions ($A = 24$, $A = 145$ to 185 and $A > 200$). For ^{180}Hf nucleus, this is illustrated in Figure 5.25. The levels of the band may be excited electromagnetically (Coulomb excitation) and decay mainly by the emission of electric quadrupole radiation. From (5.34) the ratios of the excitation energies in a rotational band are found to be

$$\frac{E_4}{E_2} = \frac{10}{3}, \ \frac{E_6}{E_2} = 7, \ \frac{E_8}{E_2} = 12.$$

independently of the moment of inertia, provided that this is independent of rotational frequency.

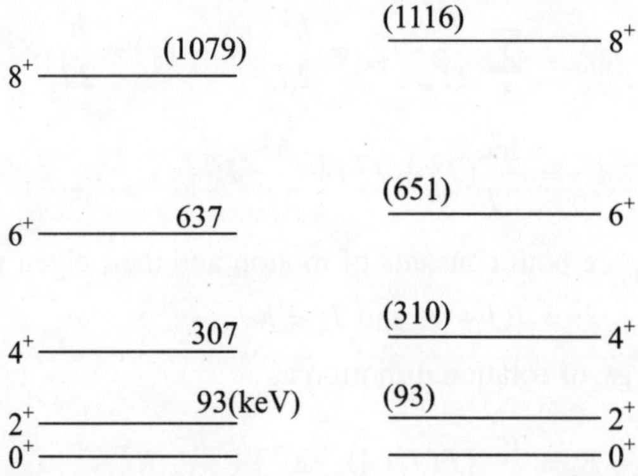

Fig. 5.25 Energy level (collective rotational levels) diagram of $^{180}_{72}$Hf: on the right, energy levels calculated using (5.34), the first excited level at 93 keV is matching with the experimental energy level.

One can calculate $\dfrac{\hbar^2}{2I}$ from the observed excitation energy for the 2^+ level. We have $E_2 = \dfrac{\hbar^2}{2I}$ $\times 2 \times 3 = 93$ keV, which gives $\dfrac{\hbar^2}{2I} = 15.5$ keV, where I is moment of inertia. Using this value of $\dfrac{\hbar^2}{2I}$, the energies of other levles are calculated and listed to the right of each level in Figure 5.25.

The agreement with the experimental values though quite satisfactory, but there is symmetric difference which increases with increasing excitation energy. This seems to be due to an increase in moment of inertia with increasing I because of the actions of the centrifugal force. This will add a term proportional to $-J^2(J+1)^2$. Equation (5.34), then becomes

$$E = \frac{\hbar^2}{2I} J(J+1) - BJ^2(J+1)^2 \qquad \ldots (5.36)$$

Now, it is found that agreement with experimental value is quite good. One can also explain the discrepancy between experimental and calculated values by other ways also.

We have seen that collective model is quite successful in interpreting the pattern of excited states of even-even nuclei. However, for *odd-even nuclei*, the situation is more complicated by the fact that the motion of the core and the motion of the odd-nucleon must be coupled. Obviously, the motion of the odd nucleon is no longer adequately described by the shell model states which correspond to a static spherical core.

It is reported that the excitation energy of the first rotational state of the deformed nuclei shows a smooth variation with A, whereas magic and near magic nuclei show no rotational spectra. Using these data and Equation (5.36), one can obtain effective moment of inertia of the various nuclei. With the help of these values, one can understand the kind of rotational motion that occurs. If R be the mean radius and ΔR the difference between the major and minor semi axes of the deformed nuclear potential well, then one can define deformation parameter with the help of relation:

$$R = R_0\left[1+\beta\left(\frac{5}{4\pi}\right)^{\frac{1}{2}}\left(\frac{3}{2}\cos^2\theta-\frac{1}{2}\right)\right] \text{ as}$$

$$\beta = \frac{4}{3}\left(\frac{\pi}{5}\right)^{\frac{1}{2}}\frac{\Delta R}{R} = 1.06\frac{\Delta R}{R_0} = 1.068 \qquad \ldots (5.37)$$

Where R_0 is the average nuclear radius, and $\delta = \frac{\Delta R}{R_0}$. The intrinsic quadrupole moment Q_0 is given by

$$Q_0 = \frac{4}{3}Z R^2 \delta \qquad \ldots (5.38)$$

$$= \frac{3}{\sqrt{5\pi}} Ze R_0^2 \beta[1+0.36\beta+\ldots] \qquad \ldots (5.39)$$

The spectroscopically observed quadrupole moment is

$$Q = Q_0 \frac{3K^2 - J(J+1)}{(J+1)(2J+3)}$$

Which becomes for the lowest state $K = J$ of the band, which may be the ground state,

$$Q = Q_0 \frac{J(2J-1)}{(J+1)(2J+1)} \qquad \ldots (5.40)$$

The difference between Q and Q_0 arises because of averaging of the direction of the nuclear axis by the rotational motion. For $J = 0$ and $\frac{1}{2}$ Q vanishes but Q_0 may still exist. The Q_0 values obtained experimentally in this way are about ten times the single particle value in the rare-earth region and suggest $\delta \approx 0.3$.

Coupling of particle and Collective Motions

Now, we consider the coupling of the motion of the odd particle with the collective motion for odd A, in the regions of deformed nuclei. The particles in this case are no longer in the same average potential and the motion of the nucleons will be altered. There may be both strong coupling and weak coupling. In case of spherical nuclei, the coupling is weak and one can treat it as a perturbation. When the particle motion and the collective parameters influence one another strongly, there will be strong coupling limit. Which method is appropriate is indicated by the observed levels. Let us assume that the single odd nucleon contributes $\Omega = K$ as the projection of the angular momentum on the axis of symmetry, the loose nucleon pairing off to $\sum \Omega_i = 0$. Obviously, K is half integer. Now, the energy of the nuclear state for the axially symmetric nucleus is as follows:

$$E_{JK} = \varepsilon_k + \frac{\hbar^2}{2I} \left[J(J+1) - K^2 + \delta_{K,\frac{1}{2}} \, a(-1)^{J+\frac{1}{2}} \left(J + \frac{1}{2} \right) \right] \quad \text{... (5.41)}$$

Where $\delta_{K,\frac{1}{2}} = 0$ for $K \neq \frac{1}{2}$ and $\delta_{K,\frac{1}{2}} = 1$ for $K = \frac{1}{2}$, and $\in_K$ is the contribution of the energy due to the particle Hamiltonian. a is the decoupling parameter.

$$a = \sum_j (-1)^{j+\frac{1}{2}} \left(j + \frac{1}{2} \right) |C_j|^2$$

Where $|C_j|^2$ is the probability that the odd particle has the angular momentum j. We may note that for $K \neq \frac{1}{2}$, the term involving a is absent.

The rotational bands are built on the particle states of different K which seems to be entirely due to the particle angular momentum and is constant throughout the band. All higher values of $J > K$ are allowed. The lowest state $K = J$ for $K = \frac{1}{2}$ and other successive states have $J = K + 1$, $K + 2$, etc. The parity is the same as for the odd particle configuration.

One can determine the level order for $K = \frac{1}{2}$ by the values of a, e.g. for $-3 < a < -2$, the level order is $\frac{3}{2}, \frac{1}{2}, \frac{7}{2}, \frac{5}{2}, \frac{11}{2}, \frac{9}{2}$ etc.

Electric Quadrupole moments for Strongly Deformed Nuclei

One can find the intrinsic quadrupole moment with the help of 5.27(a). To a first approximation 5.27(a) becomes

$$Q_0 = \frac{3}{\sqrt{5\pi}}\, Z R_0^2\, \beta = \frac{4}{5} Z R_0^2\, \frac{\delta R}{R_0} \qquad \text{... (5.42)}$$

Where δR is the difference between the semi-major and semi-minor axes of the ellipsoidal potential well. Q_0 is the quadrupole moment w.r.t. the axis of symmetry. The measured quadrupole moment is related to the symmetry of the averaged charged distribution. In the strong coupling limit, one obtains for the ground state ($K = 0$)

$$Q = Q_0\, \frac{I(2I-1)}{(I+1)(2I+3)} \qquad \text{... (5.43)}$$

$I = 0$ for a deformed even-even nucleus. This gives $Q = 0$, though Q_0 may have a large value. For $I \neq 0$, $Q < Q_0$ by a factor given by relation (5.43).

Magnetic Moments of Nuclei

One can calculate magnetic moments of nuclei with the help of collective model. The calculated values are found to be in good agreement with the experimentally observed values. Since the core of a nucleus shares the angular momentum and hence it also contributes to the magnetic moment. The magnetic moment for a single particle coupled to the surface is given by

$$\mu = <g_z\, s_z + g_l\, l_z + g_R\, R_z>_{M = J} \qquad \text{... (5.44)}$$

Where g_s, g_l, s_z, l_z refer to odd nucleon. g_R refers to the core rotation and is approximately given by $g_R = \dfrac{Z}{A}$, provided the neutron and proton densities are constant throughout the nucleus. For odd nucleon, writing values, one can show that for strongly deformed nuclei for $J = K \neq \dfrac{1}{2}$, the ground state ($K = 0$) magnetic moment is given by

$$\mu = \frac{J}{J+1}\, (J g_\Omega + g_R) \qquad \text{... (5.45)}$$

Here g_Ω refers to single particle motion. A comparison with the observed values lends support to the collective model.

We have seen that the collective model is capable of explaining all the features of measured electric quadrupole moments that are incorrectly predicted by the shell model. It leads to large enough values of Q because the core can be deformed so that the charges of many protons contribute to the total electric quadrupole moment. For nuclei between the magic numbers the core deformations become quite large, and therefore the electric quadrupole moments also become quite large. As the deformations can be due to extra nucleons of either species, the collective model explains why the observed values of Q do not depend significantly on whether the odd nucleons are neutrons or protons.

5.9 UNIFIED (NILSSON) MODEL (COUPLING OF PARTICLE AND COLLECTIVE MOTION)

In practice, we come across many such nuclei, which when excited undergoes both the collective modes of excitation as well as the particle excitation; in spherical or spheroidal nuclear potentials, depending upon the nucleus under consideration. S.G. Nilsson (1955) was the first to study the motion of a single particle in a field of force which is not spherically symmetric. We may say that this is a generalization of single particle shell model and called as unified model. Nilson considered the case of axially symmetric potential and T.D. Newton (1960) extended his work and gave the general shape.

Nilsson in his treatment introduced the axially symmetric distortion parameter δ such that

$$\delta_1 = \delta_2 = -\frac{1}{3}\delta \text{ and } \delta_3 = -\frac{2}{3}\delta \qquad \text{... (5.1)}$$

He considered the deformed potential as an anisotropic harmonic oscillator potential:

$$V(r) = \frac{M}{2}(\omega_x^2 x^2 + \omega_y^2 y^2 + \omega_z^2 z^2) \qquad \text{... (5.2)}$$

One can write for axial symmetry $\omega_x = \omega_y = \omega_\perp$. Using this (5.2) takes the form

$$V(r) = \frac{M}{2}(\omega_z^2 z^2 + \omega_\perp^2 \rho^2) \qquad \text{... (5.3)}$$

Where $\rho^2 = x^2 + y^2$. A spin-orbit coupling term $\xi_2(r) \mathbf{l} \cdot \mathbf{s}$ was added to (5.3) and further a term of the type Dl^2 was added to (5.3) to adjust the energy of the higher angular momentum terms below their oscillator values. Using the experimental data, the parameter is determined. One can treat the $\mathbf{l} \cdot \mathbf{s}$ and l^2 terms as small compared to the anisotropic oscillator potential for very strong deformation. Then calculation can be carried out with first order perturbation. However, for small and moderate deformations, the non-diagonal matrix elements of the spin-orbit coupling and l_t^2 have to be retained, where we have replaced l by l_t which is the orbital angular momentum operator constructed with the help of the dimensionless coordinates and momenta ξ_1 and $-i\left(\dfrac{\partial}{\partial \xi_i}\right)$ as in the case of the isotropic harmonic oscillator potential. Due to the symmetry about the z-axis, one can treat x and y equations together and z-equation is treated separately.

One can define the deformation parameter d by the relations:

$$\omega_z = \omega_0(\delta)\left(1 - \frac{4}{3}\delta\right)^{\frac{1}{2}} \qquad \text{... (5.4)}$$

and

$$\omega_x = \omega_y = \omega_\perp = \omega_0(\delta)\left(1 + \frac{2}{3}\delta\right)^{\frac{1}{2}} \qquad \text{... (5.5)}$$

One finds that the frequency ω_0 is related to the constant of the oscillator, which in turn related to the volume of the nuclei, i.e. $\omega_0 \propto A^{-\frac{1}{3}}$. when distortions are large, ω_0 also depends on δ and the dependence of ω_0 on δ is given by

$$\omega_0(\delta) = \tilde{\omega}_0\left(1 + \frac{2}{3}\delta\right)^{-\frac{1}{3}}\left(1 - \frac{4}{3}\delta\right)^{-\frac{1}{6}}$$

or
$$\omega_0(\delta) = \tilde{\omega}_0(0)\left[1 + \frac{4}{3}\delta^2 - \frac{16}{27}\delta^3\right]^{-\frac{1}{6}} \qquad \qquad ... (5.6)$$

Where $\tilde{\omega}_0$ is the deformation-independent constant, defined from the condition of constant volume equipotential surface as

$$\omega_x\,\omega_y\,\omega_y = \omega_0^3(\delta)\left(1 + \frac{2}{3}\delta\right)\left(1 - \frac{4}{3}\delta\right)^{\frac{1}{2}} = \text{Const.} \qquad \qquad ... (5.7)$$

δ is also related to the parameter β, i.e. $\delta = \left(\dfrac{45}{16\pi}\right)^{\frac{1}{2}}\beta$ to a first order of approximation.

One can write the Hamiltonian for the anisotropic oscillator as (due to Nilsson)

$$H = H_0 + H' = H_0 + C(\mathbf{l_t} \cdot \mathbf{s}) + D\,\mathbf{l_t} \cdot \mathbf{l_t} \qquad \qquad ... (5.8)$$

Where
$$H_0 = -\frac{\hbar^2}{2M}\nabla^2 + \frac{M}{2}(\omega_0^2 x^2 + \omega_y^2 y^2 + \omega_0^2 z^2) \qquad \qquad ... (5.8a)$$

The Hamiltonian H_0 is an isotropic oscillator Hamiltonian which can be solved by standard model. Eigen states of the Hamiltonian H_0 are specified by $N_1 = n_1 + n_2 + n_3$ which is the total number of quanta of vibration and pseudo angular momentum λ in a transformed system (dimensionless). H' is diagonal in n_1, n_2, n_3 and hence in N_1. Obviously, the eigenstates of H depends on N_1, l, λ and

$\sum$, where $\lambda^2 = l(l+1)$, $\lambda_3 + \wedge$ and the spin component $S_3 = \sum = \pm\dfrac{1}{2}$.

Nilsson and Newton obtained the solution of single particle wave equation numerically with δ given above. Their results depend upon C, D, ω_0 and δ. The results for $Z < 20$ and $N < 20$ are shown in Figure 5.26. The graphs show the energy of each state as a function of the distortion δ. In figure, the line representing each state is labelled by the value of Ω, the parity and three parameters $(N, n_3, \wedge)$, n_3 being constant for very large deformation (δ).

Using the wave function due to Nilsson, one can predict several important properties of odd A function of deformation, such as magnetic and electric quadrupole moments, reduced transition probabilities $B(L)$, etc. As mentioned earlier, the static quadrupole moment (also the quadrupole transition probability $B(E_2)$) has a part due to the single particle motion in a non-spherical potential and a part due to collective effect. We may note that for light nuclei, the two effects are comparable, where as for heavior nuclei, the collective effect ($\propto z$) will dominate.

One can write the Hamiltonian for a general unified case, i.e. including both the collective modes of excitation and particle excitation in spherical or spheroidal nuclear potentials depending on the nucleus under consideration. We have

$$H = H_c + H_p \qquad \qquad ... (5.9)$$

Where H_c represents collective motion, H_p represents the particle motion. Considering the general case for spherical or spheroidal potential, one can write H_p as

$$H_p = \sum_{i=1}^{k}[T + V(\beta,\ \gamma_i,\ r_i,\ l_i,\ s_i)] \qquad \qquad ... (5.10)$$

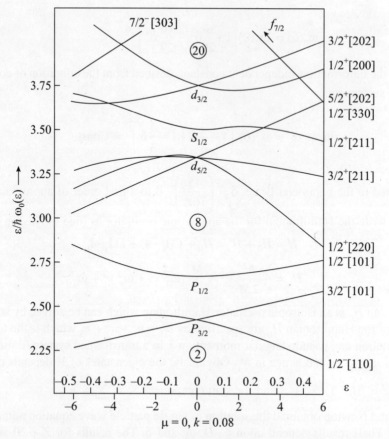

Fig. 5.26 Single particle levels as a function of nuclear deformation (δ) for $Z < 20$ and $N < 20$ due to Nilsson.

Where the average of body potential V is a function of the nuclear shape. Let us assume that the potential at a point in the nucleus is closely related to the nuclear density, and hence equipotentials are also the surface of constant nuclear density, we can write

$$r = r_0 f (\beta, r, \theta, \phi) \qquad \qquad ... (5.11)$$

Where r_0 represents a surface of constant nuclear density. Thus, we have

$$V = V_0 \left(\frac{r}{f (\beta, \gamma, \theta, \phi)}, \mathbf{l} \cdot \mathbf{s} \right) \qquad \qquad ... (5.12)$$

Where $f(0, 0, \theta, \phi) = 1$, and $V_0(\mathbf{r}, \mathbf{l}, \mathbf{s})$ are shell model potentials of a spherical nucleus. However, the exact expression for $V(\beta, \gamma, r_i, \mathbf{l}_i, \mathbf{s}_i)$ will depend on the shape of the nucleus.

One can couple the particle motion to the collective motion in two ways: (i) **weak coupling**, when β is small, and the changes in the potential shape are slow and small and the particle excitations are much larger. Then the particles move in an effective spherical potential, which in itself may undergo small vibrational changes, i.e. the coupling between particle and collective motion is weak. This corresponds to vibrational motion and applies to even-even spherical nuclei, with very small number of particles as loose nuclei, (ii) **strong coupling:** this corresponds to the situation, where a large number of 'loose' nucleons move around a closed shell 'core' and deform the core. In other words, we can say that their individual motion is strongly couples to the core, which, then, may undergo a resultant rotational motion, due to the collective effect of the 'loose' particles.

5.10 SUPERFLUID MODEL

We have seen that there appears a term δ in the semi-empirical binding energy formula in the liquid drop model. This term gives the differences in E_B between even-even, odd-odd and odd-A nuclei. None of the nuclear models discussed so far provide any explanation for the occurrence of this term on the basis of interactions between the nucleons in the different shells of the nucleus. Also, these nuclear models are unable to provide any explanation of the relatively large energy (~ 1 MeV) in the vicinity of the ground state of non-rotational levels of the even-even nuclei, the small value of the moment of inertia of the $e - e$ nuclei and the larger density (almost twice) of the single particle levels in the deformed odd-A nuclei as compared to the value obtained from Nilsson potential.

Superfluid model provides explanation of these and some other features. This model takes into account of the very short range pairing interaction different from the residual internucleon interaction considered in the models discussed so far. This pairing interaction between the identical nucleon with opposite spins in an even-even nucleus pushes down the energy considerably. In order to produce an excited state in such a nucleus, there is a necessity to break the bond. This helps to unpaired nucleon to cross the relatively large energy gap between the levels of the paired nucleon (ground state) and the first excited state.

The above situation is just similar to the pairing of electrons with equal and opposite momenta and spins (Cooper pairs) in superconductors. The pairs form structures of considerable spatial extent. The movement of these pairs in a nucleus take place in a correlated manner. In accordance with the exclusion principle, the wave functions generally change considerably, even though the force of pairing correlation may be weak.

According to this model, the individual nucleons within the nucleus retain their shell model states. To describe the pairing of nucleons, the quantum numbers l, j and $\pm m_j$ are used. We have seen that within two closed shells the nuclear shell model predicts a determinable arrangement for individual nucleon states, e.g. $2p_{\frac{3}{2}}$, $1f_{\frac{5}{2}}$, $2p_{\frac{1}{2}}$ and $1g_{\frac{9}{2}}$ states. According to this order the closed shell are predicted at $N = 28$ and that at $N = 50$. In case, if individual nucleon are non interacting (shell model), the first nucleon or two beyond a closed shell would be expected to describe by the lowest energy state configurations beyond the closed shell. However, quantum mechanics permits some admixture of levels, e.g. in nuclide $^{58}_{28}$Ni, the two neutrons would be expected to be $p_{\frac{3}{2}}$ fermions. They are in this state for most of the time, i.e. 66 %, but calculations reveal that they are expected to spend about 28 % of their time as $f_{\frac{5}{2}}$ neutrons, 3 % as $p_{\frac{1}{2}}$ neutrons and 3 % as $g_{\frac{9}{2}}$ neutrons.

A missing nucleon (hole) can create the same effect as an extra nucleon. Obviously, a nucleus with two nucleons is less than a closed shell and one can treat it in a manner analogous to a nucleus with two extra nucleons, e.g. the nuclide $^{200}_{82}$Pb has four neutron pairs (or 8 holes) less than the doubly closed shell nuclide $^{208}_{82}$Pb. The neutron levels immediately between the closed shell at $N = 126$ and that at $N = 82$ are in decending order, i.e. $3p_{\frac{1}{2}}$, $2f_{\frac{5}{2}}$, $3p_{\frac{3}{2}}$, $1i_{\frac{13}{2}}$, $1h_{\frac{9}{2}}$ and $2f_{\frac{7}{2}}$. One may expect the ground state of ^{200}Pb to appear as two $p_{\frac{1}{2}}$ holes and six $f_{\frac{5}{2}}$ holes. Thus the configuration of the state is $p_{\frac{5}{2}}^{-2} f_{\frac{5}{2}}^{-6}$. Negative numbers are indicating the number of holes. In order to break a pair from a stable nuclear system consisting solely of pairs, enough energy is required. After the pair is broken, each separate particle (hole) goes into a state that can be described by shell model configuration. Obviously, the particles of these unpaired state take on many characterstic of the non-interacting particles of the shell model. However, they are still coupled with the remaining paired particles by long range nuclear forces. By assuming that long range part of the nuclear force is a quadrupole force that can interact with particles in different orbits as well as in the same orbit, one can explain the vibrational and rotational states that are observed in energy gap.

Superfluid formalism of nuclear model gives almost the same levels as the other nuclear models. However, the different nuclear models based on assumptions which appear radically different and are often contradictory to each other. Each model explain fairly well certain limited aspects of nuclear properties. It is almost clear that they must represent different aspects of some basic principle governing the nature of nuclear matter. This basic principle is the *weak interaction* between nucleons within the nucleus. We may note that the analysis of the weak residual interaction leads to different forms of the collective models as also of the superfluid model.

A comparison of four widely used nuclear models and the ground state properties of nuclei is made in Table 5.4.

Table 5.4 Nuclear Models and the Ground State Properties of Nuclei

Name	Assumptions	Theory Used	Properties Predicted
Liquid drop model	Nuclei have similar mass densities, and binding energies nearly proportional to masses—like charged liquid drops	Classical (asymmetry and pairing terms introduced with no justification)	Accurate average masses and binding energies through semiempirical mass formula
Fermi gas model	Nucleons move independently in net nuclear potential	Quantum statistics of Fermi gas of nucleons	Depth of net nuclear potential Asymmetry term
Shell model	Nucleon move independently in net nuclear potential, with strong inverted spin-orbit coupling	Schroedinger equation solved for net nuclear potential	Magic numbers Nuclear spins Nuclear parities Pairing term
Collective model	Net nuclear potential undergoes deformations	Schroedinger equation solved for non-spherical net netnuclear potential	Magnetic dipole moments, Electric quadrupole moments

5.11 GIANT RESONANCES

This is a term generally used to describe broad resonances at energies tens of MeV above the ground state. The reason that these excitations are called 'giant' resonances comes from the fact that both their total strengths and their widths are much larger than typical resonances built upon single-particle (non-collective) excitations. In the energy region where giant resonances appear, the density of states is sufficiently high and the number of open decay channels sufficiently large that states in a given energy region cannot be very different from each other in character. As a result, one expect only smooth variations of reaction cross-sections when the nucleus is excited. The presence of large strength localized in the region of a few MeV is therefore of interest, since it must be related to some special feature of the nuclear system particular to the energy region.

For most giant resonances, the excitation strengths are found to be essentially independent of the probe used to excite the nucleus, γ-rays, electron, proton, α-particle and ^{16}O ion. Furthermore, both the width and the location of the peak of the strength distributions in different nuclei are found to very smoothly as a function of nucleon number A without any significant dependence on the detailed structure of the nucleus involved. For example, the location of isovector giant dipole resonance in different nuclei is well described by the relation

$$E_0 \approx 78 \ A^{\frac{1}{3}} \qquad\qquad \ldots (5.1)$$

Prominent dipole resonances, as well as other types of giant resonances, are found almost in all the nuclei studied, from ^{16}O to ^{208}Pb. Figure 5.27 shows the angular dependence of various giant resonances excited in the reaction ^{208}Pb(p, p').

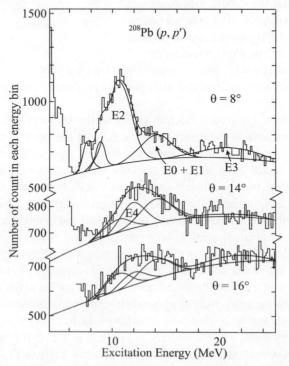

Fig. 5.27 Differential cross section of ^{208}Pb (p, p') reaction with 200 MeV protons at different scattering angles, showing the angular dependence of various giant resonances excited in the reaction.

The cause of giant resonances is the collective excitation of the nucleons. The energy gap between two adjacent major shells is roughly $41\ A^{-\frac{1}{3}}$ MeV and the parity of states produced by $1p\ 1h$ – excitations up one major shell is negative in general. *Negative parity* giant resonances are therefore predominantly caused by one-major shell excitations. For positive parity excitations there are predominantly caused by one-major shell-excitations. For positive parity excitations there are two possibilities, rearrangement of the particles in the same major shell ($0\ \hbar\omega$ – excitation) or elevating a particle up two major shells ($2\ \hbar\omega$ – excitation). Other possibilities, such as excitations by four major shells ($4\ \hbar\omega$ – excitation) for positive parity resonances and three major shells ($\hbar\omega$ – excitation) for negative parity resonances are less likely because of the larger amounts of energy involved.

(i) **Giant Dipole Resonances**   Although the low-lying excited states of the nucleus are not well described by surface oscillations there is a high lying excited stte in all nuclei which is well described as a vibrational mode. Usually these corresponds to $\lambda = 0$ and $\lambda = 1$ vibrations. This is the giant dipole resonance ($\lambda = 1$) state and arises from an oscillation of the protons with respect to the neutrons in the nucleus illustrated in Figure 5.28.

We may note that the dipole vibrations can occur only under the action of external forces. The vectors $\mathbf{R}_N$ and $\mathbf{R}_Z$ are the centre-of-mass of the neutrons and protons, respectively.

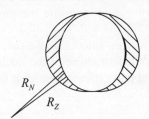

Fig. 5.28 Representation of the oscillation of protons with respect to neutrons in a nucleus. The vectors R_N and R_Z are the centre of mass of the neutrons and protons, respectively.

These giant dipole states are seen in reactions in which a photon is absorbed by the nucleus, e.g. inelastic electron scattering and from photons produced in the annihilation of positrons in flight.

The external forces due to the electric vector of the incident e.m. radiation, i.e. γ-rays, make the protons move in the direction of the field. If the neutron and proton fluids in the nucleus move in opposite phases, there is an effective separation between the centre of the charge and the centre of mass in the nucleus, which gives rise to a dipole moment to develop that is time varying. In self-conjugate ($N = Z$) even-even nuclei these states have collective states of spin-parity 1^- with an excitation energy of $\sim 20 - 25$ MeV. These states can also be described by an independent-particle model and in this model their wave functions are a superposition of many particle-hole configurations.

These giant dipole states are seen in many nuclear reactions, especially in photo-disintegration of the nuclei of both even and odd A. Examples are inelastic electron scattering and from photons produced in the annihilation of positrons in flight. In a plot of cross-section for the emission of a neutron against E_γ in photo-disintegrarion, a broad peak, called as the *giant electric dipole resonance* is observed. Similar peaks have been reported in many nuclei, ranging from very light, e.g. ^{12}C to very heavy, e.g. ^{208}Pb. It is found that the energies at which the peaks appear and their widths vary schematically with A. Due to large widths ($\sim$ several MeV) of such peaks, they are given the name giant resonances. These giant resonances are also called as *isovector resonances* since the neutrons and protons move in opposite phases and replaces each other, as it were. In this case the change in isospin is $\Delta T = 1$.

One can calculate the induced dipole moment due to the separation between the neutron and proton fluids using the *dipole sum* rule. For atoms, according to this rule, the total dipole oscillator strength summed over the discrete levels, as also those in continuum, is just equal to the total charge Z of the system. In order to take into account the relative motion between a nucleon and the recoiling nucleus, as also the effect of the nuclear force, in addition to the Coulomb force, one has to modify the dipole sum rule for atoms.

Let us consider the displacement of a proton of charge e through $\mathbf{r}_i$ relative to the centre of mass and the recoil of the charge $(Z-1)e$ and mass $(A-1)$. The induced dipole moment is

$$\mathbf{p}_i = e\mathbf{r}_i \; \frac{N}{A-1} \qquad \qquad \text{... (5.2)}$$

Where $N - A - Z$ is the number of neutrons. The energy at which the giant dipole resonance appears can be given by

$$E_0 = \left[\frac{e^2 \hbar^2}{M} \frac{NZ}{A} \frac{\beta}{\alpha_0} \right]^{\frac{1}{2}} \qquad \qquad \text{... (5.3)}$$

Where β is a correction factor due to the nuclear exchange forces ($\beta \approx 1.4$ for equal mixture of ordinary and exchange forces) and α_0 is polarizability. On the basis of liquid drop model, the estimation α_0 is given by

$$\alpha_0 = \frac{e^2 R^2 A}{40 \, a_4} \qquad \qquad \text{... (5.4)}$$

Where a_4 is the assymetry parameter in the semi-empirical mass relation and R is the nuclear radius. Using (5.4), one gets, $E_0 \sim 78 \; A^{\frac{1}{3}}$ MeV. This value of E_0 agrees fairly well with the experimental values for $A > 100$. The experimental values of E_0 for the lighter nuclei are than those obtained theoretically.

The width of the giant resonance lie between 3 to 8 MeV. The widths are narrower near closed nuclei shell and much broader for deformed nuclei. In case of deformed nuclei, there is superposition of two giant resonances, due to oscillations along the two short axes and along the long axis of an axially symmetric nucleus. In some cases, e.g. holononium and erbium, the existence of two such peakss have been reported.

We may note that the above explanation of the giant resonances is based on the collective behaviour of the nucleons. A microscopic theory based on shell model has also been proposed. This explains the appearance of these giant resonances. The parity changes involved reveals that in these cases transition between two major shells must be taken into account. In accordance with the exclusion principle, only the states near the top of Fermi sea are available to combine with the unbound states above. Dipole transitions takes place between states differing by a single quantum jump and harmonic oscillator states are favoured. Detailed calculations reveal that the main transitions take place from the closed shells which lie within one oscillator spacing above the Fermi surface. For nuclei between oxygen to calcium good agreement has been reported.

Quadrupole Resonances

In 1971, another type of giant quadrupole resonance also known as *isoscalar resonance* was observed at an energy $\approx 63\ A^{-\frac{1}{3}}$ MeV. In this type of resonance the neutrons and protons oscillate in phase, so that there is no change in the isospin ($\Delta T = 0$). In another type, the neutrons and the protons oscillate in opposite phase, so that $\Delta T = 1$, which reveals that an *isovector* type quadrupole resonance is produced.

The giant quadrupole resonances were first observed in inelastic electron scattering experiments and later in experiments on the inelastic scattering of protons. In these experiments, besides the giant electric dipole resonance peaks were observed at slightly lower energies. Angular distribution measurement revealed that these were quadrupole resonances corresponding to $\lambda = 2$. Later on, inelastic scattering of α-particles were used to study these peaks for nuclei with A ranging from 14 to 208. The energy of these peaks can be expressed by the relation

$$E_0 \approx 63\ A^{-\frac{1}{3}}\ \text{MeV} \qquad \qquad \ldots (5.5)$$

Corresponding roughly with excitations across two major shells and with a change in oscillator number $\Delta N = 2$. The resonance of course has $I^P = 2^+$ in even-even nuclei.

The width vary from 6 MeV for $A = 40$ to 3 MeV for $A = 208$. We may note that for the isovector peaks, the energy *Octapole States* ($I^P = 3^-$): These are often seen near the two-photon triplet in the even-even nuclei, e.g. as the lowest excited state in the double closed-shell nuclei ^{16}O, ^{40}Ca and ^{208}Pb, for which no $\Delta N = 0$ quadrupole vibrations are possible because shells are filled. These state appear as low-lying giant resonance, with $\Delta N = 1$, at an energy

$$E_0 \approx 32\ A^{\frac{1}{3}}\ \text{MeV}$$

A higher energy octupole mode with $\Delta N = 3$, $I^P = 3^-$ would be expected at about three times this energy.

Monopole Giant Resonances

Giant monopole resonances were first observed in experiments on the inelastic scattering of deutrons of 80 MeV energy by the nuclei ^{40}Ca, ^{90}Zr and ^{208}Pb in 1975. Subsequently, experiments on the inelastic scattering of α-particles confirmed these findings. These resonances appear at the energy given by $E_0 \approx 80\ A^{\frac{1}{3}}$ MeV. Angular distribution ($\lambda = 0$) studies revealed that these were monopole resonances. The nature of oscillations of the neutrons and protons reveal that these are isoscalar resonances. There are also evidences for isovector type giant monopole resonances ($\Delta T = 1$). We may note that γ-rays cannot excite the monopole resonances as they are transverse in nature.

5.12 OPTICAL MODEL

The existent of giant resonance is usually explained on the basis of optical model. This model of the nucleus is capable of explaining the behaviour of reaction cross-sections at both low as well high energies. At low energies the optical model deals with the energy average of the reaction cross-section. If the incident neutron energy is not sharply defined, a number of resonances may be covered in the energy spread. One can then average the cross-section and look for its energy and mass number dependence. For such purposes, we often invoke the analogy of an optical wave through a "cloudy" crystal ball. In a nuclear reaction, the scattered wave may be divided into two categories: (i) **elastic scattering** in which only the direction of wave propagation is changed and (ii) **inelastic scattering** in which the particle is scattered into an exit channel different from the incident one. The former corresponds to a refraction of the optical waves and the latter corresponds to an absorption due to the fact that the crystal ball is cloudy.

The aim of optical model is to find a potential to describe smooth variations of the scatteringg cross-section as a function of energy E and target nucleon number A. The general situation of scattering may be quite complex; however, great simplification may be obtained if we are interested only in the average properties, away from resonances and states strongly excited by direct reactions.

The optical model proposes that a nuclear reaction should be determined by the solution of the Shcrodinger equation for a one-body rather than a many body problem, namely the motion of an incident particle of given energy in a limited region of complex negative potential of the form

$$V(r) = V_0(r) - i\omega_0 \qquad \text{for } r < r_0$$
$$= 0 \qquad \text{for } r > r_0 \qquad \qquad \text{... (5.1)}$$

The real potential $V_0(r)$ is essentially that of single-particle shell model with a Woods Saxon radial dependence, i.e.

$$V = -(V_0 + i\omega_0) \left\{ 1 + \exp\left(\frac{r - R}{r_0} \right) \right\}^{-1} \qquad \text{... (5.2)}$$

It contains the Coulomb potential in the case of charged particles and a spin-orbit term connected with polarization effects. V_0, ω_0, R and r_0 are parameters for the optical madel. In order to take into account of the deformed nuclei, several workers have used more sofisticated forms of potential. Here V_0 and W_0 are assumed to be energy dependent.

We have mentioned that in optical model the cross-sections are averaged over energy range Δ which is larger than the separation D between the energy levels of the compound nucleus, the actual cross-section may fluctuate considerably with energy due to resonances. However, the average cross-section may be expected to vary smoothly with the incident energy.

A Schrodinger equation for the neutron with paten (5.1) is

$$\Delta^2 \psi + \frac{2M}{\hbar^2} [E + (V_0 + i W_0)]\psi = 0$$

for $r < R$, where $\psi = \psi(r, \theta, \phi)$. Let us write

$$\psi = \frac{u_l(r)}{r} Y_{lm}(\theta, \phi)$$

One obtains the radial equation for $l = 0$ as

$$\frac{d^2 u_0}{dr^2} + \frac{2M}{\hbar^2}[E + (V_0 + i W_0)]u_0 = 0$$

for $r < R$. For $r > R$, we have

$$\frac{d^2 u_0}{dr^2} + \frac{2ME}{\hbar^2}u_0 = 0 \qquad \ldots (5.3)$$

Let us substitute

$$k^2 = \frac{2ME}{\hbar^2} \text{ and } K^2 = \frac{2M}{\hbar^2}(E + V_0 + i W_0)$$

One obtains two radial equations as

$$\frac{d^2 u_0}{dr^2} + K^2 u_0 = 0 \qquad (r < R) \qquad \ldots (5.5)$$

and

$$\frac{d^2 u_0}{dr^2} + k^2 u_0 = 0 \qquad (r > R) \qquad \ldots (5.6)$$

Here K is complex and one can obtain its value from Equation (5.4). The complex refractive index of the nuclear matter for the entering nucleon is then given by

$$n = \sqrt{\frac{K}{k}} = \sqrt{\frac{E + V_0}{E} + \frac{i W_0}{E}} \qquad \ldots (5.7)$$

Now, writing $n = \alpha + i\beta$, one can define an attenuation coefficient for the incident wave. Obviously, it depends on β.

The elastic scattering cross-section can be calculated by determing the n_l. We have

$$\sigma_{sc}^{(l)} = \pi \lambda^2 (2l + 1) |1 - n_l|^2 \qquad \ldots (5.8)$$

We may note that the calculations using two parameter square well potential considered above is not in agreement with the experimental results.

Wood and Saxon in 1954 proposed a potential with a radial dependence as follow:

$$U(r) = -(V_c + i W_c) f(r, r_0, a)$$

$$= -\frac{(V_c - i W_c)}{1 + \exp\left[\frac{(r - R)}{a}\right]} \qquad \ldots (5.9)$$

Where $R = r_0 A^{\frac{1}{3}}$ and V_c and W_c are constants. We may note that the radial dependence is similar to Fermi charge distribution within a nucleus.

Figure 5.29 shows a plot of the real part of Woods and Saxon potential against r. The value of R is somewhat greater than for the uniform charge distribution. The parameter a (surface density) determines the skin depth. The potential falls from 90 % of the maximum magitude (at the centre) to 10 % in a distance about 4.4 a near the nuclear surface as depicted in figure.

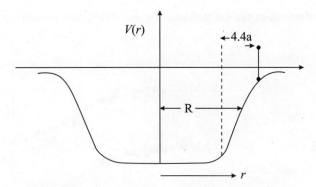

Fig. 5.29 Real part of Woods and Saxon potential

Using the above potential, the differential cross-sections obtained for elastic scattering agrees fairly well in the forward direction, while there is no agreement at the backward direction.

In addition to the theory of the neutron-nuclear reaction optical model has been applied to the scattering of protons, deutrons, α-particles and some heavier ions. Using the Bjorkund and Fernback optical potential,

$$U(\mathrm{r}) = V\rho(r) + iWq(r) + \frac{\lambda \hbar^2}{4m_0^2 c^2} \frac{V}{r} \frac{d\rho}{dr} (\mathbf{\rho} \cdot \mathbf{l})$$

Where

$$\rho(r) = \left[1 + e^{\frac{(r-R)}{a}} \right]^{-1} \qquad \qquad ...\,(5.10)$$

$q(r) = e^{-\left[\frac{(r+R)}{b} \right]^2}$, $R = r_0 A^{\frac{1}{3}}$, and m_0 the mass of electron, Albert in 1959 measured (p, n) cross-sections in some medium weight nuclei between the energy range 4 to 5.5 MeV. The experimental results are shown in Figure 5.30.

Figure shows the experimental and theoretical curves for (p, n) cross-section vs. atomic weight for protons of energies; $E_p = 5.5, 5.0, 4.5$ and 4.0 MeV. The theoretical curves obtained based on optical model is shown by solid lines. We note that there is an excellent agreement between theory and experiment. The parameters used for theoretical calculations are given in Table 5.5.

Table 5.5 $a = 0.65$ F, $r_0 = 1.25$ F, $b = 0.98$ F

E (MeV)	V (MeV)	W (MeV)
4.0	74	70
4.5	45	7.7
5.0	45	8.25
5.5	44	8.7

Shore and Coworkers has also used the optical potential functions to analyze angular distribution of 7.5 MeV protons scattered elastically by Vanadium. Corresponding to surface absorption and to volume absorption, they used two sets of parameter. Very recently Rosen and Coworkers have proposed a spin dependent optical potential by analyzing and measuring 80 separate angular distribution data.

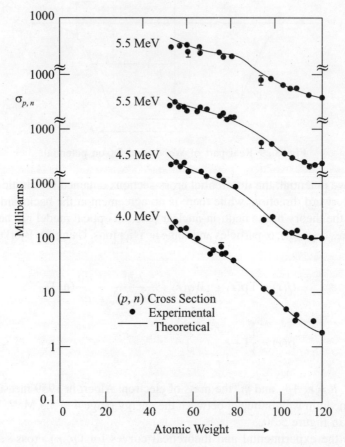

Fig. 5.30 The (p, n) cross-section vs. atomic weight for protons of energies, $E_p = 5.5, 5.0, 4.5$ and 4.0 MeV used by Albert. The solid lines are the theoretical curves obtained on the basis of optical model.

Using the following form of nuclear potentials

$$V(r) = V\left[1 + e^{(r - r_0 A^{\frac{1}{3}})/a}\right]^{-1} + iW\left[1 + e^{(r - r_w A^{\frac{1}{3}})/a_w}\right]^{-1} \qquad \dots (5.11)$$

for volume absorption and

$$V(r) = V\left[1 + e^{(r - r_0 A^{\frac{1}{3}})/a}\right]^{-1} - ia_w W_0 \frac{d}{dr}\left[1 + e^{(r - r_w A^{\frac{1}{3}})/a_w}\right]^{-1} \qquad \dots (5.12)$$

for surface absorption, Halbert and Coworkers carried out an analysis of the $d - d$ scattering data

11 MeV and 11.8 MeV. In each case they assumed uniform charge distribution of radius 1.3 $A^{\frac{1}{3}}$ F

for the Coulomb potential and obtained excellent fit for the target nuclei Ni, Zr, Ag and Sn and reasonably good fit for Ca and Ti.

The case of deuteron is slightly complicated, due to polarization or stripping of the deuteron by Coulomb interaction, non-elastic channel may couple to the elastic channels and may affect or change the optical parameters.

When the incident particle has spin other than zero, the optical potential may contain spin-orbit terms. To fit the measured polarization of the scattered particles, the strength of the spin-orbit potential is adjusted. Experiments reveal that the idea of the optical potential is not limited to the calculation of cross-sections of neutron reactions and incident energies of a few million eV but is applicable to the description of the reactions of protons and heavier particles in the region of high energies as well.

The phenomenological optical model essentially provides a highly adaptable technique for describing elastic scattering and it has been used for complex particles as well as for nucleons. The model gives no detail about the inelastic processes. The optical model potential is the potential acting on an unbound nucleon and is partly *absorptive*. The absorption represents the fact that such a nucleon has enough energy to collide with a nucleon in the nucleus, and thus be absorbed from the incident beam. (It is absorbed in the sense that it no longer has the same energy, or de Broglie wave length so there can be no interference between its wave function and the wave function for the incident nucleon). Collisions are possible since the exclusion principle does not have its usual inhibiting effect if the incident nucleon brings in enough energy that both it, and the struck nucleon, can easily find unfilled states to occupy. The incident nucleon can, of course, also scatter from the more familiar nonabsorptive part of the potential (i.e., it can also interact with the nucleus as a whole, represented by the usual attractive potential, without colliding with an individual nucleon of the nucleus). The optical model is essentially a generalization of the shell model which applies to nucleons of any energy-not just to nucleons of energy such that they are bound in a nucleus.

Example 5.1 Use the semi empirical mass formula to predict the binding energy made available if a $^{235}_{92}$U nucleus captures a neutron. This is the energy which induces fission of the $^{236}_{92}$U nucleus that is formed in the capture.

Sol. The binding energy is

$$E_n = \{[E_{92.235} + M_{0.1}] - [M_{92.236}]\}c^2 \qquad \dots (5.1)$$

The term in the first square bracket is the mass of a $^{235}_{92}$U atom plus the mass of a neutron, which are the constitutents of the $^{236}_{92}$U atom whose mass appears in the second square bracket. Since the neutron mass, $M_{0.1}$ is precisely 1.008665 amu, the first two terms from the semi empirical mass formula,

$$M_{Z,\,A} = 1.007825\,Z + 1.008665\,(A - Z) - a_1 A + a_2\,A^{\frac{2}{3}} + a_3\,Z^2\,A^{-\frac{1}{3}}$$

$$+ a_4 \left(Z - \frac{A}{2}\right)^2 A^{-1} + \begin{pmatrix} -1 \\ 0 \\ 1 \end{pmatrix} a_5\,A^{-\frac{1}{2}} \text{ (in amu)}$$

cancal out in the expression for energy, E_n (i). Then, one obtains

$$E_n = \left\{ \left[-a_1(235) + a_2(235)^{\frac{2}{3}} + a_3 \frac{(92)^2}{(235)^{\frac{1}{3}}} + a_4 \frac{\left(92 - \frac{235}{2}\right)^2}{235} \right] \right.$$

$$\left. - \left[-a_1(236) + a_2(236)^{\frac{2}{3}} + a_3 \frac{(92)^2}{(236)^{\frac{1}{3}}} + a_4 \frac{\left(92 - \frac{236}{2}\right)^2}{236} - \frac{a_5}{(236)^{\frac{1}{2}}} \right] \right\} c^2$$

$$= \left\{ \left[a_1 - a_2(236)^{\frac{2}{3}} + a_3 \frac{(92)^2}{(235)^{\frac{1}{3}}} + a_4 \frac{\left(92 - \frac{236}{2}\right)^2}{236} - \frac{a_5}{(236)^{\frac{1}{2}}} \right] \right\} c^2$$

$$= \left\{ a_1 - a_2 \left[(236)^{\frac{2}{3}} - (235)^{\frac{2}{3}} \right] + a_3 (92)^2 \left[\frac{1}{(235)^{\frac{1}{3}}} - \frac{1}{(236)^{\frac{1}{3}}} \right] \right.$$

$$\left. - a_4 \left[\frac{(26.0)^2}{236} - \frac{(25.5)^2}{235} \right] + \frac{a_5}{(236)^{\frac{1}{2}}} \right\} c^2$$

$$\approx \{ 0.0169 - 0.0191 \times 0.11 + 0.00076 \times 1.9 - 0.1018 \times 0.097 + 0.012 \times 0.065 \} c^2$$

$$= \{ 0.0169 - 0.0021 + 0.0014 - 0.0099 + 0.0008 \} c^2$$

$$= (0.0071 \text{ amu}) c^2 = 6.6 \text{ MeV} \qquad [\because 1 \text{ amu} \times c^2 = 931.5 \text{ MeV}]$$

If the neutron has negligible kinetic energy before it is captured, the $^{236}_{92}U$ nucleus is formed in a state of excitation energy equal to E_n. The excitation energy often sets the nucleus into a vibration in which it oscillates between being elongated (having a positive quadrupole moment) and being flattened (having a negative quadrupole moment). This vibration cannot take place without the excitation energy since the surface term of the semiemprical mass formula inhibits departures of the nucleus from the approximately spherical shape it has in its ground state. When the nucleus has a maximum elongation, the effect of Coulomb term can cause it to fission of great importance in nuclear reactor technology is the fact that E_n for neutron capture by a $^{238}_{92}U$ nucleus is about 1.5 MeV smaller than the value just calculated for capture by $^{235}_{92}U$. The terms in the proceeding expressions have almost the same values, except that the contribution of the pairing term (the last term) is negative instead of positive. Since all $_{92}U$ will fission only if the neutron it captures brings in more about 1 MeV of kinetic energy, in addition to its binding energy. This means $^{238}_{92}U$ is not very useful in the "Chain reaction that takes place in reactors.

Example 5.2 Use Figure 5.31 to predict the first four magnetic numbers for nuclei with potentials in the form of square wells with rounded edges (a) under the assumption that there is no spin orbit interaction, and (b) under the assumption that there is a strong inverted spin orbit interaction.

Sol.

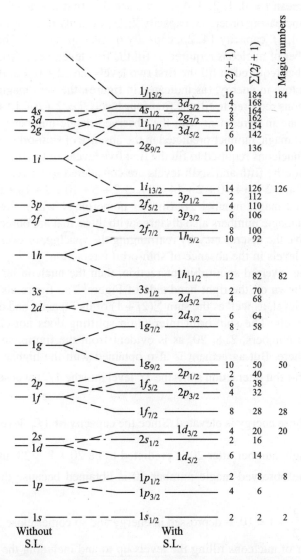

Fig. 5.31 Left: The order of filling, as the occupancy and well radius increases, of the levels of rounded edge square wells with no spin-orbit interaction.
Right: The levels that arise when a strong inverted $S \cdot L$ interaction is added.
The column marked $(2j + 1)$ shows the number of like nucleons that may occupy the corresponding level without violating the Pauli exclusion principle. The column marked $\Sigma(2j + 1)$ gives for each level the cummulative number of nucleons that lie in all levels up through that level. Significant energy gaps lie above each of the levels marked with a magic number in the last column.

(a) If there is no spin-orbit interaction then the nucleon energy levels are simply those shown on the left part of the figure. We know that the capacity of each level is $2(2j + 1)$, and the *s*, *p*, *d*, *f*, *g*, ... mean l = 0, 1, 2, 3, 4, ... , we see that first few levels, and their capacities are, in order of increasing order: 1*s*, capacity 2; 1*p*, capacity 6; 1*d*, capacity 10; 2*s*, capacity 2; 1*f*, capacity 14; 1*f*, capacity 14; 2*p*, capacity 6; 1*g*, capacity 18. The first magic number will be the number of nucleons required to fill the first level, i.e. 2. The next magic number will be the number required to fill the first two levels, i.e., 2 + 6 = 8. If the third and fourth levels are very close in energy, (as indicated in figure), the next magic number will be the number of nucleons required to fill the first four levels, i.e. 2 + 6 + 10 + 2 = 20. So far these magic numbers are in agreement with the observed magic numbers: 2, 8, 20, 28, 50, 82, 126. But the next magic number predicted in the absence of spin-orbit interaction will be the total number of nucleons required to fill the first five levels, or the first six levels, depending on whether or not the fifth and sixth levels are considered to be very close in energy. The two possibilities are, 2 + 6 + 10 + 2 + 14 = 34, or 2 + 6 + 10 + 2 + 14 + 6 = 40. Both disagree with the observed magic number 28. Similar numerology will make it apparent that the higher predicted magic numbers also disagree with those that are observed, and that there is no way to remove the discrepancy by rearranging the spacing, or even the ordering, of the nucleon energy levels in the absence of spin-orbit interaction.

(b) If there is strong inverted spin-orbit interaction, then the nucleon levels are split into the filling pattern shown on the right-hand-side of Figure. The figure also slows the capacity $(2j + 1)$ of each level, as well as the sum $\Sigma(2j + 1)$ of its capacity and the capacity of all the lower energy levels. The spin-orbit interaction splitting does not change the first three predicted magic numbers, 2, 8, 20, as is evident from the figure, so the agreement with observation is there. But agreement is also obtained with the higher magic numbers. For instance, the spin-orbit interaction splits the 1*f* level into the $1f_{\frac{7}{2}}$, whose energy is depressed,

and the $1f_{\frac{5}{2}}$, whose energy is elevated. Since the capacity of $1f_{\frac{7}{2}}$ level is $(2j + 1) = 2 \times \frac{7}{2}$ + 1 = 8, the magic number after 20 is predicted to be 20 + 8 = 28, in agreement with the observation. The observed magic number 50 is obtained because the $1g_{\frac{9}{2}}$ level, with a capacity of $2 \times \frac{9}{2}$ + 1 = 10, is depressed in energy and so comes close to the 2*p* level. Since the total number of nucleons filling the levels up to and including the 2*p* is 40, as we saw earlier, the total number filling the levels up to and including the $1g_{\frac{9}{2}}$ is 40 + 10 = 50. Figure makes the origin of remaining magic numbers apparent. We may note the fact that the *spin-orbit* splitting increases in magnitudes, with increasing, plays an important role in achieving agreement with the observation.

Example 5.3 Use Figure 5.31, and the Pauli exclusion principle argument just stated, to predict the ground state spins of the following nuclei: (a) $^{15}_{7}N$, (b) $^{17}_{8}O$, (c) $^{39}_{19}K$, (d) $^{207}_{82}Pb$, and (e) $^{209}_{83}Bi$.

Sol. (a) Figure predicts that $^{15}_{7}$N is doubly magic except for a proton hole in the $1p_{\frac{1}{2}}$ subshell.

So it should have a spin i equal to the value $j = \frac{1}{2}$ for that subshell. This prediction agrees with measurement.

(b) The Figure predicts that $^{17}_{8}$O is doubly magic except for an extra neutron in the $1d_{\frac{5}{2}}$ subshell. So it should have $i = j = \frac{5}{2}$, in agreement with measurement.

(c) $^{39}_{19}$K is predicted to be double magic except for a proton hole in the $1d_{\frac{3}{2}}$ subshell, so it should have $i = j = \frac{3}{2}$. It does.

(d) According to Figure $^{207}_{82}$Pb is doubly magic except for a neutron hole in the $1i_{\frac{3}{2}}$ subshell, the measured spin is $i = \frac{1}{2}$. However, this is not a failure of the exclusion principle, but instead is a failure of figure.

(e) The figure predicts that $^{209}_{83}$Bi is doubly magic except for an extra proton in the $1h_{\frac{9}{2}}$ subshell. So its spin should be $i = j = \frac{9}{2}$. This agrees with the measurement.

Example 5.4 Predict the ground state nuclear spin and parity for the following nuclei: (a) $^{16}_{8}$O, (b) $^{17}_{8}$O, (c) $^{16}_{8}$O, (d) $^{15}_{7}$N and (e) $^{14}_{7}$N.

Sol. (a) The $^{16}_{8}$O nucleus hav even N and even Z, and it is also doubly magic since both N and Z equal 8. It has two neutrons in the $1s_{\frac{1}{2}}$ subshell which couple together in a pair to yield zero total angular momentum. Both of these neutrons are described by even parity eigen functions, since $l = 0$, so their part of product eigen function for the nucleus is even. There are four neutrons in the $1p_{\frac{3}{2}}$ subshell, that couple into two pairs, both of which have zero total angular momentum. All four of these neutrons are described by odd parity eigen functions since $l = 1$, but the product of four odd functions is an even function, so their part of the product eigen function for the nucleus is also even. There are two neutrons in the $1p_{\frac{1}{2}}$ subshell, which form a pair of zero total angular momentum.

They contribute two odd eigen functions to the product eigen function for the nucleus, so their part of the product eigen function is also even. Exactly the same arguments apply to the protons. The net result is that the nuclear spin is zero, and the nuclear parity is even.

(b) $^{17}_{8}$O is an odd-N, even Z nucleus. Its neutrons and protons are doing the same things as the neutrons and protons in $^{16}_{8}$O, except that it has a single extra unpaired neutron in a $1d_{\frac{5}{2}}$ subshell. This gives the nucleus a spin of $i = \frac{5}{2}$. The parity of the eigen function for the unpaired neutron is even since $l = 2$, so the nuclear parity is even.

(c) $^{18}_{8}$O is an even-N, even-Z nucleus. The predicted spin and parity are $i = 0$, and even. The reasons are that there are two neutrons in the $1d_{\frac{5}{2}}$ subshell, which form a pair of zero total angular momentum, and which both have even parity eigen functions.

(d) $^{15}_{7}$N is an even N, odd-Z nucleus. Its neutrons and protons behave as an $^{16}_{8}$O, except that it has only one unpaired proton in the $1p_{\frac{1}{2}}$ subshell. This odd proton gives the nucleus a spin of $i = \frac{1}{2}$. Since the eigen function for the proton is odd because $l = 1$, the nuclear parity is odd. We may note that we predicted the nuclear spin, from a somewhat different point of view in the previous example.

(e) $^{14}_{7}$N is an odd-N, odd Z-nucleus. It has an unpaired proton in the $1p_{\frac{1}{2}}$ subshell, and also an unpaired neutron in the $1p_{\frac{1}{2}}$ subshell. Both have a total angular momentum quantum number of $j = \frac{1}{2}$. One can not say precisely what the nuclear spin should be without knowing how two different particles couple their angular momenta. But one can say that there are only two possibilities for the nuclear spins, $i = 0$, or $i = 1$. Experiments show that $i = 1$ is the correct value. One can predict unambiguously that the nuclear parity will be even, since the unpaired protons and the unpaired neutron both contribute an odd eigen function to the product eigen function for the nucleus, and the product of two odd functions is an even function. This prediction is borne out by the experiments, as are all the prediction made earlier in this example.

Example 5.5 Estimate the shell model prediction for the average electric quadrupole moment q of the nucleus $^{123}_{51}$Sb, and compare it with the measured value.

Sol. According to the shell model, the charge distribution of this nucleus is due to a spherically symmetrical core of completely filled proton subshells, plus a single odd proton in a $1g_{\frac{7}{2}}$ subshell. Since the orbital angular momentum of this proton is high ($l = 4$), to a fair approximation it can be thought of as moving in a Bohr-like orbit of radius r' (orbital motion approaches the class limit as l becomes large). Thus the spherical core makes no contribution to the nuclear quadrupole moment q. So, We take the symmetry axis perpendicular to the orbit as the Z-axis, we have

$$q = \int \rho \ [3z^2 - (x^2 + y^2 + z^2)]d\tau$$

Where ρ is approximately the charge density for a uniformly charged ring, of radius r', in a plane perpendicular to Z. This r is zero except where $x^2 + y^2 + r'^2$ and $Z = 0$. Thus

$$q = \int \rho \; [-r'^2] d\tau = -r'^2 \int \rho \, d\tau$$

This integral of ρ yields one since the ring contains the charge of one proton and ρ is measured in groups of one proton charges. Therefore, the result we obtain for an estimate of the shell model prediction of q for $^{133}_{51}$Sb is

$$q \approx -r'^2$$

Measured value of q for this nucleus is such that

$$\frac{q}{Z \, r'^2} \approx -0.09$$

or

$$q \approx -0.09 \, Z \, r'^2 \approx -0.09 \times 51 \; r'^2$$
$$\approx -5 \; r'^2$$

Obviously, the magnitude of the shell model prediction is too low, compared to the measurment by about a factor of 5.

Example 5.6 Use the single particle shell model to predict the ground state spins, parities and magnetic moments of $^{27}_{13}$Al, $^{33}_{16}$S and $^{41}_{18}$A.

Sol. We can write the shell model terms of the nuclei as

$^{27}_{13}$Al : $\left(1s_{\frac{1}{2}}\right)^2 \left(1p_{\frac{3}{2}}\right)^4 \left(1p_{\frac{1}{2}}\right)^2 \left(1d_{\frac{5}{2}}\right)^6$

Here $J = \dfrac{5}{2}$ and $l = 2$ and thus Parity $\Pi = (-1)^l = (-1)^2 = +$ even

$^{33}_{16}$S : $\left(1s_{\frac{1}{2}}\right)^2 \left(1p_{\frac{3}{2}}\right)^4 \left(1p_{\frac{1}{2}}\right)^2 \left(1d_{\frac{5}{2}}\right)^6 \left(2s_{\frac{1}{2}}\right)^2 \left(1d_{\frac{3}{2}}\right)^1$

Thus $J = \dfrac{3}{2}$ and $l = 2$. Thus parity is even.

$^{41}_{18}$A : $\left(1s_{\frac{1}{2}}\right)^2 \left(1p_{\frac{3}{2}}\right)^4 \left(1p_{\frac{1}{2}}\right)^2 \left(1d_{\frac{5}{2}}\right)^6 \left(2s_{\frac{1}{2}}\right)^2 \left(1d_{\frac{3}{2}}\right)^4 \left(1f_{\frac{7}{2}}\right)^3$

Hence $J = \dfrac{7}{2}$ and $l = 3$. Thus parity is odd.

Example 5.7 Consider a shell model in which the nucleons are in pairs in equally spaced energy levels. Considering the number of neutrons and protons same, calculate the energy needed to change n-proton pairs to neutrons and move them to neutron orbits.

Sol. Let the final nucleus is to have N neutrons and Z protons. Let us define the nucleon difference by $n = N - Z$ which is twice the number of nucleon pairs to be moved from proton levels to neutron levels. If all levels are separated by an energy δ, the total energy needed to create the final nucleus is

$$E = (2\delta)(1) + (2\delta)(3) + (2\delta)(3) + (2\delta)5 + \dots + (2\delta)(n-1)$$

$$= 2\delta[1 + 3 + 5 + \dots + (n-1)]$$

$$= 2\delta\left(\frac{n^2}{4}\right) = \frac{\delta}{2}(N-Z)^2 = \frac{\delta}{2}(A-2Z)^2$$

We can see that this term is directly related to the $a_4(A-2Z)^2 A^{-1}$ term in the semi-empirical mass formula, which is an expression for the neutron and proton excess energy.

Example 5.8 Determine the harmonic oscillator frequencies ω appropriate to the nuclei ^{17}O and ^{60}Ni.   [Kerala]

Sol. We have the total single particle energies in a nucleus of mass number A are:

$$E = M\omega^2 A(r_{rms})^2 = \frac{3}{5} M\omega^2 A R_c^2 \qquad \dots (5.1)$$

The Coulomb radius, $R_c = 1.2\ A^{\frac{1}{3}}$ fermi. Now, assuming an equal number of neutrons and protons and that all energy states upto $E_{\wedge_0}$ are occupied, one obtains

$$A = \sum_{\wedge = 0}^{\wedge_0} 2N_\wedge = \frac{2}{3}(\wedge_0 + 1)(\wedge_0 + 2)(\wedge_0 + 3)$$

$$= \frac{2}{3}(\wedge_0 + 2)^2 + \dots \text{ terms of higher order } \wedge_0 \qquad \dots (5.2)$$

and

$$\frac{E}{\hbar\omega} = \sum_{\wedge = 0}^{\wedge_0} 2N_\wedge\left(\wedge + \frac{3}{2}\right) \simeq \frac{1}{2}(\wedge_0 + 2)^4 - \frac{1}{3}(\wedge_0 + 2)^3 + \dots \qquad \dots (5.3)$$

$$\simeq \frac{1}{2}\left[\frac{3}{2}\wedge\right]^{\frac{4}{3}} \qquad \dots (5.4)$$

Solving (5.1) and (5.4), one obtains

$$\omega = \frac{5 \times 3^{\frac{1}{3}} \hbar A^{\frac{1}{3}}}{2^{\frac{7}{3}} M R_0^2} = \frac{5 \times 3^{\frac{1}{3}} \times 1.0545 \times 10^{-34}}{2^{\frac{7}{3}} \times (1.67 \times 10^{-27})(1.2 \times 10^{-15})^2} A^{-\frac{1}{3}}$$

$$= 6.3 \times 10^{23}\ A^{-\frac{1}{3}}$$

For ^{17}O nucleus, $\omega = 6.3 \times 10^{22} \times (10^{17})^{-\frac{1}{3}}$

$$= 2.45 \times 10^{22}$$

For ^{60}Ni nucleus, $\omega = 6.3 \times 10^{22} \times (60)^{-\frac{1}{3}}$

$= 1.61 \times 10^{22}$

Example 5.9 The first excited state of ^{182}W is 2^+ and is 0.1 MeV above the ground state. Estimate the energies of the lowest lying 4^+ and 6^+ states of ^{182}W? [Punjab]

Sol. The expression for energy to low lying rotational states is

$$E = I(I + 1)\, \frac{\hbar^2}{2j}$$

For 2^+ state, $0.1 = \left(\frac{\hbar^2}{2j} \right) 2 \times 3$

or $\left(\frac{\hbar^2}{2j} \right) = \frac{0.1}{6} = 0.01667$

Energies corresponding to 4^+ and 6^+ states are:

$$E(4^+) = 0.01667 \times 4 \times 5 = 0.333 \text{ MeV}$$
$$E(6^+) = 0.01667 \times 6 \times 7 = 0.700 \text{ MeV}$$

Example 5.10 What will be the characterstics of the ground state of $^{17}_{8}$O, $^{33}_{16}$S, $^{63}_{29}$Cu, and $^{209}_{83}$Bi nuclei?

Sol. We can write the shell model terms of the nuclides as

For $^{17}_{8}$O : $\left(1s_{\frac{1}{2}} \right)^2 \left(1p_{\frac{3}{2}} \right)^4 \left(1p_{\frac{1}{2}} \right)^2 \left(1d_{\frac{5}{2}} \right)^1$

Thus $I = \frac{5}{2}$, $l = 2$ and parity even

$\therefore$ Magnetic moment of this nucleus due to odd neutron $\left(1d_{\frac{5}{2}} \right) = -1.91$ and quadrupole

moment

$$Q = -\frac{3}{5} \frac{2I-1}{2I+2} R_c^2 = -\frac{3}{5} \times \frac{4}{7} \times (1.2 \times 17^{\frac{1}{3}} \times 10^{-15})^2 = 0.326 \text{ barn}$$

$^{33}_{16}$S : $\left(1s_{\frac{1}{2}} \right)^2 \left(1p_{\frac{3}{2}} \right)^4 \left(1p_{\frac{1}{2}} \right)^2 \left(1d_{\frac{5}{2}} \right)^6 \left(2s_{\frac{1}{2}} \right)^2 \left(1d_{\frac{3}{2}} \right)^1$

Thus $I = \frac{3}{2}$, $l = 2$ and parity even. The magnetic moment of this nucleus due to odd neutron

$1d_{\frac{3}{2}}$ is $1.91 \left(\frac{3}{2} \Big/ \frac{5}{2} \right) = 1.146$

The quadrupole moment

$$Q = -\frac{3}{5} \times \frac{2}{5} \times \left[1.2 \times (33)^{\frac{1}{3}} \times 10^{-15}\right]^2$$

$$= -0.0355 \text{ barn.}$$

$^{63}_{29}$Cu :   [Shells upto 28 protons]; $\left(2p_{\frac{3}{2}}\right)^1$, hence $I = \frac{3}{2}$, $l = 1$ and parity odd. The magnetic

moment of this nucleus due to odd proton $2p_{\frac{3}{2}}$ is $\frac{3}{2} + 2.29 = 3.79$. The quadrupole

moment $Q = -\frac{3}{5} \times \frac{2}{5} \times \left(1.2 \times 63^{\frac{1}{3}} \times 10^{-15}\right)^2 = -0.0547$ barn.

$^{209}_{83}$Bi: [Shells upto 82 protons]; $\left(1h_{\frac{9}{2}}\right)^1$, hence $I = \frac{9}{2}$, $l = 5$ and parity odd. The magnetic

moment due to odd proton $1h_{\frac{9}{2}} = 4.5 - 2.29 \times \frac{9}{11} = 2.63$

The quadrupole moment $Q = -\frac{3}{5} \times \frac{8}{11} \times \left(1.2 \times 209^{\frac{1}{3}} \times 10^{-15}\right)^2$

$$= 0.176 \text{ barn}$$

Example 5.11 Evaluate the Fermi energy of a typical nucleus, and use the results to determine the depth of the net nuclear potential.

Sol.   The Fermi energy, E_F, is the energy indicated in Figure 5.32 of the nucleon in the highest filled level of the system, measured from the bottom of the potential well. It is related to the nucleon mass M, and nucleon density ρ as

$$E_F = \frac{\pi^2 \hbar^2}{2M} \left(\frac{3}{\pi} \rho\right)^{\frac{2}{3}} \qquad \dots (5.1)$$

Let us consider a Fermi gas of neutrons in a uniform spherical nucleus of radius

$$r' = a \; A^{\frac{1}{3}}$$

For a typical nucleus, the number of neutrons is

$$N \simeq 0.60 \; A$$

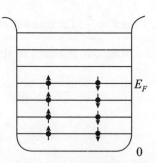

Fig. 4.32 A schematic representation of the energy levels filled by the neutrons in the ground state of a nucleus. The lowest levels are filled, according to the limitations of the Pauli exclusion principle, upto the Fermi energy, E_F.

Thus $\qquad \rho \simeq \dfrac{N}{\dfrac{4}{3}\pi a^3 A}$

$\qquad\qquad \simeq \dfrac{0.60\,A}{1.33\,\pi\,a^3\,A} = \dfrac{0.45}{\pi\,a^3}$

and the Fermi energy is

$$E_F \simeq \frac{\pi^2\,\hbar^2\,(0.26)}{2\,Ma^2} \qquad\qquad \dots (5.2)$$

Using $a \simeq 1.1$ F consistent with electron scattering measurements, and evaluating one obtains

$$E_F \simeq 43 \text{ MeV}$$

The relation between the depth of the potential V_0, the Fermi energy E_F, and the binding energy of the last neutron E_n, (Figure 5.33). $E_n \sim 7$ MeV for a typical nucleus. Thus for this nucleus the depth of the net nuclear potential acting on its neutrons is

$$V_0 = E_F + E_n \simeq 43 \text{ MeV} + 7 \text{ MeV}$$
$$\simeq 50 \text{ MeV}$$

We may note that a very similar result is obtained for the net nuclear potential for protons (of course protons also feel a net Coulomb potential exerted by the charges of other protons in the nucleus).

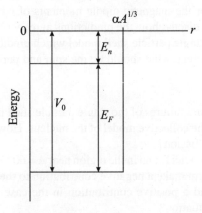

Fig. 5.33 Illustrating the relation between the depth V_0 of a nuclear square well potential of radius $r' = a\ A^{\frac{1}{3}}$, the Fermi energy E_F, and the binding energy E_n of the last neutron.

SUGGESTED READINGS

1. G. E. Brown, The Unified Theory of Nuclear Models and Forces, North Holland (1967).

2. M. A. Perston and R. K. Bhaduri, Structure of the Nucleus, Addison Wesley (1975).

3. M. G. Mayer and J. H. D. Jensen, The Elementary Theory of Nuclear Shell Structure, Wiley (1955)

4. J. P. Elliott and A. M. Lane, The Nuclear Shell Model, Encyclopdedia of Physics, Vol. 39, Springer (1957)

5. A. Bohr and B. Mottelson, Nuclear Structure, Benjamin (1975)

6. D. J. Rowe, Nuclear Collective Motion, Methuen (1957).

7. D. E. Hodgson, Superdeformed Nuclei, Contemp. Phys. 28, 365 (1987)

8. G. R. Satchler, New Giant Resonances in Nuclei, Physics Reports 14 C, 98 (1974)

9. J. O. Newton, Spinning Nuclei, Contemp, Phys. 30, 277 (1989)

10. J. Wambach, Nuclear Sound, Contemp Phys. 32, 291 (1991)

REVIEW QUESTIONS

1. Nuclei have magnetic dipole moments. Why do they not have magnetic monopole moments? What about magnetic quadrupole moments?

2. All nuclei have an electric monopole moment which measures their total charge. Some nuclei have an electric quadrupole moment which measures the departure from a spherical shape of their charge distribution. No nuclei have an electric dipole moment which would measure the departure of the centre of their charge distribution from the centre of their mass distribution. Why would we do not expect electric dipole moments for nuclei?

3. Why are the atomic magic numbers not the same as the nuclear magic numbers?

4. What fundamental law of physics is most responsible for the existence of nuclear magic numbers?

5. Why do most nuclei obey JJ coupling, whereas most atoms obey LS coupling?

6. Why there are no magic numbers that are odd?

7. What is the simplest distribution of point charges that has an electric quadrupole moment?

8. Obtain an expression for the magnetic dipole moments of odd A nuclei on the basis of single particle shell model and discuss how the predictions agree with the experimental values.

9. Give a brief account of single particle shell model which predicts the magic numbers. Assuming the shell model to be correct, what should be the spin and parity of the ground state of $^{15}_{7}N$.

$$\left[\frac{1}{2}, \text{ odd} \right]$$

10. Discuss the limitations and failures of the single particle shell model.

11. Give a brief account of the collective model of the nucleus. How does it help in understanding the phenomenon of nuclear fission?

12. Why are the most stable nuclei found in the region near $A \simeq 60$? Why do not all nuclei have $A \simeq 60$?

13. Why does the pairing term make a negative contribution to the energy liberated when a neutron is captured by $^{238}_{92}U$, and a positive contribution in the case of $^{235}_{92}U$? What are the practical consequences of this situation?

14. Explain why there can be no collisions between a typical nucleon and others in a nucleus in its ground state? If a high energy nucleon, say from a cyclotron beam, enters a nucleus in its ground state, can it collide with a nucleon in the nucleus?

PROBLEMS

1. Calculate the potential energy of a proton and an electron separated by a distance of 10^{-14} m.

[0.14 MeV]

2. Using the relativistic relation between momentum and energy find the minimum kinetic energy of (a) an electron, (b) a proton confined within a dimension of 7×10^{-15} m (assume $\Delta p \times \Delta r = \hbar$).

[28 MeV, 0.42 MeV]

3. Write down the expected odd-particle configuration of the following nuclei: $^{27}_{13}$Al; $^{29}_{14}$Si; $^{40}_{19}$K; $^{93}_{41}$Nb; $^{157}_{64}$Gd.

$$\left[d_{\frac{5}{2}}; s_{\frac{1}{2}}; d_{\frac{3}{2}} + f_{\frac{7}{2}}; g_{\frac{9}{2}}; h_{\frac{9}{2}} \right]$$

4. Predict the spin-parity of the first excited state of the nuclei $^{31}_{14}$Si, $^{41}_{19}$K and $^{49}_{21}$Sc.

$$\left[\frac{7^-}{2}, \frac{7^-}{2}, \frac{3^-}{2} \right]$$

Comment on the fact that the observed values are $\dfrac{1^+}{2}, \dfrac{1^+}{2}, \dfrac{3^+}{2}$.

5. From the equation $j = l + s$ for a single particle in an eigen state of the operators $\hat{j}, \hat{l}, \hat{s}$, obtain the value of the quantity $(l \cdot s)$ and show that the energy separation of a nucleon spin-orbit doublet is proportional to $2l + 1$. For the case $l = 1$, $j = \dfrac{3}{2}$, what is the angle between l and j?

[24°]

6. The low-lying levels of $^{39}_{20}$Ca have spin-parity values, starting from the ground state, of $\dfrac{3^+}{2}, \dfrac{1^+}{2}, \dfrac{7^-}{2}$ and $\dfrac{3^-}{2}$. Interpret these values on the basis of the SPSM.

7. Calculate the ground-state magnetic moment of ^{7}Li, ^{39}K and ^{45}Sc. The spin values are $\left[\dfrac{3}{2}, \dfrac{3}{2}, \dfrac{7}{2} \right]$ respectively.

[3.79, 0.12, 5.79 μ_N]

8. Consider two masses m bound together with an equilibrium separation R. Estimate the order of magnitude of the lowest rotational frequency of the system and compare it with the vibrational frequency due to a vibrational amplitude βR.

$$\left[\frac{\omega_v}{\omega_r} \approx \frac{1}{\beta^2} \right]$$

9. Use the empirical equation representing the measured nuclear charge densities,

$$\rho(r) = \frac{\rho(0)}{1 + e^{\frac{(r-a)}{b}}}$$

Where $a = 1.07 \ A^{\frac{1}{3}} \times 10^{-15}$ m, $b = 0.55 \times 10^{-15}$ m, determine the distance in which the nuclear charge density fall from 90 % to 10 % of their initial values.

[2.4 F = 2.4×10^{-15} m]

10. Use the shell model to predict for the ground state of $^{17}_{8}$O (a) the spin; (b) parity; (c) sign of the magnetic dipole moment; sign of the electric dipole moment.

[(a) $\dfrac{5}{2}$, (b) even (c) negative, (d) zero]

11. The rotational band based on the state $K = \dfrac{7^-}{2}$ is known in $^{179}_{74}W$. The $\dfrac{9^-}{2}$ state is at an excitation of 120 keV; predict the excitation of 120 keV; predict the excitations of $\dfrac{11^-}{2}$, $\dfrac{13^-}{2}$ and $\dfrac{15^-}{2}$ levels.

[267 keV, 440 keV, 640 keV]

12. Taking the nuclear density in the shell model to have the radial dependence given by Woods-Saxon expression,

$$\rho = \frac{\rho_0}{1 + \exp\left(\dfrac{r - R}{a}\right)}$$

obtain an expression for the spin-orbit potential V_{So}.

13. (a) The total binding energies of the nuclei $^{15}_8O$, $^{16}_8O$ and $^{17}_8O$ are 111.95 MeV, 127.62 MeV and 131.76 MeV respectively. Deduce the energies of the last occupied state and of the first unoccupied state of neutrons in $^{16}_8O$.

   (b) The total binding energy of $^{17}_9F$ is 128.22 MeV. Deduce the energy of the first unoccupied proton state in $^{16}_8O$. Why is it different from the energy of the corresponding neutron state? Account for the order of magnitude of the difference.

   (c) The first excited state $\left(J^\pi = \dfrac{7^-}{2}\right)$ of $^{17}_8O$ has an excitation energy of 0.87 MeV. The corresponding state in $^{17}_9F$ has an excitation energy of 0.5 MeV. Suggest a reason for the difference in excitation energy.   [(a) −15.67 MeV; (b) −4.14 MeV; (c) −0.60 MeV]

14. The nucleus $^{235}_{92}U$ $\left(J^\pi = \dfrac{7^-}{2}\right)$ would have closed subshell and partically filled $l = 4$ neutron subshell if the spin-orbit interation were neglected. In this approximation use the rain water model to predict the observed quadrupole moment of $^{235}_{92}U$ and compare with the measured value of $+4.1 \times 10^{-28}$ m². What value would the single particle model predict?   [−25.06]

15. The observed quadrupole moment of $^{177}_{71}Lu$ is 3.33 barns and its ground state has $J^\pi = \dfrac{7^+}{2}$. Calculate the deformation parameter d.   [0.277]

SHORT-QUESTION ANSWERS

1. Can a single nuclear model explain all the properties?

Ans. No, each model can provide satisfactory explanations of certain properties of nuclei in their ground states.

2. What informations are provided about the nucleus by liquid drop model?

Ans. The masses and stability of nuclei are understood using the liquid drop model which accounts for the gross features of the interactions between nucleons with nuclei.

3. What is single-particle shell model?

Ans. The approximate independent particle motion of nucleons leads to the single-particle shell model in which nucleons move in a spherical potential well. This explains the magic numbers and the spins and parities of many nuclear ground states.

4. Why most nuclei are deformed? How this is explained?

Ans. Consideration of the effect of deformation on the kinetic and potential energies of nucleons explains why most nuclei are deformed. The Nilsson model provides a generalisation of the single-particle shell model to one in which nucleons move in a deformed potential well.

5. What is collective rotational motion of the nucleus?

Ans. Although many levels of nuclei can be understood using the simple shell model (spherical or deformed) several states are seen which are not easily described by this model and which are reminiscent of molecular rotational levels. They correspond to collective rotational motion of the nucleus.

6. What is pairing residual interaction between the nucleons?

Ans. Many of the low-lying levels in a rotational band are described in terms of a nucleus with a constant moment of inertia which is typically about $\frac{1}{4} - \frac{3}{4}$ of the classical value. This reduction reflects the effect of the pairing residual interaction between the nucleons.

7. What is the phenomenon of backbending?

Ans. At high spin the effect of pairing is reduced and larger moments of inertia and the phenomenon of backbending are observed.

8. Is vibrational levels are also seen in nuclei?

Ans. Yes, besides rotational levels, vibrational levels are also seen in nuclei. Prominent vibrational states in all nuclei are the 1^- giant dipole state and the 3^- octupole state.

9. What are enhanced electromagnetic multipole transition rates?

Ans. All the collective states correspond in the shell model to a superposition of many simple shell model configurations. This superposition leads to enhanced electromagnetic multipole transition rates. These correspond to the collective motion of many nucleons and understood in the shell model as arising from the coherent superposition of many single-particle terms in the transition matrix elements. Examples are seen in the giant dipole resonances and the enhanced $E2$ transition rates down a rotational band.

10. What is pairing interaction?

Ans. Pairing interaction is a residual nuclear interaction, i.e. a part of the total nuclear interaction experienced by the nucleons that is not described by the spherically symmetrical net potential $V(r)$ of the shell model, or by the spin-orbit interaction.

11. Is there any similarity between the nuclear theory due to Brueckner and others and Hartree theory the atom?

Ans. The nuclear theory, which is largely the work of Brueckner, is very similar to the Hartree theory of the atom in the sence that involves self-consistent calculations for a system of fermions, but the calculations are even more complicated because of the complicated nature of nuclear forces.

12. Why it is said that the liquid drop model only gives a good account of average behaviour of nuclei?

Ans. The liquid drop model gives a good account of the average behaviour of nuclei in regard to mass, or binding energy, i.e. the average behaviour of nuclei in regard to their stability.

13. What are magic numbers? How this situation is analogous to the unusual stability of the electron shells of noble gas atoms?

Ans. Nuclei with certain values of Z and/or N show significant departures from the average behaviour by being unusually stable. These values of Z and/or N are the magic numbers

$$Z \text{ and/or } N = 2, 8, 20, 28, 50, 82, 126$$

The situation is analogous to the unusual stability of the electron shells of noble gas atoms containing $Z = 2, 10, 18, 36, 54, 86$ electrons. However, in the nuclear case the indications are not as pronounced as in the atomic case, and it is necessary to consider several of them to demonstrate the "magic" character of the number quoted above.

14. Is there any similarity between Fermi gas model of the nucleus and the free electron gas model of the conduction electrons in a metal?

Ans. Fermi gas model is essentially the same as the free electron gas model of the conduction electrons in metal. It assumes that each nucleon of the nucleus moves in an attractive *net potential*, that represents the average effect of its interactions with other nucleons in the nucleus.

15. What is the achievement of Fermi gas model?

Ans. This model establishes the validity of treating the motion of the bound nucleons in terms of the independent motion of each nucleon in a net nuclear potential.

16. Is shell model predicts only the magic numbers and all their cosequences?

Ans. No, shell model can also predict the total angular momentum of the ground states of almost all the nuclei.

17. Is shell model not so successful in predicting dipole moments of nuclei?

Ans. Yes. It says that the magnetic dipole moment of an odd-A nucleus (i.e., even N and odd-Z, or odd N and even Z) should be due entirely to that of the single odd (unpaired) nucleon. The reason is that the magnetic dipole moments of the other nucleons would be expected to cancel out in pairs, if their total angular momenta do the same. The failure in the model is due to its assumption that the nuclear magnetic dipole moment is due entirely to the single odd nucleon. It is not true that all the other nucleons are always paired off with total angular momenta and magnetic dipole moments that strictly cancel.

18. What is collective model of the nucleus?

Ans. This model is based on the assumption that the nucleons move independently in a net nuclear potential produced by the core of filled subshells, as in the shell model. This potential is capable of undergoing deformations in shape.

19. What is the achievement of the collective model?

Ans. This model explains all the quadrupole moments that are incorrectly predicted by the shell model.

20. Besides collective rotations, what other types of behaviour display by nucleii?

Ans. Collective vibrations and clustering behaviour. Most spectacular example of collective vibrations is nuclear fission.

21. Is there any analogy between a super-conductor and an unfilled subshell of a nucleus?

Ans. Yes, there is an analogy between the behaviour of two electrons moving in opposite directions with antiparallel spins in a Cooper pair of a super-conductor, and two neutrons or two protons always moving in opposite directions in an unfilled subshell of a nucleus with spins that, because of the nuclear pairing interaction, are also anti-parallel. Moreover, in both the cases the behaviour of a pair of interacting particles influences, and is influenced by, the behaviour of the other particles in the system, which move collectively.

OBJECTIVE QUESTIONS

1. Nuclear magnetic moments are of the order of
   (a) Nuclear magnetion $\left(\dfrac{e\hbar}{2m_p}\right)$
   (b) Bohr magneton $\left(\dfrac{e\hbar}{2m_e}\right)$
   (c) Plack's constant ($\hbar$)
   (d) None of the above is correct

2. The mass formula for the liquid drop model of nucleus is called
   (a) Einstein energy mass-relation
   (b) semiempirical mass formula
   (c) Bohr quantum condition
   (d) velocity-mass relation

3. Pairing energy term appearing in the semi-empirical mass formula for the liquid drop model depends only on
   (a) atomic mass of the nucleus $M(A, Z)$
   (b) only on mass number A
   (c) only on atomic number Z
   (d) none of the above

4. As the energy of the nucleon in a nucleus increases, the depth of the net nuclear potential, V_0
   (a) rapidly increases
   (b) rapidly decreases
   (c) slowly decreases
   (d) increases exponentially

5. The dependence of the depth of the net nuclear potential V_0 on the difference between the number Z of protons and number N of neutrons that the nucleus contains is a result of
   (a) Coulomb energy of the charged nucleus
   (b) Bohr quantum condition
   (c) Heisenberg's uncertainty relation
   (d) Pauli exclusion principle

6. Shell model of the nucleus is based upon
   (a) spherically symmetric potential
   (b) ellipsoidal symmetric potential
   (c) Bohr correspondence principle
   (d) none of the above

7. In a shell model it is assumed that the constitutent parts of a nucleus move
   (a) independently
   (b) collectively
   (c) opposite to each other
   (d) none of the above

8. The collective model combines certain features of the
   (a) shell and liquid drop models
   (b) shell model and Pauli exclusion principle
   (c) liquid drop model and Pauli exclusion principle
   (d) liquid drop model and Bohr atom model

9. A nuclear property which can be explained quite well in terms of the collective model is the
   (a) electric quadrupole moments only
   (b) magnetic dipole moments only
   (c) magnetic dipole moments and electric quadrupole moments
   (d) none of the above

10. Theory used in collective model is
    (a) Schrodinger equations solved for net nuclear potential
    (b) Schrodinger equation solved for non-spherical net nuclear potential
    (c) classical (asymmetry and pairing terms introduced with no justification)
    (d) quantum statistics of Fermi gas of nucleons

11. Liquid drop model predict
    (a) depth of net nuclear potential asymmetry term
    (b) magnetic numbers, nuclear spins, nuclear parities, pairing term
    (c) electric quadrupole moment
    (d) accurate average masses and binding energies through semi-empirical mass formula (4)
12. Besides collective rotations, nuclei also display
    (a) only collective vibrations
    (b) clustering behaviour
    (c) collective vibrations and clustering behaviour
    (d) none of the above

──── ANSWERS ────

1. (a)	**2.** (b)	**3.** (b)	**4.** (c)	**5.** (d)
6. (a)	**7.** (a)	**8.** (a)	**9.** (c)	**10.** (b)
11. (d)	**12.** (c)			

6

NUCLEAR REACTIONS
(INTERACTIONS)

6.1 INTRODUCTION

The study of nuclear reactions provides considerable information on the structure of nuclei as well as on the nature of their interaction. Interest in the nature of nuclear reactions is two fold, first because the dynamical process resulting in scattering or transmutation is a study in itself and secondly because such properties provide most of the tabulated information on nuclear properties. Progress in accelarator and detector technology has led to extensive and detailed knowledge of the level systems of a great many nuclei, a knowledge which often extends far beyond the limit of credible theoretical prediction. Naturally these two aspects of reaction studies are closely linked because it is necessary to understand reaction dynamics in order to extract reliable values for level parameters. In this chapter general features of nuclear reactions are described. At higher incident energies direct reactions become important. For heavy-ion direct reactions new features are seen reflecting the more classical behaviour of ions: selectivity arising through kinematic matching and characterstic bell-shaped angular distributions. At energies above the Coulomb barrier, as well as compound nucleus (fusion) reactions, deep-inelastic reactions are also observed, with sbstantial cross-sections for very heavy ions. In these reactions there is considerable transfer of the incident energy to internal excitation of the product nuclei. For even higher energies, a new phase of nuclear matter, a quark-gluon plasma, is predicted to be found in reactions with relativistic heavy ions, and it is in this area that nuclear and particle physics overlap.

A nuclear reaction is a process which occurs when a nuclear particle (e.g., a nucleon, an elementary particle or a nucleus) comes into close contact with another during which exchange of energy and momentum takes place. The final products of the nuclear reaction are again some nuclear particle or particles which leave the point of contact (reaction site) in different directions.

The changes produced in a nuclear reaction usually involve strong nuclear force. However, purely electromagnetic effects (e.g. Coulomb scattering) or processes involving weak interactions, e.g. β-decay are usually out of the purview of nuclear reactions. But the changes of nuclear states under the influence of electromagnetic interactions falls under the purview of nuclear reactions. In general, one can represent a nuclear reaction by an equation of the form

$$^A_Z X + x \rightarrow ^{A'}_{Z'} Y + y + Q \qquad \dots (6.1)$$

or simply as $A_X(x, y) A'_Y$.

Here X is the target nucleus which is bombarded by the projectile particle x. The compound nucleus formed due to this process, breaks up almost immediately by ejecting particle y, leaving a residual nucleus Y. While writing the nuclear reaction equation, atomic numbers (Z) of the atoms are often omitted as the chemical symbol of atoms indicates it. In many cases, the projectiles x and the emitted particle y are light nuclei such as protons (p), neutrons (n), deutrons ($d = ^2_1 H$), α-particles ($\alpha = ^4_2 He$), γ-rays (γ or photon $h\nu$), etc. Generally these symbols are used in nuclear reaction equations. Q is called the Q-value of reaction which is the difference of the energies in the outgoing and entrance channel of the reaction. When Q is *negative*, the reaction is an *endothermic* reaction and for Q *positive*, the reaction is called as an *exothermic reaction*.

6.2 TYPES OF NUCLEAR REACTIONS

Prior to considering the quantitative aspects of nuclear reactions, let us first consider some of the important mechanisms by which nuclear reactions takes place. One can conveniently classify them as follows:

(i) **Elastic Scattering** The ejected particle y is the same as the projectile x. The ejected particle y leaves the target nucleus without energy loss, but in general with altered direction of motion, i.e. the residual nucleus Y is the same as the target nucleus X and is left in the same state (ground state) as the latter. This process can be represented as $X(x, x)X$. A good example of this process is the scattering of α-particles by gold nucleus.

$$^4_2 He + ^{197}_{79} Au \rightarrow ^{197}_{79} Au + ^4_2 He$$

(ii) **Inelastic Scattering** In this type of reaction y is the same as x, but it has different energy and angular momentum. Obviously, the residual nucleus $Y(\equiv X)$ is left in an excited state. This process can be represented by $X(x, y)X^*$, where $*$ on X indicates that this is an excited state of X. A familiar example is

$$^7_3 Li + ^1_1 H \rightarrow (^7_3 Li)^* + ^1_1 H$$

In this example, the excess energy is latter radiated away in the form of γ-quantum.

(iii) **Radiative Capture** The projectile x is absorbed by the target nucleus X to form the excited compound nucleus (C^*). The compound nucleus subsequently return to the ground state by the emission of one or more γ-ray quanta. One can represent this process as $X(x, y)Y^*$, ($Y = C$). A familiar example of radiative capture is

$$^{26}_{12} Mg + ^1_1 H \rightarrow (^{27}_{13} Al)^* \rightarrow ^{27}_{13} Al + \gamma$$

(iv) Disintegration Process one can represent this as $X(x, y)Y$, where X, x, Y and y are all different either in Z or in A or in both. The first nuclear transmutation observed by Rutherford, $^{14}N(\alpha, p)\ ^{17}O$, i.e.

$$^{14}_{7}N + {}^{4}_{2}He \rightarrow {}^{17}_{8}O + {}^{1}_{1}H$$

is an familiar example of this process.

(v) Photo-disintegration The target nucleus is bombarded with very high energy γ-rays, so that it is raised to an excited state by the absorption of γ-rays. If the energy of γ-rays is high enough, the compound nucleus may liberate one or two particles. The reaction can be expresses as $X(\gamma, y)Y$. An example is

$$^{2}_{1}H + \gamma \rightarrow {}^{1}_{1}H + {}^{1}_{0}n.$$

(vi) Direct Reactions A collision of an incident particle with the target nucleus may immediately pull one of the nucleons out of the target nucleus by the so called direct '*pice up*' or *stripping reaction*. In the inverse process, a bombarding particle composed of more than one nucleon may lose one of them to the target by stripping reaction. This can be expressed as $X(x, y)Y$. An example is

$$^{63}_{29}Cu + {}^{2}_{1}D \rightarrow {}^{64}_{29}Cu + {}^{1}_{1}H$$

(vii) Many body reactions and Spallation Reactions When the kinetic energy of the incident particle is quite high, two or particles can come out of compound nucleus. If y_1, y_2, y_3, ... etc. represent the product particles, one can express the reaction as $X(x, y_1, y_2, y_3 ...)Y$. Few examples of such reactions are $^{16}O(p, 2p)^{15}N$, $^{16}O(p, pn)^{15}O$, $^{16}O(p, 3p)^{14}C$ etc. When the energy of the incident particle x is very high, after capture, a heavy nucleus has sufficient energy for the ejection of several particles (3 to 20). Such a reaction is known as spallation reaction e.g. the nuclear fission. The reaction is

$$^{235}_{92}U + {}^{1}_{0}n \rightarrow {}^{98}_{40}Zr + {}^{136}_{52}Te + 2\ {}_{0}n^{1}$$

(viii) Spontaneous Decay β-and α-decay processes may be considered as spontaneous decay processes. The total energy of the system is not under control of the experimenter.

(ix) Heavy ion reactions The target nucleus is bombarded by projectiles heavier than α-particles. Various types of products may be produced. Usually, the reaction takes place at fairly high energies (~ 100 MeV) of the projectile. Examples of these reactions are ^{10}B (^{16}O, ^{4}He) ^{22}Na, $^{14}N(^{14}N, {}^{15}N)\ ^{13}N$ etc.

(x) High Energy Reactions In the energy range of about 150 MeV, spallation process merges into new kinds of reactions in which new kinds of particles, e.g. mesons, strange particles are produced along with protons and neutrons. Familiar examples are:

$$p + p \rightarrow p + n + \pi^{+}$$
$$\pi^{-} + p \rightarrow \pi^{\circ} + n$$
$$p + \pi^{\circ} \rightarrow K^{\circ} + \wedge^{\circ}$$

(xi) Nuclear Fission When target nucleus is heavy nucleus and y, Y have comparable masses, the reaction is known as nuclear fission. An example is $^{238}U\ (n, f)$

$$_{92}U + n \rightarrow {}_{92}U^{*} \rightarrow {}_{56}Ba + {}_{36}Kr + Q$$

6.3 CONSERVATION LAWS

A nuclear reaction usually follow certain conservation laws:

(i) Conservation of Mass Number   The total number of neutrons and protons in the nuclei taking part in a nuclear reaction remains unchanges after the reaction. Obviously, in a reaction $X(x, y)Y$, the sum of mass numbers of X and x must be equal to the sum of mass numbers of Y and y, i.e.

$$A + a = A' + a' \qquad \qquad \text{... (6.1)}$$

In the general case of reactions involving elementary particles, one can express the law as, the number of heavy particles (baryons) remains unchanged in a reaction.

(ii) Conservation of Atomic Number   The total number of protons of nuclei taking part in a reaction remain unchanged after the reaction, i.e. the sum of atomic numbers of X and x is equal to the sum of atomic numbers of Y and y: Thus

$$Z + z = Z' + z' \qquad \qquad \text{... (6.2)}$$

In view of (6.1) and (6.2) conservation laws, *neutron number* N also remains unchanged in a nuclear reaction

(iii) Conservation of Energy and *Q*-value of a reaction   In any nuclear reaction the total energy of products, including both mass energy (in accordance with mass-energy equivalence principle) and kinetic energy of the particles plus the energy involved must be equal to the mass energy of the initial ingredients plus the kinetic energy of the bombarding particle.

Let M_X, M_x, M_Y and M_y are the rest masses of different atoms in Equation (6.1), their rest mass energies are $M_X c^2, M_x c^2, M_Y c^2$, and $M_y c^2$ respectively. Denoting the kinetic energy by E, one can write

$$M_X c^2 + M_x c^2 + E_X + E_x = M_Y c^2 + M_y c^2 + E_Y + E_y \qquad \qquad \text{... (6.3)}$$

Equation (6.3) is often expressed without the factor c^2 in mass-energy terms. This clearly means that the masses are expressed in energy units.

We may note that though nuclear masses are involved in a nuclear reaction, it is still possible for one to write the energy-balance equation in terms of the atomic masses. This is possible because the electronic masses cancel out on both sides of the equation and one can neglect electronic binding energies.

We may note that at relatively lower energies, the kinetic energy is given by the non-relativistic equations: $E = \dfrac{M v^2}{2}$. However, in the case of elementary particle reactions, usually very high energy is involved and therefore, one must use the relativistic expression $E = \sqrt{p^2 c^2 + M_0^2 c^4} - M_0 c^2$, where M_0 is the rest mass of the particle and $p = \dfrac{M_0 v}{\sqrt{1 - \beta^2}}$ is the linear momentum.

(iv) Conservation of Linear Momentum   The total linear momentum of the products must be equal to the linear momentum of the bombarding particle. The target nucleus is normally considered to be at rest. If $\mathbf{P}_X$, $\mathbf{P}_x$, $\mathbf{P}_Y$ and $\mathbf{P}_y$ represent the momentum vectors of the different nuclei taking part in the nuclear reaction, then conservation of linear momentum gives

$$\mathbf{P}_X + \mathbf{P}_x = \mathbf{P}_Y + \mathbf{P}_y \qquad \dots (6.4)$$

Equation (6.4) holds in an arbitrary frame of reference.

In the laboratory frame (*L*-system), the target nucleus is at rest, i.e. $\mathbf{P}_X = 0$, we have

$$\mathbf{P}_x = \mathbf{P}_Y + \mathbf{P}_y \qquad \dots (6.5)$$

In the centre of mass frame (*C*-system) the two particles before collision is at rest, we have $\mathbf{P}_X + \mathbf{P}_x = 0$, which gives $\mathbf{P}_Y + \mathbf{P}_y = 0$, i.e. centre of mass of the product particles is also at rest in *C*-system.

(v) **Conservation of Angular Momentum** The total angular momentum l comprising the vector sum of the intrinsic angular momentum s and relative orbital momentum l of the products must be equal to the total angular momentum of the initial particles. Let I_X, I_x, I_Y and I_y denote the nuclear spins (total angular momentum of the nuclei X, x, Y and y of a nuclear reaction $X + x \rightarrow Y + y$. Let, in the initial state l_x represent the relative angular momentum of X and x, and l_y in the final state of Y and y. Then according to this conservation law, we have

$$I_X + I_x + l_x = I_Y + I_y + l_y \qquad \dots (6.6)$$

We may note that the application of the law of conservation of the angular momentum taking into account the well-known quantum mechanical properties of the former leads to certain selection rules.

(vi) **Conservation of Nucleons** The nucleons in any nuclear reaction can neither be created nor can be destroyed, i.e. the total number nucleons minus the number of anti-nucleons in the universe remains constant.

(vii) **Conservation of spin** This law asserts that the spin character of closed system cannot change, i.e. the statistics remains same as that existed before the reaction.

(ix) **Conservation of Parity** In this chapter we will remain confined to the nuclear reactions taking place due to strong interaction. In these types of reactions parity is conserved, i.e. the parity π_i before the reaction must be equal to the parity, π_f after the reaction. In strong interaction nuclear reactions, no violation of parity has been observed so far. However, parity does not appear to be conserved in weak interaction.

Let us denote the intrinsic parities of the nuclei taking part in the reaction by π_X, π_x, π_Y and π_y, one obtain for the initial and final states of the reaction

$$\pi_i = \pi_X \pi_x (-1)^{l_x}$$
$$\pi_f = \pi_Y \pi_y (-1)^{l_y}$$

According to law of conservation of parity, we have

$$\pi_X \pi_x (-1)^{l_x} = \pi_Y \pi_y (-1)^{l_x}$$

We may note that except in the cases of elementary particle reactions, the intrinsic parity need not be taken into account, and hence we hence

$$(-1)^{l_x} = (-1)^{l_y}$$

Parity conservation results in certain selection rules. This limits the possible nuclear reactions that may occur starting from a given initial state *i*, e.g. in the case of elastic scattering *l* can change only by an even integer.

(x) Conservation of Isotopic Spin   Let us denote the isotopic spin vectors for the initial and final states by T_i and T_f respectively. One can express the invariance of the nuclear Hamiltonian function towards the charge character of the nucleons analytically as an invariance towards rotational shifts of the axes in isotopic spin space, and there should correspondingly exist a conservation law for the isotopic spin of a nuclear system. For the reaction $X + x = Y + y$, we have $T_i = T_X + T_x$ and $T_f = T_Y + T_y$. Thus for isotopic spin conservation

$$T_X + T_x = T_Y + T_y \qquad \qquad \dots (6.7)$$

We may note that isotopic spin is a characteristic of the nuclear level. Thus, one use (6.7) to identify the levels of the nuclei produced in the reaction. In particular, if $T_x = T_y = 0$, e.g. for the deutron or the α-particle, one gets $T_X = T_Y$.

Isotopic spin conservation law must be obeyed in reactions of the type (d, α), (d, d), (α, d), (α, α), etc. Moreover, the rule has been verified for the nuclei ^{6}Li, ^{10}B, ^{14}N, etc. for $T = 0$ ground states. Figure 6.1(a) illustrates the concept of isospin.

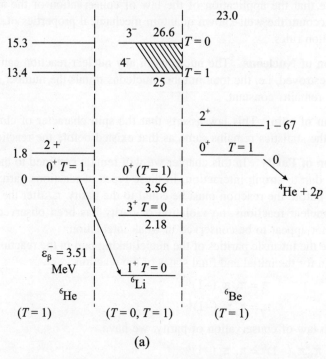

Fig. 6.1(a) The energy level diagram of $^6_2$He$_4$, $^6_3$Li$_3$ and $^6_4$Be$_2$ illustrating the concept of isotopic spin *T*

Figure 6.1(b) shows the decay of ^{12}N and ^{12}B to ^{12}C with $\Delta T = 0, +1$. Figure 6.1(c) shows the similarity of excited states of two isotopic spin doublets $^5_3$Li and $^5_2$He.

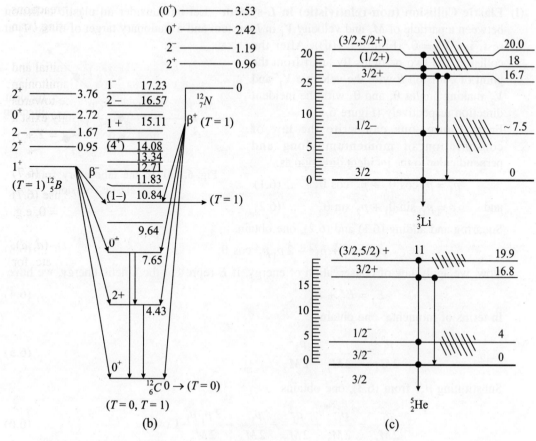

Fig. 6.1(b) The decay ^{12}N and ^{12}B to ^{12}C with $\Delta T = 0, +1$ **(c)** Similarity of excited states of two iso-spin doubles $^{5}_{3}Li$ and $^{5}_{2}He$.

6.4 COLLISION BETWEEN SUBATOMIC PARTICLES

We shall now find the energies and momenta of the particles produced in a nuclear reaction from simple kinematical considerations.

When a nuclear particle (projectile) is incident upon a nuclei (usually at rest), the particles may either suffer *elastic collision* or new particles may be produced as reaction products. No change in the internal states of colliding particles takes place in an elastic collision. Obviously, for analysing the kinematics of collision process, it is enough to apply the law of conservation of kinetic energy and momentum. However, in nuclear reactions, including *inelastic collision*, one will have to take into account the change in internal state of the particles while appalying the law of conservation of energy.

One can analyse the collision between a projectile and target nucleus, keeping in view that the observer is at rest in the laboratory frame of reference of *L*-system. Alternatively, one can also analyse the collision from the point of view of an observer at rest with respect to the centre of mass of colliding particles, know as *C*-system

(i) Elastic Collision (non-relativistic) in *L*-system   Let us consider an elastic collision between a particle of M_1 and velocity V_1 in *L*-system and a stationary target of mass M_2 at rest, i.e. $V_2 = 0$. (Figure 6.1(d)). After the collision, the two particles fly apart from the point of collision B with the velocitie V_1' and V_2' making angles θ_1 and θ_2 with the incident direction respectively (Figure 6.1).

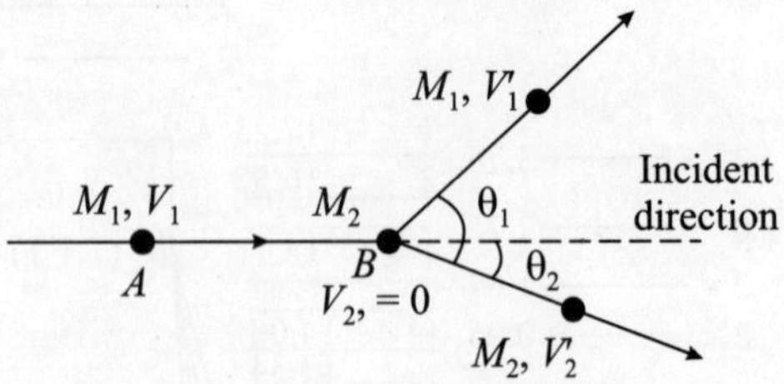

Fig. 6.1(d) In the laboratory frame the elastic collision.

From figure, one can write the law of conservation of momentum along and perpendicular to the incident direction as:

$$p_1 = p_1' \cos \theta_1 + p_2' \cos \theta_2 \quad \text{... (6.1)}$$

and

$$o = p_1' \sin \theta_1 + p_2' \sin \theta_2 \quad \text{... (6.2)}$$

Squaring and adding (6.1) and (6.2), one obtains

$$p_2' = p_1^2 + p_1'^2 - 2 p_1 p_1' \cos \theta_1 \qquad \text{... (6.3)}$$

Now, we apply law of conservation of energy. If E represent the kinetic energy, we have

$$E_1 = E_1' + E_2' \qquad \text{... (6.4)}$$

In terms of momenta, one obtains

$$\frac{p_1^2}{2M_1} = \frac{p_1'^2}{2M_1} + \frac{p_2'^2}{2M_2} \qquad \text{... (6.5)}$$

Substituting $p_2'^2$ from (6.3), one obtains

$$\frac{p_1^2}{2M_1} = \frac{p_1'^2}{2M_1} + \frac{p_1^2}{2M_2} + \frac{p_1'^2}{2M_2} - \frac{2 p_1 p_1'}{2M_2} \cos \theta_1 \qquad \text{... (6.6)}$$

or

$$p_1'^2 \left(1 + \frac{M_2}{M_1} \right) - 2 p_1 p_1' \cos \theta_1 + p_1^2 \left(1 + \frac{M_2}{M_1} \right) = 0 \qquad \text{... (6.7)}$$

Now, putting $r = \dfrac{M_2}{M_1}$, Equation (6.6) takes the form

$$p_1'^2 (1 + r) - 2 \sqrt{E_1 E_1'} \cos \theta_1 + E_1 (1 - r) = 0$$

or

$$\frac{E_1'}{E_1} (1 + r) - 2 \sqrt{\frac{E_1'}{E_1}} \cos \theta_1 + (1 - r) = 0 \qquad \text{... (6.8)}$$

We note that Equation (6.8) is quadratic in $\sqrt{\dfrac{E_1'}{E_1}}$ and therefore the solution of (6.8) is

$$\sqrt{\frac{E_1'}{E_1}} = \frac{\cos \theta_1 \pm \sqrt{\cos^2 \theta_1 + (r^2 - 1)}}{r + 1} \qquad \text{... (6.9)}$$

In case if the target particle is heavier, then $r > 1$. Since $\sqrt{\dfrac{E_1'}{E_1}} = \dfrac{p_1'}{p_1}$, one can choose the

sign in (6.9) such that $\dfrac{p_1'}{p_1} > 0$. Obviously, then we have to take + sign before the square

root symbol. Squaring (6.9), one obtains

$$\frac{E_1'}{E_1} = \frac{2\cos^2\theta_1 + (r^2 - 1) + 2\cos\theta_1 \sqrt{\cos^2\theta_1 + r^2 - 1}}{(r+1)^2} \quad \ldots (6.10)$$

For $\theta_1 = \pi$, the energy received is maximum. For $r \gg 1$, (i.e., $M_2 \gg M_1$), one obtains

$$\frac{E_1 - E_1'}{E_1} = \frac{2(r+1 - \cos^2\theta_1) - 2\cos\theta_1 \sqrt{\cos^2\theta_1 + r^2 - 1}}{(r+1)^2} \quad \ldots (6.11)$$

We may note that in this case all values of θ_1 are possible. This type of scattering is observed when β-particles are scattered by nuclei as they pass through matter. We may note that the

energy received by the nucleus in such collisions is negligibly small, i.e. $r = \dfrac{M_{nucl}}{m_e} \gg 1$.

When $r < 1$, e.g. in the collision of a nuclear particle with an electron, for real solution of (6.9), the condition is

$$\cos^2\theta_1 \geq 1 - r^2 \quad \ldots (6.12)$$

We may note that both + and – signs are possible in (6.9). Thus $\sin\theta_1 \leq r$, or $\theta_1 < \dfrac{\pi}{2}$ since

$r < 1$.

For $r \ll 1$, $\theta_1 \to 0$. This means that a heavy projectile scattered by a very light target, e.g. an electron, goes almost undeviated.

From Equation (6.9), the energy given to the target nucleus for $\theta = 0$ is

$$\frac{E_1 - E_1'}{E_1} = \frac{2r \pm 2\sqrt{r^2}}{(r+1)^2} = \frac{4r}{(r+1)^2} \quad \text{or} \quad 0 \quad \ldots (6.13)$$

Obviously, the energy given to the target nucleus is

$$E_1 - E_1' = 4rE_1 = \frac{4M_2 E_1}{M_1} \ll E_1 \quad \ldots (6.14)$$

Clearly, the energy loss by the incident particles in a collision is small compared to the incident energy.

In Equation (6.9), + and – signs before the square root show that for each θ_1 there are two possible values of E_1' except for $\sin\theta_1 = r$. In this case, we have $\cos^2\theta_1 = 1 - r^2$, and thus the term under square root is zero.

(ii) **Elastic collision (non-relativistic) in C-system** Now we consider the collision between two particles w.r.t. an observer at rest relative to the centre of mass C of the particles as shown in Figure. 6.2. Let us denote the velocity and momenta in *C*-system by the capital letters V and P and those in *L*-system by small letters v and p respectively. Let the energies and angle of scattering in *C*-system be denoted by E and θ and in *L*-system by ε and ϕ respectively.

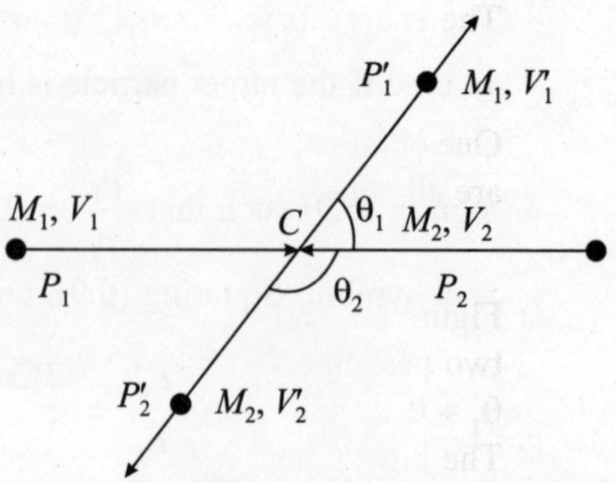

Prior to the collision, the particle M_2 is at rest in *L*-system ($v_2 = 0$). The velocity of the centre of mass in the *L*-system is

$$v_c = \frac{M_1 v_1 + M_2 v_2}{M_1 + M_2} = \frac{M_1 v_1}{M_1 + M_2} \qquad \ldots (6.15)$$

Fig. 6.2 Momentum diagram for collision between two particle in C.M. frame.

Thus the velocities of M_1 and M_2 in the *C*-system prior to collision are

$$V_1 = v_1 - v_c = \frac{M_2 v_1}{M_1 + M_2} \qquad \ldots (6.16)$$

and

$$V_2 = v_2 - v_c = \frac{M_1 v_1}{M_1 + M_2} \qquad \ldots (6.17)$$

respectively. The corresponding momenta in the *C*-system are

$$P_1 = M_1 V_1 = \frac{M_1 M_2 v_1}{M_1 + M_2} = \mu v_1 \qquad \ldots (6.18)$$

and

$$P_2 = M_2 V_2 = -\frac{M_1 M_2 v_1}{M_1 + M_2} = -\mu v_1 \qquad \ldots (6.19)$$

where $\mu = \dfrac{M_1 M_2}{(M_1 + M_2)}$ is the *reduced mass.*

Clearly, the two particles have equal and opposite momenta before collision. This means that their total momentum $P_1 + P_2 = 0$. In accordance with the conservation of momentum, the total momentum of these two particles after the collision is also zero, i.e.

$$P_1' + P_2' = P_1 + P_2 = 0$$

Where primes indicate momentum after collision. Now, the sums of the kinetic energies before and after the collision are

$$E_1 + E_2 = \frac{P_1^2}{2M_1} + \frac{P_2^2}{2M_2} = \frac{P^2}{2\mu} \quad \text{(Before collision)} \qquad \ldots (6.20)$$

$$E_1' + E_2' = \frac{P_1'^2}{2M_1} + \frac{P_2'^2}{2M_2} = \frac{P'^2}{2\mu} \qquad \ldots (6.21)$$

Where

$$|P_1| = |P_2| = |P| \quad \text{and} \quad |P_1'| = |P_2'| = |P'| \qquad \ldots (6.21a)$$

The energy conservation demands that

$$E_1' + E_2' = E_1 + E_2$$

One obtains $P' = P$, i.e. the magnitude of momenta of the particles before and after collision are all equal. This means

$$P_1 = P_2 = P_1' = P_2' \qquad \qquad ...(6.22)$$

Figure 6.2 shows the momentum diagram of the two particles. As shown in the figure the two particles fly apart from the point of collision with equal and opposite momenta. Thus $\theta_1 + \theta_2 = \pi$.

The kinetic energy of the centre of mass (C) is given by

$$E_c = \frac{1}{2}(M_1 + M_2)V_c^2 = \frac{M_1}{M_1 + M_2}\varepsilon_1 \qquad \qquad ...(6.23)$$

Where $\varepsilon_1 = \frac{1}{2} M_1 \upsilon_1^2$ is the kinetic energy of the incident particle in the L-system. This means that the energy E_c given by (6.23) is not available for the production of any inelastic effect, i.e. reaction. Clearly, the total amount of energy available for this purpose is

$$\varepsilon_1 - E_c = \varepsilon_1 - \frac{M_1}{M_1 + M_2}\varepsilon_1 = \frac{M_2}{M_1 + M_2}\varepsilon_1 = \frac{1}{2}\mu\upsilon_1^2 \qquad \qquad ...(6.24)$$

Obviously, there is no change in the kinetic energies and momenta of the particles after the collision in the C-system.

Now, we find the relationships between the angles of scattering in C and L systems.

The velocity diagram corresponding to the momentum diagram of Figure 6.2 is shown in Figure 6.3. In an elastic collision in the C-system, the magnitudes of momenta remain unchanged. The magnitudes of velocities also remain unchanged. Thus $V_1' = V_1$, $V_2' = V_2$ (Figure 6.3).

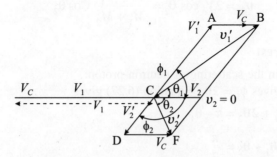

Fig. 6.3 Vector diagram exhibiting the relationship between scattering angles θ and ϕ in the C and L systems.

After collision the velocities υ_1' and υ_2' of the two particles can then be easily obtained by the vector addition of the centre of mass velocity V_C (which remains unchanged after collision) with V_1' and V_2' respectively, i.e.

$$\mathbf{v}_1' = \mathbf{V}_1' + \upsilon_c \quad \text{and} \quad \mathbf{v}_2' = \mathbf{V}_1' + \upsilon_c$$

In Figure 6.3, these are shown by *CB* and *CF* respectively. Referring to triangle *CAB*, one can write

$$\frac{V_c}{\text{Sin}(\phi_1 - \theta_1)} = \frac{V_1'}{\text{Sin}\,\theta_1} = \frac{V_1}{\text{Sin}\,\theta_1}$$

This gives the angle of scattering in the *L*-system as

$$\phi_1 = \theta_1 + \sin^{-1}\left(\frac{M_1}{M_2}\,\text{Sin}\,\theta_1\right) \qquad \dots (6.25)$$

for $\qquad M_1 = M_2$, one obtains $\phi_1 = \theta_1$

One can obtain the similar relationship between θ_2 and ϕ_2. Now, considering the $\triangle CDF$, one obtains

$$\frac{V_2'}{\text{Sin}\,\theta_2} = \frac{V_c}{\text{Sin}(\phi_1 - \theta_2)}$$

or since $\qquad V_2' = V_c$, one obtains

$$\phi_2 = 2\,\theta_2 \qquad \dots (6.26)$$

Thus, $\qquad \phi_1 + \phi_2 = \theta_1 + 2\,\theta_2 + \sin^{-1}\left(\frac{M_1}{M_2}\,\text{Sin}\,\theta_1\right) = \pi$

$$\therefore \qquad \sin(\theta_1 + 2\,\theta_2) = \frac{M_1}{M_2}\,\sin\theta_1 \qquad \dots (6.27)$$

In the *L*-system, the velocity of the struck particle is

$$\upsilon_2' = 2V_c \cos\theta_2 = \frac{2M_1\upsilon_1}{M_1 + M_2}\,\text{Cos}\,\theta_2 \qquad \dots (6.28)$$

Special cases of interest

(a) For $M_1 = M_2$ as in the scattering of neutron-proton:
Equation (6.25) gives $\phi_1 = 2\theta_1$. Equation (6.27) gives

$$\theta_1 + 2\theta_2 = \pi - \theta_1$$

or $\qquad \theta_1 + \theta_2 = \dfrac{\pi}{2} \qquad \dots (6.29)$

Obviously, in the *L*-system, the angle between the paths of the two particles of equal mass after collision is always $\dfrac{\pi}{2}$.

(b) For $M_2 \gg M_1$, i.e. scattering of light particle, e.g. electron from a very heavy particle (a nucleus). One finds,

$$\theta_1 + 2\theta_2 = \pi$$

This means

$$\theta_2 = \frac{\pi}{2} - \frac{\theta_1}{2}$$

Further, Equation (6.28) gives $v_1' \ll v_1$, i.e. the struck particle gets very little energy.

(c) For $M_2 \ll M_1$, i.e. scattering of a very heavy particle (a nucleus) by a very light particle, say electron. One obtains from Equation (6.27),

$$\sin\theta_1 = \frac{M_2}{M_1} \sin(\theta_1 + 2\theta_2) \to 0$$

Hence $\theta_1 = 0$. This means the incident particle almost goes on undeviated after collision, e.g. there is energy loss by heavy charged particle in passing through matter.

Nonelastic Collisions

Nuclear reactions, e.g. pp', nn' falls under this category. In this type of nuclear reactions, the particles produced after collision are usually different from those before collision. If M_3 and M_4 be the masses of the two particles produced and E_3 and E_4 are their energies repectively, one can write the energy conservation equation for the reaction as

$$M_1 + M_2 + E_1 + E_2 = M_3 + M_4 + E_3 + E_4 \qquad \dots (6.30)$$

Here the masses are expressed in energy units (i.e. Mc^2). Now, writing

$$Q = M_1 + M_2 - M_3 - M_4$$

One obtains ($\because E_2 = 0$),

$$Q_1 + E_1 = E_3 + E_4 \qquad \dots (6.31)$$

In order to find the energies of reaction products, we will have to use Equation (6.31) along with momentum conservation equation.

6.5 ENERGY CONSIDERATIONS IN NUCLEAR REACTIONS

Energy is either evolved or absorbed in a nuclear reaction. Those nuclear reactions in which energy is *evolved* are called *exoergic* nuclear reactions, while those nuclear reactions in which energy is *absorbed* is called *endoergic*. Energy evolved or absorbed during a nuclear reaction is usually denoted by Q and called as Q-value or Q or the reaction. Thus,

$$Q = E_Y + E_y - E_X - E_x = E_Y + E_y - E_x \qquad \dots (6.1)$$

(when the target nucleus X is at rest)

Obviously, Q is equal to the net surplus (or deficit) of the energies of the reaction products $E_Y + E_y$ over the energy supplied, i.e. E_x.

Now, expressing the atomic masses in energy units, one can write

$$M_X + M_x + E_x = M_Y + M_y + E_Y + E_y$$

Using (6.1), one obtains

$$Q = M_X + M_x - M_Y - M_y \qquad \dots (6.2)$$

Expressing (6.2) in terms of binding energies of different nuclei, one finds

$$Q = B_Y + B_y - B_X - B_x \qquad \dots (6.3)$$

We know that $Q > 0$ for an exoergic reaction and $Q < 0$ for an endoergic reaction. Since there is a net deficit of energy in the latter case, one has to supply some energy to occur. Usually this comes from the kinetic energy E_x of the projectile.

Equation (6.2) reveals that for an exoergic reaction $M_X + M_x$ is greater than $M_Y + M_y$, while for an endoergic reaction $M_X + M_x$ is less than $M_Y + M_y$.

Threshold Energy

(i) Endoergic Reaction

Equation (6.1) reveals that the Q value of a nuclear reaction can be expressed in terms of the kinetic energies of the projectile (E_x) and of the product nuclei E_y and E_Y.

One can express E_Y in terms of E_x and E_y with the help of energy and momentum conservation laws. From Figure 6.4, one obtain from the law of conservation of momentum *along* and *perpendicular* to the direction of motion of the projective (since $p = \sqrt{2ME}$)

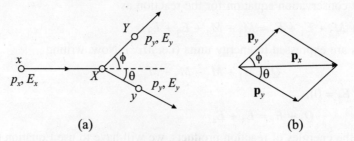

(a)   (b)

Fig. 6.4(a) Motion of x(projectile) and y and Y(product particles) in a nuclear reaction. (b) Momentum diagram.

$$\sqrt{2M_x E_x} = \sqrt{2M_y E_y}\ \cos\theta + \sqrt{2M_Y E_Y}\ \cos\phi \qquad \dots (6.4)$$

and

$$0 = \sqrt{2M_y E_y}\ \sin\theta - \sqrt{2M_Y E_Y}\ \sin\phi \qquad \dots (6.5)$$

Equation (6.1) gives the law of conservation of energy as

$$Q = E_Y + E_y - E_x \qquad \dots (6.5a)$$

Squaring and adding (6.4) and (6.5), one obtains

$$2\,M_Y E_Y = 2\,M_x E_x + 2\,M_y E_y - 4\,\sqrt{M_x M_y E_x E_y}\ \cos\theta$$

or

$$E_Y = \frac{M_x}{M_Y}E_x + \frac{M_y}{M_Y}E_y - \frac{2}{M_Y}\sqrt{M_x M_y E_x E_y}\ \cos\theta \qquad \dots (6.6)$$

Then from (6.1) and (6.6), one obtains

$$Q = E_Y \left(1 + \frac{M_y}{M_Y}\right) - E_x \left(1 - \frac{M_x}{M_Y}\right) - \frac{2}{M_Y} \sqrt{M_x M_y E_x E_y} \ \text{Cos}\ \theta \qquad \dots (6.7)$$

We see that (6.7) is quadratic in $\sqrt{Ey}$ $(= z)$. Thus, we can write

$$az^2 + bz + c = 0 \qquad \dots (6.8)$$

Here

$$a = \frac{M_y}{M_Y}, \quad b = -\left(\frac{2}{M_Y}\right)\sqrt{M_x M_y E_x} \ \cos\theta$$

and

$$c = -E_x\left(1 - \frac{M_x}{M_Y}\right) - Q \qquad \dots (6.9)$$

Writing the solution of (6.8) explicitly, one finds

$$\sqrt{E_y} = \frac{1}{M_Y - M_y}\left[(M_x M_y E_x)^{\frac{1}{2}} \text{Cos}\,\theta\right]$$

$$\pm \left[M_x M_y E_x \, \text{Cos}^2\,\theta + (M_Y + M_y)\{Q M_Y + E_x(M_Y - M_y)\}\right]^{\frac{1}{2}} \quad \dots (6.10)$$

Now, taking $Q' = -Q$, then for endoergic nuclear reactions $Q' > 0$ since $Q < 0$. Thus, in this case if $E_x = 0$, we have

$$b = 0 \quad \text{and} \quad c = -Q = Q' > 0$$

In this case the solution of z becomes

$$Z = \sqrt{E_y} = \frac{\sqrt{-4ac}}{2a} = \pm \frac{\sqrt{-Q'}}{a}$$

In this case, since both a and Q are positive, $Z = \sqrt{E_y}$ is imaginary. Clearly, the reaction is not possible. To initiate endoergic reaction, $E_x = E_{\min}$, a minimum energy is needed. Now, putting the term under square root sign in Equation (6.9) zero, one obtains for this case

$$b^2 - 4ac = 0$$

or

$$\frac{4}{M_Y^2}(M_x M_y M_{\min})\cos^2\theta = 4\left(1 + \frac{M_y}{M_Y}\right)\left[-Q - E_{\min}\left(1 - \frac{M_x}{M_Y}\right)\right]$$

This yields

$$E_{\min} = \frac{(M_y + M_Y)Q}{M_y + M_Y - M_x - \left(\dfrac{M_x M_y}{M_Y}\right)\text{Sin}^2\,\theta} \qquad \dots (6.11)$$

Since $Q < 0$, $E_{\min} > 0$. Using (6.2), one obtains

$$E_{\min} = \frac{-(M_y + M_Y)Q}{M_x - Q - \left(\dfrac{M_x M_y}{M_Y}\right)\sin^2\theta} \qquad \ldots (6.12)$$

From (6.12), we note that $E_{\min}$ depends on the angle θ at which the particle is emitted. For $\theta = 0$, y is emitted in the forward direction and $E_{\min}$ has the lowest value (E_{th}). E_{th} is known as the *threshold energy* for the endoergic reaction. From (6.12), one obtains

$$E_{\text{th}} = \frac{(M_y + M_Y)Q}{M_x - Q} \qquad \ldots (6.13)$$

We note that $Q \ll M_x$ and hence one can neglect Q in the denominator of (6.13). Further, replacing $M_y + M_Y$ by $M_x + M_X$ in the numerator of (6.13), one obtains

$$E_{\text{th}} = -Q\left(1 + \frac{M_x}{M_X}\right) \qquad \ldots (6.14)$$

Clearly, by measuring E_{th} (the minimum energy), at which an exoergic reaction is initiated, it is possible to determine Q-value of the nuclear reaction.

Reexamination of (6.10) reveals that under certain circumstances E_y will be a double valued function of the projectile energy E_x, i.e. there may be two values of E_y for a given E_x. We may note that this happens only for endoergic reactions. An example of double valued nature of E_y is revealed in the endoergic reaction $^3\text{H}(p, n)\,^3\text{He}$ which has $Q = -0.7638$ MeV. Figure 6.5 clearly shows the double values of neutron energy for this reaction. If the following condition is satisfied in (6.10), E_y is single valued. We have

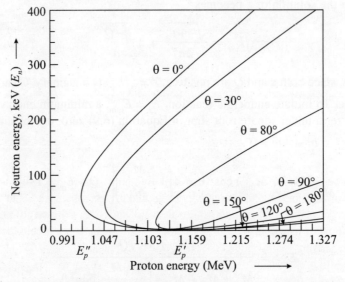

Fig. 6.5 For nuclear reaction $^3\text{H}(p, n)\,^3\text{He}$ E_n *vs* E_p graph. We can note the double valued of neutron energy.

$$Q\, M_y + E_x(M_y - M_x) \geq 0$$

or
$$E_x \geq -Q\, \frac{M_Y}{M_Y - M_x}$$

Obviously, there is a limiting energy of the projectile above which the emitted particle wll be single valued. For this, one obtains

$$E_x' = -Q\, \frac{M_Y}{M_Y - M_x} \qquad \qquad \qquad \dots (6.15)$$

For the example cited above, $E_x' = 1.145$ MeV. We further, note that for projectile energy greater than E_x', the product particle y can be emitted at all angles between $0°$ and a minimum angle θ_{max}, which can be easily found with the help of (6.10).

Endoergic Reaction   The reaction can occur for all values of E_x including $E_x = 0$. In the case of $E_\lambda = 0$, the incident momentum is zero and hence the sum of the momenta of the product particles must be zero, i.e. $\mathbf{p}_Y + \mathbf{p}_y = 0$. This clearly reveals that Y and y proceed in opposite directions and hence $\theta + \phi = \pi$. Moreover, in this case $Q = E_y + E_Y$.

In general, one finds that for $Q > 0$ only one value of E_y, i.e. the energy of the emitted particle for a given E_x and at a given angle of emission θ, is obtained. However, all values of θ are possible. Clearly, there is an energy distribution of the emitted particles between a maximum at $\theta = 0$ and a minimum at $\theta = \pi$. If we want to obtain a positive value of p_y, then in the solution of (6.7) given by (6.9), we have to chose plus sign only.

6.6   EXPERIMENTAL DETERMINATION OF Q-VALUE

By measuring the energies E_x, E_y and E_y accurately, one can determine the Q-value of reaction with the help of Equation (6.1) section 6.5. Using Equation (6.2) and knowing the precise values of atomic masses, Q-value can also be estimated. When one of the nuclei(Y) taking part in the reaction is a heavy particle, then it is very difficult to measure its kinetic energy (E_Y) accurately. However, one can estimate it with the help of Equation (5.6), knowing the masses and by measuring of E_x and E_y accurately. In this case the corresponding mass numbers in place of precise values of the masses will suffice.

One can use a scintillation counter, a proportional counter (gas filled), a solid state counter or a magnetic spectrograph, for the determination of energy if the emitted particle y is a charged particle.

When the resolving power needed is not very high, one can use scintillation spectrometers. The resolving power of scintillation spectrometers is usually low, being ~ 20 to 30. We may note that the charged particles have very small ranges in solids and hence the scintillation detector to be used can be quite thin for low energy nuclear reaction's study. One can locate the scintillator close to the target to increase the solid angle, which helps to improve the statistics of counting.

Now a days, solid state detectors with sufficiently thick layers having much better resolving power ~ 200 to 300 are available. With the help of these detectors, one can go upto fairly high energies.

Resolving powers of the order of 1000 have been achieved with *magnetic spectrometers*. These are most suitable for high resolution work. Now a days, both single focusing and double focussing magnetic spectrometers are available. In the former type, one can focus the particles emerging from a point on the meridian plane along a line perpendicular to that plane, while in the latter type, a point-object produces a point-image. One can achieve the double focussing in a number of ways. An inhomogeneous magnetic field is used in one of these, which is similar β-rays spectrometer due to Svartholm and Siegbahn.

Enge and Buechner of MIT, USA in 1963 developed a highly versatile multi gap spectrograph. Which is in fact twenty four instruments in one large vacuum chamber. Both energy and angular distributions of the particles are recorded simultaneously, whereas the energy distributions at different angles θ at the interval of 7.5° are obtained.

After proper exposure, we have to develop and scann the plates under microscopes. Then, one will have to count the number of tracks of the particles having the correct length and diffraction. We may note that at each exposure lasting about 1 to 10 hr, the number of data points obtained is 36000.

A typical spectrum obtained with multigap magnetic spectrograph for the reaction ^{45}Sc (d, p) ^{46}Sc using a deutron beam accelerated in the 8 MeV Van de Graff generator is shown in Figure 6.6. The different peaks in the spectrum corresponds to different states of the residual nucleus ^{46}Sc.

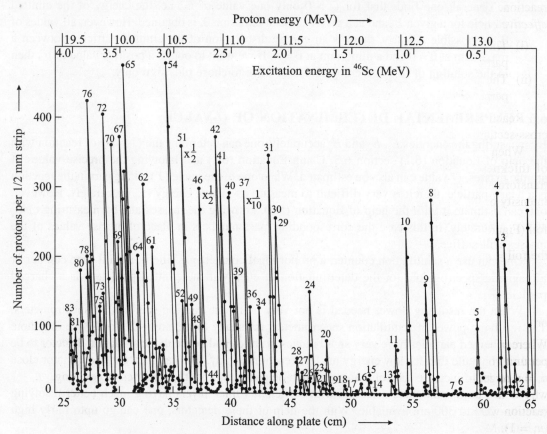

Fig. 6.6 A typical energy spectrum of protons from ^{45}Sc (d, p) ^{46}Sc reaction at θ = 37.5° obtained in a multigap magnetic spectrograph.

The determination of differential cross section $\dfrac{d\sigma}{d\Omega}$ for the peaks at different energies as function of angles of emission revealed pronounced maxima in the forward direction.

With the knowledge of orbital angular momenta of the states in which the neutron is captured in the (d, p) reaction, the positions of the maxima can be determined.

6.7 NUCLEAR REACTION CROSS-SECTION

Nuclear reaction cross-section is one of the most important quantities which one defines in a nuclear reaction. This is the probability of the occurrence of a reaction or the cross-section, which is closely related to the probability. One can define the probability of a nuclear reaction in terms of the number of particles emitted, or of nuclei undergoing transmutation for a specified number of incident particles. Usually, it is expressed in terms of effective area presented by a nucleus towards the beam of bombarding particles, i.e. the number of incident particles that would strike such an area, calculated purely on the geometrical basis. Obviously, this is the number observed to lead to the nuclear reaction in question. This effective area is termed as the cross-section of that nuclear reaction. Thus one can define the *probability of occurrence of a particular nuclear reaction as the effective cross-section for that nuclear process*. One may also define the cross-section as:

(i) The probability that an event may take place when a single nucleus is exposed to a beam of particles of total flux one particle per unit area.

(ii) The probability that an event may take place when a single particle is bombarded perpendicularly at a target having one particle per unit area.

Reaction cross-section is usually designated by the symbol, σ. One can write the nuclear cross-section of a reaction $X(x, y)Y$ as $\sigma(x, y)$.

When a parallel beam of N projectiles is incident in a given interval of time upon a target foil T of thickness Δx and surface area S normally, then the number of nuclei in T undergoing transformation due to the nuclear reaction of the type under consideration, is proportional to the intensity of the incident beam of projectiles and to the total number of target nuclei present in the foil (Figure 6.7(a)). Now, the incident particle intensity is $\dfrac{N}{S}$ and the number of nuclei present in the foil $= n.\Delta x.S$. Obviously, the nuclei transformed is

$$\Delta N \propto \left(\frac{N}{S}\right)(nS\,\Delta x)$$

or
$$\Delta N = \sigma N n\, \Delta x = \sigma N\, n_1 \qquad\qquad\qquad \dots (6.1)$$

Where $n_1 = n\,\Delta x$ is the number of target nuclei per unit area of the foil and n is the number of nuclei per unit volume. From Equation (6.1) it is obvious that since both ΔN and N are pure numbers and $n_1 = n\,\Delta x$ has the dimensions of the reciprocal of an area, σ has the dimensions of an area. This is why σ is called cross-section and measures the probability of the occurrence of the nuclear reaction when a single particle ($N = 1$) is incident on a single target nucleus present per unit area ($n_1 = 1$). Moreover, the nuclear radii are of the order of 10^{-14} to 10^{-15} m, and therefore the cross section of nuclear reaction is of the order of 10^{-28} m^2. Due to this very small value of σ, it has been found more convenient to use the unit *barn* ($= 10^{-28}$ m²) for σ. There are some special reactions such as the (n, γ) reaction induced by thermal neutrons or the neutron induced resonance reactions for which σ may be very high (several thousand barns).

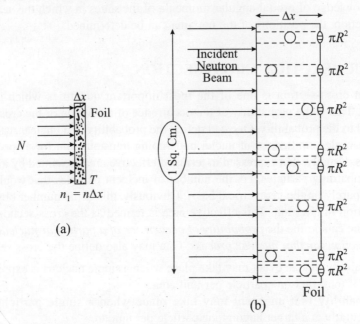

Fig. 6.7(a) A beam of particles is bombarded on a target foil, T of thickness Δx. (b) Illustration of geometrical significance of nuclear reaction cross-section.

Geometrical Significance of σ   One can understand the geometrical significance of σ with the help of Figure 6.7(b). If R is the effective radius of the target nucleus for a given nuclear reaction, then the projection of its surface area on a plane perpendicular to the direction of projectile (shown by the shaded portion in Figure 6.7(b) $= \pi R^2$). Thus the number of projectiles encountering each target nucleus is $\pi R^2 N_S$, where $N_S = \dfrac{N}{S}$ is the number of projectiles incident per unit area of the target. Here, the projectiles are considered to be mass-points. As there are n_1 nuclei per unit area of the target, and therefore the number of projectiles intercepted by the target nuclei in the foil is

$$n_1 S \, \pi R^2 N_S = \pi R^2 N n_1 \qquad \qquad \text{... (6.2)}$$

Here $N = N_S S$ is the total number of projectiles incident on the target. Thus the probability of encounter between a single projectile ($N = 1$) with one nucleus per unit area (i.e. $n_1 = 1$), in the target foil is obtained as

$$\frac{\pi R^2 N n_1}{n_1} = \pi R^2 N = \pi R^3 \qquad \qquad \text{... (6.3)}$$

Truly speaking, the probability of encounter between a single target nucleus per unit area is not determined by πR^2 alone. This probability depends on: (i) the nature of the interaction between the projectile and the target nucleus, (ii) the energy of the projectile, and (iii) several other factors. As assumed earlier, the incident particle is also not a mass-point. Clearly, the reaction cross-section σ depends on its size also. For very low energy projectiles, the de Broglie wave length $\left(\lambda = \dfrac{h}{p} \right)$ is much longer than their geometrical extension. This means that the reason over which they interact is much larger than their geometrical cross section. This is the basic reason for σ of (n, γ) reaction with thermal neutrons to be usually very large as mentioned above.

Due to strong electrostatic repulsion of the target nucleus, the cross section for charged particles is reduced considerably.

In the above discussions, we have assumed that the total projected area of all the nuclei in the foil ($= \pi R^2 n, S$) is small as compared to the area S of the foil. However, this is true only if the foil thickness is small.

6.7.1 Partial Cross-section

We have read that the total nuclear cross-section is the effective area possessed by a nucleus for removing the incident particles from a collimated beam by all possible processes. One can write the total nuclear cross-section as a sum of several partial cross-sections which represent the contributions of the various distinct, independent processes which can remove particles from the incident beam. Thus

$$\sigma_t = \sigma_s + \sigma_r \qquad \text{... (6.1)}$$

Where σ_t is the total cross-section, σ_s the cross-section that produces measurable scattering and σ_r the reaction cross-section. One can classify the scattering cross section as: *inelastic scattering and elastic scattering processes*. Thus, one can write

$$\sigma_S = \sigma_{\text{el}} + \sigma_{\text{inel}} \qquad \text{... (6.2)}$$

One can further subdivide these partial cross-sections. In the case of elastic cross-section, one cannot write separate partial cross-sections because of the possibility of interference between them. However, all inelastic scattering processes are incoherent and therefore their cross-sections are additive, i.e.

$$\sigma_{\text{Inel}} = \sigma_1 + \sigma_2 + \sigma_3 + \sigma_4 + ... \qquad \text{... (6.3)}$$

Similarly, one may subdivide the partial cross section, σ_r into component parts associated with the various possible nuclear reaction. Thus, one can write

$$\sigma_r = \sigma' + \sigma'' + \sigma''' + ... \qquad \text{... (6.4)}$$

In case of elastic scattering cross-section, the total cross-section, $\sigma_t = \sigma_r + \sigma_{\text{el}}$, determines the absorption coefficient for the beam of particles incident on the target foil. Using $\Delta N = \sigma N n_1$, one can write for a target foil having n nuclei per unit volume of infinitesimal thickness dx on which a particle beam of intensity $N_S = \dfrac{N}{S}$ is incident *perpendicularly* as

$$dn_s = -\sigma_t n_s n \, dx \qquad \text{... (6.5)}$$

Where the minus sign on the rhs of Equation (6.5) indicate the diminution in the beam intensity as it comes out of the foil. Integration gives for a foil of finite thickness x as

$$n_s = n_{so} \exp(-\mu s) \qquad \text{... (6.6)}$$

where $$\mu = \sigma_t\, n \qquad \qquad \qquad \text{... (6.7)}$$

is the total absorption coefficient n_{so} is the intensity of the beam incident on the foil and n_s is the emergent intensity. Will the help of Equation (6.6) and (7.7) and by measuring n_s, one can determine the total cross section σ_t, provided n_{so} is known.

Differential Cross-section

One can describe the distribution in angle of the emitted particles in a nuclear reaction in terms of a cross-section which is a function of the angular coordinates in the problem. The cross-section which defines a distribution of emitted particles with respect to the solid angle is called *differential cross-section* and defined as $\dfrac{d\sigma}{d\Omega}$. For a given process, the partial cross-section is given by

$$\sigma = \int \frac{d\sigma}{d\Omega}\, d\Omega \qquad \qquad \text{... (6.8)}$$

where $$d\Omega = \sin\theta\, d\theta\, d\phi \qquad \qquad \text{... (6.9)}$$

The differential cross section $\sigma(\theta, \phi)$ can be defined as the cross-section per unit solid angle, for particles going along θ and ϕ direction. The angular spreads $d\theta$ and $d\phi$ are the angles subtended by the detector over the scatterer or target.

Sometimes, one is interested to measure the energy dependence of a cross-section, as well as the angular dependence. Then one can write

$$N_r(\theta, \phi, E) = I\, N_S\, \sigma(\theta, \phi, E)\, d\Omega\, dE \qquad \qquad \text{... (6.10)}$$

Where N_r is the number of reactions per unit time, N_S is the number of particles in the target which will be offered to the incident beam and I is the number of incident particles per unit area per unit time (flux). The double dependence of the cross-section in (6.10) can also be expressed as

$$\frac{d^2 N(\theta, \phi, E)}{d\Omega\, dE} = \frac{I\, N_S}{} \frac{d^2 \sigma(\theta, \phi, E)}{d\Omega\, dE} \qquad \qquad \text{... (6.11)}$$

We may note that the measurement of differential cross-sections as a function of anlges and energies gives information about the reaction mechanism involved in a particular reaction.

One can also express the cross-sections in nuclear reactions quantum mechanically in terms of the transitions between the initial and final states.

Statistical Consideration of Nuclear Reaction Cross Section   One can calculate the probability for a particular type of reaction to take place with the help of perturbation theory. Let ψ_i and ψ_f are the wave functions of incoming and outgoing channels, the probability of transition from the state i to state f is

$$W = \frac{2\pi}{\hbar}\, |H_{if}|^2\, \frac{dn}{dE} \qquad \qquad \text{... (6.1)}$$

Where H_{if} is the matrix element for the transition and $\dfrac{dn}{dE}$ is the number of states possible for the nuclear reaction channel per unit energy. The energy density of states in a finite volume V is given as

$$\frac{dn}{dE} = \frac{V \, 4\pi \, P_b^2}{(2\pi\hbar)^3} \frac{dP_b}{dE} \qquad \ldots (6.2)$$

The momentum P_b of a particle b for a non-relativistic case is

$$\frac{P_b^2}{2m_b} = E \quad \therefore \quad dE = \frac{2P_b \, dP_b}{2m_b}$$

Thus, we have

$$\frac{dP_b}{dE} = \frac{1}{V_b} \qquad \ldots (6.3)$$

Due to spin of the particle, in the final state, the degeneracy of the state of particle b is $(2I_b + 1)$ and that of the nucleus B is $(2I_B + 1)$. Thus, we have

$$\frac{dn}{dE} = \frac{4\pi \, P_b^2 \, V}{(2\pi\hbar)^3 \, v_b} (2I_b + 1)(2I_B + 1) \qquad \ldots (6.4)$$

Now, the probability of transition from state i to state f per unit time is thus obtained as

$$W = \frac{1}{\pi\hbar^4} \frac{P_b^2}{v_b} V |H_{if}|^2 (2I_B + 1)(2I_b + 1) \qquad \ldots (6.5)$$

Now, if the projectiles 'a' have density of n_a particles per unit volume and move with velocity v_a, then the number of particles hitting the target per unit area per unit time is $n_a v_a$. Obviously, the probability of reaction taking place per unit time is

$$W = n_a v_a \sigma_r \qquad \ldots (6.6)$$

Let us consider that there is one particle in volume (V), then the density $n_a = V$, one finds

$$\sigma_r = \frac{1}{\pi\hbar^4} \frac{P_b^2}{v_a v_b} (2I_B + 1)(2I_b + 1)\} \, |V H_{if}|^2 \qquad \ldots (6.7)$$

Let U is the interaction energy between the initial state $(A + a)$ and the final state $(B + b)$ responsible for the transition. Then, one finds

$$H_{if} = \frac{1}{V} \int \psi_f^* \, U \psi_i \, d\tau \qquad \ldots (6.8)$$

The integration in (6.7) is extended over the nuclear volume V and ψ_i and ψ_f are functions of only internal coordinates.

Now, if the projectile is a positively charged particle (usually this is positively charged particle), it experiences a Coulomb repulsion in the vicinity of the nucleus. This causes in the reduction of the wave function by a factor $\exp(-G_a)$, Where

$$G_a = \frac{1}{\hbar} \int_{R_1}^{R_2} [2m_a (V_a - E_a)]^{\frac{1}{2}} \, dr$$

$$\simeq \frac{\pi Z_A Z_a e^2}{\hbar v_a} \qquad \ldots (6.9)$$

Here eZ_A and eZ_a are the charges on particles A and a respectively. This would give a factor e^{-2G_a} in $|H_{if}|^2$. Now, if the outgoing particle is also positively charged, it has to overcome a Coulomb barrier. This causes a reduction in ψ_f^* by a factor e^{-G_b}. Clearly, when the outgoing particles are both positively charged, one finds

$$H_{if} = \frac{1}{V} \int \psi_f^* \, U \psi_i \, e^{-G_a} e^{-G_b} \, d\tau \qquad \qquad ... (6.10)$$

One can write (6.10) approximately as

$$H_{if} = \frac{<U> (\text{Volume of the nucleus})}{V} e^{-G_a} e^{-G_b} \qquad ... (6.11)$$

Where $<U>$ represent the average of the interaction over the nuclear volume. Now, putting the value of H_{if} in (6.7), one obtains

$$\sigma_r = \frac{1}{\pi \hbar^4} \frac{P_b^2}{v_a \, v_b} \, [|V \, H_{if}|^2 \, (2I_B + 1)(2I_B + 1)] \qquad ... (6.12)$$

One can predict the general behaviour of the variation of reaction cross-section with projectile energy (excitation potential) from Equation (6.12). We have various cases of interest as:

(i) **Elastic Scattering; both particles are uncharge** Obviously, this concerns the neutron scattering. Since neutron being neutral does not experience any Coulomb repulsion so $G_a = G_b = 0$. In case of elastic scattering, $V_a = V_b$, thus

$$\frac{P_b^2}{v_a \, v_b} = \frac{P_b^2}{v_b} = M_n^2$$

Where M_n is the mass of the neutron. This reveals that at low projectile energies H_{if} is approximately constant and cross-section is independent of energy (Figure 6.8).

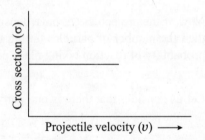

Fig. 6.8 Elastic scattering nuclear reaction cross-section vs projectile velocity for neutrons.

(ii) **Exothermic reactions in which the projectile is neutral** Such reactions can be represented as (n, p), (n, γ) and neutron induced fission (n, f) reactions. In such types of reactions Q is usually a positive and about a few MeV. The incoming neutron energy is low (< 1 eV) and thus v_a is almost constant. Similarly, for the out going particle b, v_b is constant. Moreover, G_b is also independent of projectile energy. Obviously, the reaction cross-section varies as $\frac{1}{v_a}$. This is the well known '$\frac{1}{v}$ law' for neutron induced exothermic reactions (Figure 6.9).

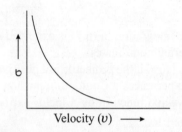

Fig. 6.9 Exothermic reaction cross-section vs. Velocity (of the incident neutron)

(iii) Exothermic Reactions with charged Projectiles The outgoing particle in reactions like (p, γ), (p, n), (α, n) and (α, γ) is charged particle and hence $G_b = 0$. Since v_b is almost constant, one finds

$$\sigma \propto v_a^{-1} e^{-G_a}$$

Figure 6.10 shows the variation of reaction cross section with projectile velocity.

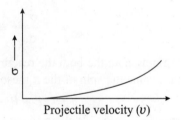

Fig. 6.10 For charged particles, reaction cross section vs velocity curve.

(iv) Inelastic Scattering (n, n') reaction In this case $G_a = G_b = 0$. The reaction is endothermic, since the residual nucleus is left in an excited state and the reaction has negative Q value. The energy of the incident projectile has to be above the threshold energy of the nuclear reaction. With variation of energy near threshold, v_a does

not change much and $\dfrac{P_b^2}{v_b} \propto \sqrt{(\text{excess energy})}$. v_b

changes rapidly near threshold and hence the reaction cross-section increases sharply above the threshold as shown in Figure 6.11.

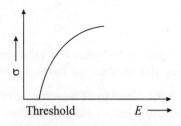

Fig. 6.11 For inelastic neutron scattering the behaviour of reaction cross-section near threshold.

(v) Endothermic reaction with outgoing charged particle Example of such type of reactions are (n, α) and (n, p) reactions. This situation is just similar to (n, n') reactions except that the factor e^{-G_b} becomes significant. Due to Coulomb barrier, the low energy outgoing particle is not able to come out. The reaction cross section varies as

$$\sigma \propto (\text{excess energy})^{\frac{1}{2}} e^{-G_b}$$

Figure 6.12 shows the variation of cross section with projectile energy. We may note that in the above discussion, we have omitted the effect of the orbital angular momentum of the incoming particle 'a' which may affect the value of the matrix element significantly.

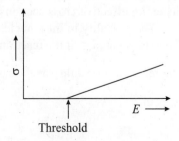

Fig. 6.12 For endothermic neutron reaction with charged particle emission, the behaviour of cross section near threshold.

Inverse Nuclear Reaction Cross-Section

Equation (6.7) allows one to write an relation between the nuclear reaction $X + x \rightleftharpoons Y + y$ going from left to right and from right to length. Let σ_{XY} be the reaction cross-section for the above reaction and σ_{YX} for the said reaction going from right to left. The matrix element H_{if} is Hermitian so that $H_{if} = H_{if}^*$. Thus, we have

$$<|H_{if}|^2> = <|H_{fi}|^2>$$

One can write Equation (6.7) for forward and inverse reactions as

$$\frac{\sigma_{XY}}{\sigma_{YX}} = \frac{P_Y^2}{P_X^2}\frac{(2I_Y+1)(2I_y+1)}{(2I_X+1)(2I_x+1))} \qquad \dots (6.13)$$

We may note the both the reactions are considered in *CM* frame. One can use Equation (6.13) to determine the spin of the π meson from the reaction

$$P + P \rightarrow \Leftrightarrow \leftarrow \pi^+ + d - 137 \text{ MeV}$$

The statistical weight $(2I_X+1)(2I_x+1)$ is equal to 4 but exclusion principle does not permit identical states of two protons and hence statistical weight is 2. On the RHS the statistical weight factor is $3(2I_\pi+1)$ and there in the *CM* mass system, one finds

$$\frac{\sigma_{pp \rightarrow \pi d}}{\sigma_{\pi d \rightarrow pp}} = \frac{3(2I_\pi^2+1)}{2}\frac{P_\pi^2}{P_p^2} \qquad \dots (6.14)$$

Experimental measured values of the cross-section of the said two reactions at known energies, the spin of π-meson has found to be zero, i.e. $I_\pi = 0$.

Determination of Reaction cross-section   One can obtain experimentally the values of σ by the measurment of the attenuation of an incident beam in a known thickness of material. Let a monoenergetic beam consisting of I particles per unit time distributed uniformly over an area A is allowed to pass through a thin sheet of target material of thickness Δx and having n nuclei per unit volume. Then the available nuclear target area which is the product of the cross-section per target nucleus (σ) and the number of target nuclei lying within the area A. Here, we have assumed that the target is thin enough so that there is no overlapping of nuclei. Obviously, every nucleus presents its entire effective cross-section to the incident beam.

The probality to hit a nuclear target for the incoming particle = nuclear target are $\sigma n\, A\, \Delta x /$ Total target area A. If the reaction produces N light product particles per unit time, then one defines the reaction probability as the ratio $\dfrac{N}{I}$. Thus, one finds

$$\frac{N}{I} = \frac{\sigma n\, A\, \Delta x}{A} \quad \text{or} \quad \sigma = \frac{N}{\left(\dfrac{I}{A}\right)(nA\,\Delta x)} \qquad \dots (6.15)$$

Obviously, one can define the nuclear cross section as the number of light product particles per unit time per unit incident flux and per target nucleus. Now, if a bombarding particle is producing a variety of light reaction products $N_1, N_2, \dots$ per unit time, then the total cross section is defined as

$$\sigma_{\text{total}} = \frac{N_1 + N_2 + \dots}{\left(\dfrac{I}{A}\right)(nA\,\Delta x)} \qquad \dots (6.16)$$

Let the incident flux (beam intensity) is I_0 and the flux after penetrating the target a distance x is I. On further penetrating the distance dx, it will become $I - dI$. This reduction in flux is attributed to the various collision events taking place within this distance dx. We have

$$-dI = N = n\sigma \, dx \, I \quad \text{or} \quad -\frac{dI}{I} = \sigma_n \, dx \qquad \text{... (6.17)}$$

The expression (6.17) contains the product of σ, the cross-section per nucleus or the *microscopic cross-section* and n, the number of nuclei per unit volume. The product is termed as *macroscopic cross section* and is denoted by Σ. Thus, one finds

$$-\frac{dI}{I} = \Sigma dx$$

or
$$I = I_0 \, e^{-\Sigma x} \qquad \text{... (6.18)}$$

The experimental arrangement consists of a source and detector between which one can place a slab of the experimental material. In the absence of the absorbing material, I_0 is the particle intensity reaching the detector and I is the intensity when the slab of thickness x, containing n atoms per unit volume is present there, one can calculate σ.

There are many nuclear reactions in which the light product particles are not product in an isotopic manner. One can then define the differential cross-section $\dfrac{d\sigma}{d\Omega}$ in terms of the number of light products dN emitted per unit time in a small solid angle $d\Omega$ at some angle θ with the direction of incident beam. From Equation (6.15), one finds

$$\frac{1}{I}\frac{d\sigma}{d\Omega} = \frac{NA\,\Delta x \left(\dfrac{d\sigma}{d\Omega}\right)}{A}$$

or
$$\frac{d\sigma}{d\Omega} = \frac{\dfrac{dN}{d\Omega}}{\left(\dfrac{I}{A}\right)(nA\,\Delta x)} \qquad \text{... (6.19)}$$

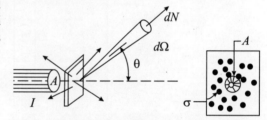

Fig. 6.13 Experimental arranged for the determination of differential cross-section, (left) side view and (right) along beam direction view.

Macroscopic Cross Section and Mean Free Path

Equation (6.18) reveals that Σ corresponds to linear absorption coefficient and has the dimensions of reciprocal length. One can define the penetration distance (λ) as tha path length over which the neutrons one travel without suffering a collision with a target. This distance is called *mean free path*. Since beam intensity is proportional to the beam particle density q, hence one can write Equation (6.18) as

$$q = q_0 \, e^{-\Sigma x}$$

Thus the average path length,

$$\lambda = \frac{\text{Total path length}}{\text{Total number of particles}} = \frac{\displaystyle\int_0^\infty x \, dq}{q_0}$$

$$= \frac{1}{\Sigma} \int_0^{\infty} (-\Sigma x)\, e^{-\Sigma x}\, d\,(\Sigma x) = \frac{1}{\Sigma} \qquad \ldots\ (6.20)$$

Obviously, the average path length λ between successive collisions is equal to the reciprocal of the macroscopic cross section Σ.

6.8 THE NUCLEAR LEVEL SPECTRUM

The experimental result to examine the level spectrum of a nucleus of middle mass say $A = 100$, by exciting it through the reaction $X(x, y)Y$ might be as shown in Figure 6.14. On the r.h.s of the figure is shown the cross-section for the reaction as a function of energy of the out going particle y and on the r.b.s. is given the level structure inferred from the observed outgoing particle energies. One can that there are two quite distinct regions, namely the low energy discrete region in which well-separated *bound* levels may be distinguished and the highest energy *continuum* region in which the levels are unbound or *virtual* and overlap considerably, although special features may still be observed.

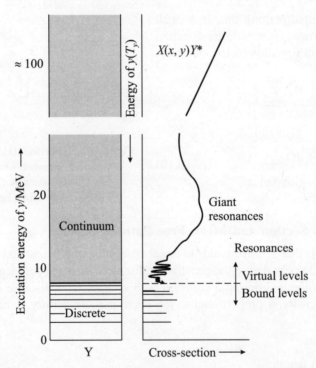

Fig. 6.14 The nuclear level spectrum for a nuclide Y of $A \approx 100$, as deduced from the energy spectrum of particles y emitted in a nuclear reaction $X(x, y)Y$.

Each excited level has its own set of the properties. For the purpose of discussing a level spectrum however, one may assume a level to be mainly characterized by an excitation energy above the ground state by a width Γ which is related to the life time τ of the level against all forms of decay as

$$\Gamma\tau = \hbar = 6.6 \times 10^{-16} \text{ eV-s} \qquad \ldots (6.1)$$

and by quantum numbers for spin, Parity P and isobaric spin, although effects exist which can cause some mixing of isobaric spin values in a state of good angular momentum. There is no abrupt change of level properties, other than level density, between the discrete and continuum regions and the boundary between these regions is near the energy of excitation at which the emission of a particle becomes possible and probable. This will be often be the neutron binding energy of about 8 MeV and above this energy the widths of the virtual levels increase rapidly with energy, although the discrete structure survives for the lowest virtual levels and may be seen as individual *resonances* in approproate reactions. Below the particle separation energy the bound levels decay by emission of radiation or even by β-particle emission if electromagnetic transitions are inhibited.

The theoretical description of nuclear level properties requires the formulation of a nuclear Hamiltonian and the solution of formidable quantum mechanical many-body problem to provide a wave function from which properties may be calculated. In the discrete region of excitation these may take the form of assumptions that the motion of most of the particles may be disregarded and that the levels arise from the independent motion of a few active nucleons. Or, alternatively that there is strong coupling between all the particles of the nucleus and a resulting coherent motion which leads to collective effects of rotation and vibration. At the higher energies of the continuum region detailed models of this type lose their validity and although the underlying microscopic structure must still exist, the motion assumes a random or chaotic character which should be treated statistically.

Special states that are found mainly, though not entirely, above the neutron binding energy are the *giant resonances* and the *isobaric analogue levels.* As states earlier, the former may be regarded as the basic electromagnetic oscillations of the nucleus as a whole and are at an excitation that varies smoothly with mass number. The latter are structurally related to the low-lying states of an isobaric neighbour nucleus, and although they are found at high excitation because they are pushed up by the Coulomb energy they still exhibit the wide spacing characterstic of low lying states.

Both nuclear decays and nuclear reactions provide information on the nuclear level spectrum and on properties of individual levels.

6.9 CLASSICAL ANALYSIS OF NUCLEAR REACTION CROSS-SECTION

Let the impact parameter (distance of closest approach) of the bombarding particle is b with respect to the nucleus. The angular momentum of the system will be normal to the relative momentum p of the two particles and having magnitude

$$L = pb = \frac{\hbar b}{\lambdabar} \qquad \ldots (6.1)$$

Let us consider the case when $b > R$. The bombarding particle should not have much effect on the target nucleus as it is outside the range of nuclear forces. Thus one expect the most important nuclear reactions to take place with those bombarding particles which have angular momentum less than or equal to the maximum value $\dfrac{\hbar R}{\lambda}$. According to quantum mechanics, the angular momentum is quantized and can have discrete values $\hbar l$ with $l = 0, 1, 2, \ldots$. We now consider the associated wave λ and we note that it may be resolved into partial waves, each corresponding to a value of l.

The significance of the angular momenta in a plane was explained by Blatt and Weisskopf. They divided the incident beam into cylindrical zones, such that each impact parameter is associated with a different impact parameter. There will only particles with an impact parameter b less than $\lambda \left(= \dfrac{\hbar}{p} \right)$ in the central zone (Figure 6.15). Due to *uncertainty* principle, it is not really possible to visualize an impact parameter less than the associated wavelength. However, the angular momentum is zero throughout this zone. The next zone contains all particles with impact parameters between λ and 2λ. Similarly, one can see that l^{th} zone contains particles with impact parameters between $l\lambda$ and $(l + 1)\lambda$. On passing from zone l to $(l + 1)$, the change in angular momentum (i.e., momentum X impact parameter) is $\hbar l$ to $\hbar(l + 1)$

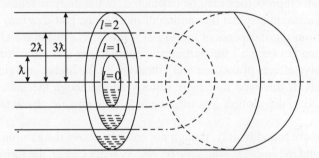

Fig. 6.15 Resolution of associated wave (λ) into partial waves

The difference between the areas due to radii $(l + 1)\lambda$ and $l\lambda$ gives the area of the lth zone. This is the cross-sectional area (σ_r^l) of the lth zone. If we assume that each particle hitting the nucleus causes a reaction then the reaction cross-section can be obtained by summing σ_r^l over all values l varying from o to lm. Thus, we have

$$\sigma_r = \sum_0^{l_m} \pi \lambda^2 (2l + 1) = \pi \lambda^2 (l_m + 1)^2 \qquad \ldots (6.2)$$

Corresponding to maximum impact parameter R, the maximum value of l is given by $l_m = \dfrac{R}{\lambda}$.

Using this value of l_m in (6.2), one obtains the maximum cross section as

$$\sigma_r = \pi (R + \lambda)^2 \qquad \ldots (6.3)$$

Here, we may recall that the incident wave may also be partially reflected and that a transmission factor T_l, depending on the value of l, should be taken into account. The quantum mechanical treatment of the problem yield

$$\sigma_r = \pi \lambda^2 \sum_{0}^{\infty} (2l + 1)T_l \qquad \dots (6.4)$$

We now consider the case of a nucleus which has a radius $R(R \gg \lambda)$, i.e. when the high energy particles strike the nucleus. The nucleus is considered as completely absorbent (i.e. perfectly black body) which corresponds to $T_l = 1$. This is for all values of l upto and including $l_m \left(= \dfrac{R}{\lambda} \right)$. The cross section obtained with these conditions is same as given by Equation (6.3). When λ is small compared to R, this is close to geometrical cross section. At very high energies, the nucleus exhibits a certain transparency to the incident particles, which may penetrate the nucleus without interaction. For the weaker energies, i.e. $\lambda \gg R$, $T_l = 0$ for large l to hold, since even for $l = 0$, $l\lambda > R$.

The *elastic cross-section* of the nuclear reaction depends upon the condition under which the incident wave is reflected. However, the interference with the incident wave gives rise to amplitudes four times that of the penetrating wave. We have

$$\sigma_{el}^{max} = 4\pi(R + \lambda)^2 \qquad \dots (6.5)$$

In contrast, when $T_l = 1$ for all values of l, in the case of nucleus behaving as black body (black nucleus) and examined previously, the minimum value is obtained as

$$\sigma_{el}^{min} = 4\pi(R + \lambda)^2 = \sigma_{inela} \qquad \dots (6.6)$$

Therefore, the effective total cross-section is equal to

$$\sigma_t = \sigma_{el} + \sigma_{inela} = 2\pi(R + \lambda)^2 \qquad \dots (6.7)$$

Obviously, the total cross-section for black nucleus is twice the geometrical cross-section. The opaque disc of the nucleus projects a shadow and diffraction occurs as in wave optics. Now, if we imagine that the nucleus is replaced by a disc of radius $(R + \lambda)$, this shadow may be shown to exist. The disc then emits particle waves of the same intensity in the direction of the incident beam but opposite in phase to those of the beam.

In the above derivation, we have not considered the effect of Coulomb's repulsion. However, when the Coulomb repulsion is taken into account, then this will bring the relative K.E. of the charged particle from T (when the particle is far away from the nucleus) to $T - B$, when the particle is in just touching position to the nucleus of radius R. Now, if z and Z are the atomic numbers of incident particle and target nucleus respectively, then we have $B = \dfrac{z Z e^2}{4\pi \varepsilon_0 R}$. Clearly, the Coulomb repulsion slows down the particle and hence the initial moment $\sqrt{2MT}$ is decreased to $\left[2M \sqrt{(T - B)} \right]$, where M is the reduced mass of the system. Now, the angular momentum is given by

$$L_m = R \left[2M \sqrt{(T - B)} \right]^{\frac{1}{2}} = R\sqrt{2MT} \left(1 - \frac{B}{T} \right)^{\frac{1}{2}}$$

$$\therefore \qquad b_m = \frac{L_m}{p} = \frac{L_m}{\sqrt{2MT}} = R\sqrt{1 - \frac{B}{T}}$$

Clearly, the Coulomb repulsion diminishes the L_m by a factor of $\sqrt{1 - \dfrac{B}{T}}$. One can estimate the upper limit for the capture of charged particles as the area of the disc of radius bm. We have

$$\sigma_r = \pi R^2 \left(1 - \frac{B}{T} \right) \qquad \text{... (6.8)}$$

6.10 PARTIAL-WAVE DESCRIPTION OF REACTION CROSS-SECTION

Let us consider a plane wave representing a particle beam incident on a nucleus located at the origin. For the sake of simplicity, let us assume that target nucleus as well as projectile are spinless, i.e. zero spin. We can assume that particles are neutrons (spin zero particles). However, the results, in principle, are the same for charged particles as well.

Let us consider that an incident plane wave of neutrons travelling parallel to the Z-axis. The spatial part of this plane can be described by $\exp(ikz)$, where $k = \dfrac{2\pi}{\lambda}$. One can express this as a superposition of spherical waves with different orbital angular momentum quantum numbers relative to the target nucleus. One can expand $\exp(ikz)$ as a sum of eigen functions of orbital angular momentum:

$$\exp(ikx) = \exp(ikr \cos\theta) = \sum_{i=0}^{\infty} i^l \sqrt{[4\pi(2l+1)]}\ j_l(kr)\ Y_{l,0}(\theta) \qquad \text{... (6.1)}$$

This involves the spherical Bessel functions $j_l(kr)$ and spherical harmonics $Y_{l,m}(\theta)$. Since there is cylindrical symmetry about the Z-axis the order m of the spherical harmonics is zero. If r, the distance of the incident particle from the target nucleus is very large, i.e. $kr \gg l$, then

$$j_l(kr) = \frac{\mathrm{Sin}\left(kr - \dfrac{1}{2}l\pi \right)}{(kr)} = \frac{i}{2kr}\left[\exp -i\left(kr - \frac{l\pi}{2} \right) - \exp i\left(kr + \frac{l\pi}{2} \right) \right] \qquad \text{... (6.2)}$$

Which is the sum of *ingoing* (first exponential term) and *outgoing* (second exponential term) spherical waves. The first wave coming towards origin with momentum $k\hbar$ and momentum $l\hbar$ and the second exponential term represents a series of a outgoing spherical waves with just the same properties. Obviously, the above expression describes the undisturbed wave. Now, if a nucleus is located at the origin, the amplitudes and phases of the outgoing spherical waves from the origin are, in general, changed. Let us consider that the outgoing part of the partial wave with the orbital angular momentum l be changed by a factor η_l. However, η_l may be a complex number. This is related to the phase shift by $\eta_l = |\eta_l|\ e^{2i\delta_l}$. When $|\eta_l| = 1$, elastic scattering will take place. Now, the wave function for the disturbed wave is

$$\psi(r) = \frac{\sqrt{\pi}}{kr} \sum_{l=0}^{\infty} \sqrt{(2l+1)}\ i^{l+1}\left[\exp -i\left(kr - \frac{l\pi}{2} \right) - \eta_l \exp i\left(kr - \frac{l\pi}{2} \right) \right] Y_{l,0}(\theta) \qquad \text{... (6.3)}$$

The wave function describing the elastically scattered particle, ψ_{sc}, which is the difference between the disturbed (final) wave and incident wave is

$$\psi_{sc} = \psi_f - \psi_i = \psi_f - \exp(ikz)$$

$$= \frac{\sqrt{\pi}}{kr} \sum_0^\infty \sqrt{(2l+1)} \; i^{l+1} (1-\eta_l) e^{i\left(kr - \frac{i\pi}{2}\right)} Y_{l,0}(\theta) \qquad \ldots (6.4)$$

Since $\psi_f \propto e^{ikz} + f(\theta) \dfrac{e^{ikr}}{r}$, one obtains

$$f(\theta) = \frac{\sqrt{\pi}}{k} \sum_{l=0}^\infty \sqrt{(2l+1)} \; i^{l+1} (1-\eta_l) e^{-\frac{il\pi}{2}} Y_{l,0}(\theta) \qquad \ldots (6.5)$$

One can obtain the scattered cross-section by dividing the number N_{sc} of scattered particles per second by the number of incident particles per unit area per second, which for the plane wave e^{ikx} is v, the velocity of the particles. Let us consider a sphere of radius r_0. Now, if r_0 is very much larger than the range of the nuclear forces, than the number of scattered particles per second, N_{sc} is equal to the flux of ψ_{sc} through this sphere of radius r_0. We have

$$N_{sc} = \frac{\hbar}{2iM} \int \left(\frac{\partial \psi_{sc}}{\partial r} \psi_{sc}^* - \frac{\partial \psi_{sc}^*}{\partial r} \psi_{sc} \right)_{r=r_0} r_0^2 \sin\theta \, d\theta \, d\phi \qquad \ldots (6.6)$$

Here M is the mass of the incident particle and the integral is extended over the surface of the sphere. Now, on substitution the values of ψ_{sc} and ψ_{sc}^* in (6.6), one obtains

$$N_{sc} = \frac{\hbar}{2iM} \int \left[f^*(\theta) \frac{e^{-ikr_0}}{r_0} f(\theta) \left(ik \frac{e^{ikr_0}}{r_0} - \frac{1}{r_0} \frac{e^{ikr_0}}{r_0} \right) \right.$$

$$\left. - f(\theta) \frac{e^{ikr_0}}{r_0} f^*(\theta) \left(-ik \frac{e^{-ikr_0}}{r_0} - \frac{1}{r_0} \frac{e^{ikr_0}}{r_0} \right) \right] r_0^2 \sin\theta \, d\theta \, d\phi$$

$$= \frac{2ik\hbar}{2iM} \int f^*(\theta) f(\theta) \frac{1}{r_0^2} r_0^2 \sin\theta \, d\theta \, d\phi$$

Since

$$\frac{\hbar k}{M} = \left(\frac{h}{2\pi} \right) \left(\frac{2\pi}{\lambda} \right) \left(\frac{1}{M} \right) = \frac{p}{M} = v$$

We have

$$N_{sc} = \frac{\hbar k}{M} \int |f(\theta)|^2 \sin\theta \, d\theta \, d\phi = v \int |f(\theta)|^2 \, d\Omega \qquad \ldots (6.7)$$

Since $Y_{l,0}(\theta)$ are orthogonal and normalized, one obtains

$$N_{sc} = v \frac{\pi}{k^2} \sum_{l=0}^\infty (2l+1)|1-\eta_l|^2 \qquad \ldots (6.8)$$

One obtains the scattering cross section as

$$\sigma_{sc} = \frac{N_{sc}}{v} = \frac{\pi}{k^2} \sum_{l=0}^\infty (2l+1)|1-\eta_l|^2 \qquad \ldots (6.9)$$

Let us define

$$\sigma_{sc,\, l} = \pi\lambdabar^2(2l + 1)\, |1 - \eta_l|^2 \qquad \ldots (6.10)$$

Using (6.10), one can write (6.9) as

$$\sigma_{sc} = \sum_{l=0}^{\infty} \sigma_{sc,\, l} \qquad \ldots (6.11)$$

Obviously, in the integral cross-section the contributions of the partial waves for different l are simply added.

Let us now derive the expression for the reaction cross-section, σ_r. This can be determined by the number N_r, the number of particles which enter a sphere of radius r_0 around the centre without leaving it again through the entrance channel. Obviously, this is equal to the net flux within this sphere as computed from the completed wave function $\psi(r)$. We have

$$N_r = -\frac{\hbar}{2iM} \int \left(\psi^* \frac{\partial \psi}{\partial r} - \psi \frac{\partial \psi^*}{\partial r} \right)_{r = r_0} r_0^2 \sin\theta \; d\theta \; d\phi \qquad \ldots (6.12)$$

Minus sign indicates that inward flux N_r is positive. Using the orthogonality and normalization of $Y_{l,\,0}(\theta)$, one obtains

$$N_r = \upsilon\, \frac{\pi}{k^2} \sum_{l=0}^{\infty} (2l + 1)(1 - |\eta_l|^2) \qquad \ldots (6.13)$$

$$\therefore \qquad \sigma_r = \frac{\pi}{k^2} \sum_{l=0}^{\infty} (2l + 1)(1 - |\eta_l|^2) = \pi\lambdabar^2 \sum_{l=0}^{\infty} (2l + 1)(1 - |\eta_l|^2) \qquad \ldots (6.14)$$

Now, making use of the definition

$$\sigma_{l,\, r} = \pi\lambdabar^2(2l + 1)(1 - |\eta_l|^2) \qquad \ldots (6.15)$$

One can write Equation (6.14) as

$$\sigma_r = \sum_{l=0}^{\infty} \sigma_{l,\, r} \qquad \ldots (6.16)$$

The total reaction cross-section for a given l is the sum of (6.10) and (6.15). One obtains

$$\begin{aligned}
\sigma_{l,\, t} &= \sigma_{l,\, sc} + \sigma_{l,\, r} \\
&= \pi\lambdabar^2(2l + 1)\, [|1 - \eta_l|^2 + |1 - \eta_l|^2] \\
&= 2\pi\lambdabar^2(2l + 1)\, [1 - \mathrm{Re}(\eta_l)] \qquad \ldots (6.17)
\end{aligned}$$

Where $\mathrm{Re}(\eta_l)$ is the real part of complex number η_l.

one can draw following conclusions:

(i) If $|\eta_l| = 1$, then $\sigma_{l,\, r} = 0$. However, $\sigma_{l,\, sc}$ may be finite for the complex η_l.

(ii) If $|\eta_l| = 1$, then $\sigma_{l,\, sc} = 0$. This shows that there are no reactions, because there is not a single example known so far in which reactions take place without elastic scattering.

(iii) If $|\eta_l| = -1$, then $\sigma_{l,\, r} = 0$. In this case the scattering cross-section $\sigma_{l,\, sc}$ is maximum having value $4\pi\lambdabar^2(2l + 1)$.

(iv) If $|\eta_l| = 0$, then $\sigma_{l,r}$ attains its maximum value $\pi \lambda^2(2l + 1)$, but the scattering cross-section takes on its finite value. We may note that the scattering and reaction cross-sections are then both of the same size.

(v) $|\eta_l|^2$ can never be negative, otherwise the outgoing wave will have a higher intensity than the incident, i.e. incoming wave. Obviously, $|\eta_l|^2 \le 1$, condition insures that $\sigma_{l,r}$ does not become negative.

We note that the maximum value of the scattering cross-section is four times the maximum value of reaction cross-section. From Equation (6.10), we see that the phase of η affects the magnitude of $\sigma_{l,sc}$, whereas only $|\eta_l|^2$ enters in $\sigma_{l,r}$. In elastic scattering, both incoming and outgoing waves are coherent. These two can interfere contructively or destructively. The factor 4 mentioned above appears from the possibility of constructive interference and is, therefore, absent in the reaction cross-section. Figure 6.16 clearly shows the upper and lower limits of elastic scattering cross-section for a given reaction cross-section. Shaded region in the figure shows the possible values.

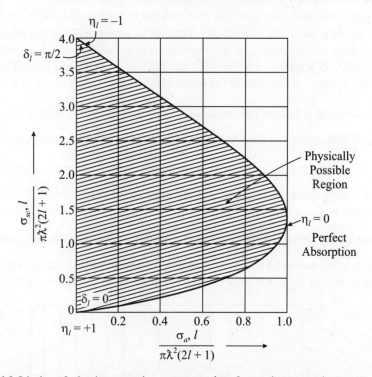

Fig. 6.16 Limits of elastic scattering cross-section for a given reaction cross-section

When the incident neutron has a high enough energy, the reduced wave length λ is small in comparison to the radius R of the nucleus. Now, if consider that nucleus is completely black to the incident neutrons, i.e. all neutrons which hit the target nucleus are absorbed by it and, therefore, lead to a reaction. All the particles with $l \le \dfrac{R}{\lambda}$ strike the nucleus. One can easily express this as

$$\eta_l = 0 \text{ if } l \le \frac{R}{\lambda} \quad \text{and} \quad \eta_l = 1, \text{ if } l > \frac{R}{\lambda} \qquad \qquad \dots (6.18)$$

Obviously, all the incident particles with $l \le \dfrac{R}{\lambda}$ react and all those with $l > \dfrac{R}{\lambda}$ pass by. Thus Equation (6.14) yields

$$\sigma_r = \sum_{l=0}^{R/\lambda} \pi\lambda^2 (2l + 1) = \pi(R + \lambda)^2 \qquad \qquad \dots (6.19)$$

When σ_r has its maximum value, one finds $\sigma_{sc} = \sigma_r$, and thus one finds that the *total cross-section is twice the geometrical cross-section of the nucleus*, apparently a paradoxial result. We have

$$\sigma_t = \sigma_{sc} + \sigma_r = 2\pi(R + \lambda)^2 \approx 2\pi R^2 \qquad \qquad \dots (6.20)$$

This may be explained to be due to the black nucleus casting its shadow behind it under the action of the neutron wave. Weisskopf confirmed the said result experimentally.

Let us consider N neutrons incident per unit area per second on the nucleus which presents an area πR^2 to the incident collimated beam. The number of neutrons absorbed by it is $N_a = N\pi R^2$, assuming localization of the impact on the nuclear surface $\lambda (\lambda \ll R)$. This gives an absorption coefficient $\sigma_a = \pi R^2$. In addition, a *shadow* of the nucleus is produced behind it. However, this shadow is not sharp due to diffraction at the edges (Figure 6.17). Moreover, it will be completely blurred at some distance L behind the nucleus ($L \sim \dfrac{R^2}{\lambda}$ or greater). This is very large for most black objects studied in optics. Obviously, the considerations of geometrical optics hold, the shadow being sharply defined. As regards nucleus, $\dfrac{R^2}{\lambda}$ is not as great, being only a few times the nuclear diameter, even for fairly energetic particles, e.g. for neutrons of energy 14 MeV, $\lambda \sim 1.25 \times 10^{-15}$ m; thus for fairly heavy nuclei R is only a few times λ). This means that the trajectories of the incident neutrons are bent at the surface of the nucleus due to diffraction through the scattering angle $\dfrac{\lambda}{R}$. In this way, the number scattered has been calculated (theoretically) and one obtains just $N\pi R^2$. This gives a cross section $\sigma_{sc} = \pi R^2$. Obviously, the total cross section is $2\pi R^2$.

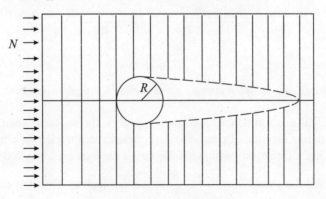

Fig. 6.17 Shadow scattering.

6.11   REACTION YIELD

The yield of the reaction is obtained from the product nuclei Y produced as a result of the reaction $X(x, y)Y$. When Y is stable, its number goes on increasing linearly with time. The number of Y nuclei produced in time dt is equal to the number of X-nuclei transmuted as a result of the reaction $X(x, y)Y$. Let $\sigma(x, y)$ denotes the cross-section. Thus, one can write

$$dN_Y = \sigma(x, y)N_0\, n_s\, dt \qquad \ldots (6.1)$$

Where N_0 is the total number of target nuclei in the foil and n_s is the number of projectiles incident on the target foil per second per unit area. Then, the number of Y nuclei produced in time t, is

$$N_Y = \sigma(x, y)N_0\, n_s\, t \qquad \ldots (6.2)$$

If the product nuclei Y are radioactive having disintegration constant λ, then the rate of change of the number of Y nuclei is equal to the difference between the rate of its production $\sigma(x, y)\, N_0 n$ and the rate of disintegration λN_Y. Thus, one finds

$$\frac{d N_Y}{dt} = \sigma(x, y)\, N_0\, n_s - \lambda\, N_Y \qquad \ldots (6.3)$$

or
$$\frac{d N_Y}{N_Y - \sigma(x, y)N_0\,\dfrac{n_s}{\lambda}} = -\lambda\, dt$$

On integration, one obtains

$$N_Y - \sigma(x, y)\, N_0\,\frac{n_s}{\lambda} = A\exp(-\lambda t)$$

Where A is the constant of integration. If at $t = 0$, $N_Y = 0$, then, one obtains

$$A = \frac{\sigma(x, y)\, N_0\, n_s}{\lambda}\,[1 - \exp(-\lambda t)]$$

One obtains finally,

$$N_Y(t) = \frac{\sigma(x, y)\, N_0\, n_s}{\lambda}\,[1 - \exp(-\lambda t)] \qquad \ldots (6.4)$$

From (6.4), we note that $N_Y(t)$ increases exponentially till it reaches a saturation value after a long time $t \to \infty$ from the start of the bombardment, i.e.

$$N_\infty = \frac{\sigma(x, y)\, N_0\, n_s}{\lambda} \qquad \ldots (6.5)$$

In practice it is found that N_Y becomes almost equal to the saturation value after 10 to 12 half lives. For larger values of $\sigma(x, y)$, N_∞ becomes larger. Interestengily, $\sigma(x, y)$ is larger for slow neutrons, the saturation yield N_∞ of the product nuclei can also be increased by using slow neutrons as projectiles. Moreover, N_∞ becomes larger for smaller values of l, i.e. for longer lived products nuclei. Finally, one can increase N_∞ by increasing the intensity of the incident beam of projectiles $\left(n_s = \dfrac{N}{S}\right)$. We may note that the saturation yield of short half-lived isotopes (τ few seconds to few days) by thermal neutron bombardment within a nuclear reactor is usually of the order of 10^{12} to 10^{17} nuclei, i.e. 10^{-4} to 10 µg. It is reported that for a long half-lived isotope, e.g. ^{239}Pu with a half life of the order of 10^4 y, the yield of reaction may be of the order of kilogramme.

Measurment of the yield of the reaction product nucleus Y gives us the value of reaction cross-section $\sigma(x, y)$.

The number of Y atoms left in the sample after the lapse of time t' from the end of the bombardment of the foil is given by

$$N_Y(t') = N_Y(t) \exp(-\lambda t')$$

$$= \frac{\sigma(x, y) N_0 n_s}{\lambda} [1 - \exp(-\lambda t)] \exp(-\lambda t') \qquad \dots (6.6)$$

At this instant its radioactivity is

$$\left| \frac{d N_Y}{dt'} \right| = \lambda N_Y(t') = \sigma(x, y) N_0 n_s \{-\exp(-\lambda t)\} \exp(-\lambda t') \qquad \dots (6.7)$$

Now, if the bombardment lasts for a very long time, i.e. $t \to \infty$, then the number of Y atoms reaches saturation value. One obtains at the time t' after the stoppage of bombardment.

$$N_Y(t') = \frac{\sigma(x, y) N_0 n_s}{\lambda} \exp(-\lambda t') \qquad \dots (6.8)$$

and

$$\left| \frac{d N_Y}{dt'} \right| = \sigma(x, y) N_0 n_s \exp(-\lambda t') \qquad \dots (6.9)$$

6.12 NUCLEAR TRANSMUTATIONS

In nuclear transmutation reactions, the particles involved are of not too high energy (<50 MeV). The first stage in the majority of instances is the formation of compound nucleus. In order to form the compound nucleus, the incident particle must overcome or penetrate the Coulomb barrier. In this section, we present specific examples of nuclear reactions with various projectiles:

(i) **Transmutation by α-particles, i.e. α-induced reactions**   When a nuclear reaction is induced by an α-particle, the compound nucleus so formed may break up by the emission of a proton, a neutron a γ-ray phonon, etc.

(α, p) Reaction   In general (α, p) reactions can be represented by the following equation:

$$^{A}_{Z}X + ^{4}_{2}\text{He} \rightarrow ^{A+4}_{Z+2}C^* \rightarrow ^{A+3}_{Z+1}Y + ^{1}_{1}\text{H}$$

One can write the Q-value of this reactions as

$$Q = B_Y - B_X - B_\alpha$$
$$= (A + 3) f_{BY} - A f_{BX} - 4 f_{B\alpha}$$

Where f_B is binding fraction. For most medium heavy nuclei, f_B is almost constant (~ 8 MeV per nucleon). Thus, one obtains ($\because f_{BY} = f_{BX}$)

$$Q(\alpha, p) \approx 3 \times 8 - 28 = -4 \text{ MeV}$$

Obviously, (α, p) reaction is *endoergic* for such nuclei. However, for some lighter nuclei, (α, p) reaction be exoergic. Examples of some (α, p) reactions are:

$$^{10}_{5}B + ^{4}_{2}He \rightarrow (^{14}_{7}N^*) \rightarrow ^{13}_{6}C + ^{1}_{1}H + 4.06 \text{ MeV}$$

$$^{14}_{7}N + ^{4}_{2}He \rightarrow (^{18}_{9}F^*) \rightarrow ^{16}_{8}O + ^{1}_{1}H + 1.2 \text{ MeV}$$

$$^{20}_{10}Ne + ^{4}_{2}He \rightarrow (^{24}_{12}Mg^*) \rightarrow ^{23}_{11}Na + ^{1}_{1}H - 2.38 \text{ MeV}$$

$$^{27}_{13}Al + ^{4}_{2}He \rightarrow (^{31}_{15}P^*) \rightarrow ^{30}_{14}Si + ^{1}_{1}H + 2.266 \text{ MeV}$$

$$^{39}_{19}K + ^{4}_{2}He \rightarrow (^{43}_{21}Sc^*) \rightarrow ^{42}_{20}Ca + ^{1}_{1}H - 0.89 \text{ MeV}$$

$$^{64}_{30}Zn + ^{4}_{2}He \rightarrow (^{68}_{32}Ge^*) \rightarrow ^{67}_{31}Ga + ^{1}_{1}H - 4 \text{ MeV}$$

There are few ($\alpha - p$) reactions lead to the formation of radioactive isotopes and that decay by β-emission, e.g.

$$^{11}_{5}B + ^{4}_{2}He \rightarrow (^{15}_{7}N^*) \rightarrow ^{14}_{6}C + ^{1}_{1}H + 0.75 \text{ MeV}$$

followed by

$$^{14}_{6}C \rightarrow {}_{7}N^{14} + \beta^-, \quad \text{(Half life } T_{\frac{1}{2}} = 5580 \text{ years)}$$

(α, n) reactions The general equation for such reaction is

$$^{A}_{Z}X + ^{4}_{2}He \rightarrow {}^{A+4}_{Z+2}C^* \rightarrow {}^{A+3}_{Z+2}Y + {}_{0}n^1$$

(α, n) reactions are also endoergic for medium heavy nuclei and for lighter nuclei these may be exoergic. Few examples of (α, n) reactions are:

$$^{7}_{3}Li + ^{4}_{2}He \rightarrow (^{11}_{5}B^*) \rightarrow ^{10}_{5}B + ^{1}_{0}n - 2.79 \text{ MeV}$$

$$^{9}_{4}Be + ^{4}_{2}He \rightarrow (^{13}_{6}C^*) \rightarrow ^{12}_{6}C + ^{1}_{0}n + 5.7 \text{ MeV}$$

$$^{11}_{5}B + ^{4}_{2}He \rightarrow (^{15}_{7}N^*) \rightarrow ^{14}_{7}N + ^{1}_{0}n + 0.15 \text{ MeV}$$

$$^{18}_{8}O + ^{4}_{2}He \rightarrow (^{22}_{10}Ne^*) \rightarrow ^{21}_{10}Ne + ^{1}_{0}n - 0.7 \text{ MeV}$$

$$^{27}_{13}Al + ^{4}_{2}He \rightarrow (^{31}_{15}P^*) \rightarrow ^{30}_{15}P + ^{1}_{0}n - 2.65 \text{ MeV}$$

$$^{65}_{29}Cu + ^{4}_{2}He \rightarrow (^{69}_{31}Ga^*) \rightarrow ^{68}_{31}Ga + ^{1}_{0}n - 5.84 \text{ MeV}$$

The product nuclei in the first four reactions are stable, while in last two reactions are unstable and radioactive. The fifth of the above reaction led to the discovery of artificial radioactivity in 1934 by I. Curie-Joliot and F. Joliot. Many of the product nuclei formed by ($\alpha - n$) reaction are unstable isotopes which then disintegrate with the emission of positron (β^+), e.g.

$$^{13}_{7}N \rightarrow ^{13}_{6}C + \beta^+, \quad (T_{\frac{1}{2}} = 10.1 \text{ min})$$

$$^{17}_{9}F \rightarrow ^{17}_{8}O + \beta^+, \quad (T_{\frac{1}{2}} = 1.2 \text{ min})$$

$$^{22}_{11}Na \rightarrow ^{22}_{10}Ne + \beta^+, \quad (T_{\frac{1}{2}} = 2.6 \text{ years})$$

$$^{26}_{13}Al \rightarrow ^{26}_{12}Mg + \beta^+, \quad (T_{\frac{1}{2}} = 7.0 \text{ sec})$$

$$^{28}_{14}\text{Si} \rightarrow {}^{27}_{13}\text{Al} + \beta^+, (T_{\frac{1}{2}} = 4.9 \text{ sec})$$

$$^{30}_{15}\text{P} \rightarrow {}^{30}_{14}\text{Si} + \beta^+, (T_{\frac{1}{2}} = 2.5 \text{ min})$$

Radiative capture (α, γ) reaction   There is a possibility for the compound nucleus formed by the capture of an α-particle to go to more stable configuration with emitting only a γ-ray photon, e.g.

$$^{7}_{3}\text{Li} + {}^{4}_{2}\text{He} \rightarrow ({}^{11}_{5}\text{B}^*) \rightarrow {}^{11}_{5}\text{B} + \gamma(h\nu)$$

General formula is

$$^{A}_{Z}X + {}^{4}_{2}\text{He} \rightarrow \left({}^{A+4}_{Z+2}C^* \right) \rightarrow {}^{A+4}_{Z+2}C + \gamma$$

These reactions are usually exoergic.

More than one Particle Emission   When the energy of the α-particle is high enough, more than one particle may be emitted from the compound nucleus, e.g. (α, $2n$), (α, np), (α, $3n$), (α, $4n$), (α, $3np$), etc.

(α, n) reaction finds wide use in the preparation of *neutron sources*.

Apart from the natural radio-elements, one can produce high-energy α-particle beams by accelerating doubly charged helium ions in particle accelerators. These α-particle beams are extensively used in the study of α-induced nuclear reactions.

(ii) **Transmutation by Protons, i.e. Proton induced Reactions**   One can obtain the high energy proton beams from particle accelerators by accelerating hydrogen ion. When such a high energy proton strikes a target nucleus, the compound nucleus so formed may disintegrate by the emission of different types of particles, e.g. (p, α), (p, n), (p, d), (p, α), (p, $2p$), (p, $3n$), etc.

(p, α) Reaction   The first artificial transmutation produced by Cockroft and Walton in 1932 on $^{7}_{3}\text{Li}$.

$$^{7}_{3}\text{Li} + {}^{1}_{1}\text{H} \rightarrow ({}^{8}_{4}\text{Be}^*) \rightarrow {}^{4}_{2}\text{He} + {}^{4}_{2}\text{He} + 17.28 \text{ MeV}$$

The general formula for (p, d) reaction is

$$^{A}_{Z}X + {}^{1}_{1}\text{H} \rightarrow \left({}^{A+1}_{Z+1}C^* \right) \rightarrow {}^{A-3}_{Z-1}Y + {}^{4}_{2}\text{He}$$

with
$$Q(p, \alpha) = (A-3)f_{BY} + 4f_{B\alpha} - Apf_{BX}$$
$$\approx 28 - 3f_{BY} \approx 4 \text{ MeV}$$

Where, we have assumed $f_{BY} \approx 8$ MeV for medium heavy nuclei. However, for target nuclei with lower values of A, Q may be much higher.

In the (α, p) reactions, we get elastic or inelastic scattering. Few examples of (α, p) reactions are:

$$^{6}_{3}\text{Li} + {}^{1}_{1}\text{H} \rightarrow ({}^{7}_{4}\text{Be}^*) \rightarrow {}^{3}_{2}\text{He} + {}^{4}_{2}\text{He} + 4 \text{ MeV}$$

$$^{7}_{3}\text{Li} + {}^{1}_{1}\text{H} \rightarrow ({}^{8}_{4}\text{Be}^*) \rightarrow {}^{4}_{2}\text{He} + {}^{4}_{2}\text{He} + 17.35 \text{ MeV}$$

$$^{11}_{5}B + ^{1}_{1}H \rightarrow (^{12}_{6}C^*) \rightarrow ^{8}_{4}Be + ^{4}_{2}He + 8.59 \text{ MeV}$$

$$^{19}_{9}F + ^{1}_{1}H \rightarrow (^{20}_{10}Ne^*) \rightarrow ^{16}_{8}O + ^{4}_{2}He + 8.12 \text{ MeV}$$

$$^{23}_{11}Na + ^{1}_{1}H \rightarrow (^{24}_{12}Mg^*) \rightarrow ^{20}_{10}Ne + ^{4}_{2}He + 2.38 \text{ MeV}$$

$$^{63}_{29}Cu + ^{1}_{1}H \rightarrow (^{64}_{30}Zn^*) \rightarrow ^{60}_{28}Ni + ^{4}_{2}He + 3.76 \text{ MeV}$$

The residual nucleus $^{8}_{4}Be$ formed in the third reaction is most unstable and breaks immediately after its production into two α-particles, i.e. $^{8}_{4}Be \rightarrow ^{4}_{2}He + ^{4}_{2}He$. Obviously, the final products of this reaction are three α-particles.

(p, n) Reaction The general formula is

$$^{A}_{Z}X + ^{1}_{1}H \rightarrow \left(^{A+1}_{Z+1}C^*\right) \rightarrow ^{A}_{Z+1}Y + ^{1}_{0}n$$

The residual nucleus Y is isobaric, i.e. having same A with the target nucleus with Z value one unit higher. We know that two isobars differing in Z by one unit cannot both be stable and hence the residual nucleus $^{A}_{Z+1}Y$ must be β-active. However, the target nucleus being necessarily stable. We have

$$^{A}_{Z+1}Y \xrightarrow[EC]{\beta^+} ^{A}_{Z}X + \beta^+ \text{ (or by electron capture)}$$

Few examples are:

$$^{11}_{5}B + ^{1}_{1}H \rightarrow (^{12}_{6}C^*) \rightarrow ^{11}_{6}C + ^{1}_{0}n - 1.763 \text{ MeV}$$

$$^{23}_{11}Na + ^{1}_{1}H \rightarrow (^{24}_{12}Mg^*) \rightarrow ^{23}_{12}Mg + ^{1}_{0}n - 4.84 \text{ MeV}$$

$$^{54}_{24}Cr + ^{1}_{1}H \rightarrow (^{55}_{25}Mn^*) \rightarrow ^{54}_{25}Mn + ^{1}_{0}n - 2.16 \text{ MeV}$$

$$^{63}_{29}Cu + ^{1}_{1}H \rightarrow (^{64}_{30}Zn^*) \rightarrow ^{63}_{30}Zn + ^{1}_{0}n - 4.15 \text{ MeV}$$

We may note that in each of the above reactions product nucleus is radioactive:

$$^{11}_{6}C \xrightarrow{\beta^+} ^{11}_{5}B \ (T_{\frac{1}{2}} = 2.5 \text{ min})$$

$$^{23}_{12}Mg \xrightarrow{\beta^+} ^{23}_{11}Na \ (T_{\frac{1}{2}} = 12.3 \text{ s})$$

$$^{54}_{25}Mn \xrightarrow{E.C.} ^{54}_{24}Cr \ (T_{\frac{1}{2}} = 310 \text{ days})$$

$$^{63}_{30}Zn \xrightarrow[E.C.]{\beta^+} ^{63}_{29}Cu \ (T_{\frac{1}{2}} = 38.5 \text{ min})$$

One can write Q of the (p, n) reaction as

$$Q(p, n) = M_X + M_H - M_Y - M_n$$

Since product nucleus Y is β^+ emitter, one finds

$$Q(\beta^+) = M_Y - M_X - 2m_e$$

$$Q(p, n) = -Q(\beta^+) - 2m_e - (M_n - M_H)$$

Now, since $M_n > M_H$, $Q(p, n) < 0$, i.e., for a stable target nucleus, the (p, n) reaction is always endoergic.

(p, γ) Reaction, i.e. Proton Capture Reaction   In some reactions, the compound nucleus formed by the absorbtion of proton by target nucleus X does not disintegrate by the emission of a particle and returns to its normal state by the emission of one or more γ-ray photons. Gereral equation is

$$\ _Z^A X + \ _1^1 H \rightarrow \left(_{Z+1}^{A+1} C^* \right) \rightarrow \ _{Z+1}^{A+1} C + \gamma$$

Few examples are:

$$\ _3^7 Li + \ _1^1 H \rightarrow (_4 Be^{8\,*}) \rightarrow \ _4^8 Be + \gamma$$

$$\ _6^{12} C + \ _1^1 H \rightarrow (_7 N^{13}\,*) \rightarrow \ _7^{13} N + \gamma$$

$$\ _7^{14} N + \ _1^1 H \rightarrow (_8^{15} O^*) \rightarrow \ _5^{15} O + \gamma$$

$$\ _{12}^{24} Mg + \ _1^1 H \rightarrow (_{13}^{25} Al^*) \rightarrow \ _{13}^{25} Al + \gamma$$

In (p, γ) reactions, the emitted γ-ray may have very high energy in some cases, e.g. in the first of the above reactions, $E_\gamma = 17.2$ MeV.

(p, d) Reaction

$$\ _Z^A X + \ _1^1 H \rightarrow \ ^{A-1}_Z Y + \ _1^2 H$$

Few examples are

$$\ _3^7 Li + \ _1^1 H \rightarrow \ _3^6 Li + \ _1^2 H$$

$$\ _4^9 Be + \ _1^1 H \rightarrow \ _4^8 Be + \ _1^2 H$$

These reactions are more common with heavier elements and protons of fairly high energy. Interestengily, these do not form any compound nucleus and they fall under the category of *pick-up* reactions.

There are examples of proton induced reactions in which if the incident proton beam has very high energy ($E_p > 20$ MeV), more than one particle may be emitted from the compound nucleus so formed, e.g. $^{63}Cu(p, pn)\ ^{62}Cu$, $^{63}Cu(p, 2d)\ ^{62}Zn$, and $^{88}Sr(p, 3n)\ ^{86}Y$

(iii)   **Transmutation by Deutrons, i.e. Deutron Induced Reactions**   Deutron is an isotope of hydrogen of mass number 2 and is present in natural hydrogen with a relative abudance of 0.015 %. One can obtain deuterium by repeated electrolysis of water. In all natural sources of water, ordinary water, H_2O is always mixed with a small fixed proportion of heavy water, D_2O in all natural sources of water. During electrolysis, in comparison to heavy water, light water is electrolyses faster. As a consequence, if one carries the electcrolysis for a very long time, the proportion of heavy water in the residue that is left becomes higher. Obviously, repeated electrolysis ultimately produce pure heavy water. However, for such separation of D_2O from ordinary water requires huge quantity of electrical energy, e.g. for obtaining 10^{-3} kg (= 1g) of pure D_2O, one requires about 30,000 ampere hours of electrical energy. D_2O is widely used in atomic reactors as moderators. In India heavy water plants are operating at Rana Pratap Sagar, Nangal, Baroda and Tutikorin. *Sp.* gravity of heavy water is 1.08. There are some chemical properties of D_2O same as of ordinary water.

When electrolysis of pure heavy water is carried, heavy hydrogen atom which when ionized gives the heavy hydrogen nucleus or deutron (D or $_1^2H$). A deutron has one proton and one neutron, bound together with a binding energy of 2.226 MeV. With the help of particle accelerators, deutrons can be accelerated to high energies and can be used as projectiles to induce various types of nuclear reactions. During these reactions p, n, α-particles etc. may be emitted. For these reactions, deutrons of 2 MeV energy are sufficient.

($d - \alpha$) Reactions   The general formula for this type of reaction is

$$^A_Z X + ^2_1 H \rightarrow \left(^{A+2}_{Z+1} C^* \right) \rightarrow ^{A+2}_{Z-1} Y + ^4_2 He$$

This reaction is usually exoergic. For a medium heavy nucleus, we have

$$Q(d, \alpha) = B_Y + B_\alpha - B_X - B_d = (A - 2) f_{BY} + 28 - A f_{BX} - 2.2$$
$$= 25.8 - 2 f_B \approx 25.8 - 16 = 9.8 \text{ MeV}$$

Obviously, $Q(d, \alpha) > 0$. Few examples of (d, α) reactions are:

$$^6_3 Li + ^2_1 H \rightarrow (^8_4 Be^*) \rightarrow ^4_2 He + ^4_2 He + 22.4 \text{ MeV}$$
$$^{14}_7 N + ^1_1 H \rightarrow (^{16}_8 O^*) \rightarrow ^{12}_6 C + ^4_2 He + 13.57 \text{ MeV}$$
$$^{23}_{11} Na + ^2_1 H \rightarrow (^{25}_{12} Mg^*) \rightarrow ^{21}_{10} Ne + ^4_2 He + 6.9 \text{ MeV}$$
$$^{24}_{12} Mg + ^2_1 H \rightarrow (^{26}_{13} Al^*) \rightarrow ^{22}_{11} Na + ^4_2 He + 1.96 \text{ MeV}$$
$$^{27}_{13} Al + ^2_1 H \rightarrow (^{29}_{14} Si^*) \rightarrow ^{25}_{12} Mg + ^4_2 He + 6.7 \text{ MeV}$$

Due to the high potential barrier to be crossed by the α-particle coming out of the compound nucleus, the (d, α) reaction is observed at fairly high energy of deutron and for low Z-target nuclei.

(d, p) Reaction

$$^A_Z X + ^2_1 H \rightarrow \left(^{A+2}_{Z+1} C^* \right) \rightarrow ^{A+1}_Z Y + ^1_1 H$$

In this case, the product nucleus is an isotope of the target nucleus having mass number A one unit higher. These reactions are usually exoergic. For medium heavy nucleus, we have

$$Q(d, p) = B_Y - B_X - B_d = (A + 1) f_{BY} - A f_{BX} - 2.2$$
$$\approx f_B - 2.2 \sim 8 - 2.2 = 5.8 \text{ MeV}$$

Obviously, $Q > 0$. However, for some light nuclei, Q may be negative, i.e. reaction endoergic. Few examples of (d, p) reactions are:

$$^7_3 Li + ^2_1 H \rightarrow (^9_4 Be^*) \rightarrow ^8_3 Li + ^1_1 H - 0.193 \text{ MeV}$$
$$^{12}_6 C + ^2_1 H \rightarrow (^{14}_7 N^*) \rightarrow ^{13}_6 C + ^1_1 H + 2.72 \text{ MeV}$$
$$^{23}_{11} Na + ^2_1 H \rightarrow (^{25}_{12} Mg^*) \rightarrow ^{24}_{11} Na + ^1_1 H + 4.74 \text{ MeV}$$
$$^{31}_{15} P + ^2_1 H \rightarrow (^{33}_{16} S^*) \rightarrow ^{32}_{15} P + ^1_1 H + 5.71 \text{ MeV}$$
$$^{109}_{47} Ag + ^2_1 H \rightarrow (^{111}_{48} Cd^*) \rightarrow ^{110}_{47} Ag + ^1_1 H + 4.6 \text{ MeV}$$

We will see that products of (d, p) reactions are the same as those of (n, γ) reactions and are usually radioactive. The product in the second example above is stable.

(d, n) Reaction

$$^A_Z X + ^2_1 H \rightarrow \left(^{A+2}_{Z-1} C^* \right) \rightarrow ^{A+1}_{Z+1} Y + _0 n^1$$

We can see that the product Y is an isotope of the compound nucleus. These reactions are usually exoergic with some exception. Few examples are

$$^7_3\text{Li} + ^2_1\text{H} \rightarrow (^9_4\text{Be}^*) \rightarrow ^8_4\text{Be} + ^1_0\text{n} + 15.024 \text{ MeV}$$

$$^9_4\text{Be} + ^2_1\text{H} \rightarrow (^{11}_5\text{B}^*) \rightarrow ^{10}_5\text{B} + ^1_0\text{n} + 4.36 \text{ MeV}$$

$$^{12}_6\text{C} + ^2_1\text{H} \rightarrow (^{14}_7\text{N}^*) \rightarrow ^{13}_7\text{N} + ^1_0\text{n} - 0.283 \text{ MeV}$$

$$^{16}_8\text{O} + ^2_1\text{H} \rightarrow (^{18}_9\text{F}^*) \rightarrow ^{17}_9\text{F} + ^1_0\text{n} - 1.625 \text{ MeV}$$

$$^{35}_{17}\text{Cl} + ^2_1\text{H} \rightarrow (^{37}_{18}\text{Ar}^*) \rightarrow ^{36}_{18}\text{Ar} + ^1_0\text{n} + 6.28 \text{ MeV}$$

We may note that when two deutrons interact, both the (d, n) and (d, p) reactions have been observed

$$^2_1\text{H} + ^2_1\text{H} \rightarrow (^4_2\text{He}^*) \Big\langle \begin{array}{l} ^3_1\text{H} + ^1_1\text{H} + 4.03 \text{ MeV} \\[6pt] ^3_2\text{He} + ^1_0\text{n} + 3.26 \text{ MeV} \end{array}$$

^3_1H is *tritium*, an isotope of hydrogen. Its nucleus is called *triton*. It is β-active having half life of about 12.4 years. We have

$$^3_1\text{H}(T) \xrightarrow{\beta^+} ^3_2\text{He} \ (T_{\frac{1}{2}} = 12.4 \text{ years})$$

In the second of the above reaction, the product ^3_2He is stable and it is an isotope of helium of mass number 3. ^3_2He is present in natural helium with relative abundance of 1.4×10^{-4} %. If we can generate sufficient quantity of tritium by the above reaction or through some other nuclear reaction, then one can bombard it with deutrons to produce (d, n) reaction as

$$^3_1\text{H}(T) + ^2_1\text{H}(D) \rightarrow (^5_2\text{He}^*) \rightarrow ^4_2\text{He} + _0n^1 + 17.6 \text{ MeV}$$

The above three reactions are known as *nuclear fusion reactions* and are of great importance in the release of energy by thermonuclear process. Research on nuclear fusion reactors is in progress.

(d, t) Reactions

$$^A_Z X + ^2_1\text{H} \rightarrow \left(^{A+2}_{Z+1} X^* \right) \rightarrow ^{A-1}_{Z} Y + ^3_1\text{H}$$

We can see that the product nucleus Y is an isotope of the target X with mass number one unit lower. The reaction cross-section of this type of reaction is rather low. Few examples are:

$$^7_3\text{Li} + ^2_1\text{H} \rightarrow (^9_4\text{Be}^*) \rightarrow ^6_3\text{Li} + ^3_1\text{H} - 0.996 \text{ MeV}$$

$$^9_4\text{Be} + ^2_1\text{H} \rightarrow (^{11}_5\text{B}^*) \rightarrow ^8_4\text{Be} + ^3_1\text{H} + 4.59 \text{ MeV}$$

We may note that in deutron induced reactions, the probability of radiative capture reaction is small because the extra nucleons in a compound nucleus are in virtual states. When the energy of the incident deutrons, i.e. $E_d > 20$ MeV or more, the reactions $(d, 2n)$, $(d, 3n)$, $(d, 2p)$, $(d, p\alpha)$, etc. in which two or more particles are emitted from the compound nucleus are possible.

(iv) Transmutation by Neutrons, i.e. Neutron-induced Reactions
Neutrons was discovered by Sir James Chadwick. Neutrons have no electric charge and can penetrate positively

charged nucleli without any experience of repulsive electrostatic forces. This is why that they have been used extensively in producing nuclear reactions. Even a zero energy neutron can penetrate into a nucleus, however high may be its atomic number. These zero energy neutrons can produce exoergic nuclear reactions. However, for endoergic nuclear reactions, the incident neutrons must have kinetic energy greater than the reaction threshold.

In order to produce neutron induced reactions, one must have high intensity neutron sources. Nuclear reactions also produce neutrons and those reactions which have high cross-sections are specially suitable as neutron sources.

Neutron induced reactions are associated with the emission of α-particles, protons, γ-rays, deutrons, etc.

(n, α) Reactions

$$^A_Z X + ^1_0 n \rightarrow ^{A+1}_Z C^* \rightarrow ^{A-3}_{Z-2} Y + ^4_2 He$$

In this types of reactions, the capture of slow neutrons causes in the emission of α-particle. Few examples are:

$$^6_3 Li + ^1_0 n \rightarrow (^7_3 Li^*) \rightarrow ^3_1 H + ^4_2 He + 4.785 \text{ MeV}$$

$$^{10}_5 B + ^1_0 n \rightarrow (^{11}_5 B^*) \rightarrow ^7_3 Li + ^4_2 He + 2.79 \text{ MeV}$$

$$^{35}_{17} Cl + ^1_0 n \rightarrow (^{36}_{17} Cl^*) \rightarrow ^{32}_{15} P + ^4_2 He + 0.935 \text{ MeV}$$

$$^{14}_7 N + ^1_0 n \, (^{15}_7 N^*) \begin{cases} \nearrow & ^{11}_5 B + ^4_2 He \\ \searrow & ^7_3 Li + ^4_2 He + ^4_2 He \end{cases}$$

The first two of the above reactions have fairly large reaction cross-section and therefore they are utilized in the construction of neutron detectors.

The capture of fast neutrons by heavier nuclei results in the emission of an α-particle, the product nucleus is usually radioactive, e.g.

$$^{23}_{11} Na + ^1_0 n \rightarrow (^{24}_{11} Na^*) \rightarrow ^{20}_9 F + ^4_2 He$$

followed by
$$^{20}_9 F \rightarrow ^{20}_{10} Ne + \beta^- \quad (T_{\frac{1}{2}} = 10.7 \text{ s})$$

$$^{27}_{13} Al + ^1_0 n \rightarrow (^{28}_{13} Al^*) \rightarrow ^{24}_{11} Na + ^4_2 He$$

followed by
$$^{24}_{11} Na \rightarrow ^{24}_{12} Mg + \beta^- \quad (T_{\frac{1}{2}} = 15 \text{ hrs})$$

(n, b) Reaction

$$^A_Z X + ^1_0 n \rightarrow (^{A+1}_Z C^*) \rightarrow _{Z-1}{}^A Y + ^1_1 H$$

Obviously, the product nucleus Y is an *isobar* of the target nucleus X with atomic number, Z one unit lower. Hence, the product nucleus Y is β^- active, decaying to the target nucleus

$$_{Z-1}{}^A Y \xrightarrow{\beta^-} {}^A_Z X$$

Now, $\qquad Q(\beta^-) = M_Y - M_X$. Thus Q of this reaction is

$$Q(n, p) = B_Y - B_X = M_X + M_n - M_Y - M_H$$
$$= (M_n - M_H) - Q(\beta^-)$$
$$= [0.782 - Q(\beta^-)] \text{ MeV}$$

If $Q(\beta^-) < 0.782$ MeV, the reaction is *exoergic* and for $Q(\beta^-) > 0.782$ MeV, the reaction is *endoergic*. We may note that in this reaction, the neutron is replacing the proton inside the nucleus. Few examples are:

$$^3_2\text{He} + ^1_0\text{n} \rightarrow (^4_2\text{He}^*) \rightarrow ^3_1\text{H} + ^1_1\text{H} + 0.764 \text{ MeV}$$

followed by $\quad _1\text{H}^3 \rightarrow ^3_2\text{He} + \beta^-$

$$^{14}_7\text{N} + ^1_0\text{n} \rightarrow (^{15}_7\text{N}^*) \rightarrow ^{14}_6\text{C} + ^1_1\text{H} + 0.627 \text{ MeV}$$

followed by $\quad _6\text{C}^{14} \rightarrow _7\text{N}^{14} + \beta^-$

$$^{27}_{13}\text{Al} + ^1_0\text{n} \rightarrow (^{28}_{13}\text{Al}^*) \rightarrow ^{27}_{12}\text{Mg} + ^1_1\text{H} - 1.83 \text{ MeV}$$

followed by $\quad ^{27}_{12}\text{Mg} \rightarrow ^{27}_{13}\text{Al} + \beta^-$

$$^{35}_{17}\text{Cl} + _0\text{n}^1 \rightarrow (^{36}_{17}\text{Cl}^*) \rightarrow ^{35}_{16}\text{S} + _1\text{H}^1 + Q$$

followed by $\quad ^{35}_{16}\text{S} \rightarrow ^{35}_{17}\text{Cl} + \beta^-$

$$^{65}_{29}\text{Cu} + ^1_0 n^* \rightarrow (^{66}_{29}\text{Cu}^*) \rightarrow ^{65}_{28}\text{Ni} + ^1_1\text{H} - Q$$

$$^{64}_{30}\text{Zn} + ^1_0\text{n} \rightarrow (^{65}_{30}\text{Zn}^*) \rightarrow ^{64}_{29}\text{Cu} + ^1_1\text{H} - Q$$

The last two along with the third reactions are performed with fast neutrons and have Q values negative. The product nuclei in the last two are also radioactive and they also produce finally nuclei identical with the initial target nuclei.

(n, γ) Reactions   Probably, the radioactive capture reaction with slow neutrons is the most common place and takes place almost with all elements.

Fermi and coworkers exposed a large number of elements in the periodic table to the action of *thermal neutrons* having energy about ~ 0.026 MeV at ordinary room temperature 300 K. They observed that in many cases radioactive product nuclei are produced. By comparing the radioactivity induced in the same elements by fast neutrons coming out of the source, they arrived at the conclusion that the production of (n, γ) reaction was greatly enhanced by the use of thermal neutrons. In some cases, they noticed that the increase was a factor 1000 or more. This discovery led to the development of the production of large scale production of radioactive substances artificially.

The general formula is

$$^A_Z X + ^1_0\text{n} \rightarrow (^{A+1}_Z C^*) \rightarrow ^{A+1}_Z Y + \gamma$$

The product is an isotope of the target nuclei with mass number larger by unity and same as the compound nucleus in the ground state, i.e. $Y = C$. The (n, γ) reactions are always exoergic ($Q > 0$) and one can induce these reactions by almost zero energy neutrons. The Q-value of the reaction is

$$Q(n, \gamma) = M_X + M_n - M_Y$$

Except for few light nuclei $Q(n, \gamma) \sim 8$ MeV. Clearly, in the (n, γ) reaction induced by zero energy neutrons, γ-rays having energy upto about 8 MeV are emitted. Few examples of these reactions are:

$$\ce{^1_1H + ^1_0n -> (^2_1H^*) -> ^2_1H + \gamma}$$

$$\ce{^2_1H + ^1_0n -> (^3_1H^*) -> ^3_1H + \gamma}$$

$$\ce{^{23}_{11}Na + ^1_0n -> (^{24}_{11}Na^*) -> ^{24}_{11}Na + \gamma}$$

$$\ce{^{63}_{29}Cu + ^1_0n -> (^{64}_{29}Cu^*) -> ^{64}_{29}Cu + \gamma}$$

$$\ce{^{103}_{45}Rh + ^1_0n -> (^{104}_{45}Rh^*) -> ^{104}_{45}Rh + \gamma}$$

$$\ce{^{115}_{45}In + ^1_0n -> (^{116}_{49}In^*) -> ^{116}_{49}In + \gamma}$$

$$\ce{^{197}_{79}Au + ^1_0n -> (^{198}_{79}Au^*) -> (^{198}_{79}Au) + \gamma}$$

We may note that except in the first case, the product nuclei in all other above reactions are radioactive. Interestingly, the capture of a neutron by the target nucleus increases the neutron-proton ratio and hence shifts the nucleus to the left above the stability line. This is why the product nucleus usually becomes β^- active, since it has an excess number of neutrons as compared to the number of protons which would make it stable. It is reported that in the case of some odd-odd product nuclei, e.g. ^{64}Cu and ^{108}Ag, both β^- and β^+ (or electron-capture) activities takes place. We have

$$^{64}_{29}\text{Cu} \xrightarrow{\quad\beta^-\quad} {}^{64}_{30}\text{Zn} \qquad (T_{\frac{1}{2}} = 12.8 \text{ hours})$$

$$\xrightarrow[\text{EC}]{\quad\beta^-\quad} {}^{64}_{28}\text{Ni}$$

$$^{108}_{47}\text{Ag} \xrightarrow{\quad\beta^+\quad} {}^{108}_{48}\text{Cd} \qquad (T_{\frac{1}{2}} = 2.3 \text{ min})$$

$$\xrightarrow[\text{EC}]{\quad\beta^+\quad} {}^{108}_{46}\text{Pd}$$

There are also few cases in which the radiative capture of neutrons exhibit exceptionally high probability at certain definite neutron kinetic energy. This is termed as *resonance capturing of neutrons*. The reaction ^{115}In (n, γ) ^{116}In exhibit a peak in the reaction cross-section curve at $E_n = 1.44$ MeV at which the cross section is 3×10^4 barns.

More than one particle emission in (n, γ) Reactions When the neutron energies are greater than 8 MeV, it is observed that more than one particle may be emitted by the compound nucleus formed by the capture of a neutron, e.g. $(n, 2n)$, $(n, 3n)$, (n, pn), $(n, p2n)$ reactions etc. For the emission of proton along with one or two neutrons, the energy of the excitation of the compound nucleus should be sufficiently high. This will enable the proton to cross the potential barrier of the nucleus and it will able to come out.

(n, d) and (n, t) Reactions These reactions are known as pick-up reactions, i.e. in these reactions compound nucleus is not formed. An example of (n, t) reaction is

$$\ce{^{14}_7N + ^1_0n -> ^{12}_6C + ^3_1H}(T)$$

The traces of tritium (T) is found in the atmosphere in the form of $^3_2$HO and mixed with the ordinary water is due to the formation of $^3_1$H in the atmosphere between cosmic ray neutrons and the nitrogen nuclei present in the atmosphere produced most likely by the above reaction.

(v) Gamma Ray Induced Reaction, i.e. Transmutation by Radiation Reactions of this type brought about by sufficiently high energy radiations are also referred as *photo-*

disintegrations or *photo nuclear reactions*. In order to produce these reactions, the energy of incident photon should exceed the binding energy of the particle to be ejected, e.g. ^1_0n, ^1_1p, ^4_2He, etc. These reactions are endoergic. If the energy of incident γ-radiation, i.e. photon is not sufficient to remove even ^1_0n, the nucleus may enter one of the excited states. The excess energy is emitted as radiation. Few examples of these types of reactions are:

$$^2_1H + \gamma \rightarrow \,^1_1H + \,^1_0n$$

This reaction is known as photo-disintegration of the deutron. The energy of γ-ray photon which can induce this reaction must be greater than the binding energy (2.226 MeV) of the deutron. This reaction has also been used in the measurment of the binding energy of the deutron accurately. The γ-rays from naturally radioactive isotope Th C″ having energy 2.62 MeV and also from the artificially radioactive ^{24}Na isotope having energy 2.76 MeV have been used for these measurments. To determine binding energy of deutron (B_d), conservation laws of for energy and momentum have been used.

Another (γ, n) reaction is

$$^9_4Be + \gamma \rightarrow (^9_4Be^*) \rightarrow \,^8_4Be + \,_0n^1$$

This reaction is used for the preparation of *photo-neutron source* and its threshold is 1.66 MeV. γ-rays from ^{24}Na or preferably from the isotope ^{124}Sb of antimony ($Z = 51$) is used for the said purpose due to relatively longer half of the latter ($T_{\frac{1}{2}} = 60$ days). The maximum energy γ-ray available from this source is 2.04 MeV.

Few other photo nuclear reactions are:

$$^{31}_{15}P + \gamma \rightarrow (^{31}_{15}P^*) \rightarrow \,^{30}_{15}P + \,^1_0n$$

$$^9_4Be + \gamma \rightarrow (^9_4Be^*) \rightarrow \,^8_3Li + \,^1_0n$$

$$^{25}_{12}Mg + \gamma \rightarrow (^{25}_{12}Mg^*) \rightarrow \,^{24}_{11}Na + \,^1_1H$$

$$^{10}_5B + \gamma \rightarrow (^{10}_5B^*) \rightarrow \,^8_4Be + \,^2_1H$$

$$^{11}_5B + \gamma \rightarrow (^{11}_5B^*) \rightarrow \,^8_4Be + \,^3_1H$$

$$^{12}_6C + \gamma \rightarrow (^{12}_6C^*) \rightarrow \,^8_4Be + \,^4_2He$$

$$^{83}_{36}Kr + \gamma \rightarrow (^{83}_{36}Kr^*) \rightarrow \,^{83}_{36}Kr + \gamma'$$

$$^{87}_{38}Sr + \gamma \rightarrow (^{87}_{38}Sr^*) \rightarrow \,^{87}_{38}Sr + \gamma'$$

When the γ-rays are of energy of the order of 10 MeV, either one neutron or proton can be emitted. At higher energies of γ-rays, reactions of the types (γ, np), $(\gamma, 2n)$, $(\gamma, n2p)$, etc., involving the ejection of two or more than two particles reactions have been observed, e.g.

$$^{16}_8O + \gamma \rightarrow (^{16}_8O)^* \rightarrow \,^{11}_6C + \,^1_0n + \,^1_0n + \,^1_0n + \,^1_1H + \,^1_1H$$

$$^{12}_6C + \gamma \rightarrow (^{12}_6C)^* \rightarrow \,^4_2He + \,^4_2He + \,^4_2He$$

(vi) Transmutation by ^3_1H and ^3_2He   Few examples are:

$$^6_3Li + \,^3_1H \rightarrow (^9_4Be^*) \rightarrow \,^7_3Li + \,^2_1H$$

$$^{59}_{27}Co + \,^3_1H \rightarrow (^{62}_{28}Ni^*) \rightarrow \,^{60}_{27}Co + \,^2_1H$$

$$^{63}_{29}\text{Cu} + {}^{3}_{1}\text{H} \rightarrow ({}^{66}_{30}\text{Zn}^*) \rightarrow {}^{64}_{29}\text{Cu} + {}^{2}_{1}\text{H}$$

$$^{2}_{1}\text{H} + {}^{3}_{1}\text{H} \rightarrow ({}^{5}_{2}\text{He}^*) \rightarrow {}^{4}_{2}\text{He} + {}^{1}_{0}\text{n}$$

$$^{32}_{15}\text{Si} + {}^{3}_{1}\text{H} \rightarrow ({}^{35}_{17}\text{Cl}^*) \rightarrow {}^{34}_{17}\text{Cl} + {}^{1}_{0}\text{n}$$

$$^{27}_{13}\text{Al} + {}^{3}_{1}\text{H} \rightarrow ({}^{30}_{14}\text{Si}^*) \rightarrow {}^{27}_{12}\text{Mg} + {}^{3}_{2}\text{He}$$

$$^{7}_{3}\text{Li} + {}^{3}_{1}\text{H} \rightarrow ({}^{10}_{5}\text{Be}^*) \rightarrow {}^{6}_{2}\text{He} + {}^{4}_{2}\text{He}$$

$$^{2}_{1}\text{H} + {}^{3}_{2}\text{He} \rightarrow ({}^{5}_{3}\text{Li}^*) \rightarrow {}^{4}_{2}\text{He} + {}^{1}_{1}\text{H}$$

$$^{3}_{1}\text{H} + {}^{3}_{2}\text{He} \rightarrow ({}^{6}_{3}\text{Li}^*) \rightarrow {}^{5}_{2}\text{He} + {}^{1}_{1}\text{H}$$

$$^{3}_{2}\text{He} + {}^{3}_{2}\text{He} \rightarrow ({}^{6}_{4}\text{Be}^*) \rightarrow {}^{5}_{3}\text{Li} + {}^{1}_{1}\text{H}$$

$$^{14}_{7}\text{N} + {}^{3}_{2}\text{He} \rightarrow ({}^{17}_{9}\text{F}^*) \rightarrow {}^{16}_{9}\text{F} + {}^{1}_{0}\text{n}$$

$$^{12}_{6}\text{C} + {}^{3}_{2}\text{He} \rightarrow ({}^{15}_{8}\text{O}^*) \rightarrow {}^{13}_{7}\text{N} + {}^{2}_{1}\text{H}$$

$$^{13}_{6}\text{C} + {}^{3}_{2}\text{He} \rightarrow ({}^{16}_{8}\text{O}^*) \rightarrow {}^{13}_{7}\text{N} + {}^{3}_{1}\text{H}$$

$$^{14}_{7}\text{N} + {}^{3}_{2}\text{He} \rightarrow ({}^{17}_{9}\text{F}^*) \rightarrow {}^{13}_{7}\text{N} + {}^{4}_{2}\text{He}$$

(vii) Heavy ions Induced Nuclear Reactions When particles heavier than α-particles are used as projectiles, nuclear reactions of this kind are placed under following categories:

(a) Transfer Reaction Such types of reactions involves the transfer of a nucleon or a few nucleons from one nucleus or a few nucleons from one nucleus to another as illustrated by the following examples:

$$^{14}_{7}\text{N} + {}^{10}_{5}\text{B} \rightarrow {}^{13}_{7}\text{N} + {}^{11}_{5}\text{B}$$

$$^{14}_{7}\text{N} + {}^{14}_{7}\text{N} \begin{cases} \rightarrow {}^{13}_{7}\text{N} + {}^{15}_{8}\text{O} \\ \rightarrow {}^{13}_{6}\text{C} + {}^{15}_{8}\text{O} \end{cases}$$

$$^{27}_{13}\text{Al} + {}^{14}_{7}\text{N} \begin{cases} \rightarrow {}^{25}_{13}\text{Al} + {}^{16}_{7}\text{N} \\ \rightarrow {}^{27}_{12}\text{Mg} + {}^{14}_{8}\text{O} \end{cases}$$

These reactions do not form any compound nucleus and hence they are also called as *direct reactions*.

(b) Accelerated Heavy ions induced Reactions In such reactions, the compound nucleus is formed and two, three, four or even more particles, mostly neutrons sometimes accompanied by one or two protons or α-particles are evaporated from the compound nucleus. It is observed that higher the energy of the projectile, the larger the number of particles ejected. Few examples of these reactions are:

$^{93}\text{Nb}({}^{12}\text{C}, 2n)^{103}\text{Ag}$, $^{197}\text{Au}({}^{12}\text{C}, 4n)^{205}\text{At}$, $^{27}\text{Al}({}^{12}\text{C}, \alpha n)^{34}\text{Cl}$, $^{238}\text{U}(14\text{N}, \alpha\ 5n)^{243}\text{Bk}$, and $^{51}\text{V}({}^{15}\text{O}, 2n)\text{Ga}^{65}$.

(c) Fission Spallation and Fragmentation Reactions (especially of the heavier elements) under this category, we can also mention the reaction wth target nuclei with moderate mass number, e.g. the products of the interaction of 125 MeV nitrogen ions with aluminium-27 (^{27}Al) range from ^{12}C to ^{38}K. We have

$$^{27}_{13}\text{Al} + ^{14}_{7}\text{N} \rightarrow (^{41}_{20}\text{Ca})^* \rightarrow ^{1}_{1}\text{H} + 2^{1}_{0}n + ^{38}_{19}\text{K}$$

Perhaps, the similar reactions involving the emission of α-particles can lead to $^{31}_{14}\text{Si}$ and other nuclei in the same region.

6.13 RECIPROCITY THEOREM (INVERSE PROCESS)

Let us consider a reversible process represented by $X + x \rightleftharpoons Y + y$, where X, x, Y and y occur in arbitrary numbers in a large box of volume V. Now, we want to find the relation between the total reaction cross-section $\sigma(x \rightarrow y)$, generally written as $\sigma(\alpha \rightarrow \beta)$ with α as entrance channel and β as reaction channel. Thus the total reaction cross-section can be written as $\sigma(\beta \rightarrow \alpha)$ for the *inverse reaction*. Here we use the fundamental theorem of statistical mechanics: *the principle of overall balance.* This theorem states that *when the system is in dynamical equilibrium all energetically permissible states are occupied with equal probability.* As mentioned above, we are interested in the reaction channels α and β. For this purpose, the theorem can be equivalently stated that in a given energy range the number of possible channels in the box is proportional to the number of possible channels into the box. The latter can be represented by

$$N_\alpha = \frac{4\pi \, p\, \alpha^2 \, V \, d\, p_\alpha}{h^3} = \frac{p\alpha^2 \, V \, d\, p_\alpha}{2\pi^2 \, \hbar^3} \qquad \ldots (6.1)$$

Now,
$$\upsilon = \frac{dE}{dp}, \quad \therefore N_\alpha = \frac{p\alpha^2 \, V \, d\, E_\alpha}{2\pi^2 \, \hbar^3 \, \upsilon_\alpha} \qquad \ldots (6.2)$$

and
$$N_\beta = \frac{p\beta^2 \, V \, d\, E_\beta}{2\pi^2 \, \hbar^3 \, \upsilon_\beta} \qquad \ldots (6.3)$$

However, the energy range for the two channels is of course the same, i.e. $dE_\alpha = dE_\beta$. Thus, one finds

$$\frac{\text{No. of channels } \alpha \text{ in the box}}{\text{No. of channels } \beta \text{ in the box}} = \frac{N_\alpha}{N_\beta} = \frac{p_\alpha^2 \, \upsilon_\beta}{p_\beta^2 \, \upsilon_\alpha} \qquad \ldots (6.4)$$

Obviously, the system is said to be in dynamical equilibrium when the number of transitions $\alpha \rightarrow \beta$ per second is equal to the number of transitions $\beta \rightarrow \alpha$ per second. This condition is found to usually holds and is termed as *principle of detailed balance.* Moreover, No. of transitions $\alpha \rightarrow \beta$ per sec. $= N_\alpha \, \omega(\alpha \rightarrow \beta)$, where $\omega(\alpha \rightarrow \beta)$ is the transition probability for the reaction $(\alpha \rightarrow \beta)$. Hence, one finds

$$p_\alpha^2 \, \upsilon_\beta \, \omega(\alpha \rightarrow \beta) = p\beta^2 \, \upsilon_\alpha \, \omega(\beta \rightarrow \alpha) \qquad \ldots (6.5)$$

The transition probability measures the chance that one particle moving with velocity υ in volume V is scattered per sec. Now, the cross-section σ corresponding to unit incident flux is given by

$$\sigma = \frac{\omega V}{\upsilon} \qquad \ldots (6.6)$$

Using $k = \dfrac{p}{\hbar}$, one obtains from (6.5) and (6.6),

$$k\alpha^2\, \sigma(\alpha \to \beta) = k\beta^2\, \sigma(\beta \to \alpha)$$

or
$$\frac{\sigma(\alpha \to \beta)}{\lambda\alpha^2} = \frac{\sigma(\beta \to \alpha)}{\lambda\beta^2} \qquad\qquad \text{... (6.7)}$$

So far we have assumed zero interinsic angular momenta for the particles. However if I is the intrinsic angular momentum of any one of the particles, then the corresponding density of states have to be multiplied by $2I + 1$.

Obviously, there are intrinsic momenta for X, x, Y and y. Now, one can write

$$(2I_X + 1)(2I_x + 1)\, k\alpha^2\, \sigma(\alpha \to \beta) = (2I_Y + 1)(2I_y + 1)\, k\beta^2\, \sigma(\beta \to \alpha) \qquad \text{... (6.8)}$$

The above equation must be employed when the initial and final states have definite angular momenta.

6.14 THEORIES OF NUCLEAR REACTIONS

In the previous sections we have studied the many types of nuclear reactions which can be broadly classified as:

(a) Particle induced reactions: (i) Compound nucleus formation reactions and (ii) Direct reactions

(b) Electromagnetic induced reactions There are various formal abstract theories based on quantum mechanics, but they are in principle inadequate to describe nuclear reactions under very general assumptions and moreover they are too complicated. Presently, there is no unified simple theory of nuclear reactions is available. Obviously, this makes necessary to use different available models and approximations depending upon the nature of the projectile and the target. In this section, we present a brief study of nuclear reaction models.

(i) Compound Nucleus Hypothesis of Nuclear Reactions The Compound nucleus theory of nuclear reactions was put forth by N. Bohr in 1936. The primary evidences on which this model was developed came after the discovery of neutron from 1935 onwards. This discovery led to major advances, because the absence of Coulomb barriers for neutron made a wide range of nuclei available for the study of nuclear reactions. For high energy neutrons the total cross section for neutron absorption and scattering was found to be of the nucleus. However, for very low energies, the reaction cross section is higher and approaches the limiting value of $\pi\lambda^2$, where λ is the reduced de-Broglie wave length of the neutrons.

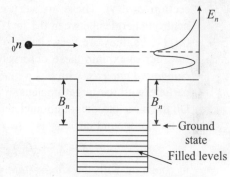

Fig. 6.18 Interaction of a neutron with a nucleus. An average single particle potential-well levels shown represent states of excitation of the system (neutron + nucleus) near and above the neutron binding energy B_n.

In order to explain the above results, it was assumed that the incident neutron moved in an average potential well due to all the nucleons within the nucleus (Figure 6.18) for an interval of time which is of the order of

$$t \sim \frac{\text{Nuclear diameter}}{\text{Nuclear velocity}} \sim \frac{10^{-14}}{10^7} = 10^{-21} \text{ s}.$$

The incident particle, i.e. neutron's brief time of association with the well indicates that it would have a large probability of escaping from the nucleus with absorption. Obviously, in this picture, the elastic scattering cross-section (σ_{sc}) should be large while capture or reaction cross-section (σ_{re}) would be quite small. However, at very low energies, σ_{re} should however be relatively large, since the incident neutron would spend a longer time near the target nucleus. Thus, in this case σ_{re} should depend on $\dfrac{1}{\upsilon}$. This means that at thermal energies σ_{sc} and σ_{re} should the comparable, as both of these approach the limiting value of $\pi \lambda^2$.

In the single particle potential well, the energy levels should be well separated from each other, the separation being of the order of 5 to 10 MeV. Above the neutron separation energy S_n (virtual levels), there should also be some levels. The level width which measures the probability of decay would be $\Gamma = \dfrac{\hbar}{\tau} = \dfrac{\hbar}{10^{-21}} \approx 1$ MeV because of the brief time of association ($\approx 10^{-21}$ s) of the neutron with the well. Variations of capture probability with energy in the thermal range were not likely to be produced by these resonances.

This picture is in sharp disagreement with observation at several points, i.e. observed results on neutron induced reactions. In several cases, the absorption cross section of neutron are reported to be very large at thermal neutron energies while elastic scattering cross-sections are much lower. We may note that at these very low energies, the absorption cross-section is due to radiative capture, i.e. (n, γ) type of reaction. Resonances are reported in most nuclei for both elastic scattering cross section (σ_{sc}) as well as for radiative capture cross section $\sigma(n, \gamma)$. These are very closely spaced, i.e. these resonances mostly appear at neutron energies between 0.1 to 10 eV. The widths of these are found to be of the order of 0.1 eV, i.e. they are very sharp.

In order to explain these observations, Niels Bohr in 1936 introduced the concept of the *compound nucleus*, which is a many body system formed by amalgamating of the incident particle with the target nucleus.

On the basis of compound nucleus hypothesis, a nuclear reaction can be described in terms of the compound nucleus as a *two stage process*

$$X + x \rightarrow (C^*) \rightarrow Y + y \qquad \qquad \text{... (6.1)}$$

in constraint with the *single-state* process

$$X + x \rightarrow Y + y \qquad \qquad \text{... (6.2)}$$

envisaged by the single-particle shell model and leading to the same products.

The composite system that is formed as a result of the incident particle x by the target nucleus X (Equation (6.1) is known as *Compound* nucleus (C^*). The compound nucleus ultimately breaks up by the emission of a particle y or of a γ-ray, it lives long enough compared to time taken by a nucleon of a few MeV energy to travel through the mean free path of collision between the nucleons within the nucleus (which is of the same order of the magnitude of nuclear radius but slightly less than the nuclear radius). The mean time between the collisions is about, $\dfrac{2\times10^{-15}}{5\times10^{-7}} \approx 10^{-22}$ s. Thus the life (t) of the compound nucleus is

$$t \sim 10^7 \times 10^{-22} \sim 10^{-15} \text{ s.}$$

Here we may note that the mean time for radiative transitions within the nuclei is $\sim 10^{-13}$ s, which is much longer than the time for decay of the compound nucleus by particle emission. Usually the particle y emitted by the decay of compound nucleus is different from the projectile x which enters X to produce the compound nucleus. In case if the particle y is identical with x, we have *inelastic scattering*. In this case $Y = X^*$ (as the target nucleus produced in a different energy state). There are very rare chances in which the residual nucleus is identical with the target nucleus and is produced exactly in the same state as the latter, one obtains what is known as the *compound elastic* or *resonance scattering*. This has been observed experimentally in the total cross-sections for interaction of slow neutrons with ^{238}U. We may note that the elastic scattering may alternatively take place by the action of the nuclear potential on the incident particle x without the entry of x into X to form the compound nucleus. This is called as the *potential* scattering and this has a much greater probability than the other.

The compound nucleus formed in a reaction is a relatively longer lived entity, and therefore the reaction actually proceeds in two steps: (i) the formation of the compound nucleus (C^*) by the absorption of projectile particle x by the target particle nucleus X, and (ii) the disintegration of C^* in a manner which is independent of the process of its formation into reaction product Y and y, Y in definite quantum states. The residual nucleus Y may sometime be left in the highly excited state which may then '*boil off*', another particle y' to leave the nucleus Y' leading to a two particle emission reaction, i.e. $X(x, y\, y')Y'$. This process may continue and further another particle y'' may be emitted from the excited nucleus Y' leaving the residual nucleus Y' in this (third) stage. Thus a series of particles, usually neutrons, may be 'boiled off' successively from a highly excited residual nucleus in each stage of the process. One can represent the two stages by which a nuclear reaction proceeds can be expressed symbolically as

$$X + x \rightarrow (C^*) \rightarrow Y + y$$

$$\rightarrow C + \gamma$$

The second stage of the nuclear reaction according to Bohr's suggestion is *independent* of the first. This means that the break-up of the compound nucleus C^* into different channels $Y + y$, $Y' + y'$, $Y'' + y''$, etc. should be determined only by the properties of the compound nucleus at its particular excitation and not by the mode of the formation.

The probability of decay of a nucleus is equal to its reciprocal of the mean-life t of the compound nucleus. If T is the width of the level, then the *uncertainty principle* gives

$$\Gamma\tau \sim \hbar \qquad\qquad\qquad \text{... (6.3)}$$

The width of the level is therefore

$$\Gamma = \frac{\hbar}{\tau} \qquad\qquad\qquad \text{... (6.4)}$$

The width of the level, Γ is obviously a measure of its decay. As stated above, the compound nucleus C^* may decay by the emission of different types of particles y', y'', y''', etc. leaving a different residual nucleus in each case. No doubt, each of these has a different probability of occurrence.

If Γ_y is partial width of the level for decay of C^* by the emission of particle y, then by considering the various possible stages of decay as mentioned above, one obtains the total width of the level as

$$\Gamma = \Gamma_y + \Gamma_{y'} + \Gamma_{y''} + \Gamma_{y'''} + \dots \Gamma_\gamma = \sum_y \Gamma_y + \Gamma_\gamma \qquad \dots (6.5)$$

The relative probabilities of different types of decay are then obtained as

$$\eta_y = \frac{\Gamma_y}{\Gamma}, \; \eta_{y'} = \frac{\Gamma_{y'}}{\Gamma}, \; \dots, \; \eta_\gamma = \frac{\Gamma_y}{\Gamma} \qquad \dots (6.6)$$

The angle-integrated cross-section for the reaction $X(x, y)Y$ near a compound nucleus resonance cen then be expressed as

$$\sigma_{xy} = \frac{\sigma_x \, \Gamma_y}{\Gamma} \qquad \dots (6.7)$$

Where σ_x is the reaction cross-section of the compound nucleus formation and Γ_y is proportional to the probability of the break up into the channel $Y + y$. The total width Γ is the sum of all partial widths (Equation 6.5), including Γ_y which measures the probability of decay back into the incident channel. Equation (6.7) clearly gives us the reaction cross-section from an entrance channel x to an exist channel y and obviously, it is then given by the product of the probabilities to form the compound nucleus C^* and for C^* to decay through y.

One can easily show that in case of s-wave scattering, the elastic scattering cross-section is obtained as

$$\sigma_{el} = \frac{\pi}{k^2} \, |1 - \eta_0|^2 = \frac{\pi}{k^2} \left| e^{2ik\,R_c} - 1 - \frac{2ik\,R_c}{\rho_0 - ik\,R_c} \right|^2 \qquad \dots (6.8)$$

and the expression for the reaction cross section

$$\sigma_{re} = \frac{\pi}{k^2} \, (1 - |\eta_0|^2) = \frac{\pi}{k^2} \frac{-4k\,R_c\,I_m\,\rho_0}{(R_e\,\rho_0)^2 + (\mathrm{Im}\rho_0 - k\,R_c)^2} \qquad \dots (6.9)$$

Where $\eta_0 = \exp 2i\delta_0$ is the *inelasticity parameter* and δ_0 is the complex phase shift for the $l = 0$ channel, R_c is the channel radius (outside which there is no interaction between the scattered particle and the residual nucleus). If r_0 is real, corresponding to the case of scattering from a real potential, σ_{re} vanishes as expected. Furthermore, since the reaction cross-section can not be negative, the absolute value of η_0 must be equal to or less than unity. This in turn, implies that the imaginary part of r_0 must be negative.

(ii) Discrete Resonances in the Compound Nucleus; Breit-Weigner One Level Relation   Let us consider a nuclear reaction produced by the absorption of the projectile x

by target nucleus, where x and X both are in the ground state. This reaction produce a compound nucleus C^* in a state of excitation near an isolated level of X, which is far removed from any of the other levels. This implies that the separation between the levels $D \gg T$, where D is the mean level spacing and T being the width of the level (Figure 6.19a). We may note the said reaction is initiated through a definite entrance channel $(X + x)$ which is characterized by a definite value of the kinetic energy of relative motion, E_x, between X and x having definite relative angular momentum. The energy of the excitation of the compound nucleus (C^*), so formed is expressed by

$$E_c = E_x + S_x = E_r \qquad \qquad ... (6.1)$$

Where E_r is the energy of the isolated level in which C^* is formed and S_x is the separation energy of x from the C^* in its ground state. S_x is given by

$$S_x = B_c - B_X - B_x \qquad \qquad ... (6.2)$$

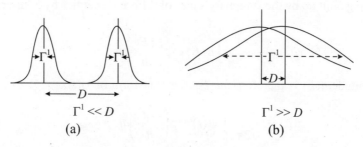

Fig. 6.19 Levels of compound nucleus (formed during the reaction $x + X \rightarrow C^* \rightarrow$) (a) Isolated levels, and (b) overlapping levels.

$B's$ in (6.2) denote the binding energies of the corresponding nuclei. Now, if x is a nucleon, then $B_x = 0$. Figure 6.20(a) shows the energy level diagram for the compound nucleus (C^*) and the subsequent break up of C^* into Y and y is shown in Figure 6.20(b). One can write, $E_c = E_y + S_y$, Where S_y is the separation energy of y from the ground state C^*, $S_y = B_c - B_Y - B_y$. Now, if y is a nucleon, $B_y = 0$. Here, we have assumed that both Y and y are formed in their ground states. However, this may not be the case and nucleus Y may be left in different excited states, which may give rise to different possible exist channels.

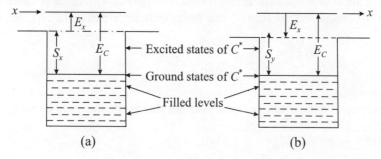

Fig. 6.20 Energy level diagram for the compound nucleus. (a) Excitation energy of C^* formed by the absorption of x by X, and (b) break of C^* into $(Y + y)$.

One can represent the state of the C^* so formed as above can be represented by a damped harmonic wave function as

$$\psi(t) = \psi_0 \, \exp\left[-\frac{i\,E_r t}{\hbar}\right] \exp\left[-\frac{\Gamma t}{2\hbar}\right]$$

$$= \psi_0 \, \exp\left[-i\left(\frac{E_r - i\Gamma}{2}\right)\frac{t}{\hbar}\right] \qquad \ldots (6.3)$$

Where $\dfrac{\Gamma}{2}$ is the half width of the level (which is actually decaying state) and its life time $\tau = \dfrac{\hbar}{T}$.

However, the wave function (6.3) does not represent a stationary state. One can think that this is being built up by the stationary states of different energies by Fourier integral method as

$$\psi(t) = -\frac{1}{\sqrt{2\pi}} \int a(E) \exp\left(\frac{i\,Et}{\hbar}\right) dt \qquad \ldots (6.4)$$

The amplitude $a(E)$ is therefore the Fourier transform of $\psi(t)$:

$$a(E) = \frac{1}{\sqrt{2\pi}} \int\limits_0^\infty \psi(t) \exp\left(\frac{i\,Et'}{\hbar}\right) dt$$

$$= \frac{1}{\sqrt{2\pi}} \int\limits_0^\infty \psi_0 \exp\left\{i\left(E - E_r + \frac{i\Gamma}{2}\right)\frac{t'}{\hbar}\right\} dt'$$

$$= \frac{\psi_0}{\sqrt{2\pi}} \left[\frac{\exp\left[i\left(E - E_r + \dfrac{i\Gamma}{2}\right)\dfrac{t'}{\hbar}\right]}{\dfrac{i\left(E - E_r + \dfrac{i\Gamma}{2}\right)}{\hbar}}\right]_0^\infty$$

The limits are 0 to ∞ because $\psi(t) = 0$ for $t < 0$. Now, considering positive values of time only (as the compound nucleus C^* can decay only after its formation), one obtains

$$a(E) = \frac{\psi_0}{\sqrt{2\pi}} \frac{i\hbar}{E - E_r + \dfrac{i\Gamma}{2}} \qquad \ldots (6.5)$$

The upper limit of the above integral vanishes due to the damping term $\exp\left(-\dfrac{\Gamma t'}{2\hbar}\right)$. Thus one obtains

$$|a(E)|^2 = \frac{|\psi_0|^2}{2\pi} \frac{\hbar^2}{(E - E_r)^2 + \dfrac{\Gamma^2}{4}} \qquad \ldots (6.6)$$

or

$$|a(E)|^2 \propto \frac{1}{(E_r - E)^2 + \dfrac{\Gamma^2}{4}} \qquad \dots (6.7)$$

The cross-section for the formation of the state E_c is by the process $X + x$ proportional to $|a(E)|^2$. Thus

$$\sigma_x = \frac{C}{(E - E_r)^2 + \dfrac{\Gamma^2}{4}} \qquad \dots (6.8)$$

Where C is a constant. One can determine C by noting the number of possible states in the incident channel, i.e.

$$dn = \frac{4\pi \, p_x^2 \Omega \, d p_x}{(2\pi\hbar)^3} \qquad \dots (6.9)$$

Where Ω is the volume of the enclosure within which the nuclear reaction is taking place. If σ_x is the reaction cross-section for the absorption of projectile x by target X. Obviously, the volume swept out by the effective collision area in one second is $\sigma_x \, v_x$, where v_x is the relative velocity of the incident particle. Thus the probability of finding the target nucleus X in volume $\dfrac{\sigma_x \, v_x}{\Omega}$ and the probability of the C^* nucleus in the given entrance channel per second is

$$\frac{\sigma_x \, v_x}{\Omega} \frac{4\pi \, p_x^2 \, \Omega}{(2\pi\hbar)^3} dp_x = \frac{\sigma_x \, p_x^2 \, v_x}{2\pi^2 \, \hbar^3} dp_x = \frac{\sigma_x \, p_x^2}{2\pi^2 \, \hbar^3} dE_x$$

Now, integrating the above over all possible energies, one obtains the total probability as

$$P = \frac{1}{2\pi^2 \hbar} \int_{-\infty}^{\infty} \frac{\sigma_x \, p_x^2}{\hbar^2} dE_x = \frac{1}{2\pi^2 \hbar} \int_{-\infty}^{\infty} \frac{\sigma_x}{\lambda^2} dE_x$$

The integrand has finite values only for within the width Γ of the level which are so narrow that one can neglect the variation of λ and write ($\because dE = dE_x$)

$$P = \frac{1}{2\pi^2 \hbar \lambda^2} \int_{-\infty}^{\infty} \sigma_x \, dE$$

$$= \frac{C}{2\pi^2 \hbar \lambda^2} \int_{-\infty}^{\infty} \frac{dE}{(E - E_r)^2 + \dfrac{\Gamma^2}{4}}$$

$$= \frac{C}{2\pi^2 \hbar \lambda^2} \frac{2\pi}{\Gamma} = \frac{C}{\pi \hbar \Gamma \lambda^2} \qquad \dots (6.10)$$

The probability of formation of C^* given by (6.10) must be equal to the probability of decay of C^* through the same channel (exit channel = entrance channel). This is a consequence of *reciprocity theorem*. Writing this probability of decay through the entrance channel as $\dfrac{T_x}{\hbar}$, one obtains

$$\frac{C}{\pi \hbar T \lambdabar^2} = \frac{T_x}{\hbar}$$

or
$$C = \pi \lambdabar^2 \, \Gamma_x \, \Gamma \qquad\qquad \text{... (6.11)}$$

Thus, one finally obtains the cross section for the formation of C^* as

$$\sigma_x = \frac{\pi \lambdabar^2 \, \Gamma_x \, \Gamma}{(E - E_r)^2 + \dfrac{\Gamma^2}{4}} \qquad\qquad \text{... (6.12)}$$

Through the exit channel $Y + y$, one can write the relative probability of decay of C^* as $\dfrac{\Gamma_y}{\Gamma}$. Then, one obtains the cross-section for the reaction $X(x, y)Y$ as

$$\sigma(x, y) = \sigma_x \frac{\Gamma_y}{\Gamma} = \pi \lambdabar^2 \frac{\Gamma_x \, \Gamma_y}{(E - E_r)^2 + \dfrac{\Gamma^2}{4}} \qquad\qquad \text{... (6.13)}$$

Equation (6.13) is the well known *Breit Wigner one level formula* for spinless nuclei at very low energies. Thus the relative angular momentum of the particles in the entrance channel is $l = 0$. However, if l is not zero, which in the case when energy is higher, one finds to take into account the *statistical factor* of the C^* state formed which, for spinless nuclei x and X, is given by

$$g = 2l_c + 1 = 2l + 1 \qquad\qquad \text{... (6.14)}$$

We know that each of $(2l + 1)$ substates can decay with equal probability. Obviously Γ_x has to be multiplied by g. One obtains

$$\sigma_x^{(l)} = \pi \lambdabar^2 (2l + 1) \frac{\Gamma_x \, \Gamma}{(E - E_r)^2 + \dfrac{\Gamma^2}{4}} \qquad\qquad \text{... (6.15)}$$

or
$$\sigma^l(x, y) = \pi \lambdabar^2 (2l + 1) \frac{\Gamma_x \, \Gamma_y}{(E - E_r)^2 + \dfrac{\Gamma^2}{4}} \qquad\qquad \text{... (6.16)}$$

For Spin Particles

Let us consider the case in which x and X have spins I_x and I_X respectively. Now, for a relatively orbital angular momentum l between x and X, the angular momentum of the compound nucleus state is

$$\mathbf{I}_C = \mathbf{I}_X + \mathbf{I}_x + \mathbf{l}$$

Let us define the *channel spin* for the entrance channel as $j = I_x + I_X$. Now the possible values of the channel spin are

$$|I_X - I_x| \le j \le (I_X + I_x)$$

Obviously, one can write

$$\mathbf{I}_C = \mathbf{j} + \mathbf{l}$$

For the compound nucleus state of spin $\mathbf{I}_C$, the statistical factor gives the probability for the occurrence of this spin by the vector addition of j and l. We have its value

$$g_C = \frac{2I_C + 1}{(2j + 1)(2l + 1)}$$

One will also to consider the probability of the occurrence of a particular channel spin j by the vector addition of $\mathbf{I}_X$ and $\mathbf{I}_x$, which is equal to the statistical factor $(2l + 1)$ for a given angular momentum l. Thus one obtains

$$g = g_c\, g_j\, g_l = \frac{2I_C + 1}{(2j + 1)(2l + 1)} \frac{(2j + 1)}{(2I_X + 1)(2I_x + 1)} (2l + 1)$$

$$= \frac{2I_C + 1}{(2I_X + 1)(2I_x + 1)} \qquad \dots (6.17)$$

Now, the cross section becomes

$$\sigma^{(l)}(x, y) = \frac{\pi\lambda^2 g\, \Gamma_x \Gamma_y}{(E - E_r)^2 + \dfrac{\Gamma^2}{4}}$$

$$= \pi\lambda^2 \frac{2I_C + 1}{(2I_X + 1)(2I_x + 1)} \left\{ \frac{\Gamma_x \Gamma_y}{(E - E_r)^2 + \dfrac{\Gamma^2}{4}} \right\} \qquad \dots (6.18)$$

Neutron Resonances at Low Energy

Very low-energy (near thermal) neutron capture resonances were seen in 1935 by Fermi and others which correspond to narrow and hence relatively long-lived states in the compound nucleus at an excitation energy of ~ 8 MeV. This was interpreted by Bohr in 1936 in his compound nucleus reaction hypotheris as evidence for the neutron rapidly sharing its energy through the strong interaction with all other nucleons.

We have for compound elastic scattering $y = x$, so that $\Gamma_y = \Gamma_x$. One obtains from Equation (6.18),

$$\sigma_{sc}^{(l)} = \sigma^{(l)}(x, x) = \pi\lambda^2 g \frac{\Gamma_x^2}{(E - E_r)^2 + \dfrac{\Gamma^2}{4}} \qquad \dots (6.19)$$

At resonance, we have $E = E_r$. This gives

$$\sigma_{sc}^{(l)} = 4\pi \lambdabar^2 g \left(\frac{\Gamma_x}{\Gamma}\right)^2 \qquad \ldots (6.20)$$

With low energy neutrons ($l = 0$), only exit channels open are (n, n) and (n, g). Hence, one finds $\Gamma = \Gamma_n + \Gamma_\gamma \approx \Gamma_\gamma$ (since $\Gamma_n << \Gamma_\gamma$). One obtains

$$\sigma(n,\ n) = \sigma_{sc}^{(0)} = \pi \lambdabar^2 g \frac{\Gamma_n^2}{(E - E_r)^2 + \frac{\Gamma^2}{4}} \qquad \ldots (6.21)$$

and

$$\sigma(n,\ \gamma) = \pi \lambdabar^2 g \frac{\Gamma_n \Gamma_\gamma}{(E - E_r)^2 + \frac{\Gamma^2}{4}} \qquad \ldots (6.22)$$

At resonance, one finds

$$\sigma_{sc}^{(r)} = 4\pi \lambdabar^2 g \left(\frac{\Gamma_n}{\Gamma}\right)^2 \qquad \ldots (6.23)$$

and

$$\sigma^{(r)}(n,\ \gamma) = 4\pi \lambdabar^2 g \left(\frac{\Gamma_n}{\Gamma_\gamma}\right) \qquad \ldots (6.24)$$

Dividing (6.23) by (6.24), one obtains

$$\frac{\sigma_{sc}(r)}{\sigma^{(r)}(n,\ \gamma)} = \frac{\Gamma_n}{\Gamma_\gamma} << 1 \qquad \ldots (6.25)$$

Using typical values of Γ_n and Γ_γ as 10^{-3} eV and 10^{-1} eV and $E_n \sim 1$ eV, one obtains from (6.25)

$$\frac{\sigma_{sc}(r)}{\sigma^{(r)}(n,\ \gamma)} = 10^{-2}$$

As mentioned earlier, we again see that the levels of the compound nucleus just above the neutron separation energy S_n are usually very sharp and well-separated from one another. In many cases this gives rise to a number of resonance peaks in the σ *vs* E plots. With the selection of proper energy. One can resolve these peaks as can be seen in Figure 6.21 for indium.

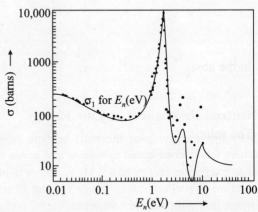

Fig. 6.21 Slow neutron cross-section for neutron

Interference Between Resonance and Potential Elastic Scattering

When we see that both resonance and potential scatterings take place, then we have to add the amplitude A_{pot} of the latter with the amplitude of resonance scattering. Then for low energy neutrons, one can write the total elastic scattering cross section as

$$\sigma_{sc} = \pi \lambdabar^2 [A_{res} + A_{pot}]$$

The resonance scattering cross section for spinless particles for $l = 0$ is given by

$$(\sigma_{sc})_{res} = \pi \lambda^2 \frac{\Gamma^2}{(E - E_r)^2 + \dfrac{\Gamma^2}{4}}$$

In case, if no other channel is present, then the width $\Gamma = \Gamma_x$. Thus, one obtains

$$(\sigma_{sc})_{res} = \pi \lambda^2 \frac{\Gamma^2}{(E - Er)^2 + \dfrac{\Gamma^2}{4}}$$

Now, writing this as $\pi \lambda^2 |A_{res}|^2$, one finds (except for a phase factor)

$$A_{res} = \sqrt{\pi} \; \lambda = \frac{\Gamma}{(E - E_r) + \dfrac{i\Gamma}{2}} \qquad \qquad \text{... (6.26)}$$

Let us now calculate A_{pot}. One can write the asymptotic form of the wave equation as

$$\Delta^2 \psi + k^2 \psi = 0$$

Where

$$k = \sqrt{\frac{2ME}{\hbar^2}} \; . \text{ Now, writing}$$

$$\psi(\mathbf{r}) = \psi(r, \theta) = \frac{u_l(r)}{r} Y_l^0(\theta)$$

Then the radial wave equation takes the form

$$\frac{d^2 u_0}{dr^2} + \frac{2M}{\hbar^2} \left\{ E - \frac{\hbar^2 l(l+1)}{2Mr^2} \right\} u_l = 0$$

In the absence of any long rang force (i.e., for neutrons), the above takes the form

$$\frac{d^2 u_0}{dr^2} + k^2 u_0 = 0 \qquad \qquad \text{... (6.27)}$$

The solution of (6.27) can be written as

$$u_0 = a \exp(-ikr) + b \exp(ikr) \qquad \qquad \text{... (6.28)}$$

From Legendre Polynomials, one obtains the value of constants a and b as

$$a = i\frac{\sqrt{\pi}}{k} \quad \text{and} \quad b = -i\eta_0 \frac{\sqrt{\pi}}{k} \qquad \qquad \text{... (6.29)}$$

(The total wave function in the presence of the absorbing nucleus can be written as

$$\psi(r, \theta) = \frac{1}{2\pi kr} \sum_{l=0}^{\infty} i^l (2l+1) \left[\eta_l \exp\left\{ i\left(kr - \frac{l\pi}{2} \right) \right\} - \exp\left\{ -i\left(kr - \frac{l\pi}{2} \right) \right\} \right] P_l \cos\theta$$

Where η_l is a complex number and it is assumed that outgoing wave is affected both in amplitude and phase. One can obtain a and b by comparing with this equation).

Now, writing

$$R\left(\frac{1}{u_0}\frac{du_0}{dr}\right)_R = f_0 \qquad \dots (6.30)$$

Where f_0 denotes the logarithmic derivative at $r = R$ for $l = 0$ neutrons. Using (6.28), one obtains

$$f_0 = -ikr\,\frac{\exp(-ikR) + \eta_0\exp(ikR)}{\exp(-ikR) - \eta_0\exp(ikR)} \qquad \dots (6.31)$$

This gives η_0 in terms of f_0 as

$$\eta_0 = \frac{f_0 + ikR}{f_0 - ikR}\,\exp(-2ikR) \qquad \dots (6.32)$$

Now, if f_0 is real, then $|\eta_0|^2 = 1$, i.e. there is no reaction. One obtains from (6.32),

$$1 - \eta_0 = \left[-\frac{2ikR}{f_0 - ikR} + \exp(2ikR) - 1\right]\exp(-2ikR)$$

$$\therefore \qquad |1 - \eta_0|^2 = \left[-\frac{2ikR}{f_0 - ikR} + \exp(2ikR) - 1\right]^2 \qquad \dots (6.33)$$

We have

$$\sigma_{sc}^{(l)} = \pi\lambda^2(2l + 1)\,|1 - \eta_l|^2$$

This gives for $l = 0$

$$\sigma_{sc} = \pi\lambda^2\,|A_{res} + A_{pot}|^2$$

One can interpret the term containing f_0 in (6.33) as the resonance scattering amplitude, i.e.

$$A_{res} = -\frac{2ikR}{f_0 - ikR} \qquad \dots (6.34)$$

and

$$A_{pot} = \exp(2ikR) - 1$$
$$= 2i\exp(ikR)\sin kR \qquad \dots (6.35)$$

In the neighbourhood of isolated resonance, (6.34) for A_{res} reduces to (6.26). Then, one obtains for the total scattering amplitude as

$$\sigma_{sc} = \pi\lambda^2\left|\frac{\Gamma}{E - E_r + \dfrac{i\Gamma}{2}} + 2\exp(ikR)\sin kR\right|^2 \qquad \dots (6.36)$$

When $(E - E_r) \sim \dfrac{\Gamma}{2}$, the resonance scattering amplitude is of the order of unity. At the resonance $E = E_r$, one finds that it is 2 and obviously much larger than the potential scattering amplitude (Equation 6.35) if $kR \ll 1$. At the resonance, the scattering cross-section attains the maximum value, i.e. $4\pi\lambda^2$. Far away from the resonance, we have $A_{res} \ll A_{pot}$. This shows that in this situation the scattering cross section is determined by A_{pot} and this is equal to that due to an impenetrable sphere if $kR \ll 1$. Thus, we have $\sigma_{sc} \sim 4\pi R^2$.

When the two amplitudes are equal, i.e. $A_{pot} = A_{res}$, there is *interference* between them (Figure 6.22). However, for $E < E_r$, A_{res} and A_{pot} both have opposite sign and they produce *destructive interference*. On the otherhand for $E > E_r$, both have the same sign and produce *constructive interference*. Figure 6.22 reveals these features.

When we are dealing with incident charged particles, we will also have to consider Coulomb scatteing. However, in the case of nuclear reactions, such coherence does not exist and there is no interference except when more than one resonance levels are present.

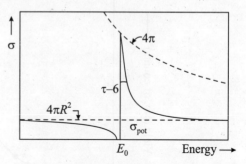

Fig. 6.22 For $l = 0$, elastic scattering cross section near resonance for a target of spin zero.

Very Low Energy Neutron Resonance

The width of the neutron T_n depends upon the density of states per unit energy interval available for it. We have

$$T_n \propto \frac{4\pi p^2 \, dp}{(2\pi\lambda)^3 \, dE} = \frac{M^2 v}{2\pi^2 \, \hbar^3} \qquad \text{... (6.37)}$$

or

$$T_n \propto v \qquad \text{... (6.38)}$$

For extremely low energies, the denominator in the Equations (6.21) or Equation (6.22) are almost constant since T_γ is independent of E_n. Clearly,

$$\sigma(n, \gamma) \propto \lambda^2 v \qquad \text{... (6.39)}$$

But $\lambda \propto v^{-1}$ and hence

$$\sigma(n, \gamma) \propto v^{-1} \qquad \text{... (6.40)}$$

Relation (6.40) is very famous $\dfrac{1}{v}$ law, one obtains

$$\sigma_{sc} \propto \lambda^2 v^2 = \text{constant} \qquad \text{... (6.41)}$$

Figure 6.21 shows the above features of the slow neutron resonance for (n, γ) reaction for indium.

Charged Particle Resonances

When charged particles are projectiles, it is found that in the case of intermediate and heavy nuclei, resonance reactions usually does not produce because of high Coulomb potential barrier. At lower energies, transmission through such high barrier is quite unlikely. On the other hand with light charged nuclei, such reactions are possible and have been reported in many cases. In the case of light nuclei Coulomb barrier is low and the energies of the resonance levels are also relatively high. Figure 6.23 shows a charged particle resonance in the case of ^{4}Be (p, γ) reaction.

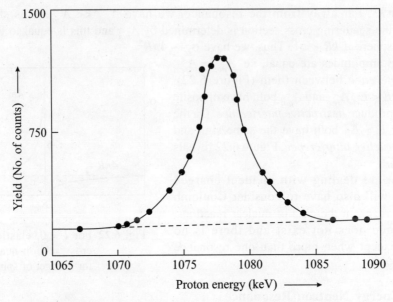

Fig. 6.23 ^{4}Be (p, γ) reaction yield. The thickness of beryllium target is 2×10^{-5} cm and deposited on copper at $E_p = 1077$ keV.

6.15 CONTINUUM STATES OF THE COMPOUND NUCLEUS

For compound nucleus excitations to the energy range at which the total width of levels become comparable with their spacing the resonances merge into a *continuum*. The cross-section for formation of the compound nucleus must now be calculated by averaging a Breit-Wigner type of cross-section over an energy interval containing many levels, and the result for spinless particles incident as an *s*-wave ($l = 0$) is

$$\bar{\sigma}_a = \pi \lambdabar^2 \frac{2\pi \, \bar{\Gamma}_a}{D} \qquad \qquad \dots (6.1)$$

Where $\bar{\Gamma}_a$ is the mean level width over the averaging interval and D is the mean spacing of levels in this interval. The quantity $\dfrac{\bar{\Gamma}_a}{D}$ is the *s*-wave *strength function* and is a measure of the complexity of internal motion in the compound state.

The concepts of the continuum region (especially that of evaporation) suggest that the emergent particles should often have an isotropic angular distribution in the *CM* system. Experiments confirm this but also shows that for transitions to the lower states of Y (i.e., the higher energy emitted particles) the angular distributions have a marked forward peak. These pearks are associated with non-compound nucleus processes known as direct interactions.

(ii) Barrier Penetration Effect on Reaction Cross-section All particles, except very low energy neutrons ($l = 0$) have to penetrate through the potential barrier surrounding the nucleus to initiate a reaction. Usually, there are two types of barrier to be penetrated:

(a) **Centrifugal Barrier** Due to orbital angular momentum $l\hbar$ of the incident particle x relative to the target nucleus X when $l = 0$. In this case, $V_l = \dfrac{\hbar^2 l(l+1)}{2Mr^2}$.

(b) **Coulomb barrier** When x is a charged particle carring a charge $Z'e$ and target nucleus having charge Ze. In this case $V_c = \dfrac{ZZ'e^2}{4\pi\varepsilon_0 r}$.

Outside the nucleus, the radial wave function obeys the wave equation

$$n\frac{d^2 u_l}{dr^2} + \frac{2M}{\hbar^2}\left\{ E - \frac{ZZ'e^2}{4\pi\varepsilon_0 r} - \frac{\hbar^2}{2M}\frac{l(l+1)}{r^2}\right\}u_l = 0 \qquad \dots (6.1)$$

We may note that in the case of wave function within the nucleus, one has to add an attractive short range nuclear potential $(-V_n)$ in place of the repulsive Coulomb potential, usually this is neglected for the inside region.

In the case of slow neutrons $(l = 0)$, both the potential terms in (6.1) become zero outside the nucleus. The above wave equation for this case, gives a solution which may be considered to be a superposition of two waves except $\exp(ikr)$ and $\exp(-ikr)$. These represent a wave incident on the nucleus near surface and a wave reflected from it for $r > R$, i.e.

$$u_l(r) = A\,\exp(ikr) + B\,\exp(-ikr) \qquad \dots (6.2)$$

Due to nuclear potential V_n, the wave number of projectile nucleus x suddenly changes to a very high value as it enters the target nucleus X. Within the nucleus, the solution is

$$u_l(r) = C\,\exp(iKr) \qquad \dots (6.3)$$

Where $K = \dfrac{2M}{\hbar^2}(E + V_n)$ is the wave number inside the nucleus $(\because K \gg k)$.

Whenever there is sudden discontinuity in the potential, the wave is partly reflected and partly transmitted at the boundary. Figure 6.24 shows a one dimensional potential discontinuity for a step function type potential. One obtains the transmission coefficient T as

$$T = \frac{4kK}{(k+K)^2} \qquad \dots (6.4)$$

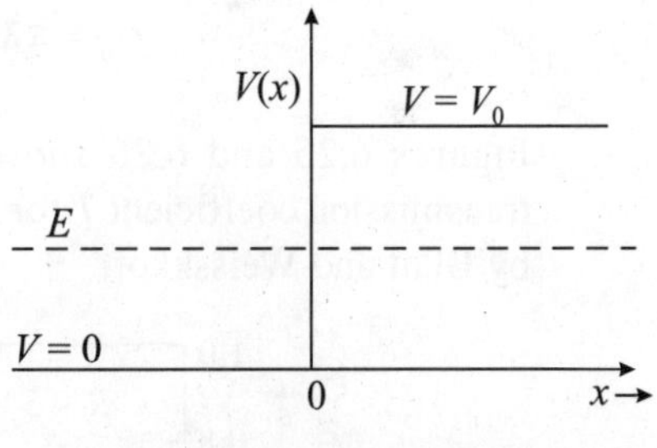

Fig. 6.24 Step function type one dimensional potential discontinuity at $x = 0$

If the incident particle while entering the nucleus (like a zero energy neutron) does not experience any repulsion, all the particles intercepted by the latter may be absorbed by the *black nucleus* approximation. In the approximation of x being a incident mass point. It we take the spatial extension of x into consideration, we have to replace the above by $\pi(R + \lambda)^2$, where λ is de Broglie wave length.

Equation (6.4) is approximately valid for the case of formation of compound nucleus by a low energy neutron. Using (6.4), one obtains the cross-section as

$$\sigma_c(n) = \pi(R + \lambda)^2\,\frac{4kK}{(K+k)^2} \qquad \dots (6.5)$$

For very low energy neutrons ($k << K$), one obtains

$$\frac{4\,k\,K}{(k+K)^2} = \frac{4k}{K}$$

$\therefore$
$$\sigma_c(n) = \pi(R + \lambdabar)^2\,\frac{4k}{K} \qquad \ldots (6.6)$$

One can deduce a more accurate relation for the transmission coefficient by introducing the quantity called as the logarithmic derivative and defined as

$$f_l = R\left[\frac{1}{u_l}\frac{du_l}{dr}\right]_R \qquad \ldots (6.7)$$

Here $u_l(r)$ is the radial wave function for the incident particle of relative orbital angular momentum $l\lambdabar$. Now, writing

$$f_l = \Delta_l + is_l \qquad \ldots (6.8)$$

One can easily calculate η_l in terms of f_l, Δ_l, s_l and phase shift ξ_l.

Assuming that the number of exit channels is so large that the probability of the disintegration of the compound nucleus through the entrance channel (resonance scattering) can be neglected, one can calculate the cross section for absorption of the incident particle x by the target nucleus X. On the basis of the continuum assumption, one can write the wave function in the entrance channel as an incoming wave only, i.e. $u_l \sim \exp(-ikr)$ for $r < R$. Then $f_l = -ikR$, which is independent of l. Then the cross section for the formation of the compound nucleus is obtained as

$$\sigma_c = \pi\lambdabar^2 \sum_{\lambda=0}^{\infty}(2l+1)\,\frac{4s_l\,KR}{\Delta_l^2 + (KR + s_l)^2} \qquad \ldots (6.9)$$

Figures 6.25 and 6.26 shows the graphical representation of the expressions for the transmission coefficient T for neutrons and charged particles (protons) respectively obtained by Blatt and Weisskkoff.

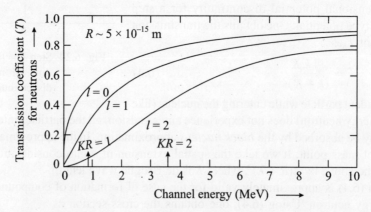

Fig. 6.25 Transmission coefficient (T) for neutrons of different l.

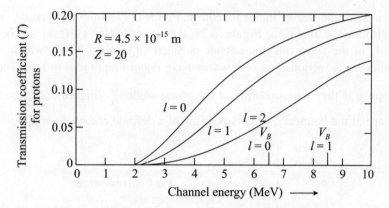

Fig. 6.26 Transmission coefficient (T) for charged particles (protons) of different l for $Z = 20$.

For charged particles of energies greater than the barrier height $E > \dfrac{Z\,Z'\,e^2}{4\pi\,\varepsilon_0\,(R + \lambda)}$, one

obtains

$$T \approx 1 - \frac{Z\,Z'\,e^2}{4\pi\,\varepsilon_0\,(R + \lambda)\,E} \qquad \qquad \ldots (6.10)$$

and

$$\sigma_c = \pi(R + \lambda)^2 \left[1 - \frac{Z\,Z'\,e^2}{4\pi\,\varepsilon_0\,(R + \lambda)\,E} \right] \qquad \ldots (6.11)$$

(iii) Statistical Model: Overlapping Levels   While studying the compound nucleus, we have seen that one of its salient features is the existence of thee resonances in the cross-section when $T << D$, where T is the level width and D is the separation between levels. We have further noticed that these resonances appear at very low incident energies (usually for neutrons) which indicates energies (usually for neutrons) which indicates that thee isolated levels exist at excitation energies close to the separation energy of the neutron (S_n) from the compound nucleus. The total energy ($E_n + S_n$) brough in by the incident particle x may be distributed in a various ways amongst the nucleons within the target nucleus X, till enough energy may be concentrated on a particular nucleon (or nucleon group) which will enable it to come out. However, even if it is energetically favourable for this to happen, the particle (or nucleon group) may not be able to come out, because of the existence of a sharp potential discontinuity at the surface of the nucleus. This causes only a part of the incident wave to be transmitted through the latter into the nucleus. Obviously, the particular configuration, in which a given particle may be able to come out with a given energy will repeat itself several times with an average period p given by $p \sim \dfrac{\lambda}{D}$, where D is the mean level spacing of the compound nucleus. After many such periods, the particle finally manages to escapes out after period τ, given by $\tau = \dfrac{\lambda}{\Gamma}$, where Γ is the width of the level. Now, since $\tau >> p$, i.e. $\Gamma << D$ which is really the condition for the formation of resonance level.

Let us consider the case of incident neutron. Figure 6.27 shows the qualitative behaviour of the neutron wave function. Figure 6.27(a) shows that away from the resonances, the amplitude of the wave function outside is much larger than inside, which means a small absorption cross-section. The two waves have equal amplitudes at resonance, which can only happen if they join smoothly at the phase angle $\dfrac{\pi}{2}$ (Figure 6.27c) corresponding to zero shape at the boundary. This is possible at a definite energy $E_n = E_r$.

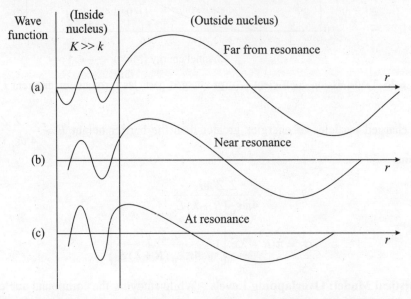

Fig. 6.27 Behaviour of wave functions for slow neutrons inside and outside a nucleus; (a) far from resonance, (b) near resonance; and (c) at resonance, when two functions join at zero slope

We may note that in this case there is large absorption cross section for the incident particles. The difference in k and K decreases as the energy of incident particle increases. This reveals that the number of attempts required to escape by the particle from the nucleus decreases because the transmission coefficient increases. This makes τ smaller and hence the width of the levels Γ becomes larger. Now, the compound nucleus can decay through many exit channels, corresponding to y and Y nuclei in the final state of the reaction and the different possible energy states in which they may be produced. This affects the width of the level and this may become comparable to or even larger than the level separation, i.e. $\Gamma > D$. When $\Gamma \gg D$, the levels are no longer well defined and one finds that there is considerable *overlapping* between levels (Figure 6.19b), and we have a *continuum of levels*. It is not obvious that Bohr independependent hypothesis applies to the continuum, but experimental evidence confirms that in some cases it does.

We have mentioned earlier that Bohr independence hypothesis holds for isolated levels, since the decay of the level formed can only depend on the property of the level and not on how the level was formed. At higher energies, levels get crowded together and thus $\Gamma \approx D$.

Clearly, the same level may be regarded as the superposition of a few states of different energies having different phases. However, in this region, there is some doubt as to the validity of the independence hypothesis. But, when there is complete overlapping due to extremely large number of exit channels being available, there can not exist any definite phase relationship between the different levels superposed and the independence hypothesis *regains validity*. This region is termed as *continuum region*. The mode of formation of the compound state is forgotten at the time of its decay in this case. This is not because of the time scales involved, but because any effect of formation will be averaged out over many overlapping states. This is *statistical assumption*. Now, we describe an experiment for the experimental confirmation of continuum region performed by S. N. Ghoshal in 1950.

Ghoshal Experiment With the help of following two processes, the same compound nucleus $_{30}^{64}$Zn was formed in the state of excitation:

$$_{28}^{60}\text{Ni} + _{2}^{4}\text{He} \rightarrow _{30}^{64}\text{Zn}^*$$

$$_{29}^{63}\text{Cu} + _{1}^{1}\text{H} \rightarrow _{30}^{64}\text{Zn}^*$$ ... (6.1a)

To measure the excitation function, i.e. the reaction yield versus α-energy curve, in the first case α-particles with energies ranging from 10 to 40 MeV and ^{60}Ni as the target nucleus were used. For similar measurments in the second case, protons with energy between about 3 to 33 MeV were used by producting the same compound nucleus using ^{63}Cu as the target nucleus.

In order to produce the compound nucleus in the same state of excitation by the two different methods cited above, it was found necessary to have the kinetic energies of the incident protons and α-particles such that $\varepsilon_\alpha = \varepsilon_p + 7$ MeV (Figure 6.28).

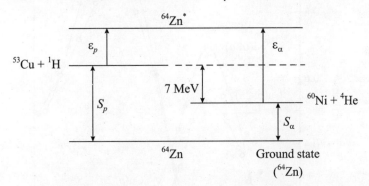

Fig. 6.28 Energy level scheme for the formation of compound nucleus ^{64}Zn* in the same state of excitation by two different processes.

We can see from figure that the absorption of proton of energy ε_p by ^{63}Cu nucleus produces the compound nucleus ^{64}Zn* in a state of excitation energy $E_c = \varepsilon_p + S_p$, where S_p is the separation energy of the proton from the nucleus ^{64}Zn in its ground state. Similarly, the absorption of the α-particle of energy ε_α by ^{60}Ni nucleus produces the same compound nucleus ^{64}Zn* in the same state of excitation, i.e. $E_c = \varepsilon_\alpha + S_\alpha$.

Comparing the two expressions for E_c, one obtains,

$$\varepsilon_\alpha = \varepsilon_p + (S_p - S_\alpha)$$

We may note that the currently accepted values of S_p and S_α gives the difference $(\varepsilon_\alpha - \varepsilon_p)$ some what lower (3.7 MeV) than the above estimate of 7 MeV. This seems to be due to the uncertainties in the energy definitions by Stacked foil method used in Ghoshal's experiment. In both the above mentioned cases, the following disintegration processes were studied:

$$^{64}Zn^* \rightarrow {}^{63}Zn + {}^1n \qquad \qquad ... (6.2a)$$

$$\rightarrow {}^{62}Zn + 2\ {}^1n \qquad \qquad ... (6.2b)$$

$$\rightarrow {}^{62}Cu + {}^1n + {}^1p \qquad \qquad ... (6.2c)$$

For the two cases, the excitation functions of above reactions were determined as functions of the proton and α-energies. Figure 6.29 shows the results. We can see from figure that the following relationships are satisfied for all proton energies ε_p between 3 to 33 MeV and corresponding α-energies $\varepsilon_\alpha = \varepsilon_p + 7$ MeV between 10 to 40 MeV:

$$\sigma(\alpha, n): \sigma(\alpha, 2n): \sigma(\alpha, pn): = \sigma(p, n): \sigma(p, 2n): \sigma(p, pn) \qquad ... (6.3)$$

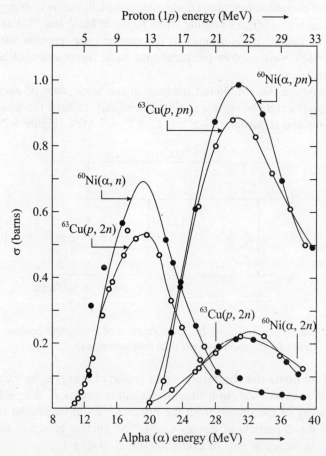

Fig. 6.29 Experimental results on the verification of the independence hypothesis due to Ghoshal

We note that relations (6.3) provide direct confirmation of independence hypothesis. One can easily see it from the following considerations.

One can write the cross-section of the reaction $X + x + (C^*) + Y + y$ as

$$\sigma(x,\, y) = \sigma_x(\varepsilon_x)\,\eta_y \qquad \ldots (6.4)$$

Where $\sigma_x(\varepsilon_x)$ denotes the cross-section for the formation of C^* in a particular state of excitation due to the absorption of x by X for the energy ε_x and η_y is the total relative probability of decay of C^* into $Y + y$ irrespective of energies of Y and y.

Let us now consider the reaction $X' + x' \rightarrow (C^*) \rightarrow Y + y$. We can see that same compound nucleus C^* is produced in the same state by the absorption of a different particle x having energy ε'_x by another nucleus X' (different from X) decay into the same products $Y + y$. Now, one can write the cross section as

$$\sigma(x',\, y) = \sigma_{x'}(\varepsilon'_x)\eta_y \qquad \ldots (6.6)$$

Thus, $$\frac{\sigma(x,\, y)}{\sigma(x',\, y)} = \frac{\sigma_x(\varepsilon_x)}{\sigma_{x'}(\varepsilon'_x)} \qquad \ldots (6.7)$$

Now, if the decay of C^* takes place through another exit channel, i.e. $C^* \rightarrow Y' + y'$, in which two other nuclei are produced in the final state. One can write the corresponding cross-sections

$$\sigma(x,\, y') = \sigma_x(\varepsilon_x)\,\eta'_y \qquad \ldots (6.7)$$
$$\sigma(x',\, y') = \sigma'_x(\varepsilon'_x)\,\eta'_y \qquad \ldots (6.8)$$

Now, taking the ratios, one obtains

$$\frac{\sigma(x,\, y')}{\sigma(x',\, y')} = \frac{\sigma_x(\varepsilon_x)}{\sigma_{x'}(\varepsilon'_x)} \qquad \ldots (6.9)$$

Which is again the result obtained by Ghoshal experiment.

6.16 DECAY (OR EMISSION) RATES OF THE COMPOUND NUCLEUS

So far we have discussed the theory of the formation of compound nucleus, which corresponds to the total reactions cross-section. Now we discuss the decay or disintegration of the compound nuclear state into various channels.

When the excitation energy E_C of the compound nucleus C^* is high, many exit channels are available for the disintegration of C^*. This is not only due to possibility of the emission of different types of particles, e.g.

$$C^* \rightarrow Y_1 + y_1 \qquad \ldots (6.1)$$
$$\rightarrow Y_2 + y_2$$
$$\rightarrow Y_3 + y_3$$

etc., but also due to the emission of any one type with different possible energies. This means, if C^* breaks up into $Y + y$, these may be emitted with different kinetic energies ε_y (in the C-system) leaving Y in different possible states of excitations E_Y. Figure 6.30 clearly show this. We can see that the excited residual nucleus Y will subsequently go down to the ground state by the emission of γ-ray (Figure 6.30b) and particle y may be emitted with different possible energies upto a maximum $(\varepsilon_y)_{max}$ leaving Y in different states of excitation. Now, if the excitation energy E_Y of Y is greater than the separation energy S_y' of another nucleon or nucleon group y' (usually a neutron), then it may disintegrate according to $Y^* \rightarrow Y' + y'$. Obviously, a different residual nucleus Y is left in this case (either in the excited or in the ground state) (Figure 6.30c).

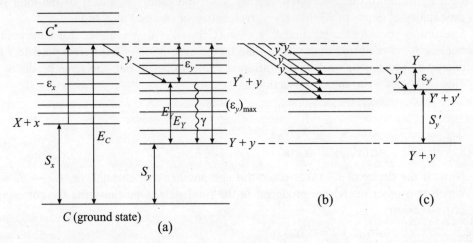

Fig. 6.30 (a) Disintegration of C^* into $Y + y$ (b) Emission of y with different energies to produces Y in different states, e.g. Y^* (c) Illustrating the further break up of Y into $Y' + y'$.

If the partial widths for the disintegration of C^* into the different channels are $\Gamma_{y_1}, \Gamma_{y_2}, \Gamma_{y_3}$, etc., then one can write the total width as

$$\Gamma = \sum_i \Gamma_{y_i} = \Gamma_{y_1} + \Gamma_{y_2} + \Gamma_{y_3} + ...$$

If one knows the relative probabilities as branching ratios of the disintegration through the different channels, $\eta_{y_i} = \dfrac{T_{y_1}}{T}$, then one can write

$$\sum_i \eta_{y_i} = \sum_i \frac{\Gamma_{y_i}}{\Gamma} = 1 \qquad\qquad ... (6.2)$$

If C^* is formed through the process $x + X \rightarrow (C^*)$. Where x having an energy ε_x is absorbed by X in its ground states, then for the different processes of disintegration of C^*, one finds the cross sections

$$\sigma(x, y_1) = \sigma_x(\varepsilon_x)\eta_{y_1}, \;\; \sigma(x, y_2) = \sigma_x(\varepsilon_x)\eta_{y_2} \qquad \text{etc.}$$

Then, we have

$$\sum_i \sigma(x, y) = \sigma(x, y_1) + \sigma(x, y_2) + \sigma(x, y_3) + ...$$

$$= \sigma_x(\varepsilon_x)\eta_{y_1} + \sigma_x(\varepsilon_x)\eta_{y_2} + \sigma_x(\varepsilon_x)\eta_{y_3} + ...$$

$$= \sigma_x(\varepsilon_x) \sum_i \eta_{y_i}$$

$$= \sigma_x(\varepsilon_x) \qquad \qquad \qquad \text{... (6.3)}$$

Obviously, the sum of the cross sections of all the different reactions which can take place when C^* is formed by the absorption of energy ε_x by X equal to the cross section of formation of C^*.

We now consider the reaction $X(x, y)Y$. Its reactions cross section may be written as

$$\sigma(x, y) = \sigma_x(\varepsilon_x) \eta_y(E_C) \qquad \qquad \text{... (6.4)}$$

Where E_C is the excitation energy of C^*. Let us specify the reactions in more detail, i.e. by further subdivision for incoming channel α and outgoing channel β, we have

$$\sigma(\alpha, \beta) = \sigma_c(\alpha) \eta_c(\beta) \qquad \qquad \text{... (6.5)}$$

where

$$\Gamma = \Gamma(E_C) = \sum_\beta \Gamma_\beta(E_C) \qquad \qquad \text{... (6.6)}$$

If $\tau(E_C)$ is a mean life of C^*, we have

$$\Gamma(E_C) = \frac{\hbar}{\tau(E_C)} \qquad \qquad \text{... (6.7)}$$

It is well known in the formal theory of cross-sections that reciprocity theorem holds good in all nuclear reactions. We have from this theorem, $p_x^2 \sigma_{XY} = p_y^2 \sigma_{YX}$ for the forward and reverse nuclear reactions $X(x, y)Y$ and $Y(y, x)X$. Now, expressing in terms of wave numbers, one can write for the reactions $\alpha \to \beta$ and $\beta \to \alpha$ as

$$k_\alpha^2 \sigma(\alpha, \beta) = k_\beta^2 \sigma(\beta, \alpha) \qquad \qquad \text{... (6.8)}$$

or

$$k_\alpha^2 \sigma_c(\alpha) \eta_c(\beta) = k_\beta^2 \sigma_c(\beta) \eta_c(\alpha) \qquad \qquad \text{... (6.9)}$$

$$\therefore \qquad k_\alpha^2 \sigma_c(\alpha) \frac{\Gamma_\beta}{\Gamma} = k_\beta^2 \sigma_c(\beta) \frac{\Gamma_\alpha}{\Gamma} \qquad \qquad \text{... (6.10)}$$

or

$$k_\alpha^2 \frac{\sigma_c(\alpha)}{\Gamma_\alpha} = k_\beta^2 \frac{\sigma_c(\beta)}{\Gamma_\beta} = f(E_C) \qquad \qquad \text{... (6.11)}$$

The r.h.s. of (6.11) is a function of the excitation energy E_C independent of channel α or β. Thus, one finds

$$\Gamma_\alpha = \frac{k_\alpha^2 \sigma_c(\alpha)}{f(E_C)}, \quad \Gamma_\beta = \frac{k_\beta^2 \sigma_c(\beta)}{f(E_C)}, \text{ etc.} \qquad \qquad \text{... (6.12)}$$

This gives

$$\Gamma = \sum_\gamma \Gamma_\gamma(E_C) = \frac{1}{f(E_C)} \sum_\gamma k_\gamma^2 \sigma_c(\gamma) \qquad \qquad \text{... (6.13)}$$

Where $k = \dfrac{1}{\lambda}$. Here, the summation is over all possible exit channels γ through which C^* can decay. One finally obtains

$$\eta_c(\beta) = \frac{\Gamma_\beta}{\Gamma} = \frac{k_\beta^2\, \sigma_c\,(\beta)}{\sum\limits_\gamma k_\gamma^2\, \sigma_c\,(\gamma)} \qquad\qquad ...\,(6.14)$$

Where the sum extends over all possible exit channels γ. If we assume the validity of Bohr's independence hypothesis and the cross section for the formation of C^* through all possible channels are known, one can calculate the reaction cross-section $\sigma(\alpha, \beta)$.

For neutrons $\sigma_c(\beta)$ is much larger than for charged particles as the latter have to penetrate through the Coulomb potential barrier. Thus, according to (6.11), the neutron width is generally larger, i.e. the probability of decay of C^* by neutron emission is far greater than for other processes of decay.

On the basis of the above assumptions, we are now in a position to write the expression for the shape of the energy spectrum of the emitted particles.

Energy spectrum of Particles emitted in the decay of the Compound Nucleus: (Evaporation Model)   Consider the reaction $X + x \rightarrow (C^*) \rightarrow Y + y$. Here the emitted particle y may be emitted with different energies ε_y, leaving behind Y in different possible energy states: $E_Y = (\varepsilon_y)_{max} - \varepsilon_y$ (Figure 6.31). We can see that each of these represents a separate exit channel β for which the ratio η_c is given by Equation (6.14). For neutrons, the variation of η_c with ε_y show a monotonic increase (Figure 6.32a). Now, assuming the levels in Y to be closely spaced, the total probability for the emission of y with energy ε_β (= ε_y) lying between energy ε and $(\varepsilon + d\varepsilon)$ is given by

$$\eta_y(\varepsilon)\,d\varepsilon = \sum_\beta \eta_c(\beta) = \sum_\beta \frac{k_\beta^2\, \sigma_c\,(\beta)}{\sum\limits_\gamma k_\gamma^2\, \sigma_c\,(\gamma)} \qquad\qquad ...\,(6.15)$$

Here the summation extends from $\varepsilon_\beta = \varepsilon$ to $\varepsilon + d\varepsilon$.

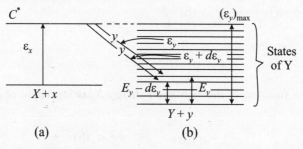

(a)   (b)

Fig. 6.31 The emission of particles y from the decay of C^* with different energies, leaving behind Y in different states.

From (6.15), we note that the denominator on the r.h.s. is constant and in the numerator, $k_\beta^2 \propto \varepsilon$.

We expect that the higher energy end of the spectrum contains evidence of energy levels, while at the low energy end there will be a continuous spectrum because of the high density of excited states of the residual nucleus.

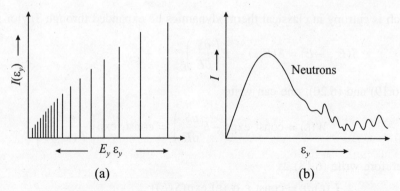

Fig. 6.32 **(a)** Energy spectrum of the emitted particle *y* due to break up of C^* (ideal resolution), **(b)** Experimentally observed energy spectrum in which overlapping of levels are clearly visible.

The shape will evidently be governed by the shape of the decay function $\eta(\varepsilon)dE$ given by

$$\eta(\varepsilon)d\varepsilon = \sum_{\varepsilon < \varepsilon_\beta < \varepsilon + d\varepsilon} \eta_c(\varepsilon_\beta) \qquad \qquad \text{... (6.16)}$$

Let the number of terms in the sum appearing on r.h.s. of (6.16) be given by the number of levels of the residual nucleus *Y* with an excitation energy between ε and $\varepsilon - d\varepsilon$. If $W(\varepsilon)$ is the level density the number of terms will be $W(\varepsilon)d\varepsilon$, Assuming that the denominator may be taken as constant, we can write

$$\eta(\varepsilon)d\varepsilon = \text{constant } \sigma_c(\beta)\, k_\beta^2\, W_Y(\varepsilon)d\varepsilon \qquad \qquad \text{... (6.7)}$$

Remembering that k_β^2 is proportional to the incident energy ε, and assuming the number of states of *Y* in the interval $d\varepsilon$ to be $W_Y(E_Y)\,d\varepsilon$, we then obtain the number of particles *y* emitted (i.e. the shape of the spectrum) with energy ε and $\varepsilon + d\varepsilon$ as

$$I_y(\varepsilon)d\varepsilon = \text{constant } \sigma_c(\beta)\, \varepsilon\, W_Y(\varepsilon_{yY}^m - \varepsilon)d\varepsilon \qquad \qquad \text{... (6.18)}$$

Where ε_{yY}^m is the maximum energy with which *y* can be emitted from C^* leaving *Y* in the ground state. Here, we have assumed that the number of states *Y* is so large that a continuous variation of the energy ε_y of the emitted particle *y* may take place. Because of the use of $W(\varepsilon)d\varepsilon$ for the number of levels in the energy range of ε and $\varepsilon + d\varepsilon$, this theory corresponds to *continuum of energy levels*. This theory is therefore termed as *continuum theory* or *statistical model of nuclear reactions*.

In Equation (6.18), $\sigma_c(\beta)$ is the cross-section of C^*, if the incident particle *x* had the energy of the outgoing channel, which is a function of channel energy $\varepsilon = \varepsilon_\beta$. Hence $\varepsilon\, \sigma_c(\beta)$ is an increasing function of ε. $W_Y(\varepsilon_{yY}^m - \varepsilon)$ is easily seen to be decreasing function of ε, because $W_Y(\varepsilon)$, the level density increases with the excitation energy. Hence $W_Y(\varepsilon_{yY}^m - \varepsilon)$ decreases with ε. Because of these two opposite trends the function $I_y(\varepsilon)$ has a maximum in the middle of the energy range.

We now define a quantity *T*, known as the *nuclear temperature* as

$$\frac{1}{T(E)} = \frac{\partial S(E)}{\partial E} \qquad \qquad \text{... (6.19)}$$

Where $\qquad\qquad S(E) = \log W(E) \qquad \qquad \text{... (6.20)}$

Let $S(E)$ which is entropy in classical thermodynamics be expanded through Taylor series, i.e.

$$S(E_{yY} - E) = S(E_{yY}) - E\left(\frac{\partial S}{\partial E}\right)_{E = E_{yY}} + \dots \qquad \dots (6.21)$$

Now, using (6.19) and (6.20), one can write

$$W(\varepsilon) = \text{const} \exp\left(-E\frac{\partial S}{\partial E}\right) = \text{const.} \exp\left(\frac{-E}{T(E_{yY})}\right) \qquad \dots (6.22)$$

One may, therefore write (6.18) as

$$I_y(\varepsilon)d\varepsilon = \text{const } \varepsilon \, \sigma_c(\varepsilon) \exp(S(E)) \qquad \dots (6.23a)$$

or $$I_y(\varepsilon)d\varepsilon = \text{const } \varepsilon \, \sigma_c(\varepsilon) \exp\left(\frac{-\varepsilon}{T(\varepsilon_{yY})}\right) \qquad \dots (6.23b)$$

Where the factor $(S(E))$ has been incorporated in the constant. It $T(\varepsilon_{yY})$ is taken as constant, then (6.23) becomes

$$I_y(\varepsilon)d\varepsilon = \text{const } \varepsilon \, \sigma_c(\varepsilon) \exp\left(-\frac{\varepsilon}{T}\right)$$

or $$\log\left[\frac{I_y(\varepsilon)}{\varepsilon\,\sigma_c(\varepsilon)}\right] = -\left(\frac{1}{T}\right)\varepsilon \qquad \dots (6.24)$$

Obviously, a plot between $\log\left[\dfrac{I_y(\varepsilon)}{\varepsilon\,\sigma_c(\varepsilon)}\right]$ versus ε will yield a *straight line* (representing antual experimental spectra). Which will determine the nuclear temperature.

In order to have a better idea about the energy distribution of y, we must know the nature of variation of σ_c with ε. For neutrons, this is found to be almost constant which means that

$$\text{In}(\varepsilon) \propto \varepsilon \exp\left(-\frac{\varepsilon}{kT}\right) \qquad \dots (6.25)$$

Clearly, for the emitted neutrons, one obtains an almost *Maxwellian distribution* for the energies of the molecules of a gas. The analogy is often used to understand and the concept of nuclear temperature. The analogy suggest that a quantity with a dimensions of energy be defined as the *nuclear temperature* and that the reaction particles y be regarded as evaporation products.

However, in the case of charged particles σ_c is strongly influenced by the energy due to potential barrier penetration factor.

Statistical Theory of Nuclear Level Densities

The concept of nuclear level density is only applicable for higher excitation energies, i.e. $\Gamma \gg D$. This is especially true for medium and heavy nuclei for energies much above the Fermi energies, i.e. above 8 MeV (excitation energy). Though the degrees of freedom available to the system is limited because, the number of nucleons, i.e. particles is not very large at these energies, the degrees of freedom available to each particle is also limited obeying Fermi gas model of single particles, one still applies the principles of statistical model as a first approximation to calculate the nuclear level densities.

One can obtain total number of particles y of a particular type in the decay of C^* by integrating (6.18) for all possible energies from 0 to ε_y^m. Let us introduce the quantity,

$$F_y = \frac{2N}{\hbar^2} \int_0^{\varepsilon_y^m} \varepsilon_\beta \, \sigma_c \, W_Y(\varepsilon_y^m - \varepsilon_\beta) \, d\varepsilon_\beta \qquad \ldots (6.26)$$

Here $\sigma_c = \sigma_c(\varepsilon_\beta)$ is the cross section for the formation of C^* by the collision of y with Y^* at the channel energy ε_β. F_y is a function of ε_y^m. Knowing the level density, one can calculate F_y for different types of emitted particles y. Clearly, F_y is a measure of the total probability for the emission of y with all possible energies. Thus, one can write

$$\sigma(x, y) = \frac{\sigma_c(x) F_\beta}{\sum_\beta F_\beta} \qquad \ldots (6.27)$$

Where the denominator in (6.27) is summed over all possible energies of y types of particles which can be emitted in the decay of C^*.

In order to obtain an expression for level density W for densely packed levles, one can apply statistical mechanics. The function $E(T)$ has a vanishing derivative, i.e. $\frac{dE}{dT} = 0$ at $T = 0$. Obviously, this corresponds to zero specific heat at $T = 0$. A Taylor series expansion of $E(T)$ about $T = 0$, gives the first term in the expansion a quadratic term $\propto T^2$. Neglecting higher terms, one can write

$$E = aT^2 \qquad \ldots (6.28)$$

Where a is constant. From 6.23(b), one obtains

$$S = \int \frac{dE}{T(E)} = \sqrt{a} \int \frac{dE}{\sqrt{E}} = \sqrt{aE} + \text{const.} \qquad \ldots (6.29)$$

One finally obtains

$$W(E) = \exp[S(E)] = C \exp\left(2\sqrt{aE}\right) \qquad \ldots (6.30)$$

Gilbert and Cameron derived through a more detailed procedure

$$W(E) = \frac{\sqrt{\pi}}{12} \frac{\exp\left(2\sqrt{aE}\right)}{a^{\frac{1}{4}} E^{\frac{3}{4}}} \qquad \ldots (6.31)$$

Where $\qquad a = \frac{\pi^2 g}{6} \quad$ and $\quad E = \varepsilon - E_0 \qquad \ldots (6.32)$

E_0 is the energy of the fully degenerate system which is constant.

Lang and LeCouteur using alternative dependence of E, obtained a relation, which fits the experimental data better. They used

$$E = aT^2 - T$$

and obtained

$$W_{L.C.}(E) = \frac{\sqrt{\pi}}{12} \frac{\exp\left(2\sqrt{aE}\right)}{a^{\frac{1}{4}} (E+T)^{\frac{5}{4}}} \qquad \text{... (6.33)}$$

A slightly different form of the level density is due to Newton according to which

$$E = aT^2 - \frac{3}{2} T$$

$$W_{Newton}(E) = \frac{\sqrt{\pi}}{12} \frac{\exp\left(2\sqrt{aE}\right)}{a^{\frac{1}{4}} \left(E+\frac{3}{2}T\right)^{\frac{5}{4}}} \qquad \text{... (6.34)}$$

In all these formulations

$$a \approx \text{constan } A \qquad \text{... (6.35)}$$

and 'a' is determined empirically. g has been taken to be independent of ε. However, pre-equilibrium studies reveal that this may not be case.

We may note that the nuclear level densities can be measured experimentally by comparing the energy distribution of the outgoing particles in the nuclear reaction with relation (6.27). A plot of $I_y(\varepsilon)$ (measured distribution function) divided by the product $\varepsilon\,\sigma_c(\varepsilon)$ against E gives us the variation nature of W_Y with energy:

$$\frac{I_y(\varepsilon)}{\varepsilon\,\sigma_c(\varepsilon)} \propto W_Y(\varepsilon_y^m - \varepsilon) \qquad \text{... (6.36)}$$

Figure 6.32(b) shows the nature of energy spectrum of the emitted particle. No doubt, the general nature corresponds to approximately Maxwellian distribution as predicted by theory. In the high energy regions, the peaks are due to direct reactions in which no compound nucleus is formed and deviations in the low energy region are due to *secondary reactions*, e.g. (n, np), (p, np), etc.

Angular Distribution of emitted Particles, i.e. Reaction Products   The angular distribution of reaction products in general, depends on the reaction mechanism, (e.g. compound, direct or pre-equilibrium) and the angular momenta-both orbital and spin-of the two states connecting the emitting particle and, of course, on the angular momenta of the emitted particles itself.

The angular distribution of the emitted particles y in a nuclear reaction $x + X \rightarrow C^* \rightarrow Y + y$, which takes place via compound nucleus formation, i.e. $x + X \rightarrow C^* \rightarrow Y^* + y$, where C^* is the compound nucleus in a highly excited state and Y^* is the residual nucleus after the emission of particle y. If Y^* is excited to an energy so that only γ-rays are emitted after emission of particle y, then the angular distribution of y is of interest.

Various experimental observations about angular distributions seems to indicate that there are two components of angular distribution, (i) Symmetric about 90° in C-system, which means that there should be fore-aft symmetry. This is due to compound nucleus formation, (ii) A forward peaked angular distribution assumed to be due to pre-equilibrium reaction mechanism.

One can understand the fore-aft symmetry in the angular distribution by remembering that the only relevant vector associated with C^* in the C-system is the total angular momentum. This is made up of the spins of all the initial particles and the relative orbital angular momenta. We may note that the spins are oriented at random while the relative orbital angular momentum $(\mathbf{V} \times \mathbf{p})$ is perpendicular to $\mathbf{p}$ of the projectile and can be oriented at random in the plane $\perp^r$ to momentum $\mathbf{p}$. This is why that any vector quantity associated with the compound nucleus C^* should have fore-aft symmetry. At the time of decay of C^*, this fore-aft symmetry is maintained, giving rise to the fore-aft symmetry in the angular distribution of the reaction products.

(iv) **Direct Reactions**   Direct reactions occur without the formation of any compound nucleus, i.e. mostly at very high energies of the bombarding particles when the nucleons of the nucleus may be considered free. These reactions are most commonly observed for incident particles with an energy of say 5 MeV or above. These energies would excite continuum states in any compound nucleus but the outgoing particles in the reaction $X(x, y)Y$ often have the following characterstics which are not explained by compound-nucleus theory:

(a) Emission of excess particles at high energy in comparison with the number expected according to statistical evaporation theory. These particles show resolvable structure since they lead to discrete, well separated states of final nucleus Y (Figure 6.14).

(b) Forward peaking of the higher energy particles of the spectrum in contrast with the symmetric angular distributions expected for evaporation particles, and in fact found for the lower energy particles y.

(c) Gradual and monotonic dependence of the yield of y on bombarding energy (excitation function)

These characterstics strongly suggest that the basic process at these energies is not an interaction with the target nucleus X as a whole, but with just a part of it, e.g. a single nucleon in the nuclear surface, whose emission excites particular states of residual nucleus Y. Few example of direct process are:

(a) direct *inelastic scattering*, such as the reaction

$$^{24}\text{Mg} + \alpha \rightarrow {}^{24}\text{Mg}^* + \alpha'$$

Which tend to pick out *collective modes* of excitation of the residual nucleus.

(b) transfer reactions, such as the deutron 'stripping'

$$^{16}\text{O} + d \rightarrow {}^{17}\text{O} + p$$

or $$\rightarrow {}^{17}\text{O}^* + p'$$

Which select *single-particle* or *single-hole* states, including the ground state, preferentially.

A good example of direct reaction is a (d, p) process in which a deutron with more than a few MeV of kinetic energy, incidents on a target nucleus and a proton is observed to emerge. Since the deutron is a loosely bound system of a proton and a neutron, it is easy to envisage that the neutron is captured into a single particle orbit of the target nucleus without disturbing the structure of the rest of the nucleons and the proton continues on to become the scattered particle. One can view this process as one in which a neutron is "stripped" from the projectile. For this reason the reaction is known as a one-neutron *stripping* reaction. States in the final nucleus that are strongly excited by such a stripping reaction are those formed predominantly by a nucleon coupled to the ground state of the target nucleus. Other stripping reactions, such as (t, p) transfer two nucleons from the projectile to the target nucleus. Even more complicated reactions, such as those involving the transfer of a cluster of nucleons, are also possible. To qualify as a direct reaction, both the target nucleus and the internal structure of the clusture transferred must be undisturbed by the reaction; the residual nucleus is simply the coupling of the cluster and the ground state of the target nucleus. However, this condition is generally difficult to meet for transfer reactions involving a large number of nucleons.

The complement of a stripping reaction is a pick-up reaction. In this case one or more nucleons are taken away from the target nucleus without changing the structure among the rest of the nucleons. A good example is the reaction ^{40}Ca (p, d) ^{39}Ca. The states in the residual nucleus, ^{39}Ca here, that are strongly excited by a pick up reaction are the one-hole states; i.e. those formed by removing one of the particles in ^{40}Ca and leaving the remaining 39 nucleons unchanged in their relative motion.

The scattering cross section in a direct reaction, stripping as well as pick up, may be derived in a straight forward way using the first Born approximation. The reaction mechanism is relatively straight forward here because of simple relation between initial and final nuclear states underlying the direct reaction assumption. One can write the Schrodinger equation for the scattering in the form of a standard second order differential equation,

$$(\nabla^2 + k^2)\, \psi(\mathbf{r}) = \frac{2\mu}{\hbar^2}\, V(\mathbf{r})\, \psi(\mathbf{r}) \qquad \qquad ...\ (6.1)$$

Where $k^2 = \dfrac{2\mu E}{\hbar^2}$. A formal solution of (6.1) for the outgoing wave function may be expressed in terms of a Green's function $G(\mathbf{r}, \mathbf{r}')$ as

$$\psi(\mathbf{r}) = e^{i\,\mathbf{k}_i\cdot\mathbf{r}} + \frac{2\mu}{\hbar^2} \int G(\mathbf{r}, \mathbf{r}')\, V(\mathbf{r}')\, \psi(\mathbf{r}')\, d^3\mathbf{r}' \qquad \qquad ...\ (6.2)$$

Where we have chosen $\mathbf{k}_i$ to be along the direction of the incident particle and the function $\exp i(\mathbf{k}_i \cdot r)$ is the solution of the homogeneous part of (6.1), i.e., for $V(r) = 0$.

We shall take the explicit form of the Green's function as

$$G(\mathbf{r}, \mathbf{r}') = -\frac{1}{4\pi} \frac{e^{ik\,|\mathbf{r} - \mathbf{r}'|}}{|\mathbf{r} - \mathbf{r}'|} \qquad \qquad ...\ (6.3)$$

It is the solution of the following equation:

$$(\nabla^2 + k^2)\, G(\mathbf{r}, \mathbf{r}') = \delta(\mathbf{r} - \mathbf{r}') \qquad \qquad ...\ (6.4)$$

Generalizing, one can include within the defining equation for $G(\mathbf{r}, \mathbf{r}')$ a part of $V(\mathbf{r})$, for e.g., the part representing the averaging effect of the target nucleons on the incident particle. This is just similar to the mean field approach.

Using (6.3), the formal solution for the scattering wave function in (6.2) may be written in the form

$$\psi(\mathbf{r}) = e^{i\,\mathbf{k}_i \cdot \mathbf{r}} - \frac{\mu}{2\pi\hbar^2} \int \frac{e^{ik|\mathbf{r}-\mathbf{r}'|}}{|\mathbf{r}-\mathbf{r}'|} V(\mathbf{r}')\,\psi(\mathbf{r}')\,d^3r' \qquad \dots (6.5)$$

Except for contributions from the Coulomb interaction, the potential $V(\mathbf{r}')$ is short-ranged, and one may approximate the argument of the exponential function in the asymptotic region by the first two terms in the following expansion:

$$ik(|\mathbf{r}-\mathbf{r}'|) = k\sqrt{r^2 - 2\mathbf{r}\cdot\mathbf{r}' + r'2} = kr - \mathbf{k}_f \cdot \mathbf{r}' + O(r'^2)$$

$$\approx kr - \mathbf{k}_f \cdot \mathbf{r}'$$

Where $\mathbf{k}_f = \dfrac{k\mathbf{r}}{r}$ is taken along the direction of the emerging particle. The formal solution of the scattering equation now takes the form

$$\psi(\mathbf{r}) \approx e^{i\,\mathbf{k}_i \cdot \mathbf{r}} - \frac{\mu}{2\pi\hbar^2} \frac{e^{ikr}}{r} \int e^{-i\,\mathbf{k}_f \cdot \mathbf{r}'} V(\mathbf{r}')\,\psi(\mathbf{r}')\,d^3\mathbf{r}' \qquad \dots (6.6)$$

Where we have taken $|\mathbf{r}-\mathbf{r}'| \simeq r$, correct in the asymptotic region where the scattered particle is observed.

We have the asymptotic form of the scattering wave function

$$\psi(\mathbf{r}) \underset{r \to \infty}{\to} e^{ikx} + f(\theta, \phi)\,\frac{e^{ikr}}{r}$$

Comparing (6.6) with the above, one obtains the scattering amplitude

$$f(\theta) = -\frac{\mu}{2\pi\hbar^2} \int e^{-i\,\mathbf{k}_f \cdot \mathbf{r}'} V(\mathbf{r}')\,\psi(\mathbf{r}')\,d^3r' \qquad \dots (6.7)$$

This is only a formal or integral equation solution for the scattering amplitude, since the expression involves the unknown function $\psi(\mathbf{r}')$, the solution to the scattering Equation (6.1). Equation (6.7) is useful in that it provides us with a starting point to expand scattering cross section in terms of a Born series. In the (first) Born approximation, the unknown function $\psi(\mathbf{r}')$ in (6.7) is replaced by the first term in (6.2). In this way, one obtains an approximate form of the scattering amplitude.

$$f(\theta) \approx \frac{\mu}{2\pi\hbar^2} \int e^{-i\,\mathbf{k}_f \cdot \mathbf{r}'} V(\mathbf{r}')\,e^{i\,\mathbf{k}_i \cdot \mathbf{r}'}\,d^3r' \qquad \dots (6.8)$$

We shall make use of this result to obtain the differential scattering cross-section for stripping and pick-up reactions.

One may simplify the expression in (6.8) further by expressing the results in terms of the momentum transfer vector,

$$\mathbf{q} = \mathbf{k}_i - \mathbf{k}_f \qquad \dots (6.9)$$

and by expanding the plane wave in terms of spherical harmonics as

$$e^{i\mathbf{q}\cdot\mathbf{r}'} = \sum_l i^l \sqrt{4\pi(2l+1)}\; j_l(qr')\,Y_{lo}(\theta') \qquad \dots (6.10)$$

The angle θ' is the angle between vectors $\mathbf{q}$ and $\mathbf{r}'$ and is therefore one of the variables of integration in (6.8); in contrast, the scattering angle θ is between $\mathbf{k}_f$ and $\mathbf{k}_i$.

Angular Distribution   So far we have ignored the internal structure of particles participating in the scattering. Since in stripping and pick up reactions we are dealing with a change in the structure of nuclei involved, the wave functions of both the initial nuclei and the final nuclei must enter into the expression for the scattering amplitude. Let us take the asymptotic forms of initial and final wave functions of the scattering system to be

$$\psi_i \to e^{i\mathbf{k}_i \cdot \mathbf{r}} \phi_i \quad \text{and} \quad \psi_f = e^{i\mathbf{k}_f \cdot \mathbf{r}_f} \phi_f \qquad \text{... (6.11)}$$

Where ϕ_i and ϕ_f are, respectively, the product of the internal wave functions of the incident particle and the target nucleus, and the product of the wave functions of the scattered particle and the residual nucleus. We consider an example ^{40}Ca(d, p) ^{41}Ca. For this reaction, we have

$$\phi_i = \{\phi(d) \times \phi(^{40}\text{Ca})\}$$

$$\phi_f = \{\phi(p) \times \phi(^{41}\text{Ca})\}$$

Where $\phi(d)$, $\phi(p)$, $\phi(^{40}\text{Ca})$ and $\phi(^{41}\text{Ca})$ are the wave functions describing the internal structure of deuteron, proton, ^{40}Ca, and ^{41}Ca, respectively. The multiplication symbols here imply that the wave functions are coupled together to some definites value in angular momentum and isospin.

In the spirit of direct reaction, the deutron wave function may be taken as (weakly) coupled state of a proton and a neutron.

$$\phi(d) = (\phi(p) \times \phi(n)) \qquad \text{... (6.12)}$$

We can make it more simplified to avoid complications due to angular momentum recoupling. Let us treat the proton as purely as a spectator in the entire scattering process. If the neutron is captured into a single particle state of the target nucleus with orbital angular momentum l_t, the wave function of the residual nucleus may be expressed in the form

$$\phi_{l_t}(^{41}\text{Ca}) \sim (\phi(n)\ \phi(^{40}\text{Ca})\ Y_{l_t, m_t}(\theta', \phi')) \qquad \text{... (6.13)}$$

Using these wave functions, the scattering amplitude for ^{40}Ca(d, p) ^{41}Ca may be written as

$$f(\theta) \approx \frac{\mu}{2\pi\hbar^2} \int e^{-i\mathbf{q}\cdot\mathbf{r}'} < \{\phi(p) \times (\phi(n)\ \phi(^{40}\text{Ca})\ Y_{l_t m_t}(\theta',\ \phi'))\}\ |V(\mathbf{r}')|$$

$$\{\phi(^{40}\text{Ca}) \times (\phi(p) \times \phi(n))\} > d^3r' \qquad \text{... (6.14)}$$

The role of the potential $V(\mathbf{r}')$ here is to strip the neutron from the deutron and put it into the residual nucleus. For our purposes, we may approximate it by a delta function at the nuclear surface

$$V(\mathbf{r}') = V_0 \delta(\mathbf{r}' - R) \qquad \text{... (6.15)}$$

so as the derivation is simplified. Here R is the radius of the residual nucleus. The meaining of this approximation is that the neutron is stripped off the incident deutron and captured by the ^{40}Ca on contact. The strength of V_0 represents the probability of such a process to take place and may be treated as a parameter related to the absolute magnitude of the scattering cross-section.

When we integrate (6.14) over the coordinates of both nucleons and ^{40}Ca, the nuclear wave function drop out. One can expand the exponential factor in terms of the spherical harmonics using (6.10), and the scattering amplitude reduces to the form

$$f(\theta) \approx -\frac{2\mu}{\hbar^2} V_0 \sum_l i^l \sqrt{\frac{2l+1}{4\pi}} \, j_l(qR) \int Y_{l_0}(\theta') \, Y_{l_t m_t}^*(\theta', \phi') \sin\theta' \, d\theta' \, d\phi'$$

$$= -\frac{2\mu}{\hbar^2} V_0 \sum_l i^l \sqrt{\frac{2l+1}{4\pi}} \, j_l(qR) \, \delta_{l\,l_t} \, \delta_{m_t 0}$$

$$= -\frac{2\mu}{\hbar^2} V_0 \, i^{l_t} \sqrt{\frac{2l+1}{4\pi}} \, j_{l_t}(qR) \qquad\qquad \dots (6.16)$$

In integrating over the angular variables, we have made use of the orthonormal condition of spherical harmonics

$$\int_0^{2\pi} \int_0^{\pi} Y_{lm}^*(\theta, \phi) \, Y_{l'm'}(\theta, \phi) \sin\theta \, d\theta \, d\phi = \delta_{ll'} \, \delta_{mm'}$$

The only angular dependence in (6.4) is the spherical Bessel function through the relation between momentum transfer q and scattering angle θ. Since we have used a plane wave to approximate the solution to the scattering equation, the result is generally known as *plane wave Born approximation* (PWBA) result. A more rigorous derivation can be found in any standard text.

From (6.16), the differential cross section for direct reaction is given by

$$\frac{d\sigma}{d\Omega} \sim |j_{l_t}(qR)|^2 \qquad\qquad \dots (6.17)$$

The momentum transfer depends on the scattering angle θ

$$q = \sqrt{k_i^2 + k_f^2 - 2k_i k_f \cos\theta} \approx 2k \sin\left(\frac{\theta}{2}\right) \qquad\qquad \dots (6.18)$$

Where we have taken $k \approx k_i \approx k_f$, valid if incident energy is sufficiently high. The angular distribution is characterized by the angular momentum transferred and given by the factor

$$\left| j_{l_t}\left(2kR \sin\left(\frac{\theta}{2}\right)\right) \right|^2 .$$

For the pick-up reaction ^{41}Ca(p, d) ^{40}Ca, the low lying states are of ^{41}Ca are expected to be very like ^{40}Ca $+ n$ and a spectcroscopic factor S_l can be defined by

$$< \psi(^{40}\text{Ca}^*) \,|\psi(^{41}\text{Ca}) > \, = S_l |\phi_n^l >$$

Where ϕ_n^l is the wave function of the neutron in an l state about ^{40}Ca and $S_l \leq 1$ is the extent to which the state in ^{41}Ca is a single particle state. In this reaction $\psi_x = \psi(^{41}\text{Ca})$, $\psi_y = \psi(^{40}\text{Ca}^*)$, $\psi_y = \psi_d$ and $\psi_x = \psi_p$. Ignoring the spins of the deuteron and proton then ψ_p is unity and

$$<Yy \,|V|Xx> \, = S_l <\psi_d |V_{pn}|\phi_n'>$$

Where V_{pn} is the interaction between the incident p and the neutron in the target. If this is taken to have zero range then:

$$<\psi_d|V_{pn} = V_0\delta(\mathbf{r}_p - \mathbf{r}_n)$$

and $$M_{if} = V_0\, S_l\, \int \exp(i\,\mathbf{q}\cdot\mathbf{r}_n)\,\phi_n'(\mathbf{r}_n)\,d^3r_n$$

Therefore the reaction cross-section is proportional to the probability (square of the probability amplitude M_{if}) that the bound state neutron has a momentum equal to q, which is the momentum transferred to the incoming proton when it picks up the neutron, and to the square of the spectroscopic factor S_l.

The bound state wave function can be written:

$$\phi_n^l(\mathbf{r}_n) = r^{-1}\, u_l(\mathbf{r}_n)\, Y_{lm}(\theta,\, \phi)$$

If the reaction is assumed localised to the surface, i.e. $\delta(r_n - R) = 0$. then:

$$\frac{d\sigma}{d\Omega} \propto S_l^2\, [j_l(qR)]^2 \qquad\qquad \dots (6.19)$$

This example shows that the magnitude of the cross-section is a measure of the spectroscopic factor S_l and hence is a test of the nuclear wave function describing the state. Much better agreement (in the case of inelastic scattering as above) is obtained with experiment if distorted waves are used instead of plane waves.

Direct reactions with heavy ions   For energies near the Coulomb barrier direct pick up and transfer reactions occur with heavy ions. However, because of the strong Coulomb interaction the angular distributions are less characterstic of the transferred angular momentum of the exchanged particle. For angles much less than the grazing angle θ_g the probability of transfer is small while for angles much greater than θ_g inelastic process predominate and a bell-shaped angular distribution is seen.

(v) Optical Model   (See chapter on Nuclear Models)

(vi) Mechanisms of Nuclear Reactions in Different Stages   So far we have read that the nuclear reaction can take place through differen mechanisms which correspond to different stages of the reaction. Figure 6.33 shows a general scheme proposed by V. F. Weisskopf. This scheme helps us to visulize the different stages of a nuclear reaction.

When a projectile is incident on a target nucleus, it first experiences the presence of the complex optical potential. This represents the *first stage,* called as *single particle stage.* The real part of the potential V leads to shape elastic scattering (Figure 6.33) whereas a part of the incident wave is absorbed due to the imaginary part W of the potential leading either to *direct interaction* or to the formation of *compound nucleus* (doorway state) which corresponds to the establishment of the statistical equilibrium of the energy imparted into the nucleus. In between, there is a region-called as the *pre-equilibrium stage* in which a limited number of collisions with the nucleon or nucleon groups within the nucleus takes place prior to attainment of statistical equilibrium. There may also be *collective effects* in which some types of excitations of collective motion, e.g. surface vibrations, may take place. All these belong to the *second stage* and termed as the *compound nucleus stage* (Figure 6.33).

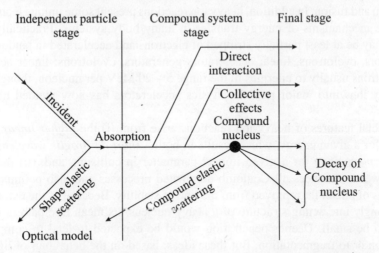

Fig. 6.33 Different stages of nuclear reaction mechanisms by Weisskopf.

Lastly, there is the *final* or *third* stage. This stage describes the production of the reaction products which proceed on a very short time duration. The two extremes of this stage represent the *direct reaction* in a nucleon or a group of nucleon comes out due to the impact of the projectile nucleon, while the compound nucleus formation is followed by the imparted energy on a group of nucleons which enable them to come out. In between, in the pre-equilibrium reaction, a few nucleon may come out of the nucleus successively, provide enough energy is received by these nucleons for emission. One can recognize the different mechanisms by the energy and angular distribution for the emitted particles.

We may note that there is some apparent lack of consistency in two extreme models, i.e. optical model and the compound nucleus model. The nucleus is treated as a perfectly body in the compound nucleus model, in which the energy imparted by the projectile particle is dissipated very rapidly amongst all the nucleus with the nucleus. On the other hand, the nucleus is considered partially transparent (grey matter) endowed both with an absorption coefficient and a refractive index, described by the imaginary (W) and real (V) parts respectively of the potential in the optical model.

Weisskopf tried to reconcile the above contradiction by invoking exclusion principle. The nucleons are fermions and they fill up all long lying states successively, like the electrons in a solid body. No single nucleon within the nucleus can occupy an excited unbound state which enables it to come out, unless sufficient energy for excitation is available. Thus, at relatively lower energy of excitation, the nucleus behaves like a perfectly black body and the imparted energy simply increase the thermal energy, i.e. energy of random motion of the nucleons which lead to the formation of the compound nucleus. One or two nucleons at higher energies, may gain sufficient energy to be excited to higher unfilled states from which they are able to come out by *direct reaction* process.

(vii) Heavy Ions (HI) Induced Nuclear Reactions Heavy ions have characterstics which emphasize particular features of the whole range of nuclear phenomena from elastic scattering to fission and fusion. In addition, heavy ion reactions present some unique features connected with the mechanisms of energy transfer in many body systems. Practically all available atoms may be at least partially stripped of electrons and accelerated in tandom electrostatic generators, cyclotrons, linear accelerators generators, cyclotrons, linear accelerators and synchrotrons usually to energies in the range 5 – 20 MeV per nucleon, though the injection of heavy ions into major particle physics accelerators has now yielded high relativistic energies.

The special features of heavy-ion reactions arise from (i) the *higher linear momentum* of the ion for a given energy, which results in both a *short de Broglie wavelength* and a high *angular momentum* for a given impact parameter in collision, and, (ii) the high nuclear charge which enhances all Coulomb-dominated processes and also permits production in reactions of nuclei far removed from the line of stability. Because of the extended, complex and strongly interacting structure of a heavy nucleus, its mean free path in nuclear matter ought to be small. Deeper penetration would be expected to lead to compound nucleus formation or to fragmentation. But these ideas, based on the behaviour of light projectiles, have had to be supplemented by new mechanisms in the case of heavy ions.

Usually heavy ions are projectiles with $A > 4$ (i.e., heavier than α-particles). Special experimental techniques are required to study reactions induced by heavy ions, because of the large variety of the reaction products, their high Z-values, short ranges and short life times. one can identify the reaction products, their high Z-values, short ranges and short life times. One can identify the reaction products by using E and ΔE counters where $E = \dfrac{M v^2}{2}$

is the non-relativistic kinetic energy. Since $\dfrac{dE}{dx} \propto \dfrac{Z^2}{v^2}$, where Ze is the charge of the product, one obtains:

$$E \Delta E \propto Z^2 M \qquad \qquad ... (6.1)$$

Obviously, by simultaneously measuring E and ΔE. One can identify the products unambiguously, specially the lighter fragments. For e.g., for the two neighbouring isotopes ^{18}O and ^{19}F, the values of $Z^2 M$ are 1152 and 1539 respectively. These values differ by 33.6 %. However, for heavier isotopes, the resolving power of the method is lower. The values of $Z^2 M$ for the neighbouring isotopes ^{42}Ca and ^{45}Sc differ by 18 %, whereas for ^{109}Ag and ^{113}Cd only by 8 %.

One can improve the resolving power of the method by using two ΔE detectors in a telescope and measuring the time of flight between them accurately. Now a days, a combination of $E - \Delta E$ telescope with a magnetic spectrometer is also used as an improved method for identification.

In many cases, *HI* induced reaction products are very short lived radioisotopes.

Figure 6.34 shows a schematic diagram of the collision of an incident *HI* with a target nucleus for different impact parameters b. For large b, usual Coulomb type scattering or Coulomb excitations occurs at low energies, i.e. when

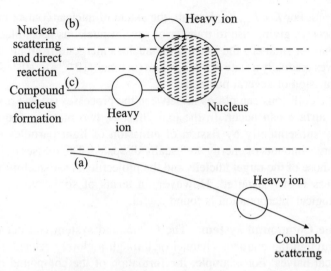

Fig. 6.34 For different parameters, the schematic diagram of collision between heavy ion (*HI*) and a nucleus.

$$E < \frac{1}{4\pi\varepsilon_0} \frac{Z_1 Z_2}{(R_{\text{ion}} + R_{\text{nuc}})} \qquad \ldots (6.2)$$

Here $Z_1 = Z_{\text{ion}}$ and $Z_2 = Z_{\text{nuc}}$. Equation (6.2) follows from the fact that for $r = b = R_{\text{ion}} + R_{\text{nuc}}$, the incident energy E is equal to the Coulomb potential energy.

When the impact parameter $b \approx R_{\text{nuc}}$ (radius of the target nucleus), scattering due to specific nuclear forces will take place (Figure 6.34b). In this case, direct reaction involving a few nucleon transfer may also take place. Fusion of two nuclei for very small b may result, causing the formation of compound nucleus (Figure 6.34c). In this case, the incident ion must of course be quite energetic so that it may overcome the potential barrier, and compound nucleus will be formed with very high excitation. In these reactions, there are many new and unusual features are observed. In this case, E must eceed the Coulomb barrier height given above.

In case of grazing incident of the ion on the nucleus, the impact parameter, b should be, $b = (R_{\text{ion}} + R_{\text{nuc}})$ and the angular momentum $L = l_{gr}\hbar$ can be determined using the relation

$$l_{gr} = (R\hbar^{-1}) \left[2M \left(\frac{E - Z_1 Z_2}{4\pi\varepsilon_0 R} \right) \right]^{\frac{1}{2}} \qquad \ldots (6.3)$$

Where $R = (R_{\text{ion}} + R_{\text{nuc}})$ is the distance between the centres of the incident ion and the nucleus at grazing incidence. Compared to the Coulomb barrier height l_{gr} if E is large, l_{gr} can be quite high. For head on grazing collision, $E = \dfrac{Z_1 Z_2}{4\pi\varepsilon_0 R}$, $l_{gr} = 0$. such type of collision usually takes place in *inelastic* processes, e.g. excitation of giant resonances, reactions involving transfer of one nucleon and other direct processes with characterstic times of

10^{-23} to 10^{-22} s. For $L < L_{gr}$, i.e. b small, the extent of nuclear contact increases and highly inelastic processes giving rise to transfer of several nucleons takes place.

It is found that for smaller l the ion and the nucleus merge to from the compound nucleus, which achieves gradually a state of equilibrium (in time $\sim 10^{-20}$ s to 10^{-18} s) with a subsequent fission or emission of several nucleons.

Deep inelastic collisions can places in between the processes (b) and (c), i.e. collision with the nuclear surface can occur. In the exit channel, two massive fragments are formed, which decay subsequently by fission or emission of light particles or γ-rays. In these processes, large amounts of energy are usually disspated. The masses of primary fragements are close to those of the target nucleus and the projectile. A satisfactory microscopic theory of the process is still awaited. However, in terms of some macroscopic parameters, phenomenological interpretation is found useful.

Decay of the Compound system   The Compound system can decay through several channels. Through the entrance channel or through a closely related channel, it can split into two equal halves. For example, the formation of the compound nucleus ^{32}S and its decay through entrance channel as

$$^{16}O + {}^{16}O \rightarrow {}^{32}S^*$$

$$^{32}S^* \rightarrow {}^{16}O + {}^{16}O$$

$^{32}S^*$ can also decay through some channel as

$$^{32}S^* \rightarrow {}^{20}Ne + {}^{12}C$$

A more probable mode of the decay of the compound nucleus is through the emission of nucleons, promarily neutrons, as fission is inhibited by Coulomb barrier. The number of neutrons emitted by decay may range from 5 to 10. One finds that this is also true even in reactions with very heavy nuclei for which the compound nucleus is highly proton rich. However, the break up by proton emission is inhibited due to Coulomb potential barrier.

In the case of lighter nuclei, the compound nucleus formed by complete fusion is neutron deficient, e.g.

$$^{40}_{20}Ca + {}^{40}_{20}Ca \rightarrow {}^{80}_{40}Zr^*$$

The lighest known stable isotope of Zr is $^{90}_{40}Zr$ so that there a deficiency of 10 neutrons. Moreover, the compound nucleus is formed in an excited state and the excited energy is removed by the emission of a few more neutrons. This means that the resulting product nucleus is still more neutron deficient. Usually, the decay of such nuclei takes place by delayed proton emission from the ground state. Two-proton radioactivity has also been reported.

The most probable mode of decay in the case of heavy nucleus fusion is fission with the formation of neutron rich fragments having $A = 100 - 150$. In some cases transuranic nuclei are also obtained.

(viii) **Measurment of Nuclear Reaction Life-times**   The nuclear reaction processes are extremely fast, e.g. the order of decay of compound nucleus is 10^{-15} s and that of direct reactions is $\sim 10^{-21}$ s to 10^{-22} s. Obviously, direct reactions are even faster than the decay of compound nucleus.

A method based on shadow effect has been developed for the measurment of these extremely short reaction life-times. This helps one to distinguish between different reaction mechanisms. Figure 6.35 shows a method of measuring nuclear reaction life times based on shadow effect. In this method a crystal is exposed to a beam of projectiles from an particle accelerator. There is an arrangement to prevent the decay products of the nuclei in the lattice sites undergoing nuclear transformation from escaping in all directions. Along the crystal axes due to much larger concentration of atoms in the lattice, these directions are forbidden for the decay products. Now, if a photographic plate is placed behind the crystal, a pattern of shadows, which is characterstic of the crystal will be observed.

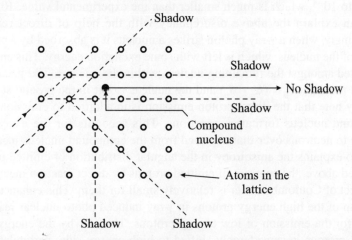

Fig. 6.35 Measurment of nuclear reaction life-times with the help of shadow effect method

In the case of direct reactions, the shadows will be formed only if the nucleus does not have time to leave the lattice site while emitting its decay products. However, if the reaction process involves the formation of the relatively longer lived compound nucleus, then the compound nucleus moves away from the lattice site prior to emitting the decay products. This is why that no shadow will be formed in this case (Figure 6.35). The compound nucleus life time $\tau = 10^{-15} - 10^{-14}$ s and its velocity is ~ 106 m/s. It moves away from the lattice site through a distance $d \sim 10^{-9}$ to 10^{-8} m. This distance is much larger than the lattice dimension $\sim 10^{-10}$ m.

We may note that this method has also been used for the measurment of very short life-times of nuclei formed in heavy ion induced reactions and also for the determination of fission time.

(ix) **Mechanism of Photo-Nuclear (γ-ray induced) Reactions**   It is observed that in gamma ray induced reactions, e.g. (γ, n) and (γ, p), the cross-section for the absorption of γ-ray increases linearly with the charge of the nucleus, i.e. $\sigma_\gamma \propto Z$.

Experimental observations on the angular distribution of the products of n or p, photo nuclear reactions reveal that these reactions do not involve the formation of compound nucleus in many cases. Thus, the neutrons and low energy protons are emitted from nuclei with $A > 100$ isotropically. The angular distribution is anisotropic for higher energy protons and the ratio $\dfrac{\sigma(\gamma,\,p)}{\sigma(\gamma,\,n)} \sim 10^{-2}$.

If we consider that the above reactions proceed through compound nucleus formation reaction mechanism, then the emitted protons and neutrons should have Maxwellian distribution. Theoretical calculations based on this mechanism and taking into account the probability of penetration through the Coulomb barrier in the case of protons yield $\dfrac{\sigma(\gamma,\,p)}{\sigma(\gamma,\,n)}$

$\sim 10^{-3}$ to 10^{-4}, which is much smaller than the experimental value, 10^{-2}.

One can explain the above discrepency with the help of direct reaction mechanism. Accordingly, when a γ-ray photon strikes a nucleus it is absorbed by a photon in the surface region of the nucleus, which is left with some excitation energy. This energy is then quickly dissipated amongst the remaining nucleus within the nucleus. The maximum energy of the emitted proton is $\varepsilon_p = (E_\gamma - S_p)$ and the nucleus is left in the ground state in this case.

We may note that the direct proton emission is in addition to emission of proton from the compound nucleus formation reactions. This accounts for the excess yield of protons relative to neutrons over that expected from the compound nucleus mechanism. Moreover, this also explains the anisotropy in the angular distribution of emitted protons.

As stated above, the energies of emitted protons by direct reaction mechanism are high and the effect of Coulomb barrier is relatively small for them. This enhances the probability of emission of the high energy protons in γ-ray induced photo-nuclear reactions as compared to that for the emission of low energy protons. This is why the energy distribution of the emitted protons is considerably shifted to high energy side, particularly for the heaviour nuclei. The high potential barrier for these reduces considerably the probability of the emission of low energy protons through compound nucleus formation process significantly.

(x) **Nuclear Molecules**   The compound nucleus system gets a large angular momentum in heavy ion nuclear reactions. In this case the resulting rotational velocity of the system is so large that the nuclear force may not be able to provide the neccessary centripetal acceleration. This is why the compound nucleus system becomes totally unstable and therefore it is unable to form as such. In this case, one finds that instead of fusion, the nuclei form a nuclear molecule much like a diatomic molecule. This nuclear molecule exists for a short time and then breaks up through the entrance channel. In this case the independence of hypothesis does not hold as the composite system retains the memory of its formation. Obviously, this is contrary to the basic assumption of the mechanism of the compound nucleus. Figure 6.36 illustrates the processes of formation (a) and break up (c) of nuclear molecule. Figure 6.36(b) also shows the composite system.

Experimental observations on the rotational and vibrational excitations provide evidence of the formation of such a composite system. These observation are just similar to those observed in ordinary molecules.

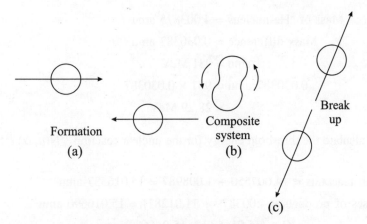

Fig. 6.36 (a) Formation of nuclear molecule, (c) break up of nuclear molecule, and (b) composite system.

Example 6.1 Calculate the energy generated in MeV when 0.1 kg of ^{7}Li is converted to ^{5}He by proton bombardment. Given: masses of ^{7}Li, ^{4}He and ^{1}H in amu are 7.0183, 4.0040 and 1.0081 respectively.

Sol. The equation for the above nuclear reaction is

$$^7_3\text{Li} + ^1_1\text{H} = 2^4_2\text{He} + Q$$

Masses of the reactants = 7.0183 + 1.0081 = 8.0264 amu

Masses of the products = 2 × 1.0040 = 8.0080 amu

∴ Difference in masses = 0.0184 amu

∴ Energy liberated when 7.0183 kg of Li is converted to He is 0.0184 kg. For 0.1 kg of Li, therefore, the amount of energy liberated is

$$E = \frac{0.1 \times 0.0184}{7.0183} \text{ kg}$$

$$= \frac{0.1 \times 0.0184}{7.0183} \times (3 \times 10^8)^2 \text{ J}$$

$$= \frac{0.1 \times 0.0184 \times 9 \times 10^{16}}{7.0183 \times 1.6 \times 10^{-13}} \text{ MeV}$$

$$= 14.74 \times 10^{25} \text{ MeV}$$

Example 6.2 Calculate the binding energy in MeV of ^{4}He from the following data: Mass of ^{4}He = 4.003875 amu; mass of ^{1}H = 1.008145 amu and mass of a neutron = 1.008986 amu.

Sol. A ^{4}He-nucleus consists of 2 protons and 2 neutrons.

Mass of (2 protons + 2 neutrons) = 2(1.008145 + 1.008986)

$$= 4.034262 \text{ amu}$$

$$\text{Mass of } {}^4\text{He-nucleus} = 4.003875 \text{ amu}$$

$$\therefore \qquad \text{Mass difference} = 0.030387 \text{ amu}$$

We have,
$$1 \text{ amu} = 931 \text{ MeV}$$

$$\therefore \qquad 0.030387 \text{ amu} = 931 \times 0.030387$$

$$= 28.29 \text{ MeV}$$

Example 6.3 Calculate the threshold energy for the nuclear reaction ${}^{14}\text{N}(n, \alpha) \, {}^{11}\text{B}$ in MeV.

[Vikaram]

Sol. Masses of reactants = 14.007550 + 1.008987 = 15.016537 amu

Masses of products = 4.003879 + 11.012811 = 15.016690 amu

$$\therefore \qquad Q = (15.016537 - 15.016690) \text{ amu}$$

$$= -0.000153 \text{ amu}$$

$$= -0.000153 \times 931 \text{ MeV} = -0.14 \text{ MeV}$$

$$\therefore \qquad E_{\text{th}} = -Q\left[1 + \frac{M(n)}{M(N)}\right] = 0.14\left(1 + \frac{1.008987}{14.007550}\right)$$

$$= 0.14\left(1 + \frac{1}{14}\right) = 0.15 \text{ MeV}$$

Example 6.4 Find the amount of energy in joules during the process in which 0.001 kg of radium is converted into lead (masses: ${}^{226}\text{Ra} = 226.0955$ amu, ${}^{206}\text{Pb} = 206.0386$ amu and α-particle = 4.003 amu).

[Raj, Punjab]

Sol. In the conversion of 1 atom of ${}^{226}\text{Ra}$ into 1 atom of ${}^{206}\text{Pb}$, 5α-particles are emitted in all.

$$\text{Initial mass of } {}^{226}\text{Ra} = 226.0955 \text{ amu}$$

$$\text{Final mass of } {}^{206}\text{Pb} = 206.0386 \text{ amu}$$

$$\therefore \qquad \text{Difference in masses} = 20.0569 \text{ amu}$$

$$\text{Mass of } 5\alpha\text{-particles} = 20.0150 \text{ amu}$$

$\therefore$ Mass converted into energy = 0.0419 amu

This is equivalent to an energy 0.0419×931 or 39 MeV (for 1 atom).

Now, 0.001 kg (= 1 g) of radium = $\dfrac{6.023 \times 10^{23}}{226}$ Ra-Atoms.

$$\therefore \text{Total energy released,} \qquad E = \frac{6.023 \times 10^{23} \times 39}{226} \text{ MeV}$$

$$= \frac{6.023 \times 10^{23} \times 39 \times 1.6 \times 10^{-13}}{226} \text{ J}$$

$$= 16.63 \times 10^9 \text{ J}$$

Example 6.5 Calculate the minimum energy of γ-rays neccessary to disintegrate a deuteron into a proton and a neutron. Given: mass of proton = 1.00759 amu, mass of neutron = 1.00898 amu and mass of deutron = 2.01471 amu.

Sol. From the nuclear reaction: ${}_1^2H + \gamma = {}_1^1H + {}_0^1n$, the minimum energy of γ-rays required to disintegrate a deuteron is:

$$1.00759 + 1.00898 - 2.01471 = 0.00186 \text{ amu}$$

$$0.00186 \text{ amu} = 0.00186 \times 931 \text{ MeV} = 1.73 \text{ MeV}$$

Example 6.6 The Q-value in MeV of the following three reactions

$$\begin{aligned} {}_1^2H + {}_1^2H &\rightarrow {}_1^3H + {}_1^1H \\ {}_1^2H + {}_1^2H &\rightarrow {}_2^3He + {}_0^1n \\ {}_1^3H &\rightarrow {}_2^3He + \beta^- \end{aligned}$$

and

are 4.031, 3.265 and 0.0185 respectively. Calculate the mass difference between the neutron and the hydrogen atoms from these data. [Punjab]

Sol. We have

$$Q_1 = 4.031 \text{ MeV} = \text{mass of } 2{}_1^2H - \text{mass } {}_1^3H - \text{mass of } {}_1^1H$$

$$Q_2 = 3.265 \text{ MeV} = \text{mass of } 2{}_1^2H - \text{mass of } {}_2^3He - \text{mass of } {}_0^1n$$

$$Q_3 = 0.0185 \text{ MeV} = \text{mass of } {}_1^3H - \text{mass of } {}_2^3He$$

The mass of β^- is too small to be taken into account and has been neglected.

$\therefore \qquad Q_1 - Q_2 + Q_3 = \text{mass of } {}_0^1n - \text{mass of } {}_1^1H.$

$\therefore$ Mass difference between neutron and H-atom is thus

$$Q_1 - Q_2 + Q_3 = (4.031 - 3.265 + 0.0185) \text{ MeV}$$

$$= 0.7845 \text{ MeV}$$

$$= \frac{0.7845}{931} \text{ amu}$$

$$= 0.000842 \text{ amu}$$

Example 6.7 The masses of the different nuclei taking part in ${}_3^7Li$ (p, n) reaction in amu are as follows: $M({}_3^7Li) = 7.01822$, $M({}_1^1H) = 1.00814$, $M({}_0^1n) = 1.00898$ and mass of the product nucleus = 7.01915. Calculate the Q-value of this reaction in MeV. Is it exoergic or endoergic? What is the threshold of this reaction? [Mumbai]

Sol. Masses of the reactants = 7.01822 + 1.00814 = 8.02636 amu

Masses of the products = 7.01915 + 1.00898 = 8.02813 amu

$\therefore$ Q-value of the reaction = (8.02636 − 8.02813) amu

$$= -0.00117 \text{ amu}$$

$$= -0.00117 \times 931 \text{ MeV}$$

$$= -1.089 \text{ MeV}$$

The negative sign of Q-value shows that the reaction is endoergic.

Threshold energy E_{th} is given by, $E_{th} = (-Q)\left[1 + \dfrac{M\,(_1^1H)}{M\,(_3^7Li)}\right]$

$$\therefore \qquad E_{th} = 0.00117 \times \left(1 + \dfrac{1.00814}{7.01822}\right) \text{amu}$$

$$= 0.001388 \text{ amu}$$

$$= 0.001338 \times 931 \text{ MeV}$$

$$= 1.2456 \text{ MeV}$$

Example 6.8 In the reaction $_5^{11}B + _2^4He \rightarrow _7^{14}N + _0^1n$, the masses of ^{11}B, ^{14}N and ^4He are 11.01280 amu, 14.00752 amu and 4.00387 amu respectively. If the incident α-particle has a kinetic energy 5.250 MeV towards ^{11}B which is at rest and the kinetic energies of product nuclei ^{14}N and ^1n anre 3.260 MeV ane 2.139 MeV respectively, compute the mass of neutron.   [Mysore]

Sol. Mass of ^{11}B = 11.01280; mass of ^{14}N = 14.00752, mass of ^4He = 4.00387 in amu.

Kinetic energy of ^4He, $K_\alpha = 5.250$ MeV

Kinetic energy of ^{14}N, $K_N = 3.260$ MeV

Kinetic energy of ^1n, $K_n = 2.139$ MeV

Kinetic energy of ^{11}B, $K_B = 0$ ($\because$ it is at rest)

From the energetics of the reactions, We have

$$M_B c^2 + M_{He} c^2 + K_\alpha = M_N c^2 + K_N + M_n c^2 + K_n$$

$$\therefore \qquad M_n c^2 = (M_B + M_{He} - M_N)c^2 + (K_\alpha - K_N - K_n)$$

$$\therefore \qquad M_n = \dfrac{(M_B + M_{He} - M_N) + (K_\alpha - K_N - K_n)}{c^2}$$

$$= (11.01280 + 4.00387 - 14.00752) + \dfrac{(5.250 - 3.260 - 2.139)}{9 \times 10^{16} \times 1.66 \times 10^{-27}} \times 1.6 \times 10^{-13}$$

$$= 1.00915 + \dfrac{-0.149 \times 1.6 \times 10^{-2}}{9 \times 1.66} \qquad (\because 1 \text{ amu} = 1.66 \times 10^{-27} \text{ kg})$$

$$= [1.00915 - (0.149 \times 0.1071 \times 10^{-2})] \text{ amu}$$

$$= (1.00915 - 0.000159) \text{ amu}$$

$$= 1.00899 \text{ amu}$$

Example 6.9 Calculate the Q-values of the following two reactions; (i) $_1^3H + _1^2H \rightarrow _2^4He + _0^1n$ and (ii) $_2^4He + _7^{14}N \rightarrow _8^{17}O + _1^1H$. Given: $M(_1^3H) = 3.0169982$, $M(_1^2H) = 2.0147361$, $M(_2^4He) = 4.0038727$, $M(_0^1n) = 1.0089932$, $M(_7^{14}N) = 14.003074$, $M(_8^{17}O) = 16.999133$ and $M(_1^1H) = 1.007825$ all in amu. Indicate also if the reaction is exoergic or endoergic.   [Kerala]

Sol. We have: $Q = $ (mass of reactants – mass of products) $\times c^2 = \Delta m \cdot c^2$

(i) Here $\Delta m = [\{M(^3H) + M(^2H)\} - \{M(^4He) + M(^1n)\}]$

$\qquad = (3.0169982 + 2.01473561) - (4.0038727 + 1.0089832)$

$\qquad = 0.0188784$ amu

$\therefore \qquad Q = 0.0188784 \times 931.48$

$\qquad = +17.57$ MeV Reaction is exoergic.

(ii) Here $\Delta m = [\{M(^4He) + M(^{14}N)\} - \{M(^{17}O) + M(^1H)\}]$

$\qquad = (4.003872 + 14.003074) - (16.999133 + 1.007825)$

$\qquad = 18.006946 - 18.006958 = -0.000012$ amu

$\qquad = -0.000012 \times 931.48 = -0.11178$ MeV.

So, the reaction is endoergic.

Example 6.10 Consider the reaction

$$_9F^{19} + _0n^1 \rightarrow _8O^{10} + _1H^1 - 7.6342 \text{ MeV}$$

Kinetic energy of the incident neutrons is 15 MeV and protons are emitted at an angle of 90° with the direction of the incident neutrons. Calculate (i) Kinetic energy of protons (ii) Threshold energy. Given m $(_8O^{19}) = 19.05862$ a.m.u., $m(_0n^1) = 1.0087$ a.m.u., $m(_1H^1) = 1.0073$ a.m.u., $m(_9F^9) = 19.0457$ a.m.u.

Sol. (i) Q-value under non-relativistic case is given by:

$$Q = \left\{\frac{M^i}{M^R} - 1\right\}K^i + \left\{\frac{M^e}{M^R} + 1\right\}K^e - \frac{2\cos\theta\sqrt{(M^i M^e K^i K^e)}}{M^R}$$

In this case, $\theta = 90°$; $\qquad \therefore \cos\theta = 0$

$$Q = \left\{\frac{M^i}{M^R} - 1\right\}K^i + \left\{\frac{M^e}{M^R} + 1\right\}K^e$$

$\therefore$ Kinetic energy of protons

$$K^e = \frac{Q - \left(\dfrac{M^i}{M^R} - 1\right)K^i}{\left(\dfrac{M^e}{M^R} + 1\right)}$$

$$= \frac{(-7.6342 \text{ MeV}) - \left(\dfrac{1.0087}{19.05862} - 1\right)(15 \text{ MeV})}{\left(\dfrac{1.00728}{19.05862} + 1\right)}$$

$$= 6.2370 \text{ MeV}$$

(ii) Threshold energy,

$$E_{\text{th}} = (-Q)\left(\frac{M^i + M^T}{M^T}\right) = 7.6342 \times \left(\frac{1.0087 + 10.0457}{19.05862}\right) \text{ MeV}$$

$$= 8.0347 \text{ MeV}.$$

Example 6.11 Calculate the Q-value for the reaction $^{11}_5\text{B}$ (α, n) $^{14}_7\text{N}$ from the following set of nuclear reaction equations:

(i) $^{14}_7\text{N} + ^2_1\text{H} \rightarrow ^{16}_7\text{N} + ^1_1\text{H} + 9.42$ MeV

(ii) $^{16}_7\text{N} + ^2_1\text{H} \rightarrow ^{13}_6\text{C} + ^4_2\text{He} + 8.2$ MeV

(iii) $^{13}_6\text{C} + ^2_1\text{H} \rightarrow ^{11}_5\text{B} + ^4_2\text{He} + 5.56$ MeV

(iv) $^1_1\text{H} + ^1_0\text{n} \rightarrow ^2_1\text{H} + 0.530$ MeV

Given that $m_\alpha = 3726.292$ MeV, $m_\alpha + 1875.069$ MeV.

Sol. Adding the first three of the above equations together,

$$^{14}_7\text{N} + ^2_1\text{H} + ^{15}_7\text{N} + ^2_1\text{H} + ^{13}_6\text{C} + ^2_1\text{H} \rightarrow ^{15}_7\text{N} + ^1_1\text{H} + 9.42 \text{ MeV}$$

$$+ ^{13}_6\text{C} + ^4_2\text{He} + 8.2 \text{ MeV} + ^{11}_5\text{B} + ^4_2\text{He} + 5.56 \text{ MeV}$$

or $^{14}_7\text{N} + 3^2_1\text{H} \rightarrow ^{11}_5\text{B} + 2^4_2\text{He} + ^1_1\text{H} + 23.18$ MeV

$^{11}_5\text{B} + ^4_2\text{He} \rightarrow ^{14}_7\text{N} + 3^2_1\text{H} - ^1_1\text{H} - ^4_2\text{He} - 23.18$ MeV

With the help of equation (iv), this can be written as

$^{11}_5\text{B} + ^4_2\text{He} \rightarrow ^{14}_7\text{N} + ^1_0\text{n} + 2^2_1\text{H} - ^4_2\text{He} - 23.18$ MeV $- 0.530$ MeV

or $^{11}_5\text{B} + ^4_2\text{He} \rightarrow ^{14}_7\text{N} + ^1_0\text{n} + Q$

where $Q = (2^2_1\text{H} - ^4_2\text{He} - 23.710$ MeV

$$= (2 \times 1875.060 - 3726.292 - 23.710 \text{ MeV}$$

$$= 0.136 \text{ MeV}.$$

Example 6.12 Calcuate the Q-value for the reaction $^{29}_{14}\text{Si}$ (d, n) $^{30}_{15}\text{P}$ from the following nuclear reaction reactions for the formation of $^{30}_{15}\text{P}$ in the ground state.

(i) $2^2_1\text{H} \rightarrow ^4_2\text{He} + 22.5$ MeV

(ii) $^{31}_{15}\text{P} + ^1_1\text{H} \rightarrow ^{28}_{14}\text{Si} + ^4_2\text{He} + 2.0$ MeV

(iii) $^{31}_{15}\text{P} + \gamma \rightarrow ^{30}_{15}\text{P} + ^1_0\text{n} - 12.5$ MeV

(iv) $^{28}_{14}\text{Si} + ^2_1\text{H} \rightarrow ^{29}_{14}\text{Si} + ^1_1\text{H} + 6.0$ MeV

Sol. We have to find out the Q-value for the reaction $^{29}_{14}\text{Si}$ (d, n) $^{30}_{15}\text{P}$, so by the combinations of the above reactions, we have to see how this reaction can be accomplished. From reaction (iv)

$$^{29}_{14}\text{Si} + ^1_1\text{H} \rightarrow ^{28}_{14}\text{Si} + ^2_1\text{H} - 6.0 \text{ MeV} \qquad \qquad \text{... (6.1)}$$

with the help of (ii) it can be expressed as

$$^{29}_{14}\text{Si} + {}^1_1\text{H} \rightarrow {}^{31}_{15}\text{P} + {}^1_1\text{H} - {}^4_2\text{He} - 2.0 \text{ MeV} + {}^2_1\text{H} - 6.0 \text{ MeV}$$

or
$$^{29}_{14}\text{Si} \rightarrow {}^{31}_{15}\text{P} - {}^4_2\text{He} + {}^2_1\text{H} - 8.0 \text{ MeV} \qquad \qquad \text{... (6.2)}$$

with the help of (iii), Equation (6.2) can be expressed as:

$$^{29}_{14}\text{si} \rightarrow {}^{30}_{15}\text{P} + {}^1_0\text{n} - \gamma - 12.5 \text{ MeV} - {}^4_2\text{He} + {}^2_1\text{H} - 8.0 \text{ MeV}$$

or
$$^{29}_{14}\text{Si} \rightarrow {}^{30}_{15}\text{P} + {}^1_0\text{n} - \gamma - {}^4_2\text{He} + {}^2_1\text{H} - 20.5 \text{ MeV} \qquad \qquad \text{... (6.3)}$$

Now with the help of (6.1), Equation (6.3) can be written as:

$$^{29}_{14}\text{Si} \rightarrow {}^{30}_{15}\text{P} + {}^1_0\text{n} - \gamma - 2{}^2_1\text{H} + 22.5 \text{ MeV} + {}^2_1\text{H} - 20.5 \text{ MeV}$$

or
$$^{29}_{11}\text{Si} + {}^2_1\text{H} \rightarrow {}^{30}_{15}\text{P} + {}^1_0\text{n} - \gamma + 2.0 \text{ MeV} \qquad \qquad \text{... (6.4)}$$

This is the desired reaction. If $^{30}_{15}\text{P}$ is formed in the ground state. Q-value for this reaction from the last equation is $Q = 2.0$ MeV

Example 6.13 Calculate the threshold energy for the $^{14}\text{N}\,(n,\,a)\,^{11}\text{B}$ reaction.

Sol.

$$M^T + M^I = 14.007550 + 1.008987 = 15.016537 \text{ a.m.u.}$$

$$M^e + M^R = 4.003879 + 11.0120811 = 15.016690 \text{ a.m.u.}$$

$$\therefore \qquad Q = [(M^T + M^I) - (M^e + M^R) \text{ a.m.u.}]$$

$$= [15.016537 - 15.016690] \text{ a.m.u.}$$

$$= -0.000153 \times 931 \text{ MeV (Since 1 a.m.u.} = 931 \text{ MeV)}$$

$$= -0.14 \text{ MeV.}$$

Thus the reaction is endoergic.

The threshold energy is

$$E_{\text{thre}} = (-Q)\left(\frac{M^i + M^T}{M^T}\right) = 0.14\left(1 + \frac{1}{14}\right) = 0.15 \text{ MeV.}$$

Example 6.14 The $^{10}_{5}\text{B}\,(\alpha,\,p)\,^{13}_{6}\text{C}$ reaction shows among others a resonance for all excitation energy of the compound nucleus of 13.23 MeV. The width of this level as found experimentally is 130 keV. Calculate the mean life of the nucleus for the excitation.

Sol. The reaction is

$$^{10}_{5}\text{B} + {}^4_2\text{He} \rightarrow ({}^{14}_{7}\text{N})^* \rightarrow {}^{13}_{6}\text{C} + {}^1_1\text{H}$$

The mean life time is given by:

$$\gamma = \frac{h}{2\pi\Gamma} = \frac{6.625 \times 10^{-34} \text{ J.sec}}{2(3.14) \times 130 \times 10^3 \text{ eV} \times 1.6 \times 10^{-19} \text{ J/eV}}$$

$$= 5 \times 10^{-21} \text{ sec.}$$

Example 6.15 Neutron capture of slow neutrons by ^{235}U shows a resonance for an excitation energy of 0.29 eV. The compound nucleus can become de-excited either through γ-emission or by fission into larger nuclear fragements. The mean life time of the compound nucleus was found to be 4.7×10^{-15} sec. and the partial width for γ-emission $\Gamma_\gamma = 3.4 \times 10^{-2}$ eV. Calculate the partial fission width Γ_f.

Sol. Total width

$$\Gamma = \frac{h}{2\pi r} = \frac{(6.625 \times 10^{-34} \text{ J.sec})}{2 \times 3.14 \times (4.7 \times 10^{-15} \text{ sec})(1.6 \times 10^{-19} \text{ J/eV})}$$

$$= 0.14 \text{ eV}$$

Now since $\Gamma = \Gamma_\gamma + \Gamma_f$

$\therefore \qquad \Gamma_f = \Gamma - \Gamma_\gamma$

$$= 0.14 \text{ eV} - 0.034 \text{ eV}$$

$$= 0.106 \text{ eV}.$$

Example 6.16 Protons of energy 4.99 MeV, are made to bombard $^{19}_{9}$F as a result of which a neutron is emitted according to the following reaction:

$$^{19}_{9}\text{F} + ^{1}_{1}\text{H} \rightarrow ^{20}_{10}\text{Ne}^* \rightarrow ^{19}_{10}\text{Ne} + ^{1}_{0}\text{n}$$

Calculate the excitation energy of the compound nucleus $^{20}_{10}\text{Ne}^*$

Sol. Kinetic energy of the proton available in the C.M. system is given by

$$= \frac{K^i}{1 + \dfrac{M^i}{M^T}}$$

$$= \frac{4.99}{1 + \dfrac{1}{19}} = \frac{(4.99) \times 19}{20} = 4.74 \text{ MeV}$$

Further

$$M^i = ^{1}_{1}\text{H} = 1.008146 \text{ a.m.u.} \qquad M_c = ^{20}_{10}\text{Ne} = 19.99872$$

$$M^T = ^{19}_{9}\text{F} = 18.004444 \text{ a.m.u.}$$

$$M^i = M^T = 20.012590$$

$\therefore \qquad (M^i + M^T) = 20.012590 - 19.998772$

$$= 0.013818 \text{ a.m.u.}$$

which is equivalent to $(0.013818 + 931)$ MeV = 12.88 MeV

∴ Excitation energy of the compound nucleus is given by

$$E_{\text{exc.}} = 4.74 + 12.88 = 17.62 \text{ MeV}.$$

SUGGESTED READINGS

1. D. F. Jackson, Nuclear Reaction Mechanism, Methuen (1970)
2. G. R. Satchler, Introduction to Nuclear Reaction, Macmillan (1980)
3. S. L. Kakani Quantum Mechanics Sultanchand, New Delhi-2 (4th ed. 2005)
4. W. Tabocman, Theory of Direct Nuclear Reactions, Oxford University Press (1961)
5. S. De Benedetti, Nuclear Interactions, John Wiley (1964)
6. W. E. Cottinghan and D. A. Greenwood, An Introduction to Nuclear Physics, Cambridge University Press (1986).
7. P. E. Nodgson, Compound Nucleus Reactions, Rep. Prog. Phys. 50, 1171 (1987)
8. P. E. Hodgson, The optical Model of Elastic Scattering, Oxford University Press (1963)
9. H. J. Specht, Rev. Mod. Phys. 46, 773 (1974)
10. J. O. Newton, Contemp Phys. 30, 277 (1989)
11. A. Bohr and B. R. Mottelson, Physics Today, 32 June (25), (1979).
12. R. Huby, Stripping Reactions, Prog. Nucl Phys 3, 177 (1953).
13. J. M. Blatt and V. F. Weisskopf, Theoretical Nuclear Physics, Wiley (1952)
14. R. J. Bilin-Stoyle, Fundamental Interactions and the Nucleus, North-Holland (1952)
15. A de Shalit and H. Feshback, Theoretical Nuclear Physics, Wiley (1974).

REVIEW QUESTIONS

1. What do you understand by Q-value and threshold energy of a reaction? Obtain expression for non-relativistic Q-value and threshold energy. How they are related together?

2. Show that in elastic collision between particles of equal masses, the kinetic energy in the centre of mass system is half that in laboratory system. [Raj]

3. A neutron of mass M_N collides with an atom M_A with kinetic energy E_0. Show that minimum energy of scattered neutron is $E_0 = \dfrac{M_A}{(M_A + M_N)}$ [Mysore]

4. Show that for maximum value of the slow neutron reaction cross-section, the scattering cross section is equal to reaction cross-section. Hence show that $(\sigma_r)_{max} = \dfrac{0.5 \times 10^{-6}}{E_r \,(\text{eV})}$ barns, where E_r is the resonance energy. [Delhi]

5. Give an outline of Bohr's compound nucleus formation hypothesis for nuclear reactions. Discuss an experiment showing that the formation and decay modes of the compound nucleus are independent of each other. [Kerala]

6. State and explain the compound nucleus hypothesis. Also explain the term excitation energy of the compound nucleus. Obtain an expression for excitation energy. [Punjab]

7. Give an outline of statistical theory of nuclear reactions. Explain the term nuclear temperature. [Kerala]

8. Explain Stripping and pick-up reactions. Give the theory of (d, p) stripping reaction.

9. Discuss the partial wave analysis for nuclear reaction cross-section and obtain expressions for scattering as well as reaction cross-sections for lth partial wave. Also show that scattering may and may not be accompanied by absorption but absorption, is always accompanied by scattering. [Bangalore]

10. Give Breit and Wigner single level relation for scattering and absorption cross-section in the vicinity of a resonance observed in neutron reaction. [Kerala]

11. Give the theory of resonances in scattering and reaction cross-sections in nuclear reactions. Mention the conditions under which these resonance are obtained? [Udaipur]

12. What is level width of a resonance in nuclear reactions? What is its relation with the life time of an excited state?

13. A particle with mass m_i is projected with a kinetic energy K in the L-frame of reference against a target nucleus of mass m_i initially at rest. Show that,

   (i) the total kinetic energy of the system in the centre of mass system is $\dfrac{K M_i}{(m_i + M_i)}$,

   (ii) total energy available for the reaction is $M_i(m_i, m_f)\, M_f$ is $Q + \dfrac{K M_i}{(m_i + M_i)}$

   (iii) Threshold energy in the L-frame of reference is $\dfrac{-Q(M_i + m_i)}{M_i}$ [Lucknow]

14. Show that in the scattering of a particle M_1 by a target nucleus M_2, the momentum transfer to the nucleus M_2 is the same in both the laboratory and CM systems of coordinates.
   [**Hint:** The transformation from L to CM system is linear in momentum coordinates and the momentum transfer vector is not altered].

PROBLEMS

1. Calculate the Q-value for the formation of ^{30}P in the ground state in the reaction ^{29}Si (d, n) ^{30}P from the following cycle of nuclear reactions:

$$^{31}P + \gamma \rightarrow {}^{30}P + n - 12.37 \text{ MeV}$$

$$^{31}P + p \rightarrow {}^{28}Si + {}^4He + 1.909 \text{ MeV}$$

$$^{28}Si + d \rightarrow {}^{29}Si + p + 6.246 \text{ MeV}$$

$$2d \rightarrow {}^4He + 23.834 \text{ MeV} \qquad [\textbf{Ans.}\ Q = 3.309 \text{ MeV}]\ [\text{Goa}]$$

2. The total cross-section of nickel for 1 MeV neutrons is 3.5 barns. What is the fractional attenuation of a beam of such neutrons on passing through a sheet of 0.01 cm in thickness? (Density of nickel is 8.9 gm/cm³). [Mysore] [**Ans.** 0.0014]

[**Hint:** Macroscopic cross-section $\Sigma = N_0 \sigma = \left(\dfrac{\rho}{M}\right) N\sigma$

$$= \left(\frac{8.9 \times 10^3}{58}\right) \times (6.02252 \times 10^{26} \times 3.5 \times 10^{-28})$$

$$= 32.37 \text{ m}^{-1}$$

Now, $\qquad I = I_0\, e^{-\Sigma x} \quad \text{or} \quad 2.3026 \log_{10}\left(\dfrac{I_0}{I}\right) = \Sigma x$

$\therefore \qquad \dfrac{I_0}{I} = \text{Antilog} \dfrac{32.37 \times 0.01 \times 10^{-2}}{2.3026} = 0.0014$

3. When the ^{12}C(α, γ) cross-section is measured, a peak is found at an α-particle energy of 7045 keV. Explain this observation and find the energy of the γ-ray emitted. Discuss whether it would be possible to confirm your explanation by measurment of the ^{15}N (p, γ) cross-section as a function of proton energy. (The mass excesses of the proton, α-particle, ^{15}N and ^{16}O are 7289, 2425, 100 and -4737 MeV, respectively). **[Ans.** 10.148 keV]

4. The cross-section for the reaction $^{10}_{5}$B(n, α) ^{7}Li is 630 b for incident neutrons of 1 eV. This reaction if often used as a means of detecting neutrons by filling an ionisation chamber with ^{10}BF$_3$ gas. Such a chamber, with the gas at STP, of rectangular cross-section is exposed to a broad of neutrons of energy 1 eV; if the chamber thickness (in a beam direction) is 0.1 m, the cross-sectional area is 2×10^{-2} m^2 and the counting rate is 10^{-3} s^{-1}. Find the incident neutron flux. **[Ans.** 2.95×10^{25}]

5. The ground state of ^{8}Be is unstable and its only decay mode is into two α-particles with a Q-value of 92 keV. Calculate the laboratory energy of an α-particle that would excite this resonance in scattering by a helium target and calculate the cross-section on resonance. [14.3 b]

6. The total energy of a degenerate Fermi gas may be written $E = a(kT)^2$ where T is the absolute temperature. The entropy of the gas, apart from a constant is given by $S(E) = k \log[\omega(E)]$. Use these formulate to show that the density of states in an excited nucleus is proportional to

$$\exp\left[2(aE)^{\frac{1}{2}}\right].$$

[HInt: The entropy of a gas may be written

$$\int_{S(0)}^{S(E)} dS = \int_{0}^{E} \frac{dE}{T}$$

Integrating the LHS becomes

$$k\{\log[\omega(E)] - \log[\omega(0)]$$

Now, using $\quad dE = 2\,ak^2T\,dT$, the r.h.s. is $2\,ak^2T$

Thus $\quad\quad\quad \omega(E) = \omega(0)\exp(2akT) = \omega(0)\exp\left[2(aE)^{\frac{1}{2}}\right]$

7. For a nucleus described by the semi-empirical mass formula, show that, neglecting symmetry terms, the energy released in fission into two fragments is a maximum for equal division of charge and mass. Calculate the value of $\dfrac{Z^2}{A}$ at which this division just becomes possible energetically.

[Hint: Excluding the asymmetry and pairing terms the semi-empirical mass formula gives

$$-B = \alpha A + \beta A^{\frac{2}{3}} + \frac{\varepsilon Z^2}{A^{\frac{1}{3}}}$$

If a nucleus A splits into $A_1 + A_2$, the energy released is given by

$$\beta\left(A^{\frac{2}{3}} - A_1^{\frac{2}{3}} - A_2^{\frac{2}{3}}\right) + \varepsilon\left(\frac{Z^2}{A^{\frac{1}{3}}} - \frac{Z_1^2}{A_1^{\frac{1}{3}}} - \frac{Z_2^2}{A_1^{\frac{1}{3}}}\right)$$

Writing $A_2 = A - A_1$ and $Z_2 = Z - Z_1$ and now differentiate w.r.t. A_1. The first term gives a maximum for $A_1 = \dfrac{A}{2}$ and the second has turning points at $A_1 = A$ and $A_1 = \dfrac{A}{2}$. Differentiation w.r.t. Z_1 gives turning points at $Z_1 = Z$ and $Z_1 = \dfrac{Z}{2}$ of these possibilities, one describes no fission at all and the other, $A_1 = A_2 = \dfrac{A}{2}$ and $Z_1 = Z_2 = \dfrac{Z}{2}$, is the required condition. We can easily check to be a maximum by a second differentiation.

Now, by setting the energy release to zero for the case of equal division and using $\beta = 17.8$ MeV and $\varepsilon = 0.71$ MeV, one finds $\dfrac{Z^2}{2} \geq 17.6$]

8. Cadmium has a resonance for neutrons of energy 0.178 MeV and the peak value of the total cross-section is about 7000 b. Estimate the contribution of scattering to the resonance.

[**Ans.** 3.37 b]

[**Hint:** From Breit-Wigner Equation,

$$\sigma_a = \pi \lambda^2 g \ \frac{T_a T}{(E_R - E) + \dfrac{T^2}{4}}$$

Where σ_a is compound nucleus formation cross-section, T is the sum of all partial widths, including T_a which measures the probability of decay back into the incident channel. E is the total energy of particle which is near to the energy E_R of a resonance level, g is statistical factor given by

$$g = \frac{2I + 1}{(2s_1 + 1)(2s_2 + 1)}$$

Where s_1 and s_2 are the spins of the bombarding particle and the target nucleus X and I is the spin of the compound state, which is obtained as a vector sum of the spins and the incident orbital angular momentum in the collision. Now, the total cross-section at resonance.

$$\sigma_n = \frac{4\pi \lambda^2 \Gamma_n}{T}$$

Where we have neglected the spin factor. If only scattering and capture contribute to the resonance then $\Gamma = \Gamma_n + \Gamma_\gamma$ and the scattering part is $\dfrac{\sigma_n \Gamma_n}{\Gamma} = \dfrac{\sigma_n^2}{4\pi \lambda^2}$. For a nucleus as heavy as cadmium the *CM* correction is negligible and one finds $4\pi\lambda^2 = 1.45 \times 10^{-21}$ m^2 where $\sigma_{nn} = 3.37$ b.

9. Estimate the relative probabilities of (n, n) and (n, γ) in indium, known to have a neutron resonance at 1.44 eV with a T of 0.1 eV and cross-section of 28000 barns. [Vikram]

[**Hint:** At resonance, i.e. $E = E_0$, we have the Breit-Wigner formula

$$\sigma(n, \gamma) = \frac{\lambda^2 \Gamma_n \Gamma_\gamma}{4\Gamma^2}$$

Now,

$$\Gamma_n \Gamma_\gamma = \frac{4\sigma \Gamma^2}{\lambda^2} = \frac{4 \times 2.8 \times 10^{-24} \times (0.1)^2}{(2.4 \times 10^{-11})^2} = 1.5 \times 10^{-4}$$

Since, $\Gamma_\gamma \approx \Gamma = 0.1,$ $\therefore \dfrac{\Gamma_n}{\Gamma_\gamma} = 0.015$]

10. The neutron capture reaction of ^{197}Au at neutron energies upto a few hundred eV is characterized by a number of resonances. The most prominents is at 4.906 eV and has $\Gamma_\gamma = 0.124$ eV and $\Gamma_n = 0.007\, E^{\frac{1}{2}}$ eV. The spin of the compound nucleus formed in this resonance absorption has value 2. Show that the peak cross section is 3.446×10^4 barns. [Sagar, Bhopal]

[**Hint:** For resonance cross-section

$$\sigma(n,\gamma) = \frac{\lambda^2}{\pi} \frac{2I_c + 1}{2(2I+1)} \frac{\Gamma_\gamma \Gamma_n}{(\Gamma_\gamma + \Gamma_n)^2}$$

$$\Gamma_n = 0.007\, E^{\frac{1}{2}} = 0.007 \times (4.906)^{\frac{1}{2}} = 0.0155 \,\text{eV}$$

Now,

$$\sigma_{(n,\gamma)} = \frac{h^2}{2\pi mE} \frac{2I_c + 1}{2(2I+1)} \frac{\Gamma_\gamma \Gamma_n}{(\Gamma_\gamma + \Gamma_n)^2}$$

$$= \frac{(6.6\times10^{-34})^2}{2\times3.14\times1.67\times10^{-27}\times4.906\times1.6\times10^{-15}} \times \frac{2(2+1)}{2(3+1)} \times \frac{0.124\times0.0155}{(0.1395)^2}$$

$$= 3.446 \,\text{barns.}$$

11. The reaction $^{13}_{6}$C (d,p) $^{14}_{6}$C has a resonance at deutron energy of 2.45 MeV. At what α-particle energy, does it allow to predict a resonance level for the reaction $^{11}_{5}$B$(\alpha,n)^{14}$N.

[**Ans.** 9.94 MeV]

12. Determine the threshold energy of the projectile in the L-frame assuming the target nucleus to be stationary:

 (a) $^{14}_{7}$N (α,p) $^{17}_{8}$O   (b) $^{16}_{8}$O (n,α) $^{13}_{6}$C

[**Ans.** (a) 1.535 MeV (b) 2.34 MeV] [Bhopal]

SHORT-QUESTION ANSWERS

1. Does at energies well below the Coulomb barrier both light and heavy ions can undergo scattering?
Ans. Yes
2. Can diffraction and interference phenomena occur in elastic and inelastic scattering? If yes, how can these be explained?
Ans. Yes, using the Born approximation.
3. Where the compound nucleus reactions are significant?
Ans. For light and heavy ions at energies above the Coulomb barrier and for low energy neutrons in particular. The occurrence of resonances in the cross-section for these reactions corresponds to the formation of long-lived states in the compound nucleus.
4. When direct reactions for both light and heavy ions become more important?
Ans. At higher incident energies

5. Is the angular distributions are characterstic of the angular momentum transfer in light-ion induced reactions but are generally more featureless for heavy-ion induced reactions?

Ans. Yes. The distributions are well described using the distorted wave Born approximation

6. What type of reactions are observed for heavy-ion reactions at energies above the Coulomb-barrier?

Ans. Deep-inelastic reactions are observed, particularly for reactions involving very heavy ions.

7. At what energy of the incident particles, direct reactions are seen?

Ans. About 5 MeV per nucleon and above.

8. Give an example of direct inelastic scattering.

Ans.
$$^{24}\text{Mg} + \alpha \rightarrow {}^{24}\text{Mg}^* + \alpha'$$

Which tends to pickout *collective modes* of excitation of the residual nucleus.

9. In general, how a nuclear reaction can be represented?

Ans.
$$X + x \rightarrow C^* \rightarrow Y + y$$

Where C^* is a compound nucleus, i.e. the incident particle becomes an integral part of the new nucleus, called compound nucleus.

10. What is a pick up reaction?

Ans. When the projectile gains nucleons from the target, the nuclear reaction is referred to as pickup reaction. An example is

$$^{16}_{8}\text{O}\,(^{2}_{1}\text{H},\ ^{3}_{1}\text{H})\ ^{15}_{8}\text{O}$$

11. Give an example of stripping reactions.

Ans.
$$^{16}_{8}\text{O}\,(^{4}_{2}\text{He},\ ^{2}_{1}\text{H})\ ^{18}_{9}\text{F}$$

12. What is the Q-equation?

Ans. The analytical relationship between the kinetic energy of the projectile and outgoing particle and the nuclear disintegration energy Q is called the Q-equation.

13. What do you understand by threshold energy?

Ans. The smallest value of projectile energy (bombarding energy) at which an endoergic reaction can take place is called the threshold energy for that reaction.

14. What is the main characterstic of compound nucleus reaction?

Ans. The absence of any dependence between formation and decay of compound nucleus.

15. What is a *doorway state*?

Ans. In practice resonances are observed at higher energies. These are due primarily to the coupling of a large number of small resonances, for example, to a state in the vicinity that is strongly excited due to some special reasons related to nuclear structure. Such a strongly excited state is often called a doorway state.

16. Why the centre of mass frame is preferred in describing the kinematics of a nuclear reaction.

Ans. Since in this frame there is no net momentum.

17. Why charged particles with very low incident energies only elastically scatter?

Ans. Due to Coulomb repulsion between the incident and target nuclei.

OBJECTIVE QUESTIONS

1. Which one of the following represents elastic scattering type nuclear reaction.
   (1) $X(x, y)Y$
   (2) $X(x, y)X$
   (3) $X(x, x)X$
   (4) $A_X(x, x)A'_Y$

2. The first nuclear transmutation observed by Rutherford, can be represented by the nuclear reaction
   (1) $^{14}N\,(\alpha, p)\,^{17}O$
   (2) $^{14}N\,(\alpha, p)\,^{16}O$
   (3) $^{14}N\,(\alpha, \alpha)\,^{14}N$
   (4) $^{14}N\,(^{14}N, ^{15}N)\,^{13}N$

3. Which one represents the conservation of isotopic spin in nuclear reactions
   (1) $I_X + I_x + l_x = I_Y + I_y + l_Y$
   (2) $T_i = T_f$
   (3) $\pi_X\,\pi_x\,(-1)^{l_x} = \pi_Y\,\pi_y\,(-1)^{l_y}$
   (4) $p_X + p_x = p_Y + p_y$

4. 1 barn is equal to
   (1) $10^{-14}\,m^2$
   (2) $10^{-19}\,m^2$
   (3) $10^{-28}\,m^2$
   (d) $10^{-34}\,m^2$

5. Which one of the following is (α, p) reaction:
   (1) $^{14}_{7}N + ^{4}_{2}He \to ^{18}_{9}F^* \to ^{17}_{9}O + ^{1}_{1}H$
   (2) $^{7}_{3}Li + ^{4}_{2}He \to ^{11}_{5}B^* \to ^{10}_{5}B + _0n^1$
   (3) $^{6}_{3}Li + ^{1}_{1}He \to ^{7}_{4}Be^* - ^{3}_{2}He + ^{4}_{2}He$
   (4) $^{11}_{5}B + ^{1}_{1}H \to ^{12}_{6}C^* \to ^{11}_{6}C + _0n^1$

6. Which one is radiative α-capture reaction:
   (1) $\alpha + ^{12}_{6}C \to ^{16}_{8}O^* \to ^{16}_{8}O + \gamma$
   (2) $p + ^{7}_{3}Li \to ^{8}_{4}Be^* \to ^{8}_{4}Be + \gamma$
   (3) $n + ^{23}_{11}Na \to ^{24}_{11}Na^* \to ^{24}_{11}Na + \gamma$
   (4) $^{2}_{1}H + \gamma \to ^{1}_{1}H + ^{1}_{0}n$

7. The net momentum of a nuclear reaction in centre of mass frame is
   (1) maximum
   (2) minimum but not zero
   (3) zero
   (4) sometimes maximum and sometimes minimum

8. The relation between the width Γ of the resonances and life time τ of the compound nucleus level is

   (1) $\Gamma = \dfrac{k}{\tau}$
   (2) $\Gamma = \dfrac{\hbar}{\tau}$
   (3) $\Gamma = \tau^2$
   (4) $\Gamma\tau = 0$

9. Isospin is a useful concept particularly for - - - - -.   [light-ion reactions]

10. The predicted energy dependence of compound nucleus formation cross-section also follows from a partial wave treatment assuming that that the nucleus acts like a strongly - - - - -.
    [absorbing sphere]

11. At low energies and small impact parameters a nuclear reaction is most likely to occur by the formation of a - - - - -. Such reactions are characterised by - - - - -. Occuring when the energy of the incident particle is such that a long-lived compound nucleus state is formed.
    [compound nucleus, resonances]

12. A useful approximation when describing compound nucleus reaction is to assume that the nucleus is strongly - - - - -. For an s-wave reaction this assumption is equivalent to taking the internal wave function as: - - - - - for $r < R$ corresponding to an ingoing wave only.
    [absorbing, $u = \exp(-kr)$ for $r < R$]

13. In a reaction at energies above the Coulomb barrier two heavy nuclei would be expected to react if they came closer than the strong interaction radius (say R_S). If the motion of the ions is described by classical mechanics then the distance of closest approach d of the heavy ions depend on the impact parameter b and hence on the incoming orbital angular momentum l. If $2a$ is the distance of closest approach of the ions in a head-on collision, then the relation between d and b is - - - - - - .

$$[d = a + (a^2 + b^2)^{\frac{1}{2}}]$$

ANSWERS

1. (c)	2. (a)	3. (b)	4. (c)	5. (a)
6. (a)	7. (c)	8. (b)		

7

NEUTRON PHYSICS AND
NUCLEAR ENERGY:
(Fission, Fusion and Reactions)

7.1 DISCOVERY OF NEUTRON

Bothe and Becker in 1930 found that when beryllium was bombarded with 5.3 MeV high speed α-particles from a radiation source (e.g. Po), a highly penetrating radiation is emitted. This radiation was found to readily penetrate several centimeters of lead and not deflected by a magnetic or electric field. They assumed that the new radiation was highly energetic γ-rays. The energy of the radiation from a measurement of the absorption coefficient was estimated to by ~ 7 MeV.

Curie and Joliot in 1932 made the historic discovery that the radiation was able to eject *protons* from substances rich in hydrogen (e.g. paraffin) and was not affected by magnetic field. From the ranges of these recoil protons, the maximum proton E proved to be about 5.3 MeV. They assumed that the protons were produced as the result of elastic collisions with the γ-rays photons. Calculations showed that each photon must have possessed energy of about 52 MeV. This they ascribed to a *Compton scattering* of the supposed electromagnetic radiation by the target protons and from the observed range of protons the energy of the radiation was calculated to be 35 – 55 MeV. These results were totally inconsistent with the results from the experiments on the absorption of these rays by Pb (~ 7 MeV).

Soon afterwards, Chadwick repeated the experiment not only with hydrogeneous substance like paraffin but also with He and N_2. Figure 7.1 shows the experimental arrangement. From the measurment of recoil particles (e.g. nitrogen) ejected by the incident penetrating radiation, Chadwick estimated that the energy of unknown radiation, if assumed to be γ-rays, is of the order of 90 MeV. He pointed out that (i) no *conceivable nuclear reaction can produce such high energy γ-rays,*

(ii) *γ-ray photons could not produce even a thousandth as many recoil particles by any process hittherto known.*

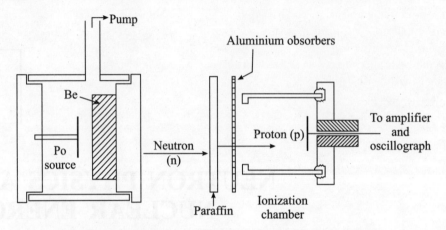

Fig. 7.1 Experimental arrangement for the discovery of neutron

Chadwick said that *these results are very difficult to explain on the assumption that the radiation from beryllium is a quantum radiation, if energy and momentum are to be conserved in the collisions.* He said that *all apparent anomalies disappear if it is assumed that the new radiation was a stream of a new type of particles of about the same mass as photon but of zero charge.*

The name *neutron* was assigned to this new particle whose existence was suggested some 12 years earlier by Rutherford but remain undetected so far. Chadwick proposed that these neutrons were formed as a result of highly exoergic nuclear reaction

$$\ {}_2^4He + {}_4^9Be \rightarrow ({}_6^{12}C) \rightarrow {}_6^{12}C + {}_0^1n + 5.6 \text{ MeV} \qquad \dots (6.1)$$

where ${}_0^1n$ denotes neutron.

Using the above reaction, calculations show that the energy released would be 0.006125 amu or 0.006125×931 MeV = 5.70 MeV.

Chadwick made careful measurments of the kinetic energy of protons and nitrogen atoms knocked out by the neutrons (unknown radiation) when they were passing through a substance containing hydrogen and nitrogen respectively. If the unknown radiation is assumed to consist of particles of atomic mass M_n and velocity u_n and if M_p and M_N are the atomic masses of the proton and nitrogen nucleus respectively, then one can easily show that

$$u_p = \frac{2M_n u_n}{M_n + M_p} \quad \text{and} \quad u_N = \frac{2M_n u_n}{M_n + M_n} \qquad \dots (7.2)$$

Where up is the initial velocity of the protons (3×10^7 m/s) deduced from their measured range and u_N the initial velocity of nitrogen recoils was found to be 4.7×10^6 m/s from a separate expansion chamber study of recoil ranges by feather. So that

$$\frac{u_p}{u_N} = \frac{3.3 \times 10^7}{4.7 \times 10^6} = \frac{M_n + M_N}{M_n + M_p}$$

For $M_N = 14$ amu and $M_p = 1$ amu, one obtains

$M_n = 1.16$ amu with an error of some 10 %. This value was soon refined by observation of other neutron-producing reactions, to $1.005 < M_n < 1.008$ amu. This completed the proof of the existence of a neutral nuclear particle of mass closely similar to that of the proton.

Chadwick's experiments were the foundation of our present model of the nucleus, a model which assumes that the nucleus is composed of protons and neutrons.

7.1.1 Neutron Sources

There are various ways in which one can extract neutrons from a nucleus in sufficient numbers in order to be called a neutron source.

(a) Radioactive Sources The most common laboratory neutron source is radium beryllium source consists of an intimate mixture of about 5 parts of fine beryllium powder (as the range of α-particles in solids is short) to 1 part of radium. The majority of neutrons is obtained according to (α, *n*) reaction (7.1). If the radium is in secular equilibrium with its decay products, we find that extra neutrons are released in accordance to the reaction

$$\,^{9}_{4}\text{Be} + h\nu \rightarrow \,^{8}_{4}\text{Be} + \,^{}_{0}\text{n}^{1} \qquad \qquad ... (7.3)$$

In reaction (7.3), the product nucleus $\,^{8}_{4}\text{Be}$ is unstable and thus decays into two α-particles within 10^{-15} s. The energy spectrum of neutrons obtained, extends upto 13 MeV and confirms that this not a monoenergetic source due to following reasons: (a) all the α-particles do not have the same initial energy, (b) several α-particles lose part of their initial energy by ionization prior to nuclear capture, (c) the energy of the neutron varies with the direction of emission having least value if emitted in the direction opposite to that of the incident α particles, and (d) $\,^{12}_{6}\text{C}$ can be left in an excited state (see reaction 7.1).

Slow neutrons are usually obtained by surrounding the source with paraffin, (Figure 7.2) water, or some other hydrogen-containing material. For a 2 MeV neutron only 25 elastic collisions with protons are needed. The energy spectrum of neutrons obtains, extends upto 13 MeV and hence this is not a monoenergetic source. Sometimes Radon is used in place of radium. Randon is a gas and it can be made as a more compact source. However, Radon has the disadvantage of decaying rapidly because of the short half life of 3.8 days. Polonium is used instead of radium when a neutron source with relatively few γ-rays is needed. This has many disadvantages, e.g.

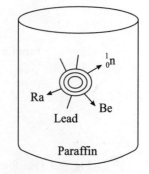

Fig. 7.2 Production of slow neutrons.

(i) the neutron output is only 3×10^6 neutrons/sec/Curie of Po whereas it is $10 - 18 \times 10^6$ neutrons/sec/Curie of Ra, i.e. in the case of Po it is much lower,

(ii) to prepare Po special chemical facilities are required,

(iii) compared to Ra-Be source, the half life is short (140 days).

Pu-Be neutron sources are also used. ^{239}Pu is the most commonly available isotope of Pu and it emits α-particles of energy 5.1 MeV. Pu-Be neutron source have some advantages over Ra-Be neutron source. Half life of Pu is 2.3×10^4 yrs as such Pu-Be sources have long life times and secondly intensity of γ-radiation is low. However, the neutron yields from Pu-Be sources is (1.2×10^6 neutrons/s) is somewhat lower than that in the case of Ra-Be sources.

(b) Photo-neutron Sources   Neutrons can be knocked out of the nuclei by high energy γ-ray photons (γ, n reactions). The energy of the γ ray photons must be greater than the neutron binding energy. Two of the following reactions have low enough γ-ray threshold for ${}_0^1n$ emission.

$${}_4^9Be + \gamma \rightarrow {}_4^8Be + {}_0^1n \ ... \ \text{threshold} = 1.6 \ \text{MeV}$$

$${}_1^2H + \gamma \rightarrow {}_1^1H + {}_0^1n \ ... \ \text{threshold} = 2.23 \ \text{MeV}$$

As far as (γ, n) reaction is concerned, one can conveniently divide the periodic table into two groups: (i) deuterium and beryllium and (ii) all the rest. The former (${}_1^2H$ and ${}_4^9Be$) have thresholds of 2.226 ± 0.003 MeV and 1.666 ± 0.002 MeV respectively. This enables natural γ-ray sources to be used. The advantage in such reactions is that the neutrons are monoenergetic. In general, it is observed that 1 gm of Ra at 1 cm from 1 gm of Be gives about 30,000 neutrons per second. We may note that this yield is smaller than (α, n) reaction yield. We may also note that radioactive nuclei capable of emitting γ-rays of such energies are not known, i.e. other than ^2H and ^9Be, the thresholds is in excess of 6 MeV. This is why radioactive (γ, n) sources employ only ^2H and ^9Be. However, γ-rays from artificially produced radio-nuclei can serve in photo-disintegration sources. Snell et al. developed a standard photoneutron source. They placed a strong mono-energetic gamma source such as Be with ^{24}Na and ^{72}Ga at the centre of a surrounding sphere of Be of heavy water (Figure 7.3). By the activation of an indium foil, neutrons were detected.

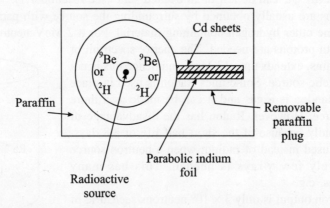

Fig. 7.3 One form of standard photoneutron source developed by Snell et at.

The yields of neutrons from various photo-neutron sources are given in Table 7.1. These sources have following advantages: (i) The reproducible neutron strength (ii) By choosing a suitable value of E_γ one can choose E_n (iii) From atomic reactors one can obtain wide variety of artificial γ-sources, and (iv) In comparision to (Be – α) sources, neutrons produced from these sources are more energetic.

Table 7.1 Yields of Radioactive (γ, n) sources

γ-ray source	Target	Half life $T_{\frac{1}{2}}$	E_γ (MeV)	E_n (MeV)	Yield (neutrons/s)
^{24}Na	Be	14.8 hrs	2.76	0.83	13×10^4
^{24}Na	D_2O	14.0 hrs	2.76	0.22	27×10^4
^{56}Mn	Be	2.59 hrs	1.81, 2.13, 2.7	0.15, 0.30	2.9×10^4
^{56}Mn	D_2O	2.59 hrs	2.7	0.22	0.31×10^4
^{124}Sb	Be	60 days	1.7	0.025	19×10^4
^{226}Ra	D_2O	1620 years	2.42	0.12	0.1×10^4
^{72}Ga	Be	14 hrs	1.81, 2.21, 2.51	0.2	5.9×10^4
^{72}Ga	D_2O	14 hrs	2.5	0.13	6.9×10^4

For the energy E_γ of the emitted neutrons, we have the following relation due to Wattenberg,

$$E_n = \frac{A-1}{A}\left[E_\gamma - Q - \frac{E_\gamma^2}{1862(A-1)}\right] + E_\gamma \cos\theta \left[\frac{2(A-1)(E_\gamma - Q)}{931\,A^3}\right]^{\frac{1}{2}} \quad \text{... (7.4)}$$

Where E_γ is the energy of γ-ray in MeV, A is the mass of the target nucleus, Q is the threshold energy in MeV for (γ, n) reaction in the target nucleus and θ is the angle between the γ-ray path and the emitted neutron direction. Since ^{24}Na source is having short half life and hence ^{124}Sb source is more suitable for intermediate energy neutrons with a relatively long half life.

Several workers have investigated the emission of photoneutrons using prompt γ-rays produced in nuclear reactions, e.g. 17 MeV γ-rays in the ^{7}Li (p, γ) reaction, More attention has been paid to the investigation of high energy X-ray induced photoneutron emission using electron accelerators.

We know that when a high energy electron beam falls on a target (usually some heavy element), X-rays are produced with an energy distribution having the maximum equal to the energy of the electron beam. These X-rays are then allowed to fall on a suitable target so that (γ, n) reaction may be produced. This method can provide neutron sources of considerable energy. In a lead target, X-rays produced 3.2 MeV electron beam from a linear accelerator, using D_2O or beryllium target an average yield of $\sim 10^9$ neutrons/s and a peak yield of $\sim 2 \times 10^{17}$ neutrons in 2 μs bursts have been obtained.

We may note that the spectrum of neutron energies is relatively broad having almost Maxwellian distribution. However, there is a high energy "tail" of neutrons ejected by direct reaction by X-rays.

(c) Neutrons from Accelerated charged Particle Reactions The invention of charged particle accelerators, e.g. Van de Graff generator machine, cyclotron, synchrotron, etc. led to a very rapid development of new neutron sources by using accelerated charged particles as projectiles to knock out neutrons from various nuclides. Protons, deutrons, tritons, α-particles and other heavy nuclides have been used as projectiles. Accelerator based neutron sources usually use (d, n) and (p, n) reactions on lighter elements. These sources produce monoenergetic neutrons in the range from a few keV to 20 MeV. The (d, n) reactions are usually exoergic, where as (p, n) reactions are always endoergic.

There are few exceptions where (d, n) reactions are endoergic, e.g. $^{12}C(d, n)\ ^{13}N$ with $Q = -0.283$. In deutron bombardment reaction the energy of the released neutron is progressively larger for most of the target nuclei, which is due to the low binding energy of the deutron.

For the purpose of neutron production, few commonly used (p, n) and (d, n) reactions are given in Table 7.2.

Table 7.2 (p, n) and (d, n) neutron producing reactions

Reaction	Q-value (MeV)	
$^3H\ (p, n)\ ^3He$	-0.764	endoergic
$^7Li\ (p, n)\ ^7Be$	-1.644	endoergic
$^2H\ (d, n)\ ^3He$	$+3.269$	endoergic
$^3H\ (d, n)\ ^4He$	$+17.6$	endoergic
$^7Li\ (d, n)\ ^8Be$	$+15.03$	endoergic
$^9Be\ (d, n)\ ^{10}B$	$+4.36$	endoergic
$^{12}C\ (d, n)\ ^{13}N$	-0.283	endoergic

In comparison to radioactive neutron production sources, charge particle accelerators possess following advantages:

(i) Neutron flux density obtained from charge particle accelerators is very high because they are capable of producing neutrons in sufficiently high numbers.

(ii) One can vary the energy of the accelerated charged particles over a wide range energy from a few MeV to several hundred BeV. Obviously, the neutrons from them will have a wide energy spectrum. By controlling the energy of the accelerated particles, one can easily control the energy of neutrons.

(iii) One can switched off the neutron beam at will by switching the particle accelerator off or on. One can obtain pulsed neutron source using synchrotron, required in the study of many neutron reaction cross-section experimentally.

The characterstics of major neutron producing sources are:

(p, n) sources   With increasing proton energy the cross-section of the $^3H\ (p, n)\ ^3He$ reaction increases. The angular distribution is quite comlex. In the (p, γ) reaction on ^3H accompanies the neutron emission with a large yield of 20 MeV γ rays.

$^7Li\ (p, n)\ ^7Be$ is widely used intermediate energy neutron source. This reaction suggests that accelerated charged particles can be used to produce nearly monoenergetic neutrons.

(d, n) sources   Fast and very fast neutrons are produced by these sources. At low bombarding energies $^2H\ (d, n)\ ^3He$ or (d, d) reaction has high yield. This makes it particularly useful as the neutron source with low voltage accelerators, e.g. Cockroft-Walton generator. Upto 10 MeV, neutrons are monoenergetic. To provide strong neutron sources, thick targets of heavy ice are frequently used.

^{3}H (d, n) ^{4}He or (t, d) reactions is the most strongly exoergic reaction. This yield mono-energetic neutrons of 12 to 20 MeV energy by bombarding deutrons of upto 3 MeV energy on tritium.

(**α**, **n**) **Sources** There is a great difficulty in obtaining intense α-particle beam from accelerators and moreover comparatively low neutron yields, (α, n) reactions have not been found very useful as neutron sources in particle accelerators.

(d) **Ultrafast Neutrons** Deutron has a low binding energy and therefore neutron and proton in it are relatively loosely bound, i.e. Deutron behaves as a loosely bound system. n and p within it are frequently outside the range of their mutual forces. This is why deutrons accelerated to about 200 MeV in a synchro-cyclotron can readily be '*stripped*' into protons and neutrons by directing the deutron beam to pass through a target nucleus. Proton may be captured by the target and neutron moves almost undeviated through the target. Thus very fast (ultrafast) neutrons are produced. Recently neutrons of energy greater than 300 MeV have been produced by stripping 650 MeV deutrons accelerated in the Birmingham synchrotron. Almost any element may be used as the target nucleus for stripping but the yield varies with its nature. We may note that the incident projectile energy of the deutrons is shared almost equally between p and n forming the deutron. Obviously, the available

neutron energy is about half of the deutron projectile energy (E_d), i.e. $\dfrac{E_d}{2}$. For very heavy

target nuclei, the angular spread is usually larger. Moreover, with ultrafast protons, the emerging neutrons are neither so well collimated and nor so nearly monochromatic.

(e) **X-ray Sources** High energy X-ray machines through (γ, n) reaction can also produce high energy neutrons. When the electrons accelerated to high energies in linear accelerators, e.g. betatrons, synchrotrons, etc. impinge on target elements of high atomic number, a flux of high energy photons is produced. Due to high flux of photons, a large number of neutrons within the target material are released. The major disadvantage of X-ray source is the wide spread in neutron energy.

(f) **Neutrons from Chain Reactors** The best source of thermal neutrons (energy ~ 1 eV) is a chain nuclear reactor or pile. In these reactors the fission process is accompanied by the emission of fast neutrons. In a reactor, the fission of each heavy nucleus of uranium by slow neutron capture is accompanied by the release on the average of about 2.5 neutrons. These neutrons are fast and can be slowed down in the moderator. Figure 7.4 shows a flux distribution for such neutrons. We may note the high energy tail, which is effectively removed from the thermal column of the reactor by the moderating graphite. At the centre of a high power reactor the thermal neutron flux can be of the order of 10^{13} neutrons per cm^2 per sec, where as in the thermal column of the same reactor, the neutron flux reduces to ~ 10^9 n/cm^2-s. The disanvantages of chain reactor source of neutrons are the wide spread in the neutron energy and also inability of directly pulsing the neutron source.

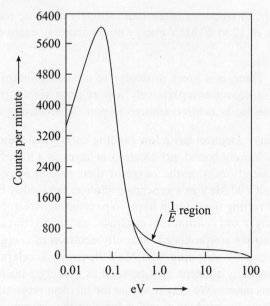

Fig. 7.4 Flux distribution for thermal neutrons from a reactor

7.1.2 NEUTRON COLLIMATORS

Neutrons obtained from the sources possess very high range of energy. Most of the neutrons have enormous energy and are called fast neutrons. One cannot use the ordinary methods used for charged particles and γ-rays for collimating a beam of fast neutrons. Dunning in 1938 devised a *neutron howitzer* for collimating neutrons. This device is based on the property of slowing down of neutron in hydrogenous materials like water or paraffin. The source S(radium beryllium) is placed at the bottom of a barrel inside a paraffin block. A slit of cylindrical size is cut in paraffin block and lined from all sides by cadmium sheet. Neutrons are emitted in all directions but their energy is negligibly small in all directions except through the barrel. The thermal neutrons resulting from the slowing down of the fast neutrons in the paraffin are absorbed effectively by the cadmium coating over the outer surface of paraffin and also on the walls of the barrel. The cadmium layer is also covered by a layer of lead to absorb γ-rays arising from neutron capture.

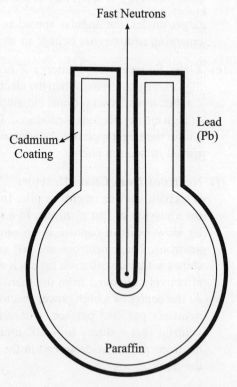

Fig. 7.5 Neutron howitzer due to Dunning.

7.1.3 FUNDAMENTAL PROPERTIES OF NEUTRON

(i) **Neutron Mass (M_n)**   While announcing the discovery of neutron, Chadwick gave a rough estimation of the mass of neutron that it need a mass very nearly equal to that of proton. Since the neutron is a chargeless particle and hence its mass could not be determined from the deflection measurments in electric and magnetic fields. Chadwick studied the collision of neutrons and protons and from the conservation of energy and conservation of momentum equations for a head on collision, and also from the lengths of accompanying tracks in the cloud chamber, he estimated that neutron mass is about $M_n = 1.15$ amu, roughly the same as that of proton.

Chadwick obtained the more precise value from the study of the following nuclear reaction

$$^{11}_{5}B + ^{4}_{2}He \rightarrow ^{14}_{7}N + ^{1}_{0}n + Q$$

He estimated the Q value from the kinetic energies of ^{11}B, ^{14}N, $^{1}_{0}n$ and α-particles. Chadwick substituted all the known values in terms of amu in the above equation and obtained $M_n = 1.0067$ amu.

A better and more precise estimate was made by Chadwick and Goldhaber in 1934 through the study of the following nuclear reaction

$$^{2}_{1}H + h\nu \rightarrow ^{1}_{1}H + ^{1}_{0}n$$

using γ-rays of Th C'', with $h\nu = 2.62 \times 106$ eV. The mass of the deutron and proton were known and the only requirement was to measure the kinetic energy of the proton and the neutron. Actually since proton and neutron will have nearly equal kinetic energies, the measurment of kinetic energy of the proton would suffice. With the help of ionization method, they measured the proton kinetic energy and found to be about 1.05 MeV. With the help of best mass-spectrographic data on atomic masses of hydrogen and deuterium and by substituting the mass equivalents for the γ-ray energy and disintegration energy (proton energy + neutron energy) and assuming all particles essentially at rest, they obtained

$$M_n = 1.0087 \pm 0.0003 \text{ amu}$$

Bell and Elliott from the study of inverse reaction to that of photo disintegration of deutron,

$$^{1}_{0}n + ^{1}_{1}H \rightarrow ^{2}_{1}H + h\nu$$

and by measuring the binding energy of the deutron to a high degree of accuracy and careful measurment of the energy of γ-rays, obtained a value of 2.230 ± 0.007 MeV (= 0.002395 amu) for the energy of these γ-ray. By substituting the best values of proton and deutron masses, they obtained neutron mass (M_n) as

$$M(^{1}_{0}n) = M(^{2}_{1}H) - M(^{1}_{1}H) + M(\gamma)$$

$$= 2.014735 - 1.008142 + 0.002394$$

$$= 1.008988 \text{ amu}$$

However, Bainbridge, through a series of measurments, obtained the binding energy of deutron = 0.002489 amu, which yield

$$M_n = 1.008986 \pm 0.0000015 \text{ amu}$$

Presently accepted value is

$$M_n = 1.008664904 \text{ amu} = 1.67493 \times 10^{-27} \text{ kg}$$

(ii) **Charge of the Neutron** Usually, it is assumed that neutron have no charge. However, the observations do not preclude the possibility of the neutron possessing a charge. Dee investigated the ionisation produced in air in a cloud chamber irradiated by fast neutrons. He concluded from the experimental data that the charge of one neutron is less than $\dfrac{1}{700}$ of the charge on proton. Rabi and Coworkers have reported the neutron charge less than 10^{-12} of charge on electron.

Despite, being the electrically neutral character of the neutron, there are observations of very small electrom-agnetic interaction between neutrons and charged particles. This has been attributed to the magnetic dipole moment of the neutron. Neutron is assumed to have a complex structure and in that it possesses an equal number of positive and negative mesons. One can not regard neutron strictly as a fundamental particle. Clearly, the structure of the neutron may lead to the existence of a purely electrical interaction although it may be very small between a neutron and an electron. Obviously, the smallness of this interaction supports the view that the neutron carries an exceedingly small charge. Fermi and Masshall set an upper limit of $\sim 10^{-18}$ electron charge for the charge on the neutron from the scattering of thermal neutrons by Xenon atoms.

(iii) **Neutron Spin and Statistics** There is enough evidence to support that neutrons obey Fermi-Dirac (FD) statistics (obeyed by half integral spin particles), i.e. neutrons are half integral spin ($\dfrac{1}{2}$ in units of $\hbar$) particles. Most important evidences in support that neutron is a half integral spin particle come (i) from the cross-section and energy dependence of the scattering of neutrons by protons taken in conjuction with the evidence concerning, the $n - p$ nuclear force, concluded from spin and binding energy of the deutron. Deutron is a very loose combination of p and n and usually considered to have a spin 1. Experimentally it is found that the magnetic moment of deutron is almost equal to the algebric sum of the magnetic moments of the n and p. (ii) the coherent scattering of neutrons by crystals containing hydrogen, (iii) polarisation of neutron produced in the scattering by ferromagnetic materials and total reflection of neutrons by magnetic mirrors studies, and (v) the values of magnetic moments of deutron, proton and neutron.

The most important argument concerning that neutron is a half integral spin particle comes from (i) the saturation of nuclear forces as manifested in the constant density of the matter and (ii) obey of B.E. statistics by light particles like deutron and α-particles. This could be understood only through the *uncertainty principle*.

Hughes and Burgy confirmed that neutrons are half integral spin particles by reflection of neutrons from magnetised mirrors.

(iv) **Decay and Half life of the Neutron** We have read that accurate determination of the neutron mass shows that it is slightly greater than the proton mass. This strengthened the hypothesis that neutron possesses a complex structure and moreover it may be considered to be composed of a proton and an electron. Outside the nucleus, neutron is unstable and expected to decay with the emission of a β-particle, plus a neutrino, leaving a proton. Energetically, this is a possible process and so outside the nucleus (neutrons are essential constituents of all nuclei except hydrogen) the expected decay process is

$$_0^1 n \rightarrow {}_1^1 H + {}_{-1}^0 e + \bar{\nu} + (0.782 \pm 0.13) \text{ MeV}$$

This was confirmed experimentally by Robson in 1951, when the decay energy and half life of the neutron was measured by him using a thermal neutrons flux from the Chalk River Reactor (pile). Robson not only observed the decay of the neutrons but also succeeded in measuring the energy spectrum of the decay electrons. Robson's plan of experimental arrangement is shown in Figure 7.6.

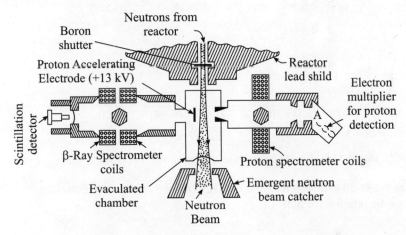

Fig. 7.6 Experimental arrangement for measuring decay energy and half life of the radioactive decay of the neutron used by Robson in 1951.

A well Collimated beam of pile neutrons from a nuclear reactor having high flux density ($\approx 1.5 \times 10^{10}$ neutrons/cm^2/sec) filtered through a bismuth plug so that γ-ray background passes, was allowed to pass into an evacuated chamber. Some of the neutrons decayed in their flight through the vacuum chamber and decay protons, with a small fraction of the disintegration energy, were deflected by the deflector (high voltage elector at a potential of +13 kV w.r.t. the earthed vacuum chamber. The evacuated chamber was provided with two lateral and opposite openings, the left one of which communicated with a β-ray spectrometer and right one with proton spectrometer. The protons were deflected by means of an electrode kept at a potential of +1 kV and situated within the vacuum chamber just opposite to the right opening, into the proton spectrometer. Proton spectrometer was provided with an *electron multiplier* to detect protons. To detect electrons, the β-ray spectrometer was provided with a *scintillation counter*. The β-rays emitted during disintegration were deflected to the left into a β-rays spectrometer and were detected with the helps of a scintillation counter using anthracene crystals as the phosphor. The two detectors were connected to a coincidence circuit which would register only when the pulses from two detectors arrived simultaneously there. We may note that due to greater mass, the proton would move more slowly through its spectrometer system and further would be delayed w.r.t. the electrons. This delay was calculated to be about 0.9 micro second.

Figure 7.7 shows a plot of the rate of proton-beta particle coincidences against the energy of the β-particles (measured in the spectrometer). The extrapolation of the curve gave a maximum energy value of 0.782 ± 0.013 MeV, which closely approximates the mass difference between the neutron and the proton.

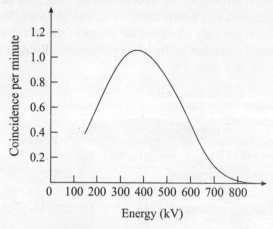

Fig. 7.7 A plot between coincidence rate and β-particle spectrum

One can make an estimation of the half-life of the neutron as follows:
We have the relation for half life

$$T_{\frac{1}{2}} = \frac{0.693}{\lambda} \frac{\varepsilon VP}{N}$$

Where ε is the detection efficiency of the proton detector which was found by callibration with a standard 6 MeV ^{212}Cm, α-particle source, V is the effective decay volume of neutrons and is found by observing the trajectories of small steel balls rolling over a rubber sheet model so that it may stimulate the potential gradient of the electrostatic field which acts on the decay protons and P denotes the density of thermal neutrons in the volume V. Lastly, N denotes the number of protons recorded by the proton spectrometer per minute. For the above quantities, Roben's observations yielded the following values: ε = 0.237, $P = 1.14 \times 10^4$ per cm^3, $V = 4.7$ cm^3 and $N = 7000$ counts/min. Therefore, the half life is obtained as

$$T_{\frac{1}{2}} = \frac{0.6931 \times 2.237 \times (4.7 \text{ cm}^3) \times (1.6 \times 10^4 \text{ cm}^3)}{7000/\text{min}}$$

$$= 12.8 \text{ min}$$

Robson's observations gave a value of $T_{\frac{1}{2}} = 12.8 \pm 2.5$ min. Recently, Sosnovski et al.

determined the accurate value of $T_{\frac{1}{2}}$ by a similar method used by D. Angelo at Argonne

National Laboratory and obtained the result: $T_{\frac{1}{2}} = 12.7 \pm 1.9$ min.

(v) Wave Properties of Neutron For $v \ll c$, i.e. for non-relativistic velocities, the wave length associated with neutrons is given by

$$\lambda = \frac{h}{p} = \frac{h}{\sqrt{2M_n E}} = \frac{2.86 \times 10^{-11}}{\sqrt{E}} \text{ m}$$

Where E is in eV. We have for thermal neutrons, $\lambda = 1.82$ A, which is comparable with atomic dimensions. Let $E = 1$ MeV, we see that $\lambda = 2.86 \times 10^{-14}$ m is approaching nuclear dimensions. However, one will have to apply relativistic correction above $E = 100$ MeV. We may note that the wave properties are of primary importance in the determination of the nature of the interaction between neutrons and nuclei.

(vi) Magnetic Moment of the Neutron In the units of nuclear magneton, the measured magnetic moments of proton and deutron are $\mu_p = 2.79270$ and $\mu_d = 0.857393$ respectively. Now, deutron contains a neutron and a proton as its constitutents. The magnetic moment of the deutron is so much different from the magnetic moment of the proton ($\mu_p > \mu_d$ by about 2 nuclear magnetons), strongly suggests that the neutron has an intrinsic magnetic moment. The strong interaction of neutrons with protons and neutrons is believed to be a consequence of meson exchange forces. Thus even a free neutron is surrounded for part of its life time by a cloud of virtual negative pions: $_0^1\text{n} \rightarrow _1^1\text{p} + \pi^-$. This indicates that the neutron has a finite negative magnetic dipole moment.

Alvarez and Bloch in 1940 using a technique similar to Rabi's molecular beam method of measuring nuclear magnetic moments, first measured the magnetic moment of the neutron (μ_n). Figure 7.8(a) shows the appratus for measuring the magnetic moment of neutron by Bloch et al. In this experimental arrangement, slabs of magnetized iron were used as polariser and analyser, in place of the inhomogeneous magnetic fields as used by Rabi in his method. These slabs of magnetized iron have the property of transmitting neutrons preferentially with a particular direction of polarization.

It is not possible to keep a sample of free neutrons in a small container and therefore resonance beam methods were used. Actually, neutrons were produced by bombarding a beryllium target by a beam of deutrons accelerated in a cyclotron (c) through induced reactions such as ^9Be (d, n) ^{10}B and then are slowed down to thermal velocities by a hydrogeneous moderator, e.g. paraffin wax. They were then collimated by passing them through a hole in the 9″ thick shielding cadmium sheets. Bloch first pointed out that a neutron beam when passed through a magnetised ferromagnetic substance gets partially polarised as a result of the interference between the nuclear scattering and magnetic scattering caused by the magnetic interaction of the nuclear magnetic moment with atomic magnetic moment. However, the iron must be magnetically highly saturated. Keeping this in view, they allowed to pass the collimated beam through a series of three magnets M_1, M_2 and M_3 in all of which the magnetic fields were directed upwards. Two iron sheets I_1 and I_3 were placed in the gaps between the pole-pieces of the first and third magnets. These sheets attained almost completely saturated magnetization due to the action of the magnetic fields in M_1 and M_3.

After passing through I_1, the neutron beam became partially polarised with net down ward polarisation, since more neutrons with up spin were scattered out by I_1 than those with down spin due to the interaction between the magnetic moment of neutron and that of the iron atoms in the plates.

In the absence of magnet (M_2), when the partially polarized beam of neutron passed through the I_3 sheet of iron, some of the neutrons in the beam were scattered out, so that the beam intensity was further reduced. The transmission of beam through I_3 depends on the degree of polarisation of the neutron beam previously attained and incident on it. Since the beam emerging out of I_1 was partially polarised, the transmission through I_3 was better than through I_1 as the neutrons with downward spins gets scattered less than those with upward spins. However, the percentage of the number of atoms of the former type being greater than that of the latter in the incident beam on I_3. Moreover, if the polarisation of beam incident on I_3 some how reduced, then the transmission through it will also be reduced.

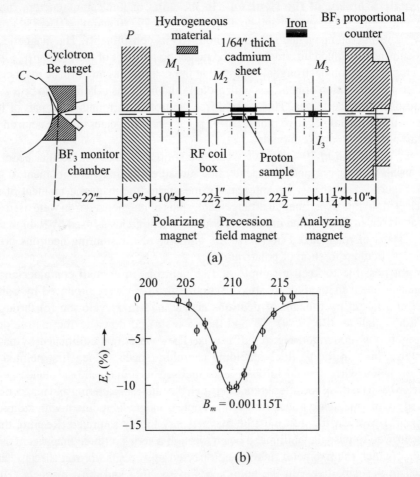

Fig. 7.8(a) Schematic experimental arrangement for the measurment of the magnetic moment of neutron due to Bloch and Coworkers. (b) Typical neutron resonance dip at B_m obtained by Bloch and Coworkers by varying the current through the magnet.

Now, if an r_f field with field direction perpendicular to the main field, i.e. steady magnetic field in M_2, was applied, a spin flipping, i.e. a change in the orientation of the neutron spin occurs preferentially when the frequency of the applied r_f field became equal to the Larmor processional frequency of the neutrons around the steady field in M_2. The effect was to destroy fully or partially the beam polarisation and the result was a reduction in intensity observed at the counter BF_3 used as the detector. Figure 7.8(b) shows a typical neutron resonance exhibiting a drop in the intensity of the neutron beam.

The condition for resonance is given by

$$\omega_L = \frac{g_n \, \mu_N \, B}{\hbar} \qquad \qquad ...\ (7.1)$$

Where ω_L is the Larmor precessional frequency of the neutron in the magnetic field B, μ_N is the nuclear magneton and g_n is the gyromagnetic ratio for neutron and determined in the experiment. B(steady field in M_2) is usually kept constant during experiment and the frequency of the applied r_f field is varied till the resonance condition is attained.

The first experimental value obtained by Alvarez and Bloch, $|\mu_N| = |g_n| \, s_n \, \mu_N = 1.9345 \, \mu_N$.

Here $s_N = \dfrac{1}{2}$ is the spin of the neutron.

More accurate value has been obtained from a improved method due to Ramsay et al. which in principle is similar to that of Bloch and Alvarez but in which a much higher degree of polarization is achieved in the neutron beam. Figure 7.9 shows the experimental arrangement.

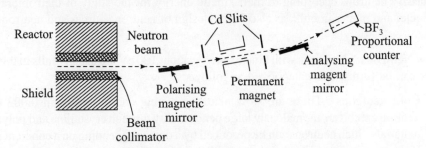

Fig. 7.9 Experimental arrangement for the determination of the magnetic moment of the neutron due to Ramsay et al.

The magnetic mirror in this method is made of an alloy containing 93 % cobalt and 7 % of iron. A well collimated beam of thermal neutrons from a nuclear reactor was allowed to incident on a magnetic mirror at a glancing angle of $17'$. After reflection from the magnetic mirror, the neutron beam gets polarised. This reflected beam has a polarisation as high as 90 %. The beam of polarised neutrons was then passed through a lit and a 1.5 m long uniform magnetic field of intensity about 8.5 kG produced by a permanent magnet. The beam is then allowed to pass through another slit on to another magnetic mirror used as a analyzer, which is magnetised in the same direction as the polarising magnetic mirror. The intensity of the doubly reflected beam (~ 7500 counts/min) is recorded by means of a BF_3 proportional neutron detector.

The 1.5 m long magnetic field produced by a permanent magnet contains two rf current loops mounted some distance (about 4211) apart produces narrower resonance lines. Resonance flipping will occur when the rf supply frequency equals the Larmour angular frequency of the neutron precession. The spin flipping reverses the polarisation of the neutron beam. Obviously, the onset of the resonance is revealed by a sharp decrease in the number of doubly reflected neutrons detected by BF_3 counter. The condition for resonance is

$$2\,\mu_N\,B_0 = \hbar\omega \quad \text{or} \quad \mu_N = \frac{\hbar\omega}{2\,B_0} \qquad \qquad \dots (7.2)$$

The magnetic sign reveals that the magnetic moment of the neutron (μ_N) is similar to that generated by rotating negative charge with its spin parallel to that of the neutron.

We have mentioned earlier that the deutron is believed to consists of a proton and a neutron having parallel spins. Obviously, the magnetic moment of the deutron should be equal to the sum of the separate nucleons moments. However, this sum differs from the moment of the deutron by 0.022 nuclear magneton. This discrepancy may be due to: (i) spending part of its time by deutron in a higher orbital momentum state, in which its magnetic moment would be definitely less than that of the ground state, and (ii) there will be a difference in the magnetic moments of the nucleons in the free and bound states.

7.1.4 Classification of Neutrons According to Energy

One can classify neutrons according to their kinetic energy for the study of their interaction with particles, nuclei and matter in bulk, as also in the design of neutron sources and neutron detectors. We have

(a) **Slow Neutrons** Neutrons with kinetic energy about 100 eV are usually falls in this category. This can be further sub categorized as follows:

(i) **Cold neutrons** These are slow neutrons having energy less than 0.002 eV. These neutron exhibit an anomalously large penetrability through crystalline and polycrystalline materials. Such neutrons can be produced by a device depending on a coherent scattering. If the neutron wavelength (λ) exceed $2d$, where d is the grating spacing of the crystal, then in accordance with Bragg's law ($n\lambda = 2d \sin \theta$), there will be no reflection. Obviously, neutrons with the lower energies, i.e. neutrons with wavelength less than $2d$ shall be reflected and removed from the beam. For graphite, $2d = 6.7$ A°, the computed maximum energy of the transmitted neutrons is obtained as 0.002 eV. This value is well below the average energy of the thermal neutrons.

(ii) **Thermal Neutrons** When fast neutrons have been diffused through matter in bulk, they are slowed down until the average energy of the neutrons is equal to the average thermal energy of atoms around them. Such neutrons are called thermal neutrons and the energy distribution of these neutrons in weakly absorbind media is the Maxwellian distribution. Accordingly, the number of neutrons with velocity between v and $v + dv$ per unit volume is given by

$$dn(v) = n_v \, dv = A v^2 \exp\left(-\frac{M_n v^2}{2 k_B T}\right) dv \qquad \text{... (7.3)}$$

Where A is constant, T is the absolute temperature and M_n is the mass of the neutron. In the distribution curve, the maximum occurs at the energy $E_m = k_B T$, which is about 0.025 eV at 20°C.

We may note that the neutron distribution is, nevertheless, very close to the Maxwellian distribution but corresponding to a higher temperature than the temperature of the medium. This 'neutron temperature' T_n is evaluated by the following relation:

$$T_n = T\left(1 + 0.92 \, A \frac{\Sigma_a}{\Sigma_s}\right) \qquad \text{... (7.4)}$$

Where T is the temperature of the medium, Σ_s and Σ_a are the macroscopic cross-sections of thermal neutron scattering and absorption at the energy $k_B T$ and A, the mass number of atoms of the moderator.

(iii) **Epithermal Neutrons**   When higher energy neutrons enter a material medium, they achieve ultimately thermal equilibrium with the molecules in the medium. However, prior to this happens, there is a preponderance of relatively higher energy slow neutrons than is permitted by Maxwellian distribution function. Such a distribution in it are called epithermal neutrons and usually have energy above ~ 0.5 eV.

(iv) **Resonance Neutrons**   Slow neutrons in the energy range 1 to 100 eV are called resonance neutrons due to the fact that large number of very sharp resonances produced in heavier nuclei by these neutrons.

(b) **Intermediate Neutrons**   Neutron having energies in the range 1 keV to 0.5 MeV are usually called intermediate neutrons and are obtained by the deceleration of fast neutrons. In comparison to other regions, neutrons in this region of energy range has been less extensively studies. For the study of these neutrons several new techniques have been developed.

(c) **Fast Neutrons**   The neutrons having energy between 0.5 to 10 MeV are called fast neutrons. In this range several nuclear reactions, which are energetically not possible at lower neutron energies, take place, of which the inelastic scattering of neutrons is most important. The inelastic neutron scattering reaction helps in the investigation of the excited states of the target nuclei.

(d) **Very fast Neutrons**   The neutrons having energy between 10 to 50 MeV are termed as very fast neutrons. This range of energy is characterised by nuclear reactions in which more than one products appear such as $(n, 2n)$ reactions. In this range, the availability of suitable neutron sources with well-defined energies is rather limited.

(e) **Ultrafast Neutrons**   These neutrons are obtained from cosmic radiations and also produced in (pn) reactions by very high energy protons accelerated in ultrahigh energy particle (proton) accelerators. These neutrons produce spallation reactions in which a nucleus breaks up into a large number of fragements by the impact of these neutrons. The energy of such neutrons is beyond 50 MeV. These neutrons have rather small probability of interactions with nuclei, resulting in *partial transparency* of nuclei to these neutrons.

Note   The energy of 99 % fission neutrons lies in the fast region. The total cross-section in this region approximately equals $\sigma_t \approx 2\pi(R + \lambdabar)^2$, where R is the radius of the nucleus and $\lambdabar$, the wavelength of the neutron of energy E. The main feature of the fast region is that since the radiative capture cross section σ_γ is very small here, the total cross-section is the scattering cross-section: $\sigma_t \approx \sigma_s = \sigma_n + \sigma_{n'}$. The thresholds of inelastic scattering lie just within this region or a little below its boundary with the intermediate region.

7.1.5 Neutron Detectors

Neutrons can not produce ionisation in matter being electrically neutral. However, during nuclear reactions, neutrons can eject charged particles, e.g. p, ${}^2_1D$, α, etc., when they have quite low energies. By elastic collisions, fast neutrons can also eject recoil nuclei. These charged particles ejected by neutrons during nuclear reactions can produce ionization in matter when they pass through it.

In general, two principles have been used in the detection of neutrons: (i) use is made of the charged particles produced in nuclear reactions by the interaction of neutrons with various nuclei of substances introduced in the counter, and (ii) advantage is taken of the recoil of the nuclei of light elements after interaction with neutrons. However, for the detection of slow neutrons, the devices related to first method is used, whereas for the detection of fast neutrons, one uses second method.

Conventional particle detectors based on the production of ionisation can be used for the detection of these charged particles, e.g. p, D, α, etc., which with some modification can therefore serve as convenient tools for the detection of the neutrons and also for different types of measurments.

Another method of detection of neutrons is based on the radioactive isotope production by nuclear reactions induced by neutron, especially at low energies of neutron.

One can construct special instruments for neutron detection based on intense ionisation produced by fragments of large nuclear charges produced during the nuclear fission of heavy nuclei induced by neutrons.

Obviously, one can put the neutron detection methods into two groups: (i) for slow neutrons, and (ii) for fast neutrons.

7.1.5.1 Slow Neutron Detection

Most widely used common reactions for neutron detections are

(n, α) reactions: ${}^{10}B\ (n, \alpha)\ {}^{7}Li + 2.79$ MeV

${}^{10}B\ (n, \alpha)\ {}^{7}Li^{*} + 2.31$ MeV

${}^{6}Li\ (n, \alpha)\ {}^{3}H + 4.79$ MeV

The second of the above nuclear reaction leads to an excited state of the product nuclei ${}^{7}Li$ with an excitation energy $= 0.48$ MeV.

(n, p) reactions: ${}^{3}He\ (n, p)\ {}^{3}H + 0.764$ MeV

The above four nuclear reactions have large thermal neutron cross-sections, which is proprotional to v^{-1}. These reactions are used for the detection of thermal neutrons. The second reaction is most slow neutron capture reactions in ${}^{10}B$ which lead to the formation of the excited state ${}^{7}Li^{*}$ after the emission of the α-particle. With the emission of γ-ray, this state then comes down to the ground state.

For thermal neutrons (0.025 eV) the cross section for ^{10}B (n, α) reactions 3837 barns. Natural boron contains 18.83 % of ^{10}B isotope and rest is ^{11}B. This has a much lower cross-section (~ 710 b). Because of its relatively large Q (= 2.78 MeV) value, the resolution is not very good for neutron energy measurment using the boron nuclear reaction.

Because of higher Q value, ^6Li (n, α) reaction having 940 barns thermal neutron cross-section is also not so good for the measurment of very low energy. However, this is more suitable for the measurment of neutron energies above several hundred keV.

The thermal neutron cross section of reaction ^3He (n, p) is 5330 barns with low Q value. At lower energies, it has better resolution for energy measurment.

We may note that at higher energies, scattering collisions become important. The scattering cross-sections must be known for neutron spectrum analysis.

(i) Boron Trifluride (BF₃) Proportional Counter

The sensitivity of BF_3 counters suffer from one major defect, i.e. their sensivity decreases very rapidly with neutron energy becoming pronihitively small in the intermediate energy range. Hanson and Mckibben in 1947 modified it to make it sensitive to fast neutrons in the energy range 10 keV to 3 MeV. Figure 7.10 shows a typical BF_3 counter, which is also called a long counter. It has a cylindrical cothode, usually made of stainless steel, aluminium or copper with an axial anode wire of tungsten or nickel having diameter ~ 0.5 mm. The counter tube is embedded in a cylinder of paraffin of diameter 8″. Around this paraffin, there is a layer of an absorber for slow neutrons which is also covered with an outer layer of paraffin.

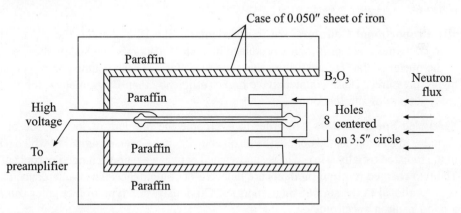

Fig. 7.10 Hanson and Mckibben type long BF_3 proportional counter embedded in paraffin

We know that only those neutrons will be detected with sufficient efficiency which follows $\dfrac{1}{v}$ law and obviously these are thermal neutrons. The neutrons having an energy of about 100 keV have a slightly better chance of detecting than neutrons having lower energy. Experiments have shown that the long counter has an efficiency for the detection of neutrons which is practically constant over a considerable range of neutron energies. If the counter and paraffin mentle are quite long, even neutrons of highest energy shall be slowed down to thermal energies and could be detected. We may note that the detector efficiency for neutrons incident on the 8″ circle on the front face is less than 1 %.

The operating voltage of BF_3 counter is usually ~ 2000 to 2500 Volts at the gas pressure in the range of 100 to 600 torr. So far BF_3 counters having lengths upto 1.8 m and diameter 5 cm have been achieved. Smaller counters with higher gap pressures have also been in use. Figure 7.11 shows the operating characterstic of a BF_3 counter.

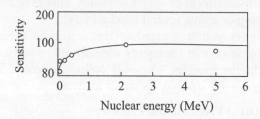

Fig. 7.11 Operating characterstic of a BF_3 counter

(ii) Boron Coated Propotional Counters
The inner walls of these counters are lined with a coating of boron (> 1 mg/cm^2) and within which the $^{10}B(n, \alpha)$ nuclear reaction is induced by slow neutrons. The coating of boron serves as a radiator. One may also use enriched ^{10}B isotope to increase the efficiency of the counter. The two charged particles, i.e. $^4_2$He and $^7_3$Li fly off in opposite directions and hence only one of them can enter the counter to produce a pulse. Due to energy loss of the charged particles in the coating of boron, these counters do not have a satisfactory counting plateau.

(iii) ^{3}He Proportional Counters
In these counters, like BF_3 counter, the pulse producing charged particles i.e., protons are produced in nuclear reactions induced by thermal neutrons in the nuclei of the ^{3}He gas filling this type of counter. These counters can be operated at higher pressures. These counters have higher efficiency than BF_3 counters due to the large cross-section of the reaction.

(iv) Sandwich Spectrometers
In order to overcome the problem of loss of one of charged particles in a boron coated counter, neutron counters have been designed in which a thin foil of the radiator, usually ^{3}He or ^{6}Li is sandwiched between a pair of charged particle detectors. The two charged reaction products are detected in coincidence. The coincidence pulse size is proportional to the sum of their energies. Obviously, this type of a counter can serve as a good neutron spectrometer.

Usually semiconductor detectors are employed for the detection of the charged particles. In the coincidence mode of operation, much of the background due to γ-rays and recoil contribution are eliminated and therefore they give a cleaner spectrum.

If ^{3}He is sandwiched between the two detectors, then compared to ^{6}Li spectrometer for the same energy resolution, its efficiency for neutron detection is about 20 to 50 times higher. Moreover, they have better energy resolution. Further, the former has an advantage that ^{3}He can be used in the elemental form whereas in the ^{6}Li spectrometer this is not possible. The efficiencies of these counters for fast neutrons are quite low $\sim 10^{-5}$ to 10^{-6}.

(v) Fission Counters

The odd-odd nuclei ^{235}U and ^{239}Pu undergo thermal neutron fission for which the reaction cross-section is quite high. Moreover, they also undergo fast neutron fission with much lower cross-section. However, the even-even nuclei ^{238}U and ^{232}Th undergo fission only with fast neutrons of energy $E_n \sim 1.5$ MeV. Because of large masses of two fission fragments produced during fission, their ranges are very short. They produce intense ionisation and as such the pulses that they give rise to can be easily distinguished from the background due to other ionizing radiation, e.g. α-particles which are emitted spontaneously by these heavy nuclei.

Usually, the fissionable material is introduced as a lining on the interior walls of the counter having thickness ranging from a few tens of $\dfrac{\mu g}{cm^2}$ to $\sim \dfrac{2\ mg}{cm^2}$. However, due to very high energy of the fission fragments relatively large pulses are produced even if one of the fragments is absorbed in the coating.

One can detect the fission fragments by a gas counter, a scintillation counter or a semiconductor detector. One can monitor very high neutron fluxes by the fission counters, as in a nuclear reactor.

For detection of slow neutrons, usually counters with internal coating of ^{235}U are employed. In the case the cross section for fission being ~ 600 b. In case if ^{238}U is used as the coating, then the counter will be only sensitive to fast neutrons.

(vi) Activation Method Fermi discovered that many elements exposed to a flux of the thermal neutrons become radioactive. This discovery has led to the detection of slow neutrons. The feasibility of detecting the induced ratio-activity depends on its life time, which cannot be much shorter than the time between the end of exposure of the target to the neutron beam and the measurment of the induced radioactivity. For these measurments life times between ~ 0.1 s to 10^4 y are found useful.

A number of elements are found to have a large activation cross-section for the (n, γ) reaction few are listed in Table 7.3. Thin foils of these elements (suitable for the purpose are exposed to a neutron flux for a time interval greater than six or seven times the half life of the radio induced activity. Then, they are removed and the extent of the activity is determined with an appropriate Geiger counter, ionisation chamber, scillation counter or other radiation detector by counting the emitted radiations. At low energy, indium has a large cross-section for the induced reaction and therefore it is very efficient detector for slow neutrons at the energy. One can estimate the thermal component of a neutron flux by measuring the induced activity in a bare indium foil and other wrapped in cadmium foil having thickness ~ 1 mm. Cadmium is nearly opaque for low energy neutrons (~ 0.3 eV) having cross section for the reaction in several thousand barns, but it does not yield a radioactive isotope. However, for epithermal neutrons, the cadmium foil is nearly transparent.

Table 7.3

Element as a target	Isotope	Abundance %	Reaction Cross-section $\sigma(n, \gamma)$ barns	Activity	τ
Manganese	^{55}Mn	100	13.3	$_{-1}e^0$ (β^-)	2.58 h
Copper	^{63}Cu	69.2	4.4	β^-, β^+, E.C.	12.7 h
	^{65}Cu	30.8	2.2	β^-	5.1 min
Rhodium	^{103}Rh	100	11	β^-, I.T.	4.35 min
Indium	^{113}In	4.3	8	I.T.	49.5 days
	^{115}In	95.7	87	I.T.	2.2 s
Silver	^{107}Ag	51.83	38	β^-	2.4 min
	^{109}Ag	48.17	4.4	β^-	252 days
Gold	^{197}Au	100	98.8	β^-	2.7 days

The radioactive isotopes produced in (n, γ) nuclear reactions are isotopes of the target nuclei. *Szilard and Chalmers* in 1934 devised a technique for separating radio-nuclei from their isotopic environment. The method is based on the fact that the emission of γ-ray photon in the (n, γ) reaction after the capture of slow neutron can provide recoil energy E_γ sufficient to break the chemical (molecular) bond between radiactive atoms and thereby to change the chemical state of the product nucleus compared to a normal nucleus in the medium. Under certain circumstances, one can separate the radioactive nuclei from the other with 100 % efficiency.

For a single γ-ray of energy E_γ (MeV) emitted from a nucleus of mass number A, the recoil energy E_r (eV) is given by

$$E_r = 536 \, \frac{E_\gamma^2}{A} \qquad \qquad \ldots (7.1)$$

Where A is the mass number of the recoiling atom. For example, for $E_\gamma = 8$ MeV, $A = 120$, one obtains the recoil energy $E_r = 286$ eV. We can see that E_r is fairly large than necessary for rupturing a molecular band ($\sim$ 1 to 5 eV). The above reaction is termed as *Szilard-Chalmers reaction*. This is used to obtain radioactive samples of high specific activity, especially when only relatively weak neutron source were available.

(vii) Scintillation Counters for slow Neutrons detection   Various types of slow neutron scintillation have been developed by investigators. Because neutrons are not capable of producing scintillation directly, and hence it is essential to incorporate some material in phosphor which interacts with neutron and produce ionising radiation on the passage of neutrons through the mixture. We may note that in crystal phosphor light pulses rise at a rapid rate. In an early method alternate thin layers of a boron compound with *ZnS* preparation was used. The α-particles released by the neutrons from the boron were detected by the ability of *ZnS* to scintillate under the action of α-particles. However, this did not have a satisfactory efficiency for the detection of slow neutrons, because α-particles were relatively insensitive to γ-rays. Moreover, not all ionizing particles from the disintegration of the boron succeeded in producing scintillations and only a fraction of the light from the scintillations cound reach the photo-multiplier and much of the light lost by reflection. These difficulties to a large extent were removed by fusing *Zn S* in a glassy layer of boric anhydride which was put directly at the window of the photomultiplier tube (Figure 7.12).

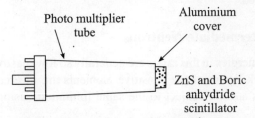

Photo multiplier tube

Aluminium cover

ZnS and Boric anhydride scintillator

Fig. 7.12 Scintillation counter using a photomultiplier tube and a scintillator containing *Zn S* and boric anhydride for the detection of slow neutrons.

The efficiency of scintillation counter as a slow neutron detector reaches 100 % when phosphor itself contains an element sensitive to neutrons. This is the main advantage. When large volume slow neutron detectors are required, liquid scintillators containing boron compounds are often used. Methyl borate dissolved in terphenyl and phenycyclohexane as main solvent is found to be an efficient detector of slow neutrons. Nicholson and Snelling used lithium iodide phosphors activated with europium. They produced scintillators by the products of the nuclear reaction ^6_3Li (*n*, α) ^3_1H having $Q = 4.79$ MeV.

One can defect slow neutrons in a wide variety of ways. 20 cms diameter solid discs of fused B_2O_3 and *Zn S* has been successfully used for slow neutrons detection.

The usual difficulty which is encountered with these detectors is to distinguish between the pulses of neutron and γ-ray. To some extent, one can overcome this by knowing the actual shape of the scintillation pulses.

(viii) Photographic emulsion and cloud chambers as slow neutron detectors

These are the most widely used devices for slow neutron detection because the secondary charged particles produced by neutrons give rise to visible tracks in both these devices. These two devices yield information is of the same sort as photographic emulsion may be regarded as a cloud chamber being operated at very high pressure, however these two devices are complimentary devices. One can make the choice between the two by keeping in mind that the photographic emulsions are much more sensitive and therefore more useful for weak sources whereas the cloud chamber have the advantage that the gas filling the chamber may be chosen as desired. There has been considerable progress in emulsion technique and so far a number of nuclear emulsions with varying degree of sensitivity over a wide range, have been developed. Most of the nuclear emulsion devices contain an appreciable amount of nitrogen which is a quite useful for the detection of slow neutrons through the reaction $^{14}_7N$ (*n*, *p*) $^{12}_6C$. Here the protons resulting from thermal neutron capture have a range of few microns and one can easily observe and count in the emulsions giving an estimate of the flux of thermal neutron.

One can make nuclear emulsions more sensitive for the detection of thermal neutrons by adding small amounts (or 1 % by weight) of boron or lithium to them. The purpose of this is to initiate (n, α) reaction in these loaded emulsions, as this is suitable for small fluxes of slow neutrons. One can load the nuclear emulsion with uranium citrate or acetate for observing tracks of fission fragments produced by slow neutron fission. Now a days, emulsions with greatly decreased sensitivity for α-particles have been developed so that the effect of a-particles is reduced.

7.1.5.2 Detection of Intermediate Neutrons

Cross sections of neutron energies in this range are generally so low that to yield adequate intensities of ionizing radiations thick layers of the sensitive elements are desired. However, the sensitive element which emit γ-rays are not subject to the same limitation. Using the soft γ-rays from the reaction

$$^{10}_{5}B\,(n,\,\alpha)\,^{7}_{3}Li^{*} \to \,^{7}_{3}Li + \gamma\,(480\text{ keV})$$

for measuring intermediate energies of neutrons, Rae and Bowey in 1953 designed a detector. Layers of $^{10}_{5}B$ opaque to neutrons with energies upto 1 keV may then be used as the detector. With the use of NaI phosphor with a rapid rate of rise of the luminuous response in a single channel scintillation spectrometer, one can substantially reduce the disturbing background from other radioactive sources and the cosmic radiation. Moreover, in addition to this, a through system of lead screening and neutron shields were also provided. This detector mainly consists of two photomultipliers, were with NaI phosphor, faced the boron detector from opposite sides. Since the amount of boron used is no longer opaque to neutrons above 1 keV and therefore a drop of about 15 % in the efficiency is expected at 5 keV and about 1 % at 50 keV.

7.1.5.3 Detection of Fast Neutrons

   (i) Proton-recoil Counters   In this method, the observation is made of the ionisation produced by the recoiling protons in the elastic scattering of neutrons by hydrogeneous materials. The cross-section of the scattering is energy dependent.

The energy of the proton depends on the angle of recoil in accordance.

$$E_p = E_n \cos^2 \theta \qquad\qquad ...\,(7.1)$$

Where θ can vary between $0°$ and $90°$ in the L-system may be filled with hydrogen or deuterium, but a better procedure is to use argon or one of the heavier inert gases as a filling gas and to place a thin sheet of a hydrogeneous material such as paraffin at one end of the chamber from which the protons are ejected by the impact of incident neutron beam. The recoil protons passing through the counter produce ionization in their paths and so can be recorded by the usual method.

To a large extent, these neutron counters are insensitive to slow and thermal neutrons and also to γ-rays, unless one operates the counter at a very high gas pressure.

For a given neutron energy, the pulse height distribution is usually broad due to the broad range of the recoil proton energy. This is why that these counters are not very suitable for the measurment of complex neutron spectra, though one may obtain some information in the case of simple neutron spectrum.

Sun and Richardson in 1954 designed a multiple-wire proportional counter containing methane at 1.2 atm and operated at 3400 V. When operated, this has been found to be almost completely insensitive to γ-radiations at an over all efficiency of 0.17 % for detecting neutrons in the energy range from 0.3 to 10 MeV.

(ii) **Fission Counter** ^{238}U has a very large cross section for fast neutrons and ^{235}U has a very large cross-section for slow neutrons. Obviously, uranium depleted of the isotope ^{235}U can be used for the detection of neutron in a fission counter which the similar to that described previously.

There are some heavy stable nuclei, e.g. $^{209}_{83}$Bi, which undergo fission by very high energy neutrons ($E_n > 50$ MeV). One can use these materials as lining within a fission counter for detection of ultrahigh energy neutrons.

(iii) **Scintillation Counter**

For fast neutron detection one can use some organic scintillators, e.g. stilbene or anthracene and various liquid and plastic scintillators. These materials are quite rich in hydrogen content and therefore protons are released when fast neutrons are incident on them. These protons produce scintillation pulses.

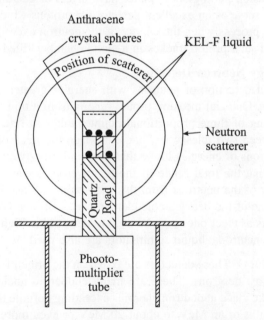

Fig. 7.13 Fast neutron scintillation counter

Hornyak in 1952 designed scintillators which were effective for fast neutrons by molding a scintillating phosphor which was sensitive to recoiling protons, such as *Zn S* into a plastic serving as a fairly dense radiator. Another ingeneous scintillation detector is due to Taylor et al. This scintillation detector uses a number of anthracence plastic scintillating liquid spheres immersed in a nonlydrogeneous non scintillating liquid such as C_6Cl_F6. This nonscintiallating liquid couples the plastic spheres through a quartz rod to the cathode of photomultiplier tube. One has to choose the size of the spheres so as to dissipate completely the energy of recoil protons by the fast neutrons. The adjacent spheres are sufficiently separated so that the neutron pulses may be well distinguished from the γ-ray pulses. The optimum diameters of these plastic spheres for detecting neutrons of K. E. 3.5, 4.5, 7.0 and 14 MeV are reported to be 1.5, 2.6, 6.0 and 10 mm respectively. The arrangement is schematically shown in Figure 7.13. For different media, the apparatus measures fast neutron elastic scattering cross-sections. The scattering of fast neutrons is in the form of a spherical shell which surrounds the detector.

Organic scintillators have good efficiency for neutron detection. One can detect recoil protons of energy as low as 10^4 eV with low noise photomultiplier tubes and the associated electronics.

One can achieve an efficiency of upto 10 % for neutrons of a few MeV energy with organic crystals of thickness few centimeters. Corrections are needed to determine the neutron energy spectrum from the observed pulse height distribution of the protons due to nonlinear response of these scintillators to the recoil proton energy.

(iv) Photographic process

The nuclear emplusion photographic plates have layers of gelatine which are quite rich in hydrogen and can serve as an excellent detector for protons which recoil from neutrons in the *n-p* scattering process when the energy of the neutron exceeds about 2 MeV. One can also photographed such proton trackes in a cloud chamber filled with water vapour.

(v) Ultra-High-Energy Neutron Detection

The efficiencies of detection of neutrons with energies greater than 50 MeV are usually reported to be low. One can measure proton recoils from a thin hydrogeneous target with telescope consisting of three proportional counter tubes in line. By introducing suitable absorbers in between two telescopes, one can made the telescope insensitive to protons generated by neutrons of energies lower than some arbitrary value. The main advantage of this telescope is that the total range of the recoil protons can be measured and one can deduce the energy of the neutrons from this measurment. Since neutrons (> 50 MeV) can induce fission some of the heavy stable elements, e.g. Bi, one can make fission chambers with such elements as electrodes may be used to detect ultra high energy neutrons. For the detection of these neutrons, liquid scintillators are also used.

(vi) Threshold Detectors   These detectors are based on the principle of production of induced radioactivity by fast neutrons. There are many endoergic nuclear reactions induced by neutrons which take place at neutron energies exceeding definite thresholds which lie in the range of a few tenths of an MeV to about 20 MeV or even more involve either (*n*, 2*n*) or charged particle reactions, e.g. $^{12}_{6}C + ^{1}_{0}n \rightarrow ^{11}_{6}C + 2^{1}_{0}n$ does not occur unless the neutron energy is > 22 MeV, the *threshold energy* of the reaction. Obviously, $^{12}_{6}C$ can be used as a threshold detector for neutrons whose energies exceed 22 MeV. Table 7.4 gives threshold energies for few such reactions.

One can use the relative activations to derive the neutron spectra, provided, if we choose a series of threshold detectors for which σ vs E_n curves are known. One uses this *spectral index method* for the study of reactor neutron spectra.

Table 7.4 Neutron Threshold Energies

Nuclear reaction	Half life of the product	Reaction threshold (MeV)
$^{31}P\ (n, p)\ ^{31}Si$	2.7 hours	0.97
$^{32}S\ (n, p)\ ^{32}P$	14.3 days	0.96
$^{27}Al\ (n, \alpha)\ ^{24}Na$	14.9 hours	2.44
$^{14}N\ (n, 2n)\ ^{13}N$	10 min	11.3
$^{16}O\ (n, 2n)\ ^{15}O$	2 min	17.3
$^{12}C\ (n, 2n)\ ^{11}C$	21 min	20.3

7.1.5.4 Neutron Detection through slowing down of fast neutrons

One can slow down fast neutrons by diffusion and elastic scattering in a moderating medium containing low mass number nuclei. Hydrogen ($A = 1$) is the most efficient slowing down medium. However, this suffers from the disadvantage of having a relatively high neutron capture cross section. One can conveniently detect the slow neutrons produced from fast neutrons in a moderator by a suitable slow neutron counter. The advantage of this method of detection of fast neutrons is relatively high efficiency which is almost constant over a considerable range of neutron energy. Based on this principle, there are two detectors:

(i) **Long Counter** First devised Hanson and Mckibben in 1947, this counters has a uniform sensitivity in the energy range ~ 2 keV to 6 MeV. This counter consists of cylindrical BF_3 counter having diameter 0.5″ and length 10″ placed along the paraffin cylinder of diameter ~ 8″, which acts as the moderating medium (see Figure 7.10). The neutrons approach parallel to the axis. To prevent the background of scattered neutrons impinging on the paraffin from all sides, a slow neutron absorber is used as a shield around the moderator in some arrangements. The absolute efficiency of this counter is about 1 % or even less. However, the use of ^3He detector in place of BF_3 detector increases the efficiency.

(ii) **4π-Detectors** In these counters, the neutron source is placed at the centre of a large moderating medium with an assembly of thermal neutron detectors. Over a broad range of energies, the arrangement has nearly constant sensitivity. First, the neutrons from the source are slowed down to thermal energies. These are detected by detectors embedded within the moderator volume. Graphite, paraffin, polyethylene and water have been used as moderators and BF_3 or ^3He counters are used as detectors of thermal neutrons. For neutron detection, moderators with low neutron capture cross sections give high efficiencies. One can determine the optimum size of the moderating assembly to give high efficiency, by slowing down length of neutrons.

To study neutron multiplication, 4π neutron detectors with organic scintillators have been used.

7.1.6 Slowing Down of Neutrons

The energy of neutrons emitted in nuclear reactions is about a few MeV. Neutrons cannot loose their energy by ionisation as they are not electrically charged. However, neutrons can be slowed down by successive scattering in matter until they reach in thermal equilibrium with their surroundings. After this neutrons diffuse like gaseous molecules until they are captured by nuclei.

(i) **Energy loss** Fermi and Collobrators showed that the neutrons are slowed down in hydrogeneous materials and these slower neutrons have a greater probability of inducing radioactivity in comparison to more energetic neutrons. One can calculate the loss of energy in a single collision by neutron with the help of equations of conservation of energy and momentum. One can also use another very simple method, which involves the use of two reference systems, i.e. *L*-system, in which the target nucleus is initially at rest and is approached by the incident neutron. The second reference system is *CM* system in which the *CM* of the colliding particles is at rest initially and also throughout the collision. Prior to collision in the *L*-system, the neutron of mass m_n moves towards a nucleus of mass M(assumed to be at rest) with speed v_0. Let E_0 and E be the energies of the neutron before and after the collision in the *L*-system. The scattering angles in the *L* and *CM* systems be θ and φ respectively. In the *CM* system, Let v be the velocity in *L*-system after collision. Now, the velocity v_c in the *CM* system is obtained as

$$v_c = \frac{m_n v_0}{m_n + M} = \frac{v_0}{A+1}$$

Where $A = \dfrac{M}{m_n}$. Now, the velocity of the target nucleus V in *CM* System will be

$$V = v_0 - v_c = \frac{A v_0}{A+1}$$

From Figure 7.14, one finds

$$v^2 = (V \sin \theta)^2 + (V \cos \theta + v_c)^2$$

$$= v_0^2 \frac{(A^2 + 2A \cos \theta + 1)}{(A+1)^2}$$

Hence, one obtains

$$\frac{E}{E_0} = \left(\frac{v}{v_0}\right)^2 = \frac{A^2 + 2A \cos \theta + 1}{(A+1)^2} \qquad ... \ (7.1)$$

Equation (7.1) clearly reveals that the energy of the neutron after collision varies with the angle of scattering θ. The limiting values of angle θ are 0 and π. One finds the maximum and minimum values of energy of the neutron after scattering are:

For $\theta = 0$ $E_{max} = E_0$ ... (7.2)

$$E_{min} = \frac{(A-1)^2}{(A+1)^2} E_0$$

$$= \alpha \, E_0 \qquad ... \ (7.3)$$

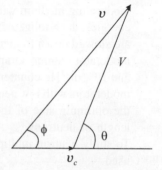

Fig. 7.14 Vector diagram exhibiting relation between neutron velocity (v) in *L*-system and velocity (v_c) in the *C*-system after collision.

Here $$\alpha = \frac{(A-1)^2}{(A+1)^2} < 1$$

Now, if the scattering nucleus is a proton, then we see that $A = 1$ and we obtain $E_{min} = 0$. Obviously, the neutron energy in this case after collision varies from 0 to E_0. One can easily see from Figure 7.14 that

$$\text{Cos } \phi = \frac{V \text{ Cos } \theta + v_c}{\sqrt{(V \text{ Cos } \theta + v_c)^2 + (V \text{ Sin } \theta)^2}}$$

$$= \frac{A \text{ Cos } \theta + 1}{\sqrt{A^2 + 1 + 2 A \text{ Cos } \theta}} \qquad \qquad ... \ (7.4)$$

Relation (7.4) gives us the relation between the scattering angles in the L-system (θ) and C.M. system (ϕ).

In case if the nucleus is heavy, i.e. $A \gg 1$, one obtains

$$\cos \phi = \frac{A \text{ Cos } \theta}{\sqrt{A^2 + 2 A \text{ Cos } \theta}} = \frac{A \text{ Cos } \theta}{A} = \cos \theta$$

or $$\phi \approx \theta \qquad \qquad ... \ (7.4a)$$

Clearly, if the nucleus is heavy, the scattering angles in the two cases are almost same.

7.1.6.1 Mean Energy Loss

After scattering, an incident neutron of Energy E_0 may have energy E between αE_0 and E_0. Thus the loss of energy in collision is $\Delta E = E_0 - E$. To find the mean value of ΔE, one must know the probability that the neutron will have the energy E after suffering an elastic collision. Since E depends on the scattering angle θ and hence the above probability is equal to the probability that the neutron will be scattered at an angle θ due to the collision. We know from the theory of partial wave scattering that the scattered amplitude $f(\theta)$ can be expresses as a sum over different l values. Of these, $l = 0$ corresponds to S-wave scattering and this is spherically symmetric, i.e. independent of θ. Clearly, for low energy neutrons, the scattering is spherically symmetric, i.e. the probability of the scattering is the same for all angles. Moreover, it depends only on solid angle $d\Omega$ at an angle θ. Now, one can express the probability of scattering into the solid angle $d\Omega$ at an angle θ as

$$dP = \frac{d\Omega}{4\pi} = \frac{2\pi \text{ Sin } \theta \, d\theta}{4\pi}$$

$$= \frac{1}{2} \sin \theta \, d\theta = -\frac{1}{2} \, d(\cos \theta) \qquad \qquad ... \ (7.5)$$

From (7.1), we obtain

$$\frac{dE}{E_0} = \frac{2A}{(A+1)^2} \, d(\text{Cos } \theta) = \frac{4A}{(A+1)^2} \, dP$$

Negative sign has been omitted.

Now, we have $\qquad (1 - \alpha) = 1 - \dfrac{(A-1)^2}{(A+1)^2} = \dfrac{4A}{(A+1)^2}$

Thus, one obtains

$$dP = \frac{dE}{E_0\,(1-\alpha)} \qquad\qquad \text{... (7.6)}$$

$\therefore$ Mean energy loss per collision is obtained as

$$<\Delta E> = \frac{\displaystyle\int_{\alpha E_0}^{E_0} (E_0 - E)\,dP}{\displaystyle\int_{\alpha E_0}^{E_0} dP}$$

One can easily that the denominator in the above is equal to unity. Thus, one obtains

$$<\Delta E> = \int_{\alpha E_0}^{E_0} \frac{(E_0 - E)\,dP}{E_0\,(1-\alpha)} = \frac{E_0\,(1-\alpha)}{2}$$

Thus the mean fractional energy per collision is

$$\frac{<\Delta E>}{E_0} = \frac{1-\alpha}{2} \qquad\qquad \text{... (7.7)}$$

We note that the above is independent of initial energy E_0. This clearly reveals that though the neutron energy is reduced by successive collisions, the mean fractional energy loss per collision remains the same for all collisions. This means, if one plot the successive energies of the scattered neutron on a logarithmic energy scale, the intervals between the points will, on the average, be equal, as $\dfrac{dE}{E} = d(\ln E)$.

This also shows that in the theory of neutron slowing down, one finds it convenient to adopt the logarithm of energy as the variable, rather than the energy itself.

7.1.6.2 Logarithmic Energy Decrement

We now find the average logarithmic energy decrement, usually denoted by ξ. For this, we first calculate the mean value of the cosine of the scattering angle in L-system. Using (7.4) and (7.5), one obtains

$$<\cos\phi> = \frac{\displaystyle\int \cos\theta\,dP}{\displaystyle\int dP}$$

$$= \frac{\int_{-1}^{1} \cos\phi \, d\,(\cos\theta)}{\int_{-1}^{1} d\,(\cos\theta)}$$

$$= \frac{1}{2} \int_{-1}^{1} \frac{(A\cos\theta + 1) \, d\,(\cos\theta)}{\sqrt{A^2 + 1 + 2A\cos\theta}}$$

Evaluating the integral, one finally obtains

$$<\cos \phi> = \frac{2}{3A} \qquad\qquad \dots (7.8)$$

This reveals that $<\cos \phi>$ decreases with increasing mass of the scattering nucleus.

The average logarithmic energy decrement is expressed as

$$\xi = <\Delta \log E> = <\log E_0 - \log E> = <\log \frac{E_0}{E}>$$

($\because$ logarithmic decrement in energy for a neutron of initial energy E_0, $\varepsilon = \ln\left(\dfrac{E_0}{E}\right)$, $\therefore$ mean lograthmic decrement in energy $\xi = <\varepsilon>$). We have

$$\xi = \int_{\alpha E_0}^{E_0} (\ln E_0 - \ln E) \frac{dE}{E_0\,(1-\alpha)}$$

$$= 1 + \frac{\alpha \ln \alpha}{1 - \alpha} \qquad\qquad \dots (7.9)$$

In terms of atomic mass number A, one obtains (7.9) as

$$\xi = 1 + \frac{(A-1)^2}{2A} \ln \frac{(A-1)}{(A+1)} \qquad\qquad \dots (7.10)$$

For $A > 10$, one finds that (7.10) reduces to

$$\xi = \frac{2}{A + \frac{2}{3}} \qquad\qquad \dots (7.11)$$

Clearly, for very large A, ξ approaches zero. We may note that even for $A = 2$, the error in (7.11) is only 3.3 %. This result is consistent with our earlier finding that *heavy elements are poor moderators.* One finds a very good agreement between theoretical values and experimental results.

Thus, in any collision with a nucleus, the neutron always loses on the average the same fraction of its energy prior collision. We may note that the above derivation is based on the assumption of collision with an initially stationary nucleus.

Knowing ξ, one can easily know the average number of collisions needed to bring about a given decrease in neutron energy. In case, if neutrons start out with $<E_1>$ and are slowed down to $<E_2>$, then the total logarithmic energy loss is $\ln\left(\dfrac{<E_1>}{<E_2>}\right)$. The average number of collisions is obtained as

$$n = \ln\frac{\left(\dfrac{<E_1>}{<E_2>}\right)}{\xi} \qquad \qquad ... \,(7.12)$$

If $<E_1> = 2$ Mev   and $<E_2> = \dfrac{1}{40}$ eV, then

$$n = \ln\frac{\left(\dfrac{2\times10^6}{1/40}\right)}{\xi} = \frac{1}{\xi}\,\ln(8\times10^7)$$

$$= \frac{18.2}{\xi}$$

Table 7.5 gives the values of n for nuclei of different mass numbers.

Table 7.5 Values of n for nuclei of different A

Element	Mass number (A)	α	ξ	n
$^1_1$H	1	0	1.000	18
$^2_1$H (D)	2	0.111	0.725	25
$^4_2$He	4	0.225	0.425	43
Be	9	0.640	0.209	86
C	12	0.716	0.158	114
U	238	0.983	0.00838	2172

7.1.6.3 Slowing-down Power and Moderating ratio

For efficient slowing down of neutrons, one not only require a large value of ξ and a large probability of cross-section for scattering and large N (the number of atoms per unit volume), but the *absorption cross-section* must also be small otherwise too many neutrons may be lost by absorption.

We have the slowing down power of the moderator = $\xi\,N\,\sigma_s$. Obviously, the slowing down power is the scattering cross-section per unit volume of the material times the mean logarithmic decrement in energy. Thus slowing down power is

$$sdp = \xi\,\sigma_s\,N = \xi\Sigma_s = \frac{\xi}{\lambda_s} = \frac{\xi\,N_0\,\rho\,\sigma_s}{A} \qquad ... \,(7.13)$$

Where N_0 is Avogadro number, ρ is density and A is the atomic weight of the material.

Obviously, *sdp* can be interpreted as the mean energy loss in logarithm of the energy per unit distance of travel in the moderator. For a good moderator, we see that it must have large values of ξ and of macroscopic cross section Σ_s. We may note that any material which is a strong neutron absorber is useless as a moderator, because the main purpose of the moderator is to slow down the fast neutrons generated in fission to thermal energies and produce more fission for the release of energy on large scale in a nuclear reactor. Obviously, if the moderator is a strong absorber of neutrons, then the purpose of slowing down of neutrons is defeated.

Equation (7.13) reveals that the material chosen for the moderator must have a large *sdp* and low cross section for absorption. The ratio of *sdp* to macroscopic absorption cross section is the quantity which indicate the better moderator, and this ratio is called '*moderating ratio*' (m_r). Thus

$$m_r = \frac{sdp}{\Sigma_\alpha} = \frac{\xi \Sigma_s}{\Sigma_\alpha} \qquad \ldots (7.14)$$

If the moderator consists of more than one element, than the approximate value of ξ for such combination is given by

$$\xi = \frac{\xi_1 \Sigma_s^1 + \xi_2 \Sigma_s^2 + \xi_3 \Sigma_s^3 + \ldots}{\Sigma_s^1 + \Sigma_s^2 + \Sigma_s^3 + \ldots} \qquad \ldots (7.15)$$

In this case, the slowing down power is given by

$$s'dp' = \xi_1 \Sigma_s^1 + \xi_2 \Sigma_s^2 + \xi_3 \Sigma_s^3 + \ldots \qquad \ldots (7.16)$$

and

$$m_r = \frac{\xi_1 \Sigma_s^1 + \xi_2 \Sigma_s^2 + \xi_3 \Sigma_s^3 + \ldots}{\Sigma_a^1 + \Sigma_a^2 + \Sigma_a^3 + \ldots} \qquad \ldots (7.17)$$

Moderating ratio for few materials is given in Table 7.6.

Table 7.6 Moderating ratio for few materials

Material	Logarithmic decrement in energy (x)	n	Moderating ratio (m_r)
Hydrogen	1.000	18	66
H_2O	0.927	19	67
Deuterium ($_1^2D$)	0.725	25	75820
D_2O (Heavy water)	0.510	35	5820
Helium	0.425	43	94
Beryllium	0.209	86	146
Carbon	0.158	114	237
Oxygen	0.120	150	487

7.1.6.4 Slowing-down Density ($q(E)$)

The rate at which the neutrons slow down post a particular energy value *E*, per unit of the moderator is termed as the slowing down density. Due to the discontinuous nature of the slowing down process and also of the presence of the absorption resonances for neutrons of discrete energies, one does not expect the variation of slowing down density with energy to be simple. One can simplify the problem by treating the slowing down as a virtually continuous process, which is possible with sufficiently heavy moderator nuclei for which loss of average energy per collision is small.

Let the rate of neutrons production with an initial high energy (E_0) is constant and equal to Q neutrons/unit volume/sec. Let prior to attaining thermal energies, no neutrons are lost either by escaping, i.e. leakage or by absorption, then the slowing down density $q(E)$ will remain constant and will be equal to the source density (Q). This is possible for an ideal moderator as one can only reduce the neutron escaping to negligible proportions by using a moderator as one can only reduce the neutron escaping to negligible proportions by using a moderator of thick walls and also the absorption is not zero, even if the moderator is infinite in extent.

The number of neutrons/unit volume whose energies lie between E and $E + \Delta E$ is equal to $n(E)\Delta E$. Now, if dt be the time needed by the neutrons to transit through the energy interval ΔE, the number of neutrons that are within this energy interval at a given time will be represented by

$$n(E)\Delta E = -q(E)\Delta t \qquad \qquad ...\,(7.18)$$

The negative sign in (7.18) has been introduced because energy decreases with the increase of time $n(E)\Delta E$ in (7.18) represents the neutron density crossing E/collision. The number of collisions makes per second by this neutron group $= \upsilon \lambda_s = \upsilon \Sigma_s$. Thus the neutron density that crosses the energy E/sec is given by

$$q(E) = [n(E)\Delta E]\, \upsilon \Sigma_s \qquad \qquad ...\,(7.19)$$

If the energy changes in collision is small, then for such event, one can write

$$\frac{\Delta E}{E} = \Delta(\ln E) = \xi$$

or
$$\Delta E = \xi E$$

$$\therefore \qquad q(E) = n(E)E\, \xi\, \upsilon\, \Sigma_s \qquad \qquad ...\,(7.20)$$

Thus in the absence of neutron absorption, i.e. $q(E) = Q$, one finds the neutron flux per unit energy as

$$\phi(E) = \upsilon\, n(E) = \frac{Q}{\xi E \Sigma_s} \qquad \qquad ...\,(7.21)$$

(7.21) reveals that the slowing down neutron flux per unit energy is inversely perportional to the energy E. The quantity $\phi(E)\Sigma_s$ is termed as the *collision density* $F(E)$ and is expressed as

$$F(E) = \phi(E)\Sigma_s = \frac{Q}{\xi E} \qquad \qquad ...\,(7.22)$$

Relation (7.22) reveals that the scattering loss, i.e. number of neutrons scattered out of the energy interval ΔE/sec/unit volume is equal to the neutron influx gain, i.e. the number of neutrons scattered into the energy interval ΔE/sec/unit volume. $q(E)$ will bear a relation with Q in the presence of absorption.

$$q(E) = Q\, P(E) \qquad \qquad ...\,(7.23)$$

Here $P(E)$ is the probability or the fraction of unabsorbed neutrons. Now, the absorption loss,

$$\Delta q = n(E) \, \Delta E \, \upsilon \, \Sigma_a = \phi(E) \Delta E \sum_a$$

Thus in the steady state within the energy interval ΔE, we have

Scattering loss + Absorption loss = Influx

or

$$\phi(E) \sum_s \Delta E + \phi(E) \sum_a \Delta E = q(E) \frac{\Delta E}{\xi E}$$

or

$$\frac{\phi(E) \sum_s \Delta E}{\phi(E) \sum_a \Delta E + \phi(E) \sum_s \Delta E} = \frac{\Delta q \, E \, \xi}{q(E) \, \Delta E}$$

or

$$\left(\frac{\Sigma_a}{\Sigma_a + \Sigma_s} \right) \left(\frac{\Delta E}{\xi E} \right) = \frac{\Delta q}{q(E)}$$

7.1.6.5 Slowing down time

The time taken by the neutrons to slow down from an initial energy E_0 to a final energy E is termed as the slowing down time. This can be obtained from Equations (7.18) and (7.20). We have

$$\Delta t = - \frac{\Delta E}{\xi E \upsilon \Sigma_s} \qquad \qquad ...\,(7.24)$$

Let us assume that Σ_s and ξ remains constant over the slowing down range of energy. One can integrate (7.24) as

$$T \text{ (slowing down time)} = \int_0^T dT = - \frac{1}{\xi E_s} \int_{E_0}^{E_f} \frac{dE}{\upsilon E}$$

$$= \frac{\sqrt{\frac{m}{2}}}{\xi \Sigma_s} \int_{E_0}^{E_f} E^{-\frac{3}{2}} \, dE$$

$$= \frac{\sqrt{2m}}{\xi \Sigma_s} \left[\frac{1}{E_f^{\frac{1}{2}}} - \frac{1}{E_0^{\frac{1}{2}}} \right] \qquad ...\,(7.25)$$

$$= \frac{\lambda_s}{\xi} \sqrt{2m} \left[E_f^{\frac{1}{2}} - E_0^{\frac{1}{2}} \right] \qquad ...\,(7.24a)$$

7.1.6.6 Resonance Escape Probability

Let us assume that there is no absorption of neutrons in the moderator as they slow down. If however, one assumes that the absorption cross-section of the moderator for neutrons is not zero but has a little value, $q(E)$ will no longer will be equal to $\Omega(E_0)$, (the rate of production of neutrons in a nuclear reactor, with an initial high energy value E_0) but will be lesser than it and its value will depend upon the neutron energy. As the neutrons pass from energy E to $E - \Delta E$, the change in $q(E)$ due to absorption is equal to the neutron absorption/cm^3, i.e.

$$\Delta q = n(E)\, \Delta E\, \upsilon \sum_a = \phi(E)\, \Delta E \sum_a \qquad \text{... (7.26)}$$

When there is also absorption, the slowing down density $q(E)$ will be only a fraction of $\Omega(E_0)$ and therefore, one finds

$$\frac{q(E)}{\Omega(E_0)} = p(E) \qquad \text{... (7.27)}$$

Where $p(E)$ is fraction, i.e. $p(E) < 1$, i.e. $p(E)$ is a measure of the fraction of neutrons that escape absorption while slowing down from E_0 to E. $q(E)$ is called the *resonance escape probability*. In the presence of neutron absorption, one can write the steady state condition as

Influx = scattering loss of neutrons + absorption loss of neutrons

$$\therefore \qquad \frac{q(E)}{\xi E}\, \Delta E = \phi(E)\, \Sigma_s\, \Delta E + \phi(E)\, \Sigma_0\, \Delta E$$

or

$$\frac{q(E)}{\xi E}\, \Delta E = (\Sigma_s + \Sigma_s)\, \phi(E)\, \Delta E$$

But we have from Equation (7.26), $\phi(E)\, \Delta E = \dfrac{\Delta p}{\Sigma_a}$, Using this, the above reduces to

$$\frac{q(E)}{\xi E}\, \Delta E = \frac{(\Sigma_s + \Sigma_a)\Delta q}{\Sigma_a}$$

or

$$\frac{\Delta q}{q(E)} = \left(\frac{\Sigma_a}{\Sigma_s + \Sigma_a} \right) \frac{\Delta E}{E\,\xi}$$

On integration, one obtains

$$\int_{q(E)}^{q(E_0)} \frac{\Delta q}{q(E)} = \int_{E}^{E_0} \left(\frac{\Sigma_a}{\Sigma_s + \Sigma_a} \right) \frac{\Delta E}{E\,\xi}$$

or

$$q(E) = q(E_0)\, \exp\left[-\int_{E_0}^{E} \frac{\Sigma_a}{\xi(\Sigma_s + \Sigma_a)} \frac{\Delta E}{E} \right] \qquad \text{... (7.28)}$$

Since $\dfrac{\Sigma_a}{\Sigma_s + \Sigma_a}$ is simply the partial probability of neutron absorption and hence the integral in the exponent simply gives the total probability of the neutron being absorbed as it to showing down from energy E_0 to E. Let us denote this probability by P_a, then

$$P_a = \int\limits_{E}^{E_0} \frac{\Sigma_a}{\xi(\Sigma_s + \Sigma_a)} \frac{\Delta E}{E} \qquad \dots (7.29)$$

Now, Equation (7.28) can be expressed as

$$q(E) = q(E_0) \exp(-P_a) \qquad \dots (7.30)$$

Since the initial value of the slowing down density $q(E_0)$ is equal to the rate of production of fast neutrons, viz $W(E_0)$, therefore, comparing Equation (7.30) and Equation (7.27), one obtains

$$p(E) = e^{-P_a} \qquad \dots (7.31)$$

In practice, the exponential is small and therefore, to a first approximation, one can write

$$p(E) = 1 - P_a \qquad \dots (7.32)$$

Since P_a is the absorption probability and therefore, $(1 - P_a)$ is the probability that the neutron escapes absorption or resonance escape probability. This is why $p(E)$ is called as *resonance escape probability*.

The integral $\int\limits_{E}^{E_0} \frac{\Sigma_a}{\xi(\Sigma_s + \Sigma_a)} \frac{\Delta E}{E}$ is called *resonance integral* and can also be expressed as

$$P_a = \int_{E}^{E_0} \frac{\Sigma_a}{(\Sigma_s + \Sigma_a)} \frac{\Delta E}{E} = \frac{1}{\xi \Sigma_s} - \int\limits_{E}^{E_0} \frac{\Sigma_a}{1 + \dfrac{\Sigma_a}{\Sigma_s}} \frac{\Delta E}{E}$$

$$= \frac{N_0}{\xi \Sigma_s} \int\limits_{E}^{E_0} \frac{\sigma_a}{1 + \left(\dfrac{N_0 \sigma_a}{\Sigma_s}\right)} \frac{\Delta E}{E}$$

Thus, one can write resonance escape probability as

$$p(E) = \exp\left[-\frac{N_0}{\xi \Sigma_s} \int\limits_{E}^{E_0} \frac{\sigma_a}{1 + \dfrac{N_0 \sigma_a}{\Sigma_s}} \frac{\Delta E}{E} \right]$$

$$= \exp\left[-\frac{N_0}{\xi \Sigma_s} \int\limits_{E}^{E_0} (\sigma_a)_{eff} \frac{\Delta E}{E} \right] \qquad \dots (7.33)$$

Where $(\sigma_a)_{eff}$ is called effective absorption.

$$(\sigma_a)_{eff} = \frac{\sigma_a}{1 + \dfrac{N_0 \sigma_a}{\Sigma_s}} \qquad \dots (7.34)$$

and the integral $\int\limits_{E}^{E_0} (\sigma_a)_{eff} \frac{\Delta E}{E}$ is called *effective resonance integral*.

7.1.7 Energy Distribution of Thermal Neutrons

We have seen that the slowing down of neutron process is due to an elastic scattering process. Obviously, the material used as moderator for slowing down process should have a large scattering cross-section and small absorption cross-section for neutrons. As a result of slowing down process, neutrons attain the state in which their energies are in equilibrium with those of the atoms or molecules of moderator in which they are moving. The energy distribution of neutrons at this time will be approximately Maxwellain, corresponding to the temperature of the medium surrounding them. The velocity distribution of neutrons in thermal equilibrium can be expressed as

$$dn = n(v)dv = 4\pi n \left(\frac{m}{2\pi kT} \right)^{\frac{3}{2}} v^2 \exp\left[\frac{-mv^2}{2 kT} \right] dv \qquad \text{... (7.1)}$$

Where m is the mass of the neutron, n the total number of neutrons per unit volume, k the Boltzmann constant and T the absolute temperature. $dn = n(v)dv$ represents the number of neutrons whose velocity lie between v and $v + dv$. Figure 7.15 shows the velocity distribution function for neutrons, which has a maximum for $v = v_p$, called the *most probable velocity*. One can determine it by differentiating the r.h.s. with respect to v of Equation (7.1) and setting equal to zero. Thus, one

obtains $v_p = \sqrt{\dfrac{2 kT}{m}}$. The energy corresponding to v_p is given by

$$E_p = \frac{1}{2} m v_p^2 = kT \qquad \text{... (7.2)}$$

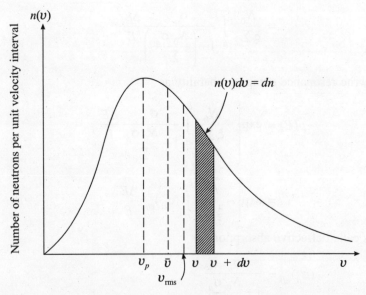

Fig. 7.15 Velocity distribution of neutrons exhibiting meximum at v_p (most probable velocity).

Now, the energy distribution of neutrons is obtained as

$$dn = n(E)dE = 2\pi n \left(\pi kT\right)^{-\frac{3}{2}} E^{\frac{1}{2}} \exp\left(-\frac{E}{kT}\right)dE \qquad \ldots (7.2)$$

Figure 7.16 shows the energy distribution of neutrons in accordance with Equation (7.2). This energy distribution leads to a most probable energy E_0. One can found E_0 by differentiating Equation (7.2) and then equating this to zero. One obtains $E_0 = \frac{1}{2} kT$, which is one half of E_p given by (7.2). Corresponding to this energy, the velocity is called as *average velocity* and is given by.

$$<v> = \bar{v} = \left(\frac{2}{\sqrt{\pi}}\right)v_p = 1.128\, v_p \qquad \ldots (7.3)$$

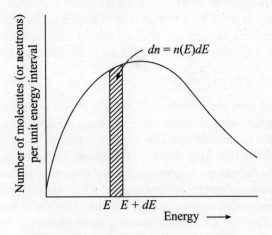

Fig. 7.16 Energy distribution of neutrons

From kinetic theory, we have the average energy ($\bar{E}$) as

$$\bar{E} = <E> = \frac{3}{2} kT = \frac{3}{2} E_0 \qquad \ldots (7.4)$$

We can see thatt $\bar{E}$ does not correspond to the average velocity but to the v_{rms} (root mean square velocity), given by

$$\bar{E} = \frac{1}{2} m\bar{v}^2 = \frac{1}{2} mv_{rms}^2$$

or
$$v_{rms} = \left(\frac{3kT}{m}\right)^{\frac{1}{2}} = \sqrt{\frac{3}{2}}\, v_p \qquad \ldots (7.5)$$

We may note that velocity distributions of thermal neutrons from different sources have been measured experimentally but it is difficult to make highly accurate measurments. However, the measured velocities do follow Maxwellian distribution within the experimental accuracy ($\sim 10\,\%$). This discrepancy seems to be due to temperature which is slighly greater than that of the material of the moderator. We see that the steady influx of high energy neutrons and at the same time a steady absorption of low energy neutrons raises the high energy and of the Maxwellian distribution and thereby depresses the low energy portion of Maxwellian distribution.

7.1.8 Continuous Slowing Down of Neutrons

In the studies of nuclear reactors, the process of slowing down of neutrons has a unique importance, e.g. the average (net vector) distance between the point at which a neutron is produced and at which it becomes thermal due to the slowing down process determines the leakage of neutrons in a finite medium. In case if this distance is large, the reactor size will be large and thus the leakage could be reduced prior the neutrons attain thermal energies. Clearly, one must know the spatial distribution of neutrons of different energies during the process of their slowing down.

Let us consider a neutron produced from a certain source or fission. This neutron will travel a certain source or fission. This neutron will travel a certain time with this energy prior to making a scattering collision with a nucleus. Now, if $\lambda(E)$ is the mean free path of collision and $v(E)$ is the velocity, then the time between two successive collisions $= \dfrac{\lambda}{v}$. The energy of the neutron after collision is decreased and then the neutron will diffuse at this constant lower energy = until it makes a collision with another nucleus of the material. Since at each collision the energy of neutron is decreased with which the neutron moves. Obviously, the time between collisions, on the average increases with decreasing energy. This process of alternate diffusion at constant energy with a collision in between when the energy is decreased, i.e. the neutron is slowed down, continues till the energy of neutron equals the energy of surrounding. This means that neutron gets thermalised. At each collision, the logarithm of neutron energy is reduced on the average by a constant amount ξ, which is independent of energy. Clearly, the slowing down process of neutron can be represented by a series of discontinuous steps of the same vertical height (ξ) in a logarithmic energy (Figure 7.17). The horizontal lines represent the time intervals between successive collisions and which increases with time, i.e. with decreasing energy. These horizontal lines also represent the constant energy at which the neutron diffuses between collisions. The plot is for a single neutron. There may be some variation for different neutrons. However, the deviations from the average are not great, specially for the heavier nuclei e.g. carbon, graphite, berrilium. Further, under these conditions, one may replace the series of steps by a continuous curve as shown in Figure 7.17. This clearly implies that a neutron slows down continuosuly in time.

We may note that the continuous slowing down model is not valid for hydrogen, deuterium or for their compounds, e.g. H_2O, D_2O since neutron may lose all its energy in a single collision, with a hydrogen nucleus. Moreover, the number of collisions required to reduce the neutron energy to thermal energy value is relatively small and therefore continuous approximation cannot hold at all in the case of hydrogen, deutron and their compounds.

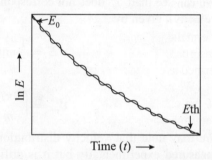

Fig. 7.27 ln E vs time plot. Continuous slowing down model.

Fermi Age Equation

Let us assume that the slowing down of neutron is taking place in a non-absorbing medium. Let the energy of neutron be E_0 at birth (due to fission process). Due to collision the energy of neutron reduced to E(velocity v). If λ_s be the mean free path for scattering collisions, then the number of scattering collisions in time dt with velocity v is $\dfrac{v\,dt}{\lambda_s}$, $v\,dt$ being the distance travelled by neutron in time dt. Since ξ is the mean log decrement at each collision, the decrease in the logarithm of neutron energy ($\ln E$) in time dt is given by

$$- d \ln(E) = -\frac{dE}{E} = \xi \frac{v}{\lambda_s} dt \qquad \qquad \text{... (7.1)}$$

or

$$\frac{dE}{E} = -\xi \frac{v E}{\lambda_s} \qquad \qquad \text{... (7.1a)}$$

The logarithmic energy decrement is called as *lethargy* and denoted by u,

$$u = \ln \left(\frac{E_0}{E} \right) \qquad \qquad \text{... (7.2)}$$

Where E_0 is an arbitrary energy corresponding to zero lethargy. With the decrease of energy of neutron, lethargy increases. From Equation (7.2), we have

$$du = -d(\ln E)$$

$$= \xi \frac{v}{\lambda_s} dt \qquad \qquad \text{... (7.3)}$$

Now, let $n(r, t)$ be the neutron density or number of neutrons per unit volume at a point r and at time t. Clearly, the number of neutron leaving a unit volume in unit time $= \dfrac{dn}{dt}$.

Apart from the slowing down of the neutrons in moderator medium, neutrons also diffuse through the medium due to scattering collisions. There is also the possibility of the capture of the neutron by the nuclei of the medium. Once the neutron is produced by the source, e.g. by fission, neutrons start slowing down in a moderator material medium. Neutron, therefore takes certain average time known as *slowing down time* to slow down to thermal energies and on attaining the thermal equilibrium with the moderator (material medium) the thermal neutrons diffuse prior being absorbed. The average time for which the thermal neutrons diffuse prior absorption/leakage is termed as *diffusion time*. Let us first consider the problem of diffusion of mono-energetic neutrons.

We may note that the neutron diffusion is somewhat simpler than the diffusion of gas, because of the extremely low density of neutrons in the medium and moreover the collisions between neutrons are rare. Obviously, we have to consider only collisions of neutrons with the nuclei which we may regard to be stationary. Clearly, the phenomenon of diffraction of neutron is quite similar to the electron diffraction in the metallic lattice.

Neutron diffuses at a constant energy in between scattering collisions. Figure 7.18 shows a typical neutron trajectory as a zig-zag path. We can see that the path elements being of varying lenghts, pointing in different directions with different speeds. These are the scattering free paths and the average of these paths is the mean free path (λ_s). After the collision, the direction of the neutron is not known exactly and governed by the laws of probability. We may remember that slowing down process is statistical in nature.

However, when we consider that there are large number of neutrons, there is a net monement of the neutrons from the regions of higher concentration to lower concentration. The rate at which this motion of neutrons takes place is known as *diffusion*. Several results of kinetic theory of gases can also be applied to neutron diffraction.

Let us assume that the absorption of neutrons in the medium is small, then one can apply the well knwon law, i.e. *Fick's law of diffusion to mono-energetic neutrons* which do not lose any energy by scattering collision. In the present case, Fick's law states that there is a net number of neutrons flowing in unit time through a unit area held normal to the direction of flow of the neutrons. We have

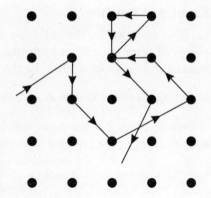

Fig. 7.18   Elastic collisions of neutrons with the nuclei in a solid medium exhibiting zig-zig trajectory.

$$\mathbf{J} = -D_0 \, \mathrm{grad} \, n(\mathbf{r}, t) = -D_0 \, \nabla n \qquad \qquad \ldots (7.4)$$

Where $\mathbf{J}$ is the neutron current density vector, representing the net number of neutrons flowing in a given direction in unit time through a unit area held normal to the direction of flow, n is the neutron density and D_0 is called diffusion constant having dimensions of $cm^2 \, sec^{-1}$.

We may note that above assumptions are applicable with a good degree of approximation, to neutrons in thermal equilibrium with the atoms in a weakly absorbing medium. Moreover, in these calculations, one must use the properly averaged values of the cross section.

We now consider the flow of neutrons through a rectangular volume element dV having dimensions dx, dy, and dz located at a point (x, y, z) as shown in Figure 7.19. We can see that the number of neutrons moving along the Z-direction and entering the lower face $\dfrac{dx \, dy}{sec}$ is $J_z \, dx \, dy$,

where J_z is the net current along the Z-direction. Now, the number leaving the upper face per sec is $J_{z+dz} \, dx \, dy$. The net rate of flow of neutrons out of the given volume dV, through the faces parallel to the x, y plane can be obtained from Equation (7.4) as

$$(J_{z+dz} - J_z)dx \, dy = -D_0 \left[\left(\frac{\partial n}{\partial z} \right)_{z+dz} - \left(\frac{\partial n}{\partial z} \right)_z \right] dx \, dy$$

$$= -D_0 \frac{\partial^2 n}{\partial z^2} \, dx \, dy \, dz = -D_0 \frac{\partial^2 n}{\partial z^2} \, dV \qquad \qquad \ldots (7.5)$$

Similarly, one can find the net rate of loss of neutrons from other two faces parallel to (y, z) and (x, z) planes as $-D_0^2 \dfrac{\partial^2 n}{\partial x^2} \, dV$ and $-D_0^2 \dfrac{\partial^2 n}{\partial y^2} \, dV$.

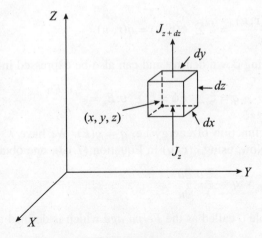

Fig. 7.19 Neutron leakage calculation

Thus the total number of neutrons leaving a unit volume/sec is obtained as

$$-D_0 \left[\frac{\partial^2 n}{\partial x^2} + \frac{\partial^2 n}{\partial y^2} + \frac{\partial^2 n}{\partial z^2} \right] = -D_0 \, \nabla^2 n(\mathbf{r}, t)$$

By definition, Equation (7.6) is equal to $\dfrac{dn}{dt}$, i.e.

$$\frac{dn}{dt} = D_0 \, \nabla^2 n(\mathbf{r}, t) \qquad \qquad \text{... (7.7)}$$

Equation (7.7) is a *neutron balance equation* in a non-absorbing medium far away from the neutron source. Equation (7.7) is also known as *diffusion equation*. Once the neutrons are produced then after leaving the source, the number of neutrons which diffuse between the time interval t and $t + dt$ at $\mathbf{r}$ in a unit volume is $n(\mathbf{r}, t)dt$ (by definition). If 'u' is lithargy, then one can transform the variable 't' to the variable 'u' such that the number of variables within the interval u and $u + du$ (lithargy) is given by

$$n(\mathbf{r}, u)du = n(\mathbf{r}, t)dt \qquad \qquad \text{... (7.8)}$$

Using Equation (7.3), one obtains

$$n(\mathbf{r}, u) = n(\mathbf{r}, t) \, \frac{\lambda_s}{\xi \upsilon} \qquad \qquad \text{... (7.9)}$$

Now, expressing (7.7) in terms of u, one obtains

$$D_0 \, \nabla^2 n(\mathbf{r}, u) \frac{\xi \upsilon}{\lambda_s} = \frac{dn}{du} \frac{du}{dt}$$

$$= \frac{du}{dt} \frac{\partial}{\partial u} \left[\frac{\xi \upsilon}{\lambda_s} \, n(\mathbf{r}, u) \right] \qquad \qquad \text{... (7.10)}$$

Now, $n(\mathbf{r}, u) = \phi(\mathbf{r}, u)$, where $\phi(\mathbf{r}, u)$ is the neutron flux at r per unit lithargy interval. Using this, one obtains

$$\frac{\xi \phi(\mathbf{r}, u)}{\lambda_s} = \xi \sum_s \phi(\mathbf{r}, u) = q(\mathbf{r}, u) \qquad \ldots (7.11)$$

Where $q(\mathbf{r}, u)$ is the slowing down density and can also be expressed in terms of n through

$$q = \xi \sum_s E\phi = \xi \sum_s \upsilon \, E_n = \frac{\xi \upsilon \, En}{\lambda_s} \qquad \ldots (7.11a)$$

We may note that q is a function of energy i.e. $q = q(E)$. We have $\lambda_s^{-1} = \sum_s$, the macroscopic scattering cross section. Now, using $q(\mathbf{r}, u)$ in Equation (7.10), one obtains

$$D_0 \, \nabla^2 q = \frac{\xi \upsilon}{\lambda_s} \frac{\partial q}{\partial u} \qquad \ldots (7.12)$$

Now, introducing a variable t, called as the *Fermi age* which is defined as

$$\tau = \int_0^t D_0 \, dt = \int_0^u \frac{D_0 \lambda_s}{\xi \upsilon} \, du \qquad \ldots (7.13)$$

Where we have used Equation (7.3) for dt. Now,

$$\frac{\partial q}{\partial u} = \frac{\partial q}{\partial \tau} \frac{\partial \tau}{\partial u} = \frac{\partial q}{\partial \tau} \frac{D_0 \lambda_s}{\xi \upsilon} \qquad \ldots (7.14)$$

Now, Equation (7.12) reduces to

$$\nabla^2 q = \frac{\partial q}{\partial \tau}$$

or
$$\nabla^2 q - \frac{\partial q}{\partial \tau} = 0 \qquad \ldots (7.15)$$

This equation is known as *Fermi age equation*. The unit of t is the square of length and obviously its dimensions is (meter)^2. We may note that τ does not in any way represent the elapsed time but it is related to the chronological age of the neutrons, i.e. the time elapsing between the production of a neutron (of energy E_0) and its slowing down to the energy E. Equation (7.15) reveals that as E decreases, τ increases, as it should be. For $E = E_0$, $\tau = 0$.

Solution of Age Equation

We can see that Fermi age equation provides a complete description of neutron in space (r) and energy (lithargy). We can further see that age equation has a form similar to the heat conduction equation in a continuous medium, under non-stationary condition. We are familiar with the solution of heat conduction equation for different forms. Obviously, we can directly borrow from these known solutions and write the solutions of age equation for an isotropic point source emitting Q neutrons/sec of energy E_0 in an infinite homogeneous non-absorbing medium as

$$q(\mathbf{r}, t) = \frac{Q}{(4\pi \, \tau)^{\frac{3}{2}}} \exp\left(-\frac{r^2}{4\tau}\right) \qquad \ldots (7.16)$$

Where $q(\mathbf{r}, t)$ is slowing down density for neutrons of age τ at a distance r from the point source. Obviously, the distribution of slowing down density is thus Gaussian. Figure 7.20 shows the shape of the curve for $q(\mathbf{r}, \tau)$ as a function of r for two different values of age τ. We see that for smaller values of τ the curve has a sharper peak and falls off rapidly to low values as r increases and for τ larger, the peak is broader and smaller, falling off slowly with the increase of r. One can easily understand these features from physical considerations. For τ small, the energy E to which the neutrons are slowed down is relatively large and therefore they would not have sufficient time to diffuse too far away from the source and hence they are confined to the average distances closer to the latter. On the other hand for τ large, the energy E decrease to a much smaller value which can happen only if sufficient time has lapsed for the neutrons to a larger number of collisions, i.e. neutrons diffuse farther out. Thus the curve should be low and broad.

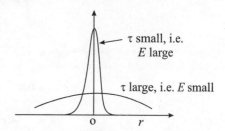

Fig. 7.20 Neutron slowing down density for a point source for two different values of τ.

Slowing Down Length

One can define the width of the Gaussian curve as the value of r at which the slowing down density $q(\mathbf{r}, \tau)$ become $\dfrac{1}{e}$ times of its maximum value. It is expressed as

$$r_0 = 2\sqrt{\tau} \qquad\qquad \ldots (7.17)$$

One can calculate from Equation (7.16) the mean distance from the source of the neutrons at which neutrons get thermalised and for $r_0 = r_{th}$ and $\tau = \tau_{th}$, this distance is given by Equation (7.17). We may note that beyond this distance r_{th} from the source, neutrons diffuse without further losing any energy.

One can also find the another important quantity, i.e. mean square distance $<r^2>$ travelled by the neutrons to attain a given value of τ, $<r^2>$ is also called as the mean square (net vector) slowing down distance about a point source. We have

$$<r^2> = \frac{\displaystyle\int_0^\infty r^2 q(\mathbf{r}, \tau)\, 4\pi\, r^2\, dr}{\displaystyle\int_0^\infty q(\mathbf{r}, \tau)\, 4\pi\, r^2\, dr} \qquad\qquad \ldots (7.18)$$

Using (7.16), (7.18) becomes

$$<r^2> = \frac{\displaystyle\int_0^\infty r^4 Q\, e^{-\frac{r^2}{4\tau}}\, dr}{\displaystyle\int_0^\infty r^2 Q\, e^{-\frac{r^2}{4\tau}}\, dr}$$

$$= \frac{\displaystyle\int_0^\infty r^4 e^{-\frac{r^2}{4\tau}}\,dr}{\displaystyle\int_0^\infty r^2 e^{-\frac{r^2}{4\tau}}\,dr}$$

Using
$$x^{2n} e^{-\alpha x^2}dx = \frac{1.3.5\,......\,(2n-1)}{2^{n+1}\,a^n}\sqrt{\frac{\pi}{a}}\,, \text{ the above yields}$$

$$<r^2> = 6\,\tau$$

or
$$\tau = \frac{1}{6}\,<r^2> \qquad\qquad ...\,(7.19)$$

Clearly, for a given energy the age τ is one sixth of the mean square distance ($<r^2>$) travelled by neutrons to attain that energy, distance between the source of neutrons where $\tau = 0$ to the point at which its Fermi age is τ. The square root or τ or τ_{th} gives the *slowing down length* of thermal neutrons. We may note that $\sqrt{\tau_{th}}$ determines the critical size of a nuclear reactor and measures the extent to which the fast neutrons leak out during slowing down to thermal energy.

Experimental Determination of Fermi Age (τ)

One can measure τ by activation method. The detector by which τ is to be measured must have high absorption for neutrons of that particular energy. If one chooses a material which has an absorption resonance at some specific energy, e.g. the indium resonance at 1.44 eV, then by placing foils of known thickness of the material at different distance from the neutron source, one may determine the variation of the slowing down density q with the distance r from the measured activities in different foils. Usually, the indium foils are covered with cadmium so that the effect of thermal neutrons may be elimanated. Thus, by measuring the saturated activities in the foil at different places, one can know the values of q as a function of r. Equation (7.16) yield

$$\ln q = \text{Constant} - \frac{r^2}{4\tau} \qquad\qquad ...\,(7.20)$$

The plot of $\ln(q)$ as a function of r^2 give a straight line from which one can calculate the age $\tau(E_r)$ at the indium resonance energy 1.44 eV. Since the indium resonance energy (1.44 eV) is some what above the thermal energy (0.025 eV), a correction is to be applied to $\tau(E_r)$ to get the value $\tau(E_{th})$. This correction (i.e. estimation of τ from 1.44 eV to 0.025 eV has to be obtained theoretically. Table 7.7 provides the values of τ for thermal neutrons in various moderators, i.e. materials. Since

Table 7.7 Fermi age (τ) and slowing down length ($\sqrt{\tau}$) for moderators

Moderator material	Fermi age (t) (cm²)	Slowing down length ($\sqrt{\tau}$) (cm)
H_2O (water)	33	5.7
D_2O (Heavy water)	120	11.0
Be (Beryllium)	98	9.9
Graphite	350	18.7

Fermi age has been defined in terms of the mean square distance travelled by the neutron while slowing down (Equation (7.19)), and hence $\sqrt{\tau}$ is slowing down length.

We may note that although one can not strictly define Fermi age (τ) for hydrogen of deuterium like moderators for neutrons, but still Equation (7.19) is applicable to theses substances.

Slowing Down Time (τ_s)

Consider that neutrons are slowing down in a non-absorbing material. Let their velocity after time t be v. Now, the average number of collisions in a time interval $dt = \dfrac{v\,dt}{\lambda_s}$. We have the average decrease in the logarithim of energy, i.e. $-d(\ln E)$, in time dt is given by (7.1). We have

$$-d(\ln E) = \frac{dE}{E} = \frac{v\xi}{\lambda_s}\,dt$$

or

$$dt = -\frac{\lambda_s}{\xi}\sqrt{\frac{m}{2}}\,\frac{dE}{E^{\frac{3}{2}}} \qquad \ldots (7.21)$$

$\therefore$ Slowing down time t_s is given by

$$t_s = \int_0^t dt = -\frac{\lambda_s}{\xi}\sqrt{\frac{m}{2}}\int_{E_0}^{E_{th}} E^{-\frac{3}{2}}\,dE$$

Here E_0 is the average energy of the neutrons at the souce and Eth is the average energy of the thermal neutrons. We have

$$t_s = \sqrt{2m}\,\frac{\lambda_s}{\xi}\left(\frac{1}{\sqrt{E_{th}}} - \frac{1}{\sqrt{E_0}}\right)$$

$$= \sqrt{2m}\,\frac{\lambda_s}{\xi}\left(E_{th}^{-\frac{1}{2}} - E_{th}^{-\frac{1}{2}}\right) \qquad \ldots (7.22)$$

For few moderating materials the values of slowing down times with $E_0 = 2$ MeV to $E_{th} = 0.25$ eV are given in Table 7.8.

Table 7.8 Slowing down time (t_s) and Diffusion time (t_d) for certain moderators (materials)

Moderator (material)	s (cm)	Slowing down time (t_s) (sec)	Average diffusion time (t_d) (sec)
H_2O (water)	1.1	10^{-5}	2.1×10^{-4}
D_2O (Heavy water)	2.6	4.6×10^{-5}	15×10^{-2}
Be (Beryllium)	1.6	6.7×10^{-5}	4.3×10^{-3}
Graphite	2.6	1.5×10^{-4}	1.2×10^{-2}

For thermal neutrons, the average diffusion time is given by

$$t_d = \frac{\lambda_a}{\upsilon_{th}} = \frac{1}{\upsilon \sum_a} \qquad \qquad \dots (7.23)$$

Where λ_a is the absorption mean free path. At ordinary temperature the υ_{th} (mean velocity) $= 2.2 \times 10^5$ cm/s.

7.1.9 Neutron Monochromators (Velocity Selectors)

We have read that neutron reaction cross-sections frequently show rapid variations with neutron energies and therefore it becomes essential to have a knowledge of energy of the neutrons involved in an interaction so that one can obtain an accurate knowledge about the energy dependence of neutron cross-section. Such a knowledge is of great importance for two basic reasons: (i) this helps in revealing the various energy levels of the compound nuclei, in terms of which one can investigate theoretically the nuclear reactions, and (ii) using this information one can get vital information useful for the design of nuclear reactors.

Unfortunately, many of neutron sources e.g. reactors, do not produce monoenergetic neutrons. Even accelertors do not provide mono-energetic neutrons for the accurate analysis of cross-section. A great deal of efforts has benn made into the fabrication of sorting out mono-energetic neutrons from a beam of hetro-energetic neutrons. A number of special devices have been designed for use in the intermediate energy range.

Earlier slow neutrons have been sorted out by introducing absorbers into a beam of hetero-energetic neutrons to remove fast neutrons. However, this technique was not much successful chiefly because most of the substances were found to possess slow neutron attenuation coefficients as large as those for fast neutrons. There was another approach which used neutron absorbers that obey υ^{-1} law. However, neutron monochromator (velocity selector) helped to select neutron beams of definite energies. Two basic factors which determine the usefulness of a neutron monochromator can be operated, and (ii) the resolution which determines the energy spread of the beam at a given energy. We now consider some of these monochromators.

(i) Crystal Monochromator

This is based on the quantum mechanical wave properties of neutron. The de Broglie wavelength of a slow neutron of kinetic energy E is given by

$$\lambda = \frac{h}{M_n \upsilon} = \frac{h}{\sqrt{2ME}} \qquad \qquad \dots (7.1)$$

Where M_n is the mass of neutron, $h = 6.62 \times 10^{-34}$ J–s is Planck's constant. For thermal neutrons of mean, velocity $= 2.2 \times 10^3$ m/s, one obtains de Broglie wave length λ (using 1), $= 1.81 \times 10^{-10}$ m $= 1.81$ A. We can see that this wave length is of the same order as the X-ray wavelength and also equal to the distance between crystal lattice planes. This holds good for neutrons with energy range 0.01 eV to about 10 eV. This reveals that when a beam of slow neutrons of wavelength λ is reflected from the surface of a crystal, the diffraction maxima are observed at angles given by the Bragg's condition

$$2d \sin \theta = n\lambda \qquad \qquad \dots (7.2)$$

Where n (=1, 2, 3, ... etc.) is an integer and represents the order of reflection and d is the separation between the adjacent lattice planes, θ is the glancing angle of the incident neutrons for n_{th} order reflection. Now, if a beam of neutrons of heterogeneous energies is allowed to fall on a set of parallel crystal planes at angle θ, the beam reflected at the same glancing angle will mostly consist of mono-energetic neutrons of velocity given by

$$v = \frac{h}{M_n \lambda} = \frac{nh}{2M_n d \sin \theta} \qquad \dots (7.3)$$

and energy

$$E = \frac{n^2 h^2}{8 M_n d^2 \sin^2 \theta} \qquad \dots (7.4)$$

Obviously, the crystal can be used as a monochromators of slow neutrons. By varying angle θ, one can vary the velocity of the neutrons. The experimental set up for crystal spectrometer (velocity selector) is as shown in Figure 7.21.

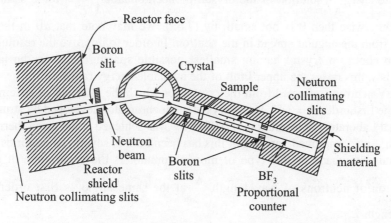

Fig. 7.21 Crystal monochromator (velocity selector) for neutrons

A well collimated beam of thermal neutrons from an intense source, e.g. fission reactor, is diffracted by a prominent set of lattice planes of a well shielded large plane crystal C. The crystals generally used are LiF, NaCl, Be and quartz. Prior to entering a well shielded BF_3 cylindrical proportional counter, the diffracted neutrons beam enter a second long collimator. Although neutrons of various speeds are present in the beam but for a particular value of θ, the great majority of neutrons reaching the detector will have a velocity given by Equation (7.3). In case if the glancing angle θ is changed, then neutrons with different velocity will now reach the detector. However, with increasing velocity, the glancing angle diminishes and ultimately the angle becomes too small to be measured. Obviously, this sets an upper limit to the speed of the neutron which can be studied in the crystal monochromator.

The energy resolving power $\left(\dfrac{dE}{E} \right)$ of the crystal monochromator can be obtained as follows:

Differentiating Equation (7.2), one obtains

$$n\, d\lambda = 2d \cos \theta \; d\theta$$

or
$$\frac{d\lambda}{\lambda} = \cot \theta \, d\theta \qquad \qquad \qquad \text{... (7.5)}$$

But
$$E = \frac{1}{2} mv^2 \quad \therefore \quad \frac{dE}{E} = 2 \frac{dv}{v} \qquad \qquad \text{... (7.6)}$$

But
$$\frac{dv}{v} = \frac{d\lambda}{\lambda}$$

$$\therefore \qquad \frac{dE}{E} = 2 \frac{dv}{v} = 2 \frac{d\lambda}{\lambda} = 2 \cot \theta \, d\theta \ (\theta \text{ small}) \qquad \text{... (7.6a)}$$

$$\therefore \qquad \frac{dE}{E} = 2 \frac{d\theta}{\theta} \qquad \qquad \qquad \text{... (7.7)}$$

Thus the energy resolution of the crystal monochromator $\left(\dfrac{dE}{E}\right)$ for θ small is given by

(7.7), otherwise then θ is not small, by (7.6a). We may note that $d\theta$ in these equations results from the angular spread in the neutron. In order to improve the resolution, one will have to choose a crystal having smallest possible spacing between the lattice planes. Obviously, this raises the upper limit of the available energy range.

For 1 eV neutrons reflected from the crystal for which the inter planar spacing is $d = 2.32$ A° in the first order, $\theta = 3.5°$. In the crystal monochromator developed by strum at Argonne National Laboratory, a resolution of 0.0015 eV was achieved at the mean thermal energy of neutron. Later, improved versions of this has been built in several laboratories of the world. The main drawback of this type of monochromator is the occurrence of second order reflection of neutrons of wavelength $\dfrac{\lambda}{2}$ at the same angle as first order neutrons of wavelength λ.

(ii) Mechanical Monochromators (velocity selector)

Looking to the range of velocities of slow neutrons, it is possible to use a mechanical device which may allow neutrons of velocities or energies lying in a narrow range to be transmitted through it. Based on this principle, Dunning et al. in 1935 designed the first slow neutron monochromator. This was based on the principle of Fizeau's toothed-wheel appratus for the measurment of velocity of light. In their deivce, two cadmium discs were mounted at the two ends of a shaft which could be rotated at high speeds (Figure 7.22). Cadmium is a strong absorber of neutrons with energies less than 0.3 eV, while other metals such as aluminium are weak absorbers of neutrons in this energy range. Both the cadmium discs had a series of uniformly spaced radial slits, which could be so aligned that the corresponding slits in the two discs faced one another of could be displaced with respect to each other. A high

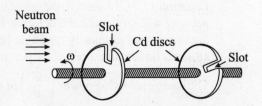

Fig. 7.22 Showing the principle of mechanical monochromator.

flux neutron beam of hetero energetic neutrons is interrupted by a rapidly rotating shutter which is normally opaque to the neutron beam. This converts the neutron beam into a series of short bursts at a fixed pulse repetition rate. One could displace the discs by an arbitrary angle with respect to each other. This shaft was rotated at large speed. With the discs so misaligned (having offset radial transmission slots), thermal neutrons moving along the shaft and passed through the slots in the first case, would normally be absorbed in the second case until of course the shaft is rotating at a speed such that neutrons of a particular pre selected velocity whose time of flight between the discs matched exactly with the time required for the second cadmium disc to rotate through the angle of misalignment. When this condition of matching is fulfilled, the neutrons of that particular velocity, i.e. energy, also pass through the second cadmium disc. Obviously, by varying either the speed of rotation of the shaft or the angular misalignment of the disc slots, neutrons of preselected velocity may be made to clear the second disc too. Then the neutron enter the detector (usually a BF_3 counter).

The two cadmium discs were misaligned to start with, so that neutrons could come out to reach the detector when the shaft was stationary. However, when the shaft was rotated at a high speed, the neutrons of a particular velocity v could come out through the slits of second cadmium disc, provided the shaft through an angle of misalignment during the time these neutrons travelled from the first to the second cadmium disc parallel to the shaft. Let the time taken by the neutron to trave from one disc to the other was $t = \dfrac{l}{v}$, where l is the distance between the two discs. Time t could be measured from the known angular speed of rotation ω of the shaft and the angle θ of misalinement, i.e. $t = \dfrac{\theta}{\omega}$. One could select the neutrons of different velocities by changing the speed of rotation of the shaft.

Although the resolution of the said initial device was not good, but it was possible to verify v^{-1} dependence of thermal neutron cross-section.

Fermi et al. in 1947 improved the initial device. They placed BF_3 neutron counter at 1.46 m beyond the rotating cylinder. Figure 7.23 shows the arrangement. They used the first nuclear reactor (pile) built by Fermi as the source of thermal neutrons.

In this device a beam of thermal neutrons from the thermal column was allowed to fall on a cylinder with axis perpendicular to the beam direction. The cylinder consists of alternate thin layers of aluminium and cadmium parallel to the length of the cylinder. Cd is opaque, whereas Al is transparent to the slow neutrons. Obviously, the cylinder thus acted as a shutter for neutron beam.

In this device neutrons could pass through the cylinder only when the layers were parallel to the direction of the neutron beam within $\pm\, 3°$ (with shutter open). When even the shutter was in open position brusts of neutrons were transmitted due to rotation of the cylinder at high speed about 15000 rpm, twice in every revolution. Number of BF_3 neutron counters placed 1.46 m beyond the rotating cylinder detected the neutrons. A mirror attached to the cylinder rotated from the mirror fell on a photocell (Figure 7.23(b)), once during each revolution of the cylinder. The time at which this happened could be adjusted by adjusting the position of the photocell. The adjustment of photocell with BF_3 counter was made in such a way that the latter incident on the photo cell. Clearly, by adjusting the position of photocell, one could vary the time t between the transmission of the neutrons through the cylinder and also the detection. Thus the neutrons detected are those with a speed $v = \dfrac{l}{t}$, where l is the distance between cylinder and counter.

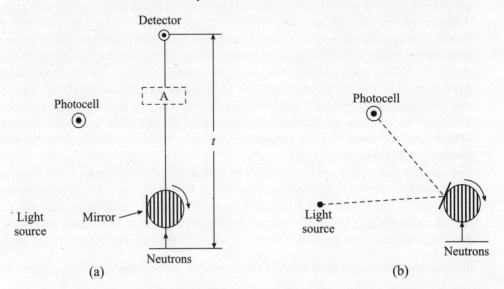

Fig. 7.23 Mechanical velocity selector (chopper) for thermal neutrons due to Fermi et al:
(a) Slow neutron chopper as a shutter, and (b) correct position of mirror.

Cadmium does not absorb the neutrons above about 0.3 eV energy and thus the velocity selector worked well for energy of neutrons, $E_n <$ 0.3 eV. A resolution of 0.005 eV was achieved at $E_n =$ 0.025 eV in an improved version of this mechanical monochromator.

Fast Chopper   These are more complicated and also difficult to operator. Selove in 1949 developed a fast chopper for neutron energies above the cadmium cut off. Selove used a long steel cylinder with thin channels cut into the cylinder parallel to the neutron beam direction as neutron shutter. The cylinder could be rotated about its axis. There was another stationary cylinder having identical channels placed in line with it. Neutrons could pass through both the sets, when the two sets of channels were aligned. However, if they were misaligned, steel thickness was sufficient to prevent neutrons of all energies from reaching the detectors.

In principle, the neutron detection and timing systems were same as those of the velocity selector of Fermi et al. In actual operation of the system, the flight path of neutrons was about 10 m and the brust duration $\Delta t =$ 6 ms and instrument has a fairly high resolution about 0.001 eV for 0.0025 eV neutrons, 0.1 eV for 1 eV neutrons and about 1.7 eV for 10 eV neutrons. It was possible for one to make the emergent neutron beam from the chopper quite narrow as the channels were of very small cross-sectional area. This means that one could use small amounts of materials in the samples for the transmission experiments. Really, this was advantageous when working with separated isotopes.

(iii) Time of Flight Velocity Selector

There is a necessity for measuring cross-sections at neutron energies greater than those covered by the crystal spectrometer and the slow chopper. One can use the time of flight method for measuring the neutron velocity with a cyclotron or linear accelerator which produces a neutron beam.

The operation of this velocity selector is based on the possibility of modulating an ion source or a beam of ions in an accelerator in such a way that the ions fall in short brusts, lasting only a few microseconds. Moreover, these bursts of ions produce neutrons through neuclear reaction and thus the neutrons are also produced in bursts. For the purpose of production of neutrons a linear accelerator of cyclotron is used.

Alvarez in 1938 first produced neutron bursts with the help of this type of velocity selector. He modulated the voltage on the 'dees'. In a latter improvement of the instrument, the arc source itself of the cyclotron was modulated. The focussing potential is modulated in some instruments so that ion bursts may be produced.

The neutron detector is placed at a distance l from the source and is also modulated. Thus it can detect the neutrons entering it only for a short interval separated from the period of neutron production by a variable time t.

Figure 7.24 shows the form of the voltage pulse which modulates the accelerator, i.e. the ion source, focussing voltage of linear accelerator or the dee voltage of the cyclotron. Pulses are usually in the microsecond range. Typical repetition rates are $\sim \dfrac{1000}{s}$.

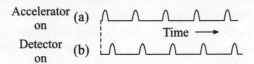

Fig. 7.24 Schematic illustration of principle of time of flight velocity selector.

Relatively high energy neutrons produced by the accelerated charged particles are slowed down by a suitable moderator, e.g. a slab of paraffin for experiments with slow neutrons. Inspite of this, neutrons of different velocities take different times for being slowed down. Thus, this sets a limit on the shortest time interval that can be used. As the slow neutrons produced in the moderator have a spread in velocities, the faster neutrons lead the slower neutrons as they travel towards the detector. Obviously, the farther away the detector from the source, the greater will be the difference in the times of arrival of the neutrons of different speeds at the detector. We have the energy of the neutrons reaching at the detector at the time t after their production is as

$$E_n = \frac{1}{2} M_n v^2 = \frac{1}{2} M_n \left(\frac{l}{t}\right)^2 \qquad \ldots (7.8)$$

Usually the distance between the moderating slab and the detector, l is usually kept as 6 to 10 m. The time required for the neutrons to travel this distance l is electronically measured. We may note that with increasing neutron energy, the resolution of the instrument decreases. However, it is somewhat better than that for fast chopper. At higher energies, i.e. $E_n > 100$

to 1000 eV, $\Delta E \sim E_n$. The resolution of the instrument is affected due to: (i) width of the voltage pulses used for modulating the source and detector, (ii) the time required for slowing down of the neutrons, (iii) variable delay between the capture of the neutron in the detector (say BF_3 counter) and the resulting neutron pulse, and (iv) detector's finite length.

By using microwave electron linear accelerators ($E_e > 20$ MeV) great improvements in the time of flight velocity selector have been achieved to product the neutron bursts having greater intensity than those obtained from cyclotrons. Moreover, the useable energy range has been pushed upto the intermediate energy region by increasing the flight path of neutron. One can also use scintillation counters in place of BF_3 counters. One can obtain the best timing resolution with plastic scintillators with nanosecond (10^{-9} s) and therefore give the best energy resolution. Organic scintillators with pulse shape discrimination against γ-rays are also often used at neutron energy above ~ 100 keV. One can use ^6Li loaded glass scintillators below 100 keV neutrons energies.

7.2 NUCLEAR FISSION

The method of releasing nuclear energy by breaking up a heavy nucleus commonly known as fission is the basis for the operation of nuclear power reactors and atom bomb.

We know that in the course of nuclear reactions, certain heavy nuclei disintegrate into two or more fragments and a few neutron. Such a disintegration process is named as fission. Fission was discovered accidently in the search for transuranic elements by neutron bombardment of natural uranium. In 1939, Hahn and Strasseman concluded from a chemical analysis of the products of such an irradiation that barium isotope ($A \approx 140$) had been produced in the $U + n$ reaction.

In 1938 Meitner and Frisch accepted these conclusions and proposed that the uranium nucleus, on absorption of a slow neutron, could assume a shape sufficiently deformed to promote fission into two approximately equal masses, e.g.

$$_{92}U + {}_0^1n \rightarrow ({}_{92}U^*) \rightarrow {}_{56}Ba + {}_{36}Kr + Q$$

Or, to many other similar pairs each with an appropriate energy release Q. They explained this phenomenon in terms of the liquid-drop model of the nucleus. They pointed out that heavy nuclei only required a little energy to become very deformed and unstable, breaking up into two nuclei of roughly equal size, similar to the division of a droplet into two, and suggested this energy was supplied in the capture of the neutron by the uranium (U) nucleus. This form of fission is called neutron-induced fission. From the semi-empirical mass formula the two product nuclei, or *fission fragments,* would each have an energy of about 75 MeV and therefore an estimated range of about 30 mm of air; they would recoil after the actual fission event (or *scission* of the intermediate nucleus U^*) in opposite directions. Their energy is converted into available heat as they slow down in matter with a yield of 1 J for 3.2×10^{10} fission events.

Spontaneous fission, which is the fission of the nucleus in its ground state, also occurs and was first observed for $_{94}^{240}Pu$ in 1940. Spontaneous fission into equal mass fragments is energetically possible for nuclei with $A \geq 90$ (for $\dfrac{A}{Z} = 2.3$). The reason that spontaneous fission is not observed for nuclei with $A \leq 240$ is that such nuclei are stable with respect to a small deformation from their equilibrium shape. Further experiments showed that lighter elements could also be fissioned by high energy particles, e.g.

$$^{63}_{29}Cu + {}^{1}_{1}p \rightarrow {}^{24}_{11}Na + {}^{39}_{19}K + {}_{0}n^1 + Q$$

In general, spontaneous fission reactions are rare events, and most of our knowledge on the subject of fission is derived from induced fission.

The proton energy threshold for this reactions ∼ 50 MeV. Such reactions in which several light particles are emitted are called *spallation reactions*, e.g. when ^{63}Cu is bombarded by energetic protons, the following spallation reactions may take place

$$^{63}Cu + {}^{1}_{1}p \rightarrow {}^{38}Cl + p + n + 6({}^{4}_{2}He)$$

or
$$^{63}Cu + {}^{1}_{1}p \rightarrow {}^{24}Na + p + 3n + 9({}^{4}_{2}He)$$

The thresholds for the above reactions comes out to be ∼ 110 MeV and ∼ 170 MeV respectively, which is much higher than the fission reaction thresholds.

7.2.1 Experimental Facts (Fission)

The fission involves the movement of many nucleons at the same time and is therefore an example of the collective degrees of freedom in nuclei. We may visualize the process in the following way.

The sequence of events leading to the production of fission fragments from uranium bombardment by neutrons is shown in Figure 7.25. One of the earliest theoretical proposals,

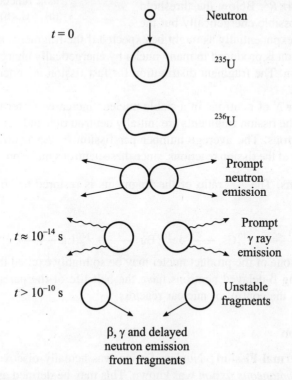

Neutron

$t = 0$

^{235}U

^{236}U

Prompt neutron emission

$t \approx 10^{-14}$ s

Prompt γ ray emission

$t > 10^{-10}$ s

Unstable fragments

β, γ and delayed neutron emission from fragments

Fig. 7.25 Schematic representation of the fission process in ^{235}U by neutron. The time scale shown gives only orders of magnitude.

rapidly verified experimentally was that the *slow neutron* fission of uranium should take place in the rare isotope ^{235}U. This followed from the observation that the complex of fission products did not show the particular resonance dependence on neutron energy known to be associated with the reaction ^{238}U + $n \rightarrow (^{239}$U$^{*})$. Thermal fission is also observed for other even-Z, odd-N nuclei such as ^{233}U and ^{239}P and the cross-section in the thermal region indeed show its own characterstic resonances (Figure 7.26).

Fast neutron fission is found for a number of even-even nuclei such as ^{232}Th and ^{238}U. There is a 'threshold' energy for the process (Figure 7.26(b)). Whereas in the thermal region the main interactions are the (n, γ) capture process with a cross-section at various resonances around $\pi \lambda^2$ and elastic scattering between resonances with the hard sphere cross-section of about $4\pi R^2$. Below the threshold energy fission is still possible energetically but the

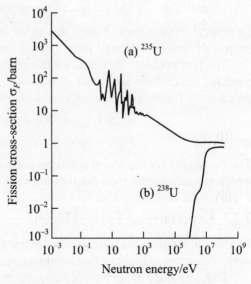

Fig. 7.26 Fission cross-section, schematic, as a function of neutron energy: (a) ^{235}U, (b) ^{238}U.

cross-section decreases exponentially as might be expected if the fragments have then to penetrate a Coulomb barrier. Fission is produced in many nuclei by energetically charged particles, e.g. *p*, *d*, α and also by γ-radiation. The fragment distribution for fast fission is generally more symmetric than for thermal fission.

Because the number N of neutrons in a stable nucleus increases faster than the number Z of protons as A increases, the fission fragments are initially neutron rich and very rapidly emit a small number of *prompt* neutrons. The average number per fission (≈ 2.5 neutrons), is an important experimental observable of the fission reactions, since these neutrons make possible the development of chain-reacting systems. The $\dfrac{N}{Z}$ ratio of the fragments is restored to normal by sequential β decay, e.g.

$$^{140}_{54}\text{Xe} \xrightarrow{\ e^{-}\ } ^{140}_{55}\text{Cs} \xrightarrow{\ e^{-}\ } ^{140}_{56}\text{Ba} \xrightarrow{\ e^{-}\ } ^{140}_{57}\text{La} \xrightarrow{\ e^{-}\ } ^{140}_{58}\text{Ce (stable)}$$

In some of these chains one of the product nuclei may be so highly excited that neutrons emission is possible. The resulting β-delayed neutrons have the half-life of the parent β-emitter, they are important technically in the control of nuclear reactors.

7.2.2 Types of Fission

(i) **Induced or Thermal Fission**   Nuclear fission was actually discovered through *induced fission* before *spontaneous fission* was known. This may be defined as any reaction of type $A(a, b)B$ with final products *b* and *B* being nuclei of comparable mass. The best-known example is the fission of $^{235}_{92}$U induced by thermal neutrons. This type of fission is also called as thermal fission. A compound nucleus ($^{236}U^*$), (where the asterik in the superscript indicates that the nucleus is in an excited state) forms the intermediate state for the reaction. Thermal neutron adds negligible energy to the fissionable nucleus, i.e. for all practical purposes we may regard the kinetic energy of the incident thermal neutron to be zero. In this limit, the excitation energy of the compound nucleus is simply the excess of the binding energy of ^{235}U, together with a "free" neutron, over that of ^{236}U. Fission of ^{235}U and ^{239}Pu by thermal neutrons are the most important reactions.

(ii) **Fast Fission** Isotopes of uranium and some other elements which form compound nuclei of even-odd structure falls into (n, a) reactions with fast neutrons (neutron energy ≥ 1 MeV). Fission of ^{238}U is an example of fast fission.

(iii) **Spontaneous Fission** Nuclear fission can take place spontaneously. However, spontaneous fission reactions are rare events. In principle spontaneous fission is possible. The only thing preventing this from happening more often is the fission barrier through which fragments must tunnel. Spontaneous fission does not become an important decay channel with $A > 240$. For e.g., the partial half-life for spontaneous fission in ^{232}U is around 10^{14} years, whereas the value for α-decay is only 2.85 years. A semi-log plot of observed spontaneous fission lifetimes versus $\dfrac{Z^2}{A}$ is shown in Figure 7.27, and can be seen that the liquid drop model accounts for the general trend, but not for the fine details of the data. The reason that spontaneous fission is not observed for nuclei with $A \leq 240$ is that such nuclei are stable with respect to a small deformation from their equilibrium shape. This is because such a change causes an increase in surface area and consequent loss in binding energy which is not offset by the decrease in Coulomb energy arising from the increase separation of charge.

(iv) **Charged Particle Fission** $Z > 90$ elements show fission with protons, deutrons and α-particles. If is found that high energy charged particles induce fission in elements even in the middle of the periodic table.

(v) **Photo fission** In heavier elements, high energy photons induce fission. γ-rays (5.1 MeV) can produce fission in ^{238}U.
Very large energy is released in all cases and also fast neutrons are emitted. Because of their exceptionally high neutrons to protons ratio, Meitner and Frish indicated that the fission fragments should be unstable, undergoing a chain of β^- disintegration, e.g.

$$^{235}_{92}U + {}^{1}_{0}n \rightarrow ({}^{236}_{92}U^*) \begin{cases} {}^{98}_{40}Zr + {}^{98}_{41}Cb \rightarrow {}^{98}_{42}Mo \\ \\ {}^{136}_{52}Te + {}^{136}_{53}I \rightarrow {}^{136}_{54}Xe \end{cases} + 2{}^{1}_{0}n$$

A typical β-decay chain is as

$$^{140}_{54}Xe \rightarrow {}^{140}_{55}Cs \rightarrow {}^{140}_{56}Ba \rightarrow {}^{140}_{57}La \rightarrow {}^{140}_{58}Ce \text{ (stable)}$$

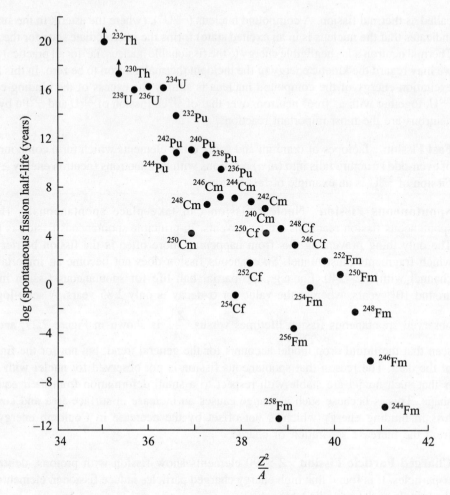

Fig. 7.27 Plot of logarithm of spontaneous fission life times versus $\dfrac{Z^2}{A}$. The significant decrease in life times for isotopes with $N > 152$ is related to a subshell closure at $N = 152$.

7.2.3 Energy Release in Fission

Nuclear fission reaction is highly exoergic reaction. Now, it is well established through various experiments that the total kinetic energy of fission fragments is about 167 MeV. Besides this, some energy is also carried by the γ-rays and few prompt neutrons emitted along with the fission fragments. To these we have to add the energies of β⁻-particles and antineutrinos emitted by fission fragments, which are usually radioactive, as also the energies of the γ-rays associated with the β-disintegrations of the fission fragments. Table 7.9 gives an idea of distribution of energies within the different components in nuclear fission. However, the total energy evolved during nuclear fission is higher than the value given in this table.

Table 7.9 Total energy shared by the different components in nuclear fission

components in nuclear fission	energy (MeV)
Kinetic energy (K. E.) of fission fragments	167
K. E. of prompt neutrons	5
Energy of prompt γ-rays	6
Energy of β⁻-particles emitted by fission fragments	8
Energy of antineutrinos emitted by fission fragments	12
Energy of the γ-rays emitted by the fission fragments	6
Total energy released	204

One can measure the energy released during nuclear fission by bombarding a piece of uranium with thermal neutrons, which is found to be heated due to the absorption of the fission fragments and some of the other products of fission. One can measure the heat thus generated by calorimetric method, which comes out to be about 186 MeV per uranium. We can see that this value is less than given in Table 7.9. This is due to the fact that γ-rays and antineutrinos produced during fission have very high penetrability and hence escape from the piece of uranium. During the process of fission about 0.9 MeV energy per nucleon is released (from B. E. consideration). Thus the total energy release for ^{238}U fission is around $238 \times 0.9 = 212$ MeV.

One can determine quantitatively the energy release from the knowledge of known atomic masses of the nuclei involved in nuclear fission. We consider following reaction and assume that three prompt neutrons are released in this reactions.

$$^{235}_{92}\text{U} + ^{1}_{0}\text{n} \rightarrow (^{236}_{92}\text{U}^*) \rightarrow ^{141}_{56}\text{Ba} + ^{92}_{36}\text{Kr} + 3^{1}_{0}\text{n} + Q \qquad \text{... (7.1)}$$

Now,

$$Q = M(^{235}\text{U}) + M(n) - M(^{141}\text{B}) - M(^{92}\text{Kr}) - 3M(n)$$
$$= 235.04278 + 1.00866 \rightarrow 140.9129 - 91.89719 - 3 \times 1.00866$$
$$= 0.21537 \text{ u} = 200.6 \text{ MeV}$$

We see that energies of the ~ 200 MeV order are released in almost fission of other heavy nuclei and comparable with the binding energy nuclei and comparable with the binding energy value.

We can see that if 1 g of ^{235}U is completely fissioned, the energy release will be about 2.29×10^4 kWH.

Number of atoms of ^{235}U in one kg,

$$n = \frac{6.025 \times 10^{23} \times 10^3}{235} = 2.564 \times 10^{24}$$

∴ Energy release per gram of ^{235}U,

$$E = \frac{nQ}{10^3} = \frac{2.564 \times 10^{24}}{10^3} \times 200.6 \times 1.6 \times 10^{-13}$$
$$= 8.229 \times 10^{10} = 2.29 \times 104 \text{ kWH}$$

We may note that the excitation energy brought along by the neutron capture sets the entire nucleus in (7.1) into vibration and many fission channels are opened as a result. The channel represented (7.1) is an typical exit channel. The three neutrons emitted in this process are called *prompted neutrons* since they are released as a part of the fission process. However, both ^{92}Kr and ^{141}B are neutron unstable. This we can see, for e.g., from the fact that the heaviest stable isotopes of Kr and Ba are ^{86}Kr and ^{138}Ba, respectively. Among the possible decays of the two nuclei, neutrons may be released either directly from ^{92}Kr and ^{141}Ba or as a result of other unstable nuclei formed from their decays. These neutrons are called as *delayed neutrons,* since they emerge after some delays-as, for e.g., due to β^- and γ-ray decays.

Delayed neutrons play an important role in maintaining a chain reaction and hence they have been studied extensively, particularly for ^{235}U fission. The results are summarized in table below. Nuclear decay scheme for fission fragments ^{87}Br and ^{137}I for delayed neutrons by Shell and Coworkers is shown in Figure 7.27(a).

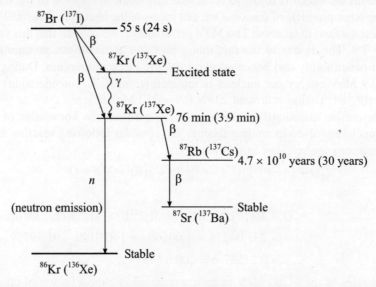

Fig. 7.27(a) Emission of delayed neutron during nuclear decay of fission fragments ^{87}Br and ^{137}I.

Table 7.10 Delayed neutrons for ^{235}U fission by thermal neutrons.

Half life (s)	Neutron energy (keV)	Fission yield relative to total neutron emission %
55.00	250	0.025
24.00	560	0.166
4.51	430	0.213
1.52	620	0.241
0.43	420	0.085
0.05	–	0.025
Total yield		0.754

7.2.4 Nature of Fission Fragments

We have mentioned above that in the nuclear fission of a heavy element like ^{235}U two fission fragements of comparable mass numbers are produced. However, the mass numbers of the fragement are not actually the same in most cases. The fragment pairs have an *asymmetric pair distribution* (Figure 7.28) in which a given fission event generally yields a high mass (H) and a low mass (L) product rather than two equal masses. A preference for asymmetric fission, i.e. fission into two fragments of unequal mass, may be understood from the large neutron excess carried by heavy nuclei. When such a nucleus undergoes fission, say into two medium weight nuclei with nucleon number roughly $\frac{1}{2}$ A each, the fragments would have to be nuclei very far away from the valley of stability. As a result, many prompt neutrons would have to be emitted and this, in turn, increases the number of independent entities in the final stage. Phase space considerations, however, strongly favour exit channels with the minimum number of products. It is therefore, more economical to have one heaviour fragment so as to reduce the number of prompt neutrons that have to be emitted. Figure 7.28 shows that the fission fragments has a bimodel distribution as a function of mass. Besides binary fission involving two final nuclei (plus prompt neutrons), *ternary fission* consisting of three final nuclei is also commonly observed. In principle larger numbers of fragments are also possible, but again, *phase space* considerations reduce their probability.

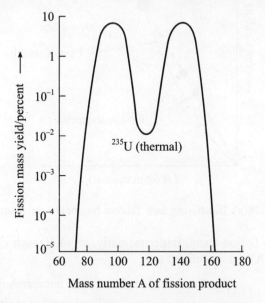

Fig. 7.28 Schematic diagram showing the distribution of fission fragments mass as a function of nucleon mass number. The reason that a bimodal distribution is favoured comes from the fact that the ratio of neutron to proton numbers $\frac{N}{Z}$ for heavy nuclei is larger than those for medium weight nuclei is larger than those for medium weight nuclei constituting the fragments.

Figure 7.28 clearly shows that probability of symmetric fission with equal mass numbers for the fragments is the least. In the case of thermal neutron fission of ^{235}U, the lighter fragments have the mass number spread between 85 to 105 with a broad peak at $A \sim 96$ and the heavier fragments have mass numbers distributed between 130 to 150 with a broad peak at $A \sim 138$.

We may note that symmetric fission becomes more probable, if the excitation energy of the compound nucleus undergoing fission is increased by high energy neutron. Moreover, symmetric fission is also more probable when fission is induced by particles other than neutrons.

As mentioned earlier, the fission fragments are highly neutron rich. The two fission fragments ^{141}Ba and ^{92}Kr (Equation 7.1) have the neutron excesses $N - Z = A - 2Z = 29$ and 20 respectively. Natural Ba and Kr have 7 and 6 stable isotopes respectively. Of these, the heaviest are ^{138}Ba with $N - Z = 26$ and ^{86}Kr with $N - Z = 14$ respectively. Clearly, these two fission fragments have neutron excesses considerably greater than are required for their-stability. Figure 7.28(a) show how fission fragments are neutron-rich. The nuclear stability condition against decay clearly reveals that these nuclei can not be stable; they must be β^--active as represented by the following β^--active chains.

$$^{141}\text{Ba} \xrightarrow{\ 18\,\text{m}\ } {}^{141}\text{La} \xrightarrow{\ 3.7\,\text{h}\ } {}^{141}\text{Ce} \xrightarrow{\ 33\,\text{d}\ } {}^{141}\text{Pr (stable)}$$

$$^{92}\text{Kr} \xrightarrow{\ 3\,\text{s}\ } {}^{92}\text{Rb} \xrightarrow{\ 80\,\text{s}\ } {}^{92}\text{Sr} \xrightarrow{\ 2.7\,\text{h}\ } {}^{92}\text{Y} \xrightarrow{\ 3.5\,\text{h}\ } {}^{92}\text{Zr (stable)}$$

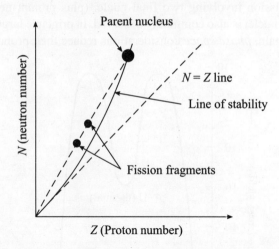

Fig. 7.28(a) Illustrating how fission fragments are neutron rich

These are examples of a fission chain. Experimentally, about 60 such chains are known.

The fission yield Y(A) can be defined as

$$Y(A) = \frac{\text{Number of nuclei of mass number A formed in fission} \times 100\,\%}{\text{Total number of fissions}}$$

$$= \frac{N_A}{N_0} \times 100\,\%$$

Figure 7.28 shows the plot of $Y(A)$ % against A for fission chains of ^{235}U.

7.2.5 Energy Distribution Between Fission Fragments

Conservation of linear momentum demands that the energy distribution of fission fragments also to be asymmetric so that the mean energies of the two groups are $\bar{E}_H = 60$ MeV and $\bar{E}_L = 95$ MeV.

Using ionization chamber, the kinetic energies of the fission fragments have been measured within which a thin foil of U or Th is placed. The fossile material is placed at one of the electrodes and the ions, which are produced when a fission fragment enters the region between the electrodes, are collected in the electrostatic field and the ionization currents measured by two electrometers calibrated by α-particles of known energy. The nucleus undergoing fission, may be considered to be initially at rest and the two fragments having masses M_1 and M_2 fly off with velocities v_1 and v_2 respectively. We have from conservation of momentum law,

$$M_1 v_1 = M_2 v_2$$

Hence the ratio of the energies is

$$\frac{E_1}{E_2} = \frac{\frac{1}{2} M_1 v_1^2}{\frac{1}{2} M_2 v_2^2} = \frac{M_2}{M_1} \qquad \dots (7.2)$$

Obviously, one can determine the ratios $\dfrac{E_1}{E_2}$ and $\dfrac{M_1}{M_2}$ and also the total energy $(E_1 + E_2)$. Figure 7.29 shows the energy distribution between the two fission fragments in the case of thermal neutron fission of ^{235}U. We see that there are two broad peaks. The higher energy peak goes with the lighter fragements and vice versa. We further see that the lower energy peak is broader than the other, such that the areas under them are equal when corrected properly for the overlapping region.

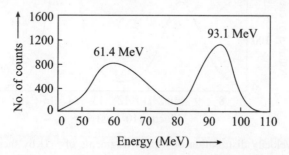

Fig. 7.29 Energy distribution of the fission fragments in ^{235}U by thermal neutron.

For the fission of ^{235}U by thermal neutron, we have $M_i \approx 95$ and $M_2 \approx 139$. Thus we have

$$\frac{E_2}{E_1} = \frac{95}{139} \approx \frac{2}{3} \qquad \dots (7.3)$$

The distribution of energy of fission fragments has been measured experimentally for several fissionable materials, e.g. ^{233}U, ^{235}U, ^{239}Pu, etc. and Equation (7.3) has been found correct in ^{235}U.

Accurate measurment of energies of fission fragments was made by Leachman in 1952. He measured the velocities of fission fragments of ^{235}U by thermal neutron, by a time-or-flight method. Figure 7.30 shows the experimental arrangement. The experimental results are displayed in Figure 7.31. From the values of the velocities, the total kinetic energy of the fission fragments works out to be ~ 166 MeV. Table 7.9 gives the complete energy distribution when a ^{235}U nucleus undergoes fission by a thermal neutron.

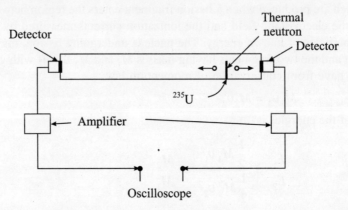

Fig. 7.30 Experimental arrangement to measure velocities of fission fragments by a time-of-flight method.

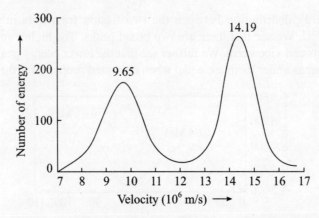

Fig. 7.31 Velocity distribution of fission fragments of ^{235}U by thermal neutron measured by a   time-of-flight-method.

Fission whether spontaneous or induced, normally leads to fragments of unequal energy and, therefore, of unequal mass. The asymmetric character of the mass distribution is most pronounced in spontaneous fission and low energy (near threshold) fission of heavy elements, becoming less pronounced and gradually giving way to symmetric fission as the excitation energy increases or the mass of the target nucleus decreases.

2.2.6 Neutron Emission in Fission Process

We have already mentioned mentioned that during the fission of heavy nuclei *prompt* and *delayed* neutrons are emitted. The emission of the prompt neutrons in nuclear fission process makes it possible to the relization of nuclear chain reaction, which is the basic requirement for the operation of a neuclear reactor. Let us write the fission reaction induced by slow neutrons in ^{235}U as

$$^{235}_{92}U + {}^{1}_{0}n \rightarrow (^{236}_{92}U^*) \rightarrow X + Y + v^{1}_{0}n + Q$$

(fast neutrons)

Where X and Y are two fission fragments and v represents the number of fast neutrons released per fission process. The knowledge of the average number of neutrons emitted per second is of great importance. Let v_{av} be the average number of neutrons released in one fission. We may note that the number of neutrons released in any one fission, i.e. v is an interger, but v_{av}, the average value of neutrons in one fission is not an integer, because the fissionable nucleus can be divided in several different ways (~ 30), i.e., there are more than 30 modes of fission in each of which a different pair of nuclei (fission fragments) is formed.

We expect n to be greater than one due to high value of neutron to proton ratio of compound nucleus ^{236}U*. The average number of fast neutrons (v_{av}) released per fission of ^{235}U by slow neutrons is 2.47 and in case of fission of ^{239}Pu, $v_{av} = 3$

These neutrons are emitted with a distribution of energy ranging from 0.05 MeV to 17 MeV; i.e. they are emitted as *fast* neutrons. The energy distribution of these neutrons is approximately Maxwellian with an average of about 2 MeV. Besides these prompt neutrons, in a few (small fraction of) cases delayed neutrons with half-lives of the order of seconds from fission-fragments are emitted. Delayed neutrons emitted are about 0.75 % of the neutrons emitted and their emission follows the radioactive decay.

We may note that not all the thermal neutrons which are absorbed, are capable of inducing nuclear fission, v_{av} is smaller than n in the same ratio as the fission cross-section σ_f to the absorption cross-section σ_a. Thus

$$\eta = v \frac{\sigma_f}{\sigma_a} \qquad \qquad ...(7.5)$$

Where η is the average number of fission neutrons released per neutron absorbed by a fissionable nuclide.

We know that all the neutrons absorbed in the fissionable material do not induce fission since some absorption may simply result in the emission of γ-ray, i.e. *radiative capture*, while others may cause fission. This means, the absorption cross-section σ_a shall be made up of two parts: (i) σ_c (the *radiative capture* cross-section) and (ii) σ_f (the fission cross-section). We have

$$\sigma_a = \sigma_c + \sigma_f \qquad \qquad ...(7.6)$$

$$\therefore \qquad \eta = v \frac{\sigma_f}{\sigma_c + \sigma_f} = v \frac{1}{1 + \frac{\sigma_c}{\sigma_f}} = \frac{v}{1 + \alpha} \qquad ...(7.7)$$

Where $\alpha = \dfrac{\sigma_c}{\sigma_f}$ is the ratio of the radiative capture cross section to fission cross section. Relation (7.7) shows that the relative probability that a compound nucleus decays by fission is thus $(1 + \alpha)^{-1}$. Table 7.11 gives the values of ν_{av}, α and η for thermal neutrons.

Table 7.11

Parameter	^{233}U	^{235}U	Uranium (natural)	^{239}Pu
σ_f	525 ± 4	577 ± 5	4.18 ± 0.06	742 ± 4
σ_c	53 ± 3	101 ± 5	3.50	286 ± 4
ν_{av}	2.51 ± 0.02	2.44 ± 0.022	2.47	2.89 ± 0.03
α	0.101 ± 0.004	0.18 ± 0.01	0.83	0.39 ± 0.03
$\eta = \dfrac{\nu}{1+\alpha}$	2.82 ± 0.02	2.07 ± 0.01	1.34 ± 0.02	2.08 ± 0.02

7.2.7 Energetics of Nuclear Fission Process: Spontaneous-Fission

In general, nuclear fission is a process of breaking up of a nucleus into two fragments nuclei having comparable masses, whether or not induced by any external agent. We have spontaneous fission in the latter case. In a spontaneous fission process, a nucleus ($^A_Z X$) under goes the spontaneous transformation and the process can be represented as

$$A^Z_X \rightarrow \,^{A_1}_{Z_1}X_1 + \,^{A_2}_{Z_2}X_2 + Q \qquad \qquad ... \ (7.8)$$

Where the two product nuclei X_1 and X_2 have mass numbers and atomic numbers of comparable values. Obviously, $A = A_1 + A_2$ and $Z = Z_1 + Z_2$ for (7.8). When $A_1 = A_2 = \dfrac{A}{2}$ and $Z_1 = Z_2 = \dfrac{Z}{2}$, we have symmetric spontaneous fission. Here, we have not included the prompt neutrons because they are not the primary products of fission.

The processes (7.8) can occur if the Q value of the transformation is positive, i.e.

$$Q = M(A, Z) - M(A_1, Z_1) - M(A_2, Z_2) > 0$$

Where the $M's$ are atomic masses expressed in the units of energy. For the symmetric case, one finds

$$Q = M(A, Z) - 2M\left(\dfrac{A}{2}, \dfrac{Z}{2}\right) > 0$$

Writing in terms of binding energies (B), one finds

$$Q = 2B\left(\dfrac{A}{2}, \dfrac{Z}{2}\right) - B(A, Z)$$

$$= 2\left(\dfrac{A}{2}\right) f'_B - A f_b = A(f'_B - f_B) = A\Delta f_B$$

Where $f'_B s$ denotes the binding fractions.

Now, for Q to be positive, Δf_B must be positive which happens if $f_B' > f_B$, i.e. the binding fraction of the product nucleus is greater than that of the parent nucleus.

Let us write the atomic masses in terms of Weizsaker's semi-empirical mass relation, neglecting the pairing interaction term, we have

$$M(A,\, Z) = ZM_H + NM_n - a_1 A + a_2 \left(\frac{A}{2}\right)^{\frac{2}{3}} + a_3 \frac{Z^2}{A^{\frac{1}{3}}} + a_4 \frac{(A-2Z)^2}{A}$$

$$M\left(\frac{A}{2},\, \frac{Z}{2}\right) = \frac{Z}{2} M_H + \frac{N}{2} M_n - a_1 \frac{A}{2} + a_2 \left(\frac{A}{2}\right)^{\frac{2}{3}} + a_3 \frac{\left(\frac{Z}{2}\right)^2}{\left(\frac{A}{2}\right)^{\frac{1}{3}}} + a_4 \frac{(A-2Z)^2}{2A}$$

Thus

$$Q = M(A,\, Z) - 2M\left(\frac{A}{2},\, \frac{Z}{2}\right)$$

$$= a_2 A^{\frac{2}{3}} \left(1 - \frac{2}{2^{\frac{2}{3}}}\right) + \frac{a_3 Z^2}{A^{\frac{1}{3}}} \left(1 - \frac{2^{\frac{1}{3}}}{2}\right)$$

$$= -0.26\, a_2\, A^{\frac{2}{3}} + 0.37 \frac{a_3 Z^2}{A^{\frac{1}{3}}} \qquad\qquad \dots (7.9)$$

For energetically possible spontaneous fission ($Q > 0$), we have

$$\frac{Z^2}{A} > \frac{0.26\, a_2}{0.39\, a_3}$$

We have $a_2 = 0.019114$ u and $a_3 = 0.0007626$ u substituting these values, one obtains

$$\frac{Z^2}{A} > 17.6 \qquad\qquad \dots (7.10)$$

This shows that the spontaneous fission should be energetically possible for $A > 90$ and $Z > 40$ (giving $\frac{Z^2}{A} = 17.8$). If fact, spontaneous fission is a very uncommon phenomenon. It is very rarely observed even amongst the nuclei of the heaviest atoms in the periodic table, e.g. uranium. For example, there is only about one spontaneous fission per hour in 1 g of ^{235}U corresponding to a half life of 2×10^{17} yr. One can understand this in terms of quantum mechanical barrier penetration problem.

Bohr and Wheeler considered the Coulomb's potential barrier of the two fragments at the instant of separation. The existence of this barrier prevents the immediate breaking of the nucleus in two fragments. Let E_b be the Coulomb barrier height and let the nucleus will be unstable and break apart into two fragments if $E_f > E_b$. Corresponding to the Coulomb potential between the two symmetric fragments when they are just in contact with each other, the barrier height is given by

$$E_b = \frac{\left(\frac{1}{2}Z\right)^2 e^2}{4\pi \varepsilon_0 (2R)} = \frac{Z^2 e^2}{32\pi \varepsilon_0 R_0 \left(\frac{1}{2}A\right)^{\frac{1}{3}}}$$

Since

$$\frac{1}{4\pi\varepsilon_0} = 9 \times 10^9 \ \frac{N-m^2}{C^2}, \quad e = 1.6 \times 10^{-19} \ C,$$

$$R_0 \sim 1.5 \times 10^{-15} \ m, \text{ and } 1 \ MeV = 1.6 \times 10^{-13} \ J$$

∴

$$E_b = \frac{9 \times 10^9 \times (1.6 \times 10^{-19})^2 \times (2)^{\frac{1}{3}}}{8 \times 1.5 \times 10^{-15} \times 1.6 \times 10^{-13}} \frac{Z^2}{A^{\frac{1}{3}}}$$

$$= 0.15 \ \frac{Z^2}{A^{\frac{1}{3}}} \ MeV \qquad \qquad \text{... (7.11)}$$

∴

$$E_b - E_f = 0.15 \ \frac{Z^2}{A^{\frac{1}{3}}} - \left[-3.42 \ A^{\frac{2}{3}} + 0.22 \ \frac{Z^2}{A^{\frac{1}{3}}} \right]$$

$$= 3.42 \ A^{\frac{2}{3}} - 0.07 \ \frac{Z^2}{A^{\frac{1}{3}}} \qquad \qquad \text{... (7.12)}$$

Therefore the condition of statibility against spontaneous fission gives

$$E_b - E_f \geq 0$$

or

$$3.38 \ A^{\frac{2}{3}} - 0.07 \ \frac{Z^2}{A^{\frac{1}{3}}} \geq 0$$

or

$$\frac{Z^2}{A} \leq \frac{3.38}{0.07} \leq 50 \qquad \qquad \text{... (7.13)}$$

A more general acceptable value for $\frac{Z^2}{A}$ parameter for spontaneous fission is 47.8. This reveals that a nucleus would undergo fission instantaneously in $\frac{Z^2}{A}$ is close to this value and if $\frac{Z^2}{A}$ is smaller than this limiting value, the nucleus will be stable against spontaneous fission. It is obvious that $\frac{Z^2}{A}$ may have value ~ 50 if for a nucleus $A = 250$. This means $A > 250$ nuclei would be too unstable to exist for more than 10^{-12} s or even less.

Figure 7.32 shows the variation of E_b and E_f with A. From graph, we note that for $A \simeq 250$, E_b becomes equal to E_f and $E_f > E_b$ for nuclei with $A > 250$. This indicates that nuclei with $A > 250$ are not expected to exist in nature. Further in the neighbourhood of $A \simeq 85$, the fission begins to becomes exoergic.

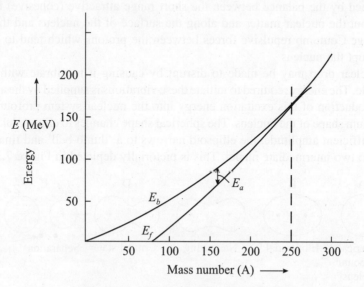

Fig. 7.32 Variation of E_b and E_f with A.

7.2.8 Theory of Nuclear Fission

(i) **The liquid drop model of fission**   If a nucleus receives sufficient amount of excitation energy by way of bombardment through energetic particles, it may undergo fission. This is true of all the nuclei from light to the heavy and the bombarding particles need not be neutrons but may be protons deuterons, α-particles or even γ-rays (in which case it is called photofission). However, we shall mainly concentrate upon the fission of heavy nuclei at the top of the periodic table by neutrons. According to classical arguments, an immediate breakup of a heavy nucleus is prevented by a fission potential barrier. Fission can then take place only if the nucleus is in energy state above the top of this fission barrier. However the existence of spontaneous fission suggests that even if the nucleus is in an energy state below the top of the fission barrier, fission may take place through barrier penetration. The potential energy curve for fission is shown in (Figure 7.34).

The first thorough theoretical treatment of the fission event based on the liquid drop model of the nucleus was given by Bohr and Wheeler in 1939. They assumed nucleus to be the drop of an incompressible, electrically charged nuclear fluid which when not under the action of any external force (in equilibrium) adopts a spherical shape. Then they pictured the fission process as a result of the vibrations induced in this liquid drop by the incident particle, thus deforming its spherical shape and ultimately leading to its break up. From our studies in chapter 4, we know that nuclear forces are short range, charge-independent nucleon-nucleon forces. The forces present within a nucleus are (i) nuclear forces and (ii) the Coulomb repulsive forces between the protons. Nuclear forces which bind the nucleons together are compared within the surface tension forces of the liquid which also hold the molecules of the drop bound together. The shape of the nucleus in an undisturbed state is determined by the balance between the short range attractive (cohesive) forces which act throughout the nuclear matter and along the surface of the nucleus and the comparatively long range Coulomb repulsive forces between the protons which tend to push them apart and disrupt the nucleus.

This nuclear drop may be made to disrupt by causing it to vibrate with a large enough amplitude. The energy required to initiate these vibrations is supplied by the absorbed neutron. The introduction of this excitation energy into the nuclear system profoundly disturbs the equilibrium shape of the nucleus. The spherical shape changes to ellipsoidal. If the vibrations are of sufficient amplitude, the ellipsoid narrows to a 'dumb-bell' and finally breaks at the neck into two intermediate nuclei. This is pictorially depicted in Figure 7.33(a).

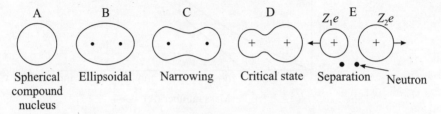

Fig. 7.33(a) Schematic pictorial depiction of fission event

In case, energy is not enough for the fission to take place, the compound nucleus which is formed as a result of absorption of the incident neutron, returns back to its spherical shape by emitting either the neutron or γ-rays. There is a critical energy called fission threshold energy or the activation energy which must be reached for complete separation. This critical energy is shown in the Figure 7.34.

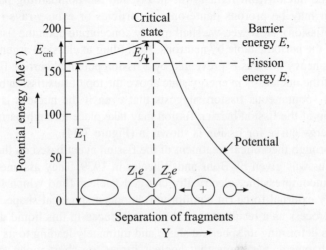

Fig. 7.34 Potential energy curve for fission

Figure 7.34 is probably an over simplification, and that the barrier actually has a double hump something like that shown in Figure 7.34(a).

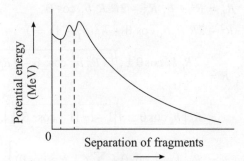

Fig. 7.34(a) A double hump fission barrier

(ii) Bohr and Wheeler's Theory of Nuclear Fission For the mathematical analysis of the problem, we consider the nucleus to be initially of spherical shape, which being made up of incompressible nuclear fluid has a constant volume. On account of this incompressibility the vibrations set up in the nucleus due to the absorption of a neutron which cannot penetrate any deep within the drop but shall remain confined to the surface of the drop only. Of course the deformation produced in the nucleus may be of any type, (the mass at which nuclei becomes unstable with respect to fission, i.e. when $E_f \gg 0$ (Figure 7.34), can be estimated by considering the change in binding energy when a spherical nucleus undergoes a small deformation to an ellipsoid of the same volume, as illustrated in 7.33(b)) but we assume for the sake of mathematical simplicity that in the lowest excitations, the drop will still retain axial symmetry. Let us choose this axis of symmetry as the polar axis of the spherical co-ordinates which we shall use here. On account of the axial symmetry, the deformation may be expressed as

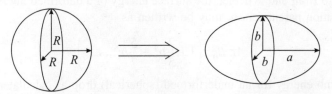

Fig. 7.33(b) Deformation of sphere of ellipsoid of same volume: $a = R(1 + \varepsilon)$; $h = \dfrac{R}{(1+\varepsilon)^{1/2}}$.
For this prolate deformation ε is positive.

$$R(\theta) = R_0\left[1 + \sum_{l=0}^{\infty} b_l\, P_l\,(\cos\theta)\right] \qquad \dots (7.1)$$

Where $R(\theta)$ is the distorted radial co-ordinate of a point on the surface of the drop, b_l are the deformation parameters and $P_l(\cos\theta)$ are the Legendre polynomials.

This formula can give both a deformation of the sphere as well as a translation of the sphere as a whole, of which latter is of no interest to us. To judge the correctness of the above statement, let us consider the sphere to be moved underformed through a distance $b_1 R_0$ along he axial line ($\theta = 0$), then we have

$$R_0^2 = R^2 + b_1^2 R_0^2 - 2R\, R_0 b_1 \cos \theta$$

or

$$R^2 - 2R\, R_0 b_1 \cos \theta + R_0^2 (b_1^2 - 1) = 0$$

$\therefore$

$$R = \frac{2R_0 b_1 \cos \theta \pm \sqrt{\left\{ 4 R_0^2 b_1^2 \cos^2 \theta - 4 R_0^2 \left(b_1^2 - 1 \right) \right\}}}{2}$$

$$= R_0 \left[b_1 \cos \theta \pm \sqrt{\left\{ 1 - b_1^2 \left(1 - \cos^2 \theta \right) \right\}} \right]$$

$$= R_0 \left[1 + b_1 \cos \theta + \sum_{\lambda = 0}^{\infty} C_{2l}\, P_{2l} (\cos \theta) \right] \qquad \text{... (7.2)}$$

Where C_{2l} arise from the expansion of the square root.

Thus the coefficient of $P_1 (\cos \theta)$ in (7.1) gives the distance through which the centre of the drop (spherical) moves and for a pure deformation without any translation we must have $b_1 = 0$. Moreover b_0 is to be chosen in such a manner that the nuclear volume remains constant, we choose $b_0 = 0$. Then equation (7.1) gives

$$R(\theta) = R_0 \{ 1 + b_2 P_2 (\cos \theta) + b_3 P_3 (\cos \theta) + \dots \} \qquad \text{... (7.3)}$$

Now the surface energy of a spherical drop is defined as

$$E_{sQ} = S\tau = 4\pi R_0^2 \tau \qquad \text{... (7.4)}$$

Where S is the surface area of the drop and τ, the surface tension.

The surface energy of a deformed drop is given by

$$E_s = \tau \int dS \qquad \text{... (7.5)}$$

According to Bohr and Wheeler, the surface energy of a deformed nuclear drop in terms of the deformation parameters b_1 may be written as

$$E_s = 4\pi R_0^2 \tau \left(1 + \frac{2}{5} b_2^2 + \frac{5}{7} b_3^2 \dots \right) \qquad \text{... (7.6)}$$

The Coulomb energy for an underformed (spherical) drop as calculated in semi-empirical mass formula is given by

$$E_{C_0} = \frac{3}{5} \frac{(Ze)^2}{4\pi \varepsilon_0 R_0} \qquad \text{... (7.7)}$$

In case of a deformed drop, the Coulomb energy in terms of the deformation parameters b_1 is given by

$$E_C = \frac{3}{5} \frac{(Ze)^2}{4\pi \varepsilon_0 R_0} \left[1 - \frac{1}{5} b_2^2 - \frac{10}{49} b_3^2 - \dots \right] \qquad \text{... (7.8)}$$

Therefore, the total deformation energy E_T is obtained by adding (7.6) and (7.8) which gives

$$E_T = E_S + E_C = 4\pi R_0^2 \tau \left(1 + \frac{2}{5}b_2^2 + \frac{5}{7}b_3^2 + \ldots \ldots \right)$$

$$+ \frac{3}{5}\frac{(Ze)^2}{4\pi\varepsilon_0 R_0}\left[1 - \frac{1}{5}b_2^2 - \frac{10}{49}b_3^2 - \ldots \right] \quad \ldots (7.9)$$

Ther total energy in the underformed state is equal to

$$E_{S_0} + E_{C_0} \quad \ldots (7.10)$$

Hence energy change due to deformation of the drop is given by

$$\Delta E = (E_S + E_C) - (E_{S_0} - E_{C_0})$$

If we consider lowest excitations only, i.e., $b_2 \neq 0$ and $b_1 = 0$ for $l > 2$, then the energy change due to the deformation of the drop is given by

$$\Delta E = (E_S + E_C - E_{S_0} + E_{C_0})$$

$$= E_{S_0}\left(1 + \frac{2}{5}b_2^2 \right) + E_{C_0}\left(1 - \frac{1}{5}b_2^2 \right) - (E_{S_0} + E_{C_0})$$

$$= \frac{2}{5}E_{S_0}b_2^2 - \frac{1}{5}E_{C_0}b_2^2$$

$$= \frac{1}{5}b_2^2\,[2E_{S_0} - E_{C_0}] \quad \ldots (7.11)$$

This shows that the surface energy appears with a positive sign and Coulomb energy with a negative sign. Therefore

(i) If $2E_{S_0} > E_{C_0}$: ΔE is positive and since surface energy prevents disruption while the Coulomb energy promotes it. The nucleus will be stable against *spontaneous decay*.

(ii) If $2E_{S_0} < E_{C_0}$: ΔE becomes negative and the nucleus will be *unstable against spontaneous decay*.

(iii) If $2E_{S_0} = E_{C_0}$, we have the *critical case*. In this case we have,

$$8\pi R_0^2 \tau = \frac{1}{4\pi\varepsilon_0}\frac{3}{5}\frac{(Ze)^2}{R_0}$$

or $\qquad 8\pi r_0^2 A^{\frac{2}{3}} \tau = \frac{1}{4\pi\varepsilon_0}\frac{3}{5}\frac{(Ze)^2 A^{-\frac{1}{3}}}{R_0}$ since $R_0 = r_0 A^{\frac{1}{3}}$

or $\qquad \dfrac{Z^2}{A} = (4\pi\varepsilon_0)\dfrac{40\pi r_0^3 \tau}{3e^2} \approx 50 \text{ or } > 45 \quad \ldots (7.12)$

Which defines a critical value of $\left[\dfrac{Z^2}{A} \right]$ and accordingly, we write it $\left[\dfrac{Z^2}{A} \right]_{cnt}$. Now we define a quantity called fissionable parameter χ as follows

$$\chi = \frac{\left[\dfrac{Z^2}{A}\right]}{\left[\dfrac{Z^2}{A}\right]_{crit}} \qquad \qquad \dots (7.13)$$

and discuss fission in terms of χ. If $\chi > 1$ the nucleus is unstable against spontaneous decay and $\chi < 1$, it is stable against spontaneous fission.

The critical energy E_f for fission defined as the energy necessary to deform a drop when it is about to split into the equal drops, then may be written as

$$E_f = 4\pi r_0 \tau A^{\frac{2}{3}} f(\chi) = E_{S_0} A^{\frac{2}{3}} f(\chi) \qquad \qquad \dots (7.14)$$

Obviously, one can predict the fission probability by comparing the critical energy with the excitation energy. The values of excitation energy calculated in this way are given in Table 7.12 for few heavy nuclei and compared with the corresponding values of critical energy.

Table 7.12 Calculated values of the excitation energy and a comparison with corresponding critical energy value

Nucleus (compound)	Excitation energy (MeV)	Critical energy (MeV)	Difference 2 – 3 (MeV)
^{233}Th	5.1	6.5	−1.4
^{236}U	6.6	5.5	1.1
^{232}Pa	5.4	5.0	0.4
^{238}Np	6.0	4.2	1.8
^{239}U	5.9	6.5	−0.6
^{240}Pu	6.4	4.0	2.4

Table 7.11 clearly reveals that for ^{238}U a critical deformation energy of 6.5 MeV is essential for fission, but it requires only 5.9 MeV when it takes up a neutron of zero K.E.. Obviously, no fission is possible with thermal neutrons with 0.03 eV.

(iv) Limitation of the Bohr-Wheeler Theory of Fission   The Bohr and Wheeler theory is based on a spherical nuclear droplet. In view of the approximate nature of the calculations, it fails to account for the fissionability of certain nuclides.

(a) **Odd-even Effect in the Binding Energy**   The values of the excitation energy of different nuclides recorded in the preceding section indicates that the binding energy of the neutron in the compound nucleus formed by the capture of a thermal neutron (e.g., ^{233}U, ^{235}U, ^{239}Pu) is greater than that for a target nucleus with an even number of neutrons (e.g., ^{232}Th, ^{238}U, ^{231}Pa, ^{237}Np). In general, the excitation energy for the nuclides with odd-N is greater than for those with even-N. Moreover, the high values of σ_f as well as the positive values of $E_{ex} - E_{ec.}$ indicate that fission with thermal neutrons occur much more readily in nuclei with odd-N than in nuclei with even-N. But the theory fails to account for the very large values of σ_f for Cm^{240} ($\sim 2 \times 10^4$ b), ^{230}U

(25b) and ^{232}U (80b) containing each even numbers of neutrons. Further, the nuclides, Pa^{231} and Np^{237} with excitation energy greaterr (of course, low corresponding to even-N) than the activation energy should undergo fission with thermal neutrons in spite of their having even-N. But, actually, they do not. Thus, it seems that the theory gives incorrect values of the activation energy for these nuclides.

(b) **Photofission Thresholds** We have seen photofission thresholds are nearly constant for several nuclides. It follows that Z and A have less influence on the fission probability. But the theory predicts the dependence of the fission probability on Z and A, which seems to be a drawback.

(c) **Asymmetric Nature of the Fission Process** As pointed out earlier, the theory predicts a greater probability for symmetrical fission. But it fails to account for the asymmetric nature of the process.

(d) **Shell Closer Effects** We saw that liquid-drop model predicts the symmetric nature of the fission process. In order to account for the highly asymmetric nature of the process, several authors have tried to explain the asymmetry on the basis of shell model as well as on the collective model of the nucleus. Wahl developed the hard core preformation theory of fission with relation to the shell structure closers and succeeded in accounting for the asymmetry in fission of ^{235}U with thermal neutrons. The compound nucleus [$^{236}_{92}$U$_{144}$] formed has an atomic number of 92, i.e., 92 protons and a mass number of 236, i.e., 236 nucleons. For fission to take place, the nucleus gets deformed following excitation, the total nucleons have a tendency to form groupings in the two end portions of the deformed nucleus, one with 50 neutrons and the other with 50 protons. Suppose, in order to form the closed shell structures, 32 protons are associated with the former group and 78 neutrons with the latter, giving mass number of 82 to 128, respectively, in the two ends. This type of mass distribution is referred to as hard core formation with closed shell structure having almost spherical shape. The total number of nucleons in the initial groupings is 210, and the rest of the nucleons, 26 (16 neutrons+ 10 protons) form a loose bridge between the two hard cores which are to be distributed between them at the moment of scission. The distribution of these bridge nucleons may be made in several different ways giving a variety of fission fragments. However, in view of the equalization of excitation energies among the two fragments, the most probable distribution of these bridge nucleons will be in equal numbers, i.e., 13 nucleons will be shared by each part. So that the mass distribution among the fission fragments would be 82 + 13 = 95 and 128 + 13 = 141, and after the emission of at least 2 neutrons, the mass numbers come to about 94 and 140, showing asymmetric fission which are close enough to the observed peaks in the mass distribution curve.

(ii) **Quantum Mechanical Barrier Penetration Theory** In the proceeding section, we have discussed fission in terms of a classical picture. However, a more refined theory of fission can be developed by introducing the idea of quantum mechanical barrier penetration which was presented by Hill and Wheeler. The fission problem is to be treated on the lines of the quantum mechanical theory of α-decay but it is more difficult than α-decay because the exact shape of the fission potential barrier is not known. Hill and Wheeler assumed a one dimensional inverted harmonic oscillator type of barrier $\left(-\dfrac{1}{2} B \omega_l^2\, \alpha_l^2\right)$ where B is inertia parameter, α_l is the deformation parameter and ω_l is the circular frequency of vibration of the deformed drop, then one can write the Hamiltonian as

$$H = \frac{\hbar^2}{2l} \frac{\partial^2}{\partial b_l^2} - \frac{1}{2} B\, \omega_l^2\, b_l^2 + E_f \qquad \text{... (7.1)}$$

The Gamow barrier penetration factor P is given by

$$P = \left[1 + e^{\frac{-2\pi(E - E_f)}{\hbar\omega_l}}\right]^{-1} \qquad \text{... (7.2)}$$

At some value of w_l, the nucleus is in the saddle point configuration for which $E = E_f$. This gives $P = \dfrac{1}{2}$ and

$$\hbar\omega_l = \hbar \left[-\frac{5}{4\pi} \frac{E_{S_0}}{B_2} \frac{\partial^2 E_f}{\partial \alpha^2}\right]^{\frac{1}{2}}$$

The kinetic energy of the deformed drop in terms of normal coordinates may be written as,

$$T = \frac{1}{2} \sum_{l=2}^{\infty} B_l\, (b_l)^2 \qquad \text{... (7.3)}$$

Where B_l is a function of the nuclear radius and mass and represent the instantaneous changes in B_l. The potential energy of the deformed drop may be written as

$$V = \frac{1}{2} \sum_{l=2}^{\infty} C_l\, b_l^2 \qquad \text{... (7.4)}$$

Where C_l = stiffness constant and is a function of the nuclear radius, charge and area. Thus when the nucleus is thrown into vibrations and is deformed, its Hamiltonian is given by

$$H = T + V = \frac{1}{2} \sum_{l=2}^{\infty} (C_l b_l^2 + B_l b_l^2) \qquad \text{... (7.5)}$$

and the characteristic (normal mode) frequencies of vibration are given by

$$\omega_l = \left(\frac{C_l}{B_l}\right)^{\frac{1}{2}} \qquad \text{... (7.6)}$$

These are the frequencies of a harmonic oscillator whose displacement is given by

$$b_l(t) = a_1 \cos(\omega_l + \phi_l)t \qquad \text{... (7.7)}$$

At some frequency ω_l the nucleus is in the critical state to break up. The penetration factor P then is equal to $\dfrac{1}{2}$ and $E = E_f$. Thus we have

$$\hbar\omega_l = \hbar \left[\frac{5 E_{S_0}}{4\pi B^2} \frac{\partial^2 E_f}{\partial b_2^2} \right] \qquad \qquad ...(7.8)$$

$\hbar\omega_l$ is ≈ 1 MeV for nuclei with $A \approx 240$. An estimation of the half life t_0 for spontaneous fission from this calculation gives

$$t_0 = \left[2\pi\,\omega_2\, e^{\frac{-2\pi(E_f - E)}{\hbar\omega_l}} \right]^{-1} \qquad \qquad ...(7.9)$$

E_f was calculated by Seaborg by drawing a graph between spontaneous fission half life and $\frac{Z^2}{A}$ and obtained E_f to be given by

$$E_f = 19.0 - 3.36 \frac{Z^2}{A} \qquad \qquad ...(7.10)$$

given by (in seconds)

$$t = 10^{-21} \times 10^{178 - 3.75 \left(\frac{Z^2}{A} \right)} \qquad \qquad ...(7.11)$$

The value obtained in (7.11) is in contrast to the expression obtained by Frenkel and Metropolis on the basis of barrier penetration probability which gives

$$t = 10^{-21} \times 10^{7.85} E_f \qquad \qquad ...(7.12)$$

Using these results, Seaborg found

$$E_f = 19.0 - 0.36 \frac{Z^2}{A} \qquad \qquad ...(7.13)$$

For a narrow range of nuclei

We may note that the zero-point energy calculated from the shape of the crater near zero deformation has a value ~ 0.4 MeV. This value is small compared to the excitation involved which justifies classical considerations.

Comparison between theory and experiment On the basis of Bohr-Wheeler theory, one obtains $E_f (\sim 15$ MeV$)$ where as the experimental value is $E_f (\sim 6$ MeV$)$. Obviously, the agreement is very poor. It shows that the activation energy $E_f = (\Delta E)_{cr}$ is not as sensitive to

change in $x = \left(\dfrac{\dfrac{Z^2}{A}}{\left(\dfrac{Z^2}{A} \right)_{crit}} \right)$. Quantum corrections discussed above also makes not much

difference. Further, the observed threshold energies are also not smooth functions of x. Moreover, liquid drop model also fails to account for asymmetric fission, which is one of its major drawbacks.

7.2.9 Charge Distribution of Fission Products

In the preceding sections, we saw that a heavier nucleus (i.e., parent fissioning nucleus) with a high $\dfrac{N}{Z}$ ratio undergoes fission with thermal neutrons into two neutron excess fragments of unequal masses, which decay by successive β^- emission producing the respective stable nuclide isobaric with the original fission fragments. For two given complementary series (i.e., the initial or primary fragments of fission) of mass numbers A_1 and A_2 of a parent fissioning nucleus of charge Z_p, let Z_1 and Z_2 be the charges (i.e. atomic numbers) of the stable isobars formed in the fission decay chains for the isobaric series of mass numbers A_1 and A_2, respectively. Let us consider that Z_1^0 and Z_2^0 are the most probable charges of the original fragments of mass numbers A_1 and A_2, respectively. For a given value of A_1 because of neutron excess, it is expected that $Z_1^0 < Z_1$ and so does $Z_2^0 < Z_2$ for the value A_2. Because of too short half-lives of the products in the series of β^- decay Z_1^0 and Z_2^0 cannot be directly deduced. The theory of the equal charge displacement (ECD) given by Glendenin, Cargell and Edwards, however, provides a convenient method for the charge distribution of heavy and light fragments of a parent fissioning nucleus. The theory assumes that the most probable charges for the heavy and the corresponding light fragment are equidistant from their final stable values for the given complementary series. Accordingly, we get

$$Z_1 - Z_1^0 = Z_2 - Z_2^0 \qquad \qquad ...\,(7.1)$$

Since the two fragments are obtained from the heavy nucleus, initial charge conservation satisfies the relation,

$$Z_p = Z_1^0 + Z_2^0 \qquad \qquad ...\,(7.2)$$

and so does the mass conservation,

$$A_p = A_1 + A_2 + \bar{v} \qquad \qquad ...\,(7.3)$$

Where $A_p = A + n$ with A as the mass number of the parent fissioning nucleus and $\bar{v}$ is the average number of neutrons emitted per fission. Combining Equation (7.1) and (7.2) we get,

$$Z_1^0 = \frac{1}{2}\,(Z_p + Z_1 - Z_2)$$

$$Z_2^0 = \frac{1}{2}\,(Z_p + Z_2 - Z_1) \qquad \qquad ...\,(7.4)$$

For example, in use of U^{236} $(= U^{235} + n)$ where $Z_p = 92$, and $\bar{v} = 2.5$ neutrons

$$Z_1^0 = \frac{1}{2}\,(92 + Z_1 - Z_{233.5 - A_1}), \text{ and similarly,}$$

$$Z_2^0 = \frac{1}{2}\,(92 + Z_2 - Z_{233.5 - A_2})$$

The values of the nuclear charges Z_1 for $A_1 = 90$ and Z_2 for $A_2 = 144$ can be calculated and hence the values for the most probable charges of the original fission fragments.

Futhermore, the theory assumes that the probability distribution around Z_p, the atomic number of fragment having the highest yield in a single chain is the same for all values of A. So the distribution around the most probable value of Z_f is given by the Gaussian equation for the fission caused by thermal neutrons,

$$P(Z) = \frac{1}{(1.5\pi)^{\frac{1}{2}}} \exp\left\{\frac{(Z - Z_f)^2}{1.5}\right\} \qquad \ldots (7.5)$$

Where Z in the atomic number of the fragment of the mass number A whose fractional yield is plotted against $(Z - Z_f)$ giving the Gaussian curve (Figure 7.35).

It must, however, be mentioned that the rule (ECD) is an empirical working hypothesis based on the experimental data for the fission reactions induced by low energy projectiles. Further, the theory suggests that the complementary chains of equal length are more probable than chains of different length.

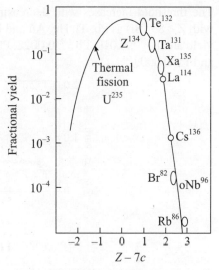

Fig. 7.35 Fractional yield of the fission fragments

7.2.10 Ionic Charge of Fission Fragments

As stated earlier, the fission of a heavy nucleus caused by the thermal neutrons leads to the rupture of the nucleus into two fragments. As a result of the rupture, a large fraction of the orbital electrons is cast off from the fissioning nucleus because of internal conversion. Consequently, the fission fragments losing orbital electrons becomes highly positively charged ions, such as Sr^{20+}, Ba^{22+} etc. Thus, the initial ionic charge of the fission fragments is high. But the charge falls off as the fragments pass through the medium. First, the charged fragments lose energy through excitation and ionization with the result that their velocity diminishes and consequently they slow down while passing through the matter. Secondly, along their track they capture electrons into their vacant orbitals, so that their ionic charge falls. Greater they slow down, the more the electron capture and hence greater the decrease in the ionic charge.

7.2.11 Fission by Fast Neutrons

So far, we have discussed the fission yield for the fission of ^{235}U with thermal neutrons ($E_n = 0.025$ eV and $v = 2200$ m/sec). As the energy of the incident neutron increases, the fission yield of U^{235} decreases according to $\frac{1}{v}$ law, where v is the neutron velocity. However, the fission of ^{238}U can be brought about only by fast neutrons of energy 1 MeV. This is also true for most other heavy nuclides. The threshold neutron energies required to effect fission in some of the heavy nuclide are recorded in Table 7.13:

Table 7.13

Nuclide	^{209}Bi	^{230}Th	^{232}Th	^{231}Pa	^{234}U	^{238}U	^{237}Np
E_n/MeV	15	~1	1.05	0.45	0.28	0.92	0.25

The threshold increases with decreasing Z. By using 84-MeV neutrons. The fission of nuclides with Z < 83 (Bi) as Pb, Tl. Hg, Au and Pt has been reported. The fission yields relative to Th (n, f) yield are: Bi^{209}: 0.019, Pb^{208}: 0.0028, Pb^{207}: 0.010, Pb^{206}: 0.007, Tl: 0.0032, Hg: 0.0023, Au: 0.002, and Pt: 0.0009

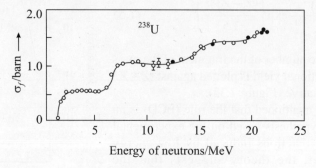

Fig. 7.36 Fission of ^{238}U

The cross sections for fast neutrons (from 0.1 to about 25 MeV) in the fission of various heavy nuclides have been reported. When the energy of the incident neutron exceeds 1 MeV, the excitation functions are divided into two categories; firstly, the nuclei, for which $B_f > S_n$, have an excitation function appearing in the range beyond thermal energies and the zone of resonances. The case of U^{238} is included in this category and its excitation function up to 12 MeV is displayed in Figure 7.36. An inspection of the figure shows that the σ_f rises rapidly beyond the threshold (0.02 MeV) and then levels off followed by a second rise which is due to the (n, n, f) reaction in which the evaporation of a neutron takes place in addition to the (n, f) reaction. In this case the nucleus is left in an excited state above the fission threshold. This situation is referred to as second chance fission. Similar curves are also obtained with other heavy nuclides, such as ^{231}Pa, ^{234}U, ^{237}Np, ^{240}Pu and ^{241}Am.

The nuclei whose σ_f is large in the region of thermal neutrons (i.e., $B_f < S_n$) are included in the second category. With increase in the neutron velocity v, the cross section varies with $\frac{1}{v}$ and reaches a value of about 1 b. The σ_f rises rapidly up to about 6 – 8 MeV followed by a second leveling off. Figure 7.37 shows the fission of different heavy nuclides.

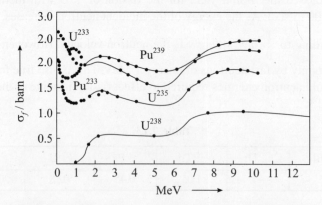

Fig. 7.37 Fission cross sections of various nuclides.

7.2.12 High Energy Fission

We have seen that the low-energy fission induced by thermal neutrons, by photons, deuterons or neutrons under about 10 MeV. Bohr compound nucleus mechanism appears to account satisfactorily for the features of nuclear reactions. However, at still higher energies new phenomena with respect to the fission process are observed which include not only much higher excitation energies but also nuclei which are lighter than ^{232}Th and whose $\dfrac{Z^2}{A}$ is as low as 30. As in the case of low-energy fission, the high energy fission results in the formation of two fragments of high kinetic energy. However, difference between these two fissions lies in two respects: first, the two fragments in the high energy fission frequently have the same mass and charge, i.e., this fission has a tendency to become symmetric, and secondly, the nucleus undergoing fission is only in rare cases a compound nucleus and may very often be a nucleus produced from different prior nuclear processes. Thus, it is apparent from these facts that the fission fragments are accompanied by a large number of particles, and it is of interest to know whether these particles are emitted before, during or after the fission phenomenon. Finally, because of high incident energy, the rupture of the target nuclei into the product nuclei of low Z is more probable. We have been concerned in this section with the high energy processes involving the nucleon evaporation as well as the emission of a large number of fragments corresponding to a very highly excited compound nucleus.

Neutron Evaporation

The characteristics of fission at high energies in the range $100 - 400$ MeV are markedly different from those of thermal neutron fission in the sense that the double hump in the low energy fission product curve is replaced by a single broad peak at about A somewhat less than half of the mass number of the target nuclide. All the primary fission products have nearly the same $\dfrac{N}{Z}$ ratio for a given target and given bombarding energy. According to Meitner, the compound nucleus because of excess of neutrons, (first evaporates a large number of neutrons in one lot forming an excited nucleus which subsequently undergoes fission into two similar fragments. These fragments in turn decay by β-emission, as in fission, giving rise to stable and isobaric nuclide. As an illustration the yield curve for the fission of ^{209}Bi by 190-MeV deuterons, which shows a broad peak at $A = 100$ (shown in Figure 7.38) follows the reaction ^{209}Bi $(d, 12n)$ ^{199}Po.

Meitner explained the symmetrical mass distribution in the case of ^{209}Bi $+ d$ reaction, assuming that after the evaporation of 12 neutrons, the compound nucleus because of insufficient neutrons does not form two fragments with 50 and 82 neutrons, respectively, rather splits into two fragments of nearly equal mass. Thus, the ^{209}Bi (d, f) reaction follows:

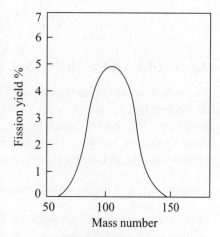

Fig. 7.38 Mass distribution of fission products of Bi^{209}.

$$^{209}\text{Bi} + d \rightarrow 12n + {}^{199}\text{Po} \xrightarrow{\text{Symmetric fission}} \begin{array}{c} ^{99}_{42}\text{Mo} \\ \downarrow 2\beta^- \\ ^{99}_{44}\text{Ru (stable)} \end{array} + {}^{100}_{42}\text{Mo} \\ \text{(stable)}$$

It may be noted that the two fission fragments (i.e., ^{99}Mo and ^{100}Mo) have each nearly the same neutron excess (i.e., 7 and 8, respectively) relative to the nearest magic number (closed shell) 50.

7.2.13 Fissile and Fertile Materials

We have read that of the three isotopes of uranium, which are fissionable with thermal neutrons, only ^{235}U is naturally occuring with an abundance of 0.71 % in natural uranium. The other two nuclides which can undergo fission with thermal neutrons are ^{233}U and ^{239}Pu do not exit in nature. These two can be produced artificially by the interaction of neutrons with ^{232}Th and ^{238}U respectively. These are called as *fissile materials*. We may note that ^{232}Th and ^{238}U are not fissile materials but can be used as raw material for the production of fissile isotopes and they are named as *fertile* or *fissionable* materials. The relative abundances of the above two fertile isotopes are quite high, i.e. 99.3 % for ^{238}U in natural uranium and 100 % for ^{232}Th.

The nuclear reactions which convert these fertile materials into fissile materials neutron capture processes with subsequent β-decay

$$^{238}_{92}\text{U} + {}^{1}_{0}\text{n} \rightarrow {}^{239}_{92}\text{U} + \gamma \qquad (\tau = 23.5 \text{ min})$$

$$^{239}_{92}\text{U} \xrightarrow{\beta^-} {}^{239}_{93}\text{Np} \qquad (\tau = 2.35 \text{ days})$$

$$^{239}_{93}\text{Np} \xrightarrow{\beta^-} {}^{239}_{94}\text{Pu} \qquad (\tau = 2.411 \times 10^4 \text{ years})$$

The nuclear reactions involved in the production of ^{233}U are:

$$^{232}_{90}\text{Th} + {}^{1}_{0}\text{n} \rightarrow {}^{233}_{90}\text{Th} + \gamma \qquad (\tau = 22.2 \text{ min})$$

$$^{233}_{90}\text{Th} \xrightarrow{\beta^-} {}^{233}_{91}\text{Pa} \qquad (\tau = 27 \text{ days})$$

$$^{233}_{91}\text{Pa} \xrightarrow{\beta^-} {}^{233}_{92}\text{U} \qquad (\tau = 1.592 \times 10^5 \text{ days})$$

7.3 NUCLEAR FISSION AS A SOURCE OF ENERGY

As pointed out earlier, the fission of ^{235}U with thermal neutrons releases about 200 MeV along with, on the average, 2.5 neutrons per fission. These neutrons may not all be capable of inducing fission with other ^{235}U nucleus since their energy may be a few MeV, and at these energies the fission cross section, σ_f is 1.44 b. However, irrespective of the energy, the fission cross section is always greater than the radiative capture cross section, i.e., $\sigma_f > \sigma(n, \gamma)$ and this eliminates the neutrons without leading to fission. But at thermal level $\dfrac{\sigma_f}{\sigma(n, \gamma)} = 5.7$ which indicates that about 80 per cent of the liberated neutrons are available for fission. Moreover, in a volume of pure U^{235} which is large enough to prevent the escape of neutrons without their first encountering a nucleus, a single fission due to a neutron would produce a chain reaction in which the liberated neutrons produce more fissions and more neutrons, and so on, till all the uranium nuclei are used up. Under some conditions, the numbers of fissions produced and neutrons liberated increase exponentially with time because each fission produces more neutrons than the one used up. In this way, the amount of energy released in a fission process can become enormous.

The fission process differs from other nuclear reactions in a sense that in the latter cases each incident particle brings about the transmutation of a single nuclide only, while in the former, a single neutron can lead to the fission of all the nuclei present. It is interesting from the point to the fission of all the U^{235} nuclei present. It is interesting from the point of view of chain reactions to think of the initial neutron as the burn which is applied to a heap of combustible materials. We know that the heat of the burner causes a part of the material to burn, and the resulting heat brings combustion on other portions, until the whole heat of materials is burnt up and its energy liberated. Thus, combustion is regarded as a thermal of heat chain process. On the other hand, neutrons initiate a branching chain of fission.

Since each fission releases about 200 MeV, the energy released by all the nuclei of ^{235}U in a kilogram of this element can be calculated. Given that 1 MeV $= 1.6 \times 10^{-6}$ ergs, approximately 200 McV released in a single fission $- 3.2 \times 10^{-4} - 3.2 \times 10^{-11}$ Watt sec, (since watt is a unit of power). Which is defined as the rate of production or expenditure of energy and equal to 107 ergs sec^{-1}). Since in 1 kg of ^{235}U there are $6.02 \times 10^{23} \times \dfrac{1000}{235}$ atoms of U^{235}, the fissioning of all these atoms releases a total energy of $3.2 \times 10^{-11} \times 6.02 \times \dfrac{10^{26}}{235} = 8.2 \times 10^{13}$ Watt-sec, i.e., the energy released by the complete fission of 1 kg of U^{235} would be 8.2×10^{10} kilo Watt-sec, this is equivalent to $8.2 \times \dfrac{10^{10}}{3600} = 2.28 \times 10^7$ kilo Watt-hours, or $2.28 \times \dfrac{10^7}{24} = 95 \times 10^4$ kilo Watt $\times$ days $= 950$ MW. Hence, the (heat) power production corresponding to the fission of 1 kg of ^{235}U per day would thus be 950 MW. To obtain this amount of power by combustion would require 2500 tons of coal per day. This shows the much larger order of magnitude of nuclear energy as compared with chemical energy. In terms of destructive capacity the release of so much of energy by the fission of 1 kg of ^{235}U is equivalent to an explosion of 20 kilotons of TNT. The fact that 1 kg of a fissioning material such as ^{235}U is equivalent to 2500 tons of coal as a source of electric power led to the use of nuclear chain reactions and to the setting up nuclear power plants.

7.3.1 The Nuclear Chain Reaction

In the preceding section, we saw that the thermal neutron induced fission of ^{235}U is accompanied by the liberation of about 200 MeV as well as by the emission of 2.5 neutrons per fission.

However, the following processes may occur:

The neutrons produced during the fission may face the following processes:

(i) The neutron, prior to being slowed down by moderator, may be absored in ^{238}U to cause fission (fast) if its energy is greater than fission threshold energy.

(ii) While slowing down, the neutron may be absorbed in ^{238}U to cause non-fission reactions, mainly radiative capture reaction.

(iii) neutron may also be absorbed in ^{235}U to cause non-fission reactions.

(iv) after being slowed down to thermal energy, the neutron may cause fission of ^{235}U.

(v) neutron may be absorbed by other materials present in the reactor, i.e. moderator, control materials, etc.

(vi) there may be leakage of neutron from the reactor assembly.

These neutrons can of course bring about further fissions in their turn, so that the number of neutrons multiplies and the overall rate of the process increases continuously. In this way, a chain reaction is possible, and the amount of energy liberated becomes enormous. Since nuclear processes take place rapidly the time interval between successive generations of fissions will be a very small fraction of a second. Hence, the release of so much energy in the small fraction of a second in the chain reaction can take the form of an explosion: the result is an "atomic bomb". Thus, a fission chain forms the basis of a powerful atomic bomb. It follows that the release of a staggering amount of energy in a fission chain within a very short-time interval leads to the explosive disappearance of the fissioning nuclide present.

The chain reaction can, however, be controlled under suitable conditions, and a steady state of self-propagated fission chains can be attained. The number of fissions per second and the rate at which energy liberated are kept constant. This condition is achieved in a device known as chain-reacting pile, i.e., nuclear reactor which can be used as a source of neutrons or of power.

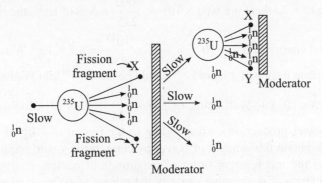

Fig. 7.39 Chain reaction

Four factor formula

Let us consider the absorption of n thermal neutrons in ^{235}U in an infinite reactor assembly of natural uranium and a moderator. These neutrons may cause fission of ^{235}U after the neutron has been slowed down to thermal energy. Now, if n be the fast neutrons emitted per fission of ^{235}U, then the total number of fast neutrons produced due to the fission of n nuclei of ^{235}U by thermal neutrons is $n\nu$. However, the number of fast neutrons produced is in fact slightly greater than $n\nu$ due to the fast fission of some of ^{238}U nuclei. The neutron, prior to being slowed down may be absorbed in ^{238}U nuclei and causes fast fission provided its energy is greater than the fission threshold energy. Let us suppose that the total number of fast neutrons produced due to fission of ^{235}U in increased by a factor, say $\varepsilon > 1$ due to fast fission of ^{238}U and this is over the number $n\nu$. Here ε is termed as *fast fission factor*. Obviously, the total number of fast neutrons produced is $n\nu\varepsilon$.

We have already mentioned that the fast neutrons produced by fission are slowed down in moderator (say graphite). Some neutrons are absorbed during the slowing down process in ^{238}U to cause non-fission reactions, i.e. radiative capture reaction, etc. and may also be absorbed in ^{235}U to cause non-fission reactions or by other materials present in the reactor, e.g. moderator, coolant, control rods, etc. However, the primary contribution comes from ^{238}U. ^{238}U absorbs neutrons in the epithermal region to produce (n, γ) reaction at some specific energies (resonance capture). Those neutrons which escape such resonance capture by finally slowed down to thermal energies by moderator. Let us suppose that a fraction p of neutrons escape resonance capture while slowing down. We call p as the *resonance escape probability*. Obviously, the number of neutrons slowed down to thermal energies is $n \vee \varepsilon p$ while the number $n\vee \varepsilon(1 - p)$ is clearly the number of neutrons captured to produce non-fission reactions while slowing down. One can estimate p from a knowledge about the non-fission capture cross-sections and their dependence on the neutrons energy.

The above number of neutrons after attaining the thermal energy, may be absorbed either by the nuclear fuel (here uranium) or in other materials present in the reactor. The fraction (say f) of thermal neutrons which are absorbed by fuel is called as the *thermal utilization factor*. Clearly, the total number of thermal neutrons absorbed in fuel is $= n\vee \varepsilon pf$ and the rest, i.e. $(1 - f) \, n\vee \varepsilon p$ is absorbed in other materials present in the reactor.

We may note that all the thermal neutrons absorbed in the fuel (^{235}U) actually do not produce fission. Some thermal neutrons absorbed produce non-fission reactions in both ^{235}U and ^{238}U. Let in the second generation, g be the fraction of thermal neutrons absorbed in the fuel to produce fission (of ^{235}U). Thus the total number of thermal neutrons available for producing fission in the second generation due to thermal neutron fissions in the first generation fission will be

$$N = n \vee \varepsilon \, p \, f \, g \qquad \qquad \ldots (7.1)$$

Thus the infinite multiplication factor can be written as

$$k_\infty = \frac{N}{n} = (\vee g)\varepsilon pf \qquad \qquad \ldots (7.2)$$

If $\eta = \vee g$, i.e. η gives the number of fast neutrons produced for each thermal neutron absorbed in the fuel (say natural uranium). One obtains

$$k_\infty = \eta \, \varepsilon pf \qquad \qquad \ldots (7.3)$$

Relation (7.3) is known as *four factor formula*.

Finite Size Correction

In the above treatment, we have assumed that there is no escapes of neutrons from the reactor. However, due to finite size of the reactor, there is always some leakage of neutrons from the reactor, both during the process of slowing down and also after the attainment of thermal energy. This means that the actual multiplication factor k becomes less than k_∞ ($k < k_\infty$). We may note that the fraction of neutrons escaping depend on the ratio of the surface area of the reactor faces to the volume of the reactor and this is larger for a smaller reactor. When the size of the reactor is too small, then there will be so much leakage of neutrons that the reactor may not become self-sustaining, i.e. $k < 1$ for such a reactor. The reactor is said to be *sub-critical* under this condition. Now, if the size of the reactor is such that inspite of escape or loss of neutrons due to leakage, the self sustained fission chain reaction goes on in the reactor so that $k = 1$. This condition for the reactor is called *critical condition*. The third situation is for relatively large size reactor. In this case to a great extent the loss of neutrons due to leakage will be minimised making $k > 1$. This condition is known as *super-critical* condition and the rate of chain reaction increases from one generation to the next.

Let the fraction of neutrons leaking out during the process of slowing down is ls, then $(1 - l_s)$ will be the fraction not leaking out during this stage of fission. Similarly, if after thermalization the fraction of neutrons which leak out is l_{th}, then the fraction of thermal neutrons not leaking out = $(1 - l_{th})$. Writing the non-leakage factor by combining the above two as, $l = (1 - l_s)(1 - l_{th})$, we obtain

$$k = k_\infty l = k_\infty (1 - l_s)(1 - l_{th}) = \eta \varepsilon p f (1 - l_s)(1 - l_{th})$$

For an reactor of infinite size there is no leakage and hence $l_s = l_{th} = 0$ and $k = k_\infty = \eta \varepsilon p f$.

Obviously, one can manipulate k by the values of p and f which depends on the geometry of the arrangement of the fuel and the moderator, i.e. homogeneous or heterogeneous.

7.3.2 Conditions for Controlled Chain Reactions

The growth of the chain reaction can be controlled by choosing appropriate conditions, so that a steady state is attained in a nuclear reactor. Several factors, with a common objective of preventing the propagation of chain, involving the suitable method either by slowing down the fission neutrons or allowing on the average, at least one of the fission neutrons to cause another fission can, however, be considered. The ratio of the number of fissions produced by a particular generation of neutrons to the number of fission giving rise to these neutrons, is called the multiplication factor, or reproduction factor, k; the condition for the maintenance of the steady state chain, on average, at least one fission neutrons induce another fission is $k = 1$. If $k > 1$ the number of neutrons and so the number of fission, increases in each generation, and the chain is said to be divergent. In such a case, most of the fissioning material is used up and a terrific amount of energy is liberated in an extremely short time which leads to explosion. If, on the other hand, $k > 1$, the reaction comes to a stand still after a limited number of generations. Because the number of neutrons available for fission gradually decreases with time and finally, the chain converges. Thus, $k = 1$ indicates a "critical" state in which fissions are maintained as opposed to a "sub critical" state when $k < 1$, and $k > 1$ refers of a "super-critical" state.

Since one neutron per fission is required to propagate the chain reaction, the number of neutrons increases in each generation by a factor of $(k - 1)$ in a super-critical mass. If N is the number of neutrons present at a time t after the fission starts, and t is the average time between successive neutron generations, the increase in the number of neutrons in a time t between two generations is $N(k - 1)$. At the end of time t, even if it is short, the rate of change of the number of neutrons in a chain reaching system is,

$$\frac{dN}{dt} = \frac{N(k-1)}{\tau}$$

The number of neutrons at time t is given by

$$N_t = \int_0^t N(k-1) \frac{dt}{\tau} = N_0 \exp\left[\frac{(k-1)t}{\tau}\right] \qquad \qquad \ldots (7.1)$$

where N_0 is the number of neutrons at $t = 0$, and $\dfrac{t}{\tau}$ represents the number of fission generations that have elapsed during the time t.

The reactivity of a mass experiencing a chain reaction is given by

$$\frac{N_t}{N_0} = \exp\frac{(k-1)t}{\tau} \qquad \qquad \dots (7.2)$$

significance of k will be discussed in the following section with relation to the thermal neutron reactors.

Resonance escape probability

Let us suppose that q_0 fast neutrons are produced uniformly throughout the reactor/volume/s. Due to scattering collisions with moderator nuclei, the neutrons are slowed down to lower energies. Thus there will be an energy distribution of neutrons from source energy (fast) to thermal energy.

Let us first assume an infinite system with no absorption. Obviously, for such a system the slowing down density q for all energies is the same, i.e. $q = q_0$. We can easily see that there exist a relation between q and neutron flux ϕ as

$$\phi = \frac{q}{\xi \, \Sigma_s \, E} \qquad \qquad \dots (7.1)$$

Where $\Sigma_s = N_s \, \sigma_s$ the macroscopic scattering cross section, ξ is the mean logarithmic decrement in energy, N_s is the number of scattering nuclei per unit volume and σ_s is the microscopic scattering cross section.

Now, taking into account of the neutron absorption, Equation (7.1) is modified as

$$\phi = \frac{q(E)}{\xi(\Sigma_s + \Sigma_a)E} \qquad \qquad \dots (7.2)$$

Where $\Sigma_a = N_a \, \sigma_a$ is the macroscopic absorption coefficient, N_a is the number absorbing nuclei/volume and σ_a is the microscopic absorption cross-section.

We may note that $q(E)$ in (7.2) is no longer constant. Because of neutron absorption in the energy interval dE, as the energy is reduced from E to $E - dE$, q is reduced to $q - dq$. Thus, we have the number of neutrons absorbed in the energy interval dE as

$$dq = \phi \, \Sigma_a \, dE \qquad \qquad \dots (7.3)$$

Using (7.1), one obtains

$$\frac{dq}{q} = \frac{\Sigma_a \, dE}{\xi(\Sigma_s + \Sigma_a)E}$$

or

$$\int_{q_0}^{q} \frac{dq}{q} = \int_{E_0}^{E} \frac{\Sigma_a}{\xi(\Sigma_s + \Sigma_a)} \frac{dE}{E} \qquad \qquad \dots (7.4)$$

The resonance escape probability is given by

$$p = \frac{q}{q_0} = \exp\left[-\int_E^{E_0} \frac{\Sigma_a}{\xi(\Sigma_a + \Sigma_s)} \frac{dE}{E}\right] \qquad \dots (7.5)$$

σ_s is almost constant in the epithermal region (~ 1 eV to 10 keV). Since most of the absorption takes place in this region and therefore, one finds

$$p = \exp\left[-\frac{N_a}{\xi \Sigma_s} \int_E^{E_0} \frac{\sigma_0}{1 + \left(\frac{N_0}{\Sigma_s}\right)\sigma_a} \frac{dE}{E}\right] \qquad \dots (7.6)$$

Now, writing

$$(\sigma_a)_{eff} = \left(\frac{\Sigma_s \sigma_a}{\Sigma_s + \Sigma_a}\right) = \left[\frac{\sigma_a}{1 + \left(\frac{N_0}{\Sigma_s}\right)\sigma_a}\right]$$

one obtains

$$p = \exp\left[-\frac{N_a}{\xi \Sigma_s} \int_E^{E_0} (\sigma_a)_{eff} \frac{dE}{E}\right] \qquad \dots (7.7)$$

The integral within (7.7) is termed as *'effective resonance integral'* having dimensions of an area.

Homogeneous reactor   In this type of a reactor, the fuel and the moderator are intimately mixed in the form of a solution or a slurry. In general, this type of reactor comprises the solution of some salt of uranium or plutonium in H_2O or D_2O.

In a highly dilute solution, i.e. $N_a \ll N_s$, one has $\Sigma_a \ll \Sigma_s$. The integral in Equation (7.5) takes the form

$$\frac{1}{\xi} \int_E^{E_0} \frac{\Sigma_a}{\Sigma_s + \Sigma_a} \frac{dE}{E} \approx \frac{1}{\xi} \int_E^{E_0} \frac{\Sigma_a}{\Sigma_s} \frac{dE}{E} \qquad \dots (7.8)$$

In a reactor in which $\Sigma_a \gg \Sigma_s$, i.e. ratio of moderator to fuel is small, then the integral in Equation (7.5) becomes quite large. In this case p becomes *small*.

From Equation (7.7), we note that the effective resonance integral depends on the ratio $\dfrac{\Sigma_s}{N_a}$ or the scattering cross section associated with each absorber atom. The value of this integral is not affected by the nature of the moderator used, i.e. there is not much difference in the resonance, absorption in reactors using different moderators, e.g. H_2O, D_2O and graphite. For natural uranium as fuel, the effective resonance integral at high dilutions has the value ~ 282 barns and ~ 70 barns for ^{232}Th as fuel.

To calculate the effective resonance integral (ERI) for a homogeneous reactor, one can use the following empirical relationship

$$\text{ERI} = 3.9 \left(\frac{\Sigma_s}{N_a}\right)^{0.415} \text{ barns} \qquad \dots (7.9)$$

Heterogeneous Reactor

One can decrease the value of ERI by using a heterogeneous arrangements in which lumps or rods of uranium are arranged in a lattice within the reactor's body.

Heterogeneous arrangement has the advantage that the resonance neutrons are almost completely absorbed in the outer layers of the uranium lumps, which is facilitated by using lumps with a small surface to volume ratio, i.e. using larger lumps. This means that as the thermal neutrons enter the latter, they undergo only fission absorption in the interior of the lumps, but no resonance absorption to any appreciable extent. The fast neutrons are mainly slowed down in the moderator. Since there are no ^{238}U nuclei are present in the neighbourhood during this process and hence they will be slowed down past the epithermal region without any appreciable absorption. Due to these reasons ERI tends to decrease and resulting in an increase of p, so that $p_{\text{hetero}} > p_{\text{homoge}}$.

Equation (7.6) for p is also applicable to a homogeneous system in which both fuel (u) and moderator are exposed to the same neutron flux. This is modified for a heterogeneous system and approximated as

$$p = \exp\left[-\frac{N_u V_u \phi_u}{\xi \sum_{sm} V_m \phi_m} \int_E^{E_0} (\sigma_{au})_{eff} \frac{dE}{E}\right] \qquad \dots (7.10)$$

Where V_m and V_u denotes the volumes of the fuel and the moderator respectively and ϕ_u and ϕ_m denote the average flux of neutrons for the two.

Based on experimental data, an empirical relationship for the effective resonance integral gives for a heterogeneous reactor

$$\text{ERI} = A\left[1 + \frac{\mu}{(r + r_0)}\right] \qquad \dots (7.11)$$

Where A, μ and r_0 are constants having values $A = 10.2$ barns, $\mu = 29.5$ kg/m^2 and $40 = 1.1$ kg/m^2 for natural uranium metal and for thorium metal the values of these constants are: $A = 8.37$ barns, $\mu = 5.5$ kg/m^2 and $r_0 = 0.8$ kg/m^2 and r is the mass to surface area ratio of the resonance absorber measured in kg/m^2.

Thermal Utilization Factor: Homogeneous and Hetrogeneous reactors

Thermal utilization factor ($+f$) is defined as the ratio of the number of thermal neutrons absorbed in the fuel to the total number absorbed in all the processes. Let us denote the fuel by subscript 1. Now, we write foa a reactor in which the different components having volumes V_1, V_2, V_3, etc. face the neutron fluxes ϕ_1, ϕ_2, ϕ_3, etc. Thus, we have

$$f = \frac{\text{Thermal neutrons absorbed in fuel}}{\text{Total number of thermal neutrons absorbed}}$$

$$= \frac{\sum_{a_1} V_1 \phi_1}{\sum_{a_1} V_1 \phi_1 + \sum_{a_2} V_2 \phi_2 + \sum_{a_3} V_3 \phi_3 + \dots} \qquad \dots (7.12)$$

For a homogeneous reactor, we have $\phi_1 = \phi_2 = \phi_3 = ...$, and $V_1 = V_2 = V_3 = ...$. Thus, one obtains

$$f = \frac{\sum_{a_1}}{\sum_{a_1} + \sum_{a_2} + \sum_{a_3} + ...}$$

$$= \frac{N_1 \, \sigma_{a_1}}{N_1 \, \sigma_{a_1} + N_2 \, \sigma_{a_2} + N_3 \, \sigma_{a_3} + ...} \qquad ... (7.13)$$

Where N_1 is the number of nuclei of the fuel per unit volume, N_i is the number of nuclei per unit volume of the ith component and $\sigma_a's$ are the microscopic absorption cross section.

Now, assuming that only fuel (u) and moderator (m) nuclei are to be present, one finds

$$f = \frac{N_u \, \sigma_{au}}{N_u \, \sigma_{au} + N_m \, \sigma_{am}} = \frac{\sum_{au}}{\sum_{au} + \sum_{am}} \qquad ... (7.14)$$

We may note that $\sum_{a_1}$, for fuel includes both fission as well as radiative capture. For uranium,

$\sum_{a_1} = \sum_{au}$ includes absorption in both ^{235}U and ^{238}U.

Heterogeneous Reactor

The moderator and fuel assembly in a graphite moderated nautural uranium reactor has a heterogeneous assembly, the thermal utilization factor f is smaller for this assembly due to the fact that the average thermal neutron flux in the fuel is less than in the moderator. We have the rate of capture of neutron in a given material is equal to the product of the flux and absorption cross section in the material. Thus, one obtains

$$f = \frac{N_1 \, \sigma_{a_1} \, \phi_1 \, V_1}{N_1 \, \sigma_{a_1} \, \phi_1 \, V_1 + N_2 \, \sigma_{a_2} \, \phi_2 \, V_2 + N_3 \, \sigma_{a_3} \, \phi_3 \, V_3 + ...} \qquad ... (7.15)$$

Now, assuming that only fuel (u) and moderator (m) to be present, one finds

$$f = \frac{N_u \, \sigma_{au}}{N_u \, \sigma_{au} + N_m \, \sigma_{am} \, F} = \frac{\sum_{au}}{\sum_{au} + \sum_{am} F} \qquad ... (7.16)$$

Here $F = \left(\dfrac{\phi_m \, V_m}{\phi_u \, V_u}\right)$. This ratio is usually called as the *thermal disadvantage factor.*

We may note that F depends on: (i) the size of the fuel element, (ii) its absorption cross-section, (iii) spacing within the fuel elements and (iv) moderator's neutron diffraction properties.

η (The number of fast neutrons produced per thermal neutron absorbed in the fuel)
One can estimate the value of η as follows. Let us denote the two types of nuclei by the subscripts 1 and 2 for ^{235}U and ^{238}U respectively for natural uranium as fuel. For thermal neutrons, we obtain

$$\eta = \nu g = \nu \frac{N_1 \sigma_{if}}{N_1 (\sigma_{if} + \sigma_{ir}) + N_2 \sigma_{2r}}$$

$$= \nu \frac{\sigma_{if}}{\sigma_{if} + \sigma_{ir} + \left(\dfrac{N_2}{N_1}\right)\sigma_{2r}} \qquad \ldots (7.17)$$

We have $\dfrac{N_2}{N_1} = 139$ for natural uranium and $\sigma_{1f} = 580$ b, $\sigma_{1r} = 112$ b and $\sigma_{2r} = 2.8$ b for thermal neutrons. For the fission of ^{235}U by thermal neutrons, let us assume $\nu = 2.5$. One obtains

$$\eta = \frac{2.5 \times 580}{(580 + 112) + 139 \times 2.8} = 1.34 \qquad \ldots (7.18)$$

We may note that the value of η is determined by the type of the fuel. Above calculations reveal that for every 100 thermal neutrons absorbed in the natural uranium fuel, 134 fast neutrons are produced. Since $k_\infty = \eta \varepsilon p f$ with ε slightly greater that 1, it is possible in principle by suitably adjusting the values of p and f, to make an assembly of natural uranium with a moderator go critical ($k = k_\infty l > 1$).

In case if the fuel used is enriched in ^{235}U content, then one finds the value of η is different given by (7.18). For pure ^{235}U fuel, one obtains

$$\eta = \frac{\nu N_1 \sigma_{if}}{N_1 (\sigma_{if} + \sigma_{ir})} = \frac{\nu \sigma_{if}}{\sigma_{1f} + \sigma_{1r}}$$

$$= \frac{2.47 \times 580}{580 + 112} = 2.07 \qquad \ldots (7.19)$$

Obviously, η is quite large in this case if the fission is induced by thermal neutrons. We may note that η is different for fission induced by fast neutrons in ^{235}U which is of importance in hydrogen bomb. Values of ν and η for the principal isotopes useful for nuclear power production are given in Table 7.14.

Table 7.14 Values of n and h

Atomic nucleus	Thermal ν	neutron η	Fast neutrons	
			ν	η
$^{233}_{92}$U	2.52	2.28	2.70	2.45
$^{235}_{92}$U	2.47	2.07	2.65	2.30
^{239}Pu	2.91	2.09	3.00	2.70

Multiplication Factor (k)

k depends on η, ε, p and f apart from non leakage factors. One can achieve the criticality condition ($k > 1$) with natural uranium as fuel in a homogeneous assembly only if D_2O is used as moderator. D_2O is the best moderator available. However, one can achieve the condition of criticality ($k > 1$) in a heterogeneous assembly using natural uranium as fuel and a moderator of inferior quality, e.g. graphite. Thus for a uranium-graphite reactor in which the molar ratio of carbon to uranium is 225, we have pf = 0.823 for a heterogeneous assembly. For this $k_\infty = \nu \varepsilon p f > 1$. For the same ratio of graphite to U nuclei in a homogeneous assembly pf = 0.595 giving $k_\infty < 1$.

7.3.3 The Principle of Nuclear Reactors

In the preceding section, we saw that the multiplication factor k finds its importance for the maintenance of a chain reaction. As pointed out earlier, such a steady state self-propagated fission chain is established in a nuclear reactor. For steady state operation k is kept equal to unity in a nuclear reactor. However a reactor must be designed in such a way that k can be made slightly greater than unity (say, 1.01 or 1.02) so as to permit an increase in neutron flux and, therefore, in principle to bring to desired level. Thus, it is apparent that the condition for the maintenance of a chain reaction is $k \geq 1$. This is realized in a reactor by adopting two methods; the first is, the fission neutrons are slowed down without loss and transformed into thermal neutrons before inducing other fission reactions, and the second, in the thermal-neutron fission U^{235}, the isotopic composition of uranium (since the total amount of U^{235} in natural uranium is very small being 0.72 %) is changed by increasing the proportion of U^{235}. We shall consider these facts in a nuclear reactor employing natural uranium with thermal neutrons.

(i) **Nuclear Reactor Employing Natural Uranium with Thermal Neutrons**   We have seen that in the fission of ^{235}U by thermal neutrons, 2.5 neutrons are liberated per fission. As pointed out earlier, all these fission neutrons are not capable of inducing fission with other ^{235}U nuclides present, for which the propagation of fission chains in natural uranium may not be possible. This is because, the fission neutrons are of high energy (0.5 – 17 MeV), i.e., fast, and the cross section for the fission of ^{235}U by fast neutrons is not very large. The $\dfrac{1}{\upsilon}$ law is observed in (n, f) reactions with ^{235}U, i.e., the fission cross section decreases with increase of neutron energy. In addition, the natural isotope ratio of the uranium mixture is 0.72 % ^{235}U to 99.28 % ^{238}U and the fast neutrons can fission ^{238}U although with a relatively small cross section, being around 0.01 b for 2 MeV neutrons. Moreover, thermal neutrons are captured by ^{238}U to give ^{239}U by (n, γ) reaction, and there is a strong resonance at neutron energies of about 6 eV. This is referred to as the resonance capture of thermal neutrons by ^{238}U leading to the formation of transuranic elements, such as Np and Pu. This results in the loss of thermal neutrons. Therefore, it would not be proper to say that the neutrons liberated in the fission of ^{235}U have only a small change of inducing other fission and the multiplication factor k will be very much less than 1, i.e., the system attains a "sub-critical" state.

Although the resonance capture by ^{238}U brings about a loss of thermal neutrons, ^{238}U nuclide is non-fissile by thermal neutrons. However, for the propagation of fission chains in natural uranium, it becomes necessary to slow down (i.e., thermalize) the fission neutrons from some MeV to less than eV range. This slowing down process is carried out by incorporating an inert substance known as moderator (discussed in (iv) below) into the nuclear reactors which also reduces markedly the neutron capture by ^{238}U.

(ii) **Sub-Critical Mixture of Natural Uranium Leading to Diverging Chain Reactions**   In spite of the handicap due to the small proportion of ^{235}U in natural uranium the propagation of the cahins in natural uranium is, however, sustained. We shall now consider how a sub-critical (i.e., $k < 1$) mixture of natural uranium leads to fission chains. In the energy range of fission neutrons, the cross section for 0.72 % of ^{235}U is 1.44 b, whereas the inelastic scattering cross section of fast neutrons is 2.1 b. Because of this, a large fraction of the fast neutrons is slowed down to an energy less than the fission threshold (1 MeV). Below this threshold they can neither give rise to (n, f) reactions with ^{238}U nor scatter inelastically. Elastic scattering, however, slows down, because of fission neutrons to an energy region between 1 and 10^5 eV ('resonance' region). In this energy range the probability of resonance capture in ^{238}U is much greater than the probability of fission in ^{235}U. Since the neutrons lose energy very slowly, the probability that they will reach the thermal energies without being captured, is very small. So that in this case, the compound nucleus [^{239}U] is formed, followed by de-excitation by gamma emission (n, γ) and the result is that the chain production of neutrons is stopped.

The capture cross section σ_c for ^{238}U is more than 10000 b for three resonance energy values around 7 eV, 20 eV, 37 eV, and approaches this height at 65 and 100 eV. The σ_f for ^{235}U may remain high, but does not exceed 600 b in the energy range of 5 to 1000 eV. So in this energy range. Indeed, the energy region of thermal neutrons (< 1 eV) favours fission.

The cross section for U^{235} increases from 60 b at 1 eV to 2000 b at 0.002 eV (the $\dfrac{1}{v}$ law).

The ratio $\sigma(n, \gamma) \ \dfrac{^{238}\text{U}}{\sigma_f \ ^{235}\text{U}}$ is about 0.18 in this energy region. The cross section for the radiative capture of U^{238} is about 10 b at the neutron energy of 1 eV. At the average thermal neutron velocity (2200 m/sec), the ratio $\sigma(n, \gamma) \ \dfrac{^{238}\text{U}}{\sigma_f \ ^{235}\text{U}}$ is 0.0047 as compared to the ratio of the abundance of the two isotopes in natural uranium is $\left(\dfrac{99.28}{0.72}\right) = 137.8$. Thus, it is apparent that the fission of U^{235} is much more probable than radiative capture of ^{238}U. However, to maintain a chain reaction with a $k \geq 1$ factor, as stated earlier either the fission neutrons must be slowed down or the isotopic composition of ^{235}U must be increased.

By adopting the slowing down process of the fission neutrons, Fermi and Szilard in 1942 constructed the fast nuclear reactor. Using natural uranium as fuel and graphite as moderator. This reactor called the "thermal" nuclear reactor. Started functioning at a low energy level of 100 W with a thermal neutron flux of 4×10^6 n cm^{-2} sec^{-1}. We shall discuss some basic principles of thermal nuclear reactors.

(iii) The Principle of Thermal Nuclear Reactors: The Four-Factor Formula It is clear from the foregoing discussion that in the fission of natural uranium with thermal neutrons, some of the neutrons are lost by the resonance capture by ^{238}U the principal component of the natural uranium. In addition, there is loss of thermal neutrons due to their capture by the atoms of the moderator itself. The loss of thermal neutrons due to their capture by other substance such as impurities in the uranium and in the moderator, and by fission products is also observed. Finally, a certain number of neutrons, both fast and slow, escapes from the surface. In view of these processes. Which bring about the loss of thermal neutrons from the system the chain reaction is not possible. If however, the fission of uranium nuclei, (i.e., 235U nuclei) gives rise to more number of neutrons than are captured and the number of fission capture, the multiplication factor becomes greater than unity and the chain reaction is possible. We shall be concerned in this discussion with the various factors for the achievement of a chain reaction in thermal nuclear. These are:

(a) **Fast fission factor ε** In a thermal nuclear reactor employing natural uranium and moderator such as graphite, the fission of a ^{235}U nucleus by a thermal neutron yields on average n fast, (i.e., fission) neutrons. Since these neutrons possess an average energy above the fission threshold for ^{238}U, some of them can cause fission of ^{238}U nucleus, before reaching the moderator and thus some additional fast neutrons are liberated from the fast fission of U^{238} nuclei. The total number of fast neutrons from fission increases from ν to $\nu \varepsilon$, where ε is the fast fission factor; it is very close to unity in a homogeneous reactor ($\varepsilon = 1.03$ for heterogeneous graphite natural uranium reactors). Since the fast fission factor is a measure of the increase in fast neutrons due to the fast fission of ^{238}U, the 3 % increase (in the fast fission of ^{238}U nuclei) in the total number of fast neutrons appears to be an important effect in the thermal neutron fission of natural uranium.

(b) **The resonance escape probability p** The $\nu \varepsilon$ fast neutrons move through the uranium and moderator. Most of them are slowed down because of elastic collisions with moderator atoms and inelastic collisions with ^{238}U nuclei. Let us consider a fraction l_f of the fast neutrons which escapes from the system, (i.e., reactor) before slowed down to thermal energies, i.e., $\nu \varepsilon\, l_f$ fast neutrons are lost. The number of fast neutrons remaining in the reactor is $\nu \varepsilon (1 - l_f)$.

The $\nu \varepsilon (1 - l_f)$ fast neutrons are slowed down by collisions with moderator atoms. But during the slowing down process, all the fast neutrons pass through the energy region between 1 and 10^5 eV (mentioned in (ii) and in this process neutrons with resonance energy values around 7 eV, 20 eV, and 37 eV, and few other values (corresponding to the peaks in the curve of capture cross section against the neutron energies, Figure 7.40, are captured by ^{238}U to form [^{239}U] with very large cross sections, which decays to ^{239}Np and then to ^{239}Pu. Of the $\nu \varepsilon (1 - l_f)$ neutrons, which start the thermalization process, let a fraction p escape resonance capture i.e., the $\nu \varepsilon (1 - l_f)\, p$ neutrons, while the fraction $(1 - p)$, i.e., the $\nu \varepsilon (1 - l_f)(1 - p)$ neutrons are captured by ^{238}U to produce ^{239}Pu. The quantity p is called the resonance escape probability; which is a measure of the probability of the fast neutrons that are thermalized before capture.

(c) **Thermal utilization factor** f From the preceding discussion, it is seen that number of fast neutrons escaping resonance capture is $v \in (1 - l_f)p$, and these neutrons are to be thermalized in the moderator. But in the course of thermalization, some of the fast neutrons, without being captured by the moderator, escape from the reactor assembly. If the fraction escaping is denoted by lt the number of neutrons that escape is given by $v \in (1 - l_f)l_t$ and that left behind in the reactor is $p(1 - l_t) \cdot v \in (1 - l_f)$. Further, these remaining neutrons are not all thermalized and so do not all participate in a new generation of U^{235} fission. A fraction f is absorbed in uranium, and the fraction $(1 - f)$ is absorbed in other materials, such as the moderator, the walls of the reactor, impurities or other substances which may be present and these neutrons are lost. The number of neutrons available for carrying on the chain reaction is then $v \in (1 - l_f)p (1 - l_t)f$.. The quantity f is called the thermal utilization factor, which is the ratio of the number of thermal neutrons absorbed in the uranium to the total

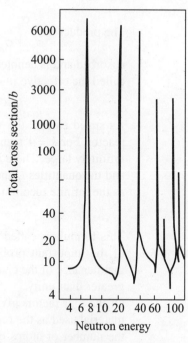

Fig. 7.40 Total capture cross section against neutron energies.

number absorbed in the whole assembly, (i.e., by the uranium moderator, impurities. etc.).

(d) **η factor** Now we see that the net number of thermal neutrons available to produce new fissions in U^{235} is $v \in (1 - l_f) p(1 - l_t)f$. But not all of these thermal neutrons cause to fission of ^{235}U nuclei. A certain proportion leads to fission with a cross section σ_f, while the other lead to radiative capture by ^{238}U nuclei with a cross section σ_r. So the fraction of the thermal neutrons absorbed in uranium that effect fission of ^{235}U, i.e., the

relative fission cross section $\dfrac{\sigma_f}{\sigma_f + \sigma_r}$, where $(\sigma_f + \sigma_r)$ is the total absorption cross section, σ_a of thermal neutrons. The $v \in (1 - l_f) p(1 - l_t)f$ thermal neutrons will induce

$v \in (1 - l_f) p(1 - l_t)f \times \dfrac{\sigma_f}{(\sigma_f + \sigma_r)}$ additional fissions. Thus, the number of fission of

U^{235} in natural uranium in the second generation per fission of u^{235} by a first generation neutron, which is the multiplication or reproduction factor k is given by

$$k = v \in (1 - l_f) p(1 - l_t) \frac{f \times \sigma_f}{(\sigma_f + \sigma_r)} \qquad \text{... (7.1)}$$

The product $\nu \dfrac{\sigma_f}{\sigma_f + \sigma_r}$ is the number of fission neutrons produced per thermal neutron absorbed and is denoted by η. Thus, for a reactor of finite size, the quantity k is usually called the effective multiplication factor, k and is written as

$$k = \eta \in p\, f(1 - l_f)(1 - l_t) \qquad \text{... (7.2)}$$

As stated earlier, the value of k is important as it determines the functioning of the reactor. For $k = 1$ signifies the chain reacting system. When the size of the reactor is infinitely large. There is no leakage of neutrons both fast and thermal from the reactor, and the quantities l_f and l_t are both zero. In such a case, the multiplication is referred to as the infinite medium multiplication factor and represented by k_∞, and is given by

$$k_\infty = \eta \in p_f \qquad \text{... (7.3)}$$

This formula is called the *Fermi's four-factor formula* and the calculation of k_∞ is one of the important problems in reactor design theory. As pointed out earlier for the maintenance of the chain reaction in a nuclear reactor it is necessary to keep k_∞ slightly greater than unity.

Of the four factors giving rise to k_∞, η depends on the nuclear properties of the fissile material used as the fuel. For natural uranium η is expressed in terms of N^{235} and N^{238}, the number of atoms of two isotopes per unit volume. In fact, $\sigma_f = N^{235}\, \sigma_f\,(^{235}U)$ and σ_a is the sum of all the absorption process by ^{235}U and ^{238}U. Thus,

$$\eta = \nu \frac{N^{235}\, \sigma_f\,(^{235}U)}{N^{235}\left[\sigma_f\,(^{235}U) + \sigma_{n,r}\,(^{235}U) + N^{238}\, \sigma_{n,r}\,(^{238}U)\right]} \qquad \text{... (7.4)}$$

Taking into account that σ_n, $\dfrac{(^{235}U)}{\sigma_f\,(^{235}U)} = 0.180$ becomes.

where $\nu = 2.44$ (Section 6.1.6) and $\dfrac{N^{238}}{N^{235}} = 137.8$. It follows that by using uranium "enriched" in ^{235}U, the ratio $\dfrac{N^{235}}{N^{238}}$ increases and the ratio $\dfrac{N^{238}}{N^{235}}$ decreases, so that η increases, i.e., more neutrons are produced per thermal neutrons absorbed in uranium. On the other hand, $\in$ depends both on the nuclear properties of the fuel as well as the size and shape of the assembly forming the reactor. For natural uranium graphite reactor $\in = 1.03$. The other two quantities, p and f, depend on the nuclear properties of the fuel, moderator, and other materials present in the reactor, and also on the relative proportion of the fuel and the moderator substance, as well as on the geometry of the structure of the reactor. It should be noted here that the change in the relative proportions of the fuel element and a particular moderator causes the oppositely directed changes in p and f. If the reactor contains a large proportion of the moderator as compared with uranium, the quantity p will increase, while f decreases, because of a smaller proportion of uranium (and hence of U^{235} nuclei). Similarly, by decreasing the ratio of the moderator of uranium, p decreases and f increases at the same time. In the design of a nuclear reactor, it is, therefore, necessary to find the particular composition of the system, which gives the largest value of the product pf.

In the special case of a reactor in which the fuel consists of a pure fissile material, i.e., ^{233}U, ^{235}U, or ^{239}Pu, both the fast fission factor ϵ and the resonance escape probability p are practically unity. In this case, k_∞ becomes,

$$k_\infty = \eta f \qquad \qquad \dots (7.5)$$

(e) **Escape factors** As pointed out earlier, Equation (7.3) is only valid for a reactor of infinitely large dimensions, which support the assumption that the loss of neutrons by escape from the reactor is neglected. But actually, fast neutrons and thermal neutrons escape from the volume of the nuclear fuel and from the moderator, so that for a given reactor one reaches a critical state for which the effective multiplication factor k is given by

$$k = (1 - P)k_\infty$$

where p is the "escape" probability of neutrons. The probability of escape and also the critical dimensions can be reduced by surrounding the reactor with a neutron reflector. The materials, such as graphite, BeO, D_2O, etc. which act as moderators, can also reflect the thermal neutrons back into the reactor.

(iv) Moderators From the preceding discussion, we know that to achieve a self-sustaining chain reacting system moderators are needed for slowing down neutrons. The choice of a moderator is considered from two points: to allow enough neutrons to reach thermal energies as rapidly as possible and in the smallest possible average path; and to present the smallest possible neutron capture cross section so as to make the f factor close to 1. The neutrons lose enough energy by elastic collisions with moderator atoms, so that they skip over many resonances for which the resonance absorption of neutrons can be made small. Heavy water, graphite and beryllium have been used successively as modertors with natural uranium. Ordinary water, with two protons which can capture neutrons, cannot be used as an efficient moderarters with natural uranium, since the absorption cross section of H for thermal neutrons is 0.31 b which is too large whereas for deuterium in heavy water the captures cross section is 0.00032 b which is a thousand times weaker than for H.

The values for the mean thickness of different moderators necessary to slow down a fast neutron are tabulaed in Table 7.15. The square of the thickness gives the area of the moderator.

Table 7.15

Moderator	Thickness in cm (t_s)	t_s^2 (area) in cm^2
H_2O	5.3	28
D_2O (density 1.1)	11.2	125
Be (density 1.85)	9.8	96
C (density 1.6)	19.2	364

(v) **Reactor Control: The Importance of Delayed Neutrons**   As mentioned earlier that 1 kg of ^{235}U can liberate 959 MW, which is equivalent to the power generated by a power station whose daily coal consumption is 2500 tons. In order to release this amount of energy in a reasonable time without explosion, it is, however, essential to keep k factor around one. In view of this, the control of a reactor must be very accurately carried out.

For maintaining the effective multiplication factor at unity, once the desired energy level has been reached, control bars containing efficient neutron absorbers, such as Cd, B^{10}, or Hf, are used. Since these three elements have very high capture cross section for thermal neutrons, the bars are penetrated more or less deeply in the reactor and regulated the neutron flux so that the condition $k = 1$ maintained.

We have, $\dfrac{N_t}{N_0} = \exp\left[(k-1)\dfrac{t}{\tau}\right]$. It can be seen that when k is greater than one, the rate of neutron production increases very rapidly and becomes explosive. The average life t of a majority of neutrons (i.e., the average interval between two successive generations of neutrons) is in fact very short, being of the order of 10^{-3} sec in the case of the slowest reactors. The number of neutrons in the chain increases exponentially. $k-1$ is the so called reactivity of the reactor. The time constant of the reactor $\dfrac{\tau}{k-1}$ is extremely small even for low reactivities. Thus, for the slowest reactor with $\tau = 10^{-3}$ sec, $k = 1.001$ and $\dfrac{N_t}{N_0} = 3$ in 1 sec and $\dfrac{N_t}{N_0}$ becomes more than 20,000 in 10 sec. It follows that the neutron density and the energy output would increase rapidly, so that mechanical control, even if automatic, might not prevent the chain from becoming divergent. Fortunately, the neutrons produced during fission (that is in 10^{-14} to 10^{-16} sec) are accompanied by delayed neutrons, has helped to overcome this difficulty. The proportion of these delayed neutrons is about 0.65 per cent of the total for ^{235}U. Their half-lives are of the order of tens of seconds. This brings the average life τ up to about 0.1 sec. The dimensions of the reactor are chosen in such a way that it may be critical in terms of the whole neutron fluex, and sub-critical only for the prompt neutrons. For such a reactor, the neutron density and the energy output increase fairly slowly; and control by means of neutron absorbers (i.e., the control bars) becomes easier.

(vi) **Reactor Coolants**   The fission energy, initially in form of kinetic energy fission fragments, neutrons, β-rays and γ-rays, is converted into heat, when these particles are stopped in the various materials present in the reactor. In order to maintain the characteristics of the reactor, it is essential to remove the heat by using some sort of cooling system. The materials used as coolant may be divided into four categories as light and heavy water, liquid metals, organic liquids (hydrocarbons) and gases.

For thermal reactors, both light and heavy water are good coolants. Reactors operating at high temperatures, water is circulated under high pressure to prevent boilling.

Liquid metals, e.g., sodium can be used as coolants at high temperatures without pressurization. Certain hydrocarbon compounds of high boiling point, such as polyphenyls have been used as moderator coolants. But these are not as efficient as water or liquid sodium.

Gaseous coolants such as air, CO_2 and hydrogen are used under high pressure. no doubt each type of coolant has its advantages and disadvantages. Nevertheless the choice of a coolant depends on the purpose of the reactor. In some cases, to prevent the corrosive action of the coolants, the fuel element is protected by a metallic (e.g., Al) or by a metalalloy (e.g., zirconium alloy, or stainless steel) casing.

A schematic self-explanatory diagram of nuclear power reactor is shown in Figure 7.41.

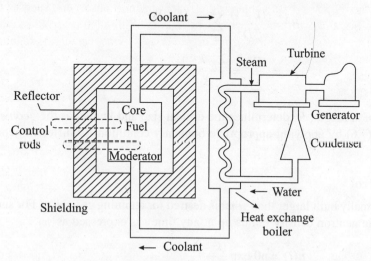

Fig. 7.41 Nuclear power reactor

7.3.4 Theory of Atomic Reactors

There is a definite and finite size of the reactor for a given arrangement of the fuel and the moderator, in which $k_\infty > 1$, at which the neutron distribution, once established will maintain itself, i.e. neither increasing nor decreasing. This size of the reactor is called *critical size*. One can determine the critical size of the reactor by the condition that the rate of neutron leakage is equal to that rate of neutron production (strength of the source) minus the neutron absorption rate.

Let us assume that neutrons are produced and absorbed at the thermal energy only. This forms the basis of *one-group theory*. We further consider that there is a homogeneous arrangement of the fuel and the moderator. If the value of k corresponding to the latter is used, then these considerations will apply to a heterogeneous arrangement also, provided the critical size is large compared to the size of the unit cell in te lattice.

We have the diffusion equation under steady state condition.

$$D \nabla^2 \phi - \phi \sum_a + Q = 0 \qquad \qquad ...(7.1)$$

Where $D = \dfrac{\lambda_{tr}}{3}$ is the diffusion coefficient, ϕ is the neutron flux, Q is the source term and $\sum_a$ is the macroscopic absorption cross section. Now, $Q = k \sum_a \phi$, where k is multiplication factor.

Thus, one obtains

$$DV^2\phi + k\sum_a \phi - \phi\sum_a = 0 \qquad \text{... (7.2)}$$

or
$$\nabla^2\phi + \frac{(k-1)}{D}\sum_a \phi = 0 \qquad \text{... (7.3)}$$

or
$$\nabla^2\phi + B^2\phi = 0 \qquad \text{... (7.4)}$$

Where
$$B^2 = \frac{(k-1)}{D}\sum_a = \frac{(k-1)}{L^2} \qquad \text{... (7.5)}$$

and
$$L^2 = \frac{D}{\sum_a} = \frac{\lambda_{tr}\,\lambda_a}{3} \qquad \text{... (7.6)}$$

Here L is diffusion length, B^2 determines the critical size and known as the *geometrical buckling*. One can solve (7.6) by applying appropriate boundary condition for different geometrical shapes of the reactor.

Reactor Control

A reactor is normally built larger than what is desired for attaining criticality. For such a reactor, the rate at which the neutron density builds up at any time t is expressed as

$$n(t) = n0 \exp\frac{(k-t)t}{t_0} \qquad \text{... (7.5)}$$

where n_0 is the initial number of neutrons and t_0 is the average life time of neutrons. Let us neglect the effect of delayed neutrons always produced in the fission of ^{235}U. Thus the life time of prompt neutrons comes out to be around a milli-second (0.001 s). In accordance to the Equation (7.5), one finds that even $k-1 = 0.01$, the number of neutrons would increase by a factor of 2×10^4 every

second. Thus $\dfrac{n(t)}{n_0} = \exp\left(\dfrac{0.01}{0.001}\right) = e^{10} \simeq 2.2 \times 10^4$. One cannot control such a reactor easily. It

is the delayed neutrons which come to the rescue although they constitute 0.75 % of the total number of fission neutrons. Delayed neutrons increase the average life time of fission neutrons from 10^{-3} s to 0.1 s. Obviously, delayed neutrons would increase the number of neutrons by a factor of 2.7 for every 10 s even with the same excess reactivity $(k-1) = 0.01$. Clearly, the delayed neutrons presence in a reactor produced during fission makes it possible to control the reactor. One controls the chain reaction in a reactor by inserting cadmium or boron rods inside. Cadimum and boron have large values for absorption cross-section of thermal neutrons. Clearly, these rods of Cd or B can effectively control the thermal neutron flux and also the reactivity. When one wants to stop the chain reaction then the rods Cd or B will have to be pushed sufficiently inside the reactor. This will make the rate of absorption greater than the rate of production of neutrons. As a result, ultimately the neutron flux goes down almost to zero.

Reactor Power

Let $\sum\limits_{f}$ and ϕ are the macroscopic fission cross section and the neutron flux of monoenergetic neutrons respectively. The *rate of fission per unit volume* is given by given by $\sum\limits_{f} \phi$, where $\sum\limits_{f} = N \sigma_f$ and $\phi = nv$, where N is fissile nuclei per unit velocity of neutrons and σ_f is the fission cross section.

We have read that on an average about 200 MW $= 3.2 \times 10^{-11}$ W-s energy is produced per fission. Clearly, 3×10^{10} fission are required to produce 1 Watt-s of energy. Now, if V is the volume of a reactor, then the number of fission occuring per second $= V \sum\limits_{f} \phi$. In case if the reactor is operating for some time, then the power (P) of the reactor is given by

$$P = \frac{V \sum\limits_{f} \phi}{3 \times 10^{10}} \text{ W}$$

The mass of ^{235}U fuel consumed in a reactor of volume V is

$$M = \frac{NV \, 235}{6 \times 10^{23}}$$

Then

$$P = \frac{VN \sigma_f \phi}{3 \times 10^{10}} = \frac{6 \times 10^{23} \, M \, \sigma_f \, \phi}{3 \times 10^{10} \times 235}$$

$$= 8 \times 10^{10} \, M \, \sigma_f \, \phi \text{ Watts}$$

Here σ_f is in barns (1 barn $= 10^{-24}$ cm^2) and M in grammes. Now, using $\sigma_f = 580 \times 10^{-24}$ cm^2, we obtain

$$P = 8 \times 10^{10} \times 580 \times 10^{-24} \, M\phi \text{ Watts}$$

$$= 4.6 \times 10^{-11} \, M\phi \text{ Watts.}$$

From the above equation, it is evident that the power output of a given reactor containing a definite quantity of fissile material is proportional to the neutron flux. Hence by measuring the appropriate thermal neutron flux, the reactor power can be determined.

It has been shown that the average flux in a uranium (U^{235}) reactor working at 1 MW/ton is about 3×10^{12} n cm^{-2} sec^{-1}. The maximum flux at the centre of the reactor is about three times higher. The total number of neutrons produced per sq cm by a material exposed to a flux nv for a time is referred to as the integrated flux and given by nvt.

7.3.5 Classification of Reactors

In the preceding section we saw the designing of self-sustaining chain reacting systems. Depending on the characteristics of the systems, numerous reactors of different categories have been constructed. These are:

(i) Reactors on the basis of fuel elements:

    (a) Natural uranium reactor containing 0.72 % ^{235}U

    (b) Enriched uranium reactor with more than 0.72 % ^{235}U

    (c) ^{239}Pu or ^{233}U

(ii) Reactors in terms of neutrons energies Reactors are classified as thermal (low energy), fast (high energy) or intermediate according to the neutron energies at which most of the fissions occur. Since the fission rate cannot be controlled, the atomic bomb is not strictly a reactor. As mentioned earlier, in thermal reactors most fissions occur due to the absorption of slow neutrons resulting from the slowing down process of the fission neutrons. In a fast reactor, the majority of fissions occur by interaction of fissile material with neutrons of high energy, i.e., fission neutrons which either escape during slowing down process or suffer relatively little slowing down. On the other hand, in an intermediate reactor, fissions are effected mainly by neutrons slowed down to energies between the fast and thermal energy range, preferably above the resonance energy region.

(iii) Reactors in terms of the moderator used As graphite, heavy water, or water (swimming-pool reactor) and beryllium or beryllium oxide moderator reactors.

  (iv) Reactors are also categorised according to the *coolant* used, such as air, CO_2, water or liquid metal (sodium) cooled reactors.

(v) Reactors in terms of fuel-moderator composition As homogeneous reactor when the fuel and the moderator are in one phase as solution, or heterogeneous reactor when the fuel and moderator form two different phases.

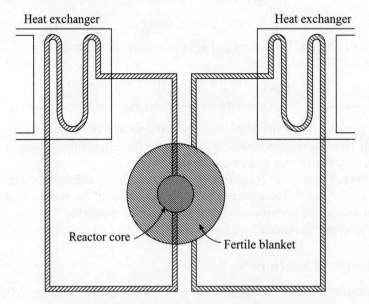

Heat exchanger   Heat exchanger

Reactor core   Fertile blanket

Fig. 7.42 Schematic Diagram of homogeeous reactor.

(vi) Reactors are also named according to the purpose for which they are intended As power generation, research, plutonium extraction of isotope production reactors, breeder reactors etc.

Sometimes reactors are named in terms of the combination of different characteristics, e.g., thermal, natural uranium, graphite moderated, heterogeneous air cooled research reactor, etc.

(vii) Reactors according to fuel and moderator Thermal reactors can also be classified as *homogeneous* or heterogeneous from the point of view of their structure. Fuel and moderator in homogeneous reactors are in the same phase, i.e. same physical state both in solid or both in liquid phase. The two are intimately mixed, whereas the fuel and the moderator in the case of heterogeneous reactors are in different phase. The fuel may be in the form of plates or rods may be dispersed uniformly in the moderator which may be a liquid (D_2O or H_2O) or a solid (BeO or graphite). Obviously, the fuel and the moderator are thus geometrically separated in this type of reactor. We may note that most of the present day reactors are of heterogeneous type.

7.3.5 Typical Reactors

In the foregoing discussion, we have seen that numerous reactors of different designs can be constructed. Considering different fuel-moderator combinations, a quanity referred to as the design parameter F is defined by

$$F = \frac{\text{Number of moderator atoms}}{\text{Number of fuel atoms}}$$

Since a reactor contains a fissile material, i.e., a fuel along with a moderator for slowing down the fission neutrons to maintain the chain reaction, the design parameter Finds its importance in the design of a self-sustaining chain reacting system with $k \geq 1$. We have been concerned with this parameter in some of the typical reactors.

(i) Homogeneous Thermal Reactors: Natural Uranium Moderated with Graphite, Ordinary Water or Heavy Water As stated earlier, in a homogeneous reactor the fuel and the moderator form a single phase (Figure 7.42). In the case of natural uranium-graphite moderated thermal reactor, very small particles of uranium are dispersed uniformly throughout a large block of graphite. In H_2O or D_2O moderated reactor, uranium is present in the form of a solution of uranyl sulphate or in form of fine suspensions of U_2O in water or in D_2O. Since the fuel is in the form of very small particles, the fast fission neutrons from U^{235} escape from the fuel (i.e., natural uranium) and enter the moderator and get slowed down before causing fission of U^{238}. The result is that the fast fission factor ϵ becomes unity, and k_∞ is then equal to *npf*.

For a chain reaction in a homogeneous reactor, it is necessary for $k_\infty > 1$ or $npf > 1$, or $pf > \dfrac{1}{\eta}$. Since η is fixed at 1.34 in natural uranium, then $pf > 0.75$. Let us consider the product *pf*. As previously, p is the resonance capture escape probability, and f, the thermal utilization, is the ratio of the thermal neutron absorption in uranium to the total thermal neutron absorption throughout the reactor.

As pointed out earlier, these two quantities p and f change in an opposite manner. For various values of the design parameter F, the values of f, p, pf and k_∞ are recorded in Table 7.16 for the three systems, U-graphite, UO_2-H_2O and UO_2-D_2O. For small values of F, f approaches unity, while P becomes small.

It is seen that $k_\infty = 1$ is never reached for the homogeneous mixtures containing natural uranium and graphite, or natural uranium and ordinary water. This follows that there can be no fission chain reaction with either ordinary water or graphite as moderator. However, the homogeneous system UO_2-D_2O leads to $k_\infty > 1$ between $F = 50$ and $F = 1000$.

Table 7.16 Values of f, p, pf and k_∞ for three homogeneous systems

F	Uranium-graphite				UO_2-H_2O			
	f	p	pf	k_∞	f	p	pf	k_∞
1	0.999	0			0.950	0.317	0.304	0.41
5	0.997	0.012			0.822	0.724	0.595	0.80
10	0.994	0.062	0.062	0.08	0.698	0.832	0.581	0.78
50	0.972	0.364	0.354	0.47	0.316	0.974	0.308	0.41
100	0.945	0.523	0.494	0.66				
200	0.896	0.641	0.574	0.77				
300	0.852	0.704	0.604	0.80				
400	0.812	0.744	0.604	0.81				
500	0.775	0.772	0.598	0.80				
1000	0.633	0.841	0.532	0.71				
DO_2-D_2O								
1	1.000	0.018	0.018	0.02				
5	0.999	0.333	0.333	0.45				
10	0.998	0.509	0.508	0.63				
50	0.991	0.787	0.780	1.04				
100	0.984	0.853	0.839	1.12				
200	0.968	0.900	0.871	1.16				
300	0.952	0.920	0.876	1.17				
400	0.938	0.932	0.874	1.17				
500	0.924	0.940	0.869	1.16				
1000	0.857	0.960	0.823	1.10				

(ii) Enriched Uranium Graphite Reactor When natural uranium enriched in ^{235}U, homogeneous thermal neutron-graphite-uranium reactors or water-UO_2 reactors can be made critical, i.e., $k_\infty \geq 1$. With enriched uranium, both f and η are increased, the increase in η being more important. Calculation of η from

$$\eta = \frac{2.44}{1.180 + 0.0047\,\dfrac{^{238}N}{^{235}N}}$$

shows that $\eta = 2.07$ may be reached with pure ^{235}U. In the preceding tabulation it is seen that the maximum value of k_∞ is 0.81 with natural uranium (0.72 % ^{235}U) graphite reactor; η being 1.34 for 0.72 % ^{235}U. It follows that an increase of about 20 % in η is necessary to bring k_∞ upto unity, or a value of η between 1.6 and 1.7, which can be attained with enrichment of the natural uranium with ^{235}U. The values of k_∞ calculated for $F = 400$ with $p = 0.744$ for natural uranium with enrichment in ^{235}U in the fuel (i.e., natural uranium) are given below. The enrichment has been expressed in terms of x, the percent of ^{235}U in the natural uranium.

Table 7.17 Variation of k_∞ with Percentage of U^{235}

$\dfrac{N^{238}}{N^{235}}$	x	f	pf	h	k_∞
137.8	0.72	0.810	0.603	1.335	0.81
120	0.83	0.824	0.613	1.399	0.86
100	0.99	0.841	0.626	1.475	0.93
90	1.10	0.852	0.634	1.523	0.97
80	1.23	0.862	0.611	1.569	1.01
70	1.41	0.874	0.650	1.618	1.05
60	1.64	0.887	0.660	1.670	1.10
50	1.96	0.901	0.670	1.726	1.16

It is seen that k_∞ increases significantly with enrichment in ^{235}U. It is evident that a chain reaction can be achieved when k_∞ increases from a sub-critical value of 0.81 for $x = 0.72$ to 1.01 for $x = 1.23$. That is, for a chain reaction to be possible in the homogeneous natural uranium-graphite reactor, the minimum enrichment of ^{235}U is taken as 1.23. Subsequent increase in k_∞ indicates that a higher degree of enrichment in ^{235}U makes it possible to consider fission with fast, non-decelerated neutrons.

Similarly, a chain reaction can be achieved with the enrichment of ^{235}U in homogeneous water-UO_2 reactors. Enriched uranium-ordinary water (swimming pool type) reactor like the Apsara developed in India went critical in August 1956.

(iii) **Natural Uranium Heterogeneous Reactors** As stated earlier, the maintenance of a chain reaction is not possible in natural uranium homogeneious reactors with ordinary water or graphite moderator because of loss of neutrons by resonance capture from the large surface area of the fuel element which are in form of suspension or dispersion in the moderator. However, Fermi and Sziland suggested the use of large blocks of uranium (or UO_2) encased in graphite to reduce the ratio of the surface area to the volume of the fuel element, so that the loss of neutrons by resonance capture which takes place mainly at the surface decreases. It may be once again stated that the absorption of neutrons by radiative capture by ^{238}U occurs when neutrons, in the course of thermalization, reach energies which, together with the binding energy, correspond to levels of compound nucleus [^{239}U]. Neutrons with such an energy when encounter a ^{238}U nucleus are essentially all captured. In a homogeneous uranium-graphite system, each particle of uranium (since uranium is dispersed in the form of single atoms or as very small particles) has the same probability of capturing a certain fraction of the resonance neutrons, but when the uranium is present in large blocks, the resonance absorption of neutrons diminishes as the number of neutrons penetrating from the surface to the centre of a large block of uranium decreases. Thus, the use of large blocks of uranium increases the p factor, the resonance escape probability. Nevertheless, there is an accompanying decrease in the thermalization factor f. The use of certain dimension of the blocks of uranium, however, compensates the decrease in f with the gain in p, so that the product pf increases leading the reactor to go critical i.e., $k_\infty \geq 1$.

It should be noted that among the neutrons, which impinge on the encased uranium, those which possess an energy greaterr than the resonance energy are not absorbed at the surface. Moreover, the slowing down of the neutrons inside the uranium block is less than in the moderator. So the neutrons pass through the uranium block without being slowed down sufficiently. After emerging out of the uranium, the neutrons re-enter the moderator (i.e., graphite) and slow down and only encounter with another block of uranium when their energy is very small. The magnitude of the effect depends on the spacings of the uranium block in the moderator.

It is apparent from the foregoing discussion that the size and shape of the uranium blocks and their suitable spacings in a moderator can lead to a chain reaction in natural uranium-graphite reactor. We shall now consider two parameters, which make the resonance escape probability high and lead the reactor to go critical.

(a) **The size and shape of the uranium block or optimum diameter of the uranium block**   In order to compensate the opposite changes in the two quantities p and f and hence to increase the product pf, it is desirable that a neutron liberated from the fission of ^{235}U should at least collide once or at best twice with the ^{238}U nucleus before leaving the surface. It follows that the diameter of the uranium block should not be too large compared with one mean free path of a thermal neutron in the uranium block. This diameter corresponding to the mean free path value is reffered to as the optimum diameter. The mean free path of a thermal neutron is given by

$$\lambda = (N\,\sigma_t)^{-1} \qquad \qquad \dots (7.1)$$

where N is the concentration of the fuel atoms (= 0.048×10^{24} atoms of U^{238}/cm^3) and σ_t is the total absorption cross section for normal uranium (= 16b),

$$\lambda = (0.048 \times 16)^{-1} \approx 1.3 \text{ cm}$$

Taking the average total cross section for fast neutrons equal to 7.2 b, the mean free path comes to 2.89 cm. Thus, the uranium blocks of diameter between 1.3 and 2.89 cm clad in aluminium or zirconium-aluminium alloy are used in reactors moderated with graphite or D_2O.

Table 7.18 Nuclear Reactors in India

Reactor	Type	Neutron	Power/MW Flux/sec	Fuel	Cladding	Moderator	Coolant	Purpose	Date undergoing critical
Apsara (First Indian Reactor)	Swimming pool	1.25×10^{13}	1.0	enriched uranium 2 %	Al	H_2O	H_2O	Research and isotope production	August 1956
Cirus (First Indian D_2O reactor)	Tank	6.7×10^{13}	40	U (Natural)	Al	D_2O	D_2O/H_2O	isotope production	July, 1960
Zerlina	Tank		0	U	Al	D_2O	D_2O/H_2O	Research	July, 1961
Purnima I	Fast		0	U-Pu-oxide	Stainless steel		air	–do–	May, 1972
Tarapur	Boiling water		2×200	Enciched U	Zr-Al	H_2O	H_2O	Power generation	1969
Kota	Pressurized heavy water under He-pressure		2×200	U	Zr-Al	D_2O	D_2O/H_2O	–do–	1973
Purnima II	Fast		0	U-Pu-Oxide	Stainless steel	–	air	Research	1976
Kalpakkam	Pressurized heavy water under He-pressure	1.3×10^{14}	2×235	U	Zr-Al	D_2O	D_2O/H_2O	Power generation	1983
Kalpakkam	–do–			U	Zr-Al	D_2O	D_2O/H_2O	–do–	August, 1985
Dhruva	Tank	1.3×10^{14}	100	U	Al	D_2O	D_2O/H_2O	Research and isotope production	–do–
Fast Breder			13	Pu C(70) + UC(30)	Stainless steel	–	Liquid Na	Research	Dec, 1985

(b) Optimum distance between the fuel elements, or the slowing down length: The distance between the uranium blocks should be such that a fast neutron leaving the surface of one block slows down the thermal energy before reaching the surface of another block. This distance is called the slowing down length of the moderator, which can be either determined experimentally or calculated theoretically. The slowing down length (t_s) recorded earlier for different moderators gives the lattice spacing of the uranium blocks incorporated in the corresponding moderator.

With the optimum arrangement of natural uranium in the graphite lattice, the value of k_∞ is found to be 1.07, and for natural uranium heavy water moderator reactors, the value of k_∞ is 1.21.

Neutron Balance in Natural Uranium Graphite Moderated Reactor

In a natural uranium graphite, heterogeneous reactor, 2.52 neutrons are produced for each fission of ^{235}U (2.46 from ^{235}U thermal fission +0.06 from ^{238}U fast fission). These are consumed as follows:

Resonance capture by ^{238}U	:	0.8
Radiative capture by ^{238}U	:	0.2
Absorption by graphite	:	0.3
Absorption by other materials of the reactor	:	0.05
Loss due to escape	:	0.09
Chain propagation	:	1.08
Total	:	2.52

Life of a Reactor

It is seen that during the slowing down process in a thermal nuclear reactor, some of the thermal neutrons may be captured by ^{238}U to form ^{239}U, which decays to ^{239}Np and then to ^{239}Pu. It follows that ^{238}U becomes indirectly fissile by slow neutrons. If each fission of ^{235}U leads to 0.8 ^{239}Pu fission, it is found that the total number of fissions per atom of ^{235}U is 5. It means that the production of ^{239}Pu helps in extending the working life of the reactor.

Indeed, plutonium, fission products and other short-lived activities are accumulated on the uranium blocks used in thermal reactors moderated with graphite, light or heavy water, This may bring about a decrease in the reactivity of the system, and consequently in the working life of the reactor. However, by periodically replacing the used uranium blocks, the working life of the reactor can be extended. The used uranium blocks are left for some time for the short-lived activites to decay away and then chemically treated to remove the fission products as well as plutonium from the surface of the uranium blocks.

7.3.7 Critical Size of a Thermal Reactor

As pointed out earlier, the condition k is close to k_∞, and that $k_\infty \geq 1$ for the maintenance of a self-sustaining fission chain is only valid for reactor of infinitely large dimensions with the assumption that no neutrons escape. But actually, fast neutrons ans slow neutrons leak out of the reactor surface, whatever large the dimensions are, and the multiplication factor k is no longer close to k_∞

rather a fraction of it. In view of the neutron leakage, it is essential to invoke a quantity which in combination with k_∞ gives information regarding the feasibility of a fission chain in a reactor of finite dimensions. This quantity is referred to as the critical size which is governed by the steady state conditions that the loss of neutrons by leakage and by capture is balanced by the rate of production in a fission process in a reactor of finite dimensions. It may be noted that the critical size is not a constant, but depends on the isotopic composition of the uranium, the proportion of moderator, fuel-moderator lattice and the presence of other materials causing the non-fission capture of neutrons. As stated earlier, for the fission chain to occur, the system should be super-critical, i.e., the size of the reactor should be equal to or larger than the critical value.

We shall consider a homogeneous reactor in form of a rectangular slab of infinite length of the sides along x and y directions but of finite thickness in the z direction. For such a reactor, since two of the sides are infinitely large to ensure no leakage of neutrons from the surface, it is assumed that the nuclear properties including k_∞ are known, and the situation drives to calculate the critical thickness, t at which the reactor will maintain a chain reaction. This implies that the steady state condition for the reactor to function is governed by the neutron balance equation as

Rate of neutron escape + Rate of neutron capture = Rate of neutron product,

i.e., $\qquad n_{es} + n_{cap} = n_{pr}$ ... (7.1)

For the present purpose, it is assumed that the neutrons are produced and captured uniformly throughout the reactor, and can escape from the surface through the boundary of the reactor. In addition, it is assumed that all the neutrons are thermal and monoenergic.

The orgin of the coordinates system is taken as the centre of the reactor, and the finite thickness is considered along the Z-axis at which the neutron balance equation applied. Adopting the boundary conditions for the state operations of the reactor, the neutron density at the outer boundary must vanish, i.e.,

$$n(z) = (n_{pr} - n_{es} - n_{cap}) = 0, \text{ at } z = \frac{1}{2t}$$ ... (7.2)

where t is the critical thickness of the reactor. For a uniform reactor of the type in the present case, the neutron density is symmetric about the centre, $z = 0$, and positive throughout the interior of the reactor.

The general solution for the equation n(z) = 0 is given by

$$n(z) = A \cos Bz + C \sin Bz$$ ... (7.3)

where A and C are the arbitrary constants and B is defined in the neutron balance equation,

$$\frac{d^2 n}{dz^2} + \frac{(k_\infty - 1)^{\frac{1}{2}}}{M} n = 0$$

as $\qquad B = \dfrac{(k_\infty - 1)^{\frac{1}{2}}}{M}$ ... (7.4)

where M is the diameter of the migration area, $M = \left(l^2 + t_s^2\right)^{\frac{1}{2}}$, where t_s has already been defined as the thickness necessary for thermalization, i.e., the slowing down length of the moderator, and

I is the diffusion length of a thermal neutron, i.e., the average distance measured as a straight line from the point at which a thermal (moderator + fuel) as a whole. Since the neutron density is symmetric about $z = 0$, $\sin B z$ is not symmetric, so C must vanish, and Equation (7.3) becomes,

$$n(z) = A \cos B z \qquad \qquad \text{... (7.5)}$$

According to the condition 2, $\quad B\dfrac{t}{2} = \dfrac{\pi}{2}$

or
$$t = \frac{\pi}{B} = \frac{\pi M}{(k_\infty - 1)^{\frac{1}{2}}} \qquad \qquad \text{... (7.6)}$$

Thus, the critical thickness of the reactor in terms of the nuclear properties including k_∞ and M, can be determined from Equation (7.6). It is seen that, if $k_\infty < 1$; t has not real value, if $k_\infty = 1$, t is infinite; if $k_\infty > 1$, t can have a finite real value, and a chain reaction is possible in a reactor of finite size.

The loss of neutrons due to leakage can be minimized by decreasing the ratio of area of volume of a reactor. Since for a given volume, a sphere has the smallest possible area, a nuclear reactor in form of a sphere is more convenient. The critical thickness derived by Equation (7.6) is based on the symmetrical distribution of neutrons, and therefore, t may be taken as the critical radius of a spherical homogeneous reactor. Thus, the critical radius R_c of a spherical reactor is

$$R_c = t = \frac{\pi M}{(k_\infty - 1)^{\frac{1}{2}}} \qquad \qquad \text{... (7.7)}$$

and the critical size of a cubical reactor is given by

$$S_c = x = y = z = \sqrt{3}\,\frac{\pi M}{(k_\infty - 1)^{\frac{1}{2}}} \qquad \qquad \text{... (7.8)}$$

For the sphere, the critical volume is

$$V_c = \frac{4}{3}\pi R_c^3 = 130\left(\frac{M^2}{k_\infty - 1}\right)^{\frac{1}{2}} \qquad \qquad \text{... (7.9)}$$

and for the cube,

$$\dot{V}_c = S_c^3 = 161\left(\frac{M^2}{k_\infty - 1}\right)^{\frac{1}{2}} \qquad \qquad \text{... (7.10)}$$

It is apparent from Equation (7.9) and (7.10) that the critical volume of a sphere is smaller than that of a cube and hence the sphere has a smaller area to volume ratio. The critical mass of uranium can be determined from the critical volume, and hence the mass of uranium per unit volume. For natural uranium, the critical mass is of the order of a few tons with heavy water and about ten times with graphite moderator. For pure ^{235}U, the critical mass is only 2 to 3 kg.

7.3.8 The Breeder Reactor

In a natural uranium thermal reactor, ^{235}U present to the extent of about 0.72 percent is only fissile which undergoes fission producing fast neutrons, whereas ^{238}U (99.28 % of natural uranium) is not fissile, but by radiative capture of fission neutrons produces ^{239}Np is not fissile, which rapidly leads to ^{239}Pu. This nuclide is fissile and the average number of neutrons produced is 2.89. Thus ^{238}U is converted into ^{239}Pu by the radiative capture of neutrons. Similarly, Th^{232} by the (n, γ) capture of slow neutrons is converted into Th^{233} which subsequently decaying by β-emission produces a fissile nuclide ^{233}U. The reactors which produce a fissile material at the consumption of an initial fissile (or fissionable) nuclide are termed as converter reactors. In a converter reactor, the rate of ^{239}Pu production is not much less than the rate of ^{235}U consumption, i.e., both the rates are nearly equal. The reactors, however, produce new fissile elements at a greater rate as compared to the rate of destruction of the initial one.

As an illustration, let us consider a natural uranium thermal reactor in which 2.5 neutrons, on the average, are liberated per ^{235}U fission. Because of competition between the radiative capture $(n, \gamma$ reaction) and fission only approximately 2.1 neutrons, on the average, are released for each neutron captured by ^{235}U. Remembering that at least one neutron is required to encounter with another ^{235}U nuclide to sustain the fission chain, there are approximately 1.1 neutron per ^{235}U nuclide consumed available to produce new fissionable material, i.e., ^{239}Pu. It means that when a neutron induces a fresh fission in order to maintain a chain reaction, more than one neutron remains available to create more ^{239}Pu in ^{238}U than the amount of ^{235}U consumed. This process is called breeding, and this is achieved better in a fast-neutron reactor. Such a reactor is called the breeder reactor. In breeder reactors, more and more fissionable material is produced than what is consumed in generating energy and power. The newly produced fuel can be used not only for the reactor itself but also for new reactors. This is the principle of the breeder reactor, which finds its importance because of the limited availability of the fissile materials in nature.

The materials consumed or converted and breed in the reactor are termed as fertile and fissile, respectively, This, natural uranium, ^{238}U and ^{232}U are called the fertile, whereas ^{239}Pu and ^{233}U fissile materials.

Conversion of ^{232}Th and ^{238}U into ^{233}U and ^{239}Pu as a result of neutron absorption is as shown below

$$_0n^1 + {}^{232}\text{Th} \rightarrow {}^{233}\text{Th} + \gamma$$
$$\text{(fast)} \qquad \downarrow \beta^-$$
$$^{233}\text{Pa}$$
$$\downarrow$$
$$^{233}\text{U} + \beta$$

and
$$_0n^1 + {}^{238}\text{U} \rightarrow {}^{239}\text{U} + \gamma$$
$$\text{(fast)} \qquad \downarrow \beta^-$$
$$^{239}\text{Np}$$
$$\downarrow$$
$$^{239}\text{Np}$$
$$\downarrow$$
$$^{239}\text{Pu} + \beta^-$$

Ovbiously, in a fast breeder reactor, more fissionable material is produced than consumed.

A reactor, generally consists of an active core in which the fission chain is maintained and in which most of the fission energy is released as heat. The core contains the reactor fuel, and also a moderator, if necessary, to thermalize the fission neutrons. The control rods, i.e., the strong neutron absorbers, such as boron steel rods or cadmium strips, if necessary are dipped into the core. The core is surrounded by a reflector of a substance of weak neutron absorber. The core is surrounded by a reflector of a substance of weak neutron absorber. The core is surrounded with a loop for the circulation of a coolant. The arrangement of core and reflector along with other materials present, must be capable of sustaining a fission chain. The reactor system including the reflector, must be housed in a shield, of concrete or stainless steel vessel. A breeder reactor with a plutonium core (^{239}Pu) is displayed in Figure 7.43.

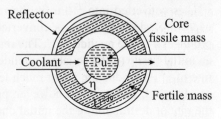

Fig. 7.43 Principle of a breeder reactor.

A sodium cooled, plutonium fuelled loop type Fast Breeder Test Reactor (FBTR) of 40 MW was set up in 1985 in India at Kalapakkam.

7.3.9 Nuclear Reactors In India

A number of different nuclear reactors have been set up in the country. These reactors are either moderated with D_2O or H_2O. D_2O has been used in certain reactors as the primary coolant, and the light water as the secondary coolant (Table 7.15).

7.3.10 Reprocessing of Spent Fuels

Reprocessing of the spent fuels from the reactor is done to recover the fuel elements and then to use once again in the reactors. Since the spent fuels are highly radioactive and toxic, reprocessing is carried out by means of very sophisticated equipments shielded by concrete walls with lead glass windows. All operations and analysis are to be carried out by remote control. The following processes are carried out for reprocessing the spent fuels:

Since fuel elements in the reactor are usually clad either with aluminium or zirconium alloy, or stainless steel. The cladding is removed by dissolving in a suitable reagent, either a suitable acid for a base. The aluminium cladding is removed by digesting with hot NaOH and the residue containing the spend fuels is dissolved in HNO_3 to form a solution containing the nitrates of fuel elements (e.g., U and Pu) and the fission product elements. The solution is then concentrated and subjected to the process of solvent extraction using the organic solvent tributyl phosphate diluted with a hydrocarbon, such as kerosene. The fission products remain in the aqueous solution and can be discarded. The nitrates of the spent fuels remain in the organic solvent. The plutonium (IV) nitrate in the organic solvent is reduced to the Pu(III) state, and so it is re-extracted into an aqueous solution, from which plutonium can be extracted by ion-exchange method using sodium oxalate as the complexing agent, followed by heating. Thus, plutonium in form of its oxide can be recovered. The uranium nitrate containing U(VI), since U(VI) is very stable and is not reduced, remains in organic solvent and can be extracted by ion-exchange method.

7.3.11 Radiation Damage and Fission Product Poisoning

Indeed, the working life of a reactor is related to the nature, shape, composition, spacings, etc., of fuel elements in the reactor. It implies that the life of a fuel element decides the fate of a reactor. Because of the high cost of recovering the fissible and fertile materials from the spent fuels, it is essential to increase the life of a fuel element for a reactor. It will be seen that the life of a fuel element is limited by a number of factors of which radiation damage and fission product poisoning are the most disturbing ones, for fuel elements, especially for the solid fuels when used in a nuclear reactor. The solid fuels when subjected to the action of high energy particles, such as neutrons, the atoms in the crystal lattice are displaced from their normal positions, as a result the physical and structural properties of the solid materials are changed. Consequently, the fuel element cannot operate satisfactorily for a longer period in a nuclear reactor. This is referred to as the radiation damage of the fuel elements.

As the reactor goes on operating, the amount of the fissile element decreases, and the non-fission capture of neutrons by the fission products accumulating on the solid fuels increases. The latter effect is called the fission product poisoning which affects the life of fuel elements.

The limitations of the operating life of the fuel elements can, however, be overcome by using the fertile material in the fuel element, which by the neutron capture is converted into the fissile material to replenish the decrease of the latter in the reactor. Since a fuel is non-affected by the radiation damage, the fuel operates satisfactorily when used in the fluid form, i.e. its solution in water, or in a molten metal, or in a molten salt mixture. In the case of decomposition of water in aqueous solution of the fuel and in the case of effects of depletion in fossile material and accumulation of fission products can be overcome by replacing some of the fluid fuel by fresh one, even during the course of operation.

Uses of atomic energy

Apart from the atomic energy for military purposes it is of very great use for the benefit of mankind. Some of them are listed below:

(i) Power production
(ii) Production of radio-active isotopes, which are used in the field of agriculture, medicine, industries etc. through a technique called as *tracer technique*.
(iii) Navigation

India's Three Stage Programme

India has plenty of thorium (reserves one of the world's largest), but uranium is scare. India's uranium resources-about 0.8 % of the world cannot contribute to any significant improvement in the situation if uranium is to be used on once-through basis and then disposed off as waste. However, with a carefully planned programme, the available uranium can be used to harness the energy contained in non-fossil thorium, of which India possesses 32 % of the world reserves.

The first stage of this programme involves using indigeneous uranium in PHWRs, which efficiently produces energy as well plutonium. In the second stage, by reprocessing the spent nuclear fuel and using recovered plutonium in FBRs, the non-fossil depleted uranium and thorium can breed additional fissile nuclear fuel, plutonium and ^{233}U, respectively. In the third stage, thorium and ^{233}U based nuclear reactors can meet long term energy demands.

Presntly India's total installed capacity of nuclear electricity generation is 2720 MW and by the end of 11th plan the projected nuclear capacity will be 10,280 MW and by the year 2020 it will be about 20,000 MW.

7.4 NUCLEAR FUSION

This is the process by which the sun and other stars generate their energy, is being developed to produce electrical power on earth. Fusion power will be safe, environmentally attractive and its fuel deuterium exists in abundance in ordinary water. The deuterium in the oceans could provide enough energy to satisfy the world's electricity requirement at the present rate of consumption for millions of years. Scientists hope to build fusion power reactor soon.

7.4.1 The Source of Stellar Energy

In 1928, Professor Knowlton wrote physics text in which a unifying theme was the question: where does the Sun energy come from? He considers the possibility that the Sun is leaking energy stored within it during some stage of astronomical evolution. He treats the Sun as a coal pile deriving its energy from chemical reaction. He estimates the energy brought to the Sun by falling meteors. He assumes that possibility the Sun is contracting, with the potential energy of its parts becoming kinetic and thermal. He concludes as follows:

"The english astronomer Eddington has calculated that if $\frac{1}{10}$ of the hydrogen now present in the Sun was to be converted into helium, the reaction would liberate a supply of that great enough to keep the Sun radiating at its present rate for a billion years. If such a process is going on anywhere, it must be at some place where atomic nuclei are forced to approach one another either as in the impact of swiftly moving atoms or by enormus external pressure. A few years ago we had not the slightest ground for believing that either atomic disintegration or atom building was possible. Perhaps there are other things happening in the great solar crucible of which we have as yet no suspicion. At any rate we have here a plausible and fascinating hypothesis as to the source of solar energy which unlike any of these previously considered, is quantitatively adequate."

H.A. Bethe in 1939 suggested that the production of steller energy is by thermonuclear reactions in which protons are continuously transformed into helium nuclei. For comparatively low steller temperatures he proposed the proton-proton cycle represented in Figure 7.44 and whose reactions are listed in Table 7.16

Table 7.16 Proton-Proton cycle reaction

$$_1^1H + _1^1H \rightarrow _1^2D + _{+1}^0e(\beta^+) + \nu + 0.42 \text{ MeV}$$

$$_1^2D + _1^1H \rightarrow _2^3He + \gamma + 5.49 \text{ MeV}$$

$$_2^3He + _1^1H \rightarrow _2^4He + _{+1}^0e(\beta^+) + \nu$$

$$_2^3He + _2^3He \rightarrow _2^4He + p + p + 12.86 \text{ MeV}$$

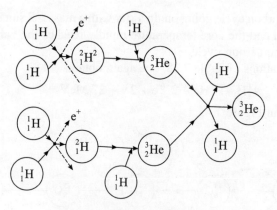

Fig. 7.44 Proton-Proton Cycle

On average neutrinos take away 0.26 MeV each. The positron annihilates with an electron and releases 1.02 MeV so each proton gives rise to 6.55 MeV. The observed luminosity of sun is 3.9×10^{26} J/s and if pp cycle is the principle source of power this implies that there are $\sim 3.7 \times 10^{38}$ protons being converted a second. Equating this rate to $n^2 \, v \, \sigma V$ gives an active volume $V = 10^{25}$ m^3. This volume is $\sim 10^{-3}$ of the solar volume so the pp cycle which starts with a weak interaction can account for the observed solar luminority.

If we add first three equations, nuclides that appear on both sides cancel out, we have

$$4\,{}^1_1H \rightarrow {}^4_2He + 2\,{}^0_{+1}e(\beta^+) + 2\nu + 27 \text{ MeV}$$

This is considered to be an important source of energy in the Sun. It predominates in states of comparatively low temperatures.

For main sequence stars (the Sun is only a small star) Bethe suggested an alternative to the proton-proton cycle the carbon-nitrogen cycle or carbon cycle. It is called a carbon cycle, since carbon serves as a sort of nuclear catalyst. The reactions of carbon cycle are listed in Table 7.18.

Table 7.18 Carbon cycle

$${}^{12}_6C + {}^1_1H \rightarrow {}^{13}_7N + \gamma + 1.94 \text{ MeV}$$
$${}^{13}_7N \rightarrow {}^{13}_6C + {}^0_{-1}e + \nu + 1.20 \text{ MeV}$$
$${}^{13}_6C + {}^1_1H \rightarrow {}^{14}_7N + \gamma + 7.55 \text{ MeV}$$
$${}^{14}_7N + {}^1_1H \rightarrow {}^{15}_8O + \gamma + 7.29 \text{ MeV}$$
$${}^{15}_8O \rightarrow {}^{15}_7N + {}^0_{-1}e + \nu + 1.73 \text{ MeV}$$
$${}^{15}_7N + {}^1_1H \rightarrow {}^{12}_6C + {}^4_2He + 4.96 \text{ MeV}$$

Counting the energy liberated in the annihilation of the two positrons, the total energy liberated in one cycle is 26.72 MeV, just as in one $p-p$ cycle. In the carbon cycle a little more than 5 % of the energy is lost from the star by the two neutrino emitted in the higher energy β-decays. The rate at which the carbon cycle occurs is much higher than the rate for the $p-p$ cycle, because no step in the carbon cycle is anywhere as near as slow as the first step in $p-p$ cycle. The sun has not yet reached the stage in its development where the carbon cycle dominates the energy production, although there is some carbon cycle going on. In a star with a mass > 2 sun masses, the gravitational contraction is very rapid and the core temperature rapidly reaches the value $\sim 10^8$ K required for carbon formation and the carbon cycle.

On addition the equations of Table 7.18, one again obtains

$$4\,{}_1^1\text{H} \rightarrow {}_2^4\text{He} + 2\,{}_{-1}^0\text{e} + 2\,\nu + 27 \text{ MeV}$$

Carbon cycle is shown in Figure 7.45.

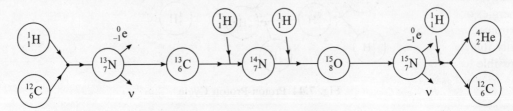

Fig. 7. 45 Carbon cycle

The energy liberated in the fusion of light nuclei into heavier ones is often called thermonuclear energy, particularly when fusion takes place under man's control. On the earth neither proton-proton nor carbon cycle offers any hope of practical application, since the reaction rates of both the above cycles would be far too low. So we have to consider the possibility of fusion of light nuclei. This requires an initial temperature of the order of 10^8 degrees Kelvin, at which the state of the gas is known as Plasma[*]. A plasma of $D - T$ at a temperature ~ 20 keV (~ 200 million K) is needed to produce an adequate reaction rate. Such a plasma must not be allowed to come into contact with any material as it would melt it.

7.4.2 The Plasma: The Fourth State of the Matter

At very high temperature, of the order of 10^8 °K the atoms of the gas are fully ionised and these ions and free electrons are moving about very rapidly. It is possible that the separation of positive ions and free electrons is never very large because of the electrostatic attraction but they do move much more independently of each other than at ordinary temperatures. The mixture is still electrically neutral of course, and the whole state of the matter is called the plasma. Plasma engineering is now very important subject for research and is occupying the time of a large section of the scientists of the world.

A plasma is classified as (i) weakly ionized. (ii) moderately ionized and (iii) fully ionized. In nature, weakly ionized plasma is found in the ionisphere. The Sun, hot stars and certain intersteller clouds are examples of fully ionized plasma, formed at very high temperatures (high temperature plasma). Artificially, plasma is produced in gas discharges and gas discharge tubes. The control of its motion is the basis for the use of plasma as the working medium in various engines and for the direct conversion of its internal energy into electrical energy (MHD, generators, plasma sources of electrical energy; etc.). Here we shall confine overselves to the study of fully ionised or hot plasmas

---

[*] Plasma is a state of matter characterized by a high, or even complete, ionization of its particles.

7.4.3 Fusion Reactions

As stated earlier fusion reactions occur when two light nuclei such as Deuterium (D), Tritium (T), or Helium (^{3}He) collide and re-arrange themselves so as to form two other nuclei of smaller mass with a consequent release of energy. The following reactions are of primary interest in CTR research.

There are several important features of these reactions. They take place only at very high energy, because of the strong Coulomb repulsion between approaching nuclei. To achieve particle energies of 10 keV the mixture must be at a temperature of about 100 million degrees kelvin. At this extremed temperature the gas atoms are fully ionized. They have all shared their electrons. Such a fully ionized gas is called a plasma, and the fact that the particles are charged particles, makes possible their containment with magnetic fields.

Table 7.19 Fusion reactions

Energy required	*Fusion reaction*	*Energy released*
10 keV	$^2_1H + ^3_1T \rightarrow ^4_2He + n$ (D – T reaction)	17.67 MeV
50 keV	$\rightarrow ^3_2He + n$ $^2_1D + ^2_1D \rightarrow$ $^3_1T + p$ (D – D reaction)	3.3 MeV 4.0 MeV
100 keV	$^2_1D + ^3_2He \rightarrow ^4_2He + p$ (D – ^{3}He reaction)	18.3 MeV

The reaction products of a fusion reaction carry away the energy of the fusion reactions. In the *DT* reactions most of the energy is carried away by 14 MeV neutrons where as in $D - {}^3$He reaction all of the energy is released determines the method of energy conversion. In the case of the *DT* reaction, the energy of the escaping neutrons must be degreaded to heat and converted through a thermal cycle to electricity. When the energy appears in the charged particles, a unique direct conversion of their energy to electrocity is possible. This has the potential for very high efficiency and low waste heat.

7.4.4 Energy Balance and Lawson criterion

A measure of success in obtaining net power from the thermonuclear reactions is the ratio of thermonuclear power generated to the power required to create and sustain the hot reaction mixture.

$$Q = \frac{\text{Thermonuclear Power}}{\text{Injected Power}} = \frac{\left(\dfrac{n}{2}\right)^2 \langle \sigma V \rangle E_f}{\left(\dfrac{3n\,KT}{\tau}\right)}$$

where

$$\left(\frac{n}{2}\right)^2 \langle \sigma V \rangle \rightarrow \text{number of fusion reaction per cubic cm per sec.}$$

$$E_f \rightarrow \text{Fusion energy released per reaction}$$

The injected power is obtained by supposing that the entire energy, $3nkT$, must be replaced every τ seconds. Thus τ is the energy replacement time, approximately the plasma confinements time.

Lawson first stated the requirement for power balance in a fusion reactor by assuming that the entire energy, both the energy from thermonuclear reactions and energy invested in initially creating the hot plasma could be reinvested, with an efficiency of $\dfrac{1}{3^{\text{rd}}}$ in creating new plasma. This leads to the requirement $Q \geq 2$. For the D-T reaction, this places a requirement on the density-containment time product.

$$n\tau > 10^{14} \text{ sec cm}^{-3} \text{ for } T = 10 \text{ keV}$$

Necessary parameters based on Lawson Criterion for a fusion reactor are given in a Table 7.20.

Table 7.20 Parameters for a fusion reactor

Temperature (kT)	$\geq 10 \text{ keV}$ ($T \geq 108$ °K)
Plasma density (nt)	$\cong 10^{15}$ per cubic cm.
Magnetic field (B)	$\cong 20$ kilo Gauss
Containment time (τ)	$\geq 1/10$ Sec D-T reaction
	≥ 10 Sec; D-D reaction

The above estimate reveal that D-T reaction can produce self sustained thermo-nuclear reaction easily compared to others. D-T reaction has an enhanced cross-section because of a resonant state in ^5He.

The two main problems of generating fusion power are:

(i) **Heating the plasma of the fusion fuel to very high kinetic temperature** ($T \sim 10^8$ K for D-T reaction) so that fusion reaction may take place, and (ii) confinement of high temperature heated plasma long enough for economic generation of fusion power.

One can use the following heating mechanisms to produce temperatures $\sim 10^8$ K or more (i) By passing a high current through plasma. This is expected to produce temperature upto $\sim 10^7$ K. Beyond this temperature the resistance of plasma drops to almost zero above $\sim 10^7$ K and therefore this method is not effective.

(ii) **Pinch effect**   one can compress the plasma by electrodynamic forces resulting from the passage of electric current through the plasma. One can regard the plasma column carrying current as made up of a large number of plasma threads, each carrying an electric current in same direction. These attract each other and as a result whole of the plasma column is compressed perpendicular to its axis. This effect is well known *Pinch effect*. The sudden adiabatic compression results in the heating of the plasma column and heat so generated is carried away by shock waves and turbulence processes.

There are other processes also for heating the plasma. These include:

(iii) Irradiation by high frequency electromagnetic field

(iv) Irradiation by high energy beam of neutral atoms;

(v) Irradiation by very frequency electromagnetic field;

(vi) Irradiation by a high intensity beam of electrons;

(vii) Irradiation by a high intensity laser radiation.

Clearly, if the power produced in the fusion reactions, e.g. 17.6 MeV per fusion in $D-t$ reaction is greater than the loss of energy from the system (mainly due to bremsstrahlung process) then it is possible in principle to achieve controlled release of fusion energy in a self-sustained chain reaction.

7.4.5(a) Cross-section of Fusion Reactions

The values of cross-sections for $D-D$ ($D-D, p$ and $D-D, n$), $D-T$ and $D-{}^3$He fusion reactions upto 100 keV energy of the incident deutrons obtained experimentally are shown in Figure 7.46. At small bombarding energies (few keV) the cross-section (σ) is very low and rises very rapidly as the energy increases and then rises rather slowly at higher energies. One can fairly well represent the experimental data at low energies by Gamow's theory of Coulomb barrier penetration in the case of $D-D$ reaction.

$$\sigma_{DD} = \frac{288}{E} \times 10^{-28} \exp\left[-\frac{45.8}{\sqrt{E}}\right] \qquad \dots (7.1)$$

where E is in keV. Equation (7.1) shows that σ_{DD} increases by a factor of about 4×10^{12} as E increases from 1 to 10 keV.

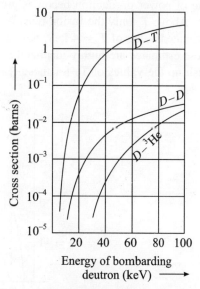

Fig. 7.46 Variation of reaction cross-section with energy for $D-D$, $D-T$ and $D-{}^3$He fusion reactions

7.4.5(b) Reaction Rates

The reaction rate (R_{12}) per unit volume for the fusion reaction occuring in a high temperature plasma is proportional to the particle densities n_1 and n_2 of the particles taking part in the fusion reactions. The reaction cross section (σ) is a function of the relative velocity (v) of the particles of the reaction, then the reaction rate can be expressed as

$$R_{12} = n_1 \, n_2 \, \langle \sigma v \rangle \qquad \dots (7.2)$$

where $\langle \sigma v \rangle$ denotes the mean value of the product σ_v computed over Maxwellian velocities.

Using the variation of σ in Figure (7.46), one obtains the curves for the variation of $\ln(\sigma v)$ against the kinetic temperature T(Figure 7.47). We may note that the kinetic temperature 1 keV $\sim$ 1.15×10^7 K.

Figure 7.47 shows that $\ln(\sigma v)$ rises very steeply with rising T at lower temperatures (σv) varies as $T^{6.3}$ at $T \sim 1$ keV, while at $T \sim 0.1$ keV, (σv) varies as T^{133}. At $T \sim 100$ keV, variation of σv is quite slow.

Let Q is the energy released per fission, then the power density (P) is given by

$$P = R_{12}Q \qquad \dots (7.3)$$

It has been estimated that for fusion to be economically viable, $P \sim 10^8$ W/m^3/s.

7.4.5(c) Critical Temperature

We have read that nuclear fusion reaction can be self-sustaining only if the rate of energy loss from the reacting system is less than the rate of energy generation. Moreover, the main energy loss is bremstrahlung. However, the rate of energy loss P_{br} increases much more slowly as compared to the rate of power production. A comparative study of two processes is projected graphically in Figure 7.48. There is a certain temperature (T_c) as shown by dashed line in figure, above which the rate of power production becomes greater that of the rate of energy loss.

Above T_c only the fusion reaction can become self-sustaining. For $D - T$ fusion reaction the estimated value of $T_c \approx 4$ keV, whereas for $D - D$ reaction it is about 10 times higher. However, these estimates are highly idealized as there may be additional losses in actual situation and may push up the values of Tc for these reactions.

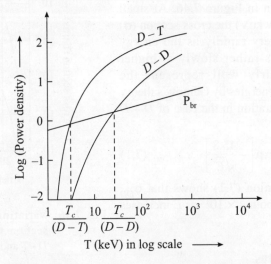

Fig. 7.48 Comparative values of log (power density) and bremstrahlung loss (P_{br}) for $D - D$ and $D - T$ fusion reactions. The dashed vertical lines give critical temperatures of these reactions.

Above studies clearly reveal that the most promising fusion reaction from the power generation point of view is $D - T$ fusion reaction. However, tritium $({}^3_1\text{H})$ does not occur naturally and it is radioactive with a half life of 12 years. We have to bred it effectively by the following reaction

$$^6_3\text{Li} + {}^1_0\text{n} \rightarrow {}^4_2\text{He} + {}^3_1\text{H}(T) + 4.8 \text{ MeV}$$

If we place a mantle of lithium around the fusion reactor, then in the fusion reactor the neutrons generated will breed tritium from ^{6}Li by the above reaction.

7.4.6 Confinement of Hot Plasma

Plasma heated to very high temperature ($\sim 10^8$ K) has to be confined within the fusion reactor for a sufficiently long time. One can easily estimate the confinement time. Let us consider a thermo-nuclear reactor having volume 10 to 100 m³ at a power level of $\sim 10^6$ kW. The power density should be about 10^4 to 10^5 kW/m³/s. Now knowing mean value $\langle \sigma v \rangle$ and energy released per fusion (Q) one finds minimum plasma particle density (n_{min}) $\sim 10^{20}$ particles/m^3. Using the previously given values of $n\tau$, one obtains $\tau \sim 1$ to 100 s. Obviously, there must be some special method for confining the plasma within the reactor at about 10^8 K as the natural escape time is $\sim 10^{-6}$ s. Such a plasma must not be allowed to come into contact with any material as it would melt it.

There are two possible ways: (i) Low density gaseous plasma magnetic confinement and (ii) Production of high density plasma in very small pallets of the fussion fuel in the frozen state, by very sudden compression using intense laser beams or by the use of highly focused beams of protons or heavy ions. Now, we discuss the two approaches:

(i) **Magnetic Confinement**   The magnetic approach to fusion has already produced megawatts of fusion energy-an amount roughly comparable to the external heating supplied to the plasma.

We know that a plasma is a mixture of positive ions and negative electrons. Thus, it behaves like an electrically conducting gas, exerting an outward pressure. Due to this, plasma develops a tendency to expand outwards which is resisted by the electromagnetic stresses produced by the strong external magnetic field. The plasma particles in a properly shaped magnetic field will move in orbits that help them to remain within the container, i.e. vacuum chamber of the reactor without coming in contact with the walls of the container.

No confined plasma can be in thermodynamic equilibrium and each departure from equilibrium provides a free energy source to drive *instability*. There are various types of instabilities of pinched plasma. The instabilities that have most affected fusion progress are driven either by plasma currents or by plasma pressure gradients.

The most serious plasma instabilities are: (i) *bottle necks* and (ii) *bends* (Figure 7.49). Once formed, these instabilities tend to grow rapidly, which make the particles of the plasma escape prematurely to the wall, prior to having sufficiently to fuse.

Today, new methods exist for confining plasma without having to impose a current, thus eliminating the first free energy source. In addition, ingenious confinement geometries-still in exploratory stages can counteract the instabilities due to plasma pressure gradients.

There are various magnetic confinement devices proposed so far. Of these the prominent are: (i) toroidal confinement device (ii) magnetic mirror and tendom mirror devices, and (ii) field reversed device etc.

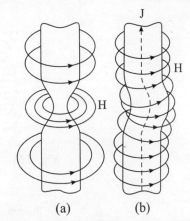

Fig. 7.49   Instabilities in a pinched plasma: (a) bottlenecks, and (b) bends

Magnetic confinement of plasmas takes advantage of the fact that in strong magnetic fields individual charged particles are confined to move along field lines in tight helical trajectories (Figure 7.50). Thus individual particles are confined to one dimensional motion along the field lines.

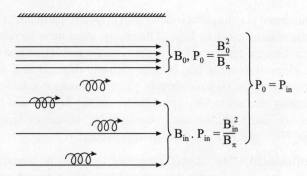

Fig. 7.50 Principle of containment of hot plasma in magnetic field

The two basically different approaches to magnetic confinement shown in Figure 7.51 differ as to whether magnetic field lines, and hence trajectories, lead directly out of the containment region as in the open systems (a), or whether magnetic lines remain largely within the containment region as in the closed toroidal system (b).

In *open system*, magnetic mirrors, or regions where the strength or the magnetic field increases, are used to reflect particles and reduce the loss from the ends.

All "magnetic bottles" are variants of these simple *open* or *closed* systems. Variations are desirable, indeed necessary, to provide the necessary equilibrium and stability of the plasma confinement.

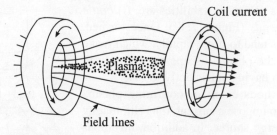

Fig. 7.51(a) Open System-Simple magnetic mirror

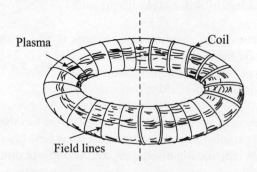

Fig. 7.51(b) Closed System-Simple torus

(i) Toridal Confinement: Tokamak One way being explored is to try and contain the plasma, using a toroidal magnetic field (Figure 7.52), for sufficient time for the fusion reactions to generate an appreciable amount of energy. The reaction chamber is in the shape of a toroid which is encircled by current carrying coils producing strong axial magnetic fields ~ 1 T or more. The field lines are directed the long way around the toroid. A weaker magnetic field called as poloidal field with the field lines circling the short way around the coil is generated by a strong toroidally directed electric current which is induced to flow in the plasma by transformer action.

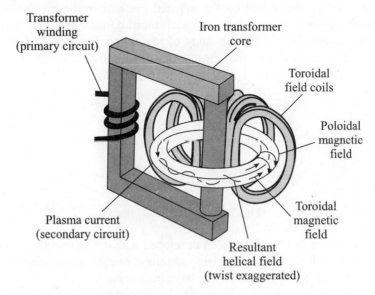

Fig. 7.52 Schematic diagram of the tokamak magnetic field configuration at the Jointt European Torus (JET) facility.

From Figure 7.52 it is clear that an iron transformer core with a primary coil wound around it is introduced through the central part of the toroidal chamber, which with the gas in it serves as the secondary. Alternating current flowing in the core winding produces the secondary current in the gas which generates the plasma and heats it. In the tokmak, the superposition of the toroidal and poloidal fields makes the resultant field helical. We may note that the circulating current in the tokmak not only provides the main confinement force for plasma but also helps in the initial Joule heating of the plasma.

The main effect the strong toroidal field in the tokmak is to stabilize the plasma against the kink instabilities of the bend and bottle neck types.

From the economical point of view, the *magnetic field efficiency or beta* (β) *factor* should be as high as possible, where β-factor is the ratio of the energy-density of the plasma to that of the externally applied field. High β-factor value ensure better utilization of the externally generated magnetic field and thus reduces the cost ratio of the magnet system relative to the power output by fusion reactor. However, the use of very strong toroidal magnetic field to minimise the plasma instabilities limits the value of β-factor. This is main disadvantage of this system.

Due to the closed nature of the magnetic field lines in the tokmak, it is found that the plasma particles can escape from the system only by diffusion across the former and the time for this to happen depends on the square of the radius of the plasma column. Obviously, one can increase considerably the plasma confinement time by increasing the dimensions of the confining magnetic system.

Spitzer suggested to twist the toroidal tube into a space like a figure of 8 (Figure 7.53). In this tube, the polodal field is generated by an array of longitudional conductors laid around the chamber with a helical pitch. This creates a field resembling that in the tokmak. A magnetic field of this type is said to possess a *rotational transform*. The confinement of plasma in it is not dependent on the induced currents in the plasma and therefore no transformer is needed. Now a days, more advanced designs which are large and complex but having flexible facility for the study of plasma confinement and heating have been developed. Russians have prepared the tokmak-10 with following datas: $T = 10^7$ K (1 keV), $n = 8 \times 10^{19}/m^3$, $\tau = 6 \times 10^{-2}$ s and $n\tau = 5 \times 10^{18}/m^3$.

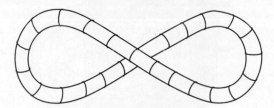

Fig. 7.53 A Stellarator tube.

The Alcator device in USA have been developed with the following parameters: $n\tau = 2 \times 10^{19}/m^3$, $T = 5 \times 10^6$ K (= 0.5 keV). Using additional heating with injected neutral atoms, the ion temperature of 2 keV (= 2×10^7 K) have been achieved.

(ii) Magnetic Mirror Plasma Confinement Device   In the Stellarator and other magnetic confinement plasma devices using closed containing tubes, the plasma can escape in the axial direction and so confinement is required only in radial direction. Magnetic mirror system is an alternate device proposed for the plasma confinement *having bottle shaped magnetic field*. Magnetic mirror system consists of a straight tube with magnetic coils wound around it in such a way as to provide a field that is considerably stronger at the ends than in the middle (Figure 7.54(a)). Obviously, it is an open ended system in which the field lines, which are essentially longitudion, leave at the ends.

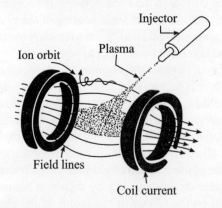

Fig. 7.54(a) Simple magnetic mirror system

The ions move in spiral orbits with radii of curvature inversely proportional to the field strength and under certain conditions are reflected when move into regions of higher field strengths. Theoretical study of the motion of an ion in a non-homogeneous steady longitudional magnetic field shows that $\dfrac{W_s}{B}$ is a constant, i.e. adiabatic invariant, where W_s $= \dfrac{1}{2}Mv_s^2$, where v_s being the velocity component of the ion to the magnetic induction B. In accordance with the shape of the magnetic field shown in Figure 7.55(a) and (b), i.e. having lower value at the middle, compared to that at the two ends, then obviously the value of v_s increases towards the ends.

The total energy $W = W_s + W_p$, where $W_p = \dfrac{m\,v_p^2}{2}$ ($v_p \to$ parallel component of ion velocity) must remain constant. Clearly, as v_s increases near the ends v_p will decrease as the ion moves towards the regions of increasing B at the two ends. Now, if it becomes zero, then the ion will be reflected, i.e. turn backwards and will retrace its path parallel to the field line to be again reflected back from the other end. Obviously, the regions of the maximum field at the two ends thus act as two magnetic mirrors. Clearly, the ions are trapped between these two ends where the fields are maximum. This is why the device acts like a magnetic bottle and ions can be confined within the bottle for an appreciable length of time. Actually the ions describe helical paths (Figure 7.50 and Figure 7.54(a)) of gradually decreasing pitch.

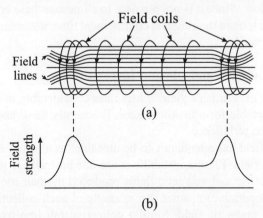

Fig. 7.55 Simple magnetic mirror system: (a) Field coils and lines of force (b) Variation in the magnetic field strength

The heating of plasma in the magnetic mirror system is accomplished by increasing the magnetic field relatively slowly such that the plasma is compressed. One can carry out this compression in several stages from one magnetic bottle to a smaller one in which it is further compressed.

Theory shows that if the direction of the ion velocity v at the canter of magnetic bottle lies within a cone of semi-vertical angle $\theta = \sin^{-1}\left(\dfrac{B_o}{B_m}\right)$, there will be no reflection at the ends and ions will escape from the magnetic bottle, where B_o is the magnetic induction field at the centre and B_m at the ends (where the field is maximum). Such a cone is termed as loss cone. However, if magnetic field lines are properly shaped, gross plasma instabilities reduces considerably.

To maintain the plasma temperature and density in a mirror system, energetic neutral atoms have to be continusously injected from a transverse direction. These energetic atoms are ionized by collision with the particles of plasma already trapped in the device and there by replenish the ions lost due to reasons mentioned earlier.

Some of the drawbacks of the original magnetic error concept have been rectified in other magnetic mirror designs, e.g. *tendem mirror* and field reversed mirror.

In the past two decades, the research in the field of magnetic confinement of hot plasma has changed the picture of fusion plasma. Several new designes have been developed, e.g. quasi-symmetric coil configuration, etc. The new configurations are descendants of early asymmetric configurations known as stellerators.

Classical plasma losses from the magnetic container

With increasing plasma density, particles occasionally collide with one another, and change their orbits somewhat. A collision may, in the case of mirror systems, allow the particle to escape through the mirror directly, or in the case of closed systems, allow the particle to take a small step towards the surface of its magnetic container. In the latter case, after many such collisions the particle is lost altogether. Thus, collisions are responsible for a slow leak from the magnetic container, referred to as a classical loss. Since it is not possible to elemenate these collisions, they set an upper limit on continement time termed the classical confinement time, a standard with which confinement times are generally compared.

Anomalous lose, Turbulence and plasma instabilities

Until recently, all plasma experiments yielded loss rates considerably in excess of classical values, loss rates clearly unacceptable for a fusion reactor. Frequently these anomalous losses seemed to be associated with plasma turbulence.

A plasma in a magnetic field has a tendency to be unstable; as a result, it can break up and escape from confinement by the field. Plasma instabilities are due basically to the presence of electrically charged particles; the electric and magnetic fields produced by their motion cause the particles to act in a collective (or cooperative) manner. An example of such collective action is the drift of a plasma in a nonuniform magnetic field. Similar collective fall into two broad categories called gross. Hydromagnetic instabilities and more localized microinsablities.

Suppose a small displacement of a plasma occurs in a magnetic field; if the system reacts in such a way as to restore the original condition, then it is stable. In the case of a hydromagnetic (or gross) instability, however, the plasma does not recover, but the displacement increase rapidly in magnitude. The whole plasma may then break up and escape even from a strong magnetic field.

Mircoinstabilities, as the name implies, are on a small scale compared with the dimensions of the plasma. As a rule, these instabilities do not lead to complete loss of confinement, but rather to an increase in the rate at which the plasma diffuses out of the magnetic field. Microinstablities apparently result from the interaction of electrically charged particles with electromagnetic waves.

They are, therefore sometimes called *wave-particle instabilities*. Under certain conditions, the energies of the charged particles can be repeatedly added to the waves so that they grow in amplitude. As a result, a high-frequency turbulence can develop in the plasma that makes its escape from the confining field easier. One of the most serious consequences of turbulence, especially in closed systems, is calle *Bohm diffusion*.

The ability to understand and overcome plasma instabilities is vital to the success of the controlled fusion research program. Consequently, the highly complex problems involved have been the subject of extensive experimental and theoretical studies. As a result of step by step interplay between theory and experiments over a period of years, considerable progress has been made.

7.3.8B The laser-fusion process

In the laser-fusion process a solid, spherical pellet of deuterium-tritium is envisioned to be irradiated by a short pulse, high energy laser beam, which accomplishes a number of tasks; (a) the leading portion of the laser beam first creates a plasma surrounding the pellet such that reminder of the beam is nearly fully absorbed, (b) the main portion of the beam then delivers energy to the pallet which not only vaporizes; ionizes, and heats the *DT* mixture but it also induces an implosion, which theoretically results in an ultimate compression of 1000-100000 times the solid density; (c) the resultant highly compressed, hot *DT* plasma then supports fusion reactions at an extremely rapid rate in the short time (of the order of picoseconds $\sim 10^{-12}$ sec) before the pallet files apart due to its high internal pressure. The length of time which the pallet remains in its highly compressed state is determined in large part by the inertia of the fuel mixture and the confinement is thereby called inertia of the fuel mixture and the confinement is thereby called inertial.

Laboratory experiments have established the fact that nuclear fusion reactions can indeed be induced by the action of lasers on solid deuterium. However, the amount of energy released is only a small fraction of that in the laser beam. More powerful lasers with shorter pulses are required and efforts are being made to achieve this objective. One possibility being studied for increasing the power is to focus several individual laser beams into the solid pallet of reacting material.

7.3.9 Fusion Reactor

There is a feeling among plasma physicists that the scientific possibility of nuclear fusion as a means of generating useful power will probably be established early. Consequently there will be a need to design and construct fusion reactor. Scientists have started to give some thought to this problem. Two conceptual designs of fusion reactors are shown in Figure 7.56. Figure 7.56(a) represents the section through a torus showing the basic principles of a possible fusion reactor, while Figure 7.56(b) represents the principles of a conceptual spherical (or approximately spherical) reactor utilizing laser for fusion of **deuterium-tritium pellets**.

It is easy to see why research on fusion continues in spite of its many discouraging aspects. Although fission presents a far greater energy source than our remaining fossil fuels (coal, gas etc.), this source could be used up in a few centuries of modern civilization. Furthermore, fission produces radioactive wastes which are certain to be an increasingly serious problem, possibly requiring that these "poisons" be projected into space. On the other hand, the product material of fusion are almost harmless, and the known fuel supply is sufficient for man's extravagant use for billions of years. No doubt, the reality of fusion power will solve the problem of power for ever.

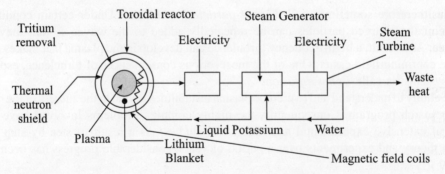

Fig. 7.56(a) Section through a torus showing the basic principles of a possible fusion reactor

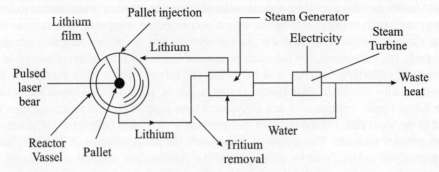

Fig. 7.56(b) Basic principle of a conceptual spherical reactor utilizing laser fusion of $D-T$ pellets

Fusion Research in India

India's fusion research program started in early 1980s at the Institute of Plasma Research in Gandhi nagar. Indian scientists have indigenously built a tokamak ADITYA which was commissioned in 1989, and on which experiments have been conducted since then. ADITYA is like a miniature artificial sun which is contained by magnetic cages for fractions of a second and in wich temperature of several million degrees have been achieved. Scientists are now in the process of assembling second superconducting tokmak SST1, in which the kea feature will be keeping the fusion fire alive for 1000 seconds.

Relative Merits of Fusion Technology

   (i) The fuel (i.e. deuterium) for this process is essentially limitless in sea water.
   (ii) The energy is very clean. There is no atmospheric pollution, no greenhouse gas emission.
   (iii) Radioactivity from waste products in fusion is negligible, and in principle, can be eliminated totally.
   (iv) Fusion reaction are inherently safe; the reaction is difficult to ignite and there is no possibility of a chain reaction or a melt-down.
   (v) There are no dangers of porliferation, no worries that some rouge nation or terrorist group will steal strategic material for a nuclear weapon.

Fusion energy seems to be the ultimate solution to the energy problems of mankind. Most people believe that the earliest that fusion reactors may be commercially deployed is by the year 2035 and latest in by the year 2050.

SUGGESTED READINGS

1. N. Feather, A history of nucleons and nuclei, Contemp Phys. 1, (191, 257), 1954; Chadwick's
2. Neutron, Contemp Phys. 15, 565 (1954)
3. E. Segre, Nuclei and Particles, Benjamin, Ch.11 (1977)
4. W. R. Phillips Rep. Prog. Phys. 40, 345 (1977)
5. N. K. glendenning, Direct nuclear Reactions, Academic Press (1983)
6. S. E. Liverhant, Elementary Introduction to Nuclear Reactor Physics, Sir, Issac John Wiley (1960)
7. C. M. H. Smith, A Text Book of Nuclear Physics, Pergamon Press (1966)
8. W. R. Phillips, Rep. Prog. Phys. 40, 345 (1977)
9. Nuclear Data Tables 1973, Neutron Cross-sections (activation), A 11 (601), 1973
10. W. E. Cottingham and D. A. Greenwood, An Introduction to Nuclear Physics, Combridge (1986)
11. N. A. Jelley, Fundamentals of Nuclear Physics Cambridge (1990).

Example 7.1 An foil of indium having an area of 4×10^{-4} m^2 and $Nx = 3.8 \times 10^{24}$ atoms/m^2 and $\sigma_{act} = 145$ barns, is irradiated for 100 minutes. The counting started 20 minutes after the irradiation is stopped and continued for 10 minutes. The total number of counts observed with a detector of efficiency 0.95, was reported to be 164000. Taking mean life of Indium 78.1 minutes, show that the neutron flux is 2140 neutrons/sec/cm^2. [Vikram]

Sol. We have the maximum counting rate

$$C_0 = \frac{I}{\tau\left[\exp\left(-\frac{t_1}{\tau}\right) - \exp\left(-\frac{t_2}{\tau}\right)\right]\left[1 - \exp\left(-\frac{t_e}{\tau}\right)\right]}$$

Where t_e is the exposure or irradiation time length, τ is average life of the radioactive product $\left(= \dfrac{1}{\lambda}\right)$ and $t = t_1$ and $t = t_2$ are indicating the time interval lying between t_1 and t_2, where t_1 and t_2 are referred to the same origin of time. I is the number of counts per unit area of the foil for an interval of time $(t_1 - t_2)$

Given:

$$I = \frac{164000}{4} = 41000 \text{ Counts/unit area}$$

$$\tau = 78.1 \text{ min}, \ t_1 = 20 \text{ min}, \ t_2 = 30 \text{ min, and}$$

$$t_e = 100 \text{ min}$$

$$\therefore \qquad C_0 = \frac{41000}{78.1\left[e^{-\frac{20}{78.1}} - e^{-\frac{30}{78.1}}\right]\left[1 - e^{-\frac{100}{78.1}}\right]}$$

$$= 6850 \text{ Counts/min}$$

∴ Capture rate of neutron

$$R = \frac{C_0}{\epsilon} = \frac{6850}{0.95} = 7100 \text{ Counts/min}$$

∴ Flux,

$$\phi = \frac{R}{\sigma_{\text{act}}} Nx = \frac{118}{145 \times 10^{-24} \times 3.8 \times 10^{20}}$$

$$= 2140 \text{ neutrons/sec/cm}^2$$

Example 7.2 How many collisions are required for neutrons to loss on the average 99 % of an initial energy of 2 MeV in ^{9}Be moderated assembly (average logarithmic energy decrement $x = 0.209$). Compare this with the total number of collisions required to attain thermal energies.

[Mysore]

Sol. If neutrons of energy E_1 are slowed down to energy E_2, then we have the number of collisions required to reduce this amount of energy is

$$n = \frac{1}{\xi} \ln_e \left(\frac{E_1}{E_2} \right) = \frac{1}{0.209} \log_e \frac{100}{1} = 22$$

Now, the number of collisions riquired to reduce the energy to 0.025 eV, i.e. thermal energy is

$$n_1 = \frac{1}{0.209} \log_e \left(\frac{2 \times 10^6}{0.025} \right) = 87$$

Example 7.3 Show that average energy loss for neutrons which have one collision with carbon nuclei is 0.1462. Given ξ for carbon is 0.158.

[Jodhpur]

Sol. We have

$$\xi = \overline{\log_e \left(\frac{E_1}{E_2} \right)}$$

or

$$\left(\frac{E_1}{E_2} \right) = e^{\xi} \quad \text{or} \quad \left(\frac{E_2}{E_1} \right) = e^{-\xi}$$

∴

$$1 - \left(\frac{E_2}{E_1} \right) = \left(\frac{E_1 - E_2}{E_1} \right) = 1 - e^{-\xi}$$

$$= 1 - e^{-0.158} = 0.1462$$

The result obtained shows that it is incorrect to use ξ which is the mean value of the log and not the log of the mean value for obtaining average energy loss per collision.

Example 7.4 Calculate the energy of the neutrons reflected from the crystal of lattice spacing 2.3 A°, if the glancing angle of the incident neutrons for first order reflection is 3.5°.

[Vikram]

Sol. We have the energy of neutrons

$$E = \frac{n^2 \, h^2}{8m \, d^2 \, \text{Sin}^2 \, \theta}$$

$$= \frac{(6.6256 \times 10^{-34})^2}{8 \times 1.6748 \times 10^{-27} \times (2.3 \times 10^{-10})^3 \times (0.061)^2}$$

$$= 1.04 \text{ eV}$$

$n = 1$
$h = 6.6256 \times 10^{-34}$ J-s
$m = 1.6748 \times 10^{-27}$ kg
$d = 2.3 \times 10^{-10}$ m
$\sin 3.5° = 0.061$

Example 7.5 Show that the energy of neutrons which have made 40 collisions with Be nuclei, starting with an initial energy of 2 MeV is 460 eV. Given x for Be = 0.209.

Sol. We have

$$E_t = E_1 \, e^{-n\xi}$$

$$= 2 \times 10^{-6} \times \exp[-40 \times 0.209]$$

$$= 2 \times 10^6 \times 0.00023 = 460 \text{ eV}.$$

Example 7.6 Calculate the slowing down time in graphite for neutrons having an initial energy of 2 MeV and final thermal energy = 0.025 eV. Given $\xi = 0.155$ for graphite and $\lambda_s = 2.6 \times 10^{-2}$ m.

[Kerala]

Sol. We have

$$t = \frac{\lambda_s}{\xi} \sqrt{(2m)} \, (E_t)^{-\frac{1}{2}}$$

$$= \frac{\lambda_s}{\xi c} \sqrt{2mc^2} \, (E_t)^{-\frac{1}{2}} \qquad \qquad \dots (7.1)$$

Since the final energy in this case is far less than the initial energy, i.e. $E_t^{-\frac{1}{2}}$ as compared to $E_0^{-\frac{1}{2}}$ in the expression without any appreciable error

$$t = \frac{\lambda_s}{\xi} \sqrt{\left(2m \left[E_t^{-\frac{1}{2}} - E_0^{-\frac{1}{2}} \right] \right)}$$

∴ Therefor from (7.1), we have

$$t = \frac{2.6}{0.155 \times 3 \times 10^8} \sqrt{\frac{2 \times 931 \times 10^6}{0.025}}$$

$$= 150 \times 10^6 \text{ s} \qquad \qquad (\because mc^2 = 931 \text{ MeV})$$

Example 7.7 Show that the average distancee $(\bar{x})$ travelled by a neutron away from an infinite planar neutron source in a moderator prior being absorbed is equal to its thermal diffusion length (L), i.e. $\bar{x} = L$.

Also show that the root mean square of the distance travelled $(x_{rms}) = \sqrt{2}\ L = \sqrt{2}\ \bar{x}$.

Sol. We have the average distance

$$\bar{x} = \frac{\int\limits_0^\infty x\, e^{-\frac{x}{L}}\, dx}{\int\limits_0^\infty e^{-\frac{x}{L}}\, dx} = L$$

Which is obviously, the *thermal diffusion length*. Similarly, one obtains

$$\bar{x^2} = \frac{\int\limits_0^\infty x^2\, e^{-\frac{x}{L}}\, dx}{\int\limits_0^\infty e^{-\frac{x}{L}}\, dx} = 2L^2$$

$$\therefore \qquad x_{r.m.s} = \sqrt{\bar{x^2}} = \sqrt{2}L = \sqrt{2}\ \bar{x}$$

Example 7.8 Calculate the diffusion length for thermal neutrons in graphite using the following values: $\sigma_a = 3.2$ milli barns, $\sigma_s = 4.8$ barns and $\rho = 1.62 \times 10^3$ kg/m³   [Kerala]

Sol. We have the expression for diffusion length,

$$L = \left[\frac{\lambda_a\, \lambda_{tr}}{\sqrt{3}} \right]^{\frac{1}{2}}$$

Here
$$\lambda_a = \sum_a^{-1} = \frac{1}{N_0\, \sigma_a} = \frac{1}{N \rho\, \sigma_a}$$

$$= \frac{12}{6.02 \times 10^{26} \times 1.62 \times 10^3 \times 3.2 \times 10^{-31}} = 38.4 \text{ m}$$

and
$$\lambda_{tr} = \frac{\lambda_s}{1 - \frac{2}{3}A} = \frac{\sum_s^{-1}}{1 - \frac{2}{3}A} = \frac{\dfrac{A}{N \rho\, \sigma_s}}{1 - \frac{2}{3}A}$$

$$= \frac{\dfrac{12}{[6.02 \times 10^{26} \times 1.62 \times 10^3 \times 4.8 \times 10^{-28}]}}{\left[1 - \frac{2}{3} \times 12 \right]}$$

$$= 0.0254 \text{ m}$$

$$\therefore \qquad L = 0.572 \text{ m}$$

Example 7.9 When a U-235 nucleus undergoes fussion, 200 MeV energy is released. How many fissions per sec are needed for generation a power of 1 Watt? How much energy in joule would be released when 1 g of U-235 is fissioned? [Kerala, MDS]

Sol. We have: 1 Watt = 1 Js^{-1} = 6.242 × 10^{12} MeV/s

$$\therefore \text{ Number of fission needed/sec} = \frac{6.242 \times 10^{12}}{200}$$

$$= 3.12 \times 10^{10}$$

No. of U-235 atoms in 1 g of U-235 = $\dfrac{6.02 \times 10^{23}}{235}$

$\therefore$ Energy released by 1 g of U-235, on fission, is given by

$$E = \frac{6.02 \times 10^{23}}{235} \times 200 \text{ MeV}$$

$$= \frac{6.02 \times 10^{23} \times 200 \times 1.6 \times 10^{-13}}{235} \text{ J}$$

$$= 8.2 \times 10^{10} \text{ J}$$

Example 7.10 Calculate the energy released in the following reaction.

$$^{235}_{92}U + {}_0n^1 \rightarrow {}^{95}_{42}Mo + {}^{139}_{57}La + 7({}^{0}_{-1}e) + 2({}^{1}_{0}n) + Q$$

[Raj, MDS]

Sol.
$$^{235}Mu = 235.0439 \text{ a.m.u.}$$

$$Mn = 1.0087 \text{ a.m.u.}$$

$$\text{Total initial mass} = 236.0526 \text{ a.m.u.}$$

$$\text{Mass of } ^{95}Mo = 94.9058 \text{ a.m.u.}$$

$$\text{Mass of } ^{139}La = 139 = 138.9061 \text{ a.m.u.}$$

$$M_s = 0.00055 \text{ a.m.u.}$$

$$\text{Total final mass} = 235.8332 \text{ a.m.u.}$$

Mass converted into energy

Energy equivalent = 0.219 × 931 MeV = 204.3 MeV

Example 7.11 If U-236 nucleus is fissioned by a neutron, two fission fragments of mass numbers 96 and 138, and two neutrons are obtained. If the masses of the nuclei and neutron are 235.1175, 95.9385, 137.9487 and 1.00898 amu, calculate the amount the amount of energy released.

[Kerala]

Sol.   Masses before fission = 235.1175 + 1.0089 = 236.1264 amu

Masses after fission = 95.9385 + 137.9487 + 2 × 1.0089

$$= 95.9385 + 137.9487 + 2.0178 = 235.9050 \text{ amu}$$

∴   Difference in mass = 236.1264 − 235.9050 = 0.2214 amu

∴   Energy released, $E = 0.2214 \times 931$ MeV = 206 MeV

Example 7.12 Assuming that energy released by the fission of a single uranium atom is 200 MeV. Calculate the number of fissions per sec. required to produce 1 Watt of power.   [Jodhpur]

Sol. Given: Energy released per fission,

$$= 200 \text{ MeV} = 2 \times 108 \times 1.6 \times 10^{-12} \text{ erg} = 3.2 \times 10^{-4} \text{ erg}$$

$$= \frac{3.2 \times 10^{-4}}{10^7} = 3.2 \times 10^{-11} \text{ Watt}$$

∴ No of fissions required to produce 1 Watt power

$$= \frac{1}{3.2 \times 10^{-11}} = 3.125 \times 10^{10} \text{ fissions}$$

Example 7.13 A power reactor is developing energy at the rate of 30000 kW, How many atoms of ^{235}U undergo fission per sec? How many kg of ^{235}U would be consumed in 1000 hours of operation, assuming that on an average 200 MeV energy is released per fission.   [Kerala, Mysore]

Sol. Given that the rate of development of energy by power reactor,

$$= 30000 \text{ kW} = 3 \times 10^7 \text{ Watt}$$

$$= 3 \times 10^{14} \text{ erg/s}$$

Energy released per fission = 200 MeV

$$= 200 \times 10^8 \times 1.6 \times 10^{-12} \text{ erg}$$

$$= 3.2 \times 10^{-4} \text{ erg}$$

No. of atoms undergoing fission/sec $= \dfrac{3 \times 10^{14}}{3.2 \times 10^{-4}}$

$$= 9.4 \times 10^{17}$$

∴ No. of atoms whose fission takes place in 1000 hours

$$= 9.4 \times 10^{17} \times 1000 \times 60 \times 60 = 3.384 \times 10^{23}$$

∴ 6.02×10^{23} atoms of ^{235}U weight 235 gm.

∴ 3.384×10^{23} " " " will weight

$$\frac{235}{6.02 \times 10^{23}} \times 3.384 \times 10^{23} = 132.1 \text{ gm} = 0.1321 \text{ kg.}$$

Example 7.14 Calculate the electrostatic potential energy between two equal nuclei produced in the fission os ^{235}U at the moment of their separation. Use the relation $R = R_0 \ A^{\frac{1}{3}}$.   [Mysore]

Sol. If R is the radii of either nuclei at the moment of fission, then

$$R = 1.3 \left(\frac{226}{2} \right)^{\frac{1}{3}} \times 10^{-15} \text{ m} = 1.3 \times 4.88 \times 10^{-15} \text{ m}$$

$$(R_0 = 1.3 \times 10^{-15} \text{ m})$$

Then electrostatic P.E. $= \dfrac{\left(\dfrac{Z}{2} \right)^2 e^2}{4\pi\, \varepsilon_0 \times 2 \times R}$

$$= \frac{(46)^2 \times (1.6 \times 10^{-19})^2}{4\pi \times 8.85 \times 10^{-12} \times 2 \times 1.3 \times 4.88 \times 10^{-15}} \text{ Joules}$$

$$= 240 \text{ MeV}$$

Example 7.15 Calculate the energy released by the fission of a gram of uranium ^{235}U. Energy released per fission is 200 MeV.

Sol. At. Wt. of Uranium = 235

∴ No. of atoms in 1 gm. Uranium

$$= \frac{6.02 \times 10^{23}}{235}$$

∴ Energy released by the fission of 1 gm. U^{235}

$$= \frac{6.02 \times 10^{23}}{235} \times 200 \text{ MeV}$$

$$= 2.8 \times 10^5 \text{ Kilo Watt Hour (kWH)}$$

Example 7.16 On fission, U-235 yields two fragments of $A = 95$ and $A = 140$ roughly. Obtain the energy distribution of the fission products. Assume that the two fragments are ejected with equal and opposite momentum.

Sol. Using the condition $M_1 v_1 = M_2 v_2$ ... (i)

$$\therefore \qquad \frac{E_2}{E_1} = \frac{\frac{1}{2} M_2 v_2^2}{\frac{1}{2} M_1 v_1^2} = \frac{\frac{1}{2}(M_2 v_2)^2}{\frac{1}{2}(M_1 v_1)^2} \cdot \frac{M_1}{M_2} \qquad \text{... (ii)}$$

From relation (i): $\qquad \dfrac{E_2}{E_1} = \dfrac{M_1}{M_2} = \dfrac{95}{140} = \dfrac{19}{28}$

$$\therefore \qquad E_2 : E_1 = 19 : 28 = 2 : 3 \text{ (approximately)}$$

Example 7.17 A nuclear power generating station has a capacity of 60,000 kW. If only 20 % of the thermal energy generated is converted into electricity, and if 200 MeV energy is produced per fussion, calculate the amount of U-235 spent per year. If natural uranium is used as fuel, how much uranium would be needed per year if the relative abunance of U-235 in natural uranium is 0.7 %. Given: 1 MeV = 1.6×10^{-13} J.

Sol. 1 kg of U-235 = $6.02 \times 10^{23} \times \dfrac{10^3}{235}$ = 2.56×10^{24} atoms of U-235

$\therefore$ Energy released by fission of 1 kg of U-235 is

$$E = 2.56 \times 10^{24} \times 200 \text{ MeV}$$

$$= 2.56 \times 10^{24} \times 200 \times 1.6 \times 10^{-13} \text{ J}$$

$$= 8.19 \times 10^{13} \text{ J}$$

$$= \dfrac{8.19 \times 10^{13}}{60 \times 60 \times 10^3} \text{ kilo-Watt hour (kWH)}$$

$$= 22.75 \times 10^6 \text{ kWH}$$

$\therefore$ Energy obtained as electricity per kg of U-235 is given by

$$E_0 = 22.75 \times 10^6 \times \dfrac{20}{100} = 4.55 \times 106 \text{ kWH}$$

$\therefore$ Amount of U-235 spent per hour = $\dfrac{60,000}{4.55 \times 10^6} = 13.1 \times 10^{-3}$ kg

$\therefore$ Consumption of U-235 per year = $13.1 \times 365 \times 24 \times 10^{-3}$ kg = 114.7 kg

If natural uranium is used, the consumption per year would be

$$114.7 \times \dfrac{100}{0.7} \text{ kg} = 16385 \text{ kg} = 16.385 \text{ metric ton}$$

Example 7.18 In a nuclear reactor, fission is produced in 1 gm of ^{235}U, (235.0439 amu) in 24 hours by a slow neutron (1.008 amu). Assuming that $^{92}_{36}$Kr (91.8973 amu) and $^{141}_{56}$Ba (140.9139 amu) are produced in all reaction and no energy is lost, write the complete reaction and calculate the total energy produced in kilowatt-Hour. Given 1 amu = 931 MeV.   [Rajasthan, Mysore]

Sol. The nuclear fission reaction is

$$^{235}_{92}U + ^{1}_{0}n \rightarrow ^{141}_{56}Ba + ^{92}_{36}Kr + 3^{1}_{0}n$$

The sum of the masses before the reaction is

235.0439 + 1.0087 = 236.0526 amu

The sum of the masses after the reaction is

140.9139 + 91.8973 + 3(1.0087) = 235.8373 amu

The mass loss in the fussion is

$$\Delta m = 236.0526 - 235.8373 = 0.2153 \text{ amu}$$

The energy equivalent of 1 amu is 931 MeV. Therefore the energy released in the fission of a U^{235} nucleus

$$= 0.2153 \times 931 = 200 \text{ MeV}$$

The number of atoms in 235 gm of ^{235}U is 6.02×10^{23} (Avogadro number). Therefore, the number of atoms of 1 gm

$$= \frac{6.02 \times 10^{23}}{235} = 2.56 \times 10^{21}$$

Hence the energy released in the fission of 1 gm U^{235} i.e. in 2.56×10^{21} fission is

$$E = 200 \times 2.56 \times 10^{21} = 5.12 \times 10^{23} \text{ MeV}$$

Now, $\qquad 1 \text{ MeV} = 1.6 \times 10^{-13} \text{ J}$

$\therefore \qquad E = 5.12 \times 10^{23} \times (1.6 \times 10^{-13}) = 8.2 \times 10^{10}$ joule

Now, 1 killowatt-hour = 1000 Watt × 3600 sec

$$= 1000 \ \frac{\text{joule}}{\text{sec}} \times 3600 \text{ sec}$$

$$= 3.6 \times 106 \text{ J}$$

$\therefore \qquad E - \frac{8.2 \times 10^{10}}{3.6 \times 10^6} - 2.28 \times 10^4 \text{ kWH.}$

Example 7.19 A reactor is developing nuclear energy at a rate of 32,000 kilowatts. How many atoms of ^{235}U undergo fission per second? How many kg of ^{235}U would be used up in 1000 hours of operation? Assume an average energy of 200 MeV released per fission. Take Avogadro's number as 6×10^{23} and 1 MeV = 1.6×10^{-13} joule. [Lucknow, Delhi]

Sol. The power developed by the reactor is 32000 kilowatt, i.e., 3.2×10^7 Watt. Therefore, the energy released by per second is

$$= 3.2 \times 10^7 \text{ joule} \qquad [\because 1 \text{ Watt} = 1 \text{ joule/sec}]$$

$$= \frac{3.2 \times 10^7}{1.6 \times 10^{-13}} \qquad [\because 1.6 \times 10^{-13} \text{ joule} = 1 \text{ MeV}]$$

$$= 2 \times 10^{20} \text{ MeV.}$$

The energy released per fission is 200 MeV. Therefore, the number of fissions occurring in the reactor per second

$$= \frac{2 \times 10^{20}}{200} = 10^{18}$$

The number of atoms (or nuclei) of U^{235} consumed in 1000 hours

$$= 10^{18} \times (1000 \times 3600)$$

$$= 36 \times 10^{23}$$

Now, 1 gm-atom (i.e. 235 gm) of U^{235} has 6×10^{23} atoms. Therefore, the mass of U^{235} consumed in 1000 hours is

$$= \frac{36 \times 10^{23}}{6 \times 10^{23}} \times 235$$

$$= 1410 \text{ gm} = 1.41 \text{ kg}$$

Example 7.20 Calculate the approximate mass of uranium which must undergo fission to produce same energy as is produced by the combustion of 10^5 kg of coal. Heat of combusion of coal is 8000 kcal/kg; the energy released per fission of U^{235} per gm-atom (1 cal = 4.2 joules).

[Rajasthan]

Sol. The energy produced by 10^5 kg. of coal

$$= 10^5 \times 8000 \text{ k/cal}$$

$$= 8 \times 10^{11} \text{ cal}$$

$$= 8 \times 10^{11} \times 4.2 = 3.36 \times 10^{12} \text{ joule} \qquad [\because 1 \text{ cal} = 4.2 \text{ joule}]$$

$$= \frac{3.36 \times 10^{12}}{1.6 \times 10^{-13}} = 2.1 \times 10^{25} \text{ MeV} \quad [\because 1 \text{ MeV} = 1.6 \times 10^{-13} \text{ joule}]$$

The energy released per fission is 200 MeV. Therefore, the number of fission required for 2.1×10^{25} MeV energy is

$$\frac{2.1 \times 10^{25}}{200} = 1.05 \times 10^{23}$$

Since 235 gm of U^{235} contains 6.02×10^{23} atoms, the mass of uranium containing 1.05×10^{23} atoms is

$$\frac{1.05 \times 10^{23}}{6.02 \times 10^{23}} \times 235 = 41 \text{ gm}$$

Example 7.21 The reproduction factor for fission of pure U-235 is 1.125. Calculate the number of secondary neutrons produced in (i) 10 generations and (ii) 100 generations.   (Kerala)

Sol. From the relation $k^n = x$, we obtain: $\ln x = n \ln k$.

$\therefore$ For (i), $\ln x_1 = 10 \times \ln 1.25 = 1.1728$ $\therefore x_1 = 3.247$

For (ii), $\ln x_2 = 100 \times \ln 1.125 = 11.778$ $\therefore x_2 = 2.18 \times 10^{11}$

$\therefore$ We get $x_1 = 3.247$; $x_2 = 2.18 \times 10^{11}$ as the number of secondary neutrons in 10 and 100 generations respectively.

Example 7.22 Certain stars obtain part of their energy by the fusion of three α-particles to form a ^{126}C nucleus. How much energy does each such reaction evolve? The mass of helium atom is 4.00260 amu while the mass of an electron is 0.00055 amu. The mass of ^{126}C atom is 12.0000 amu by definition.   (1 amu = 931 MeV)

Sol. The mass of $_2He^4$ atom is 4.00260 amu and it has 2 electrons. Therefore, the mass of its nucleus (α-particle)

$$= 4.00260 - \text{mass of 2 electrons}$$

$$= 4.00260 - (2 \times 0.00055)$$

$$= 4.00260 - 0.00110 = 4.00150 \text{ amu}$$

The mass of ^{126}C atom is 12.00000 amu and it has 6 electrons. Therefore, the mass of ^{126}C nucleus

$$= 12.00000 - (6 \times 0.00055)$$

$$= 12.00000 - 0.00330$$

$$= 11.99670 \text{ amu.}$$

when 3 α-particles fuse in a ^{126}C nucleus, the mass-loss is

$(3 \times 4.00150) - 11.99670 = 0.00780$ amu.

The equivalent energy is

$$0.00780 \times 931 = 7.26 \text{ MeV}$$

Example 7.23 Compute the mass of neutron from the following reaction:

$$\,_0^1 n + \,_1^1 H \rightarrow \,_1^2 H + \gamma$$

where the masses $M(^1H) = 1.007825$ amu, $M(^2H) = 2.014102$ amu and $E(\gamma) = 2.23$ MeV.

Sol.

$$M(n) = M(^2H) - M(^1H) + \gamma$$

$$= (2.014102 - 1.007825 + 2.23/931) \text{ amu}$$

$$= (2.014102 - 1.007825 + 0.002395) \text{ amu}$$

$$= 1.008672 \text{ amu}$$

Example 7.24 A thermonuclear device consists of a torus of 3 m dia with a tube of 1 m dia. It has in it deuterium gas at a temperature of 20°C and at a pressure of 10^{-2} mm of mercury. A capacitor of 1.2×10^3 μF is discharged through the tube at 40 kV. Find the maximum temperature attained, if only 10 % of the electrical energy is transformed to plasma kinetic energy. The energy is equally shared by deutrons and electrons in the plasma.   [Kerala, MDSA]

Sol. The electrical energy, from the discharge, is obtained from

$$E = \frac{1}{2} CV^2 = \frac{1}{2} \times 1.2 \times 10^3 \times 10^{-6} \times (40 \times 10^3)^2 = 96 \times 10^4 \text{ J}$$

∴ Energy used by plasma $= \dfrac{1}{10} \times 96 \times 10^4 = 96 \times 10^3$ J

Let T_K be the kinetic temperature of plasma particles. So the average kinetic energy is

$$E_A = \frac{3}{2} Nk T_K$$

N being the total number of deuterium particles.

Now, the pressure P os the deuterium gas is

$$P = 10^{-5} \text{ m of mercury}$$

$$= 10^{-5} \times 13.6 \times 10^3 \times 9.81$$

$$= 1.34 \; \frac{\text{N}}{\text{m}^2}$$

Volume of torus, V = cross section × circumference

$$= \left(\frac{\pi \times 1^2}{4}\right) \times (\pi \times 3) = \frac{3}{4}\pi^2 = 7.4 \text{ m}^2$$

$\therefore$ $pV = NkT$ gives: $1.34 \times 7.4 = Nk(273 + 20)$

$\therefore$ $Nk = \dfrac{1.34 - 7.4}{293} = 0.0338$

Since each deuterium molecule gives rise to two ions and two electrons, we have

$$4 \times \frac{3}{2} NkT_k = 96 \times 10^3$$

$\therefore$ $T_k = \dfrac{96 \times 10^3}{4 \times \dfrac{3}{2} \times 0.338}$

$$= 4.73 \times 10^5 \text{ K}$$

Example 7.25 In neutron-induced fission of a $^{235}_{92}U$ nucleus, unstable energy of 185 MeV is released. If a $^{236}_{92}U$ reactor is continuously operating at a power level of 100 MW, how long will it taken for 1 kg of uranium to be consumed in this reactor.

Sol. Operating power level of the reactor of 100 MW, which is equivalent to

$$10^8 \text{ J/s} = \frac{10^8}{1.6 \times 10^{-13}} \text{ MeV/s}$$

Fission of one $^{236}_{92}U$ nucleus produced 185 MeV of usable energy, hence the number of uranium nuclei fissioning to generate this power

$$= \frac{10^8}{1.6 \times 10^{-13}} \times \left(\frac{1 \text{ fission}}{185 \text{ MeV}}\right) \text{ per second}$$

$$= 3.378 \times 10^{18} \text{ fission per second}$$

$$= \left(\frac{1 \text{ kg}}{235 \text{ kg}/k \text{ mol}}\right)\left(6.02 \times 10^{26} \frac{\text{nuclei}}{k \text{ mol}}\right)$$

$$= 2.5617 \times 10^{24} \text{ nuclei}$$

$\therefore$ The time in which this number of uranium nuclei fissions is

$$= \frac{2.5617 \times 10^{34}}{3.378 \times 10^{29}} = 7.583 \times 10^4 \text{ sec}$$

$$= 8.777 \text{ days}$$

Example 7.26 In a reactor, fission is produced in 1 g of U-235 (235.0439 amu) in 24 hours by slow neutrons (1.0087 amu). Assuming that $^{92}_{36}Kr$ (91.8973 amu) and $^{141}_{56}Ba$ (140.9139 amu) are produced in all reactions and no energy is lost, write the complete reaction and calculate the total energy produced in kWh. Given: 1 amu = 931 MeV. [Raj, Mysore]

Sol. The fission reaction is:

$$^{295}_{92}U + ^{1}_{0}n \rightarrow ^{141}_{56}Ba + ^{92}_{36}Kr + 3^{1}_{0}n$$

Masses before reaction = 235.0439 + 1.0087 = 236.0526 amu

Masses after the reaction = 140.9139 + 91.8973 + 3 × 1.0087 = 235.8373 amu

∴ Mass loss in fusion, Δm = 236.0526 − 235.8373 = 0.2153 amu

∴ Energy released in fission of a U-235 nucleus is

$$\Delta E = 0.2153 \times 931 \text{ MeV} = 200 \text{ MeV}$$

No. of U-235 atoms in 1 g = $\dfrac{6.02 \times 10^{23}}{235}$ = 2.56 × 10^{21}

∴ Energy released on fission of 1 g U-235

$$= 2.56 \times 10^{21} \times 200 \text{ MeV}$$

$$= 2.56 \times 10^{21} \times 200 \times 106 \times 1.6 \times 10^{-19} \text{ J} = 8.2 \times 10^{10} \text{ J}$$

Now, 1 kWh = 1000 Wh = 1000 × 3600 J = 36 × 105 J

∴ Energy released by 1 g of fission of U-235, expressed in kWh is

$$E = \frac{8.2 \times 10^{10}}{36 \times 10^5} \text{ kWh} = 2.28 \times 10^4 \text{ kWh}$$

Example 7.27 A reactor is developing nuclear energy at a rate of 32,000 kW. How many atoms of U-235 would undergo fission per second? How many kg of U-235 would be used up in 1000 hours of operation? Assume that on an average 200 MeV is released per fission and Avogadro number = 6 × 10^{23}. [Delhi Hons.]

Sol. We have: 32,000 kW = 32,000 × 10^3 W = 3.2 × 10^7 W

∴ Energy released per sec. by the reactor, E = 3.2 × 10^7 J ($\because$ 1 W = 1 J/s)

∴ $E = \dfrac{3.2 \times 10^7}{1.6 \times 10^{-13}}$ MeV = 2 × 10^{20} MeV

∴ No. of U-235 atoms undergoing fission per sec is

$$N = \frac{2 \times 10^{20}}{200} = 10^{18}$$

∴ No. of U-atoms and up in 1000 hrs is

$$N_U = 10^{18} \times 3600 \times 1000 = 36 \times 10^{23}$$

But 235 g of U contains 6 × 10^{23} atoms.

∴ Mass of U-235 used up in 1000 hours is

$$m = \frac{36 \times 10^{23} \times 235}{6 \times 10^{23}}$$

$$= \frac{36 \times 10^{23} \times 235}{6 \times 10^{23} \times 10^3} = 1.41 \text{ kg}$$

Example 7.28 Calculate the energy liberated when a single helium nuclei is formed by the fusion of two deuterium nuclei.

Sol. The desired reaction is

$$\underset{(D)}{^2_1 H} + \underset{(D)}{^2_1 H} \rightarrow {}^4_2 He + Q$$

Mass of 2D atoms = 2×2.01478 a.m.u.

$$= 4.02956 \text{ a.m.u.}$$

Mass of the helium atom = 4.00388 a.m.u.

Mass defect = 4.2956 − 4.00388 a.m.u.

$$= 0.02568 \text{ a.m.u.}$$

Energy released = 0.02568×931.8 MeV

$$= 24 \text{ MeV (approx.)}$$

Example 7.29 The area of a lake is about 106 square miles and its average depth is roughly $\frac{1}{20}$ th mile. Calculate the amount of energy released if all the deuterium atoms in the water are used up in fusion

Sol. Volume of the water in the lake = $10^6 \times \frac{1}{20} = 5000$ cubic miles

Mass of the water = $5000 \times (1.6 \times 10^3)^3 \times 10^3$ kg

$$= 21.3 \times 10^{15} \text{ kg} = 2.13 \times 10^{16} \text{ kg.}$$

∵ Mol. wt. of the water if 18, so that the number of molecules of water in 2.13×10^{16} kg is

$$= \frac{2.13 \times 10^{16} \times 6.02 \times 10^{26}}{18} = 7.13 \times 10^{41} \text{ molecules of water}$$

The abundance of deuterium is 0.0156 %; so that the total number of deuterium atom is

$7.3 \times 10^{41} + 2 \times 0.0156 \times 10^{-2} = 2.22 \times 1038$ deuterium atoms

We know that 6 atoms of deuterium give an energy of 43 MeV = 7.17 MeV per *D* atom.

Total energy release = $7.17 \times 2.22 \times 10^{40} = 1.6 \times 10^{39}$ MeV

Example 7.30 Calculate the *Q*-value for the *D-T* fusion reaction ${}^3_1 H$ (*d*, *n*) ${}^4_2 He$. Given mass of deuterium $m_D = 2.01402$ amu mass of tritium $m_T = 3.016030$ amu, $m_n = 1.008665$ amu and mass of helium $m_{He} = 1.008665$ amu.

Sol. From the principle of conservation of mass and energy one can write

$$m_T c^2 + m_D c^2 = m_n c^2 + m_{He} c^2$$

$$\therefore \qquad Q = [(m_T + m_D) - (m_n + m_{He})]c^2$$

$$= [(3.016030 + 2.014102) - (1.008665 + 4.002603)] \text{ amu}$$

$$= 0.018864 \times 931.5 \text{ MeV}$$

$$= 17.572 \text{ MeV}$$

Example 7.31 Supposing all energy from *D-T* fusion reaction available, calculate the rate at which deuterium and tritium are consumed to produce 10 MW in a *D-T* fusion reactor.

Sol. The *Q* value for the *D-T* fusion reaction is 17.572 MeV (See example 7.30).

Hence the rate at which the *D-T* reaction should, proceed in order to produce a power level of 10 MW, is given by

$$\text{reaction rate} = \frac{10^7 \text{ J/s}}{1.6 \times 10^{-13} \text{ J/MeV}} \times \frac{1 \text{ reaction}}{17.572 \text{ MeV}}$$

$$= 3.556 \times 10^{18} \frac{\text{reactions}}{\text{sec}}$$

In each reaction one atom of deuterium and one atom of tritium is consumed. Hence the deuterium ($A = 2$)

$$-\frac{d}{dt}(m_D) = 3.556 \times 10^{18} \left(\frac{\text{atoms}}{\text{sec}}\right) \left(\frac{1\,k \text{ mol}}{6.02 \times 10^{26} \text{ atoms}}\right) \times \left(\frac{2 \text{ kg}}{1\,k \text{ mol}}\right)$$

and for tritium ($A = 3$)

$$-\frac{d}{dt}(m_T) = \frac{3}{2} (1.81 \times 10^{-8} \text{ kg/sec})$$

$$= 1.772 \times 10^{-8} \text{ kg}$$

Example 7.32 For a nucleus described by the semi-empirical mass formula

$$M(A, Z) = Z_{MH} + (A - Z)M_n - \alpha A + \beta\, A^{\frac{2}{3}} + \frac{\gamma \left(\frac{A}{2} - Z\right)^2}{A} + \frac{\epsilon\, Z^2}{A^{\frac{1}{3}}}$$

Where α, β, γ and ϵ are constants, show that, neglecting asymmetry terms, the energy released in fission into two fragments is a maximum for equal division of charge and mass-calculate the value of $\frac{Z^2}{A}$ at which this division just becomes possible energetically.

Sol. Excluding the asymmetry and pairing terms the semi-emprical mass formula gives

$$-B = -\alpha A + \beta\, A^{\frac{2}{3}} + \frac{\epsilon\, Z^2}{A} A^{\frac{1}{3}}$$

If a nucleus A splits into $A_1 + A_2$, the energy released is given by

$$\beta\left(A^{\frac{2}{3}} - A_1^{\frac{2}{3}} - A_2^{\frac{2}{3}}\right) + \in\left(\frac{Z^2}{A^{\frac{1}{3}}} - \frac{Z_1^2}{A_1^{\frac{1}{3}}} - \frac{Z_2^2}{A_2^{\frac{2}{3}}}\right)$$

writing $A_2 = A - A_1$, and $Z_2 = Z - Z_1$ and differentiating with respect to A_1. The first term gives a maximum for $A_1 = \dfrac{A}{2}$ and the second has turning points at $A_1 = A$ and $A_1 = \dfrac{A}{2}$.

Differentiating w.r.t. Z_1 gives turning points at $Z_2 = Z$ and $Z_1 = \dfrac{Z}{2}$. Of these possibilities, one describes no fission at all the other, $A_1 = A_2 = \dfrac{A}{2}$ and $Z_1 = Z_2 = \dfrac{Z}{2}$, is the required condition. We may easily check it to be a maximum by second differentiation.

By setting the energy release to zero for the case of equal division and using $\beta = 17.8$ MeV and $\varepsilon = 0.71$ MeV, one obtain $\dfrac{Z^2}{A} \geq 17.6$.

REVIEW QUESTIONS

1. Give an account of various methods of production and detection of neutrons.
2. Describe an accurate method for the determination of magnetic moments of neutrons experimentally.
3. Explain how the liquid drop model of the nucleus enables one to understand, why ^{235}U undergoes fission if it captures a thermal neutron, whereas ^{238}U does not.   [Mysore]
4. What do you understand by 'prompt' and 'delayed' neutrons? Explain the role played by delayed neutrons in nuclear reactors.
5. What is asymmetrical fission? Show the mass distribution of fission fragments for different fission chains of ^{235}U by a sketch.   [Raj]
6. Discuss th neutron cycle and possibility of a chain reaction in natural ^{238}U. Obtain a four-factor formula and explain on what factors depend each term in the formula. What is the advantage of 'lumping' in a heterogeneous reactor?   [Kerala]
7. Explain why a self-sustaining chain reaction can not be obtained with natural ^{238}U as fuel (except with heavy water as moderator) in a homogeneous assembly.   [Kerala]
8. What is a nuclear reactor? Explain criticality of a nuclear reactor.
9. What is the principle of fast breeder reactor? Explain the working of this reactor. Mention its advantages.
10. Distinguish between the processes of nuclear fission and nuclear fusion. Why both processes may lead to the evolution of energy?
11. Why do neutrons of low energy causes fission in ^{235}U, ^{233}U and ^{239}U whereas high energy neutrons are needed to cause fission in ^{238}U?
12. Describe the experiment for neutron life time and discuss its results.
13. Give an clear account of the physical principles underlying the design of a thermal nuclear reactor.
14. Discuss the four factor formula and the cycle of neutron absorption and production for a critical reactor.

15. Show that if a neutron of initial kinetic energy E_n is scattered by a heavy nucleus to mass number A, the recoil energy given to the nucleus is given by

$$E_R = 4 A E_n \operatorname{Sin}^2 \phi \, (A + 1)^2$$

where ϕ is the angle of recoil of the nucleus with respect to the direction of the incident neutron.

You may assume that $A = \dfrac{M}{m_n}$, where M is the mass of nucleus and m_n is the mass of the neutron.

[Raj, Udaipur]

16. How are neutrons produced in the laboratory? Give few illustrative nuclear reactions for their production and discuss the applications of neutrons of different energies for effecting nuclear transmissions. [Punjab]

17. Estimate the energy released in the fission of a nucleus. Explain the distribution of fragments with respect to their mass numbers. Give the basic principles of a nuclear reactor. Explain how it can be used for producing electric power? [Punjab]

18. Discuss Bethe Carbon-nitrogen and proton-proton cycles. How carbon-nitrogen cycle produces stellar energy?

19. What is a plasma. Discuss a laboratory method for its production and confinement. What are the fundamental difficulties in achieving sustainable controlled fusion reactions in the laboratory.

20. Give an brief account of nuclear fusion reactor. What are the basic difficulties in achieving it?

21. Write short notes on:

    (a) Neutron detector (b) Neutron: production and detection (c) Nuclear fission (d) Nuclear reactor (e) Fast breeder reactor (f) Nuclear fusion (g) Thermonuclear fusion reactor (h) Plasma (i) Steller energy (j) Laser fusion reactor (k) Fission and fusion

22. What is the source of energy in Sun and Stars? Give a short account of Bethe's explanation.

23. Explain, what do you understand by nuclear fusion reactions? Mention clearly the difficulties in the way of achieving self sustained fusion reactions and evaluate the progress made in this direction.

24. Compare nuclear fusion and nuclear fission processes and throw light on the importance of fusion reactions.

PROBLEMS

1. A neutron is scattered through an angle of 60° in the C.M. system in collision with $^{12}_{6}C$. Find the fractional loss in the energy of the neutrons. Also calculate the maximum fractional energy loss in this case? Compare it with that in case of collision of the neutron with $^{235}_{92}U$ nuclei.

   [**Ans.** 7 %, 28.4 %, 9.83 %] [Jodhpur]

2. Calculate the thermal diffusion length of neutrons in an aqueous solution of neutral uranium salt. Given, weight composition 5 % uranium, 95 % of light water. [**Ans.** 6.8 cm] [Kerala]

3. A point source of thermal neutrons emitting 10^3 neutrons/s is placed at the centre of graphite sphere of radius 0.1 m. Calculate the flux distribution of the thermal neutrons within the sphere, assuming it to be surrounded by a material which absorbs all the neutrons escaping through the surface of the sphere. Also calculate the fraction of neutrons which are thus absorbed.

   [Mysore]

4. Show that the age difference for neutrons slowing down in graphite between neutrons that have reduced the indium resonance energy of 1.44 eV and neutrons that have reduced thermal energies 0.025 eV is $60 \times 10^{-4} \, m^2$. [Raj, MDSA]

5. A crystal spectrometer has a beryllium crystal with a lattice spacing of 0.7323 A°. What are the Bragg angles for the first order reflection and of neutrons of energy 1, 3, 5 and 30 eV respectively?

   [**Ans.** $11°\ 15',\ 6°\ 28',\ 5°,\ 2°\ 2'$] [Mysore, Kerala]

6. A neutron causes fission in $^{235}_{92}U$ producing $^{97}_{40}Zr$ and ^{134}Te and some neutrons. What is the atomic number of Te, and how many neutrons are released?   [52, 4]

7. The fission of $^{235}_{92}U$ releases approximately 200 MeV. What percentage of the original mass of $^{235}_{92}U + {}_0n^1$ disappears?   [0.1 %]

8. Assuming that the 2 neutrons and 2 fission fragments are produced in the thermal neutron fission of ^{235}U, estimate the energy liberated in the fission. Binding energy per nucleon of the fission nucleus and fission fragments are 7.8 and 8.5 MeV respectively.   [148 MeV]

9. A nuclear reactor containing ^{235}U is operating at a power level of 5 Watt. Calculate the fission rate assuming that 200 MeV of useful energy is released in one fission.   [1.56×10^{11} per second]

10. A reactor is developoing nuclear energy at the rate of 30,000 kW. How many atoms of ^{235}U undergo fission per sec? How many kg. of ^{235}U would be used in 100 hours of operating that on an average energy of 200 MeV is released per fission.   [0.132 kg.]

11. Assuming the energy released by the fission of a single uranium atom is 200 MeV. Calculate the number of fission per sec. required to produce 1 Watt of power.   [3.12×10^{10} fission]

12. The energy reaching our earth from the Sun is 2 cal/cm^2-min on a surface normal to the Sun rays. At what rate must hydrogen be consumed in the Sun in the fusion reaction.

$$4{}_1^1H \rightarrow {}_2^4He + 2{}_{+1}^0e\ ?$$

[Distance from the earth to the sun is 1.495×10^8 km.]

[**Ans.** 5.7×10^{11} kg/sec.]

13. What is the energy released when 1 kg of nuclear fuel is consumed if the fusion reaction

$${}_1^2H + {}_1^2H \rightarrow {}_2^4He \text{ is possible?}$$

[**Ans.** 5.74×10^{14} J]

14. Certain Stars obtain part of their energy by the fusion of three α-particles to form a $^{12}_6C$ nucleus. How much energy does such reaction evolve?

15. Estimate the energy released in the Bethe Carbon-Nitrogen cycle in Stars.   [**Ans.** 25.6 MeV]

16. Calculate the energy liberated when a single helium nucleus is formed by the fusion of two deuterium nuclei. Given

   Mass of ${}_1^2H = 2.01478$ a.m.u.

   Mass of ${}_2^4He = 4.00388$ a.m.u.

17. The fusion reaction $2{}_1^2H \rightarrow {}_2^4He + Q$ is proposed to be used for the production of industrial power, assuming the efficiency of the process to be 30 % find how many grams of deuterium will be consumed in a day for an output of 50,000 kW. Given

   Mass of ${}_1^2H = 2.01478$ a.m.u.

   Mass of ${}_2^4He = 4.00388$ a.m.u.

   [25.22 gm]

18. A nuclear reactor is set up fulled up by $^{235}_{92}U$. Estimate the mass of $^{235}_{92}U$ consumed in giving a power output of 100 MW for a period of 5 years. Estimate an upper limit for the production of $^{239}_{94}Pu$ if the reactor core were blanketed with a layer of natural uranium.

   [**Ans.** ^{235}U: 2.1 kg; ^{239}Pu: 3.15 kg]

SHORT-QUESTION ANSWERS

1. Can neutrons be knocked out of nuclei by high energy gamma ray photons? If yes, give one example.

Ans. Yes. $^9\text{Be} + \gamma \rightarrow {}^8\text{Be} + {}_0\text{n}^1 + 1.6\,\text{MeV}$

2. Write an reaction, where high energy neutrons are produced.

Ans. ${}_1^3\text{H} + {}_1^2\text{H} \rightarrow {}_2^4\text{He} + {}_0^1\text{n} + 17.6\,\text{MeV}$

3. What is the best source of thermal neutrons?

Ans. Chain reactor

4. Is neutron outside the nucleus is unstable?

Ans. Yes. Outside the nucleus, a neutron is expected to decay as follows

$$ {}_0^1\text{n} \rightarrow {}_1^1\text{p} + {}_{-1}^0\text{e} + n + 782\,\text{keV}. $$

5. How you say that the value of the spin of neutron is $\dfrac{1}{2}$?

Ans. Deutron is a loose combination of p and n and has spin 1. Proton (p) has the value $\dfrac{1}{2}$. Neutrons obey FD statistics and thus have spin $\dfrac{1}{2}$. Hyghes and Burgy have confirmed it by reflection of neutrons from magnetic mirrors.

6. Does a neutron posses magnetic moment?

Ans. Yes. $\mu_n = -1.913148 \pm 0.000066\,nm$

The negative sign indicates that the magnetic moment of the neutron is similar to that generated by rotating negative charge with its spin parallel to that of the neutron.

7. What do you understand by neutron-induced fission?

Ans. Neutron Capture by heavy nuclei induces fission if the compound nucleus formed in the capture is at an excitation energy greater than the fission barrier. This depends in particular on the Z of the nucleus and for a particular on the Z of the nucleus and for a particular element on whether the target nucleus has even or odd A because of the effect of pairing. For e.g. in the slow neutron $(E_n \approx \dfrac{1}{40}\,\text{eV})$ capture by ^{235}U, the compound nucleus, ^{236}U, is formed at an excitation energy equal to the separation energy ^{236}Sn and as $^{236}\text{Sn} > E_f$, the barrier height, prompt fission occurs.

8. How you will explain the occurrence of fission isomers?

Ans. The occurrence of fission isomers is seen to be caused by the significant effect of shell corrections to the liquid drop model of fission.

9. Why the following reaction is consider to be promising for a fusion reactor?

$$ {}_1^2\text{H} + {}_1^3\text{H} \rightarrow {}_2^4\text{He} + {}_0^1\text{n} + 17.62\,\text{MeV} $$

Ans. The reaction has an enhanced cross-section because of resonant state in ^5He.

10. What are the basic requirement of a fusion reactor?

Ans. (i) A plasma of deuterium and tritium corresponding to $\sim 20\,\text{keV}$ (~ 200 million K) is needed to produce an adequate reaction rate. A confinement time $\sim 1.5\,\text{s}$ and a plasma density of $\sim 2 \times 10^{20}\,\text{m}^{-3}$ is required.

11. What is basic approach in laser fusion?

Ans. To use pellets of deuterium and tritium and implode them using very high powered lasers for controlled fussion. The implosion would raise the temperature sufficiently for the fussion reasons to occur.

12. What is difference between thermally fossile and fast-fossile nuclei?

Ans. In the former case an even-Z, and even-N compound nucleus is formed, e.g.

$$^{235}U + {}^1_0n \rightarrow {}^{236}U^*$$

and in the latter case an odd-Z or odd-N system is formed, e.g.

$$^{238}U + {}^1_0n \rightarrow {}^{239}U^*$$

13. What is the importance of the discovery of double hump in fission process?

Ans. The discovery of double hump, which is not necessarily confied to fissionable nuclei, also explains some remarkable features of fission excitation function: (i) the induced fission shows sharp compound-nucleus resonances in the electron volt (eV)-kilo electron Volt (keV) energy range, corresponding to a compound nucleus excitation approximately equal to the height of the fission barrier, (ii) the excitation function for fission of some nuclei such as ^{240}Pu shows well-marked resonance below the threshold energy for over-the-barrier fast fission.

14. Why fusion of two nuclei does not occur at room temperatures?

Ans. Due to Coulomb repulsion between the nuclei.

15. What is the advantage of fast neutron induced fission?

Ans. An alternate technique for generating power is to use quite highly enriched fuel (~ 20 % fossile nuclei) with control rods but no moderator. The increased enrichment enables the reactor to operator on fast neutron-induced fission. The principle attraction of this technique, which is technically more difficult, is the possibility of breeding $^{239}_{94}Pu$.

OBJECTIVE QUESTIONS

1. A free neutron decays into
   (a) a proton only
   (b) a neutrino only
   (c) a beta particle
   (d) a proton, a neutrino and a β^--particle.

2. The basic constitutents of a nucleus are
   (a) a proton an electron
   (b) proton and α-particle
   (c) proton and neutron
   (d) neutron and α-particle

3. When aluminium is bombarded with fast neutrons, it changes into sodium with emission of particle

$$^{27}_{13}Al + {}^1_0n \rightarrow {}^{24}_{11}Na + x$$

What is x in the reaction?
   (a) proton
   (b) electron
   (c) neutron
   (d) α-particle

4. The critical mass of a fissionable material can be reduced by
   (a) Cooling it
   (b) heating it
   (c) additing impurities to it
   (d) surrourding it with a shield that will reflect neutrons.

5. Cadmium rods are used in a nuclear reactor for
   (a) absorbing neutrons
   (b) slowing down fast neutrons
   (c) speeding up slow neutrons
   (d) absorbing neutrons

6. Graphite and D_2O are two common moderators used in a nuclear reactor. The function of the moderator is
   (a) to cool the reactor
   (b) to control the energy released in the reactor
   (c) to absorb the neutrons and stop the chain reaction
   (d) to slow down the neutrons to thermal energies

7. The kinetic energy of a 300 K thermal neutron is
   (a) 300 MeV
   (b) 0.026 eV
   (c) 300 eV
   (d) 0.026 MeV

8. The average number of neutrons released by the fission of one uranium atom is
   (a) 2.5
   (b) 3
   (c) 1
   (d) 2

9. The energy released by the fission of one uranium atom is 200 MeV. The number of fissions per second required to produce 3.2 W of power is
   (a) 1015
   (b) 10^{17}
   (c) 10^{17}
   (d) 10^{19}

10. In the fission of U-235, the percentage of mass converted into energy is about
    (a) 5 %
    (b) 1 %
    (c) 0.1 %
    (d) 15 %

11. What is the energy released in the fission reaction

$$^{236}_{92}U + ^{117}_{46}X + ^{117}_{46}Y + 2^1_0n$$

if the binding energy per nucleon of X any Y is 8.5 MeV and that of $^{236}_{92}U$ is 7.6 MeV?
   (a) 500 MeV
   (b) 100 MeV
   (c) 200 MeV
   (d) 1000 MeV

12. The binding energy of deutron (^2_1H) is 1.15 MeV per nucleon and α-particle has a binding energy of 7.1 MeV per nucleon. The energy released (Q) in the reaction is

$$^2_1H + ^2_1H \rightarrow ^4_2He + Q$$

   (a) 23.8 MeV
   (b) 13.9 MeV
   (c) 931 MeV
   (d) 1 MeV

13. What is the main source of energy of the sun?
    (a) nuclear fusions of lighter elements
    (b) nuclear fission of heavier unstable elements
    (c) combustion of pure carbon present in the sun
    (d) gravitational energy liberated during the slow contraction of the sun

14. $^{238}_{92}U$ emits on α-particle and the resultant nucleus emits a β-particle. The atomic and mass number of the final nucleus are
    (a) 91, 234
    (b) 92, 232
    (c) 90, 240
    (d) 90, 236

15. The equation

$$4\,_1^1\text{H}^+ \rightarrow \,_2^4\text{He}^+ + 2e^+ + 26\,\text{MeV}$$

represents
   (a) fusion   (b) β-decay
   (c) γ-decay   (d) fission

16. A star intially has 10^{40} deutrons. It produces energy via the processes

$$_1^2\text{H} + \,_1^2\text{H} \rightarrow \,_1^3\text{H} + p$$

   and $$_1^2\text{H} + \,_1^3\text{H} \rightarrow \,_2^4\text{He} + n$$

   Where the masses of the nuclei are: $m(_1^2\text{H}) = 2.014$ amu, $m(_1^1p) = 1.007$ amu, $m(_0^1\text{n}) = 1.008$ amu and $m(_2^4\text{He}) = 4.001$ amu. If the average power radiated by the star is 10^{16} W, the deutron supply of the star is exhausted in a time of the order of
   (a) 10^{12} s   (b) 10^8 s
   (c) 10^6 s   (d) 10^{16} s

17. Thermal neutrons can cause fission in
   (a) ^{235}U (b) ^{238}U
   (c) ^{238}Pu (d) ^{232}Th

18. Nuclear fission experiments show that the neutrons split the uranium nucleus into two fragments of about the same size. This process is accompanied by the emission of several:
   (a) neutrons   (b) α-particles
   (c) protons and positrons   (d) protons and α-particle

19. 200 MeV of energy is obtained in the fission of one nucleus of ^{235}U. A reactor is generating 1000 kW of power. The rate of nuclear fission in the reactor is:
   (a) 931   (b) 1000
   (c) 2×10^8 (d) 3.125×10^{16}

20. Which of the following is a fusion reaction

   (a) $_1^2\text{H} + \,_1^2\text{H} \rightarrow \,_2^4\text{He}$ (b) $_0^1\text{n} + \,_7^{14}\text{N} \rightarrow \,_6^{14}\text{C} + \,_1^1\text{H}$

   (c) $_0^1\text{n} + \,_{92}^{238}\text{U} \rightarrow \,_{93}^{239}\text{Np} + \beta^- + \gamma$ (d) $_1^3\text{H} \rightarrow \,_2^3\text{He} + \beta^- + \gamma$

ANSWERS

1. (d)	**2.** (b)	**3.** (d)	**4.** (d)	**5.** (d)
6. (a)	**7.** (b)	**8.** (a)	**9.** (b)	**10.** (c)
11. (c)	**12.** (a)	**13.** (a)	**14.** (a)	**15.** (a)
16. (a)	**17.** (a)	**18.** (a)	**19.** (d)	**20.** (a)

EXPERIMENTAL TECHNIQUES:
Nuclear Radiation and Particle Detectors and Particle Accelerators

8.1 INTRODUCTION

The penetration of ionizing radiations into matter has been of theoretical interestt and of practical importance for nuclear physics since the early days of the subject. The classification of the radiations from radioactive substances as α, β and γ rays was based on the ease with which their intensity could be reduced by absorbers. The range or absorption coefficient provided, and still provides, a useful method of energy determination, and the associated processes of ionization and excitation underlie the operation of nearly all present day particle and photon detectors. In the present chapter, we start by examing the basic process of the interaction of radiation with matter, with particular reference to nuclear detecting techniques. Detectors for both nuclear structure and particle physics are presented. In the latter field especially, large and complex installations are now used.

8.1.1 Passage of charged particles and radiation through matter

For energies upto a few hundred mega eVs, the main process by which a charged particle such as a proton loses energy in passing through matter is the transfer of kinetic energy to atomic electrons through the Coulomb interaction between charges. One may think of two extreme types of transfer collision the 'heavy' type in which the electrons behave as free particles and the light type in which the interaction is with the atom as a whole. *Heavy collisions* leads to ionisation of the atom and to energetic electrons which are visible along the track of a charged particle in the expansion chamber as δ-*rays*. Inhibition of this type of collision leads to the reduced energy loss observed in the phenomenon of channelling.

In special cases, e.g. for fission fragments or other heavy ions, transfer of energy to the nuclei of the medium may also be important. In such cases of high charge or low velocity or both, *large angle scattering* of the Rutherford type from nuclei may be observed, but for fast ions this is a rare event. It is consequently reasonable, at least for the matter (Figure 8.1(a)).

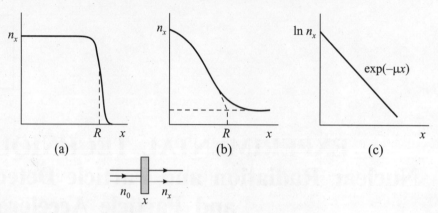

Fig. 8.1 Absorption of initially homogeneous particles by matter (schematic): (a) light ions (e.g. α, p, muons), showing mean range R; (b) electrons, showing extrapolated range R; a finite level of background, often due to γ radiation, is shown; (c) photons, $E_g < 10$ MeV. Absorber thickness x and ranges R are conveniently measured as mass per unit area of absorber and μ in (c) is then a mass absorption coefficient with units of metres squared per kilogram.

Fast electrons also lose energy by collision with other electrons and may be scattered through angles up to 90° in such collisions. They lose very little energy to nuclei because of the disparity in mass, but can suffer large deflections of more than 90°. In all collisions in which accelerations are experienced, electrons also lose energy by the radiative process known as *bremsstrahlung*. Heavy particles do not radiate appreciably in this way because their accelerations are much smaller. As a result of scattering and radiative energy loss a group of initially monoenergetic electrons may traverse considerably different thickness of matter prior being brought to rest and range can best be defined by an extrapolation process (Figure 8.1(b)).

Electromagnetic radiation interacts with matter through the processes of elastic (Rayleigh and nuclear) scattering, the photoelectric effect, the compton effect and pair production. These processes reduce the *number* of photons in a beam in proportion to the numbers incident on an absorber, thereby creating an exponential attenuation (Figure 8.(c). No definition of range is appropriate, but the attenuation coefficient is energy dependent. For radiation of energy greater than 10 MeV, the electrons and positrons from the pair production processes themselves generate bremsstrahlung and annihilation quanta and an electromagnetic shower builds up characterized by a linear dimension called the *radiation length*.

Figure 8.2 shows the relative importance of collision and radiative energy loss for electrons and protons in lead.

If a particle other than an electron, of charge Ze and non-relativistic velocity v, passes through a substance in which there are N atoms of atomic number Z per unit volume, then if it is also assumed that all the Z electrons per atom are free, the energy loss per unit path length or *stopping power* is

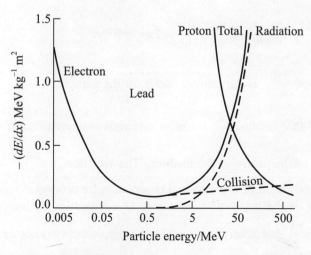

Fig. 8.2 Energy loss for electrons and protons in lead as a function of kinetic energy.

$$-\left(\frac{dE}{dx}\right)_{coll} = \frac{1}{(4\pi\varepsilon_0)^2} \frac{4\pi Z^2 e^4}{m_e v^4} NZ \ln\left(\frac{2m_e v^2}{I}\right) \text{ J/m} \qquad \dots (8.1)$$

where I is a mean excitation potential for the atoms of the substance and E is the total energy of the particle. The observed stopping power of Al for protons of energy 2.0 MeV is 11.2 MeV kg⁻¹ m². Relation (8.1) gives the value 11.8 MeV kg⁻¹ m², which is in satisfactory agreement with the observed value.

If the path x is measured as mass per unit area rather than as a length, then the number density in (8.1) is replaced by the inverse of mean mass, in kg, of an atom of the stopping material. This in turn is equal to $\dfrac{N_A}{A}$, where N_A is Avogadro's number and A is the atomic weight in kg. If A is taken, as an approximation, to be a mass number, $\dfrac{N_A}{A}$ must be multiplied by a factor of 10^3 as N_A is defined in terms of the mole rather than the kilomole.

Bethe and Bloch presented (8.1) in relativistic form. They also showed how to correct for the binding of the inner electrons in the absorber atoms. Taking x as mass per unit area the relativistic equation for a heavy particle can be written

$$\left(-\frac{dE}{dx}\right)_{coll} = \frac{KZ}{A} \frac{1}{\beta^2} \left[\ln\left\{\frac{2m_e c^2 \beta^2 E_{max}}{I^2(1-\beta^2)}\right\} - 2\beta^2\right] \text{ J-kg}^{-1} \text{ m}^2 \qquad \dots (8.2)$$

where
$$K = \frac{1}{(4\pi\varepsilon_0)^2} \frac{2\pi N_A Z^2 e^4}{m_e c^2} \qquad \dots (8.3)$$

Also $\beta = \dfrac{v}{c}$, v is the velocity of the particle and c the velocity of light. E_{max} is the maximum energy that may be transferred to an electron, i.e. the maximum δ ray energy, and is given by

$$E_{max} = \frac{2 m_e c^2 \beta^2}{1 - \beta^2} = 2 m_e c^2 \beta^2 \gamma^2 \qquad \qquad ... (8.4)$$

where $\gamma = (1 - \beta^2)^{-\frac{1}{2}} = \dfrac{E}{mc^2}$ is the Lorentz factor for the particle.

Equation (8.2) reveal that because $\dfrac{Z}{A}$ for most materials is approximately 0.5, collision loss is roughly independent of the nature of the medium. The variation of $\dfrac{dE}{dx}$ with energy in the low energy ($\gamma = 1$) region is similar to experimental observation for protons in aluminium (Figure 8.3). We may note that m_e is the mass of the electron. Over much of the energy range shown the collision loss varies as $\dfrac{1}{v^2}$ but at very low energies the simple theory is not valid for heavy particles because of the effect of capture and loss of electrons by the particle. In the $\dfrac{1}{v^2}$ region, $\dfrac{dE}{dx}$ can be used as a means of *particle identification* if the corresponding energy is known because particles of same energy but different mass will have a different velocity.

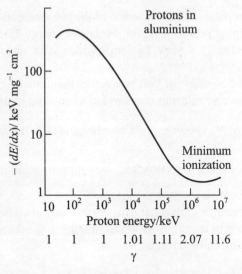

Fig. 8.3 Specific energy loss, or stopping power, of aluminium for protons. The quantity $\gamma = (1 - \beta^2)^{-\frac{1}{2}}$ gives the mass of the particle in units of its rest mass. Note that the energy loss at minimum ionization is about 2 keV mg^{-1} cm^2.

At kinetic energies comparable with and greater than the mass of the particle ($\gamma = \dfrac{E}{mc^2} \approx 3$, in Figure 8.3) a region of *minimum ionization*, is reached. All particles of this energy suffer the same collision loss of 0.1 to 0.2 MeV/kg/m^2 (for a charge particle) and particle identification by $\dfrac{dE}{dx}$ is no longer possible. At higher energies still ($\gamma \approx 10$) the logarithmic term in Equation (8.2) leads to so called *relativistic rise* in energy loss and there is a small mass dependence for a given momentum, so that identification is again possible. At very high energies ($\gamma \approx 100$) the collision loss reaches the *Fermi plateau* value because polarisation effects in the medium limit the energy transfer.

Collision loss, or ionization density, varies as Z^2 and for large Z the high density may result in lattice damage and storage of energy (in dielectrics) over long periods, until it is released by thermal annealing or chemical etching.

Observable phenomena basically governed by the collision-loss formula for a single particle are: (i) production of ion pairs in solids, liquids and gases and of electron-hole pairs in semi-conductors, (ii) production of light in scintillator materials, (iii) production of bubbles in superheated liquid, (iv) production of developable grains in a photographic emulsion or of damage trails in a dielectric. Each of these has been made the basis of an energy-sensitive particle detector.

Collision loss of energy determines the range of a heavy charged particle in matter, at least up to energies at which nuclear interactions become important. The range is the distance in which the velocity v of a particle (of mass m and charge Ze) reduces to zero and it may be estimated by changing the variable from E to V in Equation (8.1). Since in the present approximation E is the sum of a constant rest energy mc^2 and a kinetic energy $T = \dfrac{1}{2} mv^2$, thus

$$R = -\int_{v}^{0} dv \left(\frac{dv}{dx}\right)^{-1} = \frac{m}{Z^2} f(v) \qquad \text{... (8.5)}$$

In relation (8.5) the mass and charge number of the particle are shown explicitly and the integral $f(v)$ contains other constants. This permits the range to be compared with the range R_p of a reference particle such as a proton ($Z = 1$). If one chooses kinetic energies so that the velocity is the same in each case, it follows that

$$R(T) = \frac{m}{m_p Z^2} R_p \left(\frac{T m_p}{m}\right) \qquad \text{... (8.6)}$$

Relation (8.6) shows that a deutron has about twice the range of a proton of half the deutron kinetic energy E.

Since loss of energy is a statistical process, particles of initially homogeneous energy will exhibit a spread of energy after passage through a given thickness of matter. Similarly, for a given energy loss, the range has a spread and this is known as *straggling*. The distribution of energy loss in a thin absorber was calculated by Landon, and verified by others experimentally, to be asymmetri about the mean loss. These effects are enhanced for electrons, which may lose upto half of their energy in a single collision.

If a charge particle moves through a single crystal, it may follow certain directions defined by the crystal structure in which the electron density is especially low. The collision loss is then lower than energy loss in the other directions and the particles are said to be channelled.

8.1.2 Energy Loss by Radiation

Electrons of kinetic energy T and total energy $E (= T + m_e c^2)$ passing through a thin absorber give rise in each radiative collision to a *bremsstrahlung* photon of any energy between 0 and T. The process may be envisaged as shown in Figure 8.4 in which the Coulomb field existing between the incident electron and a nucleus or another electron of the absorbing medium, is represented by an emission of *virtual* photons (A virtual particle is one which is emitted only because conservation of energy may be relaxed over a time given by the uncertainty relation $\Delta E \, \Delta t \approx \hbar$. It must disappear by re-absorption or interaction within the time Δt for energy increment ΔE so that conservation is obeyed overall). The real electron makes a *Compton scattering* collision with a virtual photon, creating a virtual lepton which then gives rise to a real photon, i.e. the bremsstrahlung quantum and a real electron of reduced energy. The bremsstrahlung process $e \rightarrow e' + \gamma$ cannot take place in free space because momentum and energy cannot then be conserved. Photons can, however, be emitted by electrons moving in the magnetic field of an orbital accelerator and this emission is called *synchrotron radiation*.

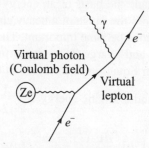

Fig. 8.4 Production of bremsstrahlung by an electron in the field of a nucleus Ze; the virtual lepton is said to be "off the mass shell" because the mass derived from the familiar expression $E^2 - p^2c^2$ is not that of a real particle. This and subsequent similar figures are not intended to display the precise order of events, and other diagrams showing a different order may be drawn. Time increases to the top of the page, and a fixed Coulomb field is shown as a wavy horizontal line.

Classically, bremsstrahlung originates in the acceleration of a charged particle as a result of Coulomb interaction, and its intensity is therefore inversely proportional to the square of the particle mass according to Maxwell's theory. The muon has 200 times the electron mass and its radiative energy loss is therefore negligible except at very high energies. This holds a *fortiori* for all other charged particles, since they are still heavier.

All charged particles independently of their mass do, however, suffer a small electromagnetic loss in matter through *Cherenkov radiation*. This depends on the gross structure of the medium through which the particle passes, and in which a macroscopic polarisation is set up. If the velocity $v \, (= \beta c)$ of the particle exceeds the velocity of light in the medium, thee polarization is longitudinally asymmetric because it can only relax to symmetry on a time scale determined by this velocity. Secondary wavelets then originate along the track *AB* of the particle (Figure 8.5) and these form a coherent wave front propagating in a direction at angle θ with the path of the particle where

$$\text{Cos } \theta = \frac{\frac{c}{n}}{\beta c} = \frac{1}{\beta n} \qquad \qquad \ldots (8.7)$$

n being the refractive index. It has the important features for nuclear detectors that coherence does not appear until $\beta = \dfrac{1}{n}$, i.e. there is a *velocity threshold* for observable radiation, and that the direction of emission is very well defined. There are counters based on this effect for the identification of particles of relativistic energy.

Cherenkov loss is only about 0.1 % of collision loss at the ionization minimum, although it increases with energy in the region of the relativistic rise. The loss for a singly charged particle is

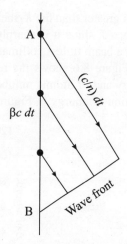

$$\left(-\frac{dE}{dx}\right)_{Ch} = \frac{\pi e^2}{c^2 \varepsilon_0} \int \left(1 - \frac{1}{\beta^2 n^2}\right) \nu \, d\nu \qquad \dots (8.8)$$

Where the integration extends over all frequencies for which $\beta n > 1$. Within these bands the spectral distribution of energy loss is proportional to $\nu \, d\nu$, in constrast with the $d\nu$ proportionality of bremsstrahlung. The difference is that bremsstrahlung originates incoherently from the incident particle itself while Cerenkov radiation arises coherently from the medium.

Fig. 8.5 Formation of a coherent wave front of Cherenkov radiation

If a charged particle moves through an *inhomogenous* medium, e.g. if it crosses a boundary between different media, a further small contribution to radiative loss known as *transition radiation* occurs. It arrises in the relativistic case because of the different extent to the electric field of the particle in the transverse plane due to different polarization effects in the two media. Transition radiation, unlike Cherenkov radiation, occurs for any velocity of the particle and it yields photons not only in the optical but also in the X-ray region.

8.1.3 Absorption of Electromagnetic Radiation

One may describe the exponential absorption of a beam of homogeneous photons shown in Figure 8.1(c) by the equation

$$n_x = n_0 \exp(-\mu x) \qquad \dots (8.9)$$

where $\mu \, (m^{-1})$ is the *linear attenuation coefficient* and n_0 is the initial number of photons in the beam. $\mu = N\sigma$, where σ is cross section ascribed to each atomic absorbing centre and N is the volume density of these centres. If x is measured in kilograms/m^2 rather than in metres, Equation (8.9) becomes

$$n_x = n_0 \exp(-\mu_m x) \qquad \dots (8.10)$$

where $\mu_m = \dfrac{\mu}{\rho} = \dfrac{N\sigma}{\rho} = \dfrac{N_A \sigma}{A}$ is *mass attenuation* coefficient, σ is the density of the absorbing material and A is the atomic weight, in kilograms, of the absorber atoms. μ_m is measured in m^2/kg.

σ can be written as

$$\sigma = \sigma_{PE} + Z\sigma_C + \sigma_{PP} \qquad \dots (8.11)$$

where σ_{PE} is the cross section due to *photoelectric effect*, σ_C is due to the *Compton effect* and σ_{PP} due to *pair production*. The factor Z embodies the assumption that all the atomic electrons contribute individually (and incoherently) to compton scattering. This will be true if the photon energy is much greater than the *K*-shell ionization energy of the atom. Elastic (Rayleigh) scattering has been neglected, since it is sharply forward peaked at mega-eV energies and does not remove photons from a beam unless collimation is very fine.

Figure 8.6 shows the relative importance of the three absorption processes as a function of energy and of atomic number an the actual mass absorption coefficient mm for lead is given as a function of energy in Figure 8.7.

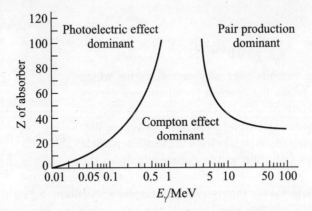

Fig. 8.6 Relative importance of the three major types of γ ray interaction.

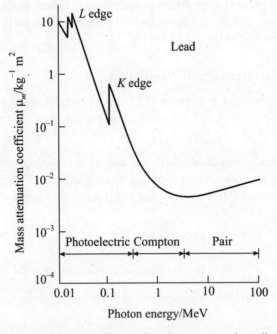

Fig. 8.7 Mass attenuation coefficient for electromagnetic radiation in lead

Figure 8.7 shows (a) sharp peaks at energies corresponding with the atomic absorption edges and indicating the onset of increased photoelectric absorption as a new electron shell becomes available, and (b) a minimum absorption coefficient due to the compensation of photoelectric and Compton cross sections, which fall with energy, by the pair production cross-section which rises. Both effects mean that the energy is not a single-valued function of μ_m throughout its range for a given absorbing material. The individual processes indicated in Equation (8.11), together with elastic scattering, are all part of the total electromagnetic interaction between radiation and matter and are therefore connected. For most practical purposes, in the energy range of interest to nuclear physics, it is possible to regard them as independent, each contributing to the total cross-section to a degree determined by the quantum energy. The salient features of the processes are:

Photoelectric effect

Because of the necessity to conserve energy and momentum, a free electron can not wholly absorb a photon. Photoelectric absorption, therefore, tends to take place most readily in the most tightly bound electronic shell of the atom available, for a particular incident photon energy, since then momentum is most easily conveyed to the atom. The kinetic energy of a photoelectron the K-shell is

$$T = h\nu - E_K \qquad \qquad \text{... (8.12)}$$

where E_K is the K-ionization energy. The cross-section depends on atomic number Z and wavelength of radiation λ approximately as

$$\sigma_{PE} \approx Z^5 \, \lambda^{\frac{7}{2}} \qquad \qquad \text{... (8.13)}$$

in the energy range $0.1 - 0.35$ eV. The vacancy in the atom created by the ejection of a photoelectron leads to the emission of characterstic X-rays or of low energy photoelectrons known as *Auger electrons* from less tightly bound shells as an process favoured for light atoms.

Campton effect

The scattering of photons by atomic electrons that may be regarded as free is a simple collision process. With the quanties shown in Figure 8.8(a), the *Compton wave length* $\left(\dfrac{h}{m_e c} \right)$ of the electron is given by

$$\frac{c}{\nu'} - \frac{c}{\nu} = \lambda' - \lambda = \frac{h}{m_e c} (1 - \cos \theta) \qquad \qquad \text{... (8.14)}$$

The Compton wavelength $\left(\dfrac{h}{m_e c} \right) = 2.43 \times 10^{-12}$ m. The energy of the recoil electron is $h(\nu - \nu')$; it is zero for $\theta = 0$ and is continuously distributed up to a maximum for $\theta = \pi$. Emission of electron leaves an excited atom.

The Compton wavelength shift is independent of wavelength and of the material of the scatterer. The differential cross-section for the process is given by a formula due to Klein and Nishina, which predicts that at a given angle the cross-section decreases with increasing photon energy. The angular variation for a range of energies is shown in Figure 8.8(b); in all cases there is a finite cross-section for $\theta = 0$ and Equation (8.14) shows that over a small angular range about zero the wavelength is essentially unmodified.

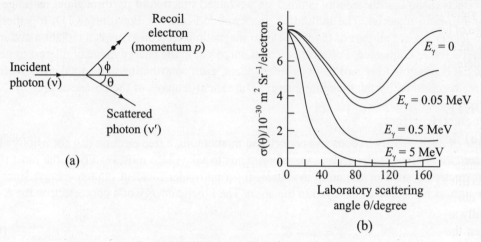

Fig. 8.8 (a) Compton scattering of a photon by a free electron. (b) differential cross section for Compton scattering giving the number of photons scattered per unit solid angle at angle θ, for energies of $0 - 5$ MeV

As $\dfrac{h\nu}{m_e c^2}$ tends to zero the angular distribution shown in Figure 8.8(b) become more symmetrical about 90° and finally reach the form characterstic of classical *Thomson scattering* from a free charge. However, the assumption of scattering from free electrons then becomes less tenable for matter in its normal state and coherent *Rayleigh scattering* from the bound electrons forming the atomic charge distribution takes over.

Pair production and annihilation

The pair production process

$$h\nu \rightarrow e^+ + e^- \qquad \qquad ... (8.15)$$

contributes increasingly to γ-ray absorption as the quantum energy rises above the threshold energy of 1.02 MeV. The process take place in free space because momentum cannot be conserved but it occurs if a third electron, or better still a nucleus, is present to absorb recoil momentum. In the latter case the cross-section is proportional to the value of Z^2 for the absorber as in the case of bremsstrahlung, which is theoretically at least a similar process. The rising cross-section for each of these phenomena as quantum energy increases is ultimately limited by screening of the nuclear charge by the atomic electrons.

The positron is a stable particle like the electron but it may disappear by annihilation with an electron. The exact converse of process (8.15) is *single-quantum annihilation* which cannot take place in free space and occurs only with fairly low probability when positrons are brought to rest in matter. The more prolific *two-quantum annihilation radiation*

$$e^+ + e^- \rightarrow 2h\nu \qquad \qquad ... (8.16)$$

is readily observed as a pair of oppositely directed 511 keV quanta, if e^- and e^+ are essentially at rest. This radiation contributes a soft component to a flux of γ-radiation of energy above the pair production threshold. Fast moving positrons may both radiate and annihilate in flight in a Coulomb field, yielding energetic quanta which contribute to the build up of *cosmic ray showers* by the passage of a charged particle between two electrodes maintained at a sufficient potential difference. These detectors are generally termed as *gas-filled detectors. Ionisation chamber* is simplest of such detectors.

8.2 DETECTORS FOR PARTICLES

8.2.1(a) Ionisation Chamber

The ionisation chamber essentially consist of a pair of conductors, well insulated from each other, in a gas filled chamber. Most common forms are using parallel plate or co-axial cylindrical electrodes.

In the chamber a number of gases may be used. But generally argon is preferred. With the electric field E, between the conductors, the chamber can be used to count single ionising particle, directed into the chamber through a window. If the particle flux is large, the ions produce ionisation more or less continuously, giving rise to steady current output. The chamber in which the ionising particle is completely absorbed is know as deep chamber. In a shallow chamber the chamber depth is small part of the range in the chamber gas.

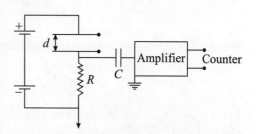

Fig. 8.9 Ionisation chamber

Let a high potential difference V_0 be connected to the electrodes of the chamber. The collector electrode at lower potential is connected to earth through a high resistance. Thus when no ion passes the chamber the electric energy stored in electric field, is given by $\frac{1}{2} CV_0^2$, where C is capacity of the electrodes. If an α-particle enters the chamber producing say n ion pairs. The electrons will drift in a direction opposite to field with an average velocity V_e. The positive ion drift velocity V_p would be much less than V_e. In a time t, the electron will move a distance $V_e t$ towards high voltage electrode and positive ion will move a distance $V_p t$ towards collector. Due to these movements the collector potential would tend to rise. The time constant RC is made so large that a very little current would flow through resistance R.

Due to the motion of electrons and positive ions there would be a drop of energy of each $Ene V_e t$ and $Ene V_p t$, for electron and positive ions respectively. Due to collision of these with the gas molecules this energy would appear as heat. But as the total energy of the system (isolated) is to remain constant, this thermal energy must have come from the energy stored in the electric field and the potential difference between the electrodes must drop from V_0 to a smaller value V, i.e., the collector electrode will acquire a potential $V' = V_0 - V$. Equating the decrease of stored energy to amount of energy converted to work, we get

$$\frac{1}{2} CV_0^2 - \frac{1}{2} CV^2 = Enet(V_e + V_p)$$

or $\qquad \dfrac{1}{2} C(V_0 + V)(V_0 - V) = nEe(V_e + V_p)t$ $\qquad$ ... (8.1)

Usually V' will be much smaller than V_0 and hence $E \approx \dfrac{V_0}{d}$

Also $\qquad\qquad\qquad V_0 + V \approx 2V_0$

substituting these in (8.1), we get

$$\dfrac{V_0}{d} CV' \, 2V_0 = n \dfrac{V_0}{d} e(V_e + V_p)t \quad \text{or} \quad V' = \dfrac{ne}{Cd}(V_e + V_p)t \qquad \text{... (8.2)}$$

Since $V_e \gg V_p$, when electrons are still moving, we can write

$$V' = \dfrac{ne}{Cd} V_e t \qquad \text{... (8.3)}$$

$$t_e = \dfrac{d - d'}{V_e} \qquad \text{... (8.4)}$$

Where d' is the distance through which proton moved to plate.

$$t_p = \dfrac{d'}{V_p} \qquad \text{... (8.5)}$$

Since $V_p \ll V_e$, when the positive ions are still moving in the field, the collector potential when electron strike the electrode is

$$V' = \dfrac{ne}{Cd} V_e \dfrac{d - d'}{V_e} = \dfrac{ne}{Cd}(d - d') = \dfrac{ne}{Cd}\left(1 - \dfrac{d'}{d}\right) \qquad \text{... (8.6)}$$

This potential still increases until the positive ions strike the collector. The expression for V' can be written as

$$V' = \dfrac{ne}{C}\left(1 - \dfrac{d'}{d}\right) + \dfrac{ne}{Cd} V_p t \qquad \text{... (8.7)}$$

The expected limiting value of V' when positive ions also strike the collector can be determined by putting $t = t_p$ from equation (8.5) viz.

$$V' = \dfrac{ne}{C}\left(1 - \dfrac{d'}{d}\right) + \dfrac{ne}{Cd} v_p \dfrac{d'}{v_p} = \dfrac{ne}{C} \qquad \text{... (8.8)}$$

After the time $t > t_p$ there were no ions left in the chamber. The values of V' are plotted against t. This has been shown in Figure (8.2). For time $t > t_e$ there is an abrupt change of the variation of V and up to $t = t_p$ it is very small. From Equation (8.8) it is clear that the change in potential V' is proportional to n and is independent of, where the ions were formed. Although the shape of the collector potential depends upon where the ions are formed. The maximum variation of V is very small and hence to detect this variation or pulse it has to be amplified. Two kinds

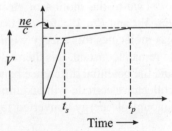

Fig. 8.10 Variation of V' with time

of amplifiers are used to detect the pulse. The shown amplifiers in which shortest time constant is made long as compared to drift time of positive ions. The other amplifier is known as fast in which the time constant RC is made short enough so that the rapid varying potentials induced by electrons, are reproduced accurately. The potential induced by positive ions are of no interest here.

The amplifiers may be proportional type of liner type. With the help of ionisation chamber the energies of α-particles and other ions can be measured accurately with derscrete spectra. These may also be used for measurement on photons and count fission rate.

Region of Multiplicative operation

In an ionisation chamber even after reaching saturation current, at high potential, the electrons receive more energy producing secondary ion. Thus the value of current increases to more than its saturated value. The current pulse depends upon initial ion pairs and applied voltage. If Figure 8.11 log of the pulse v have been plotted against the applied voltage V. The curve I, is drawn for electrons producing less ionisation, whereas the curve II is drawn for α-particles producing high ionisation.

To explain this variation, the whole region can be divided into six regions:

(1) In region A, the d.c. voltage V is less and due to recombination of positive ions and electrons, the number of ion pairs can not be detected by electronic circuit.

(2) In region B, the voltage V is sufficient and recombination of positive ion and electron is almost zero. This region is called *ion chamber region*. The detection of particle by ionisation chamber is done in this region. In this region almost all ions reach respective electrodes and pulse does not depend on V.

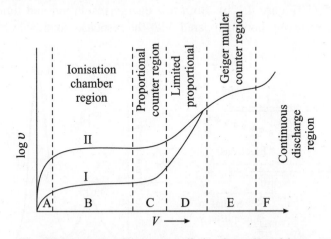

Fig. 8.11 Variation between applied voltage V and log v

(3) In region C, the potential V is still larger and the electrons reaching the electrode gains sufficient energy, producing new ion pairs due to collision of gas molecules. The new electrons and the initial electrons produce large number of ion pairs and compared to initial charge reaching to electrode, a very high charge reaches it. The process is cumulative, and produces an avalanche of electrons. The number of ion pairs produced by every electron,

before reaching to electrode, is called amplification. In this region pulse and voltage are proportional. The region is called *proportionality region* and the counter operated in this region is called proportional counter.

(4) In region D, at still higher potential the proportionality disappears. This is called *limited proportionality region*. This is not useful for measurement.

(5) In region E, even a fine ion particle produces a very high pulse, which does not depend upon initial ion pairs. This is called *Geiger-Muller region*. The counter operated in this region is called Geiger-Muller counter (G.M. Counter). In this region the ionisation extends to whole electrode.

(6) In region F, spurious untriggered breakdown of gas occurs and a continuous unquenchable discharge starts. In this region discharge does not depend upon initial particles and the region is not suitable for measurement.

An ionisation chamber is said to be, for a particular isonising radiation, deep or shallow according as it can completely absorb or not, the ionising radiation.

For measurement of the energy of a single particle which stops in the chamber, gases such as argon or argon-CO_2 mixtures which do not form negative ions are usually used. There is then a fast pulse at the anode due to electron collection and this may be rendered independent of the position of initial track in the chamber by placing a 'Frisch' grid near the electrode (Figure 8.12a) to screen off the induction effect of the positive ions. The chamber voltage must of course ensure complex electron collection but must not lead to ionisation by collision. The high mobility of electrons permits pulses of a width of approximately 1 μs to be derived from the chamber, although recovery is slower because of the low mobility of the positive ions. The resolution of an ionisation chamber for α-particle of energy about 4.5 MeV can be about 0.5 % (Figure 8.12b). Both this energy resolution and time resolution of the chamber are relatively poor compared with the response obtainable from semiconductor detectors.

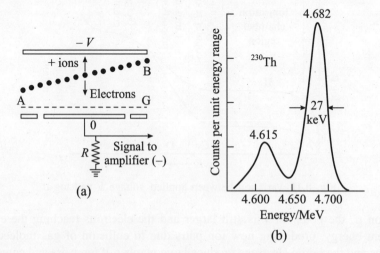

Fig. 8.12 (a) Ionization chamber with Frisch grid; *AB* is the track of a charged particle from which ion and electron collection starts; (b) spectrum of α particles from ^{230}Th in a gridded ionization chamber.

8.2.1(b) Proportional Counter

The general form of an ionisation chamber is called a counter. In these counting tubes, the ions created by the entry of an elementary particle move in such an intense electric field that ionisation by collision is produced, i.e., gas amplification results. This amplification produces a pulse of current large enough to permit the detection of the particle. In a proportional counter the amplitude of the output pulse is proportional to the number of ions produced initially and hence the name.

A proportional counter consists of an ionisation chamber in which one electrode is a cylinder of radius b and other an axial wire of radius a. The arrangement is shown in Figure 8.13.

For low potential difference V_0 between the electrodes the arrangement is just an ionisation chamber. When this is increased to about a few thousands, then with the entry of the particle gas multiplication starts. When the electrons of initial ionisation reach the high field strength region, they acquire enough kinetic energy between collision to more ions. The electrons so formed can continue the process. This is called *avalanche*.

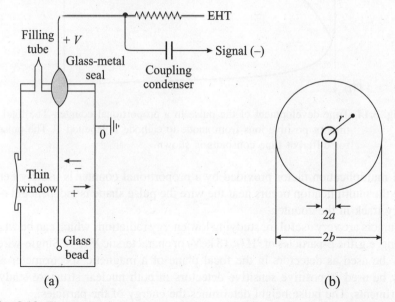

Fig. 8.13 The proportional counter: (a) construction – the glass bead prevents sparking from the end of the wire; (b) cross-section in cylindrical geometry.

The avalanche effect can be made use to detect single particle. If initially there are n ion pairs producing $m\,n$ electrons and $m\,n$ positive ions due to avalanche, then the factor m is known as multiplication factor. The multiplication factor depends upon counter potential and for greater multiplication, the avalanche is more effective. Multiplication factor m up to thousand have been used in a proportional counter. Still greater m and hence higher counter potential cannot be used as the counter looses its proportionality feature.

The electric field strength E, at a point distant r from the axis of the cylinder is given by

$$E_r = \frac{V_0}{r \log_c\left(\dfrac{b}{a}\right)}$$

The electrons formed in the avalanche are collected quickly by the wire. In addition they do not change the wire potential by any appreciable amount. It is because that by the time the electrons reach the wire, positive ions, which are still very nearby the wire, neutralize their effect. It is only as the positive ions drift away, then the wire potential drops.

The field necessary for ionisation by collision in argon is about 10 V/m at atmospheric pressure and it is convenient to realize this in the wire-cylinder geometry of the counter (Figure 8.13). In a typical argon field counters with $a = 10$ mm, $b = 8$ mm and an operating voltage of 3 kV the total drift time is 550 μs, but the pulse height increases very rapidly at the beginning and very short signals (≈ 30 ns) can be obtained by differentiation (Figure 8.14).

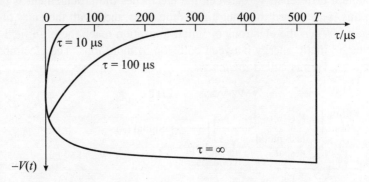

Fig. 8.14 Time development of the pulse in a proportional counter. The total drift time of positive ions from anode to cathode is labelled T. The pulse shape for different time consterstic is shown.

The total multiplication factor provided by a proportional counter is usually between 10 and 10^3 and since the multiplication occurs near the wire the pulse shape is independent of the position of the primary track in the counter.

These counters are very useful for studying low energy radiations which can be wholly absorbed in the gas filling, e.g. the β-particles of ^{3}H (< 18 keV) or characterstic X-rays. Single-wire proportional counters may be used as detectors in the focal plane of a magnetic spectrometer and multiwire counters may be used in positive sensitive detectors in both nuclear structure study and particle physics experiments. The pulse height determines the energy of the particles.

8.2.1(c) Geiger-Muller counter (G.M.counter)

If in a proportional counter the applied voltage is increased, its proportionality feature disappears. The avalanche does not form at one point but spread over the entire length of the central wire. Hence all pulses, for given voltage become of the same size and the amplification does not depend on the ionisation produced by the ionising particle. A counter so operated is called G.M. Counter. In the Geiger region with higher values of V_0 multiplication factor m is much larger and the magnitude of the current pulse in the counter output is independent of energy and nature of the particles detected. The electron avalanche is concentrated near the anode wire, quickly collected by it whereas positive ions move much slower, around anode wire make a sheath and thus increase the anode radius. The electric field therefore drops to a value below that which is capable of supporting ionisation by collision. When the positive ions reach the wall, they emit electrons from cathode by field emission, leaving the excited atoms. The excess amount of energy will be radiated in the form of photons. The new produced electrons may lead unwanted avalanche and hence output pulse. In this way the discharge would be continuous, consisting a series of pulses.

To avoid these unwanted pulses in a G.M. Counter, a suitable polyatomic vapour is used. This is called quenching, Due to presence of polyatomic quenching gas the ions reaching the cathode would be of this gas and not of the inert gas. These gas ions do not eject electrons from the cathode and hence a single pulse per particle only will be produced.

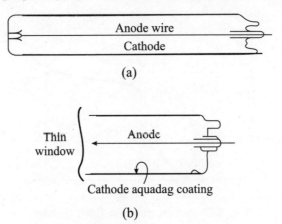

(a)

(b)

Fig. 8.15 End Window G.M. tube

A variety of designs of G.M. Counter tube are available. A particular tube shown in Figure 8.15(a) consists of a cylindrical copper cathode with a thin axial tungsten wire anode supported inside a cylindrical glass envelope. The other type (Figure 8.15(b)) is an end window counter tube, consisting of a thin window of mica sheet. In this tube the anode is tungsten wire and cathode is in the from of graphite coating inside the glass envelope. The tubes are made self quenching and are filled with a mixture of argon at a partial pressure 10 cm. of Hg and alcohol vapour at I cm. of Hg. In some tubes instead of alcohol, bromine vapour have also been used for quenching. But in this case cathode is to be made of stainless steel, or nichrome or graphite.

The time constant used with G.M. Counter is short and hence the output pulse is over before the ion sheath has drifted very for from the wire. The counter will remain dead for a certain time and then operate with a reduced pulse size, for a still further time before it makes a complete recovery. These times and recovery times are of the order of 100 m sec., setting an upper limit to the counting rate.

The G.M. Counter circuit is shown in Figure 8.16.

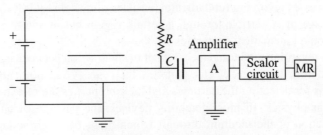

Fig. 8.16 G.M. Counter circuit

For a tube using alcohol as quenching gas the operating voltage is 1000 to 1200 V. For bromine quenched it is only about 250 V. With the pulse time 100 m-sec., a maximum counting rate of the order of 5000 per sec. can be achieved with. The tube, an amplifier, scalar circuit and mechanical register are shown. A curve showing variation counter potential and corresponding counting rate for a G.M. Counter gives the operation conditions of it. This is shown in Figure 8.17. This curve is known as plateau curve. At about 1000 Volts the pulse becomes uniform in size. This is known a G.M. threshold. The horizontal region is known as plateau region.

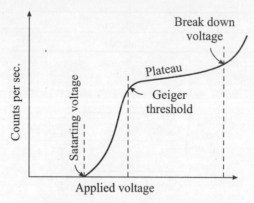

Fig. 8.17 Plateau curve for G.M. Counter

The region extends for about 300 to 400 Volts. After-wards double pulsing begins and at higher voltages continuous discharge develops.

The G.M. Counter is slow in working, but the advantage of this lies in the fact that large pulses are being produced with it. Their life is limited because of the decomposition of the organic vapour when the positive ions neutralized at cathode and dissociated into molecular fragments. With about 10 counts the organic vapour dissociates. However bromine vapour tubes have long life. Furthermore a G.M. Counter cannot discriminate between the particles of different energies.

The G.M. Counters show both a threshold for operation and uniform sized signal when operating. The filling is usually an argon-halogen mixture chosen to quench the counter discharge and terminate a signal. There are *dead-time* and *recovery-time* interval resulting from the development and quenching of the discharge which in practice limit counting rates to perhaps a few hundred per second and even then large corrections are necessary. For this region, and also because of its poor sensitivity as a γ-ray detector, the GM counter is now little used in nuclear structure study.

8.2.2 Nuclear Radiation Detectors

"The analysis of processes involving fundamental particles of nuclei requires techniques which not only reveal the passage of A particle through a certain region but in addition allow us to obtain information about some kinematical properties."

The information regarding nuclei and the study of the interaction between nuclei and elementary particles are obtained from the type of ejected particles, their energy and momentum. Hence sensitive and accurate methods for detecting the particles and their properties are of extreme importance for researches in nuclear physics. In most detecting devices, the passage of charged particles are menifested by ionisation of the medium through which these pass. These charged particles are electrons, positrons, protons, α-particles, nuclei of heavier element, mesons, hyperons etc. If the particles are uncharged viz, neutron, X-radiation etc., these impart energy to charged particles of the matter, which then cause ionisation. This ionisation is the fundamental basis of detection of charged particles and their energies. The devices used for detecting theses are called detectors and used to detect and record the number of particles emitted in different experiments of nuclear radiations, disintegration and transmutation. The detectors are classified into different categories depending upon the type of particles and their effect, visible of measurable.

One may broadly classify the nuclear radiation detectors under following three categories: (i) devices based on the detection of free charge carriers by the ionisation produced by them, e.g. *Ionisation Chambers, proportional Counters, Geiger-Muller* (GM) *Counters* and *semiconductor detectors*; (ii) devices based on visualisation of the tracks of radiation, e.g. *Wilson cloud chamber, bubble chamber nuclear emulsion plates, spark chamber,* etc. and (iii) devices based on *light sensing,* e.g. Scintillation Counter, Cerenkov Counter, etc.

8.2.2(a) Gaseous Detectors

There is an extensive class of particle detectors exists that depends for their performance on the electric pulse of current produced when ions are formed.

Self-quenching action of GM-tube

A GM tube must produce a single pulse due to the entry of a single particle. It should not give any spurious pulse and recover quickly to the passive state. Unfortunately, the positive Ar ions which eventually strike the cathode gain electrons from it and become neutral Ar-atoms which are left in an excited state. These excited Ar atoms return to the normal state with the emission of photons and these photons cause avalanches and hence spurious pulses.

These unwanted pulses are avoided by the presence of a diatomic (or polyatomic) gas, such as bromine, in the tube. The positive Ar-ions moving slowly toward the cathode would have multiple collisions with Br-molecules and transfer their charge to them. On the other-hand, now Br molecular ions in their turn reach the cathode where they gain electrons to become neutral bromine molecules and move into excited states. However, the excited Br-molecules lose their excitation energy not by the emission of photon but by dissociation into Br-atoms and thus no spurious pulses are produced. In due course of time, Br-atoms recombine back into Br-molecules. Thus Br *quenches* the discharge, and the tube is ready to receive the next particle within about 10^{-4} s.

Dead time, Recovery time and Resolving time

Dead time of a GM Counter is the time interval between the production of the initial pulse and initiation of second Geiger discharge. During this period the GM Counter is insensitive, i.e. dead to further ionisation pulses. Usually, it is $\sim 500 - 100$ μs. It arises due to slow mobility of heavier positive ions from anode region to the cathode. Around the inner electrode, the presence of positive ion sheath lowers the electric field the Geiger Muller Counter threshold, V_{th} ($= V_s$).

The field at the central electrode recovers quickly as the ion sheath moves towards the cathode since the external resistance is usually quite small.

The time for recovery to threshold is called the *dead time*. GM counter can record another ionising particle only after the field has restored to a value above V_{th} ($= V_s$). Since a finite pulse must be developed so that the counter circuit can count it, as the actual resolving time of the counter is slightly greater than the dead time. One can define the recovery time of the GM counter as the time interval for the counter to return to its original state to produce the full sized pulses again. Three time intervals are marked in Figure 8.18.

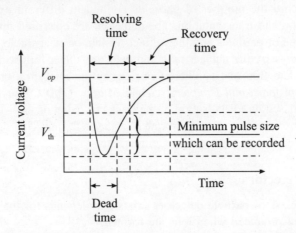

Fig. 8.18 Dead time, resolving time and recovery time associated with the operation of a GM counter.

We may note that (i) the resolving time of GM counter is often taken to be synonymous with dead time (ii) during dead time, a high flux through the GM Counter should be avoided, and (iii) during resolving time the pulses are recorded, but they are of smaller size.

Resolving time of G.M. Counter

For determining the true counting rate of a GM Counter, one must know its resolving time (τ). If n be the counting rate of GM Counter and N the actual rate of arrival of the particles in the Counter, then

$$N - n = N n \tau \qquad \qquad \text{... (8.1)}$$

Counter was insensitive for an interval $n\tau$/second and hence the missing particles per sec is given by

$$N = \frac{n}{1 - n\tau} \qquad \qquad \text{... (8.2)}$$

Using (8.2), one can obtain the true counting rate N from the observed counting rate n, provided τ is known.

τ can be determined by measuring separately the counting rates n_1' and n_2' with two radioactive sources 1 and 2 of almost equal strength. Let the total counting rate by n_4' when both sources are placed in the same position as before and n_b be the background rate, i.e. when there is no source. We have

$$n_1 = n_1' - n_b; \ n_2 = n_2' - n_b \text{ and } n_t = n_t' - n_b \qquad \qquad \text{... (8.3)}$$

Where n_1, n_2 and n_t are above counting rates corrected for the background.

Let N_1, N_2 and N_t be the true rates of arival of particles in the GM counter in the above three situations, then from (8.1), one obtains

$$N_1 = \frac{n_1}{1 - n_1\tau} \; ; \; N_2 = \frac{n_2}{1 - n_2\tau} \; \text{and} \; N_3 = \frac{n_t}{1 - n_t\tau} \qquad \ldots (8.4)$$

Also $$N_t = N_1 + N_2 \qquad \ldots (8.5)$$

Neglecting the terms involving τ_2 and solving for τ, one obtains

$$\tau = \frac{n_1 + n_2 - n_t}{2n_1 n_2} \qquad \ldots (8.6)$$

8.2.3 Scintillation Counters

The development of the modern scintillation particle-counting system from the original ZnS screen viewed by a microscope, as used by Rutherford, is first due to the application of the photomultiplier as a light detector and secondly to the discovery of scintillation materials that are transparent to their own fluorescent radiations, (e.g. anthracene by Kallman and NaI by Hofstadter).

The processes resulting in the emission of light when the charge particle passes through matter depend on the nature of the scintillating material. Among the phosphors, inorganic crystals such as NaI are especially activated with impurities (e.g. thallium) to provide luminescence centres in the band gap. So also are the organic crystals like napthalene and anthracene, and organic liquids such as few percent of diphenyloxazole or terphenyl dissolved in toluene or xylene. Gaseous phosphors such as argon, Krypton and Xenon also used.

Luminescence Radiations: Modes of Energy Transfer

We know that the atoms in a crystal are arranged in a regular array in the form of a lattice. In *ionic crystals*, the alternate lattice sites are occupied by positive and negative ions, e.g. Na^+ and I^- in NaI. Due to strong electrostatic attraction between adjacent ions, the energy levels of individual valence electrons overlap forming a broad energy band called as *valence band*, the characterstic of the crystal. Above this band, there exist *conduction band*. There two bands are separated by an energy gap giving rise to the *forbidden* zone. In the forbidden zone no electron can exist. In halogen halide crystals, the energy gap is $\sim 6 - 8$ eV. These crystals are especially activated with impurities like thallium.

When a high speed charged particle traverse through the crystal lattice of the luminescent medium, good amount of its energy is converted into heat and a small fraction goes into exciting electron from the valence band for transition into the conduction band which is normally empty in an insulator. The electron now moves through the crystal and so does the *hole* left in the valence band. Now, the trapped electron in the conduction band may subsequently return to the valence bond after remitting the excitation energy as *fluorescent* or *luminiscent radiation*. Since the energy of the radiation $(6 - 8$ eV) is too high to be in the visible region. However, the presence of an impurity or activator (NaI) shifts the emission spectrum to visible region by deexcitation process. Electrons raised from the valence band of the crystal by Coulomb interaction may themselves excite an electron in a luminescence centre. This excitation will take place, preferentially, so that the relevant interatomic spacing in the centre is left unchanged. It is followed by a radiationless transition and then by photon emission also preserving interatomic spacing. The fluorescent radiation is of a longer wavelength than that corresponding to the absorption transition and thus not usually reabsorbed by the scintillator.

The scinrtillators chiefly used for particle or photon detectors are:

(a) **Sodium Iodide (thallium activated);** *bismuth germinate*   Because of the high density NaI is an efficient scintillator for γ-ray detection. The rise time of the light pulse is about 0.06 μs and the decay time is about 0.25 μs, with a longer component of about 1.5 μs. Even higher efficiency is obtainable with bismuth germanate (BGO) because of its higher atomic numbers and density.
Other materials used scintillators are CsI(Tl), CsI(Na), Ba F_2.

(b) **Plastic and liquid scintillators**   These are readily obtainable in very large volumes and can be adapted to many different geometrical arrangements, including counting of source over a solid angle of 4π. The scintillator often contains a wavelength shifter to degrade the emitted spectrum to a wavelength that is transmitted efficiently through the crystal and matches a photomultiplier detector. The large pulse associated with an instantaneous brust of ionisation decays at first very rapidly, with a time constant of about 0.005 μs, so that these scintillators are useful for fast counting.

Apparatus

A scintillation counter is a combination of a scintillator and a photomultiplier tube. The scintillator is placed directly on the window of a photomultiplier tube (Figure 8.19), or is coupled to it via a

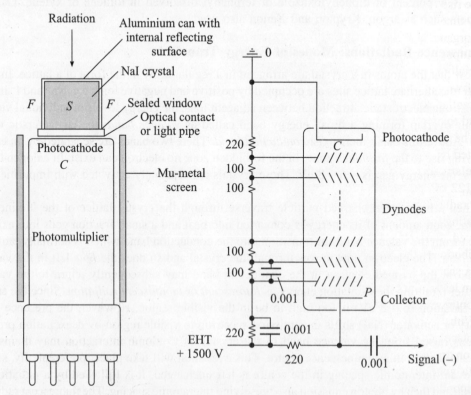

Fig. 8.19 Scintillation counter, showing the circuit arrangement for supplying dynode potentials (resistances in kilo-Ohms, capacitances in microfarads).

light guide. In the latter case, the luminiscent radiation is guided to the cathode *C* by way of total internal reflection in a suitable perspex (lucite) pipe. The window is coated on the inside with a thin antimony-caesium layer from which photoelectrons are emitted with up to about 10 % effciency for photons of wavelength 400 – 500 nm. The tube and the phosphor are to be enclosed in a light-tight case and must be air-tight, otherwise the phosphor crystals may deliquesce in contact with moist air. For different ionising particles a wide variety of phosphor is used, e.g. ZnS phosphor is used for detecting α-particles. Phosphor is deposited on a perspex plate mounted on the end of the photomultiplier. As the radiation is emitted in all directions and therefore it is covered with a thin aluminium foil *F* as reflector so that escape of radiation in the unwanted direction may be prevented charged particles of high energy find no difficulty to reach the phosphor penetrating the reflector. Since the phosphor must be transparent to the luminiscent radiation and hence the latter may reach the photocathode of the tube.

For radiations more penetrating than α-particles, e.g. β-rays, γ-rays, high energy protons etc., usually phosphors having large volume is employed. A crystal of anthracene is commonly used for β-particles and thallium activated NaI is suitable for γ-rays.

The multiplier provides an amplification determined by the inter electrode potential difference, and the output pulse shape, before external amplification, depends on the relative magnitudes of rise time τ, for the scintillator response, the decay time τ for the response and the time constant τ_a of the photomultiplier anode circuit. For $\tau_a \ll \tau_1$ the rise time of the output pulse is limited by τ_1 or by the rise time of the photomultiplier itself, while the decay time is determined by the scintillator decay time τ. This is known as the *current mode* of operation, and is used in fast timing applications with organic scintillators. If $\tau_a \gg \tau_1$, however, the current pulse is integrated and the rise time of the output pulse is determined by τ and the decay time by τ_a. This is the voltage-mode and is especially suitable for pulse-height analysis.

If we use external amplification, the pulse shape and duration may be further controlled by differentiation and integration time constants. In particular they can be used to minimise pulse pile-up and to discriminate against circuit noise.

The spectra observed with a pulse height analyser for radiations form ^{57}Co, ^{137}Cs and ^{22}Na detected by a NaI crystal are shown in Figure 8.20. Their characterstic features are determined by the relative probabilities of the basic photon interaction processes in the crystal. For ^{57}Co, $E_\gamma = 122$ keV, there is a full energy peak due to the production of photo electrons and the capture of all radiations resulting from the atomic vacancy produced. A much lower energy 'escape peak' indicating that sometimes the atomic *K* X-ray (26.8 keV for iodine) leaves the crystal may also be resolved. For ^{137}Cs with $E_\gamma = 662$ keV, a similar full energy peak is obserbed, but there is a contribution to this from Compton scattering and below the main peak is the continuous distribution with a sharp 'Compton edge' due to recoil electrons for which the associated Compton scattered photon leaves the crystal. The angle dependence of the recoil electron energy leads to a peaking of the continuous distribution as a function of energy near the *Compton edge*.

With ^{22}Na, $E_\gamma = 1.28$ MeV, both photoelectric and Compton effect and especially the latter, take place and in addition *pair production* ($\gamma \rightarrow e^+ e^-$) with a threshold of 1.02 MeV occurs. If the position produced in this last effect ultimately annihilates in the crystal the two 511 keV annihilation quanta generally produced may both be fully detected by the crystal. A pulse with height corresponding to E_γ then appears since the annihilation energy adds to the kinetic energy of the e^+ e^- pair. If one annihilation quantum escapes there is a pulse at energy $E_\gamma - m_e c^2$, and if both escape

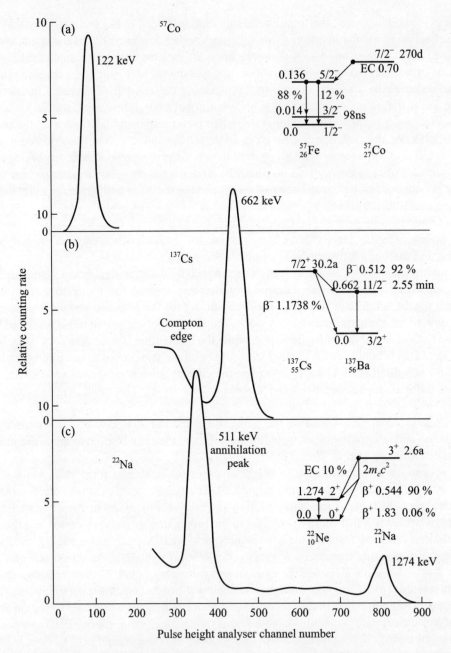

Fig. 8.20 Scintillation counter: pulse height distribution for radiations from (a) ^{57}Co, (b) ^{137}Cs and (c) ^{23}Na detected in a cylindrical NaI(Tl) crystal of dimensions about 25 mm × 40 mm. In part (a) the weak 136 keV transition is not resolved; in part (b) internal conversion electrons accompany the 0.662 keV transition but the Ba X-rays of energy 32 keV are not shown. In all spectra peaks due to back-scattering of forward-going Compton-scattered photons and due to the escape of the 28.6 keV X-ray of iodine. The lifetimes shown are halflives, $t_{\frac{1}{2}}$.

there is one at $E_\gamma - 2m_e c^2$. The resulting spectrum is some what complex even for a homogeneous photon energy. The width of the full energy peaks is seen to be approximately 10 %.

For each photoelectron released from the photocathode by incident radiation from phosphor, the number of electrons reaching the plate is r^n, where r is the secondary emission ration and n is the number of dynodes of the photomultiplier tube. Now, if Q be the charge reaching the plate P due to stopping of a charged particle of kinetic energy W_k in the phosphor, we have

$$Q = \frac{W_k}{W} \, f \, \rho\sigma \, d\tau \, r^n \, e \qquad\qquad \dots (8.1)$$

Here f is the fraction of the particle energy W_k converted to luminiscent radiation, ρ is the fraction photons not lost in phosphor, σ is the fractions of these photons reaching photocathode, τ is the fraction of photoelectrons reaching the first dynode, e is the electronic charge, W is the mean energy of a photon luminiscent radiation, and d is the fraction of photons capable of releasing photoelectrons reaching photo cathode. We may note that it is a function of wavelength range of luminiscent radiation and spectral sensitivity of photocathode.

The above factors are subject to *statistical fluctuations*. Thus, if V be the amplitude of the output voltage pulse appearing on a capacitor of capacitance C, connected to the plate P, and the earth through a high resistance, $V = \left(\dfrac{Q}{C}\right)$ will also fluctuate about a mean value of particle energy W_k. We see from Equation (8.1) that the amplitude of output pulse is proportional to the particle energy dissipated in the phosphor. We may note that the linear relation does not hold good for low energy particles with organic scintillators.

The main advantage of the scintillator detector is that a high efficiency coupled with moderate resolution for γ-ray detection may be obtained with NaI (Tl) or BGO in a reasonably small space and at a reasonably low cost. Anticoincidence shields made of these materials are used in nuclear spectroscopic work with escape-suppressed spectrometers (e.g. γ-ray spectrometers).

Organic Scintillators

These are ring compounds composed of carbon, hydrogen and oxygen. When ionization in an organic scintillator is produced by a charged particle, the electronic states along with their vibrational states of the molecules are excited and the excitation energy of electronic states is about 4 eV and the vibrational energies are about 0.1 eV. The excited vibrational states decay to their associated electronic state in a very short time. The electronic state decays by emitting characterstic wavelength light to the ground state of the molecule and its associated vibrational states. However, the coupling between both electronic and vibrational states provides a channel for dissipation of the excitation energy of a molecule. This causes in reduction of the amount of energy emitted as light. The molecules where electronic excitation is not accompanied with large configurational changes, this effect is minimum. Benzene derivatives are often used as scintillators because they fulfill this requirement. The wavelength of the emission band of these organic scintillators increases with the number of conjugated rings in the molecule. Pure anthracene and stilbene crystals are commonly used as scintillators and these have their emission bands in the blue region. The scintillation decay time in an organic scintillator is about 10 – 8 s and emission of light in these scintillators is a molecular phenomenon. We may note that the wavelength of light emitted by both anthracene as well as stilbene are same in all the three phases, i.e. solid, liquid and gas.

In bulk organic scintillators, the molecular excitation is transmitted to its different regions by resonance phenomenon. As a result, foreign molecules situated in the matrix of the phosphor and having their characterstic absorption energy lower than the excitation energy of phosphor, can absorb the energy and also reemit. In case if the transition probability for de-excitation in the foreign molecule is greater than that for the phosphor molecule, then most of the light is emitted in the wave length band characterstic of the foreign molecule. However, the decay time of light emission remains the same as that of the phosphor molecule Naphthalene emits light in the near ultraviolet region. However, 1 % addition of anthracene to naphthalene results in the emission of almost the entire excitation energy in the blue region, characterstic of anthracene. Obviously, anthracene is acting as a wavelength shifter. These binary scintillators with wavelength shifters are commonly used. By adding 1 – 4 % tetraphenyl butadiene in polystyrene and polymerising it, one gets *plastic scintillator*. Two wavelength shifters can also be used in a scintillator. One can prepare *liquid scintillator* by adding 4 gm of *p*-terphenyl and 8 mg of diphenyl hexatriene in a litre of pure toulene or phenyl cyclohexane. In this liquid scintillator the solvent emits light of vary short wavelength, which is transformed by *p*-terphenyl to a wavelength band at 3500 A. Diphenyl hexatriene molecules shifts this light to the blue region. One should add an optimum amount of wavelength shifter to a phosphor. By transferring the excitation energy to its vibrational states a larger amount reduces the light output and ultimately appears as heat. Minute quantities of other impurities in the phosphor also quench the light emission by transferring the excitation energy to the vibrational states. We may note that the organic compounds used in a organic scintillators must be of very high purity.

The amount of light emitted depends upon the specific ionization for the same amount of ionization produced in an organic scintillator. The light output diminishes with the increase in specific ionization for the same total ionization. One can explain this due to quenching of light emission by the molecules damaged due to ionization. One finds that the amount of light emitted by an ionising particle $\left(\dfrac{dL}{dE}\right)$ is given by

$$\frac{dL}{dE} = \frac{L_0}{\left(1 + a\dfrac{dE}{dx}\right)} \qquad\qquad \text{... (8.2)}$$

Where a and L_0 are constants and $\dfrac{dE}{dx}$ indicates the specific energy loss by the particle. We may note that the specific energy loss by electrons is very small and one can observe its effect at kinetic energies below 100 keV.

Anthracene and Stilbene are the commonly used scintillators among the pure materials. Anthracene has the largest scintillation efficiency, whereas stilbene has a decay time of 4.5 ns. Figure 8.21 shows the decay of scintillation pulses with time from a stilbene scintillator for different ionising particles.

Liquid scintillators often have two wavelength shifter and their light efficiency is only about half of that of an anthracene crystal. The main advantage of these scintillators is that they can be obtained in large quantities. One can load these scintillators with heavy elements for specific purposes, e.g. Pb loaded liquid scintillators are used for the detection of γ-rays. Gadolinium loaded liquid scintillators are used for the detection of thermal neutrons.

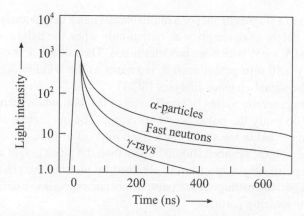

Fig. 8.21 For different ionising particles the luminiscence decay of stilbene

Plastic scintillators are used in counter telescopes which are generally used in high energy physics to define the path of a particle. These scintillators have a light output only as good as the liquid scintillators. One can obtain plastic scintillators in large size blocks and can easily machined and polished. For the detection of thermal neutrons boron and gadolinium loaded plastic scintillators are often used. They are also employed in fast timing experiments. In these scintillators, prolonged exposure to ionizing radiations, light and oxygen causes polymer degradation.

Organic scintillators provide excellent timing signals and can be built into the very large and complex arrays necessary for particle location and identification of high energy physics.

Advantages

The pulses given by scintillation counters are of extremely short duration and therefore the method is capable of rapid counting, shorther than gas filled ones.

The method has the special merit that the strength of the flash depends on the type and energy of the particle that causes it. One can sort out particles in a given range and analyse the distribution of the energy in the incident radiation by using suitably biased electronic circuits. The arrangement then behaves as a *scintillation spectrometer.*

The stopping power of solid and liquid phosphors is much higher than that of the gas used in gas-filled counters. This is why this instrument is preferred where high *stopping power* is demanded, as in the case of experiments with γ-rays. One of the most important applications of the scintillation detector is in the field of γ-ray detection and spectroscopy.

For detecting very weak signals, detectors of very large volume have to be built with liquid.

We may note that pulses due to the passage of the charged particles through the scintillator are also accompanied by smaller unwanted pulses due to small thermionic emission of the photomultiplier. However, when the desired pulses are quite large, the unwanted pulses ones can be *biassed out* very easily. However, if they are small, as with β⁻ and γ-rays, one can eliminate the unwanted pulses by operating the photomultiplier at a very low temperature.

8.2.3(a) Scintillation Spectrometer

Figure 8.22 shows a typical scintillation spectrometer. The bleeder resistance and preamplifier are mounted near the base of the photomultiplier tube. The output of preamplifier after amplification in a linear amplifier is fed to a single channel or a multichannel pulse height analyser. We may note that a single channel pulse height analyser gives an output only when the height of the input signal lies between voltages V and $V + dV$, which are predetermined. The voltage V is called *baseline* voltage and can be adjusted by a 10 turn potentiometer, is known as the window width. Electronic scalar counts the output of the signal channel analyses (SCA).

Baseline voltage is generally varies in small steps of 0.1 Volt while the window width is kept fixed at 0.05 or 0.1 V to study the pulse height spectrum. The number of output pulses for each base line voltage are recorded in the scalar for a fixed interval of time-say one minute. However, the maximum base line voltage scanned should be less than the pulse height at which the amplifier starts to saturate. The pulse height spectrum is obtained by plotting the counting rate recorded at different base line voltages. Interestingly, each pulse height corresponds to a definite energy dissipated in the scintillator by the ionizing particle.

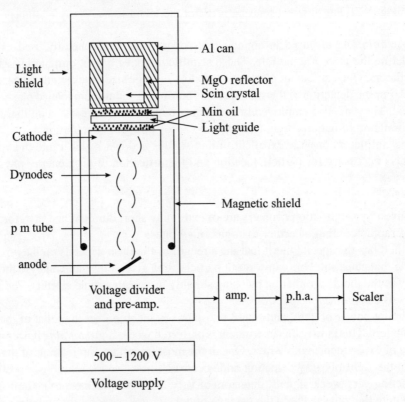

Fig. 8.22 A Scintillation Spectrometer

The pulse height spectrum of all the input pulses in a multichannel analyser is analysed simultaneously. The whole pulse height spectrum can be observed on an oscilloscope screen within few minutes and can be recorded as per requirement. Nowadays, Computer based multichannel analysers are widely used.

Energy Resolution of a Scintillation Spectrometer

This may be defined as the energy spread (due to spread in pulse height) as measured by the scintillation spectrometer to the energy dissipated by the ionizing particle in the scintillator. Resolution (R) for the Gaussion distribution of energy is expressed as

$$R = \frac{\Delta E}{E} \propto \frac{\sigma_N}{N} = \frac{1}{\sqrt{N}} = E^{-\frac{1}{2}} \qquad \qquad \dots (8.3)$$

Where N is the average number of photoelectrons emitted from the photocathode when the dissipated energy in the scintillator is E. Even if we consider the statistical distribution of secondary electrons emitted from dynodes, the energy resolution $R \propto E^{-\frac{1}{2}}$.

Figure 8.23 shows the effect of resolution on the energy spectrum in the scintillation counter.

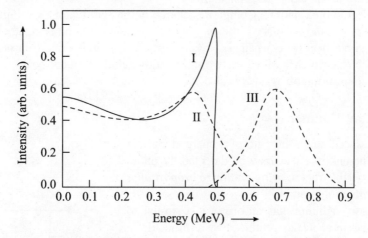

Fig. 8.23 The statistical resolution of the scintillatin counter exhibiting the modification of the spectra due to photo-electron and Compton electron. The Gaussian peak ('III') is due to photoelectrons and the Compton electron spectrum ('I') is modified to give spectrum 'II'.

For a definite energy dissipation in scintillator, the output pulse height spectrum is Gaussian and the energy resolution is usually defined as the *ratio of full width at half maximum of the peak and the peak position on the base line voltage*. By observing the pulse height spectra due to monoenergetic radiations of different energies, one can establish the relationship between the base line and the energy dissipated (calibration curve).

8.2.3(b) Scintillation Detector for γ-ray Spectroscopy

For γ-ray spectroscopy, usually a NaI(Tl) scintillation crystal is used. One centimeter thick crystal can be used for γ-rays of energies below 100 keV. A crystal having thickness 3.5 to 5 cm have a fairly good detection efficiency for higher energy γ-rays. The diameter of photo cathode in photomultiplier tubes used is about 4.8 to 5 cm. The diameter of scintillation crystal could be between 3.5 to 4.8 cm. In some special purpose experiments, scintillators of diameter greater than 4.8 cm are generally used and they are coupled to photomultiplier tube through a light guide.

When a γ-ray photon collides a scintillator detector, there is some probability that the photon may pass through it without any interaction. We have read that a photon may interact with the scintillator by photoelectric absorption or by Compton scattering. If photon energy is greater than $2\,m_0c^2$, it may also be absorbed by pair production process. An energetic electrons is produced in either of these processes, which looses its kinetic energy in the scintillator and produces ionisation. A number of photons are emitted by scintillator which is proportional to the energy deposited by the electron. The observed pulse height spectrum in scintillation spectrometer represents the spectrum of energies dissipated by the electrons in the scintillator-modified by the statistical fluctuation. The following events contribute to the pulse height spectrum when a beam of monoenergetic photons strikes a NaI(Tl) scintillation crystal:

(i) Photon with energy $h\nu_0$ may produce photoelectric effect in the high atomic number atoms of the scintillator, e.g. iodine in NaI(Tl). This may give rise to the 'full energy peak' or 'photo peak' in the pulse height spectrum.

(ii) The incident photon may be Compton scattered in the scintillator (See Figure 8.23).

(iii) Pair formation will take place if the energy of the incident photon is greater than the energy of positron-electron ($e^+\,e^-$) pair, i.e. $h\nu_0 > 2\,m_0c^2$. This gives rise to a peak in the pulse height spectrum which is called as 'two annihilation radiation escape peak'. This corresponds to energy deposition of $h\nu_0 - 2\,m_0c^2$.

8.2.4 Cherenkov Counter

The photons produced along the path of a charged particle in matter by Cherenkov effect may be detected by photo multiplies in the visible region or by *multiwire proportional* counters in the ultraviolet region.

The Cherenkov radiation has an origin just similar to that of *shock wave* in acoustics produced by a projectile in air when it moves faster than an ordinary sound wave, e.g. supersonic flight.

The Cherenkov radiation can be observed only along certain directions. Let us consider that a charged particle

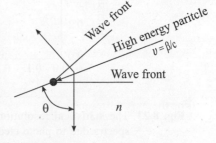

Fig. 8.24(a) Cherenkov radiation

moving with a speed υ through a transparent dielectric medium of refractive index n (Figure 8.24(a)). From Huygen's construction, it is obvious that Cherenkov radiation is observed only along the direction θ, such that the time taken by the particle to travel from A to B with speed υ (assuming that it essentially remains constant over this length) is equal to the time taken by the radiation to travel from A to point C (Figure 8.24b).

The direction of propagation of Cherenkov radiation wavefront is determined by the ratio $\dfrac{(c/n)}{\upsilon}$, where c is the velocity of light. We have

$$\frac{c/n}{\upsilon} = \cos\theta = \frac{1}{n\beta}, \text{ where } \beta = \frac{\upsilon}{c} \qquad \dots (8.1a)$$

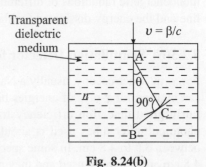

Fig. 8.24(b)

Angle θ is called as the angle of propagation of the wavefront (Figure 8.24(a) and (b)) with respect to the direction of motion of the particle.

Cherenkov radiation is emitted if $\beta < \dfrac{1}{n}$, and hence it is confined to high-energy charged particles. Further, β is always less than 1 (since $v < c$) and hence no Cherenkov radiation occurs in the X-ray region, since n is less than 1 in this region. Obviously, the Cherenkov radiation is thus confined to the surface of a cone of semiapex angle θ with its axis coinciding with the path AB of the particle in the medium. If $\cos \theta = 1$, the radiation is emitted along the direction of travel of the particle through the medium. In a given medium, the angle of emergence of the Cherenkov radiation is determined only by the velocity of the particle; this offers the posibility of measuring the velocities of high-energy particles by methods based on the detection of this radiation.

The range of the values of β is thus

$$\frac{1}{n} \le \beta \le 1 \qquad \qquad \dots (8.1b)$$

At very high particle velocities, as those of cosmic rays, $\beta \approx 1$, so that $\cos \theta = \dfrac{1}{n}$ which is a constant for a given medium.

For $\beta = \dfrac{1}{n}$, one finds $\theta = 0$ and for $\beta = 1$, $\theta = \theta_{max}$ which is expressed by the relation

$$\theta_{max} = \cos^{-1}\left(\frac{1}{n}\right) \qquad \qquad \dots (8.2)$$

If $\beta < \dfrac{1}{n}$, no Cherenkov radiation is found to be observed.

For the emission of Cherenkov radiation, the minimum (or threshold) velocity is

$$v_{min} \text{ (or } v_{th}) = c\,\beta_{min} \quad (\because \beta = \frac{v}{c})$$

which gives $n\,\beta_{min} = 1$. This occurs for $\theta = 0$.

The minimum kinetic energy T_{min} required for the particle of mass m_0 for the emission of Cherenkov radiation is given by

$$T_{min} = m_0 c^2 \left(\frac{1}{\sqrt{1 - \beta_{min}^2}} - 1\right) \qquad \qquad \dots (8.3)$$

Obviously the threshold velocity $v_{th} = \beta_{th} c$ below which no Cherenkov radiation is seen is determined by the refractive index n of the medium and, in the case of gases, can be conveniently varied by altering the pressure.

Since n of a gas is *pressure-dependent* and hence one can make detectors with varying thresholds by controlling the pressure. An interesting application of Cherenkov effect in a gas in a gas has been made in the radiation produced by cosmic rays as they penetrate the earth's atmosphere.z

Table 8.1 shows the corresponding Lorentz factor $\gamma_{th} = (1 - \beta_{th}^2)^{-1/2}$, which gives the threshold energy in terms of the particle mass for a few radiating

Table 8.1 Cherenkov radiators

Radiator	$n - 1$	γ_{th}
Plastic scintillator	0.58	1.3
Water	0.33	1.5
Pentane (STP)	1.7×10^{-3}	17
Helium (STP)	3.3×10^{-5}	123

Figure 8.25 shows the plot of the angle of emission of Cherenkov radiation given by Equation 8.1(a) vs. momentum for pions and kaons in a radiator with $n = 1.1$. We note that discrimination of particles is also possible by angle selection. This has been used for particle identification in the region of minimum ionisation where collision-loss methods fail.

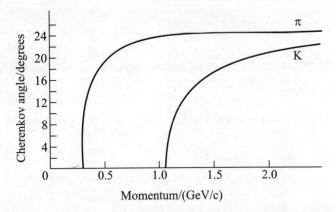

Fig. 8.25 For π and K partiles Cherenkov angle vs. momentum in a radiator with $n = 1.1$

Figure 8.26(a) and 8.26(b) shows the simple arrangements for identification of relativistic particle of different mass but the same momentum by the Cherenkov effect. In the threshold detector (Figure 8.26a) the refractive index of the detector is chosen so that the heavier particle of a possible pair has a velocity just below the threshold. The angle selection for a ring image in a

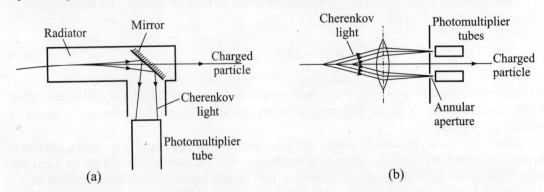

Fig. 8.26 Principle of operation of Cherenkov counters: (a) threshold, and (b) differential

differential counter is achieved using adjustable annular aperatures. Velocity resolution $\dfrac{d\beta}{\beta}$ as good as 10^{-7} has been reported in the literature providing pion, kaon and proton separation above 100 GeV/c momentum.

Like all detectors that give a response to γ radiation, Cherenkov detectors take advantage of the fact that the electric field of a relativistic particle is concentrated in the plane perpendicular to its motion by the Lorentz transformation; the field measured in the laboratory in that plane is increased by a the factor γ. As γ increases processes can take place at greater and greater impact parameters and energy loss increases until polarization of the atoms of the medium inhibits the spread of the field. This saturation effect can be pushed to higher values of γ by the use of low density radiators but the energy loss is then so small that very long radiators are required. To overcome this difficulty in the ultrarelativistic region the pehnomenon of *transition radiation* has been exploited.

The weakness of transition radiation can be augmented by allowing particles to pass through a series of foils whose contributions add together coherently as in multiple beam interferometry. In one detector a stack of 650 Li foils 50 μm thick and 250 μm apart was used, X-ray photons were detected (Figure 8.27) in a multiwire proportional counter and by selecting the frequency of the radiation it was possible to identify electrons in the energy range of a few giga-eV ($\gamma > 1000$) while rejecting heavy particles ($\gamma < 1000$).

The Cherenkov counter is an indispensable tool in high energy physics. Though the intensity of the Cherenkov radiation is very much smaller than that of the luminescent radiation from a scintillation counter, a Cherenkov counter has practically no measurable time lag between the entry of the particle and the start of the emission of the radiation; in the case of scintillation counters, this time lag is of the order of 10^{-9} s, which, though very small, is significant in recording extremely energetic particles. Cherenkov counter could define the direction of approach of the particle and therefore used as parts of trigger systems for bubble chambers, cloud chambers, etc.

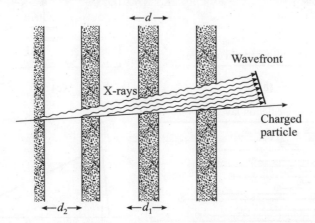

Fig. 8.27 Use of multiple-foil interference in the practical application of transition radiation.

8.2.5 Semiconductor Detectors

Solid state detectors operate essentially through the promotion of electrons from the valence band of a solid to the conduction band as a result of the entry of the particle or photon to be detected into the solid. Since the band gap is some solids is only about 1 eV, production of electron-hole (e^- e^+) pair as part of general general energy loss by collision may require only 3 – 4 eV on average. The stopping of a particle of energy 1 MeV may thus produce 3×10^5 electron-hole pairs and the associated statistical fluctuations would be only about 0.2 %, which offers a great advantage in basic resolution over the scintillation detector and indeed over the simple gas-filled ionisation chamber for which an energy expenditure of 30 eV per ion is required.

The difficulty in realizing the attractive features of solid state detectors in practice has been in obtaining materials in which residual conductivity is sufficiently low to permit conduction pulses due to single particles to be distinguished above background and in which the charged carriers are not rapidly 'trapped' by impurities. This has, however, been achieved for certain *semiconductors* in the following type of detector.

Junction Detectors for Charged Particles

We know that in an *n* type semiconductor conduction is due to the motion of *electrons* in the conduction band, and in *p*-type material the process is effectively a motion of posite *holes* resulting from the rearrangement of electrons between atoms in the crystal. The two materials are prepared from an intrinsic semiconductor such as *silicon* by the controlled addition of electron-donating or electron-accepting elements (Figure 8.28(a)). In the case of silicon these elements could be phosphorus (*P*) and indium respectively and the effect of their existence within surroundings of silicon is to produce hydrogen-like structures with their own set of energy levels, known as

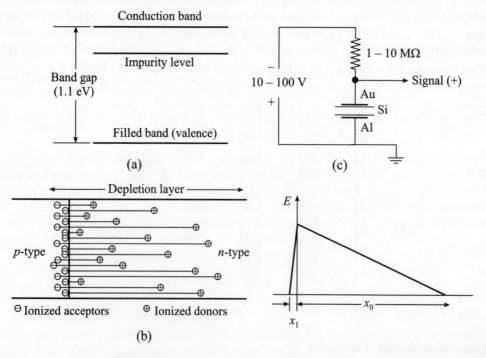

Fig. 8.28 Semiconductor detector: (a) bands and impurity levels in a semi conductor; (b) charge and electric field distributions in the depletion layer of a *p-n* junction; (c) circuit diagram.

impurity levels. These are slightly below the conduction band and slightly above the valence band respectively, so that thermal excitation of electrons to the former and from the latter is readily possible. The corresponding conduction process is then by electrons (*n*-type) or by holes (*p*-type); it is assumed that all impurities are ionized.

Usually Si and Ge intrinsic semiconductors are used for semiconductor counters. Their intrinsic properties are summarized in Table 8.2.

Table 8.2 Intrinsic Silicon and Germanium Semiconductors

Property	*Silicon* (Si)	*Germanium* (Ge)
Atomic number	14	32
Atomic weight	28.09	72.60
Density (300 K)	2.33 g/cm^3	5.32 g/cm^3
Atoms/cm^3	4.96×10^{22}	4.41×10^{22}
Stable isotope mass numbers	28 – 29 – 30	70 – 72 – 73 – 74 – 76
Forbidden energy gap (0 K)	1.165 eV	0.746 eV
Intrinsic carrier density (300 K)	1.5×10^{10} cm^{-3}	2.4×10^{13} cm^{-3}
Intrinsic resistivity (300 K)	2.3×10^3 Ω-m	47×10^{-2} Ω-m
Electron mobility : 77 K	2.1×10^4 cm^2/V-s	3.6×10^4 cm^2/V-s
: 300 K	1350 cm^2/V-s	3900 cm^2/V-s
Hole mobility : 77 K	1.1×10^4 cm^2/V-s	4.2×10^4 cm^2/V-s
: 300 K	480 cm^2/V-s	1900 cm^2/V-s
Dielectric constant	12	16
Fano factor (77 K)	0.09	0.08
Energy per electron hole pair: 77 K	3.76 eV	2.96 eV
300 K	3.62 eV	

If a junction between *p*- and *n*-type extrinsic semiconducting regions is formed in a crystal then conduction electrons predominate in the *n*-region but a few will diffuse into the *p*-region (Figure 8.28(b)) where the conduction electron density is low. Similarly, holes from the *p* region will diffuse into the *n*-region and the joint effect of those motion is to leave a positive space charge in the *n*-region and the negative space charge in the *p* region, near the boundary.

The electric field resulting due to transition from *n*-type to *p*-type is given by

$$E_{max} = \left(\frac{2V\,Ne}{\sigma} \right)^{\frac{1}{2}}$$

... (8.1)

where N is the acceptor concentration ($N_D \gg N_A$), σ is the conductivity of the semiconductor. E_{max} could be as high as $10^6 - 10^7$ V/m. This electric field from this double layer *n*- and the *p*-regions generates a potential difference between the *n*-and *p*-regions and this causes a drift current of thermally excited electrons or holes in the opposite direction to the diffusion current so that a dynamical equilibrium exists in which there is no net transfer of charge. The region of the crystal occupied by the space charges has lost charge carriers in comparison with the rest of the solid and is able to sustain an electric field. It is known as *depletion layer.*

The depth x_0 of the depletion layer depends on the density of impurity centres and on any applied potential difference V (reverse bias marking the n-region more positive with respect to the p-region). If P is the resistivity of n-type region,

$$x_0 \approx (\rho V)^{\frac{1}{2}} \qquad \qquad \dots (8.2)$$

For silicon of the highest practicable resistivity a depletion depth of about 5 mm may be obtained.

If a charged particle enters the depletion layer the electron-hole pairs produced are swept away by the field existing the layer and a pulse may be detected in an external circuit. Because of the low value of the energy required to form an electron-hole pair, and because of the existence of another favourable factor (fano factor F) which recognizes that energy losses for a particle of finite energy are not wholly independent, a resolution as good as 0.25 % is obtainable at 5 MeV. The spectrum of α-particles from ^{212}Bi observed with a silicon detector of resistivity 27 Ω-m is shown in Figure 8.29.

The semiconductor detectors are made in different configurations. Few of these are as follows:

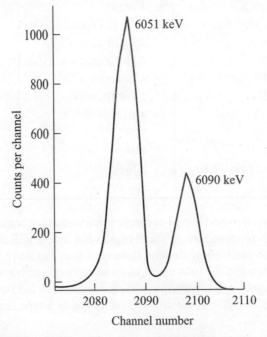

Fig. 8.29 Semiconductor Counter: pulse height distribution for 6 MeV α-particles from ^{212}Bi (Th C) detected in a silicon surface barrier 134 mm thick. The groups are of width 16 keV at half maximum intensity.

8.2.5(a) Diffused Junction Detectors

In this type of detector a donor impurity, usually phosphorus (P) is introduced into a p-type (boron-doped) silicon single crystal to form a depletion layer at the diffusion depth. The n-type layer acts as the window for the ionizing particle to enter. It does not contribute to the formation of the electron-hole pairs and acts only as a *dead layer*, whose thickness can be measured experimentally.

8.2.5(b) Surface Barrier Detectors

A *p*-type layer is formed on the surface of *n*-type silicon by oxidation. Contact is made to the detectors through a layer of gold. The counter base is often a layer of aluminium. Signals are derived from the surface barrier detector to the connections shown in Figure 8.28(c). Since the depth of the depletion layer, and consequently the interelectrode capacitance, depends on the applied voltage, a charge sensitive rather than voltage sensitive pre-amplifier is used.

8.2.5(c) Ion implanted layer Detectors

The doping impurities in these type of detectors are implanted on a *p*- or *n*-type crystal surface by exposing it to suitable ion beams accelerated to an energy ~ 10 keV. For this purpose, usually the beams of phosphorus or boron are used. In the crystal, the ion beams have a definite range, which can be increases by increasing the accelerating voltage. One can closely control the density of the added impurity by ion implantation. The crystal is annealed after implantation at a temperature of about 500 °C. As compared to surface barrier detectors these are more stable and less subject to ambient temperatures. One can have detectors with extremely thin entrace window.

8.2.5(d) Fully Depleted Detectors

When we apply a reverse bias to a *p-n* junction, the depletion layer extends in the region with smaller number of carriers. However, in fully depleted detectors, the depletion layer depth extends over the entire thickness of the semiconductor crystal wafer. For these detectors *n-p* junction is made by depositing a heavy concentration of say *n*-type impurity on the surface of middle *p*-type highly pure semiconductor. At a given reverse bias to maximise the depletion depth, the concentration of the doping impurity in the high purity side of the junction is minimised. At the surface of the waves, the heavily doped layer can be the incoming radiations.

In these detectors, the entire bias voltage drop is across the depletion region. If the reverse bias is sufficiently high then the electric field becomes uniform over the whole depletion depth, which results in a uniform collection of $e-h$ pairs formed due to incoming radiation. By blocking (heavily doped) *p*- or *n*-type contact away from a *p-n* junction, the Ohmic connection at the back of the wafter is made.

This detector can be used as a transmission detector in which the incoming radiation loses only part of its kinetic energy in the fixed thickness of the detector and then goes out so that it may enter another thick detector. In the detector material, the pulse height at the output of the detector is a measure of the specific energy loss $\left(-\dfrac{dE}{dx}\right)$ of the incoming radiation. In the thickness range of about 50 to 2000 microns, such detectors are available in the market. The incident particle is stopped in a thick detector after passing through a fully depleted $\dfrac{dE}{dx}$ detector, where the particle looses whole of its kinetic energy. Pulse height of this detector measures the total energy E of the incident particle. Usually, $\dfrac{dE}{dx}$ detector is coupled with E of a incident particle is employed for identification of the nature of the particle.

The advantages of the silicon *p-n* junction as a detector of heavy particles are its excellent *resolution* and *linearity*, its small size and consequent fast response time permitting a high counting rate and the fairly simple nature of necessary electronic circuits. The solid state counter is not actually a windowless counter because of the existence of *p*- or *n*-type insensitive layers of thickness 0.1 – 0.5 μm in surface-barrier and diffused-junction detectors respectively. This is not a serious drawback, although it adds to the line width, and its effect is outweighed by the over all simplicity of the counters, permitting the use of *multi-detector arrays*. The handling of data from such a complex system demands the availability of an on-line computer for storage of information and data processing. A special facility of high value in use of solid state detectors is the easy construction of *particle identification telescopes* based on energy loss X energy product signals. Counters of a length of a few centimeters may also be used as position sensitive indicators in the focal plane of a spectrometer, since it can be arranged that output pulse height depends on the distance of entry of the incident particle from the collecting electrode.

A disadvantage of silicon detectors for use with heavy ions and fission fragments is that nuclear scattering by the silicon nuclei worsens resolution and give rise to an energy defect because of the occurrence of non-ionizing transitions with large statistical fluctuations in the solid lattice. This is why, gas filled detectors can give a better performance in this application.

8.2.5(e) Germanium-Lithium (Ge-Li) and HP Ge Detectors

Increased sensitive volumes of semiconductor detectors can be achied by compensating the acceptor centres in a *p*-type semiconductor by controlled addition of donor impurities. The most suitable donor is lithium because Li atom has a very low ionization potential in a semiconductor and also a high mobility. The Li is 'drifted' into the bulk material from a surface layer by raising the temperature to about 420 K in the presence of an electric field. Compensated depths of about 15 mm and active volumes (in Ge) upto 100 cm^3 can be made. Large volume detectors may also be prepared from highly purified Ge (impurity concentration about 1 part in 10^{13}). The Ge (Li) detectors must be kept at a temperature below 150 K to prevent the lithium drifting away from the counting region. Ge (Li) counter consists of an $n - i - p$ junction device, where i being the intrinsic semiconductor. Li-ions are allowed, under an intense electric field, to drift radially in a *p*-type Ge-cylinder where they are neutralized to create an intrinsic region. The region has the low conductivity and forms the effective area of this detector. When a charge particle or radiation enters the region it produces electron-hole pairs which are collected by the external field. Ge (Li) counter is maintained at liquid nitrogen temperature at all stages, i.e. from operation to preservation. With the rise in temperature, the efficiency reduces as Li-ions drift outwardly and the intrinsic region is not maintained. One will have to take adequate care to maintain the low temperature as any accidental exposure to high temperature may cause permanent damage to the detector.

Ge (Li) detectors are particularly suitable for the *detection of γ-rays* since the photoelectric absorption cross section of γ-radiation is proportional to Z^5 and germanium has relatively larger Z. With this detector a high resolution of about 1 % can be achieved.

8.2.5(f) γ-ray Detectors

The major application of the lithium-drift technique is to the Ge (Li) counter, which has transformed the subject of γ-ray spectroscopy. Ge has a band gap of only 0.67 eV and must be cooled to liquid nitrogen temperature when used as a counter in order to minimise thermal excitation of electrons to the conduction band. This is an inconvenience which results in preference for silicon counters (although these are also improved by cooling) for charged particle spectrometry. For γ-ray detection, however, the higher Z of germanium (Z_{Ge} = 32, Z_{Si} = 14) leads to a marked improvement in efficiency as may be seen from the Z-dependence of the basic processes of photoelectric, Compton and pair production absorption. The difference between Ge (Li) and Si (Li) counters is especially marked in the energy ranges in which the photo effect and pair production predominate because of particular Z dependence. The Ge (Li) detector cannot yet be made as large as a sodium iodide or bismuth germanate scintillator and is less suitable when the highest efficiency is desired, but its resolution is better by more than an order of magnitude and line widths of less than 1 keV have been reported. These are narrower than those observed for charged particles because there is no counter 'window' and energy loss in atomic collisions is not important.

A spectrum of ^{24}Na-γ rays taken with Ge (Li) detector compared with one obtained with a scintillation counter is shown in Figure 8.30.

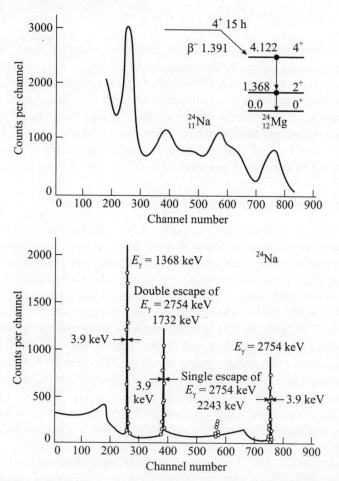

Fig. 8.30 Semiconductor Counter: spectrum of radiations from ^{24}Na-γ rays as seen (a) in an NaI (Tl) crystal and (b) in a 30 cm^3 Ge (Li) detector.

Low energy radiations, e.g. electrons of approximately 10 keV, or characterstic X-rays may be recorded with high resolution by Si (Li) counters.

If a high resolution Ge detector is surrounded by a large NaI or bismuth germanate (BGO) scintillator operated in an anticoincidence mode it is possible to veto those events in the Ge whose secondary radiations, (e.g. Compton-scattered photons) give a signal in the surrounding detector. The background in the Ge spectrum of full energy events such as photo peaks is then much reduced. Such escape-suppressed spectrometers (ESS) are sometimes arranged around central BGO detectors which sense total energy and γ ray multiplicity in nuclear cascade.

HP Ge Detectors

The thickness of the depletion layer of the usual junction detectors is a millimeter or two. This is why that these detectors are not suitable for the detection of γ-rays. The depletion depth in a junction detector is given by

$$d = \sqrt{\frac{2\,\varepsilon V}{e\,N}} \qquad\qquad ...\ (8.3)$$

Where N is the net impurity concentration in the bulk semiconductor and V is the reverse bias, which is limited (maximum value) by the break down voltage. Thus by reducing net impurity concentration, the depletion depth can be increased. Now a days, such techniques have been developed to get purified germanium having impurity level as only 1 part in 10^{12}, which corresponds to $N = 10^{10}$ impurity atoms/cm^3. These impurities could render the germanium either p-type or n-type. Using this germanium, detectors have been made and these are called as high purity germanium (HP Ge) detectors and these can have a depletion depth of a few centimeters or more. So far silicon could not be purified to this level.

8.2.6 Photographic Emulsion or Nuclear Emulsion Technique

Use of photographic emulsions directly as particle track detectors has been found useful for studying certain nuclear reactions. It has been found that the blackening of a photographic plate by the radiation is roughly proportional to the product of the intensity and the time of exposure.

When a heavy charged particle traverses a photographic emulsion produces a latent image of its track which can be developed and fixed when the deposited black silver in the negative gives a permanent record. Ordinary optical photographic emulsion are unsuitable for the quantitative studies with nuclear radiations, since the density of such emulsions is low and the developed crystal grains are too large and widely spaced to give any well-defined particle tracks. These defects are overcome in *special nuclear emulsions* in which the proportion of Ag Br to gelatine is about 4 – 5 times higher than in ordinary optical emulsions and Ag Br grains have diameters 0.1 to 0.2 mm only. Such fine grain nuclear emulsions having thickness varying from 25 – 2000 mm have been developed by Powell and Coworkers. Such fine grain nuclear emulsions have reopened this technique as a detector of nuclear radiations with renewed vigour. The nuclear emulsion in a sense, is a kind of *frozen cloud chamber.*

For different types of particles the sensitivity of nuclear emulsion depends on the grain size, and smaller the grain diameter, less is the sensitivity. Obviously, highly ionising particles would therefore require less sensitive plates for recording their tracks, e.g. the fission fragments in a nuclear fission reaction. However, high energy electrons produce very few ions, and unless the nuclear emulsion is most sensitive, its track is not recordable. For detecting α-particles, fission fragments Ilford D-1 emulsion (grain dia ~ 0.12 m) is suitable, whereas for proton and meson tracks Ilford C-1 is appropriate and Ilford G-2 (grain dia ~ 0.18 m) is sensitive to high-energy electrons.

To develop such thick emulsions uniformly one method is to soak them in the developer which is maintained at a very low temperature, till the developer diffuses throughout the entire body of the emulsion. Naturally, at such low temperatures the rate of development will be very slow. The temperature is gradually raised after the whole emulsion has been properly suffused by the developer, so that it is uniformly developed quickly.

Another development is the emulsion stack where emulsions from dozens of plates are stripped from their glass backings (Figure 8.31) and then stacked together (since the thickness of emulsions are only about 1 mm thick). After exposing the stack, the emulsion layers are separated and remounted on the glass backings and then developed. From Figure 8.31, it is clear that an incoming beam of high energy particles *AB* produces an event at *C* and one can study the resulting tracks *CD* and *CF* of the two particles produced. Different types of ionising particles demand different types of emulsions. Now, these are commercially available, including one which is sensitive enough to record the tracks of electrons at their minimum specific ionisation.

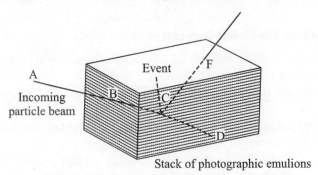

Fig. 8.31 Photographic emulsion stack.

Prior to stacking, the stripped emulsions are numbered and fiducial marks given so that the continuity of a track through the stack could be followed. Measurments by scanning microscopic observations are made of track densities (variation of grain density), range, direction of travel of the particles and the small angle of scattering of the particles due to collision with nuclei of atoms in the emulsion. The probability of an ionising particle leaving a track inclined at an angle to the surface of the emulsion. High quality microscopes with small depths of view, accurate depth guages and attached devices for measuring the track lengths and angles are used to scan the developed plates. The small depth of focus of the high magnification scanning microscopes makes it imperative to focus properly the microscope on short sections of the track and take microphotographs of these short sections. By placing together enlargements of these sections, a mosaic picture of entire track of ionising particle through the emulsion is obtained.

The above measurments provide information about the charge, mass and energy of the particle.

The variation of grain density along the track $\left(\dfrac{dn}{dx}\right)$ bears a relation with specific energy-loss of

the particle $\left(\dfrac{dE}{dx}\right)$ by an empirical relation

$$\frac{dn}{dx} = A\left[1 - \exp\left(\alpha\, z\, e\sqrt{\frac{dE}{dx}} - \beta\right)\right] \qquad \text{... (8.1)}$$

Where Ze is the charge of the particle and A, α and β are constants.

The mass m of the particle (in proton mass unit and its kinetic energy E are related through the range R by the following relation

$$E = BZ^{2\mu}\, m^{1-\mu}\, R^{\mu} \qquad \text{... (8.2)}$$

Where μ and B are constants.

One can obtain the total number of grains (n) from the following relation,

$$n = C\, Z^{2\nu}\, m^{1-\nu}\, R^{\nu} \qquad \text{... (8.3)}$$

Where ν and C are two empirical constants.

One can identify the particle from the knowledge of the range R and the grain-density $\dfrac{dn}{dx}$ when the grains are resolved. Usually, the energy of the particle is evaluated from the range of the particle in the emulsion.

Advantages of Nuclear Emulsion Detectors

  (i) This method is quite cheap and simple and there is no elaborate arrangement is required either mechanical or electronic.
 (ii) Its weight is light and hence easily portable. Thus it well-suited for high altitude cosmic ray experiments.
(iii) It is continuously sensitive, whereas bubble chambers or cloud chambers are sensitive only for a fraction of a second.
 (iv) These can give complete tracks of very high energy particles, being a solid medium.
  (v) The track of the particle can be recorded for considerable lengths as the emulsions are now available in 'stripped' form.
 (vi) One can distinguish one particle from the other as the tracks of different types of particles show different appearances and characterstics.
(vii) These can also be adapted to the study of neutrons. In the emulsion high energy neutrons can knock off protons, which in their turn develop tracks in it. By studying the direction of emission of protons with respect to that of neutrons, one can estimate the energy of the incident neutron by measuring the range of the proton. For the detection of slow neutron, a special type of the emulsion containing boron or lithium having high thermal neutron cross section is used.
(viii) The density of emulsion is high and also its stopping power is high (~ 1000 times that of standard air). The significant nuclear events take place too frequently and therefore interesting results e.g. π-μ-e decay, explosion or evaporation of nuclei, heavy nuclei in primary cosmic radiation, existence of new particles, etc. have been obtained by emulsion method with the aid of accelerators.
 (ix) Emulsions impregnated with target materials are used for studying different nuclear reactions.
  (x) Emulsions gives highest spatial resolution and 4π acceptance which is its main advantage.

Drawbacks

(i) Temperature, humidity, age of the emulsion etc. affect the sensitivity and thickness of the emulsions under which they are developed.
(ii) Unlike cloud chamber tracks, the nuclear plate tracks cannot be bent satisfactorily in a magnetic field since the scattering obscures the curvature which is small enough.
(iii) Due to the density of the emulsion the tracks are relatively short, at best few mm in length and therefore they must be studied only after enlargement or under a high power microscope.
(iv) These remain always sensitive to ionising particles and there is no way or method to trigger them by the particles one wishes to study, unlike cloud chambers.

8.2.7 Cloud Chamber

A cloud chamber is one of the most important instruments for researches on radiation. This was first used by C.T.R. Wilson in 1912 and is capable of giving visual observation of the tracks of the charged particles when theses pass through a gas vapour mixture. This instrument can be used for photographing the tracks of α-particles and that due to secondary ionisation effects accompanying the passage of X-rays and γ-ray through gases, and gives valuable information regarding cosmic radiation.

The physical principle involved is that of suddenly increasing a volume of air which is saturated with water vapour or vapour of any other liquid. Due to adiabatic expansion of this mixture, the temperature reduces. The condition of the surrounding is such that the space becomes supersaturated with vapours. The excess vapour will condense on small particles, i.e., ions or molecules, which are suitable centre for the condensation, forming a cloud. It is this droplet on the ion which is illuminated and photographed.

Consider a cylinder and piston-arrangement containing mixture of air and water vapour. Let V_0 and T_0 be volume and temperature of the mixture. If the vapour is in equilibrium with liquid surface at temperature T_0 then its pressure p_0 will be the saturated vapour pressure. The mass m_0 (in molecules) will be given by

$$P_0 V_0 = m_0 RT_0 \qquad \qquad ... (8.1)$$

Let the chamber be expanded adiabatically and let its volume become V. The resulting lower temperature T may be given by

$$T_0 V_0^{\gamma-1} = T_V^{\gamma-1} \qquad \qquad ... (8.2)$$

Where γ is specific heat ratio of the mixture. If no vapour condenses, the new pressure P would be given by

$$P'V = m_0 RT \qquad \qquad ... (8.3)$$

But at new temperature the saturated vapour pressure P would be less than P. Hence this is an unstable condition and the vapour will condense till its pressure reduces to P. If m be the mass of vapour remained, we can write

$$PV = mRT \qquad \qquad ... (8.4)$$

This condensation would take place on small particles or ion present and under suitable conditions, can have visible size. The condition may be achieved by supersaturation ratio σ, viz, the ratio of the vapour density which actually exists at stable equilibrium. Thus

$$\sigma = \frac{\dfrac{m_0}{V}}{\dfrac{m}{V}} \cdot = \frac{m_0}{m} = \frac{P'}{P} \tag{8.5}$$

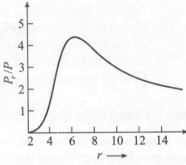

Fig. 8.32 Variation of $\dfrac{P_r}{P}$ with r

(using equations (8.3) and (8.4))

It is this expansion ratio which determines the growth condition of the condensation droplets. Further the saturated pressure for charged and uncharged drops are different. The variation of the ratio $\dfrac{P_r}{P}$, i.e., the ratio of saturated vapourr pressure for a drop of charge e and radius r to normal saturated vapour pressure, against r has been shown in Figure (8.32). It has been observed that corresponding to maximum of this curve is taken to be the supersaturation ratio, the droplet will grow in size, whatever its initial radius may be.

The schematic diagram of cloud chamber is shown in Figure (8.33). It consists of a glass chamber enclosed by glass plate on one side and piston on the other. The chamber contains air and a few drops of alcohol which readily provides vapour at the saturated pressure. To start with, the piston is pushed in, to compress the mixture. After a short interval of time the mixture and the space come to thermal equilibrium. Now on sudden release the piston produces an expansion. The distance to which the piston is released is fixed by the required supersaturation ratio. Due to cooling the supersaturated alcohol vapour condenses resulting in a cloud formation. At first the dust present in air causes the could to form all over the chamber. By maintaining a small field this is easily swept out, because dust particles and hence the water droplets carry electric charges. Now the chamber is ready for the experiment. When the ionising sample is introduced, the tracks of the ionising particle is rendered visible on the loss produced by the particle which can be illuminated and photographed. The chamber and the camera can be reset after clearing off the ions.

Cloud chamber's technique can be used to determine variation of specific ionisation along the track of charged particles and their ranges. A magnetic field can also be used along with the cloud chamber to measure sign, and momentum of a charged particle. To increase the probability of nuclear events, high pressure gas filling may also be used. Furthermore a three dimensional reconstruction is possible by using stereoscopic picture along with the cloud chamber technique.

Cloud Chambers are sensitive for a very short interval after the expansion stroke. A temperature gradient by using hot top and cold bottom, the vapour is allowed to diffuse downward continuously. With this arrangement supersaturation exists continuously and the device, which becomes continuously sensitive, has been developed. This is know as diffusion cloud chamber.

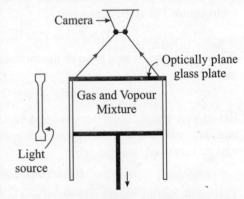

Fig. 8.33 Schematic diagram of cloud chamber

8.2.7(a) Modern Cloud Chamber

Figure 8.34 shows a modern cloud chamber. This essentially consists of large chamber A with an optically flat glass front G. For smooth and fast expansion a piston P supported by the rubber diaphragm RR is provided. There are two adjustable screws S_1 and S_2, which limit expansion and thus control the *expansion ratio*.

The space behind the piston P is closed with the help of a smooth value V and is filled with compressed air. With the release of V, air rushes into the chamber B and the piston P quickly completes the stroke. If then a charged particle passes through the chamber A during or immediately after the expansion, by condensing vapour its track is made visible. The tracks are photographed by stereo-cameras CC on being illuminated by high pressure Xenon lamps. This enables three dimensional events to be reconstructed.

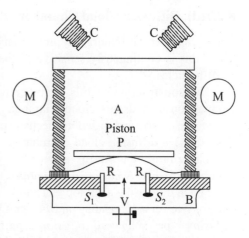

Fig. 8.34 Modern Cloud Chamber

Between the top and the floor of the chamber a small potential difference applied, immediately on completion of photographing the tracks, sweeps away any residual ions left and makes the chamber ready for the next event.

8.2.7(b) Blackett's Modification: Counter Controlled Cloud Chamber

When an expansion is made, chamber is not certain. To overcome this difficulty Blackett in 1932 provided an ingenious modification to the cloud chamber.

On the either side of chamber connected in coincidence are mounted two G.M. Counters MM (Figure 8.34). When a charged particle passes through the two GM counters, it has to pass through A and the output coincidence pulse from the GM counters, suitably modified, is made to operate the value V to expand the chamber A, switch on illumination, click the camera and wind the film ready for the next arrival. The ions are swept away from the chamber by an automatic 'switching on' of a clearing field after the photograph of the track is taken. Such an arrangement is called *counter controlled cloud chamber*.

8.2.7(c) Magnetically Controlled Cloud Chamber

Often cloud chamber are used in conjuction with magnetic fields. Such arrangement is exteremely useful in determining momenta of the particles and for their identification. This arrangement is called magnetically controlled cloud chamber.

The cloud chamber is placed in the magnetic field at right angles to the plane of the figure. The tracks get bent and the direction of bending and measurment of the curvature of the track would give the sign of the charge and the momentum of the particle respectively.

We may note that for determining the range and energy loss of particles and for observing cosmic ray showers and other such events, often, regularly spaced metal plates are introduced inside the cloud chamber.

8.2.7(d) Diffusion Cloud Chamber

In this chamber gas requires no preliminary expansion. Thus, this is continuously sensitive to the passage of swift particles.

The construction of this chamber is simpler in comparison to Wilson's cloud chamber. This has no *moving part, valve* and *re-setting* mechanism, etc. Figure 8.35 shows a diffusion cloud chamber.

The chamber contains a suitable gap-vapour (alcohol) mixture. A large vertical temperature gradient is set up in it by using a hot top plate and a cold bottom plate or vice versa, depending on whether it is *downward diffusion* type or *upward diffusion* type respectively. By keeping a vopour source (alcohol-filled trough) T, T near the hot plate a vapour density gradient is maintained.

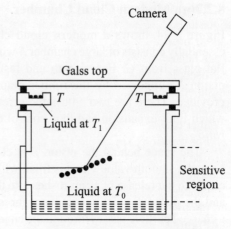

Fig. 8.35 Diffusion cloud chamber

The vapour is unsaturated in the warm region and as the vapour diffuses downward to regions of lower and lower temperatures it becomes supersatured. This means that the gas in a given region is supersaturated at all times and is continuously renewed by the diffusion of fresh vapour from other parts of the diffusion chamber. The location and the extent of the region of supersaturation depend on the geometric design as well as the temperature gradient in the chamber. The entire system is illuminated by a source of strong light and the track photographed by a camera at the top.

This is a simple and continuously sensitive. In comparison to Wilson type cloud chambers, the draw backs due to *convection* and *turbulence* are for less. For observing tracks the fraction of the total time useful is much larger than in the expansion chamber. Moreover, the *recovery time* being very small and can be further reduced by the application of an electrostatic field, the diffusion chamber becomes quite useful for use with pulsed accelerators.

This instrument cannot be operated at a background level higher than the cosmic ray level due to being continuously sensitive. It is not possible to introduce absorbers to record nuclear reactions as the sensitive region in this chamber does not extend to the surfaces placed inside it. The sensitive layer of this chamber is less than 7.5 cm thick and one can not make it vertical to detect cosmic rays.

One can operate this chamber with hydrogen gas of relatively high density which makes it possible to use it for meson production experiments.

We may note the Wilson cloud chamber has a glorious history in the field of high energy physics. This instrument has been responsible for the discovery of number of particles e.g. e^+, $\wedge$, μ^+, π^+ present in the cosmic ray showers.

Bubble Chamber

A cloud chamber is not suitable for full length track of high energy particle. Moreover theses track in photographic emulsions are so short that their magnetic curvature cannot be observed. Both of these disadvantages have been over come in bubble chamber. In a bubble chamber liquids having stopping power are used.

The principle of the bubble chamber is very simple. It is well known that a liquid can be heated above its boiling point without boiling if it is subjected to a pressure slightly greater than its saturation vapour pressure at the higher temperature. Now if the pressure is released, the liquid remains in super heated state. To this liquid if fast particles pass, they provide triggering points at which microscopic bubbles from, that starts boiling process. Thus when the fast particles pass through the super heated liquid, along its track it leaves a string of bubbles, and can be photographed. Thus instead of a liquid drop forms in a supersaturated vapour, as in cloud chamber, here in bubble chamber can easily be reset by applying the original pressure.

For working of the bubble chamber a large number of liquids can be used. The liquids preferred must be non conducting, transparent and have low surface tension and high vapour pressure. The liquids used are either pentane, liquid hydrogen propane etc.

Glacer, the inventor of the bubble chamber techniques (In 1952), used ether for the purpose and photographed the tracks of the tiny bubbles.

Figure 8.36 shows a simplified diagram of a bubble chamber. A thick walled glass chamber is filled with the liquid. The chamber is connected to a pressure system and heated to some temperature higher than its boiling point. This temperature and corresponding pressure are predetermined. High energy particle enter the liquid through a window. A release of pressure suddenly, followed by flash of light enable the photographs to be taken. A sharply defined trails of double on the path of particle will be obtained. If the bubble of double on the path of particle will be obtained. If the bubble chamber is located in a strong magnetic field, the particle charge and momentum relation can be obtained from the track curvature.

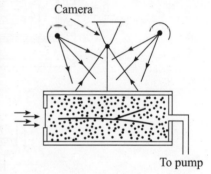

Fig. 8.36 Principle of Bubble chamber

One can show that the critical radius rc of the bubble is given by

$$r_c = \frac{2S}{p_v - p} \qquad \qquad ... (8.1)$$

Where p is external pressure acting on the bubble, σ is surface tension trying to reduce the bubble size and p_v is the vapour pressure tryingg to expand the bubble. The bubbles formed below the critical radius tend to diminish in size and disappear, while those above the critical radius continue to grow. The bubbles are initiated due to the secondary electrons (δ-rays) resulting from the primary ionisation.

A schematic diagram of a typical bubble chamber is shown in Figure 8.37(a) and 8.37(b). We may note that chamber is similar in construction to the cloud chamber. Bubble chamber are generally used in conjuction with electromagnets MM for determining charge and momentum. The magnetic field should be very high ($\sim 2T$ or more). Now a days, superconducting magnets are used.

For selecting a liquid for the bubble chamber, one should take care of the following points: (i) liquid should be *non-conducting* so that the ions retain their charges (ii) liquid should have *low surface tension* so that the force tending to collapse the bubble is weak and (iii) liquid should have *high vapour pressure* that would tend to enlarge each of the bubbles formed. Table 8.3 summarizes some important characterstics of few bubble chamber liquids.

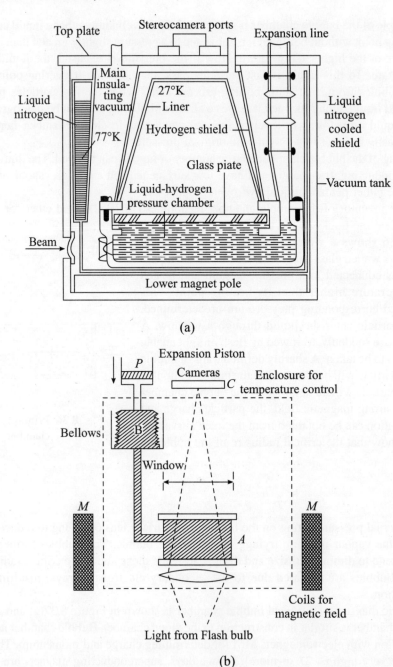

Fig. 8.37 Schematic diagram of a bubble chamber

Very large size hydrogen bubble chamber have been developed for experiments related to particle physics. Alvarez in 1959 built a bubble chamber 72″ long with a diameter 30″ containing 520 litres of liquid nitrogen at Lawrence Berkeley laboratory. Now, a days still bigger bubble chambers have been built. Modern bubble chambers are very complex and at the same time very expensive also.

Table 8.3 Important characterstics of few bubble chamber liquids

Liquid	Operating Temperature (°C)	Density (kg/m³)	Pressure (atm)	$\left(\dfrac{dE}{dx}\right)_{\min}$ (MeV/m)
Hydrogen (H₂)	−246	60	5	24
Deuterium (D₂)	−241	130	7	22
Helium (He)	−269	130	1	21
Propane	58	430	21	100
Pentane	157	500	23	–
Xenon	−20	2300	26	11,000

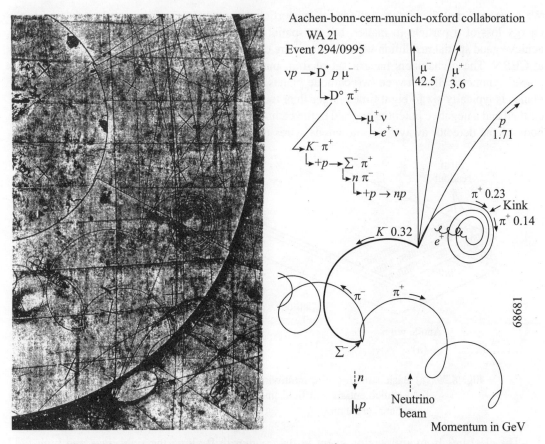

Fig. 8.38 Production of a heavy charmed meson D^* by a neutrino in the hydrogen-filled bubble chamber BEBC CERN. On the left can be seen the actual photograph with the complete interpretation of the event on the right.

In high energy physics laboratories, bubble chambers have now been replaced by purely *electron detectors* due to following reasons:

(i)   the chamber cannot be triggered,

(ii) the repetition rate of 1 – 10 Hz is low and a pulse of charged particles must be limited so that only about 30 tracks are formed,

(iii) event processing using views from three stereo cameras is inevitably slow, and

(iv) the bubble chambers are not suitable for use with colliding beam channels.

We may note that bubble chamber with its solid angle acceptance of 4π sr, it unsurpassed in its capability for track reconstruction. In addition, the bubble density along the tracks can be used for mass identification purposes over a certain range of β values. A rather special event, but one that well shows the full capabilities of the techniques, is shown in Figure 8.38.

8.2.9 Multiwire Proportional Chambers and Drift Chambers

We have read that the simple cylindrical proportional counter is extensity used to determine the energy loss of a particle in matter, but its spatial resolution is clearly limited by size. One can achieve good spatial resolution with the multiwire chambers developed by Charpak and Coworkers at CERN. These chambers have a set of thin, parallel, equally spaced anodes or 'sense' wires placed symmetrically between two cathode planes (Figure 8.39(a)). The gap between the cathode planes is typically six to eight times larger than the gap between the anode wires. With the anode earthed and a negative potential applied to the cathodes, a nearly uniform field develops in the main body of the detector away from the anode wires (Figure 8.39(b)).

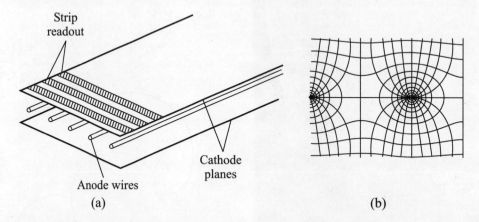

Fig. 8.39 (a) Basic structure of a multiwire proportional chamber, (b) view of the electric field equipotentials and field lines around the anode wires (wire spacing 2 mm, wire diameter 20 mm).

The electrons from an ionizing event in the counter drift along the field lines and high field region near the anode wire. The subsequent motion of the positive ion space charge in the field induces a negative pulse on the wire. Corresponding positive signals are produced on all other electrodes which largely compensate the negative signals produced by capacitative coupling. Each sense wire therefore acts as an independent counter and the chamber as a whole is continuously sensitive. Today, track-sensitive detectors are generally multiwire structures operated in proportional mode. They range from small high precision devices to large, multi-square metre chambers with tens of thousands of read-out channels. A spatial resolution of approximately 0.7 mm is possible if the anode puses are used but if the cathode plane is in strip from a determination of the centre of the cathode pulses can yield resolutions of the order of 50 mm.

The position of the primary ionization column caused by the passage of a charged particle may be determined by measuring the time for collection on the sense wire using an arrangement similar to that shown in Figure 8.40. The wider anode wire spacing in such *drift chambers* leads to a decrease in the number of channels needed for operation. Spatial resolutions of about 100 mm are possible in such chambers, but they are only useful when detection rates are low.

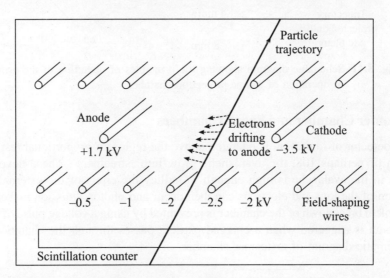

Fig. 8.40 Schematic diagram showing the principles of operation of a drift chamber. The electrode configuration creates an almost uniform drift field so that the relation between track distance from the anode wire and drift time is linear.

The *time projection chamber* (TPC) is a large drift volume with a single planar multiwire chamber. One can obtain the complete track images by drifting the trajectory ionization on to the wire plane as shown in Figure 8.41. Proportional wires and a segmented cathode provide coordinates perpendicular to the drift direction and time information determines the coordinates along the drift direction. Particle identification by $\frac{dE}{dx}$ measurment is possible, although because of Landau fluctuations many independent samples of energy loss must be taken into account.

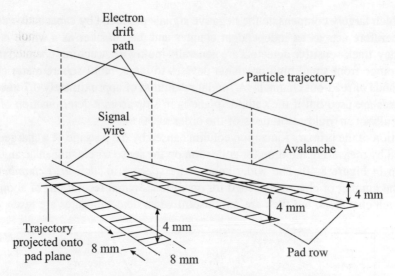

Fig. 8.41 Schematic diagram showing the principles of operation of the principles of operation of a time-projection chamber.

8.2.10 Streamer Chamber and Spark Chambers

If a gaseous detector of ionisation is operated above the region of proportional response, with an amplification of perhaps 108, the avanlanche grow into 'streamers'. These develop when the space charge in the avalanche is high enough to nullicy the effect of the external field locally. Recombination of the ions and electrons can then occur and photon emission makes the streamer visible. Complete breakdown of the chamber is prevented by using a voltage pulse of short duration ($\approx$ 10 ns) which is triggered when a charged particle passes through the chamber. Figure 8.42 shows a schematic diagram of a steamer chamber.

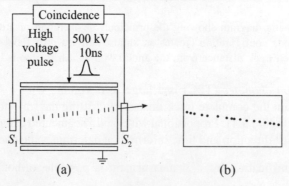

Fig. 8.42 Illustration of the principles of operation of a streamer chamber. A coincidence in scintillation counters S_1 and S_2 triggers the high-voltage pulsar causing streamer development in the chamber volume. The tracks are normally viewed transparent electrodes along a direction parallel to the electric field when they appear as a series of dots reminiscent of a bubble chamber photograph. (a) view normal to the electric field; (b) view parallel to the electric field.

If a longer pulse is applied to the chamber the streamers grow into *sparks* and such discharge between closely spaced electrodes have also been used for track recording.

Figure 8.43 shows a schematic diagram of modern spark chamber. It consists of a set of equally spaced identical plane conducting plates, usually of aluminium, mounted horizontally in an air tight chamber containing a mixture of Ne-He gas a pressure of 1 atm. Spark chamber is also provided (not shown in Figure 8.43) with a counter and a logic circuit for determining the arrival of a particular particles. The alternate plates are connected to a high d.c. voltage source (~ 20 kV).

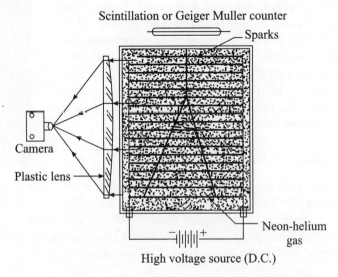

Fig. 8.43 Schematic diagram of a spark chamber

The number of plates in a typical chamber varies from 25 to 100. The area of each plate is about 1 m^2 and thickness 1 mm accurately. The spacing between any two successive plates is about 3 to 5 mm. When a high voltage is suddenly applied to alternate plates, while other plates are left at ground potential, causes a very high electric fields across the gaps. Then a spark breakdown of the gas takes place being initiated by the trails of ions produced in it by the ionising particle traversing the chamber. The trayectory of the particle is recorded by a series of sparks that jumps from one plate to the next close to the path of the particle. The light from the spark is quite intense and the spark trails are photographed spectroscopically. Thus, inside the chamber the three-dimensional path of the particle may be re-projected for measurments like those used in cloud chambers or bubble chambers.

A field (~ 10 kV/m) is then applied between the plates to sweep away the unwanted ions from the gas and the chamber is ready for the next firing.

The sensitive time of a spark chamber is very short of the order of 0.5 μs and hence the spark chamber can be operated in an intense beam (up to about 10^6 particles/s) and can still record single particles.

Like bubble chamber, the photographs of the sparks can be analysed and the track coordinates can be digitzed with the help of an image tube. Another method of determining the track coordinates to use transducers for picking up of sound of the spark at the edge of the chamber. One can also register multiple tracks (Figure 8.44) by following the actual particle trajectories.

The tracks in a spark chamber show broken appearances. However, if the interplate gaps are made relatively wider (~ 0.4 m), then the tracks may appear somewhat continuous. In such case, one may possibly identify case, one may possibly identify the vertices of different events.

Similar to bubble chamber, magnetic analysis of the tracks is also possible in the case spark chamber. This yields information about the sign of the charge of the ionizing particle and its momentum. The spark chambers although cheaper to build and simple in construction but not as accurate as the bubble chambers in yielding information. No doubt, this detector can be triggered but because of the need to clear ions after the spark the time resolution and repetition rate are low and spark chambers are now rarely used.

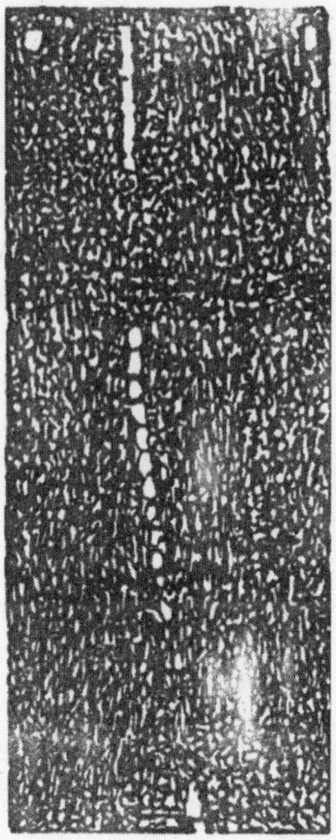

8.2.11 Total Absorption Calorimeters

Fig. 8.44 Multiple tracks in a spark chamber

In the particle detectors so far described, a particle loses only a small fraction of its total energy. In a total absorption calorimeter, however, a particle interacts and produces a cascade of shower particles in a block of sense material. If a section of the calorimeter is sensitive to scintillation light, Cherenkov radiation or ionization a signal proportional to the total energy can be received. The uncertainty in such an energy measurment

arises from statistical fluctuations in the shower development and the energy resolution $\dfrac{\Delta E}{E}$ improves

with increasing energy as $E^{-\frac{1}{2}}$.

If we assume that the rate of energy deposition is proportional to the particle energy then it is clear that the size of a calorimeter scales only logarithmically with energy, whereas a magnetic

spectrometer scales as $p^{\frac{1}{2}}$ for a given momentum resolution $\dfrac{\Delta p}{p}$. Calorimeters can be used to

detect neutral particles as well as charged particles and if the detector is segmented the position and direction of an incident particle can also be measured. In general signals are formed on a time scale between 10 and 100 ns and may be used for real-time event selection. Identification is also possible employing the specific response of the detectors to electrons, muons and hadrons.

The main types of calorimeters are:

(i) **Electromagnetic Calorimeters**   In these calorimeters, a shower is initiated by a high energy electron or photon and proceed through successive processes of pair production and bremsstrahlung. As the shower develops the number of particles increases exponentially with depth until it reaches a maximum when the particle energy is of the order of the *critical energy EC*. The scale of longitudional development is set by the *radiation length*. The transverse shower dimensions are determined by the radiation length and by the multiple

Coulomb scattering of the low energy electrons in the shower. In homogeneous electromagnetic calorimeters the whole detector is sensitive. Commonly used materials are lead glass and NaI (Tl); in the former, Cherenkov radiation from electrons and positrons is detected and in the latter, scintillation light is detected. In heterogeneous calorimeters there are alternate layers of passive absorber and active material such as plastic scintillator, in which about 10 % of the total energy is deposited.

(ii) Hadronic Calorimeters In these calorimeters, the shower development is similar in principle to the electromagnetic case but is more hadronic shower development is governed

by the nuclear absorption length $\lambda = \dfrac{A}{N_A \, \rho \sigma}$, where A is the atomic mass number, N_A the

Avogadro number, ρ the density and σ the cross section for inelastic collisions, materials commonly used are liquid argon, Fe and uranium, with λ values for nucleons of 80.9, 17.1, and 12.0 cm respectively. In contrast with the electromagnetic case about 30 % of the incident energy is dissipated in reactions which do not produce an observable signal owing to leakage of muons, neutrinos or slow neutrons for instance. These calorimeters are heterogeneous and if plastic scintillator is used as the active medium it is coupled to photo multipless via a wavelength shifting guide (Figure 8.45).

Alternatively, one may use direct charge collection read-out with or without amplification of ionization.

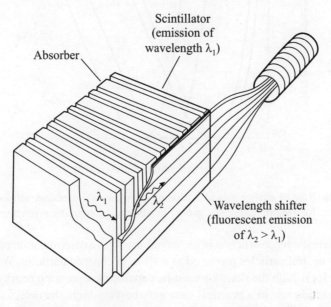

Fig. 8.45 Calorimeter readout using wavelength shifters.

In high energy collision many particles can be produced and the basic aim of the detector is to identify them and to measure their momenta. If the particles are produced in a thick target a large variety of purpose-built detectors may be used for the study of interactions with colliding beams detectors can not be so easily changed and tend to be large multipurpose installations. The term detector is applied both to such installations and to the individual particle-sensitive components from which they are assumbled.

In most experiments it is necessary to measure coordinates along the trajectories of charged particles. If a magnetic field is employed, the curvature of trajectory determines the sign of charge and the momentum of the particle. For identification by mass determination, a simultaneous measurment of momentum and velocity, or of momentum and energy, must be made. Figure 8.46 illustrates the use of relativistic rise in collision loss for this purpose.

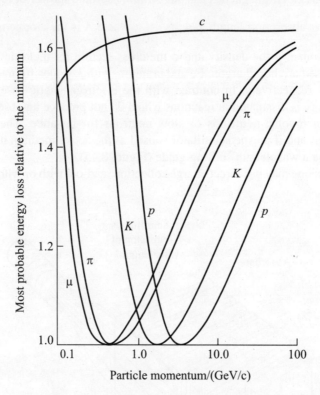

Fig. 8.46 Rate of energy loss in a mixture of argon and methane, relative to the minimum, for electrons, muons, pions, kaons and protons, as a function of momentum.

A complete kinematic reconstruction of an 'event' involves a determination of energy-momentum four-vectors of any neutral particles produced as well as of charged particles. When the multiplicity of secondary particles is high the detector must be capable of separating nearby particles in space, and, since the interaction rate in a practical case may be very high, the detectors may need a good time resolution. A typical apparatus in high energy physics experiment involves many types of component detector, each performing one or more of above tasks. We may note that particle detectors also rely on the loss of energy of a charge particle in matter. The collision-loss phenomena led to the following general categories:

(i) **Multiwire proportional counters and streamer or spark chambers**   There rely on ionization of gases. The miniature 'microstrip' silicon detectors, which offer enormous advantages in complex installations required good spatial resolution, are based on electron-hole pairs in semiconductors

(ii) **Scintillation Counters**   There detect light flashes associated with primary ionisation or excitation processes. They are mainly used to provide time-reference signals and triggers for other detectors in particle physics experiments, though the high proton detection efficiency and good energy resolution NaI(Tl) makes it ideal for use in total absorption shower counters.

(iii) **Bubble Chambers**   These rely on the deposition of energy along the track of a particle in a superheated liquid. Although now declining in importance they have had a highly significant as well as aesthetic role in particle physics.

(iv) **Nuclear Emulsions**   These record particles because of the production of developable grains along the track of the particle. They have had important applications in cosmic ray physics because of their small size and integrating property. They were used in 1947 in the discovery of the pion and of the $\pi - \mu$ decay sequence but they now have only some what special applications in particle physics. Such as the decay of short-lived particles, e.g. *B*-mesons.

Radiative processes such as *bremsstrahlung Cherenkov radiation* and *transition radiation* are the basis of the detectors. Photons themselves and heavy neutral particles are detected by observation of the charged particles that they excite as secondary radiations in passing through matter. Neutrinos can only be detected as a result of their weak interactions with nuclei.

Although traditionally the problem of particle detection has involved the characterization of stable particles, γ, e, μ, p, n, $\wedge$, Σ, ... $\wedge_c$, ..., π, K, D etc., there is now increasingly a change of emphasis. The events, particularly at colliding beam machines, have characterstic 'jet' structures caused by fragmentation of heavy quarks and gluons and often contain energetic leptons and exhibit 'missing' energy due to escaping neutrinos. The traditional magnetic moment analysis now has to be supplemented to precise global measurments of event structure involving determinations of energy and momentum flow among multiple jets of particles. Calorimeters are ideally suited to this task and play an important part, in the large hybrid detectors ALEPH, DELPHI, L3 and OPAL now in use with the large electron-positron collider LEP at CERN. The object of these installations above all is to test the predictions of the *standard model of electroweak interactions* and of *gange theories of the strong interaction*.

8.3 PARTICLE ACCELERATORS

Important informations regarding the nuclei of matter and the different particles in its constitutions can be had by studying the interaction of matter with high energy particles. With the energetic particles a large variety of nuclear reactions can be made to occur. The machines used to accelerate or energize the atomic particles are known as *particle accelerators*. Probably the most important application of the behaviour of charged particles in electric and magnetic fields lies in the construction of accelerators to produce beams of high-energy particles which are used for many types of scattering experiments. These machines are playing important role in conducting most modern research work in particle physics concerning artificial transmutation and over other particles. Now a days atomic particles of almost any desired energy, can be obtained with the suitable machines. The particles used as projectiles are: electron, proton, deutron, α-particle, cosmic rays, γ-ray, etc. In recent times the nuclei of light atoms, e.g. Li, Be, B, C and N have also been artificially accelerated. These particles have been obtained by ionisation of appropriate gas.

After the particles being produced it is required to accelerate them to high energy of the order of keV, MeV or BeV. In some accelerators particles, during their acceleraton move along linear path, whereas in some other particles are made to traverse spiral or circular paths. The former process is known as *linear acceleration* while the latter process is called the *orbital acceleration.*

The accelerators now used for nuclear structure studies may be classified broadly into those which develop a steady accelerating field (electrostatic generators) and those in which radio frequency electric fields are used (linear accelerators, cyclotrons, synchrotrons, microtrons). As interest moves into the 100 – 1000 MeV range of energies only accelerators in the second class can be considered; and for linear accelerators and cyclotrons in particular use of super conducting techniques to reduce radio frequency power input then offers great advantages.

8.3.1 Ion Source

Ion source is main part of an accelerator, in which ions are produced by ionisation of a gas. The ions are in turn accelerated. Thus the gases used for ionisation are, hydrogen for protons, deuterium for deutrons helium for α-particles etc. For obtaining electrons, hot cathode is used, in which intensity of electron is being controlled by the filament temperature i.e., inturn filament current. At the beginning positive ions were produced by gas discharge, in which perforated cathode, allowed flow of positive ions on the other side of discharge tube kept at low pressure. But this type of source was defective as regards to larger energy range, low ion current and lesser life of the electrodes. Hence the development of hot cathode are took place. Using hot cathode a copiou's supply of electron can be obtained and arc can be maintained with a small voltage between cathode and anode. By using capillary arc source and an insulated tube having fine hole, positive ion along radial or axial direction can be obtained by producing arc discharge in small region. Such a source is useful in Cockcraft Walton and Van-de-Graff machines.

M. Von Ardenne developed an efficient ion source known as Duoplamastron. This is capable of giving a proton current of 500 mA through 10^{-3} m hole. In this source anode structure consists of poles of electromagnet, producing intense magnetic field in discharge region and the source can be operated at low pressure. The electrons emitted by hot cathode are used to ionise the gas.

By using a radio-frequency oscillator and ring electrode, an electrodeless discharge also serves the purpose of a good ion source. In some accelerators and cyclotrons positive ions are produced by accelerated ions and stripping of two or more electrons.

In certain time of flight instruments instead a continuous ion beam, a pulsed beam (10^{-8} sec. pulse) is required. In such cases, a high frequency electric field is applied for ion sweeping, which gives pulsed beam.

8.3.2 The Tandem Electrostatic Generator

In the famous 1932 experiment of *Cockcroft and Walton* in which the ^{7}Li nucleus was disintegrated for the first time by laboratory-accelerated protons, a transformer-rectifier-capacitance-voltage multiplying circuit (Figure 8.47) was used to produce protons of velocity v given by

$$\frac{1}{2}m_p v^2 = eV \qquad \qquad \dots (8.1)$$

where V was the output voltage of the generator on load (≈ 500 keV). This method can be used for voltages upto 5 MeV but for general nuclear structure research the most versatile and powerful accelerator is now the electrostatic generator developed by Van-de-Graff in 1931.

8.3.2(a) Cock-Craft-Walten and voltage generator

In Cock-Craft-Walten generator the principle of rectification of alternating current is used. It consists of a high frequency transformer T, two sets of condensers of equal capacity and a set of similar high voltage rectifiers. The principle of generator is shown in Figure (8.47 *a, b, c, d*).

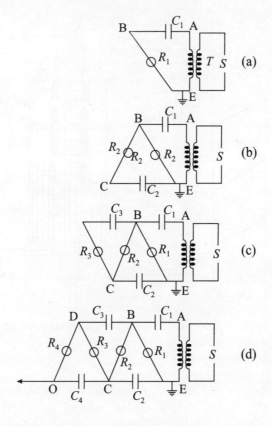

Fig. 8.47 Principle of Cock-Craft-Walten generator

Consider Figure (8.47a), in which the point A of secondary of transformer would be at alternating potential say V, w.r.t. earth. The voltage on the condenser C_1 would be peak of this voltage on secondary on transformer i.e., $\sqrt{2}\,V$. Thus the point B would be at voltage $\sqrt{2}\,V$ w.r.t. A and $2\sqrt{2}\,V$ w.r.t. earth. Now if another condenser C_2 is connected to this system (see Figure 8.47b), this would be charged to a potential $2\sqrt{2}\,V$ w.r.t. earth, and the potential of C_2 would be double of that of C_1. When the condenser C_3 connected to the system (Figure 8.47c), this would be at potential $2\sqrt{2}\,V$ w.r.t. B and $4\sqrt{2}\,V$ w.r.t. earth. This in turn charges the condenser C_4 to a potential $4\sqrt{2}\,V$. Thus the maximum voltage at the point O reached to $4\sqrt{2}\,V$, Increasing the number of condersers and the rectifiers, the voltage can further be increased. If there are n condensers on output side the voltage, which can be obtained, will be $2\sqrt{2}\,nV$. A voltage of the order of 10^6 V may be obtained with this generator.

Figure 8.48 illustrates a three stage Cockcrof Walton cascade accelerator.

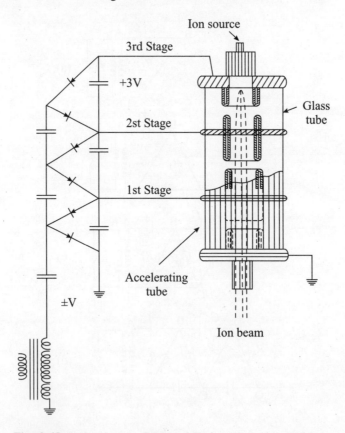

Fig. 8.48 A three stage Cockcrof Walton cascade accelerator

If there are N stages, the final high voltage terminal attains a voltage equal to 2 VN. Due to stray capacitances, and other leaks, the voltage falls significantly above 500 kV. To avoid this, one may use two transformers and two capacitors. This arrangement is called *symmetrical cascade rectifier*. The high voltage terminal is connected to the accelerating column which acts like the load. If a current I is drawn at the output stage and f be the frequency of the power supply or (a.c. voltage), there is a voltage drop across the last condenser, given by

$$\Delta V_0 = \frac{I}{Cf}\left(\frac{2}{3}N^3 + \frac{1}{4}N^2 + \frac{1}{12}N\right) \qquad \dots (8.2)$$

The terminal voltage will have a ripple R given by

$$R = \frac{IN(N+2)}{16\,fC} \qquad \dots (8.3)$$

One can decrease the ripple either by increasing the value of C or f and increasing f to RF values which is more convenient than having big capacitances which may cause the associated insulation problems.

Solid state rectifiers are used in modern Cockcroft-Walton generators since these do not require any power supply for their operation and one can operate them without difficulty within a pressure vessel housing the generator. These generators are quite suitable for delivering large ion currents (~ 1 mA) at voltages about 1 MV. However, the voltages has fluctuations of the order of 1 % which is relatively large.

8.3.2(b) Van-de-Graaff Generator

The first generator of Van-de-graaff type capable of giving 25 mA current at 1.5 MV. The principle of this generator is very simple. If a charged conductor is connected internally to an outer uncharged conductor, whole of its charge would be transferred to the outer surface of the hollow conductor, and this would not be dependent on the potential of the hollow conductor. In this way the potential of hollow conductor can be increased to a limit determined by leakage of charge by the machine.

Figure 8.49 shows the main components of a typical two stage *tendon Van de Graaff generator*. Referring first of all to the lower half of the Figure 8.49, charge is conveyed to the terminal by an insulating belt or by a 'load ertron' of conducting bars insulated from each other but forming a flexible chain. Charge is transferred to the belt or chain at the low potential and by a corona discharge from a spray points or by electrostatic induction. At the terminal end the transfer process is reversed and by allowing the terminal pulley to reach a higher potential than the terminal itself, negative charge may be conveyed to the downgoing charging track. A sectionalized evacuated accelerating tube traverses the terminal and the insulating stack structure that supports it and terminates at earth potential. The stack structure itself is a series of equipotential surfaces separated by rigid of equipotential surfaces separated by rigid insulators and connected by resistors. It provides a uniform axial field, within which the accelerating tube is contained.

Electrostatic generators are normally enclosed in a pressure vessel filled with an insulating gas such as sulphur hexafluoride to allow the axial voltage gradient to be raised to about 2 MV/m. Pressure of up to 10 atmospheres are used in large machines.

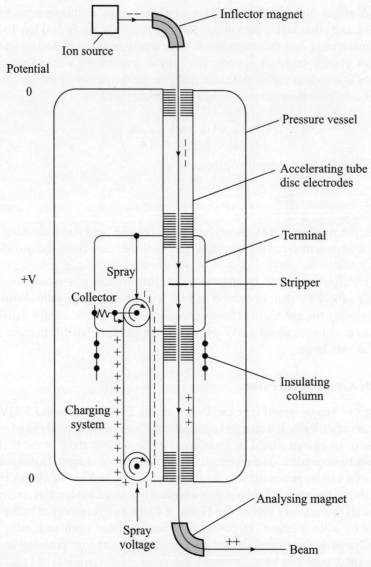

Fig. 8.49 A schematic diagram of two stage tandem Van de Graaff generator. The lower half is essentially a single ended machine and the addition of the upper half with an extra tube and stripper converts it to a tandem accelerator.

In the early electrostatic accelerators ions were injected into the accelerating tube from a gaseous discharge tube housed in terminal. In the tandem accelerator, however shown in Figure 8.49, the ion source is at earth potential and produces negative ions, e.g. H⁻, He⁻, ¹⁶O⁻. These are accelerated to the terminal through an extension of the accelerating tube already mentioned. In the terminal, H⁻ ions for instance, are moving with a velocity v given by (8.1) and passage through a thin stripper, e.g. a carbon foil of thickness about 50 μg/cm² or a tube containing gas at low pressure, removes the extra electron and a further electron as well so that a positive ion H⁺ is available for acceleration to earth potential from the terminal, yielding an energy of 2 eV. Negative ions of heavier elements may be stripped to a positive charge state qe in the terminal and the final velocity in this case corresponds to acceleration through a potential $(1 + q)V$; additional strippers in the positive ion tube can be used to increase q.

By controlling the charging current from an error signal derived from the accelerated beam itself, the energy of an electrostatic generator may be stabilized. This passes through an slit with insulated jaws, from which a difference signal is derived if the beam deviates from the axis.

Energy definition to about 0.5 keV is possible in this way with a spread about the mean energy of not more than $1 - 1.5$ keV even in the largest accelerators of, say 20 MV terminal voltage applying proton currents of, say, 10 mA. Thus type of performance is ideal for nuclear reaction experiments.

These accelerators are particularly useful in the case of heavy ion acceleration, e.g., suppose a beam of O^+ ions is first converted into a beam of O^- ions, which is accelerated to 10 MeV energy in the first stage of a 10 MV Van de Graaff generator. In an electron stripper these negative ions are then reconverted into positively charged oxygen ions.

The ions which are so produced carry different multiples of electronic charges, e.g. O^+, O^{2+}, O^{3+}, O^{4+}, O^{5+} etc. After passing through the remaining portion of the accelerator tube, ions gain additional energy of $10n$ MeV, where ne is the charge on the ion. Obviously, the final energy of the positive ion beam becomes $10(n + 1)$ MeV. O^{5+} ion attains the energy of about 60 MeV in this accelerator. We may note that these machines cannot accelerates atoms which cannot be made into negative ions.

8.3.2(c) Pelletron Accelerators

A more recently developed concept of Van de Graaff generator known as *pelletron* differs in design of the charging mechanism. The charging belt in it is replaced by a charging chain, which consists of steel cylinders joined by links of solid insulating material such as nylon. The chain is intrinsically spark protected. The metal cylinders are charged as they leave a pulley at the ground potential and as they pass over a pulley within the high voltage terminal the charge is removed.

The maximum operating voltage range beam current capabilities range from a few microamperes to about 0.8 mA. The machines are enclosed within pressure tanks filled with SF_6 above ~ 1 MV voltage.

These machines are more stable and at the same time better energy resolution. These machines accelerate most of the heavy ions except those that cannot be produced into negative ions. There are two pelletrons in India: (i) at TIFR having dome voltage of 14 MV and (ii) at Nuclear Science centre, Delhi whose dome voltage is 15 MV. New a days, many of the older Van de Graff generators have been converted into pelletrons.

8.3.2(d) Folded Tandem Accelerators

Now a days folded version of tandem accelerators have been developed so that the size of the accelerator is reduced and also cost of building and insulating gas is minimised. In this version, both acceleration columns are put in same single insulating column with a 180° magnet in the terminal to steer the beam from one tube into the other tube. This type of design not only provides a compact system but it also locates both the ion source and sample chamber near the system control panel.

8.3.3 Linear Accelerators

It was recognized early that the extension of direct voltage methods to the production of very high particle energies would ultimately encounter insulation problems. The possibility of successive re-application of the same moderate electric field to a particle beam was therefore studied and led to the development of the linear accelerator.

8.3.3(a) Drift Tube Linear Accelerators

The earliest heavy particle accelerators were of this type. Figure 8.50 illustrates the principle of Sloan and Lawerence linear accelerator developed in 1931. In this accelerator, operated at about 30 MHz, a number of field-free drift tubes of length L_1, L_2, ... L_n, separated by small accelerated gaps, were connected alternately to the output terminals of an oscillator of free-space wavelength λ. The length of the drift tube is such that the field in the gap just reverses in the time that a particle takes to pass from one gap to the next gap. If the voltage across each gap at the time of passage of the particles is V then the particle energy at entry to the drift-tube numbered n is neV (for an assumed initial injection energy of eV) and the particle velocity is

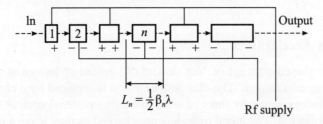

Fig. 8.50 Principle of Linear accelerator due to Sloan and Lawerence (1931).

$$v_n = \sqrt{\frac{2\,neV}{m}} \qquad \qquad ...\,(8.1)$$

where m is the mass of the particles being accelerated. The frequency of the oscillator is $\dfrac{c}{\lambda}$ and for a time of flight of a half cycle the length of the drift tube n must therefore be

$$L_n = \frac{1}{2}\frac{v_n\,V}{c} = \frac{1}{2}\beta_n\lambda \qquad \qquad ...\,(8.2)$$

so that for non-relativistic energies ($v \ll c$),

$$L_n \propto n^{\frac{1}{2}} \qquad \qquad ...\,(8.3)$$

and
$$v \propto \sqrt{V} \quad \text{(for the ion of constant charge and mass)} \qquad ...\,(8.4)$$

It also follows that if the energy gain per gap is held constant the accelerator length is directly proportional to wave length. The particles emerge in bunches corresponding closely with the times of appearance of the gap voltage V, at which resonance is possible.

The apparent requirement that the drift-tube structure should be designed for exact resonance with the accelerating beam was realized to be unnecessary following the enunciation of the principle of phase *stability* by McMillan and Veksler in 1945. In its application to linear accelerator (Figure 8.51(a)) this principle considers a particle that crosses a gap with a phase angle ϕ_0 with respect to the accelerating voltage waveform (point A). If ϕ_s corresponds to voltage V (Equation 8.1) for which the drift tube structure is designed, then the particle arrives at the next gap with the same phase angle. Let particles, however ($\phi > \phi_s$, points B) receive a larger acceleration in the gap, traverse the drift tube more quickly and move towards point A in phase at the next gap.

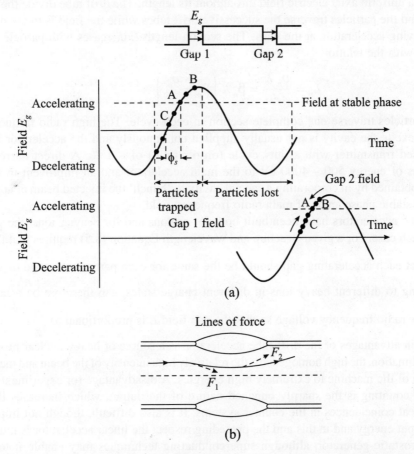

Fig. 8.51 (a) Phase stability in a drift-tube ion accelerator. The dots show the phase angles with respect to the gap field of a bunch of particles of uniform velocity arriving at gap 1. At gap 2 there is increased bunching about the stable phase ϕ_s. (b) Radial defocusing of particles passing through a cylindrical gap in a field increasing with time.

Similarly, early particles ($\phi < \phi_s$, points C) will be less accelerated and will also move towards A in phase. Particles corresponding to point A thus have stable phase, and all particles with phase angles within a certain range of ϕ_s will be trapped and will oscillate about the point of stable phase. Latitude is therefore possible in the mechanical tolerances that must be applied to the drift tube structure.

The desirable feature of axial (phase) stability leads to radial instability because the stable phase point is on the rising part of the voltage wave. Figure 8.51(b) illustrates this; because the field increases as the particle traverses the gap, the defocussing force predominates radial stability is now generally restored by means of quadrupole magnets within the drift tubes themselves.

The original mode of excitation indicated in Figure 8.50 is used only for low energy accelerators. For energies above, say, 5 MeV for protons, the availability of high powers at microwave frequencies makes it highly desirable to use the resonant cavity type of excitation introduced by Alvarez (Figure 8.52). The cavity is tuned by radius rather than length and is excited in the lowest longitudional mode, with a uniform axial electric field throughout its length. The drift tube divide the cavity into n sections and the particles traverse the successive drift tubes while the field is in the decelerating phase, receiving acceleration at the gaps. The section length L increases with particle velocity in accordance with the relation

$$ L_n = \frac{v_n \lambda}{c} = \beta_n \lambda \left(\beta_n = \frac{v_n}{c} \right) \qquad \qquad \dots (8.5)$$

since the particles traverse one complete section in each cycle. The high radio frequency power required to excite the cavity is not usually supplied continuously and the accelerator is operated from a pulsed transmitter with a duty cycle (on-off ratio) of $\sim 1\ \%$. A direct current injector supplies ions of energy 500 – 4000 eV to the main accelerator and an improvment in intensity is sometimes obtained by in corporating a special cavity to 'bunch' the injected beam at appropriately the selected stable phase angle of main radio frequency field.

Drift-tube accelerators have been built both for protons and for heavier ions; the principle is similar in each case. For a given structure and wavelength Equation (8.5) requires that the velocity increments at each accelerating gap should be the same for each particle. A range of values $\frac{ze}{m}$, corresponding to different heavy ions in different charge states, can therefore be accelerated by adjusting the radio frequency voltage so that the gap field E is proportional to $\frac{m}{z}$.

The main advantages of the drift-tube accelerator as a source of heavy nuclear projectiles are the good collimation, the high homogeneity, the relatively high intensity of the beam and the possibility of extension of the machine to extremely high energies. A disadvantage for experiments requiring coincidence counting is the sharply bunched nature of the output, which increases the ratio of random to real coincidences in the counter systems. It is also difficult, though not impossible, to vary the output energy and in this and the preceding respect the linear accelerator is much inferior to the electrostatic generator, although superconducting techniques may enable it to approach continuous operation.

Nowadays, two types of linacs are available: (i) travelling wave type and (ii) the standing wave type, depending on the coupling of the cavity and the r.f. power phase shift per cavity in operation. In the first type the coupling between adjacent cavities results in a phase shift less than 180° and in the second type the energy is strongly reflected from the end of each cavity chain and the cavity oscillations are 180° out of phase. We may note that early electron Linacs are of first type where as all proton Linacs are of second type.

8.3.4 Wave guide Accelerator

It has been stated above that electron cannot be accelerated in drift tube accelerator. A wave guide pattern can be used for the electrons. In a wave guide accelerator electric field is established by the use of wave guide which simply is a pipe of conducting material. In free space electromagnetic waves propagate in such a way that the electric and magnetic fields are always in the plane perpendicular to the direction of propagation, whereas in a wave guide these waves propagate with a transverse magnetic field and an axial component of electric field. It is this axial electric field which is used to accelerate the electromagnetic waves in wave guide. The phase of electromagnetic wave in a simple wave guide is usually greater than the velocity of light. Hence to match these velocities a smooth wave guide cannot be used for the purpose. But along the length of the wave guide a series of metal discs, with holes in the centre are placed as shown in Figure 8.53(a) and (b). The electron beam is injected at one end of the disc loaded guide and passes through the hole of the metal discs (Figure 8.54). At the initial stage the electron velocity is adjusted of the order of half the velocity of light i.e. $\approx \dfrac{c}{2}$. This corresponds to electron energy 79 keV. In travelling a short length, the beam acquires the velocity of light. The ratio-frequency signal is produced by oscillator and amplified by Klystron amplifier.

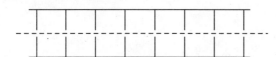

Fig. 8.53 (a) Loaded wave guide

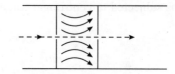

Fig. 8.53 (b) Electric field lines

First wave guide linear accelerator had been developed by D.W Fry and his collaborators in England in 1946. Hausan produced electron beam of energy 6 MeV. At Stanford University such an accelerator with a length 220 ft. and diameter 3 inches has been divided into 21 sections. Using Klystron amplifier with high frequency power (3×10^9 cycle per sec.), the electron obtained with this accelerator can also be used to accelerate proton. It should also be noted that a given machine can only be used to accelerate on type of particles only.

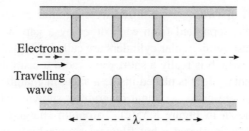

Fig. 8.54 Disc-loaded circular wave guide

Wave guide accelerators are especially suitable for electrons since these particles have a velocity 0.98c for an energy of only 2 MeV, which may easily be provided at injection by an electrostatic accelerator.

8.3.5 Cyclotron

The cyclotron is a magnetic-resonance orbital accelerator developed by Lawrence and his Collaborators at Berkely in 1930 s. After Cockcroft and Walton's success in disintegration experiments with high velocity protons in 1932 the cyclotron rapidly became a powerful competitor in the field, with the particular advantage that it did not require generation of a high voltage.

Cyclotron is first orbital type of accelerator in which charged atomic particles move under the action of a strong magnetic field, in circular paths. The important feature of the machine is that, when the particles are to be accelerated, the phase of the accelerating field is always such as to speed them up. The machine is capable of giving energetic ions of larger intensity than that of linear accelerator. Because the principle of this accelerator depends upon resonance of electric and magnetic field, it is called as magnetic resonance accelerator. But the name cyclotron is much more popular as the acceleration of particles involves the cyclic motion.

To understand principle of cyclotron, consider a charged particle of charge q and mass m, moving with uniform velocity v, in a transverse magnetic field B. This transverse magnetic field will exert a force on the particle perpendicular to plane of v and B both. The magnitude of this force is $q v B$. Under the influence of this force, the particle will move along a circular path; and will be subjected to centrifugal force given by $\dfrac{mv^2}{r}$, where r is radius of curvature of the path. Thus

$$\frac{mv^2}{r} = q v B \quad \text{or} \quad \frac{v}{r} = \omega = \frac{q}{m} B \qquad \text{... (8.1)}$$

Where ω is angular velocity of rotation of particle.

The time T required by the particle in completing one revolution will be given by

$$T = \frac{2\pi r}{v} = \frac{2\pi m}{eB} \qquad \text{... (8.2)}$$

This equation shown that irrespective of the velocity of the particle and for given value of B and $\dfrac{e}{m}$, the time of revolution remains constant.

The cyclotron consists of two short, hollow, semi-circular cylinders D_1 and D_2, mounted rigidly along their diameter, separated from each other by a gap as shown in Figure 8.55(a). Because of their shapes, these semicircular cylinders are called "Dees". The dees are connected to a high frequency oscillator, which is really a short wave radio transmitter supplying energy to the dees. The whole arrangement of dees is placed inside a vacuum chamber within the pole pieces of a strong magnet. The complete cross sectional diagram is shown in Figure 8.55(b). Schematic diagram of cyclotron is shown in Figure 8.55(c). To obtain charged particles, say protons, say protons, at the centre of dees is placed a hot filament-from which, by passing electric-current, thermions (electrons) can be obtained. Now when a trace of hydrogen gas is admitted to the evacuated chamber, these electrons ionise some of the hydrogen atoms by collision, producing protons.

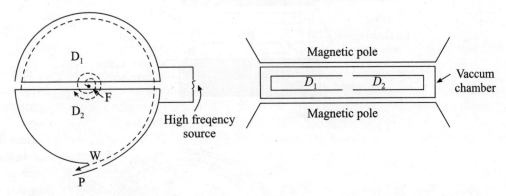

Fig. 8.55 (a) Position of dees of a cyclotron (b) Cross-sectional diagram of cyclotron

Consider protons in the neighbourhood of F, and the dees are connected to high frequency oscillator. At the instant when D_1 is negative and D_2 is positive, the protons will be accelerated towards D_1. As soon as the protons enter the space inside D_1 the electric field would not interact its motion and under the influence of strong magnetic field their path would be circular. If after making half revolution, the potential difference between the dees D_1 and D_2 is reversed so that D_1 becomes positive and D_2 negative and the protons reach in the gap between D_1 and D_2, these would be accelerated towards D_2, (as D_2 negative) and repelled by D_1 causing the protons to increase in their speed. Hence inside the space of D_2 the particles will move in a circular path of larger radius. Now again after completing half revolution in D_2 as the protons reach in the gap, again the polarity of dees reverses and the protons would be accelerated towards D_1.

In this way every time when the protons reach in the gap, these would be accelerated every time because of the electric field. When the radius of the path reaches to maximum, with the help of negatively charged plates, these protons can be made to pass through a narrow opening W. Due to negative charge the path of the protons straightens and emerge from cyclotron in pulses because of alternating potential difference which is inevitable.

In the cyclotron alternating potential difference and magnetic field are so adjusted that by the time the proton moves along semi-circular path inside the dees, their polarity reverses. Thus

$$\text{frequency of electric field } f = \frac{\omega}{2\pi} = \text{cyclotron frequency} = \frac{qB}{2\pi m} \qquad \dots (8.3)$$

As the protons emerge from the cyclotron, their velocity and hence energy would be maximum. At the same time the radius of curvature of the path equals that of radius R of the dees. Thus

$$B q \upsilon_{max} = \frac{m\upsilon_{max}^2}{R} \quad \text{or} \quad \upsilon_{max} = \frac{BqR}{m} \qquad \dots (8.4)$$

Hence maximum energy E_{max} is given by

$$E_{max} = \frac{1}{2}m\upsilon_{max}^2 = \frac{1}{2}\frac{B^2 q^2 R^2}{m} \qquad \dots (8.5)$$

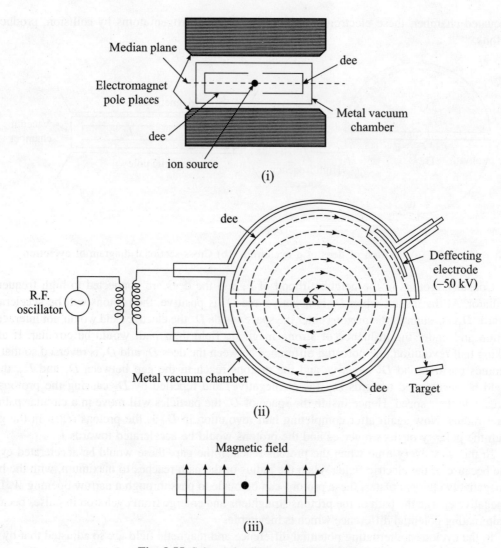

Median plane

dee

Electromagnet
pole places

Metal vacuum
chamber

dee

ion source

(i)

dee

Deffecting
electrode
(–50 kV)

R.F.
oscillator

S

Metal vacuum chamber

dee

Target

(ii)

Magnetic field

(iii)

Fig. 8.55 Schematic diagram of Cyclotron

Thus we see that in cyclotron the energy can be increased to any amount by increasing cyclotron radius. However, it is not possible. Because, as the velocity of the particles increases, the mass of the particles also increases according to the relation.

$$m = m_0 \left(1 - \frac{v^2}{c^2}\right)^{-\frac{1}{2}} = \gamma m_0 \qquad \qquad \dots (8.6)$$

where

$$\gamma = \left(1 - \beta^2\right)^{-\frac{1}{2}} \text{ with } \beta = \frac{v}{c}$$

The frequency of rotation

$$f = \frac{qB}{2\pi\gamma m}$$

Relativistic mass increases with velocity and limits the maximum energy is obtainable, it is because of this reason, electrons cannot be accelerated by cyclotron.

Lawrence's first cyclotron with magnetic pole diameter 2.5 inches provided proton of energy 80 keV. At Berkeley California U.S.A., so many cyclotrons of different diameters of pole pieces have been built. A 60 inch pole piece cyclotron is capable of giving deutrons at 24 MeV and Helium atom to about 40 MeV. The magnetic field applied in 60 inch cyclotron was 1.6 Webers/meter^2 or 16000 gauss.

If the field is uniform azimuthally, but non uniform radially, the axial component at radius r may be given by

$$B_z = B_0 \left(\frac{r_0}{r}\right)^n \qquad \ldots (8.7)$$

where B_0 is the flux density at a reference radius r_0, for which it is reasonable to assume that $\gamma \left(= \dfrac{v}{c}\right) = 1$ and the field index is

$$n = -\frac{r}{B_z} \frac{\partial B_z}{\partial r} \qquad \ldots (8.8)$$

The frequencies of radial oscillation about an equilibrium orbit of radius r_0 are

$$f_z = f_0 \ (n)^{\frac{1}{2}}$$

$$f_r = f_0 \ (1-n)^{\frac{1}{2}} \qquad \ldots (8.9)$$

For the orbit to be stable n must be between 0 (radially uniform field) and 1.

The total energy of a particle occuping such a stable orbit is (setting $c = 1$), we have

$$E = \gamma m = (p^2 + m^2)^{\frac{1}{2}}$$

$$= \{(qBr)^2 + m^2\}^{\frac{1}{2}} \qquad \ldots (8.10)$$

but for extreme relativistic energies

$$T = E = p \qquad \ldots (8.11)$$

To increase the particle energy it may be given a series of properly timed pulses from an electric field. Lawrence noticed that for non-relativistic motion ($\gamma = 1$) the cyclotron frequency f is given by the Equation (8.3) is constant and equal to f_0, i.e. the motion is *isochronous*. The electrical impulses can therefore be derived from an oscillating field of frequency f_0, in resonance with the cyclotron frequency. This is achieved in practice by the use of D shaped electrode in the vacuum system in which the particles move.

In a uniform field field cyclotron, bunching of ions from the central source occurs early in the acceleration process and is maintained, at about a 10 % duty cycle during the passage to maximum radius, the stability condition $0 < n < 1$ requires a radially *decreasing* magnetic field (Figure 8.55(e)). Unfortunately this destroys *isochronism* which is in any case prejudiced by the gradual decrease of γ above 1 as the particle velocity increases. The ion bunch therefore gradually loses phase with respect to the accelerating field and would in the end begin to suffer deceleration. The conventional azimuthally uniform field cyclotron is therefore limited in output energy-in particle to about 30 MeV for protons.

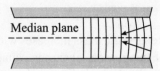

Fig. 8.55 **(e)** Axial forces in a radial decreasing cyclotron field.

Magnetic Focussing in Cyclotrons

The basic cyclotron resonance condition applies strictly to the ions moving in an orbit helix situated in the median plane (central plane) of the cyclotron chamber and crossing the gaps in perfect synchronism with the R.F. electric field applied to the dees. Such particles experience only an inward radial force from the magnetic field. However, the ions, being accelerated, deviate from this ideal condition and move up and down the median plane and so have all chances of being drifted to the dees and be lost in case of an uniform magnetic field. Thus the probability of ions reaching the target shall be vanishingly small. This is a very undesirable effect. Obviously it can be remedied by supplying restoring forces to bring the ions back to the median plane. These restoring forces can be provided by the main magnet itself if the magnetic field, produced by the magnet, decreases in magnitude from the centre outwards. This is called magnetic focussing. In such a field the magnetic lines of force will be concave inward (see Figure 8.55d).

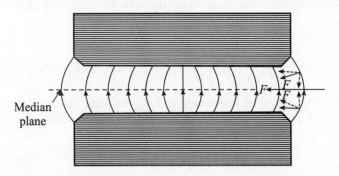

Fig. 8.55(d) Shimmed pole faces for magnetic focussing

An ion moving in the median plane in such an outwardly decreasing magnetic field, will experience only an inward force confined wholly to median plane. But the ions moving above or below the median plane experience forces that have radial as well as vertical components, of course the vertical component is always directed towards the median plane. Larger the deviation from the median plane, larger is this force, i.e. the force is proportional to the displacement of the ion from the median plane – a true characterstic of all types of restoring forces. The ion under the action of this restoring force excutes small amplitude operations in a vertical plane. These are known as '*betatron oscillations*'. Thus magnetic focussing results from a small decrease in the magnetic field with increasing radius. For small cyclotrons which use relatively high 'dee voltage' the decrease of magnetic field at the periphery is about 3 to 4 % of the field magnitude at the centre. For median energy cyclotron (upto 20 MeV), it may be 2 % and it should be still smaller (about 1 %) for cyclotrons, reaching relativistic limit of energy.

8.3.6 Synchro Cyclotron (Frequency Modulated Cyclotron)

The most important concept, upon which the operation of Syncro cyclotron and also of synchrotrons and linear accelerators, for accelerating the ions in the relativistic energy range depends is the principle of *phase stability*. Because of the principle of phase stability it is possible to design a synchrocyclotron in which synchronism at an accelerating phase can be preserved by a programmed decrease of the frequency f of the electric field, it is preferable to use a magnetic field shape that permits the radial *increase* required for isochronism, but provides some extra focusing forces.

The radial variation of the mean field required is given by

$$B = \gamma B_0 = B_0 \left(1 - \beta^2\right)^{-\frac{1}{2}} = B_0 \left(1 - \frac{v^2}{c^2}\right)^{-\frac{1}{2}} \qquad \text{... (8.1)}$$

and approximately this gives

$$B = B_0 \left(1 + \frac{1}{2}\beta^2\right) = B_0 \left(1 + \frac{r^2 \omega_0^2}{2c^2}\right)$$

$$= B_0 \left(1 + \frac{2\pi^2 f_0^2 r^2}{c^2}\right) \qquad \text{... (8.2)}$$

The extra focusing is obtained, following a suggestion of L.H. Thomas by introducing *azimuthal variation* into the field, also known as *sector focusing*. In such a field there are alternate high and low field regions. A closed orbit in this field is non-circular and as a particle in such an orbit crosses a sector boundary (either high → low or low → high) radial components of velocity arise. The field variation at the boundaries gives rise to *azimuthal* components and the new $v \times \mathbf{B}$ force is axially focusing in both types of transition region. The sector focusing may be improved by using spiral instead of radial sector boundaries.

The phase stability in cyclotron for certain orbits may be explained as follows:

It is clear form above equation that as kinetic energy of particles increases the frequency of rotation decreases. Thus for a particle in phase stable orbit, the motion will be for specific energy. At this value of energy only its frequency of rotation would be equal to frequency of alternating voltage applied. The corresponding orbit would be phase stable orbit for this value of energy, and the ion enter the D gap, when the instantaneous field in it is zero and just at the verge of decelerating. Under the condition the ion continue moving for long in the orbit. This condition is shown in Figure (8.56) at points O, T, 2T. etc.

If the ion enter the gap prior to zero phase at A with positive phase, it would be accelerated and its frequency would decrease. That being so it would take long to return in the gap as has been shown at B and C, in next cycles. Thus is a few cycle it would return in state of zero phase with greater energy. When the ion enters the gap in decelerating field, the decrease in energy in it would be accompanied by increase in frequency, returning it to zero phase state. Hence about equilibrium please at $\theta = 0$, the ion would oscillate on the sides of the stable orbit.

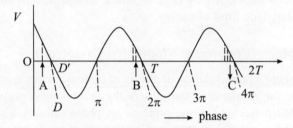

Fig. 8.56 Principle of phase stability.

Thus when the ion moves along stable orbit, its energy remains constant. In synchro cyclotron the dee voltage is decreased and by constant magnetic field the radius of stable orbit is increased to get more energetic ions. This is why, the machine is also called frequency modulated cyclotron.

For maintaining phase stability the frequency variation in synchro cyclotron must be slow and continuous. Hence the frequency modulated source is used. In addition as high electric field is not necessary, the small gap between the dees is also not required. Hence only one dee is provided. The electric field is applied between this dee and the earth. The Figure (8.57) shows the arrangement of synchro cyclotron.

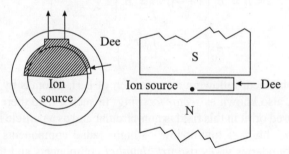

Fig. 8.57 Synchrocycolotron

A synchro cyclotron has following advantages over a cyclotron:
 (i) Since there is no necessity of providing high electric field, hence there is no difficulty in electrical insulation.
 (ii) The machine is simpler mechanically and there is sufficient space for putting ion source, target etc. inside the machine.
 (iii) By minimizing the distance between poles of electromagnet, B can be increased.
 (iv) For better focussing of accelerated ions the surface of poles may be cut in such a way that magnetic flux density is less from centre towards out side.
 (v) The efficiency of high frequency value oscillator is greater.

Due to frequency modulation only a portion of ions is brought to phase stability. Hence the out beam of ions is limited and an accelerated ion pulse is obtained.

The protons, the deutrons, α-particles etc. can be accelerated by synchro cyclotron. In Berkeley California University with 184" poles, Synchro cyclotron 200 MeV deutron, and 400 MeV α-particles have been obtained. For this a frequency modulation from 11.5 million C/s to 9.8 million C/s was done. The frequency modulation from 23 million C/s to 15.6 million C/s had been used for producing protons of energy 300 MeV. In the later case dee voltage and magnetic field were 15 kV and 15,000 gauss respectively. In Columbia University, with 164" Synchro cyclotron, nuclear charge distribution had been studied, using high energy μ-meson. Using high energy ions from synchro cyclotron, important results have been obtained during researches on atomic nuclei.

We may note that due to low mass and therefore rapidly increasing γ under acceleration, electrons cannot be accelerated in synchro-cyclotron too. Phase stable machines with a varible magnetic field, however, can accelerate either electrons or protons, and following the development of powerful focusing techniques these *synchrotrons* have become of outstanding importance as particle accelerators.

8.3.7 Microtron or Electron Cyclotron

The relativistic limitation of the ordinary cyclotron may also be overcome if the increase in energy at each acceleration is so large that the revolution time increases by one radio frequency period $\dfrac{1}{f_0}$. From relation

$$f = \frac{\omega}{2\pi} = \frac{qB}{2\pi \gamma m} \qquad \qquad \dots (8.1)$$

We can see that this requires increments of γ, i.e. an energy increase of mc^2, per turn. This is practicable for electrons ($m_e c^2 = 511$ keV) and microtron uses this principle (Figure 8.58).

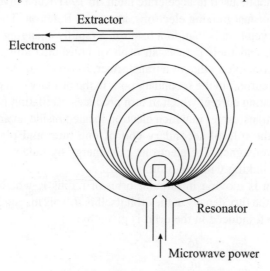

Fig. 8.58 Successive electron orbits in microtron.

Electrons pass through a cavity excited from a pulsed radio frequency source of frequency about 3000 MHz. The corresponding magnetic field from Equation (8.1) with $\gamma = 1$, is $= 0.107\ T$ and this field is maintained over a large is maintained over a large area bounded by cavity. After N transits the total energy has become $(N + 1)\ m_e c^2$ and the orbit radius is given by

$$evB = \frac{\gamma m v^2}{r}$$

as
$$R_N = \frac{(N+1) m_e c^2}{e B c} = \frac{(N+1) m_e c}{eB} \qquad \qquad \text{... (8.2)}$$

since the electron velocity rapidly approaches c.

The orbits are widely spaced and extraction is thus busy. There is phase stability during the acceleration and the beam is monochromatic to about 0.1 %.

Microtron built with a 'race-track' geometry use a linear accelerator as the accelerating section and with superconducting techniques continuous operation is possible. Such a facility is very suitable for electron scattering experiments at energies upto 1 GeV as an alternate to a large linear accelerator since the continuous mode of operation allows the use of coincidence counting techniques.

If the electron beam from microtron is allowed to strike a heavy target bremsstrahlung photons will be produced, with each photon there is a reduced energy electron and if the energy of this electron is determined by magnetic analysis, the energy of the associated photon is known. Thus a spectrum of '*tagged*' *photons* is provided and can be used, in coincidence with a signal from the tagging electron, for photo disintegration 3.8.

8.3.8 Betatron: The electromagnetic induction accelerator

It has been stated above that due to large relativistic increase in mass of the electrons at low energy, a cyclotron can not be used to accelerate them. In 1941 Kerst invented another accelerator to accelerate β-particles or fast moving electrons, known as Betatron. The fundamental difference between a betatron and cyclotron is that, in a betatron a rapidly changing magnetic field is used to accelerate the electrons, and their path is an orbit of constant radius. The action of betatron depends upon the same principle as that of a transformer, in which an alternating current applied to a primary coil, induces a current in the secondary coil. In the primary coil oscillating magnetic field is produced due to alternating current which in turn produces oscillating potential in the secondary. In this way betatron is a transformer in which electron located inside an annular, doughnut-shapted vacuum chamber play the role of secondary coil of one tum, and placed within the poles of laminated steel of elecrtromagnet. The electrons gain energy by induction, because of the change of the flux with ϕ, time, linking with the orbit.

Consider an electron is moving in circular orbit of radius r, where the total magnetic flux through the orbit is f and the flux density at the orbit itself is B. This magnetic field B is perpendicular to orbital plane. The e.m.f. induced in the orbit is given by:

$$\varepsilon = -\frac{d\phi}{dt} \qquad \qquad \text{... (8.1)}$$

The work done on the electron of charge-e, in one revolution is given by $-\varepsilon e$ which equals $e\frac{d\phi}{dt}$.

If F is tangential force on the orbit, the work can also be given by $F\, 2\pi r$. Hence

$$F = \frac{e}{2\pi r} \frac{d\phi}{dt} \qquad \qquad \text{... (8.2)}$$

Due to influence of this applied force the energy of the electron will increase and hence it would try to move to an orbit of larger radius. It an orbit of constant radius is to be maintained, this increase in radius of curvature of the orbit, be resisted. A radial force due to magnetic field and acting perpendicular to the direction of motion of the electron keeps it moving in the orbit of constant radius. This inward radial force is BeV, and must be equal to the centrifugal force $\dfrac{mv^2}{r}$. Thus

$$Bev = \frac{mv^2}{r}$$

or
$$mv = Ber \qquad \qquad \qquad ... (8.3)$$

The tangential force on the electron is for the orbit of constant radius, is given by:

$$F = dt(mv) = \frac{d}{dt}(Ber) = er\frac{dB}{dt} \qquad \qquad ... (8.4)$$

Hence from equation (8.2) and (8.4)

$$\frac{e}{2\pi r}\frac{d\phi}{dt} = er\frac{dB}{dt} \quad \text{or} \quad d\phi = 2\pi r^2\, dB$$

Integrating
$$\int_0^\phi d\phi = 2\pi r^2 \int_0^B dB \quad \text{or} \quad \phi = 2\pi r^2 B \qquad \qquad ... (8.5)$$

This is know as *betatron condition*. The result shows that the total flux ϕ within the radius of the orbit r, must be twice the value which would be obtained if flux density B were uniform over whole area of the orbit and also this should be proportional to B.

This condition holds for relativistic energies as well as non-relativistic energies. Such a flux distribution is obtained by specially shaped countered pole-piece faces, where the flux density at the centre of the orbit is greater than it is at the circumference of the orbit.

The construction of betatron is illustrated in Figure 8.59(a) and (b). The electrons are produced from a hot filament and are given primarily acceleration by the application of electric field of the order of a few thousand volts. These are then injected in the highly evacuated doughnut tube DD' made of glass or ceramic at G. In the coils of electromagnet an alternating current from mains supply of normal frequency say 50 c/sec., is passed. Thus increasing flux in a given direction is obtained for quarter cycle only. This is shown in Figure 8.59(c).

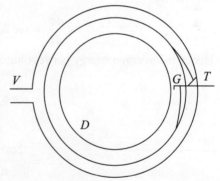

Fig. 8.59(a) Formation of a coherent wave front of Cherenkov radiation

At an instant when magnetic field is just rising from zero in first quarter cycle, induces a potential in the doughnut tube, increasing the energy of the electrons. When the strength of the field passes its maximum and starts decreasing in the next quarter cycle, the direction of induced e.m.f. reverses and the electrons slow down. To avoid this effect the electrons are removed from the stable orbit. This is done by discharging a capacitor through the primary coils or through auxiliary coils around primary, at the instant when electron received the desired amount of energy. Due to this extra current momentarily the total flux φ increases rapidly, but flux density *B* increases less rapidly, and causes orbital radius *r* to increase. Thus when the radius of final orbit increases, the beam can be made to hit the target *T*, producing hard *X*-ray or γ-rays depending upon their energies, in pulses.

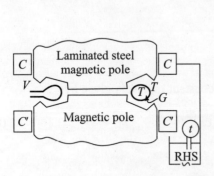

Fig. 8.59(b) Position of D.T. in the pole
pieces of electro-magnet

Fig. 8.59(c) Acceleration cycle in a between

The energy of the electron beam can be calculated from the average induced e.m.f. and the total number of revolutions made by electron. Let the flux variation be given by

$$\phi = \phi_0 \sin \omega t$$

Where ϕ_0 is amplitude and ω is angular frequency of changing flux. Hence the energy per revolution is given by

$$e \frac{d\phi}{dt} = \omega e \, \phi_0 \cos \omega t \qquad ... (8.6)$$

Hence the average energy per revolution

$$= \omega e \, \phi_0 \, \frac{1}{\dfrac{T}{4}} \int_0^{T/4} \cos \omega t \, dt$$

$$= \frac{2 \omega e \, \phi_0}{\pi} \qquad ... (8.7)$$

This represents energy per revolution. To calculate number of revolutions, we can approximately consider these to travel with the velocity of light *c*. Because electron of energy 1 MeV moves with velocity 0.94c. Furthermore these accelerate only during quarter of alternating cycle i.e., for time

$$\frac{1}{4} \ T \ \text{viz,} \ \frac{1}{4} \cdot \frac{2\pi}{\omega} = \frac{\pi}{2\omega} \ .$$

$$\text{Distance traversed in this time} = \frac{c\pi}{2\omega}$$

$$\text{Distance traversed in one revolution} = 2\pi r$$

$$\therefore \qquad \text{Number of revolution} = \frac{c\pi}{2\omega} \cdot \frac{1}{2\pi r} = \frac{c}{4\omega r} \qquad \qquad \text{... (8.8)}$$

Hence total average energy acquired by the electrons in accelerating quarter cycle is given by

$$E = (\text{No. of revolution Energy per revolution})$$

$$= \frac{c}{4\omega r} \frac{2\omega e \phi_0}{\pi} = \frac{ec\phi}{2\pi r} \qquad \qquad \text{... (8.9)}$$

With betatrons electrons at 300 MeV have been obtained. The 100 MeV betatron of General Electric Research Laboratories has 66 inches orbit of electron and 76 inches pole face diameter. This is operated with 60 C/sec. alternating current. At the University of Illinois electrons with energies more than 300 MeV have been obtained.

Any accelerated charge particle loses energy by radiations. Hence an electron also loses energy by radiation and the rate of Loss of radiating energy depends upon four the power of the energy, it is acquiring. This in turn sets limitations on the energies to which electrons can be accelerated in a betatron. The maximum energy is reached when the energy lost per turn by radiation equals the maximum energy acquired by the electron per turn under practical conditions maximum electrons energy can be reached to about 1000 MeV. To obtain higher energies one uses a linear accelerator or a synchrotron.

Magnetic Focussing and Orbit Stability in Betatron We have had a qualitative discussion of the magnetic focussing in cyclotron which is similar in many respects in all the orbital accelerators and so also in a betatron. As we have seen there that magnetic focussing can be achieved by an outward radial decrease in the magnetic field. We shall develop here the mathematical theory.

The equilibrium orbit radius r_0 of the electron in a betatron is held constant despite the fact that the electron-momentum continuously increases. The equilibrium orbit is confined to the median plane and the magnetic flux is supposed to have axial symmetry.

For vertical restoration of the electron to the equilibrium orbit, the magnetic field must decrease in the radial direction along the median plane. The vertical component (z-component) of the magnetic field in the neighbourhood of the equilibrium orbit is say given by the relation.

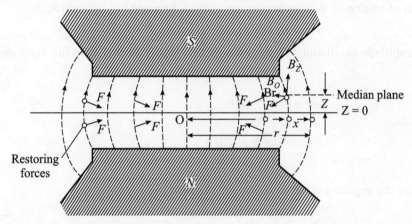

Fig. 8.59(d) Restoring forces acting on the particles above or below the median plane in a radially decreasing magnetic field.

$$B_z = B_0 \left(\frac{r_0}{r} \right)^n \qquad \qquad \ldots (8.10)$$

Where B_0 is the value of B_z at the equilibrium orbit ($z = 0$) and a positive constant called the field index. The field at the orbit has two component- a radial component B_r and a vertical component B.

For small variations r from r_0, i.e. $\frac{r - r_0}{r_0} \ll 1$, we can have Taylor series expansion of equation (8.10) and to a first approximation, obtain

$$B_z = \left(1 - n \frac{r - r_0}{r_0} + \ldots \right) \qquad \qquad \ldots (8.11)$$

B_r can now be determined from the Maxwell equation, $\nabla \times \mathbf{B} = 0$, which in our case give

$$\frac{\partial B_r}{\partial z} = \frac{\partial B_z}{\partial r} \qquad \qquad \ldots (8.12)$$

Differentiating Equation (8.11) with respect to r, we get

$$\frac{\partial B_z}{\partial r} = -n \frac{B_0}{r_0} \qquad \qquad \ldots (8.13)$$

Combining Equation (8.13) with Equation (8.14),

$$\frac{\partial B_r}{\partial z} = -n \frac{B_0}{r_0} \qquad \qquad \ldots (8.14)$$

Integrating this equation with respect to z and remembering that $B_r = 0$ at $z = 0$, we get

$$B_r = -n \frac{B_0}{r_0} z \qquad \qquad \ldots (8.15)$$

The equation of motion of the electron in the vertical direction i.e. z-direction, is

$$m z = B_r \, ev \qquad \qquad \ldots (8.16)$$

For small amplitude oscillations B_r is given (8.15). Hence in this case the equation of motion becomes

$$m z = -n \frac{B_0}{r_0} z e v$$

or

$$z = -n \left(\frac{B_0 e}{m} \right) \left(\frac{v}{r_0} \right) z \qquad \qquad \ldots (8.17)$$

We know that the angular cyclotron frequency is given by

$$\omega_0 = \frac{B_0 e}{m} = \frac{v}{r_0} \qquad \qquad \ldots (8.18)$$

Hence equation of motion (8.17) reduces to

$$z = n \, \omega_0^2 z \qquad (8.19)$$

Which represents vertical oscillations of angular frequency

$$\omega_z = \omega_0 \sqrt{(n)} \qquad \ldots (8.20)$$

Next we write down the equation of motion in the radial direction. Let the electron be displaced radially in the medium plane so that its instantaneous displacement 'r' is given by $r_0 + x$, i.e. $r - r_0 = x$, where $x \ll r_0$, r_0 being the radius of the equilibrium orbit. In terms of x, equation (B) becomes

$$B_z = B_0 \left(1 - \frac{nx}{r_0} \right) \qquad \ldots (8.21)$$

The equation of motion in the radial direction may be written as

$$mx = \frac{mv^2}{r_0 + x} - Bzev$$

or

$$x = \frac{v^2}{r_0 + x} - \frac{B_z ev}{m}$$

$$= \frac{v^2}{r_0} \left(1 + \frac{x}{r_0} \right)^{-1} - \frac{B_0 ev}{m} \left(1 - \frac{nx}{r_0} \right)$$

$$= \frac{v^2}{r_0} \left(1 - \frac{x}{r_0} \right) - \frac{B_0 ev}{m} \left(1 - \frac{nx}{r_0} \right)$$

or to first order appoximation.

$$= \left[\omega_0 \left(1 - \frac{x}{r_0} \right) - \omega_0 \left(1 - \frac{nx}{r_0} \right) \right]$$

From equation (8.18)

$$= -v[1 - n]x \, \frac{\omega_0}{r_0}$$

$$= -\omega_0^2 [1 - n]x, \text{ since } v = \omega_0 r_0 \qquad \ldots (8.22)$$

which represents radial oscillations of angular frequency

$$\omega_r = \omega_0 \, (1 - n)^{\frac{1}{2}} \qquad \ldots (8.23)$$

The Equations (8.19) and (8.22) represent oscillatory motion in vertical and radial directions respectively, of frequencies given by Equation (8.20) and (8.23) provided the coefficients to z and X are positive, i.e. $1 > n > 0$. This means that the field must decrease from the centre to periphery. These oscillations are called the *betatron oscillation*. The frequencies of both the oscillations are always less than ω_0. n is farther from zero, stronger is the vertical restoring force and n is farther from 1 stronger is the radial restoring force. In the cyclotron radial osciallations are relatively unimportant, but in a betatron both the vertical and radial oscillations must be small, since if the oscillations exceed the dimensions of the cavity, the electron beam will strike the walls and be lost.

8.3.9 Electron Synchrotron

The betatron has the disadvantage that a large magnet is needed to supply variable flux to the accelerating electrons. In a synchrotron, first proposed by McMillan and Veksler, this has been avoided by using radio-frequency electric field to accelerate the electron, when they have already been accelerated to energies about 2 MeV. To achieve 2 MeV energy betatron principle is used. At this energy, the speed of electrons is about 0.98c, and hence they travel practically at constant speed but the energy is imparted to them in terms of increase in mass.

For the electron moving in circular orbit of radius r with a velocity v, we have

$$\frac{mv^2}{r} = Bev \quad \text{or} \quad \omega = \frac{v}{r} = \frac{Be}{m} \qquad \qquad ... (8.1)$$

Where ω is angular velocity and B is magnetic flux density at the orbit in superimposed perpendicular magnetic field.

From the equation (8.1) it is clear that if ω is to remain constant at relativistic speed, then B, must be increased in the same ration as m, the mass of the electron.

Figure 8.10(a) and (b) shows constructional details of electron-synchrotron. Instead of massive magnets in betatron, in an electron synchrotron, a ring shaped magnet with flux bars are used. This maintains the stable orbit of the electron as in the betatron. In between the poles of the electromagnet is placed the doughnut vacuum chamber. When the electron have been accelerated by the betatron action to an energy of about 2 MeV the flux bars become magnetically saturated ceasing the betatron action. The acceleration of the electrons beyond the energy 2 MeV is achieved by a radio frequency oscillator. The radio frequency electric field is provided between electrodes on the inner walls of doughnut tube. In one of the design, the electrodes consists of a silver coating on the inner walls of small angular sector of doughnut tube with a short accelerating gap across which radio frequency oscillator is connected (Figure 8.60). Thus by selecting proper phase for the accelerating voltage with its proper frequency matching with the frequency of revolution of electron in the stable orbit, it can be made to cross the gap at right moment, to gain the same amount of energy in each revolution, showing phase stability. This demands that the time of one revolution, showing phase stability. This demands that the time of one revolution of electron in the orbit must be equal to the period of radio-frequency oscillator. Furthermore when the electrons have attended the maximum output energy and the magnetic field reaches maximum, the radio-frequency supply is switched-off. As the magnetic field still continues a little, the radius of the orbit decreases and the beam can be made to hit the target, T, producing hard X-rays, in pulses.

The electrons are injected into the doughnut tube of the electron synchrotron in same way as that used in betatron. The preliminary acceleration in an electrostatic field given in a Van de Graff generator or a linear accelerator. It is important to note that the same final energy can be attained by the electrons than that in the betatron. Thus loss of energy due to radiation is less as compared to the betatron.

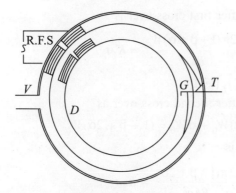

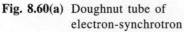

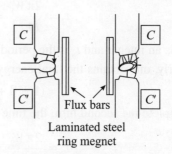

Fig. 8.60(a) Doughnut tube of electron-synchrotron

Fig. 8.60(b) E*l*ectron synchrotron

The 70 MeV electro-synchrotron of the General Electric Company U.S.A has magnet to 8 tons in weight. The electron injected at energy 50 keV, have the orbit diameter 59.2 cm. At 2 MeV synchrotron action starts. The frequency of radio-frequency oscillator, at which it is operated is 163 Mc/sec. with a peak of 1000 V. The operating frequency of the electromagnet is 60 c/sec. The electron of energies more than 300 MeV have been obtained. At Massachusetts Institute of Technology has an electro-synchrotron of 50 ton magnet and radio frequency oscillator of 46.5 Mc/sec. The orbits is 40 inches. Electrons are injected at 80 keV accelerated to 7 MeV by betatron action and finally to 330 MeV by synchrotron action.

Above 100 MeV, acceleration of electrons is usually done in linear accelerators. These accelerators have been used up to an energy of 600 MeV. For accelerating protons, synchrotron machine are also used. One injects protons in these machines after accelerating them to energy ~ 20 MeV.

Energy attained

Let the electrons are injected with energy βW_0, where $W_0 = m_0 c^2$ (rest mass energy of the electron). Each time the electrons are passing through the resonator and the gain in kinetic energy $\Delta W_k = \alpha W_0$, a constant fraction of the rest energy.

Now, the total energy in first cross-over of resonator is

$$W_1 = W_0 + \beta W_0 + \alpha W_0 = (1 + \beta + \alpha)W_0 \qquad \text{... (8.2)}$$

In the magnetic field B, the radius of the orbit is

$$r = \frac{mv}{Be} = \frac{Wv}{Bec^2} \qquad (\because W = mc^2) \qquad \text{... (8.3)}$$

Where v is the velocity of the electron.

To traverse a complete cycle, the time t for electrons is

$$t = \frac{2\pi r}{v} = \frac{2\pi}{v} \frac{Wv}{Bec^2} = \frac{2\pi W}{Bec^2}$$

Therefore, the time t_1 to complete a circle after first cross-over is

$$t_1 = \frac{2\pi W_1}{Be c^2} = \frac{2\pi(1+\beta+\alpha)W_0}{Be c^2} = K_1 t_e \qquad \text{... (8.4)}$$

Where K_1 is an integer and t_e is the period of r_f field.

Similarly, one obtains the total energy after second cross-over as

$$W_2 = (1 + \beta + \alpha)W_0 + \alpha W_0 = (1 + \beta + 2\alpha)W_0$$

and to complete the second turn, the time t_2 is

$$t_2 = \frac{2\pi W_2}{Be c^2} = \frac{2\pi(1+\beta+2\alpha)}{Be c^2} = K_2 t_e \qquad \text{... (8.5)}$$

Where K_2 is an integer

$$\therefore \qquad t_2 - t_1 = \frac{2\pi\alpha W_0}{Be c^2} = (K_2 - K_1)t_e = K t_e \qquad \text{... (8.6)}$$

Where $K = K_1 - K_2 = 1, 2, 3, 4, \ldots$

This leads to

$$\frac{t_2 - t_1}{t_1} = \frac{\alpha W_0}{(1+\beta+\alpha)W_0} = \frac{K}{K_1}$$

or

$$\alpha = \left(\frac{K}{K_1 - K}\right)(1 + \beta) \qquad \text{... (8.7)}$$

thus the total kinetic energy of the electron after n turns is given by

$$W_{nk} = \beta W_0 + n\,\alpha W_0 = (\beta + n\alpha)W_0 \qquad \text{... (8.8)}$$

In microtrones, the electrons are accelerated to a few ten of MeV, where the maximum energy being limited by the size of the *rf* cavity.

There is a improvement in microtron. The *race-track microtron* uses a magnetic field split into two semicircular sectors. They are at a sufficient drift space apart into which a linac is inserted. One can achieve few hundred MeV of maximum energy with this type of microtron.

8.3.10 Proton-Synchrotron or Bevatron

The principles on which the proton-synchrotron is designed and operated are basically the same as those of the electron synchrotron. However a proton-synchrotron differs from an electron synchrotron in following two fundamental differences and design of the machine is different to the of an electron synchrotron. These differences are:

(1) The rest mass of a proton is nearly two thousand times, that of an electron. Hence whereas an electron has velocity 0.98 times the velocity of light at an energy 2 MeV only, the proton does not approach this velocity until its energy increases to 1000 MeV. It is, therefore, necessary that the frequency of radio frequency supply across the gap can no longer be kept constant but must be varied considerably, So that the proton may more in stable orbit.

(2) The protons are not accelerated initially by betatron method. Because in the betatron method to accelerate proton an excessively large magnet is required which would be uneconomical. Hence for the proton into stable orbit of doughnut chamber preliminary acceleration is provided by a radio-frequency voltage with a rising magnetic field.

The doughnut chamber is placed between the poles of ring shaped magnet, which produces a magnetic field normal to the chamber. The protons at low energy are injected in the form of periodic pulses into synchrotron orbit. They are accelerated by oscillating magnetic field, which is in resonance with them, while the field increases to maximum. The phase stability principle is utilized and the frequency of radio frequency supply is made exactly equal to that of the revolving protons. When the proton crosses the gap; proper acceleration is provided. The magnet is excited periodically and accelerate the proton during the time when the magnetic field is increasing. Finally when the protons reach to maximum energy, the frequency is distorted, so that the orbital path of the proton beam changes and either these may be made to strike the target fitted inside the doughnut vacuum chamber or can be taken out by an ejection device at the periphery.

The proton synchrotron at Brookhaven National Laboratory U.S.A. is known as *cosmotron*. This has an orbit of 30 ft, with a magnetic field at injection of 300 gauss and maximum magnetic field of 14000 gauses. The diameter of the magnet is 75 ft. and it weigh about 200 tons. Figure 8.61 shows it schematically.

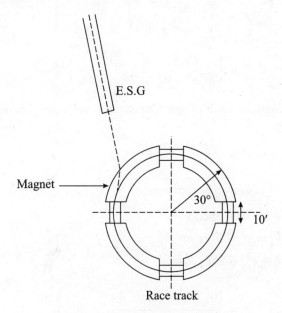

Fig. 8.61 Brookhaven cosmotron

It is built in four quadrants separated by 10 ft. gaps for allowing straight sections of vacuum chamber. These gaps are free from magnetic field, used for injection, acceleration and ejection of the protons. Proton in the form of pulses obtained by electrostatic generator of energy about 3.5 MeV, are injected in doughnut tube of the machine. The acceleration interval in the cosmotron is one second and about 800 eV energy is added per revolution. During the accelerating period, the radio/frequency changes from 0.37 Mc/sec to 4 Mc/sec, applied at one of the straight section made of ferrite tube, a non/metallic material of low conductivity and high permeability. A single turn of wire around the ferrite tube carrying radio frequency supply. The proton beam within the ferrite section form secondary of a transformer with ferrite tube as core and outer turn of wire as primary. This produces protons of energy 32 BeV.

The other proton-synchrotron are at Berkely, U.S.A., known as *Bevatron*, has race track 100 ft. The Bevatron produces proton of energy 6.2 BeV and has been used to create proton-antiproton and neutron-antineutron pairs which gave valuable understanding of fundamental particles.

Frequency, Energy orbit radius

The circular frequency of protons of energy W_k in equilibrium orbit of radius r in a magnetic field B in the absence of circular section is given by

$$\frac{Bec^2}{2\pi(m_p c^2 + W_k)}$$

Where m_p and e are mass of the proton and charge on it respectively.

The circulation frequency f' due to straight section is modified as

$$f' = \frac{Bec^2}{2\pi(m_p c^2 + W_k)}\left(\frac{2\pi r_0}{2\pi r_0 + 4L}\right)$$

$$= \left(\frac{Bec^2}{m_p c^2 + W_k}\right)\left(\frac{r_0}{2\pi r_0 + 4L}\right) \qquad \dots (8.1)$$

Where L denotes the length of each of the four straight sections. The K.E., W_k of a proton of momentum **p** is given by

$$E = W_k + m_p c^2 = \sqrt{p^2 c^2 + m_p^2 c^4}$$

$$= \sqrt{Be\, r_0\, c^2 + m_p^2 c^4}$$

Where $p = Be\, r_0$ and E is the total energy.

$\therefore \qquad W_k^2 + 2\, W_k\, m_p c^2 = (Be\, r_0 c)^2$

or $\qquad W_k(W_k + 2\, m_p c^2) = (Be\, r_0 c)^2 \qquad \dots (8.2)$

or $\qquad r_0 = \dfrac{[W_k\,(W_k + 2\, m_p c^2)]^{\frac{1}{2}}}{Bec} \qquad \dots (8.3)$

r will be in meter, if B is in tesla, and W_k and $m_p c^2$ in GeV. Now, from Equation (8.3), we have

$$r_0 = \frac{[W_k\,(W_k + 2\, m_p c^2)]^{\frac{1}{2}}}{0.3\, B} \qquad \dots (8.4)$$

and
$$B = \frac{[W_k (W_k + 2m_p c^2)]^{\frac{1}{2}}}{e\, r_0 c} \qquad \text{... (8.5)}$$

This reveals how to increase the magnetic field B at the equilibrium orbit with increase in proton energy W_k as it circulates in the orbit. Now, using (8.5), Equation (8.1) gives

$$f' = \frac{c[W_k (W_k + 2m_p c^2)]^{\frac{1}{2}}}{(m_p c^2 + W_k)(2\pi\, r_0 + 4L)} \qquad \text{... (8.6)}$$

If the accelerating voltage is in phase with f' then its *rf* value should always match f' and vary in accordance with the variation of f' with W_k.

This accelerator can energise protons to about 10 BeV energy. This machine is also called as *Cosmotron*.

8.3.11 Phase Stable Synchrotron

Nowadays these machines are used to produce the synchrotron radiations. These machines are ring shaped accelerators and use the principle of synchronous nature of acceleration process termed as *phase stability*. The high energy particle beam obtained from it are used in medical application and treatment. The very high energy versions of synchrotrons are called as *storage rings*. Figure 8.69 shows the main components of such a synchrotron.

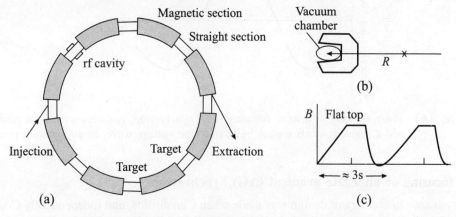

Fig. 8.62 Schematic diagram of a weak focusing proton synchrotron: (a) magnet ring (b) Cross-section through a magnetic sector and (c) magnetic field cycle exhibiting the flat top to give a long 'spill time' of a extracted time.

The particles are injected into the magnet ring from a linear accelerator and the magnetic field at this time is that required to maintain them on the stable orbit. An accelerating electric field is provided by radio frequency (*rf*) cavities through which particles pass and as their momentum increases the magnetic field is increased synchronously to maintain them on an orbit of radius R. Figure (8.62(c)) shows a typical magnetic field cycle. One the particles have reached full energy they can be ejected by a pulsed extraction magnet to form an external beam, or directed to an internal target from which secondary beams of product particle may be obtained. The frequency necessary at total energy E is given by

$$f = \frac{qB}{2\pi E} = \frac{qB}{2\pi (m^2 + q^2 B^2 R^2)^{\frac{1}{2}}}$$

$$\Rightarrow \frac{1}{2\pi R} \text{ for } E \gg m.$$

The particle motion is phase stable in synchrotrons. In weak focusing machines a particle arriving early (B) at the accelerating gap (Figure 8.63) receive a greater energy than it should and moves to a slightly greater radius, at which an orbit takes *longer* to describe. At the next transit of the accelerating cavity it has therefore moved towards the phase stable point A. Similarly, late particles (C) also move towards *A*, which is on the falling part of cavity voltage waveform, in contrast with the situation for linear accelerators as mentioned earlier.

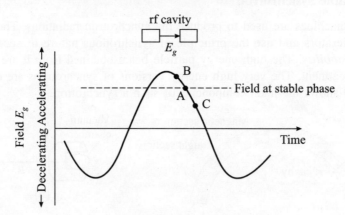

Fig. 8.63 Phase stability in a weak focusing (CG) synchrotron; particles arriving at times *B* and *C* move towards time *A*, relative to the voltage wave, as acceleration proceeds

Strong focusing or alternate gradient (AG) Synchrotron

A major advance in accelerator design was made when Christofilos, and independently Courant et al. realized that it is not necessary to render axial and radial oscillations stable simultaneously. In the AG synchrotron alternate magnetic field sectors have reversed gradients so that there is either axial or radial focussing in a given sector but not together. There is, however, net focusing in both directions after passage through two sectors (e.g. achromatic lens combination). A typical value of |n| in *AG* synchrotrons is 300, and this means a shorter free oscillation period the amplitude of these oscillations is reduced and a considerable spread of momentum can be accomodated in a small radial space, with consequent saving of magnet and vacuum chamber costs. In recent *AG* synchrotrons the bending and focusing take place in separated sectors.

In these machines, because of the large value of n, the change of radius with momentum is rather small and the orbits for late particles at low energy may actually take a shorter time to describe. The phase stable point is then on the rising part of the voltage wave, as with the linear accelerator. Above a certain point known as the *transition energy*, the orbit for late particles begins to take a longer time to describe than total for the synchronous beam and the phase-stable point then moves to the *falling* part of voltage wave as for *CG* machines.

The first accelerator of *AG* type was designed and commissioned by CERN near Geneva in 1959, which has a maximum proton energy of 29 GeV. A slightly later accelerator, the alternating gradient synchrotron (AGS) was built at Brookhaven National Laboratory in the USA and commissioned in 1960. The 1970s saw a ten fold increase in beam energy. In 1972 a 500 GeV machine was commissioned at Fermilab near chicago and in 1976 the SPS, which uses CPS as an injector, became active at CERN. This accelerator now provides protons not only for fixed target experiments but also for colliding-beam experiments, heavy ions have also been accelerated to relativistic energies.

The highest energy orbital accelerators, the 7 GeV machine DESY in Hamburg and the Cornell 12 GeV synchrotron, are both AG machines. However, extension of orbital acceleration to higher electron energies meets the severe and unaviodable problem of radiation loss. A charged particle of

mass m and energy E, moving in an orbit of radius R, suffers a loss proportional to $\left(\dfrac{E}{m}\right)^4\left(\dfrac{1}{R}\right)$.

This is negligible for proton machines due to the factor m^{-4}, but becomes prohibitive for high energy electron machines, unless they have a very large radius, because the necessary radio frequency power cannot be supplied. The *synchrotron radiation* accompanying a circulating electron beam can be used as an intense spectroscopic source from the *X*-ray region to the infra-red.

Both electron and proton machines produce a beam in which there is bunching within pulses of a repetition rate governed by the modulation cycle.

8.3.12 Storage Rings and Collider Rings

The high energy charged particles from particle accelerators are made to strike target materials in experiments designed to study the new particle generated due to such collisions. Momentum has to be conserved in all such collisions, and the useful kinetic energy available for the production of new particles from the head on collision of two particles is readily obtained by calculating its value in a frame of reference in which the resultant momentum of the colliding particles is zero.

The synchrotrons which are used to collide the particles are termed as collider rings. Two counter rotating beams are brought into collision at one or more points in the collider ring. The two-counter-rotating beams can be anti particles of each other, e.g. e^- and e^+ or protons and anti protons. If the energy of both the beams of particles is same then one syncrotron ring can accelerate and store both beams in the same vacuum chamber.

The first head on collision between high energy protons was achieved in a accelerator named as '*Intersecting Storage Rings*' (ISR). This consisted of two interlaced rings of magnet, with two beam pipes that crossed at eight places. CERN, Geneva has a 400 GeV proton-anti proton storage ring. Fermi Lab; Chicago also has a 1 TeV proton-anti proton collider. Several proton synchrotrons have been built to make proton-proton collision to study nuclear reactions and 30 GeV interacting storage ring at CERN. 26 GeV protons from a proton synchrotron were fed into the machine and the two proton beams made to collide head on after being accelerated to 31.5 GeV. This is equivalent to a stationary target being struck by protons of energy over 2000 GeV. The concentration of protons in the colliding beams was enhanced using *stochastic colling* method due to Van der Meer. The average position of a sample of protons in a slice through the beam is sensed at one point in the orbit and this information is used to send a signal across the ring to a *kicker* which generates a right amount of electric field to nudge the protons, on average, towards the path designed for this purpose with the help of sophisticated electronic circuits. By repeating the said path, the entire beam is slowly concentrated close to the designed path. We may note that relative advantage in this method increases as the square root of beam energy.

8.3.13 TeV Accelerators

The accelerating fields in present day accelerating machines are set up by radio waves in hollow copper cavities. These waves are coupled oscillating electric and magnetic fields. These waves form a standing wave that varies between regions of positive and negative electric field when pumped into a cavity of correct dimensions, i.e. shape and size. The charged particles entering a cavity absorb energy from the radio waves with the right direction of the field as small accelerating kicks while they are steered on many thousands of circuits by magnetic fields.

By probing progressively smaller space-time regions of matter, the sub nuclear realm is studied. In order to study the finer details of the structure of matter requires the giant particle accelerators and colliders with sophisticated detector systems.

The superconducting magnet technology that has worked satisfactorily in the *tevatron* is employed in the *Superconducting Super collider* (SSC) in USA to accelerate protons and anti-protons to design energy of 20,000 GeV. When the proton-anti proton beams collide with a total energy of 40,000 GeV, new particles in the mass range of 1000 – 4000 GeV are expected to be created.

We may note that in all high energy machines, heavy shielding against harmful radiations is required and this forms a massive structure.

We list some of the accelerating machines either in operations or under construction in Table 8.4.

Table 8.4 Some Accelerating Machines

Machine	Date of Commissioning	Radius (Approximate) (m)	Particles	Beam energy (GeV)
• AG Synchrotron, Brookhaven, Upton, New york, USA	1961	125	Protons	33
• SLAC (Stanford Linear Accelerator, California	1961	–	Electrons	22
• Cornell Electron Synchrotron, New york	1967	–	Electrons	12
• Serpukhov Proton Synchrotron, Russia	1967	235	Protons	76
• Mainring, Fermi lab* Batavia, Illinois	1972	1000	Protons	400

(Contd.)

Machine	Date of Commissioning	Radius (Approximate) (m)	Particles	Beam energy (GeV)
• Deutsches Elektronen Synchrotron (DESY) Germany	1974	–	electrons	22
• CERN super proton Synchrotron (SPS) Switchrotron-France	1976	1100	Protons	400
• Fermi lab Tevatron Illinois, USA	1984	920	Protons	1000
• Japanese National Laboratory (KEX)	1986	–	Electrons	30
• CERN large electron-positron storage ring (LEP)	1988	–	Electron	60
• Serpukhov UNK Russia	1990(?)	3060	Protons	3,000
• Superconducting Super Collider (SSC), USA	1994(?)	–	Protons, 20,000	

* This is now used mainly as injectors for the SPS and tevatron respectively.

The principal heavy ion synchrotrons are listed in Table 8.5.

Table 8.5 Principal Heavy Ion Synchrotrons

Facility	Maximum beam energy (GeV/nucleon)	Maximum ion mass
• Saturne II, Saclay (France)	1.2	40
• Bevalac, Berkeley, California	2.1	238
• Synchrophastron, Dubna, Russia	4.6	20
• AGS, Brook haven, Upton, New york	15.6	32
• SPS, Cern Switzerland	200	16

8.3.4 Particle Accelerators in India

The first 37″ fixed frequency cyclotron that produced proton beams of 3.7 MeV at Palit Laboratory of physics, University of Calcutta culminated in commissioning in 1960. This was used for studies in γ-spectroscopy and biomedical investigation. Later it was de-commissioned.

A 66 cm fixed frequency cyclotron with variable energy was gifted by university of Rochester, USA in 1971. This started operation at the Punjab University, Chandigarh and produced a proton beam of 5 MeV.

224 cm AVF type cyclotron at variable energy cyclotron centre (VECC) in calcutta is the biggest particle accelerator of India. This was designed to deliver proton beams of 6 – 60 MeV and α-particles of 24 – 130 MeV. Presently, it delivers α-beam of 80 MeV.

Other accelerators are: K-14 tandem type pelletron at TIFR, Mumbai which delivers 28 MeV protons, 42 MeV α-particles and heavier ions of still higher energies, and K-15 tandem type pelletron at Nuclear Science centre, Delhi. K-15 is capable to deliver energy 8 – 30 MeV.

There are also low energy electrostatic accelerators in operation at various laboratories, e.g. 5.5 MeV Van de Graaff accelerator at TIFR, 2 MeV Tandem Van de Graaff accelerators at BARC; at IGCAR, Kalapakkam; Institute of Physics, Bhubaneswar and IIT, Kanpur. There are also a number of Cockroft-Walton generators in operation, e.g. 1 MeV machine at the university of Calicut, a 400 kV machine at TIFR and a 250 kV machine at Bose Institute, Calcutta.

There are also number of electron accelerators which are exclusively used for biomedical purposes. In addition to these, there also exists facilities at other centres.

Example 8.1 An ionisation chamber exposed to a beam of α-particle registers a current of 4.8×10^{-13} ampere. On the average 20 α-particles enter the chamber per sec. Assuming that in producing ion pairs 35 eV per ion pair energy is needed. Calculate the energy of the α-particle.

Sol. Let 'E' eV be the energy of α-particles.

The amount of energy of 20 α-particles = 20 EeV

$\therefore$ number of ion pairs produced per sec. $\dfrac{20E}{35}$

$\therefore$ Charge possessed by these ion pairs/sec.

$$= \frac{20E}{35} \times 1.6 \times 10^{-19} \text{ Coulomb/sec} = \text{current}$$

or

$$\frac{20E}{35} \times 1.6 \times 10^{-19} = 1.8 \times 10^{-13}$$

$\therefore$

$$E = \frac{4.8 \times 10^{-13} \times 35}{20 \times 1.6 \times 10^{-19}} = 5.25 \times 10^6 \text{ eV} = 5.25 \text{ MeV}$$

Example 8.2 An ionisation chamber is charged to a potential 1000 volts. If its capacity be 50 pf, by what percentage its charge would reduce in passing an α-particle producing 2×10^5 ion pairs.

Sol. The fall in potential is given by

$$V = \frac{ne}{C} = \frac{2 \times 10^5 \times 1.6 \times 10^{-10}}{50 \times 10^{-12}} = 6.4 \times 10^{-4} \text{ Volt.}$$

$\therefore$ Percentage of voltage drop $= \dfrac{6.5 \times 10^{-4} \times 100}{1000}$

$$= 6.4 \times 10^{-5} \text{ \%}$$

Hence percentage of charge reduction is also equal to 6.4×10^{-5} %.

Example 8.3 An ionisation chamber is connected to an electrometer of capacity 0.5 picofarad and voltage sensitivity of 4 divisions per volt. A beam of α-particles causes a deflection of 0.8 division. If 35 eV is required to produce one ion-pair, find the number of ion-pairs produced by the α-beam and the energy of the source of α-particles.

Sol. The signal voltage $\Delta V = \dfrac{\Delta Q}{C}$. Therefore

$$\Delta Q = \Delta V \times C$$

$$= \frac{0.8}{4} \times 0.5 \times 10^{-12}$$

$$= 10^{-13} \text{ C}$$

Example 8.4 Find the number of ion pairs by 10 MeV proton. If in the proportionality region the amplification is 10^3, current pulse time is $10\,\mu s$ and resistance between electrodes is 10^4, find voltage pulse height. The amount of energy required to produce one ion pair is 34 eV.

Sol. no. of ion pairs by 10 MeV protons

$$n = \frac{10 \times 10^6}{34}$$

$$= 2.94 \times 10^5$$

Total ion pairs in proportionality region

$$= 2.94 \times 10^5 \times 10^3$$

$$= 2.94 \times 10^8$$

Hence charge on electrode due to these pair

$$q = 2.94 \times 10^8 \times 1.6 \times 10^{-19} \text{ Coulomb}$$

or $$q = 4.704 \times 10^{-11}$$

$\therefore$ $$\text{Current} = \frac{q}{t} = \frac{4.704 \times 10^{-11}}{10 \times 10^{-6}}$$

$$= 4.704 \times 10^{-6} \text{ ampere}$$

$\therefore$ Voltage pulse higher $= i \times R$

$$= 4.704 \times 10^{-6} \times 10^4$$

$$= 4.704 \times 10^{-2}$$

$$= 0.047 \text{ Volts}$$

Example 8.5 A halogen quenched G.M. Counter works at 1 kV. Anode wire has radius 0.2 mm and radius of cathode is 20 mm. The guarranted period for the counter is 10^9 counts. Find maximum radial field. If the counter is operated 30 hrs a weak and count 3000 counts/min, find the life of the counter (take 1 year = 50 weeks)

Sol. The field at an axial distance r is given by

$$E_r = \frac{V_0}{r \log_e \left(\dfrac{b}{a} \right)}$$

This would be maximum near the surface of wire

$$\therefore \qquad (E_r)_{max} = \frac{1 \times 10^3}{0.2 \times \log_e 100}$$

Here $r = 20$ mm, $a = r = 0.2$ mm

or $\qquad (E_r)_{max} = \frac{10^3}{0.2 \times 2.3026 \times 2}$

$$= 1.087 \times 10^3 \text{ V/mm}$$

If its life is x, we have

$$10^9 = x \times 50 \times 30 \times 60 \times 3000$$

$$x = \frac{10^{9-6}}{5 \times 6 \times 3 \times 3}$$

$$= \frac{1000}{5 \times 6 \times 3 \times 3}$$

$$= \frac{100}{27} = 3.7 \text{ years}$$

Example 8.6 A GM-counter has a 'dead time' 400 μs. What are the true counting rates when the observed rates are (i) 100 per minute, (ii) 1000 per minute?

Sol. The relation connecting the true rate N_t, observed counting rate N_0 and the dead time t is given by

$$N_t = \frac{N_0}{1 - N_0 t}$$

$$\therefore \quad \text{(i) } N_t = \frac{100}{1 - 100 \times \left(400 \times \dfrac{10^{-6}}{60}\right)} = 100.07/\text{min}$$

$$\text{(ii) } N_t = \frac{1000}{1 - 100 \times \left(400 \times \dfrac{10^{-6}}{60}\right)} = 1006.71/\text{min.}$$

Example 8.7 A γ-photon of energy 5 MeV produces an electron-positron pair (pair-production). The two particles generated move in opposite directions with equal speeds. The electron loses all its kinetic energy in producing ion-pairs on entering a helium field detector. Calculate the number of ion pairs produced and the pulse height, the system having a capacitance 10 μμF. Assume 42.6 eV average energy is required to produce an ion-pair in helium.

Sol. Let E^+ and E^- be the K.E. of the positron and electron respectively and E_γ the energy of γ-photon

$$\therefore \qquad E_\gamma = 2 m_0 c^2 + E^+ + E^-$$

where $2m_0c^2$ is the total rest mass of the particles.

Now, $\qquad\qquad\qquad E^+ = E^- = E$, say

$\therefore \qquad\qquad\qquad E_\gamma = 2m_0c^2 + 2E$

$\therefore \qquad\qquad E = \dfrac{E_\gamma - 2m_0c^2}{2} = \dfrac{5 - 2 \times (0.51)}{2} = 1.99 \text{ MeV}$

$\therefore$ No. of ion-pairs produced in He is

$$n = \frac{1.99 \times 10^6}{42.6} = 4.67 \times 10^4$$

$\therefore$ Output pulse height,

$$V = \frac{q}{C} = \frac{ne}{C} = \frac{4.67 \times 10^4 \times 1.6 \times 10^{-19}}{10 \times 10^{-12}} \text{ volt}$$

$$= 75 \times 10^{-5} \text{ V}$$

Example 8.8 Calculate the frequency of the electric field that must be applied between the dees of a cyclotron in which (i) proton (ii) deutron (iii) α-particles are accelerated. The applied magnetic flux is 3.5 Wb/m^2.

Sol. Frequency of electric field

$$n = \frac{Bq}{2\pi m}$$

$\qquad\qquad\qquad q = 1.6 \times 10^{-19} \text{ Coulomb}$

Here $\qquad\qquad m = 1.67 \times 10^{-27} \text{ kg}$

$\qquad\qquad\qquad B = 3.5 \text{ Wb/m}^2$

$$n = \frac{3.5 \times 1.6 \times 10^{-19}}{2 \times 22 \times 1.67 \times 10^{-27}}$$

$$= 53.4 \times 10^6 \text{ Hz} = 53.4 \text{ MHz}$$

For deutrons

$$q = 1.6 \times 10^{-19}, \ m = 2 \times 1.6 \times 10^{-27} \text{ kg}.$$

$\therefore \qquad n_{deu} = \dfrac{3.5 \times 1.6 \times 10^{-19}}{2 \times 3.14 \times 2 \times 1.67 \times 10^{-27}} = 26.7 \text{ MHz}$

For α-particles

$$n_\alpha = \frac{3.5 \times 1.67 \times 10^{-19} \times 2}{2 \times 3.14 \times 4 \times 1.67 \times 10^{-27}}$$

$$= 26.7 \text{ MHz}$$

Example 8.9 In a particular betatron the magnetic induction at the orbit is 0.4 Wb/m^2 and its frequency is 60 Hz. If the radius of the orbit is 1.65 m, estimate the final energy which can be obtained. How many revolutions will the electron make in gaining this energy?

Sol. In a betatron the velocity of accelerated electrons is of the order of c and hence energy gain can be written as

$$E = e\,c\,R\,B$$

$$B = 0.4 \text{ Wb/m}^2$$

$$\therefore \qquad E = 1.6 \times 10^{-19} \times 3 \times 10^8 \times 1.68 \times 0.4$$

$$= 3.17 \times 10 *$$

$$c = 3 \times 10^8 \text{ m/sec}^2$$

$$R = 1.65 \text{ m}$$

$$Q = 1.6 \times 10^{-19} \text{ Coulomb}$$

$$= \frac{3.17 \times 10^{-11}}{1.6 \times 10^{-19}} \text{ eV}$$

$$\approx 200 \text{ MeV}$$

Example 8.10 In certain proton-synchrotron, a proton with an initial energy of 9 MeV is accelerated to 10 GeV, gaining on the average an energy of 2.2 keV/tum. The total length of the stable orbit inclusive of the straight sections is 100 m. The acceleration time 3.3 sec.

(a) Determine the number of revolutions and paths length covered by protons in the whole period of acceleration.

(b) What is final velocity of protons?

Sol. (a) No. of rotations = $\dfrac{\text{Total energy gain}}{\text{Average gain of energy per cycle}}$

$$= \frac{10 \times 10^9 \times 1.6 \times 10^{-19} - 9 \times 10^6 \times 1.6 \times 10^{-19}}{2.2 \times 10^3 \times 1.6 \times 10^{-19}}$$

$$= \frac{(10 \times 10^9 - 9 \times 10^6)}{2.2 \times 10^3} = \frac{(10^4 - 9) \times 10^6}{2.2 \times 10^3}$$

$$= \frac{9991 \times 10^3}{2.3}$$

$$= 4.5 \times 10^6 \text{ rotations}$$

Distance travelled by the particle

$$= 4.5 \times 10^6 \times 100 \times 2 \text{ m}$$

$$= 9 \times 10^8 \text{ m}$$

Example 8.11 What radius is needed in proton synchrotron to attain particle energies of 10 GeV, assuming that guide field of 2.0 Wb/m^2 is available?

Sol. For proton synchrotron after applying the relativistic correction

$$B = \frac{[T(T + 2m_0c^2)]^{\frac{1}{2}}}{ecR}$$

$$T = 10 \text{ GeV} = 10 \times 10^9 \times 1.6 \times 10^{-19} \text{ Joule}$$

$$= 1.6 \times 10^{-9} \text{ Joule}$$

$$m_p = 1.67 \times 10^{-27} \text{ kg}$$

$$B = 21.0 \text{ Wb/m}^2$$

$$\therefore \quad R = \frac{[(T + 2m_0c^2)]^{\frac{1}{2}}}{Bec}$$

$$= \frac{\left\{1.6 \times 10^{-9}\left[1.6 \times 10^{-9} + 2 \times 1.67 \times 10^{-27} \times \left(3 \times 10^8\right)^2\right]\right\}^{\frac{1}{2}}}{2 \times 1.6 \times 10^{-19} \times 3 \times 10^8}$$

$$= 18.2 \text{ m}$$

Example 8.12 A halogen quenched GM tube operates at 1 kV and has a wire diameter 0.2 mm. The radius of the cathode is 20 mm and the tube has a guaranteed life of 10^9 counts. What is the maximum radial field and how long will the counter last, if it is used on an average of 30 hours per week at 3000 counts/min?

Sol. The field at a distance r from the axis of a cyclindrical axis is given by

$$E = \frac{V_0}{r \log_e \frac{b}{a}}$$

Its value will be maximum near the wire

$$\therefore \quad E_{max} = \frac{10^3}{0.01 \times \log_e \frac{2}{0.02}}$$

$$= \frac{10^3}{0.1 \times 2.303 \times 2}$$

$$= \frac{10^6}{4.6} \frac{\text{Volt}}{\text{cm}}$$

$$= 2.1 \times 10^5 \frac{\text{Volt}}{\text{cm}}$$

Suppose its life be t years, then

$$10^9 = t \times 50 \times 30 \times 60 \times 3000$$

$\therefore$ $$t = \frac{100}{27} = 3.7 \text{ years}$$

Example 8.13 Calculate (i) the minimum velocity for the emission of Cerenkov radiation for a medium of refractive index 1.5 and (ii) the minimum kinetic energy for protons for such emission.

Sol. The minimum velocity, $v_{\min} = \dfrac{c}{n} = \dfrac{3 \times 10^8}{1.5} = 2 \times 10^8 \text{ m/s}$

The minimum kinetic energy, $T_{\min} = Mc^2 \left(\dfrac{1}{\sqrt{1 - \dfrac{v_m^2}{c^2}}} - 1 \right)$

Now, $$\frac{v_m}{c} = \frac{1}{n} = (1.5)^{-1}$$

$\therefore$ $$T_{\min} = 938.3 \left(\frac{1}{\sqrt{1 - \left(\dfrac{1}{1.5} \right)^2}} - 1 \right) = 321 \text{ MeV}$$

Example 8.14 An energetic particle of unit charge has a range 340 μm in a nuclear emulsion detector. Another elementary particle of unit charge has a range 258 μm. The mass of the first particle is 273 m_e, what is the mass of the second particle?

Sol. The two particle have the same charge, So the ratio of their masses M is equal to the ratio of their ranges, R.

$\therefore$ $$\frac{M_1}{M_2} = \frac{R_1}{R_2}$$

$\therefore$ $$M_2 = M_1 \times \frac{R_2}{R_1} = 273 \, m_e \times \frac{258}{340} = 207 \, m_e.$$

$\therefore$ The mass of the second particle is 207 m_e.

Example 8.15 An organic quenched tube has slope of 4 % per 100 V. If the operating voltage is 1250 V, calculate the maximum voltage-fluctuation in the supply in order that the maximum error in the rate of count is 0.05 %.

Sol. Percentage slope $= \dfrac{\dfrac{(n_2 - n_1)}{n_{av}}}{\dfrac{(V_2 - V_1)}{100}} \times 100 \%$ ... (i)

Now, $\dfrac{n_2 - n_1}{n_{av}} \times 100$ = percentage change in count rate with respect to n_{av}.

$\therefore$ $\qquad \dfrac{n_2 - n_1}{n_{av}} \times 100 = 0.05$ $\qquad\qquad\qquad\qquad\qquad$ ... (ii)

Here slope per 100 V = 4 % = $\dfrac{4}{100}$.

$\therefore$ $\qquad\qquad \dfrac{4}{100} = \dfrac{0.05 \times 100}{V_2 - V_1}$, using (i) and (ii). $\qquad\qquad$ ... (iii)

$\therefore$ Percentage change in operating voltage

$$- \dfrac{V_2 - V_1}{1250} \times 100$$

$$= \dfrac{0.05 \times 100 \times 100}{4 \times 1250} \times 100, \text{ using (iii)},$$

$$= 10 \%$$

Example 8.16 A neutron fission counter is lined with U-235 to detect slow neutrons by ionising fission fragments that produce an average energy of 200 MeV. Find the pulse height produced by a capacitor of capacitance 40 pF connected to the collecting electrode. Assume that an average energy of 35 eV is needed to produce an ion pair.

Sol. Total number of ion pairs produced = $\dfrac{200 \times 10^6}{35}$ = 5.71 × 106

$\qquad \therefore$ Total charged produced across collector

$$= 5.71 \times 10^6 \times 1.6 \times 10^{-19} = 9.136 \times 10^{-13} \text{ C}$$

Capacitance of the capacitor = 40 × 10⁻¹² F.

$\therefore$ $\qquad\qquad$ Pulse height = $\dfrac{\text{Charge}}{\text{Capacitance}} = \dfrac{9.136 \times 10^{-13}}{40 \times 10^{-12}} = 2.284 \times 10^{-2}$ V

Example 8.17 In a scintillation counter, a γ-ray peak of 20 keV energy is observed at a pulse height of 32 V. The full width at half maxima (FWHM) is 4 keV. Evaluate the percentage resolution of the counter.

Sol. Peak energy of 20 keV corresponds to 32 V.

$\qquad \therefore$ Full width of peak at half maxima corresponds to $\dfrac{32 \times 4}{20} = 6.4$ V

$\therefore$ $\qquad\qquad$ FWHM = 6.4 V

$\therefore$ Percentage resolution = $\dfrac{\text{FWHM}}{\text{Peak pulse height}} = \dfrac{6.4}{32} \times 100 = 20 \%$

Example 8.18 A cyclotron with dees of radians 1.1 m has magnetic field of 0.7 Weber/cm². Calculate the energies to which (i) protons (ii) deutrons (iii) α-particle, are accelerated.

Sol.

$$E_{max} = \frac{1}{2} \frac{B^2 e^2 R^2}{m}$$

For *proton*, $m_p = 1.67 \times 10^{-27}$ kg, $R = 1.8$ m, $B = 0.7$ Wb/m²

and $e = 1.6 \times 10^{-19}$ Coulomb

$$E_{proton} = \frac{1}{2} \frac{(0.7)^2 \times (1.6 \times 10^{-19})^2 \times (1.8)^2}{1.67 \times 10^{-27}} \text{ Joule}$$

Similarly we can calculate for deutrons and a-particles taking

$$m_{max} = 2 m_p$$

$$\therefore \quad E_{max} \text{ (deutron)} = \frac{76}{2} \times 10^6 \text{ eV}$$

$$= 38 \text{ MeV}$$

(iii) $m_\alpha = 4 m_p$ but charge is $2e$

$$\therefore \quad E_{max}(\alpha) = \frac{1}{2} \frac{B^2 (2e)^2 R^2}{4 m_p} = \frac{1}{2} \frac{B^2 e^2 R^2}{m_p} = 76 \times 10^6 \text{ eV}$$

Example 8.19 Protons are accelerated in a cyclotron applying an electric field of frequency 10^5 c/sec. and of amplitude 5000 Volts. Find the energy acquired by the proton. The cyclotron radius is 50 cm.

Sol.

$$E = \frac{1}{2} \frac{e^2 B^2 R^2}{mc^2} \quad \text{and} \quad \frac{eE}{2\pi mc} = f$$

$$E = \frac{1}{2} \frac{(2\pi m f c)^2 R^2}{mc^2} = \frac{1}{2} 4\pi^2 m f^2 R^2$$

$$= 2\pi^2 \times 1.67 \times 10^{-24} \times 10^{10} \times 50 \times 50 \text{ ergs}$$

$$= \frac{2\pi^2 \times 167 \times 25 \times 10^{-12}}{1.6 \times 10^{-12}} \text{ eV}$$

$$B = 514.4 \text{ eV}$$

Example 8.20 In a cyclotron to accelerate proton the magnetic field is 15000 gauss and radius of the dee is 15 cm. Calculate the energy of proton beam and frequency of oscillator. Take $\frac{e}{m}$ for proton to be 9600 e.m. u/gm. and charge on proton = 1.6×10^{-20} a.m.u.

Sol.

$$E = \frac{1}{2} \frac{e^2 B^2 R^2}{m} \qquad\qquad [e \text{ in e.m.u.}]$$

$$= \frac{1}{2}\left(\frac{e}{m}\right)^2 mB^2 R^2$$

Substituting

$$E = \frac{1}{2} \times (9600)^2 \times \frac{1.6\times10^{-20}}{9600} (15000)^2 \times 15^2$$

$$= 48 \times 1.6 \times 225 \times 225 \times 10^{-20+2+6} \text{ ergs}$$

$$= \frac{48\times1.6\times225\times225\times10^{-12}}{1.6\times10^{-12}} \text{ eV} = 2.43 \times 10^6 \text{ eV}$$

$$E = 2.43 \text{ MeV}$$

Frequency of oscillator = cyclotron frequency

Period of cyclotron $= T = \dfrac{2\pi m}{v}$ But $v = \dfrac{BeR}{m}$

$$\therefore \qquad T = \frac{2\pi m}{Be}$$

$$\therefore \qquad \text{Frequency} = \frac{1}{T} = \frac{B}{2\pi}\left(\frac{e}{m}\right)$$

$$= \frac{15000}{2\pi} \times \frac{9600}{1} = \frac{144\times10^6}{6.28} = 22.9 \times 10^6 \text{ cycle/sec.}$$

Example 8.21 A cyclotron is used to obtain a 2 MeV beam of protons. The alternating potential difference applied between the dees has a peak value of 20 kV and its frequency is 5 mega Hertz. What is the intensity of magnetic field applied for resonance? Also calculate the radius of the dees of the cyclotron. The specific charge $\left(\dfrac{q}{m}\right)$ for the protons is 96×10^6 Coulomb/kg.

Sol. For response the frequency of alternating voltage is given by

$$n = \frac{Bq}{2\pi m}$$

$$n = 5 \times 10^6 \text{ Hz} \qquad\qquad \therefore \quad B = \frac{2\pi n}{\left(\dfrac{q}{m}\right)}$$

$$E = 2 \times 1.6 \times 10^{-13} \text{ J} \qquad\qquad = \frac{2\pi\times5\times10^6}{96\times10^6} = \frac{31.4}{96}$$

$$\frac{q}{m} = 96 \times 10^6 \text{ C/Kg} \qquad\qquad = 0.327 \text{ Wb/m}^2 \text{ (Tesla)}$$

The radius of the dees will be equal to the maximum radius of the orbit when the ions are drawn out of the machine. If R be the radius of this orbit and energy required by the ions is E then

$$E = \frac{1}{2} m v^2_{\text{max}}$$

$$= \frac{1}{2} m \left(\frac{B^2 q^2 R^2}{m^2} \right)$$

or

$$R^2 = \frac{2E}{mB^2 \left(\dfrac{q}{m} \right)^2}$$

or

$$R = \left(\frac{2E}{m} \right)^{\frac{1}{2}} \frac{1}{B \left(\dfrac{q}{m} \right)}$$

$$= \frac{2 \times 2 \times 1.6 \times 10^{-13}}{(1.667 \times 10^{-27})} \times \frac{1}{0.327 \times 96 \times 10^6}$$

$$= 0.6236 \text{ m}$$

Obviously, the radius of the dees will be slightly more than 62.4 cm.

Example 8.22 Calculate the frequency of alternating electric field and the strength of the magnetic field in a 50 cm cyclotron, for the production of beams of (a) 30 MeV α-particles (b) 20 MeV deutrons (c) 10 MeV protons. Mass of a proton = 1.67×10^{-27} Kg and electronic charge = 1.6×10^{-19} C.

Sol. The energy of the accelerated particles is given by

$$E = \frac{1}{2} m \left(\frac{B^2 q^2 R^2}{m^2} \right) = \frac{B^2 q^2 R^2}{m^2}$$

so that

$$B^2 = \frac{2mE}{q^2 R^2}$$

or

$$B = \left(\frac{2mE}{qR} \right)^{\frac{1}{2}}$$

For resonance condition the frequency of the applied field is related to B as

$$n = \frac{Bq}{2\pi m}$$

(i) 30 MeV α particles
Given

$$E = 30 \times 10^6 \times 1.6 \times 10^{-19} \text{ J}$$

$$48 \times 10^{-13} \text{ J}$$

$$q = 2 \times 1.6 \times 10^{-19} \text{ C}$$

$$m = 4 \times 1.67 \times 10^{-27} \text{ Kg}$$

$$R = 0.5 \text{ m}$$

$$B = \frac{(2 \times 4 \times 1.67 \times 10^{-27} \times 48 \times 10^{-13})^{\frac{1}{2}}}{2 \times 1.6 \times 10^{-19} \times 0.5} \text{ Tesla}$$

$$= 1.582 \text{ Tesla}$$

∴

$$n = \frac{1.582 \times 2 \times 1.6 \times 10^{-19}}{2 \times 3.14 \times 4 \times 1.67 \times 10^{-27}} \text{ Hz}$$

$$= 1.207 \times 10^7 \text{ Hz}$$

$$= 12.07 \text{ mega Hz}$$

Similarly one obtains

(ii) 10 MeV Deutrons

$$B = 1.77 \text{ Tesla}$$

and

$$n = 13.5 \text{ mega Hz}$$

(iii) 10 MeV protons

$$B = 0.913 \text{ Tesla}$$

$$n = 13.94 \text{ mega Hz}$$

Example 8.23 In a particle betatron the maximum magnetic field at the orbit is 0.4 Tesla and its frequency is 60 cycles/sec. If a radius of the stable orbit is 1.65 m, estimate the final energy which can be obtained? How many turns will the electrons turns will the electrons make in gaining this energy Assume sinusoidal variation of magnetic field with time.

Sol. The accelerated electrons in a betatron have velocities quite close to the velocities of light c. The energy gained is

$$E = e c R_0 B_0$$

Where B_0 is the peak value of magnetic field and R_0 is the radius of stable orbit.

$$B_0 = 0.4 \text{ Tesla} \qquad \therefore \ E = 1.6 \times 10^{-19} \times 3 \times 10^8 \times 1.65 \times 0.4 \text{ J}$$

$$e = 1.6 \times 10^{-19} \text{ C} \qquad = 3.168 \times 10^{-11} \text{ J}$$

$$R_0 = 1.65 \text{ m} \qquad = \frac{3.168 \times 10^{-11}}{1.6 \times 10^{-13}} \text{ MeV}$$

$$C = 3 \times 10^8 \text{ m/s} \qquad 200 \text{ MeV}$$

The number of revolutions. The number of revolutions made by the electron during quarter of a cycle when magnetic field increases and electrons gain enregy is

$$N = \frac{c\left(\dfrac{\pi}{2\omega}\right)}{2\pi R_0} = \frac{c}{8\pi\,f\,R_0}$$

$$= \frac{3\times10^8}{8\times3.14\times60\times1.65}$$

$$= 1.2\times10^5$$

Example 8.24 The r.f. potential difference applied between the dees of a cyclotron is 50 kilovolts and a magnetic field of 1.6 Tesla. Dee radius is 0.55 meters. Calculate

(i)   the energy acquired by protons in this cyclotron.

(ii) the number of revolutions these protons make in attaining this energy.

(iii) oscillator frequency and wave length.

Sol.

(i) Energy acquired by a proton in a cyclotron is given by

$$E = \frac{R^2\,B^2\,e^2}{2m}$$

$$= \frac{(0.55)^2\,(1.6\text{ Tesla})^2\times(1.6\times10^{-19})^2}{2\times1.67\times10^{-27}}$$

$$8.3\times10^{-12}\text{ J} = 51.8\text{ MeV}$$

(ii) When energy is known, number of revolutions can be found by calculating the number of accelerations. Since the number of revolutions will be equal to half the number of accelerations.

Now, if N is the number of accelerations received by the particle in acquiring the energy E, then $E = NeV$ since eV is the energy gained by the particle in one acceleration, e being the charge on proton and V is dee potential.

$\therefore$ Number of accelerations $= \dfrac{E}{eV}$

$\therefore$ Number of revolutions $= \dfrac{E}{2\,eV} = \dfrac{\dfrac{E}{e}}{2\,V}$

$$= \frac{E\,(\text{in electron-Volts})}{(\text{in Volts})}$$

$$= \frac{51.8\times10^6\text{ eV}}{50\times100\text{ V}} = 10^{36}$$

(iii) $\nu_o = \nu_e = \dfrac{Be}{2\pi m} = \dfrac{(1.6\,\text{Tesla})(1.6\times10^{-19}\,C)}{2\times3.14\times1.67\times10^{-27}}$

$$= 24.41\ \text{MHz}$$

and wavelength

$$\lambda = \frac{c}{\nu} = \frac{3\times10^8\ m_c}{24.41\times10^6\ \text{Hz}}$$

$$= 12.29\ \text{m}$$

Example 8.25 A 200 MeV betatron has a stable orbit radius of 0.8 m. The electron acquire energy at the rate of 500 eV per revolution. The magnet of this betatron is energised at 180 Hz and produces pulses of 1 m sec duration, each containing 10^{-19} electrons. Calculate (i) distance travelled by electrons in attaining maximum energy (ii) peak beam current

Sol. (i) Final electron energy = 200 MeV

Energy gain or the electrons per revolutions = 500 eV

∴ Total number of revolutions in attaining this energy $= \dfrac{200\times10^6}{500} = 4\times10^5$

Distance travelled by electrons in 4×10^5 revolutions

$$= 2\,\pi r_0 N$$

$$= 2\times3.14\times0.8\times4\times10^5$$

$$= 2.01\times10^6\ \text{m} = 2.01\times10\ \text{km}$$

(ii) Each pulse contains 10^{10} electrons and each electron carries a charge $e = 1.6\times10^{-19}$ C

∴ Each pulse carries a charge $= 10^{19}\times1.6\times10^{-19} = 1.6\times10^{-9}$ C

Pulse duration $= 1\mu$ sec.

∴ Peak value of current $= 1.6\times\dfrac{10^{-9}}{10^{-6}}$

$$= 1.6\times10^{-3}\ \text{Amp.}$$

Example 8.26 The parameters of a synchrocyclotron, meant for accelerating deutrons are;

Magnetic flux density at the orbit = 1.431 Wb/m^2

Magnetic flux density at the centre = 1.5 Wb/m^2

calculate: (i) The maximum frequency of the dee-voltage

(ii) Energy gained by deutrons if the dee voltage frequency is modulated between this maximum and a minimum of 10×10^6 Hz.

Sol. (i) The frequency of dee voltage in a synchro cyclotron is given by

$$\nu = \frac{e\,Bc^2}{2\pi(K + m_0c^2)}$$

Maximum frequency of the dee-voltage (V_{max}) corresponds to zero value of K. E. (K), i.e.

$$\nu_{max} = \frac{eB}{2\pi m_0}$$

$$= \frac{1.6 \times 10^{-19} \times 1.5}{2 \times 3.14 \times 3.34 \times 10^{-27}}$$

$$= 11.44 \times 10^6 \text{ Hz} = 11.44 \text{ MHz}$$

(ii) $\nu_{max} = \dfrac{eBc^2}{2\pi m_0 c^2}$ and $\nu_{min} = \dfrac{eBc^2}{2\pi (K + m_0 c^2)}$

$$\therefore \qquad K = \frac{\nu_{max} - \nu_{min}}{\nu_{max}\, \nu_{min}} \cdot \frac{eBc^2}{2\pi}$$

$$= \frac{[(11.44 - 10) \times 10^6] \times 1.6 \times 10^{-19} \times 1.43 \times (3.10^8)^2}{11.44 \times 10 \times 10^{12} \times 2 \times 3.14 \times 10^6 \times 10^{-13}} \text{ MeV}$$

$$= \frac{1.44 \times 1.43 \times 9}{11.44 \times 10 \times 2 \times 3.14} \times 10^4 \text{ MeV}$$

$$= 258.1 \text{ MeV}$$

Example 8.27 An electron-synchrotron has a stable orbit radius of 50 cms. and uses a magnetic field of 15 kilo-gauss at the orbit. Calculate (i) Energy gained by the electrons in it (ii) Frequency of the applied electric field.

Sol. (i) Final energy of electrons accelerated in an electron synchrotron

$$E = e\, Br_0 c = 300\, Br_0 \text{ MeV}, \quad \text{if } B \text{ is in } \frac{W}{m^2} \text{ and } r \text{ is in meters.}$$

$$= 300 \times 1.5 \times 0.5 = 225 \text{ MeV}$$

(ii) Frequency of the r.f. field $\nu = \dfrac{c}{2\pi r_0}$

$$\therefore \qquad \nu = \frac{3 \times 10^8}{2 \times 3.14 \times 0.5} = 95.54 \text{ MHz}$$

Example 8.28 A mixed beam of protons and deutrons which were accelerated to a potential of 10^5 volt is allowed to pass through a uniform magnetic field of 1.5 T in a direction at right angles to the field. Calculate the linear separation of the deuteron beam from the proton beam when each has described a semicircle.

Sol. Energy of the beam, $\dfrac{1}{2} m\upsilon^2 = qV = 10^5 \times 1.6 \times 10^{-19} = 1.6 \times 10^{-14}$ J

$$\therefore \qquad \text{Velocity}, v = \sqrt{\frac{2\times1.6\times10^{-14}}{m}} = \sqrt{\frac{3.2\times10^{-14}}{m}}$$

$$\therefore \qquad Bqv = \frac{mv^2}{r} = \frac{3.2\times10^{-14}}{Bqv}$$

$$\therefore \qquad r_1 = r\text{-value for proton beam} = \frac{3.2\times10^{-14}}{Bqv}$$

$$\therefore \qquad r_1 = \frac{3.2\times10^{-14}}{1.5\times1.6\times10^{-19}} \times \sqrt{\frac{1.67\times10^{-27}}{3.2\times10^{-14}}} = 0.0305 \text{ m}$$

$$r_2 = r\text{-value for deuteron beam} = \sqrt{2}\, r_1 = 0.0431 \text{ m}$$

$$\therefore \text{ linear separation,} \quad d = 2(r_2 - r_1) = 2(0.0431 - 0.305)\text{m} = 0.0252 \text{ m}$$

$$= 2.52 \text{ cm}$$

Example 8.29 What are the design parameters for a cyclotron that would accelerate α-particles to a maximum energy of 20 MeV? The dees are to have a diameter of 1 m. [Mysore]

Sol. Magnetic field, $B = \sqrt{\dfrac{2mE}{q^2 r^2}}$, where m = mass, E = energy, q = charge of α-particles and r = radius of dees.

$$\therefore \qquad B = \left\{\frac{2\times4\times1.67\times10^{-27}\times20\times1.6\times10^{-13}}{(2\times1.6\times10^{-19})^2\times(0.5)^2}\right\}^{\frac{1}{2}}$$

$$= 1.3 \text{ Wb/m}^2$$

Frequency of rf-oscillator, $\quad v = \dfrac{Bq}{2\pi m}$

$$\therefore \qquad v = \frac{q}{m}\times\frac{B}{2\pi} = \frac{2\times1.6\times10^{-19}}{4\times1.67\times10^{-27}}\times\frac{1.3}{2\times3.14}$$

$$= 9.9 \times 106 \text{ Hz} = 9.9 \text{ MHz}$$

Example 8.30 A linear accelerator (linac), accelerating protons, has a length 50 cm for the first drift tube and the energy attained by protons is 100 MeV. The frequency of the rf-field applied and the peak accelerating potential are 500 MHz and 1500 kV respectively. The time taken by the proton to travel the distance between two consecutive taps is equal to the period of the rf-field. Find (i) the length of the last tube, (ii) energy of protons at injection into first tube, and (iii) the total number of acceleration received to attain the maximum energy. [Ujjain]

Sol. Let v be the frequency of the r-field, L_n the length of the nth drift tube and v_n the velocity of proton while passing the nth tube.

(i) Since the protons in passing over the distance *between* two consecutive gaps take a time equal to the period of the rf-field.

$$\frac{1}{\nu} = \frac{L_n}{\upsilon_n} \quad \text{or} \quad L_n = \frac{\upsilon_n}{\nu}$$

But, we have,

$$\upsilon_n = \sqrt{\frac{2E}{m}} \quad \text{or} \quad L_n = \frac{1}{\nu}\sqrt{\frac{2E}{m}}$$

$\therefore$ Length of nth drift tube $= \dfrac{1}{500 \times 10^6} \left\{ \dfrac{2 \times 100 \times 1.6 \times 10^{-13}}{1.67 \times 10^{-27}} \right\}^{\frac{1}{2}}$

$$= \frac{1.38 \times 10^8}{500 \times 10^6} = 0.276 \text{ m}$$

(ii) If E_0 be the initial energy of the protons

$$E_0 = \frac{1}{2} m\upsilon_0^2 = \frac{1}{2} m\upsilon^2 L_1^2$$

Here $\qquad L_1 = 50 \text{ cm} = 0.05 \text{ m}$

$\therefore \qquad E_0 = 1.67 \times 10^{-27} \times (500 \times 10^6)^2 \times \dfrac{(0.05)^2}{2} \text{ J}$

$$= \frac{1.67 \times 10^{-27} \times (500 \times 10^6)^2 \times (0.05)^2}{2 \times 1.60 \times 10^{-13}} \text{ MeV}$$

$$= 3.26 \text{ MeV}$$

(iii) Net increase in energy of protons

$$= \text{Final energy attained} - \text{initial energy}$$

$$= 100 \text{ MeV} - 3.26 \text{ MeV} = 96.74 \text{ MeV}.$$

Increase in energy of protons at each gap

$$= qV = (1.6 \times 10^{-19}) \times (1500 \times 10^3) \text{ J}$$

$$= 1.6 \times 15 \times 10^{-14} \text{ J}$$

$$= \frac{1.6 \times 15 \times 10^{-14}}{1.60 \times 10^{-13}} \text{ MeV}$$

$$= 1.5 \text{ MeV}$$

$\therefore$ Total number of accelerations received to attain the maximum energy is $\dfrac{96.74}{1.5} = 65$

Example 8.31 A 250 MeV betatron has a stable orbit of radius 0.8 m. The electrons acquire energy at the rate of 500 eV per revolution. The magnet of the betatron is energised at 180 Hz and produces pulses of 1 μs duration, each containing 10^{10} electrons. Calculate (i) the distance covered by electrons to attain the maximum energy and (ii) the peak beam current. [Kerala]

Sol. Final electron energy = 250 MeV; Energy gained per revolution = 500 eV

∴ No. of revolution required to attain the final energy is

$$n = \frac{250 \times 10^6}{500} = 5 \times 10^5$$

∴ Distance travelled in 5×10^5 revolutions is

$$d = n.2\pi r_0 = 5 \times 10^5 \times 2 \times 3.14 \times 0.8$$
$$= 25.12 \times 10^5 \text{ m} = 25.12 \times 10^2 \text{ km}$$

Charge carried by each pulse, $q = Ne$, where N = no. of electrons in the pulse and e = electronic charge.

∴ $$q = 10^{10} \times 1.6 \times 10^{-19} \text{ C} = 1.6 \times 10^{-9} \text{ C}$$

Pulse duration = 1 μs = 1×10^{-6} s.

∴ Peak current $= \dfrac{1.6 \times 10^{-9}}{1 \times 10^{-6}} = 1.6 \times 10^{-3}$ A.

Example 8.32 An electrostatic separator has a uniform electric field E between the plates. If a paraxial beam of negatively charged p and k mesons with momentum p passes between the plates of the separator show that, on emerging, the angular separation between the p and K-mesons is $\Delta\theta$

$= \dfrac{eEl}{pc}\left(\dfrac{1}{\beta_k} - \dfrac{1}{\beta_\pi}\right)$ where l is the length of the plates and β_k and β_π are the relativistic velocities of

the k and π mesons respectively.

For practical calculations the angular separation can be expressed as $\Delta\theta = 1.0 \times 10^{-6} \dfrac{el}{p}\left(\dfrac{1}{\beta_k} - \dfrac{1}{\beta_\pi}\right)$

radiations with l in metres, E in $\dfrac{V}{m}$ and p in MeV/c. Calculate the angular separation in this case.

Take the masses of the p and k mesons to be 140 MeV/c^2 and 494 MeV/c^2 respectively.

Sol. The transverse momentum imparted to a particle in traversing the length l of separator is p_T

$= \int F \, dt = \dfrac{e\beta l}{\beta c}$. Hence

$$\Delta\theta = \frac{\Delta p_T}{p} = \frac{eEl}{pc}\left(\frac{1}{\beta_k} - \frac{1}{\beta_\pi}\right)$$

For a momentum p, $\beta = \dfrac{p}{E}$, hence $\beta_k^{-1} = 1.0135$ and $\beta_\pi^{-1} = 1.0011$.

Thus $$\Delta\theta = \frac{1 \times 10^{-6} \times 5 \times 5 \times 10^{-6} \times 0.0124}{3 \times 10^3}$$

$$= 0.1 \text{ m rad}$$

Example 8.33 A beam consists of positively charged pions, kaons and protons with momentum 10 GeV/c. Design a system of N_2-filled threshold Cherenkov Counters that would positively identify the particles if the refractive index of N_2 gas varies with pressure p (in atmospheres) as $n = 1 + 3 \times 10^{-4} \, p$. Take the masses of the pion, kaon and proton to be 0.14, 0.494 and 0.938 GeV/c² respectively.

Sol. Threshold Cherenkov counters discriminate between two particles of the same momentum and different masses simply by whether radiation is emitted or not (choose $\dfrac{1}{n}$ between β_1 and β_2). One need to compare values of β with values of $\dfrac{1}{n}$, or equivalently $\dfrac{1}{\beta}$ with n. We have $\beta = \dfrac{p}{E}$ or $\beta^{-1} = \dfrac{E}{p} = (p^2 + m^2)^{\frac{1}{2}}$; hence $\beta^{-1} \approx \dfrac{1 + \dfrac{1}{2} m^2}{p^2}$

Particle	$\dfrac{Mass}{GeV}$	$\dfrac{p}{(GeV/c)}$	$\beta^{-1} - 1$	$\dfrac{Pressure}{atom}$	$n - 1$
π	0.140	10	9.8×10^{-5}	1	3×10^{-4}
k	0.494	10	1.22×10^{-3}	10	3×10^{-3}
p	0.938	10	4.4×10^{-3}	20	6×10^{-3}

Positive identification may be achieved with the three Cherenkov counters with pressures chosen.

Example 8.34 In a simple heavy charged particle identification system signals from a thin counter (giving $\dfrac{dE}{dx}$) and stopping counter (giving $T = E - mc^2$) are multiplied. Show that the product $T \left(\dfrac{dE}{dx} \right)$ is to a first approximation independent of velocity and find the ratio of this quantity for deutron and ³He particles.

Sol. We have $\dfrac{dE}{dx} \approx \dfrac{z^2}{v^2}$ (From Equation 8.1 Section 1)

neglecting the logarithmic term. For a non-relativistic particle $T = \dfrac{1}{2} m v^2$ so that $T \dfrac{dE}{dx} \propto mz^2$. For deutrons $m \propto 1$, $Z = 1$ and for ³He $m \propto 3$, $Z = 2$

$$\therefore \qquad T \frac{dE}{dx} = \frac{1}{6}.$$

Example 8.35 A belt system of total width 0.3 m charges the electrode of an electrostatic generator at a speed of 20 m/s. If the break down strength of the gas surrounding the belts is 3 MV/m.

Calculate: (i) the maximum charging current, (ii) the maximum rate of rise of electrode potential, assuming a capacitance of 111 pF and no load current.

Sol. Let the surface density of charge be σ. Then by Gauss's theorem the electric field near the surface element is given by $2\,EdS = \dfrac{4\pi\,\sigma\,ds}{4\pi\,\varepsilon_0}$ or $E = \dfrac{\sigma}{2\,\varepsilon_0}$. The maximum charge carried to the terminal/s is then $\sigma\,v\,b$ where v is the speed of the belt and b is its width. Now, inserting values we get the current as

$$i = 2 \times 8.85 \times 10^{-12} \times 3 \times 10^6 \times 20 \times 0.3 = 319 \times 10^{-6} \text{ V}$$

$$= 919 \ \mu A.$$

The electrode potential rises at the rate

$$\frac{i}{C} = \frac{(319 \times 10^{-6})}{(111 \times 10^{-12})} \text{ V/s} = 2.9 \text{ MV/s}$$

Example 8.36 Calculate the energy difference between the photo peak and the high energy side of the Compton electron distribution in the pulse height spectrum from a scintillator detecting γ-radiation of energy $m_e c^2$. What would the energy of the back scattered radiation be?

Sol. From Equation 8.1, section 1, the maximum loss of energy to an electron is when $\theta = \pi$. We then have $\dfrac{hc}{E'} - \dfrac{hc}{E} = \dfrac{2h}{m_e c}$ and the high energy Compton edge is at the energy $E - E' =$

$\dfrac{2E^2}{(2E + m_e c^2)} = \dfrac{2}{3}\,m_e c^2$ for $E = m_e c^2$. This is separated from the photo peak at $E = m_e c^2$ by

$\dfrac{1}{3}\,m_e c^2 = 170$ keV. The energy of the back scattered radiation is $E' = \dfrac{1}{3}\,m_e c^2 = 170$ keV.

Example 8.37 Deutrons of energy 15 MeV are extracted from a cyclotron at a radius of 0.51 m by applying an electric field of 6 MV/m over an orbit arc of 90°. Calculate the equivalent reduction of magnetic field and the resulting increase in orbit radius Δr.

Sol. Since the magnetic for velocity v is BeV and the electric force is E_e the effective reduction of magnetic field is $\Delta B = \dfrac{E}{v}$. The deutron velocity v is given by $v^2 = \dfrac{(2 \times 15 \times 1.6 \times 10^{-13})}{(2 \times 1.67 \times 10^{-27})}$

whence $v = 3.79 \times 10^7$ m/s and for $E = 6 \times 10^6$ V/m, $\Delta B = 0.16\ T$. We have

$$B = \frac{mv}{er} = \frac{(2mT)^{\frac{1}{2}}}{er}$$

Where T is K.E. and r the orbit radius. Now, substituting the values, one obtains $B = 1.54\ T$ and Δr is the then given by $r\,\dfrac{\Delta B}{B} = 0.053$ m.

SUGGESTED READINGS

1. *E*. Segre, 'Nuclei and Particles', Benjamin (1977)
2. A.B.J. England, 'Techniques in Nuclear Structure Physics', Mcmillian (1974)
3. K. Siegbalim (ed.), 'Alpha, Beta and Gamma Ray Spectroscopy', North-Holland (1965)
4. D. A. Bromley (ed), 'Detectors in Nuclear Science', Nucl Instrum Meth, 162(1979)
5. G. Hall, 'Modern charged particle detectors Contemp Phys 33(1) (1992).
6. K. W. Allen, 'Electrostatic Accelerators' in J. Cerny (ed) Nuclear Spectroscopy and Reactions (Part A) Academic Press (1974)
7. W. Scharf, Particle Accelerators and their uses, Harwood (1991)
8. S. Meer Van der, IEEE Trans Nucl. Sci NS-28 (1994) (1991).

REVIEW QUESTIONS

1. Giva an account of methods of producing high energy charged particles.
2. Describe construction, working and necessary theory of the cyclotron. What are the limitations of a cyclotron?
3. Describe and explain linear acceleration for accelerating (i) protons (ii) Electrons.
4. Explain focussing action in linear acceleration. Hence describe the working of a wave guide accelerator. How this is advantageous over drift tube accelerator?
5. Explain the theory and working of an accelerator which is used to accelerate particles.
6. Explain the principle of synchrotron. In what ways a proton synchrotron is different than an electron synchrotron. Explain the working of proton-synchrotron.
7. Describe and explain the working of the proton synchrotron.
8. Explain the principle of working and construction of a betatron. Derive the condition for its operation. How it is achieved?
9. Describe a betatron, obtain and expression for the total energy of electron accelerated by a betatron. What are its limitations?
10. Why is cyclotron not used for producing high energy electrons?
11. (a) Describe the principle, construction and working of a cyclotron. Derive the required relations.
    (b) How does a synchrocyclotron differ from cyclotron? What is its advantage over a cyclotron?
12. Explain the principle of synchrotron. In what ways is a proton synchrotron different from an electron-synchrotron? Explain the working of proton-synchrotron.
13. Give theory and explain working of an accelerator which is used to accelerate β-particle.
14. Give theory, working and construction of a cyclotron. What are limitations of a cyclotron?
15. Discuss the working principle of a betatron. What difficulties does one encounter while accelerating electrons beyond 100 MeV?
16. State the principle of linear accelerator. Describe its construction and working.
17. Describe the construction of a betatron. How energy is gained in it? Obtain 'Betatron condition'. How is it achieved? Derive an expression for estimating the total energy of electrons accelerated by a Betatron.
18. Give an account of the detection of particles having low energies.
19. Explain and distinguish between operation of proportional and Geiger region. Describe and explain the working of a Geiger-Muller tube.
20. Give an account of proportional counter for detection charged particle.

21. Explain the meaning of avalanche. How this effect has been made use of detecting a single particle. Explain a counter working on this principle.
22. Describe and explain the function of the Wilson Cloud chamber. Summarise its application in physics.
23. How the bubble chamber is a valuable in nuclear technique? What advantages it has over the Wilson's cloud chamber. Describe and explain with necessary theory, working of a bubble chamber.
24. Write an essay on "Detection of charged particles, and their relative advantage."
25. Give the theory and working of an ionisation chamber.
26. What do you understand by self quenching. Explain the construction and working of a self quenching G.M. Counter.
27. Give a brief account of different types of particle detectors.
28. Make a comparative study of ionisation chambers, proportional and Geiger-Muller counters, explaining the differences in their mechanism of working. Explain 'avalanche' 'quenching' and 'dead time' in a self quenched Geiger Muller counter. Explain its mode of operation. Draw its voltage characteristic and lable the threshold voltage, plateau region, working voltage and continuous discharge region.
29. Describe a cloud chamber and explain its working. What types of tracks do you obtain in it for α and β particles. How can it be used to measure the energy of a charged particle? What is an automatic cloud chamber?
30. Describe a scintillaton counter and ins mode of operation. What advantages has it over other types of counter?
31. Write short notes on
    (a) Bubble chamber
    (b) Photographic emulsion
32. Make a comparative study of ionisation chambers, proportional and Geiger-Muller counters, explaining the difference in their machanism of working.
33. (a) Describe the principle, construction, working and speciality of Bubble chamber.
    (b) Which detector will you use and why
        (i) Locating radioactive source
        (ii) Determination of energies of particles
        (iii) Distinguish between particles and protons.
    (iv) Count α particles in the presence of β particles.
    (v) Photograph meson tracks?
34. (a) Explain avalanche, quenching and dead time.
    (b) Make a comparative study of ionisation chamber, proportional counter and G.M. Counter.
35. Give the theory and working of ionization chamber for detecting charged particles.
36. Give the theory and working of proportional counter detecting charged particles.

PROBLEMS

1. In a proton cyclotron the frequency of radio source is 15×10^6 cycle/sec. of amplitude 5000 V. Find (a) final velocity of proton (b) the energy of protons (c) number of revolutions made by protons and (d) time for which they remained inside dee. Considering radius of cyclotron to be 40 cms.
2. A cyclotron of 50 cm. radius has a magnetic field of 1 weber/meter^2. Find the energy acquired by protons and what frequency must be applied to the dees. If the voltage of the dees, when the protons cross gap, is 50 KV, how long the acceleration lasts?

3. Find the maximum energy of deutrons in a 60 inch diameter cyclotron operating with 12 Mc/sec. source when the magnetic field is 16000 gauss. **(Ans. 26 MeV)**

4. A proton synchroton produces protons of 3 BeV and uses maximum magnetic field 12,000 gausses. Find radius of orbit. What frequency of radio-frequency field will be required when protons have maximum energy? If the protons are injected at an initial energy 4 MeV, find the magnetic field radio frequency at the time of injection.

   Ans. [10.6 m, 4.4×10^6 c/s, 270 gauss, 0.42×10^6 C/S]

5. If the frequency of applied voltage is 1.2×10^7 c/sec, find values of field strength of resonance when (a) protons (b) deutrons and (c) α-particles, are to be accelerated. The radius of orbit is 50 cm. Find also their energies.

   Ans. [7.86×10^3, 1.572×10^4, 1.56×10^4, 7.4 MeV 14.8 MeV, 14.7 MeV]

6. In a 60 inch diameter betatron, maximum magnetic field is 5000 gauss operating at 50 C/sec. Find average energy gained per cycle and the total energy gained by the electron.

7. In a proton synchrotron, maximum energy output is 10 BeV, with an orbital radius 100 meters, if the protons are injected at 10 MeV find the frequency of revolution. Also find their frequency when they acquire maximum energy.

8. In an electron synchrotron radius of stable orbit is 25 cm. and the maximum magnetic field is 10000 gauss. Find the frequency of applied electric field and maximum energy of the electron.

   (Ans. 1.9×10^8 C/s, 75 MeV)

9. Calculate the frequency of the electric field that must be applied between the dees of a cyclotron in which deutrons are accelerated. The strength of magnetic field is 3 Wb/m². Given 1 amu = 1.66×10^{-27} kg and $e = 1.6 \times 10^{-19}$ Coulomb. **[Ans. 23 mega Hz]**

10. Find the energy of deutrons accelerated in a 60 inch diameter cyclotron. The frequency of alternating voltage applied between the dees is 12 mega Hz. Find the strength of magnetic field.

    [Ans. 35.6 MeV, 1.57 Tesla]

11. Protons are accelerated in a cyclotron. The diameter of the dees is 1 meter, the alternating voltage applied between the dees has a frequency 10 mega Hz and its peak value is 5000 Volts. Calculate (i) The magnetic field needed for resonance (ii) The energy of the emergent protons and (iii) the final velocity of protons.

12. In a betatron the magnetic field had a peak value of 4000 gauss and its was varied at a frequency of 60 cycle/sec. The diameter D of the stable orbit was 66 inches. Calculate the maximum energy acquired by the electrons. **(Ans. 100 MeV)**

13. Calculate the radius of the circular path in a 20 BeV proton synchroton. The available magnetic field has a maximum strength of 1.5 Tesla. **(Ans. 46.6 metres)**

    [Hint: rest mass = 22.49 amu and $R = \dfrac{Mv}{Be}$]

14. In a certain betratron experiments, the maximum magnetic field at the orbit is 0.4 Tesla, operating at 50 cycles/second with a stable orbit diameter of 1.5 m. Calculated the average energy gained per revolution and the final maximum energy of electrons. Assume that the electron moves with nearly the velocity of light.

15. In a certain proton synchrotron, a proton with an initial energy of 9 MeV is accelerated to 10 GeV gaining on the average an energy of 2.2 keV per turn. The total length of the stable orbit inclusive of the straight section is 100 m. The acceleration time is 3.3 sec. (i) Determine the number of

revolutions and paths length covered by protons in the whole period of acceleration. (ii) What is final velocity of protons?

16. Calculate average energy per revolution and final energy gained by electrons in a betatron to which is applied a maximum magnetic field of 5000 gauss operating at 60 cps in a stable orbit of diameter 2 meters.

17. A cyclotron of diameter 1 meter is used to accelerate protons. It is applied alternating voltage of frequency 10 mega cycles/sec of peak value 5000 volts. Calculate the magnetic field needed and energy of emergent blam.

18. Find the maximum energy of a deutron in a 60″ diameter cyclotron operating at MHz source. The magnetic field is 16000 gauss.

19. In a 50″ diameter betatron, maximum magnetic field is 5000 gauss operating at 50 Hz. Find the total energy gained by the electron.

20. The diameter of a cyclotron D is 1.8 m and strength of magnetic field is 0.7 Wb/m^2. Calculate the energies to which (i) a proton, (ii) a deutron can be accelerated.

21. A 10 MeV α-particle loses all its energy in a proportional counter. One electron pair is produced for each 35 eV energy loss. The proportional counter has a multiplication $M = 1000$, and the total capacitance between wire and ground is 30 pf. Find voltage of output pulse.

22. Find the ionisation current due to a beam of 10 MeV α-particles entering in an ionisation chamber at the rate of 10 required to produce one ion pair.

23. The vapour pressure of air water mixture at 20°C is 17.5 mm of Hg. If the specific heat ratio is 1.41, find supersaturation ratio of the mixture.

24. In an ionisation chamber $C = 30\ \mu - \mu f$ and $V_0 = 300$ Volts. How much energy is stored in electrostatic field. If an α-particle forms 10,000 ion pairs in its active volume, what change would be than in its potential. Also find the fraction of stored energy, converted to heat.

25. Find the ionization current due to a beam of 5 MeV particle entering in an ionization chamber at the rate of 15 particles per sec. Assume that on the average 35 eV energy required to produce one ion pair.

26. An 11 MeV α-particle looses all its energy in proportional counter. One electron ion pair is produced for each 35 eV energy loss. The proportional counter has multiplication $M = 1000$ and total capacitance between wire and ground is 30 pf. Find the voltage of output pulse.

27. Find the number of ion pairs produced by a 10 MeV proton. If in the proportionality counter the amplification is 10^3, current pulse time is 10 μ sec and the resistance is 10^4 Ohms then find the height of the voltage pulse. Energy for producing one ion pair is 34 eV.

SHORT QUESTION ANSWERS

1. What are nuclear radiation or particle detectors?

Ans. Nuclear radiation (or particles) such as α, β and γ-rays cannot be observed directly. They are detected by the secondary effects which they produce in the materials through which they pass. There are three such effects: ionisation, photographic and fluorescence. Number of detecting instruments are based upon them

2. What is ionisation? How ionisation produced by charged particle can be used to detect the particle?

Ans. A charged particle like α, β, p, etc. passing through the material produces ions by collisions with the molecules of the material. This forms the basis of many detecting instruments such as electroscope, spark counter, ionisation chamber, Geiger-Muller counter, cloud chamber and bubble chamber. There are three principal ways to detect the particle by these instruments.

   (i) The ionisation can be made visibly observable. Then the tracks can be seen or photographed. Cloud chamber, diffusion chamber, bubble chamber and spark chamber uses this method.

  (ii) When the electrons and positive ions formed by ionisation recombine and light is given out. One can measure this light by photosensitive devices, e.g. scintillation counter.

 (iii) By applying an electric field, the electrons and ions can be collected. This will result in an electrical signal. Ionization chamber, GM counter and proportional counters make use of this fact.

  3. What are the uses of ionisation instruments?

Ans. These are used for the detection and also for the measurments of radioactive radiations. The ionisation chamber is used for the detection of α and p. One can use the proportional counter for the rough measurment of the intensity of the α-particles. GM counter is suitable for detecting β and γ-rays. GM counters are used in uranium prospecting and health protection operations. One can trace the path taken by radioactive iodine taken into the stomach in reaching the thyroid gland with a G.M. Counter.

  4. What is an particle accelerator?

Ans. A particle accelerator is a device built to accelerate (i.e., to increase the kinetic energy) the charged particles so as to give them desired energy. The maximum energy that can be acquired by a particle depends upon the type of accelerator.

  5. What are the different classes of particle accelerators?

Ans. (i) linear or lineac   (ii) static or statitron, and   (iii) circlar

  6. What is an omnitron?

Ans. This is an particle accelerator which gives strong beams of accelerated particles of all masses from hydrogen to uranium.

  7. What is the basic principle of a betatron?

Ans. Betatron is a device for speeding up electrons to extremely high energies with the help of expanding magnetic field. One can accelerate electrons with Betatron upto energies 300 MeV.

  8. What are the advantages of the silicon p-n junction as a detector of heavy particles?

Ans. Its excellent resolution and linearity, its small size and consequent fast response time permitting a high counting rate and the fairly simple nature of the necessary electronic circuits.

  9. What do you understand by 'anti-coincidence circuit'?

Ans. In several experiments, it becomes necessary to record the arrival of one or more pulses and to exclude those in which all the Rossi circuit tubes receive input pulses in their grids. An arrangement that makes this possible is called anti-coincidence circuit.

  10. What is a Cerenkov counter?

Ans. This is essentially a scintillation counter coupled with a mirror system.

OBJECTIVE QUESTIONS

  1. The threshold energy for the photoproducting of an electron-positron pair in the field of a free electron is

(a) $m_e c^2$ (b) $2\,m_e c^2$
(c) $4\,m_e c^2$ (d) $8\,m_e c^2$

2. A triton ($^3_1$H) of energy 5000 MeV passes through a transparent medium of refractive index $n = 1.5$. The angle of emission of Cherenkov light is
  (a) 44°   (b) 22°
  (c) 11°   (d) 5°

3. The energy of 100 keV photon scattered backwards ($\theta = 180°$) from an electron of energy 100 keV moving towards the photon is
  (a) 315 keV   (b) 515 keV
  (c) 715 keV   (d) 1015 keV

4. A Geiger counter is operated in a circuit which imposes a paralysis time of 300 μs after each count. Corresponding with an observed rate of 10000 per minute, the true rate of counting is
  (a) 5526/minute   (b) 7526/minute
  (c) 2526/minute   (d) 10526/minute

5. If the terminal of a Van de Graaff generator may be regarded as a cylinder of radius r and the pressure vessel as a coaxial cylinder of radius R, the radial voltage gradient is given by $E_r = \left[\dfrac{V}{r}\ln\left(\dfrac{R}{r}\right)\right]$, where V is the terminal potential. For any given E_r, V is a maximum, when
  (a) $\dfrac{R}{r} = e$ (b) $\dfrac{R}{r} = \dfrac{1}{e}$
  (c) $\dfrac{R}{r} = e^2$ (d) $\dfrac{R}{r} = e^{-2}$

6. In the SLA electrons reach an energy of 20 GeV in a distance of 3 km. Assuming that increments of energy are uniform and that the injection energy is 10 MeV, the effective length of the accelerator to an electron is
  (a) 0.17 m   (b) 0.37 m
  (c) 0.47 m   (d) 0.57 m

7. A pulse of 10^{10} particles of single charge is injected into a cyclic accelerator and is kept circulating in a stable orbit by the application of a radio frequency field. When the radio frequency is 7 MHz, the mean current is
  (a) 3.2 mA   (b) 5.2 mA
  (c) 7.2 mA   (d) 11.2 mA

8. The extracted proton beam from a proton synchrotron has an energy of 28 GeV. The CM energy in collisions with a fixed target of hydrogen is
  (a) 5.25 GeV   (b) 7.25 GeV
  (c) 2.25 GeV   (d) 3.25 GeV

ANSWERS

1. (c)   **2.** (a)   **3.** (d)   **4.** (d)   **5.** (a)
6. (d)   **7.** (d)   **8.** (b)

HINTS

1. Let the γ-ray energy be E_γ. At threshold the particles are at rest in the CM system conservation of four momentum gives, for a final system of three electron masses,

   $$p_\gamma + p_e = p_{3m}$$

   Taking the scalar product

   $$(p_\gamma)^2 + (p_e)^2 + 2\,\mathbf{p}_\gamma \cdot \mathbf{p}_e = (p_{3m})^2$$

   and inserting values for an electron initially at rest and $c = 1$

   $$0 + m_e^2 + 2\,E_\gamma m_e = 9\,m_e^2$$

   $$\therefore \qquad E_\gamma = 4\,m_e$$

2. 1 mass unit = 93 MeV/c^2, the Lorentz force for the triton is

   $$\gamma = \frac{5000 + 5 \times 931}{3 \times 93} = 2.79 = \frac{1}{(1 - \beta^2)^{\frac{1}{2}}}$$

   Whence $\qquad \beta = 0.93$ and $\theta = \cos^{-1}\dfrac{1}{\beta n} = 44°$

3. Let the initial and final electron energy E_1 and E_2 and the initial and final photon energy be $h\nu_1$, and $h\nu_2$. The initial momenta are for the electron $p_1 = [(1511)^2 - (511)^2]^{\frac{1}{2}} = 1422\ \mathrm{keV}/c$ and for the photon 100 keV/c. Equations for conservation of momentum and energy are then

   $$1422 - 100 = p_2 + h\nu_2 \quad (\mathrm{keV}/c)$$
   $$1511 + 100 = E_2 + h\nu_2 \quad (\mathrm{keV})$$

   Using $E_2^2 - p_2^2 = m_c^2$ and eliminating E_2 and p_2, one finds

   $$(1611 - h\nu_2)^2 - (1322 - h\nu_2) = (511)^2.$$

   $$\therefore \qquad h\nu_2 = 1015\ \mathrm{keV}$$

4. Counts are lost during a period of $\dfrac{300 \times 10000}{60}$ second during each second. This is 0.05 s so the true rate will be $\dfrac{10000}{0.95} = 10526$

5. Express V in terms of r and set $\dfrac{dV}{dr} = 0$.

6. Since $E = \gamma_m$, uniform increments of energy mean that γ increases uniformly with distance along the accelerator, so that

   $$\gamma = \gamma_0 + kl \quad \text{or} \quad dl = k^{-1}\,d\gamma$$

   From the point of view of electron, a Lorentz transformation gives $dl_e = \dfrac{dl}{\gamma} = \dfrac{d\gamma}{k\gamma}$ so that $l_e = k^{-1}$

   $\ln\left(\dfrac{\gamma_r}{\gamma_0}\right)$, where γ_r refers to the final energy. Substituting the values given, one obtains $l_e = 0.57$ m

7. Since 10^{10} particles of charge a appears at a point 7×10^6 times/s, the current is $i = 10^{10} \times 1.6 \times 10^{-19} \times 7 \times 10^6$ A $= 11.2 \times 10^{-3}$ A

8. In natural units $E^2 - p^2 = E'^2$, where E is the total energy in L-system, p is the beam momentum in the laboratory and E' is the total of CM energy. If W is the beam energy, $E = W + m_p$ and $p^2 = W^2 - m_p^2$, where m_p is the mass of the proton. Hence, $E' = (2 \, m_p w)^{\frac{1}{2}} = 7.25$ GeV.

ELEMENTARY PARTICLES AND PARTICLE PHYSICS

9.1 INTRODUCTION

The discovery of the electron by J. J. Thomson in 1898 and that of atomic nucleus and the proton by Rutherford in 1914 made it clear that the atoms themself had an internal structure. The discovery of the neutron by Chadwick in 1932 in a way completed the picture of atom. On this picture the atom consists of protons and neutrons surrounded by electrons to compensate for the positive charge on the nucleus. At the begining of 20th century, these were the elementary particles of the nature. However, by the 1960's with the advent of huge accelerating machines a plethora of elementary particles hyperons, kaons, leptons, and so on had been discovered. Several hundred 'elementary' particles are known today. Almost all these particles (other than electron, proton and neutron) decay rapidly after being created in high energy collisions between other particles. It has become clear that some of these elementary particles are more elementary than others, and that the latter are probably composites of a far smaller number of rather unusual particles 'quarks' that have not as yet been detected in isolation. A time table for the experimental discovery of the fundamental particles is shown in Figure 9.1. (Some particles were predicted on theoretical grounds prior to their experimental discovery and others were observed unexpectedly).

Remarkably, the field of elementary particles is completely different today from what it was about twenty years ago. Today quarks and leptons are the fundamental objects; they interact via the gauge bosons. The forces that significantly affect them are the unified electroweak force, whose gauge bosons are the photon and the $W^\pm$ and Z° bosons, and the strong force. The theory of the strong force is called quantum chromodynamics (QCD). The gauge bosons of the strong force are the (eight) gluons. In the present chapter we shall study about these elementary particles.

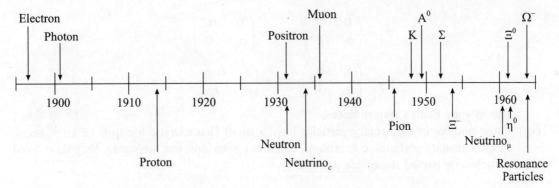

Fig. 9.1 Time table for the discovery of elementary particles.

9.2 SALIENT FEATURES

The elementary particles discovered so far form a long list. Some salient features about these particles are

(i) The elementary particles are analysed in terms of their mass, intrinsic spin, magnetic moment and interaction properties. The elementary particles having the same properties are put in the same group.

(ii) The elementary particles are confined to some relevant quantum numbers. Every particle has intrinsic angular momentum either $\frac{1}{2}\hbar$, $\frac{3}{2}\hbar$ or integral 0, 1, 2, 3 etc. in the units of $\hbar$.

(iii) The elementary particles of spins $\frac{1}{2}$ or $\frac{3}{2}$ obeying Fermi Dirac (FD) statistics are called *Fermions,* e.g. electrons, positrons, neutrinos, antineutrinos, m mesons, neutrons and many hyperons.

(iv) The elementary particles of spin 0 or 1 obeying BE statistics are called bosons, e.g. $\pi^\pm$, π^0, $K^\pm$, η^0 mesons having spin 1.

(v) Spinless particles are studied by Klein and Gordan (K.G.) equation

$$\left(\Box^2 - \frac{m^2c^2}{\hbar^2}\right)\psi = 0 \qquad \qquad \ldots (8.1)$$

where $\Box^2 = \left(\nabla^2 - \frac{1}{c^2}\frac{\partial^2}{\partial t^2}\right)$ is known as de Alemberation operator.

(vi) Spin $\frac{1}{2}$ particles are studied by Dirac's relativistic wave equation

$$i\hbar\,\frac{\partial\psi}{\partial t} = (C\,\boldsymbol{\alpha}\cdot\mathbf{p} + \beta mc^2)\psi = 0 \qquad \qquad \ldots (8.2)$$

Where α and β are Dirac matrices

$$\alpha_1 = \begin{pmatrix} 0 & \sigma_1 \\ \sigma_1 & 0 \end{pmatrix}, \quad \alpha_2 = \begin{pmatrix} 0 & \sigma_2 \\ \sigma_2 & 0 \end{pmatrix}$$

$$\alpha_2 = \begin{pmatrix} 0 & \sigma_3 \\ \sigma_3 & 0 \end{pmatrix}, \quad \beta = \begin{pmatrix} I & 0 \\ 0 & -I \end{pmatrix}$$

with σ's are Pauli's spin matrices.

(vii) Large number of elementary particles have a small characterstic-life time ($\geq 10^{-16}$ secs). Few elementary particles, e.g. protons, electrons, neutrinos, etc. are stable. Very short lived particles are named resonance particles.

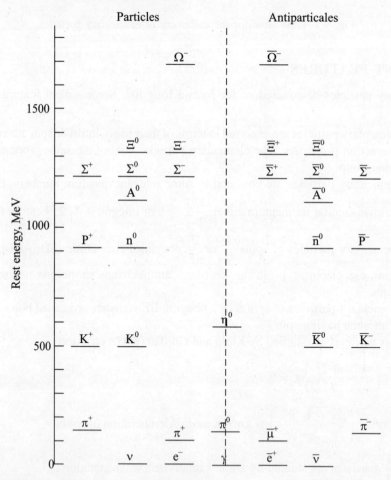

Fig. 9.2 Particles and antiparticles arranged according to their charge and rest energy.

(viii) Each particle has an associated *anti particle*. An antiparticle is designated by the same symbol as the particle, but with a bar over it. The symbol is needed and is consistently used for neutral anti-particles. However, for many others the sign of the charge is sufficient indication, e.g. e^+, p^-, μ^+, π^+ etc. antiparticles have the same mass and spin as the particles but opposite electromagnetic properties, such as charge and magnetic moment (other properties, to be mentioned later on, are also opposite). Four particles, the photon, the graviton, and π°, η° mesons are their own antiparticles. The existence of antiparticle is a requirement of relativity and quantum mechanics. In Figure 9.2 particles and antiparticles are arranged in a systematic way. A particle and its antiparticle may combine, disappear or annihilate each other, e.g.

$$e^- + e^+ \rightarrow 2\gamma \text{ (photon)}$$

The electron was the first elementary particle for which a statisfactory theory was developed by Dirac-in 1928. Relation between particles and their antiparticles is summarized in Table 9.1.

Table 9.1 Relation Between Particle and Anti Practicle

Property	Relation
Mass	Equal
Spin	Equal
Charge	Equal in magnitude but opposite in nature
Magnetic Moment	Equal in magnitude but opposite insign
Mean Life at the time of free decay	Equal
Production	In pair
Destruction	In pair
Total isotopical spin	Equal
Third component of Isotopic spin	Equal in magnitude but opposite in sign
Internal parity	Same for bosons Opposite for fermions
Stangeness Quantum number	Equal in magnitude but opposite in sign.

(ix) The unstable particles decay into simpler particles or annihilate with their antiparticles in due course of time after their production.

Now, we shall study these features in detail.

9.3 CLASSIFICATION OF ELEMENTARY PARTICLES

On the basis of the property of elementary particles called "spin", one can grouped them into two broad categories, e.g. (i) **Fermions**: obeying FD statistics and (ii) **Bosons**: obeying BE statistics.

The most important difference between these two groups of particles is that there is no conservation law controlling the total number of bosons in universe, where as the total number of fermions is strictly conserved.

The fermions fall mainly into two categories:

(i) Leptons (Weakly interacting Fermions) The lightest family of fermions is known as the family of leptons and includes four particles and four antiparticles. The lighest of the four are the neutrinos, which have rest mass zero. The two kinds of neutrino known are electron-neutrino (ν_e) and muon neutrino (ν_μ). Each is associated with its corresponding anti particle. A neutrino is said to have *negative helicity* equal to -1, and an anti neutrino is said to have positive helicity, equal to $+1$ (Figure 9.3). Both types of neutrino have spin $\dfrac{1}{2}$, and neither has an electric charge. The two are differentiated from one another by their couplings, one to the electron field and other to the muon field. These couplings are through the weak interaction. When an electron is involved in a weak interaction, so is also the corresponding electron-neutrino. When a muon is involved in a weak interaction, the corresponding muon neutrino is also involved.

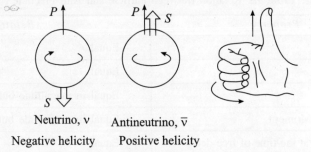

Neutrino, ν Antineutrino, $\bar{\nu}$

Negative helicity   Positive helicity

Fig. 9.3 The neutrino and antineutrino have opposite helicity.

A typical reaction involving electrons and neutrinos is that of β-decay:

$$\,_0^1 n \rightarrow \,_1^1 p + \,_{-1}^0 e + \bar{\nu}_e \qquad \qquad \text{... (8.3)}$$

and

$$\,_1^1 p \rightarrow \,_0^1 n + \,_{+1}^0 e + \nu_e \qquad \qquad \text{... (8.4)}$$

and the inverse of above are

$$\bar{\nu}_e + \,_1^1 p \rightarrow \,_0^1 n + \,_{+1}^0 e \qquad \qquad \text{... (9.5)}$$

and

$$\nu_e + \,_0 n^1 \rightarrow \,_1^1 p + \,_{-1}^0 e \qquad \qquad \text{... (9.6)}$$

Typical reactions involving muons and neutrinos are those of pion decay

$$\pi^- \rightarrow \mu^- + \bar{\nu}_\mu \qquad \qquad \text{... (9.7)}$$

and

$$\pi^+ \rightarrow \mu^+ + \nu_\mu \qquad \qquad \text{... (9.8)}$$

The muon production reaction is

$$\bar{\nu}_\mu + \,_1^1 p \rightarrow \,_0^1 n + \mu^+ \qquad \qquad \text{... (9.9)}$$

We note that in all these above reactions an electron is associated with its antineutrino when both appear on the same side of arrow, but associated with its neutrino when they appear on opposite sides. The same holds good for muon and its neutrino. This is a consequence of the law of conservation of particle number and in this case the law is the conservation of lepton number (L). Lepton number has a value $L = +1$ for leptons ($\,_{-1}^0 e$, μ, ν_e, ν_μ), $L = -1$ for anti leptons ($\,_{+1}^0 e$, μ^+, $\bar{\nu}_e$, $\bar{\nu}_\mu$) and $L = 0$ for all other Leptons.

Neutrinos carry only the weak interaction and so interact only very weakly with other particles. This is expressed through an exceedindly small interaction cross section and correspondingly long mean free path through matter. A typical neutrino interaction cross section is of the order of 10^{-49} m^2 or less, while the maximum known is about 10^{-42} m^2. A typical neutrino mean free path is thousands of light years long through condensed matter such as lead.

Neutrinos are the only known particles that appear to be absolutely polarized in spin with reference to their momentum vectors. Neutrinos spin counter clockwise (with negative helicity), and antineutrinos spin clockwise (with positive helicity) relative to their, directions of travel. Neutrinos are stable against decay into other particles.

We must note that the neutrinos involved in pion decays are not the same as those involved in β-decay. The existence of two classes of neutrinos was established in 1962.

Electrons and Muons   The two charged members of the lepton family are the electron (e) and muon (μ). These two particles act very much alike, although they differ in rest mass. The rest mass of electron (m_0) is 0.510 MeV where as the rest mass of muon is about 200 times m_0, i.e. 105.66 MeV. Both these particles have spin $\dfrac{1}{2}$ and each has negative electric charge. The anti particle of each is positively charged. Figure 9.4 illustrates the production of an electron-positron pair by a photon entering a cloud chamber from the left, i.e.

$$\gamma \rightarrow {}_{-1}^{0}e + {}_{+1}^{0}e \qquad \qquad ... (9.10)$$

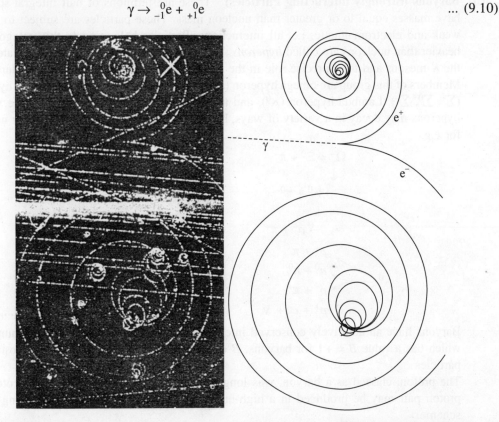

Fig. 9.4 Electron-positron pair production (Countery: Brook haven National Laboratory).

The track of photon is not visible. Obviously for an electron pair to be produced, the photon must have an energy equal to at least $2m_e c^2 = 1.022$ MeV. In order for energy and momentum to be conserved in Equation (9.10), the process must occur near a nucleus.

The electron is stable against decay into other particles, while muon decays into an electron and two neutrinos:

$$\mu^- \to {}_{-1}^{0}e + \bar{v}_e + v_\mu \qquad \qquad ... (9.11)$$

The positive muon (the anti particle) decays into the charge conjugate states of the same three products:

$$\mu^+ \to {}_{+1}^{0}e + v_e + \bar{v}_\mu \qquad \qquad ... (9.12)$$

In these decays, the electron-neutrino is associated with the presence of an electron and the muon-neutrino with the muon as in the preceding reactions. The mean life time of the muon in these decays is 2.2×10^{-6} sec.

The electron was the first of the elementary particles to be found in nature by J. J. Thomson, while the muon was the firstt cosmic ray particle heavier than the electron was found. The great majority of cosmic-ray secondary particles at sea level are muons. The muon life time is long enough for a negative muon sometimes to temporarily replace an atomic electron to form a muonic atom.

(ii) **Baryons (Strongly interacting Particles)**   These are fermions of half integral spin and have masses equal to or greater than nucleon mass. These particles are subject to strong, weak and electromagnetic, i.e. all interactions. Particles belonging to this category and heavier than nucleons are called *hyperons*. The hyperons are usually found associated with the *K*-mesons and their precise role in the construction of matter is not fully understood. Members of this group are omega hyperon (Ω^-), Cascade hyperons ($\Xi^0 \ \Xi^-$), Sigma hyperons (Σ^+, Σ^0, Σ^-), Lambda hyperon ($\wedge^0$), and the nucleons (proton and neutron). The various hyperons may decay in a variety of ways, but the end result is always a proton or neutron, for e.g.

$$\Omega^- = \Xi^0 + \pi^-$$
$$\downarrow$$
$$\to \wedge^0 + \pi^0 \qquad \qquad ... (9.13)$$
$$\downarrow$$
$$\to p^+ + \pi^-$$

$$\wedge^0 \begin{cases} p^+ + \pi^- \\ n^0 + \pi^0 \\ p^- + \pi^+ \\ p^+ + e^- + \bar{v} \end{cases} \qquad \qquad ... (9.14)$$

Baryons have an additively conserved internal quantum number called Baryon number B which has a value $B = +1$ for baryons, $B = -1$ for anti baryons and $B = 0$ for rest of the particles.

The proton, classed as a baryon, was long thought be only stable hadron. A proton-anti proton pair may be produced in a high-energy proton-proton collision according to the scheme.

$$p^+ + p^+ \rightarrow p^+ + p^+ + p^+ + \bar{p}^- \qquad \dots (9.15)$$

(This is not the only process that may take place in a proton-proton collision). Proton-antiproton annihilation is a more complex process, involving the production of several particles, most of them pions. One such process is

$$p^+ + p^- \rightarrow 4\pi^+ + 4\pi^- + x\pi^0 \qquad \dots (9.16)$$

Fig. 9.5 Proton-antiproton annhilation (Courtesy: Brookhaven Nat. Lab.)

Proton-antiproton annihilation according to (9.16) took place in a bubble chamber is shown in Figure 9.5. The experimental arrangement to observe an antiproton is shown in Figure 9.6. Protons accelerated up to 6.2 GeV, hit a suitable target which produced several reactions involving $K's$, $\pi's$ and $\mu's$ as well as π-particles. A deflecting magnet M_1 selected only negative particles, which were allowed to pass through an opening in the shielding. S_1, S_2 and S_3 are scintillation counters and C_1 and C_2 are Cerenkov detectors. These instruments are so designed that they are sensitive only to particles in a certain energy range. Between S_1

and S_2 there was placed a second deflecting magnet M_2, which acted as a momentum selector, since the radius of the path was fixed by the position of the detectors, and only particles with momentum $p = eBr$ were properly deflected toward S_2. Because it offered certain experimental advantages, the magnetic field was chosen to correspond to a momentum p = 1.19 GeV/c, which is slightly less than the momentum at which most protons should have produced (1.75 GeV/c).

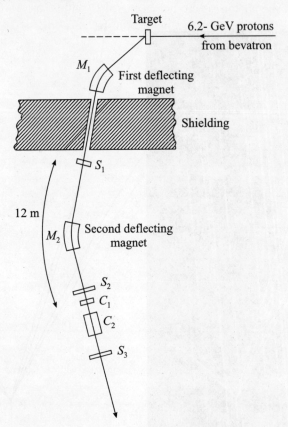

Fig. 9.6 Experimental arrangement for the observation of the antiproton.

Current theories of fundamental interactions require that protons should decay into leptons, but since the expected mean life is years, proton instability will not be easy to confirm.

Like fermions, bosons also fall into two categories:

(i) Massless Bosons (Subject to electromagnetic interaction only) This group consists of integral spin bosons, i.e. with spin 1, 2 or so and rest mass zero. Photon and graviton belongs to this group. the rest mass (m_0) of photon is zero and hence the general equation for the energy of a particle

$$E^2 = m_0^2 c^4 + p^2 c^2 = m^2 c^4 \qquad \qquad \dots (9.17)$$

reduces to

$$E = pc = mc^2 \qquad \ldots (9.18)$$

Which yields a momentum for the photon as

$$p = \frac{E}{c} = mc \qquad \ldots (9.19)$$

The rest mass of the photon is zero but its energy is finite. Therefore the velocity of photon is c(velocity of light in vacuum) and momentum $p = mc$. The energy of photon is also given by

$$E = h\nu \qquad \ldots (9.20)$$

$$\therefore \quad \text{mass of the photon} = \frac{h\nu}{c^2} \qquad \ldots (9.21)$$

The photon, given the symbol γ is the quantum of electromagnetic field. As such it is a *boson*. The photon has spin 1. The photon is itself uncharged, although it carries the Coulomb force field between electrical charges, either as 'virtual' particle in the static field or as a real particle in the radiation field.

The photon is stable and therefore, does not decay spontaneously, into any other particle. Obviously, the life time of photon is infinite so long as it does not interact with other particles, and so photons continuously reach the earth from the farthest distances of the universe. The photon may by thought of as the most commonly experienced elementary particle in its free state. The photon is its *own antiparticle*. Photon carries only the electromagnetic interaction, and so interacts with other particles only through this and through the gravitational field.

Another massless boson, so called '*graviton*' with a probable spin of 2 units, has been postulated as a field particle of gravity. Graviton has eluded detection so far.

These bosons, created by the electromagnetic field, are essentially of one kind.

(ii) **Mesons (Strongly interacting particles)** This group contains bosons of spin O and having masses intermediate between the leptons and the nucleons. Bosons of this group formed in the strong interactions are of two distinct kinds: (a) pions or π mesons (π^+, π^- and π^0) (b) Kaons or K-mesons (K^+, K^- and K^0). The decay schemes for pion or kaon are as follows:

$$\left.\begin{aligned} \pi^+ &\to \mu^+ + \nu_\mu \\ \pi^- &\to \mu^- + \bar{\nu}_\mu \\ \pi^0 &\to \gamma + \gamma \end{aligned}\right\} \qquad \ldots (9.22(a))$$

$$\left.\begin{aligned} K^+ &\to \mu^+ + \nu_\mu \\ &\to \pi^+ + \pi^0 \\ &\to \pi^+ + \pi^+ + \pi^- \\ &\to \pi^0 + e^+ + \nu_e \\ &\to \pi^0 + \mu^+ + \nu_\mu \\ &\to \pi^+ + \pi^0 + \pi^0 \end{aligned}\right\} \qquad \ldots (9.22(b))$$

Table 9.2 Particle Decay Modes[*]

	Particle	Decay mode	Relative probability, %	Half-life, s
Leptons	Photon	Stable		
	Neutrino	Stable		
	Electron	Stable		
	Muon	$\mu^- \rightarrow e^- + \nu + \bar{\nu}$		1.52×10^{-6}
Mesons	Pion	$\pi^+ \rightarrow \mu^+ + \nu$	100	1.80×10^{-8}
		$\rightarrow e^+ + \nu$	$\sim 10^{-4}$	
		$\pi^0 \rightarrow \gamma + \gamma$	99	6×10^{-17}
		$\rightarrow \gamma + e^+ + e^-$	1	
	Kaon	$K^+ \rightarrow \mu^+ + \nu$	63	
		$\rightarrow \mu^+ + \pi^0$	21	
		$\rightarrow 2\pi^+ + \pi^-$	5.6	8.56×10^{-9}
		$\rightarrow \pi^0 + e^+ + \nu$	4.8	
		$\rightarrow \pi^0 + \mu^+ + \nu$	3.4	
		$\rightarrow \pi^+ + 2\pi^0$	1.7	
		$K^0 \rightarrow \pi^{\pm} + e^{\mp} + \nu$	18	
		$\rightarrow \pi^{\pm} + \mu^{\mp} + \nu$	14	4×10^{-8}
		$\rightarrow \pi^+ + \pi^- + \pi^0$	6.3	
		$\rightarrow 3\pi^0$	11.3	
		$K^0 \rightarrow \pi^+ + \pi^-$	35	6.0×10^{-11}
		$\rightarrow 2\pi^0$	15	
	Eta	$\eta^0 \rightarrow \gamma + \gamma$	33	
		$\rightarrow \pi^+ + \gamma + \gamma$	20	
		$\rightarrow 3\pi^0$	20	$< 10^{-16}$
		$\rightarrow \pi^+ + \pi^- + \pi^0$	22	
		$\rightarrow \pi^+ + \pi^- + \gamma$	5	
Baryons	Proton	Stable		
	Neutron	$n^0 \rightarrow p^+ + e^- + \bar{\nu}$		7.0×10^2
	Lambda	$\wedge^0 \rightarrow \pi^+ + \pi^-$	66	1.76×10^{-10}
		$\rightarrow n^0 + \pi^0$	34	
	Sigma	$\Sigma^+ \rightarrow p^+ + \pi^0$	53	5.6×10^{-11}
		$\rightarrow n^0 + \pi^+$	47	
		$\Sigma^0 \rightarrow \wedge^0 + \gamma$		$< 7 \times 10^{-15}$
		$\Sigma^- \rightarrow n^0 + \pi^-$		1.1×10^{-10}
	Xi	$\Xi^- \rightarrow \wedge^0 + \pi^0$		2.0×10^{-10}
		$\Xi^0 \rightarrow \wedge^0 + \pi^-$		1.2×10^{-10}
	Omega	$\Omega^- \rightarrow \wedge^0 + K^-$	50	10^{-10}
		$\rightarrow \Xi^0 + \pi^-$	50	

[*] To obtain the decay of antiparticles, change all particles into antiparticles on both sides of the equations.

Table 9.3 Elementary Particles Stable Against Decay by the Strong Interaction Mesons are Jointly Considered Hadrons and Believed to be Composed of Quarks.

Class	Name	Particle	Anti-particle	Rest mass, MeV	Mean life, s	Spin	L_e	L_μ	L_τ	B	S	Y	I	I_z
Massless Boson	Photon	γ	$(\bar\gamma)$	0	Stable									
Lepton	e-neutrion	ν_e	$\bar\nu_e$	0	Stable	1/2	+1	0	0	0				
	μ-neutron	ν_μ	$\bar\nu_\mu$	0	Stable	1/2	0	+1	0	0				
	τ-neutrion	ν_τ	$\bar\nu_\tau$	0	Stable	1/2	0	0	+1	0				
	Electron	e^-	e^+	0.51	Stable	1/2	+1	0	0	0				
	Muon	μ^-	μ^+	106	2.2×10^{-6}	1/2	0	+1	0	0				
	Tau	τ	$\bar\tau$	1784	5×10^{-13}	1/2	0	0	+1	0				
Meson	Pion	π^+	π^-	140	2.6×10^{-8}	0	0	0	0	0	0	0		+1
		π^0	(π^0)	135	8.3×10^{-17}								1	0
		π^-	π^+	140	2.6×10^{-8}									-1
	Kaon	K^+	K^-	494	1.2×10^{-8}	0	0	0	0	0	+1	+1		+1/2
		K^0	$\bar K^0$	498	9×10^{-11} 5×10^{-8}								1/2	-1/2
	Eta Meson	η^0	η^0	549	7×10^{-19}	0	0	0	0	0	0	0	0	0
Baryon	Proton	p	$\bar p$	938.3	Stable	1/2	0	0	0	+1	0	+1		+1/2
	Neutron	n	$\bar n$	939.6	932	1/2	0	0	0	+1	0	+1	1/2	-1/2
	Lambda hyperon	Λ^0	$\bar\Lambda^0$	1116	2.5×10^{-10}	1/2	0	0	0	+1	-1	0	0	0
	Sigma hyperon	Σ^+	$\bar\Sigma^-$	1189	8.0×10^{-11}	1/2	0	0	0	+1	-1	0		+1
		Σ^0	$\bar\Sigma^0$	1192	10^{-14}								1	0
		Σ^-	$\bar\Sigma^+$	1197	105×10^{-10}									-1
	Xi hyperon	Ξ^0	$\bar\Xi^0$	1315	3.0×10^{-10}	1/2	0	0	0	+1	-2	-1		+1/2
		Ξ^-	$\bar\Xi^+$	1321	1.7×10^{-10}								1/2	-1/2
	Omega Hyperon	Ω^-	Ω^+	1672	1.3×10^{-10}	3/2	0	0	0	+1	-3	-2	0	0

Curiously enough, two distinct varities of the neutral kaon are known. One K^0 has a mean life time of 8.8×10^{-11} sec and decays either into $\pi^+ + \pi^-$ or into $\pi^0 + \pi^0$. The life time of K^0 is 5.2×10^{-8} sec. and it may decay into a pion, an electron, and a neutrino; into a pion, a muon, and a mu-neutrino or into three pions.

Baryons and mesons together are called as *hadrons*. The name hadron means strongly interacting particle. Hadrons carry the electromagnetic, the weak, and the strong interactions. The decay modes of leptons, Baryons are summarized in Table 9.2.

The classification scheme of the elementary particles is presented in Table 9.3. The particles in table 9.3 all exist long enough to travel as distinct entities along paths of measurable length and their modes of decay can be observed in such devices as bubble chambers. A considerable number of experimental evidence also points to the existence of several hundred different hadrons whose life times are only about 10^{-23} sec. These ultrashort-lived particles cannot be detected by observing their creation and subsequent decay in a bubble chamber or other instruments because the distance they cover in 10^{-23} sec. is only about $\approx 3 \times 10^{-15}$ m, even if they move at the speed of light - a length characterstic of hadron dimensions. Instead such particles appear as *resonant states* in the interactions of less remanescent (and hence more readily observable) particles. Resonant states occur in atoms as energy as revealed in Franck-Hertz experiment. We shall discuss resonance later in this chapter.

9.4 FUNDAMENTAL INTERACTIONS IN NATURE

The number of elementary particles so far discovered is very large, exceeding 200. The interactions amongst these particles may be considered as the direct source of the natural phenomena. Only four fundamental interactions known in the nature, e.g. strong, electromagnetic, weak and gravitational are apparently enough to account for all the physical processes and structures in the universe on all scale of size from hadrons to galaxies of stars. Table 9.4 summarizes salient freatures of these basic interactions.

(i) **Gravitational Interaction**   Gravitation was the first interaction to be known and is usually discussed in terms of the forces between macroscopic objects or in terms of the general relativity. It operates quite certainly on elementary particle level, but it is so weak that there if any other interactions are present, it is over whelmed and (so far) is then undetectable. The strength of gravitational interaction is commonly stated in terms of gravitational constant $G = 6.7 \times 10^{-11}$ N–m^2/kg^2. The dimension less quantity $\dfrac{G \, m_e \, M_p}{\hbar c} \sim 3 \times 10^{-42}$ is usually taken as a constant, characterizing this interaction. The gravitational interaction between two bodies is believed to be mediated through the quantum of the this interaction called the *graviton*. Graviton has not been observed experimentally as a unique particle. The extremely small value of the gravitational interaction has so far kept it out of the detailed theory of the elementary particles, although any complete theory obtained in future must include it.

Table 9.4 Fundamental Interactions in Nature

Interaction	Gravitational	Electromagnetic	Strong	Weak
Relative magnitude	10^{-39}	10^{-3}	1	$10^{-13} - 10^{-11}$
Range	infinite	infinite	Short $\sim 10^{-15}$ m	Short ($\sim 10^{-16}$ m)
Characteristic Time	10^{-16} sec	10^{-20} sec	10^{-23} sec	10^{-10} sec
Associated particles	Graviton	Photon	Pion, Kaon; Gluons	Intermediate boson, $W^{\pm}$, Z^0
Particles acted upon	Every particle having mass	Charged particles	Hadrons	Hadrons, Leptons
Examples	Astronomical forces	Atomic forces	Nuclear forces	Forces involved in β-decay
Strength (Natural units)	G (Newton) $= 6 \times 10^{-39}$	$g_e = \dfrac{1}{137}$	$g^2 \approx 1$	G Fermi $= 5 \times 10^{-17}$
Role in universe	Assembles matter into planets, stars and galaxies	Determines structure of atoms, molecules, solids and liquids. An important factor in astronomical universe	(i) Gluons holds quarks together to form nucleons (ii) Pions holds nucleons together to form atomic nuclei	Mediates transformation of quarks and leptons; helps determine composition of atomic nuclei
Conservation	—	Isospin violation	all	Flavour, CPT violation
Spin	2	1	1	1
Mass	0	0	0	90 GeV

(ii) Electromagnetic interaction This is much stronger than gravitational interaction. This is the classical interaction of Maxwell-Lorentz field between electrically charged particles. The quantum of the field is photon. It is a long range square type interaction, whose strength is determined by Sommerfeld's fine structure constant $\alpha = \dfrac{e^2}{4\pi \varepsilon_0 \hbar} \sim \dfrac{1}{137}$. This electromagnetic interaction between a proton and an electron is about 10^{37} times stronger than gravitational force between them at the same distance.

This interaction is charge dependent. In terms of isobaric spin, this interaction depends on T and is governed by the isospin rule $\Delta T = 0, \pm 1$. All other quantities, e.g. charge, baryon number, lepton number, hypercharge, parity, strangeness numbers are conserved.

The electromagnetic interaction is menifested in the chemical behaviours of atoms and molecules, Rutherford scattering and so on.

Photon capture can affect the production of mesons or hyperons by an electromagnetic interaction

$$\gamma + p \nwarrow\searrow \begin{matrix} \pi^0 + p \text{ (photo-pion production)} \\ \pi^0 + K^+ \text{ (low strange particles)} \end{matrix} \qquad \dots (9.23)$$

These decay processes illustrates particularly a important point which at first sight may appear paradoxical namely that uncharged particles can also be subjected to electromagnetic interaction, e.g

$$\left. \begin{matrix} \pi^0 \rightarrow \gamma + \gamma \\ \eta^0 \rightarrow \pi^+ + \pi^- + \pi^+, \quad \eta^0 \rightarrow \gamma + \gamma \\ \Sigma^0 \rightarrow \wedge^0 + \gamma \end{matrix} \right\} \qquad \dots (9.24)$$

In these processes no change of strangeness (S) is involved. The decay process such as $\Sigma^+ \rightarrow p^+ + \gamma$ are forbidden because the change $\Delta S = 1$ is required. The paradox that the decay of neutral particle is by electromagnetic interaction is resolved by introducing an intermediate step in the overall nuclear reaction i.e. virtual production of nucleon-anti nucleon.. We have

$$\pi^0 \longrightarrow (N + \bar{N}) \longrightarrow \gamma + \gamma \qquad \dots (9.25)$$
$$\text{Strong} \qquad \text{Virtual} \qquad \text{electromagnetic}$$

An example of the electromagnetic interaction is the process of mutual annihilation of particles and anti particles.

Unlike the gravitational force the electromagnetic interaction may be either attractive or replusive.

(iii) **Weak Interaction** This interaction is responsible for the nuclear β-decay and the weak decay of certain elementary particles, e.g. muons, pions, K-mesons and some hyperons. This interaction is about 10^{10} times weaker than the electromagnetic interaction and ulike gravitational or e.m. interaction, this is a very short range force. The range of weak interaction is given by $\dfrac{\hbar}{m_w c} \sim 2.4 \times 10^{-18}$ m, where m_w is the mass of the W boson, mediating the weak force (half life $\sim 10^{-17}$ s). The strength of the weak interaction may also be expressed in terms of the universal weak interaction coupling constant g_w, although this constant is not dimensionless as are others. It is value is 1.4×10^{-62} J-m^3 (from Fermi theory of β-decay). When expressed in dimensionless form, g_w is found to be small compared to the fine structural constant $\alpha \sim \dfrac{1}{137}$. Here $g_w = \left(\dfrac{g}{\hbar c} \right)(\lambda^2) \simeq 3 \times 10^{-12}$; $\lambda_c = \dfrac{\lambda_c}{2\pi}$, λ_c being the Compton wavelength for the electron.

The weak interaction is mediated through the heavy vector bosons, known as $W^\pm$ and Z^0 bosons. They have the masses $m_w c^2 = 80.9$ GeV and $m_z c^2 = 93$ GeV. When one compares the characterstic times for the operation of the weak ($\sim 10^{-10}$ s) and strong ($\sim 10^{-23}$ s) interactions, one finds the ratio of the former to the latter $\sim 10^{-13}$.

Few examples of weak interaction are

$$\bar{\mu} + p \rightarrow n + \nu_\mu$$

$$\mu \rightarrow \bar{e} + \nu_\mu + \bar{\nu}_e \quad \text{(Muon decay)}$$

$$\bar{\nu}_e + p \rightarrow n + e^+ \qquad \text{(proton capture of anti neutrino)} \qquad \dots (9.26)$$

Other types of weak interactions which require coupling between other pairs of fermions,

$$\wedge^0(S = -1) \rightarrow \pi(S = 0) + p(S = 0) \qquad \dots (9.27)$$

We must note that ν and $\bar{\nu}$ are associated exclusively with weak interactions just as photons are associated only with e.m. interaction.

An important characterstic of weak interaction which distinguishes it from other fundamental interactions, is that parity is violated (i.e. parity is not conserved) in weak interaction.

Recently, weak and electromagnetic interactions have been unified and the unified interaction is known as *electro-weak* interaction.

(iv) Strong Interaction   This is the dominant nuclear interaction responsible for most nuclear phenomena, nuclear energy levels, and so on. It is the interaction between protons and neutrons, having short range like the weak nuclear interaction, the range being $\dfrac{\hbar}{m_\pi c}$ ~

10^{-15} m. Within this range it predominates over all the other fundamental interactions between the neutron and the proton-with a characterstic strength parameter $\alpha \sim 0.3$. Obviously, the value of α of strong interactions is much larger than the strength parameter $\alpha \sim \dfrac{1}{137}$ of e.m. interaction. The time for the operation of the strong interaction is about $\sim 10^{-23}$ s, which is also the order of the characterstic nuclear time.

The quantum of the strong interaction is the *pi* meson or pion ($\pi^\pm$, π^0).

The nuclear strong interaction is responsible for the formation of the so called meson and baryon resonances, or excited states of the mesons and baryon particles. It is also the coupling that drives their decay, and for this reason their life times are short. A term usually used for the strongly interacting particles is hadron. The quantitative connection between the strong interactions and the structure of the nucleus is not yet fully clear.

Strong interaction is a short range force ($\sim 10^{-15}$ m) conserves baryon number (B), charge (Q), hypercharge (Y), Parity (π), isospin (T) and its component T_x. The following is an example of a strong interaction.

$$\pi^- + p \rightarrow \pi^0 + n \qquad (9.28)$$

One can investigate the nature of the strong interaction by studying the properties of the *hypernuclei* in which a nucleon in a nucleus is replaced by a hyperon. The first hypernucleus discovered was a boron nucleus in which a neutron was replaced by a $\wedge^0$-hyperon. So far several hypernuclei have been produced and their properties investigated. An example is $\wedge^0$-nucleus of tritium. This decays according to $\wedge^3\text{H} \rightarrow {}^3\text{He} + \pi^-$. The Q value for this is 41.5 MeV.

One can also obtain information about these systems through the study of the exotic atoms in which Σ^- hyperon is captured by the electric field of the nucleus. Sigma has the large mass and hence its orbit round the nucleus has a small radius. One obtain important information about the strong interaction by the measurment of the energies of the X-rays emitted during their transition from one orbit to another. This interaction between elementary particles is responsible for the total cross-section is a function of energy.

Recent studies reveal that protons and neutrons are made up of more fundamental entities known as quarks. It is believed that strong interaction is mediated through the exchange of massless quanta of spin $\hbar$ known as *gluons* between the quarks. Quarks and gluons have not been observed in the free state so far.

Examples of particles and their methods of interaction are shown in Table 9.5

Table 9.5 Elementary Particles and Corresponding Interaction modes.

Weak	Electromagnetic	Strong
e	e	π
μ	μ	K
ν	π	p
π	K	n
K	p	$\wedge$
p	n	
n	$\wedge$	
$\wedge$		

(v) Fifth Interaction   There is an indication of the existence of fifth interaction besides four fundamental interactions discussed so far. It has been reported that the force between two masses be better represented by

$$F = G_\infty \frac{m_1 m_2}{r^2}\left[1 + \left(1 + \frac{r}{\lambda}\right)\alpha e^{-\frac{r}{\lambda}}\right] \qquad \ldots (9.29)$$

with $\alpha = -0.007$ and $\lambda = 200$ m. As α is negative, the second term in the square bracket represents a repulsive force. For $r \gg 200$ m,

$$F = G\frac{m_1 m_2}{r^2} \qquad \ldots (9.30)$$

Which is the force operative between two masses. For $r \ll 200$ m,

$$F = G_\infty \frac{m_1 m_2 (1+\alpha)}{r^2} = G'\frac{m_1 m_2}{r^2} \qquad \ldots (9.31)$$

where $G' = G_\infty(1 + \alpha)$

This is the force one measures in a Cavendish experiment. The value of G for small distances is about 1 % less than the value of G for large distances.

Unified Approach   There have been efforts to unify the various interactions and deduce them as different manifestations of the same underlying theory. Maxwell in 1864 first achieved such a unification for electric and magnetic forces. Salam and Weinberg have now succeeded in unifying electromagnetic and weak forces and further efforts are being made to bring in strong and gravitational interactions. We now present the ideas which led to the unification of electromagnetic and weak interactions.

The electron-antineutron pair in the β-decay is produced in the same angular momentum state as the photon. This similarity between weak and electromagnetic processes can be further extended by postulating that the neutron changes into a proton by emitting a massive W^- (which carries spin) and this then decays into an electron-anti neutrino pair. The basic weak process here is a vector boson W^- interacting with a fermion pair similar to a photon interacting with a fermion pair. The important differences are (i) the W^- is charged and is accompanied by its antiparticle the W^+ whereas the photon is neutral and is its own antiparticle. (ii) the W^- is massive. Since weak interaction is a shortrange force. From the analysis of the weak processes. It is estimated that

$$m_w > 7.5 \times 10^4 \text{ MeV}$$

(iii) The interaction of the $W^\pm$ violates parity invariance whereas the interaction of the photon conserves parity.

inspite of the above differences, the interaction of the W bosons and the photon with matter can be combined. This is done by allowing a triplet and a singlet of vector bosons to interact with the lepton and quark doublets. Demanding that the photon does not interact directly with the neutrinos and that its interaction conserves parity identifies the photon and relates the strengths of the weak and the electromagnetic interactions. Around 1967

68, Salam, Weinberg and Glashow showed that electromagnetism and weak interactions are two faces of a unified electroweak force. At this step the number of basic forces or interactions in nature came down to three. The problem though is that this unit can be seen only in processes at energies of about one hundred GeV or higher, i.e., at and above one hundred proton masses. The model developed by Weinberg and Salam has the following interesting features:

(i) There are three massive bosons W^+, W^-, Z (which is neutral), with masses

$$m_w \geq 3.7 \times 10^4 \text{ MeV}$$
$$m_z \geq 7.3 \times 10^4 \text{ MeV}$$

and have parity violating interaction. It is large mass of these bosons which suppresses the usual weak interaction. The photon has the usual electromagnetic interaction. In high energy processes (energies comparable with $m_w c^2$) the two interactions would become comparable. These bosons with $m_w \approx 8.1 \times 10^4$ MeV and $m_z \approx 9.5 \times 10^4$ MeV been observed recently at CERN.

(ii) The neutrinos and leptons have another interaction with matter, in addition to the electromagnetic and β-decay type of interactions. This interaction has been observed in 1973 in processes of the type

$$\nu_\mu + N \to \nu_\mu + \text{hadrons}$$

$$\bar{\nu}_\mu + N \to \bar{\nu}_\mu + \text{hadrons}$$

where N is a nucleus. It has also been observed in 1978 in the scattering of polarized electrons by a deuteron target. It was reported that there was a difference in the scattering of left-handed and right-handed electrons, indicating violation of the right-left symmetry or the violation of parity invariance in the interaction of electrons. The amount of parity violation is in agreement with the prediction of Weinberg-Salam model.

With the successful unification of weak and electromagnetic interactions, efforts have been directed towards unifying strong weak and electromagnetic interactions. Soon after came a theory of the strong interaction among quarks *quantum chromodynamics*

(*QCD*), and the two together make up the so-called *Standard model* of particle physics. In most of the theories, leptons and quarks are put in the same multiplet. Therefore, there is possibility of quarks transforming into leptons which means that the protons would be unstable against decay into leptons. Search for such decays (the life time of protons expected in these theories is greater than 10^{30} years) is being pursued vigorously by different groups of experimentalists. Everything except gravity seems to be satisfactorily handled by standard model, but yet it cannot by the last word since there are some twenty independent parameters in it to be taken from experiment. So thee are constant efforts to test and see what may lie beyond the standard model.

One such attempt is to unify the *QCD* theory with the electroweak one into a *Grand unified Theory* (GUT). If this happens, the theoretical indication is that this unity will be seen only at energies of 10^{15} GeV and higher. To get a feeling for this, recall that the proton mass is about 1 GeV. Present accelerators and those expected in the next decade or so are in the range of a few thousand GeV. But grand unification is only expected at 10^{15} GeV – so no hope of reaching such scales in laboratories, and in nature too such events must be extremely rare. How one can test such ideas at all?

This brings one to cosmology. The problem of unifying quantum theory and gravity is still open. The question arises, under what conditions might such a combination be relevant? Quantum theory involves Planck's constant h, while general relativity, being a relativistic theory of gravitation, involves both the Newtonian constant G and the speed of light c.

$$\text{Planck time} \quad \left(\frac{hG}{c^5}\right)^{\frac{1}{2}} \approx 5 \times 10^{-44} \text{ s}$$

$$\text{Planck mass} \quad \left(\frac{hC}{G}\right)^{\frac{1}{2}} \approx 1.2 \times 10^{19} \text{ GeV/}c^2$$

$$\text{Plank temperature} \quad \frac{1}{k}\left(\frac{hC}{G}\right)^{\frac{1}{2}} \approx 1.4 \times 10^{32} \text{ K}$$

In a process, characterized by these scales, quantum gravity would be relevant.

9.5 PARTICLE INSTABILITY

The creation and annihilation processes, β decay etc. are manifestations of a more general property of elementary particles: their instability. In other words given the proper conditions, the fundamental particles can transform into the other particles as a result of their interaction. Particles which are unstable undergo spontaneous decay with a well defined half life. We must note that only four particles (and their anti particles) are stable against spontaneous decay: the photon, the neutrino, the electron and the proton of all the unstable particles, the one having the longest half-life is the neutron. Mesons have a half life of the order of 10^{-8} s (π^0 and η^0 are exceptions), while the baryon half-lives are of the order of 10^{-10} s.

The necessary condition for spontaneous decay is the condition of conservation of energy, i.e. the rest mass of the parent-particle be larger than the sum of the rest masses of the daughter conserved. In addition to energy, momentum as well as angular momentum are supposedly conserved. Particle decay modes have been summarized in Table 9.2. Few examples are

(i) proton-antiproton annihilation

$$p^+ + \bar{p}^- \rightarrow \bar{\Xi}^+ + \Xi^-$$

$$\pi^- + \wedge^0$$

$$\pi^+ + \pi^0$$

$$\pi^+ + \bar{p}^- \qquad \dots (9.32)$$

(ii) proton-proton collison

$$p^+ + p^+ \begin{cases} p^+ + p^+ + \pi^0 \\ p^+ + n + \pi^+ \end{cases} \qquad \dots (9.33)$$

(iii) K^- and p^+ collison

$$K^- + p^+ \rightarrow \Xi + K^0 + \pi^+$$

$$\pi^+ + \pi^-$$

$$\pi^- + \wedge^0$$

$$\pi^- + \pi^+ \qquad \dots(9.34)$$

Obviously, at high energies the collisions between particles are not elastic. The particles can no longer be considered as billiard balls, which is useful approximation at low energies.

9.6 CONSERVATION LAWS

Elementary particles follow a set of conservation rules that give clues as to their nature and which, incidently, sum up much of the knowledge of physics.

(i) **Conservation of energy** In common with all isolated physical phenomena, the reactions of fundamental particles conserve energy distributed in various forms including mass, e.g.

$$n \rightarrow p + e^- + \bar{v}_e + 0.782 \text{ MeV}$$

However the following process

$$K^0 \neq \pi^+ + \pi^- + \pi^e + \pi^0$$

is not observed, because the rest energy of the K^0 is not great enough to make four pions even if they all could memade at rest. However, the conservation of energy principle allows the following process

$$K^0 \Rightarrow \pi^+ + \pi^- + \pi^0$$

(ii) **Conservation of linear momentum** Linear momentum is always conserved in all elementary particle reactions. It is related to the invariance of the physical laws under translation in space. The laws of interaction do not depend on the place of measurment. Obviously, the space is homogeneous. This demands that the annihilation of a slow positron and an electron cannot result in one photon, but must produce atleast two, preceeding in opposite direction.

(iii) **Conservation of angular momentum**   This is also followed by all types of interactions. This is related to the invariance of the physical laws under rotation (isotropy of space). The conservation of angular momentum includes both types (orbital as well as spin) of angular momentum together. It should be noted that total angular momentum is conserved. The orbital and spin angular momentum may not be separately conserved. The orbital angular momentum is given by the motion of the object as a whole about any chosen external axis of rotation. The spin is the intrinsic angular momentum of each object about an axis through its centre of mass. Conservation of angular momentum places restrictions on the spins that the products of the reaction can have. Strongly interacting fermions have half integer spin $(S = \frac{1}{2}$ for $\Xi, \Sigma, \wedge, n, p$ and $S = \frac{3}{2}$ for $\Omega)$, strongly interacting bosons (η, K, π) have $S = 0$; weakly interacting fermions, so called leptons (μ, e, ν_e, ν_μ) have $S = \frac{1}{2}$, massless bosons (photon) have $S = 1$ and gravitons have $S = 2$. The total spins of the reaction products may only differ by 0 or 1 from the total spins of the initial components. In other words, during a reaction $\Delta S_{total} = 0$ or 1. The following reaction exhibit the spin balance

$$n \rightarrow p + e^- + \bar{\nu}_e + Q$$

$$\left(\frac{1}{2}\right) \quad \left(\frac{1}{2}\right) \left(\frac{1}{2}\right) \left(-\frac{1}{2}\right)$$

Thus $\Delta S = 0$. Spin angular momentum is given by $L = \sqrt{S(S+1)}\,\hbar$.

(iv) **Conservation of Parity**   Parity holds for strong nuclear and electromagnetic interactions, but is violated in weak interaction. Parity is related to the invariance of the physical laws under inversion of the space coordinates, i.e. x, y, z are to be replaced by $-x, -y, -z$. Obviously, this is equivalent to combined reflection and rotation. We must note that the physical laws do not depend on the left or right handed ness of the coordinate system.

The spatial parity depends on the orbital angular momentum and is given by $P = (-1)^l$ where l is the azimutal quantum number. Obviously, the states of even l have parity $(P = +1)$ while states of odd l have odd parity $(P = -1)$. Since the two successive parity operations bring the system back to its original state and hence $P^2 = 1$. This means P have two possible eigen values ± 1:

Let $\psi(x, y, z) = \psi(r)$ represents the wave function of the state of a system, then we have

$$\psi(-r) = +\psi(r) \text{ for even parity states}$$

$$\psi(-r) = -\psi(r) \text{ for odd parity states}$$

Parity is a good quantum number as it is conserved in strong and electromagnetic interactions. We further note that an energy eigenstate is also an eigenstate of parity.

We must remember that parity operation symmetry represents discrete symmetry (reflection and rotation through 180°) unlike the symmetries under infinitesimal changes implied in conservation of energy, conservation of linear momentum and conservation of angular momentum.

Parity has a broder definition, when one is dealing with elementary particles. Every elementary particle with a non-zero mass has an *intrinsic* parity π. The states with $P_i = +1$ are called even intrinsic parity states and the states with $\pi = -1$ are said to be *odd intrinsic* parity states. Obviously, the total parity of a system of n particles is the product of their intrinsic parities and the orbital parity $(-1)^l$ and can be expressed as $\pi_1 \pi_2 \pi_3 \pi_4 \dots \pi_n (-1)^l$.

The intrinsic parities of electron, proton and neutron are taken to be even. Then one finds that the intrinsic parities of other particles are fixed relative to the intrinsic parities of nucleons. We must note that the intrinsic parity of the pions ($\pi^\pm$, π^0) is odd.

Usually the parity of a particle is designated by the symbol (+) for even and (–) for odd and expressed as subscripts above the total angular momentum, i.e. (J^P). For nucleons one can write $(J^P) = \left(\dfrac{1}{2} + \right)$ and for pions ($J^P = 0$).

If space inversion symmetry exists, i.e. there is no distinction between the coordinate frame and the inverted coordinated frame, then

$$[H, P] = 0$$

i.e. P commutes with Hamiltonian H.

In a nuclear reaction the product of the parities of the initial components was always thought to be equal to the product of the parities of the final components. This conservation of parity law was shown to be invalid by the theoretical work of Tsung Lee and Chen Yang, and the experimental verification by mrs. Chien-Shiung Wu at Columbia university in 1956. Lee and Yang questioned the validity of invariance under space reflection for processes due to the weak interaction. The Lee and Yang proposal was motivated by what at that time was called the $\tau - \theta$ puzzle: Each particle, described by its corresponding field or wave function is supposed to have a certain intrinsic parity. Mesons have negative parity Kaons disintegrate in several ways: one resulting in three pions, another in two pions. The decay modes are $K^\pm \rightarrow 2\pi^\pm + \pi^\mp$ and $K^\pm \rightarrow \pi^\pm + \pi^0$. Now the parity of two pions (two 'odds') is odd. If parity were always conserved, then the two pions would have come from a particle of even parity state (it was tentatively called the "theta" meson) while three pions would have come from a particle of odd parity state (it was called the "tau" meson). However, the "tau-theta puzzle" arose over the fact that except for parity, they were identical. Lee and Yang decided that they were identical, and that a single particle disintegrated in two ways, parity not being conserved. The theta and tau mesons are now realized to be the same thing-the kaon.

(v) Conservation of Charge   We note that without exception all of the elementary particles have either no electric charge or the precise charge $\pm e$. There is no variety, as there is of masses. The positive and negative electric charges are exactly equal. The net electric charge of a system is always conserved, e.g.

$$n \rightarrow p + {}_{-1}e^0 + 0.782 \text{ MeV}$$

Nowadays, the word charge is used in a broader sense in fundamental particle physics. Apart from electric charge Q, one define the baryonic charge B for each baryon and leptonic charge l for each lepton. In addition, strangeness, charm and beauty have similar properties and are considered as charges. $B = +1$ for the baryons and -1 for the antibaryons. Similarly $l = \pm 1$ for the leptons and antileptons, while for baryons, mesons $l = 0$. We must note that there are actually three different types of leptonic charges l_e, l_μ and l_τ associated with the three different group of leptons, e.g. electron-positron $(e^+ e^-)$, $\mu^+ \mu^-$ and $\tau^+ \tau^-$ (τ-lepton or tauon). Each type of these leptons are associated with a neutrino, viz. the electron-neutrino (ν_e), the muon neutrino (ν_μ) and the taun-neutrino (ν_τ). The association of the charged leptons with their neutrinos (neutral) is expected to arise from the empirical law of leptonic charge (lepton number) conservation. All the neutrinos associated with leptons are probably mass zero particles.

It is observed that leptons and baryons disappear in pairs. This means that the net number of baryons and each kind of leptons remains unchanged in the universe. The lepton conservation law is probably related to the gauge invariance of the weak nuclear field.

With the help of conservation laws, one can understand why some reactions do not take place, e.g.

$$n \to \pi^0 + \gamma$$
$$(B = 1 \quad 0 \quad 0)$$

does not occur, because baryonic charge B is not conserved. The another example is

$$\mu^+ \to e^+ + \gamma$$
$$(l_\mu = -1 \quad 0 \quad 0)$$
$$(l_e = 0 \quad -1 \quad 0)$$

The above reaction does not occur because of the non conservation of the leptonic charges. However, conservation of the boryonic and leptonic charges are probably not satisfied exactly. The two processes stated above may actually occur, though with very small probabilities. However, they have not yet been observed.

The violation of baryon number, if any, indicates that the lightest baryon, i.e., proton, may not be absolutely stable, but may show the decay $p \to e^+ \pi^0$. This is the prediction of grand unification theory (GUT).

(vi) Conservation of Charge (c)-parity   This is related to the invariance under charge conjugation operation (c). This operation consists in replacing all particles by their antiparticles without changing any other physical property, such as momentum or spin. A system is invariant with regard to charge conjugation when, if a process is possible in the original system, the corresponding process is also possible in the charge conjugate system. Charge conjugation means reversal of the signs of all types of charge (electric, baryonic and leptonic) of the particles. We must note that the principle of charge conjugation applies not only to charged particles, but also to neutral particles. Obviously, the neutrino (electrically neutral), has an antiparticle (the antineutrino) which is different from the former, the leptonic charges of the two being equal and opposite. This reveals that they interact differently with matter.

There are self conjugate particles, e.g. π^0 meson, the photon, the η^0-meson, the $\dfrac{J}{\psi}$ particle and γ-meson for which all the different charges are zero ($Q = 0$, $B = 0$, $l = 0$). These particles must be identical with their anti particles. Obviously, these particles constitute the truly neutral particles.

Positronium is another example of a truly neutral system. In a positronium, an electron and a positron are bound together to form a hydrogen like structure. Q, B and l are all zero for it.

C parity is definite, i.e. even (+) or odd (–) for the really neutral particles. When all electrons are replaced by positrons, the e.m. field changes sign. The wave function of photon must change sign under charge conjugation. This means C is odd for photons. Similarly, one finds that the C-parities of the $\dfrac{J}{\psi}$ and γ-meson are *odd*, i.e.

$$C|\gamma> = -|\gamma>, \ C|v> = -|v> \ \text{and} \ C|\frac{J}{\psi}> = -|\frac{J}{\psi}>$$

However, C parity for π^0 must be even because it breaks up into two photons. For η^0 meson also C-parity is even:

$$C|\pi^0> = |\pi^0>, \ C|\eta^0> = |\eta^0>$$

The following system from the operation of C-parity also occurs in nature, $\pi^- \to \mu^- + \bar{v}$. This shows that weak interactions are invariant under charge-parity operation. We must note that the electromagnetic and strong nuclear interactions are invariant under C, where as the weak interaction is not invariant under C, which is evident from the asymmetry in the β-decay of ^{60}Co.

We must note that the quantum number C has no meaning for the charged particles like $\pi^\pm$: $C|\pi^+> \propto |\pi^->$. Also for a state $B \neq 0$, the quantum number C has no meaning.

(vii) Conservation of T The operation of T, which is related to the time-reversal invariance, means changing the time t to $-t$ in all equations of motion i.e., reflection of the time axis at the origin of the time coordinate (in relativistic space-time continuum). Under time reversal the quantities velocity, momentum, and angular momentum are reversed and in a collision the initial and final states are exchanged. Obviously, T operation is also a discrete change like the parity operation.

$\dfrac{d\mathbf{p}}{dt}$ is invariant under time reversal, since it implies the replacements $p \to -p$ and $t \to -t$.

Also the electric field ε is invariant under time reversal. However, the magnetic field suffers the transformation $B \to -B$ because time reversal implies reversing the velocities and therefore also the currents. Then $V \times B$ is invariant under time reversal, because both factors have changed sign, and the equation of motion of charged particle, $\dfrac{d\mathbf{p}}{dt} = -q(\mathbf{E} + \boldsymbol{v} \times \mathbf{B})$, is invariant w.r.t. time reversal, T operation causes the reversal of sign of the angular momenta and transforms the wave function to its complex conjugate. One can easily verify that the time reversed wave function $\psi_{rev}(t) = \psi^*(-t)$ is a solution of the Schrodinger equation, which is obviously time-reversal invariant.

Strong and weak interactions are also considered invariant under time reversal, although lately the problem has been open to discussion. Recent (1966) experiments regarding a special form of decay of neutral Kaon, which occurs only in 0.3 % of the cases, seem to indicate that weak interactions may violate the principle of invariance under time reversal.

(viii) CPT Theorem   All physical laws are considered to be invariant under the CPT operation, i.e., under the combined operations of time reversal T, Parity P and charge conjugation T. This theorem also known as Luders-Pauli theorem is true even in cases in which the laws may not be invariant under the individual operations. The orders of the three operations are immaterial. So far, all evidences point to the universal validity of this theorem. The CPT invariance implies that if an interaction is not invariant under any one operations C, P or T, its effect must be compensated by the combined effect of the remaining two operations. It is believed that probable CPT violation, if any can be observed in the properties of the neutral K^0 and $\bar{K}_0$ mesons.

(ix) Conservation of Isospin   If we consider that the neutron and the proton are the two charge states of the same entity known as nucleon then one finds the charge independence or isotopic invariance of nuclear force. This leads to the idea of isospin. One can take the isospin of the nucleons to be $I = \dfrac{1}{2}$ with the two components $I_3 = \dfrac{1}{2}$ for the proton and $I_3 = -\dfrac{1}{2}$ for the neutron. The charge being given by

$$Q = \frac{1}{2}\ (1 + 2I_3) \qquad \qquad ...\ (9.35)$$

We must note that some of the books have used T for the isospin.
The isospin multiplicity is obtained by

$$M = 2I + 1 \qquad \qquad ...\ (9.36)$$

In the presence of strong nuclear interaction, the two states of the nucleon with different I_3 are degenerate, i.e. symmetric. However, the symmetry is broken by the electromagnetic interaction. The electromagnetic interaction makes the energies of the two states, i.e. the proton and the neutron masses, slightly different. For both the states, the intrinsic spin and parity $\left(\dfrac{1}{2}^+\right)$ are the same

We must note that the I is conserved in strong interaction, but not in electromagnetic interaction. One can see that $[H, I] = 0$, i.e. H commute with I. This implies that conservation of the isospin is related to the invariance of H under rotation of the axes in the hypothethical isospin space, which is assumed to be isotropic.

One can also introduce the concept of the isospin for the other strongly interacting particle multiplets having nearly same masses. Because of the symmetry breaking property of the electromagnetic interaction one finds that the energies, i.e. masses of the different charge states with different I_3 (same I) are different. Few examples of isospin multiplets are the pions (π^+, π^0, π^-) with $I = 1$, $I_3 = +1, 0, -1$ respectively ($M = 3$), Sigma hyperons (Σ^+, Σ^0, Σ^-) with $I = 1$, $I_3 = +1, 0, -1$ respectively ($M = 3$) and Kaons (K^+, K^0) with $I = \dfrac{1}{2}$, $I_3 = +\dfrac{1}{2}, -\dfrac{1}{2}$ respectively ($M = 2$). $\wedge^0$ hyperon and η^0 meson are examples of a particle multiplet having only one particle corresponding $I = 0$ and $I_3 = 0$.

The components of a particular isospin multiplet exhibit the similarity in observed properties e.g., the mass spin, intrinsic parity and life time. This leads to the hypothesis of isotopic invariance in strong interactions involving the particles in the different cases states above. We shall see that the resonance particles (excited states of baryons and mesons) also form isospin multiplets. These particles have the same isospin multiplicity as the corresponding particles in their ground state. Δ-resonance is an example of an isospin quadruplet ($M = 4$) with $I = \dfrac{3}{2}$. This can exist in the following four charge states, e.g.

Δ^{++}, Δ^+, Δ^0, Δ^- with $I_3 = +\dfrac{3}{2}, +\dfrac{1}{2}, -\dfrac{1}{2}$ and $-\dfrac{3}{2}$ respectively.

(x) Conservation of Strangeness To describe the strange behaviour of certain groups of particles (strange particles) with respect to their production by strong interaction and decay by weak interaction a strangeness quantum number S is used. The law of conservation of strangeness requires that *the total strangeness S must remain the same in process due to strong or electromagnetic interactions.*

Particles and anti particles have opposite strangeness. The following relationship holds between the change Q of a particle and its strangeness

$$Q = I_3 + \left(\frac{B + S}{2}\right) \qquad \qquad ... (9.37)$$

Where B is the baryon number and I_3 is the third component of the isospin I.

One can see that the process $\pi^- + p^+ \to \eta^0 + \wedge^0$, which complies with all the other conservation laws, is not observed because the total strangeness on the left is zero, while on the right it is -1. However, processes such as $\pi^+ + p^+ \to \Sigma^+ + K^+$ and $\pi^- + p^+ \to \wedge^0 + K^0$ do occur. We can easily see that in both cases the total strangeness remains equal to zero. We have for the K^+ meson $B = 0$, $S = 1$ and $Q = +1$. This gives $I_3 = +\dfrac{1}{2}$. On the other hand we have $Q = 0$, $I = -\dfrac{1}{2}$. Obviously, K^+ and K^0 form an isospin doublet with $I = \dfrac{1}{2}$. One finds that for the non-strange baryon, proton (or neutron) $B = 1$, $S = 0$ and hence $Q = I_3 + \dfrac{B}{2} = I_3 + \dfrac{1}{2}$. This requires one to put $I_3 = +\dfrac{1}{2}\left(\text{Or } -\dfrac{1}{2}\right)$ to make $Q = +1$ (or 0). Now, we define the hypercharge Y as

$$Y = B + S \qquad\qquad\qquad\qquad \text{... (9.38)}$$

This gives $\qquad Q = I_3 + \dfrac{Y}{2} \qquad\qquad\qquad\qquad \text{... (9.39)}$

Since B and S are both conserved in strong interaction and hence Y is also conserved.
We must note that the conservation of strangeness is not a rigorous law as strangeness is not conserved in weak interaction. Conservation of strangeness law can be violated by weak interactions which allow a change of strangeness of ±1. In most baryon decays, the conservation of strangeness is violated, indicating that the decays take place by means of weak interactions, e.g.

$$\Xi^- \rightarrow \wedge^0 + \pi^- \quad (\Delta S = +1)$$
$$\qquad\qquad\qquad \downarrow \pi^- + \pi^+ \quad (\Delta S = -1)$$

Hyperons cannot decay into baryons through the strong interaction with conservation of strangeness. This explains why the hyperons decay so slowly (about 10^{-10} s) compared with the time involved in their production or annihilation in collisions by means of strong interactions, which is about 10^{-23} s. The selection rule $\Delta S = 0, \pm 1$ is obeyed in weak decay of semi-stable strange particles. Few examples are

$$\Omega^- \rightarrow \wedge^0 + K^- \qquad (\Delta S = +1)$$
$$\Sigma^0 \rightarrow \wedge^0 + \gamma \qquad (\Delta S = 0)$$
$$\Sigma^+ \rightarrow p + \pi^0 \qquad (\Delta S = +1)$$
$$\Xi^0 \rightarrow \wedge^0 + \pi^0 \qquad (\Delta S = +1)$$

The decay $\Xi^- \rightarrow n + p^- \,(\Delta S = 2)$ is not observed.
Table 9.6 exhibit the behaviour of various quantities in different types of interactions.

Table 9.6

Quantity	Symbol	Weak	Electromagnetic	Strong
Energy	E	Yes	Yes	yes
Momentum	p	Yes	Yes	Yes
Parity	P	No	Yes	Yes
Isospin	I	No	No	Yes
Third Component of I	I_3	No	Yes	Yes
Charge Conjugation	C	No	Yes	Yes
Time reversal	T	No	Yes	Yes
CP	–	No	Yes	Yes
CPT	–	Yes	Yes	Yes
Baryon Number	B	Yes	Yes	Yes
Lepton Number	l	Yes	Yes	Yes
Charge	Q	Yes	Yes	Yes
Strangeness	S	No	Yes	Yes
Hypercharge	Y	No	Yes	Yes
Charm	c	No	Yes	Yes

9.7 RESONANCES

Besides longe-lived fundamental particles, in the past few years experimental evidence has accumulated which points toward the existence of very short lived particles, called *resonance particles*. Their lifetimes are so short ($\sim 10^{-29}$ s or less) that they do not leave any recognizable track in bubble or spark chambers. Resonance particles can be classified according to baryon number, hypercharge and isotopic spin, and are designated by the same symbols as those baryons and mesons which have the same quantum numbers as the resonances, but with a subscript 1, 2, 3, ... according to increasing mass.

Particle resonances have been discovered both amongst the hyperons and mesons. The number of such resonances is about 80 if one includes all charge states and antiparticles; their number is increasing continually. Figures 9.7 and 9.8 shows some meson resonances and some of baryon resonances respectively. The masses (in MeV), spins and particles are shown.

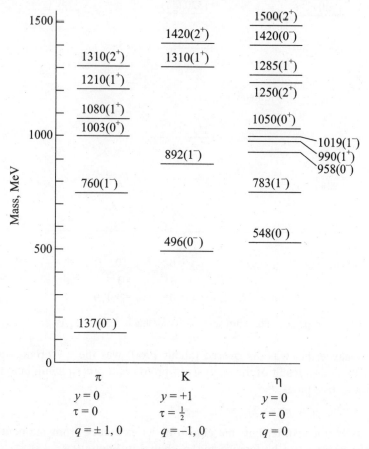

Fig. 9.7 Meson resonances. The spin and parity of each resonance is given in parentheses.

These particles are produced in high energy collisions between hadrons by strong interactions and are observed during scattering or reactions initiated by very high energy elementary particles obtained from particle accelerators.

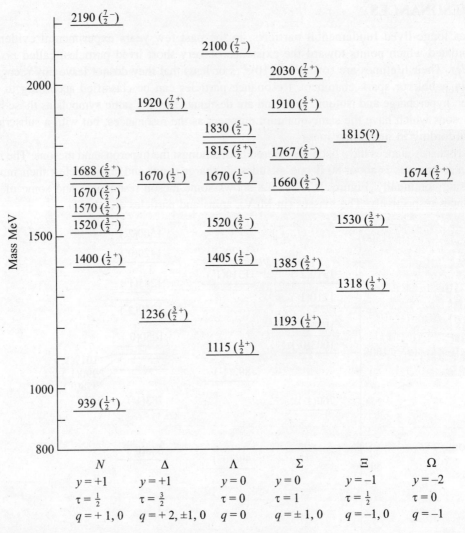

$2190 \left(\frac{7}{2}^{-}\right)$

$2100 \left(\frac{7}{2}^{-}\right)$

$2030 \left(\frac{7}{2}^{+}\right)$

$1920 \left(\frac{7}{2}^{+}\right)$ $1910 \left(\frac{5}{2}^{+}\right)$

$1830 \left(\frac{5}{2}^{-}\right)$ $1815(?)$

$1815 \left(\frac{5}{2}^{+}\right)$ $1767 \left(\frac{5}{2}^{-}\right)$

$1688 \left(\frac{5}{2}^{+}\right)$ $1670 \left(\frac{1}{2}^{-}\right)$ $1670 \left(\frac{1}{2}^{-}\right)$ $1660 \left(\frac{3}{2}^{-}\right)$ $1674 \left(\frac{3}{2}^{+}\right)$

$1670 \left(\frac{5}{2}^{-}\right)$

$1570 \left(\frac{3}{2}^{-}\right)$

$1520 \left(\frac{3}{2}^{-}\right)$ $1520 \left(\frac{3}{2}^{-}\right)$ $1530 \left(\frac{3}{2}^{+}\right)$

$1400 \left(\frac{1}{2}^{+}\right)$ $1405 \left(\frac{1}{2}^{-}\right)$ $1385 \left(\frac{3}{2}^{+}\right)$

$1318 \left(\frac{1}{2}^{+}\right)$

$1236 \left(\frac{3}{2}^{+}\right)$ $1193 \left(\frac{1}{2}^{+}\right)$

$1115 \left(\frac{1}{2}^{+}\right)$

$939 \left(\frac{1}{2}^{+}\right)$

Mass MeV — 2000, 1500, 1000, 800

N	Δ	Λ	Σ	Ξ	Ω
$y = +1$	$y = +1$	$y = 0$	$y = 0$	$y = -1$	$y = -2$
$\tau = \frac{1}{2}$	$\tau = \frac{3}{2}$	$\tau = 0$	$\tau = 1$	$\tau = \frac{1}{2}$	$\tau = 0$
$q = +1, 0$	$q = +2, \pm 1, 0$	$q = 0$	$q = \pm 1, 0$	$q = -1, 0$	$q = -1$

Fig. 9.8 Baryon resonances. The spin and parity of each resonance is given in parentheses.

The first resonance that was discovered (about 1960) was the η^0 (originally called ω^0). Its discovery came about as a result of the analysis of proton-anti proton annihilation. It corresponds to the production of five pions,

$$p^+ + p^- \rightarrow 2\pi^+ + 2\pi^- + \pi^0$$

of cource, the π^0 is not visible, but one can infer its existence from momentum and energy conservation. The process can be represented as shown in Figure 9.9.

By analysing the energies of the mesons by means of this and many other similar processes, one finds that three of the mesons (marked 1, 3, 5) proceed from the decay of a short-lived particle having a rest mass ~ 548 MeV, and thus is one of the η-mesons, designated η^0. Obviously, instead of the above scheme, one must write

$$p^+ + \bar{p}^- \rightarrow \eta^0 + \pi^+ + \pi^-$$
$$\qquad\qquad\qquad\longrightarrow \pi^+ + \pi^- + \pi^0$$

We must note that even if the η^0-meson is moving at the velocity of light, in its lifetime ($\sim 10^{-20}$ s) it can not move more than 3×10^{-12} m, a quantity impossible to measure in a bubble-chamber photograph.

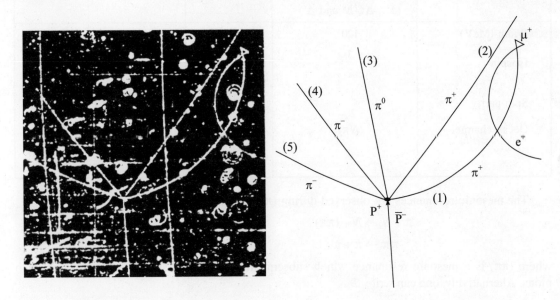

Fig. 9.9 Production of the η^0 resonance in a proton-antiproton annihilation.

Some observed resonance decays by means of strong interactions are

Mesons	Baryons	
$\pi_1^+ \rightarrow \pi^+ + \pi^0$	$N_1^+ \rightarrow \pi^+ + \pi^0$	$N_2^+ \rightarrow \eta^0 + \pi^0$
$\eta_1^0 \rightarrow \pi^+ + \pi^- + \pi^0$	$\wedge_1^0 \rightarrow \Sigma^0 + \pi^0$	$\wedge_2^0 \rightarrow \Sigma^+ + \pi^-$
$K_1^+ \rightarrow K^+ + \pi^0$	$\Sigma_1^+ \rightarrow \wedge^0 + \pi^+$	$\Delta^{++} \rightarrow p^+ + \pi^+$

Baryonic resonances are summarized in the Table 9.7. $N^*(1515)$, $N^*(1690)$ and $\Delta(1950)$ are π-N resonances. Three resonances at pion kinetic energies of 600, 900 and 1400 MeV have been reported in (πN) scattering. It is found that 600 and 900 MeV $(\pi^- p)$ resonances have isospin $\frac{1}{2}$ (charge doublet) and spin parity $\frac{3}{2}^-$ and $\frac{5}{2}^+$ respectively. The corresponding masses are 1515 and 1690 MeV respectively. These resonances are designated as $N^*(1515)$ and $N^*(1690)$ respectively and constitute the $d_{\frac{3}{2}}$ and $f_{\frac{5}{2}}$ excited states of the nucleon.

The 1400 MeV ($\pi^+ p$) resonance has spin parity $\frac{7^+}{2}$ and an isospin $\frac{3}{2}$ and constitute the $\Delta(1950)$ resonance.

Table 9.7 Baryonic Resonances

Resonance	$\Delta(1232)$ (Δ^{++}, Δ^+, Δ^0 and Δ^-)	$N^*(1515)$	$N^*(1960)$	$\Delta(1950)$
Width (MeV)	120	120	100	200
Isospin	$\frac{3}{2}$	$\frac{1}{2}$	$\frac{1}{2}$	$\frac{3}{2}$
Spin-parity	$\frac{3^+}{2}$	$\frac{3^-}{2}$	$\frac{5^+}{2}$	$\frac{7^+}{2}$
Decay channel	$(N\pi)$	$(N\pi)$ $(N\pi\pi)$ $(N\eta)$	$(N\pi)$ $(N\pi\pi)$	$(N\pi)$ $(N\pi\pi)$

The mesonic resonances are observed during (πN) collisions in inelastic processes, e.g.

$$\pi + N \rightarrow N + (\pi\pi)$$

$$\pi\pi \rightarrow \pi + \pi$$

where ($\pi\pi$) is a mesonic resonance which subsequently decays by strong interaction into two pions. Alternatively, one can write

$$\pi + N \rightarrow N + \pi + \pi$$

Obviously, in the above process three particles are produced simultaneously in the final state and hence the product particles have a continuous energy distribution. On the otherhand if the ineleastic process results in the formation of the ($\pi\pi$) or pionic resonance in the final state then it is essentially a two body process and hence the energy distribution should show a peak at a definite energy.

We must note that apart from the pionic resonance ($\pi\pi$), the baryonic resonance (πN) may also be observed. One can express the final state as

$$\pi + N \rightarrow (N\pi) + \pi$$

($\pi\pi$) and (πN) both resonances have been observed in these inelastic processes.

ρ-meson   It is a meson, which breaks up into two pions: ($\pi^+ \pi^0$), ($\pi^- \pi^+$) or ($\pi^- \pi^0$). Its mass is 769 MeV and width is 124 MeV. Its decay time is $\sim 10^{-23}$ s. Obviously, it decays by strong interaction. The charge states are $+e$, 0, $-e$ and isospin $I = 1$. One obtains $J = 1$ from a study of the angular distribution of the decay products. The pions have zero spin and odd parity each and therefore the total spin and parity of ρ-meson are 0 and even. Obviously, it is produced in $l = 1$ state and with odd parity. One finds $J^P = (1^-)$.

η-meson   Following three pion resonance have been reported in the pion-deutron reaction

$$\pi^+ + d \rightarrow p + p + (3\pi)$$

$$3\pi \rightarrow \pi^+ + \pi^- + \pi^0$$

The resonances occur in the neutral state at the masses 783 MeV and 549 MeV. The isospin $I = 0$ for both. The 549 MeV particle is called as the η-meson. It is an electromagnetic decay having $J^P = 0^-$. The electromagnetic decay $\eta \rightarrow 2\gamma$ (31 %) competes with the other stronger decay modes, e.g. $\eta \rightarrow (3\pi^0)$, $n \rightarrow (\pi^+ \pi^- \pi^0)$ etc. It is believed that this is due to the violation of G-parity conservation in the 3π decay.

G-parity The Combined operation of Charge-conjugation C and a rotation through 180° about 2-axis in the isospin sapce is known as the G-parity operation. This is conserved only in strong interaction. The rotation about the 2-axis in the isospin space changes the sign of I_3 (third component of isospin). Now, if $B + S = 0$, the sign of the charge Q does not change by this operation. This follows from the fact that the operation C changes the sige of I_3, i.e. of the charge. Obviously, the combined operation leaves the sign of I_3 unchanged. Obviously, one can assign a definite G-parity to the pion which turns out to be negative. This indicates that strong interactions can not transform an even number of pions into an odd number. The decay of η^0 meson into three pions violates the conservation of G as η^0-meson has even G-parity. Obviously, the two photon decay of η^0 competes with its decay into three pions. This clearly reveal that the most plausible cause of the three pion mode of the decay of η^0, inspite of G-parity violation, is the virtual electromagnetic effect.

ω-meson This was discovered in an analysis of the complicated reaction

$$\bar{p} + p \rightarrow \pi^+ + \pi^+ + \pi^- + \pi^- + \pi^0.$$

The tracks of all the charged particles are observed by preliminary scanning of the photographs of the events caused by the incidence of the $\bar{p}$ on the chamber. One can not observe π^0 directly and hence the reaction is identified from kinematical considerations.

Analysis of the results on the distribution of the effective mass of the product $(\pi^+ \pi^0 \pi^-)$ reveals a prominent peak at 783 MeV with a width of 9.9 MeV. This has been named as ω^0 (783) meson having $I = 0$ (isospin singlet) and $J^P = (1^-)$. The production and decay of ω can be expressed as

$$\bar{p} + p \rightarrow \pi^+ + \pi^- + \omega$$

$$\omega \rightarrow \pi^+ + \pi^- + \pi^0 \quad (89.9 \text{ %})$$

The life time of decay is $\sim 7 \times 10^{-23}$ s. This is rather long compared to the characterstic nuclear time.

Other decay modes of ω are $\pi^+ \pi^-$ (1.3 %), $\pi^0 \gamma$(8.8 %). ω(783) is also produced in $(\pi^+ d)$ reaction.

ϕ meson Mass 1020 MeV, $I = 0$ and $J^P = (1^-)$. We note that last two are the same as those of ω-meson. ϕ(1020) decays primarily into $K\bar{K}$ with a peak width of 4.2 MeV. One finds that modes of its decay include $K^+ K^-$ (48.6 %), $K_L K_S$ (35.1 %), $\pi^+ \pi^- \pi^0$ (14.7 %), $\eta^0 \gamma$ (1.6 %), $\pi^0 \gamma$ (0.14 %) and some minor leptonic decays.

K* meson Excited states of K-meson have been observed. Obviously, K^* which is a vector meson with hypercharge $Y = 1$ is a $(K\pi)$ resonance with mass 892 MeV and width 50 meV, $J^P = (1^-)$ and $I = \dfrac{1}{2}$. Its decay by $(K\gamma)$ mode is also reported.

ρ(769), ω(783), ϕ(1020) and K^*(892) are the lightest vector mesons. Many other mesonic resonances have been reported. A recent and important discovery is that of the so called $\dfrac{J}{\psi}$

particle. The discovery of this particle lends support to the idea of charmed quark c apart from three quarks u, s and d.

9.8 HIGHER BARYONIC STATES

The excited states of N, Δ, $\wedge$, Σ and Ξ^-, which are higher baryonic states have been discovered. Table 9.8 provides a list of well established of these states.

Regge recurrences Systematization of excited baryonic states exhibit that the density of excited states per unit energy interval rapidly increases with increase in mass. One finds that there are different mechanisms of excitation for the higher excited stated of baryons discovered so far. These are: (i) Rotational excitation of the low lying multiplets (ii) Vibratation excitation of the ground state multiplets of given type and (iii) Excitations involving internal orbital angular momentum (expected from the quark model of the hadrons).

The hadrons are not point particles like the leptons and have structure. Obviously, the hadrons could have rotational and other excited states like the molecules. One finds that such excited states would reveal themselves through a sequence of hadrons (resonances) with increasing energies (masses) and spins (J). It is reported that except for parities P, C, G other quantum numbers would be the same. These are termed as *Regge recurrences*. The relation between their energies and spins is named as *Regge trajectory*. Figure 9.10 shows Regge trajectory for some prominent rotational nuclear states with isospin $I = \dfrac{3}{2}$. This Figure exhibit the variation of $m^2(J)$ against J differing by two units. One can represent the excellent straight line fit of the points by the empirical relation $m^2(J) = b(J - \alpha_0)$ where $b = 1$ GeV^2. Similar Regge trajectories have been reported in the literature for other meson resonances.

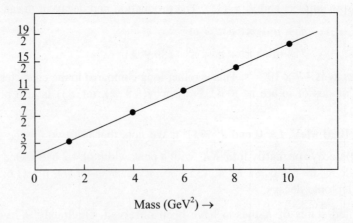

Mass (GeV^2) $\rightarrow$

Fig. 9.10 Regge Trajectory exhibition the plot of the square of the mass of the $I = \dfrac{3}{2}$ nucleonic states versus the spin J of the states.

Table 9.8 Higher Baryonic States

N States J^P $(I=\frac{1}{2}, Y=+1)$	Δ States J^P $(I=\frac{3}{2}, Y=+1)$	$\wedge$ States J^P $(I=0, Y=0)$	Σ States J^P $(I=1, Y=0)$	Ξ States J^P $(I=\frac{1}{2}, Y=1)$	Ω^{-1} States J^P $(I=0, Y=-2)$
$N(939)\,\frac{1}{2}^+$	—	$\wedge(1115)\,\frac{1}{2}^+$	$\Sigma(1193)\,\frac{1}{2}^+$	$\Xi(1315)\,\frac{1}{2}^+$	—
$N(1450)\,\frac{1}{2}^+$	$\Delta(1232)\,\frac{3}{2}^+$	$\wedge(1405)\,\frac{1}{2}^-$	$\Sigma(1385)\,\frac{3}{2}^+$	$\Xi(1530)\,\frac{3}{2}^+$	$\Omega(1672)\,\frac{3}{2}^+$
$N(1520)\,\frac{3}{2}^-$	$\Delta(1620)\,\frac{1}{2}^-$	$\wedge(1520)\,\frac{3}{2}^-$	$\Sigma(1660)\,\frac{1}{2}^+$	$\Xi(1885)\,\frac{3}{2}^-$	
$N(1535)\,\frac{1}{2}^-$	$\Delta(1640)\,\frac{3}{2}^+$	$\wedge(1670)\,\frac{1}{2}^-$	$\Sigma(1670)\,\frac{3}{2}^-$	$\Xi(1930)\,(?)$	
$N(1650)\,\frac{1}{2}^-$	$\Delta(1700)\,\frac{3}{2}^-$	$\wedge(1690)\,\frac{3}{2}^-$	$\Sigma(1750)\,\frac{1}{2}^-$	$\Xi(2030)\,(?)$	
$N(1670)\,\frac{3}{2}^-$	$\Delta(1900)\,\frac{1}{2}^-$	$\wedge(1800)\,\frac{3}{2}^+$	$\Sigma(1765)\,\frac{5}{2}^-$		
$N(1680)\,\frac{5}{2}^-$	$\Delta(1910)\,\frac{1}{2}^+$	$\wedge(1815)\,\frac{5}{2}^+$	$\Sigma(1915)\,\frac{5}{2}^+$		
$N(1680)\,\frac{5}{2}^+$	$\Delta(1920)\,\frac{5}{2}^-$	$\wedge(1830)\,\frac{5}{2}^-$	$\Sigma(1940)\,\frac{3}{2}^-$		
$N(1700)\,\frac{3}{2}^-$	$\Delta(1920)\,\frac{5}{2}^+$	$\wedge(1870)\,\frac{3}{2}^+$	$\Sigma(2030)\,\frac{7}{2}^+$		
$N(1710)\,\frac{1}{2}^+$	$\Delta(1950)\,\frac{7}{2}^+$	$\wedge(2100)\,\frac{7}{2}^-$	$\Sigma(2250)\,(?)$		
$N(1740)\,\frac{3}{2}^+$	$\Delta(1960)\,\frac{3}{2}^+$	$\wedge(2110)\,\frac{5}{2}^+$	$\Sigma(2455)\,(?)$		
$N(1830)\,\frac{3}{2}^-$	$\Delta(2010)\,\frac{3}{2}^-$	$\wedge(2350)\,\frac{9}{2}^+$	$\Sigma(2595)\,(?)$		
$N(1880)\,\frac{3}{2}^+$	$\Delta(2420)\,\frac{11}{2}^+$				
$N(1990)\,\frac{7}{2}^+$	$\Delta(2850)\,\frac{15}{2}^+$				
$N(2190)\,\frac{7}{2}^-$	$\Delta(3230)\,(?)$				
$N(2200)\,\frac{9}{2}^-$					
$N(2220)\,\frac{9}{2}^+$					
$N(2650)\,\frac{11}{2}^-$					

9.9 SYMMETRY CLASSIFICATION OF ELEMENTARY PARTICLES

Large number of elementary particles have been discovered in recent years. We have seen that these can be classified on the basis of their interactions. All the charged particles are subject to electromagnetic interaction and hence the interactions which form the basis of classification of these particles are primarily the strong nuclear and weak interactions. We have seen that elementary particles which are subject to the strong nuclear interactions are grouped as hadrons while those subject to weak nuclear interactions are grouped as leptons. Hadrons have been divided into two groups: (i) baryons and (ii) mesons. We have seen that baryons obey Fermi-Dirac statistics and are fermions whereas the mesons obey Bose-Einstein statistics and are bosons. The mesons constitute the quantum of the inter nucleon field.

One can classify Hadrons of each of the above two groups on the basis of their isospin (I). The isospin determines the multiplicity $M = 2I + 1$. We must note that for each group of particles of given I, the constituent particles have different values of I_3 (the third componenet of isospin), which can have the values $I, I - 1, \ldots -I$. For a system of more than one particle, the isospin vectors can be compounded by vector addition method where as the values of their I_3 can be added algebraically. As an example, consider two isospin doublets each with $I = \dfrac{1}{2}, I_3 = \pm \dfrac{1}{2}$.

Using the above method of compounding, one can obtain the isospin multiplicities of particular groups made up from the above two isospin doublets. The process of forming different isospin multiplets from a number of isospin doublets, e.g. as above, can be achieved with the help SU_2 (special unitary group of rank 2). The generators of such transformations are the 2×2 Hermitian matrices. One can see that the members of a doublet, e.g. p, and n are transformed into each other by the SU_2 transformations.

M. Gell-Mann and Y. Ne'eman in 1962 independently proposed an extension of the scheme of classification of the elementary particles based on the values of I_3 and hypercharge $Y = B + S$, known as SU_3 symmetry. This is also known as octet symmetry or eight fold symmetry. $SU-(3)$ group is generated by a set of 3×3 unitary matrices with unit determinant. With this scheme, called Eight fold path, the regularities of the hadron spectrum could be understood. On the basis of the $SU(3)$ symmetry, Gell-Mann predicted the existence of omega baryon which was detected within a year of its prediction. The $SU(3)$ flavour symmetry group, in addition to generating the multiplet structure of the hadrons, also related many properties, e.g. masses, moments, and decays of hadrons. For example, it yielded the following mass formula

$$\frac{(N_+ \equiv)}{2} = \frac{(3\wedge + \Sigma)}{4}$$

$$1129 \text{ MeV} \qquad 1135 \text{ MeV}$$

Later the higher symmetries, described by $SU(4)$, $SU(5)$, $SU(6)$ and such other groups, were also employed to study the hadronic behaviour.

Gell-Mann and Ne'eman proposed the existence of a group of eight baryons in a super multiplet in the SU_3 scheme, in place of the simple isotopic invariance assumed in SU_2.

These baryons are p, n, $\wedge$, Σ^+, Σ^0, Σ^-, Ξ^- and Ξ and all have $J^P = \left(\dfrac{1}{2}^+\right)$. The different baryon groups have different values either of I (isospin) or of S (strangeness) or of both. Although they have different masses but the mass differences between the different groups are within 15 %.

One can assume that the above eight baryons formed as a result of a common very strong interaction have all equal masses (eight fold degeneracy in strangeness and charm). This octet symmetry is broken due to the action of a moderately strong interaction, which depends on strangeness. This helps in removing the strangeness degeneracy between the groups of different values of S within the super multiplet. Finally, one finds that in each group of S, the charge degeneracy is removed by the electromagnetic interaction. Obviously, the components of same I with different I_3 have slightly different masses. It is reported that the first splitting (due to the moderately strong interaction) is of the order $\dfrac{\Delta M}{M} \sim 10$ % to 20 % whereas the second splitting due to electromagnetic interaction within each group of the same S is of the order $\dfrac{\Delta M}{M} \sim 1$ %.

Figure 9.11 exhibit the scheme of classification on the basis of octet symmetry. This is achieved by a graphical plot (Weight-diagram) of the baryon octet in the $I_3 - Y$ plane. As evident from Figure 9.11, that the members of the octet super multiplet form a symmetric hexagon with one baryon at each corner and two at the centre of the hexagon. It is also evident from figure that the two nucleons ($S = 0$) with $I_3 = \pm\dfrac{1}{2}$ fall on the line $Y = B + S = 1$; the three Σ^- hyperons ($S = -1$) with $I_3 = \pm 1$, 0 fall on the line $Y = 0$; and the two Ξ hyperons ($S = -2$) with $I_3 = \pm 1$, 0 fall on the line $\pm Y = -1$. We also note that the single $\wedge$-hyperon ($S = -1$) with $I_3 = 0$ is at the centre of the figure with $Y = 0$ along with Σ_0.

One can show a similar-multiplet for mesons (Figure 9.12). This consists of following members: the two kaons ($S = 1$) with $I_3 = \pm\dfrac{1}{2}$ on the line $Y = 1$; the three pions ($S = 0$) with $I_3 = \pm 1$, 0 on the line $Y = 0$; and the two antikaons ($S = -1$) with $I_3 = \pm\dfrac{1}{2}$ on the line $Y = -1$. From figure it is evident that the eighth members of the octet is the isospin signlet η^0-meson

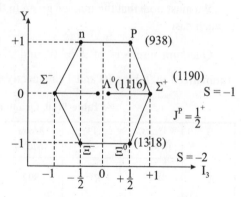

Fig. 9.11 $I_3 - Y$ plot (weight diagram) for the baryon octet.

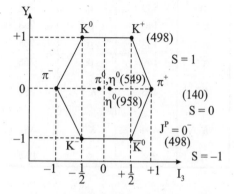

Fig. 9.12 Weight diagram ($I_3 - Y$ plot) for the meson octet.

$(I_3 = 0, Y = 0)$. It is to be noted that in Figure 9.12, we have included the resonance particle η^0 (958) with $J^P = (0^-)$ thereby making it essentially a nonet. We should also note that all the mesons in this diagram are pseudoscalar. The properties of baryonic and mesonic octets are summarized in Table 9.9.

Table 9.9 Relevant Properties of the components of baryonic and mesonic octets

I	*Y*	*Baryon* $\left(J^P = \dfrac{1}{2}^+ \right)$	*Mass* (MeV)	*Meson* ($J^0 = 0^-$)	*Mass* (MeV)
1	0	$\Sigma^+ \ \Sigma^0 \ \Sigma^-$	1194	$\pi^+ \ \pi^0 \ \pi^-$	137
$\dfrac{1}{2}$	1	$p \ n$	939	$K^+ \ K^0$	496
$\dfrac{1}{2}$	−1	$\Xi^0 \ \Xi^-$	1317	$K^- \ K^0$	496
0	0	$\wedge^0$	1116	η^0	549

We must note that the masses given in the table are the average masses for the isospin multiplet in each case.

Quantum numbers of $J^P = 0^-$ and $J^P = 1^-$ mesons and $J^0 = \dfrac{1}{2}^+$ and $J = \dfrac{3}{2}^+$ baryons are summarized in Tables 10 and 11 respectively

Table 9.10 Quantum numbers of $J^P = 0^-$ and $J^P = 1^-$ mesons

0^-	Mass (MeV)	1^-	Mass (MeV)	Charge Q	Isospin I_3	Strangeness S	Hyper charge Y
K^+	494	K'^+	892	1	$+\dfrac{1}{2}$	1	1
K^0	498	K'^0	896	0	$-\dfrac{1}{2}$	1	1
η	549	ω	782	0	0	0	0
π^+	140	ρ^+	768	1	+1	0	0
π^0	135	ρ^0	768	0	0	0	0
π^-	140	ρ^-	768	−1	−1	0	0
$\bar{K}^0$	498	$\bar{K}'^0$	896	0	$+\dfrac{1}{2}$	−1	−1
K^-	494	$\bar{K}'^-$	892	−1	$-\dfrac{1}{2}$	−1	−1
η^0	968	ϕ	1020	0	0	0	0

Table 9.11 Quantum Number of $J^P = \dfrac{1^+}{2}$ and $J = \dfrac{3^+}{2}$ Baryons

$\dfrac{1^+}{2}$	Mass (MeV)	$\dfrac{3^+}{2}$	Mass (MeV)	Charge (Q)	Isospin (I_3)	Strangeness S	Hyper charge Y
		Δ^{++}	1232	2	$+\dfrac{3}{2}$	0	1
p	938	Δ^{+}	1232	1	$+\dfrac{1}{2}$	0	1
n	940	Δ^{0}	1232	0	$-\dfrac{1}{2}$	0	1
		Δ^{-}	1232	−1	$-\dfrac{3}{2}$	0	1
Λ	1116			0	0	−1	0
Σ^{+}	1189	Σ'^{+}	1383	+1	+1	−1	0
Σ^{0}	1193	Σ'^{0}	1384	0	0	−1	0
Σ^{-}	1197	Σ'^{-}	1387	−1	−1	−1	0
Ξ^{0}	1315	Ξ^{*0}	1532	0	$+\dfrac{1}{2}$	−2	−1
Ξ^{-}	1321	Ξ^{*-}	1535	−1	$-\dfrac{1}{2}$	−2	−1
		Ω^{-}	1672	−1	0	−3	−2

One can also construct similar weight diagrams in $Y - I_3$ plane for 14 exhibit two examples. Figure 9.13 is the mesonic resonance octet with $J^P = (1^-)$ whereas Figure 9.14 is a decuplet with $J^P = \left(\dfrac{3^+}{2}\right)$.

We can see that in the case of meson octet discussed in Figure 9.12, the octet represented in Figure 9.13 is essentially a nonet due to the inclusion of the $\phi(1020)$, which has $J^P = (1^-)$.

The weight diagram for the baryon resonance decuplet predicted by the $SU(3)$ symmetry *is* an equilateral triangle as shown in Figure 9.14. The diagram comprises an isotropic quadruplet $\Delta(1232)$ with $S = 0$, $Y = 1$ and $I = \dfrac{3}{2}$; a triplet Σ^0 (1385) with $S = -1$, $Y = 0$ and $I = 1$; a doublet $\Xi^{0,-1}$ (1530) with $S = -2$, $Y = -1$ and $I = \dfrac{1}{2}$ and a singlet Ω^- (1632) with $S = -3$, $Y = -2$ and $I = 0$.

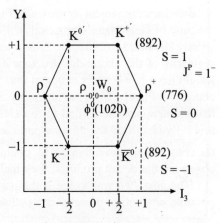

Fig. 9.13 Weight diagram for the mesonic resonance octet with $J^P = (1^-)$.

As mentioned earlier that with the help of this diagram the existence of the Ω^- was predicted and discovered subsequently. This marked a great trimpth of the $SU(3)$ symmetry theory. The anti particle of Ω^- has also been discovered. Gell-Mann received the Noble prize in 1964 for the $SU(3)$ scheme of classification.

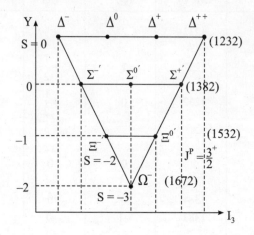

Fig. 9.14 Weight diagram for the mesonic resonance decuplet with $J^P = \left(\dfrac{3^+}{2}\right)$.

Discovery of Ω^- Hyperon and experiment

According to "eight fold way" theory of Gell-Mann, particles should be grouped in families, with all members of a family having the same spin and parity. If we look at the resonances shown in Figure 9.8 and plot the family of ten lighter particles having spin $\dfrac{3}{2}$ (and positive parity), one obtains the arrangement as shown in Figure 9.15, which exhibits a nice geometrical symmetry. At the time of Gell-Mann's proposal in 1961, only the Δ-quartet and the Σ-triplet were known. But shortly afterward in 1962, the Ξ-doublet was reported, and it fitted well with the scheme. From the regularity of the phramidal structure it was easy to predict that the remaining particle at the top should have $\tau_z = 0$ and $\tau = 0$, and thus be a singlet. Also its strangeness should be $= -3$, giving a charge $-e$. Finally, from the regularity in the mass difference $\Delta - \Sigma$ and $\Sigma - \Xi$, it called be inferred that the new particle (called Ω^- by Gell-Mann) should have a mass about 1675 MeV. Obviously, this fully identified the missing particle. From these properties experimenters expected in view of the conservation laws, that the Ω^- could decay into $\Xi^0 + \pi^-$, $\Xi^- + \pi^0$ or $\wedge^0 + \bar{K}^-$; this meant that by observing the decay products, they had a clue by which they could identify this particle.

The next step was to see if the particle could be produced and observed in the laboratory. A possible production process, compatible with the conservation law, would be

$$\bar{K}^- + P^+ \rightarrow \Omega^- + K^+ + K^-$$
$$S = -1 \qquad 0 \qquad -3 \qquad +1 \qquad -1$$

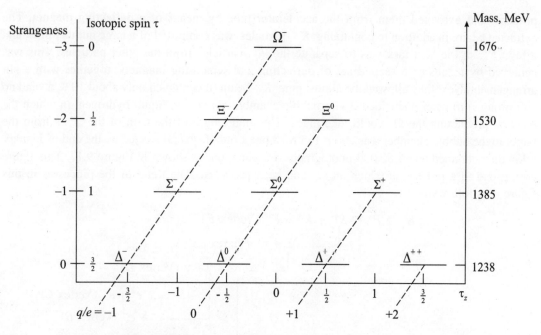

Fig. 9.15 Arrangement of lighter hyperons with spin $\frac{3}{2}$ and positive parity (Gell-Mann's "eight fold way" theory).

The experimental set up used at the Brookhaven National Laboratory in 1963 is shown in Figure 9.16. A 33 GeV proton beam generated by the accelerator hit a tungenston target, producing $\bar{K}^-$-mesons,

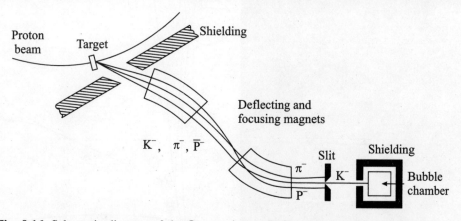

Fig. 9.16 Schematic diagram of the Ω^- experiment. Before the separation there were approximately 800 π^- and 10 $\bar{p}^-$ for every 10 $\bar{K}^-$ particles. After separation, there was only one π^- and no $\bar{p}^-$ for every 10 $\bar{K}^-$-particles.

together with pions, antiprotons and other particles. For experimental convenience, they choose $\bar{K}^-$'s with a kinetic energy of 5 GeV (which is well above the threshold energy of 3.2 MeV of the process) and extracted them from the accelerator tube by means of a deflecting magnet. The extracted beam, in addition to containing $\bar{K}^-$-particles, was composed of a large number of π^- and a few $\bar{p}^-$, so the next task was to separate the $\bar{K}^-$-particles from the other particles. This was achieved by means of a second set of deflecting and separating magnets, together with a slit arrangement. Next they allowed the almost pure $\bar{K}^-$-beam (it contained only about 10 % π^- and no anti protons) to enter a shielded $2 - m$ bubble chamber containing liquid hydrogen, in which the reaction producing the Ω^- could take place. The total length of the path of the $\bar{K}^-$'s from the target to the bubble chamber was about 135 m. After a run of several weeks, by the end of January 1964 they obtained about 50,000 photographs, of which one is shown in Figure 9.17. This figure correspond to a process in which an Ω^- had been produced. The trend of the processes in this figure is as follows:

$$\bar{K}^- + P^+ \rightarrow \Omega^- + K^+ + K^0 \text{ (Vertex } F)$$
$$\hookrightarrow \Xi + \pi^- \text{ (Vertex } E)$$
$$\hookrightarrow \Lambda^0 + \gamma + \gamma \text{ (Vertex } D)$$
$$\hookrightarrow e^+ + e^- \text{ (Vertex } C)$$
$$\hookrightarrow e^+ + e^- \text{ (Vertex } B)$$
$$\hookrightarrow P^+ + \pi^- \text{ (Vertex } A)$$

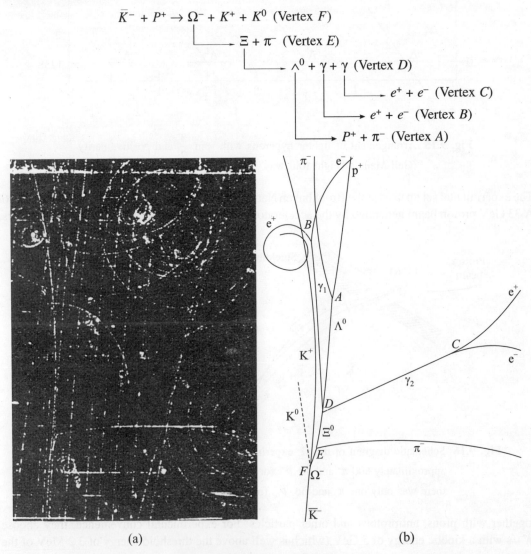

(a)   (b)

Fig. 9.17 Discovery of Ω^- hyperon

Of cource the neutral particles do not leave any track in the Chamber. Figure 9.17(b) represent the analysis. The analysis gave a mass value of Ω^- between 1668 MeV and 1686 MeV. The second photograph, obtained after a few weeks later, narrowed the mass to the range between 1671 and 1677 MeV, in excellent agreement with the theoretical prediction. The analysis provided the Ω^--life time about $\dfrac{2.5 \times 10^{-2} \text{ m}}{3 \times 10^8 \text{ m s}^{1-}} \sim 10^{-10}$ s. The broken lines in the diagram denote the unobserved paths of the neutral particles deduced from measurments on the visible particle tracks.

Important predictions of *SU*(3) symmetry

(i) The masses of the super multiplet should be connected by the following formula

$$M_n + M(\Xi)^0 = \frac{\{M(\Sigma^0) + 3M(\wedge^0)\}}{2}$$

On substitution of the mass value one finds that the above relationship holds within an accuracy of 1 %.

(ii) The second important prediction is

$$M_n - M_p + M(\Xi^-) - M(\Xi^0) + M(\Sigma^+) - M(\Sigma^-) = 0$$

This relationship is found to hold well within the limits of experimental errors.

We must note that some predictions of *SU*(3) are not satisfied and unitary symmetry is much broader in scope than the isotope invariance based on *SU*(2) symmetry.

One obtains the mathematical description of unitary symmetry with the help of *SU*(3) group for three row matrices. We have seen that the simplest *SU*(3) is a triplet while the simplest *SU*(2) multiplet is a doublet. The next more complicated representation of the *SU*(3) group is the baryon octet.

9.10 QUARK MODEL

The symmetries discussed above were only to act as intermediate step. In fact, the symmetries in turn raised several new questions in this field. For example, why nature realizes only 1, 8, 10 representations for the hadrons while *SU*(3) group does possess lower representation like 3, 3^*, 6, 6^*, etc. This led scientists to the proposal that the baryons are composed of three quarks (or aces), while mesons are quark-anti quark bound states. The hadrons upto strange sector could be composed of three types (named as 'up', 'down' and 'strange') of quarks (q) carrying fractional charges.

The two of these quarks should have charges of $-\dfrac{1}{3}e$ and the third should have a charge $+\dfrac{2}{3}e$.

The original three quarks were labeled u(for 'up'), d(for 'down') and s(for 'strange'), and they and their anti particles u, d, and s were assigned charges as follows:

$$u: \quad +\frac{2}{3}e \qquad\qquad \bar{u}: \quad -\frac{2}{3}e$$

$$d: \quad -\frac{1}{3}e \qquad\qquad \bar{d}: \quad +\frac{1}{3}e$$

$$s: \quad -\frac{1}{3}e \qquad\qquad \bar{s}: \quad +\frac{1}{3}e$$

Quarks are thought to be elementary in the same sense as leptons, essentially point particles with no internal structures, but unlike leptons-indeed, unlike anything else in nature-are supposed to fractional electric charges as shown above.

These six kinds of quarks are called six quark *"flavors"*. The quarks are called for historical reasons: *up, down, strong, charmed, bottom* and *top*. They are denoted by the first letters of their names as indicated below

$$\begin{bmatrix} u \\ d \end{bmatrix}, \begin{bmatrix} c \\ s \end{bmatrix}, \begin{bmatrix} t \\ b \end{bmatrix} \qquad\qquad \dots (9.40)$$

It is not understood why there are six or whether more will be found as machines become available to look at higher energies. We can see that the top row in (9.40) has electric charge $q = \frac{2}{3}e$ and the bottom row has $q = -\frac{1}{3}e$, where e is the magnitude of the electron's electric charge.

Each quark has a baryon number of $B = \frac{1}{3}$, each antiquark a baryon number of three antiquarks, so that $B = -1$. A meson is made up of one quark and one half-integral spins of baryons and the 0 or 1 spins of mesons. The properties of the various quarks are listed in Table 9.12.

Table 9.12 Properties of Quarks

Quark	Isospin I, I_3		Strangeness S	Bayron B	Hyper charge Y	Charge Q	Charm C	Bottom b	top t	Mass (GeV)
u	$\frac{1}{2}$	$+\frac{1}{2}$	0	$\frac{1}{3}$	$\frac{1}{3}$	$\frac{2}{3}$	0	0	0	0.39
d	$\frac{1}{2}$	$-\frac{1}{2}$	0	$\frac{1}{3}$	$\frac{1}{3}$	$-\frac{1}{3}$	0	0	0	0.39
s	0	0	-1	$\frac{1}{3}$	$-\frac{2}{3}$	$-\frac{1}{3}$	0	0	0	0.51
c	0	0	0	$\frac{1}{3}$	$\frac{1}{3}$	$\frac{2}{3}$	1	0	0	1.55
b	0	0	0	$\frac{1}{3}$	$\frac{1}{3}$	$-\frac{1}{3}$	0	$+1$	0	4.72
t	(0)	(0)	(0)	$\left(+\frac{1}{3}\right)$	$\left(+\frac{1}{3}\right)$	$\left(+\frac{2}{3}\right)$	0	0	1	$(30-50)$

We must note that the quantum numbers I_3, S, B, Y, Q and C of the antiquarks are the negatives of those of the quarks.

The stable matter, we see around us, involves only the first generation quarks and leptons. All quarks and leptons have spin $\frac{1}{2}$. Quarks have the baryon number $B = \frac{1}{3}$ and may be "red", "green" or "blue". The strong force between the quarks arises from their "color" in the same way that the electromagnetic force between charged particles arises from their charge. For each lepton and quark there is an antilepton and antiquark with opposite properties, including anticolor in the case of quarks. Baryons consists of three quarks of different colors, mesons consist of a quark of one color and an anti quark of the corresponding anticolor. The reason for the existence of other generations of quarks are not known, except that to allow CP-violation minimum number of the generations required is three. Some theoretical particle physicists have even suggested the possibility of the fourth generation, but with no experimental indication so far.

All known hadrons can be explained in terms of various quarks and antiquarks (Table 9.13).

Table 9.13 Compositions of some hadrons based on the quark model

Hadron	Quark Content	Baryon Number	Charge (e)	Spin	Strangeness
π^+	$u\bar{d}$	$\dfrac{1}{3}-\dfrac{1}{3}=0$	$+\dfrac{2}{3}+\dfrac{1}{3}=+1$	$\uparrow\downarrow=0$	$0+0=0$
K^+	$u\bar{s}$	$\dfrac{1}{3}-\dfrac{1}{3}=0$	$+\dfrac{2}{3}+\dfrac{1}{3}=+1$	$\uparrow\downarrow=0$	$0+1=+1$
p^+	uud	$\dfrac{1}{3}+\dfrac{1}{3}+\dfrac{1}{3}=+1$	$+\dfrac{2}{3}+\dfrac{2}{3}-\dfrac{1}{3}=+1$	$\uparrow\uparrow\downarrow=\dfrac{1}{2}$	$0+0+0=0$
n^0	ddu	$\dfrac{1}{3}+\dfrac{1}{3}+\dfrac{1}{3}=+1$	$-\dfrac{1}{3}-\dfrac{1}{3}+\dfrac{2}{3}=0$	$\downarrow\downarrow\uparrow=\dfrac{1}{2}$	$0+0+0=0$
Ω^-	sss	$\dfrac{1}{3}+\dfrac{1}{3}+\dfrac{1}{3}=+1$	$-\dfrac{1}{3}-\dfrac{1}{3}-\dfrac{1}{3}=-1$	$\uparrow\uparrow\uparrow=\dfrac{3}{2}$	$-1-1-1=-3$

A serious problem with the idea that baryons are composed of quarks was that the presence of two or three quarks of the same kind in a particular particle (for instance, two u quarks in a proton, three s quarks in a Ω^- hyperon) violates the exclusion principle, to which quarks ought to be subject since they are fermions with spins of $\dfrac{1}{2}$. To get around this problem, it was suggested that quarks and antiquarks have an additional property of some kind that can be manifested in a total of six different ways, rather as electric charge is a property that can be manifested in the two different ways that have come to be called positive and negative. In the case of quarks, this property became known as "Color", and its three posibilities were called red, green and blue. The antiquark colors are antired, antigreen and antiblue.

According to the color hypothesis, all three quarks in a baryon have different colors, which satisfies the exclusion principles since all are then in different states even if two or three are otherwise identical. Such a combination can be thought of as white by analogy with the way red, green, and blue light combine to make white light (but there is no connection whatever except on this metaphorical level between quark colors and actual visual colors). A meson is supposed to consist of a quark of one color and an antiquark of the corresponding anticolor, which has the effect of canceling out the color. The result is that both baryons and mesons are always colorless. We must note that quark color is a property that was significance within hadrons but is never directly observable in the outside world.

Though the quark picture has successfully explained the hadron interactions, free quarks have never been observed. Now, it is believed that quarks may never be observed since they can not exist in free state. Quarks can only form combinations which are observed in different types of hadrons. In spite of that, one can infer some of the properties of the observed properties of combinations of quarks.

If we assume that the binding energies of quarks are small compared to their masses, then one finds $m_u = m_d = 0.39$ GeV and $m_s = 0.51$ GeV. The expected masses of other three quarks are $m_c = 1.55$ GeV, $m_b = 4.72$ GeV and $m_t = 30$ to 50 GeV.

Quark Structure of Mesons and Baryons

We have metioned that the observed particles are formed by the combination of more than one quark, i.e. quarks have to be combined. However, the combination like qq is ruled out since the charge will be non integral for such a combination.

The combination of a quark and an antiquark ($q\bar{q}$) can produce a real observable particle. The method of obtaining this is illustrated in Figure 9.18(c) with the help of weight diagrams of the quarks and antiquarks.

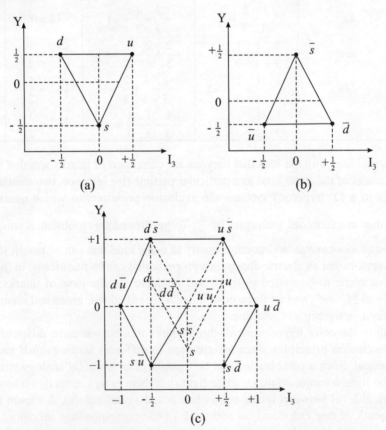

Fig. 9.18 The weight diagrams of the u, d, s quarks and their antiquarks.

From Figure 9.18(c) it is evident that the centre of symmetry of the weight diagram of the antiquarks is successively superimposed on the positions of the three quarks u, d, s, and the positions of the antiquarks $\bar{u}$, $\bar{d}$, $\bar{s}$ are shown in each case in the $Y - I_3$ plane. Let us consider that if the weight diagram of the antiquark is superimposed on the position of the u quark then the positions of the three antiquarks give corresponding values of Y and I_3 of the combinations $u\bar{u}$, $u\bar{d}$ and $u\bar{s}$. We can see from Figure 9.18(c) that these fall on the three corner points of the triangle on the upper right hand side of the hexagon. Following the same procedure, one can successively superimpose the weight diagram of the antiquarks on the positions of d and s to yield the values of Y and I_3 for the combonations $d\bar{u}$, $d\bar{d}$, $d\bar{s}$, $s\bar{u}$, $s\bar{d}$ and $s\bar{s}$.

One can see from Figure 9.18(c) that in the hexagonal weight diagram of the combination $q\bar{q}$, the different corner points are occupied by the combinations $u\bar{s}$, $u\bar{d}$, $d\bar{s}$, $d\bar{u}$, $s\bar{u}$ and $s\bar{d}$. We further note that the three other combinations $u\bar{u}$, $d\bar{d}$ and $s\bar{s}$ are located at the centre of the hexagon. Obviously, one obtains 3×3 or 9 combinations which comprise an $SU(3)$ octet and SU_3 singlet. One can express this in the group theoretical language as $3 \otimes 3 = 8 \oplus 1$.

We further note that two of the combinations at the centre of the hexagon are parts of the $SU(3)$ octet. One of these combinations belongs to an isospin triplet and the other to an isospin singlet. The third combination belongs to $SU(3)$ singlet. This has been reported as the $\eta(958)$ resonance.

We have read that quarks and antiquarks have opposite parity with spin $\frac{1}{2}$ each, and hence the combination $q\bar{q}$ has $J^P = (0^-)$ or (1^-) in the $l = 0$ state. The $J^P = (0^-)$ $q\bar{q}$ combinations have been identified with the eight mesons (Table 9.9), whereas $q\bar{q}$ combinations with $J^P = (-1)$ have been identified with the meson resonances.

Table 9.14 Quantum numbers of different qq combinations

$q\bar{q}$ combinations	I_3	S	B	Y	Q	Particle
$u\bar{s}$	$\frac{1}{2}$	1	0	1	1	K^+
$d\bar{s}$	$-\frac{1}{2}$	1	0	1	1	K^0
$s\bar{u}$	$-\frac{1}{2}$	-1	0	-1	-1	K^-
$s\bar{d}$	$\frac{1}{2}$	-1	0	-1	0	$\bar{K}^0$
$u\bar{d}$	1	0	0	0	1	π^+
$d\bar{u}$	-1	0	0	0	-1	π^-
$u\bar{u}$	0	0	0	0	0	
$d\bar{d}$	0	0	0	0	0	π^0, η^0, η^*
$s\bar{s}$	0	0	0	0	0	

The quantum numbers of the different possible $q\bar{q}$ combinations are summarized in Table 9.14. It is evident from the table that out of these, six can be identified with the six known particles listed in the last column. We further note that all the quantum numbers are equal for the remaining three ($u\bar{u}$, $d\bar{d}$, $s\bar{s}$). Theoretically, it has been shown that the wave functions of π^0, η^0 and η^0 (958) can be expressed as linear combinations of the above three combinations. This is why, all these three particles are shown associated with the above three combinations.

Three quark combinations Now, we consider the 27 possible three quark combinations. One can represent them symbolically as

$$3 \otimes 3 \otimes 3 = 1 \oplus 8 \oplus 8 \oplus 10$$

Obviously, one can have integral baryon number $B = 1$ and integral charge Q with three quarks. One can also represent qqq combinations with the help of weight diagrams for the quarks. One can strat with the weight diagram for the first quark and the weight diagram for the second is superimposed on the positions of the three quarks u, d, s of the first diagram successively, giving the nine points u, u, ud, us etc.

Now, the weight diagram for the third quark is successively superimposed on the nine points in the $Y - I_3$ plane obtained previously. In this way, one obtains the twenty seven points representing the different possible qqq combinations are obtained. These include the octet of baryons with $J^P = \left(\dfrac{1}{2}^+\right)$ and octet of corresponding antibaryons. In addition, one finds that there is decuplet containing the ten baryons (Figure 9.14) for which $J^P = \left(\dfrac{3}{2}^+\right)$. One can obtain the wave functions of the different observed particles as linear combinations of the wave functions of the qqq combinations.

Baryon Octet One can write the quark structure of the nucleons as

$$p = (uud) \quad \text{and} \quad n = (udd)$$

This can be easily verified by referring to the values of quantum numbers of the quarks (Table 9.12). One can properly antisymmetrized the wave functions for p and n by introducing the colour wave function in the quantum chromodynamic theory (QCD) of the quark structure of the hadrons. One finds that the replacement of a d quark in the nucleon by an s quark produces a baryon state with $J^P = \left(\dfrac{1}{2}^+\right)$ and $S = -1$. This procedure generates new states, an isotriplet and an isosinglet as follows:

$$(\Sigma^+, \Sigma^0, \Sigma^-) = \left\{ uus, \frac{(ud + du)s}{\sqrt{2}}, ds \right\}$$

$$\wedge = \left\{ \frac{(ud - du)s}{\sqrt{2}} \right\}$$

However, when a quark and a d quark are each replaced by an s quark the isodoublet Ξ states are obtained as follows:

$$(\Xi^0\ \Xi^-) = (uss, dss)$$

Obviously, Σ and $\wedge$ states generated have $S = -1$ and Ξ states have $S = -2$ as required. We must note that the flavour wave function (sss) is necessarily symmetric and can not occur with a total spin $J = \dfrac{1}{2}$.

Thus the eight baryonic states stated above, all have $J^P = \left(\dfrac{1^+}{2} \right)$ and for baryonic octet of Figure 9.11.

$\dfrac{3^+}{2}$ **Baryon decuplet** It is found that the three quark state (uuu) with spin $\dfrac{3}{2}$ is flavour symmetric and spin symmetric and has $L = 0$ space function which makes the total wave function which makes the total wave function including colour function antisymmetric. These are called non strange states. Out of these four are identified as particles with $I = \dfrac{3}{2}$ whose quark structures are as follows:

$$\Delta^{++} = (uuu) \qquad\qquad \Delta^{+} = (uud)$$
$$\Delta^{0} = (udd) \qquad\qquad \Delta^{-} = (ddd)$$

The isotriplet Σ^*, the isodoublet Ξ^* and isosinglet Ω^- together complete the decuplet and their quark structures are as follows

$$\Sigma^{*-} = (dds) \qquad \Sigma^{*0} = (uds) \qquad \Sigma^{*+} = (uus)$$
$$\Xi = (dds) \qquad \Xi^{*0} = (uss)$$
$$\Omega^- = (sss)$$

The above baryon decuplet is shown in Figure 9.14.

Finally, one finds that there is an $SU(3)$ singlet formed by the three quark combination which has been identified with $\wedge(1405)$. The wave function of this has slight admixture of with that of $\wedge(1675)$.

On the basis of three quark model, it is possible of fit the observed resonances in the scheme of classification of the particles. On the basis of this model it is also possible to predict some of the important properties of the hadrons from a knowledge about the quark wave functions.

It is found that the mass differences between the particles of the different isospins in a supermultiplet owe their origin in a type of strong interaction of intermediate strength. However, the origin of this symmetry breaking interaction is not clear, but it is possible to show on the basis of the weak bond model between the quarks that the masses of the u and d quarks should be heaviour. The estimated masses of m_u, m_d and m_s have been reported earlier. According to the quark model, the masses of the Δ^- (1232), Σ^- (1385), Ξ (1530) and Ω^- (1672) should be equispaced. Measurment yield $\Sigma^{--} = \Delta^- = 149$ MeV, $\Xi^- - \Sigma^- = 145$ MeV and $\Omega^- - \Xi^- = 142$ MeV. These observations are in accordance with the above conclusions.

Experimental Supports for the Quark Model

The qualitative estimates on the basis of quark model of strong processes in pion-nucleon $(\pi - N)$ scattering, the electromagnetic process of the annihilation of an electron-positron pair in hadrons and the two weak processes in the β-decay and some few other processes are in good agreement with the experimental results.

However, the observations on the inelastic scattering of high energy electrons by protons provide the most convincing proof of the correctness of the quark model. Feynman on the basis of the *Parton-Model* interpreted the results on the inelastic scattering cross section measurment for electrons with energies upto 10 GeV both with protons and neutrons exhibiting slow decrease in the cross section with energy. Feynman concluded that a nucleon at rest is a complex particle consisting of virtual point particles called as *Protons*. These partons are believed to be same as quarks. It is remarked that the partons are concentrated as mass-points, within the finite radius of the nucleons. However, this behaviour is characterstic of the hadrons. On the other hand, the leptons do not seem to have finite extension and are considered as point particles.

The measurment of the total cross sections for $N - N$, $N - \bar{N}$ and $\pi - N$ processes provides the important confirmation of the quark model. The estimated ratio $\dfrac{(\sigma_{NN} + \sigma_{N\bar{N}})}{\sigma_{\pi N}} = 3$ from the quark model agrees well with the measured value.

The ratio of the magnetic moments of the nucleons based on the weak-bond model, i.e. $\dfrac{\mu_n}{\mu_p} = -\dfrac{2}{3}$ is in very good agreement with the experimental value, i.e. $\dfrac{\mu_n}{\mu_p} = -0.68$.

Non-Observance of Free Quarks

Though the quark model has successfully explained the hadron interactions, free quarks have never been isolated in an experiment. Some experiments claim to observe free quarks. However, these observations have not been confirmed.

Scientists have made attempts to produce quarks in the high energy collisions using proton beam from high energy accelerators. The following reactions have been investigated.

$$p + p \rightarrow p + p + q + \bar{q}$$
$$p + p \rightarrow p + 2u + d$$

Although, some of these experimental efforts reported initial success, but on further careful scrutiny these predictions were found to be incorrect.

We have read that quarks carry fractional electric and baryonic charges. Therefore it is expected that the lighest of them must be stable. Obviously, if quarks are produced in very high energy collisions (in Cosmic rays) there should be some accumulation of these in the earth's crust or in sea water or in meteorities (cosmic bodies). However, all attempts by scientists to detect their presence in these bodies reveal that there can not be more than one quark per 10^{19} protons in water or 1 quark in 10^{15} meteorites.

It is argued that the absence of any evidence for the existence of free quarks is probably due to their very large mass, i.e. the energy available from the accelerators is not high enough to produce quarks. The another view supports the weak-bond model which is supported by various experiments. The weak-bond model consider that the quarks are rather loosely bound within the hadrons (parton-model). This is why that they do not come out of the latter in high energy collisions and hence it is difficult to see them.

9.11 QUANTUM CHROMODYNAMICS (QCD)

The theory of how quarks interact with each other is known is *quantum chronodynamics* since it is modeled on quantum electrodynamics the theory of how charged particles interact with each other, with colour taking the place of charge. The theory was first propounded in 1973.

We have read that one of the main problems in formulating the quark structure of the hadrons is the apparent failure of Pauli's exclusion principle.

Quarks are spin $\frac{1}{2}$ particles and obey Fermi-Dirac statistics. The proposed structure of the baryon Δ^{++} requires that the three identical spin $\frac{1}{2}$ particles are to be combined to produce a completely symmetric ground state (uuu) which reproduces all the known properties of Δ. This combination violates Pauli's exclusion principle. We should also note that the same is also true for the proposed work structures of the proton $p(uud)$ and the neutron $n(udd)$.

Although the observed sequence of baryons, antibaryons and mesons are correctly reproduced by the quark structures qqq, $\bar{q}\,\bar{q}\,\bar{q}$ and $\bar{q}\,\bar{q}$, but the other possible combinations such as qq, $\bar{q}\,\bar{q}$ etc. or single quarks are not observed.

To overcome these difficulties, Gell-Mann and Zweig in 1963 proposed a new property (or quantum number) of the quarks, known as the colour. They assumed that the quarks can have three primary colours, red (R), green (G) and blue (B) respectively, such as q_R, q_G, q_B. In this proposal the Δ-particle can be represented as

$$\Delta^{++} = u_R\, u_G\, u_B$$

We must note that the colour is a conserved additive quantum number.

At this stage, the question arises that why the various possible combinations (two on three) of colours are not observed in nature. Although the representations $u_R\, u_G\, u_B$, $u_R\, u_G\, d_G$, $u_B\, u_R\, d_R$ etc. ... can reproduce the observed properties of the protons, it is only the first which is regarded as the correct state of the proton. To answer this, they assumed that only those particles or quark bound states exist in nature which are *colourless*. The analogy with the concept of colour is shown in Figure 9.19.

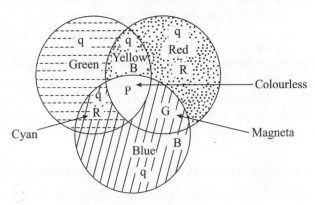

Fig. 9.19 Colour composition of hadron

As stated earlier, only the colourless combinations are manifested in nature. One finds

Proton p = R G B

Antiproton $\bar{p}$ $\bar{R}$ $\bar{G}$ $\bar{G}$

pion π $R\bar{R} \pm G\bar{G} + B\bar{B}$ and so on.

Here R, G, B denotes red, green and blue colours of light.

We must note that the combination $R\bar{R} + G\bar{G} + B\bar{B}$ remain unchanged by rotation in the R, G, B colour space and is a singlet representation of the colour group. Such colour singlet representations which provide the wave functions of the observed particles are known as colourless.

Although the combinations $R\bar{R}$, $G\bar{G}$ and $B\bar{B}$ are individually colourless but they do not represent the wave functions of any observed particle. We must note that besides, any combination which is coloured or the individual quarks which are coloured do not represent the wave function of any observed particle, then they are hidden from us.

The *QCD* theory has been developed, following the formalism of the theory of quantum electrodynamics (*QED*). *QED* and *QCD*, both theories are special versions of relativistic field theory. The most recent studies of elementary particles class of these theories, known as *gauge theories*.

QED is the simplest of gauge theories and describes the interaction for electrically charged particles like the electrons in terms of the exchange of virtual photons. An electron is supposed to emit and reabsorb such a photon with a time determined by the well known Heisenberg uncertainty principle. Feynman diagram help to describe the Coulomb attraction or repulsion between charged particles. There are several testaments, like anomalous magnetic moment of leptons, Lamb shift etc. to the success of the *QED*.

QED, as $U(1)$ abelian guage theory, has acted as a guiding theory for constructing the gauge field theories of other basic interactions. For examples, in midsixties electroweak interactions theory has been constructed by generalizing the gauge invariance to non-abelian group $SU(2)$. It introduces three W-bosons as mediator of the work force. Similarly, first field theory of strong interactions, called Quantum chromodynamics (*QCD*), has been developed in which the colour quantum number acts as the strong charge. The strong interaction (attraction) between two quarks can not be described by the *QED* theory because two quarks carrying positive charges would be subjected to the repulsive Coulomb force. The strong attractive force overcomes this repulsive force to bind quarks in hadrons. The so called colour charge of the quarks as proposed in *QCD* is responsible for the colour field just as the electric charge is responsible for the e.m. field. However, the three colours require a bigger unitary nonabelian group in three dimension-called colour $SU(3)$. As a consequence of that, the strong interactions require eight-quanta called *gluons*, which unlike the photon, carry strong charge, i.e. colour quantum number. Obviously, the gluons are the quanta of the colour field and bind the quarks in hadrons (e.g. nucleons) and also the nucleons in the nuclei.

The gluon is itself a coloured particle being depicted by the symbols $R\bar{R}$, $R\bar{B}$, $R\bar{G}$, $G\bar{R}$, $G\bar{B}$, $G\bar{G}$, $B\bar{R}$, $B\bar{B}$, $B\bar{G}$. Obviously, gluons are bicoloured objects. As is evident that there are nine possible such combinations. The combination $R\bar{R} + G\bar{G} + B\bar{B}$ is a colour singlet and is does not have any net colour. Obviously, this combination can not be regarded as a gluon which must carry colour charges between two quarks. This means the *QCD* is described in terms of eight gluons. This consist of all the six nondiagonal elements $R\bar{B}$, $R\bar{G}$, etc., besides the two diagonal elements $R\bar{R} - G\bar{G}$ and $R\bar{R} + G\bar{G} - 2B\bar{B}$. Obviously, each of these eight carry colour-charges, and hence there will be colour interaction between them, unlike in the case of photons in the e.m. theory which do not interact amongst themself. We must note that due to this property, strong interactions gets weaker at short distances (asymptotic freedom). This effect is similar to the modification of Coulomb law in *QED* due to vacuum polarization.

We can now see that with the introduction of the colour quantum, every quark is characterized by two properties *flavour* and *colour*. Thus there are four different flavours (*u, d, s, c*) and three different colours (*R, B, G*). This means there are altogether 12 types of quarks and correspondingly 12 antiquarks.

9.12 CHARMED QUARK

Not only do quarks come in three colours, but additional varieties (or "flavours") of quarks have had to be included in the scheme to supplement the original *u*, *d* and *s* trio. The first of the new ones, the "Charm" quark *C*, was proposed largely by analogy with the existence of four leptons: if quarks are elementary particles in the same sense as leptons, then there ought to be four of them, too. Apart from the usual properties (Table 9.12), this quark is endowed with an additional charm quantum number $C = 1$. The antiquark $\bar{c}$ has $C = -1$. Such a quark has charge $+\frac{2}{3}e$ and a mass 1.55 GeV and is heavier than u, d or s quarks.

Leptons and quarks differ in a number of important properties, and they are generally classified in separate families. One of the most conspicuous differences is in electric charge (Table 9.15). The lepton charges are integers (neutrino do not carry electric charge) where as the quark charges are fractions. Furthermore, the leptons exist as free particles, whereas the quarks are found only as constituents of composite particles called hadrons. It is customary to divide the leptons and the quarks into three generations. The kinds of quarks are generally called flavours, in addition the quarks have another property called colour (we must remember that flavour and colour are arbitrary labels, which have nothing to do with the sensation of taste or sight). Each flavour of quark comes in three colours; red, green and blue. The property of colour marks a major difference between the leptons and quarks.

To explain the noobservance of the leptonic decays $K^0 \to \mu^+\mu^-$, $\wedge \to n\, e^+e^-$ etc. Glashow et. al. suggested the existence of a quark of fourth flavour, known as the charmed quark (c). Its presence in intermediate states would cancel the unobserved neutral current in *K*-decay, provided the mass of the *c* quark was not very different from that of the *u*-quark. The generalization of Equation (9.37) to include *C* leads to

$$Q = I_3 + \frac{B + S + C}{2} \qquad \dots (9.41)$$

Like the strangeness *S*, charm *C* is conserved in strong and electromagnetic interactions and changes by unity in weak interactions.

Table 9.15

Family	Flavour (f)	Quarks q_{fn} Colour: n = 1, 2, 3 red green blue	Charge of q_{fn}	Lepton l_f	Charge of lepton l_f
First generation	1 up	u_1 u_2 u_3	$\frac{2}{3}$	ν_e	0
	2 down	d_1 d_2 d_3	$-\frac{1}{3}$	e	-1
Second generation	3 Charm	c_1 c_2 c_3	$\frac{2}{3}$	ν_μ	0
	4 Strange	s_1 s_2 s_3	$-\frac{1}{3}$	μ	-1
Third generation	5 top	t_1 t_2 t_3	$\frac{2}{3}$	ν_τ	0
	6 bottom	b_1 b_2 b_3	$-\frac{1}{3}$	τ	-1

The discovery of $\frac{J}{\psi}$ particles in 1974 is the first experimental evidence in support of the c-quark.

Mesons and baryons containing c quarks have been found and extensively studied. One can extend the two dimensional diagram of Figure 9.12 by adding a third dimension corresponding to the charm quantum number, to give a polyhedron whose vertices correspond to the predicted charmed mesons.

The charmed particles family members fall into two classes: (i) *hidden* charm and (ii) *bare* charm. The former class includes the $\frac{J}{\psi}$ particle which is considered to be a combination $(C\bar{C})$ of charmed and anticharmed quarks, known as *charmonium*. The bare charm class include D^+ mesons $(c\bar{d})$ with $J^P = (0^-)$ and $J^P = (1^+)$; Σ^{2+} baryon (cuu); $\wedge^+$ (cud) baryon. The life time of these bare charm states are to be of the order 10^{-13} s.

9.13 BEAUTY AND TRUTH

Ledermann et al. in 1977 discovered a new resonance state at mass 9.460 GeV which was found to decay into $\mu^+ \mu^-$ pair. Detailed study revealed that these new particles, known as the γ(upsilon) particles are made up of a new type of quark of charge $-\frac{1}{3}$. These were named as the bottom or b quark. A new quantum number b (beauty) is associated with them, $b = +1$ for the beauty quark and $b = -1$ for antiquark b. These beauty (b) quarks have the mass 4.72 GeV. Higher states of the γ(upsilon) meson have also been reported. It is found that the lower states have resonance widths ranging from 44 to 18 keV whereas the higher states have width in the tens of MeV range.

In addition to the known five quarks (u, d, s, c and b), a six quark, known as the top or t quark predicted theoretically. The recently discovered vector bosons ($W^\pm$, Z^0) experiments provided the indication of this quark. The mass of t quark is expected ~ 30 to 50 GeV. The quantum number t is assigned to it. Obviously, $t = 1$ for the top quark and $t = -1$ for its antiparticle.

Now, one can express the generalised expression for Q which includes b and t is

$$Q = I_3 + \frac{B + S + C - b + t}{2} \qquad \qquad \dots (9.42)$$

9.14 WEAK INTERACTIONS

These interactions are mainly responsible for the decay of semi-stable elementary particles and nuclei. The carriers of this interaction are called *intermediate vector bosons,* of which there are two kinds. Because the weak interaction has so short a range, the rest masses of such particles are large more than 30 times the proton mass. One kind called W, has a spin of 1 and a charge of $\pm e$ and is responsible for ordinary beta decays. The other kind, called Z, also has a spin 1 but is electrically neutral and heavier than the W; its effects seem confined to certain high energy events. Although the W particle is a natural concomitant of the weak interaction and was proposed many years ago, the idea of the Z-particle originated more recently in a theory that unites the weak and electromagnetic interactions, and its discovery helped confirm the theory.

A decay is caused by weak interactions only if it can not be caused by strong or electromagnetic interactions. A decay is said to be weak if it satisfies at least one of the following conditions: (i) Strangeness conservation law is not satisfied (ii) charm conservation law is not satisfied (iii) A neutrino is associated with the decay. We note that the strangeness is not conserved in the decay $K^+ \to (\pi^+ \pi^0)$. Its probability is 21 %. The decay $K^+ \to (\mu^+ \nu_\mu)$ is associated with both strangeness conservation violation and emission of a neutrino. Its probability is 64 %. The decay $\mu^\pm \to (e^\pm \nu_e \nu_\mu)$ is associated with the emission of neutrinos but there is no violation of strangeness conservation. The decay of the charmed meson $D^0 \to (K^+\pi)$ with $\tau \sim 10^{-13}$ s takes place due to violation of charm conservation law. One finds that $C = 1$ in the initial state while $C = 0$ in the final state in this charmed meson decay.

Nuclear β-decay is an familiar example of weak interaction decay. The coupling constant in this decay process is almost identical with the coupling constant in the purely leptonic muon decay which is governed by the $V - A$ (vector minus axial vector) interaction. Non conservation of parity in weak decay is intimately associated with the role of the neutrino. The important selection rules governing weak decays are summarized as: (i) $|\Delta S| = 0$, 1, i.e. the change in strangeness can not exceed unity. This is satisfied in the decay processes of the strange particles $\wedge$, $\Sigma^\pm$, Ξ^0 and Ω^-. No violation of this rule has ever been reported. We see that $\Xi^- \to (n\pi^-)$ and $\Omega^- \to (n, \pi^-)$, $S = -2$ for Ξ^- and $S = -3$ for Ω^-; $S = 0$ in the final states in both cases. These are unobserved decay processes. (ii) $|\Delta I| = \dfrac{1}{2}$, i.e. the change in the isotopic spin I in a decay involving a change in S is always $\dfrac{1}{2}$ and (iii) $\Delta Q = \Delta S$, i.e. in decay involving leptons, the change in charge Q is given by this rule.

Weak interaction is also refereed as universal Fermi interaction and can be described in terms of the Pauli triangle. It is assumed that the strength of the interactions in all the three processes, e.g. $n \rightarrow (pe^- \bar{\nu}_e)$, $\bar{\mu} + p \rightarrow (n \nu_\mu)$ and $\mu^\pm \rightarrow (e^\pm \nu_e \nu_\mu)$ are equal. However, more precise determinations of g_β, g_μ and $g_{\bar{\mu}}$ exhibit that the strength of various weak interactions responsible for these three processes differ slightly from one another. Considering this fact Cabbio connected all the constants through a single parameter called as the Cabbio angle ($\theta_c \sim 14°$).

The universal theory implies that along with the weak charged currents, there should also exist neutral currents. Neutral currents have been discovered. However, some of these were not observed in nature, e.g. $K^+(\mu^+ \mu^-)$. This requires an improvement of theory, i.e. one should introduce the fourth quark (c-quark).

9.15   UNIFICATION OF ELECTROMAGNETIC AND WEAK (ELECTROWEAK) INTERACTIONS

The connection between the weak and electromagnetic interactions was independently developed in the 1960 s by Steven Weinberg and Abdus Salam. The key problem to be overcome in constructing unified theory was that the carriers of the weak force have mass whereas the carriers of electromagnetic force, namely photons are massless. What Weinberg and Salam did was to show that, at a certain primitive level, both forces are aspects of a single interaction mediated by four massless bosons. Through a subtle process called spontaneous symmetry breaking, three of the bosons acquired mass and become the W and Z particles, with a consequent reduction in the range of what now appears as the weak part of the total interaction. One way to look at the situation is to regard the masses of the W and Z bosons as being attributes of the states they happen to occupy rather than as intrinsic attributes. The fourth boson, the photon, remained massless and the range of the electromagnetic part of the total interaction accordingly stayed infinite.

Weinberg and Salam's unified electroweak theory is based on the principle of local gauge symmetry. The existence of local symmetry is connected with the presence of an additional field. Yang and Mills in 1954 investigated the problem and concluded that the local invariance of the theory must always lead to the assumption of a compensating field with new quanta (gauge bosons).

Yang and Mills, on the basis of local isotopic invariance developed a theory with three compensating fields with three gauge bosons, all of them massless. Obviously, this was unacceptable for the cases of strong or weak nuclear interactions. To resolve this problem the theory of *QCD* was evolved. In the case of weak interactions, Weinberg and Salam employed the idea of spontaneous symmetry breaking by introducing the so-called spinless Higgs bosons of non-zero mass. This field takes care of the symmetry violation. The interaction with the Higgs field provides the gauge bosons and their mass. Hooft in 1971 has shown the theory as renormalizable and hence calculable as in electrodynamics. This motivated the prediction of Weinberg-Salam model.

The discovery of the $W^\pm$ and Z^0 bosons (1983) provided a great boost to the Weinberg-Salam electroweak model. The decay modes of these particles are $W^+ \rightarrow (e\nu^-)$, $W^- \rightarrow (e^- \bar{\nu})$ and $Z^0 \rightarrow (\mu^+ \mu^-)$. The masses are 80.9 and 93 GeV respectively. The spin of W is 1 as revealed by the preferred directions of the decay products. We must note that the leptonic weak decays are probably only a few percent of the total number of decays. Weinberg-Salam electroweak theory predicts the following expressions for the masses of W and Z particles.

$$m_W = \frac{37.4}{\text{Sin } \theta_W} \text{ GeV, and}$$

$$m_Z = 37.4 \text{ Sin } \theta_W \cos \theta_W \text{ GeV}$$

where θ_W, the Weinberg angle is the mixing parameter and can be determined from experiments:

$$\text{Sin}^2 \theta_W = 0.218 \pm 0.010$$

There is excellent agreement between the theoretically predicted and experimentally determined masses of W and Z particles.

An experiment at Stanford (1978) exhibiting interference effects between the photons (γ) and the Z^0 particles in the scattering of polarized electrons in the reaction $e^- + p \rightarrow e^- + p$ provided indirect but quantitative support for Weinberg-Salam electroweak model.

Neutral Currents The discovery of neutral current weak interactions and the excellent agreement of their properties with the predictions of electro-weak theory were the most important developments.

The decay $K^+ \rightarrow (\pi^+ \pi^0)$ in which the neutral π^0 is formed from the charged K^+, is a strangeness violating decay. Few examples of neutral weak interactions are

Lepton-nucleon scattering: $\nu_\mu N \rightarrow \nu_\mu N; \ eN \rightarrow eN$

Neutrino-lepton scattering: $\nu_\mu e \rightarrow \nu_\mu e$

Lepton-lepton scattering: $e^+ e^- \rightarrow \mu^+ \mu^-$

Hadron-hadron scattering: $NN \rightarrow NN$

We can see that there is no change of charge in any of the above neutral weak interactions. The results of the interference between the electromagnetic and neutral current weak interactions in the scattering of polarized electrons on deuterium has revealed valuable information about the properties of neutral currents. Parity violating effects of neutral currents in atomic transitions have also been reported.

Neutral current are found to be a mixture of vector (V) and axial vector (A) currents. This means that neutral currents transform like vectors and axial vectors under rotations of the spatial coordinates. Neutral currents are also reported to be a mixture of isoscalar ($I = 0$) and isovector ($I = 1$). The coupling constants which determine the relative strengths of the vector, axial vector, isoscalar and isovector have been measured and found to be in good agreement with the predictions of Weinberg-Salam electroweak theory.

The final confirmation of electroweak model must however await the discovery of Higgs scalar bosons with mass $\sim 10^2$ GeV. This will definitely suit the purpose which treats the quarks and leptons on equal footing.

Obviously, the electro-weak model also known as *Quantum Flavourdynamics* (*QFD*) now stands on solid ground. Glashow-Weinberg and Salam were jointly awarded Noble prize in 1979 for their outstanding contribution.

Standard Model The Weinberg-Salam electroweak model together with Glashow-Ilipoulos-Malani Charm scheme is called Standard model. This model forms the basis of the current understanding of particle physics. The fundamental constitutents of the standard model are the six quarks (u, d, s, c, b, t) and the six leptons (e, ν_e, μ, ν_μ, τ, ν_τ). Obviously, the standard model of the strong, weak and electromagnetic interactions is just the combination of Glashow-Weinberg-Salam model and

QCD. The gauge group is the direct product $G_s = SU^c(3) \times SU(2) \times U(1)$ with couplings g_s, g and g' for the three factors.

The quarks interact both by strong and electroweak interactions, where the leptons have only electro-weak interactions. The quanta of the strong interaction are bosons of different colours known as gluons, where as the quanta of electro-weak interactions are the photons (γ) and the intermediate bosons ($W^{\pm}$, Z^0) and the theory is based on Weinberg-Salam model. We must note that the photons are also vector bosons.

9.16 GRAND UNIFIED THEORY

The theoreticians have further developed grand unification theory (GUT) of the strong, electromagnetic and weak forces. In such a GUT the strong, weak and electromagnetic interactions appear as different manifestations of one basic phenomenon, with leptons and quarks finding natural places within the scheme. Obviously, GUT provides a super structure in which $SU(3)$ and $SU(2) \times U(1)$ theories can be embedded. The super structure would take the form of a larger symmetry in which quarks and leptons would be treated on equal footing. The search of large symmetry leads to a larger group $SU(5)$, the smallest simple group that can accomodate the constitutent $SU(3)$ and $SU(2) \times U(1)$ symmetries. It is believed that $SU(5)$ to be a likely symmetry group-of nature.

A starling consequence of GUT is the disappearance of the distinction between quarks and leptons, i.e. a transition from a quark to a lepton or vice versa can take place. The interconversion of quarks and leptons is possible only in GUT. Obviously, the protons would then not be stable particles. The present theoretical estimated life of proton decay to be $\sim 10^{30}$ years for the decay mode $p \rightarrow (e^+ \pi^0)$. Proton decay requires that the baryon number violating interactions may be able to explain the asymmetry between matter and anti matter. There is as yet no experimental evidence in support of proton decay. Obviously, if proton can decay, the atom itself is unstable, and all matter is impermanent.

Another important consequence of the GUT is the existence of very massive magnetic monopoles (mass $\sim 10^{16}$ GeV). The other consequences of GUT are

(i) GUT suggests the explanation for the quantization of electric charge, i.e. the experimental observation that charge always comes in discrete multiples of a fundamental smallest charge e ($e \rightarrow$ the magnitude of the charge on electron).

(ii) GUT yields a reasonable value for the weak angle θ_w, $\sin^2 \theta_w = \dfrac{3}{8}$

(iii) GUT reveals that all the three interactions (electromagnetic, weak and strong) are to be manifestations of the different facets of a single universal gauge force with a primitive universal strength of $\dfrac{\alpha}{\sin^2 \theta_w}$.

(iv) GUT may possibly provide an explanation for the dominance of matter over antimatter in the universe.

GUT has wide implications in cosmology. If GUT finds an experimental basis, e.g. the discovery of proton radioactivity or of the heavy magnetic monopoles, it will provide a good support for the big-bang theory of the creation of the universe.

Now, scientists are trying to develop a theory of everything (TOE), i.e. a super unification theory which involves gravity. Since gravity is not well understood at the quantum level, it poses the real challenge to such efforts. Presently, efforts are made in developing the so-called supergravity and *superstring field theories*. However, these efforts have important role to play in understanding the origin and history of early universe, matter generation and the inflation problems.

We now have a good understanding of the natural phenomena ranging from microcosm to macrocosm. Obviously, we have a unified view of the cosmos. However, the question about the ultimate structure of matter is still unanswered. It may have more surprises in its store. Having knowledge of the many elementary particles (6 leptons, 18 quarks and 13 quanta), some scientists have proposed *preon models,* i.e., next layer of matter which may be approach at the TeV energy scale. Obviously, there may be many more layers of the structure of the matter leading to the problem of infinite divisibility. This means ultimate constitutes of matter simply do not exist. On the contrary, there are scientists believe that there should be an end to the sequence of ever finer levels of the structure and the colour confinement-provides an clear indication to that. We will have to wait for future experiments for deciding these exciting crucial issues.

SOLVED PROBLEMS

Example 9.1 If a pion decays from rest to give a muon of 4 MeV energy, what is the kinetic energy of the accompanying neutrino? What is the mass of the neutrino in this process?

Sol. The pion decay is

$$\pi^+ \rightarrow \mu^+ + \nu_\mu + E$$

Here
$$E = M_\pi - M_\mu \text{ (neutrino has zero 'rest' mass)}$$
$$= 273\, m_e - 207\, m_e = 66\, m_e$$
$$= 66 \times 0.51 \text{ MeV} \quad (\because m_e = 0.51 \text{ MeV})$$
$$= 33.7 \text{ MeV}$$

Since the muon takes 4.0 MeV of energy, the energy shared by the neutrino is 33.7 − 4.0 = 29.7 MeV.

$$\text{Mass of this neutrino} = \frac{29.7}{0.51}\, m_e$$
$$= 58.2\, m_e$$

Example 9.2 Find the maximum kinetic energy of the electron emitted in the beta decay of the free electron. The neutron-proton mass different is 1.30 MeV.

Sol. The decay equation is

$$n^\circ \rightarrow p^+ + e^- + \bar{\nu}_e + E$$

From this, we have

$$E = M_n - M_p - m_e$$

$$= (1.30 \text{ MeV}) - m_e$$
$$= 1.30 \text{ MeV} - 0.51 \text{ MeV}$$
$$= 0.79 \text{ MeV}$$

This energy release goes into the kinetic energy of the electron plus the kinetic energy of the antineutrino. The two particles can share the 0.97 MeV of energy in anyway they choose. Thus the electron may have a maximum kinetic energy of 0.79 MeV.

Example 9.3 Which of the following reactions are allowed and forbidden through the strong interactions?

(a) $\Sigma^+ + {}^1_0 n \rightarrow \Sigma^- + {}^1_1 p$

(b) $\pi^+ + {}^1_0 n \rightarrow K^+ + \Sigma^0$

(c) $\wedge^0 \rightarrow \Sigma^+ + \pi^-$

(d) $\wedge^0 + {}^1_0 n \rightarrow \Sigma^- + {}^1_1 p$

(e) $K^+ \rightarrow \pi^+ + \pi^\circ + \pi^0$

(f) $K^- \rightarrow \pi^- + \pi^\circ$

(g) $K^- + {}^1_0 p \rightarrow K^\circ + {}^1_0 n$

Sol. (a) Forbidden since charge is not conserved.

   (b) Allowed by strong interaction as strangeness is not conserved.

   (c) Forbidden as energy is not conserved.

   (d) Allowed, strangeness is conserved

   (e) Forbidden as strangeness is not conserved

   (f) Allowed, as strangeness is conserved.

Example 9.4 Classify the following processes as strong, electromagnetic, weak of totally forbidden and stable whether the following are conserved in each process: (i) parity (*P*), strangeness (*S*), Isospin (*I*), third component os isospin (I_3).

   (i) $\pi^- + p \rightarrow \wedge^0 + K^0$

   (ii) $\pi^- + p \rightarrow n + \pi^\circ$

   (iii) $p + \gamma \rightarrow p + \pi^\circ$

   (iv) $\Sigma^0 \rightarrow \wedge^0 + \gamma$

   (v) $\wedge^0 \rightarrow p + \pi^-$

   (vi) $\Xi^- \rightarrow \wedge^0 + \pi^-$

   (vii) $K^0 \rightarrow \pi^+ + \pi^-$

Sol. (i) Strong: parity, isospin, strangeness and I_3 are conserved

   (ii) Strong: parity, isospin, strangeness and I_3 are conserved.

   (iii) Electromagnetic-Parity, strangeness I_3 are conserved but isospin not conserved.

   (iv) Electromagnetic: Parity, strangeness, I_3 conserved but isospin not conserved.

       (v), (vi) and (vii): weak, Parity, strangeness, isospin and I_3 not conserved.

Example 9.5 With the help of conservation laws determine which of the following reactions are allowed or forbidden?

(i) $\pi^+ + n \rightarrow \wedge^0 + K^+$

(ii) $\pi^+ + n \rightarrow K^0 + K^+$

(iii) $\pi^0 + n \rightarrow \bar{K}^0 + \Sigma^0$

(iv) $p \rightarrow e^+ + \gamma$

(v) $\pi^0 + \pi^- \rightarrow \bar{n} + p$

(vi) $n \rightarrow p + e^+ + \bar{\nu}$

(vii) $\pi^+ + n \rightarrow \pi^- + p$

Sol.

(i)

		π^+	+	n	$\rightarrow$	$\wedge^0$	+	K^+	(strong interaction).
Charge	q	1		0		0		1	
Baryon number	B	0		1		1		0	
Lepton number	L	0		0		0		0	
Isospin	I_3	+1		$-\dfrac{1}{2}$		0		$+\dfrac{1}{2}$	
Strangeness	S	0		0		-1		1	

Obviously, in the above strong interacting reactions all the conservation laws are obeyed and hence it is allowed.

(ii)

		π^+	+	n	$\rightarrow$	K^0	+	K^+	(Strong interaction)
Charge	q	+1		0		0		+1	
	B	0		1		0		0	
	L	0		0		0		0	
	I_3	+1		$-\dfrac{1}{2}$		$-\dfrac{1}{2}$		$+\dfrac{1}{2}$	
	S	0		0		1		1	

Since B, I_3 and S quantum numbers are not conserved and hence this reaction is not allowed.

(iii)

		π^+	+	n	$\rightarrow$	$\bar{K}^0$	+	Σ^0	(Strong interaction)
	q	0		0		0		0	
	B	0		1		0		1	
	L	0		0		0		0	
	I_3	0		$-\dfrac{1}{2}$		$+\dfrac{1}{2}$		0	
	S	0		0		-1		-1	

Since I_3 and S are not conserved and hence this reaction is not allowed.

(iv) $p \rightarrow e^+ + \gamma$ (weak interaction)

q	1	1	0
B	1	0	0
L	0	−1	0

Since B and L are not conserved and hence this reaction is not allowed.

(v) $\pi^+ + \pi^- \rightarrow \bar{n} + p$

q	0	−1	0	1
B	0	0	−1	+1
L	0	0	0	0

Obviously, q is not conserved and hence this is not allowed.

(vi) $n \rightarrow p + e^+ + n$

q	0	1	1	0
B	1	1	0	0
L	0	0	−1	+1

Obviously, q is not conserved and hence this is not allowed.

(vii) $\pi^+ + n \rightarrow \pi + p$

q	1	0	−1	1
B	0	1	0	1
L	0	0	0	0

Obviously, q is not conserved and hence this is not allowed.

Example 9.6 Construct a Feynman diagram for neutron-proton scattering resulting from the exchange of a π meson.

Initially the neutron and proton approach each other, so lines are needed which start apart and converge. It does not matter whether the neutron or the proton is at the top, nor does it matter what kind of line is used. However, solid lines are most frequently used for baryons, just as the wavy line is almost always used for the photon. For the exchanged pion a dashed line will be used to distinguish it more clearly from the baryons.

Now we have a choice. The first possibility is that the proton emits a π^+, turning into a neutron and the π^+ is a absorbed by the initial neutron, turning it into a proton, as shown in Figure 9.20. The second possibility is that the neutron emits a π^-, turning into a proton, and the π^- is absorbed by the initial proton, turning it into a neutron, as shown in Figure 9.20(b). Note that the dashed lines for the pions have appropriately different slopes in the two cases, indicating the two different origins and time progressions. However, these two diagrams are completely equivalent. The reason is that these virtual pions exist for too short a time to permit, even in principle, any measurement which could distinguish Figure 9.20(a) from 9.20(b). Since the π^+ and π^- are antiparticles of each other, this illustrates the principles that an anti-particle is equivalent to a particle going backward in time. (That is, the emission of an antiparticle is equivalent to the absorption of a particle). Because the distinction between (a) and (b) is meaningless, we shall frequently draw vertical lines for the extremely short-lived exchanged virtural particle. (The infinite slope of a vertical line does not imply that the particle travels with infinite speed).

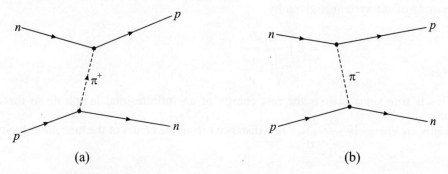

(a)   (b)

Fig. 9.20 Feynman diagrams for proton-neutron scattering through the exchange of a virtual π meson. In (a) the proton emits a π^+ meson, becoming a neutron, and the neutron absorbs the π^+, becoming a proton. In (b) exactly the same process is described as the neutron emitting a π^- to become a proton, and the proton absorbing the π^- to become a neutron.

Example 9.7 Show that the quark quantum numbers give the corresponding quantities for the $\Sigma^0(1385)$ particle.

Sol. The $\Sigma^0(1385)$ has $Q = 0$, $B = 1$, $S = -1$ (hence $Y = 0$), $T = 1$, and $T_z = 0$. It is made up one of each kind of quark, or uds. Taking the u, d, and s properties in order, we have

$$Q = +\frac{2}{3} - \frac{1}{3} - \frac{1}{3} = 0$$

$$B = \frac{1}{3} + \frac{1}{3} - \frac{1}{3} = 1$$

$$S = 0 + 0 - 1 = -1$$

$$T = \frac{1}{2} + \frac{1}{2} + 0 = 1$$

$$T_z = +\frac{1}{2} - \frac{1}{2} + 0 = 0$$

Example 9.8 Determine k_2 in Equation (9.1) (k_1 and k_2 are constants) from the energy in the color lines of force between a quark and an antiquark by determine the angular momentum of this meson.

Sol. Suppose the color lines of force have been pulled together until they form a tube, and the interaction energy is then so high that the masses of the quark can be neglected in comparison

$$V_c = \frac{k_1}{r} + k_2 r \qquad \ldots (9.1)$$

to it. If this system is now considered to be rotating, we have a crude model for a meson with angular momentum. We can use this to deduce k_2, which will be the energy per unit length of the force tube, and also the second constant in (9.1). For definiteness, assume the ends of the force tube rotate at velocity c and that the tube has a half length of ρ. The total mass M of the system is given by

$$Mc^2 = 2 \int_0^\rho \frac{k_2 \, dr}{\sqrt{1 - \frac{v^2}{c^2}}} \qquad \qquad \ldots (9.2)$$

This is true since $k_2 \, dr$ is the rest energy of an infinitesimal length dr so that its total relativstic energy is $\dfrac{k_2 \, dr}{\sqrt{1 - \dfrac{v^2}{c^2}}}$. At a distance r from the center of the tube the velocity will be

$v = \dfrac{ce}{\rho}$. Making this substitution in (9.2) gives

$$Mc^2 = 2 k_2 \int_0^\rho \frac{dr}{\sqrt{1 - \frac{r^2}{\rho^2}}} = \pi k_2 \rho \qquad \qquad (9.3)$$

Now, the angular momentum of the infinitesimal mass at the distance r from the center where the velocity is v is $vrk^2 \dfrac{dr}{c^2 \sqrt{1 - \dfrac{v^2}{c^2}}}$. Thus the total angular momentum of the tube in units of $\hbar$ is

$$J = \frac{2}{\hbar} \int_0^\rho \frac{vrk_2 \, dr}{c^2 \sqrt{1 - \frac{v^2}{c^2}}} = \frac{2k_2}{\hbar} \int_0^\rho \frac{r^2 \, dr}{c\rho \sqrt{1 - \frac{r^2}{\rho^2}}} = \frac{rk_2 \rho^2}{2\hbar c} = \frac{(Mc^2)^2}{2\pi k_2 \hbar c} \qquad (9.4)$$

Although this is a crude model, the result that $J \propto M^2$ is in agreement with experiment. If the mass squared of mesons of the same structure but differing in angular momentum is plotted against that quantity, a straight line is obtained with the slope $\dfrac{dJ}{d(Mc^2)^2} = 0.9 \text{ GeV}^{-2}$.

A similar plot, Figure 9.21, for baryons is more spectacular because there are more known examples. Again, straight lines and the same slope are obtained. According to the model this slope has the value

$$\frac{dJ}{d(Mc^2)^2} = 0.9 \text{ GeV}^{-2} = (2\pi k_2 \hbar c)^{-1} \qquad \qquad \ldots (9.5)$$

Solving (9.5) for k_2 gives

$$k_2 = [2\pi (0.9 \text{ GeV}^{-2}) (0.2 \text{ GeV-F})]^{-1} = 1 \text{ GeV-F}^{-1}$$

where we have used the convenient value $\hbar c = 197 \text{ MeV-F}$.

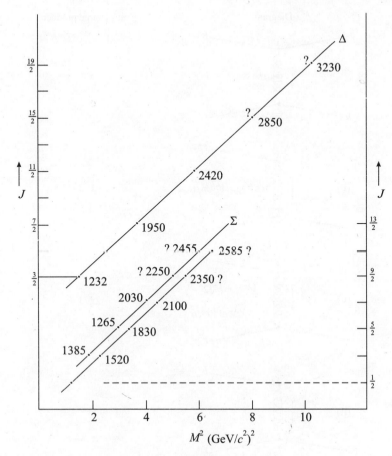

Fig. 9.21 Baryon spins versus the square of their masses for three sequences: Δ has $T = \frac{3}{2}$, $S = 0$, and spin J and sign of parity, P, expressed as $J^P = \frac{3^+}{2}, \frac{7^+}{2}, \frac{11^+}{2}$; Λ has $T = 0$, $S = -1$, $J^P = \frac{1^+}{2}, \frac{3^-}{2}, \frac{5^+}{2}, \ldots$; and Σ has $T = 1$, $S = -1$, $J^P = \frac{3^+}{2}, \frac{5^-}{2}, \frac{7^+}{2} \ldots$. Particles for which the spin-parity is not well established at the time of writing have a question mark with their mass value in MeV/c^2.

Is this result reasonable? Since the proton has a rest mass energy of about 1 GeV and a radius of about 1 F, this is indeed a correct order of magnitude energy density for a hadron. Accepting this value, we then find that at a distance of a typical hadron radius of 1 F the confinement energy of the quark is about 1 GeV, which is a hundred times nuclear binding energies. Put another way, the force, which is constant with distance, is 10^{15} GeV/m ($\sim 10^2$ newtons), or about 10 tons on each pointlike quark.

Example 9.9 Show that a baryon made of a colorless combination of three quarks does bind.

Sol. Since a baryon will have to have a totally antismmetric color eigenfunction for its three quarks, it will be of the form

Diagram   Color charge product

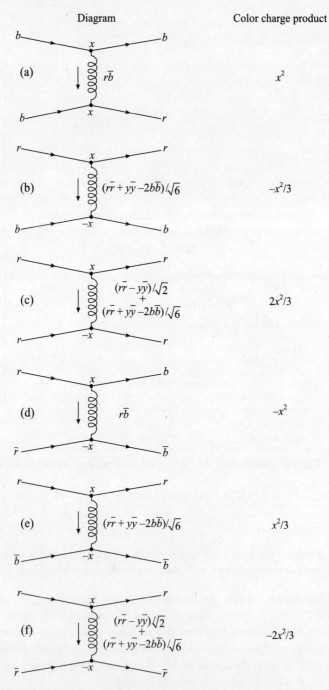

Fig. 9.22 Gluon coupling between quarks. All possible types of gluon exchange are
represented by these six diagrams. That is, all other exchanges just involve
a permutation of color labels. The color eigenfucntion is given for each
exchanged gluon. The relative probability for each type of exchange is given
by the "color charge product." where z is the color charge.

$$\frac{[(rb - br)y + (by - yr)r + (yr - ry)b]}{\sqrt{6}} \qquad \dots (9.1)$$

Its antisymmetry can be seen by interchanging any two color labels. This eigenfunction is to be used to determine the interaction between quarks, which occurs by gluon exchange. Any one interaction must be between the two quarks exchanging the gluon, with the third quark not participating, but all possible two-quark interactions must be considered. The mathematical form expressing such an interaction involves the product of the initial state eigenfunction, the final state eigenfunction, and the interaction potential (it is a matrix element;). The part of the interaction potential relevant here is the gluon exchange color charge product, given in Figure 9.22 Equation (9.1) is the form of both the initial and final state eigenfunctions.

Example 9.10 Show that the gluon couplings give binding also for a colorless quark and antiquark.

Sol. Since the quark-antiquark pair, if bound, form a meson (which is a boson), it will have a totally symmetric color part to its eigenfunction

$$\frac{(r\bar{r} + y\bar{y} + b\bar{b})}{\sqrt{3}} \qquad (9.1)$$

The first term, $r\bar{r} \rightarrow r\bar{r}$ contributes $\left(\frac{1}{\sqrt{3}}\right)^2 \left(-\frac{2\chi^2}{3}\right) = -\frac{2\chi^2}{9}$, but each of the other two terms are identical in form with different color labels. All three then give a total of $3\left(-\frac{2\chi^2}{9}\right) = -\frac{2\chi^2}{3}$. Also $r\bar{r} \rightarrow b\bar{b}$ or $y\bar{y}$, each giving $\left(\frac{1}{\sqrt{3}}\right)^2 (-\chi^2)$ for a total of $-\frac{2\chi^2}{3}$. However, $y\bar{y} \rightarrow r\bar{r}$ or $b\bar{b}$ and $b\bar{b} \rightarrow r\bar{r}$ or $y\bar{y}$, giving the same contributions as the $r\bar{r}$. So the total is $-2\chi^2$. The net coupling strength is $-\frac{2\chi^2}{3} - 2\chi^2 = -\frac{8\chi^2}{3}$ giving a potential of $-\frac{8\chi^2}{3r}$. Again, the minus sign indicates binding. But other quark combinations give positive signs and nonbinding potentials.

Example 9.11 Suppose a quark-antiquark pair posses color. Then it would have color-anticolor like a gluon. For definiteness, say it is $r\bar{b}$. Find the form of the potential.

Sol. The gluon exchange between r and $\bar{b}$ cannot involve swapping colors since $r \rightarrow \bar{b}$ is not possible because a quark cannot become an antiquark. Thus only a non-color-changing gluon can be involved. Of the two available, only one has both r and $\bar{b}$ color; it is $\frac{(r\bar{r} + y\bar{y} - 2b\bar{b})}{\sqrt{6}}$. The red part of the gluon couples at the upper vertex with color charge

$\dfrac{\chi}{\sqrt{6}}$. The antiblue part of the gluon couples at the lower vertex with color charge $(-\chi)\left(\dfrac{-2}{\sqrt{6}}\right)$

$= \dfrac{2\chi}{\sqrt{6}}$. The color charge product is then $\left(\dfrac{\chi}{\sqrt{6}}\right)\left(\dfrac{2\chi}{\sqrt{6}}\right) = \dfrac{\chi^2}{3}$. This gives the positive, non-

binding potential $\dfrac{\chi^2}{3r}$.

Example 9.12 If in the following reaction, the incident kaon has a kinetic energy of 1.63 GeV, calculate the total energy to be divided between the four recoiling particles.   (Mysore)

$$\bar{K}^- + p^+ \rightarrow \Sigma^- + \pi^+ + \pi^- + \pi^+$$

The mass energy of p-mesons are 139.6 MeV, $\Sigma^- = 1197.3$ MeV, proton = 938.3 MeV and $\bar{K}^- = 493.8$ MeV.

Sol. Energy of particles taken together are as follows:

$$\bar{K}^- = 0.4938 \text{ GeV} \qquad (\because 1 \text{ GeV} = 10^3 \text{ MeV})$$
$$p^+ = 0.9383 \text{ GeV}$$
$$\underline{E_{\bar{K}^-} = 1.63 \text{ GeV}}$$

$\therefore$ Total energy, $E = 3.0621$ GeV

Energy of the four recoiling particles are:

$$\Sigma^- = 1.1973 \text{ GeV}$$
$$\pi^+ = 0.1396 \text{ GeV}$$
$$\pi^- = 0.1396 \text{ GeV}$$
$$\underline{\pi^+ = 0.1396 \text{ GeV}}$$

Total energy = 1.6161 GeV

$\therefore$ Excess energy = $3.0621 - 1.6161 = 1.446$ GeV

$\therefore$ Average energy per particle = $\dfrac{1.446}{4} = 0.3615$ GeV $= 361.5$ MeV

Example 9.13 If a pion decays from rest to give a muon of 4.0 MeV energy, what is the kinetic energy of the accompanying neutrino? What is the mass of the neutrino in the process?   (MDS)

Sol. The decay mode of pion is given by

$$\pi^+ \rightarrow \mu^+ + \nu_\mu + E$$

$\therefore$ Energy, $E = m_\pi - m_\mu$ (neutrino has zero 'rest' mass)

$$= 273\,m_e - 207\,m_e = 66\,m_e$$
$$= 66 \times 0.51 \text{ MeV} \qquad (\because m_e = 0.51 \text{ MeV})$$
$$= 33.7 \text{ MeV}$$

Since muon takes 4.0 MeV of energy, the kinetic energy of the accompanying neutrino is

$$33.7 - 4.0 = 29.7 \text{ MeV}$$

$$\therefore \quad \text{Mass of neutrino} = \left(\frac{29.7}{0.51}\right)m_e = 58.23 \ m_e$$

Example 9.14 Find the maximum kinetic energy of the electron emitted in the β-decay of the free neutron. The neutron-proton mass difference is 1.30 MeV. (Jodhpur)

Sol. The decay scheme of a free neutron is

$$n^0 \rightarrow p^+ + e^- + \bar{\nu}_e + E$$

$$\therefore \text{ Energy,} \qquad E = m_n - m_p - m_e$$

$$= (1.30 \text{ MeV}) - m_e$$

$$= (1.30 - 0.51) \text{ MeV} \qquad (\because \ m_e = 0.51 \text{ MeV})$$

This energy goes into the kinetic energy of the electron and antineutrino. The sharing of energy may be made in any manner so that the maximum kinetic energy of the electron is 0.70 MeV.

Example 9.15 A μ^- meson decays into an electron e^- and a pair of neutrons. Calculate the maximum available energy for the process and the average electron energy. [Punjab]

Sol. The decay mode of μ^- meson is given by

$$\mu^- \rightarrow e^- + \nu_\mu + \bar{\nu}_e$$

Taking $m_\mu = 207 \ m_e$, the total available energy for the process is given by

$$207 \ m_e \rightarrow m_e + 0 + 0 + E$$

($\because$ neutrino has zero 'rest' mass)

$$\therefore \qquad E = (207 - 1)m_e = 206 \ m_e$$

$$= 206 \times 0.51 \text{ MeV} \qquad (\because \ m_e = 0.51 \text{ MeV})$$

$$= 105 \text{ MeV}$$

If the above energy is equally shared by the particles, the average energy of the electron is

$$E_e = \frac{105}{3} = 35 \text{ MeV}$$

Example 9.16 A π^- is captured by a stable krypton atom, $^{84}_{36}$Kr. Calculate (a) the radius of the first and second Bohr orbits, (b) the energy radiated from transition $n = 2$ to $n = 1$, (c) the frequency of corresponding radiation and (d) the wavelength. (Bangalore)

Sol. We have $Z = 36$, $m = 273 \ m_e$, $n = 1, 2$. We have also:

$$r_1 = \text{radius of the first Bohr orbit for H-atom}$$

$$= 5.29 \times 10^{-11} \text{ m}$$

$$h = \text{Planck's constant} = 6.62 \times 10^{-34} \text{ Js}.$$

(a) By direct substitution in the relation, we can find the r-values. For

$$n = 1, \quad r = r_1 \frac{n^2}{mZ} = 5.29 \times 10^{-11} \times \frac{1^2}{273 \times 36} = 5.38 \times 10^{-15} \text{ m}$$

$$n = 2, \quad r = r_1 \frac{2^2}{mZ} = 5.29 \times 10^{-11} \times \frac{4}{273 \times 36} = 21.52 \times 10^{-15} \text{ m}$$

(b) The energy value for $n = 1$ state is, also by direct substitution,

$$E_1 = R \frac{mZ^2}{n^2} = -2.18 \times 10^{-18} \times \frac{273 \times (36)^2}{1}$$

$$= -7.71 \times 10^{-13} \text{ J}$$

Similarly, for $n = 2$, $E_2 = \frac{1}{4} E_1 = -1.92 \times 10^{-13}$ J

∴ Energy radiated in transition is: $E = E_2 - E_1 = 5.79 \times 10^{-13}$ J

(c) Frequency of radiation, n is given by

$$n = \frac{E}{h} = \frac{5.79 \times 10^{-13} \text{ J}}{6.62 \times 10^{-34} \text{ Js}}$$

$$= 8.74 \times 10^{20} \text{ Hz}$$

(d) Wavelength of radiation λ is given by

$$\lambda = \frac{c}{v} = \frac{3 \times 10^8 \text{ ms}^{-1}}{8.74 \times 10^{20} \text{ s}^{-1}}$$

$$= 3.44 \times 10^{-13} \text{ m} = 0.00344 \text{ A}$$

Example 9.17 Find the value of the third component of isotopic spirt of Ξ^- in the following strong insteraction:

$$\pi^+ + n \rightarrow \Xi^- + K^+ + K^+$$

Sol. The interaction is $\quad \pi^+ + n \rightarrow \Xi^- + K^+ + K^+$

∴ Isotopic spin, I: $\qquad\qquad 1 + \frac{1}{2} \rightarrow \frac{1}{2} + \frac{1}{2} + \frac{1}{2}$

Third component of isospin, I_3: $\quad +1 - \frac{1}{2} \rightarrow I_3 + \frac{1}{2} + \frac{1}{2}$

∴ $\qquad\qquad\qquad I_3 \text{ for } \Xi^- = -\frac{1}{2}$

Example 9.18 Identify the type of the following interaction from the conservation laws:

$$\Sigma^0 \rightarrow \wedge^0 + \gamma \text{ (life time} \leq 10^{-14} \text{ s)}$$

Sol. We have: $\qquad\qquad\qquad \Sigma^0 \rightarrow \wedge^0 + \gamma$

Charge, Q: $\qquad\qquad\qquad 0 \rightarrow 0 + 0 \qquad\qquad \therefore \Delta Q = 0$

Baryon number, B: $\qquad\qquad +1 \rightarrow +1 + 0 \qquad \therefore \Delta B = 0$

Lepton number, L: $\qquad\qquad 0 \rightarrow 0 + 0 \qquad\qquad \therefore \Delta L = 0$

Strangeness no., S: $\qquad\quad -1 \rightarrow -1 + 0 \qquad \therefore \Delta S = 0$

Hypercharge, Y: $\qquad\qquad 0 \rightarrow 0 + 0 \qquad\qquad \therefore \Delta Y = 0$

Since the strangeness number is conserved, the interaction is either a strong interaction or an electromagnetic one. Its half-life is $\leq 10^{-14}$ s which points to the fact that it cannot be a strong interaction, but is a weak decay. As S is conserved, it cannot be a weak interaction. So it is an electromagnetic interaction and a γ-photon is produced.

Example 9.19 Identify the unknown particle in the reactions given below, using the conservation laws. $\hfill$ [Jodhpur]

(i) $\qquad\qquad\qquad \mu^- + p \rightarrow {}_0^1 n + \dots$

(ii) $\qquad\qquad\qquad \pi^- + p \rightarrow K^0 + \dots$

Sol. (i) The reaction is: $\quad \mu^{-1} + p \rightarrow {}_0^1 n + \dots$

The unknown particle must be of zero charge and mass, spin $\dfrac{1}{2}$ and lepton number 1 so that the charge, mass and lepton numbers are conserved. Since the interacting particle is μ^- meson, the unknown particle is identified as mu-neutrino, ν_μ.

(ii) The reaction is $\qquad\qquad\qquad \pi^- + p \rightarrow K^0 + \dots$

For charge conservation, Q: $\qquad -1 + 1 \rightarrow 0 + Q \qquad \therefore Q = 0$

Conservation of baryon no. B: $\qquad 0 + 1 \rightarrow 0 + B \qquad \therefore B = +1$

Strangeness conservation, S: $\qquad 0 + 0 \rightarrow -1 + S \qquad \therefore S = +1$

Third component of isospin, I_3: $\quad -1 + \dfrac{1}{2} \rightarrow -\dfrac{1}{2} + I_3 \qquad \therefore I_3 = 0$

Thus the unknown particle has charge zero, baryon number $+1$, strangeness number $+1$ and third component of isospin 0. So the particle could be $\wedge^0$ or Σ^0.

We may note that Σ^0 has a rest mass greater than $\wedge^0$. So Σ^0 will be produced only if the kinetic energy imparted to π^- is greater than that required for the production of $\wedge^0$.

Example 9.20 Check if the following reactions are allowed or forbidden. $\hfill$ [Ujjain]

(i) $\qquad\qquad\qquad \pi^- + p \rightarrow \wedge^0 + \pi^0$

(ii) $\qquad\qquad\qquad p + \bar{p} \rightarrow 2\pi^+ + 2\pi^- + 2\pi^0$

Sol. (i) The reaction is: $\quad \pi^- + p \rightarrow \wedge^0 + \pi^0$

Q:	$-1 + 1 \rightarrow 0 + 0$	$\therefore \Delta Q = 0,$	$Q \rightarrow$ conserved
B:	$0 + 1 \rightarrow +1 + 0$	$\therefore \Delta B = 0,$	$B \rightarrow$ conserved
S:	$0 + 1 \rightarrow -1 + 0$	$\therefore \Delta S \neq 0,$	$S \rightarrow$ not conserved

It is a strong interaction where the charge and baryon number are conserved. But since the strangeness number is not conserved, the reaction is forbidden.

(ii) The reaction is: $p + \bar{p} \rightarrow 2\pi^+ + 2\pi^- + 2\pi^0$

Q:	$+1 - 1 \rightarrow +2 - 2 + 0;$	$\Delta Q = 0$
B:	$+1 - 1 \rightarrow 0 + 0 + 0;$	$\Delta B = 0$
S:	$0 + 0 \rightarrow 0 + 0 + 0;$	$\Delta S = 0$
Y:	$+1 - 1 \rightarrow 0 + 0 + 0;$	$\Delta Y = 0$

Clearly, the above reaction is an allowed reaction.

Example 9.22 A μ^- meson decays at rest and emits an electron. Which other particle (s) will be emitted in the above decay? Use the conservation laws in support of your answer.   [Punjab]

Sol. Since μ^- decay from rest, the conservation of energy and linear momentum demands that two or more particles are emitted. Since one of the ejected particle is an electron, the charge conservation demands that the other particle (s) must be neutral. Since its rest mass must be less than μ^-, it could be a neutrino. Now μ^- meson is a spin $\frac{1}{2}$ particle, and the vector sum of spins of $e^- \left(\frac{1}{2} \right)$ and that of $\nu \left(\frac{1}{2} \right)$ is either 0 or $\frac{1}{2}$. The conservation of spin requires that another neutral and spin $\frac{1}{2}$ particle should be emitted in the opposite direction.

This could be an antineutrino, $\bar{\nu}$ and the decay scheme would be: $\mu^- \rightarrow e^- + \nu + \bar{\nu}$. However, if ν and $\bar{\nu}$ are simultaneously emitted they would annihilate to produce photons. Since photons are not observed, a particle and its antiparticle are not possibly emitted. So there should be two different neutrinos. Since μ^- and e^- are involved in the decay, the neutrinos could be ν_e and ν_μ. To conserve spin, one will be anti-mu-neutrino, $\bar{\nu}_\mu$. The other particle is electron neutrino, ν_e. We have

$$\mu^- \rightarrow e^- + \nu_e + \bar{\nu}_\mu$$

Example 9.23 An ultra-relativistic proton moves in a magnetic field. Can it radiate π^+, π^- and π^0 mesons, electrons and positrons?   [Kerala]

Sol. If the energy of the proton is sufficiently large, it can radiate π^0 and π^+ mesons, and also positrons. The reactions are:

$$p \rightarrow p + \pi^0; \, p \rightarrow n + \pi^+; \, p \rightarrow n + e^+ + \nu$$

But π^- mesons and electrons cannot be radiated.

Example 9.24 Which of the following reactions are permissible with regard to strangeness.

(a) $\pi^- + p \rightarrow \wedge^0 + K^0$; (b) $\pi^- + p \rightarrow \wedge^0 + \pi^0$;

(c) $\wedge \rightarrow p + \pi^-$; (d) $\Xi^- \rightarrow \wedge + \pi^-$;

(e) $\Xi^- \rightarrow 2\pi^- + p$; (f) $\pi^+ + \bar{p} \rightarrow \Sigma^- + K^-$;

(g) $\pi^+ + \bar{p} \rightarrow \bar{\Sigma}^- + \pi^-$; (h) $\pi^+ + n \rightarrow \wedge^0 + K^+$;

(i) $K^- + p \rightarrow \Sigma^- + \pi^-$; (j) $p + \bar{\Sigma}^- \rightarrow K^+ + \pi^+$;

(k) $p + \bar{\Sigma} \rightarrow \pi^+ + \pi^+$; (l) $\pi^- + p \rightarrow \bar{\Xi} + X^- + \pi$;

(m) $\pi^- + p \rightarrow \Sigma^+ + K^-$; (n) $\pi^- + \bar{p} \rightarrow \Xi^- + K^+ + K^0$.

Sol. Reaction (a), (f), (h), (j), (l) and (n) gives $|\Delta S| = 0$ and they proceed through strong interaction. In reaction (b), (c), (d), (g), (i) and (k), $|\Delta S| = 1$. Hence disintegrations. (c), (d) have relatively small probabilities, while reactions (b), (g), (i) and (k) are not observed in practice. In (e) and (m), $|\Delta S| = 2$. They are not observed. Cascade hyperon decays in accordance with reaction (d) and afterward $\wedge$ decays according to (c).

SUGGESTED READINGS

1. S. Beneditti, Nuclear interactions, John Wiley and Sons (1964)
2. W. R. Frazer, Elementary Particles, Prentice Hall Inc. (1966)
3. S. R. Parker (Ed.) Nuclear and Particle Physics, Source book, McGraw Hill Co. (1987)
4. F. Halzen and A. D. Martin, Quarks and Leptons, John Wiley and Sons (1984)
5. K. N. Mukkin, Experimental Nuclear Physics, Vol. II, Mir Publishers, Moscow (1987)
6. E. B. Paul, Nuclear and Particle Physics, North Holland Pub. Co. (1969)
7. M. Gell-Mann, Phys-Rev. 92, 833 (1953)
8. J. C. Pati and A. Salam, Phys. Rev. Lett. 31, 661 (1973)
9. Y. Nambu, Scientific American 235(5), 48 (1976)
10. R. P. Crease and C. C. Mann, The second Creation, Affiliated East West Pvt. Ltd. (1986)

REVIEW QUESTIONS

1. Classify the elementary particles. What do you know about leptons and mesons? What conservation laws are obeyed in the case of production and annihilation of particles?
2. What do yor understand by antiparticles. Explain their characterstic properties taking examples of positron and antiproton.
3. Compare the properties of pions, muons and kaons.
4. Discuss and classify the different types of fundamental interactions known in nature. What attempts have been made to unify them?
5. Illustrate by taking examples the different conservation laws followed by the elementary particles.
6. Define Isospin, Strangeness and baryon number. Is there any relation connecting them? What conservation laws and selection rules are obeyed for the creation of strange particles?
7. Classify the elementary particles. How are these particles assigned (i) isospin, Baryon number, (iii) strangeness number (iv) hypercharge.
   What conservation laws are obeyed in the case of annihilation of particles.

8. What are quarks? Give an account of quark model.
9. Discuss Salam-Weinberg model and mention salient features.
10. Mention Salient features of GUT.
11. Describe the discovery of Ω^- Hyperson with experimental details.
12. What are weight diagrams? Draw weight diagrams for mesonic resonance octet with $J^P = (1^-)$.
13. What are resonance particles? Give their salient features.
14. Mention important predictions of $SU(3)$ symmetry.
15. Discuss the composition of hadrons based on quark model.
16. Give a brief account of quark structure of mesons and baryons.
17. Why free quarks could not be observed?
18. What is Quantum Chromodynamics? How it explains interaction of examples?
19. What are neutral currents? Explain with the help of few examples.
20. Mention Salient fratures of standard model.
21. Write short notes on
(i) Baryon number (ii) Strangeness (iii) Isospin, Strangeness and Baryon number (iv) Parity, (v) Quantum numbers and conservation laws of elementary particles (vi) Quark model (vii) QCD (viii) GUT.

SHORT QUESTION ANSWERS

1. Write the names of fundamental interactions of nature with their relative strength, taking strength of strong interaction as 1.

(**Ans.** e.m. 10^{-1}, weak 10^{-13}, Grav. 10^{-37})

2. Write the exchange quanta in case of fundamental interactions

(**Ans.** (i) Strong: Pions and in Quark model: gluons; (ii) e.m. : Photons (iii) Weak: $W^\pm$, Z^0 bosons and (iv) Gravitational: Graviton).

3. Why the decay $n \to \pi^0 + \gamma$ is forbidden?

(**Ans.** B is not conserved, $B = 1$ for neutron, but $B = 0$ for both π^0 and γ)

4. What is CPT theorem?

5. Write quark structure of the nucleons

(**Ans.** $p = (uud)$, $n = (udd)$?

6. Write the generalized expression for Q to include b and t.

(**Ans.** $Q = I_3 + \dfrac{B + S + C - b + t}{2}$)

PROBLEMS

1. Which of the following reactions can occur? State the conservation laws followed and violated:
   (i) $p + p \to n + p + \pi^+$ (ii) $p + p \to p + \wedge^0 + e^+$
   (iii) $e^+ + e \to \mu^+ + \pi^-$ (iv) $\wedge^0 \to \pi^+ + \pi^-$
   (v) $\pi^- + p \to n + \pi^0$

2. Discuss whether the following particle reactions are allowed or forbidden under conservation of charge Q, Baryon number B and strangeness S?

(i) $\pi^+ + n \to \wedge^0 + K^+$ (ii) $\pi^+ + p \to \wedge^0 + \pi^0$

3. Which of the following reaction can occur? State the conservation principles violated by the others.

  (i) $\wedge^0 \to \pi^+ + \pi^-$ (ii) $\pi^- + p \to n + \pi^0$

  (iii) $\pi^+ + p \to \pi^+ + p + \pi^- + \pi^0$

[**Ans.** (i) B not conserved (ii) can occur (iii) charge not conserved]

4. Which of the following reactions can occur? State the conservation principles violated by the others

  (i) $p + p \to n + p + \pi^+$ (ii) $p + p \to p + \wedge^0 + \Sigma^+$

  (iii) $e^+ + e^+ \to \mu^+ + \pi^-$

5. Mention conservation laws and test the occurrence of following reactions.

  (i) $\mu^+ \to e^+ + \bar{\nu} + \gamma$ (ii) $n \to p + e^- + \bar{\nu}$

  (iii) $p \to n + e^- + \bar{\nu}$ (iv) $\gamma \to e^+ + e^-$

  (v) $\pi^0 \to \pi^+ + \pi^-$ (vi) $\pi^+ \to \mu^+ + \nu + \bar{\nu}$

  (vii) $\mu^+ \to e^+ + e^- + \nu$ (viii) $p + \bar{n} \to n + \bar{p}$

6. Classify the following reactions with respective interactions and give reasons in support of your answer.

  (i) $\pi^- \to p \to \pi^0 + n$ (ii) $\pi^- + p \to \wedge^0 + K^0$

  (iii) $\Sigma^0 \to \wedge^0 + \gamma$ (iv) $p + \gamma \to p + \pi^0$

  (v) $\pi^0 \to \gamma + \gamma$ (vi) $\wedge^0 \to p + \pi^-$

  (vii) $K^0 \to \pi^+ + \pi^-$ (viii) $\wedge^0 \to p + e^- + \bar{\nu}$

  (ix) $\Xi^- \to \wedge^0 + \pi^-$

[**Ans.** Short range weak: (vi), (vii), (ix), (viii)
Short range: (i), (ii)
Electromagnetic: (iii), (iv), (v)]

7. Which of the following processes are allowed or forbidden? Give reasons in support of your answers.

  (i) $\pi^- + p \to \wedge^0 + \pi^0$ (ii) $\pi^- + n \to \Sigma^- + K^0$

  (iii) $n \to \pi^0 + \nu$ (iv) $n \to p + \mu^- + \nu_\mu$

  (v) $n \to p + e^-$ (vi) $p + p \to p + \bar{p}$

  (vii) $p + p \to p + \bar{p} + p + p$ (viii) $\pi^+ \to \mu^+ + \gamma$

  (ix) $\pi^+ \to \mu^+ + \nu_\mu$

8. A negative pion may be captured into a stable oribit around the nucleus, constituting a mesic atom. Assume that one can compute the energy and the radius of the mesonic orbit by the same formulas used for electronic orbits in the hydrogen atom. (i) What energy is released when a free pion at rest is captured into the ground state around a proton? (ii) Compute the radius of the ground-state orbit of a pion moving around a pion. (iii) Estimate the nucleus for which the radius of the ground-state orbit of pion is equal to the radius of the nucleus.

[**Ans.** (i) 2.53 keV, (ii) 2.85×10^{-13} m (iii) $^{110}_{47}$Ag]

9. Consider the decay of a π^0 meson into two photons. Given that the photon energies are E_1 and E_2 show that they are related by

$$E_1 E_2 = \frac{1}{2} \frac{(m_\pi c^2)^2}{(1 - \cos\theta)}$$

Where θ is the angle between photon directions as measured in the *L*-frame.

10. Analyse the process $\bar{K}^- + p^+ \to \Omega^- + K^+ + K^0$ from the point of view of conservation laws. Calculate the threshold kinetic energy of the *K*-particle.

$(I_3, I$ and S are conserved; 2710 MeV)

11. Find the energy of photon emitted in the decay $\Sigma^0 \to \wedge^0 + \gamma$. [**Ans.** 73.6 MeV]

12. A proton of kinetic energy *K* collides with a stationary proton, and a proton antiproton pair is produced. If the momentum of the bombarding proton is shared equally by the four particles that emerge form the collision, find the minimum value of *K*. [**Ans.** $6\, m_p c^2$]

14. Why does a free neutron not decay into an electron and a positron? Into a proton-antiproton pair?

15. According to the theory of the continuous creation of matter (which has turned out to be inconsistent with astronomical observations), the evolution of the universe can be traced to the spontaneous appearance of neutrons and antineutrons in free space. Which conservation law(s) would violate this process? [**Ans.** energy]

16. A μ^- meson collides with a proton, and a neutron plus another particle are created. What is the other particle? [**Ans.** ν_μ (mu neutrino)]

17. All resonance particles have very short life times. Why does this suggest they must be hadrons? [**Ans.** Only the strong interaction can produce such rapid decays).

Table 9.16 Elementary Particles and their Properties

Particles and their types	Symbol	$I(J^P)C$	Charge	Mass (MeV)	Mean life (s)	Main decay modes	Strangeness (S)	Antiparticles
Photons	γ	$0(1^+)$	0	0	∞	—	—	γ
Hadrons								
Baryons (B = 1)								
Proton (p)	p	$\frac{1}{2}\left(\frac{1}{2}^+\right)$	$+e$	938.280	Stable	—	0	$\bar{p}$
Neutron (n)	n	$\frac{1}{2}\left(\frac{1}{2}^+\right)$	0	939.573	898 ± 16	$pe\,\nu_e$	0	$\bar{n}$
Hyperons:								
Lambda	Λ^0	$0\left(\frac{1}{2}^+\right)$	0	1115.60	2.63×10^{-10}	$p\pi^-$ (64 %); $\pi\pi^0$ (36 %); $pe^-\bar{\nu}_e$ (0.84 %)	-1	$\bar{\Lambda}^0$
Sigma	Σ^+	$1\left(\frac{1}{2}^+\right)$	$+e$	1189.36	8×10^{-11}	$p\pi^0$ (52 %), $n\pi^+$ (48 %); p, γ(0.12 %)	-1	$\bar{\Sigma}^-$
	Σ^0	$1\left(\frac{1}{2}^+\right)$	0	1192.46	6×10^{-20}	$\Lambda^0\gamma$	-1	$\bar{\Sigma}^0$
	Σ^-	$1\left(\frac{1}{2}^+\right)$	$-e$	1197.34	1.48×10^{-10}	$n, \pi^-; n\,\bar{e}\,\bar{\nu}_e$ (0.11 %)	-1	$\bar{\Sigma}^+$
Xi	Ξ^0	$\frac{1}{2}\left(\frac{1}{2}^+\right)$	0	1314.9	2.9×10^{-10}	$\Lambda^0\pi^0$	-2	$\bar{\Xi}^0$
	Ξ^-	$\frac{1}{2}\left(\frac{1}{2}^+\right)$	$-e$	1321.3	1.64×10^{-10}	$\Lambda^0\pi^-$; $\Lambda^0\,\bar{e}\,\bar{\nu}_e$ (0.028 %)	-2	$\bar{\Xi}^+$

Particles and their types	Symbol	$I(J^P)C$	Charge	Mass (MeV)	Mean life (s)	Main decay modes	Strangeness (S)	Antiparticles
Omega	Ω^-	$0\left(\frac{3}{2}^+\right)$	$-e$	1672.5	0.82×10^{-10}	$\Lambda^0 K^-$ (66 %) $\Xi^0\pi^-$ (23 %) $\Xi^-\pi^0$ (8 %)	-3	$\bar{\Omega}^+$
Charmed baryons								
Lambda-C	Λ_c^+ (C=1)	$0\left(\frac{1}{2}^+\right)$	$+e$	2282	2.3×10^{-13}	$p\pi^+K^-; \Lambda^0\pi^+$	0	$\bar{\Lambda}_c^-(C=-1)$
Sigma-C	Σ_c^+ (C=+1)	—	$+e$	2457	2.3×10^{-13}	$\Lambda_c^+\pi^0$	0	$\bar{\Sigma}_c^-(C=-1)$
	Σ_c^{++} (C=+1)	—	$+2e$	2457	—	—	0	$\bar{\Sigma}_c^-(C=-1)$
Beautiful baryons:								
Lambda-b	Λ_b (b=+1)	—	—	5425	—	—	0	$\Lambda_b(b=-1)$
Mesons (B=0)								
Pion	π^+	$1(0^-)$	$+e$	139.57	2.6×10^{-8}	$\mu^+\nu_\mu$	0	π^-
	π^0	$1(0^-)$	0	134.96	0.83×10^{-16}	$\gamma\gamma$	0	π^0
	π^-	$1(0^-)$	$-e$	139.57	2.6×10^{-8}	$\mu^-\bar{\nu}_\mu$	0	π^+
Kaon	K^+	$\frac{1}{2}(0^-)$	$+e$	493.67	1.24×10^{-8}	$\mu^+\nu_\mu(63.5\%)$ $\pi^0\mu^+\nu_\mu$ (3.2 %) $\pi^0 e^+\nu_e$ (4.8 %) $\pi^+\pi^+\pi^-$ (5.6 %) $\pi^+\pi^0\pi^0$ (1.7 %)	$+1$	K^-
Kaon	K^0	$\frac{1}{2}(0^-)$	0	497.67	$K_s^0(0.89\times10^{-10})$ $K_L^0(5.18\times10^{-8})$	$\pi^+\pi^-$ (69 %) $\pi^0\pi^0$ (31 %) $\pi^+\pi^-\pi^0$ $3\pi^0$	$+1$	$\bar{K}^0$

Particles and their types	Symbol	$I(J^P)C$	Charge	Mass (MeV)	Mean life (s)	Main decay modes	Strangeness (S)	Antiparticles
Eta	η^0	$0(0^-)$	0	548.8	7.5×10^{-9}	$\pi^\pm \mu^\mp \bar{\nu}_\mu\,(\nu_\mu)$ $\pi^\pm e^\mp \bar{\nu}_e\,(\nu_e)$ $3\pi^0\,(32\%)$ $\pi^+ \pi^- \pi^0\,(24\%)$ $\pi^+ \pi^- \gamma\,(5\%)$ $\gamma\gamma\,(39\%)$	0	$\bar{\eta}^0$
Charmed mesons Dee	D^+ $(C=+1)$	$\tfrac{1}{2}(0^-)$	$+e$	1869.4	9×10^{-13}	$\bar{K}^0 \tau^+\,(2\%)$	0	$\bar{D}^-$ $(C=-1)$
	D^0 $(C=+1)$	$\tfrac{1}{2}(0^-)$	0	1864.7	4.5×10^{-13}	$K^- \pi^+ \pi^+\,(5\%)$ $K^- \pi^+ \pi^+ \pi^0\,(3\%)$ $\bar{K}^0 \pi^+ \pi^0\,(13\%)$ $\bar{K}^0 \pi^+ \pi^+ \pi^-\,(8\%)$ All $e^+ \nu_e\,(19\%)$ $K^- \tau^+\,(2\%)$	0	$\bar{D}^0$ $(c=-1)$
	F^+ $(C=+1)$	$0(0^-)$	$+e$	1971	3×10^{-13}	$K^- + \pi^+ \pi^0\,(9\%)$ $K^- \pi^+ \pi^+ \pi^-\,(5\%)$ All $F^-\,(44\%)$ All $K^0_s\,(17\%)$ All $e^+ \nu_e\,(5\%)$ $\varphi\pi^+$	$+1$	$\bar{F}^-$ $(C=-1)$
Beautiful mesons	B^{+1} $(b=+1)$	$\tfrac{1}{2}$	$+e$	5270.8	15×10^{-12}	$\varphi\pi^+$ $\eta\,\pi^+ \pi^+ \pi^-$ All $e^+ \nu\,(13\%)$	0	B^- $(b=-1)$

Particles and their types	Symbol	$I(J^P)C$	Charge	Mass (MeV)	Mean life (s)	Main decay modes	Strangeness (S)	Antiparticles
	B^0 (b=−1)	$\frac{1}{2}$	0	5274	1.5×10^{-12}	All $\mu^+\nu$ (12%) $\bar{D}^0\pi^+$ (4%) $D^{*-}\pi^+\pi^+$ (5%) All $e^+\nu$ (13%) All $\mu^+\nu$ (12%) $D^0\pi^+\pi^-$ (13%) $\bar{D}^*\pi^-$ (3%)	0	$\bar{B}^0$ (b=−1)
Leptons $(l=1)$								
Electron	e^-	$\left(\frac{1}{2}\right)$	$-e$	0.511	Stable	—	—	e^+
Muon	μ^-	$\left(\frac{1}{2}\right)$	$-e$	105.66	2.2×10^{-6}	$e^-\,\bar{\nu}_e\,\nu_\mu$	—	μ^+
Tauon	τ^-	$\left(\frac{1}{2}\right)$	$-e$	1784.2	4.6×10^{-13}	$\mu^-\,\bar{\nu}_\mu\,\nu_\tau$	—	τ^+
Electron neutrino	ν_e	$\left(\frac{1}{2}\right)$	0	<35 eV	Stable	—	—	$\bar{\nu}_e$
Muon-neutrino	ν_μ	$\left(\frac{1}{2}\right)$	0	<0.52 MeV (?)	Stable	—	—	$\bar{\nu}_\mu$
Tauon-neutri-no	ν_τ	$\left(\frac{1}{2}\right)$	0	<150 MeV (?)	Stable	—	—	$\bar{\nu}_\tau$
Photon	γ	(1)	0	0	Stable	—	—	γ (Same)
Graviton (not yet observed)		(2^+)	0	0	Stable	—	—	?

COSMIC RAYS

10.1 INTRODUCTION

Today the cosmic rays are considered to be one of the most active field of research. The study of cosmic rays has led to the discovery of several elementary particles of nature, nemely mesons, hyperons, positron etc. The study of cosmic rays has revealed some of the most striking and complex phenomena, has brought to light particle energies of the order 10^{10} GeV which no accelerating machine on earth can produce in the foreseeable future. It has also raised several fundamental and perplexing questions about nuclear structure and nuclear interactions.

Cosmic rays are highly penetrating radiations consisting of high energy atomic nuclei which are continuously coming from outer space. About 10^{18} of them reach the earth surface per second. Cosmic ray particles have a wide range of energy from tens of MeV to about 10^{20} GeV.

10.2 DISCOVERY OF A COSMIC RAYS

It has long been known that a charged electroscope, if left standing for some little time, will discharge regardless of how well the gold leaf is insulated, i.e., the air has a slight conductivity. When the properties of radioactive radiations were better known Rutherford showed that the rate of leakage was considerably reduced by shielding the electroscope with thick slabs of lead, but there was always a residual leakage of charge which could not be eliminated.

To explore the origin of this ionising agent further and to investigate whether or not an extra terrestrial radiation was indeed present, experiments were later conducted on the measurment of the small leak of an electrometer maintained at a high altitude. In 1909, Wulf made such measurment at the top of the Eiffel Tower in Paris, but detected no significant difference from results obtained on the ground. The famous experiment of Hess in 1912, in which he sent up an ionization chamber in a balloon and found that the intensity of ionisation actually increased up to a height of 5000 m and then decreased again, showed beyond, that these ionizing radiations travel down to earth through the air. A further observation showed that the intensities were the same for night or day indicating that the origin of these radiations was not solar. Hess suggested therefore that these rays were of cosmic origin, and they were finally called "cosmic rays" by Millikan in 1925. The experiments of Hess and Kolhorster may be regarded as constituting the discovery of cosmic rays. For this reason Hess was granted Noble prize in Physics for the year 1936.

From 1926 to 1928, Millikan and Cameron conducted experiments on the penetration of cosmic rays through water by lowering the electroscope in deep lakes. They observed that the cosmic rays reaching the earth consisted of two components: (i) soft and (ii) hard of very different penetrating powers. The soft component (now known to consist of electrons, positrons and photons) was absorbed at a depth of about 1 m of water, but the more penetrating hard component (now known to consist of mesons) was not fully absorbed even at the bottom of the 500 m deep lake. This showed that the energy of the cosmic rays was higher than any radiation known at that time.

The most exciting and fruitful field of work with comic ray is the search for new fundamental particles. These particles have thrown much light on the details of nuclear and nucleonic structure, and on the nature of nuclear interactions. In this chapter we shall study the salient features of cosmic rays.

10.3 NATURE OF COSMIC RAYS

From large number of experiments, now it is firmly established that cosmic rays are of two kinds (i) Primary cosmic rays and (ii) Secondary cosmic rays.

(i) Primary Cosmic rays   Primary cosmic rays have their origin some where out in space. They travel with speeds almost as great as the speed of light and can be deflected by planetary or inter galactic magnetic fields.

The composition of cosmic rays entering the earth's atmosphere, is fairly well known from balloon experiments and it is found that these primary cosmic rays consists of mainly of fast protons. There are very few positrons, electrons or photons, and the 'particle' composition is mainly 92 % protons, 7 % d-particles and 1 % heavy nuclei, carbon, nitrogen, oxygen, neon, magnesium, silicon, iron, cobalt and nickel stripped of their electrons. The average energy of cosmic ray flux is 6 GeV (1 GeV = 109 eV), with a maximum of about 10^{10} GeV. The radiation reaching the earth is almost completely *isotropic*. Composition of cosmic rays entering earth's atmosphere is shown in Table 10.1.

Table 10.1 Composition of primary cosmic rays entering earth's atmosphere

Group of nuclei	Charge Z	Flux density	Percentage of total flux
Protons (P)	1	1300	92.9
Helium nuclei (α)	2	88	6.3
Light nuclei (Li, Be, B)	3 to 5	1.9	0.13
Medium nuclei (C, N, O, F)	6 to 9	5.6	0.4
Heavy nuclei (beyond Ne)	≥ 10	2.5	0.18
Super heavy nuclei	≥ 20	0.7	0.05

As soon as the primary cosmic rays enter the earth's atmosphere multiple collisions readily take place with atmospheric atoms producing a large number of secondary particles in showers. Thus when a primary proton strikes an oxygen or nitrogen nucleus a nuclear *cascade* results.

The collision cross sections for the primary component of cosmic rays are of the order of 10^{-1} barns and the mean free path for a collision process at the top of the atmosphere may be as high as several kilometers. The new particles produced after primary collision give in their turn more secondary radiations by further collisions until a cascade of particles have developed, increasing in intensity towards the earth.

The energy spectrum of the primary cosmic rays ranges from 10^9 eV to about 10^{19} eV and can be expresses as

$$\frac{dN}{dE} = K(E + m_0 c^2)^{-\gamma} \qquad \qquad ...\,(10.1)$$

Where N is the number of nuclei with a kinetic energy per nucleon $> E$ (in GeV) and $m_0 c^2$ is the nucleonic rest energy. K and γ are constants for a given cosmic ray component. This is expressed in Table 10.2.

Table 10.2

Nucleus	% Composition	Energy range (GeV/nucleon)	Flux, i.e. no. of particles $m^{-2}\,S^{-1}$ per unit solid angle
H	92	2 – 20	$400\,E^{-\frac{8}{7}}$
He	7	1.5 – 8	$460\,E^{-\frac{7}{4}}$
Li, Be, B	0.18	—	$12\,E^{-\frac{7}{4}}$
C, N, O, F	0.36	3 – 8	$24\,E^{-\frac{7}{4}}$
Ne and beyond	0.15	3 – 8	$16E^{-2}$

About 1.5 GeV per nucleon the cosmic ray intensity is fairly steady with time, with flux values in space of about

Protons	1500 nuclei/m²-S unit-solid angle
α-particles	90 nuclei/m²-S unit solid angle
Heavies	10 nuclei/m²-S unit solid angle

At lower energies the cosmic ray intensity is not constant with time but depends on the activity of the sun.

The differential energy spectrum of primary cosmic rays is shown in Figure 10.1(a). The primary spectrum exhibit a maximum at about 5×10^8 eV below which the intensity falls off rapidly. From Figure it is evident that there is a minimum at about 20 MeV and below this the intensity again rises. One obtains the differential energies between E and $E + dE$ incident per unit area per unit solid angle from Figure 10.1(a). On integrating it for all energies above E, one obtains the integral energy spectrum as shown in Figure 10.1(b).

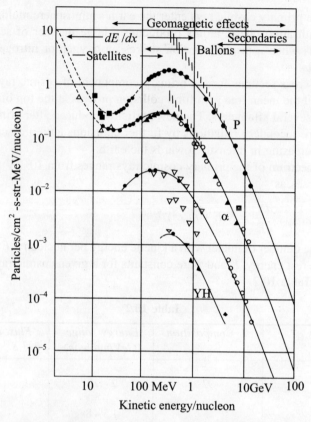

Fig. 10.1(a) Differential energy spectrum of primary cosmic rays.

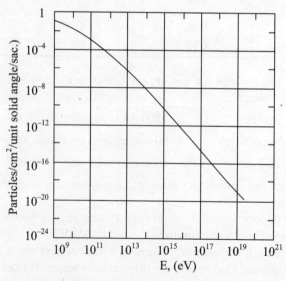

Fig. 10.1(b) Integral energy spectrum of primary cosmic rays.

(ii) Secondary cosmic rays   These are formed as a result of inelastic collisions of primary rays with the nuclei of the oxygen and nitrogen atoms of the air in the upper layers of the atmosphere. Due to these inelastic collisions in addition to showers of mesons (positive negative and neutral) few hyperons are also produced. These are called secondary cosmic rays. Below an altitude of 20 km all cosmic radiations are secondary. At high energy of primary particles (greater than 5×10^9 eV), their collisions with the atoms of air lead, as a rule to the initation of electron-nuclear showers. The result of interaction between the particles and the nuclei of atoms of the air is the disrupting of the latter into separate nucleons or larger fragements, and the formation of unstable particles ($\pi^\pm$ and π^0 mesons). The subsequent decays

$$\pi^+ \to \mu^+ + \nu \text{ (neutrino)}$$
$$\pi^- \to \mu^- + \nu$$
$$\pi^0 \to \gamma + \gamma$$
$$\gamma \to e^+ + e^-$$

lead to the formation of a soft electron-photon component of the shower. This component is then intensively multiplied due to the consecutive (cascade) formation of new $e^+ - e^-$ pairs and to bremsstrahlung of new gamma quanta by the particles (electron cascade process) as shown in Figure 10.1(c). the totality of nuclear high energy interactions brings about *extensive air showers* also called (Auger showers). At energies of the primary particles over 10^{13} eV, these showers may contain millions of particles (mainly $e^\pm$) with the transverse dimensions of the shower exceeding 1 km^2. These showers were first observed in cloud chamber photographs in 1933 by Blackette and Occhialini.

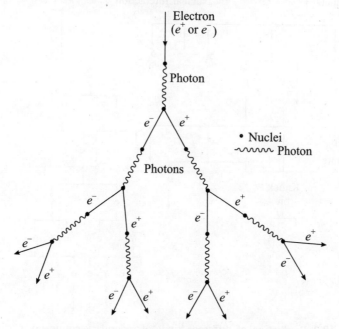

Fig. 10.1(c) The principle of shower formation.

At this stage, one can summarize the four principal components of rays in the atmosphere:

(i) A nuclear component containing particles capable of producing nuclear collisions. It is attenuated strongly in passing through the atmosphere. It consists mostly of nucleons; the fraction of pions in it increases with increasing depth in the atmosphere.

(ii) An electron-phonon component which owes its origin mainly to the production of neutral π mesons in nuclear collisions and multiplies by pair-production and bremsstrahlung process. It reaches a broad maximum at about 13 km altitude; most of it dies out, however, before reaching the earth's surface.

(iii) A muon component which owes its origin mainly to the production of $\pi^{\pm}$ mesons in nuclear collision. Muons do not multiply like electrons. They produce fast electrons, both by collisions with atomic electrons and decay. At sea level muons constitute about 80 percent and electrons about 20 percent of the charged particles incident on the earth's surface.

(iv) A neutrino component which arises from the decay of unstable particles in the atmosphere, mostly pions and muons. Neutrinos have only weak interaction with matter and are capable of passing through the earth without and sensible attenuation.

Figure 10.2 is a schematic diagram exhibiting the principal generic relations between various types of secondary cosmic ray particles.

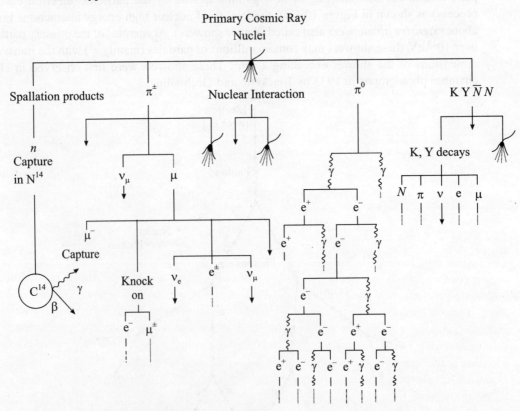

Fig. 10.2 Production of different kinds of secondary comic ray particles

It is found that the intensity of the nuclear active component (N) rapidly goes down with the atmospheric depth below the top of the atmosphere due to their loss of energy by ionization and absorption. The neutrons (n), on the other hand come down to lower altitudes. It is also reported that the intensities of the electronic component (s) and muon component (μ), after reaching maxima at the atmospheric depth $\sim$ 2000 kg/m^2 go down (Figure 10.3). Figure 10.3 also exhibit the total intensity (t).

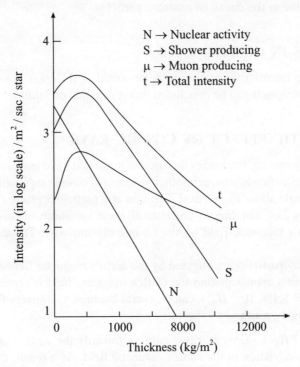

Fig. 10.3 Variation of the intensities of the different components of the secondary cosmic rays with the atmospheric depth. Total intensity (t) is also shown.

10.4 SOFT AND HARD COMPONENTS

The cosmic rays are not a homogeneous beam. Their absorption in matter exhibit that they consist of two components (i) one being completely stopped by lead of 0.1 m thickness. This component of cosmic rays is called the *soft component* (ii) the other which passes through 0.1 m of lead or more. This component of cosmic rays is called *hard* component. Such distinction between hard and soft components, based upon their absorption in 0.1 m of lead is quite arbitrary but has proved very convenient. The soft component of cosmic rays constitute about 20 % of the total cosmic ray particles at the sea level at 50° geomagnetic latitude. The proportion increases with altitude upto about 15 km beyond which there is a marked decrease in its intensity. Such a change in the intensity of the soft component with altitude is well evident if we associate this component with the secondary cosmic rays. Since these are low energy particles produced at the height about 15 km, their intensity is quite remarkable in that region but as these are regularly absorbed in atmosphere on their way to the earth, their intensity decreases with decrease of altitude. The hard component can well be related to primary cosmic rays which are highly energetic.

The soft component consists of electrons, positrons and photons. Muons, formed upon the decay of pi(π) mesons and weakly interacting with the nuclei of the atoms in the atmosphere, make up the hard component. The ratio of the intensities of the soft and hard components varies substantially with the altitude because of the unequal absorption of different particles by the atmosphere and also due to the decay of unstable particles.

10.5 VARIATIONS IN COSMIC RAYS

It is observed that the intensity of cosmic rays measured at different places of the earth is not same. From the observations it can be concluded that it depends on latitude, altitude and direction.

10.6 GEOMAGNETIC EFFECT OF COSMIC RAYS

Measurments of the cosmic ray intensities throughout the world have revealed variations both with the lattitude and longitude. In addition, azimuthal variation of cosmic ray intensity has been reported at a particular polar angle about the vertical direction at a particular place, called as the east-west and north-south effects. One can directly correlate all these variations in cosmic ray intensity with the effect of the earth's magnetic field on the cosmic ray intensity. These effects are known as *geomagnetic effects*.

Charged cosmic-ray particles are affected by the earth's magnetic field long before they enter the atmosphere. To a first approximation the earth's magnetic field is represented by a dipole of magnetic moment $M = 8.1 \times 10^{22}$ J/T located several hundred kilometers from the centre of the earth and pointing roughly south.

(i) **The Latitude Effect**   Cosmic rays, coming towards the earth isotropically from outer space. Undergo deviation in the earth's magnetic field. As a result, the intensity of cosmic rays that reaches the earth's surface depends upon the latitude. Near the equator, the deviating effect of the earth's magnetic field is more strongly manifested and the greater part of thee particles, subject to a large deviation, never enter the atmosphere. This effect is called latitude effect.

The Dutch physicist J. Clay in 1931 reported variations in the cosmic ray intensities with latitude. The measurment of cosmic ray intensities by clay at various latitudes revealed that starting from the latitude of 90°, the cosmic ray intensity remained constant until a latitude of about 50° was reached (Figure 10.4).

Therefore, the intensity began to drop appreciably till it was minimum at the equator. Same symmetrical rise in intensity was observed in the southern hemisphere. There was some uncertainty whether intensity of

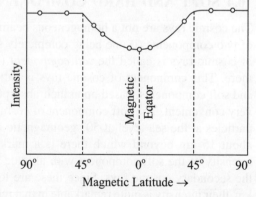

Fig. 10.4 Latitude effect in cosmic rays

cosmic rays is constant between the latitudes of 50° and 90°, which would mean that no primary cosmic particles with energy less than about 1500 MeV can reach the earth's atmosphere. Later investigations by A. H. Compton, R. A. Millikan and H. V. Neher and others confirmed the existence of the latitude effect 55° N to 45° S. Figure 10.5 shows the results of these surveys at different altitudes. It is reported that the intensity of cosmic rays increases from geomagnetic equator to about 40° latitude by nearly 14 % at sea level. Further, the intensity remains almost constant for latitudes above 40°.

It is also reported that the intensity at high altitudes increases from 0° to 55° latitude above which the intensity of cosmic rays is found to be constant. The increase in the intensity of cosmic rays at higher latitudes is reported to be much greater than at the sea level.

Cause of Latitude Effect The latitude effect can only be explained if the cosmic rays are primarily in the form of charged particles. The discovery of latitude effect is thus important in establishing that the majority of cosmic ray particles must indeed be charged.

The variation in the cosmic ray intensity at different geomagnetic latitudes can be understood by considering the effect of earth magnetic field on cosmic rays. The presence of such effect

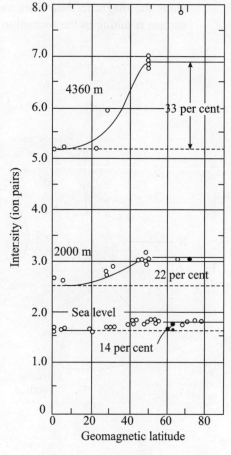

Fig. 10.5 Latitude effect exhibited by cosmic rays.

clearly reveals that the cosmic rays are charged particles, which otherwise would not be effected. At the equator, the earth's magnetic field is perpendicular to the direction of motion of cosmic ray particles. Therefore, it exerts maximum force upon the particles, which thus suffer maximum deflection. Obviously, their intensity is minimum at the equator. At poles, however, the particles approach parallel to the earth's field and suffer minimum deflection, which implies maximum intensity of cosmic rays at the poles (Figure 10.6).

Because of the magnetic forces, many charged particles from sun and outer space are trapped by the field in two doughnut shaped belts called *Van Allen belts*. These belts surround the earth except at the regions of magnetic poles. The outer belt constitutes the low energy particles, protons and electrons from the sun and attains maximum intensity roughly at 16,000 km from the earth. The inner belt is caused by high energy protons from the outerspace and is situated at a distance of about 5000 km from the earth's surface. The trapping of charged particles by the earth's magnetic field is not surprising. The earth's magnetic field being stronger near the earth's surface behaves like a giant magnetic mirror system. The charged particles such as electrons and protons entering the earth's magnetic field sprial along the magnetic field lines and are repeatedly reflected at the mirror points near the surface resulting in the formation of these belts.

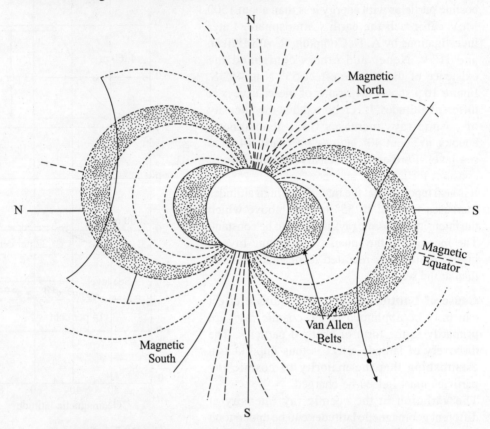

Fig. 10.6 Behaviour of cosmic rays in the earth's magnetic field.

An extensive series of measurments by Compton and Turner made on a series of sea Voyages between the latitudes 55° N and 45° S provides some of the best evidence on the latitude effect. Figure 10.7 shows the results of these experiments. The intensity at the magnetic equator is 90 % of that at a latitude of 50° N. The curve is approximately symmetrical as the intensity at a magnetic latitude 1° N is very nearly the same as at 1° S. The change over latitudes between 40° and the poles is small.

The latitude effect is considerable more marked if observation are taken at high altitudes. The results of Bowen, Millikan and Nehr are shown in Figure 10.8. This work revealed the dependence of cosmic ray intensity upon geometrical latitude. One can express the latitude effect by quantity r, and its dependence upon the relative change of intensity between geometric equator and pole, i.e.

$$r = \frac{(I_{90} - I_0)}{I_{90}} \qquad \ldots (10.2)$$

It was reported to be 0.1 for sea level and 0.33 at 4360 m altitude.

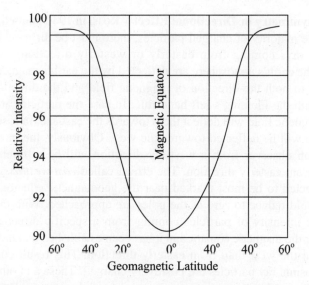

Fig. 10.7 Variation of cosmic ray intensity with geomagnetic latitude.

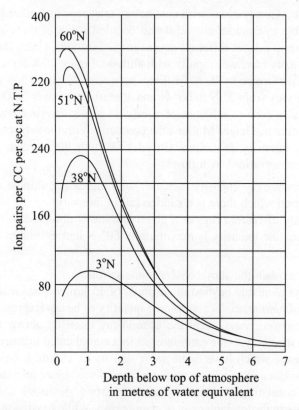

Fig. 10.8 Variation of the cosmic ray intensity with the altitude at four difference geomagnetic latitudes. (Altitude expressed in equivalent metres of water below top of atmosphere).

(ii) East-West Asymmetry or Directional Effect   Rossi in 1930 pointed out that if the cosmic rays constitute majority of charged particles, there must be difference in the intensities of these cosmic rays coming from easterly or westerly direction. The charged particles approaching the earth's atmosphere are deflected by the earth's magnetic field in a direction perpendicular to both the direction of magnetic field and the direction of their motion in accordance with the Fleming's left hand rule. In case the particles are positively charged, the earth's magnetic field will deflect them towards the east and, in case these are negatively charged, these will be deflected towards the west. Obviously, the intensity of the particles approaching an observer from a westerly direction will be different from that of those approaching from easterly direction. The effect, called *east-west assymetry,* or directional effect, is expected to be most marked near the geomagnetic equator, where the excess is 14 % and should reduce to zero as the poles are approached. With the help of cosmic ray telescope, the intensity of particles coming from a specific direction has been actually measured near the magnetic equator and it has been established that more particles approach the earth from the west than from easterly direction. This result confirms the view that majority of cosmic ray particles are positively charged. These and other considerations led Johnson to suggest that primary cosmic rays are mainly protons. On the basis of this one can easily explain east-west assymetry effect.

(iii) Altitude Effect   Observations with cosmic ray telescope and other improved apparatus, it has been possible to conduct a careful and detailed study of the effect of altitude on the intensity of cosmic rays at different geomagnetic latitudes. Observations indicate that the cosmic ray intensity increases rapidly with altitude. Figure 10.8 depicts the results of series of experiments performed by R. A. Millikan between 1933 and 1938 at five different places whose latitudes vary from 3° N to 60° N and at altitudes of over 20,000 m. In 1935 Stevens and Anderson also rose to a height of nearly 14 miles carrying with them, among other scientific instruments, Geiger Muller tube counters. With these instruments they measured the cosmic ray intensity at various altitudes on both their ascent and decent. One can summarize the observations as follows:

(i) The total cosmic ray intensity increases with increasing altitude upto a height of about 15 km beyond which there is a sudden fall in intensity.

(ii) The intensity of cosmic rays at a given latitude is observed to rise with geomagnetic lattitude, i.e., the intensity is maximum at 60° N and minimum at 3° N for any given latitude.

One can easily explain the above observations as follows:
The cosmic rays available beyond the altitude of 15 km consists mainly of protons with a small fraction of *helium nuclei* and minute quantity of heavy fragments. These particles, so called *primary cosmic rays,* form the cosmic ray intensity above 15 km. these primary cosmic rays on their way to the earth interact in a complicated manner with the nuclei of the atmospheric gases, which have appreciable density at 15 km or below. Such interactions results in the production of quite a good number of secondary particles with comparatively low energies, consisting of muons, pions, electrons, positrons and gamma rays which cause an increase in the total intensity of cosmic rays. While passing through the atmosphere, many of these secondary particles are slowed down until these no longer produce ionisation

and hence are not recorded. Thus the intensity of cosmic rays decrease with the decrease of altitude. At sea level, only a few particles higher energies are left to be recorded.

One can easily understood the difference in intensities at different latitudes (for a given altitude) by considering the action of earth's magnetic field on the cosmic ray particles. At lower latitudes, the magnetic force due to earth's field is strong enough to deflect back most of the cosmic ray particles approaching earth. Only highly energetic particles succeed in breaking through the magnetic field and are able to reach the earth's surface. At higher latitudes, the magnetic force is comparatively weak and can not provide an effective check on the entry of cosmic rays to the earth's surface. Thus the intensity of cosmic rays at higher latitudes is much greater than that found at lower latitudes.

Explanation of Geomagnetic Effect

Stoermer in 1917 developed a theory of the origin of *aurora borealis,* observed on the polar regions of the earth. According to Stoermer's theory, the high energy electrons coming from the sun during solar activity strike the atoms and molecules in the upper layers of the atmospheric air and excite the orbital electrons, resulting in the emission of visible radiation which is seen as the aurorae in the polar regions. Due to the action of earth's magnetic field, the electrons coming from the sun are mostly deflected towards the poles and hence the aurorae are seen in the polar regions. According to Stormer's theory, the motion of the charged particle in the moving meridian plane (the plane containing the moving particle and the magnetic dipole axis) is given by

$$\cos \omega = \frac{2I}{r \cos \lambda} - \frac{\cos \lambda}{r^2} \qquad \text{... (10.3)}$$

Where ω is the angle between the orbit and the circle λ; r is a constant; I is a constant of motion (I is defined as the angular momentum of the particle at infinity w.r.t. the magnetic dipole of the earth) and λ is the geomagnetic latitude.

One can split the motion of the particle into two components:

(i) the motion in the meridian plane, i.e. r, Z plane and (ii) the rotation of meridian plane round the Z-axis. We must note that all the points with $r < 1$ for orbits $I > 1$ are forbidden. Obviously, directions are forbidden for which

$$\cos \omega > \frac{2}{r \cos \lambda} - \frac{\cos \lambda}{r^2} \quad (r < 1) \qquad \text{... (10.4)}$$

These forbidden directions lie outside a circular cone, known as stormer cone. The axis of the cone is in the east-west direction. The semivertical angle ω_s of the cone is given by

$$\cos \omega_s = \frac{2}{r \cos \lambda} - \frac{\cos \lambda}{r^2} \qquad \text{... (10.5)}$$

One can define a critical value of I, say such that all directions corresponding to directions with

$$\cos \omega < \cos \omega_s = \frac{2I_c}{r \cos \lambda} - \frac{\cos \lambda}{r^2} \quad (r < 1) \qquad \text{... (10.6)}$$

are forbidden with all other directions are allowed. This relation yields

$$r = \frac{\cos^2 \lambda}{(I_C^2 - \cos \omega \cos^3 \lambda)^{\frac{1}{2}} + I_C} \qquad \text{... (10.7)}$$

Thus the smallest momentum which can reach the earth from outside at a latitude λ through an angle ω to the east-west direction is given by

$$p(\lambda, \omega) = p_{max} \frac{\cos^4 \lambda}{\left[(I_C^2 - \cos \omega \cos^3 \lambda)^{\frac{1}{2}} + I_C \right]^2} \qquad \text{... (10.8)}$$

Now, we consider the following special cases

(i) $\omega = 0$, i.e. particles with momenta

$$p > p\,(\lambda, \pi) = p_{max} \frac{\cos^4 \lambda}{\left[(I_C^2 - \cos^3 \lambda)^{\frac{1}{2}} + I_C \right]^2} \qquad \text{... (10.9)}$$

all allowed from are directions at the latitude λ.

(ii) $\omega = \pi$, i.e. particles with momenta

$$p < p\,(\lambda, \pi) = p_{max} \frac{\cos^4 \lambda}{\left[(I_C^2 - \cos^3 \lambda)^{\frac{1}{2}} + I_C \right]^2} \qquad \text{... (10.10)}$$

are forbidden in all directions at the latitude λ.

(iii) $\omega = \dfrac{\pi}{2}$, i.e. the case for which the momentum is called the limiting momentum incident in the vertical direction

$$p_{vert}(\lambda) = p_{max} \cos^4 \frac{\lambda}{4} \; I_C^2 \qquad \text{... (10.11)}$$

$p(\lambda, 0)$, $p(\lambda, \pi)$ and $p\left(\lambda, \dfrac{\lambda}{2}\right)$ momenta are called the limiting momenta. Figure 10.9 exhibit the plot of these momenta against λ for $I_C = 1$.

Relation (10.7) reveals that the particles with momentum p are partly forbidden in a belt of latitudes λ:

$$\lambda_1 > \lambda > \lambda_2$$

One finds that the momentum $p(= p_{max}\; r^2)$ is allowed in all directions in the region round the pole above λ_1 and is forbidden in all directions in the equatorial belt $\pm \lambda_2$. Now we refer back to Stormer cone. We have defined this as a cone outside of which the particles are forbidden. Now, we shall see which directions inside this cone are allowed. Inside Stormer's cone a main cone has been defined by Vallarta. All directions inside the main cone are allowed, whereas all directions are forbidden in a simple shadow cone. The simple shadow cone is due to particles that interact the earth. This cone lies in the region near the horizon toward the north or south pole and is of the most importance at large Zenith angles. In between the main cone and the simple shadow cone, there is a region called the *penumbra*. The penumbra is a complex region containing an infinite number of allowed and forbidden sub-regions. We must note that for the equator there is no penumbra and the allowed cone coincides with the main cone.

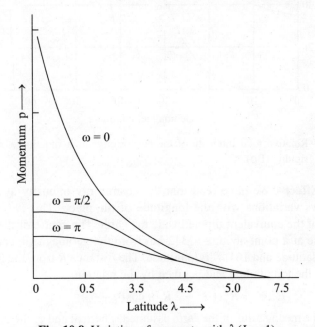

Fig. 10.9 Variation of momenta with $\lambda (I_c = 1)$.

Now, we shall try to explain the geomagnetic effects on the basis of Stormer's theory.

(i) **Latitude Effect** One can understand this effect on the basis of allowed cones described earlier. We have seen that the intensity of the cosmic ray incident particles increases with geomagnetic latitude as the cone formed by the allowed directions opens gradually when one moves away from the equator towards the pole. It increases upto 50° and remains constant for higher latitudes. Obviously, the lattitude effect arises from a primary spectrum containing various energies. Observations clearly reveals that the intensity above 50° remains constant. This clearly indicates no particles which are latitude sensitive above 50° and obviously with momentum below 3,000 MeV/c.

A plot of the relative total intensity of the cosmics rays as a function of the geomagnetic latitude (λ) for particles of different momenta expressed in terms of magnetic rigidity ($B \rho C$) can help to make quantitative predictions about the lattitude effect for different magnetic rigidities (Figure 10.10)

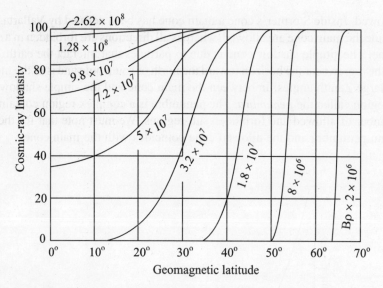

Fig. 10.10 Relative total intensity of the cosmic rays as a function λ for different magnetic rigidity (Bρ).

(ii) Longitude Effect We have read that the cosmic ray intensity at a given latitude and altitude shows variations with the longitude of the place. This was explained as due to eccentricity of the equivalent dipole inside the earth. The centre of the equivalent dipole is reported to lie at a point about $\varepsilon \simeq 342$ km from the geomagnetic centre of the earth to a point 6.5° N latitude and 161.8° E longitude. The distance R from the geomagnetic centre is a function of the longitude ϕ and is given by the relation

$$R^2(\phi) = R_e^2 + \varepsilon^2 - 2\,R_e\,\varepsilon\cos(\phi - \phi_0) \qquad \ldots (10.12)$$

Where R_e is the mean radius of the earth assumed spherical and ϕ_0 the geomagnetic longitude of the point of nearest approach to the geomagnetic centre. Corresponding to this variation in distance, it is found that the limiting momentum also varies with ϕ. One finds that the limiting momentum in the vertical direction at the equator is given by the relation

$$p_{\text{vert.}}\,(\lambda = 0,\,\phi) = \frac{15000\ \text{MeV/c}}{\left(1 + \dfrac{\varepsilon}{R_e}\right)^2 - 2\left(\dfrac{\varepsilon}{R_e}\right)\cos(\phi - \phi_0)} \qquad \ldots (10.13)$$

Making use of proper numerical values, one obtains that $p_{\text{vert.}}$ varies along the geomagnetic equator between 13,700 MeV/c and 16,500 MeV/c. Theoretical predictions do not agree very well with observed longitude variation. It is found that the point of minimum intensity occurs about 65° west of the predicted value close to a point where the earth's field is maximum. Moreover, the magnitude of the effect requires higher strength of the earth's field than observed at the position of the maximum field strength.

The total flow of energy has a variation of 10 % for ϕ_0 and $\phi_0 + \pi$. It is reported that the longitude effect is about 7 % of the total intensity at sea level. The difference between equator and high latitudes is found to be about 7 % in USA and about 14 % in the far east.

(iii) East-west asymmetry   This can be accounted for on the assumption that the primary cosmic rays contain an excess of particles of one sign. For the positive cosmic ray primaries, the forbidden cones point towards the east and one expect the greater intensity from the west. Similarly for the negative cosmic ray primaries the forbidden cones points towards the west and one expect greater intensity from the east. Obviously, the difference of the intensities of the positive and negative primaries determines the geomagnetic effect of considerable significance, i.e. *east-west asymmetry.*

The measurments of the rate of particles arriving from a certain direction from the Zenith angle were made using Geiger Muller counter telescope as shown in Figure 10.11. The telescope is pointed to the east and west of the Zenith and the intensities for the same Zenith angle for the two directions ($\theta_W = \theta_E$) are compared.

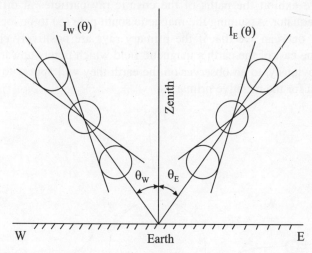

Fig. 10.11 Geiger Muller Counter telescope used for east west asymmetry.

If I_0 is the flux of particles coming from vertical directions and $I(\theta)$ is the flux of particles coming from an angle θ from the Zenith, then one finds

$$I(\theta) = I_0 \cos^2 \theta \qquad \text{... (10.14)}$$

If I_W and I_E are the two intensities at a particular point of observation then the east-west effect is defined as

$$a(\theta) = \frac{I_W(\theta) - I_E(\theta)}{[I_W(\theta) - I_E(\theta)]^{\frac{1}{2}}} \qquad \text{... (10.15)}$$

To compare the latitude effect of the vertical intensity at sea level with the asymmetry at an elevation above sea level, one uses the relation

$$\psi = \psi_0 \cos \theta \qquad \text{... (10.16)}$$

where ψ_0 is the mass equivalent of the atmosphere above sea level and ψ is that above the observer of the east-west effect.

The observations were taken at a height of 4.3 km above the sea level with $\psi = 6.24$ m water equivalent. For comparison with the vertical intensity, one can define the indication as

$$\cos \theta = \frac{\psi}{\psi_0} = \frac{6.24}{10.3} \quad \text{and} \quad \theta = 53° \qquad \qquad \ldots (10.17)$$

Assuming, $P_{\text{vert.}} = 13700$ MeV/c, the cut-off energies towards the west and east are found to be $P_W = 10^4$ MeV/c and $P_E = 2.54 \times 10^4$ MeV/c. Using these values, one obtains large east-west asymmetry.

Measurments made by Johnson and others show that the east-west asymmetry is maximum at the equator and decreases as on goes towards higher latitudes.

One can explain the east-west effect qualitatively with the help of Figure 10.12. We note that the figure exhibit the paths of the cosmic ray particles at different points on the geomagnetic equator. Assuming the magnetic south pole (S) to be located above the plane of the figure, one can see that if the primary rays are positively charged, they will be deflected to the east of the earth's magnetic field which points upwards from the plane of the figure. Obviously, to an observer on the earth they will appear to come from the west and vice versa for the negative primaries.

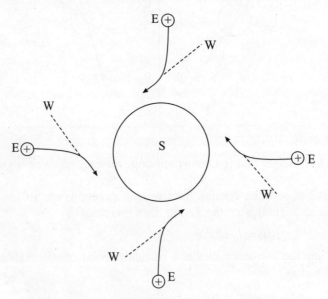

Fig. 10.12 Explanation of east-west asymmetry.

10.7 INSTRUMENTS AND APPARATUS USED IN RESEARCH ON COSMIC RAYS

Most of the work on cosmic rays prior to 1930 was undertaken using ionisation chambers in conjunction with an electrometer. Though valuable results were obtained, these devices were only able to record the total effect of the ions produced by cosmic rays during an interval of time. Furthermore, the study of directional phenomena demanded the use of cumbersome lead shields which themselves gave rise to secondary phenomena not always easy to elucidate. The further studies in this fascinating field of research demanded instruments able to record single particles, study collision processes, delineate particle tracks, and measure particle energies. With rapid progress in electronics, instruments for these purposes become increasingly available after 1930. The devices used in recording cosmic ray phenomena are as follows:

1. **Ionization Chambers**
2. **Geiger Muller Counting tubes**: (i) Geiger-Muller tubes in coincidence circuits and (ii) Geiger Muller tubes in anti-coincidence circuits
3. **Cloud Chamber**: The cloud chamber, especially the Geiger-Muller Counter controlled cloud chamber. Modern developments from the cloud chamber include the high-pressure cloud chamber, the diffusion chamber, and the bubble chamber.
4. The **nuclear research emulsion**.

Now, we present a brief discussion of these devices:

   (i) Ionization Chambers Ionization Chambers were valuable in the early studies of cosmic ray phenomena. Millikan and Coworkers used a spherical ionization chamber of 15 cm diameter with 3 mm thick steel well surrounded (during most measurments) by 10 cm of lead and a 0.5 inch thick cast iron jacket, equivalent to a total thickness of 11 cm of lead for the survey of the latitude effect of the cosmic rays. This arrangement eliminated most of the soft component and only the hard component i.e. muons was recorded. The said ionisation chamber was used with an automatically recording electroscope. Compton and Turner also used similar apparatus in their latitude effect measurments.

The ionization chambers used in the early stages were the integrating uncompensated type. Later on, with the improvised techniques, the average ionization currents in the chambers is usually compensated by some method and only the deviations from the average are measured. Nowadays, such ionisation chambers are in increasing use in which the technique of rapid electron collection is utilized. The use of good amplifiers help to reproduce the pulse shape as a function of time. This makes it possible to obtain considerable information about the events causing the total ionization.

   (ii) G.M. Counter Telescopes Coincidence and Anti-Coincidence Circuits A useful technique, originally due to Bothe and Kolhorster and later on improved by Ross is the use of two or more counting tubes in coincidence. Figure 10.13 exhibit two or three G. M. Counters arranged in a line, one above the others. This arrangements usually called as a counter telescope.

The G.M. Counters used for the study of cosmic rays are usually of the glass-walled type. High energy particles which are present in both the soft and hard components of the cosmic ray secondaries can penetrate through the glass wall of the G.M. and produce an ionization pulse at the of the Counter.

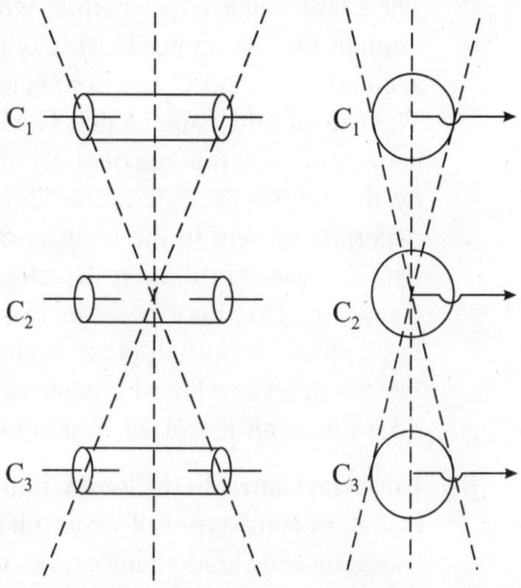

Fig. 10.13 G.M. Counter telescope.

The G.M. Counters are connected in coincidence. The object is to record an ionizing particle (e.g. a fast electron or a meson in the secondaries). Such an arrangement of G.M. Counters permits the detection of only particles which arrive within a small cone about the axis of telescope as shown in Figure 10.13. It is found that with three G.M. Counters (Figure 10.13) the magnitude is sufficient to operate the amplifier, i.e. the definition of the direction of arrival is better than with only two or less of G.M. tubes.

A high energy single cosmic-ray particle passes through all the G.M. tubes C_1, C_2 and C_3 in succession only if they are arranged one above the other in a line (Figure 10.13). When the distance between the two extreme G.M. tubes ≈ 1 m, then a cosmic ray particle travelling through them with a velocity $\sim 10^8$ m/s will pass through the three G.M. Counters within 10^{-8} s. Each G.M. Counter produces a negative electric pulse at its anode. These pulses provided by G.M. tubes are fed to the input stages of three trides (or pentodes) of the Rossi coincidence circuit. As stated above, the pulses are negative with respect to earth at the value grids. A pulse is obtained at any one of the value grids only if its corresponding G.M. tubes has been triggered by an ionising particle. If the resolving times of the G.M. tubes and the coincidence circuit are large as compared to the time of transit of the cosmic ray particle through the three G.M. tubes then the pulses from the three G.M. Counters will cut off the currents through the three tubes at the input stage of the coincidence circuit. Obviously, a positive coincidence output pulse will be obtained which can be amplified and recorded. If the incoming particle passes through the upper two G.M. tubes C_1 and C_2 but not through the third G.M. tube C_3, the output pulse at the common plate point of three G.M. tubes of the coincidence circuit will be extremely small to be recorded. This is why, when the particle passes through all the three G.M. tubes, C_1, C_2 and C_3, it is detected as a true coincidence. One will have to correct the recorded coincidence rate for the accidental coincidences.

One can direct the G.M. Counter telescope described above along different directions to determine the cosmic ray intensity at different Zenith angles. Many interesting experiments have been done with cosmic ray telescope.

There are some experiments, where it may be necessary to detect the particles passing through the two upper G.M. Counters C_1 and C_2 but not through third counter C_3. In such a situation C_1 and C_2 are connected in coincidence and C_3 is connected in anti coincidence. The use of anti coincidence G.M. Counters around a G.M. Counter cosmic ray telescope helps to reduce the rate of spurious counts due to side showers during coincidence counting by the telescope. Such side showers are produced by stray particles incident on some materials present in the vicinity of cosmic ray telescope.

One can use scintillation detectors or semiconductor detectors in an array instead of G.M. Counters. This may be used to perform coincidence experiments of different types, e.g. large plastic scintillators are widely used in extensive air shower investigations.

For the investigation of cosmic ray showers, the G.M. Counters are arranged in a geometry different from that used in a counter telescope, described above.

(iii) Counter-Controlled Cloud Chamber   The disadvantage of a simple form of cloud chamber is that, in some types of experiment, the expansion may well be made when no particles are travesing the cloud chamber, i.e. the majority of the photographs are therefore wasted. This is especially the case in experiments on cosmic rays. To overcome this difficulty Blackett introduced the G.M. Counter controlled cloud chamber (Figure 10.14).

There is also a different technique connected with the cloud chamber experiments on cosmic rays. One can introduce a delay between the production of the coincidence pulse which actuates expansion of the chamber and the photographing of the track. Due to such delay, it is observed that the ionizing droplets along the track of the ionizing particle diffuse out to some extent. Obviously, when the track is photographed, the individual droplets become visible. The number of such droplets per unit path length of the track is proportional to the specific ionization of the particle. Therefore this method permits identification of the particle by comparison with the density of the droplets along the track of some known particle.

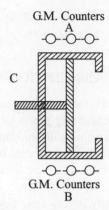

Fig. 10.14 G.M. Counter controlled cloud chamber

Usually the cloud chambers used for cosmic ray investigations are placed in magnetic fields perpendicular to the chamber. One can determine the momentum of the particle, knowing the curvature of the track in magnetic field (B) and making use of the relation $\rho = \dfrac{Bq}{mv} = \dfrac{Bq}{\rho}$, where ρ is the radius of curvature of the track and q is the charge of the ionizing particle. Whenever very high magnetic fields are required, superconducting magnets have been used. For the study of very high energy cosmic ray particles intense magnetic fields are required. Cloud chamber can measure both the momentum and the specific ionisation of the particle. These measurments helps for the identification of the cosmic ray particle.

To measure precisely the radius of curvature of the tracks, one needs the spectroscopic photographs. Scientists have developed various methods for this purpose.

One can measure the range and energy loss of the particles using cloud chambers with regularly spaced metal plates.

(iv) Nuclear Emulsion Technique   For studying certain nuclear reactions, it is found suitable to use photographic emulsions directly as particle track detectors. A heavy cosmic ray particle traversing a photographic emulsion produces a latent image of its track which can be developed and fixed when the deposited black silver in the negative gives a permanent record. Some outstanding discoveries, e.g. π-mesons have been made using nuclear emulsion techniques.

Ordinary optical photographic emulsions are not suitable for quantitative studies with nuclear radiations, since the density of such emusions is low and the developed crystal grains are too large and widely spaced to provide any well-defined particle tracks. These defects are overcome by having emulsions in which the proportion of silver bromide to gelatine is higher than that in optical emulsions (about ten times) and by using silver halide grains having diameters of the order of 0.1 to 0.2 mm only. Such fine grain nuclear emulsions with thickness varying 25 to 1000 mm have been developed by Powell et al. Thick emulsion plates upto 2000 microns thickness made by Ilford Co., of UK have been used in many investigations.

10.8 ABSORPTION OF COSMIC RAYS

Several studies have been made on the absorption of cosmic rays and the measurments reported on the penetrating power of these radiations in different materials. Hoffmann and others performed experiments with ionization chambers shielded with lead absorbers. Using three G.M. Counters, Rossi observed the rate of coincidences (Figure 10.15(a). Figure 10.15(b) exhibit the variation of the total cosmic ray intensity with lead thickness. From curve it is evident that the intensity first decreases rapidly upto 5 cm of lead and then slowly up to 10 cm. We can notice that the change in intensity is very slow beyond 10 cm. The cosmic ray particles absorbed in 5 cm of lead are about 50 % at 3500 altitude and have small penetrating power. This group of cosmic ray particles are electrons and photons and is called *soft component*.

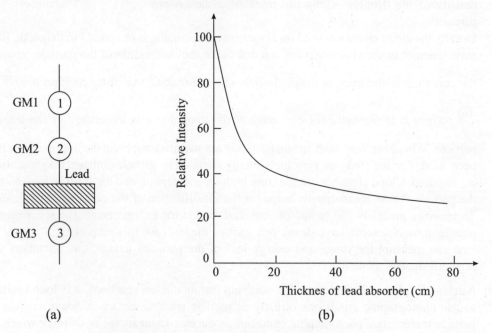

(a)   (b)

Fig. 10.15 Absorption curve of the cosmic rays in lead

The particles which are not absorbed upto 10 cm of lead, constitute the *hard component* of cosmic rays. This group of particles consists of μ-mesons and protons. It is reported that the soft component is absorbed more strongly in heavy elements than in light elements. Rossi and others reported that the absorption was proportional to $\dfrac{Z^2}{A}$. The intensity of the soft component of cosmic rays is the difference between the total intensity and the intensity of the hard component extrapolated to the zero thickness.

Myssowsky and Tuwin studied the variation of cosmic ray intensity with atmospheric pressure and interpreted this variation as being due to the absorption of cosmic rays in air. Gockel in 1915 reported the directional variation of cosmic ray intensity. It is reported that the cosmic ray intensity is larger in the vertical direction, than in inclined directions. It is found that for the angle of inclination θ to the vertical the intensity is nearly proportional to $\cos^2 \theta$. One can explain this as due to the variation of air distance travelled by the cosmic ray particles to reach sea level.

Measurments of the total vertical intensity by Pfotzer revealed that the intensity of the soft component of cosmic-rays increases rapidly upto the height of about 80 gm/cm^2 below the top of the atmosphere and decreases again for the greater heights. The studies of the intensity of the hard component by Sachein revealed that the intensity increases upto the greatest heights observed without forming any peak.

Measurments on cosmic rays have also been made in mines and under great thickness of water. It is reported that the cosmic rays are much more penetrating in sea water. With the help of G.M. Counters, observations have been carried out on cosmic rays even at the depth of 1400 m water equivalent. It is found that the relative soft intensity (ratio of soft to hard components) decrease for the few meters of water equivalent underground. The minimum value reported to be of the order of 5 %. For larger depths, it is found that this intensity increases, being of order of 30 %. One can explain this by assuming that the soft component underground is secondary to the hard Component.

Energy and Mass of Secondary Cosmic ray particles.

The energy and mass of cosmic ray particles are calculated by measuring tracks in a cloud chamber or a nuclear emulsion plate. It is observed that the tracks are thin for the particles of low density of ionisation. The density of ionization produced by a cosmic-ray particle of velocity v and charge Ze is proportional to $\left(\dfrac{Z}{v}\right)^2$ and to a factor which increases logarithmically with energy. One can determine the rest mass of the a particle from the values of velocity and momentum. The momentum can be determined from the curvature of the track in a magnetic field and the velocity from the density of ionization.

The energy of the cosmic ray particles is given by

$$E = \left[p^2 c^2 + m_0^2 c^4 \right]^{\frac{1}{2}}$$

The momentum is given by

$$p = \frac{E}{c}\left[1 - m_0^2 \frac{c^4}{E} \right]^{\frac{1}{2}}$$

For a particles of charge $q(= Ze)$ moving with the radius of curvature R in a magnetic field B, we have $p = qBR$ and hence

$$qBR = \frac{E}{c}\left[1 - \frac{m_0^2 c^4}{E^4} \right]^{\frac{1}{2}}$$

For the particles with $E \gg m_0c^2$, $E = qBRc$. Obviously, the energy is directly measured from the measurement of the curvature. Particles of all masses at relativistic energies provide similar specific ionization but the difference appears at low energies. It is observed that the specific ionization first decreases, reaches to minimum at the K.E. equal to m_0c^2 of the particles and then increases with the energy.

Secondary particles in cosmic rays were identified on the basis of these measurments. It was reported that the hard components are μ-mesons and soft particles are electrons.

On the basis of the study of the energy spectrum of cosmic ray particles, Blackett and Brode suggested that the occurrence of the particles $n(E)dE$ in the energy range E and $E + dE$ can be expressed as

$$n(E)dE = K\,E^{-2.5}\,dE \qquad\qquad\qquad ... (10.18)$$

where K is constant. We must note that this energy spectrum is for the hard components as the soft component could not be studied with good degree of accuracy.

10.9 DISCOVERY OF POSITRONS

The positron, an antiparticle of the electron having the same mass, charge (sign positive), spin and magnetic moment (sign opposite of the latter) was predicted by the relativistic quantum mechanics of Dirac and discovered few years latter by Anderson in cosmic rays.

Dirac's relativistic quantum mechanical equation for an electron gives not only positive, but also negative values for the total energy of a free electron. At a given momentum of a particle p, relativistic equation has solutions corresponding to energies

$$E = \pm\sqrt{c^2p^2 + m_e^2c^4}$$

When a positron and an electron meet, they annihilate.

Anderson in 1932 detected a positron in the composition of cosmic rays. Anderson obtained the photographs of the tracks of some cosmic ray particles produced in a cloud chamber placed between the poles of an strong electromagnet capable of producing magnetic field ~ 1.5 T. He observed that, once in a while, pairs of tracks having equal but opposite radii of curvature and appearing to originate in the metal casing around the chamber, are obtained. To find out whether these tracks are produced by particles carrying the same charge but moving in opposite directions with their paths meeting at a common point, Anderson placed a lead sheet 6 mm thick across the diameter of the chamber and found out the difference in the curvatures of the tracks on the either side of the lead sheet (Figure 10.16).

Since the particles get slowed down on traversing the lead sheet, the curvature of their paths should be greater on the side of the sheet from which they emerge than on the side of the sheet they enter. Figure 10.16 shows a diagram based on a photograph obtained by Anderson in which the positron originated near the bottom of the cloud chamber. Before traversing the lead plate the energy of this particle was computed from its curvature in a known

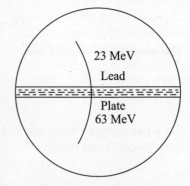

Fig. 10.14 Discovery of positrons.

magnetic field to be 63 MeV. After traversing the lead plate, its energy was found to be 23 MeV. The loss of energy of 40 MeV was calculated to be that expected on traversing 6 mm of lead plate. From the density of the tracks and the change in their curvatures on travelling the lead sheet, Anderson concluded that they were produced by particles having the same mass as the electron moving in the same direction, one carrying the negative electronic charge and the other an equivalent posite charge. That it was a positively charged particle was readily deduced as the magnetic field direction was known.

Could this positive particle be a proton? Anderson proved that the particle could not be a proton. A proton which could produce a track of the measured curvature would need to have a much lower energy than a positron. The ionisation density along the track revealed that the particle carried an electronic charge (e). The kinetic energy of the particle can be estimated from the measured momentum of the particle above the plate (23 MeV/c) from the following relation,

$$E_k = \sqrt{p^2 c^2 + m_e^2 c^4} - m_0 c^2$$

Where m_0 is the rest mass of the particle. If it is a proton then $m_0 c^2 = 938$ MeV and hence $E_k = 0.28$ MeV. Such a low kinetic energy proton would produce many more ions per centimeter of the path than an electron, and, correspondingly, a much denser track. The length of the track should only be about 5 mm. The measured length of the track was greater than 50 cm. Moreover, a proton of this energy could not traverse 6 mm of lead sheet. Thus it was definitely established that the observed particle was a positively charge particle of electronic mass. Anderson's conclusion was confirmed by Blackett and Occhialini who obtained similar tracks with their automatic cloud chamber. Anderson was awarded Noble prize in physics in 1936 for this very significant discovery. We must not that positrons are emitted by certain radionuclides as well.

10.10 COSMIC RAY SHOWERS

In section 3 we have read about cosmic ray showers. The cosmic rays showers were found to be produced by secondary cosmic rays.

Rossi observed the coincidences between three G.M. Counters arranged in a triangle, as shown in Figure 10.17(a). Obviously no coincidence could be caused by a single particle. Rossi found that the rate of coincidences in a triangular coincidence arrangement increases when a thin lead sheet is placed close above the counter (Figure 10.17(a)). It is observed that the rate of coincidence changes by changing the thickness of the lead plate (Figure 10.17(b)). From Figure 10.17(b) it is evident that the counting rate at first increases with increasing lead thickness and then decreases. It attains a maximum at a thickness between 1 to 2 cm of lead sheet. At larger thickness ~ 10 cm, the coincidence counting rate becomes almost constant because the soft particles and their secondary components are absorbed completely. The variation of the coincidence rate with the thickness of the lead sheet as shown in Figure 10.17 is called the *Ross transition curve*.

Rossi, from the above observations concluded that cosmic rays produced a large number of secondary particles in the lead plate. Obviously, with increasing thickness of the lead sheet, the probability of production of the secondary particle increases and when the thickness becomes larger (~ 10 cm), the secondary particles are completely absorbed in the lead sheet. This explains the nature of variation of the coincidence rate as shown in Figure 10.17(b). The phenomenon is named as *cosmic ray shower.*

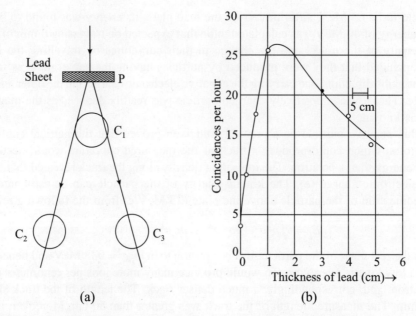

Fig. 10.17 (a) The experiments of Rossi on cosmic ray showers (b) Rossi transition curve

Blackett and Occhialini in 1933 provided direct evidence of the production of cosmic ray shower in metal plates by means of cloud chamber photographs. The cloud chamber photographs exhibit that a great majority of all showers consist of 2 – 20 thin tracks with energy $10^7 - 10^8$ V. These first multiply rapidly and then get absorbed. It is reported that sometimes a simultaneous passage of a large number of particles through the ionization chamber is observed. This phenomenon is termed as *brusts*. Brusts exhibit transition effect similar to those of showers, except that the maxima are obtained at large thickness. The following relation for the size distribution of showers is reported by Montgomery. The rate of showers containing n particles is given by

$$S(n) \sim \frac{A}{n^{z+1}} \qquad \qquad ...\,(10.19)$$

where $Z \sim 2.3$ and A is a constant.

The type of cosmic ray showers described above is usually called as the *electronic or cascade shower.*

The mechanism responsible for shower formation was the subject of much speculation until 1937, when the explanation based on pair production was put forward by Bhaba and Heitler and also carlson and openheimer. Both the theories gave more or less similar conclusions.

According to Bhabha and Heitler theory, two different physical mechanisms are responsible for the production of cosmic ray showers: (i) electron-positron pair production and (ii) the emission of electromagnetic radiation by charged particles by bremsstrahlung.

When high energy electromagnetic radiation, i.e. photon ($E >> 2\,m_0c^2 = 1.02$ MeV) passes through matter, it undergoes absorption mainly by electron-positron pair creation, i.e.

$$\gamma + X \rightarrow X + e^- + e^+$$

where X is the nucleus in whose force field the pair is born. We can see that the energy of the photon is equally distributed between the two particles created, i.e. electron and positron. However, when a high energy electron (or positron) loses energy while passing through matter by ionization and radiation. We know that when a high energy charged particle is accelerated or decelerated it emits electromagnetic radiation, there by losing a fraction of its energy. This is known as bremsstrahlung.

Theory of cosmic ray Shower

Let us suppose that a high energy photon $(E \gg 2\, m_0 c^2)$ produces an electron-positron pair in the nuclear electric field of the atom of the matter through which it passes. These particles (e^+ and e^-) are decelerated due to collisions with the atoms of the medium, which causes them to emit electromagnetic radiation of high energy by bremsstrahlung. The photons thus emitted are also of high energy and hence they create e^- and e^+ pairs. Each of the particles (e^- and e^+) so created again undergoes bremsstrahlung process and creates high energy e.m-radiations. These two processes go on repeating till a large number of electrons, positrons and photons are generated in a cascade (Figure 10.1(c)). Ultimately the energies of these particles (electrons, positrons and photons) created by the above processes are reduced to such an extent that they are unable to create any further photons or particles. The low energy photons thus created now lose energy by Compton scattering process whereas the low energy electrons and positrons lose energy by ionization. Obviously, the size of the cascade shower at first increases with increasing thickness of the material traversed, attains a maximum and then begins to decrease due to the absorption of the photons and the electrons in the material, when the thickness is high.

Now, we present the main results of the cascade theory. The complete mathematical theory of this process is bit difficult. The cascade theory is based on the assumption that the angular divergence of the shower particles can be neglected.

(i) **Radiative Collision of Electrons** When a photon with energy between E_γ and $(E_\gamma + dE_\gamma)$ is emitted by an electron of energy E passing by a nucleus of charge Ze, then the cross-section of the emission is given by

$$d\sigma_{br} = 4Z^2\, \alpha\, r_0^2\, \frac{dE_\gamma}{E_\gamma} \left[\left\{ \frac{4}{3}\left(1 + \frac{E_\gamma}{E}\right) + \left(\frac{E_\gamma}{E}\right)^2 \right\} \ln\left(\frac{183}{Z^{1/3}}\right) + \dots \text{Smaller terms} \dots \right] \quad \dots (10.19)$$

where $r_0 = \dfrac{e^2}{4\pi\,\varepsilon_0\, m_0 c^2}$ is the classical electron radius and $\alpha = \dfrac{e^2}{4\pi\,\varepsilon_p\, \hbar c}$ is the Sommerfeld's

fine structure constant. The quantity within parenthesis, i.e. $\left\{ \dfrac{4}{3}\left(1 + \dfrac{E_\gamma}{E}\right) + \left(\dfrac{E_\gamma}{E}\right)^2 \right\}$ is of

the order of unity. Obviously, (10.19) reduces to

$$d\sigma_{br} = 4Z^2\, \alpha\, r_0^2\, \frac{dE_\gamma}{E_\gamma} \ln\left(\frac{183}{Z^{\frac{1}{3}}}\right) \qquad \dots (10.20)$$

The number of photons in the energy internal E_γ to $(E_\gamma + dE_\gamma)$ which an electron $(E >> m_0 c^2)$ produces in a path length dx is obtained from

$$dn(E_\gamma) = N dx \, d\sigma_{br} = \frac{dx}{x_0} \frac{dE_\gamma}{E_\gamma} \qquad \qquad \dots (10.21)$$

Here x_0 is known as *Cascade* and *radiation* unit and is expressed by

$$x_0 = \left[4 N Z^2 \, \alpha \, r_0^2 \, \ln\left(\frac{183}{Z^{\frac{1}{3}}} \right) \right]^{-1} \qquad \qquad \dots (10.22)$$

Where N is the number of nuclei per unit volume. Thus $N dx$ represent the number of nuclei present per unit area in the thickness dx of the absorber. If the lengths are measured in radiation units, then one can write $\dfrac{dx}{x_0} = dl$. This means

$$dn(E_\gamma) = dl \, \frac{dE_\gamma}{E_\gamma} \qquad \qquad \dots (10.23)$$

We have $x_0 = 330$ m for air and $x_0 = 0.52$ cm for lead.

The cascade or radiation length is the distance in which the energy of an electron is reduced to $\dfrac{1}{e}$ of its original value. Obviously, the electron loses its energy on the average $\left(1 - \dfrac{1}{e}\right)$ or 0.632 of its initial energy. Thus one obtains the mean energy loss per unit path length as

$$\frac{dE}{dx} = -\frac{E}{x_0} \qquad \qquad \dots (10.24)$$

This reveals that all energy-losses in a collision are equally probable. This is an important conclusion

(ii) Pair Production Now we consider the pair production by a high energy photon. The cross-section of the process in which one of the particles (electron or positron) has an energy E and the other has an energy $(E_\gamma - E)$ is given by

$$d\sigma_p = 4 Z^2 \, \alpha \, r_0^2 \, \frac{dE_\gamma}{E_\gamma} \left[\left\{ \left(\frac{E}{E_\gamma} \right)^2 + \left(1 - \frac{E}{E_\gamma} \right)^2 + \frac{2}{3} \frac{E}{E_\gamma} \left(1 - \frac{E}{E_\gamma} \right) \right\} \ln \frac{183}{Z^{1/3}} + \text{Smaller terms} \right] \qquad \dots (10.25)$$

We must note that the above expression is for complete screen and very high energies. Integrating over energy (10.24) yields

$$\sigma_p = Z^2 \, \alpha \, r_0^2 \left(\frac{28}{9} \ln \frac{183}{Z^{\frac{1}{3}}} - \frac{2}{27} \right) \qquad \qquad \dots (10.26)$$

We should note that quantity within parenthesis in (10.24) is order of unity and hence $d\sigma_p$ assumes the same form as Equation (10.20) for $d\sigma_{br}$. Obviously, the primary energy is divided with equal probability between the two particles. One can obtain the probability that one pair is formed in path length dx $\left(\text{or } dl = \dfrac{dx}{x_0}\right)$ by integrating (10.25) over E and multiplying by N,

$$dp = \frac{7}{9}\frac{dx}{x_0} = \frac{7}{9}dt \qquad\qquad \text{... (10.27)}$$

One obtains the decrease in intensity of a beam of photon due to production in dx as

$$\frac{dn}{n} = -\frac{dx}{\lambda_p}$$

or

$$n = n_0 \exp\left[-\frac{x}{\lambda_0}\right] \qquad\qquad \text{... (10.28)}$$

Where $\lambda_p = \left[\dfrac{28}{9}\left(Z^2 n \,\alpha\, r_0^2\right)\ln\dfrac{183}{Z^{\frac{1}{3}}}\right]^{-1}$ is the mean free path for the pair production.

Obviously

$$\lambda_0 = \frac{7}{9}x_0 \qquad\qquad \text{... (10.29)}$$

(iii) **Ionization** We have read that the size of the cascade shower grows until the energies of the component particles become too small. After achieving this the electrons and positrons begin to lose energy by ionization. The limiting critical energy (E_c) at which the cascading process stops is equal to the energy lost by the electron in one radiation unit of distance and is given by

$$E_c = x_0 \left(\frac{dE}{dx}\right)_{ion} = \left(\frac{dE}{dx}\right)_{ion} \qquad\qquad \text{... (10.30)}$$

However, the ionization loss per unit path $\left(\dfrac{dE}{dx}\right)_{ion} \alpha\, Z$ and $x_0 \,\alpha\, \dfrac{1}{Z^2}$. Obviously, $E_c \,\alpha\, \dfrac{1}{Z}$. One obtains the an approximate value of the critical energy as

$$E_c = 1600\,\frac{m_0 c^2}{Z} \qquad\qquad \text{... (10.31)}$$

We must note that E_c given above is of the same order of magnitude (slightly less) as the energy ε_c at which the rates of radiation loss and ionization loss become equal. For some materials, the values of E_c, ε_c and x_0 are given in Table 10.3.

Table 10.3 Values of E_c, ε_c and x_0 for some materials

Substance	Al	Fe	Cu	Pb	Air	Water
E_c (MeV)	52	25	22.4	7.0	98	111
ε_c (MeV0	60	30	25	10	120	150
x_0 (m)	0.097	0.0182	0.0147	0.0052	330	0.43

Air Showers   Several workers in different countries found the existence of cascade showers in the atmosphere, usually known as air showers. These were observed in 1938 by observing coincidences between two parallel horizontal counters.

Auger and others observed such coincidences even with the counter distance upto 300 m. Scientists believe that these extensive or large air showers are produced by very high energy particles, e.g. electrons, positrons and photons, which then falls down towards sea level with a particle density as high as $\Delta \sim 10^3$ per m^2.

It is shown theoretically that the denser showers with $\Delta \sim 10^3$ per sq.m. are created by electrons with an energy $\sim 10^{15}$ to 10^{16} eV, where as the dense showers ($\Delta \sim 10$ per sq.m) are generated by electrons of energy $\sim 10^{12}$ eV.

It is believed that the extensive air showers consist mostly of electrons (e^+ and e^-) and photons. Auger and others suggested that a small fraction of particles in these air showers may contain heavier penetrating particles, but no direct experimental evidence in support of this suggestion is there.

We must note that the cascade theory does not take into account the angular spread of shower particles. Cascade theory assumes that these shower particles proceed in the direction of the primary particles, which initiated the shower. It is felt that while considering the extensive air showers, the cascade theory needs modification to account the angular and spatial distribution of shower particles. It is believed that the angular spread is mainly due to Rutherford type elastic scattering of the particles (e^- and e^+). One finds that the root mean squared lateral spread $<X(E)>$ in radiation units and the root mean squared angular spread $<q(E)>$ are expressed by

$$<X(E)> = 0.80 \ \frac{E_s}{E} \qquad \qquad ... (10.32)$$

$$<\theta(E)> = 0.73 \ \frac{E_s}{E} \qquad \qquad ... (10.33)$$

Where E_s is a constant ~ 21 MeV and $E > E_s$. It is reported that the above expressions roughly agrees with the experimental observations.

Using the following empirical relation

$$n(E) = AE^x \qquad \qquad ... (10.34)$$

One can estimate the energy E (in eV) of the primary proton from total number n of shower electrons. Here $A = 1.35 \times 10^{-12}$ and $x = 1.14$.

Extensive Air Shower Production Mechanism   When a high energy cosmic ray particle, e.g. a proton or a heavy nucleus enter the atmosphere, it give rise to different types of nuclear events. This causes to produce high energy protons, neutrons, charged and neutral pions. Many of these particles created (e.g., p, $\pi^{\pm}$, nuclei) produce further nuclear interactions and give rise to many more such particles. These particles form a narrow bundle of particles known as the core of the shower. The core is about a few metres in diameter at the sea level.

The neutral pions decay very fast into two high energy photons and these photons in the shower case then start cascade electronic showers and give rise to a very large number of electrons and positrons. It is reported that some charged pions also decays into muons which thus originate from the shower core.

10.11 DISCOVERY OF MUONS

The study of the absorption curve of the cosmic rays in matter reveals that at the sea level the cosmic rays consist of a soft component and a hard or penetrating component. It is found that the soft component is more easily absorbed in matter (about 10 cm thick sheet of lead absorbs it completely). Soft component is mainly made up of electrons, positrons and photons produced in cascade showers produced in air by high energy electrons or photons. As stated earlier that a part of soft component may also consist of low energy heavier particles.

The penetrating or hard component is made up of new particles, known as muons. Muons are heavier than the electrons but lighter than protons. Anderson and Neddermeyer in 1938 established the identity of the hard component in cosmic rays. It was reported that the penetrating or hard components of the cosmic rays are not high energy electrons, positrons, or photons. This can be seen from the fact that they do not initiate any cascade electronic shower. Cloud chamber photographs reveal that the particles in the hard component are usually single, when they pass through a high Z-absorber, i.e. lead.

We have read that a high energy charged particle loses energy in matter either by ionization or by radiation. It is observed that at lower energies, the rate of ionization loss is relatively high and goes down rapidly with increasing energy. After attaining a maximum at an energy $\approx 2m_0c^2$, it again rises slowly. However, the rate of radiation loss (due to bremsstrahlung) becomes significant only at very high energies, much above the minimum of the ionization loss curve. This is important for the e^-'s and e^+'s only. One can neglect it for the heavier particles.

Now, we consider the ionization loss of three type of particles viz. electrons (mass m_e), singly charged particles (mass m_e) and protons (mass 2000 m_e). Figure 10.18 exhibit that the slow rise in the rates of ionization loss after the minima for these particles occur above the energies 2 MeV, 300 MeV and 3 GeV respectively. One can easily see that the rates of energy loss in the region (minimum loss) are more or less the same for all these three particles. Obviously, the thickness of the tracks of the particles in cloud chamber photographs for all of them should be almost the same. This means that one can not distinguish them from one another from the appearances of their tracks near the minimum ionization.

Out of the above three types of particles the rates of radiation loss can be neglected for the protons and for the particles of mass 200 m_e. Obviously, it is of importance only for electrons at these high energies, i.e. when the ionization loss rate is near the minimum value. This means the two heavier particles will be much more penetrating then the electrons at energies above the minima in the ionization energy loss curve.

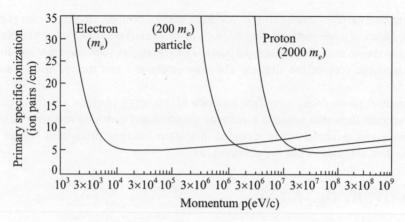

Fig. 10.18 Ionization loss of three type of particles.

Anderson and Neddermeyer in 1936 established that particles with masses 100 and 300 times m_e did indeed exist in the secondary cosmic rays. These particles were at first called *mesotrons*. They are now known as *mesons*, and, in fact, are the most abundant kind of *meson* found in cosmic rays at ground level the *mu-mesons*.

The specific ionisation produced by a very energetic charged particle is practically independent of its mass. Consequently, it is not easy to distinguish in cloud chamber studies between the types of track left by two very energetic particles of quite different masses. However, this difficulty can be overcome by a study of the specific ionisation produced by a particle which has been slowed down in an absorber in the chamber, in conjunction with measurments of the radius of curvature of its track in a strong magnetic field.

In several cloud chamber photographs produced to confirm the existence of the meson, only one given by Anderson and Neddermeyer is considered here. A lead plate 0.35 mm thick was placed in the chamber along its diameter and the chamber was operated vertically between the poles of a powerful electromagnet at the top of Pike's Peak in california. It was observed that after passing through the lead plate its range was about 4 cm in the chamber gas and the radius of curvature of its track was about 7 cm. Since this did not lose much energy in the lead plate it could not be an electron and must be heavier than the latter. If it is assumed to be a proton, then its energy could be determined from the measured range of 4 cm from which its momentum P and its magnetic rigidity $B\rho = \dfrac{pc}{e}$ could also be determined. Knowing the value of the magnetic field ($B = 0.79$ T), the radius of curvature determined was 0.2 m and not 0.07 m as observed. The direction of the track showed that the particle was positively charged. The particle, could not, however, be a proton because a proton which gave the value of $B\rho$ obtained would have had a much shorter range than that observed for the particle in question.

Soon after the discovery of meson, Street and Stevenson in 1937 showed conclusively that both positive and negative mesons of mass intermediate between that of an electron and a proton exist by arranging coincidence counters in conjunction, with a cloud chamber so that the mesons tracks were photographed near the ends of their ranges. As shown in Figure 10.19, a lead block 10 cm. thick was placed above the cloud chamber to absorb the unwanted soft component. A further lead plate was placed horizontally inside the cloud chamber along its diameter to increase the probability of particles being stopped within the chamber. Above the 10 cm block of lead, G.M. Counters C_1 and C_2 formed a coincidence telescope. G.M. Counters C_3 between thick block and the chamber detected those penetrating particles able to traverse 10 cm. of lead. Below the chamber was a triangular arrangement C_4 consisting of a band of counters in parallel. The expansion of the chamber was triggered by coincidence pulses from C_1, C_2 and C_3, but if C_4 also produced a pulse, i.e., the particle had emerged below the chamber, then this pulse was used to forbid an expansion. In this way, photographs were only taken of particles in the hard component near the ends of their ranges after traversing the lead plate in the cloud chamber. The expansion was delayed about one second to enable better estimates of the number of droplets formed to be made and so of the number of ions produced per centimetre of path.

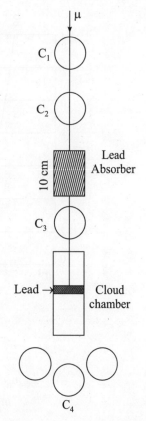

Fig. 10.14 Street and Stevenson's experiment to study a μ-meson track near the end of its range.

Slow-moving particles near the ends of their ranges will produce marked differences of specific ionisation depending on their masses. The particles photographed in Street and Stevenson's experiment had a specific ionisation in air of about six times that for an electron. The particles could not therefore be electrons. That they were not so massive as the proton was established by measuring their ranges and their radii of curvature in a known magnetic field. In one of the photographs obtained by them, the ionisation density was reported to be six times the minimum and the magnetic rigidity was 9.6×10^{-2} T-m. When they compared these values with those for particles of different masses (singly charged), they found that these fit well with the values for a particle of mass $\sim 200\ m_e$. One can understand the principle by referring to Figure 10.20. Figure 10.20 exhibit the plot of the ratio $\dfrac{I}{I_0}$ of the ionization density I to the maximum density I_0 versus BR for particles of different masses. One can easily see that the curves are well resolved at relatively lower values of BR. One finds the values of $\dfrac{I}{I_0}$ for two particles of masses 200 m_e and 300 m_e for magnetic rigidity $B\rho = 10^{-1}$ T-m as 7 and 13 respectively. Obviously, one can easily establish the identity of the particle without any ambiguity at such low values of BR.

Recent determinations of the rest mass of μ-meson by Fretter and others by careful measurments using cloud chambers gave the value $m_\mu = 207\ m_e$ for both $\mu^\pm$. The muonic mass has been determined fairly accurately by measuring the energy of the transition $2D_{\frac{5}{2}}$ to $2P_{\frac{1}{2}}$ in a muonic atom. The widely accepted values of muon rest mass in energy units is

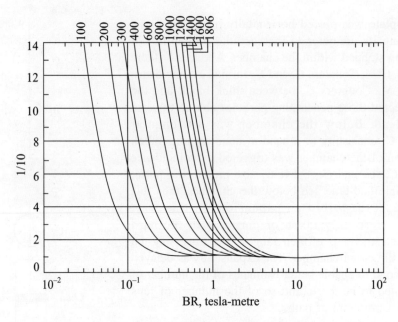

Fig. 10.20 $\dfrac{I}{I_0}$ as a function of $B\rho$.

$$m_\mu c^2 = 105.65943 \pm 18 \times 10^{-5}\ \text{MeV}$$

$$m_\mu = (206.7686 \pm 0.0005)\ m_e$$

10.12 PROPERTIES OF THE μ-MESON

We have seen that both positive and negative μ-mesons exist (μ^+ and μ^-). These are most stable particles. A muon decays by the emission of an electron (or a positron) and two neutrinos (one neutrino and one antineutrino). The mean life of decay is about 10^{-6} s. The decay scheme for the μ-meson is represented by

$$\mu^+ \rightarrow e^+ + \nu + \bar{\nu}$$

$$\mu^- \rightarrow e^- + \nu + \bar{\nu}$$

Mean life (τ_μ)   The mean life of the μ-meson has been measured in a number of experiments. A typical arrangement used by Rossi and Nereson in 1942 is shown in Figure 10.21. They measured the intensity of the hard component of the cosmic rays in the vertical direction with the help of a counter telescope and also at various places at different altitudes above the sea level. Considering that muon were stable, there would be a diminution in the intensity from the higher altitudes to the sea level due to the absorption of muons in the air mass between two places. Let us consider that a solid absorber (i.e. graphite) with a thickness equivalent to the air mass between the place of higher latitude and the sea level is introduced above the counter telescope at the higher altitude, then the intensities should be equal at the two places if the muons were stable. However, actual experimental measurments revealed that even with such modification, the intensities were not equal at the two places. They found that the intensity at the sea level was lower than at the higher altitude place with the graphite absorber placed above the counter telescope.

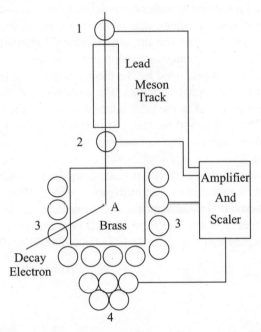

Fig. 10.21 Measurment of the mean life of the μ-meson by Rossi and Nereson (1942).

Rossi and coworkers interpreted the above result as being due to the radioactive decay of the muons as they come down from the places of higher altitude to the sea level. They found that the vertical intensity was reduced by about 10 % with the graphite absorber (thickness 870 kg/m^2) placed above the counter telescope. On the other hand they found that the intensity diminises near sea level by about 20 % for the same air mass between the two places having an equivalent thickness of 870 kg/m^2 of graphite. From these observations they estimated $\tau_\mu \sim 2\,\mu s$ (2×10^{-6} s).

We must note that the measured values of tm need relativistic correction, because of the very high velocity with which the muon travel. In accordance with the special theory of relativity, we have

$$\tau_\mu' = \frac{\tau_\mu}{\sqrt{1-\beta^2}} \qquad\qquad \ldots (10.35)$$

Where τ_μ is the mean life of the muon in the rest frame and $\beta^2 = \dfrac{v^2}{c^2}$. Obviously, the mean life of muon in the rest frame quoted above is thus

$$\tau_\mu = \frac{\tau_\mu'}{\sqrt{1-\beta^2}} \qquad\qquad \ldots (10.36)$$

Measurment of τ_μ by Gupta and Bhattacharya at an altitude 2330 m and near sea level yield a value of 2.5 μs.

Nereson and Rossi using more accurate electronic methods measured τ_μ by a coincidence-anti coincidence device using arrays of G.M. Counters above and below a brass plate within which the low energy cosmic ray muons suffered radioactive decay to produce e^-'s (or e^+'s) which entered arrays of G.M. Counters placed on either side of the brass plate. It is reported that the higher energy muons which did not decay in the brass plate entered another gruop of G.M. Counters placed below the top counters to produce anti coincidence pulses.

The delayed electrical pulses produced by the electrons give rise to coincidence pulses along with the pulses generated by the decaying muons. The amplitudes of these pulses obtained from a time circuit were found proportional to the time intervals between the pulses from the muons and the delayed pulses from the muons and the delayed pulses from the decay electrons. In this way they obtained the time-distribution of the radioactive decay of the muons in the brass plate and from which they obtained τ_μ.

Very recently Astbury et al. using the $\pi - \mu$ beam from an accelerator which after collimation produced coincidence pulses in the plastic scintillators S_1 and S_2 (Figure 10.22). The arrival of μ mesons is signalled by these coincidence pulses. The scintillator S_3 detects the subsequent decay of the μ meson in the target. The average time interval between the arrival and decay of muons provide τ_μ. Presently accepted value of τ_μ is $(2.19714 \pm 0.00007) \times 10^{-6}$ s.

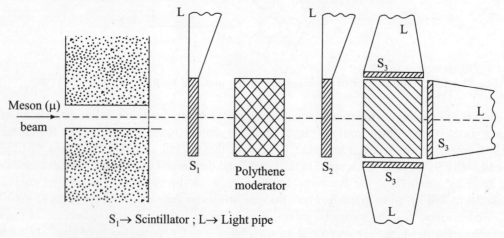

$S_1 \rightarrow$ Scintillator ; $L \rightarrow$ Light pipe

Fig. 10.22 Measurment of the mean life of the μ-meson by Astbury et al.

Spin and Magnetic moment

The studies on the frequency of cosmic ray brusts at high energies indicated that s_μ was probably $\dfrac{1}{2}$ in the unit of $\hbar$. This has been confirmed by the observations on muon production in π-meson (pion) decay. The direct determination of the pion spin yield its value to be zero. The decay scheme for pions are

$$\pi^+ \rightarrow \mu^+ + 2\nu$$
$$\pi^- \rightarrow \mu^- + 2\nu$$

As the spin of the neutrino is $\dfrac{1}{2}$ and hence the spin of the muon (s_μ) is also $\dfrac{1}{2}$.

We know that Dirac's electron theory predicts the magnetic moment to be equal to the Bohr magneton (μ_B), i.e.

$$\mu_e = \mu_B = \frac{e\hbar}{2m_e} \qquad \qquad \text{... (10.37)}$$

Taking into consideration the interaction of the electron with its own radiation field (virtual emission and reabsorption of photons in the electric field of the electron) the above value of m_e is slightly modified as

$$m_e = 1.001160\ \mu_B \qquad \qquad \text{... (10.38)}$$

This has been confirmed experimentally by Lamb and Rutherford.

If the muon is a Dirac particle like the electron, then its magnetic moment should be

$$\mu_\mu = \frac{e\hbar}{2m_\mu} \qquad \qquad \text{... (10.39)}$$

Obviously, electron mass m_e has been replaced by muon mass. This is less than μ_B by the factor $\dfrac{m_e}{m_\mu} \sim \dfrac{1}{207}$. Let us suppose that muon does not obey Dirac equation, then μ_μ should be different from the value given by Equation (10.39). Obviously, the difference, if any, should reflect the effect of some new type of interaction (other than radiation reaction) involving the muon just as the anomalous magnetic moments of the proton and the neutron from the expected value of a nuclear magneton $(\mu_N = \dfrac{e\hbar}{2M_P})$ reflects the effect of the interaction involving the virtual emission and reabsorption of the pions in the strong field of the nucleon.

Theoretical calculations considering the effect of radiation reaction on the muon magnetic ratio $g_\mu = 2 \times 1.0011654$. Using this value of g_μ, one obtains muon magnetic moment as

$$\mu_\mu = 1.00111654\ \frac{e\hbar}{2m_\mu}$$

μ_μ has been determined precisely by a method based on the quantity $\dfrac{(g_\mu - 2)}{2}$. Under this method the change in the angle between s_μ and p_μ (muon momentum) is measured when the muon passes through a magnetic field B. The muon's cyclotron frequency in this field is given by

$$\omega_c = \frac{Be}{m_\mu}$$

The Larmor frequency of precession of a particle of magnetic moment $\mu = g_{\mu \cdot s}$ in this field is obtained as

$$\omega_L = \frac{g_\mu\, Be}{2m_\mu}$$

If it is a Dirac particle, then one finds $q_\mu = 2$ and hence $\omega_C = \omega_L$.

When the above particle travels through a magnetic field, one finds that the orientation of s w.r.t. p (momentum in the orbit) remains unchanged. Now, for $q_\mu > 2$, $\omega_L > \omega_C$, it is obvious that the angle between s_μ and p_μ will change continuously. This is obtained as

$$\phi = (\omega_L - \omega_C)t = \left(\frac{g_\mu - 2}{2}\right)\frac{Bet}{m_\mu} \qquad \qquad \text{... (10.40)}$$

We have $t = \dfrac{2\pi n}{\omega_c} = \dfrac{2\pi m_\mu n}{Be}$, where t is the time for n cyclotron revolutions. Obviously, the value

$$\text{... (10.41)}$$

The value of $\dfrac{(g_\mu - 2)}{2}$ was determined in an experiment by noting the change in the orientation of s_μ for a given number of revolutions n in a magnetic field.

Recent experiments on the measurment of μ_μ yield

$$m_\mu = (1.00116616 \pm 31 \times 10^{-8})\frac{e\hbar}{2m_\mu}$$

This value is in good agreement with the theoretically expected value.

One obtains the precise value of muon mass as

$$m_\mu = 206.767 \, m_e$$

10.13 INTERACTION OF MUON WITH MATTER

The highly penetrability of the cosmic ray muons exhibit that these must interact very weakly with the matter. This observations on cosmic ray muons contradicts the proposition of Yakawa that the muons are the quanta of the internucleon field. To investigate this point further Conversi et al. performed an experiment in which radioactive disintegration of positive and negative muons in different substances were studied.

These scientists separated the positive and negative muons with the help of a magnetic lens and observed their behaviours in iron ($Z = 26$) and carbon ($Z = 6$). They reported that almost all the negative muons were stopped in iron while none in carbon. Obviously, in the latter case they mostly undergo radioactive decay. Such type of difference in behaviour in the latter case is expected because of the Z^4 dependence of the muon capture probability by the nuclei. No such difference in the case of positive muons in the two absorbers was reported.

These observations clearly reveal that there was no strong interaction between the muons and the nuclei of the atoms. In case, if there were any such strong interaction between the muons, then μ^- would be captured in the nuclei of both higher Z(Fe) and lower Z(C) elements with in $\sim 10^{-18}$ s. Moreover, no difference in the behaviours of μ^- in the two absorbers would take place.

Experimentally, it was observed that μ^- actually rotates in the K-orbit of the μ-atom at C at least for 10^{-6} s, obviously, the probability of strong interaction between the muon and the nucleus is less than 10^{-12} of that expected from Yukawa theory.

The above experimental evidence clearly reveal that mesons are different from the Yukawa particles, which being the quanta of strong internucleon field and interact strongly with nuclei.

10.14 DISCOVERY OF THE PI-MESON

The evidence summarised in 10.12 shows that the μ-meson and the Yukawa meson are not the same. This problem was resolved by the discovery of a different type of meson- the π-meson in the upper atmosphere in 1947.

Using the sensitive nuclear research emulsions produced by Ilford Ltd, Perkins (1947), and also Occhialini and Powell (1947), observed in exposures at high altitudes, the tracks of charged mesons which came to rest in the emulsion. From counts of the number of silver grains produced and a study of the deviations in the trajectories of these particles, they were estimated to have masses in the range 200 to 300 m_0. Furthermore, about 10 % of these mesons disintegrate nuclei in the emulsion to produce protons, a particles, etc. Soon afterwards Lattes et al. in 1947 showed that about 10 % of the mesons, when stopped in the emulsion, emitted a second meson. This second meson was always ejected with constant velocity. The explanation was that a parent meson the π-meson lost part of its rest mass in decaying spontaneously into a μ-meson

Further investigations established the existence of both posite and negative π-mesons. They are products of highly energetic nuclear reactions in the upper atmosphere and decay in flight, the π^+-meson producing a μ^+-meson and the π^--meson a μ^--meson. Recent work on the π-meson gives the mass of both kinds as $(273 \pm 0.2)\ m_0$, where m_0 is the rest mass of the electron, and the mean life as $(2.6 \pm 0.1) \times 10^{-8}$ s. The decay schemes are

$$\pi^+ \to \mu^+ + 2\nu$$
$$\pi^- \to \mu^- + 2\nu$$

The energy released in each decay is 33.7 MeV of this energy, a constant amount of 4 MeV appears as the kinetic energy of the μ-meson, which has a corresponding constant range in a nuclear emulsion of 600 microns approximately. The emission of two neutrinos, which do not produce a track, is postulated to conserve energy and momentum.

Using the specially prepared electron-sensitive emulsions originally produced by Berriman in 1948, the successive decay of a π^+-meson into-firstly, a μ^+-meson, and, finally a e^+, has been recorded (Figure 10.23). The mean life of the decay of the π-meson was estimated to be between 10^{-6} to 10^{-11} s. Later measurments yields a

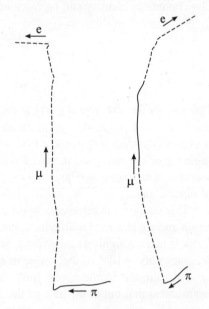

Fig. 10.23 Examples of the successive decay $\pi \to \mu \to e$ (a diagrammatic presentation).

value of about 2.6×10^{-8} s. The muon produced in the decay of a pion ($\pi^{\pm}$) has a definite range which corresponds to a kinetic energy of about 4.2 MeV.

We must note that pions are usually not observed in cosmic rays at sea level. It is reported that they are produced in nucleon-nucleon collisions in the upper regions of the atmosphere. Their mean life is much shorter and hence they disintegrate in flight, producing a muon while down through the air. It is these muons which have been detected as the hard component of the cosmic rays.

10.15 ORIGIN OF COSMIC RAYS

We have seen that cosmic rays are reported to come to the earth from all directions with the same intensity. The intensity of cosmic rays remains practically constant in time. Minor periodic variations associated with the solar day and year and with the reasons have been reported.

Various attempts to understand the origin of cosmic radiation have led to a realization that this phenomenon is by no means a minor extraneous occurrence in the scheme of things but plays an essential role in the dynamics of the galaxy and perhaps the universe.

It is reported that cosmic rays of upto 10^{14} eV are *isotropic* to at least within 1 part in 10^3. An obvious way of ensuing this would be to assume that the cosmic rays fill the whole universe with uniform density. However, one is immediately faced with a serious problem. However one is immediately faced with a serious problem. One finds that the kinetic energy of the nuclear component of cosmic rays in out neighbourhood is ~ 1 eV/cm^3, which is of the same order as the energy density of starlight.

The energy density can be calculated from the following relation

$$e \simeq \frac{4\pi \times 2.35}{c} \left[\int_1^{\infty} \frac{dE}{E^{2.67}} \times E - \int_1^{\infty} \frac{dE}{E^{2.67}} \right]$$

$$\simeq 10^{-9} \text{ BeV/cm}^3 \qquad \qquad \dots (10.42)$$

This energy density over the whole universe would require that the total energy in cosmic rays be many orders of magnitude greater than the total energy in thermal photons. However, the situation is even worse than it appears at first sight. If the high-energy cosmic ray gas has existed since the begining of the universe, it must have been subject to adiabatic cooling due to expansion of the universe, thus requiring a large fraction of the initial energy of the universe to have been in the form of cosmic rays.

This situation is relieved a good deal if one assumes that cosmic rays are mainly galactic in origin and are confined within its volume by large scale magnetic fields in the galaxy. If the storage factor is taken roughly as the inverse of the degree of anisotropy, the power going into cosmic rays becomes only $\sim 10^{-4}$ of the power in starlight.

An estimate of the lower limit on the magnetic field strength may be obtained from the requirement that out in the halo of the galaxy, where it is not well anchored in thermal plasma, the field be able to support the pressure of the cosmic-ray gas. Thus setting $\dfrac{B^2}{8\pi} \simeq \dfrac{\varepsilon}{3}$, where $\varepsilon = 1$ eV/cm$^3 = 1.6 \times 10^{-19}$ J/cm^2. One obtains $B = 3.5 \times 10^{-10}$ W/m^2.

Existence of magnetic fields of the order of 1 gamma was originally reported by the observation that the light from many red stars is partially plane polarized. The reddening is produced by scattering from dust clouds in the way. In the presence of magnetic fields the dust grains may be partially aligned, in which case the component of light whose electric vector is parallel to the longer axis of the grain would be preferentially weakened.

Radio astronomy provides a significant information about the magnetic field structure. General radio emission from the galaxy has the character of synchrotron radiation produced by high-energy electrons circulating in magnetic fields. Actually this observation led to the discovery of the radio halo of the galaxy, indicating that magnetically the galaxy is almost spherical. This is of great importance in extending the energy range upto which cosmic-ray particles may contained in the galaxy. Assuming a reasonable source strength and uniform spacial distribution for cosmic ray electrons, it is estimated that the field is ~ 2 gamma in the spiral arms and ~ 0.3 gamma in the halo. Magnetic fields of this order are also indicated by measurments of Faraday rotation of the plane of polarisation of radiowaves from extragalactic radio sources and upper limit of 0.5 gamma is obtained from the null result of an attempt to measure the Zeeman splitting of the 0.21 m line of the neutral hydrogen.

The intersteller space is also filled with plasma in which the magnetic fields are presumably embedded. We know that there exist probably a generic relationship between the plasma and the magnetic fields. It is obvious that wherever there are turbulent plasma, on the surface of the sun for e.g., magnetic fields are generated which are then frozen in the plasma and move with it with all their inhomogeneties. Ultimately, there should arise an equipartition of energy between the plasma and magnetic fields. This must also hold for the interstellar plasma. Obviously, the kinetic energy density of the plasma is also of the order of ~ 1 eV/cm^3. This reveals that the cosmic-ray energy density is also of the same order may not be accidental but may indicate a slow process in which the energy of the clouds is imparted to a few preferred particles in a statistical manner. Fermi in 1949 reported the existence of such a process. In principle this process can provide an effective mechanism for damping the energy of plasma clouds; a fraction of high-energy cosmic ray particles, after soaking up some of their energy, may leak out of the galaxy.

Possible Sources of Cosmic rays It has been known for a long time that the sun can accelerates particles, and sometimes it has been seriously suggested that most cosmic rays may be of solar origin and their intensity may be built up over long periods by storage around the planetary system. However, now one can give several obvious arguments against such a possibility. Solar particle spectra have been found to very steep, the diffusion times are not very large, solar-particle events are associated mostly with visible solar flares (including high anisotropy in diffusion), and the chemical composition of solar rays is different from that of galactic cosmic rays. Actually one can go further and say that even if all the sunlike stars in the galaxy were to produce radiation like the sun, they would not account for the observed galactic radiation. If there is subsequent acceleration in the intersteller space, the energy density of radiation would be too small by several orders of magnitude, and the spectrum would be far too steep. If this were patched up by some acceleration process subsequent to emission, the chemical composition would still remain a problem. As it is, the relative abundance of heavy elements in solar rays is less than observed for galactic cosmic radiation; further acceleration and traversal through additional amount of matter would make this discrepancy even wider. A few years ago ordinary stars could not be ruled out as injectors, because there was the possibility that during acceleration a fractional may occur favouring the emission of heavier elements. This has now been excluded.

Therefore for the supply of initial "ions" to be accelerated one has to look to sources that are comparatively richer in heavy elements and, in order to dominate injection of material from ordinary stars, extraordinary active. Such sources are noval and supernoval, red giant stars, and perhaps the magnetic stars of Babcock which are relatively newly condensed from intersteller gas and hence may be richer in heavy elements.

Particles from these sources may be injected into intersteller space after a short primary acceleration when they may gain energy by a slow statistical process, or they may be accelerated by almost their full energy in the source region itself, through a combination of inductive, statistical, and shock processes.

Acceleration Mechanisms   It is believed that the charged particles in primary rays acquire high energies due to the acceleration imparted to them by the electromagnetic fields of the stars and the sun. This acceleration of the charged particles must take place gradually otherwise, if the energies upto 10^{12} eV possessed by the heavy and superheavy nuclei in the primary radiation obtained at once, the nuclei would disintegrate into their constitutent nucleons because the binding energy of the nuclei would be insufficient to hold them together under such conditions. Gradual acceleration of the primary cosmic rays may take place similar to the acceleration of particles in the Betatron. However, we have no means of knowing whether such magnetic fields do exist in space or not.

In 1949 Fermi suggested that acceleration of cosmic rays may occur by collisions of particles with randomly moving magnetic clouds in inter steller space. Some collisions with approaching fields are more frequent than overtaking collisions, there would be a slow gain in energy. It is obvious that the total energy gain would depend on the time a particle had spent in the galaxy, as well as on the time rate of gain of energy. Obviously,

$$\frac{dE}{dt} = \bar{\alpha}(E)E \qquad\qquad \ldots (10.43)$$

and the number of particles of total energy E decreases with a mean life $\tau(E)$,

$$\frac{dN(E)}{dt} = -\frac{N(E)}{\tau(E)} \qquad\qquad \ldots (10.44)$$

Then the integral spectrum would have a slope

$$\gamma(E) = \frac{E\left[\dfrac{dN(E)}{dE}\right]}{N(E)} = \frac{1}{\bar{\alpha}(E)\,\tau(E)} \qquad\qquad \ldots (10.45)$$

If $\bar{\alpha}(E)$ and $\tau(E)$ have a weak dependence on E, the spectrum is a power law. The energy gain being proportional to the total energy, this process discriminates strongly against electrons electrons and protons are injected at comparable kinetic energies. Recent values of gas-cloud velocities and fields do not seem to yield a value of exponent in agreement with experiment. The mechanism would not be effective for heavy nuclei unless they are injected with high enough energies, because ionization loss dominates over energy gain. Obviously, the Fermi mechanism cannot operate fro any particle unless there is a minimum starting energy; this may either be provided by a nonstistical process or more likely a form like a supernova.

Now it is clear that the whole of the cosmic-ray acceleration need not occur in the intersteller space. The actual acceleration may be a hierachical process starting, for example, with an explosion in supernova, followed by a statistical and inductive process in its envelope, and later by the statistical process in the galactic space. The relative importance of these in different stages and some of the other related problems have been discussed recently by Hayakawa et al (1964).

Energetics of Supernovae as Cosmic ray Sources   Most of our present knowledge about supernova explosions and their capacity for generating high-energy particles come from the study of the Crab nebula (Taurus A) which was formed by a supernova. It is reported that light from the Crab nebula is strongly polarized indicating magnetic fields of the order 10^{-8} T to 10^{-7} T. In addition, it is found to be a strong source of highly polarized radio waves and X-rays. Radio and light spectrum has the character of synchrotron radiation emitted by electrons whose energies extend to $\sim 10^{12}$ eV. It is suggested that a system that can accelerate electrons should be even more efficient in accelerating nuclei. However, Hayakawa et al. (1964) have wondered whether the rather high injection energies needed near the centre of the star in the explosion stage would not lead to nuclear interactions and fragmentation of a large number of heavy nuclei, thus partly invalidating the very arguments that suggested a preferred role for supernovae as cosmic-ray sources. So far there is no report for direct evidence for the presence of high-energy neutrons in the Crab nebula. This may be provided by observation of ~ 100 MeV γ-rays or of high-energy neutrinos which can only arise from interactions of nuclear particles.

Total cosmic ray energy in the galaxy is 3×10^{68} cm$^3 \times 1$ eV/cm^3) ; 5×10^{49} J. If life time is considered as 2×10^6 years, the power of cosmic ray source is $\sim 10^{36}$ J/s. This power may be reduced by about a factor of 100 (to $\sim 10^{34}$ J/s) if it is assumed that most of the cosmic-ray nuclei are stored in the halo region where the gas density is $\sim 3 \times 10^2$ p/cm^3, thus increasing the cosmic-ray life time to 10^8 years. The power of Crab nebula for generating relativistic electrons is estimated as 10^{29} to 10^{32} Joules/s; hence that for generating protons may be 10^{32} to 10^{35} J/s. The life time of supernova is ~ 3000 years. Thus the total cosmic ray power of all supernovae may be 10^{33} to 10^{36} J/s. This seems to be adequate when compared with the observed power of 10^{34} to 10^{36} J/s. However, it is found that the total cosmic ray energy release required of a supernova comes uncomfortably close to its total energy release in all forms of radiation. On the other hand, all these arguments have only a qualitative significance, particularly because of the increasing evidence in recent years suggesting that the Crab is not a typical supernova.

Cosmic rays from Extragalactic Region   All radio stars and radio galaxies must have relativistic electrons and hence be capable of producing cosmic rays. Following this argument, Fujimoto et al. (1964) have estimated that there may be an extragalactic component. Since the hirerachy of acceleration processes is likely to be the same in all turbulent plasma, the shape of the extragalactic energy spectrum processes is likely to be the same in all turbulent plasma, the shape of the extragalactic energy spectrum may be similar to that of the galactic spectrum. However, while the galactic spectrum may be limited at the high-energy end essentially by the size of the accelerating and/or containment volumes, to about 10^{15} eV, the extragalactic spectrum may extend to much higher energies.

Peters (1959) first suggested this general picture of primary cosmic-ray spectrum. He showed that the more or less continuous shape for the number of air showers up to energies of the order of 10^{17} eV does not exclude the possibility that galactic radiation is subject to a rigidity cutoff at $\sim$

10^{15} eV for protons, because for the same rigidity a heavy nucleus of mass number A has $\dfrac{A}{2}$ times the total energy of a proton. Beyond 10^{17} eV a flatter extragalactic spectrum may come in.

Bray et al (1964) supported the above proposal and said that a large majority of showers of size greater than 10^6 are produced by heavy nuclei. Pal and Tandon (1965) has also shown that the muon spectrum upto 10^{13} eV and γ-ray spectra at various altitudes upto 10^{13} eV can all be understood in terms of a rigidity steepening starting around 2×10^{14} eV, if in addition to the main spectrum there exists another one, with an intensity of about 5 %, having the same slope as the first one ($\gamma \sim 1.7$) and continuing upto much higher energies. This second spectrum which may or may not extend down into energies below $\sim 10^{12}$ eV, could then have an extragalactic origin and may have a different chemical composition.

Concluding one can say that the primary cosmic rays acquire their energies in the vicinity of magnetically active stars, especially supernovae. This is supported by the observations on radio stars which show intense radio noise due to very fast electrons moving in magnetic fields suggesting that cosmic rays may also be associated with steller events of great violence. Since the cosmic rays are pushed about in all directions by these great belts of steller magnetic fields, in which they undergo multiple reflections and changes of direction, they surround the earth isotropically so that the earth can be regarded as a simple body in a whole sea of cosmic rays.

10.16 FUTURE OF COSMIC RAY RESEARCH

The world wide programme of rocket and setellite now being carried out will enable to explore fully the cosmic ray intensity variations in the space immediately surrounding the earth and perhaps to explain some of the variations in the earth's magnetic field.

The very high energies associated with cosmic rays at high altitudes and the case with which the atomic nucleus is penetrated enables cosmic ray research workers to make valuable contributions to our knowledge of the structure of the nucleus.

Perhaps the most exciting sphere of work with cosmic rays is the research for new particles. Since 1932 when the positron was discovered by Anderson in cosmic rays many strange particles often with surprising properties have been discovered. These elementary particles are playing a large part in our search for the details of nuclear and nucleonic structure.

Example 10.1 From the information given in Table 10.2, calculate the intensity of various primary particles in microwatts/m^2 unit solid angle at 10 GeV per nucleon. How would you expect this to vary with latitude (protons 0.46 microwatts)?

Sol. Considering proton only, from table 10.2; the flux is given by $\dfrac{4000}{E^{\frac{8}{7}}}$ paritcles/m^2-S unit solde angle where E is in GeV.

$$\therefore \qquad \text{Flux} = \frac{4000}{10^{\frac{8}{7}}} = 290 \text{ particles per unit solid angle with 10 GeV energy}$$

$\therefore$ Energy intensity $= 290 \times 10^4 \times 1.6 \times 10^{-13}$ J/m²S unit solid angle

$= 0.46$ micro Watts/m² unit solid angle

One can make similar calculations for other particles.

Example 10.2 A 1 MeV positron encounters a 1 MeV electron travelling exactly in opposite directions. Determine the wave lengths of the photons produced? Given, rest mass of electron or positron = 0.512 MeV.

Sol. Total K.E. of electron and positron = 2 MeV

Rest mass of electron + positron = 0.512 + 0.512 = 1.024 MeV

Total energy = 2 + 1.024 = 3.024 MeV

The total energy is equally shared by the two photons. Therefore energy of one photon $= h\nu$

$$= \frac{3.024}{2} \text{ MeV} = \frac{3.024 \times 10^6 \times 1.6 \times 10^{-19}}{2} \text{ J}$$

or $\qquad\qquad \nu = \dfrac{3.024 \times 10^6 \times 1.6 \times 10^{-19}}{2 \times 6.625 \times 10^{-34}}$ Cycles/s

Now $\qquad\qquad \lambda = \dfrac{c}{\nu} = \dfrac{3 \times 10^8 \times 2 \times 6.625 \times 10^{-34}}{3.024 \times 10^6 \times 1.6 \times 10^{-19}}$ m

$$= 8.2 \times 10^{-13} \text{ m}$$

Example 10.3 Prove that in Anderson's experiment on the discovery of positron, if the particle emerging from the lead plate be a proton having a momentum 23 MeV/c, then its kinetic energy should be 0.3 MeV. If the emergino particle be a positron calculate its kinetic energy.

[Kerala]

Sol. The kinetic energy E_k is related to the momentum p by the relation

$$E_k = \sqrt{p^2 c^2 + m_c^2 c^4} - m_0 c^2$$

Here, $m_0 c^2$ for proton = 938 MeV and pc = 23 MeV.

$\therefore \qquad\qquad E_k = \sqrt{23^2 + 938^2} - 938$

$$\simeq 938.3 - 938$$

$$= 0.3 \text{ MeV}$$

For positron $m_0 c^2 = 0.51$ MeV. So the kinetic energy E_k is given by

$$E_k = \sqrt{23^2 + (0.51)^2} - 0.51$$

$$\simeq 23 - 0.5$$

$$= 22.5 \text{ MeV}$$

Example 10.4 The average amount of intersteller matter through which cosmic ray nuclei pass from the point of origin to that of observation is 2.5 g cm⁻². If the density of intersteller space between the above two points be 1 hydrogen atom/cc, find the life time of cosmic ray nuclei. Treat the nuclei relativistically.   [Mysore]

Sol. The density of intersteller space, $\rho = 1$ H-atom/cc $= 1.67 \times 10^{-24}$ g/cc

Velocity of cosmle ray nuclei, $c = 3 \times 10^{10}$ cm/s

Avergae amount of matter traversed, $x = 2.5$ g $\times$ cm^2

Let t be the required lime time. Then, one obtains, $x = \rho c t$

$$\therefore \quad t = \frac{x}{\rho c} = \frac{2.5}{1.6 \times 10^{-24} \times 3 \times 10^{10}} \text{ s}$$

$$= \frac{2.5}{4.8 \times 10^{-14} \times 3600 \times 24 \times 365} \text{ yr}$$

$$= 1.58 \times 10^6 \text{ years}$$

Example 10.7 Muon decays into an electron according to the following reaction:

$$\mu^- \rightarrow \beta^- + \nu_\mu + \nu_e$$

The electron is ejected with relativistic energy. Compute the maximum energy that may be available. What is the average electron energy?

Sol. Mass of the muon $= 207\ m_e$, where $m_e \rightarrow$ electron mass.

$\therefore$ Maximum available energy, $Q = (207 - 1)m_e = 206\ m_e$,

$$= 206 \times 0.51 = 105 \text{ MeV}$$

$$(\because\ m_e = 0.51 \text{ MeV})$$

Average electron energy, assuming that Q is divided equally amongst the particles, is

$$E_e = \frac{Q}{3} = \frac{105}{3} = 35 \text{ MeV}$$

Example 10.8 A pion decays from rest to give a muon of 3.57 MeV energy. Find the kinetic energy of the associated neutrino. Also find its mass.   [Mysore]

Sol. The decay of pion is the following: $\pi^\pm \rightarrow \mu^\pm + \nu_\mu + Q$

$$\therefore \qquad Q = \text{mass of } \pi - \text{mass of } \mu$$

$$= (273 - 207)m_e;\ \text{where } m_e = \text{electronic mass}$$

$$= 66\ m_e = 66 \times 0.51 \text{ MeV} \quad (\because\ m_e = 0.51 \text{ MeV})$$

$$= 33.66 \text{ MeV}$$

Now the energy of muon, $E_\mu = 3.57$ MeV

$\therefore$ Energy of neutrino, $E_\nu = (33.66 - 3.57) = 30.09$ MeV

$$\therefore \qquad \text{Mass of neutrino} = \left(\frac{30.09}{0.51}\right)m_e = 59\ m_e$$

Example 10.9 A positron collides head-on with an electron and both are annihilated. Each particle had a kinetic energy of 1 MeV. Find the wavelength of each of the resulting photons. [Bangalore]

Sol. The collision process may be expressed as: $e^- + e^+ \rightarrow \gamma + \gamma$

Now;   rest mass of e^- = 0.51 MeV

rest mass of e^+ = 0.51 MeV

kinetic energy of e^- = 1.00 MeV

kinetic energy of e^+ = 1.00 MeV

$\therefore$ Total energy = 3.02 MeV

As γ-photons have no rest energy, energy of each photon is $E = \dfrac{1}{2} \times 3.02 = 1.51$ MeV. But

$E = h\nu = \dfrac{hc}{\lambda}$, where ν = frequency and λ = wavelength.

$\therefore$
$$\lambda = \frac{hc}{E} = \frac{6.62 \times 10^{-34} \text{ J.s} \times 3 \times 10^8 \text{ m.s}^{-1}}{1.51 \times 10^6 \times 1.6 \times 10^{-19} \text{ J}} 1$$
$$= 8.22 \times 10^{-13} \text{ m}$$

Example 10.10 Calculate the energy of neutron produced when a slow negative pion is captured by a proton. Treat the neutron non-relativistically. [Punjab]

Sol. The process may be expressed as:
$$\pi^- + \pi^+ \rightarrow n^\circ + \gamma + Q$$

Substituting the masses of different particles in MeV unit, one obtains
$$Q = M_\pi + M_p - M_n$$
$$= (273 \times 0.51 + 938 - 939) \text{ MeV}$$
$$= (139 + 938 - 939) \text{ MeV} = 138 \text{ MeV}$$

$\therefore \qquad E_\gamma + E_n = Q = 138 \text{ MeV}$

For a non-relativistic neutron, we have, from the conservation of momentum
$$M_n \upsilon_n = \frac{E_\gamma}{c}$$

$\therefore$
$$\frac{E_n}{E_\gamma} = \frac{\dfrac{1}{2} M_n \upsilon_n^2}{c M_n \upsilon_n} = \frac{\upsilon_n}{2c} = \frac{E_\gamma}{2M_n c^2}, \text{ using the above relation.}$$

$\therefore$
$$\frac{E_n}{E_n + E_\gamma} = \frac{E_\gamma}{2M_n c^2 + E_\gamma}$$

But $2M_n c^2 = 2 \times 939 = 1879$ MeV; $E_\gamma = (138 - E_n)$ MeV and therefore

$$\frac{E_n}{138} = \frac{138 - E_n}{1878 + 138 - E_n} = \frac{138 - E_n}{2016 - E_n}$$

$$\therefore \qquad 2016E_n - E_n^2 = 138^2 - 138E_n$$

or, $\qquad E_n^2 - 2154E_n + 1382 = 0$

$$\therefore \qquad E_n = \frac{2154 \pm \sqrt{2154^2 - 4 \times 138^2}}{2} = \frac{2154 \pm 2136}{2}$$

$$\therefore \qquad E_n = 9 \text{ MeV}$$

SUGGESTED READINGS

1. B. Ross: Cosmic Rays, McGraw Hill (1964)
2. T. H. Johnson, Rev. Mod. Phys. 10, 219 (1938)
3. P. Auger, Nature 135, 820 (1935)
4. W. R. Webber Physics Today, Oct. 1974
5. W. Ifendale Cosmic Rays, Newnes (1963)
6. A. M. Hillas Cosmic Rays, Pergman-Press (1972)
7. E. Segre Nuclei and Particles (2nd ed.), Benjamin (1977)
8. L. Janossy Cosmic Rays (1948)
9. C. M. G. Lattes et al., Nature 116, 453 and 486 (1947)
10. P. Freier et al., Phys. Rev. 74, 213 (1948)
11. B. Rossi and K. Greisen, Cosmic Ray Theory, Rev. Mod. Phys. 23, 240 (1941)
12. H. J. Bhabha and W. Heitler, Proc. Roy. Soc. A149, 432 (1937).

REVIEW QUESTIONS

1. What are cosmic rays? State the main features known about them as regards the particles involved and their energies.
2. Distinguish between primary and secondary cosmic rays.
3. How does the intensity of cosmic ray very with latitude, direction (east-west asymmetry) and altitude?
4. What are cosmic ray showers?
5. Give a brief account of the origin of cosmic rays.
6. Write an essay on cosmic rays, covering the nature of primary and secondary cosmic rays and the effect of earth's magnetic field, latitude and altitude on them.
7. Give a brief account of the earth's magnetic field on incoming cosmic rays.
8. Give an account of the latitude effect and east-west asymmetry in the intensity of cosmic rays. Clarify their significance in regard to the nature of constitutents of these rays.
9. Explain geomanetic effect on cosmic rays.
10. Discuss energy distribution of the primary cosmic rays.
11. What is cosmic ray shower. Mention the main results of the theory of the cascade process.
12. If you are travelling in an interplanetary space ship, what short of apparatus you need to measure the total intensity of cosmic rays?
13. Discuss Stoermer's theory. How it is applied to explain Geomagnetic effects?

SHORT QUESTIONS

1. Cosmic rays are . . . . particle and are extremely . . . .
Ans. (high energy, penetrating)
2. Cosmic ray intensity is maximum at the . . . . and minimum at the . . . .
Ans. (magnetic poles, magnetic equator)
3. Cosmic rays were named as such on account of their . . . . origin and they were discovered by . . . .
Ans. (extra terrestrial, Hess)
4. The variation in cosmic ray intensity at different geomagnetic latitudes is called . . . .
Ans. (latitude effect)
5. The formation of cosmic ray showers is based upon the phenomenon of . . . .
Ans. (pair production)
6. What is a *brust*?
Ans. (A brust is a sudden local increase in ionisation observed in an ionisation chamber)
7. The cosmic rays particle energies range from . . . .
Ans. (10^{-11} J to 10 J or 10^8 eV to 10^{20} eV).
8. From 10^{10} to $\sim 10^{15}$ eV the cosmic rays particle energy may be represented fairly well by the power-law expression. Write the expression.

Ans. $(N(> \text{E}) = (2.08 \times 10^4) \left(\dfrac{E}{1 \, \text{BeV}} \right)^{-1.67}$ per m^2/stes/sec.)

PROBLEMS

1. Estimate the dose rate at sea level at 45° N from cosmic ray secondary particles.
2. In a cloud chamber using a field of $2T$ the radius of an electron track in a cosmic ray shower is 20 m. Determine the energy of the electron?
3. Calculate the meson fluxdensity at sea level if the ionisation chamber used is a cylinder of 400 mm diameter and 400 mm long with air 4 atmospheric pressure, and gives a current of 56 FA.
4. A positron and an electron with negligible kinetic energy annihilate one another to produce two photons. Calculate their frequencies? (**Ans.** 1.237×10^{-20} Cycles/s, each)
5. The radius of the solar system is 1.2×10^{13} m and the magnetic induction within it is 10^{-9} T. Show that the maximum energy of protons which can be confined within the solar system is about 3.6×10^3 GeV.
6. Show that the maximum kinetic energy of the electrons in the decay of μ^- is about 53 MeV.
7. Calculate the kinetic energy of muon emitted in the decay of a pion at rest by taking $m_\pi = 139.58$ MeV and $m_\mu = 105.66$ MeV.

  (**Ans.** 4.2 MeV)
8. A neutral pion whose kinetic energy is equal to its rest energy decays in flight. Find the angle between two gamma-ray photons that are produced.
  (**Ans.** 120°, use relativistic expression for kinetic energy to find p_π).
9. Show that the radiation unit of length in the electronic cascade shower theory has the values 0.52×10^{-2} m and 330 m respectively in air and lead.

Appendix 1

FUNDAMENTAL CONSTANTS

Quantity	Symbol	Value
General		
Gravitational constant	G (or G_N)	6.6726×10^{-11} m^3 kg^{-1} s^{-2}
Velocity of light	c	299792458 m s^{-1}
Avogadro's number	N_A	6.02214×10^{23} mol^{-1}
Elementary charge	e	1.60218×10^{-19} C
Faraday constant	$F = N_A e$	9.64853×10^4 C mol^{-1}
Electron mass	m_e	9.10939×10^{-31} kg
Electron charge to mass ratio	$-\dfrac{e}{m_e}$	-1.75882×10^{11} C kg^{-1}
Classical electron radius	$r_e = \dfrac{\mu_0 e^2}{4\pi m_e}$	2.81794×10^{-15} m
Thomson cross-section	$\sigma_T = \dfrac{8\pi r_e^2}{3}$	6.65246×10^{-29} m^2
Gas constant	R	8.31451 J mol^{-1} K^{-1}
Boltzmann constant	$k = \dfrac{R}{N_A}$	1.38066×10^{-23} J K^{-1}
Planck constant	h	6.62608×10^{-34} J s
Planck constant, reduced	$\hbar$	1.05457×10^{-34} J s

Quantity	Symbol	Value
Compton wavelength		
of electron	$\lambda_C = \dfrac{h}{m_e c}$	$2.42631 \times 10^{-12}\,\mathrm{m}$
of proton	$\lambda_{C_p} = \dfrac{h}{m_p c}$	$1.32141 \times 10^{-15}\,\mathrm{m}$
Planck mass	$m_p = \left(\dfrac{\hbar c}{G}\right)^{\frac{1}{2}}$	$2.17671 \times 10^{-8}\,\mathrm{kg}$
Fermi constant	G (or G_F)	$1.4355 \times 10^{-62}\,\mathrm{J\,m^3}$
Atomic and nuclear masses		
Atomic mass unit	amu or u	$1.66054 \times 10^{-27}\,\mathrm{kg}$
Atomic mass		
of electron	M_e	$5.48580 \times 10^{-4}\,u$
of proton	M_p	$1.007276\,u$
of hydrogen atom	M_H	$1.007826\,u$
of neutron	M_n	$1.008665\,u$
Ratio of proton to electron mass	$\dfrac{m_p}{m_e}$	1836.15
Mass		
of proton	m_p	$1.67262 \times 10^{-27}\,\mathrm{kg}$
of neutron	m_n	$1.67493 \times 10^{-27}\,\mathrm{kg}$
Neutron–H-atom mass difference	$M_n - M_H$	$0.781\,\mathrm{MeV/c^2}$
Spectroscopic constants		
Rydberg constant	$R_\infty = \dfrac{m_e e^4}{8 h^3 \varepsilon_0^2 c}$	$1.09737 \times 10^7\,\mathrm{m^{-1}}$
Bohr radius	$a_0 = \dfrac{h^2 \varepsilon_0}{\pi m_e e^2}$	$5.29177 \times 10^{-11}\,\mathrm{m}$
Fine-structure constant	$\alpha = \dfrac{\mu_0 e^2 c}{2h}$	7.29735×10^{-3}
Magnetic quantities		
Bohr magneton	$\mu_B = \dfrac{eh}{4\pi m_e}$	$9.27402 \times 10^{-24}\,\mathrm{J\,T^{-1}}$
Nuclear magneton	$\mu_N = \dfrac{eh}{4\pi m_p}$	$5.05079 \times 10^{-27}\,\mathrm{J\,T^{-1}}$
Magnetic moment		
of electron	μ_e	$9.28477 \times 10^{-24}\,\mathrm{J\,T^{-1}}$
of muon	μ_μ	$4.49045 \times 10^{-26}\,\mathrm{J\,T^{-1}}$

Quantity	Symbol	Value
of proton	μ_p	$\begin{cases} 1.41061 \times 10^{-26} \text{ J T}^{-1} \\ 2.793\,\mu_N \end{cases}$
of neutron	μ_n	$\begin{cases} -0.96624 \times 10^{-26} \text{ J T}^{-1} \\ -1.913\,\mu_N \end{cases}$
Gyromagnetic ratio of proton (corrected for diagmagnetization of water)	γ_p	$2.67522 \times 10^8 \text{ rad s}^{-1} \text{ T}^{-1}$
g factors		
of electron	$\begin{cases} g_l \\ g_s \end{cases}$	-1 -2.0023
of proton	$\begin{cases} g_l \\ g_s \end{cases}$	1 $5.586 \left(= \dfrac{2\mu_p}{\mu_N} \right)$
of neutron	$\begin{cases} g_l \\ g_s \end{cases}$	0 $-3.826 \left(= \dfrac{2\mu_n}{\mu_N} \right)$
Inversion factors		
1 kg		$5.60959 \times 10^{29} \text{ MeV/c}^2$
1 MeV/c^2		$1.78266 \times 10^{-30} \text{ kg}$
1 amu (u)		$\begin{cases} 1.66054 \times 10^{-27} \text{ kg} \\ 931.494 \text{ MeV/c}^2 \end{cases}$
1 electron mass (m_e)		0.510999 MeV/c^2
1 proton mass (m_p)		938.272 MeV/c^2
1 eV		$1.60218 \times 10^{-19} \text{ J}$
Wavelength of eV/c particle $\left(\dfrac{hc}{e} \right)$	$1.23984 \times 10^{-6} \text{ m}$	
Width-lifetime produce ($\hbar$)		$6.58212 \times 10^{-16} \text{ eV s}$
$\hbar c$		197.327 MeV fm

Appendix 2

LIST OF ELEMENTS

Z	Symbol	Name	Z	Symbol	Name
1	H	hydrogen	2	He	helium
3	Li	lithium	4	Be	beryllium
5	B	boron	6	C	carbon
7	N	nitrogen	8	O	oxygen
9	F	fluorine	10	Ne	neon
11	Na	sodium	12	Mg	magnesium
13	Al	aluminum	14	Si	silicon
15	P	phosphorus	16	S	sulfur
17	Cl	chlorine	18	Ar	argon
19	K	potassium	20	Ca	calcium
21	Sc	scandium	22	Ti	titanium
23	V	vanadium	24	Cr	chromium
25	Mn	manganese	26	Fe	iron
27	Co	cobalt	28	Ni	nickel
29	Cu	copper	30	Zn	zinc
31	Ga	gallium	32	Ge	germanium
33	As	arsenic	34	Se	selenium
35	Br	bromine	36	Kr	krypton
37	Rb	rubidium	38	Sr	strontium
39	Y	yttrium	40	Zr	zirconium
41	Nb	niobium	42	Mo	molybdenum
43	Tc	technetium	44	Ru	ruthenium
45	Rh	rhodium	46	Pd	palladium
47	Ag	silver	48	Cd	cadmium
49	In	indium	50	Sn	tin
51	Sb	antimony	52	Te	tellurium

Z	Symbol	Name	Z	Symbol	Name
53	I	iodine	54	Xe	xenon
55	Cs	cesium	56	Ba	barium
57	La	lanthanum	58	Ce	cerium
59	Pr	praseodymium	60	Nd	neodymium
61	pin	promethium	62	Sm	samarium
63	Eu	europium	64	Gd	gadolinium
65	Tb	terbium	66	Dy	dysprosium
67	Ho	holmium	68	Er	erbium
69	TM	thulium	70	Yb	ytterbium
71	Lu	lutetium	72	Hf	hafnium
73	Ta	tantalum	74	W	tungsten
75	Re	rhenium	76	Os	osmium
77	Ir	iridium	78	Pt	platinum
79	An	gold	80	Hg	mercury
81	TI	thallium	82	Pb	lead
83	Bi	bismuth	84	PO	polonium
85	At	astatine	86	Rn	radon
87	Fr	francium	88	Ra	radium
89	**	actinium	90	Th	thorium
91	Pa	protactinium	92	U	uranium
93	Np	neptunium	94	Pu	plutonium
95	Am	americium	96	Cm	curium
97	Bk	berkelium	98	Cf	californium
99	Es	einsteinium	100	Fm	fermium
101	Md	mendelevium	102	No	nobelium
103	Lr	lawrencium	104	Rf	rutherfordium
105	Db	Dubnium	106	Sg	Seaborgium
107	Ns	Nielsbohrium	108	Hs	Hassium
109	Mt	Meitnerium			

* Elements with atomic numbers 110, 111, 112, 114 and 116 have also been created but not yet named.

Appendix 3

MASSES OF NEUTRAL ATOMS

In the fifth column of the table an asterisk on the mass number indicates a radioactive isotope, the half-life of which is given in the seventh column.

Z	Element	Symbol	Chemical Atomic Weight	A	Mass (u)	Half life $\left(T_{\frac{1}{2}}\right)$
0	(Neutron)	n		1*	1.008665	12 min
1	Hydrogen	H	1.0079	1	1.007825	
	Deuterium	D		2	2.014102	
	Tritium	T		3*	3.016030	12.26 y
2	Helium	He	4.0026	3	3.016030	
				4	4.002603	
				6*	6.018892	0.802 s
3	Lithium	Li	6.939	6	6.015125	
				7	7.016004	
4	Beryllium	Be	9.0122	7*	7.016929	53.4 d
				9	9.012186	
				10*	10.013534	2.7×10^6 y
5	Boron	B	10.811	10	10.012939	
				11	11.009305	
6	Carbon	C	12.01115	12	12.000000	
				13	13.003354	
				14*	14.003242	5730 y
7	Nitrogen	N	14.0067	14	14.003074	
				15	15.000108	
8	Oxygen	O	15.9994	15*	15.003070	122 s
				16	15.994915	

Z	Element	Symbol	Chemical Atomic Weight	A	Mass (u)	Half life $\left(T_{\frac{1}{2}}\right)$
				17	16.999133	
				18	17.999160	
9	Fluorine	F	18.9984	19	18.998405	
10	Neon	Ne	20.183	20	19.992440	
				21	20.993849	
				22	21.991385	
11	Sodium	Na	22.9898	22*	21.994437	2.60 y
				23	22.989771	
12	Magnesium	Mg	24.312	23*	22.994125	12 s
				24	23.985042	
				25	24.986809	
				26	25.982593	
13	Aluminum	Al	26.9815	26*	25.986892	7.4×10^5 y
				27	26.981539	
14	Silicon	Si	28.086	28	27.976929	
				29	28.976496	
				30	29.973763	
				32*	31.974020	≈ 700 y
15	Phosphorus	P	30.9738	31	30.973765	
16	Sulfur	S	32.064	32	31.972074	
				33	32.971462	
				34	33.967865	
				36	35.967089	
17	Chlorine	Cl	35.453	35	34.968851	
				36*	35.968309	3×10^5 y
				37	36.965898	
18	Argon	A	39.948	36	35.967544	
				38	37.962728	
				39*	38.964317	270 y
				40	39.962384	
				42*	41.963048	33 y
19	Potassium	K	39.102	39	38.963710	
				40*	39.964000	13×10^9 y
				41	40.961832	
20	Calcium	Ca	40.08	39*	38.970691	9.877 s
				40	39.962589	
				41*	40.962275	7.7×10^4 y
				42	41.958625	
				43	42.958780	
				44	43.955492	
				46	45.953689	
				48	47.952531	
21	Scandium	Sc	44.956	45	44.955920	
				50*	49.951730	1.73 min

Z	Element	Symbol	Chemical Atomic Weight	A	Mass (u)	Half life $\left(T_{\frac{1}{2}}\right)$
22	Titanium	Ti	47.90	44*	43.959572	47 y
				46	45.952632	
				47	46.951768	
				48	47.947950	
				49	48.947870	
				50	49.944786	
23	Vanadium	V	50.942	50*	49.947164	$= 6 \times 10^{11}$ y
				51	50.943961	
24	Chromium	Cr	51.996	50	49.946055	
				52	51.940513	
				53	52.940653	
				54	53.938882	
25	Manganese	Mn	54.9380	50*	49.954215	0.29 s
				55	54.938050	
26	Iron	Fe	55.847	54	53.939616	
				55*	54.938299	2.4 y
				56	55.939395	
				57	56.935398	
				58	57.933282	
				60*	59.933964	$= 10^5$ y
27	Cobalt	Co	58.9332	59	58.933189	
				60*	59.933813	5.24 y
28	Nickel	Ni	58.71	58	57.935342	
				59*	58.934342	8×10^4 y
				60	59.930787	
				61	60.931056	
				62	61.928342	
				63*	62.929664	92 y
				64	61.927958	
29	Copper	Cu	63.54	63	62.929592	
				65	64.927786	
30	Zinc	Zn	65.37	64	63.929145	
				66	65.926052	
				67	66.927145	
				68	67.924857	
				70	69.925334	
31	Gallium	Ga	69.72	69	68.925574	
(31)	(Gallium)			71	70.924706	
32	Germanium	Ge	72.59	70	69.924252	
				72	71.922082	
				73	72.923462	
				74	73.921181	
				76	75.921405	
33	Arsenic	As	74.9216	75	74.921596	

Z	Element	Symbol	Chemical Atomic Weight	A	Mass (u)	Half life $\left(T_{\frac{1}{2}}\right)$
34	Selenium	Se	78.96	74	73.922476	
				76	75.919207	
				77	76.919911	
				78	77.917314	
				79*	78.918494	7×10^4 y
				80	79.916527	
				82	81.916707	
35	Bromine	Br	79.909	79	78.918329	
				81	80.916292	
36	Krypton	Kr	83.80	78	77.920403	
				80	79.916380	
				81*	80.916610	2.1×10^5 y
				82	81.913482	
				83	82.914131	
				84	83.911503	
				85*	84.912523	10.76 y
				86	85.910616	
37	Rubidium	Rb	85.47	85	84.911800	
				87*	86.909186	5.2×10^{10} y
38	Strontium	Sr	87.62	84	83.913430	
				86	85.909285	
				87	86.908892	
				88	87.905641	
				90*	89.907747	28.8 y
39	Yttrium	Y	88.905	89	88.905872	
40	Zirconium	Zr	91.22	90	89.904700	
				91	90.905642	
				92	91.905031	
				93*	92.906450	9.5×10^5 y
				94	93.906313	
				96	95.908286	
41	Niobium	Nb	92.906	91*	90.906860	(long)
				92*	91.907211	= 10 y
				93	92.906382	
				94*	93.907303	2×10^4 y
42	Molybdenum	Mo	95.94	92	91.906810	
				93*	92.906830	= 10^4 y
				94	93.905090	
				95	94.905839	
				96	95.904674	
				97	96.906021	
				98	97.905409	
				100	99.907475	
43	Technetium	Te		97*	96.906340	2.6×10^6 y

Z	Element	Symbol	Chemical Atomic Weight	A	Mass (u)	Half life ($T_{\frac{1}{2}}$)
				98*	97.907110	1.5×10^6 y
				99*	98.906249	2.1×10^5 y
44	Ruthenium'	Ru	101.07	96	95.907598	
				98	97.905289	
				99	98.905936	
				100	99.904218	
				101	100.905577	
(44)	(Ruthenium)			102	101.904348	
				104	103.905430	
45	Rhodium	Rh	102.905	103	102.905511	
46	Palladium	Pd	106.4	102	101.905609	
				104	103.904011	
				105	104.905064	
				106	105.903479	
				107*	106.905132	7×10^6 y
				108	107.903891	
				110	109.905164	
47	Silver	Ag	107.870	107	106.905094	
				109	108.904756	
48	Cadmium	Cd	112.40	106	105.906463	
				108	107.904187	
				109*	108.904928	453 d
				110	109.903012	
				111	110.904188	
				112	111.902762	
				113	112.904408	
				114	113.903360	
				116	115.904462	
49	Indium	In	114.82	113	112.904089	
				115*	114.903871	6×10^{14} y
50	Tin	Sn	118.69	112	111.904835	
				114	113.902773	
				115	114.903346	
				116	115.901745	
				117	116.902958	
				118	117.901606	
				119	118.903313	
				120	119.902198	
				121*	120.904227	25 y
				122	121.903441	
				124	123.905272	
51	Antimony	Sb	121.75	121	120.903816	
				123	122.904213	
				125*	124.905232	2.7 y

Z	Element	Symbol	Chemical Atomic Weight	A	Mass (u)	Half life $\left(T_{\frac{1}{2}}\right)$
52	Tellurium	Te	127.60	120	119.904023	
				122	121.903064	
				123*	122.904277	12×10^{13} y
				124	123.902842	
				125	124.904418	
				126	125.903322	
				128	127.904476	
				130	129.906238	
53	Iodine	1	126.9044	127	126.904070	
				129*	128.904987	1.6×10^7 y
54	Xenon	Xe	131.30	124	123.906120	
				126	125.904288	
				128	127.903540	
				129	128.904784	
				130	129.903509	
				131	130:905085	
				132	131.904161	
				134	133.905815	
				136	135.907221	
55	Cesium	Cs	132.905	133	132.905355	
				134*	133.906823	2.1 y
(55)	(Cesium)			135*	134.905770	2×10^6 y
				137*	133.906770	30 y
56	Barium	Ba	137.34	130	129.906245	
				132	131.905120	
				133*	132.905879	7.2 y
				134	133.904612	
				135	134.905550	
				136	135.904300	
				137	136.905500	
				138	137.905000	
57	Lanthanum	La	138.91	137*	136.906040	6×10^4 y
				138*	137.906910	1.1×10^{11} y
				139	138.906140	
58	Cerium	Cc	140.12	136	135.907100	
				138	137.905830	
				140	139.905392	
				142*	141.909140	5×10^{15} y
59	Praseodymiu	Pr	140.907	141	140.907596	
60	Neodymium	Nd	144.24	142	141.907663	
				143	142.909779	
				144*	143.910039	2.1×10^{15} y
				145	144.912538	

Z	Element	Symbol	Chemical Atomic Weight	A	Mass (u)	Half life $\left(T_{\frac{1}{2}}\right)$
				146	145.913086	
				148	147.916869	
				150	149.920960	
61	Promethium	PM		145*	144.912691	18 y
				146*	145.914632	1600 d
				147*	146.915108	2.6 y
62	Samarium	Sm	150.35	144	143.911989	
				146*	145.912992	1.2×10^8 y
				147*	146.914867	1.08×10^{11} y
				148*	147.914791	1.2×10^{13} y
				149*	148.917180	4×10^{14} y
				150	149.917276	
				151*	150.919919	90 y
				152	151.919756	
				154	153.922282	
63	Europium	Eu	151.96	151	150.919838	
				152*	151.921749	12.4 y
				153	152.921242	
				154*	153.923053	16 y
				155*	154.922930	1.8 y
64	Gadolinium	Gd	157.25	148*	147.918101	85 y
				150*	149.918605	1.8×10^6 y
				152*	151.919794	1.1×10^{14} y
				154	153.920929	
				155	154.922664	
				156	155.922175	
				157	156.924025	
				158	157.924178	
				160	159.927115	
65	Terbium	Th	158.925	159	158.925351	
66	Dysprosium	Dy	162.50	156*	155.923930	1.2×10^3 y
				158	157.924449	
				160	159.925202	
				161	160.926945	
				162	161.926803	
				163	162.928755	
				164	163.929200	
67	Holmium	Ho	164.930	165	164.930421	
				166*	165.932289	1.2×10^3 y
68	Erbium	Er	167.26	162	161.928740	
				164	163.929287	
				166	165.930907	
				167	166.932060	
				168	167.932383	

Z	Element	Symbol	Chemical Atomic Weight	A	Mass (u)	Half life $\left(T_{\frac{1}{2}}\right)$
69	Thulium	Tm	168.934	170	169.935560	
				169	168.934245	
				171*	170.936530	1.9 y
70	Ytterbium	Yb	173.04	168	167.934160	
				170	169.935020	
				171	170.936430	
				172	171.936360	
				173	172.938060	
				174	173.938740	
				176	175.942680	
71	Lutecium	Lu	174.97	173*	172.938800	1.4 y
				175	174.940640	
				176*	175.942660	22×10^{10} y
72	Hafnium	Hf	178.49	174*	173.940360	2.0×10^{15} y
				176	175.941570	
				177	176.943400	
				178	177.943880	
				179	178.946030	
				180	179.946820	
73	Tantalum	Ta	180.948	180	179.947544	
				181	180.948007	
74	Wolfram (Tungsten)	W	183.85	180	179.947000	
				182	181.948301	
				183	182.950324	
				184	183.951025	
				186	185.954440	
75	Rhenium	Re	186.2	185	184.953059	
				187*	186.955833	5×10^{10} y
76	Osmium	Os	190.2	184	183.952750	
				186	185.953870	
				187	186.955832	
				188	187.956081	
				189	188.958300	
				190	189.958630	
				192	191.961450	
				194*	193.965229	6.0 y
77	Iridium	Ir	192.2	191	190.960640	
				193	192.963012	
78	Platinum	Pt	195.09	190*	189.959950	7×10^{11} y
				192	191.961150	
				194	193.962725	
				195	194.964813	
				196	195.964967	
				198	197.967895	

Z	Element	Symbol	Chemical Atomic Weight	A	Mass (u)	Half life $\left(T_{\frac{1}{2}}\right)$
79	Gold	Au	196.967	197	196.966541	
80	Mercury	Hg	200.59	196	195.965820	
				198	197.966756	
				199	198.968279	
				200	199.968327	
				201	200.970308	
				202	201.970642	
				204	203.973495	
81	Thallium	TI	204.19	203	202.972353	
				204*	203.973865	3.75 y
				205	204.974442	
(81)	(Thallium)	RaE"		206*	205.976104	4.3 min
		Ac	C"	207*	206.977450	4.78 min
		Th	C"	208*	207.982013	3.1 min
		Ra	C"	210*	209.990054	1.3 min
82	Lead	Pb	207.19	202*	201.927997	3×10^5 y
				204*	203.973044	1.4×10^{17} y
				205*	204.974480	3×10^7 y
				206	205.974468	
				207	206.975903	
				208	207.976650	
		Ra	D	210*	209.984187	22 y
		Ac	B	211*	210.988742	36.1 min
		Th	B	212*	211.991905	10.64 h
		Ra	B	214*	213.999764	26.8 min
83	Bismuth	Bi	209.980	207*	206.978438	30 y
				208*	207.979731	3.7×10^5 y
				209	208.980394	
		Ra	E	210*	209.984121	5.1 d
		Th	C	211*	210.987300	2.15 min
				212*	211.991876	60.6 min
		Ra	C	214*	213.998686	19.7 min
				215*	215.001830	8 min
84	Polonium	PO		209*	208.982426	103 y
		RaF¢		210*	209.982876	138.4 d
		Ac	C'	211*	210.996657	0.52 s
		Th	C'	212*	211.989629	0.30 ms
		Ra	C	214*	213.995201	164 ms
		Ac	A	215*	214.999423	0.0018 s
		Th	A	216*	216.001790	0.15 s
		Ra	A	218*	218.008930	3.05 min
85	Astatine	At		215*	214.998663	= 100 ms
				218*	218.008607	1.3 s
				219*	219.011290	0.9 min

Z	Element	Symbol	Chemical Atomic Weight	A	Mass (u)	Half life $\left(T_{\frac{1}{2}}\right)$
86	Radon	Rn				
		An		219*	219.009481	4.0 s
		Tn		220*	220.011401	56 s
		Rn		222*	222.017531	3.823 d
87	Francium	Fr				
		Ac K		223*	223.019736	22 min
88	Radium	Ra	226.0			
		Ac X	232.08	223*	223.018501	11.4 d
		Th	X	224*	224.020218	3.64 d
		Ra		226*	226.025360	1620 y
		MsTh,		228*	228.031139	5.7 y
89	Actinium	Ac		227*	227.027753	21.2 y
		MsTh2		228*	228.031080	6.13 h
90	Thorium	Th	232.038			
		RdAc		227*	227.027706	18.17 d
		RdTh		228*	228.028750	1.91 y
				229*	229.031652	7300 y
		Io		230*	230.033087	76000 y
		UY		231*	231.036291	25.6 h
		Th		232*	232.038124	1.39×10^{10} y
		UX,	231.0359	234*	234.043583	24.1 d
91	Protoactinium	Pa	238.03	231*	231.035877	32480 y
		UZ		234*	234.043298	6.66 h
92	Uranium	U		230*	230.033937	20.8 d
(92)	(Uranium)			231*	231.036264	4.3 d
				232*	232.037168	72 y
				233*	233.039522	1.62×10^5 y
				234*	234.040904	2.48×10^5 y
		Ac	U	235*	235.043915	7.13×10^8 y
				236*	236.045637	2.39×10^7 y
		U1		238*	238.048608	4.51×10^9 y
93	Neptunium	Np	237.0480	235*	235.044049	410 d
				236*	236.046624	5000 y
				237*	237.048056	2.14×10^6 y
94	Plutonium	Pu	239.0522	236*	236.046071	2.85 y
				238*	238.049511	89 y
				239*	239.052146	24360 y
				240*	240.053882	6700 y
				241*	241.056737	13 y
				242*	242.058725	3.79×10^5 y
				244*	244.064100	7.6×10^7 y

INDEX